GEOTECHNICAL SPECIAL PUBLICATION NO. 118

A HISTORY OF PROGRESS

Selected U.S. Papers in Geotechnical Engineering

VOLUME ONE

SPONSORED BY
The Geo-Institute of the American Society of Civil Engineers

EDITED BY
W. Allen Marr

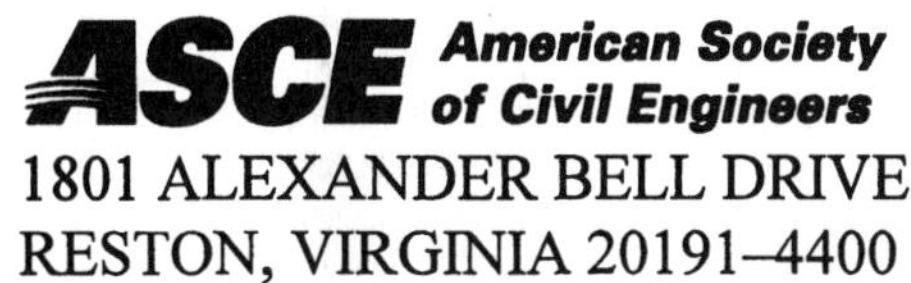

1801 ALEXANDER BELL DRIVE
RESTON, VIRGINIA 20191–4400

ISBN 0-7844-0656-1

ISBN 0-7844-0656-1

Manufactured in the United States of America.

Library of Congress Cataloguing-in-Publication data on file.

Preface

The geotechnical engineering profession has inherited a rich body of literature that documents the development of fundamental concepts and their application to practical problems. Many of these publications contain information as useful today as when they were first written. Others provide great insight into the origins and development of our field and the thought processes of its leaders.

The Geo-Institute decided to reprint some of these papers in a Geotechnical Special Publication to make them more readily available and hence promote their use by the geotechnical engineering profession. The assignment was given to the Geo-Institute's Technical Publications Committee, chaired by Professor Robert Koerner, who delegated the task of organizing this effort to a committee. A working committee was formed to select which papers to include in the GSP. The committee was comprised of the following individuals:

- Clyde N. Baker Jr., Senior Principal Engineer, STS Consultants, Ltd., Vernon Hills, IL
- John T. Christian, Consulting Engineer, Newton, Massachusetts
- Barry R. Christopher, Geotechnical Engineering Consultant, Roswell, GA
- Charles Dowding, Professor of Geotechnics, Northwestern University, Evanston, IL
- J. Michael Duncan, University Distinguished Professor, Virginia Tech, Blacksburg, VA
- Richard J. Finno, Professor of Civil Engineering, Northwestern University, Evanston, IL and Editor-in-Chief of JGEE
- John E. Garlanger, Principal, Ardaman and Associates, Orlando, FL
- Roman D. Hryciw, Professor, University of Michigan, Ann Arbor, MI
- W.F. Marcuson III, Consulting Engineer, W.F. Marcuson III and Associates, Inc., Vicksburg, MS
- W. Allen Marr, Chief Engineer, GEOCOMP/GeoTesting Express, Boxborough, MA
- Paul Mayne, Professor, Georgia Tech, Atlanta, GA
- N.R. Morgenstern, University Professor (Emeritus) of Civil Engineering, University of Alberta
- Juan M. Pestana, Professor, University of California, Berkeley, CA
- Greg Richardson, Principal, Greg Richardson and Associates, Raleigh, NC
- James L. Withiam, Principal, D'Appolonia, Monroeville, PA and Editorial Board Chair of Geo-Strata
- Stephen G. Wright, Brunswick-Abernathy Regents Professor, The University of Texas at Austin and representative from Geo-Institute Board of Governors

These people were selected because they collectively represent academia and practice in various segments of our profession, they came from different academic backgrounds, and they agreed to help.

The selection of papers considered several criteria. The paper had to have appeared in a US publication to minimize problems with ASCE obtaining a copyright release. The publication could not be too long. The publication must have made a significant contribution to the geotechnical engineering literature. The content should have something that geotechnical engineers could benefit from reading today.

The selected papers provide a wide view of the development of geotechnical engineering from a broad spectrum of contributors. They vary from the highly theoretical to the elegantly practical. Many outstanding papers could not be included due to space limitations, while many others appeared in non-US publications. Doubtless, every reader would have included other papers. We solicited nominations of papers with announcements in ASCE publications and broadcast emails. All of these nominations were considered.

Some of the most outstanding publications in geotechnical engineering were the early textbooks. Those of special note include:

Prentis, E.A. and L. White, *Underpinning-Its Practice and Applications*, New York, 1931.
Muskat, M., *Flow of Homogeneous Fluids Through Porous Media*, McGraw-Hill, 1937.
Terzaghi, K., *Theoretical Soil Mechanics*, John Wiley and Sons, New York, 1943.
Taylor, D.W., *Fundamentals of Soil Mechanics*, Wiley, 1948.
Terzaghi, K. and R.B. Peck, *Soil Mechanics in Engineering Practice*, John Wiley and Sons, New York, 1948.
Lambe, T.W., *Soil Testing for Engineers*, John Wiley and Sons, New York, 1952.

These books continue to enlighten any engineer who takes the time to consider their contents. Obviously, it was not practical to include them or many other excellent publications in this GSP.

On a personal note, it was an honor to work with this group of eminent engineers on something I consider to be of great importance to our profession. Thanks also go to Charles Ladd, John Christian, Steve Poulos, Lewis Edgars, Jorg Oosterberg, Paul Mayne, Steve Wright, Richard Woods and Harvey Wahls who loaned original materials from their libraries for this effort. This collection stands as a testament to the grand efforts of our forefathers and colleagues. I hope it inspires our members, especially our younger members, to continue in the further development of our proud heritage.

W. Allen Marr, Editor

Contents

Volume One

Volume Two

A Science of Soil Action

NOWHERE does the civil engineer encounter so many baffling problems as in connection with earthwork, his largest activity. Clay and sand in particular give him endless trouble, mainly because he knows little about them. When he excavates, when he builds dams or embankments, constructs foundations or digs tunnels, he is each moment working at the verge of engineering knowledge. Land slips and caving banks give him frequent reminders of his inability to foretell or prevent destructive soil action. This applies also to foundation settlement, which almost daily supplies an indictment of his ignorance of soils.

The need for developing such knowledge has been felt by many. That soils, representing the oldest constructive material, had no science of stresses, no mechanics other than the granular theory of the Coulomb-Rankine time, has long been recognized as a serious lack. But it has been widely believed that soil action is subject to so many erratic influences as to make a science hardly possible. This view is now proved wrong. A brilliant series of articles, the first of which we publish in this issue, shows that the behavior of clays and sands is just as definite, just as constant and subject to calculation and analysis, as the behavior of a piece of steel or concrete under load. The articles present a new science in embryo, the science of soil action. They establish convincingly that future engineering in the field of soils will be built up on the substratum of the facts here developed.

As so far worked out, the new knowledge relates to the physical behavior of sand and clay. In its further extension, no doubt, other soils also will be covered. Even in its present form, however, it is of far-reaching value to the practical man.

The background of these articles is interesting. Six or seven years ago Charles Terzaghi, an engineering teacher in Constantinople, began some laboratory experiments on sand to discover the nature and laws of its internal friction and volume changes under load and moisture. He found the task exasperatingly slow, but persevered with patience and originality, in spite of having to depend on improvised apparatus of most simple kind, and in time obtained a wholly new picture of the phenomena of sand behavior. Some of the results which that early work yielded appeared in an article published in *Engineering News-Record*, five years ago (Sept. 30, 1920). Subsequently he extended his work very greatly, and in particular made searching studies of clay. Last year the work reached a stage where its results began to group themselves into a full and highly enlightening explanation of how both plastic and granular soils behave.

The series of articles now beginning represents a somewhat sketchy outline of the principal facts so far developed.

Most interesting of the facts in the entire series, and most immediately fruitful, are those which relate to clay, brought out in the opening articles. They explain why clay coheres, why it compresses greatly under load and shrinks on drying, and why sand and clay differ. They help us also toward an understanding of what happens when a soil breaks down into a semi-liquid state, a phenomenon involved in many of those instabilities of soil that give constructors trouble and consume labor and expense.

Not all soils are clays and sands. Soils of mixed grain texture or structure are fully as important and only a first step will have been made until the facts concerning these soils also are developed. Very much further investigation needs to be made to survey the whole field that Dr. Terzaghi has reconnoitered and successfully outlined.

A great deal also remains to be done in bringing the new facts and teachings to bear upon the problems of practice with least delay. Such utilization can be brought about only through an educational process of large proportions. The present articles represent one initial stage of the process. Fortunately, Massachusetts Institute of Technology has taken up an active share of the educational work, by engaging Dr. Terzaghi to come to this country and make his work and results available to students. This enterprising action gives promise of soon making the young science an integral part of our current engineering knowledge.

Because much work remains to be done in exploring the field of soil phenomena, there is a call for the assistance of many trained investigators. It is desirable that full development of the field come about as early as possible, and it may be hoped that the present series of articles will not only bring enlightenment and practical help to the working engineer, but will also inspire laboratory investigators to co-operate in advancing our soil knowledge still further.

Principles of Soil Mechanics: I—Phenomena of Cohesion of Clay

Relation Between Load and Moisture-Content—Hydrostatic Stress in Clay—Shrinkage and Negative Hydrostatic Pressure—Surface Tension the Cause of Cohesion

BY DR. CHARLES TERZAGHI
Professor of Civil Engineering, American Robert College, Constantinople, Turkey
Temporary Lecturer at Massachusetts Institute of Technology, Cambridge

This article begins a series in which the nature and behavior of soil are studied and analyzed, with fruitful results. The first four articles discuss clay, explaining its cohesion, shrinkage, elastic behavior and settlement. Later, sand is taken up, and the characteristics of sand and clay are compared. Mixed soils are not dealt with; nevertheless the facts developed in the author's investigation bear intimately on problems arising in practical engineering work wherever clay or sand soils are concerned.—Editor.

A NUMBER of years ago the author became impressed with the perplexing character of the problems which the behavior of sand and clay soils presents to the engineer. It was evident that our knowledge of the underlying laws of action of these materials is not sufficient for proper solution of these problems, and that the first step of progress would be a careful experimental study of their physical behavior. At Robert College the author at last found opportunity to carry on such a study, which began with sands and later took up clays. The work has yielded results of interest and (it is believed) importance. Some of the early results of the study of sands were reported in *Engineering News-Record*, Sept. 30, 1920, pp. 632-637. A fuller statement of results is now possible, as the later investigations included clays, the physical properties of which had remained essentially unexplored. The term clay indicates mixed-grained, plastic soils consisting of particles from 1 mm. down to 0.006 mm. in diameter and a very small percentage of "ultra-clay" (colloidal particles in the colloid-chemical sense of the word). Hence the term includes almost all the soils which are commonly known as "clays."

Prime factors in the behavior of clay are (1) cohesion, (2) the action of the moisture contained in the clay, and (3) permeability. The present and several succeeding articles will summarize the facts concerning them as developed by the experimental investigation. Cohesion will be considered first.

Cohesion is commonly looked on as a characteristic property of certain soils. No serious attempt has thus far been made to analyze the physical causes of the property, but it has apparently been accepted as an elementary fact, unsuitable for further analysis. As a consequence the mechanics of clays has remained in a rudimentary state. We know practically nothing about the part played by the factors of time and permeability in the processes called "settlement," nor have we any clear conception of what causes the swelling of clays. Under these circumstances it is timely to investigate these phenomena systematically. The author has carried out a long series of laboratory studies in this effort.

Load and Moisture Content—The simplest process a layer of clay can undergo is compression under a uniformly distributed load when the layer is restrained from expanding laterally and when surface tension of the capillary water is excluded. In order to accomplish the latter, the author's experiments were made with a mixture of clay and water (free from air), and during the experiments the surface of the clay was kept covered with water; thus there was no external force acting on the clay particles except the surcharge. As the voids of the clay are completely filled with water, a certain quantity of the capillary water must escape during the compression of the clay by the surcharge.

The experiment was made as follows: The lower part of a glass cylinder (Fig. 1) was filled with a liquid but very viscous mixture of clay and water. The bottom of the cylinder was covered with a sheet of thin filter paper (not shown), and on this rested a bronze ring completely surrounded by and immersed in the clay solution. The surface of the mixture was covered with a layer of filter-paper, and on top of the paper was placed a filter composed of quartz sand of grain size ¼ to ½ mm. Within 24 hr. the compression produced by the weight of the filter was completed, and the filter was then loaded by a brass cup. One day later further load was added by filling the lower half of the cup with lead shot, and after the lapse of two more days the upper half of it was filled too (total load at this time 0.1 kg./cm.^{2}). For measuring the compression produced by the load, the side of the brass cup was provided with a scale. Additional surcharge was applied by means of a lever bearing on the cup through a steel ball. Thus the pressure was increased at about two-day intervals from 0.1 to 0.2, 0.3, 0.6 and 1.2 kg./cm.^{2}

Higher pressures were applied in a testing machine. For this purpose, the lever, cup and sand filter were removed, the bottom plate of the cylinder unscrewed, and the clay forced out by a piston. Then the ring was freed from the surrounding clay in such a manner that it contained a layer of clay of thickness equal to the height of the ring (1 cm.). A sample of the removed part of the clay served for determining its water content. The ring with its clay core was weighed, and then the experiment was continued by means of the apparatus Fig. 2. The change from one apparatus to the other was made rapidly, to minimize loss of water by evaporation.

The apparatus consisted of a square vessel the bottom of which was covered with two layers of thick and one layer of fine filter paper, two superimposed bronze rings, bronze loading plates, and a steel ball. The two rings fitted together with a ground conical joint. First the lower ring with its content was placed on a moist sheet of thin filter paper, the upper ring was applied and lined with filter paper (see detail in Fig. 2), the upper surface of the clay was covered with a circular piece of moistened filter paper, a sand filter was applied, and finally the whole set with the bronze plates and ball was transferred into the enclosing vessel, which was filled with water and placed in a testing machine.

In the machine the pressure was raised within twenty minutes from zero to 2 kg./cm.[2] and then left constant. Under this pressure the thickness of the

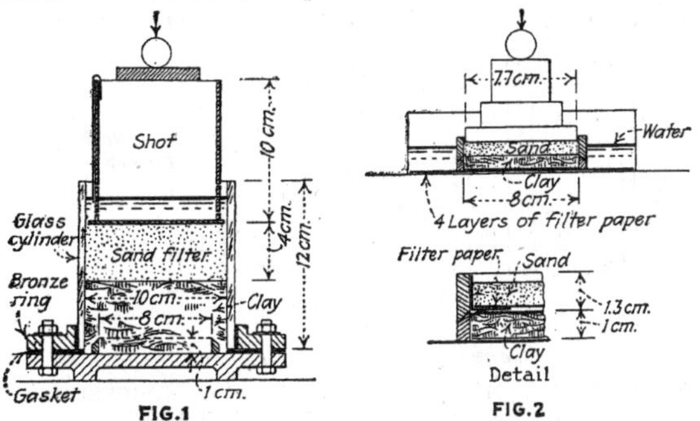

FIGS. 1 AND 2—DETERMINING RELATION BETWEEN PRESSURE AND MOISTURE-CONTENT OF CLAY
Fig. 1—Low-pressure tests. Fig. 2—High-pressure tests.

layer decreased, at first rapidly and then more slowly, and within about two days became constant, indicating hydrostatic equilibrium. Now the water was siphoned out of the vessel, the piston was rapidly lifted, the bronze plates and the filter and upper ring were removed, the upper surface of the clay was carefully dried off by means of absorbent paper, the weight of the ring together with its content was determined, the clay particles projecting beyond the upper edge of the ring were scraped away, and the weight was determined once more. Operations performed, the set was reassembled and the experiment was continued.

Thus the pressure was raised at intervals of about two days, from zero to 2, 4, 8, 14 and 20 kg./cm.[2]. The measurements furnished the data for computing the co-ordinates of corresponding points of the pressure-moisture curve. Then followed a complete cycle, in which the pressure was successively kept constant for two days at the following values: 8, 4, 2, 1, 0 (four to six days, to allow complete resaturation), 1, 2, 5, 10 and 20 kg. per cm.[2]

At the outset of the investigation the author feared that the total removal of the pressure in each of the consecutive steps of the test might represent an important source of errors of observation. As a matter of fact no results at all could be obtained by such method in investigating the compressibility of sands or of other highly permeable granular materials. But in the case of clay experience has shown that the fear was not justified. Due to low degree of permeability even of sandy clays the effect of resaturation is negligible. Thus, a layer of clay 1 cm. thick and carrying no surcharge at all was brought into contact with water, after it had been previously compressed by a load of 18.9 kg./cm.[2], and in an hour its moisture content increased only 0.15 per cent, while the time during which the clay has a chance to take up water between two consecutive tests never exceeds two minutes.

After having completed the cycle, the ring with its clay core was weighed once more. Then a sample of the clay was set apart for determining the final moisture content of the layer. The figure thus obtained served as a basis for calculating the moisture content which corresponds to the successive stages of the experiment. The initial moisture content (corresponding to a pressure of 1.2 kg. per sq.cm.) being known, the results of the calculation could be checked by theoretically calculating the subsequent moisture content. Any difference between the actual and the theoretical moisture content (in no case large) was then distributed over the intermediate values.

Fig. 3 shows typical results. Abscissas represent intensity of loading and ordinates moisture content in percentage of the space occupied by the solid matter of the clay. A yellow pottery clay and a blue marine clay are represented in the lower diagram, and a gray sandy delta mud in the upper. Each test lasted about eight weeks. The main branch of each curve portrays the effect of gradually increasing load (combined compression), while loading cycles are represented by curved hysteresis loops (elastic expansion).

Lag of Compression—As already mentioned, the compression produced by the application of a load never occurred at once, but increased for some time while the load remained constant. Equilibrium was never reached in less than about 24 hr., in spite of the fact that the layers were very thin. This fact is very significant; it indicates that the capillary water took time to drain from the interior of the clay towards the free surfaces and escape through the filters. Since this flow of water implies a hydrostatic head, and the sur-

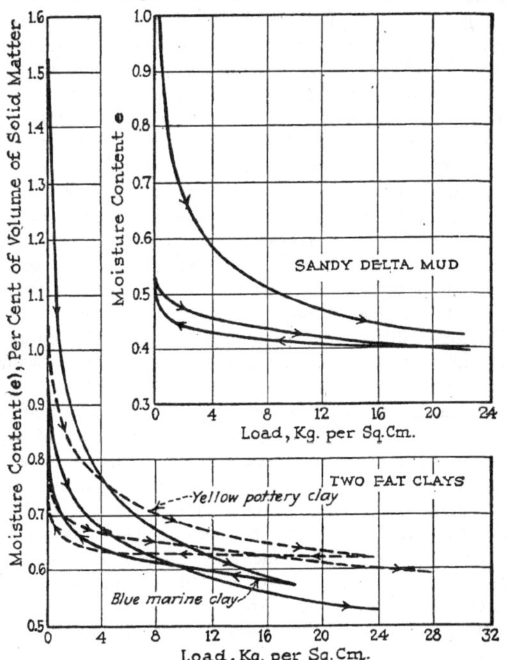

FIG. 3—PRESSURE-MOISTURE CURVES FOR THREE CLAYS

face of the clay (being covered with water) was constantly at zero hydrostatic pressure, the application of the load must have produced a positive hydrostatic pressure in the capillary water of the central parts of the clay. The *hydrostatic stress difference* which produces the outflow of the capillary water is consumed by the resistance which the water encounters in the narrow voids of the clay on its way to the surface.

The author later proved by other experiments (see a subsequent article) that the flow of water through clay follows Darcy's law as closely as does the flow of water through sand. Hence there exists simple proportionality between the value of the hydrostatic stress difference and the amount that the thickness of the layer decreases per unit of time at constant surcharge, because the decrease of clay volume is identical with the volume of the water that escapes through the surface of the clay. Immediately after the load is applied the speed with which the clay volume decreases is greatest, hence the hydrostatic stress difference must be a maximum; in the final state the hydrostatic pressure has become zero in every part of the layer and the thickness of the layer remains constant (hydrostatic equilibrium of the clay).

From the results of the tests described we learn that pressure and moisture content of a clay are as definitely related as stress and strain in solid bodies. The only difference between the two relations is that the compression of the clay develops gradually, while the application of a pressure on a solid body is almost immediately followed by the corresponding strain. In addition we have learned that a change of the volume occupied by a mixture of clay and water involves a flow of capillary water from the central parts of the clay towards the surface, which in turn requires the presence of a hydrostatic stress difference. The strain does not become constant until the hydrostatic stress difference has disappeared.

Shrinkage Test—Keeping these simple but important facts in mind, we may proceed to a second series of tests. We fill the glass cylinder of Fig. 1 with a liquid yet very viscous mixture of clay and water, just as was done before. But instead of covering the surface of the clay with a filter and loading it, we leave it as it is, in contact with the atmosphere. Precisely the same phenomenon occurs without load which we previously produced by means of loading the clay surface: the capillary water flows from the central part of the clay towards the surface, where it evaporates; the moisture content of the clay becomes smaller and the thickness of the layer decreases. *The clay shrinks.*

The only difference between the artificial compression and the natural shrinkage is that compression under load can be carried as far as we want, while in shrinkage due to evaporation a point is reached beyond which the volume of the clay remains constant. At this shrinkage limit the clay passes from the semi-solid into the solid state. Before the limit is reached, the layer shows symptoms of lateral contraction; it shrinks away from the ring to which it is confined, and cracks. At the shrinkage limit the clay usually changes its color. Yet, from the outset of the shrinkage test to a point beyond the limit of the plastic state, the process of compression and the process of shrinkage are in every respect identical.

It is astonishing that nobody seems to have felt the necessity of investigating the physical causes of this identity. As a matter of fact, analysis of the phenomenon leads to new and very important conclusions. In order to perform it we have merely to keep in mind that the elementary laws of mechanics which apply to solids and liquids in general are valid also for the constituents of a mixture of clay and water.

Negative Hydrostatic Pressure—By direct measurement it was found that the water content of a layer of clay slowly shrinking at normal temperature remains fairly uniform throughout the thickness of the layer, if this thickness amounts to not more than a few centimeters. Hence there must be a continuous flow of water from the interior of the layer towards the surface of evaporation. The flow indicates the presence of a hydrostatic stress difference between the interior and the surface.

During the loading test the hydrostatic pressure acting in the central part of the layer of clay was positive.

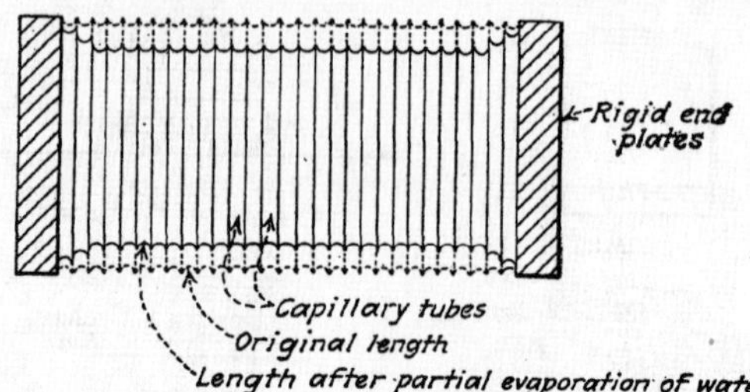

FIG. 4—BUNDLE OF CAPILLARY TUBES COMPRESSED BY EVAPORATION OF CONTAINED WATER

This positive hydrostatic pressure was balanced by the external load pressing upon the particles between which the capillary water flowed toward the free surface. During the shrinkage test the external load was zero, and as the dead weight of the clay can be considered negligible compared with the intensity of the forces we have to deal with, the hydrostatic pressure in the interior of the clay is zero. But as the flow of the capillary water toward the surface of evaporation requires the existence of a hydrostatic stress difference, the hydrostatic pressure acting in the vicinity of the surface of evaporation must have been negative.

During the compression test, surface tension of the water was not active, because the free surfaces of the clay were covered with water; only the external load forced the clay particles down. During the shrinkage test, there was no external load, but surface tension acted all over the top surface of the clay. Therefore the hydrostatic stress difference and the negative hydrostatic pressure could not be produced by other than the surface tension of the capillary water.

In a vertical capillary, water may rise to great height. The surface tension at its surface, amounting to T dynes per centimeter of circumference, exerts a lifting force of $\pi dT/g$, where d is the diameter of the capillary in centimeters and g is the acceleration of gravity in cm./sec.^{2}. Equating this force to the weight of the column of water in the capillary, of height h cm. and specific gravity w, we obtain $h = 4T/wdg$. The lifting force at the surface carries the weight of the column by a tensile stress in the water column, as water has no shearing strength. This tension, or negative hydrostatic pressure, has its maximum amount hw just under the upper surface of the column. and at any other point, at height h' above the level of water in the vessel below, it has the value $h'w$. Precisely similar tension, or negative pressure, exists in the clay during its shrinkage, and explains the phenomena which occur during the drying process.

A layer of clay surrounded by a stiff ring represents a bundle of capillary tubes whose ends are located within the free surfaces of the layer (Fig. 4). At the outset the tubes are filled with water. Evaporation at the free surface tends to cause the water to retire into the interior, but it is prevented from doing so by the same forces which drive the water up a capillary, and therefore it continues to occupy the whole void space within the clay. The volume of the water decreasing by evaporation, the clay is compressed by the surface tension of the water; in other words it shrinks, and at the same time the water comes under tension.

The analogy between compression by load and shrinkage by evaporation is set forth diagrammatically in Fig. 5. The direction of water flow and the nature of the internal pressures are indicated.

The intensity of the forces increases in direct proportion to the quantity of water evaporated, provided the solid matter follows Hooke's law. They reach their maxima when the full capillary force of the water is developed (*hw*, above); this marks the shrinkage limit. Further evaporation causes the water to retire into the interior of the bundle, and the free surfaces change from dark to light color.

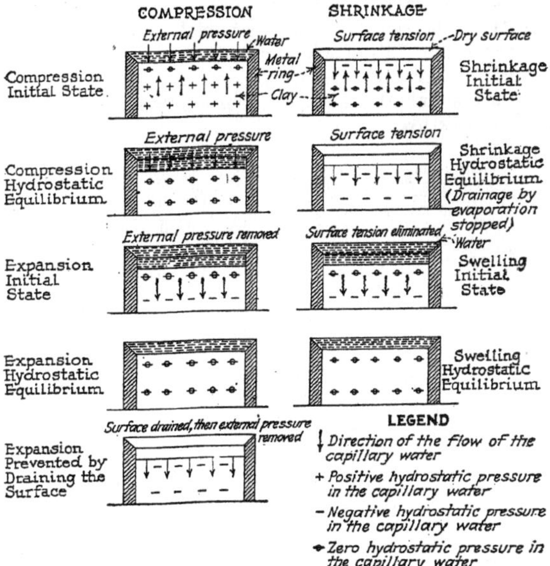

FIG. 5—DIAGRAMMATIC COMPARISON OF COMPRESSION AND SHRINKAGE OF CLAY

The pressure exerted by the surface tension will be called the *capillary pressure*, and its maximum value the *transition pressure*, which is the limit at which the clay passes from the semi-solid to the solid state. It has been proved that capillary action involves a tension in the capillary water equal to the capillary pressure. This tension cannot possibly be greater than the "intrinsic pressure" of the water. However, since the intrinsic pressure of water amounts to about 21,000 atmospheres, the transition pressure is limited only by the size of the voids of the clay.

The intensity of the negative hydrostatic pressure is evidently identical with the intensity of the external pressure required to produce the same degree of compression as the one brought forth by shrinkage. Owing to this fact, the transition pressure may be determined from the moisture content at which the clay ceases to shrink (shrinkage limit) and the external pressure required to reduce the moisture content of the clay from its initial value down to that at the shrinkage limit. Following this method, the author has found the transition pressure of a yellow, residual clay to be 171 kg./cm.^{2} (cube compressive strength 52 kg./cm.^{2}), and that of a blue, marine clay to be 339 kg./cm.^{2} (compressive strength 86 kg./cm.^{2}). Thus the capillary pressure may be a force of enormous intensity. Yet the very existence of this force has never been suspected up to this time.

Swelling of Clay—If the free surface of a layer of plastic or of semi-solid clay is covered with water, the surface tension at once becomes zero and the clay expands. This swelling is identical with the expansion of the clay produced by the removal of an external load (recurrent branches of the hysteresis loops in Fig. 3), and represents elastic expansion of the clay caused by the elimination of the surface tension of the capillary water. Since the increase in bulk means increase of water content, it is apparent that water enters through the free surface and flows into the interior, impelled by the hydrostatic stress difference between the surface (where the pressure is zero, since the surface is covered with water) and the interior (where negative hydrostatic pressure exists).

If, during a compression test, one drains the water completely off the surface and then removes the external load, the volume of the clay remains nevertheless unchanged, because expansion would mean increase of the moisture content and there is no free water available. On the other hand, if the volume remains unchanged, the inward pressure cannot possibly decrease; the surface tension of the capillary water takes the place of the external load. Like a rubber skin, it opposes any tendency to expansion.

Thus all phenomena associated with the cohesion of clays are capable of being explained by the single factor of surface tension. Cohesion is the internal frictional resistance produced by the capillary pressure. As the cause of the capillary pressure—the surface tension of the capillary water—is an external one, merely acting on the surface of the clay, the cohesion due to the capillary pressure may be called the *apparent cohesion*, in opposition to the *true cohesion* produced by initial friction*. As the initial friction was found to amount to not more than about 20 g./cm.^{2}, the true cohesion is very small compared with the apparent cohesion.

The shearing strength of a mass of clay is equal to the product of the capillary pressure and the coefficient of internal resistance. This relation, however, proves to be valid for the plastic state only, while in the semi-solid and solid states the shearing strength is smaller, just as the shearing strength of solid bodies is very much smaller than the product of the intrinsic pressure times the coefficient of internal resistance. The latter phenomenon seems to be due to important secondary stresses (unequal internal stress-distribution), and it is on the point of being thoroughly studied by various English investigators. (A. A. *Griffith*[1] Prof. B. P. *Haigh*[2] and others.)

Properties of Capillary Water—Since shrinkage is

NOTE—The term "initial friction" as used by the author denotes the shearing strength of clay when not under pressure, either external or capillary. Extended investigation of initial friction showed it to be far too small in amount to account for any of the properties commonly grouped under the general term "cohesion of clay." The internal friction which acts in the interior of a clay subjected to either load or capillary pressure is the sum of (1) initial friction, which is practically independent of the intensity of the pressure, and (2) the frictional resistance set up by the pressure, which is proportional to the intensity of the pressure. Even at small pressures, however, the first item is almost negligible compared to the second.

due to the surface tension of the capillary water, it ought to be possible to calculate the intensity of the transition pressure from the average width of the voids of the clay. As the specific gravity of water is unity, the maximum capillary pressure in grams per square centimeter is equal to the maximum capillary height h in centimeters. The data obtained by a wet mechanical analysis allowed the author to estimate for the yellow residual clay mentioned above a maximum capillary height of 306 m., or 30,600 cm. This value corresponds to the width the voids would have at the shrinkage limit. Hence the intensity of the transition pressure ought to be of the order 30,600 g. or 30.6 kg./cm.². But, as previously noted, determination of the transition pressure for this clay gave the value 171 kg./cm.². The cause of the obvious disagreement between the two figures mentioned resides in the fact that water enclosed in voids of width less than about 0.1 micron ($= 0.1\mu = 0.0001$ mm.) has properties different from those of water contained in wider vessels. The viscosity of the water as well as the rate of evaporation rapidly increases below this limit, as has been proved by permeability measurements.

A very important consequence hereof is that the capillary water of clays is but partially evaporable. If the capillary water contained in clays could completely evaporate, as does the capillary water of sands, clay upon drying would lose its cohesion. The cohesion of clay dried at normal temperatures is however almost twice as great as the cohesion at the shrinkage limit. As the cohesion of a reversible colloid is due wholly to the surface tension of the capillary water, it follows that part of the capillary water remains in the voids of air-dried clay forever without evaporating.

According to the definition of the shrinkage limit, the width of the voids of the clay remains constant below the shrinkage limit, although the moisture content continues to decrease. Evidently the surface of the capillary water retreats from the outer surface of the clay into the narrowest parts of the capillary system. This process decreases the total length of the contour line along which the surface tension acts; nevertheless the cohesion continues to increase. Hence we must conclude that the surface tension per unit of length of the contour line increases just as does the viscosity as the size of these very small capillary channels decreases. This conclusion confirms what has been assumed concerning the cause of the disagreement between the actual and the theoretical value of the capillary pressure.

References.

[1]"The Phenomena of Rupture and Flow in Solids," A. A. Griffith, *Phil. Trans.* Royal Society (London), 1920, pp. 163-198.

[2]"Theory of Rupture in Fatigue," B. P. Haigh, Proceedings of the International Congress of Applied Mechanics, Delft, April, 1924.

[3]"Die Viskosität des Wassers in engen Durchgangsquerschnitten," K. Terzaghi, *Zeitschrift für angewandte Mathematik und Mechanik*, 1924.

Principles of Soil Mechanics: II—Compressive Strength of Clay

Modulus of Elasticity Determined from Cube Tests at Different Moisture Contents Shows Constant Ratio to Capillary Pressure—Poisson's Ratio for Clay—Analogies with Metals—The Iowa Experiments

By Dr. Charles Terzaghi

Professor of Civil Engineering, American Robert College, Constantinople, Turkey
Temporary Lecturer at Massachusetts Institute of Technology, Cambridge

FOLLOWING the tests of confined samples of clay reported in the preceding article (*Engineering News-Record,* Nov. 5, 1925, p. 742), cubes of clay were tested by methods the same as used for ordinary solids. These tests brought somewhat different phenomena into view. In particular they threw light on resaturation action, on lateral effects, and on the analogies between clay and solids. The careful attention paid to the elastic properties of clay cubes was warranted by the fact that the behavior of clay cubes under load was found to be closely related to what is called the bearing capacity of clay deposits. As a consequence a clear understanding of the effect of load on clay cubes represents the key for understanding the more complicated effect of a load placed on the surface of a clay deposit.

Elastic Behavior of Clay

Cubes 2 cm. and 4 cm. on a side were used. They were molded within a lining of filter paper in a prismatic mold fitted with a plunger. According to the purpose of the test, the cube was either tested at once after molding or it was allowed to dry until the moisture content decreased to that desired. During the test the cube was surrounded by a tin cylinder inclosing an inner one of brass wire mesh, with watersoaked cotton between, to maintain an atmosphere saturated with water vapor. In such an inclosure the moisture content of the clay remained sensibly constant for several days, while a test lasted only two hours.

To measure the compressive shortening of the cubes measurements of great precision were required, and the accuracy of a micrometer screw with electric contact indicator proved to be insufficient. The difficulty was overcome by developing an interference contact indicator, consisting of a pair of thin glass sheets with a residual film of water between, a very sensitive micrometer screw, and an eyepiece. By noting the change in color of the Newtonian rings due to contact between the upper glass sheet (thickness 0.1 to 0.2 mm.) and the point of the micrometer screw, the position of the screw point can be determined to less than 0.00001 mm., which is far beyond the degree of accuracy of the micrometer screw.

Cube Test Results—The main diagrams in Fig. 1 may serve as an example of the results obtained by the tests. They represent the stress-strain diagrams for a set of cubes of yellow residual pottery clay with different percentages of moisture. The diagrams are very similar to compression diagrams of concrete and natural stone. Abscissas represent pressures (kg. / cm.^{2}), while ordinates give reduced strain, i.e., the ratio between the total compression and the reduced height—the height which the cubes would have if the volume of voids were reduced to zero at constant horizontal cross-section. Operating with reduced instead of ordinary strains is necessary in order to have a common basis for comparing with each other the strains produced by loads in cubes of different moisture content.

Strain Lag—In testing Cubes 1 and 2 the load was first gradually increased and then left constant for a certain time. During this time the strain increased, at decreasing rate (strain-time curves, under the stress-strain diagrams). This change of strain at constant load is unlike the change of moisture content at constant load in the tests previously described, as the latter process is due to the great resistance against the flow of water through the voids of the clay, while the former one goes on at constant moisture content, i.e., with the capillary water remaining in a state of equilibrium. The increase of the strain of clay cubes at a constant load seems to be identical with the elastic after-effects associated with the deformation of solid elastic bodies. The author has found that in clay it is due to the gradual compensation of unbalanced frictional resistances which at first take up a part of the load. Simple mathematical relations exist between the capillary pressure acting within the cube, the load and the value of the coefficients which determine the relation between the time and the increase of the strain.[1]

Cubes 3 and 4 in Fig. 1 were tested on a screw testing machine, in which the compression (strain) remains constant when the machine is stopped, while the pressure may change. As the lower diagrams show, when the compression is thus held constant the pressure decreases for a time. This phenomenon was found to be due to the same cause as the increase of strain at constant load, and it follows similar simple laws.

The more slowly the load was applied, the less marked were the elastic after-effects, but at the same time the steeper was the slope of the main branch of the stress-strain diagram. The main branch, which corresponds to infinitely slow application of the load, is called the reduced stress-strain curve (dash-line curves in Fig. 1). It can easily be constructed by means of the data furnished by the time-strain or the time-pressure curves.

Cyclic Loading—The next operation performed during the cube tests consisted in reducing the load to zero and re-applying it (complete cycle). The effect is represented by the hysteresis-loops. These loops differ from those shown in the pressure-moisture diagrams previously described by being straight instead of curved. When the load again reaches the value it had before release, the stress-strain curve asymptotically approaches the curve of continuous increase of load. Phenomena similar to these came in evidence when plotting the results of loading tests performed on the bottom of test pits in the field.

Modulus of Elasticity—The term "modulus of elasticity" should be confined to the reversible part of the deformation of the cube, represented by the hysteresis loops. It is equal to the tangent of the angle between the strain axis and the axis of the hysteresis loop. In general this angle differs but very little from that of the initial tangent to the continuous-loading curve.

A very important relation was found to exist between the intensity of the capillary pressure and the value of the modulus of elasticity. The abscissas of the main branches of the pressure-moisture curves (see Fig. 3, p. 743, *Engineering News-Record,* Nov. 5, 1925) indicate the pressures required to reduce the moisture content of confined clay from its initial value (for zero pressure) down to the moisture content represented by the ordinates, provided surface tension is not active. It has been shown that the capillary pressure acting within a clay of definite moisture content is equal to the external pressure which in the pressure-moisture diagram corresponds to that moisture content. Using these pressure equivalents of the moisture content the curves in Fig. 2 were plotted. Here abscissas represent capillary pressures and ordinates represent moduli of elasticity. The values for Cubes 1, 2, 3 and a fourth cube intermediate in moisture content between 2 and 3 are plotted on the line marked "Yellow Pottery Clay." As the points mark out a straight line, it appears that the ratio between modulus of elasticity and capillary pressure of a given kind of clay is constant and independent of moisture content, provided that no resaturation of the sample has occurred. The lower line is the corresponding curve for a blue marine clay and the two upper lines for two sands. The simple relation which exists between the capillary pressure and the modulus of elasticity is of considerable practical importance inasmuch as it was found that the bearing capacity of a clay increases in simple proportion with its modulus of elasticity.

Analogy of Clay and Solids—Cohesion of clay depends upon the capillary pressure. In a similar way the cohesion of solid bodies is the result of the "intrinsic pressure," i.e., of the pressure per unit of area produced by the mutual attraction of the molecules. As the elastic properties of the clay cubes are very similar to those of solid cubes, the author suspected that the ratios between the moduli of elasticity of solids and their intrinsic pressures might also be a constant. This conclusion was found to be correct. In the upper diagram in Fig. 2, abscissas represent the intrinsic pressures of various metals (according to Traube, computed from thermodynamic data), while ordinates are moduli of elasticity for compression within rigid enclosures, computed from the ordinary moduli of elasticity and the

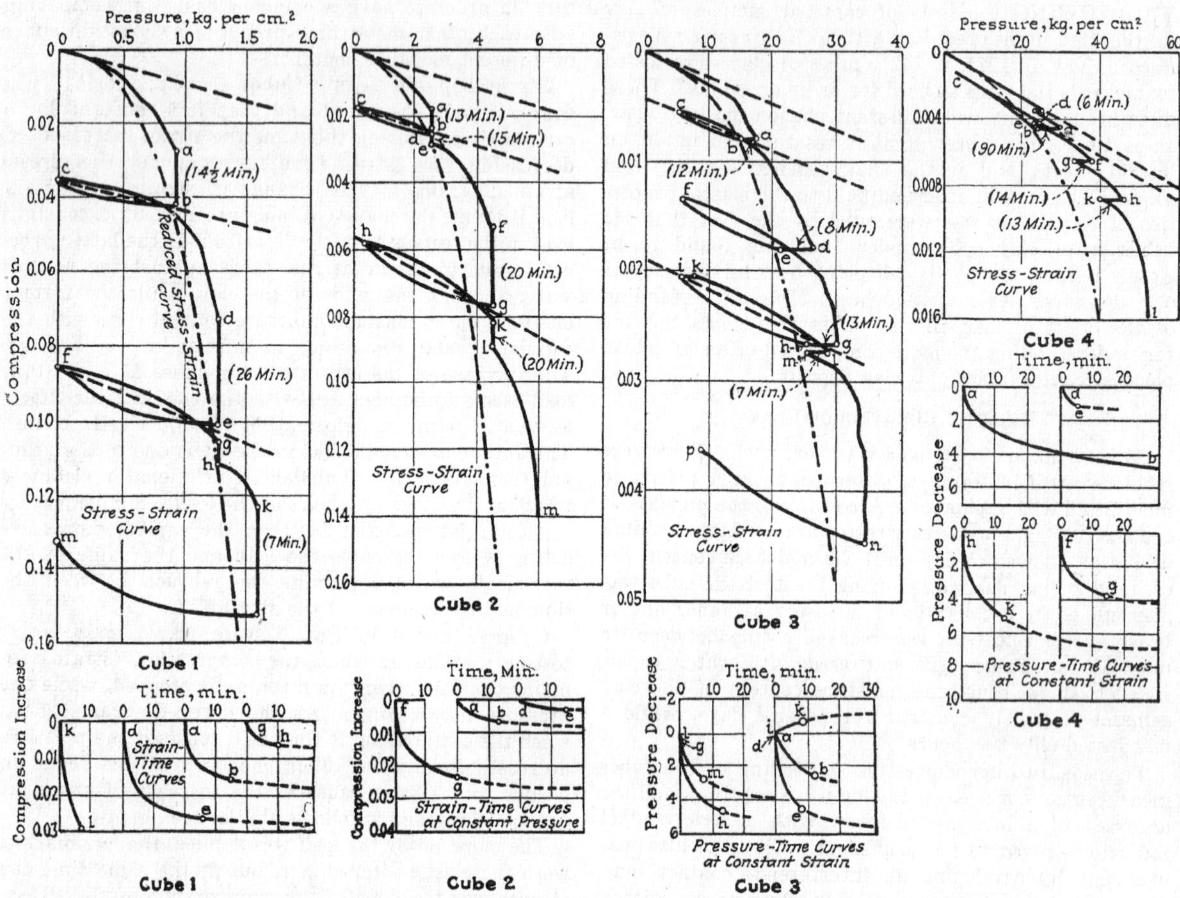

FIG. 1—COMPRESSION TESTS OF CLAY CUBES

Yellow residual pottery clay, specific gravity of solid matter 2.93, lower limit of plastic state 24.2, lower limit of liquid state 58.0 per cent, coefficient of plasticity 33.8.

	Cube 1	Cube 2	Cube 3	Cube 4
Moisture-Ratio e, (per cent of volume of solid matter)	0.792	0.681	0.490	0.482
Modulus of Elasticity (kg. per cm².)	115 to 76	310 to 195	3,760 to 3,460	7,300

Poisson constants. All the points are located along a straight line.

If the moduli of elasticity were perfectly constant for each clay cube, the hysteresis loops ought to be strictly parallel to each other, while in the diagrams Fig. 1 the slopes of the loops somewhat increase with the intensity of the pressure at which the cycle was started. In order to understand this we must keep in mind the fact that in case of a solid the intrinsic pressure remains constant during a compression test, while the volume decreases. It would be possible to maintain the volume constant by reducing the intrinsic pressure as the load is increased. In the case of a clay cube the volume cannot possibly change as long as the moisture content is unchanged, and therefore the capillary pressure decreases with increasing load, which in turn causes a decrease of the modulus of elasticity and a corresponding increase of the slope of the hysteresis loops. As a matter of fact, after having determined Poisson's ratio for several clays, the author was able to compute the relative position of their hysteresis loops, and the results checked fairly closely with the results of tests.

Poisson's Ratio; Resaturation—Poisson's ratio represents the ratio between the linear compression and the corresponding linear lateral expansion for a loaded cube. If one knows this ratio for any homogeneous material one can calculate the lateral pressure which the material will exert under load against a rigid enclosure. In turn, knowing the lateral pressure, one can compute Poisson's ratio. This method was used for the clays.

In order to find the lateral pressure exerted by a loaded layer of clay (mixture of clay and water, free from air and the clay surfaces covered with water so as to exclude the surface tension of the water), two sets of apparatus were used similar to the one represented by Fig. 2 (p. 743, *Engineering News-Record*, Nov. 5, 1925). The rings were larger and higher, and the clay was enclosed between two filters of equal dimensions and quality. At half the height of each layer of clay there was a horizontal steel tape, which passed out through slits in the enclosing ring. In one of the layers the flat side of the tape was horizontal, in the other one it was vertical. In order to keep the clay from being squeezed out through the slits, each slit was sealed on the inside by a small shield of thick filter paper. The ratio K between the horizontal and the vertical pressure acting within the clay is equal to the ratio between the forces required to overcome the friction between the clay and the two tapes. The frictional resistances were not measured until three or more days after the load was applied, to allow the hydrostatic stress differences to disappear. Computation from the results furnished the values of Poisson's ratio:

Yellow residual clay (K = 0.70)	2.73
Blue marine clay (K = 0.75)	2.55
(Water	2.00)
(Lead	2.24)
(Silver	2.63)

If both Poisson's ratio and the ratio between modulus of elasticity and capillary pressure are known, one can compute the reversible part of the deformation produced by a load on a layer of clay confined within a rigid ring. This reversible part is represented by the curved hysteresis loops of the pressure-moisture diagram (Fig. 3, p. 743, *Engineering News-Record*, Nov. 5, 1925). The computation was carried out for different kinds of clays. In every case the curve thus obtained coincided very nearly with the recurrent branches of the hysteresis loops of the corresponding pressure-moisture diagrams. This furnished a check on the computation of Poisson's ratio.

The recurrent branches of the hysteresis loops of the pressure-moisture-content diagrams are called the *resaturation curves*. They are simple logarithmic lines. The greater the resaturation coefficient, the greater is the increase in volume due to resaturation and the smaller is the ratio between modulus of elasticity and capillary pressure. The coefficient is greatest for clays rich in colloids. Experience seems to show that a clay is the more likely to slide the more it swells when brought into contact with water. Hence the value of the coefficient of resaturation may furnish a valuable indication of the stability of a natural clay deposit.

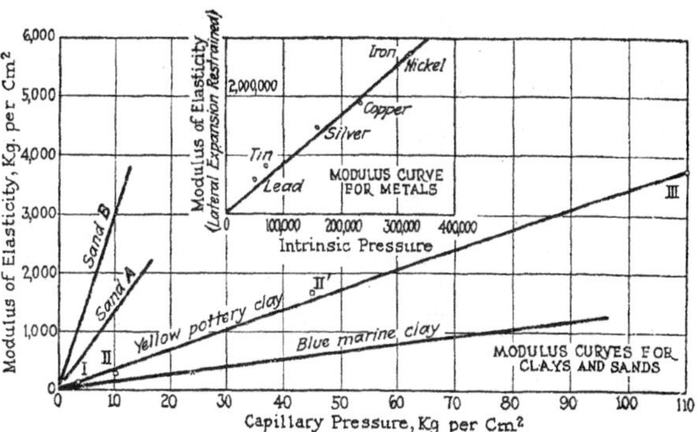

FIG. 2—RELATION BETWEEN MODULUS OF ELASTICITY AND CAPILLARY PRESSURE

Non-homogeneous Clay Soils—In 1921 and 1922 the engineering experiment station of Iowa State College determined the compressive, tensile and shearing strength of different kinds of clay exposed on the bottom of an old gravel pit[2]. The samples were taken from a depth of 1.5 to 1.8 m. below the bottom of the pit and were tested in undisturbed condition. The stress-strain diagrams seem to resemble closely those presented in Fig 1. Nevertheless, a thorough study of these diagrams leads to the conclusion that the two sets of tests differ in vital respects.

For the Iowa cubes the reversible part of the deformation is very small compared with the total deformation, and the high moduli of elasticity of the cubes seem quite out of proportion to their low compressive strengths. Calculation indicates unusually high coefficients of internal friction. One of the causes of these abnormalities seems to reside in the fact that the Iowa clay apparently had undergone repeated resaturation and shrinkage since its deposition, while the cubes of Fig. 1 were compacted under the influence of direct compression by capillary pressure. According to our experience, resaturation substantially affects the relation which exists between the moisture content and the capillary pressure. However, up to the present time, no exhaustive investigation of the effect of resaturation has yet been made.

The second and equally important cause of difference lies in the fact that the cubes of Fig. 1 were homogeneous, while the Iowa cubes were not. Thus, the paper states, "The yellow clay . . . was not infrequently permeated by roots, wormholes and fibers or small crevices, and the strength would be correspondingly decreased." "One of the prominent factors appeared to be the friability or brittleness of the blue clay, as induced by the infiltration of water and air previously mentioned, which oxidizes the ferrous matter and causes the deposition of a brownish-yellow film along the planes of separation which honeycombed the structure. A slight shock as a result of this film or matrix was often sufficient to leave the blue clay a coarse granular mass" (p. 563). Clays which include planes of separation and seams of oxidized matter cannot be considered homogeneous; they represent an intermediate type between homogeneous clay and the crumb soils. In a homogeneous clay, the hydrostatic pressure of the capillary water is the same throughout the mass (provided the clay is in hydrostatic equilibrium); in a soil composed of crumbs the particles of each crumb are bound together by a very intense capillary pressure, while the crumbs themselves are as independent as are the individual grains of a mass of sand. As a matter of fact, accumulations of dry crumbs prove to have the same elastic properties as dry sands except for the fact that they are more compressible and less elastic than sands, on account of the brittleness of the crumbs. If water penetrates the voids of such a crumb mass, the crumbs absorb the water very slowly, because their permeability is small. Water flows from the voids between the crumbs towards the centers of important negative hydrostatic pressure, i. e. penetrates the crumbs. However, even after equilibrium is reached the soil represents a honeycombed mass which can by no means be compared to a homogeneous clay and if the water later on evaporates, the mass breaks up into crumbs again, on account of the non-uniform resistance of the material against the effect of secondary stresses.

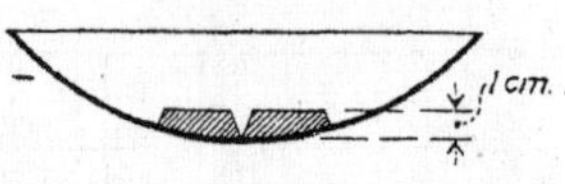

FIG. 3—DETERMINING LIMIT OF LIQUID STATE (ATTERBERG)

Due to these facts there is a great difference between the elastic properties of crumb soils and of homogeneous clays. The former still need thorough experimental investigation. Fortunately the worst types of foundations (mud deposits, soft clays and the like) belong almost without exception to the homogeneous type of clay, whose physical properties are now known.

It is to be regretted that the Iowa investigations could not be supplemented by contemporaneous loading tests performed with the soil-testing apparatus proposed by the Foundation Committee.[3] The results would have furnished a most valuable contribution to our knowledge of the resistance of clays, because the relation which exists between the cube strength of a resaturated clay and the bearing capacity of the same material is not yet known.

Data Required for Describing Clays—Engineering literature contains hardly a single description of a clay that would allow us to identify the clay with those from other localities, or a description which gives us a clear notion of what the material was like. At best the describer contents himself with mentioning the color of the clay, its moisture content and its chemical composition. Experience, however, has shown that the chemical composition of a clay has but little to do with its physical properties, and, as for moisture content, some clays with 25 per cent of water are almost liquid while others containing 30 per cent are very stiff.

An exhaustive investigation of test boring samples requires too much time and labor to be justified except for scientific purposes. Study of the problem of investigating the properties of clays for *engineering purposes* has led the author to the conclusion that it would be advisable to modify the test program according to the character of the work affected. A detailed account of the data required for describing clays extracted from the sites of future floating or pile foundations, clays for dam construction purposes, clays which threaten to slide, etc., has already been published (in the author's book "Erdbaumechanik," Vienna, 1925). Some data however are required regardless of the special purpose of the description. These data concern the relation between consistency and moisture content.

The simplest and most reliable way of expressing this relation was devised by the late Prof. A. Atterberg, of Kalmar (Sweden), for agrogeological purposes, but it suits engineering purposes as well.[4] It requires a knowledge of the following data: Moisture content of the clay, specific gravity of the dry matter, lower limit of the plastic state and lower limit of the liquid state.

Moisture content is determined in the ordinary way. The sample is taken out of the test-pit or out of the well-boring auger and transferred into a wide-necked bottle with glass stopper. The bottle is filled completely so that there remains no air between the clay and the stopper, and the joint between the neck and the stopper is sealed with wax. In the laboratory a sample of the content is placed between two watchglasses and weighed, dried at 100 deg. C. and again weighed.

The specific gravity of the dry matter has to be determined by pycnometer. In order to get fairly accurate results the clay solution should be heated to the boiling point to drive out the adsorbed air. The test is made after the solution has cooled.

The lower limit of the plastic state is determined as follows: Mix a sample of the clay with water until it becomes very plastic or, if it is too soft in its natural state, let it dry until it reaches the consistency required. Then work the sample into several thin threads (diameter about 3 mm.), put the threads together and work them out to threads again by rolling them with the palm on a smooth, clean sheet of paper. Repeat the process until no more threads can be formed, the material becoming brittle and the threads breaking to pieces while worked. Then the moisture content is determined; this value, expressed in per cent of the weight of the dried sample, was called by Atterberg the *lower limit of the plastic state*.

In order to determine the lower limit of the liquid state, take a flat porcelain cup (Fig. 3), mix a sample of the clay with water until it becomes very soft, make out of the mixture a cake about 4 cm. in diameter and 0.8 cm. thick, cut this cake with a nickel scoop into two equal parts and shake the cup. If the two lower edges of the cut do not flow together, add some water and repeat the test. The moisture content at which the lower rims join along a strip of a height of about 1 mm. is called the *lower limit of the liquid state.* Both limits should be determined at least twice.

The difference between the two limits is called the

coefficient of plasticity. A clay is plastic between the two limits, else it is either liquid or semi-solid. The greater the coefficient of plasticity, the more plastic the clay is supposed to be. For materials without any plasticity (typical quicksands or very fine quartz dust) the coefficient is zero, i.e. the two limits are identical. For equal coefficients of plasticity the limit of plasticity may be low or high, depending on the shape of the grains and the percentage of humus constituents present.

The degree of plasticity of two clays with equal coefficients of plasticity may be equal or different according to whether the specific gravities of their grains are equal or different. In order to establish a common basis for comparison, the author expresses the limiting moisture contents in per cent of the space occupied by the solid matter. Care should be taken to use samples which have not previously been dried, as the very great capillary pressures which develop during the process of shrinkage may crush many of the grains and decrease the average size of grains.

The data mentioned represent the minimum requirement for characterizing a clay. Without containing these data, the description of a clay is practically worthless and the recorded phenomena are merely curiosities.

Summary—From the experimental studies described in this and the precding article we have derived facts which may be summarized as follows:

The cohesion of clay is due to two factors. One of these is the pressure exerted by the surface tension of the capillary water, a force whose intensity exceeds all the other forces the earthwork engineer has to deal with. It may amount to several hundred atmospheres: it compacts loose, colloidal sediments more thoroughly than can be done by artificial means except in the laboratory by using a high-power testing machine. Swelling of clay is nothing more or less than the purely elastic expansion produced by the elimination of the surface tension of the capillary water. Local evaporation of the capillary water or local flooding of the surface of clay deposits produces secondary stresses the intensity of which is far greater than the weight of the heaviest structures and which were found to be the primary cause of many vast soil displacements, known as *earth slips*.

The second one of the factors mentioned consists in the fact that the properties of the water contained in voids of width less than 0.0001 mm. are no longer identical with those of ordinary water. In such voids, viscosity and surface tension are increased (in inverse proportion to the diameter of the voids), and the water loses its ability to evaporate in contact with the air. Thus the capillary water of the clays is to a certain degree solidified by the influence of the forces exerted by the molecules of the solid matter. Due to this fact the capillary pressure assumes far greater values than it would if the surface tension of the capillary water had its normal value.

Capillary pressure plays the same part in the physics of clays as does intrinsic pressure in the physics of solids. Therefore the elastic properties of the clays are qualitatively identical with those of granular solids (rocks, concrete, etc.).

The minimum requirement for describing a clay consists in presenting the following data: Water content, specific gravity of the solid matter, lower limit of the plastic and the lower limit of the liquid state of the clay.

References.

[1]"Die Beziehungen zwischen Elastizität und Innendruck," K. Terzaghi, *Sitzungsberichte der Wiener Akademie der Wissenschaften*, 1923.

[2]"Progress report of tests on undisturbed clays," J. W. Griffith, *Proceedings, Am. Soc. C. E.*, March, 1922, pp. 557-579.

[3]*Proceedings*, Am. Soc. C. E., August 1920, Plate XI and XII; January, 1922, Plate VII.

[4]"Die Plastizität der Tone," A. Atterberg, *Internationale Mitteilungen für Bodenkunde*, 1911. Atterberg's method of measuring plasticity was severely criticized by W. E. Emley five years ago (Technologic Paper 169, U. S. Bureau of Standards, "Measurement of Plasticity of Mortars and Plasters"). The arguments advanced by Mr. Emley are justified so far as the plasticity of mortars and plasters for industrial uses is concerned. But they do not seem to affect the value of the method for purposes of soil study.

Principles of Soil Mechanics: III—Determination of Permeability of Clay

Validity of Darcy's Law—Slichter's Formula and Hazen's Observations—Permeability of Sand—Experiments on Clay—Darcy's Law Valid Even for Semi-Solid State

BY DR. CHARLES TERZAGHI

Professor of Civil Engineering, American Robert College, Constantinople, Turkey
Temporary Lecturer at Massachusetts Institute of Technology, Cambridge

WHEN water percolates through a permeable body of prismatic form in parallel flow, its rate of flow is expressed by Darcy's formula, $Q = kFs$, where Q is the volume of flow per second, F is the cross-section of the body, and s is the hydraulic gradient, equal to loss of head divided by length, or h/l. The coefficient k (dimensions cm./sec.) has been called the *coefficient of permeability*, and represents the velocity of percolation under a hydraulic gradient of unity.

The law is valid for laminar flow only, and therefore does not apply to the (turbulent) flow of water through coarse gravel. Some have claimed also that it is not valid for flow through clay. But it is generally accepted as applicable to flow through fine and medium sand under moderate head, as it agrees fairly well with experimental data.

Void-Ratio and Permeability—The voids of a permeable body constitute in effect a system of capillary tubes. Percolation velocity depends on the size of the tubes. A sand of definite grain size may have wide or narrow voids, according to whether the structure is loose or dense, and the coefficient of permeability thus depends on density of structure as well as on grain size.

Many years ago Prof. C. S. Slichter[1] made an attempt to compute the coefficient of permeability of sand, assuming uniform globular grains; according to the arrangement of the spheres, the volume of voids ranges between limits of 26.0 and 47.6 per cent of the total volume of the mass. His formula was $k = 771 d^2/c$, where d is the diameter of the spheres in centimeters and the factor c has for a temperature of 10 deg. C. values as follows:

$$\text{Volume of Voids} = \frac{\text{Voids}}{\text{Total Volume}} =$$

$$\text{Void-Ratio} = \frac{\text{Voids}}{\text{Solid Volume}} =$$

n,	0.26	0.28	0.30	0.34	0.38	0.42	0.46
e,	0.352	0.388	0.428	0.515	0.612	0.723	0.850
Factor c,	84.30	65.90	52.50	34.70	24.10	17.30	12.80

Thus, the coefficient of permeability of a mass with a volume of voids of 46 per cent (very loose arrangement of the spheres) becomes $60.3 d^2$ at 10 deg. C. But Allen Hazen has found by experiment that the coefficient of permeability of a very clean and loose sand amounts to $116 d_w^2$ (d_w is effective size of the grains according to his well-known definition), and, if the grains are of approximately the same size, even $150 d^2$.

The disagreement between the theoretical value obtained by the formula of Slichter and Hazen's experimental results is due to the fact that the voids of Slichter's filter are supposed to be equal among themselves, while in practice each filter contains wide and narrow voids. It can be proved that at equal volume of voids and equal size of grains the coefficient of permeability decreases with increasing uniformity of voids. On the other hand, the relation which exists for any one kind of sand between the density and coefficient of permeability agrees fairly well with Slichter's theory.

In order to represent the theoretical relation, the void-permeability curve S in Fig. 1 has been plotted. It corresponds to a mass consisting of microscopically small spheres of equal diameters. The upper and lower ends of the middle section of the curve (void-volume 47.6 per cent, void-ratio 0.905; void-volume 26 per cent, void-ratio 0.352) indicate the extreme limits within which Slichter's formula is valid. The geometrical part of the theory of Slichter has been further developed by L. Darapsky and L. Burmester[2]; but, as the shape of the grains of natural sands is irregular rather than globular, elaborate theoretical investigations of such a kind are at best of academic value.

Permeability of Sand—As Slichter's formula for the permeability coefficient is rather complicated, I derived a semi-empirical formula based on the following facts: The widest parts of the capillary channels through which the water flows have at least five times the cross-section of the narrowest ones. Hence, if a definite quantity of water percolates through one of the capillary channels, the loss of head per unit of length of the narrowest sections of the channel is at least 25 times greater than the loss per unit of length of the widest ones. Due to this, the percolation of water through sand can be compared to the flow of water through a set of sieves in series: the resistance to percolation is confined to the sieves while in the spaces between sieves the resistance is negligible. Let d_w be the effective size of the grains (cm.), n the void-volume, v_o and v_t the coefficients of viscosity of the water at 10 deg. C. and at temperature t respectively, and C an empirical coefficient found by experiment to range from $800\,v_o$ to $460\,v_o$ depending both on the shape of the grains and on the uniformity of the sand; then

$$k = \left(\frac{C}{v_o}\right)\left(\frac{v_o}{v_t}\right)\left(\frac{n - 0.13}{\sqrt[3]{1-n}}\right)^2 d^2$$

$$= (800 \text{ to } 460)\,\frac{v_o}{v_t}\left(\frac{n - 0.13}{\sqrt[3]{1-n}}\right)^2 d^2$$

The value 800 was derived from tests on sands whose grains were well polished and rounded, while the value 460 was from tests with sands of irregular, rough grains. The influence of the uniformity of the sand upon the value of C was far less marked than the influence of the quality of the grains. Within the limits of void-ratio 0.352 and 0.905 (void-volume 26 0 to 47.6 per cent) the curve of permeability on void-ratio plotted by means of the formula just given coincides closely with the corresponding curve plotted from Slichter's formula (Fig. 1), *provided* they are drawn so as to have one point in common. It may surprise the reader to learn that two apparently very different methods for calculating the relation between the volume of voids and the coefficient of permeability furnish similar results. But it has been found by experience that any simplifying assumptions concerning the shape and the arrangement of the capillary openings lead to results agreeing fairly well with each other and with the observations, provided the nature of the material has been taken into consideration appropriately. This fact considerably facilitates the interpretation of the results of tests.

Basing calculations on the formula, one can reduce any coefficient of permeability to a volume of voids of 50 per cent (void-ratio 1.0) and to a normal temperature of 10 deg. C., regardless of the actual density of structure of the sample. The coefficient of permeability thus obtained is called the *reduced coefficient of permeability* k_r. By introducing into the formula the special values $n = 0.5$ and $v_o = v_t$, one arrives at $k_r = (174$ to $100)\ d^2_w$. As the value k_r is independent of both temperature and voids, it is a value which may serve as a basis for investigating the influence of the quality of the grains and of the uniformity of the mass upon the permeability of the material. The values 174 and 100 seem thus far to represent the extreme limits for these influences, provided the sands are perfectly clean, without any traces of clay. Even a very small quantity of clay was found to upset the validity of the formula. For checking the formula obtained by theory, and for determining the value of the empirical coefficients, the relation between void-ratio and coefficient of permeability had to be investigated by experiment.

Tests of Permeability of Sand—In order to get accurate results care must be taken that neither the sand nor the water system contains air. I use a cylindrical filter (Fig. 2), the sand resting on a diaphragm of brass wire mesh. Provision is made for sending water through the sand either upward or downward. Before the sand is introduced the water is allowed to rise above the wire mesh, and bubbles adhering to the mesh are removed by exhaustion. The sand is thoroughly wetted before it is placed in the cylinder. If, during the test the water flows upward, the rim of the cylinder serves as overflow. However, most of the tests were made with the water percolating downward. The loss of head is measured by a graduated standpipe, communicating with the space below the wire mesh. The data required for making a complete report on the results of the test are surface area and thickness of the layer, method of preparing the layer (sand loosely poured in, stirred, compacted) volume of voids, average specific gravity of the grains, temperature, sieve analysis, effective size of grains, shape of grains, values of percolation and loss of head. Basing on the results of numerous tests, the formula expressing the relation between void-ratio and permeability has been checked and verified.

Permeability of Clay—If the grains of clay differed from those of sand only in size, our formula could be used for computing the coefficient of permeability of both pure sand and pure clay. But it was found that the finer constituents of clay consist of very thin, flexible mineral scales, while the grains of sand are bulky and rigid. The capillary channels of sand accordingly have a rather compact cross-section, while those of clay resemble narrow slits. This difference has an important influence on the mechanics of the percolation process. A theoretical study of the flow of water through powders with scale-like grains[3] led to the formula

$$k = \left(\frac{C}{v_o}\right)\left(\frac{v_o}{v_t}\right)(e - 0.15)^3(1 + e)\,d^2$$

where e is the void-ratio, or ratio of volume of voids to volume of solid matter. Considering the verified accuracy of the formula for sand, the present formula might also be expected to be adequate. However, before empirical data for verifying the formula could be obtained, it was necessary to develop a method for determining experimentally the coefficient of permeability of clay.

Clay Tests—Measuring the permeability of clay is more difficult. The first systematic attempts seem to have been made by American engineers (testing materials for the Lahontan dam, Cold Springs dam, Palouse project[4]). In most of these tests the material was put into a cylindrical vessel and compacted by tamping. The height of the layer was usually about 100 cm. Below the layer of soil there was a filter. The water percolated vertically downward. In principal the apparatus was identical with that shown in Fig. 2, and it would fully serve for determining the coefficient of sand. But when applied to clay the method is open to serious objection.

Experience has shown that the coefficient of permeability of clay varies within extraordinarily wide limits, according to the volume of voids. As soon as a soil mass composed of crumbs of clay comes into contact with water, the crumbs swell and the volume of voids increases. Within the test cylinder this swelling is restrained to an unknown extent by friction along the sides of the vessel and it is doubtful if the volume of voids of the sample is identical with that of the soil in the dam. Furthermore, the samples unavoidably contain much air. Even if one shakes a sample of powdered clay with water in a bottle, a considerable quantity of air continues to stick to the clay particles and one cannot drive it out except by boiling the solution. But in the dam the air contained in the soil is gradually carried away by the percolating water in solution, while the laboratory test never lasts long enough even partially to exhaust the air out of a sample 100 cm. thick. Last but not least, one must consider that a crumb soil, as used by the investigators, contains voids much larger than the coarsest individual particles of the soil. Water percolating through such a mass carries in suspension the fine particles detached from the crumbs located near the upper surface of the sample, and the formation of a "filter skin" becomes inevitable. As soon as a filter skin is formed, no more conclusions can be drawn from the results of the tests regarding the permeability of the samples. It is very doubtful whether the coefficient of permeability of clay crumb soils can be determined at all except by indirect experimental methods.

In order to exclude these sources of error I constructed the apparatus shown in Fig. 3. Air can easily be excluded from the interior of the apparatus; the diameter of the clay sample is much greater than its thickness, hence the friction along the sides of the sample has practically no influence on the volume of voids; and the moisture content of the sample is maintained constant by keeping the sample under a constant pressure.

The apparatus consists of a vessel in which an internal rim rests on a perforated bronze plate, whose upper side is covered with a fine brass wire mesh. The plate supports a sand filter covered with a circular sheet of thin filter paper. In order to perform the test, one fills the vessel with water, introduces the bronze plate, removes the air bubbles from the wire mesh, fills the space above the mesh with clean quartz sand (size of grains 0.5 mm.), and covers the surface of the filter with filter paper. In the meantime the ring shown seated on the vessel has been partially filled with a plastic, homogeneous mixture of water and clay, about 2.5 cm. thick. This ring is applied to the top of the vessel in such a way that the air may escape when the ring is lowered, and is then clamped down by a gasketed bronze ring. The upper surface of the clay sample is covered with paper and a sand filter. This is loaded by a lever. The water-filled space below the perforated bronze plate communicates with a standpipe 100 cm. high, 0.6 cm. in diameter.

Under the influence of the load on the clay, the moisture content of the sample decreases, until equilibrium is reached. Then one fills the standpipe with water and the water percolates upward through the clay. As a rule more water evaporates in the small annular space around the loading plate than comes up through the clay. Hence it is necessary to keep the water-level in the annular space constant by occasionally adding water. In the standpipe the water-level falls during the test at a decreasing rate. I make readings three times a day and compute the coefficient of permeability from the rate at which the level sinks.

Three tests are run simultaneously, at loads of 60, 125 and 190 kg. (0.75, 1.6 and 2.4 kg./cm.^{2}). As a rule four to six weeks are required for the contents of the standpipe to filter through a fairly fat clay. After each reading, the temperature of the water is read from a thermometer whose bulb is enclosed within a brass pipe (not shown on the drawing) which passes across the vessel below the perforated bronze plate. As changes of temperature have marked effect on the elevation of the water level in the standpipe, the tests must be made in a cellar with a very constant temperature. A fourth simultaneous test is run in an apparatus of somewhat different type, containing a sample at about the lower limit of the liquid state, not loaded, the water filtering through the clay in downward direction from a standpipe over an opening in a bronze plate clamped down on the gasketed upper rim of the cylinder which contains the clay. After the test the clay in each apparatus is sampled and its moisture content determined.

Homogeneous mixtures of clay and water were used only. As the width of the voids of such samples is equal to or less than the diameter of the smaller individual particles, no filter skin can be formed, provided the water used for the test is perfectly clean.

In order to investigate whether a filter skin forms in these tests, parallel tests were run with samples of different thickness. If a filter skin formed, a thin layer should show a smaller coefficient of permeability than a thick one, since the passage of water through the filter skin is not affected by the clay below the skin. But the tests gave the same coefficient of permeability for a thin as for a thick layer. Again, after a test on a liquid-plastic sample, the surface was scraped off and the test was repeated, without any change in result.

In the moisture-permeability curves in Fig. 1, those points whose ordinates were determined by means of the apparatus shown in Fig. 3 are indicated by circles. For moisture contents approaching the lower limit of the plastic state the coefficient of permeability became exceedingly small, so that it could no longer be reliably determined in a direct way. In order to overcome this difficulty and to make it possible to investigate the permeability for the semi-solid state of the clays also, I was obliged to work out an indirect method, which enables the permeability of almost impermeable samples to be measured comparatively rapidly. This method will be described in a subsequent article. It is distinguished by the fact that the water merely circulates in the interior of the sample, without entering or leaving it, since during such a process no filter skin can possibly be formed. The close check between the points *a* and 4, Curve *A*, Fig. 1, furnishes a conclusive additional proof for the results obtained by means of the apparatus Fig. 3 to be reliable.

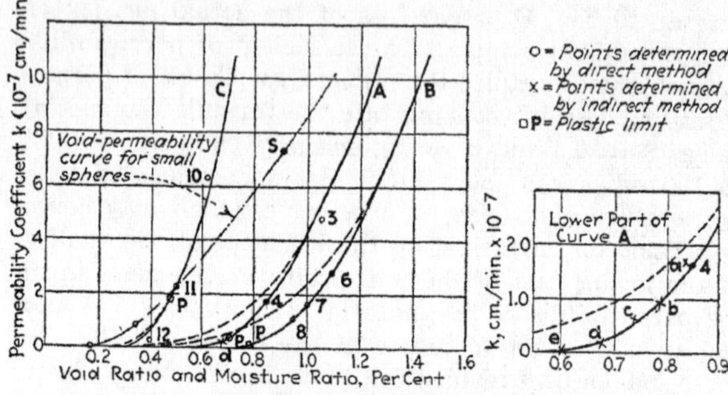

FIG. 1—PERMEABILITY-MOISTURE CURVES FOR THREE CLAYS

Darcy's Law Valid—Tests with the apparatus just described proved that Darcy's law holds for percolation through clay of plastic consistency. In addition it may be inferred from certain of the results that it is also valid for the flow of water through semi-solid clays. Distinct departure from Darcy's law could be noticed only for semi-liquid clays. Reducing the hydraulic grade from 50 down to 10 or 15 caused rapid decrease of the coefficient of permeability, and only at low heads was the coefficient fairly constant. This phenomenon, however, may be explained by the fact that the structure of a semi-liquid clay is honey-combed. The void-ratio of the loosest possible aggregation of equal spheres amounts to 0.91 (volume of voids 47.6 per cent), while for a semi-liquid mass of clay it is approximately 2. The structure of such a mass is stable merely because initial friction* keeps the particles in relative position, and the average width of the voids is far greater than the average diameter of the clay particles. When water percolates through such a mass under a considerable head it produces elastic and non-elastic deformations and grain displacements similar to the deformations produced by a stream of water forced through a system of very expansible rubber tubes, but at lower hydrostatic pressure the elastic deformations disappear and the coefficient of permeability changes accordingly.

Hence we can state that Darcy's law is valid at least for the flow of water through clays with a medium or a low moisture content (plastic or semi-solid consist-

*NOTE—The term initial friction has been used by the author for indicating the shearing strength the clay has, if it does not stand under the influence of any pressure (neither external nor capillary). It amounts to at least 10^{-6} dynes per point of contact between each pair of clay particles.

ency). It remains to be examined whether or not the permeability coefficients agree with the formula given above.

Departure from Theory—Fig. 1 contains the results of three sets of permeability tests. In order to compare these results with the formula, I determined first such values of $\frac{C}{v_o}$ and of d_w as would make the curve fit the test points 1 to 4 (curve A, points 1 and 2 outside of the figure). Then, with the same value of $\frac{C}{v_o}$, I determined two other values for d_w, to fit the respective groups 5 to 8 and 9 to 12 (curves B and C). If the

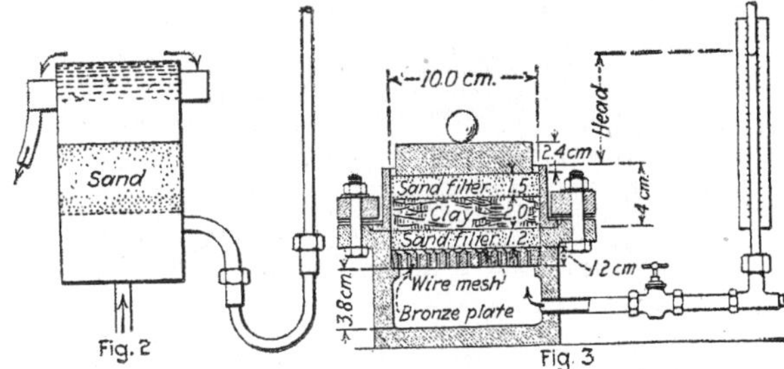

FIGS. 2 AND 3—MEASURING PERMEABILITY OF SAND AND CLAY

Fig. 2—Apparatus for sand. Fig. 3—Apparatus for clay.

formula is correct, the points of each set ought to be close to the corresponding theoretical curve. For medium moisture contents the curves and the points agree fairly well, but for moisture contents approaching the plastic limit the disagreement increases rapidly, and for the semi-solid state the coefficient is only a small fraction of the theoretical value. The plastic limit (limit between the plastic and the semi-solid state) is marked in the figure by the letter P. The anomaly becomes particularly obvious when comparing the relation between the void-ratio and the coefficient of permeability of sands with the results of the clay tests. Increasing the void-ratio of a sand from 0.6 to 0.8 increases the value of k by 75 per cent, whereas the same increase of the void-ratio of the clay increases the coefficient of permeability by several thousand per cent.

For reasons which have already been discussed, it seemed inadmissible to blame such important disagreement merely upon the approximate character of the assumptions. On the other hand there was reason for suspecting that the disagreement is due to a physical factor, one not present in flow of water through sand but appearing only when the voids are very small. If the disagreement were due to error of the formula, it ought to manifest itself whenever the void-ratio drops below some certain value but instead, it is invariably associated with the semi-solid state of a clay, regardless of whether this state (plastic limit) is reached at void-ratio 0.3 or 1.0. Experience has shown that the moisture content of a clay corresponding to the plastic limit increases with increasing fineness of the grains, provided the grains are of similar kind, so that it seemed likely that the plastic limit is reached as soon as the width of the voids drops below a definite critical value, regardless of what the effective size of the grains may be. If that be true, the observed anomaly concerning the permeability too should be in some way related to the width of the voids dropping below a certain critical value.

Change in Viscosity—From this reasoning it seemed that the physical constants of the water might change when the capillary channels of a shrinking clay become reduced to a certain size. This supposition was strengthened by anomalous results obtained in certain evaporation tests.

Prisms made of different clays were allowed to dry at normal temperature (15 to 25 deg. C.), and their loss of weight was determined three times a day, together with the shrinkage of the prisms (distance between gage-points on their surfaces); at the same time the loss of weight of a cylindrical vessel containing distilled water with a wide, free surface was measured. Basing my calculations on the data thus obtained, I computed for each one of the prisms the ratio between the speed of evaporation of the capillary water and the corresponding speed of evaporation of the water contained in the vessel (relative speed of evaporation). The results of the computation were plotted in a digram (Fig. 4) the abscissas of which represent water content (void-ratio) and ordinates represent the relative speed of evaporation. The letter P indicates the plastic limit, N the point at which the surface of the capillary water retires into the interior, and S the shrinkage limit. Down to the plastic limit the relative speed of evaporation was constant, at about 1.1. But between P and N the speed of evaporation decreased, though the surface of the capillary water remained at the surface of the prisms.

Thus, one of the physical constants, the relative speed of evaporation, changes after the state P has been reached. It has already been remarked that the surface tension of the water changes its value at this same point. Under these conditions I felt warranted in concluding that the physical constant which determines the speed of percolation, i.e., the viscosity of the water, also changes its value as the clay approaches the lower limit of the plastic state.

Deducing from the observed facts the law which connects width of voids and coefficient of viscosity, we obtained the law

$$v_o' = v_o \left(1 + \frac{R}{s^8}\right)$$

where v_o denotes the normal value of the coefficient of viscosity and v_o' the viscosity of the water flowing through a capillary channel $2s$ (cm.) wide; R is a constant of value between the limits 6×10^{-42} and 2×10^{-40} (its exact value has not been determined, but knowledge of the limits is sufficient for our purposes). In voids smaller than 0.1μ (0.0001 mm.) the value of the coefficient of viscosity increases rapidly.

Revised Permeability Formula—Because of this change in viscosity the formula for permeability of clay had to be modified, the result being,

$$k = \left(\frac{C}{v_o}\right)\left(\frac{v_o}{v_t}\right)\frac{(e - 0.15)^{11}(1 + e)}{(e - 0.15)^8 + \frac{c}{d_w^8}} d_w^2$$

It is obviously difficult to determine the effective size d_w of the grains of clay because it requires wet mechanical analysis of the material. But as the formula expresses the relation between water content and permeability it can be used even when d_w is not known. The only requirement is that we know the result of a single permeability test on the sample of clay under study. This one result, the value k of the sample at one particular moisture content, may be used to determine the value k at any other moisture content, the two values being connected by the law expressed by the formula given.

Thus suppose we have two clays, for one of which the effective grain size d_w is m times larger (m being unknown) than the grain size d_w of the other one. In addition, we suppose that the permeability of the clay with the grain size d_w be given by the formula

$$k_1 = 3.34 \times 10^{-7} \left(\frac{v_o}{v_t}\right) \frac{(e-0.15)^{11}(1+e)}{(e-0.15)^{8}+0.0166} \text{(cm./min.)}.$$

According to what precedes, the permeability of the clay with the grain size md_w should be

$$k_2 = 3.34 \times 10^{-7} \left(\frac{v_o}{v_t}\right) \frac{(e-0.15)^{11}(1+e)}{(e-0.15)^{8}+\frac{0.0166}{m^{8}}} m^{2} \text{(cm./min.)}.$$

By making a single permeability test with the second clay, one point of its moisture-permeability curve is obtained, and this point permits of solving the above equation for m. Knowing the value of m, the complete curve can be drawn from the formula and subsequent permeability tests should plot close to this curve. Curve B in Fig. 1 is an example of what agreement may be

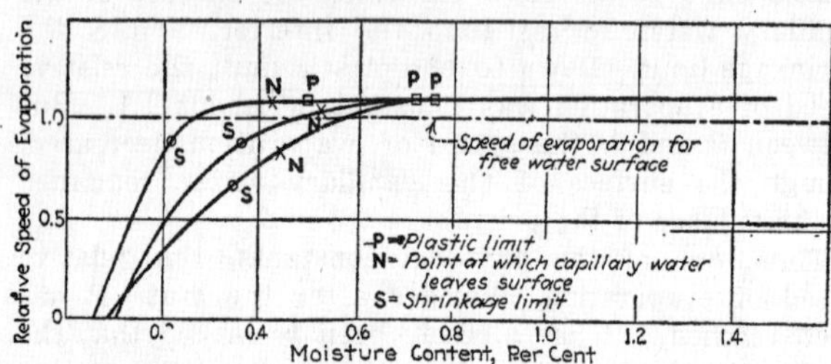

FIG. 4—SPEED OF EVAPORATION OF CAPILLARY WATER OF CLAY

expected. For one point of this curve the test furnished the value $k = 40 \times 10^{-7}$ cm./min., while the formula gave the value 35×10^{-7} cm./min. In the case of curve C, Fig. 1, the agreement between theory and test results was less satisfactory, on account of the high sand content of this material, amounting to 59.1 per cent. Such a soil may be called a sandy mud. Nevertheless its permeability also proved to vary between extraordinary wide limits as its density varies. This further illustrates that one cannot possibly describe the permeability of a clayey material except by a moisture permeability curve; a single figure means nothing.

Individual Character of Different Clays—For clays having equal grain size, the plastic limit may be high or low according to the shape of the grains and to the degree of adsorptive saturation. Therefore the values of the constants C and c in the general formula for the permeability coefficient depend on the nature of the grains as well as on their size, and an expression of the numerical type as just quoted is not valid except for clays whose grains are fairly similar in character. If the water which percolates through the clay contains substances in solution, these substances are adsorbed by the clay, and the solute accumulates along the surface of the clay particles. The concentration of the solution influences considerably the viscosity of the solvent, and in turn the viscosity of the solvent has a marked effect upon the permeability of the medium.

Attention should be called once more to the important fact that the preceding discussions apply only to homogeneous mixtures of clay and water. Crumb clay soils do not belong to this class. Percolation through such soils inevitably leads first to the formation of a filter skin and then to the filling up of the widest voids with spongy colloidal matter, detached from the top layers of the crumbs. Hence percolation through crumb soils is associated with grain displacements, internal erosion and deposition. It is a very complex phenomenon, which still needs careful investigation. It is altogether inadmissible to simply apply to the hydraulics of crumb soils the conceptions derived from sand filter tests or those applicable to homogeneous clays.

---

[1]C. S. Slichter, Annual Report of U. S. Geological Survey, 1899, p. 311.

[2]"Filtergeometrice," by L. Darapsky, *Ztschr. für Mathematik u. Physik*, 1912, p. 170; "Geometrische Untersuchung der Bewegung des Grundwassers im Gerölle und der Wasserfilterung durch Sand," by L. Burmester, *Ztschr, für angewandte Mathematik u. Mechanik*, 1924, p. 33.

[3]Erdbaumechanik, by K. Terzaghi (1925).

[4]Verbal Communication of R. R. Coghlan, chemist in charge; "Experiments on Material for Cold Springs Dam, Umatilla Project," by D. C. Henny and E. G. Hopson, *Eng. News*, 1907, p. 250; "Experiments on Cold Springs Dam and for a Projected Dam Which Was Abandoned," by T. A. Noble, *Eng. News*, 1907, p. 490.

Principles of Soil Mechanics: IV—Settlement and Consolidation of Clay

Consolidation the Result of Decrease of Moisture Content Under Load—Change in Hydrodynamic Stress—Typical Computations—Application to Permeability Determination

By Dr. Charles Terzaghi

Professor of Civil Engineering, American Robert College, Constantinople, Turkey
Temporary Lecturer at Massachusetts Institute of Technology, Cambridge

When a homogeneous layer of clay is loaded, the deformation of the clay increases for some time after the load is applied. This phenomenon is of great practical importance, being involved in all the numerous cases of settlement and subsidence met with in practice, as well as in the gradual consolidation of clap deposits.

The physical causes of this phenomenon have already been explained: The load produces a hydrostatic stress within the clay, which stress in turn causes the water in the clay to flow toward the surface, until stress equilibrium is reached, when the hydrostatic stress has decreased to zero. Basing calculations on this set of conditions, and on what has been developed in the preceding article on permeability, we can derive the relation between pressure, moisture content and time. That is to say, we obtain a solution of the time-settlement problem.

In Fig. 1 the pressure-moisture and moisture-permeability curves for a particular clay are given, as obtained by experiment. Suppose that a layer of this clay is enclosed in a rigid ring, and a load of intensity p_o per unit of area is applied to its upper surface, which is covered with water and is protected by a sand filter. What change will the layer undergo if the pressure is suddenly increased to p_1 and then kept constant?

According to the diagram, the moisture equivalent of pressure p_o is e_o, and the coefficient of permeability at moisture content e_o is k_o. When the pressure is increased these quantities ought to change to the values e_1 and k_1. But as the increased compression involves the escape of capillary water it is a rather slow process, and in every stage of this process the distribution of the hydrodynamic stresses must provide for the hydraulic gradient required to drive the capillary water out. This fact represents the point of departure for the subsequent considerations.

The diagram Fig. 2 represents a cross-section through the layer; h is the thickness the layer would have if its volume of voids were reduced to zero (reduced thickness). The origin of co-ordinates is taken at the lower surface of the layer and abscissa zero represents pressure p_o. At the upper surface of the layer the pressure is constant and equal to P_1, and the hydrodynamic stress is zero. According to the physical facts as set forth in the curves of Fig. 1, no change of water content is possible without change of pressure in the solid skeleton of the clay. But as we observe that the water content of the layer decreases while the external load remains constant, we are obliged to conclude that the pressure acting within the solid skeleton gradually increases. Let dp be the increase of the pressure p per time element dt, at height z above the base of the layer. The sum of the pressure p, within the solid skeleton and the hydrostatic pressure w at the same point must be equal to p_1, or $p_1 = p + w$. Differentiating, $dp = —dw$.

To determine how the pressure changes with time, assume a straight-line relation between pressure change and the corresponding moisture or void-ratio change. This assumption is permissible so long as the total change from p_o to p_1 is not large. The relation is expressed by the slope a of the pressure-moisture curve in Fig. 1. Then $de = —a\ dp = a\ dw$. The quantity of water which drains per unit of time through unit area of a horizontal cross-section at height z above the base is $Q = ks$, where k is the coefficient of permeability and s is the hydraulic gradient, equal to $—dw/dz$). Hence $Q = —k\ dw/dz$. But the hydraulic pressure w changes also with the time; i.e., while the water flows from elevation z to elevation $z + dz$, in the time element dt, the clay pressure p increases by dp, which equals $—dw$, and this change of the pressure involves a change of the water content according to the relation $de = —adp$, given above. Hence the quantity Q increases on its way from elevation z to $z + dz$, and

we have, $\frac{dQ}{dz} = -\frac{de}{dt} = a\frac{dp}{dt} = -a\frac{dw}{dt}$

Inserting the value of Q from the previous expression we obtain,

$$-\frac{k}{a}\frac{d^2w}{dz^2} = \frac{dw}{dt}$$

This equation represents the distribution of hydrodynamic stresses in the clay. It is mathematically identical with the equation for linear flow of heat through a plate of isotropic material of thickness $2h$ and of uniform temperature, insulated on its lateral faces, which is transferred suddenly into a space of lower temperature. The rate at which cooling proceeds from upper and lower faces toward the interior corresponds to the progress of consolidation of the clay from its upper (loaded) surface downward into its interior. This thermodynamic analogy makes it possible to transform any time-settlement problem into a thermodynamic one, noting that heat content replaces moisture content e, specific heat replaces modulus of compression a, temperature replaces hydrodynamic stress w, and coefficient of heat conductivity replaces coefficient of permeability k. There are the further relationships that (1) in thermodynamics specific heat decreases with increasing heat content, and in the hydraulics of clays the modulus of compression decreases with increasing moisture content; (2) loss of heat causes a body to contract, and loss of water causes clay to shrink.

Similar analogies play a very important part in modern applied mechanics. Thus, there is a thermodynamic analogy for the flow of water towards a series of wells, and a hydrodynamic analogy for the torsional stresses in solid bodies. Another analogy is utilized in the soap film method of solving stress problems, recently proposed by A. A. Griffith. According to this method, one can determine the distribution of torsional stresses over the cross-section of a twisted bar by simply measuring the deflection of a soap film stretched across an opening whose outlines are identical with those of the section.

The partial differential equation covers the whole field of time-settlement problems for watersoaked clay soils, and the special character of the problem merely influences the mathematical method to be employed for solving the equation. In dealing with the problem of gradual consolidation of a layer of clay under uniform load, at lateral confinement, the equation was solved by means of Fourier's series. From the results, numerical values have been computed, based on the following assumptions: thickness of the layer, 20 m., reduced thickness of the layer = 13 m., initial pressure 4 kg./cm.^{2}, final pressure 5 kg./cm.^{2}, modulus of compression 0.00002 g^{-1} cm.^{5}, and coefficient of permeability 0.100 cm. year^{-1}. The values

creases, as time passes, but the sum of these two pressures constantly equals the external load, which, in our case, is represented by the weight of the mud located above the cross-section for which p and w should be determined. For a shrinking clay cube the external pressure is equal to zero, hence the hydrodynamic stress w is negative and the pressure p (in this case called the capillary pressure), is equal to it and of opposite sign. The internal friction is equal to the product of the static pressure p times the coefficient of internal friction, while the hydrodynamic stress w cannot produce any internal friction. Hence a decrease of w at a constant external load produces not only an increase of p but a corresponding increase of the internal friction. These fundamental facts ought to become as familiar to the foundation engineer as Hooke's law to the structural engineer.

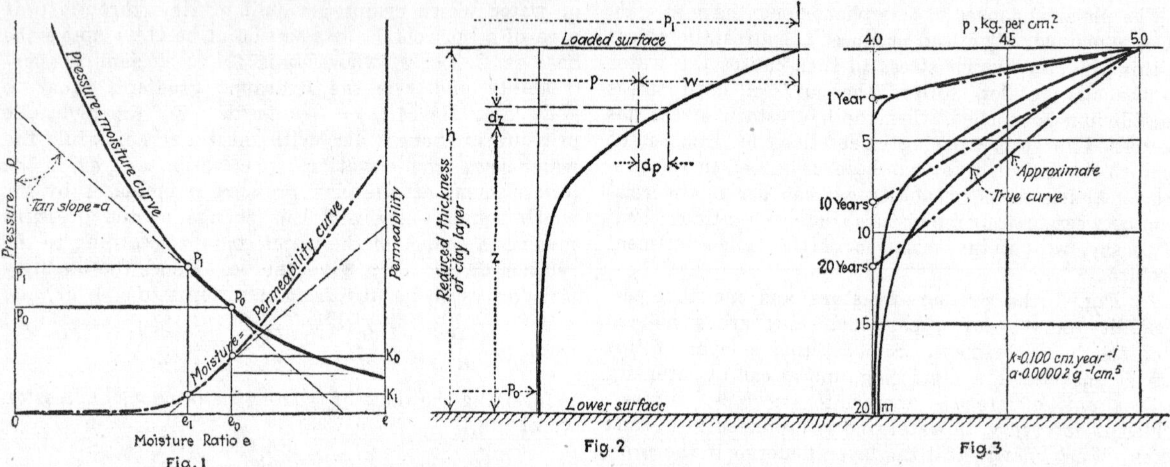

FIGS. 1 TO 3—CONSOLIDATION OF A CLAY DEPOSIT UNDER LOAD
Fig. 1—Test curves for a clay sample. Fig. 2—Distribution of pressure through the layer. Fig. 3—Pressure distribution after different periods

of the two latter constants (a and k) correspond to the properties of a fairly fat clay in a plastic state. The results of the computation are graphically represented in Fig. 3. The curves show how exceedingly slowly the compression proceeds from the surface of the layer toward the interior. This explains the gradual increase of the settlement of structures resting on the surface of strata of plastic clay. The dotted lines represent the results obtained by means of an approximate method, based on the assumption that the distribution of the stresses follows a straight-line law; they agree well with those of the accurate method.

Some Applications

A few simple examples may indicate what information the theory of hydrodynamic stresses is able to furnish.

Natural Settlement of Mud Deposits—Suppose a basin (lake or bay) is fed by a muddy stream so that mud is deposited at a uniform rate per unit of area and per unit of time. After a certain time the process of sedimentation stops. We inquire what will be the distribution of the stresses in the deposit at any definite time, and what will be the annual rate of settlement of the surface of the deposit.

We must be careful to distinguish between the hydrodynamic stress w acting in the capillary water, and the static pressure p acting in the mud. The ratio $w : p$ de-

In Fig. 4, diagrams b and c represent graphically the results of the calculation for a mud deposit formed at the rate of 5 cu.cm. of solid matter per square centimeter of bottom area per year; b corresponds to a sandy mud and c to a mud with a high percentage of colloids. The dotted lines passing through the upper

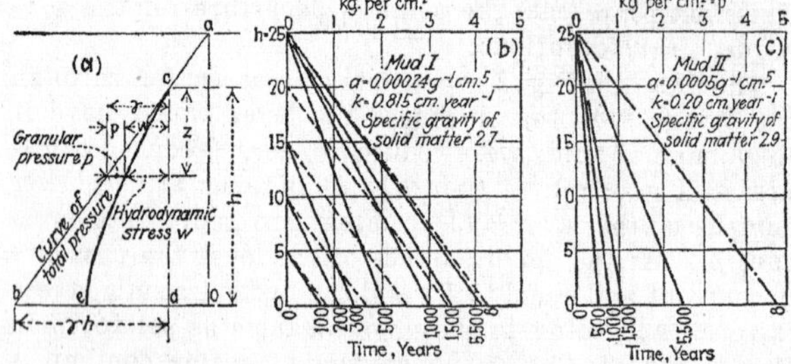

FIG. 4—HYDRODYNAMIC STRESS AS RELATED TO AGE OF DEPOSIT

corners of the graphs b and c represent the stress distribution for hydrostatic equilibrium, reached after infinite time. The full lines divide the abscissas of the dotted lines into two unequal parts, the right-hand part corresponding to the hydrodynamic stress and the left-hand part to the static pressure. At the lower end of each full line (pressure-distribution line) is written the time which elapsed between beginning of sedimentation and state represented by pressure distribution.

From the diagram *c* in Fig. 4 (mud rich in colloids) it may be seen that 5,000 years after the deposit was formed the hydrodynamic stresses still amount to about one-half of the pressure exerted by the dead weight of the deposit. This fact indicates in turn that the internal friction of the mud amounts to only one quarter of what it will be after the state of hydraulic equilibrium has been reached. As the deposit represented by *b* is more permeable and less compressible than the deposit *c*, it is at that time (5,000 years after it was formed) almost in hydrostatic equilibrium.

A short time ago I had an opportunity to examine fresh samples extracted out of a young and very colloidal mud deposit by means of test borings. The tests clearly evidenced that the pressure corresponding to both the moisture content and the internal friction of the mud amounted to a small fraction only of what would correspond to the state of hydrostatic equilibrium. On account of the excessive hydrodynamic stresses, the bearing capacity of the deposit was so small that the plan of loading its surface with factory buildings had to be abandoned. After a few thousand years more, the same deposit would undoubtedly present a fairly good foundation.

Impervious and Porous Layers—Suppose, now, that the deposit just discussed, instead of being homogeneous, included a mud layer less permeable than the remainder of the material. Such a layer has precisely the same effect on the drainage process as a layer of poor heat conductor would have on the cooling of a deep stratum of hot substance. Above the layer the mud will drain more rapidly, while below the layer it will remain almost liquid for a long time, and there will be a considerable difference of hydrodynamic stress between the two sides of the intermediate layer. If, on the other hand, the deposit contained a sand layer instead of the colloidal layer, the hydrodynamic stresses above and below the layer would be equal or nearly so. A hole drilled down to the sand layer would furnish artesian water, and drainage would proceed more rapidly than before, with correspondingly accelerated settlement.

On two occasions I had an opportunity to observe such a phenomenon in practice. In the first case the mud deposit had a thickness of about 15 m. Immediately after the sand stratum was opened up by means of a caisson well, some buildings located in the vicinity of the well started to settle badly. During the first year following the construction of the well, the settlements amounted to several inches and their rate of increase decreased from year to year, as theory demands. The well furnished artesian water. In the other case the mud deposit had a thickness of 50 m. While drilling a test hole through the mud we struck a layer of sand at a depth of about 15 m. To the surprise of the drill men, we got a rush of artesian water into the casing, although no artesian conditions were known to exist in the vicinity of the site.

Evaporation from Mud Deposits—Deposits of soft mud whose surface is exposed to the air gradually dry out. Experience shows that this drainage by evaporation leads to the formation of a semi-solid or solid crust, whose thickness increases as time goes on. I have previously proved that the solidification of a shrinking clay is caused by the capillary pressure, which in turn is associated with a negative hydrodynamic stress of equal intensity.

Consideration of the physics of the evaporation process leads to the conclusion that during the drying process two successive stages must be distinguished, the first stage lasting from the beginning of evaporation to the time when the capillary pressure at the upper surface of the deposit reaches the value of the transition pressure. At equal speed of evaporation and equal value of transition pressure, the thickness of the crust increases during the first stage in direct proportion to the permeability of the material. If the materials of two deposits are identical, the thickness of the crust at the outset of the second stage will be in inverse proportion to the speed of evaporation. During the second stage the rate of increase of the thickness of the crust becomes exceedingly small and, in opposition to what is true for the first stage, almost independent of the temperature of the atmosphere.

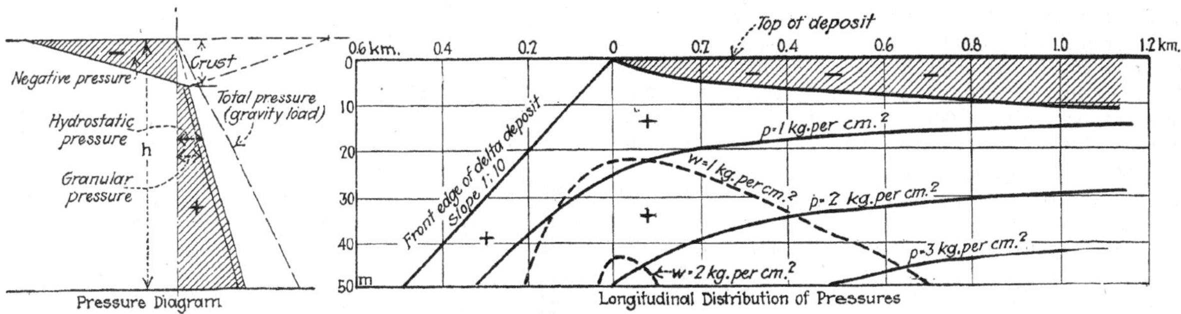

FIG. 5—STRESS DISTRIBUTION IN A DELTA DEPOSIT
Drainage by gravity and by evaporation. Negative hydro-dynamic pressures indicated by minus signs.

Basing my computations on the results graphically represented in diagram *b* of Fig. 4, and on the deductions just stated, I surveyed the stress distribution existing in a delta deposit whose outer edge advances toward the ocean at the rate of 1 m. per year and whose capillary water evaporates along the horizontal top surface of the deposit. The outcome of the investigation is represented graphically in Fig. 5. The hydrostatic pressure within the capillary water of the crust is negative, that below the crust positive. The moisture content of the deposit just below the crust must be a maximum. If a canal is dug in the deposit to a greater depth its bottom will discharge water, while those parts of the slopes which are located within the crust will absorb water.

The positive hydrostatic pressure in the capillary water of the core of the mud deposit is the greater, the more rapidly the deposit was formed and the less permeable the deposit is. Due to the assistance of this excess pressure, natural deposits of highly colloidal mud are apt to flow out spontaneously, apparently without any external cause. Such phenomena have been observed and they are known as "submarine landslides."

Conditions similar to those represented in Fig. 5 exist in hydraulic fill dams, for some time after they have been deposited. The less permeable the core material, the more important are the hydrostatic pressures in the capillary water of the core and the more imminent is the danger of a dam failure by lateral eruption of the core material.

Indirect Determination of Permeability of Clays—The theory of hydrodynamic stresses led to a very simple method for determining the coefficient of permeability of plastic or semi-solid clays. Suppose a layer of clay enclosed within a rigid ring and its surface covered with water is in hydrostatic equilibrium. If a surcharge is applied to its upper surface, there results in the capillary water a positive hydrostatic pressure whose intensity increases from the upper surface towards the bottom, while the simultaneous static pressure within the solid skeleton of the mass decreases from the top towards the bottom. If the load is kept constant, water will slowly be squeezed out. But if, instead of keeping the load constant, we keep the compression constant, by fixing the head of the testing machine in its new position, the moisture content of the layer can not change. Immediately after the head of the testing machine has assumed its new position, the hydrostatic pressure will be a maximuum near the bottom and a minimum near the top surface of the sample. As a consequence the water will move in an upward direction, causing the top part to swell, and the pressure acting against the head of the machine will decrease. Since the speed of the decrease of the pressure depends on the coefficient of permeability of the clay, the value of k can be calculated from rate of decrease of pressure.

Tests involving this action were made with an apparatus similar to that used for making load tests on saturated clay in confined condition. It consists of a zinc vessel, two bronze rings and a sand filter. The clay sample had a diameter of 8 cm. and a thickness of 4 cm. Pressure was applied by a screw testing machine. Fig. 6 gives the results. First a pressure of 14.1 kg./cm.^{2} was applied and then the head of the machine was held fixed. During the following hours the pressure decreased and gradually approached the value 1.72 kg./cm.^{2} (curve I). At the same time the rate of decrease of the pressure became slower, approaching zero. Let h be the reduced thickness of the layer, e_1 the moisture content, a_1 the modulus of compression (derived from the main branch of the pressure-moisture curve), a_2 the modulus of expansion (derived from the resaturation line of the pressure-moisture diagram), and k_1 the coefficient of permeability, all taken at the pressure p_1, and c, a coefficient whose value can be calculated from the speed with which the pressure acting against the head of the testing machine decreases; then

$$k_1 = c_1 \frac{a_1 h^2 \, (1 + e_1)}{2\left(1 + \sqrt{\frac{a_1}{a_2}}\right)^2}$$

Two days after the test was started, the pressure was raised from 1.72 kg./cm.^{2} to 27.3 kg./cm.^{2}, and the machine again held fixed. For this second stage of the test, the relation between pressure and time is represented by curve II. Similarly curves III and IV were derived, for still higher pressures.

By means of these tests it was possible to investigate with accuracy the flow of water through clays compressed by a pressure so high that their coefficient of permeability amounted to not more than 0.000,000,007 cm./min. Hence the sensibility of the method is practically unlimited.

The results of these tests proved conclusively that the flow of water through clays even of semi-solid consistency follows Darcy's law as closely as does the flow of water through sands.

Value of the Theory—Suppose an engineer observes that a structure founded on a bed of clay has suffered important settlements, and that the settlements increase in course of time at a definite rate; he publishes his observations and adds according to traditional practice some data concerning color and moisture content of the clay. A few years later another engineer wants to find

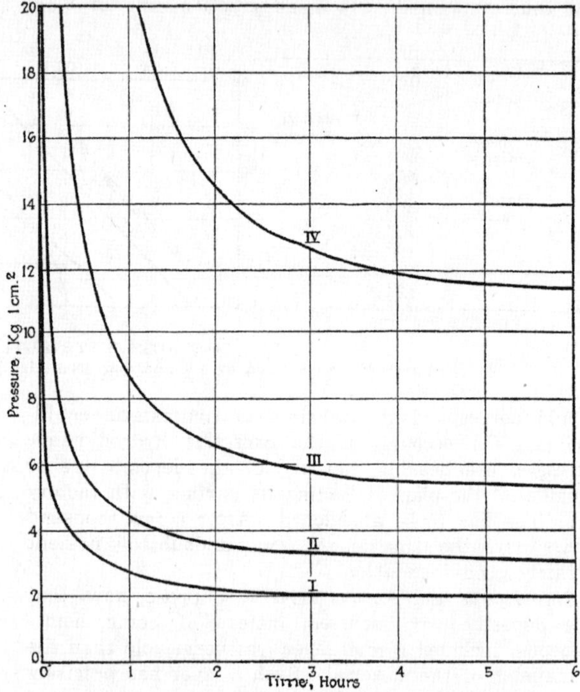

FIG. 6—PRESSURE-TIME CURVES FOR LOADED CLAY LAYER

At constant moisture content. The four curves represent different initial applied pressures. In each of the four tests the deformation was maintained constant after the load had been applied.

out whether a clay stratum at some other place will behave under load in a similar way. He will discover that he cannot. Or, if he is innocent enough to believe he can, with no other basis for conclusions than the data published by his predecessor, he will probably have later opportunity to revise his optimistic ideas.

But with the theory of hydrodynamic stresses at our disposal to explain the time settlement phenomena on a physical basis we are better equipped. The fundamental differential equation does not involve color or moisture content of the clay, but reveals the fact that the quantitative side of the time-settlement process depends only on the ratio a/k; everything else is non-essential for this particular process. Two clays may behave altogether differently in spite of their having the same blue color and the same moisture content, while two other clays of different color and different moisture contents may behave identically provided their a/k ratios are identical.

In order to compare two clays as to settlement, we must submit a sample of each to a test of such kind that the outcome of the test depends on nothing but the value of a/k of the clay. Several test arrangements are possible which satisfy this condition and, as a matter of course, we will adopt that one for which our laboratory equipment is best fitted. By comparing the results of the tests made with the two clays, we are in a position to foretell with a fair degree of accuracy how similarly or how differently the two clays will behave as to the relation between time and settlement.

The time-settlement relation, however, is only one of the many facts which may interest the earthwork-engineer. If he wants to investigate the tendency of clays to slide or the quality of core materials for hydraulic-fill dams he must execute other tests. These other tests must be planned according to the physical theory of earth slides and of the stability of dams, and with appropriate consideration of the constants which are essential to the processes in question.

No theory is required for *making* the tests. But on the other hand it is not probable that the tests will be properly planned unless the theory of the action is understood. That is the reason why the author felt obliged to start with the causative relations between the elementary facts of soil mechanics and the general laws of physics. However, the mental inertia which besets work in this field seems to be enormous, and it will still be a long time before engineers generally will see more clearly in that respect. Though more and more time and money are being spent in experimenting in the field of soil mechanics, few experimenters yet realize the fundamental difference between basing the analysis of test results on vague traditional concepts such as "internal friction" of sands or "cohesion" of clays, on the one hand, and on the other hand rigorously tracing the observed phenomena back to their physical sources —pure surface friction, structural factors, surface tension of the capillary water, and hydrodynamic stress.

Principles of Soil Mechanics: V—Physical Differences Between Sand and Clay

Commonly Noted Differences in Behavior—Grain Size and a New Uniformity Diagram—Shape of Grains and Structure—How Shrinkage, Cohesion, Plasticity and Settlement Are Affected

By Dr. Charles Terzaghi

Professor of Civil Engineering, American Robert College, Constantinople, Turkey
Temporary Lecturer, Massachusetts Institute of Technology, Cambridge

On superficial examination sand and clay seem to be essentially different materials. Their salient differences are summed up as follows:

Differences Between Sand and Clay

Clay	*Sand*
(1) Volume of voids may be as high as 98 per cent of the total volume.	(1) Volume of voids is about 50 per cent at the maximum.
(2) Shrinks in drying.	(2) Does not shrink in drying.
(3) Has a very marked cohesion, depending on the moisture content.	(3) Has negligible cohesion when clean.
(4) Is plastic.	(4) Is not plastic.
(5) Compresses very slowly when load is applied to the surface.	(5) Compresses almost immediately when load is applied to the surface.
(6) Is very compressible.	(6) Is far less compressible than clay.

Each of these properties represents the combined effect of several causes. Therefore, though the statements of the list are typical of what is found in current literature on soil mechanics. from a physical point of view they are no more satisfactory than, for example, the statement that two kinds of rock differ in brittleness, or two kinds of paraffin in tenacity and consistency. None but crude empirical conclusions can be drawn from such vague and general allegations, at the risk of confusing properties which physically may be altogether different. Atterberg, for example, confused the compressive strength of semi-solid clay with its resistance to penetration, and after assembling such heterogeneous data in a single diagram he was led to erroneous conclusions (see his "Konsistenz und Bindigkeit der Böden," *Internat. Mitteilungen für Bodenkunde*, 1912).

To characterize the properties of soils accurately we are obliged to trace these properties back to their physical sources. For the properties of clay this has been done in the preceding articles. It remains to analyze the properties of sand in similar manner and compare them with those of clay.

Grain Size and Uniformity—Considered as grain aggregates, clay and sand are characterized by their size of grain and uniformity. These factors are best expressed by a graphical representation of the results of mechanical analyses. The type of graph used should be such that the uniformity curves for materials of similar degree of uniformity will be of similar shape even though their fineness differs—the shape of the curve representing any material should be independent of the "effective size" of the grains. This can be accomplished by making the abscissas of the diagram proportionate to the logarithm of grain diameter. Thus, in Fig. 1, $x = a \log d$. Accordingly, a grain diameter d_2 which is n times as large as a grain diameter d_1 will be represented by a point at a fixed distance ($a \log n$) to the right of the point d_1, regardless of the absolute values of d_1 and d_2. Such a diagram presents exceedingly instructive uniformity curves.

Uniformity Curves—Fig. 2 represents the mechanical analyses of nine different materials. Abscissas represent logarithms of grain diameter, and ordinates represent the total weight of the grains of diameter smaller than this value, in per cent of the weight of the total mass. Points corresponding to Hazen's *effective size* are located on a horizontal line having the ordinate 10 per cent. All curves representing materials with an equal *uniformity coefficient* (Hazen) have an equal

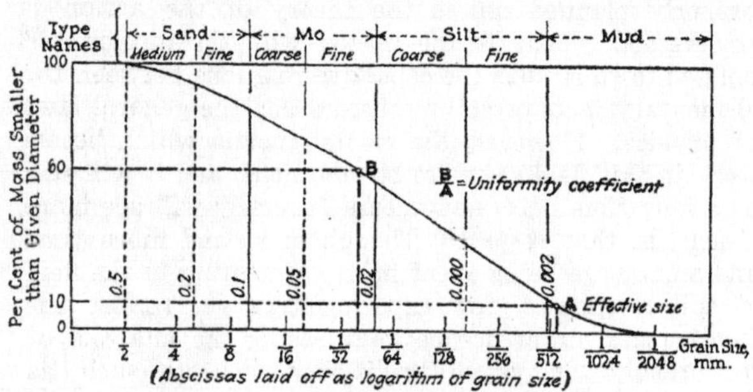

FIG. 1—METHOD OF REPRESENTING UNIFORMITY OF SOILS
(Form of curve is independent of fineness and depends only on uniformity.)

difference between the abscissas of their 10 per cent and 60 per cent points (see Fig. 1), regardless of whether the material is a coarse gravel or a colloidal powder.

The more uniform a material, the steeper is its uniformity curve. Hence two materials of different fineness but identical uniformity are represented in the diagram by curves of identical shape, and the fineness of the material (effective size) merely determines their position along the length of the diagram. Due to this property of the uniformity curves Fig. 2 reveals at a glance that in general *clays are far less uniform than are sands*. The loess from Virginia City represents an intermediate type in both uniformity and effective size.

The terms "mo," "silt" and "mud" used in the diagrams were originally proposed by Atterberg, and they seem to serve their purpose better than any other.

Shape of Grains — The grains of sand may be rounded or angular, with smooth or rough surfaces. Though these properties have a marked influence on internal friction, elasticity and permeability, yet all sands are alike (and differ from clay) in that their grains are bulky and rather rigid. Scalelike particles, as mica scales, represent only a minor accessory constituent. In contrast, clays are chiefly composed of flexible, scale-like particles. To be sure, the constituents of a "micro-mud" are so small that their shape can not be discerned even under a high-power microscope. Yet the assumption that clay particles are scale-like is supported by many proofs, of which a few may be mentioned.

A previous article described an apparatus called the interference contact indicator, consisting essentially of two glass strips enclosing a thin film of *residual* water. The film thickness, estimated by the Newtonian colors surrounding the water spots, amounts to about 0.1μ. In making several such contact couples I used (instead of distilled water) a very dilute suspension of mud, with particles of about 2μ size, and the water spots accordingly contained numerous clay particles of diameters 1 to 2μ. As the thickness of the films was not more than 0.1μ, the clay particles could not possibly be other than scale-like.

This direct evidence is strongly confirmed by indirect evidence drawn from permeability tests. The semi-empirical formula for the coefficient of permeability of clays (previously given) leads to the expression for the reduced coefficient of permeability, $k_r = 1.9\ d^2$ (cm./sec.), while the formula applicable to sand is $k_r =$ (174 to 100) d^3, in both of which d_{10} is the effective size. It thus appears that the coefficient of permeability of sand is 50 to 80 times greater than that of clay for the same effective size of grain. Both formulas refer to percolation at a standard temperature of 10 deg. C. through a material with a standard volume of voids of 50 per cent. In deriving the formula for clay it was further assumed that the water percolating through the clay has normal viscosity in spite of the small size of the voids. Hence the enormous difference between the formulas cannot be due to anything but a difference in the shape of the particles of the materials.

It is to be remembered that the term "diameter of a clay particle" by no means indicates the diameter of a globular grain of equal volume, but rather the diameter of a globular grain which sinks through the water at the same speed as the clay particle ("equivalent diameter"). As scale-like particles always sink with the flat side approximately horizontal, the volume of such a particle amounts to only a small fraction of the volume of a sphere of the same equivalent diameter. Therefore, even though sand be reduced to such fineness as to have the same equivalent grain diameter as a clay, yet a given volume of the sand would contain only a small fraction of the number of particles in an equal volume of clay, a fact which fully explains the difference between the two permeability formulas.

In chemistry a powder is not regarded as colloidal unless its particles are smaller than about 0.1μ. However, clay particles of equivalent diameter as much as 2μ have, on account of their scale-like shape, as small a volume as the volume of bulky particles of diameter approximately 0.1μ. For this reason a clay whose largest grains have an equivalent diameter of 2μ, or 0.002 mm., possesses all the properties of a colloidal substance, and Atterberg was fully justified in calling all the clay particles below 2μ colloidal mud.

It is interesting also to recall that in examining colloidal powders produced by crushing and grinding various minerals, Atterberg found that only minerals of laminated or scaly structure furnish plastic mud; and all clays are plastic.

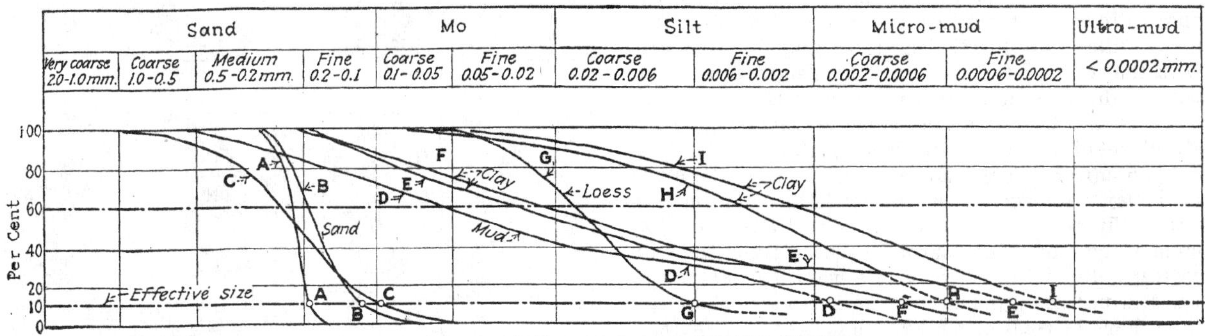

FIG. 2—NINE TYPICAL SOIL UNIFORMITY CURVES

A—Coastal dune sand, Beklemé, Thrace.
B—Beach sand, Rumeli Kawak, Thrace.
C—Glacial sand, lower end of Ankogel Glacier, Austria.
D—Mud, upper part of Golden Horn.
E—Yellow residual clay, east shore of Bosporus.
F—Yellow marine clay, from earth slide in Thrace.
G—Loess, Virginia City, Ill.
H—Blue marine clay, from earth slide on Black Sea Coast, Thrace.
I—Yellow marine clay, washed out of clay (*F*).

Sketch *D* in Fig. 3 shows a microscopic picture of the grains of coarse silt (0.02 to 0.006 mm.) extracted from a blue marine clay. A dilute suspension of this silt, examined under the microscope, was found to contain no colloidal particles at all. When the cover glass was shifted back and forth, numerous colloidal particles with a distinct Brownian movement appeared, showing how fragile are the scale-like particles. This fragility undoubtedly represents one reason why the uniformity curves of clays are so characteristically different from those of sands.

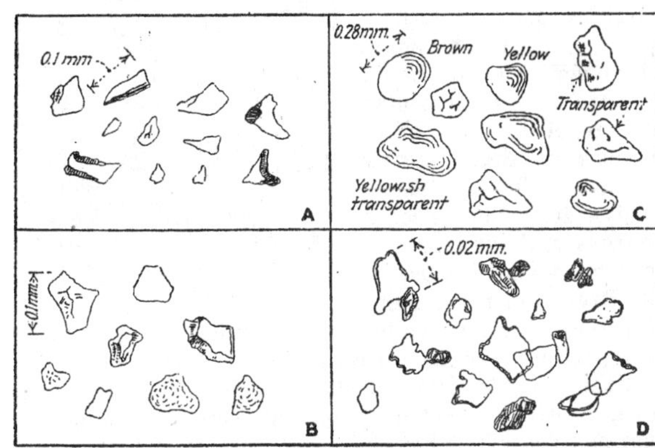

FIG. 3—CHARACTERISTIC SOIL PARTICLES (ENLARGED)
A—Crushed quartz. *C*—Dune sand (*A*, Fig. 2).
B—Glacial sand (*C*, Fig. 2). *D*—Coarse silt (from *F*, Fig. 2).

In conclusion it may be stated that throughout the author's studies no essential difference was found to exist between sands and clays other than the difference in grain size and shape. It will be shown, however, that these two differences are fully adequate to explain the more obvious distinguishing features of the two materials.

Loose and Dense Structure of Aggregate—When a grain aggregate (sand or clay) in suspension is allowed to settle, sedimentation proceeds gradually, because the particles floating next to the bottom of the basin reach the bottom first. Suppose, now, that part of the sediment has been deposited, and a newly-arriving particle settles on its uneven surface, Fig. 4. Its movement, hitherto vertical, is disturbed by two new forces, the resistance of the sediment to penetration, and initial friction* acting at the point of contact between particle and sediment. If the grain has the size of a sand grain, the initial friction is negligible compared with the weight of the grain. The weight and the reaction, then, form a couple which rolls the particle down to the bottom of the adjacent depression. The structure of the sediment thus formed will be loose (Fig. 5, upper right). By jarring the vessel the structure may be made dense (upper left). As there is practically no adhesion between the particles, the maximum volume of voids cannot be much larger than the corresponding volume of voids of a mass of equal spheres, i.e. 47.6 per cent, and in fact measurements show that the volume of voids of a cohesionless sand never exceeds about 50 per cent (void-ratio 1.0). On the other hand, if the arriving particles are very small the bond produced by initial friction may be stronger than the rolling tendency, so that the particles remain in the position of first contact, and the structure of the sediment becomes spongy (lower left in Fig. 5), with a ~~maximum~~ void-ratio (as found by measurements) of about 4.

Let us finally assume the particles to be of colloidal size. Such particles remain in suspension forever unless they are precipitated by adding a few drops of an electrolyte. Before the electrolyte was added the particles moved with a considerable speed through the liquid (Brownian movements) and repelled each other because of their electric charges. The electrolyte neutralizes the electric charges without eliminating the physical cause of the Brownian movements. Hence the particles collide in the liquid, and as soon as two particles collide initial friction comes into action and the particles stick together. Thus spongy flocs are formed, which build up a spongy sediment similar to that just described, but with a floc in place of each grain. The resulting structure may be called a spongy structure of the second order (lower right, Fig. 5). Maximum void-ratio corresponding to a simple spongy structure being 4, the maximum void-ratio of this last type ought to be about 4^2, or 16, corresponding to a volume of voids of 94 per cent.

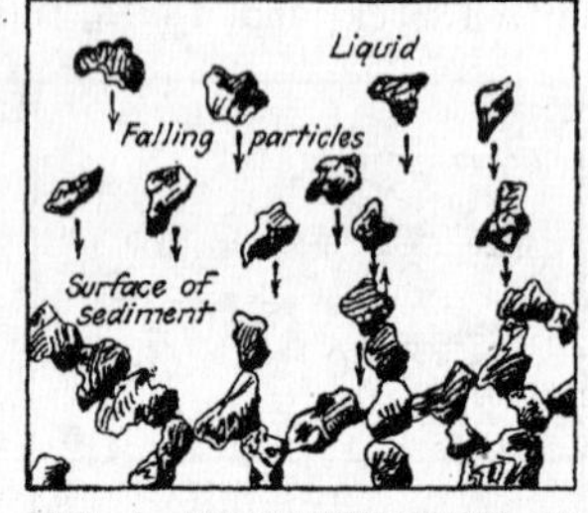

FIG. 4—SEDIMENTATION OF FINE POWDER

Thus we can explain item (1) of the list of differences between sand and clay merely by the difference in grain size.

Sedimentation tests with powders of different fineness (extracted from mixed-grain aggregates by wet mechanical analysis) gave support to this explanation. As diagrammed in Fig. 6, fine powders even when thoroughly consolidated by jarring could not be reduced to

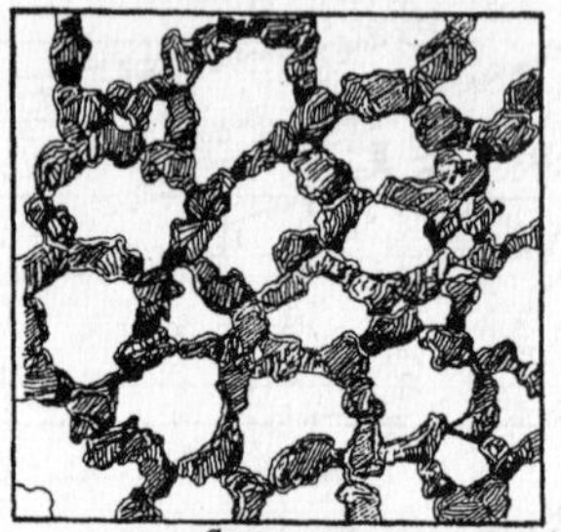

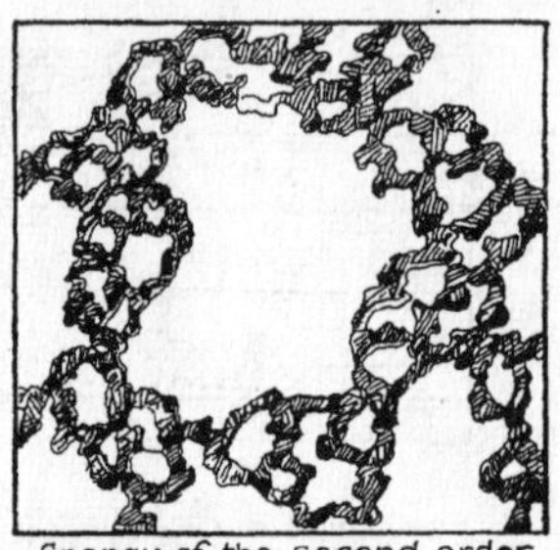

FIG. 5—MAIN STRUCTURE TYPES OF HOMOGENEOUS SOILS

the denseness of coarser sediments loosely deposited. This fact gives us an idea of what an important factor initial friction is, provided the number of points of contact per unit of volume of the sediment is very great. These sedimentation tests were made in test tubes 1.6 cm. in diameter. Slow sedimentation in wider vessels would undoubtedly lead to smaller values, yet the test results can be taken as confirming the explanation.

Difference in Shrinkage—If a piece of clay dries, it shrinks; the voids of the clay remain filled with water until the shrinkage limit is reached. In a previous article it was shown that the shrinkage represents the compression produced by the capillary pressure, and that the maximum intensity of the capillary pressure (*transition pressure*) depends on the grain size. In terms of the maximum height h in meters to which water can rise in the mass by capillarity, the transition pressure in kilograms per square centimeter is $0.1h$. For a very fine sand h is approximately 0.05m. hence the transition pressure would be 0.005 kg./cm.^{2}, while fat clays have transition pressures of 200 kg./cm.^{2} and more. A pressure of 0.005 kg. per cm.^{2} cannot possibly produce any noticeable compression of the material. This fact accounts for item (2) of the list of differences between sands and clays.

Cohesion—Item (3) is also readily accounted for, since the maximum cohesive (shearing) strength of a grain aggregate is equal to the product of the transition pressure by the coefficient of internal friction of the material, so that sand cannot exhibit much cohesion.

Plasticity—A fact rather more difficult to understand is that colloidal quartz powders (microscopic sands) have no plasticity, while clays of equal fineness are very plastic. A body is plastic if its shape can be changed while its volume remains constant. To perceive the reason why microscopic sands are not plastic, the reader must refer back to the well-known sandbag experiment performed by O. Reynolds (*Phil. Mag.*, 1885). Rey-

*The term initial friction as used by the author denotes the shearing strength of the clay when not under pressure, either external or capillary.

nolds observed that a rubber bag full of dry sand opposes small resistance to change of shape, while if the bag is sealed after the voids of the sand are first filled with water it becomes as hard as stone and can be loaded heavily without changing shape. This fact is obviously due to the bulky shape of the sand grains; the arrangement of such grains cannot possibly change unless the grains roll over each other, and these grain movements are possible only when the structure temporarily becomes very loose, which would involve increase of the volume of voids. In a dry sand there is no obstacle to change of volume of voids, but when the voids are completely filled with water, and in addition air is prevented from entering, the volume of voids must remain constant; hence the grains cannot move and the content of the rubber bag makes the impression of being solid.

The nature of this phenomenon does not depend on grain size, provided the structure of the sand is not spongy but granular. A sediment consisting of a colloidal quartz powder is originally spongy and in this state is mobile, just the same as a spongy clay sediment;

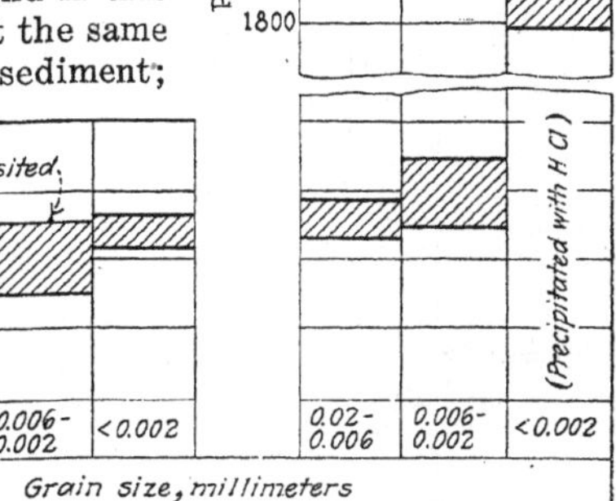

FIG. 6—VOID-RATIOS OF CLAY AND SAND SEDIMENTS

but when the capillary water of the quartz powder is allowed to evaporate, capillary pressure compresses it to granular structure. In this state, because of its bulky grains, it cannot change shape without changing volume, as shown by Reynolds' experiment. Hence neither sands nor colloidal powders with bulky grains are plastic.

In the case of clay, though held by the surface tension of the capillary water just as firmly as if enclosed in Reynolds' rubber bag, the particles are scale-like, and therefore can slide over each other while the volume of voids remains practically constant. Therefore clays are plastic. Item (4) of the differences between sands and clays is thus accounted for as the effect of size and shape of particles.

Naturally, as difference of grain size and form is one of degree, we may expect to find materials of various degrees of plasticity between the extremes of the non-plastic and the highly plastic. The most suitable method for expressing the degree of plasticity is that proposed by Prof. A. Atterberg, as already explained. Table I contains a list of the characteristic limits for colloidal powders produced by crushing and grinding different minerals. The degree of fineness of all the powders used in the tests was approximately the same. The table shows clearly the intermediate stages between high plasticity (talc, biotite) and complete lack of plasticity (colloidal quartz powders). For powders derived from the same material, plasticity was found to increase with a decrease in average size of grains.

Settlement—Coming to item (5), we again find grain size and shape to furnish a complete explanation. Loading the horizontal surface of a water-soaked layer of sand produces almost instantaneous settlement because the excess water finds small resistance to its escape from the compressed material; but in clay the low permeability causes an enormous resistance to escape of the capillary water, and therefore settlement under load proceeds slowly.

TABLE I—PLASTICITY OF FINE MINERAL POWDERS

(Grains smaller than 0.002 mm. Figures from Atterberg's "Die Plastizität und Bindigkeit der Tone," *Internat. Mitt. für Bodenkunde*, 1913. The tabulated figures are moisture content in per cent of weight of solid matter.)

Mineral	Lower Limit of Plastic State	Lower Limit of Liquid State	Coefficient of Plasticity (=Difference of Two Preceding Columns)
Biotite	44	87	43
Talcum	48	76	28
Chlorite	47	72	25
Kaolinite	43	63	20
Hematite	20	36	16
Limonite	27	36	9
Quartz	35	35	0

Principles of Soil Mechanics: VI—Elastic Behavior of Sand and Clay

Testing Sand for Compressibility and Elasticity—Expansion and Resaturation—The Ideal Sand Cube and the Effect of Lateral Expansion—Comparison with Solids

BY DR. CHARLES TERZAGHI
Professor of Civil Engineering, American Robert College, Constantinople, Turkey
Temporary Lecturer, Massachusetts Institute of Technology, Cambridge

IN STUDYING the difference in compressibility of sand and clay, which constitutes the last item of the observed characteristics as listed in the preceding article (*Engineering News-Record,* Dec. 3, 1925, p. 912), the elastic properties of sand were investigated. For this purpose a method was used similar to that previously applied to tests of the elastic properties of clay. The sand was poured into a steel ring resting on top of a polished cast-iron slab, and either left as deposited or compacted by hammering the outside of the ring. The surface was leveled with the ring, covered with a circular cast-iron plate, and loaded either by the lever of a home-made loading device (low-pressure tests) or by the piston of a screw testing machine (medium- and high-pressure tests). The diameter of the ring used for the medium- and high-pressure tests was 15 cm. and its height 4 cm. Compressions were measured by micrometer with the help of the interference contact indicator. Tests on dry sand and on sand completely immersed in water showed no difference.

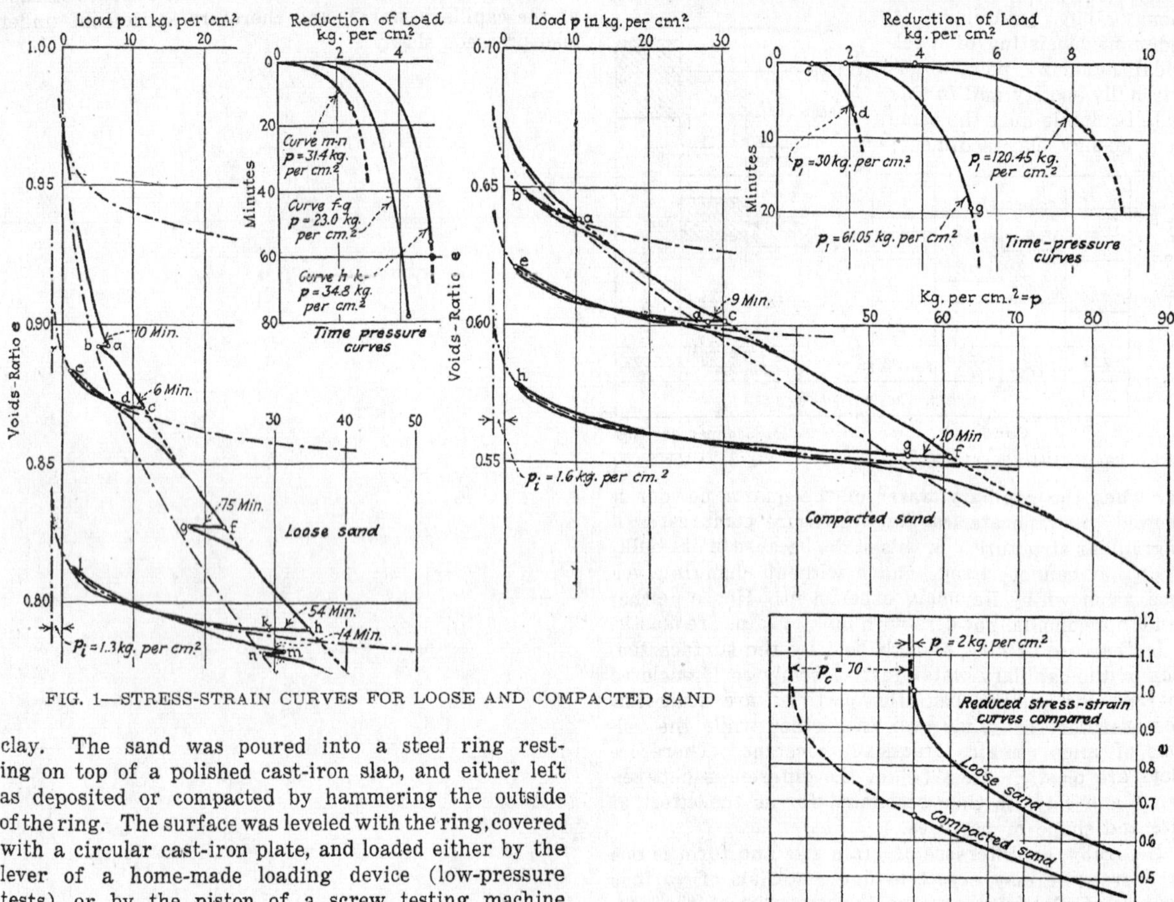

FIG. 1—STRESS-STRAIN CURVES FOR LOOSE AND COMPACTED SAND

Fig. 1 gives in separate diagrams the results of tests on loose and thoroughly compacted sand. In both cases the sand was a pure quartz sand produced by crushing white quartz pebbles, the product being sifted and washed; the grain diameter was 0.25 to 1.00 mm.

In the test on loose sand the initial volume of voids was 49.75 per cent (void-ratio 0.99). By means of a screw testing machine, pressure was increased at the rate of about 1 kg./cm². per minute to about 7 kg./cm². Then the machine was stopped for 10 min. (horizontal line *ab*) during which time the pressure exerted by the sand against the stationary piston decreased to about 5.5 kg./cm². Thereafter the machine was again started (curve *bc*), the load increasing to about 12 kg./cm². Again the machine was stopped, for 6 min. (*cd*), and then the load was released down to about 1.5 kg./cm². and reapplied (*ef*), forming a complete hysteresis loop and continuing the original stress-strain curve. A second load cycle was tried at higher load (loop *klm*).

The time-pressure diagrams (at right and above in Fig. 1) show the pressure-decrease while the machine was stopped. These curves are very similar to the time-pressure curves for clay cubes, and in fact they have the same simple differential equation. The gradual decrease of the pressure is due to a gradually proceeding mutual compensation of unbalanced frictional resistances which at first take up part of the load.

If, instead of keeping the compression constant, we keep the load constant, the compression increases at decreasing rate, and the relation between time and compression is in every respect similar to the relation between time and compression for clay cubes under constant load.

Expansion of Sand—In the test on compacted sand, Fig. 1, the initial volume of voids was 40.2 per cent (void-ratio 0.673). The stress-strain curve is less steep than that for loose sand, while the hysteresis loops of both diagrams are almost identical. The close relationship between these diagrams and the corresponding diagrams for clay is obvious. In both, the loading cycles form curved hysteresis loops. In the sand diagrams, Fig. 1, the recurrent branches of these loops are expansion lines, but they correspond in every respect to the resaturation lines of the clay diagrams, and like the latter they are logarithmic lines, $e = A \log (p + p_i) + C$, where p is the external load (kg./cm.^{2}), p_i is an initial constant having the dimensions of a pressure, A is the *expansion coefficient* of the sand, and C is a constant depending on the initial density of the sand. The expansion coefficient of sand corresponds to the resaturation coefficient of clay.

The expansion coefficient is almost independent of the density of structure of the sand. It was found to be equal to 1/100 for sand with very smooth grains and 1/176 for a sand with very rough grains. A very fat clay showed a resaturation coefficient of 1/22.3, a leaner clay 1/52.7, and a sandy mud 1/73. Thus sand is far less elastic than clay. The more sand a clayey soil contains the nearer its resaturation coefficient approaches the expansion coefficient of clean sand.

On the other hand the "initial pressure" p_i depends not only on the original density of the sand but also on the pressure at which expansion begins. This effect is due in part to the sand grains wedging together, in part to breaking of grains by the pressure, which in turn affects the structure. As illustrating the latter action it may be mentioned that a sand which initially was clean and dust-free was found, after its surface had been loaded to 50 kg./cm.^{2}, to contain 4.6 per cent of dust particles. The average value of p_i for sand (1.5 kg./cm.^{2} for medium to high pressures) is very large, compared to the value for clay (approximately 0.002 kg./cm.^{2}).

The effect of the unbalanced internal stresses on the stress-strain curve depends on the rate of loading. Under indefinitely slow application of load we would obtain the *reduced* stress-strain curve, shown by a dash line. How this line depends on the original density of the sand is represented by the third diagram in Fig. 1.

Fig. 2 shows typical stress-strain diagrams for a fat clay, a loose sand and a dense sand. The figure shows plainly that the difference between the elastic properties represented by the three curves is one of degree only. This difference, item (6) of the differences listed in the preceding article, Dec. 3, is a self-evident consequence of the fact that clay particles are scale-like. Sand can be compared to a pile of broken stone, while clay resembles a mass of flakes of paper. The higher the sand content of a clay, the less compressible and the less elastic it is.

Poisson's ratio for clay was computed from the lateral pressure in a loaded layer whose lateral expansion

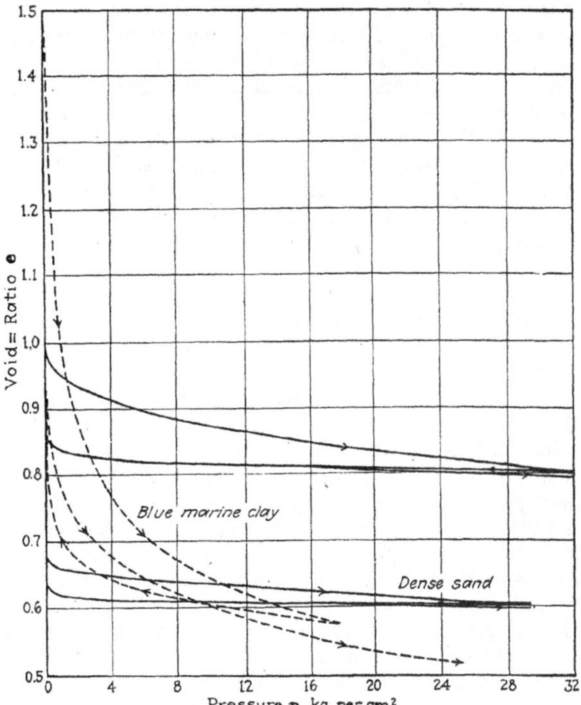

FIG. 2—PRESSURE-VOID (STRESS-STRAIN) CURVES FOR CLAY AND SAND

was prevented by a rigid enclosing ring. In a similar way Poisson's ratio for sand was determined. It has the value of 5.0 (as compared with 11 for Cooper sandstone, 5.1 for Troy granite, and 4.5 for Tuckahoe marble). Thus, while Poisson's ratio for clay is approximately identical with that of metals, its value for sand corresponds to the average Poisson's ratio for crystalline rocks.

Modulus of Elasticity of Sands—Clay is capable of being tested in the form of an unconfined cube. The particles of such a cube, before loading, are subject to no force except capillary pressure, acting like a hydrostatic pressure. Suppose the clay particles are replaced by sand particles without changing the intensity of the internal pressure; it is readily inferred that, as sand and clay have quite similar stress-strain diagrams for compression when confined laterally, the stress-strain curve of the sand cube would be of the same type as that of a clay cube, which in turn was found to resemble closely the stress-strain diagrams for a concrete cube.

Of course it is not possible to test sand cubes directly. On account of the large voids of sand the capillary pressure is very small, too small to hold the particles together. But on the other hand each cube-shaped element of the backfilling of a retaining wall is comparable to our ideal sand cube; capillary pressure is replaced by the confining pressure of the surrounding material. When the retaining wall is forced horizontally against the backfilling, a cubical element of the backfilling is subjected to forces quite like those acting on a compressed clay cube. Hence the stress-strain diagram ought to be of similar form. Experiment fully confirms this.

When a retaining wall yields under the pressure of the backfilling, any cubical element of the backfilling undergoes horizontal elongation. This corresponds precisely to the strain effect of a pull exerted on the ends of a clay prism. In such a clay prism the principal stresses at right angles to the pull are constant and equal to the capillary pressure, while the principal stress acting in the direction of the pull is a variable compressive stress, equal to the difference between the capillary pressure and the pull. This residual stress must obviously be a compression, not a tension, because the pull cannot possibly exceed the intensity of the capillary pressure. But this residual compressive stress decreases as the elongation of the prism increases. In the same way, the lateral earth pressure exerted by a cohesionless mass of sand against a yielding retaining wall must evidently decrease as the yield of the wall increases.

A confirmatory experiment is represented in Fig. 3. Two rigid vessels A and B, of similar form, are filled with sand as shown. Close to the side of each vessel and just above midheight there is within the sand a horizontal steel tape set on edge, between two sheets of smooth paper. In vessel A the tape is pressed against the wall by what I have called the *earth pressure at rest* of the sand, since near the tape the sand undergoes no lateral expansion after its deposition. In vessel B, on the other hand, the upper half of the sand body rests on the surface of the compressible lower half of the sand body, and this lower half compresses during the filling process, each layer moving downward and (because of the splay of the sides of the vessel) expanding laterally as it moves down. According to the theory above suggested this expansion ought to decrease the lateral pressure of the sand. Individual measurements of the force required to pull the tape are plotted in the figure separately for vessels A and B; in each case the lower group of points refers to the unloaded sand, while the upper set was obtained when a weight of 25 kg. rested on the sand surface. The figure shows that the lateral earth pressure was 43 per cent smaller in vessel B than in vessel A—obviously due to the lateral expansion of the sand.

A different effect resulted when the sand was subjected to a violent disturbance. The bottom of vessel B was placed on removable supports and, after a first series of pulling tests was made, this bottom was

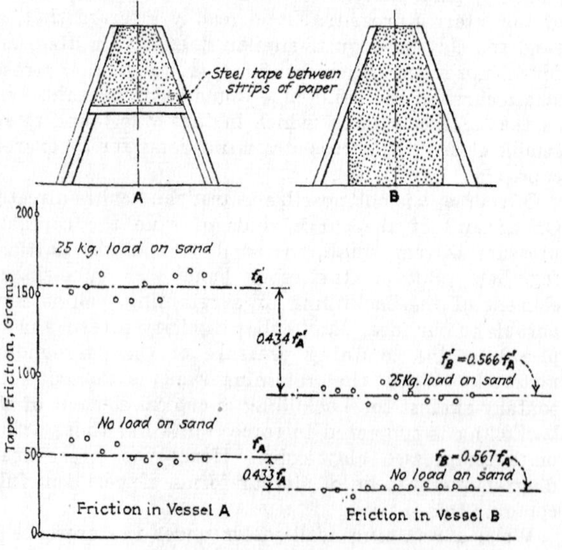

FIG. 3—LATERAL PRESSURE OF SAND (Influence of yield and settlement).

lowered suddenly a distance of 2 mm. Immediately thereafter the lateral pressure was somewhat smaller than before, but in time it increased, went beyond its original value, and after 24 hr. became as great as the lateral pressure in vessel A. The increase of the pressure was obviously due to the gradual mutual compensation of internal frictional resistances similar to those which cause the pressure in a clay cube or a layer of sand to increase when its loading is decreased and the height of the cube or layer is then held constant.

Further evidence for the pressure-relieving effect of lateral expansion of a mass of cohesionless sand has been furnished by various earth-pressure tests of G. H. Darwin, A. D. Donath, and the author.

Some interesting relations may be deduced from the stress-strain curves of sand drawn in Fig. 4. Compressing a mass of sand held in a rigid ring gives a stress-strain curve like the full line. The lateral pressure is smaller than the vertical pressure (load pressure) within the sand, and corresponds to earth pressure at rest. From the full-line curve can be derived (by the use of Poisson's ratio) the curve of cubical compression, representing the strain-effect of pressure equal in all directions (and therefore comparable to a capillary pressure); this is shown by a dash line. Suppose, now, the compression process represented by this dash curve be stopped at pressure p_1. Each element of the mass is now in the same condition as a clay cube subject to capillary pressure of the same amount p_1. Hence, if the sand specimen is further loaded in the vertical direction only, holding the lateral pressure constant at amount p_1 and allowing free lateral expansion to take place under the action of increased vertical load, we are bound to obtain a stress-strain curve identical with that of a vertically loaded clay cube. Two such curves are shown in Fig. 4, beginning at the respective cubical pressures p_1 and p_2; as the tests on clay cubes showed that the ratio between modulus of elasticity and internal pressure is constant, we may conclude that the slope of the hysteresis loops in these two curves will have the ratio $p_2:p_1$. In other words, all stress-strain curves of the type shown for the two free cubes in Fig. 4 will differ only in the scale of their abscissas; so that if curve p_1 was obtained in a test, the curve p_2 can be drawn by simply increasing the abscissas of p_1 in the proportion $p_2:p_1$. The truth of this statement was repeatedly checked by experiments on clay.

Clay and Sand Compared With Solids—In discussing the cohesion of clays I have previously shown that the ratio between the modulus of elasticity and the internal (intrinsic or molecular) pressure is a constant not only for clay but also for metals. Therefore *the laws expressed by Fig. 4 are approximately valid for the whole field of materials,* regardless of whether the particles of the material are kept together by molecular attraction (solids), by a capillary pressure (clays), or by the pressure due to the weight of the material itself (cohesionless sands).

Let us imagine for instance a cubical body consisting of cohesionless sand. From the tests already made we may calculate that an internal pressure of 1226 kg./cm.^{2} will bring the modulus of elasticity up to 200,000 kg./cm^2. This value is approximately equal to the value of the modulus of elasticity of ordinary concrete. The compressive strength of the concrete, however, amounts to not more than 300 kg./cm.^{2}, while the cube of sand can carry a maximum load of about 10,000 kg./cm.^{2}, provided the internal pressure remains unchanged. For loads ranging between zero and 300 kg./cm.^{2} the relations between stress and strain for sand agree far better with Hooke's law than do those for concrete. This statement may serve as one of the many examples of the fact that in solid bodies the ratio between compressive strength and intrinsic pressure is exceedingly small, compared to the corresponding ratio for aggregates of individual grains. Investigation of the physical causes of the low value of this ratio for solids represents one of the most attractive problems of modern research on resistance of materials.

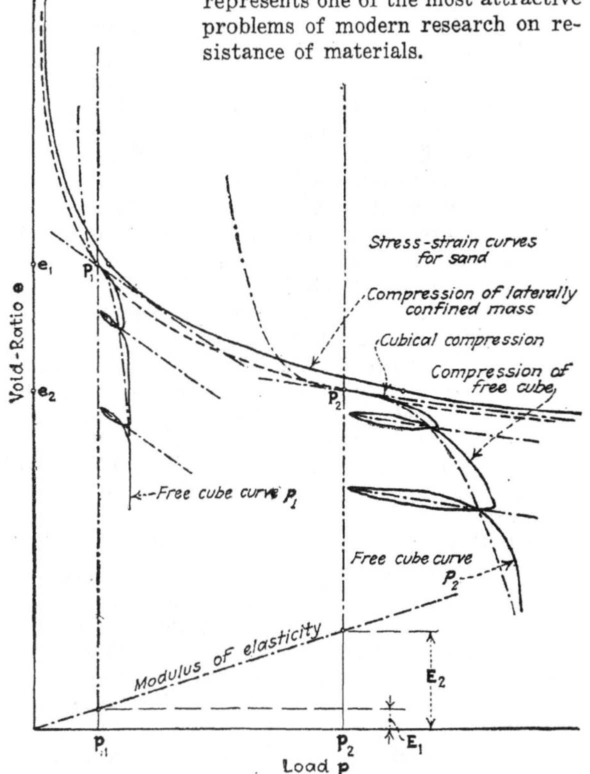

FIG. 4—STRESS-STRAIN DIAGRAM OF SAND

Due to the high ratio between modulus of elasticity and internal pressure for sands (from 238 to 419 for sands, against 31 for fat clays and about 10 for metals), a very small lateral expansion of the sand causes a considerable decrease of the lateral sand pressure. For this reason it is difficult to determine the value of earth pressure at rest, except by an indirect method such as used by the author (see "Old Earth-Pressure Theories," *Engineering News-Record*, Sept. 20, 1920, p. 632). Retaining-wall tests made by the hinged-gate method, on the other hand, are almost unavoidably affected by error because of the effect of lateral expansion.

SAND AND CLAY COMPARED AS TO PHYSICAL PHENOMENA AND THEIR CAUSES

Cause	Physical Factors Affected	Visible Consequence
Grain size.	Aggregate initial friction per unit of volume.	(1) Difference in maximum volume of voids: About 50 per cent for sand, about 98 per cent for clay.
	Capillary pressure and surface tension of the capillary water.	(2) Difference in shrinkage: Sand does not shrink in drying; clay shrinks.
		(3) Difference in cohesion: Clean sand is devoid of cohesion, clay has high cohesion.
Grain size and shape (grains bulky or scale-like).	Capillary pressure; nature of intergranular movements during deformation of the mass.	(4) Difference in plasticity: Sand is not plastic, clay is very plastic.
	Degree of permeability.	(5) Difference in speed of adjustment to loads: Sand when loaded settles to its final volume almost at once, while settlement of clay foundations proceeds very slowly.
Character of grains.	Flexibility of the particles.	(6) Difference in compressibility of the aggregate: Sand is far less compressible than clay.

Conclusions—The results of our comparison between the physical properties of sand and of clay are presented in the table. The facts stated in the first and last column of the table have long been known, but knowledge of the causative connection however between these two groups of facts is new. It is a much simpler connection than might have been expected. This table, the curves of Fig. 4, and the partial differential equation

$$\frac{K}{a} \cdot \frac{\delta^2 w}{\delta z^2} = \frac{\delta w}{\delta t}$$

represent the fundamental principles of soil mechanics in a nutshell. They are the key to the physical explanation of whatever properties a soil may display, in the laboratory or in the field.

Principles of Soil Mechanics: VII—Friction in Sand and in Clay

Complex Nature of Friction in Granular Masses as Compared with Solid Friction—Measuring Friction in Sand and Clay—Two Kinds of Frictional Motion in Sand—Hydrodynamic Effect on Clay Friction

BY DR. CHARLES TERZAGHI
Professor of Civil Engineering, American Robert College, Constantinople, Turkey
Temporary Lecturer, Massachusetts Institute of Technology, Cambridge

SOIL movements are largely determined by the resistance of the soil to sliding motions within the mass. Soil friction, therefore, is of peculiar importance in soil mechanics.

It is common, and quite natural, to approach the study of friction in sand and clay by way of conceptions derived from current views on friction between solids. This line of approach is apt to lead into error, however. Granular friction is radically different from solid friction in almost every phase, and the two subjects require separate and quite distinct modes of attack.

Largely through the work of W. B. Hardy and collaborators in very recent years, remarkable insight into the phenomena of solid friction has been gained, and the action is now known to be far more complex than had been thought during the last century or two. The coefficient of friction between perfectly clean and smooth surfaces was found to be very high and constant. However, in practice no perfectly clean surfaces exist because contaminating matter spreads over every surface either by creep or by condensation and the presence of contaminating matter considerably reduces the value of the coefficient of friction. Water is called by Hardy an anti-lubricant because it has the property of increasing the value of the coefficient of friction between contaminated, smooth surfaces by partially compensating the lubricating effect of contamination. In no case was it found to act as a lubricant. Among lubricating oils of similar chemical character the power of reducing the coefficient of friction increases with increasing molecular weight. For uneven surfaces the coefficient of friction depends in addition on whether or not the pressure acting per unit of area of the surface of actual contact is high enough to permit the minute projections to cut through the lubricating film and produce incipient abrasion.

Although it thus involves a number of factors, solid friction is nevertheless a relatively simple phenomenon, from the physical point of view. The contacting bodies are constrained to move parallel to their interface, and conditions remain fairly constant as the motion progresses, provided the motion is very slow.

In contrast, motion between masses of individual grains is a very complex process, involving not merely surface friction but also translation and rotation of the grains. The complexity of the process betrays itself in the fact that the actual movement is preceded by an incipient one, caused by minor grain displacements. In short, frictional resistance in sand depends not only on the pressure and the nature of the contacting surfaces but also on the density of structure of the sand and on the thickness of the zone to which the grain movements are confined.

Internal Friction of Sand—Extensive experiments were made by the Foundation Soils committee of the American Society of Civil Engineers to determine the coefficient of friction of Ottawa standard sand (see *Transactions,* Am. Soc. C. E., 1917, 1920), but no simple and general conclusions have resulted thus far. A rather elaborate disk apparatus was used, in which the sand was confined within radially arranged compartments. The coefficient of friction was computed from the force required to produce a rotation of the mobile part of the apparatus. The coefficient of internal friction thus obtained was remarkably low, but varied considerably with the number of preceding runs. The coefficient for the friction between standard sand and the surface of standard sand mortar was found to be considerably higher than the coefficient for sand on sand. In each series of tests the coefficient sensibly decreased (at low and at medium pressure) with increasing pressure. The results show that the coefficient of friction even of standard sand is likely to vary between surprisingly wide limits, depending on the test conditions.

In fact, there is no definite single coefficient of friction. The value of the coefficient depends both on the manner in which slip is produced and on the changes in structure of the sand prior to the slip. On theoretical considerations as well as by the results of experimental work, the author has been led to distinguish between two fundamentally different cases:

(a) Separation of a mass of sand along a plane: Within the zone of slip there is complete rearrangement of the grains. Since Reynolds' classical experiment with the sand-bag we know that the grains of a mass of sand cannot change partners unless the volume of voids of the sand temporarily increases. Hence the separation along a plane requires a gradual loosening up of the structure of the sand in the vicinity of the plane of separation, i.e., the separation involves a tendency toward increasing the volume of voids. The coefficient of friction along a plane of separation is called the *coefficient of internal friction.*

(b) Continuous deformation of a mass of sand: If a mass of sand is uniformly stretched or compressed so that the displacement of the grains occurs throughout the whole mass, the stability of the structure increases and finally assumes a maximum. In this limiting stage the frictional resistance which determines the ratio between the extreme principal stresses is the *coefficient of internal resistance.* Its value represents the maximum value which the coefficient of friction can assume within a mass of sand of a given density. It may be considerably greater than the slope of repose.

Cases (a) and (b) are limiting cases; there is an infinite number of intermediate possibilities, each of which corresponds to another type of internal displacement, involving another coefficient of friction, limiting the state of equilibrium for that particular type of displacement. Thus the value of the coefficient of internal friction depends on whether the slip is confined to a zone only a few grains in thickness or whether it occurs within a layer ten times as thick. This seems to be the chief reason for the disagreement between the results of large-scale retaining-wall tests and miniature tests.

The friction tests of the Foundation Soils committee are a notable example of the utter complexity of the frictional resistance in sands. It is timely to conclude from these and similar experiments that there can be no hope of exactly determining the value of the coefficient of friction required for calculating the outcome of more complicated earth pressure phenomena in sands. We can only guess at that value by choosing some intermediate value between the extreme ones or by indirectly computing it from what has been observed.

Internal Friction in Clay—In the case of sand, then, popular conceptions of the nature of frictional resistance are widely at variance with the physical facts. The same thing is even more true for clay.

Be it stated in advance: Experiments to determine water has completely escaped from the layer of clay.

Herein resides the physical cause of the slipperiness of the surface of fat clays. If one steps suddenly on a very slightly inclined clay-surface the foot slips, although even a very fat clay has a fairly large internal friction (friction angle at least 11 deg.). In the rapid application of the pressure of the foot, the greatest part of the weight of the body is compensated for by hydrostatic pressure, and the friction produced by the remaining weight is not sufficient to prevent the slip.

Certain types of landslides involve a similar phenomenon. The slide is preceded by the formation of shrinkage cracks or other open spaces, in which water accumulates. Due to some spontaneous displacement in the fissured material part of the weight acting in the trapped layer of water compensates part of the weight of the settling material with a speed corresponding to almost zero friction on the sliding plane. This phenomenon, in which the coefficient of friction may temporarily go down to 0.05 or less, is commonly accounted for by the lubricating action of the water. In fact

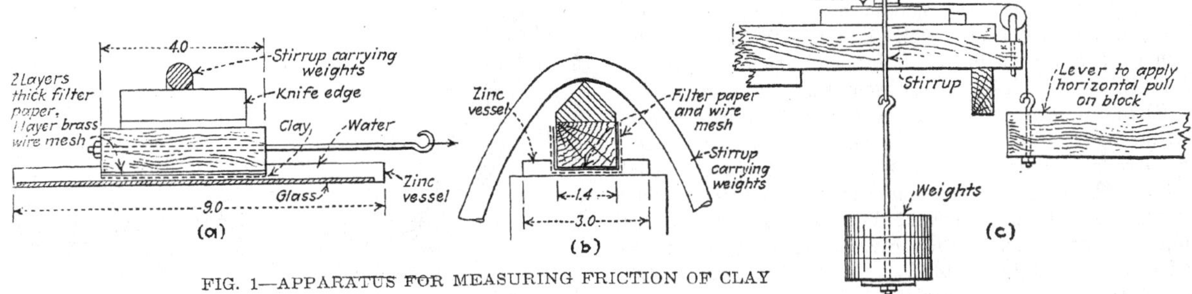

FIG. 1—APPARATUS FOR MEASURING FRICTION OF CLAY

the coefficient of internal friction of clay must (a) exclude the surface tension of the capillary water, because this tension produces internal pressures in addition to the pressure applied externally, and (b) take account of the fact that the frictional resistance depends not only on the intensity of the pressure but (within limits) also on the length of time during which the pressure acts.

Suppose the plane surface of a mass of air-free wet clay is under a definite external pressure. When the pressure is increased the clay body becomes smaller, like any other elastic body. As the voids of the clay are filled with water, the increase in pressure involves escape of the excess water. But as the permeability of the clay is very small, the escape of the water proceeds very slowly. During the drainage process part of the excess pressure is compensated for by a corresponding hydrostatic pressure of the water in the voids; the surcharge partly floats on the surface, so to say, and only the surplus is carried by the solid parts of the mass. Immediately after the surcharge is applied, the compression of the clay is practically equal to zero, hence the hydrostatic pressure at this time is almost equal to the surcharge. While the excess water gradually escapes, the hydrostatic pressure becomes smaller and approaches the value zero. The speed of decrease evidently depends both on the coefficient of permeability of the clay and on the thickness of the clay layer. A hydrostatic pressure does not produce any static friction. Hence it is merely the remainder of the surcharge which counts. The frictional resistance increases with decreasing hydrostatic pressure, and does not assume its normal value until the excess it is merely due to a spontaneous partial compensation of the weight of the overburden by the hydrostatic excess pressure acting in a layer of trapped water.

Friction Apparatus—In order to fulfill the requirements essential for a successful test, the author investigated the internal friction of clays by means of the apparatus represented in Fig. 1. A pan of zinc has its bottom covered with a sheet of glass. For determining the coefficient of friction between glass and clay, the glass sheet is covered with a thin layer of plastic mixture of clay and water (free of air). On the layer of clay is placed a prismatic block whose base and sides are lined with two layers of thick filter paper and one layer of No. 20 brass wire mesh. The brass mesh acts as a rough surface, which takes the clay with it, and the filter paper serves as a drain for the excess water. Load is applied by weights hung from a stirrup resting on a knife-edge above the loading block. The load usually reduced the thickness of the layer of clay to about 3 mm. Finally all the clay fragments projecting beyond the edges of the base of the loading block were scraped away and the zinc vessel was filled with water. Horizontal pull was applied to overcome the frictional resistance, by means of a cord leading over a pulley to the end of a loading lever.

The friction tests were never started earlier than 24 hours after the load was applied, a time sufficient for draining the thin layer of clay or allowing it to become resaturated. When it was desired to measure the coefficient of frictional resistance of clay on clay, the glass plate was replaced by a double layer of filter paper covered with a brass wire mesh.

This test arrangement is far from being ideal,

because the tests require much care, time and experience. However, for the time being it is the only one which fulfills the above-mentioned first condition for a successful test. Further efforts will be required for developing a more convenient and perfect device.

Friction Test Results—The results showed the coefficient of friction of clay to be remarkably constant for pressures of more than about 1 kg./cm.², and to be remarkably independent of many of the factors which distinctly influence the coefficient of friction of sands. This seems to be due to the fact that fat clays consist of thin and very flexible flakes, while sand is composed of bulky, fairly rigid grains. The structure of sand does not allow change of shape without change in volume of voids (sandbag experiment of Reynolds), while clay is capable of plastic deformation (deformation at a constant volume of voids).

The coefficient of friction between glass and clay is about 0.18 to 0.22 for colloidal mud, 0.23 to 0.30 for fat clays, and 0.30 to 0.32 for sandy clay. The respective coefficients of internal friction of the clays are 0.23 to 0.28, 0.25 to 0.40, and 0.40 to 0.50. The coefficients of internal friction and of internal resistance seem to be almost identical. Hence each kind of clay has its definite coefficient of internal friction. Herein lies an essential difference between sand and clay.

The sliding of clay on the smooth surface of the glass sheet occurred suddenly, as soon as the pull exceeded a threshold value. Motion along an interface located within the clay, on the other hand, proceeded rather gently, resembling somewhat the flow in a very viscous liquid. The initial friction, corresponding to zero pressure, amounted to about 20 g./cm.² The smaller the pressure, the greater was the influence of initial friction on the value of the coefficient, as may be seen from the high values at the left-hand end of curves, Fig. 2.

In each test, actual motion was preceded by an incipient motion, usually starting as soon as the pull became equal to about six-tenths of the threshold value. The incipient motion seems to be caused by the structure of the stretched clay adapting itself to the changing state of stress. If the pull was left constant before it had attained the threshold value, the incipient motion continued with decreasing speed and ceased completely within twenty-four hours. After this a further increase of pull produced no detectable incipient motion whatever.

For clays the coefficients of internal friction and of internal resistance seem to be almost identical. In a preceding paper I have shown that the shearing strength of a plastic sample of clay is equal to the product of capillary pressure times the coefficient of internal resistance. Hence there is a contradiction between the low coefficient of internal resistance and the great cohesion of certain fat clays, explained by the great intensity of the stress acting in the capillary water of these clays.

The Foundation Soils committee, in investigating the internal friction of clays, used a device similar to the one described for sand tests at the opening of this article. But the essential requirements for obtaining normal values for the coefficient of internal friction were not fulfilled. The Kentucky ball clay with 10 per cent and 25 per cent water, respectively, undoubtedly contained considerable quantities of air, or else the cohesion effect would have been far greater than that actually measured. The coefficient of friction was of the order 0.25 (0.23 to 0.27). The tests made with a mixture of clay with 39.54 per cent of water furnished an exceedingly low value of friction coefficient which clearly proves that the greatest part of the surcharge was compensated for by hydrostatic pressure, i.e., the test was made a long time before the excess water had completely escaped. In a test of such a kind the clay is not yet in hydrostatic equilibrium, and the test may furnish any friction value between zero and the normal value. Such values are not the coefficients of static friction, but coefficients of what may be called the momentary hydrodynamic friction; they depend on the pressure which acted on the clay prior to the time when the surcharge was applied, on the time during which the surcharge was allowed to act, on the thickness of the layer, and on various other factors.

Conclusions—The quantitative side of every earth-pressure phenomenon depends on the intensity of the frictional resistance acting within the soil. Through decades of years the development of soil mechanics has been handicapped by the traditional habit of identifying the laws which govern the frictional resistance in granular masses with those derived from friction tests performed with solid bodies. In the preceding the attempt has been made to characterize the phenomena of sand and clay friction and to indicate their quantitative features.

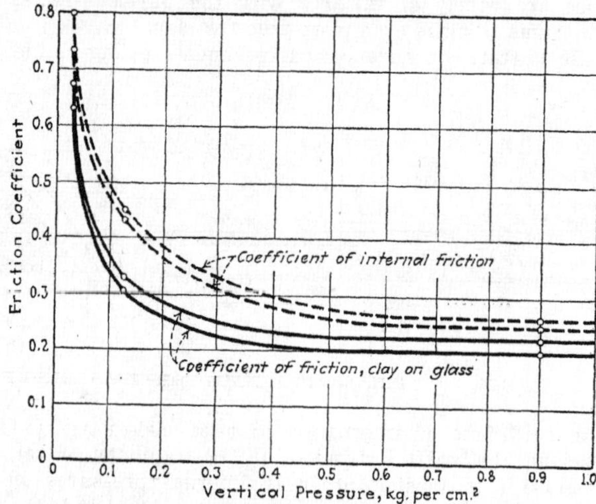

FIG. 2—STATIC FRICTION OF YELLOW CLAY

On the basis of these considerations taken in conjunction with the elaborate studies of Hardy and others on friction between solids with and without lubricants, the following statements may be made:

(*A*) Friction between smooth and absolutely clean surfaces of solid bodies is a purely physico-chemical process and is caused by direct molecular interaction.

(*B*) Friction between imperfectly smooth surfaces of solid bodies involves not only these physico-chemical causes but also a file-like action of each surface on the other. Nevertheless, from a physical point of view the phenomenon is a simple one.

(*C*) In sand the friction coefficient depends not only on the properties of the grains and the structure of the sand, but also on the nature of the process which causes the slip, and on the nature of the process which preceded the slip. It has no definite value, but may be anywhere between two limiting values, (a) the coefficient of internal friction (friction acting along a plane of separation), and (b) the coefficient of internal resistance (resistance which develops while the entire mass suffers a uniform deformation).

(*D*) In clay the friction coefficient for medium and for high pressures is remarkably constant. For low pressures, however, the value of the coefficient increases with decreasing pressure, because initial friction plays an important part, amounting to about 20 g./cm.^{2} Rapid change of pressure produces in the liquid component of the clay a positive or a negative hydrostatic pressure. The coefficient of friction does not assume its normal value (coefficient of static friction) until the hydrostatic pressure has become zero throughout the whole mass. In the preceding stage of the process the coefficient of friction (coefficient of hydrodynamic friction) may have any positive value, and is a function of the time.

Principles of Soil Mechanics: VIII—Future Development and Problems

Origin and History of Experimental Study of Soils—Future Objectives—A Theory of Models—Soil Classification—The Method of Equivalents for Engineering Practice

By Dr. Charles Terzaghi
Formerly Professor of Civil Engineering, American Robert College, Constantinople, Turkey
Temporary Lecturer, Massachusetts Institute of Technology, Cambridge

IN CONTRAST to the rapid advance of engineering science in the fields of statics, dynamics and hydraulics within the last century there has been complete stagnation in the study of soil mechanics. Since Coulomb and Rankine formulated their classical theories of granular earth pressure practically no progress has been made, and the vast field of the mechanics of plastic soils has hardly been touched. Earthwork engineering is in consequence about on that level of efficiency in design at which mechanical and structural engineering found themselves a couple of centuries ago. The reason for this lack of progress lies mainly in the great difficulty of investigation of soil phenomena. In recent years, however, various investigators have turned their attention to the subject. A summary of experimental results obtained by the author has been presented in the preceding articles of this series. Further activities along similar lines are planned, in several institutions of this country, and it may be said (speaking somewhat optimistically) that a general attack upon the unsolved problems of soil mechanics has begun. It will therefore be of value to review the situation briefly.

Engineering Geology—The first efforts were made in the direction of developing engineering geology. A text-book on the subject was written as early as half a century ago (by Dr. D. Braun), though from the present standpoint the work was of little value. Within the last generation or two, many engineers have recognized the need for a better bond between geology and engineering, and have worked to bring it about. Italy and Belgium created special departments for engineering geology. Other states sent their geologists to every important construction enterprise, in order to bring geology and engineering into constructive contact. Meritorious works on engineering geology have been written. Yet the results of the movement are on the whole disappointing.

The more one penetrates the field of engineering geology, the more its innate limitations become evident. In its present state, the science is of service in such matters as locating the sites of structures, or selecting the number and location of test borings, but this is after all only a small part of what is needed. We obtain no exact information concerning the strength and physical properties of the materials composing the several geological strata, and we do not yet find in the teachings of geology any suitable basis for identifying and classifying soils.

Experimental Soil Studies—It thus came to be realized in the course of time that soils must be studied physically as well as geologically, if we are to have a substantial basis for earthwork engineering. It is to the credit of the American Society of Civil Engineers to have made the first conscious step in this direction. In 1913 the society appointed a foundation committee, "to codify present practice on the bearing value of soils for foundations and report upon the physical characteristics of soils in their relation to engineering structures." The committee evidently realized the importance of its task; it has displayed very fruitful activity throughout its existence. By an interesting coincidence, in the same year the Swedish Government Railways appointed a Geotechnical Commission for the purpose of investigating the physical characteristics of materials in unstable slopes and devising remedial measures for slides. Thus, the need for soil investigation found simultaneous recognition on two continents.

During its ten years of existence, the Foundation committee has made many interesting reports. It has conducted friction tests, investigated the compressibility of various soils, and measured the compressive, tensile and shearing strength of undisturbed clay samples, among other things. No general conclusions have yet been reached, and in view of the lack of knowledge concerning the basic relations between the properties of soils it is impossible to make much practical use of the data obtained. The reason is obvious: Each one of the so-called physical properties of soil—internal friction, cohesion, plasticity, compressive strength, and the like—is in itself so utterly complex a phenomenon that it cannot be correlated with any of the others. It would be necessary first to discover the elementary physical factors which determine these properties. Lacking a knowledge of the elementary factors, experiments merely build up a collection of unusable data and give no knowledge of relations.

Investigation in a New Direction—Realizing the necessity for a different kind of physical study of soils, the author in 1917 undertook a program of experimentation in a new direction, beginning with a study of the elastic properties of cohesionless sand. This work was practically isolated, and at the outset found little encouragement, most of the author's friends considering the efforts to be unpromising.

The method which finally led to a considerable degree of success consisted of reducing each problem to its simplest possible terms, and using the result of one test for establishing a tentative hypothesis as to the causes of the phenomena observed, which hypothesis could then be verified or disproved by further tests. No investigation was begun until the preceding one had been carried to some definite conclusion.

Proceeding step by step in this way, the author was able to trace the many physical properties of soils to four underlying factors: (1) Friction between the grain surfaces, including initial friction; (2) viscosity of the capillary water; (3) surface tension of the capillary water; and (4) the influence of the width of voids on the physical properties of the water itself. It was found that the striking differences between cohesive and cohesionless soils could be fully explained in terms of these four physical factors, the intensity of their effects essentially depending upon the size and the shape of the soil particles.

Beginning in 1920 and continued within the last year, the author presented a summary of the action of these four physical factors and how in combination they produce the manifold physical properties of soils, in this journal.[1] A fuller account of the results of the investigation was published in book form a year ago.[2]

Application of Soil Mechanics—For practical application, soil mechanics must be applied in the first instance to the following objectives:

(A) *A theory of models* must be developed. Without such a theory no valid conclusions can be drawn from the results of loading tests on soils or from model tests (as on dams), with respect to full-sized structures.

(B) *Classification of soils* must be accomplished. It should be possible to determine and express quantitatively the relation between two apparently similar soils found in different localities.

(C) *Adequate design data* must be developed. The new science must guide the way toward obtaining all the required data for the economical design of structures consisting of soils or in contact with soils.

A Theory of Models

In modern engineering, theories of models are among the most important and most indispensable tools. To illustrate the necessity of such theories in earthwork engineering, the following example may serve:

The Foundation committee has designed a convenient apparatus for making loading tests on foundation strata, by means of which accurate load-settlement diagrams may be obtained. So far, however, no hint has been given as to how the resulting diagrams can be used. The ultimate bearing capacity of a clean, thoroughly compacted sand amounts to 0.29, 0.42, or 0.61 kg./cm.^{2} (Strohschneider), according to whether the diameter of the loaded area is 0.8, 1.25, or 1.78 sq.cm., and it is known that a circular footing 1 m. in diameter resting on such sand could carry a load several times as great without objectionable settlement, so that it is clear that the bearing capacity of sand increases rapidly with size of loaded area. In the case of plastic clay, on the other hand, the bearing capacity is almost independent of the area of the loaded surface. Soils intermediate between sand and clay will have intermediate properties. Thus, in the absence of a theory of models no conclusions can be drawn from the results of a standard loading test unless a rule of interpretation is established for each of the many kinds of soil encountered.

Even with such rules we would still have the difficulty of identifying the soil at a particular foundation site with one of the standard soils for which the rules were derived. Non-homogeneous or stratified soil deposits further complicate the matter, and make a purely theoretical approach seem hopeless. Therefore the author believed it necessary to approach the theory of models by combined empirical and theoretical attack. In order to do this, each experiment, being in effect a model test, was analyzed theoretically to the point of making clear what physical factors were concerned and developing a formula expressing the law of the result as closely as possible. These formulas were then applied to deriving coefficients having known values for definite limiting conditions (as for plastic clay and for cohesionless sand) and intermediate values for intermediate conditions.

Bearing Capacity of Soil—Some of the results obtained by this course of procedure may be stated as follows: The bearing capacity of a soil depends on its specific gravity and density of structure, the smoothness of the grain surfaces, and the intensity of the capillary pressure, in other words its cohesion. The formulas derived for the bearing capacity of the ground contain as factors the specific gravity w of the soil, the intensity of the capillary pressure p_k, the radius of the loaded surface r, and the ratio t/r between depth of foundation t and radius of loaded surface.

In wet sand, capillary pressure is negligible compared with load stresses, and the ultimate bearing capacity increases directly as the radius r and also increases very rapidly with the depth ratio t/r. In plastic or semi-solid clay, however, the capillary pressure is very large, and the load stresses are negligible in comparison. Application of theory led to the realization that under such conditions the bearing capacity is almost independent of both radius and depth-ratio. Most actual cases will lie between these two limiting cases. For any given ratio p_k/w increasing the size of the loaded area will cause the case to approach the limiting condition of cohesionless sand.

While the theory so far developed accomplishes this much, and also is of assistance in devising special tests for cases of variable soil strata, it is by no means perfect, nor have the laboratory test methods yet been developed to their best form. The results demonstrate, however, that the combined theoretical and empirical method is effective. An interpretation of the results of loading tests is now possible, however much it may still need refining. This could not have been accomplished without laboratory discovery of the fact that the cohesion of clay is due to negative hydrostatic pressure of the capillary water; that the physical properties of capillary water contained in sub-microscopic voids differ from those of water contained in coarser materials; and that the internal friction of soils is a composite of several independent actions.

A theory of models also has other applications. For example, the test loading apparatus proposed by the Foundation committee[3] is intended to be applied to a loaded area 34.7 cm. in diameter, the surrounding ground being covered by a layer of sand 70 cm. deep. Here the ratio t/r is about 4. Looked at empirically, the sand fill is only a protecting layer, and it was actually so intended. Considering the situation theoretically, however, we see that the fill may be without effect on the result of the load test or it may multiply the bearing capacity several times, depending on whether the subsoil has large or very small cohesion. Because of this relation, we are bound to conclude that a loading test intended as a guide for the design of a sand foundation of shallow depth (t/r very small) should be so arranged that the ratio t/r in the test is also small, otherwise the test results may be misleading. The disturbing influence of the protective layer becomes insignificant, however, when the soil has high cohesion.

The lack of a theory of models is responsible for the fact that engineers hold curious and highly arbitrary views as to the interpretation of small-scale experiments on soil or soil structures. Investigators who undertake large-scale experiments and find their results differing widely from those of small-scale experiments commonly explain the disagreement by charging the small-scale experimenter with being unable to appreciate the importance of errors of observation. Such charges are manifestly unfair; the real trouble is the lack of a theory of models, which in turn is due to incomplete knowledge of the physical character of the phenomena involved. Every apparent contradiction between the results of small and of large-scale tests is a symptom of a specific deficiency of our insight. No problem in soil mechanics can be considered solved unless the investigator is in a position to explain fully the cause of such disagreements. For this reason small-scale experiments should be regarded an essential part of and a supplement to every large-scale investigation. Figures alone are practically worthless. What we need is relations.

Soil Classification

Every piece of actual construction work represents an expensive full-scale test. If we are in a position to identify accurately the soil encountered at construction, the value of the test results is inestimable. If, on the other hand, the soil remains unidentified, the same test is practically without any value, so far as future utilization of the test results is concerned. Hence, a method for soil identification is just as important as a theory of models.

It has been shown that variations in character of soils are merely the visible effect of corresponding variations in size and shape of the soil particles, in water content and in structure. If these four factors were known, it would be easy to identify soils. That the importance of these factors was well appreciated is shown by the fact that the Foundation committee's classification system takes them duly into account, though unfortunately in a form not applicable to practical purposes. The properties of most soils depend on those of their smallest (so-called colloidal) particles, and one of the most important characteristics of these particles, the shape, is not susceptible of quantitative expression. Similar difficulties are met in dealing with another factor of outstanding importance, the chemical character of the adsorbed constituents. Due to the utter complexity of colloidal behavior the problem of comparing soils is in some respects like that of comparing the quality of two apples. To solve this problem according to the proposals of the Foundation committee would mean basing the comparison of the apples on the results of a physicochemical analysis of their organic constituents. Realizing the doubtful benefit to be derived from so elaborate a procedure, the author proposed, symbolically speaking, the more direct method of biting into the apples and tasting them.

Basis of Classification—In the realm of soils, the taste of the apples corresponds to the common physical properties of soil (elasticity, cohesion, permeability and the like). Although these properties are numerous and each represents the combined effect of several causes, they offer the decisive advantage that they are susceptible of accurate numerical expression by coefficients and diagrams.

The accompanying table gives a complete list of these properties. In their totality these properties express every detail of the intricate properties of the soil constituents, somewhat as the elastic properties of a metal express the nature of its intricate molecular structure, or as a man's actions express his character.

Due to the great number of items, an exhaustive study of a soil sample cannot possibly be made except for scientific purposes. However, theory and experience show that the several properties listed in sections *B* and *C* of the table are all interrelated, although the precise nature of this relationship cannot yet be stated. A systematic experimental investigation is required to establish it. If 50 or more soils of widely different character should be tested according to the full schedule of the table, it might be expected that the assembled results would soon reveal the nature of the interdependence of the physical properties. With this knowledge the tests required for furnishing complete data on the character of a particular soil could be so simplified as to make them applicable to every earthwork reconnaissance—a future result likely to be of the utmost value. By a partial unraveling of these relationships I recently succeeded in simplifying the method for estimating the intensity of the capillary pressure acting at different depths of a mud deposit, to such an extent as to save nine-tenths of the time required for testing the drill samples.

This illustrates the gain that may be expected to result from the study of the relations between soil properties. Until these relations have been worked out, however, and test programs framed accordingly, it will be necessary to adapt any phase of soil tests to the particular practical objective of the work in hand.

Adequate Design Data

Design Equivalents—In soil mechanics as in other fields, ready calculation for purposes of approximate analysis and designing requires making use of simplified empirical assumptions. In hydraulics we assume water to be incompressible, and in mechanics of materials we use Hooke's law as if strictly true. Similarly, for soil mechanics we may assume as closely correct the following principles: (1) That the ratio between modulus of elasticity and internal pressure is constant; (2) that the relation between stress and strain, for direct compression with free lateral expansion is parabolic; and (3) that Darcy's law is valid. These assumptions are extremely simple, yet most practical problems are of such nature as to lead to intricate problems in the calculus of variations if an exact solution were attempted. Even if the stress-strain law were linear (Hooke's law) it would be exceedingly difficult to deal with a simple case such as stress distribution in a trapezoidal earth embankment, while with the parabolic law the problem becomes impossible. Obstacles of this kind make necessary a further process of simplification, using *design equivalents*, i.e., approximations similar to those utilized in other fields—as when a steel bridge is assumed to have frictionless joints—and subsequently making separate allowance for the approximation if necessary.

This method of equivalents may be applied as follows: In studying, say, the effect of a load on the supporting soil, or the swelling of clay after driving a tunnel through it, or the effect of time on pile friction, the first step is to study what physical factors are involved in the process, which is readily done if the physics of soils is understood. Then an equivalent is devised, so as to represent an interaction of physical factors closely similar to the actual one, but yet simple enough for mathematical analysis. Formulas based on this equivalent give a clear conception of the relative influence of the several physical factors on the actual process, although they are strictly approximate ones.

To take an example, the equivalent for a circular foundation is a cylindrical soil core surrounded by a cylindrical soil ring. The top of the soil core corresponds to the loaded area. The loaded core presses laterally against the inside of the ring. The proper ratio between radius of core, height of core, and width of ring was estimated from the results obtained by photographing through a glass window the grain movements produced in soil by a superincumbent load. Formulas derived from such an equivalent agree better with test results than the formulas furnished by any other known method, and in addition give other valuable information, such as the influence of rate of load application on bearing capacity (in the case of wet clay), or the influence of the shape of the base (circular or rectangular) on the bearing capacity.

DATA REQUIRED FOR DESCRIBING PHYSICAL PROPERTIES OF HOMOGENEOUS SOILS

A—Material in Original Condition

1. Volume of voids, in per cent of total volume. For clayey soils in addition: microscopic structure (homogeneous, fissured, crumbly, etc.).
2. Moisture content, in per cent of weight of solid matter.

B—Modified Condition

3. Shape of grains (by microscope; include sketches where possible).
4. Specific gravity of grains (by pycnometer).

5a. (For plastic soils.) Limiting moisture content of liquid, plastic, semi-solid and solid consistencies.

5b. (For sand and granular soils.) Limiting void ratio of loosest and densest structure.

6. Coefficient of internal friction (minimum and maximum values determined at zero capillary pressure).
7. Lateral-pressure ratio in confined material (determined at zero capillary pressure).
8. (For plastic soils only.) Compressive strength of cube (dried at 100 deg. C.).

C—Diagram Characteristics

9. Uniformity (grain-size platted on semi-logarithmic chart).
10. Pressure-void diagram under continuous and cyclic loading.
11. (For plastic soils only.) Load-compression diagram for cubes of known water content under continuous and cyclic loading. (The diagrams 10 and 11 combined make it possible to compute the constants of elastic behavior.)
12. Permeability-void diagram.

Aside from its direct use in design, the method of equivalents will be of service in developing a classification of soils for purposes of practice. If the kind and relative importance of the physical factors governing a particular field phenomenon (as stability of slopes, or bearing capacity of ground) is known, a sample of soil may be tested in the laboratory for the particular coefficients which control these physical factors; then it may be concluded that every other soil having approximately the same values for these particular coefficients will behave similarly, so far as this phenomenon is concerned. The other physical factors can be disregarded, without affecting the value of the result.

Proper use of the method also helps to avoid unwarranted generalizations. Such generalizations have been particularly common in connection with the retaining-wall problem. Our classic earth-pressure theories are themselves mere design equivalents, in the sense just discussed; the sand is assumed to have a definite coefficient of internal friction, the effect of soil deformation on pressure distribution is neglected, etc., all of which assumptions are tolerable only for the purpose of the equivalent. Yet most of those who have dealt with retaining-wall experiments and analysis look upon these assumptions as rigidly correct, and in some cases resort to highly advanced mathematics to calculate the curved sliding plane of least resistance, or other details whose importance is negligible compared with the errors involved in neglecting the elastic effects. Again, the attempt has been made many times to apply Rankine's theory to determining bearing capacities, completely overlooking the fact that arching effects prevent the soil particles near the base of the load from yielding laterally.

Or, again, consider the striking remark made by a well-known authority in a recent discussion: "When an experimenter finds the angle of friction greater than the angle of repose, it at once condemns his results and shows that he is not well grounded in theory," which overlooks the radical difference between a change from motion to rest (which determines the angle of repose) or a change from rest to motion (which determines the internal friction). The fact is that the inclination of the surface on which a moving body comes to rest is invariably smaller than the angle at which a body at rest starts to move. If existing theories do not agree with such facts, the theories, not the facts, must be modified. But these examples are sufficient to show the danger of unwarranted generalization in soil mechanics. A proper study of equivalents will help to avoid this danger.

Further Developments

Engineering practice concerned with soils still depends, as did structural engineering a century or two ago, on the instinctive genius of a few individuals, and on the chance discovery and development of such individuals. Efficient and economical design is still largely a matter of hopeful anticipation rather than actual accomplishment. Moreover, whatever is accomplished in this field by the strength of individual genius remains merely an admirable monument and is of no lasting benefit because we remain powerless to identify the soils available for similar operations in two different localities, and lack methods for interpreting and applying the results of tests.

To improve this condition we need first of all a new generation of earthwork engineers, thoroughly familiar with the essentials of soil physics and trained from the very start for exhaustive analysis of the phenomena which are apt to occur in actual earthwork practice. Considering the amount and the quality of knowledge required for such activities, the courses offered by the colleges in engineering geology and in foundation engineering are obviously inadequate. On the other hand, it does not seem feasible to extend the scope of these courses without simultaneously reducing the time provided for other important subjects. Hence, the only appropriate measure would be to increase the number of traditional civil engineering options (sanitary engineering, structural engineering, etc.) by one. This new option (earthwork and foundation engineering) should include among others a more elaborate course in engineering geology and an elementary course in applied colloid chemistry.

The second essential requirement concerns the activities of engineers engaged in earthwork and foundation engineering. No satisfactory improvement is possible unless at least some of these engineers keep in permanent contact with the science of soil behavior, and try to apply the available information to their observations in the field. Progress in engineering essentially depends on the quality of the data published in professional papers, and thus far, due to inadequate discrimination between essentials and non-essentials, most of the published data have but little value.

Finally the profession needs at least one institute for scientific earthwork engineering, equipped with suitable laboratories and with a trained research staff, preferably incorporated in the organization of the department of civil engineering of some leading university. Besides serving educational and research purposes, such an institute would have two essential functions: to analyze and digest the results of important observations made and communicated by outside engineers, and to develop suitable field methods for preliminary work, based on scientific principles.

If development plans of this kind are followed out, soil mechanics may within a short time become an important factor of engineering science. In the absence of such conscious progress, the experiences of the present generation will be as valueless as those of the past, because of unintelligible records of the properties of soils.

[1]See an article by the author, "Old Earth Pressure Theories and New Test Results," published in *Engineering News-Record*, Sept. 30, 1920, p. 632; and the preceding articles of the present series, "Principles of Soil Mechanics," as follows: I—Phenomena of Cohesion of Clay, Nov. 5, 1925, p. 742; II—Compressive Strength of Clay, Nov. 12, p. 796; III—Determination of Permeability of Clay, Nov. 19, p. 832; IV—Settlement and Consolidation of Clay, Nov. 26, p. 874; V—Physical Differences Between Sand and Clay, Dec. 3, p. 912; VI—Elastic Behavior of Sand and Clay, Dec. 17, p. 987; and VII—Friction in Sand and Clay, Dec. 24, p. 1026.

[2]"Erdbaumechanik," by Charles Terzaghi (1925).

[3]Transactions, Am. Soc. C. E., 1920-1922.

[4]Transactions, Am. Soc. C. E., 1916, p. 351, Table II.

Effect of Minor Geologic Details on the Safety of Dams

By Charles Terzaghi,* Cambridge, Mass.

(New York Meeting, February, 1929)

"Minor geologic details" refer to features that can be predicted neither from the results of careful investigations of a dam site nor by means of a reasonable amount of test borings. They include such items as the exact position and the variations in width of fissures passing through the rock beneath a dam foundation, the shape and the local variations of the permeability of minor seams of coarse sand and gravel contained in fine-grained alluvial valley fills, and similar features of minor geologic importance.

Let us assume that the investigation of two different dam sites, *A* and *B*, has led to identical results—that is, the difference between the two sites consists exclusively of minor geologic details. Dam *A* was constructed and proved to be perfectly successful. According to current engineering reasoning, the success obtained by constructing dam *A* should be sufficient for being sure that dam *B* would also be a success, provided that dam *B* is a copy of dam *A*. As a matter of fact, in the field of foundation engineering almost all decisions and expert testimony is based on such analogies because no theory is supposed to be available which will yield more reliable information.

To what extent can an opinion based on analogy be depended on in the field of dam foundations? Is such an opinion justified or is it liable, in certain cases, to be misleading? In order to answer this question we must start with a clear conception of the physical factors which are likely to endanger the dam. Then we must translate the terms of the geologist into terms of physics, and finally we must draw practical conclusions. Two examples should serve to illustrate the procedure.

Dam Site of Limestone with Horizontal Bedding Planes

The first one concerns two dam sites, *A* and *B*, consisting of limestone with practically horizontal bedding planes. According to the testimony of the geologist, the individual rock layers are in both cases approximately equally thick and in both they are crossed at right angles by a few rather irregular fissures. By testing the drill-holes under pressure we also find that the fissure system has in both cases approximately the same

* Professor of Foundation Engineering, Massachusetts Institute of Technology.

flow capacity. Hence, the difference between the two sites can reside only in the position of the vertical fissures relative to the base of the dam and the local variations of the width of the fissures. As the geologist can not give information regarding these details, we have to consider the extreme mechanical possibilities connected with the existing geologic situation. Are these details of any practical consequence, and if so what is their importance?

Fig. 1 is a simplified cross-section of the proposed dam and of its rock foundation. $a'b'$ is supposed to represent one of the horizontal joints and aa' and bb' two vertical joints. The water enters the ground at a and leaves it at b. The hydrostatic uplift which acts on the roof of the joint is represented by the trapezoid 1-2-3-4. The dam is stable. However, if the vertical downstream fissure is not at bb' but at $(b)(b')$, the hydrostatic uplift will be as indicated by 1-2-(3)-(4), whereby section $b(b)$ of the rock will stand under an appreciable upward pressure. Finally if the vertical downstream fissure is much narrower than the horizontal joint or else completely absent, the hydrostatic upward pressure may be sufficient to lift section $b(b)$ of the ledge and the dam would fail. Accidents of that type seem to have happened at weir No. 26[1] on the Ohio River and at the Jumbo dam. In the first case a section of the weir together with the layer of hard shale on which it rested traveled about 150 ft. downstream; in the second case sections of horizontal layers of sandstone were forced up, downstream from the dam, and the failure of the rock strata was followed by a failure of the dam.

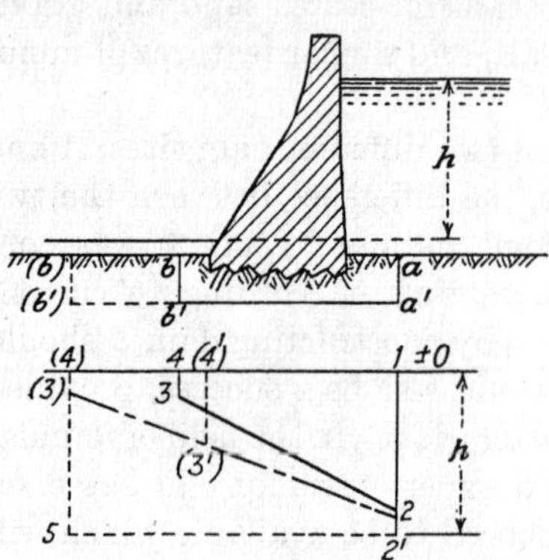

FIG. 1.—IDEAL CROSS-SECTION THROUGH MASONRY DAM WITH HORIZONTALLY STRATIFIED FOUNDATION.

Hence, it is obvious that dam A may be a perfect success, while dam B may collapse, although both were supposed to be identical and both were supposed to be on geologically identical sites. The difference resided only in the chance factor of the arrangement and the relative flow capacity of the joints and fissures as shown in Fig. 1. Yet, according to current engineering conceptions, dam site B should be fully satisfactory because A was a success. Moreover, the customary practice of reasoning on the basis of precedent is by no means limited to geologically identical sites. The mere fact that standard rules exist for rock foundations in general, concerning the intensity of the hydrostatic upward pressure, irrespective of the type of stratification and irrespective of the type of treatment by

[1] Failure of Ohio dam No. 26, Ohio River. *Eng. News* (1912) 366.

grouting, indicates a dangerous tendency to generalize. The effect thereof is an intolerably wide margin in the factor of safety of dams on rock foundations, ranging between a little more than unity and high figures. The remedy would be to abolish the standard rules and replace them by specific rules covering different geologic situations. Rocks with nearly horizontal stratification and few vertical fissures should represent the class of maximum danger, as far as hydrostatic uplift and stability against sliding is concerned, while rocks with irregular, groutable joint systems present the smallest risk.

Weirs on Deep Unconsolidated Alluvial Fills

The second case which will be considered concerns weirs on unconsolidated alluvial fills of great depth. We assume again two dam sites *A* and *B*. In both the geologist is supposed to have found that the deposit consists of fine sand extending to a depth of 200 ft. below the

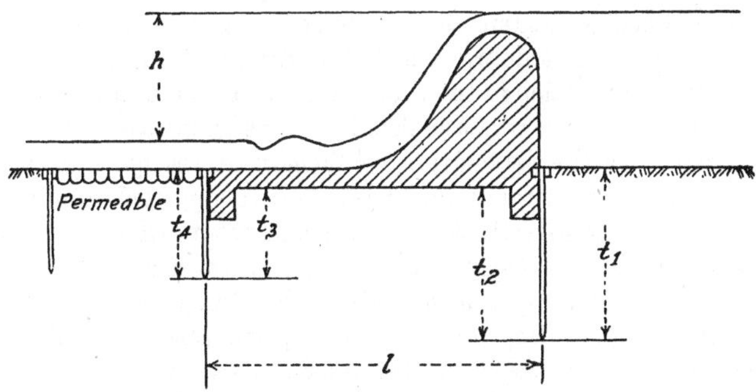

Fig. 2.—Foundation of weir designed to keep percolation coefficient within safe limits.

bottom of the valley. Furthermore, at both sites, at a depth of 30 ft. below the surface a layer of exceptionally clean, coarse sand with a thickness of 3 ft. was found. With these statements and some additional information on the geologic history of the deposit the geologist has fulfilled his obligations. He does not and can not tell whether the coarse-grained layer continues up and downstream toward the bottom or the surface of the valley fill, and to what extent the permeability of this layer varies in horizontal directions. The engineers, when dealing with unconsolidated deposits of great depth are obliged to design a weir with a shallow foundation. The most important danger which threatens the existence of such weirs is the danger of "piping," whereby the water forces its way from the storage reservoir through the underground towards the tail race. To avoid this danger it is customary to design the foundation of the weir so as to keep the "percolation coefficient" *c* within safe limits (Fig. 2.) The coefficient apparently was introduced into engineering practice by

W. P. Bligh.[2] It represents the inverse ratio between the hydraulic head h and the width of the impermeable part of the base b of the weir plus twice the sum t of the depth of all the sheet piles, or

$$c = \frac{2t + b}{h}$$

Thus for the weir shown in Fig. 2, the percolation factor would be equal to

$$c = \frac{t_1 + t_2 + t_3 + t_4 + b}{h}$$

According to Bligh, the percolation coefficient c should not be less than the following values:

Fine sand and silt	18
Fine micaceous sand	15
Coarse sand	12
Gravel and sand	9
Boulders with gravel and sand	4 to 6

If the design conforms to this rule, the engineer is usually satisfied and even inclined to assume that his weir has a definite factor of safety, similar to a concrete structure which was designed on a basis of 500 lb. per sq. in. for compression in the concrete and 14,000 lb. per sq. in. tension in the steel. However, we have decided to be skeptical in our proposed investigation. As the percolation coefficient represents a rather crude empirical conception without direct relation to physical standard units, we try first of all to investigate the physical factors which lead to piping and form our opinion on the percolation coefficient afterwards.

Fig. 3 shows a cross-section through an imaginary dam with a flat base and no sheet piles, resting on a uniform, permeable material; sand, for example. If we store the water behind the dam, it will flow along the lines indicated beneath the base of the structure. While it flows through the sand, it exerts thereon a pressure which is known to act at every point of the underground exactly in the direction of the flow. From theoretical hydrodynamics we also know the following facts:

If we select the lines of flow in such a manner that all spaces between lines pass the same quantity of water, both the velocity of the flow and the pressure which acts on the sand in the direction of the flow are in every point inversely proportional to the distance between the lines of flow. This statement, which is mathematically correct, makes it possible to compute for any point of the underground the force which acts on the sand under the influence of both the force of gravity and the pressure exerted by the flowing water. Fig. 3 shows

[2] W. P. Bligh: Dams, Barrages, and Weirs on Porous Foundations. *Eng. News* (1910) 708.

the result of the computation for different heads of impounded water. On the upstream side, where the water enters the ground, the pressure exerted by the flowing water acts in the direction of the force of

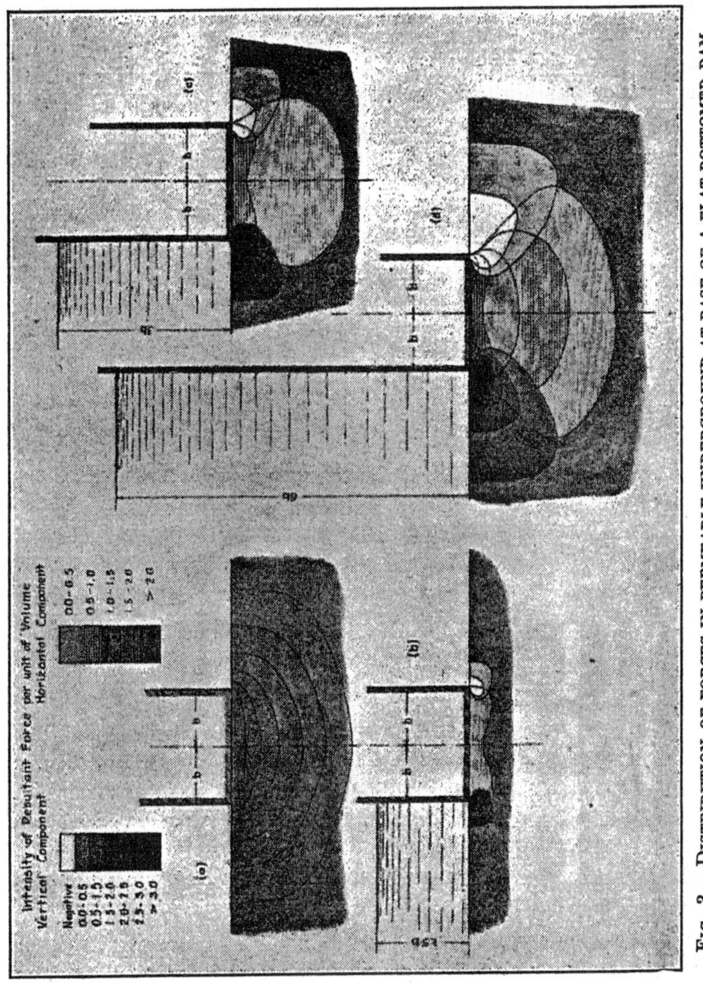

FIG. 3.—DISTRIBUTION OF FORCES IN PERMEABLE UNDERGROUND AT BASE OF A FLAT-BOTTOMED DAM.

gravity. Beneath the dam it acts more or less horizontally, while at the downstream end it counteracts the force of gravity. It is on this latter section that we have to concentrate our attention. The greater the head the more important becomes the upward pull exerted by the escaping water on the soil near the down-stream toe until, at a certain head, this upward pull of the water over-balances the downward pull exerted by the force of gravity.

The physical effect of this event was investigated by experiment and the result of the investigation is shown in Fig. 4. Before this event occurs, the quantity of water which percolates through the sand increases in direct proportion to the hydraulic head. However, the moment that the upward pull becomes equal to the force of gravity, the quantity of discharge increases abruptly as shown by the upward bend of the full

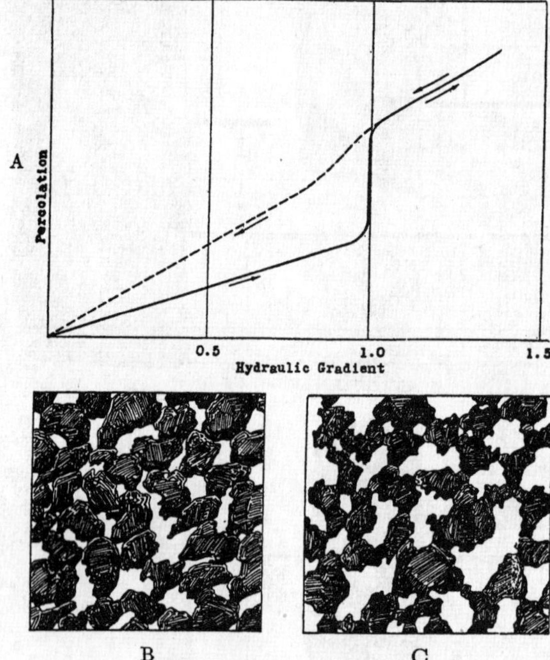

B C

FIG. 4.—EFFECT OF AN UPWARD CURRENT OF WATER (A) ON THE PERMEABILITY, (B AND C) ON THE STRUCTURE OF A SAND.

drawn line in Fig. 4*A*; at the same time the structure of the sand passes in the vicinity of the downstream toe from the dense state, Fig. 4*B*, into the loosest state at which it is still stable (Fig. 4*C*).

To investigate the effect of such a disturbance on the stability of the foundation of a weir we first of all designed a weir and computed the head at which the upward pull of the flowing water exceeds the downward pull of the force of gravity. Then we constructed a model of the weir and allowed the impounded water gradually to rise up to the computed critical height. As soon as this point was reached, piping occurred and water rushed with great violence through the sand beneath the weir.

A channel was formed and the rigid model bridged the channel, as often happens in practice.

Physical Causes of Piping

Thus we learned to understand the physical causes of piping. The fundamental requirement is that the upward pull exerted by the seepage water overcomes at some point on the bottom of the tail race the downward pull exerted by the force of gravity. As soon as this event occurs

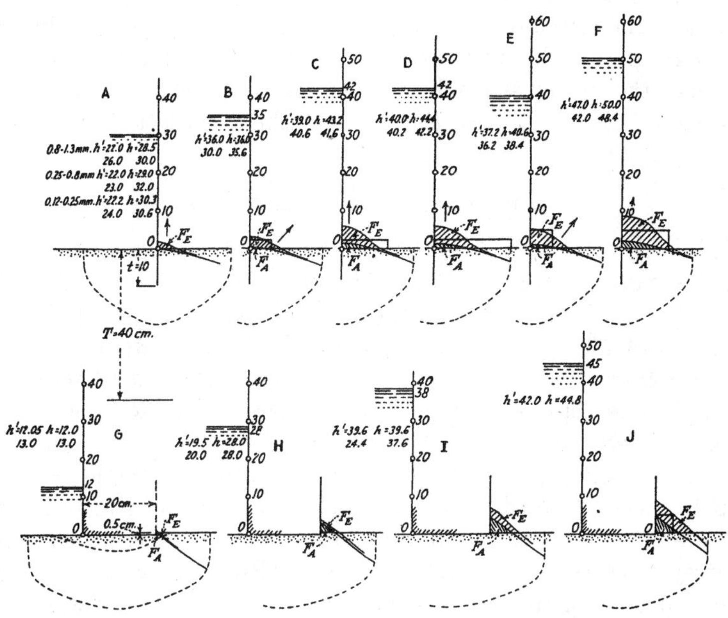

FIG. 5.—EFFECT OF FILTER LOADS (RECTANGULAR AREAS) ON THE HYDRAULIC HEAD *h* AT WHICH PIPING OCCURS. (FROM AUTHOR'S BOOK, ERDBAUMECHANIK.)

The ordinates of the shaded areas represent the computed uplifting effect which corresponds to the head *h*.

the dam is lost, no matter where it takes place. After this fact was clearly understood we were in a position to compute, for different types of foundations and for different materials, the critical head and the tests which we made for the purpose of checking the theory invariably confirmed what was predicted. Fig. 5 shows the results of some of these tests. In cases *A* to *F* the retaining structure consisted of an impermeable diaphragm; in cases *G* to *J* it consisted of a massive body with a flat base. According to current opinion the degree of safety of a dam depends on the value of the percolation coefficient. In contrast to this, both theory and the test results of Fig. 5 show nothing of that kind. For instance, the percolation coefficient at which the water piped in case *G* was almost three times greater than the percolation coefficient at which it piped in case *A*. The other figures show the effect of coarse-grained surcharges (plain rectangles) on the head at which the water piped. By properly selecting and placing the surcharges it was possible to bring the percolation coefficient at which piping occurs up to almost any value.[3] Thus it was conclusively demonstrated that the percolation coefficient represents only one of several equally important factors on which the safety against piping depends. The ultimate criterion for piping—namely, the condition that the upward pull of the water at some point of the surface of the tail race overcomes the downward pull of the force of gravity—may, in certain cases, have little to do with the value of the percolation coefficient.

Composition of Unconsolidated Deposits

Following this analysis of the piping effect we may return to our example—namely, two apparently identical dam sites, *A* and *B*, whose unconsolidated, fine-grained underground contains a coarse-grained layer at a depth of 30 ft. below the surface. Fig. 6*E* shows a cross-section through a similar dam site without any coarse-grained material and *F* and *G* are cross-sections through the dam sites referred to in our example. The only difference between the two dam sites *F* and *G* resides in the fact that the coarse-grained layer is in case *F* almost straight, whereas in case *G* it curves upward. The percolation factor is in all three cases, *E* to *G*, the same. Hence, according to current engineering conceptions the safety of these three structures should also be practically the same.

To find out whether this conception is justified we constructed, for each one of the three structures, a system of lines of flow which, according to the laws of hydrodynamics, intersect at right angles with the curves of equal pressure. Such systems of curves are known as "isothermal families" and to construct such families is a problem not of hydrodynamics but of geometry. In each of the figures the set of flow lines consists of seven curves. According to our assumption, the quantity of water which flows between each pair of curves is the same and equal to one-eighth of the total discharge, whereas for two different figures it may be different, depending on the flow capacity of the underground. In each one of the figures the danger spot—that is, the spot where piping would start—is on the downstream side of the retaining structure, at the point where the distance between the ends of the flow lines is a minimum. The upward pull exerted by the water at the danger spot is inversely proportional to the distance between the ends of the flow lines and directly proportional to the quantity of water which flows between two lines.

---

[3] C. Terzaghi: Erdbaumechanik. Vienna, 1925.

If we compare cross-sections *E* to *G* we learn the following facts: In *E* and *F* the minimum distance between the flow lines is approximately the same, while, due to the great flow capacity of the coarse-grained layer in the underground of *F*, the discharge per pair of flow lines in case *F* is greater than in case *E*. Hence, the safety of *F* is inferior to the

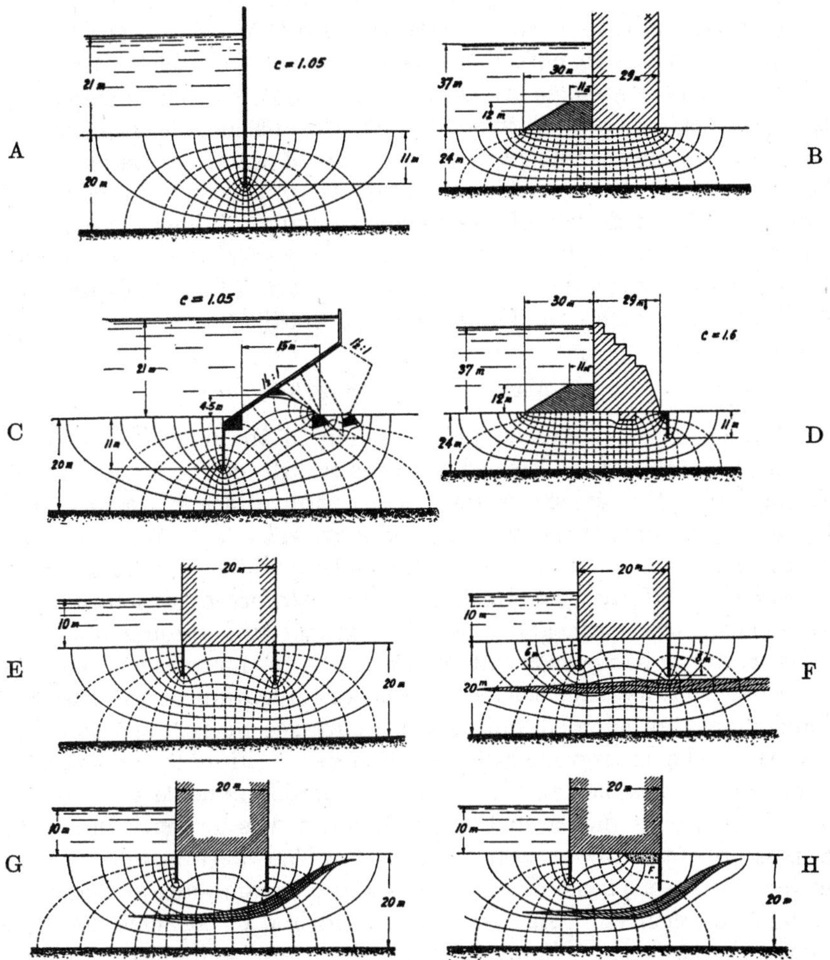

Fig. 6.—Effect of veins of permeable materials or of construction details on the shape A, distribution of the lines of flow of the seepage water. Piping starts where the distance between the downstream ends of the flow lines is a minimum.

safety of *E*, and dam *F* could fail while *E* would subsist, although the percolation coefficients are identical. In *F* and *G* the minimum distance between the flow lines is also practically identical. The danger spot—that is, the spot where the ends of the flow lines are closest together—is at *F* near the downstream toe and at *G* near the point where the coarse-grained stratum flares out. Because of the coarse-grained stratum approaching in case G the bottom of the river, it will carry much more water than it does, at equal head, in case F. Finally, due to the great velocity with which the water flows through the coarse-grained layer in G, intense erosion may take place at the boundary between the coarse and the fine sand and the material could be carried away into the tail race. No such transportation of material is possible in case F. Hence, dam G may fail, while F may be amply safe, although both dams have identical percolation coefficients and although the geologic profiles are apparently identical, except for the minor geologic detail of the shape of the permeable stratum. Yet engineers are used to comparing the safety of weirs on the basis of the value of the percolation coefficient, regardless of the type of structure and regardless of the details of the geologic profile. Hence, the factor of safety of two dams which, according to current standards, are considered to be equally safe could, in fact, be as different as 1 and 20, depending on the design and on major and minor geologic details.

Mechanics of Piping

To what extent a clearer understanding of the mechanics of piping would be helpful can be learned from the failure of the Hauserlake and Elwha dam. Fig. 5C shows the cross-section of the Hauserlake dam.[4] The underground of this dam consisted of gravel. The failure obviously started in the immediate vicinity of the footing which supported the steel construction. If the gravel backfill of the reinforced concrete skin had been much more permeable than the natural underground and if, in addition, a filter had been constructed beneath the footings with an outlet toward the downstream side, the local concentration of the flow lines would have been avoided and as a consequence no failure would have occurred. The underground of the Elwha dam[5] (Fig. 5D) consisted of a coarse and permeable gravel. Piping occurred immediately after the row of sheet piles shown on the cross-section was driven. Because of the presence of this row of sheet piles, the flow lines were crowded together between the sheet piles and the body of the dam and, as a consequence, the blowout occurred. Fig. 5H shows a method of construction which was used by the speaker in connection with several weirs in the Austrian Alps.[6] A graded filter F beneath the body of the weir diverts part of the

---

[4] F. L. Sizer: Break in the Hauser Lake Dam, Montana. *Eng. News* (1908) **59**, 491.

[5] B. L. Campbell: Accident to the Dam at Port Angeles, Washington. *Eng. News* (1912) **68**, 1232.

V. H. Reineking: Reconstruction of the Elwha River Dam. *Eng. Rec.* (1914) **69**, 373

[6] Owners of the patent for Austria: J. Pfletschinger & Co., hydraulic consulting engineers, Vienna.

flow lines towards the bottom of the weir, where the upward pull exerted by the flowing water is more than compensated by the dead weight of the structure.

Effect of Grain Size on Percolation

There is one more point concerning the percolation coefficient which deserves careful attention: The theory of piping brings out the fact that the grain size has much less influence on the value of the percolation coefficient at which piping occurs than has the type of foundation; in other words, with the same underground conditions and equal percolation coefficients, the factor of safety of two dams may be very different, depending on the general design of the cross-section of the weir. In contrast to this, the rules of Bligh previously referred to completely disregard the effect of the design of the dam on the critical percolation coefficient and seem, in addition, to indicate that the danger rapidly increases with decreasing grain size. The theory already quoted refers exclusively to perfectly uniform materials. For such materials, however, it was conclusively confirmed by experiment. Hence, within its range of validity—that is, for uniform materials—it must be strictly correct. On the other hand, Bligh's values were derived from a great number of actual dam failures and therefore must also express at least part of the truth.

Theory and Experiment versus Statistical Study

A study of this apparent contradiction between theory and experiment on one side and the results of Bligh's statistical investigations on the other, has led to rather important conclusions concerning the degree of safety of dams with a sand or gravel foundation. Without going into the details of the investigations, the conclusions can be summarized as follows:

If piping occurs in homogeneous materials as, for example, in uniform sand, no sand grain within the underground changes its position until the hydraulic head reaches almost the critical value. Failure occurs spontaneously almost without warning. Hence, if the head remains permanently below the critical value, the factor of safety remains practically constant. In contrast to this, for stratified undergrounds, the percolation of water through the substrata invariably leads to minor changes due to local erosion along boundaries between different strata, and due to deposition at those points where coarser strata flare out. Some physical aspects of these minor changes were recently investigated by A. Läufer in Vienna,[7] and the effect of these changes on the intensity and distribution of the hydrostatic uplift on the bases of the weirs became apparent from

[7] A. Läufer: Lassenbildung und deren Verheilung. Die Wasserwirtschaft (1927) **24.**

some data recently published by Hinds.[8] The nature of the changes depends to a large extent on what were called minor geologic details and their practical effect is a gradual change of the factor of safety towards either higher or lower values. A stationary state is not reached for several years. The finer and the more uniform the material of the individual strata, and the greater the difference between the permeability of adjoining strata, the more important the gradual changes are likely to be. For this reason the failure of weirs resting on stratified material sometimes occurs several years after the water is stored up, whereas, according to the theory for uniform materials, and according to laboratory evidence obtained for uniform materials, failure occurs either at once after the maximum head is brought against the structure, or not at all. Furthermore, for the same reason, the safety against piping through stratified ground beneath weirs with a definite percolation coefficient should rapidly increase with decreasing grain size, strictly conforming to Bligh's conclusions, whereas for perfectly homogeneous underground, the safety against piping should be practically independent of this factor. Thus we have learned that the difference between theory and practice is due to the possibility of erosion and deposition along geologic boundaries within the underground, a factor which theory disregards. The nature of these important changes depends essentially on what we have called minor geologic details.

The defects of the theory mentioned are matched by other and no less important shortcomings of Bligh's interpretation of the facts. As Bligh, or anybody else who would attempt a similar purely empirical investigation, was of necessity limited by the small amount and fragmentary character of observed or published evidence concerning dam failures, he could not possibly succeed in making a complete analysis of the causes of the failures. He selected among the several factors on which the safety depends, a single one, the percolation coefficient, and disregarded all the others. For this reason the factor of safety of two weirs, both resting on coarse sand, and both designed on the basis of a percolation coefficient of 12 may range between extremely wide limits. According to our present experience, these limits are more than 20 (most favorable case) and less than two, depending on the type of foundation, as well as on major and minor geologic details. Finally, Bligh's figures, based exclusively on statistics, merely express the fact that no weir with a percolation coefficient of more than 6, founded on coarse sand, has failed to date. This by no means precludes the possibility that a dam of such type may fail tomorrow. As a matter of fact, with our present knowledge of the physics of the process, we could easily succeed in

---

[8] J. Hinds: Upward Pressure under Dams. *Proc.* Amer. Soc. Civil Engrs. (1928) 686. Discussion by C. Terzaghi: *Ibid.*, 1953.

building up an underground and selecting a type of foundation in such a way that piping would occur in spite of a percolation coefficient of as much as 12.

Dams on Permeable Ground and on Rock

Thus we face in the field of dam foundations on permeable ground a similar situation as in the field of rock foundations. Bligh, some 20 years ago made with his figures one of the most valuable contributions on record in this particular field. Yet, on account of the unfortunate tendency on the part of the engineering profession to generalize, the blessing turned into a curse, inasmuch as engineers got used to the idea that the value of the percolation coefficient is a positive and reliable criterion for the safety of a dam on permeable ground and that the design based on Bligh's figures leads to structures which are both safe and economical. In contrast to this assumption, it has been shown that the factor of safety of two weirs with equal percolation coefficients and supported by apparently identical systems of strata, may be widely different, depending on the type of structure or on geologic details. The rules established by Bligh are based on the most unfavorable geologic and engineering possibilities which ever led to failure and take no account of the favorable factors which are likely to increase considerably the factor of safety of the weir foundations. Hence, many of the structures which are built according to his rules are far from being economical and strict adherence to the rules would prevent further progress in this field. To remedy this undesirable condition would require abolishing the standard rules and replacing them by specific rules whereby distinction should be made according to the type of foundation and the general geologic characteristics of the site. The investigation should be based both on theory, which furnishes information on the effect of the type of foundations, and on the results of Bligh's statistics which makes possible an estimate of the overall effect of all those factors which theory inevitably has to disregard.

Conclusions

Thus, our investigation has led to the following conclusions: If two dams of identical type are constructed under apparently identical geologic conditions, one dam may be safe and the other may fail, depending on minor geologic details. Present practice of design is based on precedent, whereby not only minor but also certain major geologic details are disregarded. Such practice means blindly trusting in purely statistical relations with an extraordinarily wide range of deviation to both sides from the average. As most of the textbooks fail to call the attention of the readers to the great uncertainty associated with the rules of design based on this practice, many engineers engaged in dam design have an exaggerated conception of the degree of reliability of their methods of procedure and as a consequence, progress in this field came practically to a standstill.

To avoid the shortcomings associated with present practice requires first of all expert translation of the findings of the geologist into physical and mechanical terms. Next it requires the evaluation of the most unfavorable mechanical possibilities which could be expected under the existing geologic conditions; and finally to assume for the design of the structure the most unfavorable possibilities. These mental operations represent by far the most important, most difficult, and most neglected tasks in the field of dam foundations. To perform them successfully requires above all a far more thorough training in general physics and in the hydraulics of seepage than most engineers receive while at school or while later engaged in practising their profession.

DISCUSSION

O. E. Meinzer, Washington, D. C.—In the tests of permeability of incoherent materials that are made in the hydrologic laboratory of the U. S. Geological Survey, water is passed upward through the material to be tested under slight head. The success of these experiments is believed to be due largely to the slight heads, hence, the slight hydraulic gradients that are used. It has been observed that if the head is progressively increased, in working with a given sample, the effluent remains clear and the rate of flow generally increases approximately in proportion to the increase in head, until a certain head is reached when at least the finest part of the material is lifted, the water becomes roily, and the rate of flow increases disproportionately. I raised the question as to whether this critical hydraulic gradient might be taken as a measure of the lifting phenomenon that was discussed by Professor Terzaghi.

G. H. Matthes, Los Angeles, Calif. (written discussion).—Professor Terzaghi has made a noteworthy contribution to the science of dam foundations and has cleared up a number of important points. It would be valuable, in connection with this paper, to have a statement from Professor Terzaghi regarding his findings on the subject of relieving upward pressure under dams built on rock foundations, by means of drill-holes and drain pipes.

It has been the custom of many engineers in this country to provide drainage in dam foundations in order to avert the very condition depicted by Professor Terzaghi in his Fig. 1. The practice in this respect varies considerably with individual judgment. In the main, it consists in drilling holes into the foundation rock under the downstream portion of the dam, at suitable intervals, to intercept water accumulating at the base of the structure and also in the foundation rock; for instance, in such a plane as is represented by the line b'a' in Fig. 1. Some engineers content themselves with placing drain pipes in the base of dams and omit the drill-holes in the rock. In either event, the assumption commonly made is to the effect that drill-holes and/or drain pipes reduce uplift pressure in their immediate neighborhood so as not to exceed the hydrostatic pressure due to the tailwater, and this is usually accompanied by the further assumption that this reduction in pressure may confidently be expected to extend continuously and undiminished from hole to hole.

Whatever merit there may be in draining foundations of dams as an added safeguard to their stability, the writer questions the propriety of taking into account such assumed reductions in uplift as factors in design. An expression from Professor Terzaghi on this point would, I believe, be welcomed by engineers and geologists.

I. B. Crosby, Boston, Mass. (written discussion).—Features which Professor Terzaghi classifies as minor geologic details not detectable by a geological investigation can nevertheless often be predicted by a geologist. Furthermore, where exact information is impossible on account of lack of exposures the geologist can foresee what unfavorable conditions may be brought about by these unknown details and thus warn the engineer so that he may design his structure to meet the worst possible conditions. In many cases, when the possibility of unfavorable conditions has been foreseen, exact information about them may be secured by a few drill-holes strategically placed in the light of the geological information already obtained.

The case given by Terzaghi of a dam site on limestone is a good example of the possibility of predicting the nature of these minor geologic details. Joint cracks in horizontally bedded limestone are often remarkably regular and if the rock is well exposed at the dam site the geologist can map out the system of joint cracks and show what effect they will have upon the dam. With that information the engineer can meet the conditions. Such studies have been made and safe dams have resulted, as at the State Line dam on the Cheat River, West Virginia, where a very detailed geologic study enabled the engineers to meet the serious difficulties presented by flat-lying, soft shales with an interbedded pervious layer. If the rock at the dam site is so concealed that it is not possible to obtain this detailed information the geologist should point out the most unfavorable possibilities, against which the engineer may then provide. In the example given, that of the failure of Dam 26 on the Ohio River, no geological investigation had been made.

In regard to the second case, that of weirs on unconsolidated alluvial fills, Terzaghi states that the geologist has found that

"The deposit consists of fine sand extending to a depth of 200 ft. below the bottom of the valley. Furthermore, at both sites, at a depth of 30 ft. below the surface a layer of exceptionally clean, coarse sand with a thickness of 3 ft. was found. With these statements and some additional information on the geologic history of the deposit the geologist has fulfilled his obligations. He does not and can not tell, whether the coarse-grained layer continues up and down stream toward the bottom or the surface of the valley fill, and to what extent the permeability of this layer varies in horizontal directions."

However, from experience it is evident that the geologist has not fulfilled his obligation with the statement quoted above, has not given all the information possible and that he frequently does give more detailed information. If a layer of coarse sand such as is shown in Fig. 6*G* existed under the dam site and rose towards the river below the dam this upward tilt of the layer would be reflected in the overlying strata and could be detected. Furthermore, it is often possible to learn the direction of the currents which deposited the sand and then it is known in what direction the sand gets finer and in which it becomes coarser. In this way the geologist can indicate the variation of permeability of the layer in horizontal directions. Where sufficient evidence cannot be obtained to give this detailed information the geologist can foresee these possibilities and with these in mind make a few borings which will show whether the layer of coarse sand remains horizontal or whether it dips upstream or downstream.

In the two examples of dam failure given by Terzaghi to illustrate "piping" the conditions are not similar to those quoted above, no geological examinations were made and at the Elwha River dam no borings or subsurface exploration of any kind had been made.

Exact information is extremely desirable but where it can not be obtained approximate information is valuable if interpreted and handled in the proper way with the realization that it is approximate merely. With this in mind the possible variations can be outlined and the worst conditions provided for in the design. The geologist must warn in what directions these variations may be, if he cannot determine which of them exist and to what extent.

C. Terzaghi (written discussion).—Mr. Meinzer raises the question whether the head at which the water starts to become roily could be considered a measure for the uplifting effect produced by the current. According to my observations, this is by no means the case. The head at which the water becomes roily essentially depends on the grain size characteristics of the material. If the material consists of coarser constituents with a small percentage of very fine material while grains of intermediate size are absent (for instance loamy sands) the water becomes roily at a very much lower hydraulic gradient than it does for more uniformly graded materials, although the uplifting effect would be practically the same for both materials provided the hydraulic gradient and the density of the materials are identical.

Mr. Matthes directs attention to a practice which may or may not be beneficial, depending on the method of application and on the interpretation of its effect. Drain pipes located at the base of the dam certainly do not serve their purpose unless the rock fissures are close together and fairly continuous. The effect of drill holes depends as much on chance as does the prospecting for water veins in fissured rock. That is, the drill-holes may or may not intersect the seats of important hydrostatic pressures, depending on the more minute details of the fissure system. The drill-holes merely reduce the probability of pressure effects in excess of what can be expected on the basis of statistical laws. Furthermore, due to the relatively small area over which the individual fissures are opened up, the relieving effect of the drill-holes is strictly local. For these reasons the author does not advise making any allowance for the effect of the presence of the drill-holes on the intensity of the hydrostatic uplift. The drill-holes should merely be considered a partial compensation for the uncertainty associated with standardized assumptions concerning the hydrostatic uplift in rock foundations. This is particularly true for foundations on horizontally stratified rocks where the drill-holes should never be omitted in spite of their not being absolutely reliable. The drill-holes should be drilled to a depth at least equal to the height of the dam and their spacing should not be smaller than about 20 feet.

With increasing experience the author becomes more and more convinced of the necessity to check the hydrostatic pressure conditions which exist after the construction of a dam is finished against the assumptions on which the design was based and to keep a record of the gradual changes in these conditions during a period of at least two years. The measuring instruments can easily be connected with the drill-holes.

Mr. Crosby is inclined to consider geological investigations a universal remedy for the present unsatisfactory state of our knowledge of dam foundations. As long as neither the average geologist nor the average engineer receives the thorough training in physics and mathematics required for properly appreciating the hydraulic importance of various geologic details, such attitude is apt to divert our attention from some of the widest gaps in our present knowledge and to breed self-sufficiency instead of stimulating much needed progress. Considering present practice, the author seriously doubts whether a geological investigation would have prevented the failure of either the Ohio Dam No. 26 or the Elwha Dam. To outline the possible variations according to the last paragraphs in Mr. Crosby's discussion requires knowledge and insight considerably beyond our present resources.

Engineering News-Record

Vol. 111 New York, August 31, 1933 No. 9

First of Four Articles on the Design and Construction of Rolled-Earth Dams

Fundamental Principles of Soil Compaction

By R. R. Proctor
Field Engineer, Bureau of Waterworks and Supply, Los Angeles, Calif.

THIS IS the first of four articles describing the application to the design and construction of rolled-earth dams of recently developed methods for controlling the compaction of soils. Under the new procedure a series of simple laboratory and field tests determines the suitability of the available soils, supplies essential data for the design of the dam, serves as a basis for the control of construction operations and furnishes a definite check on the effectiveness of the work as it proceeds. Methods are described for compacting soils so that they will be sufficiently watertight and will not become soft and unstable if completely saturated with water. The basic principles of the control of soil compaction that are described apply to all types of earthfills and to foundation design as well as to earth dams. —EDITOR.

THE NEED of a more precise procedure for the design and construction of earthfill dams initiated an extensive investigation by the Bureau of Waterworks and Supply of the city of Los Angeles into the suitability of various soils for this use, and a search for new methods of construction control that would be more certain in their application to the work than any that were available. It was particularly desired to find some method for determining and limiting the softening of an earthfill dam that may occur, under certain conditions, if the soils contained in it become completely saturated with percolating water.

The investigation resulted in the development of an entirely new method of procedure in the engineering control of earth-dam design and construction. The procedure has been further perfected in connection with the construction of four dams in the past two years, three of which are completed and in service. It is now in constant use in the construction of the Bouquet Canyon Dam No. 1, a compacted earthfill structure that will have a depth of fill of 221 ft.

As developed, the new procedure is a system of field and laboratory controls designed to secure the compaction of the soils to the density which, by actual test, is determined essential to obtain the required watertightness and stability in the dam. The required soil density is obtained by controlling the design and use of the compacting equipment and by placing the correct moisture content in the soil before it is compacted. Precise methods are provided for testing the compacted soils, as the work progresses, to determine their watertightness and stability under conditions similar to those in the finished structure. Similar tests may be made on foundation soils upon which the dam is to be constructed, thereby enabling the engineer to direct the removal of those soils found unsuitable. The methods described also apply to any earthfill, for any purpose, that is to be compacted to reduce settlement or softening from partial or complete water saturation.

Fig. 1—Placing compacted earthfill in Los Angeles' Bouquet Dam No. 1. Five sheepsfoot rollers of special design are used to compact the soil, and one is used to remove concentrated truck tracks. The soil is spread in loose layers of sufficient thickness to result in a 6-in. compacted layer.

Soils are composed of many shapes and sizes of sand, silt or clay particles that do not fill all of the space occupied by the soil. The space between the particles is known as the voids or porosity; both are expressed as their percentage of the total soil volume. The extent to which the smaller particles occupy the spaces between the larges ones determines the total quantity of soil particles that occupy a given volume, which determines the soil density and the voids. Soils are compacted by forcing the smaller particles to move into the spaces between the larger ones, thus increasing the soil density and decreasing the voids. The compacting force must overcome the frictional resistance between the soil particles in order to move them. The effectiveness of any method for compacting a soil is therefore limited by the friction between the particles. This will be shown to be determined by the moisture content of the soil.

Plasticity of soils

The plasticity of a soil is determined by the frictional resistance between its particles. It is known that the addition of water to a soil that is not already saturated makes it more plastic. The water added to a soil displaces air in the voids and has an effect similar to that of lubrication. The moisture contained in a very dry soil surrounds each particle as a thin film held in place by the force of surface tension. Where the films come in contact, the capillary force caused by the surface tension of the joined films draws the particles firmly together, causing a high frictional resistance between them. This effect may be seen by placing a few fine soil particles in water suspension on a glass slide and observing them under the microscope as the water evaporates. Small angular particles, composed of such materials as quartz, may be seen firmly attached to the glass slide by the force of capillarity, when the water has evaporated to the fullest extent possible without the application of heat. A full discussion of the effect of the surface tension of the water in soils may be found in the article "Principles of Soil Mechanics" by Dr. Charles Terzaghi *(ENR,* Nov. 5, 1925, p. 742).

The addition of water to a very dry soil reduces the capillary force, which has two effects; the particles spring apart slightly, due to the reduction of the capillary force, and the frictional resistance between them is reduced. The soil is then more plastic and has expanded slightly in volume. This effect continues with the addition of water to the soil until the forces of capillarity are almost entirely released and no further soil expansion occurs. The quantity of water in the soil when this condition is reached is not sufficient to fill the voids. The addition of still more water tends to lubricate the soil further, but no more expansion occurs. The limit of lubrication of the particles is reached when the

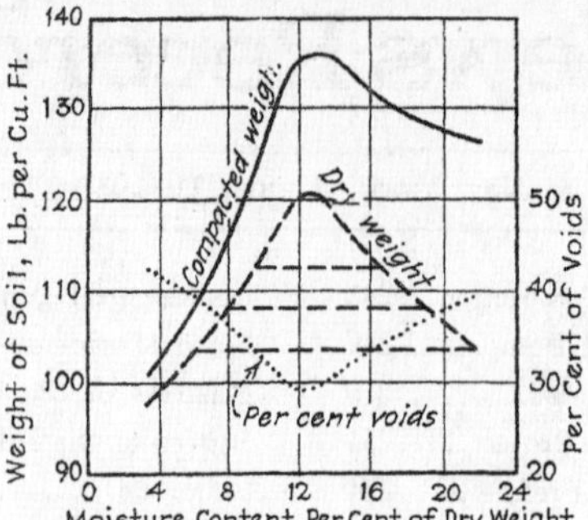

Fig. 2—Graph showing the effect of moisture content on the compacted weight, dry weight and voids of a soil when compacted by a particular method.

voids become filled with water; the friction between the particles cannot then be further reduced without rearranging them so as to increase the voids and permit the soil to absorb more water.

The limit of softening in a confined soil is therefore reached when the voids become filled with water and the soil is fully saturated. A particular soil may be compacted to many different densities and percentages of voids; obviously, there are then many degrees of softening that can take place from saturation due to this variation in water content necessary for complete saturation. Therefore, the extent to which the voids in a soil are reduced by compaction determines the softening that can take place from water saturation.

The soils in an earth dam may become fully saturated with percolating water; therefore, it is essential for the engineer to determine the density that is required to obtain the necessary stability in a particular soil before it is compacted in a dam. After the required soil density has been ascertained, an adequate process for securing it must be used, and facilities should be available for precise determination of the effectiveness of the work as it proceeds.

Moisture effect on compacted density

The effect of the moisture content of a soil upon the density to which it may be compacted is the most important principle of soil compaction. The discussion and conclusions that follow are based on data obtained from laboratory compaction tests on more than 200 different soils, each of which was compacted at several moisture contents by the same process. The laboratory methods are known to duplicate the results obtained by compacting the soil with certain types of construction equipment. The soils varied in grading from a silty gravel, 8 per cent of which passed the 200-mesh sieve, to a fine clay, 92 per cent of which passed that sieve. The results of the laboratory tests have been verified by many field tests during the construction of the dams previously mentioned.

The compaction of a soil at a very low moisture content may result in a hard and firm fill having practically no plasticity. Compacting the soil by the same method but with a slightly higher moisture content causes a greater rearrangement of the various-sized soil particles with the same effort, due to the increased lubrication furnished by the additional water. The result is a soil of greater density that contains less voids and is more plastic. This effect continues until the moisture content, plus a small amount of contained air that the compaction process cannot remove, becomes just sufficient to fill the voids when the compaction process is completed.

The soil now has the greatest density and the least voids that this method of compaction can obtain; further reduction of the voids is impossible without forcing out some of the water or contained air. It has not been found possible to do this with any construction equipment available. A higher moisture content limits the compaction of the soil to a point at which the voids equal the volume of the contained air and water, resulting in a compacted soil with more voids, less density and increased plasticity. This effect continues with the addition of more water until the soil becomes too plastic to sustain compacting equipment.

It is now evident that the moisture content of a soil, when it is being compacted, controls the soil density and consequent voids that can be attained by a particular method of compaction. The compacted soil containing the least voids will require less water to saturate it; therefore, when it is saturated it will have less lubrication between the soil particles and consequently less plasticity and greater stability. Therefore, the most stable dam that can be constructed, by use of a particular method of compacting the soils, will be obtained if the soils contain the moisture content that results in the least voids when they are compacted.

The voids are the channels of flow for percolating water; reducing them to a minimum, by compacting the soil at the proper moisture content, reduces the flow of water to the minimum attainable with the process of compaction used. Therefore, the use of the soil-moisture content that obtains the least voids with a particular method of compaction results in the most watertight and stable dam that the method of compaction can secure.

The use of still greater compacting forces and a correspondingly lower soil-moisture content results in a dam with increased soil density, decreased voids, increased watertightness, less capacity for water absorption, less plasticity and increased stability; the soil particles are arranged in closer contact by the greater force, with less water for a lubricant. Increased watertightness and stability are thus obtained by increasing the weight of construction equipment used for compacting the soil. The proper type of construction equipment, as well as the correct moisture in the soil, must therefore be used to obtain a desired degree of watertightness and stability in a dam.

The dry-weight curve

The curves of Fig. 2 were plotted from data secured by compacting a sandy clay soil at thirteen different moisture contents. The same method of compaction was used in each case. This compaction method is known to be equivalent to the use of a certain type of construction equipment. The compacted-weight curve shows the weight of soil per cubic foot after it was compacted at the indicated moisture contents. The dry-weight curve was obtained by deducting the weight of the contained moisture from the compacted weight per cubic foot. It therefore shows the actual weight of the soil particles in a cubic foot of soil that was compacted at the indicated moisture contents. The per-cent-of-voids curve, determined by the dry weight and specific gravity of the soil, shows the voids in the soil after it was compacted at the indicated moisture contents. The dry-weight curve is customarily used to show the characteristics of compacted soils, the other curves are given in order to show their relation to the dry-weight curve.

The plasticity of the thirteen compacted-soil specimens plotted in Fig. 2 was determined by measuring their resistance to penetration with an instrument designed for this purpose. Its design and method of use will be discussed later; for the present the penetrating pressures will be expressed as the pressure required to force a lead pencil, with an end area of 0.1 sq.in., to penetrate the compacted soil at a rate of about ½ in. per second.

The extent that the voids in this soil may be reduced by compacting at the proper moisture content is clearly shown. The compacted soil that contains the least voids is necessarily the most watertight. The compacted soil that contains the least voids cannot absorb as much water as the others; therefore, this soil will soften less from water absorption.

The soil had the same dry weight and voids when compacted at 6.2 and 22.0 per cent moisture contents. The plasticity of the two compacted soils was very different: the one containing 6.2 per cent moisture was very hard; and the other, containing 22 per cent moisture, was of about the consistency of soft putty. The pressure required for a pencil to penetrate the soil compacted at 6.2 per cent moisture content was 1,000 lb., the pressure required for the one compacted at 22 per cent moisture content was 0.8 lb. The same dry weight and voids in the two compacted soils determine that they can absorb the same quantity of water; therefore the soil that was compacted at 6.2 per cent moisture content can become similar to that which was compacted at 22 per cent moisture content by water absorption. This is the explanation of the extreme softening of compacted soils that sometimes occurs when they become saturated with water.

The soil in an earth dam is subject to partial or complete saturation by percolating water; for this reason the soil that was compacted at 6.2 per cent moisture content should be considered similar in every respect to that which was compacted at 22 per cent moisture content. The first consideration of any compacted specimen of this soil should be made by determining its dry weight and noting the position of this value on the dry-weight curve to the right of the point of maximum dry weight. It is definitely known that it can absorb at least this quantity of water. The remaining capacity for water absorption can be computed from the voids. A direct method for obtaining the capacity of compacted soils to absorb water will be discussed in connection with Fig. 6—in the second article.

Fig. 3—Placing compacted earthfill at Upper Hollywood Reservoir, recently completed by the city of Los Angeles. The soil shown in the center of the picture was successfully compacted by the methods described, after removing the rocks larger than 5 in. in diameter.

Compacting at low moistures

A moisture content of about 8 per cent permits rolling the soil used in preparation of Fig. 2 so dry that it will not spring under the roller. The resulting fill appears to be well compacted; it is firm and hard. The soil, when compacted at 8 per cent moisture, weighs 116.6 lb. per cubic foot, has a dry weight of 108 lb. and contains 36.5 per cent voids. It requires about 500-lb. pressure to force a lead pencil into this compacted soil. This same dry weight may be obtained by compacting the soil at 19.2 per cent moisture. The compacted weight is then 128.6 lb. per cubic foot. The increase in compacted weight of 12 lb. is caused by the increased water content. This quantity of water could be absorbed in the soil that was compacted at 8 per cent moisture without any change in the soil volume or voids. A pressure of 2 lb. would be required to force a lead pencil into this soil when compacted at 19.2 per cent moisture content. The pressure required for a pencil to penetrate the soil compacted at 8 per cent moisture should therefore be considered subject to a reduction from 500 lb. to 2 lb., or less, if the soil becomes saturated. This is not a sufficiently stable soil condition in a dam.

Compacting at high moistures

It may now be seen that it is possible to compact a soil so firm and hard as to appear entirely suitable for a dam and for this same soil to become very soft and unstable when percolating water saturates it. A compacted soil firm enough to require a pressure of about 15 lb. to force a lead pencil into it is required to prevent the wheels of a loaded truck from sinking into it. This condition is obtained in this soil at 16.5 per cent moisture content, when it is compacted by the method used in preparation of Fig. 2. Therefore, if a dam is to be built by hauling the soil in trucks and compacting it with construction equipment of the type represented by Fig. 2, the moisture content cannot be more than 16.5 per cent, which establishes a minimum dry weight of 112.5 lb.

It is now seen in this case that better results can be obtained by compacting this soil so wet and soft that it will barely support trucks than by compacting it so hard that it will not spring under the roller. The benefit would be an increase in dry weight from 108 to 112.5 lb. A dry weight of 112.5 lb. may also be obtained by using a moisture content of 9.5 per cent. About

200-lb. pressure would be required to force a pencil into the soil if compacted at this moisture. As the final results of compacting at the two moistures would be the same dry weight of soil, it would be best to compact the more plastic soil at 16.5 per cent moisture; the resulting fill would probably be more uniform in density. If more moisture than 16.5 per cent were placed in the soil, it could not be compacted sufficiently, because of this excess moisture, to permit truck travel over it, and the work could not proceed. If the moisture content were slightly less, the fill would be denser and more suitable, as shown in Fig. 2.

This shows that a soil may be compacted so as to be very plastic and have the appearance of being too soft for use in a dam, and still it may be superior to that which is compacted very firm and hard with a moisture content low enough to prevent springing under the roller. More than 200 compaction curves that have been experimentally determined, together with many field measurements, verify this conclusion.

The compaction of this soil at a moisture content of 12.5 per cent is shown by Fig. 2 to result in a dry weight of 121 lb. per cubic foot; this is the maximum compaction attainable with this type of equipment. The compacted soil contains 28.5 per cent voids. A pressure of approximately 100 lb. would be required to force a pencil into the compacted soil. Although this compacted soil is much more plastic than that which was compacted at 8 per cent moisture, one-fifth of the pressure being required for penetration, it contains 13 lb. more soil to the cubic foot. The voids are thereby reduced from 36.5 to 28.5 per cent, greatly increasing the watertightness and stability.

Determination of saturated plasticity

The saturated moisture content of this soil at its maximum dry weight of 121 lb. is 15 per cent, representing an increase in moisture content of 2.5 per cent over the 12.5 per cent moisture required for maximum compaction. This quantity of water is that necessary to replace the air which compaction does not remove from the soil particles. This contained air, for this particular soil, is equivalent to 2.5 per cent moisture at the peak of the curve and to 1 per cent moisture at 22 per cent moisture content. A specimen of this soil, which was compacted at 15.7 per cent moisture and contained the same quantity of water as the 121-lb. maximum dry-weight specimen when saturated but with a dry weight that was 6 lb. less, required 24 lb. pressure to force a pencil into it. Obviously the saturated soil of 121-lb. dry weight, containing the same quantity of water and 6 lb. more of soil particles, would be much firmer. The same computation establishes a required force of 1 lb. to force a pencil into the specimen of this soil that was compacted with 8 per cent moisture of 108 lb. dry weight, when it becomes saturated. It is thus seen that this soil when properly compacted may require 24 times more pressure for penetration when saturated than will be required when it is compacted so dry as not to spring under the roller.

Experiments have determined that no measurable softening from saturation occurs at depths greater than 2 to 5 ft. below the surface of this soil when compacted at the moisture content at the peak of the compaction curve. Precise methods for determining this are available in connection with percolation and consolidation tests to be described in a subsequent article.

Second of Four Articles on Design and Construction of Rolled-Earth Dams

Description of Field and Laboratory Methods

Air-void and dry-weight curves — Soil-plasticity needle determinations of moisture content and compaction—General field and laboratory methods

By R. R. Proctor

Field Engineer, Bureau of Waterworks and Supply, Los Angeles, Calif.

CONSIDERING the sandy-clay soil discussed in the first article of this series (*ENR,* Aug. 31, p. 000) and charted here in Fig. 2, the graph, Fig. 3, shows the results obtained from compacting the soil by three methods, as discussed in connection with Fig. 2 of the first article. Curve *A* was obtained by actual field tests in which the soil was compacted to the fullest possible extent by a sheepsfoot roller similar to many that are in use. Curve *B* is identical with the dry-weight curve in Fig. 2. Curve *C* was obtained by the laboratory standard method of compaction, which will be fully described later. Sheepsfoot rollers have been designed and constructed to duplicate the results shown on this curve.

Fig. 1—Measurement of the plasticity-needle penetration resistance in a fill compacted for a test of sheepsfoot rollers. The uncompacted surface is removed to permit full penetration of the needle.

The peaks of the dry-weight curves of Fig. 3 occur at lower moisture contents as the peak dry weight increases. This is in accord with the theoretical discussion of soil compaction. The effect of increasing the weight of compaction equipment is clearly shown. The use of a moisture content of 8 per cent has been discussed in connection with Fig. 2 of the first article.

It may be seen that the heavier equipment of curve *C* produces a dry weight of 116 lb. at this moisture content, an increase of 8 lb. over that given in Fig. 2. However, the lighter equipment of curve *A* produces a peak dry weight of 117 lb. at 13.5 per cent moisture content, 1 lb. more than that secured by the heaviest equipment with 8 per cent moisture content, as shown in curve *C*. It is thus seen that proper moisture control with light equipment may give results superior to those obtained from very heavy equipment without proper moisture control.

The zero air-voids curve

The zero air-voids curve of Fig. 3 presents combinations of dry weights and moisture contents of this soil that are the conditions of complete saturation. The dry weight and specific gravity of the soil particles fix the volume of the soil solids. The remainder of the space is the voids, which become filled with water upon complete saturation. The percentage of voids in the soil corresponding to the various dry weights is shown along the zero air-voids curve. The ultimate capacity for water absorption in this soil, when compacted at various moisture contents and by the different methods, can be determined from Fig. 3 by noting the moisture content at the intersection of the zero air-voids curve and the dry weight of the compacted soil. It will be noted that none of the compaction methods are able to attain a condition of complete saturation in the compacted soil; this is attributed to the contained air previously mentioned. The condition is similar to that found in all soils, withut exception, that have been experimentally compacted and in tests conducted of the compaction accomplished in the construction of dams.

It has now been shown that soils that are compacted at too low moisture contents, regardless of the care used in construction and of how hard they are at that time, may have a capacity for considerable more water in the voids, which causes them to become much softer when saturation occurs. The use of too light construction equipment and too little moisture in the soil may result in a dangerous condition if the dam should become fully saturated by percolating water.

Dry-weight curves

Fig. 4 shows the compacted dry-weight curves for several soils compacted by the same method. These soils varied in specific gravity from 2.68 to 2.76. The curves are plotted to show the dry weights that would have been obtained if they had been of the same

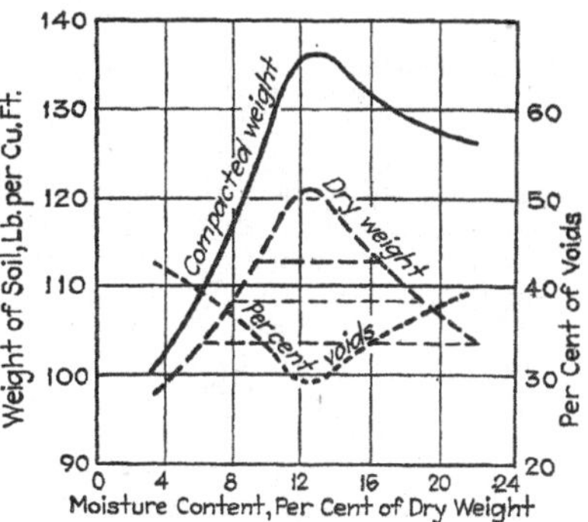

Fig. 2—Graph showing the effect of moisture content on the compacted weight, dry weight and voids of a sandy-clay soil when compacted by a particular method.

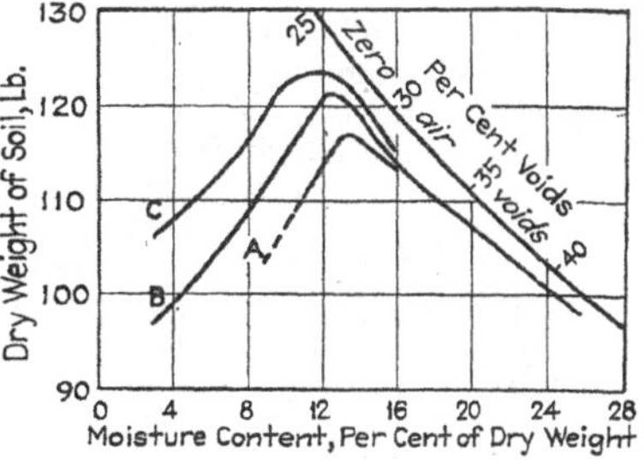

Fig. 3—Curves showing the effect of light, medium and heavy rollers on the dry weight, voids and saturated moisture content of a soil when compacted at various moisture contents.

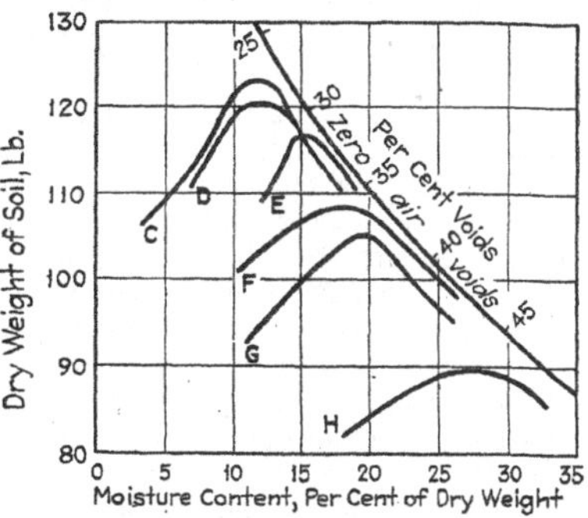

Fig. 4—Compacted dry weights, voids and saturated moisture contents for soils of different types, compacted by the same method at various moisture contents. The screen analyses of these soils are shown in Fig. 5.

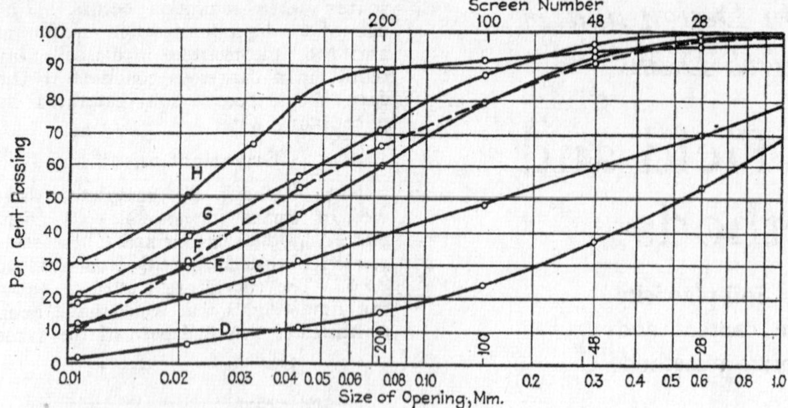

Fig. 5—Screen analyses of the soils used in preparation of Fig. 4.

specific gravity, in order properly to show the relation of the dry-weight curves to the zero air voids and per cent of voids curve. Curve *C* is from Fig. 3, curve *D* was obtained with a silty gravel, curves *E* and *G* were obtained with clayey soils, and curves *F* and *H* were obtained from California adobes. Fig. 5 shows the screen analyses of these soils. It will be noted that the peak of the compaction curve for the adobe of curve *H* is reached with a much greater moisture content and lower dry weight than the sandy clay of curve *C*. This is attributed to the larger proportion of fine particles in the soil, which require more water to lubricate their larger total surface area sufficiently for the compaction process to be most efficient. This results in a greater proportion of voids in the compacted soil, which offsets to a considerable extent the increased watertightness caused by the necessity for percolating water to flow between the finer soil particles. A given amount of compacting produces a more stable soil from the sandy clay than from the adobe, although the adobe is entirely suitable and slightly more watertight.

Soil plasticity needle

The plasticity of compacted soils has been referred to as varying with the moisture content. An instrument known as the soil-plasticity needle was devised to measure soil plasticity in terms of the pressure required to force a rod with a slightly enlarged bearing surface, to penetrate the soil at a rate of about ½ in. per second. Rods of various sizes, usually referred to as needles, are used to keep the applied pressures between 5 and 100 lb. for convenience in hand operation. The pressures are expressed in pounds per square inch on the penerating area and are known as the plasticity-needle penetration resistances or the plasticity-needle readings. Fig. 6 shows the results obtained from the application of the plasticity needle to the compacted specimens prepared to derive curve *C*, Fig. 3. The pres-

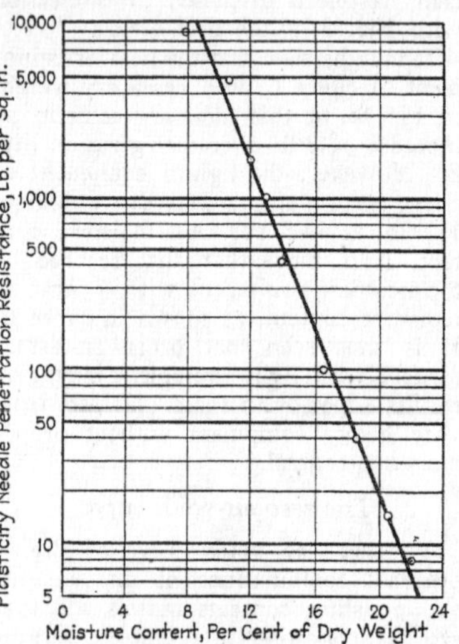

Fig. 6—Graph showing the plasticity-needle penetration resistances of a soil when compacted at various moisture contents by a particular method.

sures that have been given for a pencil to penetrate are based on the assumption that it has an area of 0.1 sq.in. It is interesting to note that Fig. 6 was obtained by use of needles that varied in size from 0.01 sq.in., with a pressure of 90 lb., to 1 sq.in., with a pressure of 8 lb. The complete use of the plasticity needle is best explained in connection with the discussion of the methods for applying the principles of soil compaction, which have been set forth, to the construction of dams.

The method of compaction required to produce a given saturated plasticity in the soil used in preparation of Fig. 2 may now be approximately determined. It is assumed that a saturated plasticity-needle reading of not less than 300 lb. per sq.in. is desired. This soil would then have twice the penetration resistance required to permit loaded-truck travel when fully saturated. An inspection of Fig. 6 shows that a moisture content not in excess of 15 per cent is required. The dry weight of the compacted soil that contains this moisture when fully saturated is found to be 121 lb. (Fig. 3). Such a condition is not attainable with the equipment of curve *A*. A dry weight of 121 lb. is secured from curve *B* at a moisture range from 12 to 13 per cent. Curve *C* fulfills this condition from 9.7 to 14 per cent moisture contents. The proper limits should be from 12 to 14 per cent, which permits compacting the soil at a plasticity needle range of 1,500 to 500 lb. per sq.in., as shown in Fig. 6. This condition of saturated plasticity applies only near the surface of the compacted fill; at depths of a few feet below the surface the weight of the soil above causes consolidation, which increases the dry weight and makes the soil less plastic and more stable.

It has been pointed out in the discussion of Fig. 4 that the very fine adobe soil of curve *G* had no particular advantage of watertightness over the much coarser but well-graded sandy clay used for curve *C*. The usual opinion that the finer clays are much more watertight is accounted for in the shape of their dry-weight curves, which show a considerable range of moisture content without much change in the compacted dry weight. The quantity of water that will flow through a soil under a given pressure increases rapidly with a decrease in the dry weight from that of the peak of the dry-weight curve. For this reason, clayey soils can be compacted at a fairly large variation of moisture content without much effect on their watertightness.

The sandy clays, similar to that shown on curve *C*, do not permit such a wide variation; moisture variation from the peak of the curve similar to that permitted by the finer clays, with only a slight variation in dry weight and watertightness, causes a decrease in dry weight of as much as 8 per cent.

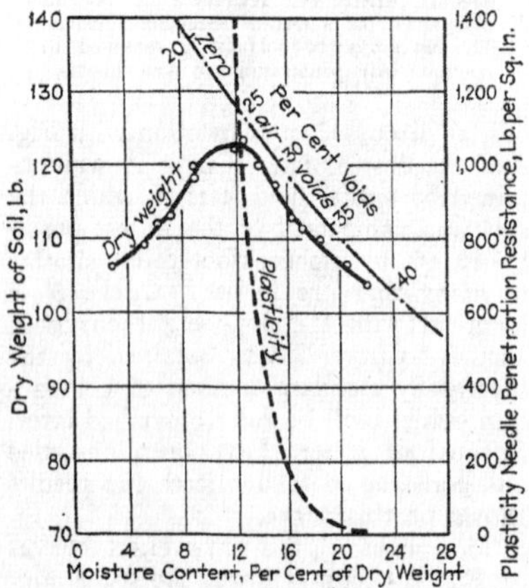

Fig. 7—Soil-characteristic curves for a sandy-clay soil showing dry weights, saturated moisture contents, voids and plasticity-needle penetration resistances when compacted by the standard laboratory method at the indicated moisture contents.

This increases the voids as much as 5 per cent, which permits a much more rapid flow of water and also, because of the larger amount of water that is absorbed, permits a considerable softening from saturation. The percolation rates that have been deermined for the soils shown in Fig. 4 do not indicate any important difference in watertightness when they are compacted to dry weights that are within 2 lb. of the peak of the dry-weight curve.

Soils suitable for dams

Any soil may be used in the construction of a dam if the velocity of flow is determined and found to be too slow to erode any of the soil particles, and the compacted and consolidated dry weight limits the softening from saturation, as determined by the plasticity needle, to that at which the soil is sufficiently stable for the particular dam. Many soils that have been thought very undesirable for use in a dam have been found to be quite suitable for this use when placed under careful engineering supervision, using the principles that have been set forth. The engineer is enabled to study the characteristics of the soils that are available by actual test before the dam is designed, and to plan their disposition in the structure. The proper type of compacting equipment and the method of its use may be determined at the same time. Methods and equipment for obtaining a precise control of the moistening and compacting of the soils will be discussed next. A method for determining the dry weight of the compacted soil after it is placed in the dam will be discussed, and a procedure will be described for testing this soil, at the dry weight of placing, for softening or consolidation and watertightness, under pressures that are fixed by its position in the finished structure.

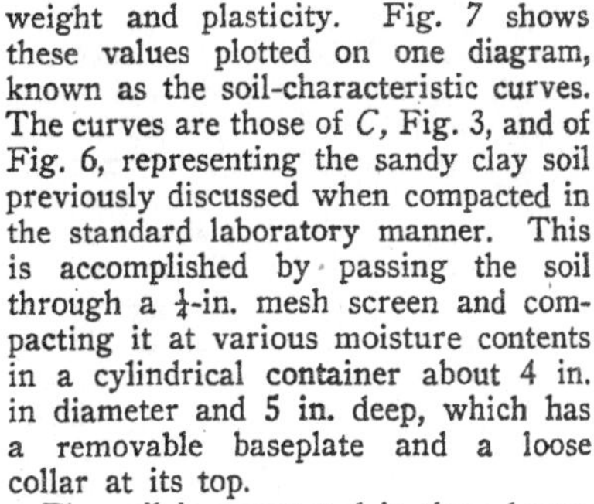

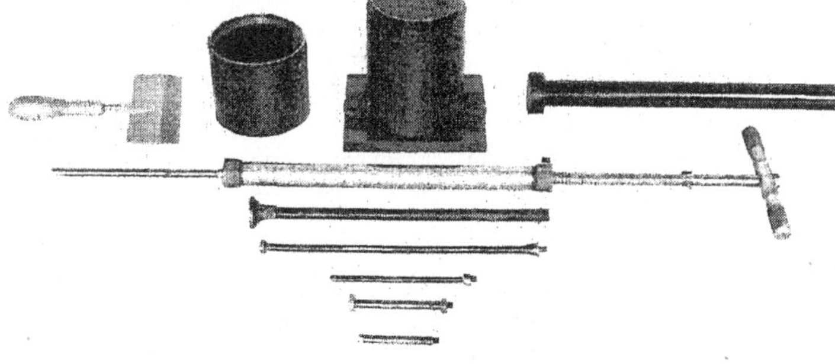

Fig. 8—The compaction cylinder, sleeve, 5½-lb. rammer, cutting tool and plasticity needle with points from 1/20 to 1½ sq.in. in area.

Laboratory methods

Compacted specimens of a soil that are prepared at different moisture contents have been shown to vary in dry weight and plasticity. Fig. 7 shows these values plotted on one diagram, known as the soil-characteristic curves. The curves are those of *C*, Fig. 3, and of Fig. 6, representing the sandy clay soil previously discussed when compacted in the standard laboratory manner. This is accomplished by passing the soil through a ¼-in. mesh screen and compacting it at various moisture contents in a cylindrical container about 4 in. in diameter and 5 in. deep, which has a removable baseplate and a loose collar at its top.

The soil is compacted in three layers of sufficient depth to permit removing the top collar and cutting the compacted soil at the top to the exact cylinder dimension. Each layer is subjected to 25 firm 12-in. strokes, using a rammer of 5½-lb. weight with a striking area 2 in. in diameter. The compacted weight of soil per cubic foot is then calculated. The moisture content is determined by drying a small sample at 250 deg. F. until no further loss in weight occurs; it is expressed as the ratio of the loss in weight by evaporation to the oven-dried weight of the sample. The dry weight is determined by deducting the weight of the contained moisture from the compacted weight of soil per cubic foot. It is the weight of soil solids contained in a cubic foot of compacted soil.

The plasticity-needle reading is taken of each compacted specimen at three points immediately after the surface is cut to the cylinder dimensions. The operator judges the average penetration pressure required for a depth of 3 in. and takes the mean of the penetrations as the reading for the specimen. The compacted-soil specimens should vary in needle readings from 100 to 3,000 lb. per sq.in., with sufficient intermediate values to determine properly the dry weight and plasticity-needle curves. A precise mechanical method could be easily devised to compact the soils in the laboratory, but it could not be conveniently used in the field. It is necessary to make frequent compaction and plasticity-needle tests in the field. The methods that have been outlined are of sufficient accuracy for use in the laboratory, and the equipment is sufficiently portable to allow quick and accurate field tests to be made wherever desired.

The specific gravity of the soil particles is determined in order to establish the position of the zero air-voids curve. The method of obtaining this is important, due to its effect on the calculations of the quantity of water that the compacted soil may absorb. Air is trapped between the smaller soil particles and is difficult to remove. This is part of the contained air not removed

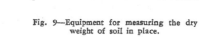

Fig. 9—Equipment for measuring the dry weight of soil in place.

by compaction processes. A weighed sample of soil that has been oven-dried for determining the moisture content is placed in a known volume of water. The volume of the mixture, after thorough shaking, is often used to determine the volume of the soil particles. The application of a partial vacuum of about 4 cm. absolute pressure at this time will produce a surprising result. Air leaves the soil very rapidly and, because of the low pressure with consequent enlarged bubbles, it appears as if a volume of air about equal to the soil was leaving. The application of this air exhaustion process is required for about a half hour. A specific gravity of soil particles of 2.68, determined by thorough shaking in water, was increased to 2.74 by using the air-exhaustion process. This error reduces the moisture content for saturation (Fig. 7) at the peak of the dry-weight curve from 14.8 to 13.6 per cent Measurements of the dry weight of saturated soils in place often indicate the presence of water in excess of the calculated voids, if the specific gravity of soil particles is determined in the usual manner.

Field and laboratory use of needle

The plasticity needle, being hand-operated, is adapted for use in the field or laboratory. It is used in the laboratory to prepare moisture-plasticity curves of the various soils, for use in the field. It is usually possible to determine the moisture content of a soil, for which a plasticity-needle penetration-resistance curve is available, within a few minutes by compacting the soil in the easily portable compaction cylinder similar to that used in the laboratory, and taking the average of three penetration readings. The moisture content of the soil is then obtained from the moisture-plasticity curve, similar to Fig. 7, which has been prepared in the laboratory for that particular soil. The readings in most cases may be taken much closer than the variation shown on the curve for 1 per cent moisture. Fig. 7 shows a plasticity-needle reading of 1,400 lb. per sq.in. for 12 per cent and 850 lb. per sq.in. for 13 per cent moistures. The use of a needle of 1/20-sq.in. area makes the pressures required 70 and 42.5 lb., respectively.

Readings are readily taken to the nearest 5-lb. pressure. The indicated range in pressure for 1 per cent moisture change is 27.5 lb. This indicates an accuracy of about 0.2 per cent in moisture content, much more accurate than the 1 per cent allowable error. When used in this manner the plasticity needle enables the inspector to make a frequent check of the moisture content of the soil as it is placed on the dam. This is absolutely essential to the proper control of the compaction process.

The completeness of compaction may be closely determined with the plasticity needle. This is accomplished by determining the reading in the compacted soil on the dam. Some of the soil is then removed and compacted in the cylinder. The needle readings are taken in the cylinder and should agree with those in the dam if the soil is sufficiently compacted. This method is adapted for frequent use and should be controlled by direct measurements of the dry weight of the soil in place. The principle of the plasticity needle may be used to measure the plasticity at various points in constructed dams, if the soil is known to be saturated, by determining the penetration resistance in auger holes or shafts. Its use in percolation and consolidation tests will be described in their connection.

Measurement of dry weight in place

It is necessary to determine the dry weight of the soil in place after compaction, definitely to check the effectiveness of the soil-compacting process. The surface soil is removed from a small area to such a depth that no uncompacted soil remains. A hole about 8 in. in diameter and 1 ft. in depth is dug, and the removed soil is weighed. The volume of the hole is determined by filling it with a fine dry sand, the weight per cubic foot of which is found by pouring it into a container of known capacity, of about the same dimensions as the hole. The weight of sand used determines the volume of the hole. No attempt is made to pack this sand; it is carefully calibrated as to its weight per cubic foot when poured in loose. It is necessary to level the top of the sand carefully to the exact position from which the soil has been removed. The soil removed is sampled for oven-drying to determine the moisture content. The weight per cubic foot of the compacted soil in place is calculated, and the dry weight is determined by deducting the weight of the contained moisture from this figure.

This method enables the dry weight to be determined within about ½ lb. per cubic foot. Some of the soil removed in making this test may be compacted in the compaction cylinder, the weight per cubic foot obtained, and in this way compared with the compacted weight found in the dam. The two should be equal, if the moisture content is not allowed to change while the test is being made. It is necessary to perform this test quickly, and in the field, particularly in hot dry climates, to prevent loss of water by evaporation. Figs. 8 and 9 show the equipment and methods used to determine the dry weight and plasticity of soil after compaction.

Placing soil in test equipment

The soil-characteristic curves of this soil, similar to Fig. 7, may now be obtained in the laboratory and a comparison made with the dry weight in place. This precisely determines the effectiveness of the compacting process. By compacting the soil removed in any desired test apparatus at the correct moisture, as shown by the dry-weight curve to give the dry weight in place, the soil is ready for any test under the exact conditions at which it was placed in the dam. The dry-weight curve can also be used to determine the moisture content required in order to place the soil at any other dry weight for any desired test. It is necessary to use the same weight of rammer as used in preparation of the compaction curve and the same number of blows on a square inch of surface as used in the compaction cylinder. The thickness of layers must be kept the same.

Third of Four Articles on the Design and Construction of Rolled-Earth Dams

Field and Laboratory Verification of Soil Suitability

Consolidation and percolation tests and test results —Saturated plasticity—Swelling and bearing power

By R. R. Proctor
Field Engineer, Bureau of Waterworks and Supply, Los Angeles, Calif.

IT IS advisable to determine the dry weight, plasticity and watertightness of a soil when under a load corresponding to the weight of soil and water that will be placed above it in the finished dam. The determination of the percolation rate of a compacted soil not under load is of little value, except to determine the relative watertightness of several soils; this comparison is best made when the soils are under a load equivalent to about 10 ft. of fill. As all soils consolidate somewhat when placed under load, it was decided to measure the soil consolidation and saturated plasticity with the same test equipment that is used for the percolation rate.

A cylinder about 8 in. in diameter, which is closed at the bottom except for a water inlet, is used. A rigid but porous disk is placed in the bottom to insure even distribution of the water for the percolation test. A little more than 3 in. of compacted soil is placed on top of the disk. It is compacted in the standard manner and at the moisture content shown by the compaction curve to result in the desired dry weight. Soil removed from a dam may be placed at the dry weight of removal in this manner. The top of the compacted soil is then cut away until a depth of 3 in. remains. The weight and volume of the compacted soil is used to check the compacted dry weight. Another porous disk similar to the first is placed on top of the soil. A loose-fitting plunger is then placed over the disk. It has projections that extend over and beyond the edges of the cylinder at opposite sides, so that a dial gage, reading to one thousandth part of an inch, may be fitted between the projections and the base plate, permitting the accurate measurement of any settlement or consolidation that occurs. A coil spring of the proper capacity to give ample travel for the load desired is placed on the plunger and is compressed with a plate on its top, which is held to the base plate by nuts on long threaded rods.

Any desired load may be placed on the spring and soil in a compression machine, after which the nuts are tightened sufficiently to relieve the machine of the load. The spring then continues to exert this force for as long a time as desired, if the soil consolidation is not enough to enable the spring to stretch enough to reduce the load materially. The load on the spring may be verified at any time by placing the apparatus in the compression machine and noting the pressure required to relieve the tension in the rods sufficiently for the nuts to be turned with the fingers. The entire assembly is shown in Fig. 2.

Fig. 1—The upstream face and left abutment of Bouquet Dam No. 1 of the bureau of water works and supply of the city of Los Angeles, showing the compacting of the earthfill and preparations for placing the reinforced-concrete apron.

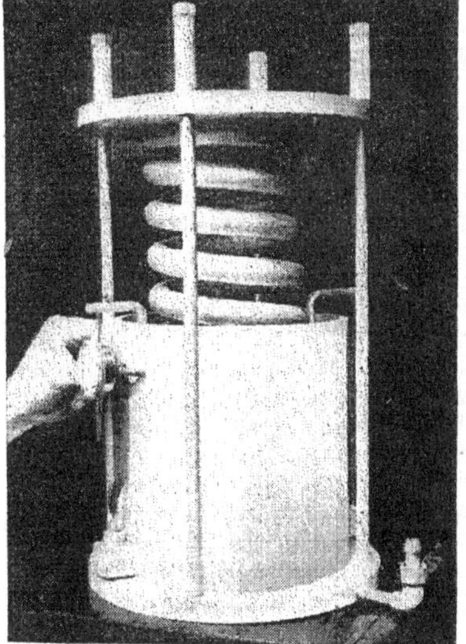

Fig. 2—The consolidation-percolation cylinder, showing the methods of measuring settlement, applying load, and the water inlet for the percolation test.

The compacted-soil specimen may be subjected to any desired load and the resulting settlement or consolidation observed until it ceases. The load may then be increased and additional consolidation data obtained. The change in dry weight, which will take place in any part of the dam during the construction period due to consolidation caused by the weight of the soil to be placed above, may be determined by this test. Fig. 4 shows the consolidation that resulted after compacting the sandy clay soil of Fig. 7 (*ENR,* Sept. 7, 1933, p. 287), to four dry weights and placing it under loads equivalent to the depth of fills shown. These results indicate that a depth of fill of at least 200 ft. is required to correct fully, by soil consolidation, an error of 6 to 8 lb. in the dry weight of placing.

Plasticity and percolation

The plasticity-needle reading of the soil at the time of placing in the cylinder may be determined after compacting a portion of the soil, as prepared for the consolidation cylinder, in the compaction cylinder. The plasticity-needle reading of the consolidated soil may be taken upon removal of the load. This value, taken immediately after the consolidating load is released and after the completion of the percolation test, represents the plasticity-needle penetration resistance of the soil in place after the dam is completed and in service. It therefore determines the extreme softening that can take place in the dam at the point from which the soil sample was taken.

The percolation rate is measured, after the consolidation is complete and with the soil still under pressure, until a constant rate of flow is obtained. The quantity of water required to maintain the desired pressure head is measured. A pressure head of about 12 ft. is usually used in order to determine the results in a few days. The water pressure acts upward, therefore the soil must be under a downward load greater than the upward force exerted by the water, equivalent to about 5 ft. of compacted soil in this case. Consolidation-percolation tests under less load require lower heads and very much longer time to complete.

It is best to determine the percolation rates of all soils under a load equal to that of 10 ft. of fill. The applied downward load is then about twice the upward water pressure, which insures the soil specimen from being moved. Many well-compacted soils require only a few drops of water to be added each day to maintain a 12-ft. head on a 3-in. compacted-soil layer. Most compacted soils have a percolation rate that is very much less than the surface evaporation; for this reason it is best to place enough water on top of the specimen to cover it and prevent surface drying. The rate of flow is reduced by direct-head ratio to that at which the pressure head over the water standing on the top of the specimen equals the thickness of the soil sample. This is a hydraulic gradient of 1. The percolation rate is determined by dividing the total quantity of water that would flow through a soil specimen in a year at a hydraulic gradient of 1 by the area of the specimen through which the water passed. It is expressed in feet per year.

Fig. 3—Laboratory equipment for consolidation and percolation tests. The method of applying equivalent fill loads is shown at the right.

The percolation rate expressed in this way is readily adapted to computing flows through dams and thin layers of compacted soil; it is not necessary to compute actual velocities through the voids. If the area exposed is one acre and the percolation rate is 1 ft. per year with a hydraulic gradient of 1, the total yearly flow is 1 acre-ft. of water. This is 0.00274 acre-ft. per day, 0.00138 sec.-ft., 0.62 gal. per minute, or 893 gal. per day. A dam having a height of 200 ft. and an average length of 650 ft. has a sectional area of 3 acres exposed to water flow at a hydraulic gradient of not more than 1. If it were constructed of compacted soil with a percolation rate of 1 ft. per year, it would have a maximum leakage of 1.86 gal. of water per minute. This quantity of leakage is extremely small for such a dam.

The percolation rates determined from nine soil samples similar to that of Fig. 5 (*ENR*, Sept. 7, 1933, p. 287) averaged 0.04 ft. per year at an average dry weight of 126 lb., resulting from consolidation at loads representing 75 to 180 ft. of fill. The maximum rate was 0.120 ft. per year; the minimum was 0.001 ft. per year. A dam of the dimensions previously mentioned would then pass not more than 1/25 of 1.86 gal., or 1/15 gal. per minute, if constructed throughout of soils similar to the nine samples. Such a dam might be classed as watertight. It would take many years for the water to flow through. The consolidated soils, when fully saturated, would have a plasticity-needle reading in exces of 1,000 lb. per sq.in., a large factor of safety against failure by softening from saturation.

The accompanying table shows the results of 27 consolidation and percolation tests on several types of soils when under loads equivalent to fills having depths of 2 to 200 ft. The dry weights, plasticity-needle penetration resistances and moisture contents are given for the following conditions: the peak of the dry-weight curve obtained by the standard laboratory compaction method, the conditions of placing in the consolidation cylinder, and the conditions found in the compaction cylinder after the load was released and the specimen exposed at its top. The percolation rate through the specimens when they were under the various loads is shown. The percentage of the weight of the soil in the specimen that could be washed through the 200-mesh screen, 0.074-mm. clear opening, is shown.

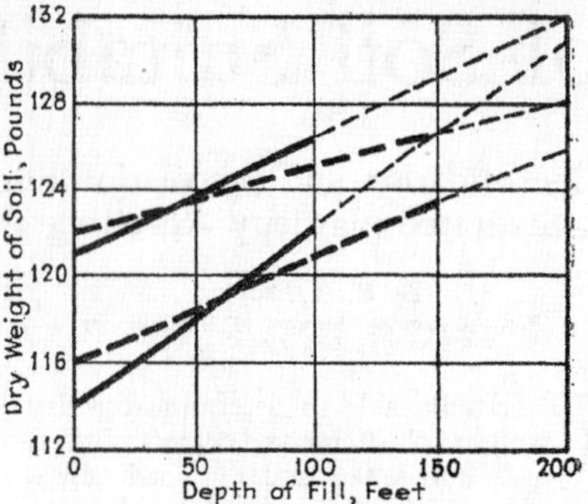

Fig. 4—Curves showing the increase in dry weight of a sandy clay soil caused by soil consolidation, under pressures equivalent to various depths of fill.

No swelling from saturation

Nine of the soil specimens were placed under 10 ft. of fill load or less. All had a dry weight, after consolidation and saturation by percolation, in excess of that to which they were compacted in the test apparatus. It may therefore be assumed that these soils will retain their watertightness and stability in a dam; they will not swell when saturated, thereby decreasing in watertightness and stability. The light load applied forced the specimens to consolidate; if any force of swelling existed in sufficient amount to exceed the weight of the indicated depths of fill, it would have forced the spring upward, permitting the specimen to swell. If a soil should be found in which this occurs, it is a simple matter to increase the load on it until such a force, if it exists, is overcome. The extent of any swelling at any intermediate load could easily be determined. Further tests on very fine soils may indicate this tendency, although no soil has been found to swell at this time.

Importance of compaction

An inspection of the percolation rates and the percentage of the soils finer than the 200-mesh screen shows that no definite relation exists between them. It is shown, however, that the extent to which the soil is compacted is much more important for obtaining watertightness than the percentage of fines. Specimen No. 17 is an interesting example. This soil, containing 8 per cent finer than the 200-mesh screen, appears to be a gravel. When the rock above ½-in. size is removed, the remainder appears to be a sand suitable for low-grade concrete. It has none of the characteristic feel of a clayey sand; it cannot be compacted in the hand to a firm ball. This soil was very watertight after consolidating under a 50-ft. fill load to a dry weight 6 lb. more than the maximum of the laboratory compaction method. This shows the possibility of compacting soils very low in fines to a condition suitable for dam construction.

Specimen No. 3 shows the effect of saturation on a soil that is insufficiently compacted at a low moisture content. The dry weight of the soil as it was placed in the consolidation cylinder was 19 lb. below that obtained by the laboratory compaction method with a moisture content 2.8 per cent higher. The plasticity-needle reading was 1,250 lb. per sq.in. When the soil was saturated under a 4-ft. fill load, the plasticity-needle reading decreased to 90 lb. per sq.in. This increased plasticity was caused by the increase in moisture content from 15.6 to 30.2 per cent. The dry weight of the specimen increased from 89 to 92 lb. This relatively low dry weight accounts for the high percolation rate. The percolation rate of this soil, if compacted to a dry weight of about 108 lb., would probably be not more than 0.010 ft. per year. It is interesting to note that the percolation rate of this clayey soil as it was compacted is many times greater than the very sandy specimen previously discussed.

A great many comparisons can be made from the accompanying table. There are many variations in placing and loading conditions. None of the soils show an excessive percolation rate, and most of the rates shown are so low that the soil may well be classed as impervious to water. The percolation rates through all soils that have been tested are negligible when they are compacted to the condition of a minimum plasticity-needle penetration resistance of 300 lb. per sq.in. when saturated. Any soil that meets this condition could probably be used in a dam; it is unlikely that the percolation velocity would be sufficiently high to erode any of the fines from the soil.

The minimum dry weight to which a soil must be compacted is easily found by the consolidation-percolation test. It is only necessary to compact the soil to lower dry weights and observe the results of the test. The lower limit of the dry weight to which the soil must be compacted is determined when the minimum allowable saturated plasticity-needle pentration resistance or the maximum allowable percolation rate is reached.

Testing foundation soils

Earthfill dams are often desirable at locations where the cost of stripping the foundation to bedrock prohibits their construction. The soils that are to be the foundation for an earth dam may be subjected to the same analysis as a compacted fill. The watertightness and stability of such soils is of equal importance to those used in the fill, particularly on the abutments, where the consolidation resulting from the lesser depth of fill is not so great.

The dry weight of foundation soils is easily determined at any depth by excavating a shaft or side-hill cut to sufficient dimensions to permit the measurement of the dry weight in place, as outlined for compacted fills. The soil may then be tested for consolidation and percolation after being placed in the test apparatus at the dry weight in place and under such load as the contemplated dam will produce at the point of sampling. The stability of the soil under these conditions can be determined by use of the plasticity needle at the conclusion of the test. The probable settlement of the foundation soils can be determined very closely by a series of tests of this nature.

It is necessary carefully to consider the rate at which the settlement takes place in the consolidation cylinder. A slow settlement of the small specimen indicates that the total foundation settlement may not take place for considerable time after the dam is completed. If this settlement can occur only by forcing some of the water from the soil, it is possible for the weight of the dam to force the saturated and quite plastic foundation soil to move outward, causing at least a partial failure of the dam before the full consolidation has taken place.

Bearing power of saturated soils

The capacity for water absorption in a soil has been shown to be determined by its dry weight. The soil used in the preparation of Fig. 7 (*ENR*, Sept. 7, p. 287) can be found in place at a dry weight of 110 lb. with a moisture content as low as 6 per cent. This soil has a plasticity-needle penetration resistance of about 10,000 lb. per sq.in. in this condition. The moisture content for complete saturation at this dry weight is 20 per cent, and the plasticity-needle penetration resistance then decreases to 20 lb. per sq.in. A standard bearing test on this soil in the drier condition would indicate a capacity to support a high unit load, and a similar test on the soil in its saturated condition would indicate a capacity to support only a very light unit load.

The determination of the dry weight

RESULTS OF CONSOLIDATION AND PERCOLATION TESTS OF VARIOUS SOILS

	Soil Characteristics at Maximum Density Obtained by Standard Laboratory Compaction			Soil Characteristics as Placed in Consolidation Cylinder			Soil Characteristics as Determined After Saturation by Percolation-Consolidating Load Released					
Test No.	Dry Weight, Lb.	Plasticity-Needle Reading, Lb. per Sq. In.	Moisture Content, Per Cent	Dry Weight, Lb.	Plasticity-Needle Reading, Lb. per Sq. In.	Moisture Content, Per Cent	Consolidating Load Equivalent Depth of Fill, Ft.	Dry Weight, Lb.	Plasticity-Needle Reading, Lb. per Sq. In.	Moisture Content, Per Cent	Percolation Rate, Gradient of 1 Ft. /Year	Soil Passing 200-Mesh Screen, Per Cent
1	121	1,200	13.0	111	40	20.2	2	118	75	17.0	0.0710	21.7
2	112	500	15.5	109	425	16.8	3	111	400	19.3	0.0154	51.2
3	108	425	18.4	89	1,250	15.6	4	92	90	30.2	2.000	68.0
4	116	1,150	14.0	116	450	15.6	5	118	475	15.8	0.0154	41.0
5	110	850	13.5	117	1,500	14.2	10	118	1,250	15.0	0.790	29.5
6	124	3,000	11.5	123	500	14.5	10	125	1,200	13.5	0.4700	18.6
7	109	525	18.5	108	300	18.8	10	111	460	19.1	0.0011	52.7
8	122	1,400	12.5	125	1,550	11.8	10	127	1,350	12.5	0.0047	34.2
9	112	825	14.8	114	550	16.4	10	115	550	17.5	0.0028	33.4
10	102	650	22.5	82	5	36.0	18	88	50	31.8	0.1090	38.0
11	89	430	27.5	91	250	32.5	20	93	300	31.6	0.00016	82.0
12	120	950	13.0	117	375	15.0	20	123	450	14.5	0.0035	37.6
13	113	650	16.5	111	425	17.9	25	112	575	18.5	0.0008	50.6
14	115	750	15.2	111	375	17.5	30	113	400	17.6	0.0010	46.2
15	110	650	18.2	111	550	17.2	35	118	900	16.4	0.0014	39.6
16	102	650	22.5	82	5	36.0	45	93	120	29.3	0.0280	43.0
17	121	700	12.5	117	475	15.2	50	127	950	12.9	0.0156	8.0
18	123	1,000	12.0	118	450	14.2	50	125	900	13.5	0.0007	38.3
19	128	1,450	11.2	126	1,225	11.2	50	131	1,800	11.2	0.0050	29.4
20	128	2,500	11.0	122	500	12.8	50	128	1,300	12.6	0.0096	31.7
21	124	1,500	12.0	114	500	13.5	100	119	700	14.7	0.8000	30.6
22	124	1,500	12.0	121	600	13.5	100	125	1,410	14.4	0.0100	30.6
23	123	2,000	11.4	122	675	13.5	140	126	700	12.4	0.0018	37.0
24	122	1,100	12.0	114	1,130	11.7	150	121	500	14.8	0.0450	32.4
25	122	800	13.5	123	1,050	12.8	150	127	1,700	12.4	0.0167	36.4
26	123	1,200	12.0	120	350	15.0	180	124	1,800	13.1	0.0010	46.4
27	123	750	12.5	119	450	14.7	200	127	1,700	12.4	0.0450	35.2

of a soil in place can therefore furnish data of much value in determining the bearing power of soils, particularly if the moisture content of the soil is low but is subject to considerable increase after the load is applied. The soil can be placed in the consolidation-percolation cylinder at the dry weight in place, and a load can be applied equal to the contemplated foundation pressure. The observed settlement would be that to be expected immediately under the foundation. The application of other lighter loads determines the settlement to be expected at various depths below the foundation, depending on the manner in which the foundation pressure is assumed to be spread in the soil below. The plasticity needle can be used directly on foundation soils in place if they are known to be saturated. Further study and experiments will undoubtedly establish a relation between the bearing power of a soil and its plasticity-needle penetration resistance when saturated.

Required depth of stripping

The dry weight of natural soils in place has been found to increase as much as 10 lb. at a depth of 15 ft. below the surface. A soil that was wholly unsuitable for foundation purposes at the surface had a dry weight equal to that to which it was necessary to compact it in the dam at depths of 12 ft. below the surface. The depth to which foundation soils should be stripped is that to which the dry weight in place is the same as required if the soil were to be placed in the fill. If the foundation soils are too deep for this to be accomplished, the cross-section of the dam should be carefully designed to meet the conditions of stability that are thus imposed. The compacted fill should then be placed at the same saturated plasticity as the foundation in order to result in a fill of sufficient plasticity to conform to such settlement as will take place in the foundation. The careful use of the principles that have been set forth should permit the design and construction of compacted-earth dams of known safety at locations that have been considered unsuitable for this purpose.

Stability of existing dams

The principles that have been set forth may be used in an analysis of the safety of existing earthfill dams. The determination of the dry weight of fill and foundation soils at critical points and the preparation of the soil-characteristic curves, similar to Fig. 7 (*ENR,* Sept. 7, p. 287), from the removed samples can result in a definite conclusion as to the margin of safety which exists in the structure. The determination of the dry weight in place for soils below the line of saturation in an earth dam is easily made. A sample of the saturated soil is removed by means of a core barrel so arranged that it will bring the sample to the surface in its original condition. The determination of the moisture content and specific gravity of the saturated soil determines its dry weight, as shown in Fig. 7 (*ENR,* Sept. 7, 1933, p. 287). The removel of sufficient soil for the compaction and plasticity-needle test enables the preparation of curves similar to those of Fig. 7, in the previous article.

Sufficient information is now at hand to proceed with the consolidation and percolation tests that have been described. This information is of particular value when the problem of increasing the height of an existing dam is being studied.

Last of Four Articles on the Design and Construction of Rolled-Earth Dams

New Principles Applied to Actual Dam-Building

Application of principles of soil compaction, as shown by several dams constructed, is entirely practicable and does not involve increased cost

By R. R. Proctor
Field Engineer, Bureau of Waterworks and Supply, Los Angeles, Calif.

THE DESIGN of an earthfill dam should be based on the findings of a thorough investigation of the watertightness and saturated stability of its foundation, particularly at the abutments, as well as a complete study of the characteristics of the soils that are available for its construction. The importance of watertightness in the abutments is often overlooked. Many engineers consider the water-percolation path under the maximum cross-section of the dam only, overlooking the possibility for water percolation from a point at the water surface near the top of the abutment and thence through the abutment to a point near the base of the dam. This path for percolation is often considerably shorter than that under the base of the dam, and particularly so if the abutments are steep. The available head is about the same—the depth of water in the reservoir. Considerable trouble with earth dams has resulted from the neglect of this point.

If the abutments of a dam cannot be established as sufficiently watertight, the cross-section should be increased at these points. This can be accomplished by extending the fill upstream from the dam along the abutment to such distance as may be required to establish a sufficiently long path of travel for percolating water in order to reduce its velocity to that which will be safe. The soil that is to be used for this portion of the dam requires more careful selection and construction than that in the dam proper; its thickness will be somewhat less than that in the main portion of the dam. The careful installation of such blankets or upstream extensions of the dam will be of great benefit at certain locations, and will usually be far more effective than the construction of concrete or other types of cutoff walls. A consideration of most cutoff-wall locations shows that the length of the percolation path through the foundation or abutment is increased by only a slight amount, not enough to be of material value.

Fig. 1—Sheepsfoot roller of extra heavy design. The appearance of the surface of the soil and the penetration of the teeth should be as shown at the completion of the compacting process.

Cross-section

The cross-section of the main portion of the dam should be designed after careful consideration of its foundation. The data which have been secured from consolidation-percolation tests of properly compacted fills indicate that only a moderate cross-section is required if the foundation is on impervious rock. There is no doubt but that side slopes of 1 on 2 or less would have a considerable margin of safety if soils were available such as were used in the preparation of Fig. 7 (*ENR*, Sept. 7, 1933, p. 287), and were compacted to the dry weight at the peak of the curve.

The computed time required for the soil of the above figure to become saturated to a depth of 20 ft. under a head of 100 ft. of water is 2½ years. This saturation is accomplished by an increase in moisture content of 2½ per cent. The thickness required for watertightness under such conditions is not great. If the saturated and consolidated soil in this structure had a minimum plasticity-needle reading of 300 lb. per sq.in., 30 lb. being required to force a pencil into it, side slopes of 1 on 2 should provide ample security. If the dam is to be built upon a soil foundation of great depth, the cross-section should be much more liberal. Many such sites would probably require side slopes of at least 1 on 4 or more, to result in the same security as would be obtained by side slopes of 1 on 2 on a rock foundation. The cross-section of each dam, particularly if it is of great height, should therefore be very carefully selected after a thorough investigation of its foundation.

The usual custom of providing a porous section on the downstream side of an earthfill dam need not be followed. The practice in the past has been to place materials of doubtful watertightness on the downstream side of the dam to maintain its stability, in case percolating waters should cause the impervious upstream section to soften. Such materials usually are loosely placed and are supposed to pass water readily and not soften from saturation, thus lending support to the softened upstream section. This practice shortens the possible downward-percolation paths through the abutments that have been mentioned. It would be much better practice to design the compacted fill to accomplish the entire purpose of the dam than to depend on doubtful pervious material for stability. The entire purpose of the procedure that has been described is to insure the stability of compacted fills when saturated with water to the fullest extent. It is not necessary to provide drainage for such fills—in fact, it is better that they retain all of the water possible, to minimize shrinkage from drying.

The materials stripped from the

foundation and abutments that are unsuitable for use in the compacted fill may be placed on the downstream side of the dam, outside the cross-section that is required for watertightness and stability. This is probably the cheapest place at which they may be wasted. They should be useful for such purposes as terracing, landscaping and providing roadways. When so placed, the waste materials can provide a loose-soil cover to prevent excessive drying of the compacted fill if the water surface in the reservoir is to be low for long periods, or if the percolation rate of the compacted fill is less than the evaporation rate from the soil surface.

Upstream slope

The protection of the slope of the dam from erosion by the water in the reservoir is very important. This may be accomplished by a properly designed concrete facing or by a rock riprap of sufficient depth. The choice of method should be governed by cost. The required depth of riprap can be determined only from consideration of the size of the reservoir and the nature of the available rock. A concrete facing of 6-in. thickness should be ample for any exposure. If made of plain concrete, there is danger that shrinkage cracks will develop with sufficient opening to permit a slow undercutting of the slab by wave action. It is best to provide steel reinforcement of the slab in sufficient quantity so that its tensile strength, without permanent elongation, is in excess of the tensile strength of the concrete, in any direction. This will not prevent shrinkage cracks, but will insure against any opening more than about 0.02 in. A concrete slab of this design is proof against undercutting by wave action. The absence of large unsightly cracks greatly improves the appearance of the structure and adds much to the layman's impression of the security of the dam. A specimen of such a concrete slab, 20 ft. in length, was stretched 6 in. before final failure at one of fifteen cracks that had opened. A slab of such flexibility, probably not adding to the watertightness of the dam in normal use, should prove of great benefit in checking a large flow of water that might take place in case of a seismic disturbance of sufficient intensity to open cracks through the dam. The ends of the slab should be anchored to a wall of such dimensions that it will not be moved by the stresses in the slab and so located that it will not be exposed to undermining by wave action.

Selection of soils

The selection of the soils that are to be used in the compacted fill should be governed by the cost of their use as well as their suitability. It is unnecessary to select those of high clay content—in fact, this should not be done. The cost of handling such soils is slightly higher than in the case of a lower clay content. The soils that are available in the best quantity and location for economical construction should be selected, if the compaction-consolidation-percolation tests indicate them to be suitable. Soils that will be found unsuitable are not numerous, and the methods that have been outlined permit the use of most natural soils, including some that appear to be streambed gravels. Soils that are poorly graded and almost wholly lacking in fine particles cannot, of course, be used.

Fig. 2—Measurement of plasticity-needle penetration resistance and dry weight of test fill compacted to determine required number of trips and weight of sheepsfoot-type rollers for the construction of Bouquet Dam No. 1.

The examination of proposed borrowpits is best made by the use of auger holes that must be so spaced as to insure samples from all types of soil that will be encountered. Representative samples of all the various types of samples secured should be selected and the complete soil-characteristic curves, such as Fig. 7 in the issue of Sept. 7, p. 287, should be obtained for each. A study of these curves indicates any necessity for further testing of this kind. If there is a wide variation in the required moisture contents for plasticity-needle readings of some intermediate value, such as 500 lb. per sq.in. in the compacted soil, sufficient additional samples should be tested so that all different types of soil will be discovered. The most porous or sandy-appearing samples should then be selected for consolidation-percolation tests. If the tests on these are satisfactory, it will not be necessary to conduct a large number of these tests on the other soils, as they can be established as satisfactory with a few tests.

When the possible borrowpits have been explored in this way the soils for use can be selected. The selection of those that are most uniform throughout the depth of the proposed cut will lower the construction costs. They may be more accurately moistened in the pit, thus requiring a minimum of moistening on the dam. The output of the construction equipment will be more uniform, resulting in greater production from a particular construction plant. If the volume of the dam is large in comparison with the volume of the reservoir, careful consideration should be given to borrowpits that are within the operating storage level. A computation of the contemplated expenditure for a cubic yard of water storage may show credit for increased storage, which will justify a slightly higher construction expense to obtain soils from within the reservoir at a point that does not affect the watertightness of the dam.

Rolling equipment

The selection of the type of construction equipment should be made after the desired density of the compacted fill has been determined. Rollers of the sheepsfoot type, shown in Fig. 1, are far superior to the flat-wheel roller. The pressure on the teeth of the sheepsfoot roller can be adjusted to give any desired degree of compaction. The efficiency of modern tractors has made possible the economic use of compacting equipment of two or three times the weight that has previously been used.

Fig. 3—Sampling of proposed borrowpit for Bouquet Dam No. 1.

Fig. 4—Placing of 6-in. reinforced-concrete slab to protect the compacted fill from wave action at Chatsworth Dam No. 2.

Pressures as high as 200 lb. per sq.in. may be obtained easily with a final penetration of 3 in. by the teeth of the roller. The action of widely spaced teeth on these rollers gives a kneading action in the soil that is necessary to force the various sizes of soil particles into the proper relation to obtain the required soil density.

Sheepsfoot rollers of the type usually manufactured are sufficiently heavy for soils that are to be compacted to a plasticity-needle reading of 200 to 400 lb. Where higher plasticity-needle readings are required, the weight of the roller must be increased in approximately the same ratio as the increase in plasticity-needle reading. The teeth on most sheepsfoot rollers available are closely spaced, but wider spacing is preferable for efficient compacting, as the soil is thereby given more of the required kneading action.

Preparation of soils

The soils in the proposed borrowpits should receive their proper moisture content well in advance of the time that they are to be placed on the dam. The proper moisture content can be determined from the soil characteristic curves, similar to Fig. 7 in our issue of Sept. 7, p. 287. The desired moisture content for each type of soil below a given point in the borrowpit can be determined in this way. The difference between this moisture content and that which is in the soil makes possible a computation of the quantity of water that must be added. The soil may be sampled for the purpose of checking this addition of water with a standard 1-in. soil sampler or by an auger hole. The auger hole is preferable, as it furnishes sufficient soil to prepare a composite sample for a determination of the soil-characteristic curves, if desired. A second computation of the required moisture to be added may be made after this sampling and should result in the proper soil-moisture content. The soil should arrive at the dam with the proper moisture content for the conditions of minimum evaporation. It is necessary to add moisture on the dam to correct for the conditions of higher evaporation.

The moisture content of the soil may be checked by use of the compaction cylinder and plasticity needle when it arrives at the dam. The inspector can then direct the addition of such water as may be required as the soil is being leveled in preparation for compaction. The moisture content should again be checked before the compacting process is started. The number of trips of a sheepsfoot roller required to compact a soil properly is very important. Extensive experiments have indicated that a minimum of 16 and as high as 25 are required, depending on the requirements for the soil density. It is very important that this work be carried on in a systematic manner. When the required number of trips has been made, the inspector should check the completeness of the compaction with the compaction cylinder and plasticity needle, as has been outlined. If the plasticity-needle reading in the fill is as much as 25 per cent lower than that in the compaction cylinder, more trips of the rollers should be required, or the dry weight in place should be determined to establish definitely the thoroughness of compaction. The inspector should be particularly alert to discover any lack of overlapping in the travel of the rollers.

Testing the compacted soil

The dry weight in place of the compacted soil should usually be determined at least once for every 1,000 cu.yd. placed on the dam. The results obtained from this test furnish a basis for keeping the inspectors informed as to the plasticity-needle requirements for the control of the moisture content and compaction of the soil. They also furnish a basis for determining the number of trips which the rollers must make in order to compact the soil to the required dry weight. The number of trips of the rollers should be determiner from the results of the dry-weight tests of the soil after it is placed in the dam.

Consolidation and percolation tests should be made at such intervals as are required to control the construction operations and to furnish sufficient records of the quality of the work performed. These should be kept with the same regard for future records as is done in the best concrete-placing practice. A close study of the results of these tests furnishes the engineer with a final check on the watertightness and stability of the structure. The construction methods can be varied, as the work progresses, to remedy any deficiencies found by the tests or to effect economies in the construction cost when it is established that the safety of the structure will not be affected.

Application to construction

Dam No. 1 of the Bouquet Canyon project is now being built by the construction forces of the Bureau of Water Works and Supply of the city of Los Angeles. The methods of construction control that have just been described are now in use on this project. The dam is to be 221 ft. high when completed, with a crest width of 50 ft. and side slopes of 3 to 1. The upstream two-thirds of the cross-section requires about 2,000,000 cu.yd. of compacted earthfill, and the downstream third requires 700,000 cu.yd. of more pervious fill, most of which is obtained from stripping operations. The entire foundation for the compacted fill is stripped to firm bedrock; dry-weight tests of the natural soils in place determined that they should be removed and compacted to a higher dry weight. They were therefore removed and compacted in other portions of the dam.

The available soils for the compacted section of the dam were explored with a 16-in. power auger, shown in Fig. 3. The auger holes were spaced about 200 ft. apart, varying in depth from 10 to 50 ft. A composite sample of each 5-ft. depth of hole was secured and removed to the laboratory. Soil-characteristic curves, similar to Fig. 7 (*ENR*, Sept. 7, 1933, p. 287, were prepared for 145 representative soil samples secured from 109 auger holes. The mean percolation rate for 40 samples tested was 0.04 ft. per year. The most economical location of the borrowpit was then selected, considering the value of the 1,250 additional acre-ft. of storage within the operating levels of the reservoir and the cost of truck haul. A minimum plasticity-needle penetration resistance (when saturated) of 300 lb. per sq.in. was established, and the percolation rate limited to 0.30 ft. per year.

A test was then made of the compaction results that could be secured with the available sheepsfoot rollers (see Fig. 2, Aug. 31, p. 246). Tests were also made to determine the compacting effect of 10 to 30 trips of the rollers, resulting in the selection of a minimum of sixteen trips. The rollers were found to be too light to accomplish the desired degree of compaction, so that rollers were then designed (Fig. 1) having a weight of about 8,000 lb. on a single drum. The effect of this change is shown in Fig. 7 of the issue of Sept. 7, p. 287.

The moisture contents of the first soils to be used were determined by sampling with 1-in. sampler tubes two months before the start of the work. The required quantity of water was added and checked, resulting in the correct soil-moisture content for proper compaction. Additional samples were taken from the borrowpit as the various soils were uncovered, so that soil-characteristic curves and consolidation-percolation results would be available at least two weeks before the soil was to be used. Sixty of these tests disclosed some 5 ft. of depth samples that had percolation rates above the maximum. Composite samples for a 20-ft. depth of these holes were then tested, resulting in percolation rates well below the maximum. The operation of the construction equipment was found thoroughly

Fig. 5—Chatsworth Dam No. 2, raised and enlarged in connection with the increase of storage facilities of the Bureau of Waterworks and Supply of the city of Los Angles. The extension of the dam along the right abutment to lengthen the path of percolation through it appears in the foreground.

to mix the soils from a 20-ft. bank in all cases. A small amount of extremely sandy soil was diverted to the porous fill for two days, until consolidation-percolation tests had determined it to be satisfactory. The consolidation-percolation tests of the compacted fill have disclosed no percolation rates above the established maximum.

Twenty trips of the rollers were used to compact the first 500,000 cu.yd. of fill, and the number of trips was then reduced to sixteen without an appreciable effect on the compaction results. A compacted plasticity-needle penetration resistance of about 900 lb. per sq. in. has been found most economical of construction, resulting in a fill with an average penetration resistance well over the minimum established. Small quantities of the compacted soil have a penetration resistance as low as 400 lb. per sq.in. The maximum allowable penetration resistance is 2,000 lb. per sq.in. Lower moisture contents than required for this value are liable to result in an undesirably smooth surface after the soil is compacted. The required surface after compaction for proper bond between soil layers is shown in Fig. 1. The dry-weight tests of the compacted soil have disclosed a few cases where additional compacting was required. The dry weight has been raised to the desired value after making additional trips of the rollers. Each area of low dry weight is tested after rerolling.

The compacted fill is being placed at a monthly average of 9,000 cu.yd. per day with two shifts. There are six in-

SUMMARY OF FIELD AND LABORATORY TESTS OF COMPACT FILL

Date and Identification Number	Fill Conditions							Laboratory Conditions										Remarks
	At Fill Moisture					At Fill Dry Wgt. (Saturated)		Characteristics At Maximum Dry Wgt.				Consolidation-Percolation Results						
	Dry Weight, Lb.		Plasticity Needle (Lb./Sq.In.)												At Removal			
	Fill	Compaction Cylinder	Fill	Compaction Cylinder	Moisture Content, Per Cent	Moisture Content, Per Cent	Indicated Plasticity Needle Lb. Per Sq. In.	Dry Weight, Lb.	Moisture Content, Per Cent	Plasticity Needle, Lb. Per Sq. In	Specific Gravity	Dry Weight as Placed, Lb.	Applied Pressure Equivalent, Ft. of Fill	Resulting Dry Weight, Lb.	Plasticity Needle, Lb. Per Sq. In.	Moisture Content, Per Cent	Final Percolation Rate, Ft. Per Year	
1-9-33 A	126.0	125.0	690	950	13.0	....	....	....	....	....	...	....	...	....	....	....	....	100′ So. of Sta. 5+75 El. 2864
1-9-33 B	122.0	124.0	950	975	13.0	....	....	....	....	....	...	....	....	....	...	....	....	520′ No. of Sta. 7+75 El. 2837
1-9-33 X	121.0	125.0	725	950	13.0	....	....	....	....	....	...	....	....	....	....	....	....	200′ No. of Sta. 7+10 El. 2851
1-10-33 A	127.0	123.0	550	750	14.0	....	....	....	....	....	...	....	....	....	....	....	....	50′ No. of Sta. 6+65 El. 2862
1-10-33A *	128.2	123.3	550	900	13.2	12.5	1,400	124.1	12.0	2,000	2.77	....	....	....	....	....	....	60′ No. of Sta. 6+65 El. 2862
1-10-33B	122.0	125.0	1,150	1,300	13.0	....	....	....	....	....	....	....	....	....	..	....	....	150′ No. of Sta. 8+50 El. 2858
1-10-33B *	123.7	124.4	1,125	1,500	11.7	14.5	300	124.3	11.5	1,500	2.78	122.0	100	127.8	1,800	13.3	0.0082	150′ No. of Sta. 8+50 El. 2858
1-10-33X	121.7	124.8	775	900	13.3	....	....	....	....	....	....	....	....	....	....	....	....	460′ No of Sta. 10+00 El. 2826
1-12-33A	128.0	126.0	1,100	1,600	12.0	....	....	....	....	....	....	....	....	....	....	....	....	300′ No. of Sta. 7+25 El. 2853
1-12-33A *	129.8	124.5	1,100	1,475	11.4	12.0	1,000	124.4	11.5	1,400	2 78	124.1	65	126.9	1,500	13.4	0.145	300′ No. of Sta. 7+25 El. 2853
1-12-33B	125.0	126.0	1,225	1,500	12 0	....	....	....	....	....	....	....	....	....	....	...	....	450′ No. of Sta. 8+00 El. 2846
1-12-33B *	125.8	126.7	1,350	2,000	11.2	13.6	800	126.5	11.4	2,000	2.76	....	....	....	....	....	....	450′ No. of Sta. 8+00 El. 2846
1-12-33X	122.0	124.8	1,325	950	13.0	....	...	...	....	....	....	....	....	....	....	....	..	510′ No. of Sta. 6+50 El. 2840
1-12-33Y	126.0	125.8	1,500	1,700	12.0	....	..	....	....	....	....	....	....	....	....	....	....	100 No. of Sta. 9+50El. 2837
1-13-33A	123.0	126.0	975	1,500	12.0	....	..	....	....	....	....	....	....	....	....	....	....	410′ No. of Sta. 10+30El. 2825
1-13-33B	127.0	128.0	1,480	2,000	11.0	....	....	....	....	....	....	....	....	....	....	....	....	40′ No. of Sta. 7+70 El. 2866
1-13-33X	126.0	123.0	1,325	1,900	12.0	....	....	....	....	....	....	....	....	....	....	....	....	350′ No. of Sta. 6+40
1-13-33Y	126.0	121.0	950	1,000	13.0	....	....	....	....	....	....	....	....	....	....	....	....	510′ No. of Sta. 7+95 El. 2840

*Check of field measurements by laboratory forces.

spectors of this operation at the work; one is in charge, one supervises the moistening of the borrowpit, and two are employed on the dam for each shift, supervising the moistening of the soil on the dam when required, testing the compaction of the soil with the plasticity needle and making dry-weight measurements of the compacted soil in place, following the methods that have been outlined. Two men are used continuously in the laboratory to prepare soil-characteristic curves and conduct consolidation-percolation tests of borrowpit samples sent from the field. This laboratory personnel also performs about four dry-weight tests of the compacted fill per week, thus insuring a uniformity of method in the laboratory and field. Soil-characteristic curves are prepared for the compacted-soil samples, and consolidation-percolation tests are made of the soil when placed in the apparatus at the compacted dry weight. These data furnish the field force with complete information to plan changes in method that may be required from time to time to secure the required results. Fifty consecutive consolidation-percolation tests, covering the placing of 700,000 cu.yd. of fill and duplicating field conditions, had an average percolation rate of 0.0136 ft. per year; the maximum rate was 0.1450 ft. per year, and the minimum was 0.0014 ft. per year. The minimum plasticity-needle penetration resistance upon completion of the test was 400 lb. per sq.in., resulting from consolidation under a load equivalent to 5 ft. of fill. The maximum penetration resistance was 2,000 lb. per sq.in., resulting from consolidation under a load equivalent to 150 ft. of fill. A summary of eighteen field and laboratory tests of the compacted soil at Bouquet Dam No. 1 is shown in the table. It is especially interesting to note the close agreement between the compacted dry weights in the fill and in the compaction cylinder.

The careful use of the methods that have been described furnishes a very close control over the soil-placing operations. The tests can be carried to any desired extent of thoroughness and do not involve any great item of expense. When proper engineering supervision over the entire excavating and compacting process is exercised, the output from a group of men and equipment is increased. When the soil in the borrowpit is properly moistened, the excavation equipment has a uniform output through the day, and there are no delays caused by avoiding areas that are too dry or are unexpectedly pronounced unsuitable by the engineer. A steady output at the borrowpits makes possible the efficient use of hauling, placing and compacting equipment. An efficient inspection organization can easily effect savings in construction costs that are several times the cost of the entire inspection and testing. The final result of such supervision should be a dam with a predetermined degree of watertightness and stability.

This procedure is, as far as the writer is informed, new to the engineering profession. It has been developed in the past three years during the construction of the Chatsworth No. 2 and No. 3, Upper Hollywood and Bouquet No. 1 dams. An extensive program of careful laboratory investigation has been undertaken and considerable effort expended to make the new ideas that have been developed thoroughly practical in their application to the work. The methods may appear rather complicated, but it must be remembered that soils are probably the most widely variable construction material with which the engineer has to deal. It is hoped that this discussion will clear up many of the uncertainties of earth-dam construction in particular and soil compaction in general that have been very difficult to overcome in the past.

Acknowledgments

The methods of procedure that have been outlined were developed in the laboratory and field by the field engineering division of the Bureau of Water Works and Supply of the city of Los Angeles, of which H. A. Van Norman is chief engineer and general manager and W. W. Hurlbut is engineer of waterworks. The late Lester L. Meyer was assistant field engineer and contributed greatly to all phases of the development of the procedure. N. M. Imbertson, H. B. Hemborg and N. F. Crossley have acted as resident engineers on the various projects and have contributed to the efficient field application of the methods. R. O. Ridenour has contributed to the design of the various equipment, and S. E. Smith directed most of the laboratory work.

LARGE RETAINING-WALL TESTS

A series of five papers reporting fundamental results

I—Pressure of Dry Sand

Tests at Massachusetts Institute of Technology furnish for first time data covering the effect on earth pressure of wall movement through the entire range up to a yield sufficient to produce slip

By Karl Terzaghi
Professor at the Technische Hochschule, Vienna, Austria

TO PROVIDE data bearing on the design of a retaining wall 170 ft. high, intended (and later built) as part of the Fifteen-Mile Falls Dam of the New England Power Association on the Connecticut River, a large earth-pressure test apparatus was built at Massachusetts Institute of Technology, as described by the author in *Engineering News-Record* of Sept. 29, 1932, p. 365. This apparatus has a bin 14x14 ft. in ground plan and 7 ft. deep, which is closed at the front by a movable retaining wall 14 ft. long and 7 ft. high so mounted as to permit of weighing the pressures against the wall; the wall also can be moved outward or inward or tilted around its lower edge.

The main purpose of constructing this apparatus was to investigate the lateral pressure of boulder clay (modified till) in different states of saturation. However, to furnish the experience required for making and for interpreting the tests with boulder clay, several tests were made with cohesionless sand. The following article deals with the results of the simplest ones of these tests, those on dry sand. The purpose of these tests was to determine how any yielding of the retaining wall affects the direction and intensity of the sand pressure.

Test material and procedure

The tests were performed with Plum Island sand (grains essentially angular, fresh quartz and feldspar with a trace of mica; effective size, 0.54 mm., uniformity coeffiicent 1.70). This material was a uniform, medium, angular sand. In the densest state its volume of voids was 38 per cent, in a loose state 46 per cent. Before use the sand was dried by steam coils on the bottom of the storage bins; due to this and to handling the sand while filling the bin it was practically dry during the tests.

The rigidity of the retaining wall and the construction of the measuring devices made it possible to determine both the intensity and the direction of the resultant earth pressure for any position of the wall up to a distance of more than 1 cm. outward or inward from the original position. The program of the tests with dry sand included the following series (Fig. 1):

Test 1—Compacted sand, wall tilting around its lower edge.

Test 2—Compacted sand, wall yielding parallel to its original position.

Test 3—Loose sand, wall tilting outward around its lower edge and then forced back again into its original position.

Test 4—Compacted sand, wall tilting around its lower edge, first advancing toward the fill, then moving outward and finally returning again to original position.

Prior to Test 1 a trial run was made for the purpose of familiarizing the operators with the testing device. For the trial run the sand was dumped into the bin and compacted in 6-in. layers with concrete tampers to a depth of 4.85 ft. (equal to about one-third the length of the wall). After the trial run, the material was excavated to an average distance of 3 ft. from the wall and the trench refilled for Test 1 by placing and compacting the sand in 6-in. layers. For Test 2 the backfill was prepared in the same way; the average width of the trench was 4 ft. For Test 3 the bin was emptied to an average distance of 8 ft. from the wall, then sand was dumped in and spread with a shovel without any attempt to compact it. For Test 4 the deposit of loose sand was entirely removed and replaced by sand compacted in 6-in. layers.

To determine the average volume of voids of the content of the bin, prismatic boxes were introduced into the bin at different stages of the filling operations. After filling had proceeded to a height of about 1 ft. above the upper rim of a box, the box was carefully excavated for the purpose of determining its volume of voids. It is obvious that these samples could not furnish more than approximate values for the average condition that prevailed in the backfill.

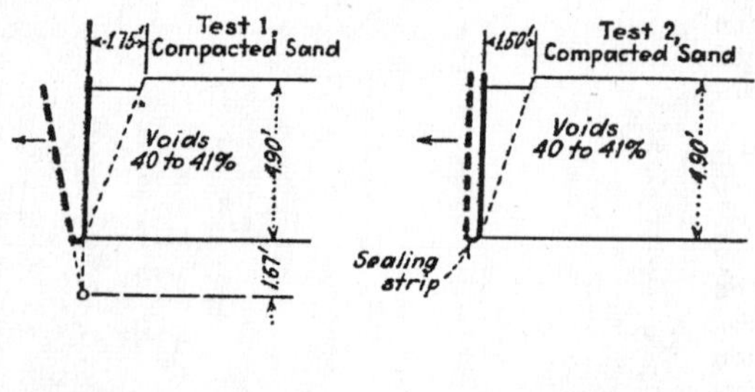

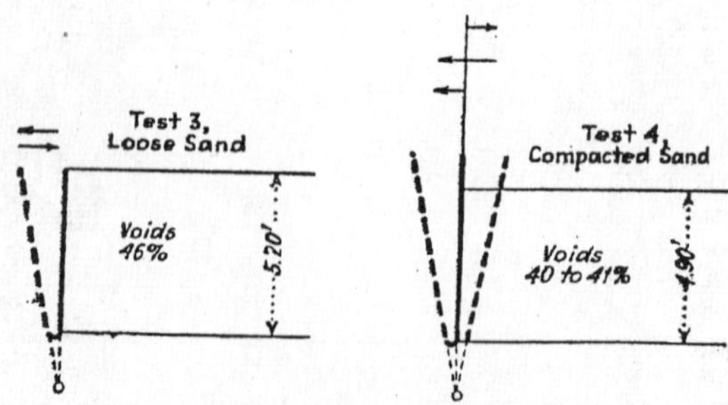

FIG. 1—FOUR CONDITIONS of sand fill and of wall movement were represented in the program of tests to determine pressure of dry sand against retaining walls.

Test results

The results of Tests 1 to 3 are shown in Figs. 2 to 5 on the following pages.

In Fig. 2, abscissas represent the wall movement in millimeters at an elevation equal to one-half of the height of the fill. Ordinates represent the hydrostatic-pressure ratio k of the earth pressure—that is, the horizontal component H of the earth pressure divided by the horizontal pressure W exerted by a liquid whose unit weight is equal to the unit weight of the backfill and whose depth is equal to the depth of the backfill.

Each test run is represented by three diagrams shown side by side. The third, (c), comprises the entire test result, with abscissas plotted to small scale, while (a) and (b) show the results of the first parts of the runs to scales fifty and five times

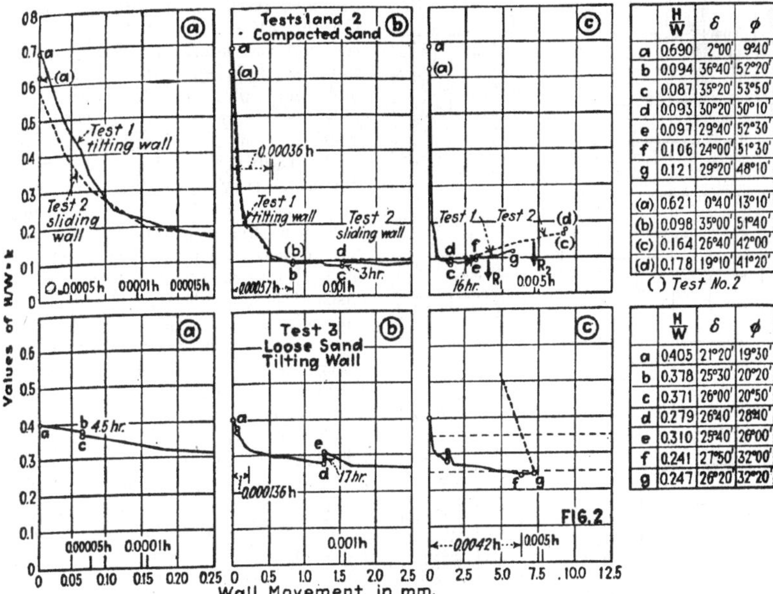

	H/W	δ	φ
a	0.690	2°00'	9°40'
b	0.094	36°40'	52°20'
c	0.087	35°20'	53°50'
d	0.093	30°20'	50°10'
e	0.097	29°40'	52°30'
f	0.106	24°00'	51°30'
g	0.121	29°20'	48°10'
(a)	0.621	0°40'	13°10'
(b)	0.098	35°00'	51°40'
(c)	0.164	26°40'	42°00'
(d)	0.178	19°10'	41°20'

( ) Test No. 2

	H/W	δ	φ
a	0.405	21°20'	19°30'
b	0.378	25°30'	20°20'
c	0.371	26°00'	20°50'
d	0.279	26°40'	28°40'
e	0.310	25°40'	26°00'
f	0.241	27°50'	32°00'
g	0.247	26°20'	32°20'

FIG. 2—HYDROSTATIC RATIO of dry sand as affected by yield of wall and compaction of fill.

larger. The position of the wall at which the first slip had occurred is marked in diagrams (c) by arrows and the letter *R*.

According to the diagrams of Fig. 2, there is almost no difference between the *k* curves for a wall which yields by tilting (Test 1) and a wall which yields parallel to its original position (Test 2), provided both walls are backfilled with compacted sand. On the other hand, there is no resemblance between the *k* curve for a tilting wall backfilled with compacted sand (Test 1) and a tilting wall backfilled with loose sand (Test 3).

In each test an intermission of several hours caused a marked increase of the intensity of the earth pressure in spite of the fact that the position of the wall did not change (vertical parts of the *k* curves). In Test 3, after the wall was moved out an average distance of 7 mm. (0.0045 *h*), it was forced back to its original position (dash line), which caused *k* to increase from 0.25 to 1.10. (All of increase not shown.)

Fig. 3 shows in similar arrangement the coefficient of wall friction, tan δ, plotted against the yield of the wall. The value of the coefficient of wall friction was obtained by dividing the measured vertical component of earth pressure, *V*, by the measured horizontal component *H*. An intermission of several hours almost invariably caused a decrease of the coefficient of wall friction, marked in the diagram by a thick vertical line.

Similarly, Fig. 4 shows the relation between the height of the center of pressure (point of application of the resultant earth pressure) and the yield of the wall. According to Coulomb's wedge theory, the center of pressure for level backfill should be one-third of the depth above the base. There is a marked difference between the curves for the two walls backfilled with compacted sand, one of which yields by tilting (Test 1) and the other by moving parallel to its original position (Test 2). But in neither case did the center of pressure descend to the lower third except in an advanced state of the test, while for loose sand it remained fairly consistently somewhat below the middle third of the depth.

In Fig. 5 the diagrams (a), at the left, show the relation between sand settlement and yield of the wall. The abscissa of the point *R* gives the average distance through which the wall had yielded at the instant when the first slip occurred within the backfill. The curves located above point *R* seem to have a horizontal asymptote, those below this point a vertical asymptote. In the test with loose sand, Test 3, no slip occurred. It seemed that the occurrence of slip in the loose backfill required a larger movement of the wall than the apparatus permitted.

The diagrams (b), at the right in Fig. 5, represent cross-sections of the surface of the fill adjoining the face of the wall, for different values of wall movement. Attention is called to the conspicuous difference between the wall which yielded by tilting (Test 1) and the wall which yielded by sliding (Test 2), both walls being backfilled by compacted sand.

Fig. 6 is an assembly drawing of all the results of test 4. This test was made for the purpose of investigating the effect of a movement of a wall toward a compacted backfill, thus mobilizing the lateral resistance of the backfill (part of the "passive earth pressure"). Starting at point *a* in diagrams *A*, *B* and *C* of Fig. 6, the wall was forced inward an average distance of 2 mm. to point *b*, tilting around an axis located somewhat below its lower rim. During this movement the coefficient of wall friction decreased from its initial value of +0.05 and asymptotically approached a value of —0.08 against a maximum value of +0.78 in the course of the outward movement of the wall during Tests 1 and 2 (Fig. 3). On reversal of the wall movement (diagram *C* of Fig. 6, branch *b e g*), the coefficient of wall friction first of all decreased to a minimum of —0.19, then

FIG. 3—WALL FRICTION of dry sand in relation to yield of wall and compaction of fill.

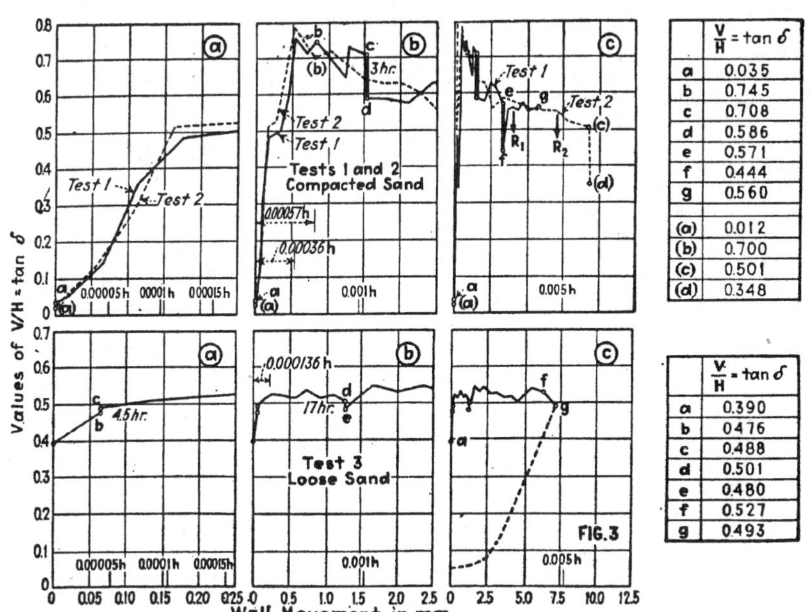

	V/H = tan δ
a	0.035
b	0.745
c	0.708
d	0.586
e	0.571
f	0.444
g	0.560
(a)	0.012
(b)	0.700
(c)	0.501
(d)	0.348

	V/H = tan δ
a	0.390
b	0476
c	0.488
d	0.501
e	0.480
f	0.527
g	0.493

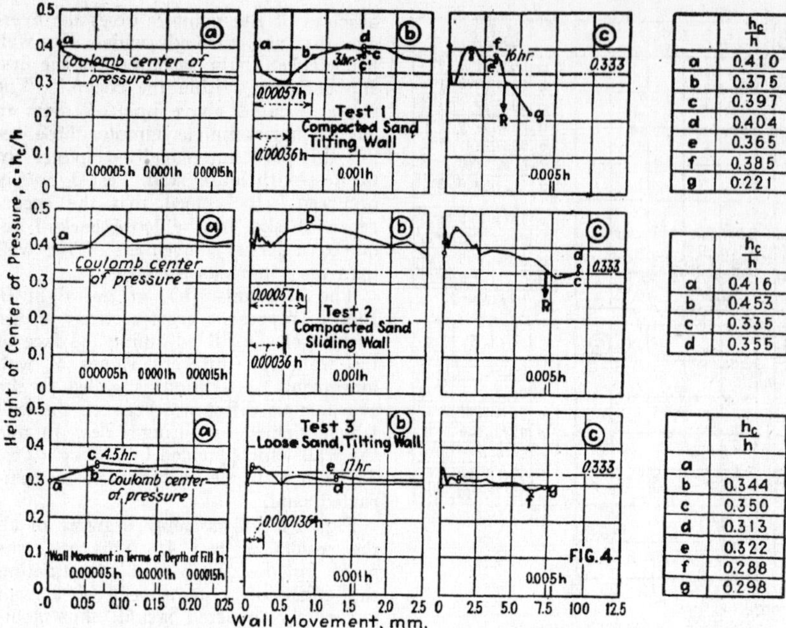

	h_c/h
a	0.410
b	0.376
c	0.397
d	0.404
e	0.365
f	0.385
g	0.221

	h_c/h
a	0.416
b	0.453
c	0.335
d	0.355

	h_c/h
a	
b	0.344
c	0.350
d	0.313
e	0.322
f	0.288
g	0.298

FIG. 4—HEIGHT of center of pressure of dry sand in relation to yield of wall and compaction of fill.

increased, became zero at *f*, when the value of the hydrostatic pressure ratio $k = H/W$ had dropped to 0.30, and finally aproached a maximum of +0.63 at point *g*.

Discussion of the test results

The most striking fact disclosed by the experiments concerns the important effect of the density of the backfill on the pressure phenomena.

With compacted sand backfill, a movement of the wall over an insignificant distance (equal to one-ten-thousandth of the depth of the backfill) decreases the hydrostatic pressure ratio k down to 0.20 or increases it up to 1.00, depending on the direction of the movement. With a loose sand backfill a like movement of the wall changes the value of k merely between the limits of 0.3 and 0.5. This contrast is indicated by Fig. 7. For the compacted backfill the value k corresponding to the initial position of the wall undoubtedly depends to a considerable extent on the method of compacting. According to the results of investigations carried out independently of the earth-pressure tests, it may have any value between about 0.35 and 0.7, but knowledge of the exact figure is without any practical importance because an insignificant outward movement of the wall reduces the figure to about 0.20, irrespective of the initial value. For loose backfill the initial value of k is approximately 0.40.

In the compacted backfill, slip occurred after the wall had moved through an average distance of 0.0027 h for tilting and 0.005 h for sliding movement. On the other hand, the movement that reduced the lateral pressure to a minimum was independent of the type of movement, about 0.0007 h (Fig. 2). To produce a slip in loose backfill, a wall movement of more than 0.008 h seems to be required; during this movement the pressure steadily decreases.

For compacted backfill, prior to the first slip, the position of the center of the pressure depends on whether the wall tilts or whether it slides. In both cases it is considerably above one-third the height of the backfill. When slip occurs, the center of pressure moves down to the middle third. For loose backfills it consistently remains at or slightly below the middle third.

The quoted differences between the behavior of dense and loose backfills are associated with corresponding differences in the changes of the structure of the sand within the sliding wedge. Based on the data of Fig. 5 we were able to compute the change of the volume of the sand contained in the wedge for a 5-mm. yield of the wall. The results in per cent of original volume are: Test 1, compacted backfill, tilting wall, 1.2 per cent expansion; Test 2, compacted backfill, sliding wall, 1.2 per cent expansion; Test 3, loose backfill, tilting wall, 0.45 per cent contraction.

A second group of interesting facts concerns the changes of the wall friction which are associated with the movement of the wall and with an intermission at stationary position of the wall. For compacted backfill the initial value of the wall friction is almost zero; a slight outward movement of the wall is sufficient to bring the coefficient of wall friction up to a maximum, but further movement causes the friction to decrease and to approach the value that corresponds to loose sand. For loose backfill a slight outward movement of the wall brings the wall friction up to an almost constant value.

Internal friction of sand

Finally, the data shown in Figs. 2 and 6 were used for an indirect determination of the angle of internal friction at the various stages of the movement of the wall. Previous experimental investigations by Müller-Breslau, Jacob Feld and many others have conclusively demonstrated that the earth-pressure theory of Coulomb reasonably agrees with reality as far as the magnitude of the lateral pressure is con-

FIG. 5—SAND SETTLEMENT in dry sand tests as affected by yield of wall and compaction of fill.

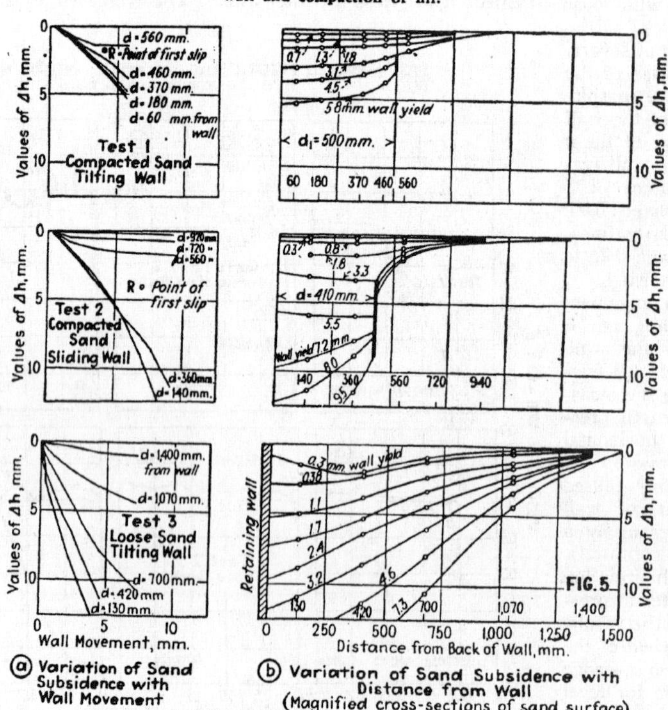

(a) Variation of Sand Subsidence with Wall Movement

(b) Variation of Sand Subsidence with Distance from Wall (Magnified cross-sections of sand surface)

cerned. Considering the approximate character of related computations, the fact that the slip surface is slightly curved rather than plane may be neglected. Hence the computations of the author were based on the well-known formula of Coulomb. Let h be the depth of the backfill; s the unit weight of the backfill; ϕ the angle of internal friction; H the horizontal component of the pressure; V the vertical component of the pressure, and $\tan \delta = \frac{V}{H}$ the tangent of the angle of wall friction. The horizontal component H of the lateral pressure acting against the back of a vertical wall is, according to Coulomb's theory,

$$H = \tfrac{1}{2} sh^2 \left\{ \frac{\cos \phi}{1 + \sqrt{\sin \phi \, (\sin \phi \cos \phi \tan \delta)}} \right\}^2$$

All the quantities of this formula are known from direct observation made during the test except ϕ, the angle of internal friction. Hence it was possible to compute the value ϕ for different stages of each test run. The results are shown in tables next to the diagrams in Figs. 2 and 6, together with the corresponding values of the angle of wall friction, δ. The tables disclose the following facts:

For compacted backfill the angle of internal friction assumes a temporary maximum of about 54 deg. prior to the first slip and decreases with further outward movement. This is obviously due to a progressive disintegration of the structure of the sand associated with the expansion mentioned before. A similar disturbing effect was apparently produced in Test 4 by starting the test with a movement of the wall toward the backfill. Therefore those parts of the diagrams in Fig. 6 that correspond to the outward movement of the wall have a character intermediate between the diagrams for dense and for loose sand, and the angle of internal friction never exceeded a value of 47 deg. For loose backfill the value of the angle consistently increased, due to progressive stabilization of the sand contained within the wedge, with a tendency to approach the "angle of repose" of 34 deg.

In connection with the preceding analysis it seemed interesting to compare the theoretical and the observed distance between the wall and the upper rim of the sliding surface. The computation was made by means of Coulomb's theory, based on the values of ϕ and δ corresponding to the stage at which the slip occurred. The results are shown in the accompanying table.

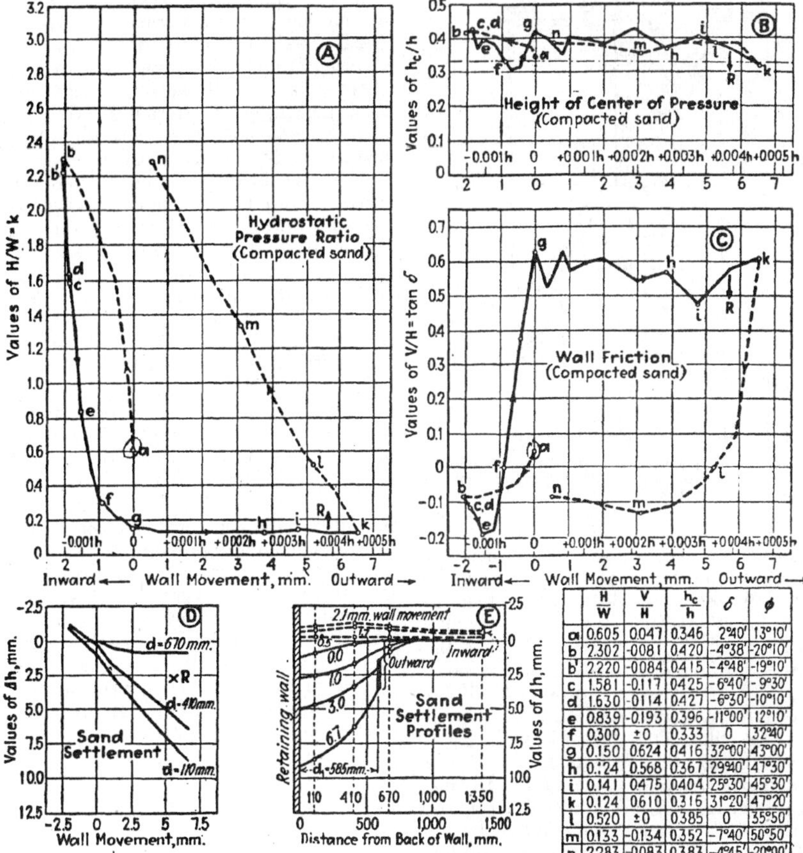

	$\frac{H}{W}$	$\frac{V}{H}$	$\frac{h_c}{h}$	δ	ϕ
a	0.605	0.047	0.346	2°40′	13°10′
b	2.302	-0.081	0.420	-4°38′	-20°10′
b′	2.220	-0.084	0.415	-4°48′	-19°10′
c	1.581	-0.117	0.425	-6°40′	-9°30′
d	1.630	-0.114	0.427	-6°30′	-10°10′
e	0.839	-0.193	0.396	-11°00′	12°10′
f	0.300	±0	0.333	0	32°40′
g	0.150	0.624	0.416	32°00′	43°00′
h	0.[illegible]24	0.568	0.367	29°40′	47°30′
i	0.141	0.475	0.404	25°30′	45°30′
k	0.124	0.610	0.316	31°20′	47°20′
l	0.520	±0	0.385	0	35°50′
m	0.133	-0.134	0.352	-7°40′	50°50′
n	2.283	-0.083	0.383	-4°45′	-20°00′

FIG. 6—WALL-PRESSURE EFFECTS of moving wall toward backfill and then reversing the movement; compacted sand.

For Tests 1 and 4 the agreement is very good, which indicates that the theoretical assumptions for slip in cohesionless material were nearly fulfilled. For Test 2 the slip plane intersected the surface of the fill at a considerably smaller distance than Coulomb's theory would lead us to expect. The discrepancy is not surprising, because the wall yielded parallel to its original position. In this type of wall movement the ratio between the movement of the wall and the corresponding width of the wedge increased from the top of the wall toward the bottom. The excessive expansion and subsidence within the lower part of the wedge caused arching in the upper part, and the slip occurred under abnormal conditions. This interpretation is substantiated by the abnormal movement of the center of pressure in Test 2 (see Fig. 4).

A conspicuous feature of all the test results appears in the fact that an intermission when the wall was stationary almost invariably caused a decrease of the coefficient of wall friction, an increase of the hydrostatic-pressure ratio k, and in general a slight rise of the center of the pressure. Similar facts had already been noticed by Jacob Feld during his own earth-pressure tests. They represent a phenomenon that is associated with any deformation of a body of sand. As an example, the results of an older experiment of the author are recalled. A layer of sand was placed in a ring and compressed by the application of a uniformly distributed surface pressure p. After the pressure was raised to p_1, the corresponding compression was kept constant for a certain period; during this intermission, the pressure which was required to keep the compression constant decreased at first rapidly, then slowly, and finally assumed a constant value p_2.

From these observations the author concluded that the stress required to *produce* a definite state of strain in the sand is invariably greater than required to *maintain* this state. Or, if a definite state of strain is established in a body of sand, the stresses decrease during the first few hours and gradually assume their final values. During these first

TOP THICKNESS OF WEDGE OF SLIP

	Computed, Ft.	Observed, Ft.	Ratio Observed to Computed
Test 1	1.80	1.75	0.98
Test 2	2.30	1.50	0.65
Test 3	3.40	....	....
Test 4	2.04	1.92	0.94

Note: The computed figure for Test 3 is added merely for comparison, in spite of the fact that during the test the wall did not move far enough to permit the occurrence of a slip. It is based on the assumption that ϕ = 34 deg. and $\tan \delta$ = 0.50.

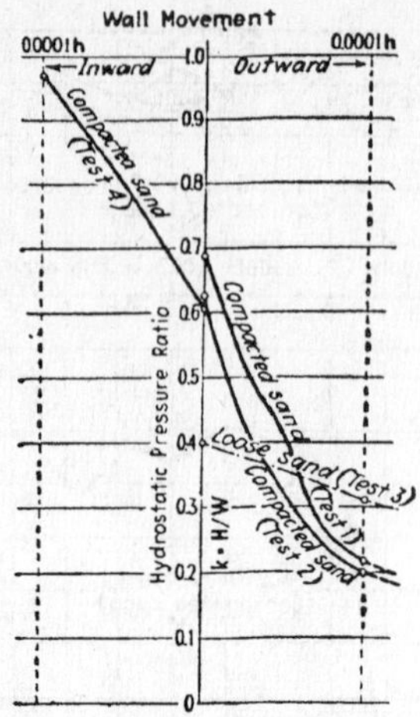

FIG. 7—CHANGE of hydrostatic ratio with movement of wall; compacted and loose backfill compared.

hours it seems that the grains slip slightly and turn into new positions, so that the forces required to keep them in this position decrease. The test results shown in Figs. 2 and 3 disclose the same fundamental property of the sand. The lateral pressure ratio $k = H/W$ depends on the intensity of the frictional stresses that act within the wedge, and the greater these stresses the smaller is the corresponding value of k. During an intermission the state of strain of the backfill remains unchanged. Hence the frictional stresses that act within the backfill and along the wall decrease, which causes an increase of the lateral pressure ratio. The gradual adjustment of the grains produces a slight subsidence of the sand within the wedge-shaped zone. Subsidence of a mass of sand inclosed between a rough wall and the stationary part of the backfill causes incipient arching, which in turn produces a slight upward movement of the center of pressure.

Conclusions on dry sand

The tests here reported furnish for the first time a set of data covering the effect of the wall movement on the earth pressure over the entire range between the original position of the wall and the yield required to produce slip. They lead to the following conclusions:

1. A striking difference exists between the behavior of dense and of loose backfill.

2. For compacted backfill the hydrostatic pressure ratio k corresponding to the pressure acting on the wall in its original position may have any value between 0.35 and 0.7, depending on how the fill was made. Inward movement of the wall through a distance of about 0.001 h (one-thousandth of the depth of the backfill) increases k up to values of 2.0 to 2.5, while outward movement of the same amount leads to the smallest values k can assume (about 0.1). Any further outward movement again causes an increase of the lateral pressure.

3. For compacted backfill the angle of internal friction assumes a well-defined maximum after the wall has yielded through an average distance of about 0.0007 h. If the wall yields further, the angle of internal friction increases again and approaches the maximum value for the loose backfill.

4. The coefficient of wall friction with dense backfill is extremely variable. If the wall moves inward toward the fill, the wall friction remains negligible. During outward movement the wall friction first assumes a maximum, then it gradually decreases toward the value that corresponds to the coefficient of wall friction for loose backfill.

5. For loose backfill, prior to any movement of the wall, the hydrostatic pressure ratio is about 0.4. Movement of the wall inward toward the fill through a distance of 0.001 h raises the value of k up to less than unity. On the other hand, outward movement decreases the lateral pressure until, after a movement through a distance of more than 0.008 h, slip occurs. At this point the hydrostatic pressure ratio assumes the Coulomb value approximately corresponding to the angle of repose. For the entire range of the movement, the coefficient of wall friction remains practically constant.

6. For both dense and loose backfill an intermission in the process of outward movement of the wall causes a slight decrease in wall friction, an increase in the value k, and in most cases a slight upward movement of the point of application of the resultant pressure.

Hence, in spite of the fact that the tests were made under the simplest conceivable conditions (dry sand, no movement of the wall in a vertical sense, etc.), the earth-pressure phenomenon was found to be more complicated than it is generally assumed to be.

Acknowledgments

These tests and those to be described in the following four papers were made in partial fulfillment of a cooperative agreement between the New Hampshire Public Service Commission, the New England Power Construction Co. and the Massachusetts Institute of Technology, the author acting as consulting engineer to the Public Service Commission. The tests were executed in the spring and summer of 1929 by G. A. Orrok, Jr., and R. G. Stafford, engineers of the New England Power Construction Co., under the supervision of the author. Their success was essentially due to the skill and initiative of these gentlemen.

LARGE RETAINING-WALL TESTS

A series of five papers reporting fundamental results

II—Pressure of Saturated Sand

Lateral pressure of submerged sand fill is the full water pressure plus the lateral pressure of the solid fraction, allowing for the reduction of effective weight by buoyancy

By Karl Terzaghi
Professor, Technische Hochschule, Vienna, Austria

IN A preceding article (*ENR*, Feb. 1, 1934, p. 136) the author described the results of earth-pressure tests with dry cohesionless sand. The present paper deals with the results of similar investigations with cohesionless sand in a state of complete submergence. Equipment and testing material (Plum Island sand) were identical in both series of experiments.

Prior to filling the bin (14x14 ft.) the rear wall of the bin was lined with a gravel filter (Fig. 1). The fill was placed in 6-in. horizontal layers and compacted by concrete tampers. After the wall was allowed to yield through an average distance of 1.4 mm., or 0.055 in., by tilting as shown in Fig. 1, an intermission of 18 hours was introduced. Then the water was slowly admitted through the rear filter, gradually rising within the fill until it appeared at the surface. Twenty-one hours later the fill was drained and the wall was allowed to yield further, up to an average distance of 4.2 mm., or 0.17 in. After an intermission of 23 hours the state of saturation was reestablished without moving the wall. Forty-seven hours later the wall was moved out to a total average distance of 6.9 mm., or 0.27 in., from its original position. It remained in this position for a period of 23 hours while the fill was in a state of complete saturation, and for 5 hours more while the fill was in a drained state. The intermissions more than covered the period required for all the elements of the lateral pressures to attain practically constant values.

When draining the sand, almost the entire water contained in the voids of the sand flowed out, only thin films remaining on the surface of the grains; the water content of the drained fill was about 2.5 per cent of the dry weight, against 25 per cent for the saturated state.

Co-existence of sand and water pressure

The term "submerged" is confined to that state in which the horizontal surface of the fill coincides with a free water surface, i.e. when the water in a standpipe connected with the sand

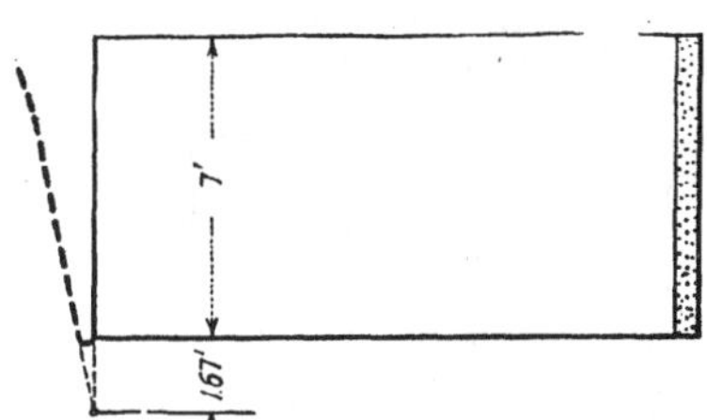

FIG. 1—GRAVEL FILTER at back of fill received water for submerging sand.

rises to the level of the surface of the sand. In such a state each sand particle is subject to the full hydrostatic uplift, which considerably reduces the active weight of the sand. Let

- s_0 be the unit weight of the water,
- s the average unit weight of the sand grains,
- n the volume of voids of the sand, expressed as a fraction of the total volume,
- s_1 the specific gravity of the dry sand, and
- s_2 the unit weight of the sand reduced by the hydrostatic uplift.

The values s_1 and s_2 are determined by the known formulas

$$s_1 = s\ (1 - n) \quad (1)$$

$$\text{and } s_2 = (s - s_0)\ (1 - n) \quad (2)$$

Hence at any depth y below the surface of a dry backfill the load carried by the grains of the backfill per unit of horizontal section is $y{\cdot}s_1$, while in a submerged backfill the load carried by the grains at the same depth is equal to $y{\cdot}s_2$, which is considerably smaller. On the other hand, in a dry backfill the back of the wall merely receives the lateral pressure exerted by the grains of the backfill, while in submerged backfill the pressure exerted by the grains combines with the full water pressure. The presence of the sand has no influence whatsoever on the intensity of the water pressure. For the sand this entire independence between the action of the water and the sand was conclusively demonstrated for a horizontal cellar floor by the tests of H. de B. Parsons[1] and many others. Therefore, when plotting the results of the tests with submerged sand a clean-cut separation was made between the water pressure and the sand pressure.

Let s_0 be the unit weight of the water. If the depth and width of the backfill are h and b respectively, the water pressure against the wall therefore is $W_0 = \frac{1}{2}\ s_0\ h^2b$. Deducting this amount from the total horizontal pressure H_s the horizontal pressure component due to the sand fraction of the submerged backfill become $H_2 = H_s - W_0$. Finally, if W_2 is the horizontal pressure that would be exerted by a liquid whose specific gravity is s_2 (see Eq. 2), the pressure coefficient of the sand fraction of the saturated fill is

$$k_s = \frac{H_2}{W_2} = \frac{H_s - \frac{1}{2}\ s_0h^2b}{\frac{1}{2}\ h^2b\ (s - s_0)\ (1 - n)}. \quad (3)$$

The drained backfill, on the other hand, has a hydrostatic pressure coefficient of

$$k_d = \frac{H_d}{\frac{1}{2}\ s_1'\ h^2b} \quad (4)$$

wherein H_d denotes the horizontal component of the pressure and s_1' the specific gravity of the sand as moistened by the preceding saturation. In our case, owing to the fact that the sand retained but very little water, we obtained an average $s_1' = 1.025\ s_1$.

In similar manner the coefficient of wall friction was computed by means of the formulas

$$\text{saturated, } \tan \delta_s = \frac{V}{H - W_0} = \frac{V}{H_s - \frac{1}{2}\ s_0h^2b} \quad (5)$$

$$\text{drained, } \tan \delta_d = \frac{V}{H_d}. \quad (6)$$

Test results

The results of the tests are plotted in Figs. 2 and 3. In all the figures the solid lines correspond to the drained state and the dotted lines to the submerged state of the backfill. The letters that refer to the submerged state are

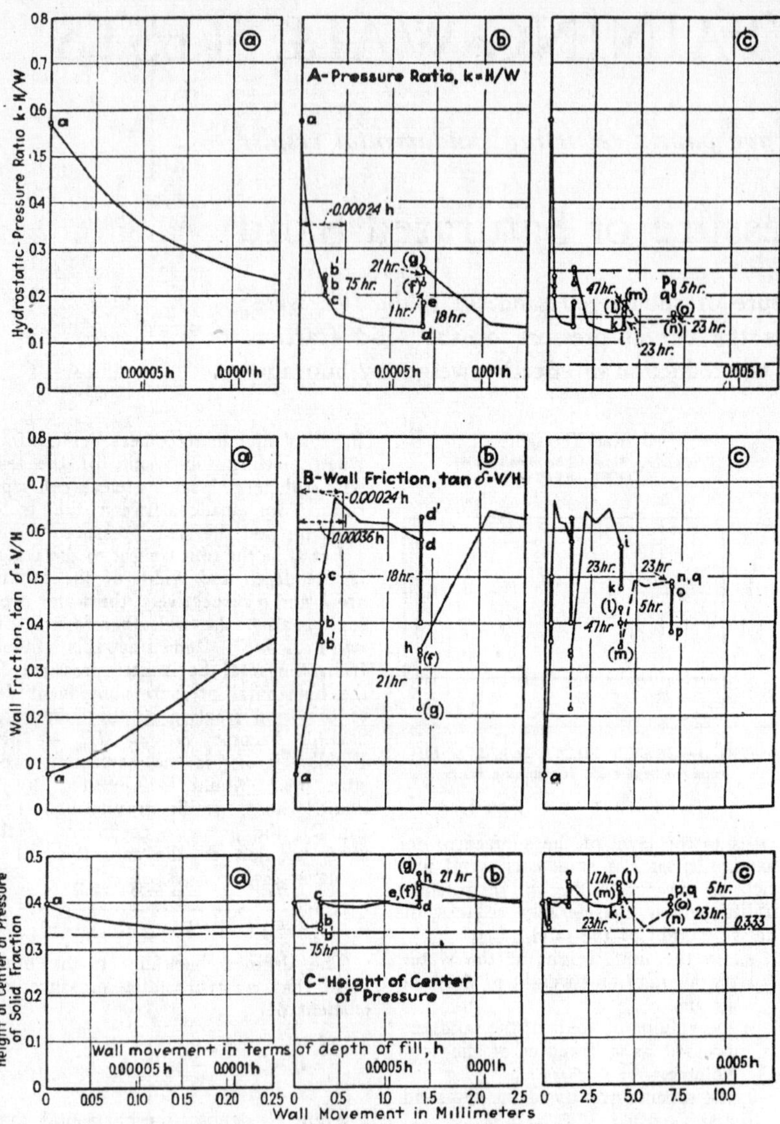

FIG. 2—SUBMERGED, SAND test results—Effect of wall movement on hydrostatic pressure ratio, wall friction and height of center of pressure of solid fraction of backfill.

shown in brackets. The changes that occur at a stationary position of the wall appear as thick vertical solid lines for the drained state and as dotted lines for the submerged state. The transition from one state to another is represented by gaps between vertical solid and dotted lines.

Fig. 2A shows the effect of the movement of the wall and of the processes of drainage and submergence on the value of the hydrostatic pressure ratio k. Abscissas represent the average distance through which the wall yielded. In Fig. 2B, the ordinates represent the tangent of the angle of wall friction (tan δ). In Fig. 2C the indicated position of the center of pressure is that of the solid fraction of the backfill only.

The diagrams of Fig. 3 correspond in every respect to the diagram of Fig. 5 in the preceding article. It should be noticed that the topmost curve in the diagram (a) of Fig. 3 of the present article is curved upward, while all the other ones are curved downward. According to the preceding article, the downward bend of the subsidence curve indicates that the corresponding point is located within the boundaries of the top surface of the wedge. Therefore, in the test now described the width of the wedge at the top must have been considerably greater than 710 mm., or 28 in.

Discussion of test results

Effect of Water on Internal Friction—In the preceding article the angle of internal friction which corresponds to the different stages of the test with dry sand was computed by means of Coulomb's formula. It was found that tan ϕ assumed its maximum value only after the wall was moved to a definite distance from its original position prior to the first slip. The relation between this maximum value of tan ϕ and the smaller values obtained in the test is the same as the relation between the ultimate shearing strength of a solid body and the shearing stresses associated with deformation with the limits of the ultimate resistance. Strictly speaking, therefore, the term "coefficient of internal friction" should be confined to the maximum value of tan ϕ, and it is used here merely for convenience. A better term would be "angle of active friction" or "angle of friction stress." If, for any stage of the earth-pressure test prior to the first slip the value of ϕ was found to be smaller than ϕ_{max}, we know that the strain in the backfill was less than that required to mobilize the entire frictional resistance of the material.

Table I contains a list of the computed values of ϕ for the different stages of the test in Fig. 2, together with the values of pressure ratio, wall friction, height of center of pressure, and elapsed time between consecutive stages. According to this table, the maximum value of the angle of internal friction is 47° for the drained state against 45°20′ for the saturated state, confirming the results of the friction tests, which showed that the presence of the water has practically no effect on the angle of internal friction. There is no contradiction between this fact and the well-known phenomenon of the "flowing out" of submerged sand fills, for the latter is a hydrodynamic phenomenon confined to fairly loose sand deposits and has nothing to do with the static internal friction. A discussion of the causes of the spontaneous spreading and flowing of submerged sand deposits can be found in one of the author's previous publications[3].

Effect of Depth of Fill on Internal Friction—Before the water was allowed to enter the dry backfill in the test diagrammed in Fig. 2, the wall had already moved through a distance of 0.00065h. According to the tests with dry sand described in the preceding paper, a movement through such a distance should be sufficient to mobilize the entire internal friction of the sand. For Test 1 with dry sand (preceding article) the value ϕ for a yield of the wall through a distance of 0.00065h was found equal to 52°20′, and the maximum value was 53°50′, against 45°10′ and 47°, respectively, found in the test described in this paper. Similarly the maximum value of the coefficient of wall friction was 0.75 for the dry sand against 0.66 for the saturated sand in Fig. 2, and the initial value of the hydrostatic pressure coefficient was 0.70 against 0.58.

These figures consistently demonstrate that both the internal friction and the wall friction for the present test (Fig. 2) were considerably lower

TABLE I — RELATION OF PRESSURE CONSTANTS TO YIELD OF RETAINING WALL

Distance of Yield of Top of Wall, Mm.	Condition of Backfill	Hydrostatic Pressure Coefficient k	Angle of Wall Friction $\tan\delta = V/H$	Angle of Internal Friction ϕ	Height of Resultant Pressure of Solid Part
	Dry (a)	0.575	0.08	14° 10′	0.396h
0.28	Dry (b)	0.219	0.40	36° 00′	0.349
	75 hr. later (c)	0.200	0.50	37° 20′	0.397
1.4	Dry (d)	0.133	0.58	45° 10′	0.387
	18 hr. later (e)	0.184	0.40	40° 00′	0.432
	Saturated (f)	0.225	0.34	36° 00′	0.433
	21 hr. later (g)	0.254	0.22	34° 00′	0.460
	Drained	0.234	0.42	34° 00′	0.420
	23 hr. later (h)	0.258	0.34	32° 20′	0.437
4.2	Drained (i)	0.127	0.57	47° 00′	0.394
	33 hr. later (k)	0.160	0.48	42° 40′	0.392
	Saturated (l),	0.163	0.44	42° 30′	0.436
	47 hr. later (m)	0.182	0.35	41° 00′	0.428
6.9	Saturated (n)	0.143	0.49	45° 20′	0.374
	23 hr. later (o)	0.161	0.48	42° 50′	0.388
	Drained (p)	0.225	0.38	35° 30′	0.411
	5 hr. later (q)	0.202	0.49	37° 00′	0.410

The letters refer to the letters in Fig. 2.

than those for the tests with dry sand. Both fills were made in the same manner (by compacting the material in 6-in. layers), and the average values of the volume of voids was found to be the same. The only difference between the fills resided in their depth, inasmuch as the depth of the present backfill was 7.0 ft. against 4.9 ft. in the dry sand test (Test 1) of the preceding article. This fact calls attention to the following important phenomenon:

For compacted sand under feeble pressure (smaller than about 0.1 kg. per cm²) the maximum angle of internal friction rapidly decreases with increasing surface pressures. For feeble surface pressures we lack a reliable method for directly determining the coefficient of internal friction. For the Plum Island sand in a compacted state tested under higher pressure, A. Casagrande obtained values of ϕ_{max} which ranged between 35° and 37°; in the tests of Fig. 2, the average normal pressure which acted on the sliding plane was 0.045 kg/cm² and ϕ_{max} was 47°; in Test 1 of the preceding article the normal pressure was 0.025 kg/cm² and ϕ_{max} was 53°50′; in previous earth-pressure tests on a very small scale with compacted sand 4 in. deep the writer obtained values of more than 65°. Hence the considerable depth of the backfill alone may account for the relatively low value of ϕ_{max} in Fig. 2.

In every other respect the fill had all the characteristic properties of a compacted backfill in contrast to those of a loose one— high initial value of k, a negligible value of the initial coefficient of wall friction, rapid decrease of k during the first part of the outward movement of the wall, a well-defined maximum for the coefficient of wall friction, and location of the center of pressure well above the middle third.

Effect of Wall Movement and of Intermissions—According to Fig. 2, in either the drained or the submerged state an outward movement of the wall increases the coefficient of wall friction and decreases the pressure ratio. During an intermission in the wall movement, the wall friction becomes smaller and the pressure ratio somewhat greater. In this respect there is no difference between the dry and the submerged sand.

Effect of Drainage and Submergence—The diagrams of Fig. 2 show that each transition from the drained to the submerged state, or vice versa, breaks the continuity of the curve and produces a rather substantial change of the stresses that previously existed in the fill. The accompanying table contains a list of related data. According to this table, after the wall yielded through an average distance of 1.4 mm., or 0.00065h, drainage of the fill caused a slight decrease of the pressure ratio combined with an increase of wall friction. As time went on, k slightly increased again, while the value of tan δ went down. In contrast to this, for a yield distance of 6.9 mm., or 0.0033h, the drainage process led to a substantial increase of k and a decrease of tan δ; but five hours later the value k had somewhat decreased, while the value of tan δ became greater than before. Hence the effect of drainage and submergence on the lateral pressure seems to depend on the distance through which the wall previously yielded.

Effect of Drainage and Saturation on Density—The effect of drainage and submergence on the state of stress in the backfill is obviously due to minor structural changes essentially produced by the flow of water through the voids of the fill. This explanation is substantiated by the vertical movement of different points located on the surface of the fill. By means of such observations it was found that submergence invariably caused the surface of the fill to rise through a distance of about 0.5 mm., irrespective of the position of the wall. This was to be expected, because the hydrostatic uplift associated with submergence decreases the vertical pressure on the grains of the sand, and hence the fill expands upward. During this expansion the fill slides along the wall in an upward direction, which relieves the shearing stresses acting in the fill along the wall and causes a decrease of the wall friction.

On the other hand, drainage produced a subsidence larger than the preceding rise due to submergence, and the distance through which the surface subsided was different, depending on the distance through which the wall had previously yielded. Thus, when the wall was at an average distance of 1.4 mm. from the original position, drainage caused a subsidence of the top surface of the wedge through a distance of 1.0 mm.; after the yield of the wall was increased to 6.9 mm., drainage caused the top surface of the wedge to subside through a vertical distance of 5.0 mm.

This apparently insignificant fact throws light on the relation that was found to exist between the position of the wall with respect to its original position and the effect of drainage and submergence on the lateral pressure. With the wall close to its original position, the grains are still in stable relation, and drainage merely causes a minute subsidence of the fill in itself, which increases the coefficient of wall friction and reduces the hydrostatic pressure ratio k. But after the wall has approached the position corresponding to the first slip, the sand in the vicinity of the future sliding plane is stressed almost to the limit of what it can stand, and in this state any external cause, such as vibrations or the flow of water through the interstices, induces the grains to settle into more stable positions. The ensuing movement of the grains has the character of a spontaneous collapse, which in turn has the effect of substantially relieving the intensity of the stresses.

Distribution of Pressures Over the Wall—According to Fig. 2C, the center of pressure reached an average height of 0.4 h and was apt to rise as high as 0.46 h. An equally high position of the center of pressure was observed during Test 2 of the preceding paper,

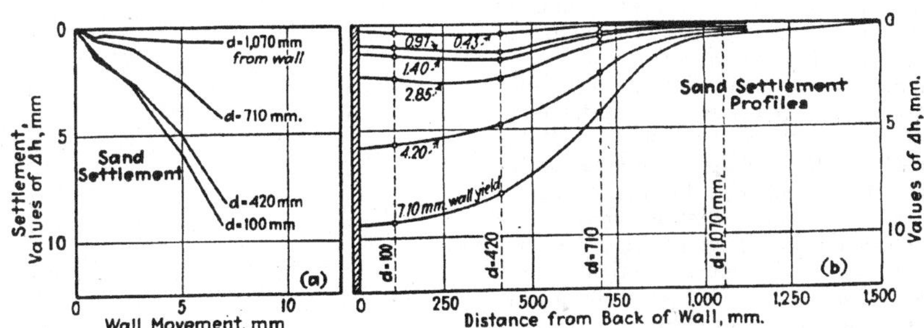

FIG. 3—EFFECT of submergence and drainage on subsidence of fill.

where the wall was allowed to yield parallel to its original position. According to Coulomb, the center of pressure should be located at a height of 0.33 *h*, and, according to the advanced theories of curved sliding surfaces, at about 0.35 *h*. Yet none of the theories consider the important fact that the grain movements associated with the outward movement of the wall give rise to "arching" between the back of the wall and the inclined zone of maximum frictional stress. Arching relieves the stresses that act on the lower part of the wall and increases those acting on the middle part, as it does in the hopper bottom of a grain bin.

Unfortunately, due to the feeble pressures involved, the Goldbeck cells failed to inform us on the distribution of the pressures over the back of the wall. Nevertheless there can be little doubt that the distribution of pressures prior to the first slip is somewhat as represented in Fig. 4. This figure shows two mechanically and physically possible pressure distributions, for height of center of pressure at 0.40 *h*, (a) of Fig. 4, and at 0.46 *h*. The distribution at (a) in Fig. 4 approximately corresponds to state *d*, Fig. 2C and the straight dotted line shows the distribution of pressures according to the current Coulomb conception. At the upper edge of the wall the pressure curve is tangent to the Coulomb line, and the area included between the curve and the back of the wall is equal to the area of the Coulomb triangle. As a result of the grain movements during an intermission and subsequent submersion, the distribution (a) passed into (b) of Fig. 4, without the wall changing its position.

In order to account for the high position of the center of pressure without admitting "arching," one would be obliged to assume that the sliding surface has a far more important curvature than ever observed in tests with cohesionless materials, provided an explanation was possible at all. And if arching occurs, the lower part of the back of the wall cannot get more than a very low pressure. The static effects of arching on the lateral pressure were recognized as early as 1920 by H. G. Moulton[3] and by J. C. Meem[4] for cohesive materials, and our test results seem to confirm the conception also for cohesionless backfill.

These conclusions by no means deny the approximate validity of Coulomb's formula concerning the magnitude of the lateral pressure, because the equations for the equilibrium of the wedge are valid irrespective of the distribution of the pressures over the back of the wall and the sliding surface. In order to adapt Coulomb's theory to reality, one merely has to eliminate the conception of the triangular pressure distribution and the hypothesis that the angle of internal friction is necessarily identical with the angle of repose, a conception which was introduced into the theory by Coulomb's successors. Since there is no way to compute the arching effect on the position of the center of pressure, we are obliged to determine it by direct observation, as was done in the earth-pressure tests.

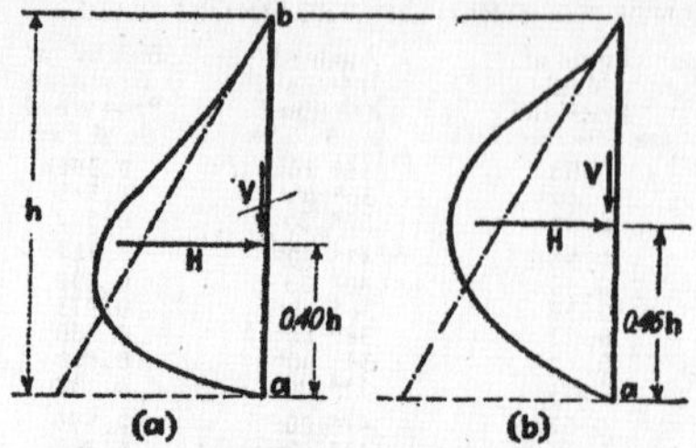

FIG. 4—DISTRIBUTION of pressure on back of wall for two positions of the center of pressure.

The preceding remarks concerning distribution of the pressures over the back of the wall merely refer to the conditions prior to slip. According to Tests 1 and 2 of the preceding article, slip causes the center of pressure to descend to a depth equal to or greater than two-thirds the depth of the backfill. The reason is obvious: at slip, the grains settle into more stable positions and the arches collapse, so that the Coulomb pressure distribution is reestablished at least approximately and temporarily.

In Test 1 of the preceding paper, the slip occurred after the wall yielded through a distance of 0.0027 *h*. In the present test no slip took place, although the wall yielded 0.0033 *h*. This failure of slip to occur seems to be connected with the fact that slip requires the material to be previously strained to the limit of its capacity, and as both drainage and submergence relieve the stresses in the backfill, the test represented by Fig. 2 gave no opportunity for stress to accumulate sufficiently for inducing a slip.

1. The lateral pressure exerted by the submerged backfill is equal to the sum of the full water pressure and the lateral pressure of the solid fraction of the fill, whose effective weight is reduced by buoyancy.
2. The presence of the water has practically no effect on the coefficients of internal friction and wall friction.
3. For a position of the wall in the immediate vicinity of the original one, the drainage of the fill produced a slight decrease of the hydrostatic pressure ratio *k* and an increase of the coefficient of wall friction tan δ. During a subsequent intermission, *k* increased and tan δ became smaller.
4. In contrast to this, after the wall moved beyond an average distance of one-thousandth the depth of the backfill, drainage produced an important increase of *k* and a decrease of the wall-friction tan δ. During the subsequent intermission *k* became smaller and tan δ greater.
5. Submergence causes a minute expansion, and drainage a more important subsidence of the fill. If the wall is close to its original position, the subsidence due to drainage is smaller than it is for more advanced positions of the wall.
6. The tests furnished indirect evidence for assuming that, prior to slip, the lower part of the back of the wall experiences lower pressure than the middle part, provided the wall is backfilled with compacted material.

References

[1] H. de B. Parsons, "Hydrostatic Uplift in Pervious Soils," *Trans.*, Am.Soc.C.E., Vol. 93, p. 1,317 (1929).

[2] K. Terzaghi, "Erdbaumechanik" (1925), p. 344.

[3] H. G. Moulton, "Earth and Rock Pressures," *Trans.*, Amer. Inst. of Min. and Met. Eng., February, 1920.

[4] J. C. Meem, "Pressure and Resistance of Soil," Brooklyn Engineers Club, February, 1920.

LARGE RETAINING-WALL TESTS

III—Action of Water Pressure on Fine-Grained Soils

A series of five papers reporting fundamental results

As with submerged sand, the retaining wall backfilled with fine-grained soil such as till or clay receives full water pressure plus the pressure of the solid fraction of the fill

By Karl Terzaghi
Professor, Technische Hochschule, Vienna, Austria

FOR coarse-grained materials, such as sands, the tests of H. de B. Parsons[1] and of many others have conclusively demonstrated that hydrostatic uplift is fully active. As a consequence, within a submerged mass of such material the water acts on every plane as if the sand were not present. This result of the Parsons tests was to be expected, yet to many engineers it came as a surprise and was accepted only with reluctance by some. Therefore the question of hydrostatic uplift in fine-grained materials, such as clay or cement-mortar, must be considered a delicate and controversial topic.

No water can penetrate into bodies that have no voids, such as metals or glass, and in such materials no hydrostatic uplift can occur. On the other hand, in coarse sand the area of contact between the grains is negligible compared with the total surface of the grains. Therefore the weight of each grain is reduced by practically the full hydrostatic uplift. In materials intermediate between these two extremes, where the grains are cemented together, the cement forms solid bridges from grain to grain, and hydrostatic forces cannot act at those parts of the surfaces of the grains that are occupied by the cement. Hence the total effect of the hydrostatic pressure on the granular material depends on the relation between the area of the cement spots and the total internal surface of the substance.

Coefficient of grain contact

Fig. 1 shows a magnified section through a submerged mass of very fine-grained material, in intimate contact with the horizontal floor and the vertical side of a solid body. The areas at which the grains are cemented together or cemented to the surface of the solid body are shown black. The curved line $S_1' S_2'$ is a section which follows an arbitrary plane section $S_1 S_2$ as closely as possible without cutting the grains. Such a section as $S_1 S_2$ will be designated as a "contact section." Concerning the areas of contact, we make the following assumption: The projection of the areas where the curved surface

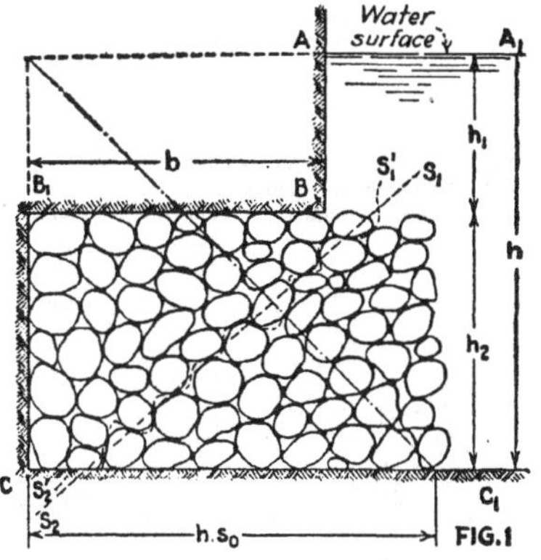

FIG. 1—DIAGRAM representing the condition of partial cementation which affects the computation of hydrostatic pressure and uplift in fine-grained soils.

$S_1' S_2'$ cuts through the films of cement upon the coordinated plane $S_1 S_2$ is equal to m per units of area of this plane. The inclosing surfaces $B\ B_1$ and $B_1\ C$ are also supposed to be cemented to the adjoining granular mass over the area m per unit of surface.

If the granular material was absent, or if it consisted of sand, the hydrostatic uplift acting on $B\ B_1$ would be $V = h_1 b s_0$ and the hydrostatic side pressure on $B_1\ C$ would be $H = \frac{h + h_1}{2} h_2 s_0$ where s_0 denotes the unit weight of the water. Due to partial cementation (Fig. 1) the pressures assume the values:

$$V_1 = h_1 b s_0 (1 - m) \tag{1}$$

$$\text{and } H_1 = \frac{h + h_1}{2} h_2 s_0 (1 - m) \tag{2}$$

Based on similar considerations, the unit weight of the submerged material reduced by the hydrostatic uplift is found to be:

$$\begin{aligned} s_2 &= (1 - n)[s - (1 - m) s_0] \\ &= (1 - n)(s - s_0 + m s_0) \end{aligned} \tag{3}$$

where n denotes the volume of voids and s the unit weight of the dry substance. The quantity m will be called the "coefficient of grain contact." The value $1 - m$ corresponds to Parsons' "effective area." For sands, according to previous test results, $m = 0$, $V_1 = V$, $H_1 = H$, and $s_2 = (1 - n)(s - s_0)$.

Thus the problem consists merely of determining the value m. The method used by Parsons and other workers in this field cannot be used for very fine-grained materials because the disturbing effect of the time element rapidly increases with decreasing grain size. Therefore the writer adopted an indirect method based on a generalized interpretation of the well-known rupture diagram of Mohr.

Interpretation of rupture diagrams

In Fig. 2, sketch (a) represents a cross-section through a cylindrical specimen of a granular material, as, for instance, mortar of portland cement. The thickened lines indicate those spots where the grains are cemented together. The specimen is supposed to be dry and completely covered with an impermeable skin. It is submitted to a compression test in the following manner: First, we apply to the surface of the specimen a hydrostatic pressure p per unit of area. Then we add to the constant hydrostatic pressure a linear pressure of increasing intensity acting only in the direction of the vertical axis. Upon reaching the intensity C this supplementary pressure causes failure. The value C represents the compressive strength which is associated with the hydrostatic pressure p.

According to Mohr, the influence of

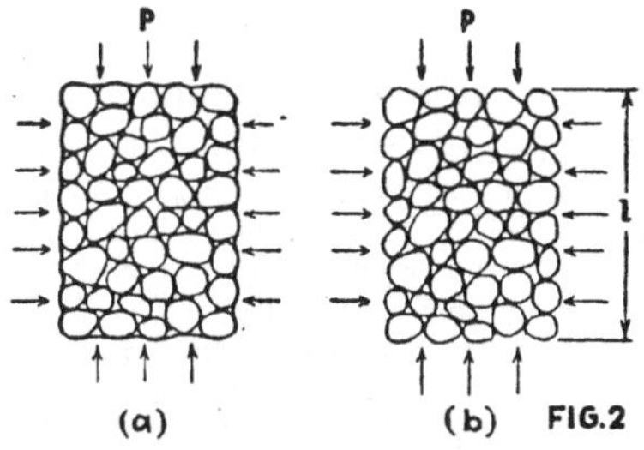

FIG. 2—SUBMERGED SPECIMENS of granular material under hydrostatic pressure with and without impermeable skin.

p on the compressive strength C of the specimen can be represented by means of the diagram Fig. 3(a). In this diagram the result of each test is represented by a circle passing through two points with the abscissas p and $(p + C)$ respectively. According to experience with granular materials, the diameter of the circles increases rapidly from the origin O toward the right. Rupture as a rule occurs along shearing planes which form an angle $(90-\alpha)$ with the direction of the supplementary pressure C.

If we draw a straight line from the center of any circle to the point of contact B_1 between this circle and the envelope A_1 A_2 of all the circles of rupture, the angle between this line and the axis is equal to 2α. The ordinate B B_1 of B_1 is the shearing stress, and the abscissa OB is the normal stress that acted on the plane of rupture at the instant of failure.

In general, the envelopes A_1 A_2 are somewhat curved. However, for pressures p up to about 100 atmospheres the curves A_1 A_2 may be replaced by straight lines as in Fig. 3(b). According to this diagram, the resistance acting along the plane of rupture consists of two parts, viz. the cohesion c, independent of the intensity of the normal stress, and a fractional resistance OB $\tan\phi$, whose intensity increases in direct proportion to the normal stress. For marble, Kármán[2] obtained for ϕ at low pressures a value of about 32°; and Brandtzaeg[3] for concrete at low pressures a value of about 40°. Both values are of the same order of magnitude as the angle of internal friction for sand.

Since an increase in the hydrostatic pressure p does not change the strength of the grains of a clay or of a mortar, the seat of the frictional resistance OB $\tan\phi$ must reside in the films of solid substance at the points of contact between the grains. Hence the influence of p on the compressive strength of the materials depends exclusively on the intensity of the pressures which act at the contacts. Based on the knowledge of this fact, the coefficient of contact m can easily be determined. We simply modify the compression test by eliminating the impervious skin of the specimen and submerge the specimen, (b) in Fig. 2, in a liquid that can be brought under a pressure p. In this case one part m p of the liquid pressure will be transmitted from grain to grain through the points of contact, while the remainder will merely be transmitted through the liquid. As a consequence, the pressure that acts on the films of cement will merely be equal to m times the pressure that acts under the test arrangement (a) of Fig. 2. Elimination of the watertight skin, therefore, should have the same effect on the compressive strength of the sample as if we reduced the intensity of the hydrostatic pressure from p to m p, and at equal

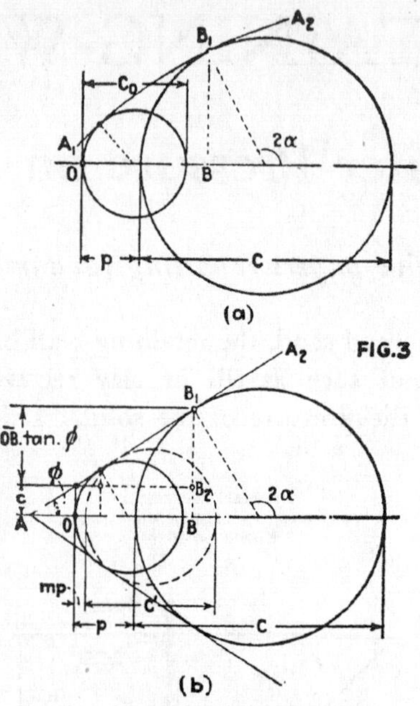

FIG. 3—MOHR'S REPRESENTATION of the influence of hydrostatic pressure on compressive strength of granular material.

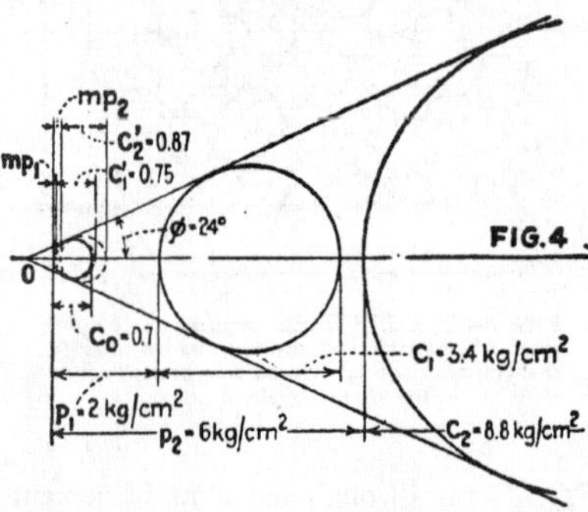

FIG. 4—TEST RESULTS with plastic marine clay, showing free transmission of hydrostatic pressure.

pressures its compressive strength C' should be smaller.

According to Mohr's diagram, no rupture can occur unless the corresponding circle makes contact with the envelope A_1 A_2, Fig. 3. Hence if we represent the result of the test with the bare specimen by a dotted circle of diameter C' tangent to A_1 A_2, the abscissa of the left point of intersection of the circle with the horizontal axis is equal to the unknown quantity m p.

Results of tests and conclusions

In 1928 some tests of this kind were made with a very plastic marine clay[4], and in 1933 with concrete. The clay had a water content of 28 per cent of the dry weight, a plasticity index of 23, and an angle of internal friction of 24 deg. The concrete represented a mixture of 236 kg. of portland cement to 1 cu.m. of the finished product.

The test results for the clay, shown in Fig. 4, furnished for m a value of about 0.02. For the concrete we obtained m practically equal to zero, and a similar interpretation of the results of older tests with cement mortar[5] also leads to m almost zero. Hence, for all these materials we are obliged to assume that hydrostatic pressures are as freely transmitted through them as through coarse-grained sand. *Any surface of a solid substance in intimate contact with submerged clay or with concrete is subject to a hydrostatic pressure of the same intensity as if the adjoining porous materials were not present.*

The reason why these important facts escaped the attention of practicing engineers is to be found in the slowness with which hydrostatic uplift manifests itself in the case of fine-grained materials because of their low permeability. Suppose, for instance, that the concrete floor of a basement located below groundwater level rested on a stratum of stiff clay. In order to lift the floor, water must percolate from the sand through the clay toward the base of the floor. Since the speed of percolation of the water through a stiff clay at moderate hydraulic gradients amounts to a small fraction of an inch per year, the process would require many years in order to produce the first visible effects. Furthermore, in the majority of cases, the permeability of the concrete is considerably greater than that of the clay. Consequently the water would permanently continue to flow, the greatest part of the hydraulic head being used up within the clay on its way from the sand toward the basement. In spite of the permanent flow, the surface of the concrete in the basement can permanently appear to be dry, because even in a cellar the speed of evaporation considerably exceeds the speed of percolation through a stiff clay.

Hence empirical evidence leads to the erroneous opinion that in all such cases hydrostatic uplift is non-existent. Yet there are conditions in engineering practice under which hydrostatic uplift in clay may lead to catastrophic consequences. Consider a masonry dam resting on a stratified shale whose layers are separated from each other by thin seams of clay. According to current conceptions, the force that resists the sliding along the seams of the substrata should be equal to $G \tan\phi$, where G is the weight of the dam and ϕ is the coefficient of internal friction of the clay. However, due to the full hydrostatic uplift in the seams of clay, the resisting force is only

$$(G - hbs_0) \tan\phi$$

where s_0 denotes the unit weight of water. The difference between the real and the assumed value of the resistance against sliding may be responsible for more than one accident of the type of the failures of Ohio River dam 26 or of the reservoir at Nashville, Tenn.

In connection with the interpretation of the results of large-scale earth-pressure tests with submerged backfill the quoted facts lead to the conclusion that, if the container of the apparatus is perfectly watertight, the back of the retaining wall in contact with a submerged backfill receives full water pressure in addition to the pressure exerted by the solid fraction of the backfill. This conclusion is valid irrespective of the grain size of the backfill material.

Previous articles in this series appeared Feb. 1, p. 136, and Feb. 22, p. 259.

References

[1]H. de B. Parsons, "Hydrostatic Uplift in Pervious Soils," *Trans.* Am.Soc.C.E., Vol. 93, p. 1,317 (1929).

[2]Frank E. Richart, Anton Brandtzaeg and Rex L. Brown, "A Study of the Failure of Concrete Under Combined Compressive Stresses," Univ. of Illinois *Bulletin*, Vol. XXVI, No. 12, 1928.

[3]Th. v. Kármán, "Festigkeitsversuche unter allseitigem Druck," Ztschr. Ver. deutsch. Ingenieure, 1911, p. 1,749.

[4]K. Terzaghi, "Tragfähigkeit der Flachgründungen," *Proc.*, First Congress Int. Assoc. for Bridge and Struct. Engineering.

[5]P. Fillunger, "Versuche über die Zugfestigkeit bei allseitigem Wasserdruck," Österr. Wochenschr. für den öff. Baudienst, Wien, 1915.

LARGE RETAINING-WALL TESTS

A series of five papers reporting fundamental results

IV—Effect of Capillary Forces in Partly Saturated Fill

Analysis of capillary forces as they complicate pressure phenomena and affect the results of tests on drained backfill of fine material

By Karl Terzaghi
Professor, Technische Hochschule, Vienna, Austria

In a preceding paper (*ENR*, Feb. 22, 1934, p. 259) the results of tests with coarse submerged sand were described. When this sand was drained the voids of the material became practically empty; water remained only in the form of thin films covering the surfaces of the individual grains, and the weight of these films did not exceed 2.5 per cent of the total weight of the sand. In a subsequent series of tests the backfill consisted of a modified till of glacial origin, whose effective size was less than one-thirtieth the effective size of the sand. When this till was drained, its voids remained practically filled with water, due to the small size of the grains. The difference between the average weight of the saturated and the drained till was less than 1 lb. per cubic foot. Most of the water was retained within the fill by capillary forces. These forces have a marked influence on the test results, and they complicate the pressure phenomena to a considerable degree. Hence, before presenting in the following paper the results of the earth-pressure tests with "modified till" an attempt will be made to analyze the various factors involved.

Drained backfill in capillary saturation

We will assume a retaining wall backfilled with fine but cohesionless sand to a depth h less than the height h_c through which water could rise in the material by capillary action. The symbols to be used in the following computation of the forces that act on the back of the wall are: h and b = depth and width of the fill; s_0 = unit weight of the water; s_1 = unit weight of the fill, excluding the weight of the water contained in the voids; s_2=unit weight of the fill with the voids completely filled with water (complete capillary saturation); s_3=average unit weight of the fill with the voids partly filled with water (partial capillary saturation); s_4=unit weight of the fill reduced by full hydrostatic uplift; $W_0 = \frac{1}{2} \cdot s_0 bh^2$ = the water pressure that acts on the wall when the backfill is replaced by water; W_1, W_2, W_3, W_4=the pressures exerted on the wall by liquids of unit weights of s_1, s_2, s_3, s_4 respectively; H=the horizontal component of the pressure that acts on the wall; V=the vertical component of the pressure that acts on the wall; tan δ=the coefficient of wall friction; k = the hydrostatic pressure ratio for the lateral earth pressure—that is, the ratio between the horizontal component of the pressure of the solid fraction of the backfill and the corresponding liquid pressure.

FIG. 1—PRESSURES on wall with the backfill at three different stages of saturation.

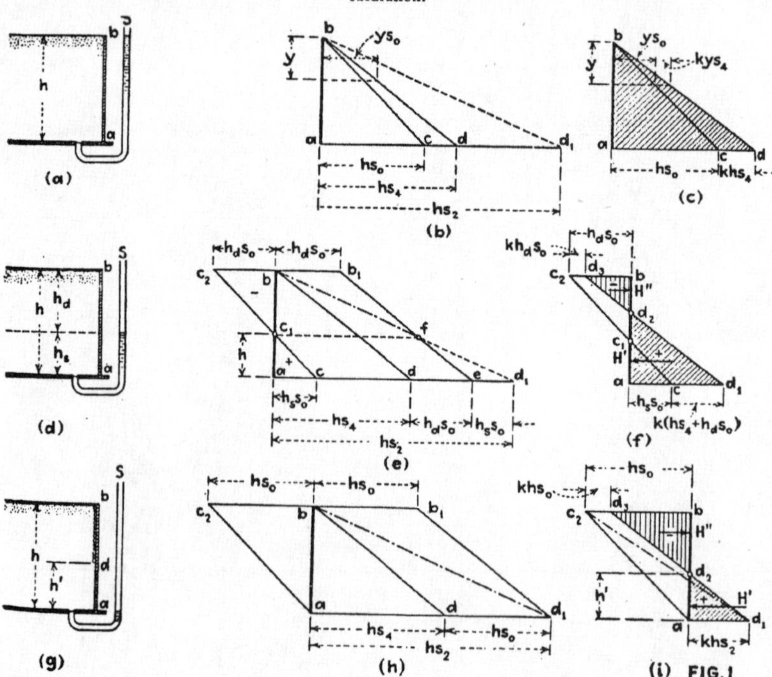

Drawings (a), (d) and (g) in Fig. 1 show sections through the wall with the backfill in three different stages of saturation: (a) completely submerged, the water level in the standpipe located at the elevation of the surface of the fill; (d) partly submerged and partly in a state of capillary saturation, the water level in the standpipe at a depth h_d below the surface of the fill; and (g) in a state of complete capillary saturation, with the level in the standpipe at the base of the fill. The depth below the surface of the fill to which the water was lowered in the standpipe will be called the "suction head," and that part of the fill that is contained within this depth is the "suction zone."

In the diagrams (b), (e) and (h) of Fig. 1 are plotted the vertical normal pressures which act at different depths y below the surface. According to the conclusions presented in a preceding paper, the presence of the backfill has practically no reducing influence on the forces that act in the liquid content of the voids of the fill. Therefore, in every case the pressures which act in the water are represented by the abscissas of a straight line (line bc in Fig. 1b, line cc_1 in Fig. 1c and line ac_1 in Fig. 1h). According to the laws of hydraulics, the standpipe level represents the boundary between negative and positive hydrostatic pressures. That part of the vertical pressure which is transmitted from grain to grain through the solid fraction of the fill is also represented, in every case, by a straight line bd. Since the grains of the backfill are almost completely surrounded with water, the effective unit weight of the sand is equal to s_4, and the base of the pressure triangles abd is equal to hs_4.

At a suction head h_d (Fig. 1d) the surface tension of the water acts on the surface of the fill like a surcharge with an intensity $h_d s_0$ per unit of area. Since this force is transmitted through the solid part of the backfill to the base of the container, the quantity $h_d s_0$ must be added to the abscissas of the pressure lines bd for the solid part. Thus we obtain, for the solid part, the pressure lines b_1c in Fig. 1e and b_1d_1 in Fig. 1h.

The same result can be obtained from the following consideration: At any point of depth y the total vertical pressure must be equal to the sum ys_2 of the weight of the sand and the water above this point. This total weight is represented in diagrams (b), (e) and (h) of Fig. 1 by the straight line bd_1. To obtain the pressure line for the solid fraction of the fill, we simply reduce the length of the abscissa of bd by the length of the corresponding abscissa of the line of the water pressure bc in Fig. 1b, cc_2 in Fig. 1e, ac_2 in Fig. 1h. Thus we obtain the pressure lines bd in Fig. 1b, b_1c in Fig. 1e, and b_1d_1 in Fig. 1h respectively.

The determination of the pressure which acts on the vertical back of a retaining wall is based on the following consideration: At any depth the horizontal pressure exerted by the liquid content of the fill is equal to the vertical pressure that acts in the liquid at the same elevation. On the other hand, the horizontal pressure exerted by the solid part at any depth y is equal to the corresponding vertical pressure multiplied by the hydrostatic pressure ratio k. This reduction expresses the influence of the frictional resistances acting at the points of contact between the grains, which increase in simple proportion to the pressure transmitted within the solid. The resultant lateral pressure is equal to the algebraic sum of the (positive or negative) pressure exerted by the liquid and the pressure of the solid. By performing this simple operation we obtained the pressure lines bd in Fig. 1c and d_1d_2 in Fig. 1f and Fig. 1i. The shaded areas inclosed between these lines and the vertical axis ab represent the total pressure on the wall. The areas located above point d_2 (Fig. 1f and i) are negative pressures (capillary suction), and those below this point are positive.

The diagrams (c), (f) and (i) of Fig. 1 clearly demonstrate the effect of the capillary forces on the pressure phenomena: When the water level in the standpipe is lowered, the lateral pressure rapidly decreases and can even assume a negative value. If capillary forces do not exist, the horizontal component of the lateral pressure of the backfill (Fig. 1g) should be equal to $\frac{1}{2}kh^2s_2$. The capillary forces reduce this pressure to $\frac{1}{2}kh^2(s_2 + s_0) - \frac{1}{2}h^2s_0$. Another effect of the capillary forces consists in shifting the center of the pressure from the middle third toward the base of the wall. Finally, if the upper part of the wall in Fig. 1g should be removed, the fill would stand unsupported with a vertical slope.

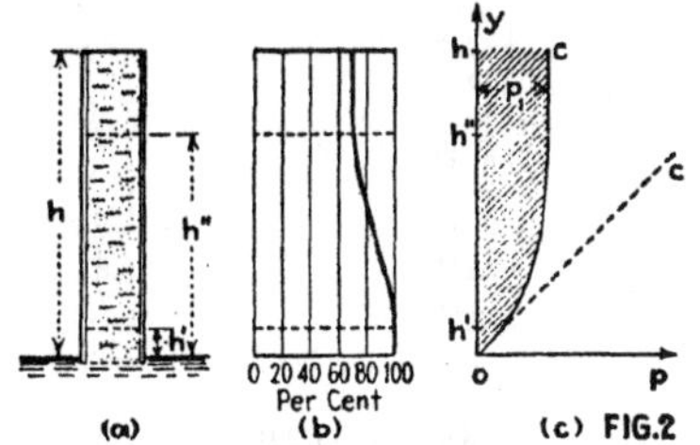

FIG. 2—DIAGRAM illustrating conditions existing in a material of partial capillary saturation.

This effect of the capillary forces can easily be demonstrated by means of the following experiment: Fill a cylindrical sieve with very fine, saturated sand; dump the content of the sieve on a table after the excess water has escaped through the perforated bottom of the vessel; the upper part of the body of sand will now stand at a vertical slope. On the other hand, if the vessel be dumped under water, the sides of the body of sand will be sloped over their full height regardless of how carefully the operation was performed.

If an attempt is made to measure the intensity of the lateral pressure by means of scales, the following difficulty is likely to arise: An insignificant impetus may suffice to cause the formation of an open fissure between the wall and the fill, extending from the surface of the fill to the depth $h - h'$ in Fig. 1g, and then the capillary suction will no longer act on the upper part of the wall, the scales recording only the positive pressure H' (Fig. 1f and 1i), while the negative presure H'' remains unknown.

Drained backfill in partial capillary saturation

In the preceding analysis we assumed that the voids of the backfill in both the submerged and the suction zone are completely filled with water. But this will rarely be the case in cohesionless backfills. To understand what conditions prevail in a material of partial capillary saturation, let us suppose a slender cylindrical vessel, Fig. 2a, completely filled with very fine sand and completely immersed in water. The height of the vessel is supposed to be considerably greater than the height h_c to which the water would rise within its content by capillary attraction. Hence, if we lift the vessel vertically, so that only the lowest part of the vessel remains immersed, part of the water will drain out. In the upper part of the column, above a height $h'' > h_c$, the amount of water replaced by air will be independent of the height above the free water level and will depend essentially on the size of the grains; in the lower part, below a height $h' < h_c$, the voids will remain completely filled; and between these parts there will be a zone of transition.

The height h_c of capillary rise is smaller than h'' but considerably greater than h', because when water rises by capillarity into a column of dry sand the upper part of the wetted section of the sand is always in a state of partial capillary saturation only. Fig 2b shows a diagram of the percentage of the voids which will be filled with water at different elevations above the free water

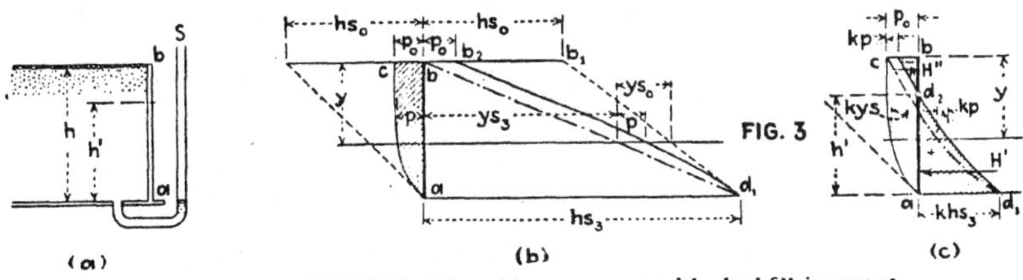

FIG. 3—COMPUTATION of lateral pressures exerted by backfill in partial capillary saturation.

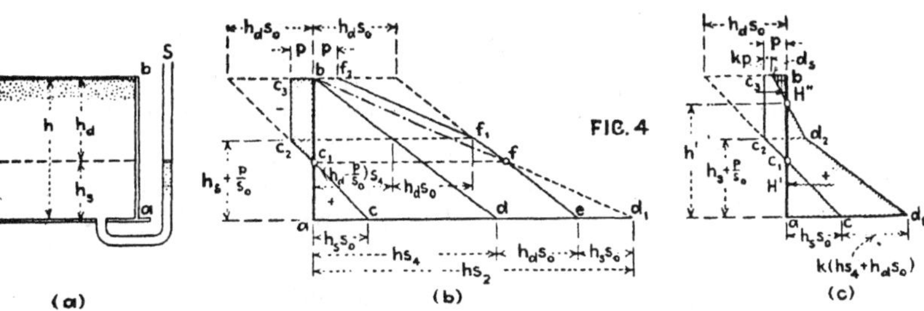

FIG. 4—INFLUENCE of partial capillary saturation limited to upper part of backfill.

level. In numerous measurements on fine silty sands from the flood plains of the lower Connecticut River as much as 90 per cent of the voids remained permanently filled with water at a height many feet above the upper level of capillary saturation. For such materials the value s_2 and s_3 are practically identical. In the upper part of the vessel the average curvature of the surface of the films of water will be everywhere the same and equal to the smallest curvature consistent with the grain size, because the water is confined to the narrowest parts of the voids. Therefore, above the zone of transition the negative hydrostatic pressure will assume a constant value p independent of the elevation and merely depending on the grain size and density of the granular material. In the diagram Fig. 2c the negative hydrostatic pressure is represented, for any elevation, by the abscissas of a curve OC with vertical asymptote.

From these facts computation of the lateral pressure exerted by backfill of partial capillary saturation may be attempted. Fig. 3 shows a cross-section through the back of the wall. Prior to drainage the conditions were identical with those shown in Fig. 1a to Fig. 1c. After drainage, the water level in the standpipe sinks to the level of the base of the fill, Fig. 3a. Since the depth of the fill is greater than the height h_c of the capillary rise, the vertical negative hydrostatic pressures which act within the water are determined by the abscissas of curve ac in Fig. 3b, while for the fill in a state of complete capillary saturation (Fig. 1g) we obtained for the negative hydrostatic pressures a straight line ac_2. The vertical pressures transmitted through the grains of the sand are determined in the same manner as in Fig. 1h. At every depth y (Fig. 3b) below the surface of the fill, the abscissas ys_3 of the straight line bd_1 represent the weight of both the sand and the water located above this elevation. In order to get the total vertical force which acts on the sand at depth y we must add to the pressure ys_3 a pressure p which is equal and opposite to the negative hydrostatic pressure p acting at the same depth in the water. Thus we obtain the pressure curve d_1b_2. The lateral pressure at depth y is equal to the algebraic sum of the negative hydrostatic pressure p and the product of the hydrostatic pressure ratio k multiplied by the vertical pressure $(ys_3 + p)$ acting on the grains of the sand, or lateral pressure $= k(ys_3 + p) - p$. Thus we obtain the pressure diagram Fig. 3c.

The shape of the curve ac in Fig. 3b can only be guessed at, and any computation based on an assumption involving a variable value of p would be complicated. In order to derive approximate formulas for estimating the effect of the capillary forces on the intensity of the lateral pressure, we replace the curve ac of Fig. 3b by the broken line cc_2c_3 as shown in Fig. 4b for a backfill under the influence of a suction head h_d which is smaller than the depth h of the backfill but greater than the height h_c of capillary rise. When dealing with this case we assume that the unit weight s_3 of the partly saturated material is practically equal to the unit weight s_2 of the saturated material.

Application of the principles evolved in connection with the preceding example leads to the following results: The vertical pressures that are transmitted from grain to grain are represented by the abscissas of the broken line ef_1f_2 of Fig. 4b, and the horizontal pressures which act on the back of the wall by the abscissas of a broken line $d_1d_2d_3$, Fig. 4c. By comparing Fig. 4b and Fig. 4c with Fig. 1e and Fig. 1f we realize that the influence of the absence of complete capillary saturation is limited to the upper part of the backfill between the surface and a depth $h_d - p/s_0$ (h_d = suction head and p = average intensity of the capillary pressure within the upper part of the zone of suction). Within this upper part the vertical pressures transmitted from grain to grain may be computed as though the weight of the sand were equal to the full weight of the sand plus the full weight of the water contained in the voids, and furthermore as though the surface of the fill carried a surcharge p per unit of area.

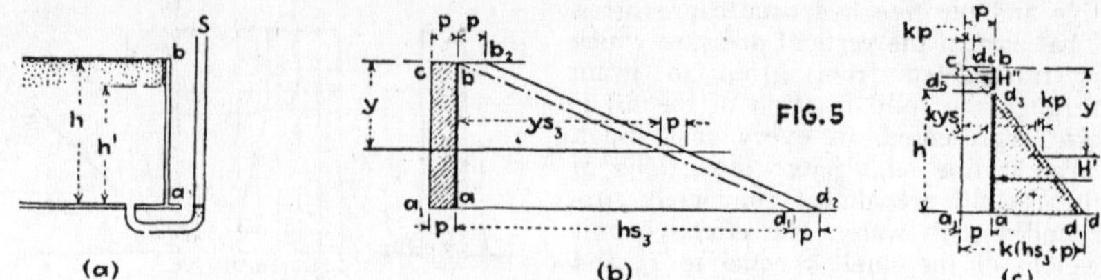

FIG. 5—VERTICAL PRESSURE is combined weight of water and sand plus capillary pressure.

For the case in which the suction head h_d is equal to the depth h of the backfill, Fig. 5a, we may further simplify the analysis by assuming the negative hydrostatic pressures to vary as the straight line a_1c of Fig. 5b, instead of as the broken line $c_1c_2c_3$ of Fig. 4b. In other words, we disregard the fact that the negative hydrostatic pressures decrease to zero in the lowest part of the suction zone. The vertical pressures which act in the solid part of the backfill are represented by the abscissas of b_2d_2, Fig. 5b, and the horizontal pressures exerted by the solid and the liquid part of the backfill jointly are given by the abscissas of the straight line d_2d_4, Fig. 5c. The procedure for determining the position of these pressure lines is identical with that described in connection with Figs. 3 and 4. According to Fig. 5b, the vertical pressure which acts at any depth y below the surface is equal to the sum of the combined weight ys_3 of water and sand and capillary pressure p. Above point d_3, Fig. 5c, the resultant horizontal pressures are negative, while below this point they are positive.

The data presented in Fig. 5c can be used for the interpretation of the results of earth-pressure tests. The analysis is based on the following assumptions:

The horizontal and vertical components of the lateral pressure are measured by scales, first for the fill in a completely submerged state and then for the fill in the state represented by Fig. 5a. With the wall in stationary position, both drainage and saturation cause the hydrostatic pressure-ratio to increase from k to $mk > k$, due to rearrangement of grains during the period of transition from one state to the other. According to the conclusions presented in Article I, such increase occurs even in dry backkfill during a period of rest.

During the test with drained fill, the upper part of the backfill (above point d_3, Fig. 5c) usually pulls loose from the wall. Hence the scales merely measure the positive part of the lateral pressure.

Based on these assumptions, we now wish to determine the average intensity p of the negative hydrostatic pressure (capillary suction) and the coefficient of wall-friction $\tan \delta$ for the drained state.

In the computations the symbols are used which were quoted at the beginning of the paper. In addition, the letter H_s is used for the measured horizontal pressure component in the submerged state and H_d in the drained state. The value of the hydrostatic pressure ratio is obtained from the results of the test on the submerged fill by using the equation $H_s = W_o + kW_4$ so that $k = (H_s - W_o)/W_4$.

For the drained state the measured pressure is, according to our assumptions and to Fig. 5c,

$$H_d = \tfrac{1}{2} h' mkhs_3 \frac{h}{h} b = \tfrac{1}{2} h'^2 mks_3 b \quad (1)$$

$$\text{and } h' = h \frac{mk(hs_3 + p) - p}{mkhs_3} = h\left[1 - \frac{p(1 - mk)}{mkhs_3}\right] \quad (2)$$

Since $\tfrac{1}{2} h^2 s_3 b = W_3$, we obtain

$$H_d = mk W_3 \left[1 - \frac{p(1 - mk)}{mkhs_2}\right]^2 \quad (3)$$

$$\text{and } p = \left[1 - \sqrt{\frac{H_d}{mk W_3}}\right] \cdot \frac{mkhs_3}{s - mk} \quad (4)$$

The measured vertical component of the lateral pressure of both the submerged and the drained backfill is equal to the product of the coefficient of wall friction $\tan \delta$ times that part of the lateral pressure which is transmitted to the wall through the grains. For the drained backfill there is no contact between the wall and the fill above point d_3 of Fig. 5c. Therefore the vertical component is equal to

$$V = \tan \delta \,.\, \text{area } (a_1\, d_2\, d_3\, d_5)\; b \text{ (width of the fill)} = \tan \delta\, (H' + ph')\, b$$

$$= \tan \delta \left(\frac{1}{b} H_d + ph^1\right) b$$

$$\text{or } \tan \delta = \frac{V}{H_d + bph^1}$$

From eq. (1) and (3) we obtain

$$h' = \sqrt{\frac{2\, H_d}{mks_3b}} = h.\ \sqrt{\frac{H_d}{m\, k\, W_3}}$$

$$\text{Hence } \tan \delta = \frac{V}{H_d + bph\sqrt{\dfrac{H_d}{m\, k\, W'_3}}} \qquad (5)$$

If we drop the assumption that the backfill pulls away from the wall, the measured pressure becomes $(H' - H'')b$, and we obtain from Fig. 5c

$$H_d = b\left(\tfrac{1}{2}\, mkh^2 s_3 + m\, k\, h\, p - h\, p\right)$$

Whence $p =$

$$\frac{\tfrac{1}{2}\, m\, k\, h^2\, s_3 - H_d/b}{h\,(1 - mk)} = \frac{m\, kW_3 - H_d}{bh(1 - mk)} \qquad (6)$$

$$\text{and } V = b \tan \delta \left(\tfrac{1}{2}\, m\, k\, h^2\, s_3 + hp\right) =$$

$$b \tan \delta \left(m\, k\, \frac{W_3}{6} + hp\right)$$

$$\text{or } \tan \delta = \frac{V}{m\, k\, W_3 + b\, h\, p} \qquad (7)$$

Conclusions

1. When draining a fine-grained, saturated backfill equipped with a standpipe, a very small discharge may lower the water in the standpipe through a great vertical distance. This vertical distance is the suction head, and that part of the fill located above standpipe level is the suction zone.

2. Within the suction zone the hydrostatic pressures in the water content of the fill are negative. These negative hydrostatic pressures are maintained by the surface tension of the water, which in turn acts on the solid part of the backfill like a surcharge.

3. The effect of the surface tension (capillary forces) on the backfill depends on whether the depth of the suction zone is smaller or greater than the height h_c of capillary rise in a dry column of backfill material.

4. If the suction head is considerably smaller than h_c (state of complete capillary saturation) the surface tension of the water acts exclusively at the top surface of the fill. In this case the vertical pressures which are carried at any depth by the grains of the backfill may be computed on the assumption that the unit weight of the solid part of the fill is reduced by the full hydrostatic uplift, and that the surface carries a surcharge equal to the suction head times the unit weight of the water.

5. If the suction head is considerably greater than h_c (state of partial capillary saturation), the places where the surface tension acts on the grains are scattered throughout the major part of the suction zone, and within the zone the intensity p of the capillary pressure is practically independent of the depth below the surface. Within this zone the vertical pressures carried at any depth by the grains may be computed on the assumption that the weight of the fill is equal to the full weight of both the sand and the water and that the surface of the fill carries a surcharge equal to p. Below the suction zone the vertical pressures carried by the grains have exactly the same intensity as if the suction zone were in a state of complete capillary saturation (see conclusion 4).

6. In every case an increase of the suction head causes a substantial decrease of the lateral pressure and a downward movement of the center of pressure.

7. The horizontal unit pressure which acts at any depth on a vertical surface is equal to the sum of the full (positive or negative) pressure which acts at this depth in the liquid part of the backfill and the product of the vertical pressure which is carried at the same depth by the grains of the backfill times the hydrostatic pressure ratio.

8. For a backfill in a state of partial capillary saturation the various elements of the lateral pressure and the intensity of the capillary pressures may be computed by Eq. (4), (5), (6) and (7).

Previous articles in this series appeared in our issues of Feb. 1, p. 136, Feb. 22, p. 259, and March 8, p. 316.

LARGE RETAINING WALL TESTS

A series of five papers reporting fundamental results

V—Pressure of Glacial Till

Results of final tests on actual soil from Fifteen-Mile Falls Dam indicate close conformity to the principles deduced in preparatory studies of sand and fine-grained soils

By Karl Terzaghi
Professor, Technische Hochschule, Vienna, Austria

FOLLOWING the wall-pressure tests with dry and saturated sand, tests were conducted with a backfill of a natural modified till, or glacial drift. The sand tests, described in the first and second article of this series, had already led to a broader knowledge of the effect of submergence and of wall movements on lateral pressure. Without this empirical knowledge, no interpretation of the results of the tests with modified till could have been attempted.

The testing material

The modified till used in the tests was a glacial material of mixed grain size. Since it was formed by erosion and subsequent redeposition of a boulder clay, it differed from the boulder clays of the same locality merely by somewhat greater effective size and a smaller uniformity coefficient. A sample of 60 cu.yd. of this material was excavated at El. + 580 ft. near the center line of the Fifteen-Mile Falls Dam across the Connecticut River, on the New Hampshire side of the river.

Fig. 1 shows a typical grain-size curve for this material. The effective size was of the order of 0.15 mm., the uniformity coefficient about 20, and the coefficient of permeability variable between 0.01 and 0.10 $\times$ 10^{-4} cm./sec. Due to the high average specific gravity of the grains (2.8) and its mixed-grained character, the average weight of the saturated, compacted till (volume of voids 28 per cent) was as high as 143.5 lb. per cu.ft.

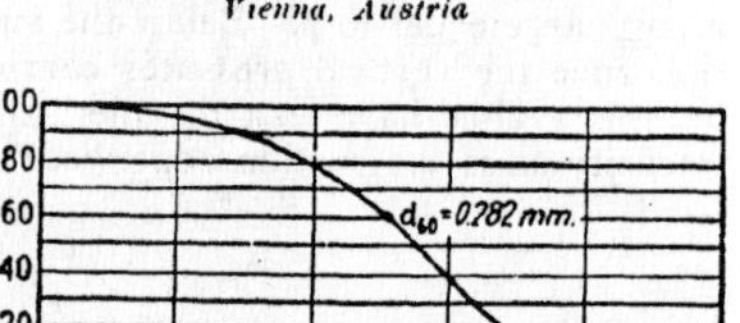

FIG. 1—GRAIN-SIZE curve of modified till selected from site of Fifteen-Mile Falls Dam.

Outline of test procedure

Before the bin was filled, its bottom and rear wall were lined with a filter of Plum Island sand (Fig. 2). Later the objection was raised that the presence of a layer of foreign material at the rear wall of the bin might affect the test result. Therefore, while the test was going on the rear part of the filter together with the adjoining part of the fill was excavated and replaced by till. This operation had no effect whatever on the results. It did not even change the readings of the pressure dials at the retaining wall beyond the normal range of variation.

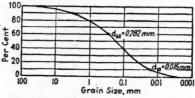

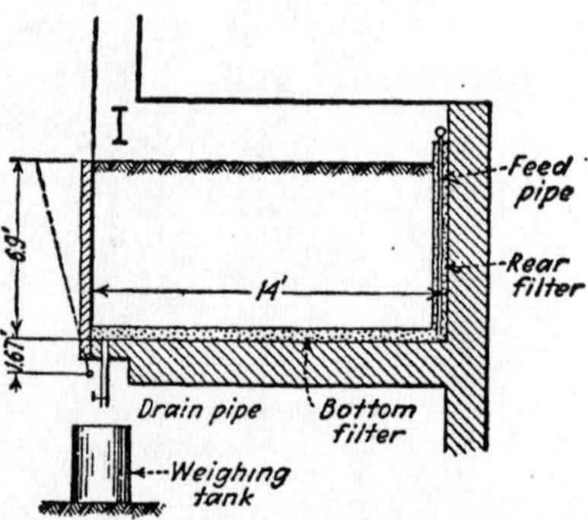

FIG. 2—ARRANGEMENT of bin for test of modified till from Fifteen-Mile Falls Dam.

The till was placed in the bin in 6-in. layers and compacted with concrete tampers. In its initial state the lower part had a moisture content of 8 per cent and the upper part of 11 per cent (per cent of dry weight); however, after the first submergence, this figure never dropped below an average of 12.6 per cent, which indicated that even in a drained state the fill was practically saturated, with an average unit weight of 142.9 lb. per cu.ft.

The test lasted from June 22 to Aug. 3, 1929, a period of 42 days. From its original position the wall was allowed to yield by increments, tilting around a horizontal axis located in the plane of contact between wall and fill (Fig. 2). The test comprised the following successive operations:

a—After the bin was filled, the saturation line (standpipe level) was allowed to rise to the surface of the fill, the wall remaining in its original position.

b—The fill was drained without moving the wall.

c—The wall was moved out an average distance of 0.15 mm. while the fill was in a drained state.

d—The fill was submerged and drained again.

e—The wall was moved out to an average distance of 0.45 mm. while the fill was in a drained state.

f—The fill was submerged.

g—The wall was moved out a distance of 0.85 mm. while the fill was submerged. Then the fill was drained.

h—The wall was moved out to a distance of 1.3 mm. while the fill was drained.

i—The wall was moved out to a distance of 4.2 mm. while the fill was in a drained state.

j—The plane of saturation (standpipe level) was raised and lowered through the full depth of the fill at stationary position of the wall four times at about two-day intervals, and the time change of the earth pressure was observed.

During the tests regular readings were made of the deformations of the machine, room temperatures and rates of evaporation, but none of these items was found to have an influence on the pressure phenomena sufficiently marked to be worth recording.

Results of the tests

The results of the tests are graphically represented in Figs. 3 and 4. In view of the facts presented in Article III we were entitled to make as clean-cut a separation between the pressure of the water and that of the solid fraction of the fine-grained backfill as was done when representing the results of the test with coarse sand in Article II. Since drainage produced little change in the water content of the fill, very marked capillary phenomena were to be expected. A brief description of the influence of capillary forces on the lateral pressure was presented in the conclusions of Article IV. However, in order to bring the results of this influence clearly into evidence, the observations on drained backfill were plotted *as if* capillary force did not exist. Hence the method of computing hydrostatic pressure ratio k, coefficient of wall friction $\tan \delta$, and height c of the center of pressure, was exactly identical with that described in Article II for the coarse sand submerged.

Diagrams A to C in Fig. 3 represent the relation between the average distance through which the wall was allowed to yield (abscissas) and the hydrostatic pressure ratio k (Fig. 3 A), the coefficient of wall friction $\tan \delta$ (Fig. 3 B), and the elevation c of the point of application of the resultant earth pressure (Fig. 3 C). In these diagrams the points corresponding to the submerged state are connected by dotted lines, and the corresponding letters are shown in brackets. The transition between drainage and saturation is marked by gaps between vertical dotted and solid lines. The time intervals which elapsed between the successive stages of the test are shown in a table on the right side of the diagrams.

The last phase of the tests, marked in Fig. 3 by capital letters, consisted of repeated submergence and drainage at stationary position of the wall. The ordinates of the ends of the vertical lines which represent this state in Fig. 3 give the extreme values observed during the operations. The vertical solid line shows the range of scatter of the values for the drained, and the dotted line for the submerged state.

Fig. 4 represents those tests which in Fig. 3 are marked with capital letters, plotted on a time scale. In the diagram at the top of Fig. 4, abscissas represent the time which passed after the wall arrived at an average distance of 4.2 mm. from its original position, and ordinates the position of the water level in the standpipes which were connected with the bin. The ordinates in the diagrams A, B and C of Fig. 4 show the hydrostatic pressure ratio, the coefficient of wall friction and the height of the center of pressure. Finally, at the bottom is plotted the vertical movement of a point of the surface of the backfill 4.5 ft. back of the wall. The data contained in this diagram were also plotted *as if* no negative hydrostatic pressures existed in the drained backfill.

FIG. 3—RELATION of wall yield, hydrostatic-pressure ratio, coefficient of wall friction, and height of center of earth pressure.

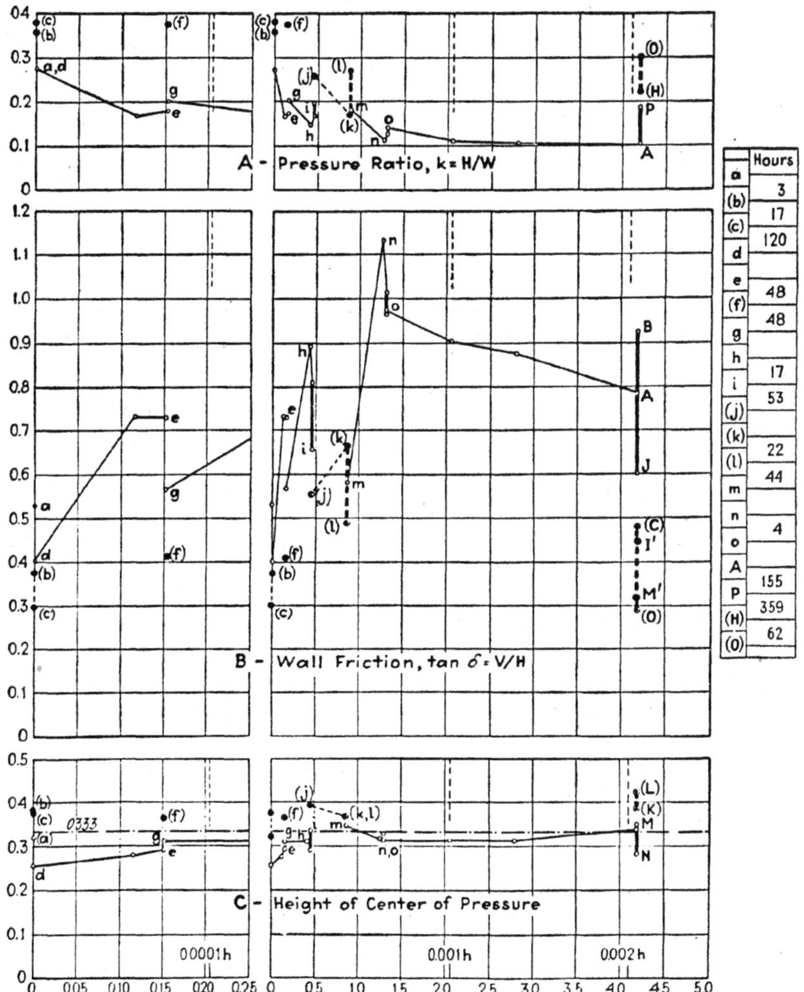

Analysis of test results

Comparing the test results shown in Fig. 3 with those described in Article II for submerged sands, the following differences will be noticed:

For compacted dry sand the hydrostatic pressure ratio at the outset of the test was 0.58, the coefficient of wall friction 0.08, and the height of the center of pressure 0.4 h. In striking contrast to these findings the till showed at the outset, and prior to the first saturation (Fig. 3), a hydrostatic pressure ratio of 0.28, coefficient of wall friction as high as 0.53, and center of pressure somewhat below the middle third, at 0.32 h.

Another conspicuous difference appeared in the effect of submerging the drained backfill. In the tests with sand, after the wall was moved out to a distance of 4.2 mm. $= 0.002\,h$, water was admitted into the fill, and the hydrostatic pressure ratio went up from 0.160 to 0.163 (increase of 2 per cent), the wall friction decreased by 9 per cent, from 0.476 to 0.435, and the center of pressure rose from 0.392 to 0.436 h. Hence the effect of saturation on the conditions of stress and strain which existed in the backfill was in these states very small. For this reason it was surprising to find that saturation of the till after the wall had been moved out to the same distance (Fig. 4, points B-C) caused very large changes, as follows: k went up from 0.11 to 0.252 (increase of 129 per cent), $\tan \delta$ decreased by 50 per cent, from 0.927 to 0.463, and the center of pressure rose from 0.337 h to 0.385 h. At the same position of the wall repeated saturation and drainage caused the center of pressure to move between the extreme positions of 0.26 h and 0.42 h.

These differences between the pressure effects of sand and till are precisely what they should be according to the conclusions of Article IV. As a matter of fact, when the tests were made these differences came as a surprise, and the theory which explains them was worked out afterwards. Since the depth of the backfill was considerably greater than the capillary height for a column of dry till, the drained backfill was in a state of partial capillary saturation, with an average unit weight of 142.9 lb. per cu.ft. (water content of 12.6 per cent) against 143.5 for the state of complete saturation (water content 13.9 per cent). Hence in order to obtain the real values of k and $\tan \delta$ for the drained state, the same separation must be made between the forces exerted by the water and by the solid fraction as was made from the beginning for the submerged state.

Capillary Forces—The computation of the negative hydrostatic pressure (capillary suction) is made by means of the equations derived in Article IV. Two extreme cases can exist, viz:

1. An open fissure is formed between the backfill and the wall through the full depth of the suction zone. For this case we obtained:

Capillary suction, $p =$

$$\left(1 - \sqrt{\frac{H_a}{m\,k\,W_3}}\right) \frac{m\,k\,h\,s_3}{1 - m\,k} \qquad (1)$$

Depth of suction zone, $h - h' =$

$$h\left(1 - \sqrt{\frac{H_a}{m\,k\,W_3}}\right) \qquad (2)$$

Coefficient of wall friction, $\tan\delta =$

$$\frac{V}{H_a + b\,p\,h'} \qquad (3)$$

2. The backfill adheres to the wall over the full depth of the fill, and no tension cracks are formed in the fill:

$$p = \frac{m\,k\,W_3 - H_a}{b\,h\,(1 - m\,k)} \qquad (4)$$

$$h - h' = \frac{p\,(1 - m\,k)}{m\,k\,s_3} \qquad (5)$$

$$\tan\delta = \frac{V}{m\,k\,W_3 + b\,h\,p} \qquad (6)$$

The coefficient m in these formulas expresses the increase of the hydrostatic pressure ratio k of the solid fraction of the backfill due to time effect and to the rearrangement of the grains produced by the processes of submergence or drainage at a stationary position of the wall. As admission or withdrawal of water was always performed after a long intermission while the wall remained stationary, we may for a rough estimate assume that $m = 1$. This assumption involves the possibility of an error of 15 per cent, or 20 per cent in the value of k.

Another source of uncertainty resides in the arbitrary character of our assumptions concerning the depth to which the open fissure between the fill and the wall extends. Both open fissures between the wall and the fill and open tension cracks within the fill were observed, but it was never possible to measure their depth except in one case of a tension crack which appeared about 3 ft. back of the wall after the wall arrived in its extreme position. Through a window in the wall of the bins it could be seen to extend almost vertically downward to a depth of about 1.84 ft. below the surface. For the first transition $B\ C$ (Fig. 4) from the drained into the submerged state, which was performed at this position of the wall, we obtained by means of Eq. (1) to (3) the following values: $p = 125.5$ lb., $h - h' = 2.60$ ft., and $\tan\delta = 0.368$. Hence the depth of the fissure was only 70 per cent of the depth of the zone of suction, and the corresponding state was very likely intermediate between what was previously called state (a) and (b).

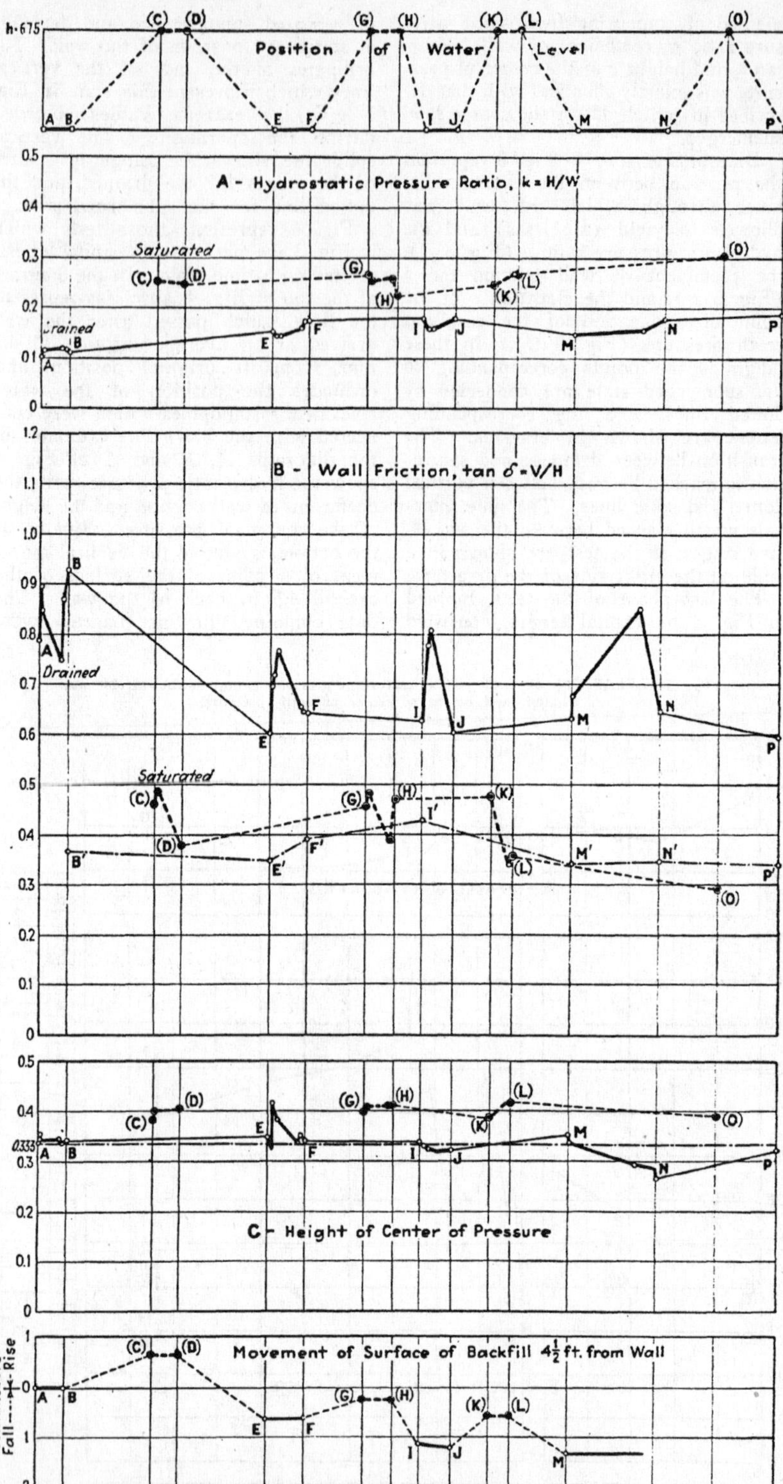

FIG. 4—TESTS indicated by capital letters in Fig. 3 plotted on a time scale.

Since the fissures closed at each submergence, the depth to which fissures opened was by no means the same for all the stages of the test. Nevertheless, for our computations we were obliged to assume that the open fissure extended in every case through the full depth of the suction zone. Under these circumstances the computation can be expected to indicate the magnitude of the quantities only approximately. It was made by means of Eq. (1) to (3), assuming $m = 1$. The results are assembled in Table I. The abnormally high values of the capillary suction p for stages c and g (figures in parentheses) seem to indicate that in these stages almost no separation occurred between the fill and the wall.

From the remaining twelve figures we obtain an average value $p = 76$ lb. per sq.ft., with a scatter of $\mp$ 50 per cent. The total negative pressure W_s, which acted on the back of the wall during the drained state, therefore averaged pbh = 7,300 lb., against a value of W_o = 20,700 lb. for the positive water pressure which acted during the state of submergence. Hence $W_s = 0.35\ W_o$, against $W_s = W_o$ for a state of complete capillary saturation.

By introducing the values of p, Table I, into Eq. (3), one can compute tan δ for any of the drained states listed in the table, because all the other quantities contained in the equation are determined by direct measurement. The results of this computation will be found in Table II, which gives the average distance of the wall from its original position, the duration of the intermissions, the values of k and tan δ for the submerged states (letters in brackets), the corresponding values of the angle of internal friction ϕ computed by means of Coulomb's formula as described in Article I, and the corrected values of tan δ for the drained state computed by means of Eq. (3). The table shows that the correction eliminated the apparent contradiction between the values of tan δ for the submerged and the drained state.

To bring this fact clearly into evidence, the corrected values of tan δ were plotted in diagram B of Fig. 4, and the points B′, E′, ... thus obtained were connected by a dot-and-dash line. According to the figure, for the submerged state the values tan δ range for the part AP of the test between the limits 0.29 and 0.48, the uncorrected drained values between the limits 0.59 and 0.93, and the corrected values between the limits 0.32 and 0.44. In diagram B of Fig. 3 the range of scatter of the corrected values of tan δ for the last part of the test (AP in Fig. 4) is shown by the vertical distance between the points M′ and P′. The range is entirely within the scatter of the values tan δ for the submerged state. Hence the high values of tan δ for the drained state were merely due to an illusion, produced by ignoring at the outset the action of the capillary forces.

For the same reason the low position of the center of the lateral pressure for the drained state in diagrams C of Figs. 3 and 4 has nothing to do with the height of the center of the pressure exerted by the solid fraction of the backfill, because it is due to the combined effect of two different causes. There is no reason for assuming that the center of the pressure exerted by the solid fraction is lower for the drained state than for the submerged state. The latter consistently gave values close to 0.4 h. The same value was found for the height of the center of pressure for a tilting wall backfilled with dry sand, before first slip (Test 1 of Article I).

TABLE I—CAPILLARY SUCTION IN TILL BACKFILL

State		p, lb./ft.^{2}	State		p, lb./ft.^{2}
a	of Fig. 3	78	B	of Fig. 4	126
d		93	E		70
e		(185)	F		66
g		(160)	I		46
i		45	J		43
m		64	M		95
			N		98
			P		88

It should be noticed that state a, Table I, corresponds to the backfill prior to first saturation. At that state the water content of the fill ranged between 9 and 11 per cent, against 13.9 per cent for the saturated and 12.6 per cent for the drained state. Nevertheless, while the fill was being placed and compacted in a moist state, a negative hydrostatic pressure developed of the same intensity as the pressure which acted in the drained fill in the later stages of the test.

Effect of repeated drainage and saturation on angle of internal friction—The values of ϕ in Table II indicate that the outward movement of the wall had but little effect on the angle of effective internal friction, which is in striking contrast to what was observed during the tests with backfills consisting of sand. In a single case (at the end of movement j-k in Fig. 3) the value of the angle of internal friction temporarily went up to 39° 40′, but during the following 21 hours it returned to 29° 20′. We suspected that the apparently abnormal behavior of the till was due to the thorough readjustment of the grains and other structural changes produced by repeated drainage and saturation. Any such readjustment reduces the stresses in the fill and prevents them from assuming their maximum value.

In order to find out whether this explanation is correct, we repeated the test with the following modification: While filling the bin in 6-in. layers we kept the water in the standpipe about 1 ft. below the surface of the fill, but immediately after the bin was filled we raised the water level to the surface. After two days we drained the fill and left it in this state of partial capillary saturation until the end of the test. During the test the wall was moved away from the fill by tilting around a horizontal axis, in the same manner as in the preceding test, to an average distance of 10 mm. or 0.005 h from its original position. In diagrams A to C of Fig. 5 the results of the test are plotted in the same manner as in Fig 3 of the preceding test—that is, the existence of the suction effect was disregarded. When the wall arrived at an average distance of 2.7 mm. from its original position, a crack appeared on the surface of the fill about 2.2 ft. back of the wall. Through the window one could see that this crack dipped away from the wall at an angle of about 70° with the horizontal and extended to a depth of about 0.8 ft.

To obtain the real values of pressure ratio and wall friction in Fig. 5, correction must be made for the capillary forces (negative hydrostatic pressure) which acted on the wall. If in the formula for capillary pressure, Eq. (1), we introduce the values H_d = 9,890 lb., W'_s = 42,900 lb., k = 0.304, s_s = 142.9 lb. per cu.ft., h = 6.55 ft., and $m = 1$, which correspond to the conditions existing at the beginning of the test, prior to draining the fill, we obtain as the value of the negative hydrostatic pressure p = 54 lb. per sq.ft. But Eq. (4), applicable when wall and fill remain in contact, gives p = 51 lb. per

TABLE II—RESULTS OF RETAINING-WALL TESTS ON TILL BACKFILL

Outward Movement of Retaining Wall from Original Position, mm.	State of Backfill	Hydrostatic Pressure Coefficient, k	Coefficient of Wall Friction for Submerged State, tanδ	Angle of Internal Friction for Submerged State, ϕ	Coefficient of Wall Friction for Drained State (corrected) tanδ	Intermission, Hours
0	a, Fig 3	0	0	0	0.07	
0	(b) Fig. 3	0.351	0.38	23° 0′	0	17
0	(c) Fig. 3	0.381	0.30	22° 20′	0	
0	d, Fig. 3	0	0	0	0.26	
0.15	e, Fig. 3	0	0	0	0.30	
0.15	(f) Fig. 3	0.376	0.41	21° 20′	0	
0.15	g, Fig. 3	0	0	0	0.26	
0.45	i, Fig. 3	0	0	0	0.47	
0.45	(j) Fig. 3	0.258	0.55	30° 10′	0	
0.85	(k) Fig. 3	0.171	0.67	39° 40′	0	22
0.85	(l) Fig. 3	0.273	0.49	29° 20′	0	
0.85	m, Fig. 3	0	0	0	0.37	
4.20	A, Fig. 4	0	0	0	0	21
4.20	B, Fig. 4	0	0	0	0.37	
4.20	(C) Fig. 4	0.252	0.46	31° 40′	0	20
4.20	(D) Fig. 4	0.247	0.38	33° 00′	0	
4.20	C, Fig. 4	0	0	0	0.35	30
4.20	F, Fig. 4	0	0	0	0.39	
4.20	(G) Fig. 4	0.262	0.46	30° 50′	0	24
4.20	(H) Fig. 4	0.222	0.47	34° 50′	0	
4.20	I, Fig. 4	0	0	0	0.43	28
4.20	J, Fig. 4	0	0	0	0.44	
4.20	(K) Fig. 4	0.245	0.48	32° 20′	0	17
4.20	(L) Fig. 4	0.271	0.35	31° 00′	0	
4.20	M, Fig. 4	0	0	0	0.32	72
4.20	N, Fig. 4	0	0	0	0.35	
4.20	(O) Fig. 4	0.302	0.29	28° 50′	0	
4.20	P, Fig. 4	0	0	0	0.34	

NOTE: Parentheses indicate that the point corresponds to the submerged state of the fill. The intermission gives the time in hours during which the fill remained in drained or saturated state at stationary position of the wall.

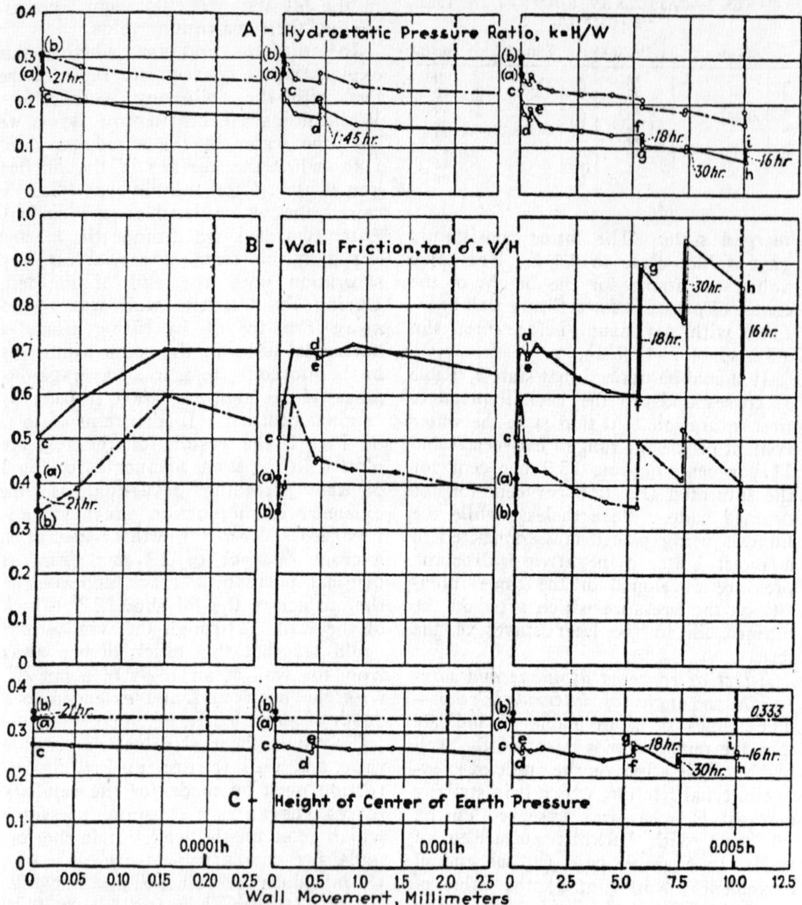

FIG. 5—RESULTS of test of till in drained state showed that a high value of internal friction was developed by wall movement.

sq.ft. For the computations the higher value was retained. It is somewhat smaller than the average value, which was obtained for p in the preceding test. Considering the scatter of the values assembled in Table I, the value $p = 54$ lb. merely represents an approximation, yet it may be used for a crude estimate of the values k and tan δ in the test Fig. 5.

Using Eq. (1) with the value $m = 1$, solving for k, and introducing the numerical values $W_2 = 42{,}900$ lb., $h = 6.55$ ft., $s_3 = 142.9$ lb. per cu.ft., and $p = 54$ lb. per sq.ft., as well as the measured values of horizontal pressure, results were obtained as plotted in A of Fig. 5 by a dot-and-dash line. Increasing the value of p from 54 to 76 lb. per sq.ft. (average value for the preceding test) would increase the computed values of k by only 0.026; hence an appreciable error in estimating the value of p has a rather small effect on the results of the computations.

Table III contains the corrected values of k, tan δ and ϕ for the test Fig. 5. This table demonstrates that the value ϕ consistently increases as the wall moves out, finally reaching a value as high as 43° 30′. Thus, when the fill is kept in drained state throughout the test, the outward movement of the wall mobilizes the internal frictional resistance of the fill in the same fashion as it does for a backfill consisting of dry sand. There is no doubt that the same result would be obtained if the fill were maintained throughout the test in a saturated state.

Subsidence due to repeated drainage and saturation—As often as the fill passes from the submerged into the drained state, the vertical pressure transmitted through the grains of the backfill increases considerably, due to the elimination of hydrostatic uplift and due to the surcharge produced by the surface tension of the water. This increase of the pressures should cause a compression of the fill and subsidence of its surface. On the other hand, a submergence of the drained fill should cause a rise due to elastic expansion of the fill. These conclusions are confirmed by the diagram at the bottom of Fig. 4c, which shows the upward and downward movement of the surface while the wall was stationary at an average distance out of 4.2 mm. According to the diagram, the rise of the surface due to submergence is always a trifle smaller than the preceding drop due to drainage, which indicates that each drainage produced a slight permanent increase of the density of the fill.

At the outset of the test, when the wall was still in its original position, a submergence-drainage cycle caused a practically irreversible lowering of the surface of the fill through a distance of 0.5 in. The total permanent subsidence prior to state A, Fig. 4, was about 0.9 in. However, at each repetition of the process of drainage, the difference between rise and fall became smaller, so that probably after many repetitions (with wall held stationary) the rise would become equal to the preceding drop.

Comparing the subsidence of the till in the first cycle with that of the sand (Article II), we notice the following facts:

	Rise and Fall of Surface	
	Till (Test Fig. 3)	Sand (Art. II)
First saturation ..	—0.5 in.	+0.003 in.
First drainage	—0.25 in.	—0.02 in.

Thus, though the initial volume of voids of the till was 28 per cent against 40 per cent for the sand, yet the till was much more affected in its structure by the circulation of the water through its voids than the sand. Further testimony for the deficient stability of the structure of the till can be found in Fig. 4, which shows the variations of pressure ratio, wall friction and height of center of pressure while the wall was held at an average distance of 4.2 mm. from its original position. Each one of the consecutive stages AB, CD, etc., represents an intermission during which the external conditions of the fill remained unchanged; nevertheless none of the quantities plotted remained constant during the intermissions, which indicates the occurrence of minor grain movements within the fill.

Due to the fact that the fill had been subjected to repeated cycles of drainage and saturation prior to state A, Fig. 4, these changes merely represent more or less erratic deviations from average values, without any tendency toward a

TABLE III—EFFECT OF WALL MOVEMENT ON MODIFIED-TILL BACKFILL

Outward Movement of Retaining Wall from Original Position, mm.	State of Backfill	Hydrostatic Pressure Coefficient, k	Coefficient of Wall Friction, tan δ	Angle of Internal Friction, ϕ	Intermission, Hours
0	c	0.304	0.33	28° 0′	
0.42	d	0.247	0.44	32° 30′	
0.44	e	0.267	0.46	30° 10′	2
5.39	f	0.209	0.36	37° 20′	
5.40	g	0.194	0.50	38° 0′	19
10.20	h	0.157	0.41	43° 30′	
10.20	i	0.182	0.35	40° 40′	16

permanent deviation. Yet these average values are very much smaller than the possible maximum values. In Table II the average values are tan δ=0.39, δ=21° 20′, and ϕ=31° 50′, while Table III gives the maximum values as tan δ=0.50, δ=27° 0′, and ϕ=43° 30′. Since friction tests merely inform us about the maximum values of δ and ϕ, it is obvious that with fine-grained backfills ample allowance should be made for the effect of subsequent structural change on the coefficient of friction.

The first drainage of the backfill in the test Fig. 3 caused a subsidence of 0.25 in. against 0.14 in. for the test Fig. 5. Hence wet placing of the fill (test Fig. 5) was found to produce a more stable arrangement of the grains than dry placing as used in the test of Fig. 3.

The following conclusions were reached as a result of these studies:

1. When the backfill was drained, a negative hydrostatic pressure developed in the water retained by the fill. This pressure pulled the wall toward the fill with a force ranging between 50 and 120 lb. per sq.ft. A similar negative pressure (suction) developed while placing and compacting the fill in 6-in. layers.

2. When draining the backfill, an open fissure is likely to be formed between the upper part of the fill and the back of the wall. In this case the lateral pressure of the upper part of the fill is taken up by the surface tension of the water, and the wall is subject to pressure from the lower part of the fill only.

3. For both the submerged and the drained state, the lateral pressure exerted by the water and by the solid fraction of the fill must be considered independently.

4. Repeated drainage and submergence partly release the stresses which are set up in the backfill by the outward movement of the wall and consequently increase the lateral pressure exerted by the solid fraction of the fill. For the final state of equilibrium to which this process leads at a stationary position of the wall, the coefficients of wall friction and of internal friction were found to be approximately equal to two-thirds the maximum values which they can assume.

5. A till that is in saturated condition at time of placing seems to have more stable structure than till that is placed and compacted in a moist state.

Previous articles in this series appeared in our issues of Feb. 1 (p. 136), Feb. 22 (p. 259), March 8 (p. 316) and March 29 (p. 403). A later article will deal with application of the results to actual design.

General Theory of Three-Dimensional Consolidation*

MAURICE A. BIOT
Columbia University, New York, New York

(Received October 25, 1940)

The settlement of soils under load is caused by a phenomenon called consolidation, whose mechanism is known to be in many cases identical with the process of squeezing water out of an elastic porous medium. The mathematical physical consequences of this viewpoint are established in the present paper. The number of physical constants necessary to determine the properties of the soil is derived along with the general equations for the prediction of settlements and stresses in three-dimensional problems. Simple applications are treated as examples. The operational calculus is shown to be a powerful method of solution of consolidation problems.

INTRODUCTION

It is well known to engineering practice that a soil under load does not assume an instantaneous deflection under that load, but settles gradually at a variable rate. Such settlement is very apparent in clays and sands saturated with water. The settlement is caused by a gradual adaptation of the soil to the load variation. This process is known as *soil consolidation*. A simple mechanism to explain this phenomenon was first proposed by K. Terzaghi.[1] He assumes that the grains or particles constituting the soil are more or less bound together by certain molecular forces and constitute a porous material with elastic properties. The voids of the elastic skeleton are filled with water. A good example of such a model is a rubber sponge saturated with water. A load applied to this system will produce a gradual settlement, depending on the rate at which the water is being squeezed out of the voids. Terzaghi applied these concepts to the analysis of the settlement of a column of soil under a constant load and prevented from lateral expansion. The remarkable success of this theory in predicting the settlement for many types of soils has been one of the strongest incentives in the creation of a science of soil mechanics.

Terzaghi's treatment, however, is restricted to the one-dimensional problem of a column under a constant load. From the viewpoint of mathematical physics two generalizations of this are possible: the extension to the three-dimensional case, and the establishment of equations valid for any arbitrary load variable with time. The theory was first presented by the author in rather abstract form in a previous publication.[2] The present paper gives a more rigorous and complete treatment of the theory which leads to results more general than those obtained in the previous paper.

The following basic properties of the soil are assumed: (1) isotropy of the material, (2) reversibility of stress-strain relations under final equilibrium conditions, (3) linearity of stress-strain relations, (4) small strains, (5) the water contained in the pores is incompressible, (6) the water may contain air bubbles, (7) the water flows through the porous skeleton according to Darcy's law.

Of these basic assumptions (2) and (3) are most subject to criticism. However, we should keep in mind that they also constitute the basis of Terzaghi's theory, which has been found quite satisfactory for the practical requirements of engineering. In fact it can be imagined that the grains composing the soil are held together in a certain pattern by surface tension forces and tend to assume a configuration of minimum potential energy. This would especially be true for the colloidal particles constituting clay. It seems reasonable to assume that for small strains, when the grain pattern is not too much disturbed, the assumption of reversibility will be applicable.

The assumption of isotropy is not essential and

* Publication assisted by the Ernest Kempton Adams Fund for Physical Research of Columbia University.

[1] K. Terzaghi, *Erdbaumechanik auf Bodenphysikalischer Grundlage* (Leipzig F. Deuticke, 1925); "Principle of soil mechanics," Eng. News Record (1925), a series of articles.

[2] M. A. Biot, "Le problème de la Consolidation des Matières argileuses sous une charge," Ann. Soc. Sci. Bruxelles B55, 110–113 (1935).

anisotropy can easily be introduced as a refinement. Another refinement which might be of practical importance is the influence, upon the stress distribution and the settlement, of the state of initial stress in the soil before application of the load. It was shown by the present author[3] that this influence is greater for materials of low elastic modulus. Both refinements will be left out of the present theory in order to avoid undue heaviness of presentation.

The first and second sections deal mainly with the mathematical formulation of the physical properties of the soil and the number of constants necessary to describe these properties. The number of these constants including Darcy's permeability coefficient is found equal to five in the most general case. Section 3 gives a discussion of the physical interpretation of these various constants. In Sections 4 and 5 are established the fundamental equations for the consolidation and an application is made to the one-dimensional problem corresponding to a standard soil test. Section 6 gives the simplified theory for the case most important in practice of a soil completely saturated with water. The equations for this case coincide with those of the previous publication.[2] In the last section is shown how the mathematical tool known as the *operational calculus* can be applied most conveniently for the calculation of the settlement without having to calculate any stress or water pressure distribution inside the soil. This method of attack constitutes a major simplification and proves to be of high value in the solution of the more complex two- and three-dimensional problems. In the present paper applications are restricted to one-dimensional examples. A series of applications to practical cases of two-dimensional consolidation will be the object of subsequent papers.

1. Soil Stresses

Consider a small cubic element of the consolidating soil, its sides being parallel with the coordinate axes. This element is taken to be large enough compared to the size of the pores so that it may be treated as homogeneous, and at the same time small enough, compared to the scale of the macroscopic phenomena in which we are interested, so that it may be considered as infinitesimal in the mathematical treatment.

The average stress condition in the soil is then represented by forces distributed uniformly on the faces of this cubic element. The corresponding stress components are denoted by

$$\begin{matrix} \sigma_x & \tau_z & \tau_y \\ \tau_z & \sigma_y & \tau_x \\ \tau_y & \tau_x & \sigma_z. \end{matrix} \tag{1.1}$$

They must satisfy the well-known equilibrium conditions of a stress field.

$$\begin{aligned} \frac{\partial \sigma_x}{\partial x}+\frac{\partial \tau_z}{\partial y}+\frac{\partial \tau_y}{\partial z}&=0, \\ \frac{\partial \tau_z}{\partial x}+\frac{\partial \sigma_y}{\partial y}+\frac{\partial \tau_x}{\partial z}&=0, \\ \frac{\partial \tau_y}{\partial x}+\frac{\partial \tau_x}{\partial y}+\frac{\partial \sigma_z}{\partial z}&=0. \end{aligned} \tag{1.2}$$

Physically we may think of these stresses as composed of two parts; one which is caused by the hydrostatic pressure of the water filling the pores, the other caused by the average stress in the skeleton. In this sense the stresses in the soil are said to be carried partly by the water and partly by the solid constituent.

2. Strain Related to Stress and Water Pressure

We now call our attention to the strain in the soil. Denoting by u, v, w the components of the displacement of the soil and assuming the strain to be small, the values of the strain components are

$$\begin{aligned} e_x&=\frac{\partial u}{\partial x}, & \gamma_x&=\frac{\partial w}{\partial y}+\frac{\partial v}{\partial z}, \\ e_y&=\frac{\partial v}{\partial y}, & \gamma_y&=\frac{\partial u}{\partial z}+\frac{\partial w}{\partial x}, \\ e_z&=\frac{\partial w}{\partial z}, & \gamma_z&=\frac{\partial v}{\partial x}+\frac{\partial u}{\partial y}. \end{aligned} \tag{2.1}$$

In order to describe completely the macroscopic condition of the soil we must consider an addi-

[3] M. A. Biot, "Nonlinear theory of elasticity and the linearized case for a body under initial stress."

tional variable giving the amount of water in the pores. We therefore denote by θ the increment of water volume per unit volume of soil and call this quantity the *variation in water content.* The *increment of water pressure* will be denoted by σ.

Let us consider a cubic element of soil. The water pressure in the pores may be considered as uniform throughout, provided either the size of the element is small enough or, if this is not the case, provided the changes occur at sufficiently slow rate to render the pressure differences negligible.

It is clear that if we assume the changes in the soil to occur by reversible processes the macroscopic condition of the soil must be a definite function of the stresses and the water pressure i.e., the seven variables

$$e_x \quad e_y \quad e_z \qquad \gamma_x \quad \gamma_y \quad \gamma_z \quad \theta$$

must be definite functions of the variables:

$$\sigma_x \quad \sigma_y \quad \sigma_z \qquad \tau_x \quad \tau_y \quad \tau_z \quad \sigma.$$

Furthermore if we assume the strains and the variations in water content to be small quantities, the relation between these two sets of variables may be taken as linear in first approximation. We first consider these functional relations for the particular case where $\sigma=0$. The six components of strain are then functions only of the six stress components $\sigma_x\ \sigma_y\ \sigma_z\ \tau_x\ \tau_y\ \tau_z$. Assuming the soil to have isotropic properties these relations must reduce to the well-known expressions of Hooke's law for an isotropic elastic body in the theory of elasticity; we have

$$\begin{aligned} e_x&=\frac{\sigma_x}{E}-\frac{\nu}{E}(\sigma_y+\sigma_z),\\ e_y&=\frac{\sigma_y}{E}-\frac{\nu}{E}(\sigma_z+\sigma_x),\\ e_z&=\frac{\sigma_z}{E}-\frac{\nu}{E}(\sigma_x+\sigma_y),\\ \gamma_x&=\tau_x/G,\\ \gamma_y&=\tau_y/G,\\ \gamma_z&=\tau_z/G. \end{aligned} \tag{2.2}$$

In these relations the constants E, G, ν may be interpreted, respectively, as Young's modulus, the shear modulus and Poisson's ratio for the solid skeleton. There are only two distinct constants because of the relation

$$G=\frac{E}{2(1+\nu)}. \tag{2.3}$$

Suppose now that the effect of the water pressure σ is introduced. First it cannot produce any shearing strain by reason of the assumed isotropy of the soil; second for the same reason its effect must be the same on all three components of strain $e_x\ e_y\ e_z$. Hence taking into account the influence of σ relations (2.2) become

$$\begin{aligned} e_x&=\frac{\sigma_x}{E}-\frac{\nu}{E}(\sigma_y+\sigma_z)+\frac{\sigma}{3H},\\ e_y&=\frac{\sigma_y}{E}-\frac{\nu}{E}(\sigma_z+\sigma_x)+\frac{\sigma}{3H},\\ e_z&=\frac{\sigma_z}{E}-\frac{\nu}{E}(\sigma_x+\sigma_y)+\frac{\sigma}{3H},\\ \gamma_x&=\tau_x/G,\\ \gamma_y&=\tau_y/G,\\ \gamma_z&=\tau_z/G, \end{aligned} \tag{2.4}$$

where H is an additional physical constant. These relations express the six strain components of the soil as a function of the stresses in the soil and the pressure of the water in the pores. We still have to consider the dependence of the increment of water content θ on these same variables. The most general relation is

$$\theta=a_1\sigma_x+a_2\sigma_y+a_3\sigma_z+a_4\tau_x+a_5\tau_y+a_6\tau_z+a_7\sigma. \tag{2.5}$$

Now because of the isotropy of the material a change in sign of $\tau_x\ \tau_y\ \tau_z$ cannot affect the water content, therefore $a_4=a_5=a_6=0$ and the effect of the shear stress components on θ vanishes. Furthermore all three directions x, y, z must have equivalent properties $a_1=a_2=a_3$. Therefore relation (2.5) may be written in the form

$$\theta=\frac{1}{3H_1}(\sigma_x+\sigma_y+\sigma_z)+\frac{\sigma}{R}, \tag{2.6}$$

where H_1 and R are two physical constants.

Relations (2.4) and (2.6) contain five distinct physical constants. We are now going to prove that this number may be reduced to four; in fact that $H=H_1$ if we introduce the assumption of the existence of a potential energy of the soil. This assumption means that if the changes occur at an infinitely slow rate, the work done to bring the soil from the initial condition to its final state of strain and water content, is independent of the way by which the final state is reached and is a definite function of the six strain components and the water content. This assumption follows quite naturally from that of reversibility introduced above, since the absence of a potential energy would then imply that an indefinite amount of energy could be drawn out of the soil by loading and unloading along a closed cycle.

The potential energy of the soil per unit volume is

$$U=\tfrac{1}{2}(\sigma_x e_x+\sigma_y e_y+\sigma_z e_z+\tau_x\gamma_x+\tau_y\gamma_y+\tau_z\gamma_z+\sigma\theta). \quad (2.7)$$

In order to prove that $H=H_1$ let us consider a particular condition of stress such that

$$\sigma_x=\sigma_y=\sigma_z=\sigma_1,$$
$$\tau_x=\tau_y=\tau_z=0.$$

Then the potential energy becomes

$$U=\tfrac{1}{2}(\sigma_1\epsilon+\sigma\theta) \quad \text{with} \quad \epsilon=e_x+e_y+e_z$$

and Eqs. (2.4) and (2.6)

$$\epsilon=\frac{3(1-2\nu)}{E}\sigma_1+\frac{\sigma}{H}, \quad \theta=\sigma_1/H_1+\sigma/R. \quad (2.8)$$

The quantity ϵ represents the volume increase of the soil per unit initial volume. Solving for σ_1 and σ

$$\sigma_1=\frac{\epsilon}{R\Delta}-\frac{\theta}{H\Delta},$$

$$\sigma=\frac{-\epsilon}{H_1\Delta}+\frac{3(1-2\nu)\theta}{E\Delta}, \quad (2.9)$$

$$\Delta=\frac{3(1-2\nu)}{ER}-\frac{1}{HH_1}.$$

The potential energy in this case may be considered as a function of the two variables ϵ, θ. Now we must have

$$\frac{\partial U}{\partial \epsilon}=\sigma_1, \quad \frac{\partial U}{\partial \theta}=\sigma.$$

Hence

$$\frac{\partial \sigma_1}{\partial \theta}=\frac{\partial \sigma}{\partial \epsilon}$$

or

$$\frac{1}{H\Delta}=\frac{1}{H_1\Delta}.$$

We have thus proved that $H=H_1$ and we may write

$$\theta=\frac{1}{3H}(\sigma_x+\sigma_y+\sigma_z)+\frac{\sigma}{R}. \quad (2.10)$$

Relations (2.4) and (2.10) are the fundamental relations describing completely in first approximation the properties of the soil, for strain and water content, under equilibrium conditions. They contain four distinct physical constants G, ν, H and R. For further use it is convenient to express the stresses as functions of the strain and the water pressure σ. Solving Eq. (2.4) with respect to the stresses we find

$$\sigma_x=2G\left(e_x+\frac{\nu\epsilon}{1-2\nu}\right)-\alpha\sigma,$$

$$\sigma_y=2G\left(e_y+\frac{\nu\epsilon}{1-2\nu}\right)-\alpha\sigma,$$

$$\sigma_z=2G\left(e_z+\frac{\nu\epsilon}{1-2\nu}\right)-\alpha\sigma, \quad (2.11)$$

$$\tau_x=G\gamma_x,$$
$$\tau_y=G\gamma_y,$$
$$\tau_z=G\gamma_z$$

with

$$\alpha=\frac{2(1+\nu)}{3(1-2\nu)}\frac{G}{H}.$$

In the same way we may express the variation in water content as

$$\theta=\alpha\epsilon+\sigma/Q, \quad (2.12)$$

where

$$\frac{1}{Q}=\frac{1}{R}-\frac{\alpha}{H}.$$

3. Physical Interpretation of the Soil Constants

The constants E, G and ν have the same meaning as Young's modulus the shear modulus and the Poisson ratio in the theory of elasticity provided time has been allowed for the excess water to squeeze out. These quantities may be considered as the average elastic constants of the solid skeleton. There are only two distinct such constants since they must satisfy relation (2.3). Assume, for example, that a column of soil supports an axial load $p_0=-\sigma_z$ while allowed to expand freely laterally. If the load has been applied long enough so that a final state of settlement is reached, i.e., all the excess water has been squeezed out and $\sigma=0$ then the axial strain is, according to (2.4),

$$e_z=-\frac{p_0}{E} \tag{3.1}$$

and the lateral strain

$$e_x=e_y=\frac{\nu p_0}{E}=-\nu e_z. \tag{3.2}$$

The coefficient ν measures the ratio of the lateral bulging to the vertical strain under final equilibrium conditions.

To interpret the constants H and R consider a sample of soil enclosed in a thin rubber bag so that the stresses applied to the soil be zero. Let us drain the water from this soil through a thin tube passing through the walls of the bag. If a negative pressure $-\sigma$ is applied to the tube a certain amount of water will be sucked out. This amount is given by (2.10)

$$\theta=-\frac{\sigma}{R}. \tag{3.3}$$

The corresponding volume change of the soil is given by (2.4)

$$\epsilon=-\frac{\sigma}{H}. \tag{3.4}$$

The coefficient $1/H$ is a measure of the compressibility of the soil for a change in water pressure, while $1/R$ measures the change in water content for a given change in water pressure. The two elastic constants and the constants H and R are the four distinct constants which under our assumption define completely the physical proportions of an isotropic soil in the equilibrium conditions.

Other constants have been derived from these four. For instance α is a coefficient defined as

$$\alpha=\frac{2(1+\nu)}{3(1-2\nu)}\frac{G}{H}. \tag{3.5}$$

According to (2.12) it measures the ratio of the water volume squeezed out to the volume change of the soil if the latter is compressed while allowing the water to escape ($\sigma=0$). The coefficient $1/Q$ defined as

$$\frac{1}{Q}=\frac{1}{R}-\frac{\alpha}{H} \tag{3.6}$$

is a measure of the amount of water which can be forced into the soil under pressure while the volume of the soil is kept constant. It is quite obvious that the constants α and Q will be of significance for a soil not completely saturated with water and containing air bubbles. In that case the constants α and Q can take values depending on the degree of saturation of the soil.

The standard soil test suggests the derivation of additional constants. A column of soil supports a load $p_0=-\sigma_z$ and is confined laterally in a rigid sheath so that no lateral expansion can occur. The water is allowed to escape for instance by applying the load through a porous slab. When all the excess water has been squeezed out the axial strain is given by relations (2.11) in which we put $\sigma=0$. We write

$$e_z=-p_0a. \tag{3.7}$$

The coefficient

$$a=\frac{1-2\nu}{2G(1-\nu)} \tag{3.8}$$

will be called the *final compressibility*.

If we measure the axial strain just after the load has been applied so that the water has not had time to flow out, we must put $\theta=0$ in relation (2.12). We deduce the value of the water pressure

$$\sigma=-\alpha Qe_z. \tag{3.9}$$

substituting this value in (2.11) we write

$$e_z = -p_0 a_i. \tag{3.10}$$

The coefficient

$$a_i = \frac{a}{1+\alpha^2 a Q} \tag{3.11}$$

will be called the *instantaneous compressibility*.

The physical constants considered above refer to the properties of the soil for the state of equilibrium when the water pressure is uniform throughout. We shall see hereafter that in order to study the transient state we must add to the four distinct constants above the so-called *coefficient of permeability* of the soil.

4. General Equations Governing Consolidation

We now proceed to establish the differential equations for the transient phenomenon of consolidation, i.e., those equations governing the distribution of stress, water content, and settlement as a function of time in a soil under given loads.

Substituting expression (2.11) for the stresses into the equilibrium conditions (1.2) we find

$$\left.\begin{aligned} G\nabla^2 u + \frac{G}{1-2\nu}\frac{\partial \epsilon}{\partial x} - \alpha\frac{\partial \sigma}{\partial x} &= 0,\\ G\nabla^2 v + \frac{G}{1-2\nu}\frac{\partial \epsilon}{\partial y} - \alpha\frac{\partial \sigma}{\partial y} &= 0,\\ G\nabla^2 w + \frac{G}{1-2\nu}\frac{\partial \epsilon}{\partial z} - \alpha\frac{\partial \sigma}{\partial z} &= 0,\end{aligned}\right\} \tag{4.1}$$

$$\nabla^2 = \partial^2/\partial x^2 + \partial^2/\partial y^2 + \partial^2/\partial z^2.$$

There are three equations with four unknowns u, v, w, σ. In order to have a complete system we need one more equation. This is done by introducing Darcy's law governing the flow of water in a porous medium. We consider again an elementary cube of soil and call V_x the volume of water flowing per second and unit area through the face of this cube perpendicular to the x axis. In the same way we define V_y and V_z. According to Darcy's law these three components of the rate of flow are related to the water pressure by the relations

$$V_x = -k\frac{\partial \sigma}{\partial x}, \quad V_y = -k\frac{\partial \sigma}{\partial y}, \quad V_z = -k\frac{\partial \sigma}{\partial z}. \tag{4.2}$$

The physical constant k is called the *coefficient of permeability* of the soil. On the other hand, if we assume the water to be incompressible the rate of water content of an element of soil must be equal to the volume of water entering per second through the surface of the element, hence

$$\frac{\partial \theta}{\partial t} = -\frac{\partial V_x}{\partial x} - \frac{\partial V_y}{\partial y} - \frac{\partial V_z}{\partial z}. \tag{4.3}$$

Combining Eqs. (2.2) (4.2) and (4.3) we obtain

$$k\nabla^2\sigma = \alpha\frac{\partial \epsilon}{\partial t} + \frac{1}{Q}\frac{\partial \sigma}{\partial t}. \tag{4.4}$$

The four differential Eqs. (4.1) and (4.4) are the basic equations satisfied by the four unknowns u, v, w, σ.

5. Application to a Standard Soil Test

Let us examine the particular case of a column of soil supporting a load $p_0 = -\sigma_z$ and confined laterally in a rigid sheath so that no lateral expansion can occur. It is assumed also that no water can escape laterally or through the bottom while it is free to escape at the upper surface by applying the load through a very porous slab.

Take the z axis positive downward; the only component of displacement in this case will be w. Both w and the water pressure σ will depend only on the coordinate z and the time t. The differential Eqs. (4.1) and (4.4) become

$$\frac{1}{a}\frac{\partial^2 w}{\partial z^2} - \alpha\frac{\partial w}{\partial z} = 0, \tag{5.1}$$

$$k\frac{\partial^2 \sigma}{\partial z^2} = \alpha\frac{\partial^2 w}{\partial z \partial t} + \frac{1}{Q}\frac{\partial \sigma}{\partial t}, \tag{5.2}$$

where a is the final compressibility defined by (3.8). The stress σ_z throughout the loaded column is a constant. From (2.11) we have

$$p_0 = -\sigma_z = -\frac{1}{a}\frac{\partial w}{\partial z} + \alpha\sigma \tag{5.3}$$

and from (2.12)

$$\theta = \alpha\frac{\partial w}{\partial z} + \frac{\sigma}{Q}.$$

Note that Eq. (5.3) implies (5.1) and that

$$-\frac{1}{a}\frac{\partial^2 w}{\partial z \partial t} = \alpha\frac{\partial \sigma}{\partial t}.$$

This relation carried into (5.2) gives

$$\frac{\partial^2 \sigma}{\partial z^2} = \frac{1}{c}\frac{\partial \sigma}{\partial t}, \tag{5.4}$$

with

$$\frac{1}{c} = \alpha^2\frac{a}{k} + \frac{1}{Qk}. \tag{5.5}$$

The constant c is called the *consolidation constant*. Equation (5.4) shows the important result that the water pressure satisfies the well-known equation of heat conduction. This equation along with the boundary and the initial conditions leads to a complete solution of the problem of consolidation.

Taking the height of the soil column to be h and $z=0$ at the top we have the boundary conditions

$$\begin{aligned} \sigma &= 0 \quad \text{for } z=0, \\ \frac{\partial \sigma}{\partial z} &= 0 \quad \text{for } z=h. \end{aligned} \tag{5.6}$$

The first condition expresses that the pressure of the water under the load is zero because the permeability of the slab through which the load is applied is assumed to be large with respect to that of the soil. The second condition expresses that no water escapes through the bottom.

The initial condition is that the change of water content is zero when the load is applied because the water must escape with a finite velocity. Hence from (2.12)

$$\theta = \alpha\frac{\partial w}{\partial z} + \frac{\sigma}{Q} = 0 \quad \text{for } t=0.$$

Carrying this into (5.3) we derive the initial value of the water pressure

$$\sigma = p_0 \Big/ \left(\frac{1}{\alpha a Q} + \alpha\right) \quad \text{for } t=0 \quad \text{or} \quad \sigma = \frac{a - a_i}{\alpha a} p_0, \tag{5.7}$$

where a_i and a are the instantaneous and final compressibility coefficients defined by (3.8) and (3.11).

The solution of the differential equation (5.4) with the boundary conditions (5.6) and the initial condition (5.7) may be written in the form of a series

$$\sigma = \frac{4}{\pi}\frac{a - a_i}{\alpha a} p_0 \left\{ \exp\left[-\left(\frac{\pi}{2h}\right)^2 ct\right] \sin\frac{\pi z}{2h} + \frac{1}{3}\exp\left[-\left(\frac{3\pi}{2h}\right)^2 ct\right] \sin\frac{3\pi z}{2h} + \cdots \right\}. \tag{5.8}$$

The settlement may be found from relation (5.3). We have

$$\frac{\partial w}{\partial z} = \alpha a \sigma - a p_0. \tag{5.9}$$

The total settlement is

$$w_0=-\int_0^h \frac{\partial w}{\partial z}dz=-\frac{8}{\pi^2}(a-a_i)hp_0\sum_0^\infty \frac{1}{(2n+1)^2}\exp\left\{-\left[\frac{(2n+1)\pi}{2h}\right]^2 ct\right\}+ahp_0. \quad (5.10)$$

Immediately after loading ($t=0$), the deflection is

$$w_i=-\frac{8}{\pi^2}(a-a_i)hp_0\sum_0^\infty \frac{1}{(2n+1)^2}+ahp.$$

Taking into account that

$$\sum_0^\infty \frac{1}{(2n+1)^2}=\frac{\pi^2}{8}, \quad w_i=a_ihp_0, \quad (5.11)$$

which checks with the result (3.10) above. The final deflection for $t=\infty$ is

$$w_\infty=ahp_0. \quad (5.12)$$

It is of interest to find a simplified expression for the law of settlement in the period of time immediately after loading. To do this we first eliminate the initial deflection w_i by considering

$$w_s=w_0-w_i=\frac{8}{\pi^2}(a-a_i)hp_0\sum_0^\infty \frac{1}{(2n+1)^2}\left\{1-\exp\left[-\left(\frac{(2n+1)\pi}{2h}\right)^2 ct\right]\right\}. \quad (5.13)$$

This expresses that part of the deflection which is caused by consolidation. We then consider the rate of settlement.

$$\frac{dw_s}{dt}=\frac{2c(a-a_i)}{h}p_0\sum_0^\infty \exp\left\{-\left[\frac{(2n+1)\pi}{2h}\right]^2 ct\right\}. \quad (5.14)$$

For $t=0$ this series does not converge; which means that at the first instant of loading the rate of settlement is infinite. Hence the curve representing the settlement w_s as a function of time starts with a vertical slope and tends asymptotically toward the value $(a-a_i)hp_0$ as shown in Fig. 1 (curve 1). It is obvious that during the initial period of settlement the height h of the column cannot have any influence on the phenomenon because the water pressure at the depth $z=h$ has not yet had time to change. Therefore in order to find the nature of the settlement curve in the vicinity of $t=0$ it is enough to consider the case where $h=\infty$. In this case we put

$$n/h=\xi, \quad 1/h=\Delta\xi$$

and write (5.14) as

$$\frac{dw_s}{dt}=2c(a-a_i)p_0\sum_0^\infty \exp\left[-\pi^2(\xi+\tfrac{1}{2}\Delta\xi)^2ct\right]\Delta\xi$$

for $h=\infty$. The rate of settlement becomes the integral

$$\frac{dw_s}{dt}=2c(a-a_i)p_0\int_0^\infty \exp(-\pi^2\xi^2ct)d\xi=\frac{c(a-a_i)p_0}{(\pi ct)^{\frac{1}{2}}}. \quad (5.15)$$

The value of the settlement is obtained by integration

$$w_s=\int_0^t \frac{dw_s}{dt}dt=2(a-a_i)p_0\left(\frac{ct}{\pi}\right)^{\frac{1}{2}}. \quad (5.16)$$

It follows a parabolic curve as a function of time (curve 2 in Fig. 1).

6. Simplified Theory for a Saturated Clay

For a completely saturated clay the standard test shows that the initial compressibility a_i may be taken equal to zero compared to the final compressibility a, and that the volume change of the soil is equal to the amount of water squeezed out. According to (2.12) and (3.11) this implies

$$Q=\infty, \quad \alpha=1. \tag{6.1}$$

This reduces the number of physical constants of the soil to the two elastic constants and the permeability. From relations (3.5) and (3.6) we deduce

$$H=R=\frac{2G(1+\nu)}{3(1-2\nu)} \tag{6.2}$$

and from (5.5) the value of the consolidation constant takes the simple form

$$c=k/a. \tag{6.3}$$

Relation (2.12) becomes

$$\theta=\epsilon. \tag{6.4}$$

The general differential equations (4.1) and (4.4) are simplified,

$$G\nabla^2 u+\frac{G}{1-2\nu}\frac{\partial\epsilon}{\partial x}-\frac{\partial\sigma}{\partial x}=0,$$
$$G\nabla^2 v+\frac{G}{1-2\nu}\frac{\partial\epsilon}{\partial y}-\frac{\partial\sigma}{\partial y}=0, \tag{6.5}$$
$$G\nabla^2 w+\frac{G}{1-2\nu}\frac{\partial\epsilon}{\partial z}-\frac{\partial\sigma}{\partial z}=0,$$

$$k\nabla\sigma^2=\frac{\partial\epsilon}{\partial t}. \tag{6.6}$$

By adding the derivatives with respect to x, y, z of Eqs. (6.5), respectively, we find

$$\nabla\epsilon^2=a\nabla\sigma^2, \tag{6.7}$$

where a is the final compressibility given by (3.8).

From (6.6) and (6.7) we derive

$$\nabla\epsilon^2=\frac{1}{c}\frac{\partial\epsilon}{\partial t}. \tag{6.8}$$

Hence the volume change of the soil satisfies the equation of heat conduction.

Equations (6.5) and (6.8) are the fundamental equations governing the consolidation of a completely saturated clay. Because of (6.4) the initial condition $\theta=0$ becomes $\epsilon=0$, i.e., at the instant of loading no volume change of the soil occurs. This condition introduced in Eq. (6.7) shows that at the instant of loading the water pressure in the pores also satisfies Laplace's equation.

$$\nabla\sigma^2=0. \tag{6.9}$$

The settlement for the standard test of a column of clay of height h under the load p_0 is given by (5.13) by putting $a_i=0$.

$$w_s=\frac{8}{\pi^2}ahp_0\sum_0^\infty\frac{1}{(2n+1)^2}\times\left\{1-\exp\left[-\left(\frac{(2n+1)\pi}{2h}\right)^2 ct\right]\right\}. \tag{6.10}$$

From (5.16) the settlement for an infinitely high column is

$$w_s=2ap_0\left(\frac{ct}{\pi}\right)^{\frac{1}{2}} \tag{6.11}$$

It is easy to imagine a mechanical model having the properties implied in these equations. Consider a system made of a great number of small rigid particles held together by tiny helical springs. This system will be elastically deformable and will possess average elastic constants. If we fill completely with water the voids between the

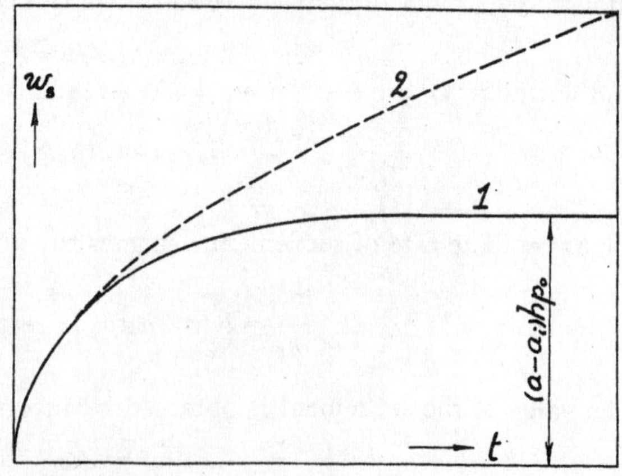

Fig. 1. Settlement caused by consolidation as a function of time. Curve 1 represents the settlement of a column of height h under a load p_0. Curve 2 represents the settlement for an infinitely high column.

particles, we shall have a model of a completely saturated clay.

Obviously such a system is incompressible if no water is allowed to be squeezed out (this corresponds to the condition $Q=\infty$) and the change of volume is equal to the volume of water squeezed out (this corresponds to the condition $\alpha=1$). If the systems contained air bubbles this would not be the case and we would have to consider the general case where Q is finite and $\alpha\neq 1$.

Whether this model represents schematically the actual constitution of soils is uncertain. It is quite possible, however, that the soil particles are held together by capillary forces which behave in pretty much the same way as the springs of the model.

7. Operational Calculus Applied to Consolidation

The calculation of settlement under a suddenly applied load leads naturally to the application of operational methods, developed by Heaviside for the analysis of transients in electric circuits. As an illustration of the power and simplicity introduced by the operational calculus in the treatment of consolidation problem we shall derive by this procedure the settlement of a completely saturated clay column already calculated in the previous section. In subsequent articles the operational method will be used extensively for the solution of various consolidation problems. We consider the case of a clay column infinitely high and take as before the top to be the origin of the vertical coordinate z. For a completely saturated clay $\alpha=1$, $Q=\infty$ and with the operational notations, replacing $\partial/\partial t$ by p, Eqs. (5.1) become

$$\frac{1}{a}\frac{\partial^2 w}{\partial z^2}=\frac{\partial\sigma}{\partial z},\quad k\frac{\partial^2\sigma}{\partial z^2}=p\frac{\partial w}{\partial z}. \tag{7.1}$$

A solution of these equations which vanishes at infinity is

$$w=C_1e^{-z(p/c)^{\frac{1}{2}}},$$
$$\sigma=C_2-\frac{1}{a}\left(\frac{p}{c}\right)^{\frac{1}{2}}C_1e^{-z(p/c)^{\frac{1}{2}}}. \tag{7.2}$$

The boundary conditions are for $z=0$

$$\sigma_z=-1=\frac{1}{a}\frac{\partial w}{\partial z},\quad \sigma=0.$$

Hence

$$C_1=a\left(\frac{c}{p}\right)^{\frac{1}{2}},\quad C_2=1.$$

The settlement w_s at the top ($z=0$) caused by the sudden application of a unit load is

$$w_s=a\left(\frac{c}{p}\right)^{\frac{1}{2}}\cdot 1(t).$$

The meaning of this symbolic expression is derived from the operational equation[4]

$$\frac{1}{p^{\frac{1}{2}}}1(t)=2\left(\frac{t}{\pi}\right)^{\frac{1}{2}}. \tag{7.3}$$

The settlement as a function of time under the load p_0 is therefore

$$w_s=2ap_0\left(\frac{ct}{\pi}\right)^{\frac{1}{2}}. \tag{7.4}$$

This coincides with the value (6.11) above.

[4] V. Bush, *Operational Circuit Analysis* (John Wiley, New York, 1929), p. 192.

Originally published in
Journal of the Boston Society of Civil Engineers
Jan., 1936

CHARACTERISTICS OF COHESIONLESS SOILS AFFECTING THE STABILITY OF SLOPES AND EARTH FILLS

BY ARTHUR CASAGRANDE, MEMBER *

(Presented at a meeting of the Designers Section of the Boston Society of Civil Engineers, November 13, 1935.)

INTRODUCTION

IT is not an exaggeration to state that throughout history and into the present day faulty designs have caused more loss of life and property in the field of earth and foundation engineering than in any other branch of engineering. In this field the largest share of these losses have been caused by failures of dams and dikes. Many times larger than the property losses due to failures has been the waste of money due to excessive over-designing.

History is full of examples which show that experience accumulated during many centuries will often lead to satisfactory solutions of problems long before they are properly understood by science. These solutions sometimes are so excellent that they can hardly be bettered by a modern, scientific approach. Therefore one may well ask why it is that just in earth and foundation engineering our empirical knowledge, derived from the experience of many hundreds, even thousands, of years, has remained so unreliable.

On closer examination one finds, in the cases where experience has crystallized into empirical knowledge that can be relied upon, that certain conditions repeat themselves with little variation, or that the number of variable factors is relatively small. In the case of soils we find just the contrary to be true. Soils exist in an almost infinite variety. In addition, the properties of an individual soil are so complex that the development of a scientific approach to the mechanics of soils could not take place until other sciences, particularly certain border fields between physics and physical chemistry, had reached a sufficiently advanced stage of development.

I do not wish to convey the impression that I deny the value of experience in foundation engineering in general. But I do wish to differ-

* Graduate School of Engineering, Harvard University, Cambridge, Mass.

entiate between experience which has proven its merit and which creates in the possessor a strong feeling that he can rely upon it, and such empirical methods which create a conscious or unconscious feeling of uncertainty, perhaps even a suspicion that one's design is, after all, not far from guesswork. Of all the empirical methods used in the field of earth and foundation engineering I consider those which have been used in the design of earth dams and dikes to be the least reliable. It is this subject, more particularly the *stability* of earth fills, consisting in part or entirely of *cohesionless* materials, which I wish to discuss in this paper.

The Meaning of the Term Stability

In mechanics a body is considered stable, or in stable equilibrium, when the forces acting upon it are in equilibrium and when small additional forces, for example, disturbances to which this body is commonly exposed, will not cause loss of equilibrium. If a slight change in the magnitude of the forces will result in an appreciable permanent displacement or movement of the body, then it is in an unstable equilibrium. Thus, a pencil standing on end, which loses its equilibrium when slightly disturbed, is, from a practical, or we may say, engineering standpoint, unstable. Naturally, such a meaning is not absolute. The limit between stable and unstable equilibrium is often subject to arbitrary definition; for example, if local conditions should make it practically impossible that a disturbance of sufficient intensity will act on the pencil to cause it to tumble, then we may say that the pencil is in a stable condition. However, if the table on which the pencil is standing is exposed to ordinary accidental disturbances, as tables usually are, then we should consider the pencil in an unstable condition.

A definitely stable condition is obtained by suspending the pencil from its upper end, because then it will always return to its original position. As another example of stability, let us consider a flexible board resting, like a beam, on two supports. In this condition the board is immune to ordinary disturbing factors. Even if I place a heavy weight upon it, causing considerable bending, its stability is not impaired; if anything, it is improved.

The pencil, when standing on end, has lost its equilibrium under a very slight disturbing force which, in itself, has not caused any noticeable deformation of the pencil. On the other hand, in the second example the beam deformed a great deal under the additional force, but its stability was not affected. Obviously, the stability of a body and its deformations due to the forces to which it is exposed are two independent effects. Thus, the gradual settlement of buildings due to consolidation of underlying clay layers is not a stability problem, so far as the soil is concerned. If the building should collapse due to the settlements, it is the building which has lost its stability, not the underlying masses of soil. As long as the stresses in the soil do not reach the shearing strength of the material, it will undergo only deformations and volume changes of a sort which are comparable to the changes that structural materials, like wood or concrete, undergo when exposed to stresses which are smaller than the strength of the material. However, if the external forces become so large that the ultimate shearing strength of the soil will be exceeded in some region, then plastic flow or rupture will result and will initiate movement in a portion of the soil mass.

As long as the shearing stresses in the soil are considerably below its ultimate shearing resistance, and as long as possible disturbances will not seriously affect the magnitude of the ultimate shearing resistance, the mass of soil is in a stable condition, regardless of the magnitude of the deformations and volume changes which are caused by the stresses in the soil. However, if the external loads create stresses within the mass of soil that approach closely its ultimate shearing resistance, or if disturbances result in a more or less sudden drop in the ultimate shearing strength of the soil to a point where it approaches closely the existing shearing stresses due to the external forces, then it would be correct to say that this mass of soil is not in a stable condition, or that it is unstable. The various types of earth movements and slides, failures of earth dams, failures of foundations of dams* and other water-retaining structures due to "piping," etc., are examples of the loss of stability of a mass of soil.

Summarizing this discussion on stability, we may say that the instinctive conception which the layman has, when using this term, the definitions used in mechanics, and also those conditions which we term stable and unstable when dealing with a mass of soil, are all inherently the same. Stability of a mass of soil refers to the equilibrium of all external and internal forces with the resistance of the soil, including the force of gravity, seepage pressures, and any possible artificial disturbances due to construction activities, etc., as well as the effects of earthquakes. Stability does not refer to the amount of deformation which these forces produce, as long as the shearing resistance of the soil is not utilized to its ultimate limit.

The stability of a mass of soil is not an individual property of the material like the specific gravity, permeability, compressibility or angle

* See "Application of Theories of Elasticity and Plasticity to Foundation Problems," by L. Jürgenson, JOURNAL, Boston Society of Civil Engineers, July, 1934.

of internal friction, which can be measured on a sample of the soil and expressed by a single quantity. It is a combined effect of one or several of such individual properties and of numerous other factors, particularly the character of the forces to which the soil mass may be exposed, its dimensions, various local conditions, and possibly other factors which are not sufficiently known; for example, the same cohesionless sand and gravel which forms a very stable and desirable foundation material for buildings or roads will yield excessively under wheel loads if it is used as a road surfacing. It forms an unstable road surface because its shearing resistance under such conditions is easily exceeded. To "stabilize" this material for such purposes one has to increase its shearing resistance by an admixture of cohesive soil, tar or asphalt. This simple example should suffice to show that one should not speak of "stability" of a soil as if it were a definite soil constant, like its permeability or compressibility.

The Shearing Resistance of Cohesionless Soils

The shearing resistance of a cohesionless material is due entirely to the force with which the grains are pressed together, which, in general, is due to the weight of the overlying soil and the load of any superimposed structures. Therefore a deposit of dry sand has on its surface very little shearing resistance; yet only a few feet beneath the surface the weight of the overlying material mobilizes sufficient friction to make the sand very resistant to deformation. This can be convincingly demonstrated by placing dry sand in a rubber bag and evacuating the air. At first the bag feels like a soft cushion, but the moment some of the air is withdrawn the bag becomes as hard and rigid as a stone. The difference in air pressure between the outside and inside of the bag presses the sand grains together, duplicating the effect of the weight of a few feet of overlying soil, and the friction thus mobilized between the grains changes the soft cushion into a hard, solid block, conveying the impression that the grains were suddenly cemented together.

From this simple experiment we can see that the resistance of sand to deformation increases rapidly with the pressure between the grains. By means of shearing tests it is possible to determine the relationship between pressure and ultimate shearing resistance or shearing strength, which can be expressed for cohesionless soils by Coulomb's law —

$$s = f \cdot p = p \cdot \tan \phi$$

in which s designates the shearing strength, p the pressure normal to the plane of shear, and f or $\tan \phi$ the coefficient of internal friction. The angle ϕ is known as the angle of internal friction, and is identical with the obliquity of the resultant force of the normal pressure and the ultimate shearing resistance.

It is known that the angle of internal friction is dependent on the density of a sand and on its grain size, shape and uniformity. However, accurate information on the magnitude of this angle for various types of cohesionless soils is still surprisingly incomplete, and that in spite of the fact that more research work has been expended in the course of time on this simplest of all mechanical soil properties than on all other mechanical properties.

Several factors, which apparently are of minor importance and have therefore been neglected, seriously influence the results of shearing tests, so that today the numerous investigations which are available in published form disagree widely. The most important of these factors is the volume change associated with the deformation of soil, particularly of cohesionless materials. An accurate observation of such volume changes is of utmost importance, not only for the purpose of proper interpretation of shearing tests, but also for the analysis of the stability of large masses of cohesionless soils.

Volume Changes associated with the Deformation of Soils

If we observe carefully the volume changes of samples of sand during shearing tests we find that *dense sand expands* and *very loose sand reduces its volume.*

In a dense sand, as in Fig. 1*a*, the grains are so closely interlocked that deformation is not possible unless accompanied by a loosening up of the structure, as indicated by Fig. 1*b*. If dense sand is so confined that it cannot expand, then the shearing strength is determined by the resistance of the grains to crushing, and therefore it acts essentially like a rigid stone. This was observed fifty years ago by Osborne Reynolds, who demonstrated it by filling a water-tight bag first with dry sand, in which condition the bag could be deformed easily, and then replacing the air in the voids of the sand completely with water, after which the bag turned rigid because the sand could no longer change its volume.

When the horizontal displacement and the volume change during a *shearing test on dense sand* are plotted against the corresponding shearing stress (Figs. 1*e* and 1*f*), it is noticed that the shearing stress reaches a maximum S_D, — corresponding to the point B on the curve, — and if the deformation is continued, the shearing stress drops again to a smaller value, S_L, at which value it remains constant for all further displacement.*

* This refers to the conditions in a large mass of sand. In a test the displacement is limited by the design of the shearing apparatus. See "Recent Developments in Soil Testing Apparatus," by P. C. Rutledge, Journal, Boston Society of Civil Engineers, October, 1935.

During this drop in shearing stress the sand continues to expand, as shown in Fig. 1*f*, curve *E–G*, finally reaching a critical density at which continuous deformation is possible at the constant shearing stress S_L. This critical density corresponds for very coarse, well-graded sand and for gravel approximately to the loose state of the material. For medium and finer grained sands it lies between the loosest and densest state. In addition, it depends to a large extent on the uniformity of the material. The more uniform a soil the lower the critical density.

The above also furnishes the explanation why most sands in their loose state have a tendency to reduce their volume when subjected to a

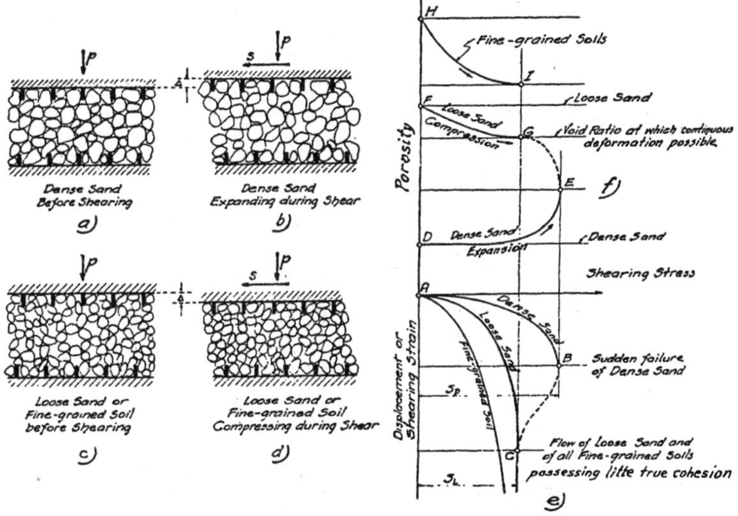

FIG. 1. — EFFECTS OF SHEARING ON THE VOLUME OF SOILS

shearing test under constant normal pressure. The shearing stress simply increases until it reaches the shearing strength S_L, and if the displacement is continued beyond this point the resistance remains unchanged. Obviously, the volume of the sand in this state must correspond to the critical density which we had finally reached when performing a test on the same material in the dense state. Therefore the curves representing the volume changes during shearing tests on material in the dense and the loose state must meet at the critical density when the stationary condition is established.

For very fine-grained soils, like rock flour, inorganic silts, and clays, the volume usually decreases during a shearing test under constant normal stress, regardless whether the material was, at the start, in a loose state or was compacted by vibrations, tamping, or static pressures of ordinary intensity. In other words, for such materials volume decrease is obtained much more efficiently by a combination of loading and deformation than by ordinary methods of compaction.

The most efficient method of reducing the volume of a cohesionless material is by means of vibrations. This is known not only to every one

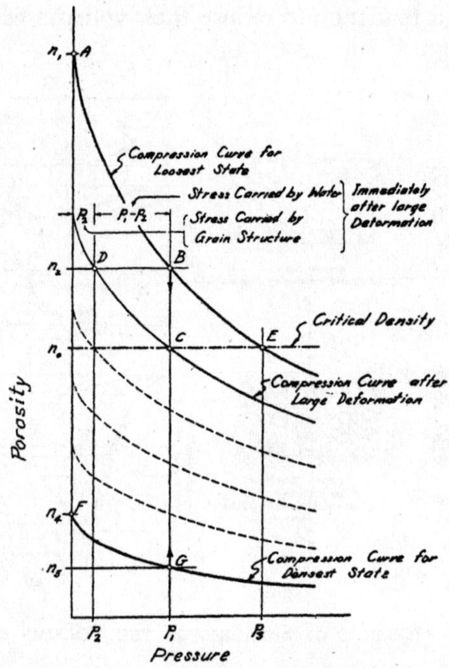

FIG. 2. — PRESSURE-DENSITY RELATIONSHIPS FOR SAND

who has worked with such materials in the laboratory, but also to engineers who have observed the behavior of deposits of sand under the effect of pile-driving or vibrations caused by heavy machinery.

Static pressure is relatively ineffective in reducing the volume of a sand; for example, it is not possible to change a loose sand into a dense sand by static pressure alone. Professor Terzaghi has designated this property of retaining the essential characteristics of a loose structure, the "conservative structure" of sand.

By combining static pressure with simultaneous deformation, one can reduce the density of a loose sand very effectively. However, even this procedure will not make the structure of a loose sand as dense as is possible by means of vibrations.

We may compare the effectiveness of these three methods for compacting soil with the help of Fig. 2. The volume decrease under static pressure is determined by the compression curve which passes through the initial density. If the material was, at the start of compression, in its loosest state, n_1, then the static pressure p_1 will reduce the volume along curve *A–B*. If the material was at the start in its densest state, then the same pressure will reduce the volume along the much flatter curve *F–G*. The several compression curves, shown dotted, correspond to intermediate densities. Note that even high pressures, for example, 20 tons per square foot, will not reduce the volume of a loose sand sufficiently to come near that of the same sand in the dense state without load.

Suppose that a loose sand has been compressed under the pressure p_1 from *A* to *B*. If, in that condition, it is desired to reduce the volume further, this can be done by applying strong vibrations which will eventually reduce the volume to point *G*. Or we can apply a deformation to the soil, while keeping the load constant at p_1, which will reduce the volume until it reaches the critical density n_o at point *C*, in which state continuous deformation is possible without further change in density. However, no change in volume will take place if, at the beginning of deformation, the sand was at its critical density, and if it was below that density, deformation will result in expansion of the sand.

Volume Change during Deformation affecting Stability

As long as the sand can freely follow the tendency to change its volume during deformation, as is the case for dry sand, the stresses in a mass of sand are, at all times, fully transmitted from grain to grain. Therefore the shearing resistance, which is directly dependent on the normal stresses acting between the grains, is always fully active. Consequently, if the mass of sand was stable before deformation started, it will also be stable during the process of deformation.

If the voids in a mass of loose sand are completely filled with water, then the reduction in volume during deformation must be accompanied by an outflow of an equivalent quantity of water. If this volume decrease lags noticeably behind that which would take place if it were not hindered by the presence of water, the pressure between the grains will temporarily be transferred in part or entirely on to the water. It is this stress in the water, known in soil mechanics as hydrodynamic stress (see Terzaghi, "Erdbaumechanik," 1925), which forces the excess water to flow out. The time-lag of volume decrease is dependent on the permeability of the material, on the amount of reduction in volume, and especially on the dimensions of the mass thus affected.

While the normal stresses in the soil are partially or fully carried by the water, the pressures acting between the individual grains are reduced by a corresponding amount, since the total stress must remain equal to the overlying loads. Simultaneously with this reduction, the frictional resistance between the grains is reduced in the same proportion. The amount of this reduction can be analyzed with the help of Fig. 2. Let us consider a volume element in the mass of sand in which the stress conditions correspond to point B. This would indicate that the sand has been deposited originally in a loose state and was subsequently compressed by a static pressure equal to p_1. Now the mass of sand shall be exposed to horizontal forces which tend to deform it without, however, changing the vertical pressures in the mass. That portion of the sand which is affected by the deformation will tend to reduce its volume. If the deformation is sufficiently large, a new state of internal equilibrium will be established after the critical density n_o is reached in point C. The compression curve through point C reflects the change in the structure of the sand which was produced by the deformation.

If the quantity of water which could escape during deformation is negligible, then the change in the structure without change in density causes a drop in the pressure acting between the grains from p_1 to p_2, the latter being determined by the intersection D between the abscissæ through B and the compression curve through C. The difference (p_1-p_2) is now carried temporarily by the water and does not produce frictional resistance, since the shearing strength of water is zero. The shearing resistance in the zone of deformation, being proportional to the pressure actually transferred between the grains, is now reduced in the ratio p_2/p_1. The pressure p_2 can be a small fraction of p_1, and may even be equal to zero, so that temporarily the soil can lose a large portion of or its entire shearing strength.

For very large pressures, which are not normally encountered in problems of earth and foundation engineering, even a loose sand may be compressed to its critical void ratio, as, for example, point E, and then deformation will not cause any reduction in shearing resistance.

If the density at the beginning of the deformation was below the critical density, as is always the case with dense sand, then deformation of a saturated mass will temporarily create tension in the water and a corresponding increase in the pressure between the grains. Hence deformation of a mass of dense, saturated sand may result in an increase of the shearing resistance beyond the normal value. In other words, the mass seems to be bracing itself, to become temporarily more stable.

Comparison of Effects of Vibrations and Deformations

Loose sand can be efficiently compacted below the critical density only by means of vibrations. Sometimes it is not easy to distinguish between deformation and vibrations, and then one is forced to resort to an arbitrary definition. For our purpose we can utilize the volume change of sand for this differentiation. We may define deformation as any disturbance which, when continued long enough, will cause compression or expansion leading to the critical density. Vibrations may be defined as such oscillatory disturbances which, when continued long enough, will cause a decrease in volume below the critical density, and which, if applied with sufficient intensity, will compact the material into the densest state.

From the standpoint of stability it would appear at first that vibrations are more dangerous than a deformation. However, we must consider that large deformations can be produced in a very short time during which excess water may not be able to drain away, while vibrations must continue for long periods in order to produce an appreciable volume decrease. And during this time interval most of the excess water can escape without impairing the stability of the mass. Therefore it is my opinion, which would have to be substantiated by further investigations, that the stability of fills of loose, cohesionless sands, excepting very fine sands, may be affected more by quick deformations than by vibrations.

The most important natural disturbance to which a fill may be subjected is an earthquake. Whether an earthquake produces chiefly vibrations or deformations in the ground is, from the standpoint of the stability of saturated fills consisting of cohesionless soils, of great importance and should be investigated. It is my opinion that the vibrations themselves are not sufficiently effective and are not of sufficient duration to cause appreciable compaction.* However, the swaying movements which are amplified in large masses of sedimentary deposits, especially clay deposits, and also artificial deposits, like large earth dams, particularly when they rest on a clay foundation, produce quite substantial deformations in such masses, frequently leading to permanent displacements,

* This could be investigated by determining the density of natural deposits of cohesionless sands in regions subjected to frequent earthquakes.

shearing cracks, etc. And since these deformations take place in a few seconds, they may be accompanied by the formation of large hydrodynamic stresses and a corresponding reduction in shearing resistance which may lead to loss of stability.

It is my belief that only artificial vibrations continued during a considerable period, as, for example, pile-driving, or the application of special vibration machines, are capable of compacting cohesionless sands into a dense state. Foundation engineers are familiar with the fact that pile-driving frequently results in settlements of neighboring structures resting on deposits of loose sand. Pile-driving has been used for the purpose of compacting such deposits to improve their ability to carry loads without undue settlements. During the past few years methods of artificial vibration with special machines have been developed in Germany for the purpose of compacting artificial fills of sandy soils.

Effect of Disturbances on the Stability of Cohesionless Soils Illustrated by Means of Experiments

In the tank shown in Fig. 3 is deposited a fine quartz sand in a loose, saturated state, with free water standing on its surface, and a weight is placed on the surface. Then a stick is thrust into the sand, and suddenly the weight sinks below the surface. The slight but rapid deformation produced by the penetration of the stick results in a change of the sand structure and the formation of hydrodynamic stresses which quickly spread through the entire mass and so decrease the internal friction that the weight can no longer be supported.

This experiment can be made even more striking by letting water percolate in an upward direction through the sand, which changes it into an exceedingly loose state, and then draining away all water from the surface, so that capillary forces will be mobilized. In this state the surface can support a heavy weight without any noticeable subsidence. Upon driving the stick into the sand under these conditions the whole mass seems to liquefy suddenly and the weight disappears completely. In this case the liberation of water due to deformation produces, as a secondary effect, the disappearance of the capillary forces which helped materially in carrying the load.

Both experiments are imitations of conditions encountered in construction practice. The only difference is that natural sand deposits are hardly ever as loose as the sand in these experiments, and therefore the effects of any disturbance in nature are not so striking.

Of utmost importance is the question of the stability of sandy soils in dikes and dams. Contrary to the belief of many engineers that a mass of sand will always be stable if the slope is less than the angle of repose, there are many examples on record of embankments, dikes, etc., consisting of fine, saturated sands, being destroyed by flowing out of the entire mass as if it had been suddenly liquefied. Such flow slides have for the first time been thoroughly investigated and explained by Terzaghi. (See "Erdbaumechanik," pp. 344–352; also "Ingenieurgeologie," by Redlich-Terzaghi-Kampe.)

The essential difference in the stability of a dense and a loose sand, when subjected to severe lateral forces, like the waves of an earthquake, can be demonstrated by experiments such as those illustrated in Fig. 4. In two identical tanks, mounted on casters, model dam sections are built of ordinary beach sand, with slopes 1 on 2. One of the sections is built of sand in a very loose state, and the other of the same sand deposited in

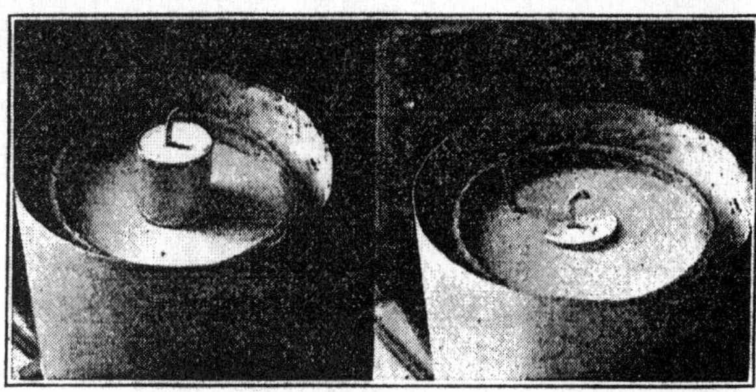

FIG. 3. — LOSS OF STABILITY IN FINE-GRAINED, COHESIONLESS, SATURATED SOILS

layers and tamped by hand into a dense state. On the downstream side the base is covered with a pervious filter blanket to keep the line of seepage away from the downstream face, which prevents sloughing and erosion.

By shaking the tank containing the loose dam three or four times with a horizontal oscillation of about one inch, one causes the dam to flow out until the surface of the sand is nearly level. The other dam, consisting of dense sand, does not change its shape in the slightest, even if its tank is shaken much more violently and more frequently than the other tank.

After our preceding discussion a further explanation of these experiments is not necessary. However, it remains to analyze the possibility that one or the other factor is reduced or exaggerated in these experiments in its relative importance, as is the case in most model experiments. The four principal factors which determine the extent of reduction in shearing resistance are the amount of deformation, the intensity of volume decrease with deformation, the permeability, and the dimensions of the mass. If only the dimensions of the mass are reduced and the other factors kept constant, then the model will display a much smaller drop

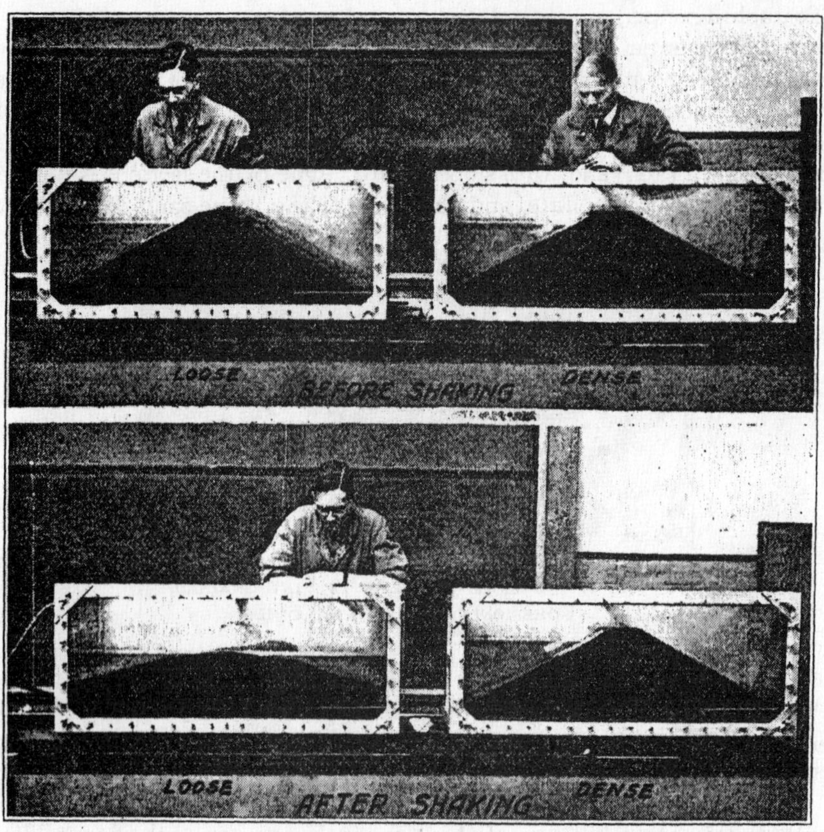

Fig. 4.— Effect of Horizontal Oscillations on Model Dam Sections Built of Sand

in shearing resistance and consequently a much larger stability than the prototype in nature. A measure for this difference is, first, the time required for the excess water to drain out, and second, the kinetic energy of the mass of soil accumulated during the motion for equal intensity of deformation. The time required for drainage increases, for the same soil, with the square of the height of the dam. The kinetic energy increases in direct proportion to the mass, that is, the square of the height, and furthermore, it increases with the square of the velocity. Since the average velocity is either independent of the height, for the same total displacement, or proportional to the height for equal angular displacement, it follows that the driving force increases at least with the fourth and possibly with the sixth power of the height. However, the resisting force increases only in direct proportion with the height. Therefore we arrive at the conclusion that, with the same type of sand, the amount of reduction in stability increases with at least the third power of the height of the dam.

If we wish to reproduce a model correctly we should reduce the permeability, and therefore the grain size of the material of which the model is built, to a very small fraction of that in the prototype, as Terzaghi has already pointed out in his book "Erdbaumechanik," page 348. However, such reduction leads to other difficulties, especially the presence of cohesion, and also the doubts which would probably be raised by engineers against the validity of test results obtained with such a very different material. The only way to make the tests with ordinary sand, perhaps of the same type as that in the prototype, and still be able to show the effect of deformation on the stability of the structure, is by exaggerating some other factor, for example, the amount of volume decrease which will occur during deformation. This recourse was taken in the model tests shown in Fig. 4, by starting out with a model in a much looser state than would normally be found in the prototype.

An interesting conclusion can be drawn regarding the amount of volume decrease necessary to produce a flow slide. The larger a structure is the smaller is the necessary amount of volume reduction to endanger the stability. With the same type and density of sand the larger structure would require a smaller deformation, or at the same deformation the larger structure must be denser than the small structure, to be equally stable. In other words, for a very large dam a density only slightly larger than the critical density may result in a flow slide.

Such experiments as shown in Fig. 4, are to me especially interesting, not on account of the behavior of the loose material, which I had anticipated before making the first tests, but on account of the behavior of the dense material, which indicates to me that a well compacted, cohesionless, saturated material is foolproof even against such disturbances as only one among a hundred dams may actually experience. Yet dams should be made safe even against such eventualities, if it is economically possible to do so.

Examples of Earth Movements caused by Formation of Hydrodynamic Stresses due to Reduction of Volume

In addition to the examples described by Terzaghi in the books "Erdbaumechanik" and "Ingenieurgeologie," the following examples may be of interest:

Recently a hydraulic fill dam in Russia failed, due to a flow slide which was initiated by the blasting of a near-by cofferdam. Russian specialists in soil mechanics who investigated this failure arrived at the conclusion that the failure was probably due to the development of hydrodynamic stresses which were created by the shock. Obviously, a large portion of the shoulders of the dam, which have the purpose of providing stability, must have consisted of loosely deposited sand, which when disturbed had the tendency to reduce its volume. It is probable that in this case the disturbance had more the character of very strong vibrations rather than that of a deformation, although this question must remain open. Certainly, it cannot be denied that the density of a large portion of the dam was above the critical density, and that an earthquake might have led to a similar failure. Furthermore, it seems to me beyond doubt that a thorough compaction of the shoulders during construction would have prevented such a failure.

Many disastrous land slides in deposits of hard or stiff clay, of ample strength to prevent slides, as long as they exist in a homogeneous mass, have their cause in the formation of hydrodynamic stresses within layers of sand or rock flour which are contained in the clay. Some natural or artificial disturbance causes the tendency to reduce the volume, and due to the slow rate at which the excess water drains away, the overlying mass of clay slides out as if suddenly placed on roller bearings.

The lack of stability of very fine-grained, saturated materials must also be considered in the placing of spoil banks of chemical wastes, which often consist of extremely fine powders. Excepting for a dry surface crust, the voids of such materials are filled with water. The large capillary pressures render it seemingly very stable, capable of standing on high vertical banks. Yet disturbances of any sort, or the formation of shrinkage or tension cracks with subsequent infiltration of surface water, resulting in deformations of the mass, will result in local liquefication of the material which may spread quickly over large areas and lead to a slide. An example of such a slide, shown in Fig. 5, had disastrous consequences. In the background of this picture one can see still standing the undisturbed banks of the deposit, 100 feet high.

Of great importance is the stability of the foundation materials on which dams or weirs are resting. It is known from experience that fine-grained sands are more liable to cause "piping" than coarse-grained sand or gravel. Professor Terzaghi* has offered an explanation why, especially in fine-grained, stratified, sandy soils, the danger exists of underground erosion and local subsidences which may lead to piping.

In my opinion it is quite possible that even in a homogeneous material piping may occur if the density is above the critical density. A serious disturbance, for example, an earthquake, or possibly even large seepage pressures, may result in a tendency to reduce the volume due to

Fig. 5. — Flow Slide in Chemical Wastes Deposit

an adjustment in the structure, reducing the shearing resistance to such a point that the whole mass can flow like a liquid under the effect of the seepage pressures. The tendency for this kind of piping is again dependent on the permeability, density and the dimensions of the soil mass. Fine-grained materials are much more apt to exist above the critical density, and, in addition, their lower permeability results in more time being required for adjustments to take place so that a movement, once it has started, will inevitably lead to piping. The reproduction of such conditions by model experiments is rendered difficult for the same reasons as discussed above for earth dams.

* "Effect of Minor Geologic Details on the Safety of Dams," American Institute of Mining and Metallurgical Engineers, Technical Publication No. 215.

Whether, in such cases, indiscriminate increase in the coefficient of percolation over that required from the standpoint of the magnitude of the discharge gradient* represents a sufficient precaution is in my opinion an open question.

Methods for Compacting Cohesionless Soils

The question arises whether there are methods available which permit the compaction of cohesionless soils below the critical density. Natural deposits of loose sands can be effectively compacted by pile-driving. At various times it has been tried to create the vibrations by other means. These attempts were not very successful because the vibrations produced on the surface did not penetrate with sufficient intensity to greater depths. However, when compacting artificial fills, this latter objection does not hold, because one can build up the fill in layers, each of which can be compacted separately. During the past three years efficient vibration machines for such purpose have been developed in Germany for the construction of embankments of fine sands, in connection with the development of a system of superhighways. Investigations† have shown that, for *cohesionless soils*, such vibration machines are more effective than any of the other methods which were commonly used for compaction. Furthermore, it was found that the thickness of the layers in which the material can be sufficiently compacted can be greatly increased if compared with other methods. Layers up to eight feet in thickness are compacted satisfactorily.

It is interesting to note that simultaneously in the United States new and efficient methods have been developed for the compaction of *cohesive soils,* — for example, the sheeps-foot roller and the Proctor‡ method for controlling the moisture content.

In my opinion it is only a matter of a few years until all fine and medium-grained cohesionless soils used in the construction of dams, dikes, etc., when exposed to saturation, will be compacted by means of vibration machinery, in order to eliminate any possibility of loss of stability due to unforeseen disturbances, and, last but not least, to permit greater economy by permitting the use of steeper slopes.

Whether stabilization by vibrations is necessary or not will often be a question that can only be decided by investigations. Particularly

* Transactions of American Society of Civil Engineers, 1935, pp. 1289–1294.

† "Verdichtung geschuetteter Daemme," by R. Mueller and A. Ramspeck, Die Strasse, No. 18, 1935.

‡ See "Fundamental Principles of Soil Compaction," by R. R. Proctor, Engineering News-Record, Vol. III, 1933.

coarse sands and gravel may assume a density below the critical density without special measures for compaction. On the other hand, whenever large masses of fine, uniform sands are used in a fill, for example, for the pervious outer sections of large earth dams, our present knowledge is sufficient to indicate the necessity of stabilization by means of vibration.*

If a portion of a dam consists of uniform, fine sand, and is subject to saturation, then its shearing resistance should not be taken into account in the design. This refers, for example, to the fine-grained, interior portions of the shoulders of a hydraulic-fill dam, when applying Professor Gilboy's stability analysis.†

Remarks on the Stability of Cohesive Soils

Most very fine-grained soils exist in their natural state above the critical density.‡

Thus the question suggests itself whether all these soils are liable to fail if subjected to disturbances. Here we must distinguish between such fine-grained soils which are not plastic and have only very little cohesion, for example, inorganic silts and rock flour, and plastic soils. The non-plastic, fine-grained soils, particularly if very uniform, are frequently of such character that all common methods for compaction, including vibrations, are not sufficient to compact the material below the critical density. There can be hardly any doubt that such materials are unsuitable in the construction of those structural sections of earth dams and dikes which are exposed to saturation. In foundation work such soils are also exceedingly treacherous, and it is certainly no coincidence that experienced foundation engineers have a sincere dislike for them.

However, as soon as we get into the plastic range, the effect of disturbances upon the stability of the soil is very much reduced. Vibrations have very little effect upon such soils and the effect of deformations is relatively small, due to their larger compressibility and their ability to undergo relatively large elastic deformations. The deformation required to reach the critical density of a plastic soil is many times larger than the deformations to which such soils may be exposed in earth and foundation

---

* After finishing this paper a publication appeared in Engineering News-Record of November 14, 1935, on "Man-made Earthquakes," by Franklin P. Ulrich, in which reference is made to investigations of the effect of earthquakes on masonry dams, and to the use of vibration machinery for various purposes, including the compaction of earth fills. The author would be very much interested to learn if earthquake experiments have also been conducted on models of earth dams.

† See "Mechanics of Hydraulic-Fill Dams," by G. Gilboy, JOURNAL, Boston Society of Civil Engineers, July, 1934.

‡ See "New Facts from the Research Laboratories," Engineering News-Record, September 5, 1935.

engineering. Deformations of a magnitude which would cause, in a sand, the development of the critical density, will cause in a plastic soil only a slight drop in the compression curve, so that in Fig. 2 the quantity (p_1-p_2) is small in relation to p_1, and only a slight reduction in stability will develop.

Recently in the Harvard Soil Mechanics Laboratory a cohesive and plastic soil was investigated, which was exceedingly well graded, containing large quantities of sand, and giving the impression that it was exceptionally well suited for the construction of a rolled earth dam for which it was intended. Subsequent investigations showed that the critical density of this soil corresponded to a higher water content than that at which it was intended to construct the dam. Hence, this material will be placed in the dam in a state below the critical density, and therefore it will be exceptionally stable. It is interesting to note that the good impression which this soil conveyed to experienced dam engineers was substantiated by laboratory tests. However, this example is not cited for the purpose of suggesting that only such cohesive soils should be used in dam construction for which the optimum water content* lies below the critical density. On the contrary, I wish to emphasize that we do not know enough about the behavior of plastic soils under shearing stress to be able to make any reliable statements.

Conclusions

1. Every cohesionless soil has a certain critical density, in which state it can undergo any amount of deformation or actual flow without volume change.

2. The density in the loose state of many cohesionless soils, particularly medium and fine, uniform sands, is considerably above their critical density. Such materials in their loose state tend to reduce their volume if exposed to continuous deformation. If the voids are filled with water and the water cannot escape as quickly as the deformation is produced, then a temporary transfer of load on to the water takes place, and the resulting reduction in friction impairs the stability of the mass, which can lead, in extreme cases, to a flow slide.

3. If a cohesionless soil is below the critical density, then it can stand any disturbance without danger of a flow slide. Whenever there is any tendency for the mass to deform, the water in the voids has a restraining influence.

* See "Fundamental Principles of Soil Compaction," by R. R. Proctor, Engineering News-Record, Vol. III, 1933.

4. Many coarse-grained, and very well-graded mixtures of cohesionless soils are in their loose state approximately at the critical density. This fact, combined with their large permeability, renders them relatively stable against any disturbances, even in the loose state.

5. Cohesionless soils in a state above the critical density can be efficiently compacted, and thereby stabilized against any disturbances, by means of special vibration machinery.

AUTHOR'S NOTE: *Confusion resulted from incorrect use of the term "density". The correct meaning is obtained by replacing in this paper the term "critical density" by "critical porosity" or "critical void ratio".*

No. D-34

DISCUSSION (By Letter)

THE DETERMINATION OF THE PRE-CONSOLIDATION LOAD AND ITS PRACTICAL SIGNIFICANCE

Dr. Arthur Casagrande, Graduate School of Engineering, Harvard University, Cambridge, Mass.

In reply to numerous questions on this subject which were addressed to the writer from Members of the Conference, the following notes were written.

Determination of the Pre-Consolidation Load. Professor Terzaghi's early investigations on the mechanics of consolidation of fine-grained soils led him to the conclusion that the relationship between void ratio and pressure for the primary or virgin branch of the compression curve could be expressed by a logarithmic curve. Extensive testing of undisturbed clay samples during the past five years have shown that such a logarithmic relation holds true at least up to 20 kg/sq cm, that is for the entire load range in which the civil engineer is interested. Any important deviations from the virgin compression curve of an undisturbed clay sample seem to be caused by the variations in loading which the soil underwent during its geologic history and by its removal from the ground. The reason for this can be understood from the shape of a rebound and re-compression curve obtained by loading a sample in increments well beyond the stress under which it was consolidated in the ground, then decreasing the load to zero and again gradually increasing it to an even larger load. The compression diagram for such a test is shown in Fig. 1. The left diagram is plotted to an arithmetic scale and in the diagram to the right the pressures are plotted on a logarithmic scale. The semi-log plot lends itself readily to an analysis of the history of the sample. The first portion (II) of the compression curve is in reality a re-compression curve which meets the virgin branch (I_b) and then continues along that branch as a straight line. At the arbitrary load, corresponding to point A, the load is again reduced in the same increments to zero, whereby a rebound curve (III) is obtained. The renewed application of the load follows the re-compression curve (IV) which meets the virgin branch (I_c) at a load higher than point A, to continue along that line.

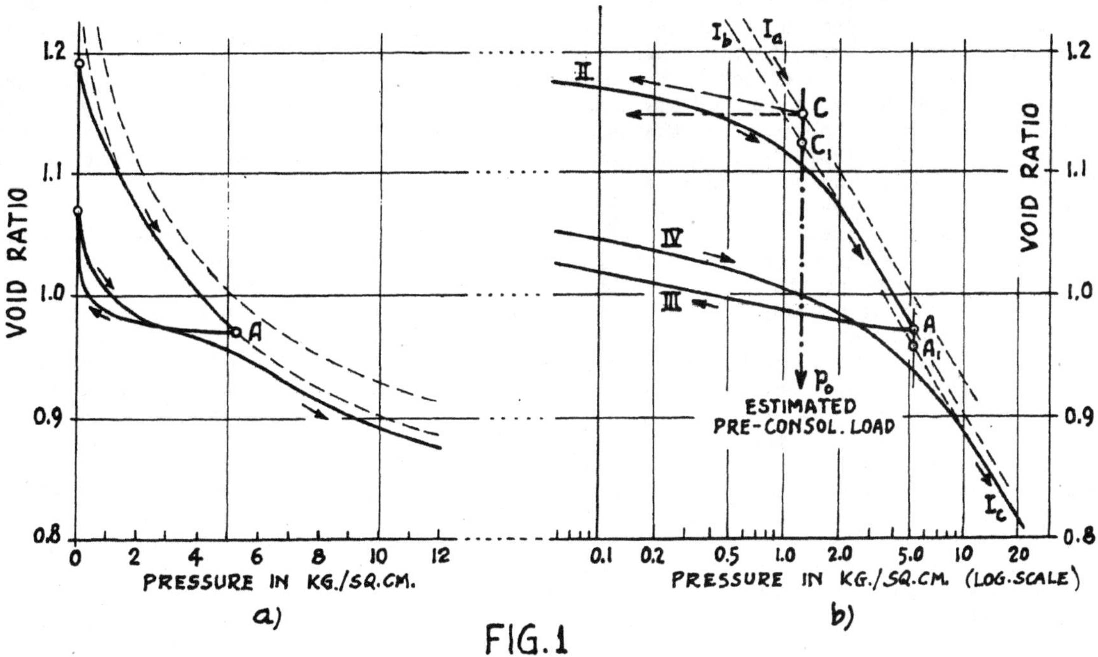

FIG. 1

The diagram shown in Fig. 1b is typical for all very fine-grained soils. The magnitude of the drop in the position of the virgin branch after each rebound loop depends chiefly on the structural characteristics of the soil. For many glacial clays this drop is very small. The exact position of the virgin branch obtained from a laboratory test, depends also on the time-increments which were allowed for each load-increment, and on the temperature.

The close similarity in the shape of branch (II) and the relative position of branch (II) and (I_b), with the shape of the recompression curve (IV) and the relative position of (IV) and (I_c), suggests that it should be possible to estimate the load p_o under which the soil was consolidated in the ground, the so-called pre-consolidation load, from a properly conducted consolidation test. Since the theoretical shape of the rebound and recompression curves is not sufficiently investigated, such estimates must be based on experience. From a large number of tests on different types of soils it was found that for the majority of clays the pre-consolidation load can be derived with a satisfactory degree of accuracy by means of the empirical method shown in Fig. 2. One determines first the position of the virgin compression line (I) with a sufficient number of points. Then one determines on the preceding branch (II) that point (T) which corresponds to the smallest radius of curvature, and draws through this point a tangent (t) to the curve, and a horizontal line (h). The angle α between these two lines is then bisected, and the point of intersection (C) of this bisecting line (c) with the virgin line (I) determined. Point (C) corresponds approximately to the pre-consolidation load (p_o) of the soil in the ground.

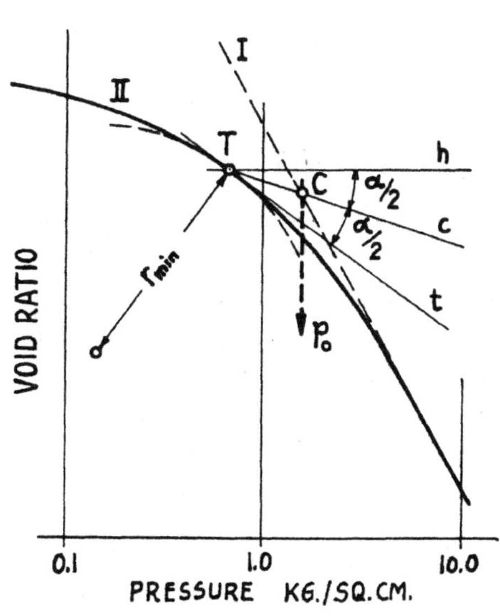

FIG. 2

The question immediately arises whether the drop in the position of the virgin compression line, due to the temporary removal of the load and the inevitable deformation of the sample during its removal from the ground and preparation for the test, affects the magnitude of the estimated pre-consolidation load. So far, experience seems to indicate that this is not the case. In other words, the partial break-down in the internal structure of the soil, due to a small amount of deformation, or alternate swelling and compression, does not obliterate or seriously distort the impression created in the material by the largest previous load.

Theoretical considerations, based on the writer's hypothesis of the structure of clays (Journal of the Boston Soc. of Civ. Eng., April 1932), lead to the same conclusion. The slight increase in compressibility is probably due to the breakdown of a small percentage of soil arches. However, in the major portion of the soil the structure is still intact and, therefore, the impression produced by the pre-consolidation load can be assumed to be unchanged.

It may be objected that the re-compression curve (II) which we obtain by removing a soil sample from the ground and then loading it again in the laboratory, is not identical with a recompression curve (IV) obtained in a consolidation apparatus without removing the sample during the test. Curve (II) is, indeed, the combined effect of the deformation of the sample during the sampling and testing operations and of an unknown amount of swelling. Even with the most careful procedure for obtaining undisturbed samples one cannot prevent the sample from being slightly deformed, because the principal stresses in the ground, particularly their ratio, differs from that in the sample after it is removed from its contact with the surrounding mass of soil.

A certain amount of swelling will take place in samples removed from drill holes, even if no water is present in the hole. This is due to the fact that the outer shell of the cylindrical samples is always disturbed and has a smaller intrinsic pressure than the undisturbed core. As a result the core will absorb some of the water from the outer shell to equalize the internal stresses. The amount of swelling is, of course, more important when free water is present in the hole, particularly when sampling operations are interrupted. While years ago we have tried to prevent such swelling by removing the water prior to sampling, we realize to-day, that the unbalanced head of water may under certain conditions cause such large deformations of the soil at the bottom of the hole that of the two evils the swelling due to a water-filled hole is to be preferred, provided the boring and sampling operations proceed without interruption.

While in the majority of tests which were performed in the Harvard Soil Mechanics laboratory, the shape of curve (II) is similar to re-compression curves of type (IV) obtained by a rebound loop in the consolidation apparatus, one observes in some cases that the initial re-compression curve (II) is steeper or less steep than curve (IV). For example, in paper C-6, Fig. 5 (Vol. II, p. 110) the first curve from the top shows a very flat initial re-compression curve with a perfect approach to the virgin compression line. This sample was taken by hand and the amount of swelling and disturbance was exceptionally small. The slope of the rebound curve indicates that a re-compression curve determined in the consolidation apparatus after a complete rebound, would be much steeper than the initial re-compression curve.

Undisturbed samples obtained from drill holes often display relatively steep initial re-compression curves which approach the virgin compression line more gradually. To what extent this is caused by partial disturbance of the internal structure of the clay and to what extent by swelling, cannot at present be decided from the shape of the curve. Often irregularities in the compression curves are due to irregular load or time increments. This may lead to serious deviations in the position of point (T) in Fig. 2. An experienced observer can easily correct the shape of curve (II) such that it will correspond in character to the ordinary shape and then apply the graphical procedure shown in Fig. 2.

In general, the most satisfactory load increments are such that each increment is exactly twice the preceding one. Each increment should remain in action for the same time period. Small variations can be corrected by extrapolation from the time-compression curves.

Irregularities in the temperature of the sample during the test are also reflected by irregularities in the compression curves and the time curves. It is not only advisable to keep the sample at a constant temperature, but preferably at a temperature which corresponds to that of the soil in the ground. Higher temperatures than the ground temperature causes the virgin compression line to move to the left (Fig. 1b) and the estimated pre-consolidation load to drop. This effect is particularly noticeable for organic silts and organic clays, but has not yet been sufficiently investigated to permit more detailed statements.

Whenever the slope of the secondary time effect on a semi-log plot (see H. Gray, Paper No. D-14, Vol. II) has been observed accurately for each load increment and at about the same temperature, it is possible to utilize the fact that this slope is considerably smaller (about one-third) for the re-compression curves, than for the virgin compression curve, for an approximate determination of the pre-consolidation load. This method is at the present state of development rather inaccurate, but permits a desirable check, particularly in such instances when the compression curves have an unusual shape.

It is evident from the foregoing discussion that a satisfactory interpretation of a consolidation test for the determination of the preconsolidation load requires experience. Refinements permitting greater accuracy undoubtedly result from further systematic studies of this question. Investigators are advised to use every instance in which the pre-consolidation load of clay samples is definitely known, for checking it against laboratory determinations. Among other things the effect of deformation of the samples require further studies. This could be done by consolidating samples in the laboratory, deforming them to various degrees after removal from the consolidation device and then subjecting them to a second consolidation test.

<u>Practical Significance of the Pre-Consolidation Load.</u> The most important practical application of the pre-consolidation load is in connection with settlement analyses. In addition, knowledge of this load may be of interest in connection with geological investigations.

It is a fortunate fact that the slope of the virgin compression branch on a semi-log plot is not noticeably affected by the swelling and minor deformation of the sample. Hence, if the soil is completely consolidated under the present overburden (p_1) the compression due to any additional stress Δp is easily obtained from Fig. 3 by multiplying the ratio $\frac{e_1 - e_2}{1 + e_1}$ with the thickness of

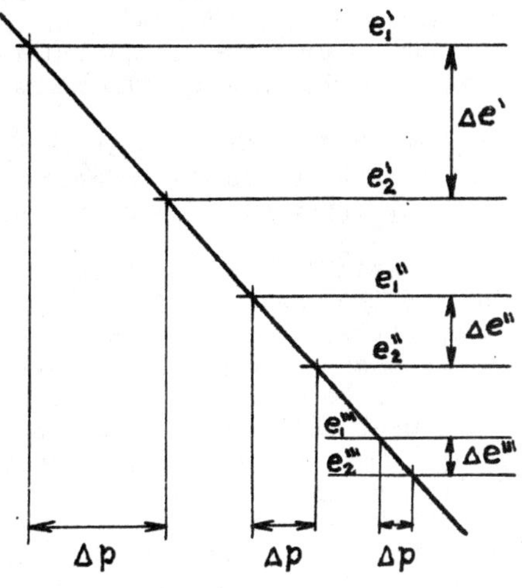

$$\text{COMPRESSION} = \frac{e_1 - e_2}{1 + e_1} \cdot 100\ \%$$

FIG.3

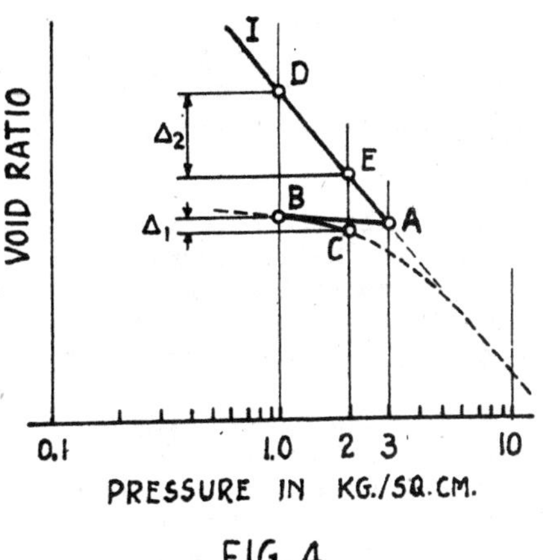

FIG.4

FIG.5

the soil layer. One can readily see from Fig. 3 that on account of the logarithmic pressure scale the amount of the compression for a given increase in stress Δp is decreasing with increasing overburden.

If the largest overburden to which the soil has been exposed during its geologic history has later been partially eroded, the compression due to the same load increase Δp will be very much smaller, in spite of the fact that the virgin compression curve remains unchanged. For example, (Fig. 4), if a clay layer has been compressed at one time by an overburden of 3 kg per sq cm which was later reduced by erosion to 1 kg per sq cm, and then again increased by a building load to 2 kg per sq cm, the compression under the building load will take place along the very flat re-compression curve from point B to C, equal to Δ_1. However, if the clay were consolidated only under the present overburden of 1 kg per sq cm, that is along the virgin compression curve to point D, then the additional stress would result in a compression from D to E, equal to Δ_2, which is between five and ten times Δ_1. This example should suffice to illustrate the fundamental importance of a careful study of the preconsolidation load in connection with settlement analysis.

In this connection it may be of interest to cite an observation the writer has recently made, which may lead to a better understanding of the stress conditions in massive clay deposits. When investigating the character of a deposit of glacial clay, about 100 ft thick, overlying rock, it was found that the preconsolidation load in the lower 50 ft was nearly constant; or in other words, that the bottom of the clay stratum was consolidated only under a stress corresponding to about one half of the overburden. While it is quite possible that recent deposits are still consolidating under their own weight, this could definitely not have been true in the case under consideration on account of the age of this deposit. Besides, the stress distribution would have to appear quite different if the clay is incompletely consolidated.

The writer's tentative explanation for such stress conditions is illustrated in Fig. 5. It is assumed that the water in the underlying rock is under a pressure p_W in excess of the hydrostatic pressure, that means it would rise in a piezometer tube to a height p_W (in metric system) above the free ground water surface. Consequently, the pressure in the solid portion of the clay would be reduced at the rock surface by the amount p_W, while at the upper surface of the clay stratum there would be no reduction, with a linear distribution of the reduction between the lower and upper clay surfaces.

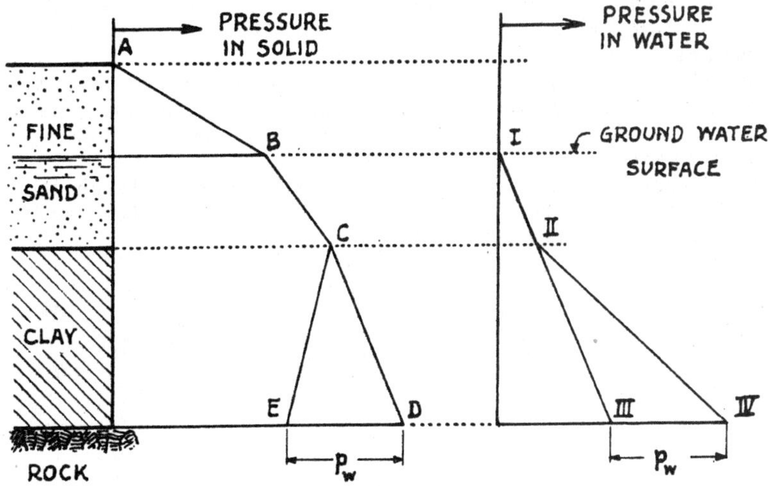

A comparison of the normal stress distribution, assuming the rock surface impervious, and the abnormal stress distribution as produced by an excess-hydrostatic pressure in a pervious rock, will clarify this paradoxical effect. If the clay is consolidated under the existing overburden, the stress distribution in the solid portion is represented in Fig. 5 by A-B-C-D, and in the water by I-II-III. A water pressure in the rock of p_W in excess of the hydrostatic pressure will change the stress distribution in the final state of consolidation in the liquid portion to I-II-IV and in the solid portion to A-B-C-E. In this case, there will be a continuous (steady) flow of water in an upward direction through the clay stratum. (This flow is not related to the flow of water in a partially consolidated clay stratum.)

In the example shown in Fig. 5, the pressure p_w is assumed so large that the condition of equilibrium would result in a clay stratum which is getting softer with increasing depth. The value of p_w could be chosen such that the clay would be consolidated to the same pressure over the entire thickness, or by a further reduction in p_w the clay should become stiffer with depth, although the normal increase in consistency can only be reached by reducing p_w to zero.

If a clay stratum is divided by a sand layer into two separate layers, and if the sand layer is somehow in communication with free surface water, then excess-hydrostatic pressure in the underlying rock may lead to a condition in which the pre-consolidation pressure of the upper clay layer increases at its normal rate with depth, while it decreases with depth for the lower layer.

Shall we consider such conditions as rare exceptions or is there a possibility that, at least in a mild degree, they are a frequent occurrence? In the writer's opinion the presence of considerable water pressures in rock at greater depths should be considered the rule, rather than the exception. However, only under certain conditions will the quantity of this flow and the pressures become very noticeable on the surface (hot springs, geysers). The normal condition will be a gradual reduction in this pressure, until it drops to atmospheric pressure in some open veins near the surface. Only if we cover the rock surface over a large area with an impervious or nearly impervious blanket, would the pressure have a chance to build up. This is precisely the result of clay overlying rock. The amount and uniformity of pressure which may develop depends to a large extent on the vicinity of rock outcrops, or of very pervious rockstrata at greater depths (large fissures) which may drain the flow to some point where it connects with free water (e.g. a subaqueous spring). It is frequently observed that drill holes extending through a clay stratum into rock, will show a water level in the drill hole considerably higher than the outside water level (either ground water or the surface of an open body of water). But even then when the water level in the drill hole does not rise higher than the free water surface, or when it remains considerably lower, it is no indication that the water in the rock is not under larger pressures. In a homogeneous mass of rock the quantities of water which seep through the rock may be so small that they cannot readily be measured; and yet, if the rock is covered with an almost impervious blanket of clay, this pressure can be fully active. The opening up of a hole would then cause locally a considerable drop in pressure, or may even show hardly any water, so that a casual inspection would lead to the conclusion that there is definitely no excess-hydrostatic pressure in the rock. And yet, if a pipe with water tight joints were sunk through the clay into the rock, filled with water and connected to a manometer, the pressure would gradually increase until it would indicate the actual pressure conditions in the rock.

<u>Summary</u>. (1) Most fine-grained, compressible soils, particularly all clays, seem to have their geologic history recorded like a photograph in an undeveloped negative. One of these factors, the largest overburden beneath which the soil has once been consolidated, the pre-consolidation load, is of importance in connection with settlement analyses. The method outlined in this report makes it possible for an experienced investigator to determine approximately the pre-consolidation load from a test on an undisturbed soil sample.

(2) The slope of the virgin compression curve seems to be very little affected by swelling and minor disturbances due to the sampling and testing operations. This is a fortunate fact because it permits a fairly reliable and simple determination of the compressibility of a soil for such cases in which it is known from geological evidence that the present overburden is practically identical with the pre-consolidation load. For such cases the entire increase in load should be considered to take place along the virgin compression curve and the shape of the approach to the virgin curve can be disregarded. Thus, for example, the amount of swelling of an otherwise undisturbed sample has practically no influence on the result.

(3) In contrast to the slope, the position of the pressure-void ratio curve for a given undisturbed soil sample may vary between considerable limits, depending on the sensitivity of the soil to swelling, alternate loading and removal of load, process of loading, the time which each increment is permitted to rest on the sample, and finally the temperature.

(4) Investigators are advised to use every opportunity to study more in detail the interrelationship between pre-consolidation load and the results of laboratory consolidation tests; further, the effects of alternate loading and removal of load, minor deformation of the sample, process of loading and temperature on the position of the virgin compression curve.

(5) A hypothesis is suggested to explain the fact that in some clay deposits which are old enough to be consolidated under their own weight, the consistency does not increase in proportion with depth, as one should expect, but is about constant or may even decrease with depth.

SECOND CONGRESS
ON LARGE DAMS
WASHINGTON, D. C., 1936

CALCULATION OF THE STABILITY OF EARTH DAMS*

WOLMAR FELLENIUS

Professor in Hydraulic Structures at the Royal Technical University, Stockholm; Lieutenant Colonel in the Royal Corps of Engineers

Sweden

To the first Congress on Large Dams, held in Stockholm in 1933, the writer contributed a paper on question IIa, viz: Research Methods for Ascertaining Whether a Given Material is Suitable for Use in the Construction of an Earth Dam (Report No. 23). As the question was not formulated to include calculations on the stability of dams, the writer confined himself to the observation that such calculations were often founded upon the basis of curved sliding surfaces, and a reference to a published work of his (1926–27) concerning such calculations.[1]

As some other reports submitted to the same Congress dealt more fully with calculations, the ensuing discussion on question IIa was diverted from the research methods of the materials to the subject of calculations, and in the first of the resolutions on this question the Congress requested that questions concerning the calculation of the stability of earth dams should be accorded special attention in the discussions at a forthcoming session.

Notwithstanding that some discussion took place at the first Congress concerning calculations, the writer considers it advisable to refer here to some of the fundamental principles presented in the aforementioned work.[1]

**Calculs de stabilité des barrages en terre.*
Berechnung der Stabilität von Erddämmen.
Cálculo de estabilidad de las presas de tierra.

[1] *Fellenius, W.: Jordstatiska beräkningar med friktion och kohesion för cirkulärcylindriska glidytor. Kungl. Väg- och Vattenbyggnadskårens 75-årsskrift p. 79–127. Stockholm 1926. Translated to German: Erdstatische Berechnungen mit Reibung and Kohäsion (adhäsion) und unter Annahme kreiszylindrischer Gleitflächen. Berlin 1927. (Earth statical calculations with friction and cohesion and upon supposition of circular cylindrical sliding surfaces.)*

All calculations concerning stability are founded on the assumption that the sliding surfaces are circular-cylindrical. The forces working in these sliding surfaces are assumed to be cohesion alone—the tangential force independent of the normal pressure—or combined cohesion and friction.

In the latter case we have the Coulomb equation

$$T=k+fN; \tag{1}$$

where

T=the tangential sliding force per unit of the sliding surface;

k=the cohesion per unit of the sliding surface;

f=the friction coefficient (=tgφ; φ being the angle of internal friction);

N=the normal pressure per unit of the surface.

The first case—pure cohesion—is a special instance of the latter, with $f=0$.

Calculations generally have for their object the determination of the properties of earth necessary for equilibrium in different given sliding surfaces. That sliding surface which needs the greatest cohesion plus the greatest angle of friction for equilibrium is the most dangerous of all, and the degree of security varies as to the ratio of the existing to the needed strength of the earth.

I. Pure Cohesion Earth

In the case where only cohesion exists in the earth the calculations are very much simplified. The equation of equilibrium for the turning movements round the axis of the cylindrical sliding surface ADC is (see fig. 1):

$$Pa=k\ s\ r \tag{2}$$

and

$$k=\frac{Pa}{sr} \tag{3}$$

where (see fig. 1):

k=the needed cohesion per unit of the sliding surface;

P=the weight of the mass of earth over the sliding surface per unit of length (of the axis);

a=the horizontal distance between the axis (M) of the sliding surface and the centre of gravity (E) of the mass of earth over the sliding surface;

r=the radius of the sliding surface;

s=the length of the sliding surface (the curve ADC);

(See Fellenius, W.: Kaj- och jordskreden i Göteborg (Quay- and landslides in Gothenburg), Teknisk Tidskrift 1918, V. U. n. 2, p. 17–19, Stockholm 1918.)

For a plane slope the calculation can be worked out numerically and, only with respect to sliding surfaces passing through the foot point of the slope, the needed cohesion in a homogeneous earth may be expressed by:

$$k=\frac{\gamma h}{4}c; \tag{4}$$

where

γ=the unit weight of the earth;

h=the height of the slope (=the difference in height between the horizontal planes limiting the slope).

c=a factor of cohesion that is a somewhat complicated trigonometrical expression (ϕ (θ, α, ω) on fig. 2), the value of which is arrived at by trial for different angles of slope, θ.

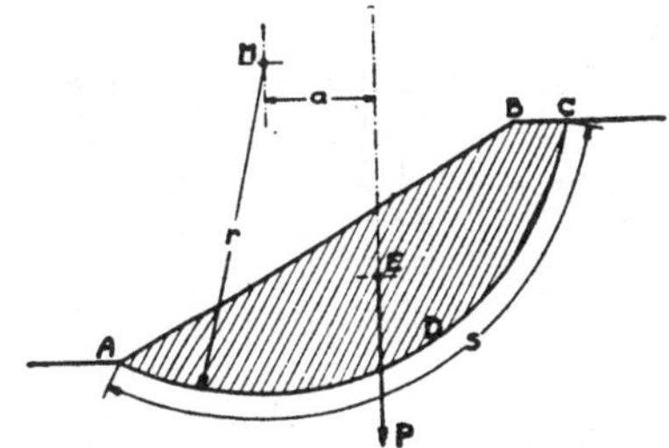

FIGURE 1.—**Stability of slope for cohesion alone for curved sliding surface through the toe of the slope.** — Stabilité du talus, cohésion simple, surface cylindrique de glissement traversant la base du talus. — Stabilität einer Böschung, nur Kohäsion, krumme Gleitfläche durch den Fusspunkt der Böschung. — Estabilidad del talud, cohesión simple, superficie cilíndrica de deslizamiento que atraviesa la base del talud.

The permissible cohesion, k, being given, the possible height, h, of the slope at different angles of slope is obtained from (4):

$$h=\frac{4k}{\gamma}\cdot\frac{1}{c} \tag{5}$$

On figure 2 the factor of cohesion, c, is shown by a full-drawn line for different angles of slope. The left part of this line (for $\theta<10°$)

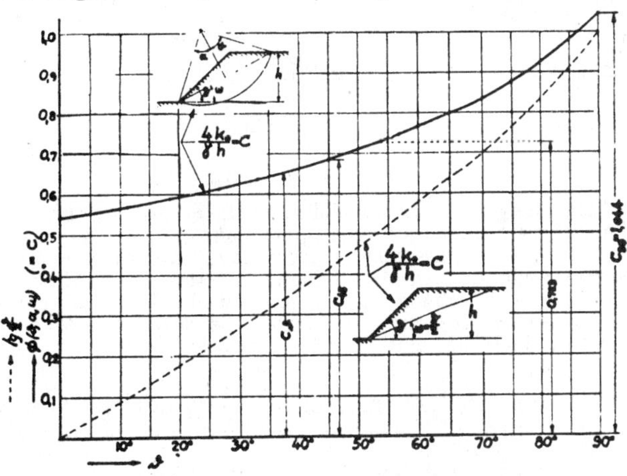

FIGURE 2.—**Diagram showing the factor of cohesion at different angles of slope: Pure cohesion, sliding surface passing through the toe of the slope.** — Diagramme indiquant le facteur de cohésion à des angles différents de pente; cohésion simple, surface de glissement traversant la base du talus. — Diagramm der Grösse des Kohäsionsfaktors bei verschiedenen Böschungswinkeln, nur Kohäsion, Gleitfläche durch den Fusspunkt der Böschung. — Diagrama indicando el factor de cohesión en diferentes ángulos de la inclinación: Cohesión simple, superficie de deslizamiento que atraviesa la base del talud.

has here been adjusted in accordance with an observation recently made to the writer by Mr. Ohde, Berlin. On the same figure the corresponding factor of cohesion, based upon calculations assuming plane sliding surfaces $\left(c=\text{tg}\frac{\theta}{2}\right)$ is shown by a dotted line.

The difference between the results of the two methods of calculation is clearly shown in this figure. One sees that the factor of cohesion is always greater for curved than for plane sliding surfaces. The increase for a vertical wall ($\theta=90°$) is 4.4 percent, but this rapidly increases at diminishing angle of slope, and is for slope 1 : 1 ($\theta=45°$) about 65 percent, and for slope 1 : 2 ($\theta=26°34'$) about 160 percent.

If, for example, one wishes to know the possible height of a slope in $\theta=30°$ and the earth has a given unit weight $\gamma=1.6$ t/m³ and a cohesion $k=1.5$ t/m², one obtains as $c_{30}=0.62$ (from fig. 2) the following height:

$$h=\frac{4\cdot1.5}{1.6}\cdot\frac{1}{0.62}=\sim6 \text{ m}$$

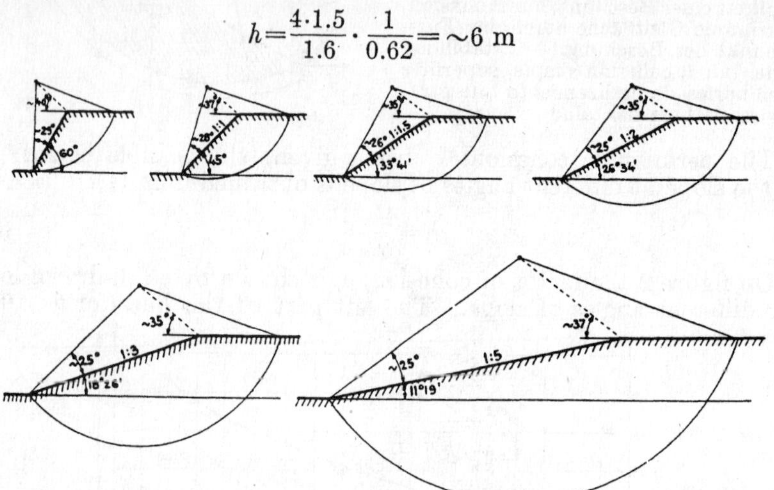

Figure 3.—Most dangerous curved sliding surfaces passing through the toe of the slope at different angles of slope; pure cohesion. — Les plus dangereuses surfaces cylindriques de glissement traversant la base du talus à des angles de pente différents; cohésion simple. — Gefährlichste krumme Gleitflächen durch den Fusspunkt der Böschung bei einigen verschiedenen Böschungswinkeln; nur Kohäsion. — Las superficies cilíndricas de deslizamiento más peligrosas que atraviesan la base del talud en diferentes ángulos de inclinación; cohesión simple.

On figure 3 the most dangerous circular-cylindrical sliding surfaces, drawn through the toe of the slope, are shown for the following slopes:

$\sqrt{3}$: 1 (60°); 1 : 1 (45°); 1 : 1.5 (33°41′);

1 : 2 (26°34′); 1 : 3 (18°26′); and 1 : 5 (11°19′).

It will be found, in nearly all these cases, that the sliding surfaces project below the horizontal plane through the toe of the slope. Naturally, entirely homogeneous material has here been assumed, and the results are therefore not valid in those most common cases where the ground on which the dam rests differs in character from the dam fill. In such cases one must determine (1) the most dangerous sliding surface that does not touch the underlying ground, and (2) the most dangerous of the sliding surfaces that pass through both the dam fill and the underlying ground, and then try to ascertain which of these two is the more dangerous.

Still more dangerous sliding surfaces may—if the underlying ground is of similar or of looser material than the dam fill—be drawn through points beyond the toe of the slope. (See fig. 4.)

It has been shown that in a homogeneous, pure cohesion material, the most dangerous sliding surface at angles of slope less than

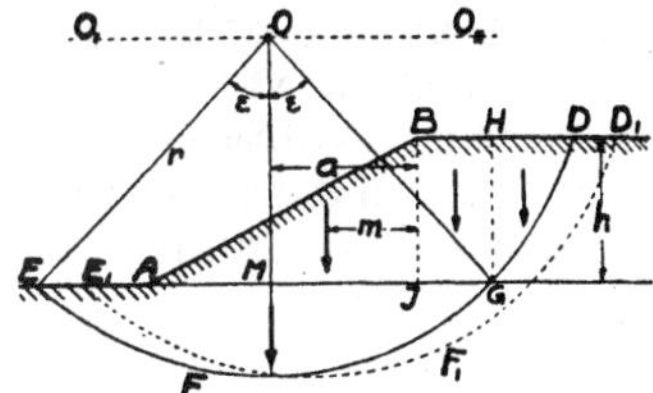

FIGURE 4.—**Stability of slope; pure cohesion, curved sliding surfaces passing outside the toe of the slope.** — Stabilité du talus; cohésion simple, surfaces cylindriques de glissement traversant des points en dehors de la base du talus. — Stabilität einer Böschung, nur Kohäsion, krumme Gleitfläche ausserhalb des Fusspunktes der Böschung. — Estabilidad del talud; cohesión simple, superficies cilíndricas de deslizamiento que atraviesan puntos fuera de la base del talud.

$\theta=53°$ theoretically is situated in infinity, that is, as deep as possible, and has a center angle$=\sim 133\frac{1}{2}°=\sim 1\frac{1}{2}$ right angle.

The cohesion corresponding to this is:

$$k=\frac{\gamma h}{4}\sin 133°34'=\frac{\gamma h}{4}0.723 \tag{6}$$

From this, for a given cohesion, k, a limit value is obtained of the possible height of the slope:

$$h=\frac{k}{0.18\gamma} \tag{7}$$

II. Friction and Cohesion Combined

In this case the calculations for curved sliding surfaces are very complicated and can scarcely be carried out otherwise than graphically. One may write approximately—especially at small angles of slope—(see fig. 1 and (1) and (2)):

$$Pa\cong(ks+fP)r \tag{8}$$

but often this approximation is not sufficiently close to be permissible for other than preliminary calculations.

The graphic calculation is shown on figure 5. The method is as follows:

The mass of earth over the supposed sliding surface is divided into vertical elements, having weights P_1, P_2, P_3, etc. The forces of cohesion acting at the sliding surface K_1, K_2, K_3, etc., are calculated upon a certain assumption of the amount of the working or permissible cohesion. With the P and K forces a continuous force polygon is drawn. Assuming that the other forces acting at the sliding surface have directions at all points making the angle of friction with the normal to the sliding surface at corresponding places and, therefore, are tangents to a circle having the same center as the sliding surface and a radius$=\sin\varphi\times$[the radius of the sliding surface], a plan of forces and a corresponding line polygon is then drawn.

The loci and the directions of the forces working in the surfaces between the different elements are not previously known, but in drawing the plan of forces and the line polygon one may, within given limits, make certain assumptions concerning them. If the calculation is based upon reasonably similar assumptions it will not influence the result to any great extent.

In the case of equilibrium, the line polygon must be reasonably situated and the force polygon must close itself. Should the force polygon not close itself, new drawings must be made and the originally assumed values of φ or k, or both, must be changed until equilibrium is obtained.

In order to determine the premises for equilibrium of a given slope, one can either assume that the cohesion is known and seek the angle

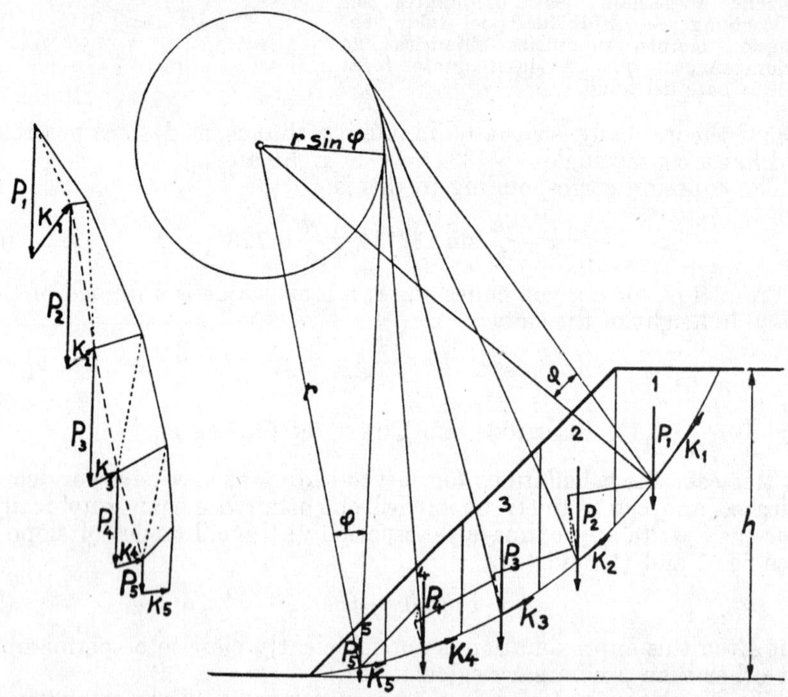

Figure 5.—Stability of slope; friction and cohesion, curved sliding surface through the toe of the slope. — Stabilité du talus, frottement et cohésion, surfaces cylindriques de glissement traversant la base du talus. — Stabilität einer Böschung; Reibung und Kohäsion, krumme Gleitfläche durch den Fusspunkt der Böschung. — Estabilidad del talud, rozamiento y cohesión, superficies cilíndricas de deslizamiento que atraviesan la base del talud.

of friction necessary for equilibrium, or that the angle of friction is known and seek the amount of cohesion necessary for equilibrium. This generally gives an infinite number of combinations and possibilities.

With the object of subjecting the problem to a systematic investigation and of obtaining a general result, the writer has chosen certain slopes and has treated them according to the following method. This is founded on the fact that the constructions at a certain angle of slope in slopes of different heights are exactly uniform, if one assumes values of cohesion proportional to the corresponding heights of the slopes. In all the cases considered herein it is assumed that the slope is just in equilibrium.

For a certain angle of slope, one must first assume that the earth material is a purely cohesive, and calculate the corresponding value of the cohesion k_o, necessary for equilibrium. k_o is calculated as in the preceding chapter (fig. 2 and (4)):

$$k_o=\frac{\gamma h}{4}c \tag{9}$$

(valid for a circular-cylindrical sliding surface through the toe of the slope).

The next step is to try to find, for the same slope, the most dangerous sliding surfaces for the cohesion=0.75 k_o, 0.50 k_o and 0.25 k_o and to determine at the same time the corresponding values of the angle of friction, $\varphi_{0.75}$, $\varphi_{0.50}$ and $\varphi_{0.25}$ just necessary to assure the equi-

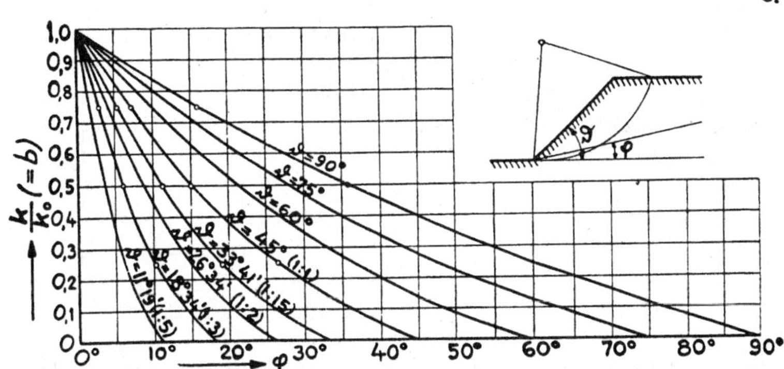

Figure 6.—Diagram showing the connection between angle of slope, angle of friction and the relative value of cohesion; curved sliding surface through the toe of the slope. — Diagramme démontrant le rapport entre l'angle de pente, l'angle de frottement et la valeur correspondante de cohésion; surfaces cylindriques de glissement traversant la base du talus. — Diagramm des Zusammanhanges zwischen Böschungswinkel, Reibungswinkel und Relativwert der Kohäsion; krumme Gleitfläche durch den Fusspunkt der Böschung. — Diagrama demostrando la relación entre el ángulo de inclinación, el ángulo de rozamiento y el valor de cohesión correspondiente; superficies cilíndricas de deslizamiento que atraviesan la base del talud.

librium of the slope. How this is done is shown in figure 5. For a certain value of k, that sliding surface is the most dangerous which needs the greatest angle of friction.

If one draws a curve for the greatest values of friction obtained, $\varphi_{1.00}=0$, $\varphi_{0.75}$, $\varphi_{0.50}$, and $\varphi_{0.25}$ as abscissas, and the corresponding relative values of cohesion, 1.00, 0.75, 0.50, and 0.25, as ordinates, this curve will be typical for the angle of slope in question. Owing to the uniformity already mentioned, this curve is evidently quite independent of the height of the slope and of the gravity of the earth and is valid for any slope of the chosen angle which is just in equilibrium.

Figure 6 illustrates some curves obtained in this way for different angles of slope.

Figure 7 shows the loci of the centers of some of the different sliding surfaces, varying according to changes of angle of slope and of relative value of cohesion. In this figure the toe of the slopes, A, is assumed to remain fixed. The center loci radiate from the centers of the most dangerous sliding surfaces in case of pure cohesion. These sliding surfaces are drawn in the figure. These curves of center loci are asymptotical to the normals to the slopes at their middle points. For $k=0$, that is, pure friction earth, the most dangerous sliding surface evidently is plane and lies in the slope itself, that is $\varphi=\theta$

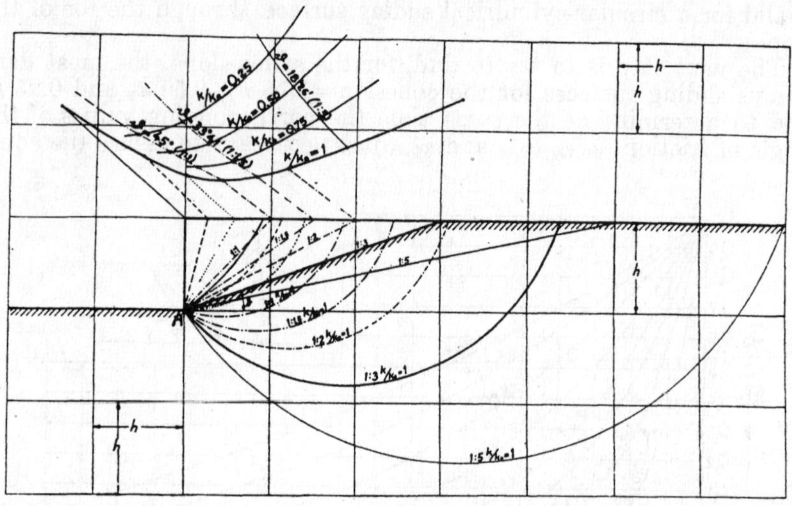

FIGURE 7.—Centers of the sliding surfaces at different angles of slope and different relative values of cohesion; sliding surfaces passing through the toe of the slope. — Les centres des surfaces de glissement à des angles de pente différents et des valeurs de cohésion différentes correspondantes; surfaces de glissement traversant la base du talus. — Die Mittelpunktlagen der Gleitfläche bei verschiedenen Böschungswinkeln und verschiedenen Relativwerten der Kohäsion; Gleitflächen durch den Fusspunkt der Böschung. — Centros de las superficies de deslizamiento en diferentes ángulos de inclinación y de los diferentes valores de cohesión correspondientes; superficies de deslizamiento que atraviesan la base del talud.

Figure 8 shows for the slope 1:3 the loci of the different sliding surfaces intersecting the toe of the slope for the relative values of cohesion: 1, 0.75, 0.50, 0.25, and 0, and the coordinates of the corresponding centers of the sliding surfaces.

We now return to figure 6, in order to show the use of the drawn curves for the calculations of different cases. If in a given case, the angle of friction, φ, and the height, h, up to which the earth is just standing in the angle of slope, θ, are known, and the value of the cohesion of that earth must be ascertained, the proportion between the desired value of cohesion and the value of k_o, valid for the case in question, may be obtained from the corresponding θ curve in figure 6 and from the known value of φ.

Ex. $\varphi = 10°$; $h = 10$ m; $\theta = 33°41'$; $\gamma = 1.6$ t/m³.
From (9) and figure 2 one obtains:

$$k_o = \frac{1.6 \cdot 10}{4} 0.636 = 2.54$$

and from figure 6:

$$k{:}k_o = 0.54 \because k = 0.54 \cdot 2.54 = 1.37 \text{ t/m}^2$$

If, however, φ is not known, but it is known that the earth is standing at the angles of slope$=\theta_1$, and θ_2, up to the respective heights h_1

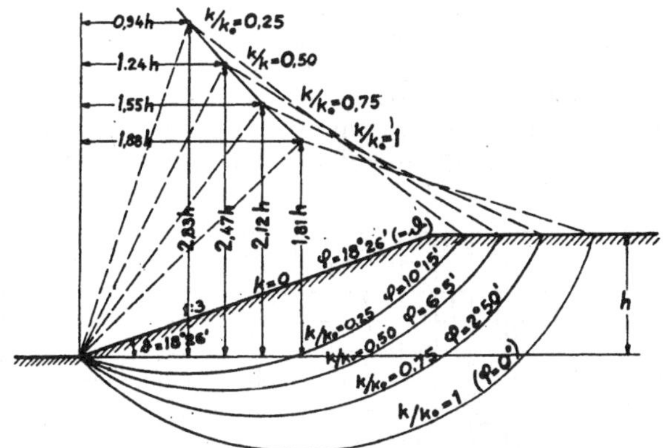

FIGURE 8.—**Displacement of the sliding surface by changing the relative value of cohesion (slope 1:3); sliding surfaces passing through the toe of the slope.** — Le déplacement de la surface de glissement par changement de la valeur de cohésion correspondante (pente 1:3); surfaces de glissement traversant la base du talus. — Die Verschiebung der Gleitfläche bei Veränderung des Relativwertes der Kohäsion (Böschung 1:3); Gleitflächen durch den Fusspunkt der Böschung. — Desplazamiento de la superficie de deslizamiento al cambiar el valor de cohesión correspondiente (inclinación 1:3); superficies de deslizamiento que atraviesan la base del talud.

and h_2, the maximum height h_3 at the angle of slope θ_3 may be ascertained in the following manner:

From figure 2 are obtained for θ_1, θ_2, and θ_3 the values of $\frac{4k_o}{\gamma h} =$ c_1, c_2, and c_3;

$$\because k_{o_1} = \frac{\gamma h_1}{4} c_1;\ k_{o_2} = \frac{\gamma h_2}{4} c_2 \text{ and } k_{o_3} = \frac{\gamma h_3}{4} c_3 \tag{10}$$

In figure 6 the θ_1, θ_2, and θ_3 curves may now be sought or drawn by interpolation. If, now, the unknown value of φ the still unknown ordinates of the θ_1, θ_2, and θ_3 curves are named b_1, b_2 *and* b_3, the cohesion of the earth is evidently:

$$k = b_1 k_{o_1} = b_2 k_{o_2} = b_3 k_{o_3} \tag{11}$$

from which, as a beginning, with the aid of (10) one obtains:

$$\frac{b_1}{b_2} = \frac{k_{o_2}}{k_{o_1}} = \frac{h_2 c_2}{h_1 c_1} \text{ and } b_1 = \frac{h_2 c_2}{h_1 c_1} b_2 \tag{12}$$

If now, in figure 6, in the vicinity of the place where φ can be supposed to be located, an aid curve is drawn the ordinates of which at every point have the relation to the corresponding ordinates of the θ_2 curve$=h_2c_2 : h_1c_1$ this curve will evidently cross the θ_1 curve at the desired φ value, and the ordinate of the crossing point will be$=b_1$. By drawing this ordinate line, b_2 and b_3 are obtained as the ordinates for the intersection of that line and the θ_2 and θ_3 curves.

From (11) and (10) are obtained:

$$\frac{b_3}{b_1}=\frac{h_1}{h_3}\frac{c_1}{c_3},$$

and from this the desired height,

$$h_3=\frac{b_1c_1}{b_3c_3}h_1$$

(b_1, b_3, c_1, and c_3 now being known).

It is certainly possible to determine the greatest angle of slope, θ_3, that can be reached at a given height, h_3, in the same way, but it will

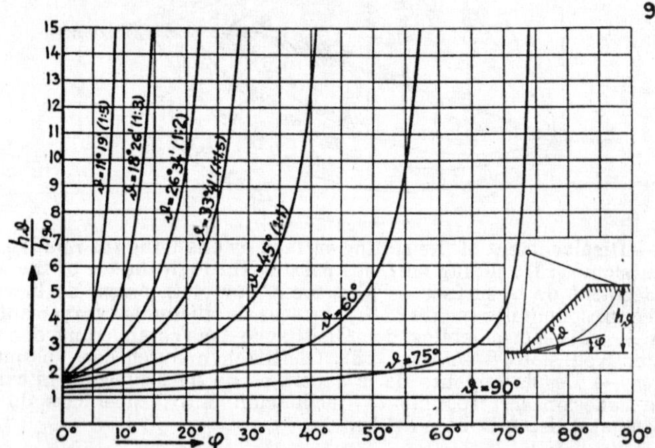

FIGURE 9.—**Diagram showing the relation between permissible heights of the slope for different angles of slope, θ, and for a vertical slope; friction and cohesion combined; circular-cylindrical sliding surface intersecting the toe of the slope.** — Diagramme démontrant le rapport entre les hauteurs de talus tolérées pour de divers angles de pente, θ, et pour le talus vertical; combinaison de cohésion et de frottement, surface cylindrique circulaire de glissement traversant la base du talus. — Diagramm des Zusammenhanges zwischen möglichen Böschungshöhen bei verschiedenen Böschungswinkeln, θ, und bei senkrechter Wand unter Annahme zusammenwirkender Reibung und Kohäsion nebst kreiszylindrischer Gleitfläche durch den Fusspunkt der Böschung. — Diagrama mostrando la relación entre las alturas de los taludes permitidas para diversos ángulos de inclinación, θ, y para el talud vertical; combinación de cohesión y de rozamiento, superficie cilíndrica circular de deslizamiento que atraviesa la base del talud.

require many tests involving a great expenditure of time. For this reason, in figure 9 a diagram has been plotted over the proportion $h\theta : h_{90}$ for different angles of slope, θ, as this proportion has been calculated upon the assumption of circular-cylindrical sliding surfaces intersecting the toe of the slope.

The ordinates in this diagram have been calculated by the aid of figure 2 and figure 6 in the following manner:

For a certain value of θ, c_θ is obtained from figure 2, and from the same figure $c_{90}=1.044$. For the same value of θ and a certain value of φ, one gets $b_{\theta,\varphi}$ from figure 6 and from the same figure $b_{90\varphi}$.

While now

$$k_{o\theta}=\frac{\gamma h_\theta}{4}c_\theta;\ k_{o_{90}}=\frac{\gamma h_{90}}{4}c_{90};$$

and further

$$k=b_{\theta,\varphi}\cdot k_{o\theta}=b_{90,\varphi}\cdot k_{o_{90}},$$

one obtains:

$$\frac{h_\theta}{h_{90}}=\frac{k_{o\theta}}{c_\theta}\cdot\frac{c_{90}}{k_{o_{90}}}=\frac{c_{90}b_{90,\varphi}}{c_\theta b_{\theta,\varphi}}=\frac{1.044\ b_{90,\varphi}}{c_\theta\ b_{\theta,\varphi}}. \qquad (13)$$

Ex.

What angle of slope, θ, should be given to a 10-meter high quite stable earth slope $k=0.5$ t/m^2, $\varphi=10°$, $\gamma=1.6$ t/m^3?

From figure 6 we obtain (for $\varphi=10°$ and $\theta=90°$):

$$\frac{k}{k_o}=0.84;$$

$$k_o=\frac{k}{0.84}=\frac{0.5}{0.84}=0.595\ \text{t/m}^2.$$

From (5) we then obtain

$$h_{90}=\frac{4\cdot 0.595}{1.6}\ 0.958=1.425\ \text{m};\ \because\ \frac{h_\theta}{h_{90}}=\frac{10}{1.425}=7.02,$$

from which (from fig. 9, for $\varphi=10°$) we obtain $\theta=17°$.

This angle of slope is only a little greater than the half of the angle of slope obtainable by a corresponding calculation based upon plane sliding surfaces.

Up to now we have assumed that, in the case of combined cohesion and friction, the sliding surface intersects the toe of the slope. In the case of dams imposed upon fairly good underlying ground this will generally be the case, but in special cases sliding surfaces situated deeper must also be considered.

For pure cohesion earth we have already found that the most dangerous sliding surface is situated very deep (theoretically infinitely deep). However, the cohesion needed in the most dangerous sliding surface through a point a little under the toe of the slope is only slightly less than that needed in the infinitely deeply situated sliding surface ($k=\frac{\gamma h}{4}\ 0.723$; see (6)). As soon as there is only a very little friction in the earth, the most dangerous sliding surface is located relatively high and generally as high as it can be developed, considering support construction such as sheet piling, etc. Thus, while in pure cohesion earth nothing is gained by pressing the sliding surface downwards, it is in friction earth, or in earth with cohesion and friction collaborating, that stability can be increased in just that way.

For the further study of this and allied conditions the writer would refer again to his work "Jordstatiska beräkningar etc.",[2] as well as to an article in Teknisk Tidskrift 1929 V. o. V. n. 5 and 6.[3]

[2] *See p. 445.*

[3] *"Jordstatiska beräkningar för vertikal belastning på horisontal mark under antagande av cirkulärcylindriska glidytor."* (*Earth statical calculations for vertical load on horizontal earth upon supposition of circular cylindrical sliding surfaces.*)

455

The method of calculation here outlined based upon circular-cylindrical sliding surfaces is, of course, approximate, but it gives a good insight into the conditions of stability. Other methods, more exact and more theoretical, have been employed (for example, by Mr. Frontard in report no. 28 at the first Congress and by Mr. F. Jonson at the second Congress on Large Dams). These methods may have certain advantages for entirely homogeneous earths, but to heterogeneous earths, which are of frequent occurrence, such methods are scarcely applicable.

The properties of the fill in earth dams are very often different from those of the underlying ground, and the dam fill itself often

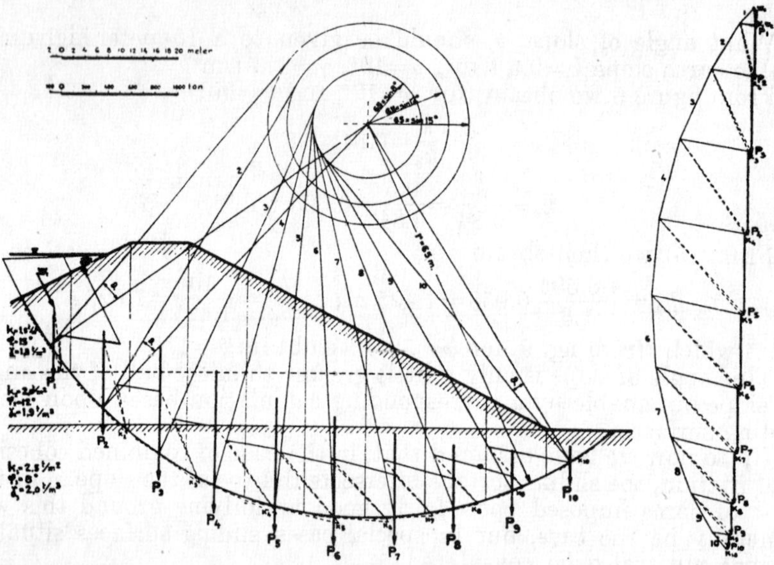

FIGURE 10.—Calculation of a dam with layers having different properties; the upstream side is assumed to be waterproof; cohesion and friction combined. — Calcul d'un barrage dont des couches possèdent des caractéristiques différentes; la face amont supposée impermé able; combinaison de cohésion et de frottement. — Berechnung eines Dammes mit Erdlagern verschiedener Eigenschaften; die Wasserseite ist wasserdicht angenommen; zusammenwirkende Kohäsion und Reibung. — Cálculo de una presa cuyas capas tienen propiedades diferentes; la cara aguas arriba supuesta impermeable; combinación de cohesión y de rozamiento.

consists of layers possessing different properties. When the dam fill is partially dry and partially saturated with water—practically a general condition with earth dams—such methods are completely unsuitable. On the other hand, the method assuming circular-cylindrical sliding surfaces is equally applicable in such cases as when the material is homogeneous. However, the more theoretical methods are recommended for comparison.

Figure 10 shows a graphical calculation of a dam, where the earth is of different kinds, not only in the dam fill itself, but also in the underlying ground. For the sliding surface here shown the water pressure against the upstream slope of the dam is considered. This slope is supposed to be waterproof.

In this case, equilibrium is realized for:

$k_1=1.5$ t/m^2; $\varphi_1=15°$; (upper part of dam fill).
$k_2=2.0$ t/m^2; $\varphi_2=12°$; (lower part of dam fill).
$k_3=2.5$ t/m^2; $\varphi_3=8°$; (underlying ground).

Figure 11 shows a corresponding calculation for the same dam, and the same sliding surface, upon the assumption that the dam fill is saturated up to the line *DEF*. The earth weights of the elements are calculated without taking into consideration the uplift, and the water pressures in the sliding surface are assumed to be working in directions perpendicular to the elements of the sliding surface.

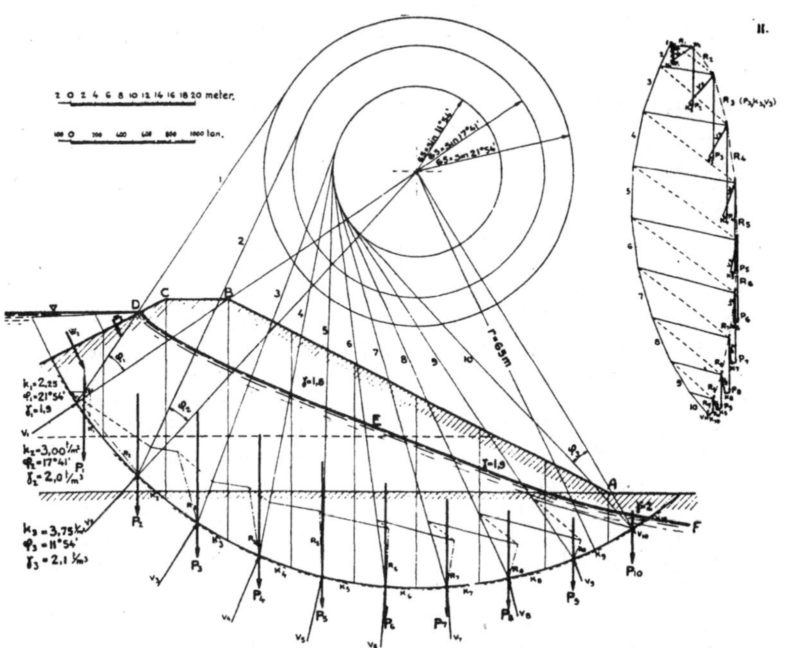

FIGURE 11.—**Calculation of the same dam as in figure 10, but with the fill saturated to the line *DEF*; the needed cohesion and friction about 50 percent greater.** — Calcul pour le barrage de figure 10, mais sans parement étanche, ligne de saturation DEF; la cohésion et le frottement nécessaires sont supérieures d'environ 50% à ceux de la figure 10. — Berechnung deselben Dammes wie in Abbildung 10; die Erde wird bis Linie DEF mit Wasser gesättigt angenommen. Die erforderliche Kohäsion und Reibung sind rd. 50% grösser als in Abbildung 10. — Cálculo para la presa de figura 10 pero sin paramento impermeable, línea de saturación DEF; la cohesión y el rozamiento necesarios son aproximadamente 50% superiores que los de la figura 10.

In this latter case, cohesion and friction needed for equilibrium are about 50 percent greater than in the former case (fig. 10).

The writer has not had the opportunity of ascertaining the *most* dangerous sliding surfaces in these cases, but it is evident that the same sliding surface cannot be the *most* dangerous in both cases. He is, however, convinced that the relation between the cohesion and friction really needed in both cases will be about the same as stated above.

III. Security

For determining the degree of security two somewhat different methods might be used.

One of these is to ascertain, by calculation, the cohesion, k_n, and the friction coefficient, f_n, respectively, needed at the boundary of equilibrium, and compare them with the existing cohesion, k_e, and friction coefficient, f_e, respectively.

The degree of security is then:

$$s = \frac{k_e}{k_n}, \text{ resp.} \cdot \frac{f_e}{f_n}$$

The second method is to choose in advance a certain required degree of security, s, and then to test the sliding surfaces upon the assumption of a permissible cohesion $= \frac{k_e}{s}$ and a permissible friction coefficient $= \frac{f_e}{s}$, respectively, and make sure that these values are in no case being exceeded.

In case of a homogeneous material it does not matter which of these methods is used, but in the case of sliding surfaces passing through layers having different properties—especially with regard to cohesion and friction combined, the second method is more convenient.

Prof. Terzaghi, in his interesting report (no. 18) at the first Congress on Large Dams, introduced the expression "instability factor" (in German: Rutschneigung). This could be defined as the proportion between the values of cohesion and friction, as obtained and verified by laboratory test, and the ones really existing in the material.

The writer would suggest that such an "instability factor" would evidently depend upon the methods of testing employed. As it must be hoped that it will be possible in due time to determine successfully the real coefficients by laboratory tests, this factor ought not to be confined to the different kinds of earth, except in connection with a certain method of testing. For that reason one may trust that this factor will gradually disappear.

By exact calculations, for example, on landslide which has occurred, one can obtain relatively exact values of the coefficients. Compare, for example, the calculations in "Erdstatische Berechnungen" concerning landslides in Gothenburg Harbor. In such cases the "instability factor" is to be considered as included in the calculated coefficients.

But it must also be noted that in laboratory tests the difference between undisturbed and disturbed samples of the earth must be taken into consideration, as well as whether the testing (that is, the determination of the resistance to shear) has been carried out rapidly, before the percentage of water has had time to adjust itself to the normal pressure, or slowly, that is, only after the percentage of water has diminished in response to the normal pressure.

For the dam fill, evidently the results of tests on disturbed samples from the earth fill are to be used, but for the underlying ground the results from tests on undisturbed samples are indicated. In the latter case it is generally best to use results obtained from rapid testings as—especially in the case of clays—one must suppose that the alterations in the loads will occur before the water percentage has adjusted itself to the normal pressure. Whether the results from rapid or slow testings are to be employed for the dam fill depends upon how rapidly the fill will be executed, and upon the percentage of water it is expected to contain at different stages of loading.

If one works along the above lines, no special "instability factor" seems to be needed. Nevertheless, because the test results are somewhat uncertain, it is convenient to claim a somewhat higher degree of security (>1.5 at the most inconvenient load). But if there is reason to regard the test results as accurate and decisive in a particular case, one can confine oneself to a somewhat lower degree of security; although generally it should be 1.2 for the most inconvenient load.

SUMMARY

The writer refers to the methods of calculation with the assumption of circular-cylindrical sliding surfaces by I, pure cohesion, and II, combination of cohesion and friction.

I. The calculation with the aid of the equation of equilibrium (2) for the turning movement is shown (fig. 1). Further, a diagram for ascertaining the degree of cohesion necessary for different angles of slope (fig. 2) and some typical sliding surfaces (figs. 3 and 4) are given. In (7) the limit of height possible for a given slope is obtained.

II. An approximative formula (equation 8) for this case is given and a graphical method of calculation (fig. 5) is shown. Moreover, a systematic investigation intended to arrive at a general result is referred to. This is based on the fact that the graphical constructions at a certain angle of slope in slopes of different heights are exactly uniform, on the assumption that values of cohesion are proportional to the corresponding heights of the slopes. In all cases herein considered it is assumed that the slope is on the border of equilibrium. The results are given by figures 6 to 9.

In figures 10 and 11, the two cases of an earth dam, with layers having different properties, are treated on the basis of different assumptions concerning saturation. It is demonstrated that for this dam the cohesion and friction needed are about 50 percent greater when the dam fill is saturated than when the upstream side of the dam is watertight.

Finally, methods and calculations for determining the degree of security are dealt with. In this connection, some reflections are made concerning the "instability factor" which is supposed to be gradually becoming unnecessary.

Special consideration must be given to the difference between undisturbed and disturbed samples of earth, also to the differences in results of tests made rapidly or slowly for the shearing resistance to various normal pressures.

RESUME

L'auteur expose des méthodes de calcul en supposant des surfaces cylindriques et circulaires de glissement pour les deux cas:

I. Cohésion simple, et II. Cohésion et frottement combinés.[4]

[4] *Fellenius, W. Erdstatische Berechnungen mit Reibung und Kohäsion (Adhäsion) und unter Annahme kreiszylindrischer Gleitflächen. Berlin, 1927.*

Pour le cas I on présente le calcul fait à l'aide de l'équation d'équilibre (2) pour le mouvement basculant (fig. 1). En outre, on éclaircit à l'aide d'un facteur de cohésion le calcul de la cohésion nécessaire (fig. 2). Quelques surfaces typiques de glissement sont montrées (figs. 3 et 4). Par l'équation (7) on arrive à la limite de hauteur permise pour une pente donnée.

Pour le cas II on donne une formule approximative, équation (8) suivie d'un calcul plus exact en forme graphique (fig. 5). Ensuite on présente une étude systématique du problème faite dans le but d'arriver à un résultat général. Cette étude est basée sur le fait que les constructions graphiques pour des angles de pente donnés et des différentes hauteurs de pente sont uniformes, si l'on se base sur des valeurs de cohésion qui sont proportionelles aux hauteurs de pente respectives. Dans tous les cas envisagés on suppose que le talus touche à l'équilibre.

Les résultats de ces calculs sont reproduits (figs. 6 à 9). L'emploi de ces diagrammes est expliqué d'une façon plus détaillée dans le texte.

On donne ensuite deux calculs graphiques d'un barrage, dont les matériaux de terre de diverses sortes ne se trouvent pas seulement dans le corps même du barrage, mais aussi dans le sous-sol (figs. 10 et 11).

On demontre que dans ce cas la cohésion et la friction doivent avoir des valeurs au moins 50 pour-cent plus grandes lorsque la barrage est saturé d'eau que lorsque la face amont a un parement étanche.

En fin, on expose les méthodes de calcul du facteur de sécurité. À ce sujet on présente quelques réflexions sur le "facteur d'instabilité." On croit qu'un facteur spécial d'instabilité deviendra de moins en moins nécessaire à la suite du développement des méthodes d'investigation des échantillons de terre. Il faut apporter une considération tout spéciale à la différence entre la terre sans détérioration de la matière (à son état naturel) et avec détérioration (travaillée), ainsi qu'aux différences entre les résultats donnés par des essais de détermination de la résistance au cisaillement sous diverses pressions normales, selon que les essais soient faits rapidement ou len tement

ZUSAMMENFASSUNG

Der Verfasser berichtet über die Berechnungsmethoden der Stabilität von Erddämmen unter Annahme kreiszylindrischer Gleitflächen in folgenden 2 Fällen:

(I) Kohäsion. (II) Zusammenhängende Kohäsion und Reibung.[5]

(I) Die einfache Berechnung mit der Gleichgewichtsbedingung (2) der Drehmomente wird in diesem Abschnitt gezeigt; dabei ist eine krumme Gleitfläche durch den Fusspunkt der Böschung angenommen. Weiter wird die Berechnung der erforderlichen Kohäsion mit Hilfe eines Kohäsions-Faktors erklärt; die Grösse des Kohäsionsfaktors ist diagrammässig bei verschiedenen Böschungswinkeln dargestellt; einige typische Gleitflächen werden gezeigt. Weitere Abbildungen zeigen gefährlichste krumme Gleitflächen durch den Fusspunkt der Böschung bei einigen verschiedenen Böschungswinkeln (nur Kohäsion) und die

[5] *Fellenius, W. Erdstatische Berechnungen mit Reibung und Kohäsion (Adhäsion) und unter Annahme kreiszylindrischer Gleitflächen. Berlin 1927.*

Stabilität einer Böschung (nur Kohäsion), wobei krumme Gleitflächen ausserhalb des Fusses der Böschung angenommen sind.

Unter II wird erst eine Annäherungsberechnung, Gleichung (8), angeführt und dann eine genaue graphische Berechnung gezeigt. (Krumme Gleitflächen durch den Fusspunkt der Böschung.) Danach wird eine systematische Behandlung des Problems zur Ermöglichung eines allgemein gültigen Ergebnisses angegeben. Diese gründet sich darauf, dass die graphischen Konstruktionen bei festgestellten Böschungswinkeln und verschiedenartig hohen Böschungen vollkommen gleichförmig sind, wenn man mit Kohäsionswerten arbeitet, die proportional zu den entsprechenden Böschungshöhen sind. In allen hier behandelten Fällen wird vorausgesetzt, dass die Böschung sich im Gleichgewicht befindet.

Die Resultate dieser Berechnungen werden in den Abbildungen 6–9 angegeben, die den Zusammenhang zwischen Böschung und Reibungswinkel und Relativwert der Kohäsion darstellen, weiter die Mittelpunktslagen der Gleitfläche bei verschiedenen Böschungswinkeln und verschiedenen Relativwerten der Kohäsion, ferner die Verschiebung der Gleitfläche bei Veränderung des Relativwertes der Kohäsion. (Böschung 1:3.) Ausserdem wird der Zusammenhang zwischen möglichen Böschungshöhen bei verschiedenen Böschungswinkeln, θ, und bei senkrechter Wand unter Annahme zusammenwirkender Reibung und Kohäsion nebst kreiszylindrischer Gleitflächen durch den Fusspunkt der Böschung veranschaulicht.

Die Anwendung dieser Diagramme wird näher erläutert.

Danach folgen 2 graphische Berechnungen eines Dammes, unter der Annahme, dass die Erde nicht nur in dem Dammkörper selbst, sondern auch in dem Untergrund verschiedenartig ist und dass die Wasserseite wasserdicht ist.

Schliesslich werden Methoden und Berechnungen für die Bestimmung des Sicherheitsgrades behandelt. Hierbei werden einige Ansichten betreffend die "Rutschneigung" ausgesprochen. Es wird behauptet, dass ein besonderer Faktor für die Rutschneigung mit der Zeit bei entwickelten Untersuchungsmethoden für die Erdproben unnötig werden wird. Der Unterschied zwischen ungestörten (gewachsenen) und gestörten (ausgebreiteten) Erdproben muss besonders beachtet werden und ebenso auch der Unterschied zwischen den Resultaten bei schnell und langsam ausgeführten Prüfungen des Scherwiderstandes bei verschiedenen Normaldrücken.

RESUMEN

El autor expone los métodos de cálculo asumiendo superficies cilíndricas y circulares de deslizamiento para los dos casos: I. Cohesión simple y, II. cohesión y rozamiento combinados.[6]

Para el caso I se presenta el cálculo hecho usando la ecuación de equilibrio (2) para el movimiento de oscilación (fig. 1). Por otro lado el cálculo de la cohesión necesaria queda esclarecido usando un factor de cohesión (fig. 2). Algunas superficies típicas de deslizamiento se muestran en las figuras 3 y 4. Por la ecuación (7) se llega al límite de altura permitido para una pendiente dada. Para el caso II se da

[6] *Fellenius, W., Erdstatische Berechnungen mit Reibung und Kohäsion (Adhäsion) und unter Annahme kreiszylindrischer Gleitflächen. Berlin 1927.*

una fórmula aproximativa, ecuación (8), seguida de un cálculo más exacto en forma gráfica (fig. 5). Seguidamente se presenta un estudio sistemático del problema hecho con el objeto de llegar a un resultado general. Este estudio se basa en el hecho de que las construcciones gráficas para *los ángulos de inclinación dados* y *las diferentes alturas de pendiente* son uniformes, si uno se basa en los valores de cohesión *que son proporcionales a las alturas de las pendientes respectivas.* En todos los casos previstos se supone que el talud llega al equilibrio.

Los resultados del cálculo se dan (figs. 6 a 9). El empleo de estos diagramas se explica más detalladamente en el texto.

A continuación se dan dos cálculos gráficos de una presa, cuyos materiales de tierra de diversas clases se encuentran no sólo en el cuerpo de la presa sino también en el subsuelo (figs. 10 y 11). Se demuestra que en este caso la cohesión y la fricción deben tener un valor cuando menos 50% mayor cuando la presa está saturada de agua que cuando el paramento aguas arriba es impermeable.

En fin se exponen los métodos de cálculo del *factor de seguridad.* Sobre este sujeto se presentan algunas consideraciones respecto del "factor de inestabilidad." Se cree que un factor especial de inestabilidad cada vez será menos necesario como corolario de los adelantos en los métodos de investigar las muestras de tierra. Se debe dar una consideración muy especial a la diferencia entre la tierra *sin deterioro de la materia* (en su estado natural) y *con deterioro* (trabajada), así como también a las diferencias entre los resultados dados por los ensayos de determinación de la resistencia a la cortadura bajo diversas presiones normales, según sean hechos los ensayos *rápidamente* o *lentamente.*

Force at a Point in the Interior of a Semi-Infinite Solid

RAYMOND D. MINDLIN, *Department of Civil Engineering, Columbia University*
(Received March 13, 1936)

A solution of the three-dimensional elasticity equations for a homogeneous isotropic solid is given for the case of a concentrated force acting in the interior of a semi-infinite solid. This represents the fundamental solution having a singular point in a solid bounded by a plane. From it may be derived, by a known method of synthesis, the solutions for the semi-infinite solid which correspond to the solutions known as nuclei of strain in the solid of indefinite extent.

1. INTRODUCTION

ONE of the fundamental results in the theory of elasticity is the Kelvin solution[1] for a force applied at a point in a solid of indefinite extent. It is well known that stress distributions for a number of problems of practical importance may be obtained from the Kelvin solution by methods of synthesis and superposition.[2] For example, the problem, first solved by Lamé, of the stresses in a spherical container under uniform internal or external pressure may be treated by superposing stresses derived from Kelvin's results. First we form, from the Kelvin single force solution, the stresses for a double force, that is, a pair of equal and opposite forces acting at neighboring points. By superposing three mutually perpendicular double forces, we arrive at the solution known as the center of compression. Finally, the addition of a uniform tension or pressure to the center of compression leads to the solution of Lamé's problem.

The classical problem of Boussinesq dealing with a normal force applied at the plane boundary of a semi-infinite solid[3] has found practical application in the study of the distribution of foundation pressures, contact stresses, and in certain problems of soil mechanics. It is known that Boussinesq's problem may also be solved by superposing solutions derived from Kelvin's results. To do this we form a line of centers of compression by integrating the stresses for a single center of compression. By combining this result with the original Kelvin single-force solution, the Boussinesq formulas are obtained.

The Kelvin solution may be used in studying stresses due to a force applied at a great distance from a boundary while the Boussinesq solution is applicable in the case where the force acts at the surface. The solutions described in this paper fill in the gap between the two by giving the stresses for the case where the force is applied near the surface. Such a condition is approached in a number of practical problems such, for example, as in the case of a guy wire anchor. Again, by integrating these new solutions along a line, we may approximate the conditions produced by a friction pile or an anchor rod.

Solutions of the elasticity equations such as the single force, double force, center of compression, and line of centers of compression are referred to as nuclei of strain.[2] The number of nuclei of strain which may be obtained by synthesis from the Kelvin solution is unlimited and further combinations lead to the solutions for such important cases as Southwell's problem of the spherical cavity in an unlimited solid under simple tension[4] and Cerruti's problem of a force applied tangentially at the plane boundary of a semi-infinite solid.[5]

Each of the nuclei of strain obtained from Kelvin's results is a solution applicable to a solid of indefinite extent. With the solution for a force in the interior of a semi-infinite solid, as described in this paper, we may apply the method of synthesis to develop a new series of nuclei of strain for which the stresses vanish on the plane boundary of the semi-infinite solid. Such solutions may be termed half-space nuclei of strain.

[1] Sir W. Thomson, Cambridge and Dublin Math. J. (1848). See also A. E. H. Love, *Mathematical Theory of Elasticity*, fourth edition (Cambridge, 1927), p. 183; and S. Timoshenko, *Theory of Elasticity* (New York, 1934), p. 321.

[2] Love, p. 186, and Timoshenko, p. 323.

[3] J. Boussinesq, *Applications des Potentials* . . . (Paris, 1885). Also Love, p. 191, and Timoshenko, p. 331.

[4] R. V. Southwell and H. J. Gough, Phil. Mag. **1**, 71 (1926).

[5] V. Cerruti, Acc. Lincei. Mem. fis. mat., Roma **13**, 81 (1882). Also Love, p. 241.

The notion of such half-space nuclei of strain suggests some interesting possibilities. By employing these half-space nuclei as unit solutions, it may be possible to find solutions analogous to those obtained from the Kelvin nuclei but with the additional feature that a new plane boundary is introduced. For example, the solution analogous to Southwell's problem would pertain to a spherical cavity near the surface of a semi-infinite solid; the solution analogous to the problem of Boussinesq would be that for a force applied at the surface of a body bounded by two perpendicular planes (i.e., the transmission of force through "quarter-space"); the solution analogous to the hollow sphere under uniform pressure would apply to an eccentrically hollow sphere.

The present paper deals with the fundamental solution for the single force in the interior of a semi-infinite solid. The problem is divided into two parts: (1) force normal to the boundary; (2) force parallel to the boundary.

The discovery of the solutions followed a study of H. M. Westergaard's interpretation[6] of the problems of Kelvin, Boussinesq and Cerruti in terms of Galerkin vectors and the results obtained are shown to include these problems for certain limiting conditions.

2. The Galerkin Vector

The components of stress in an elastic isotropic solid were expressed by B. Galerkin[7] in terms of partial differential coefficients of three functions satisfying a fourth-order equation. P. F. Papkovitch[8] and H. M. Westergaard[9] have expressed the displacements and stresses in terms of a vector stress function whose scalar coefficients are the Galerkin functions. The displacements are given, in Westergaard's form of the Galerkin vector, by[9]

$$\begin{aligned} u&=(1/2G)[2(1-\mu)\Delta X-(\partial/\partial x)\,\mathrm{div}\,\mathbf{F}],\\ v&=(1/2G)[2(1-\mu)\Delta Y-(\partial/\partial y)\,\mathrm{div}\,\mathbf{F}],\\ w&=(1/2G)[2(1-\mu)\Delta Z-(\partial/\partial z)\,\mathrm{div}\,\mathbf{F}],\end{aligned}\qquad(1)$$

where: X, Y, Z are the Galerkin functions,
$\mathbf{i}$, $\mathbf{j}$, $\mathbf{k}$ constitute an orthogonal system of unit vectors
$\mathbf{F}=\mathbf{i}X+\mathbf{j}Y+\mathbf{k}Z$ is the Galerkin vector,
Δ is Laplace's operator,
G is the modulus of rigidity,
μ is Poisson's ratio.

The stresses are given by[9]

$$\begin{aligned}\sigma_x&=2(1-\mu)\,\partial\Delta X/\partial x+(\mu\Delta-\partial^2/\partial x^2)\,\mathrm{div}\,\mathbf{F},\\ \sigma_y&=2(1-\mu)\,\partial\Delta Y/\partial y+(\mu\Delta-\partial^2/\partial y^2)\,\mathrm{div}\,\mathbf{F},\\ \sigma_z&=2(1-\mu)\,\partial\Delta Z/\partial z+(\mu\Delta-\partial^2/\partial z^2)\,\mathrm{div}\,\mathbf{F},\\ \tau_{yz}&=(1-\mu)(\partial\Delta Y/\partial z+\partial\Delta Z/\partial y)-(\partial^2/\partial y\partial z)\,\mathrm{div}\,\mathbf{F},\\ \tau_{zx}&=(1-\mu)(\partial\Delta Z/\partial x+\partial\Delta X/\partial z)-(\partial^2/\partial z\partial x)\,\mathrm{div}\,\mathbf{F},\\ \tau_{xy}&=(1-\mu)(\partial\Delta X/\partial y+\partial\Delta Y/\partial x)-(\partial^2/\partial x\partial y)\,\mathrm{div}\,\mathbf{F}.\end{aligned}\qquad(2)$$

For the case of zero body force, the Galerkin vector must satisfy the biharmonic equation,

$$\Delta^2\mathbf{F}=0.\qquad(3)$$

It was observed by Professor Westergaard[6] that the coefficient of a single axially symmetrical component of the Galerkin vector is identical with Love's function[10] for strain symmetrical about the corresponding axis.

3. Nuclei of Strain in Terms of Galerkin Vectors

The solutions given in sections 4 and 5 were obtained by superposition of nuclei of strain for an unlimited solid. It has been found convenient, in applying the method of superposition, to express the nuclei of strain in terms of Galerkin vectors.

In the following list of functions, A, B, C, $\cdots$ H are constants and $R=(x^2+y^2+z^2)^{\frac{1}{2}}$. It may be observed that each of the functions satisfies the biharmonic equation and that the stresses derived from them vanish at infinity.

[6] H. M. Westergaard, *Lectures on Elasticity while Visiting Professor at the University of Michigan* (Summer Session, 1934).

[7] B. Galerkin, Comptes rendus 190, 1047 (1930).

[8] P. F. Papkovitch, Comptes rendus 195, 513, 754 (1932).

[9] H. M. Westergaard, Bull. Am. Math. Soc. 41, 695 (1935).

[10] Love, p. 274; Timoshenko, p. 309.

Galerkin Vectors for Nuclei of Strain in an Unlimited Solid

A. Single force (Kelvin's problem).

(1) $\mathbf{i}AR$ (single force in x direction)
(2) $\mathbf{j}AR$ (single force in y direction)
(3) $\mathbf{k}AR$ (single force in z direction)

B. Double force

(1) $\mathbf{i}Bx/R$ (double force in x direction)
(2) $\mathbf{j}By/R$ (double force in y direction)
(3) $\mathbf{k}Bz/R$ (double force in z direction)

C. Double force with moment

(1) $\mathbf{i}Cz/R$ (double force in x direction with moment about y axis)
(2) $\mathbf{j}Cx/R$ (double force in y direction with moment about z axis)
(3) $\mathbf{k}Cy/R$ (double force in z direction with moment about x axis)
(4) $\mathbf{i}Cy/R$ (double force in x direction with moment about z axis)
(5) $\mathbf{j}Cz/R$ (double force in y direction with moment about x axis)
(6) $\mathbf{k}Cx/R$ (double force in z direction with moment about y axis)

D. Line of double forces with moment

(1) $\mathbf{i}Dz \log (R+x)$ (type $C(1)$ along x axis from $x=0$ to $x=-\infty$)
(2) $\mathbf{j}Dx \log (R+y)$ (type $C(2)$ along y axis from $y=0$ to $y=-\infty$)
(3) $\mathbf{k}Dy \log (R+z)$ (type $C(3)$ along z axis from $z=0$ to $z=-\infty$)
(4) $\mathbf{i}Dy \log (R+x)$ (type $C(4)$ along x axis from $x=0$ to $x=-\infty$)
(5) $\mathbf{j}Dz \log (R+y)$ (type $C(5)$ along y axis from $y=0$ to $y=-\infty$)
(6) $\mathbf{k}Dx \log (R+z)$ (type $C(6)$ along z axis from $z=0$ to $z=-\infty$)

E. Center of compression or dilatation

(1) $\mathbf{i}E \log (R+x)$
(2) $\mathbf{j}E \log (R+y)$ (All three yield the same stresses)
(3) $\mathbf{k}E \log (R+z)$

F. Line of compression or dilatation of constant strength

(1) $\mathbf{i}F[x \log (R+x)-R]$ (along x axis from $x=0$ to $x=-\infty$)
(2) $\mathbf{j}F[y \log (R+y)-R]$ (along y axis from $y=0$ to $y=-\infty$)
(3) $\mathbf{k}F[z \log (R+z)-R]$ (along z axis from $z=0$ to $z=-\infty$)

G. Doublet (double center of compression-dilatation)

(1) $\mathbf{i}G(1/R)$ (axis of doublet parallel to x axis)
(2) $\mathbf{j}G(1/R)$ (axis of doublet parallel to y axis)
(3) $\mathbf{k}G(1/R)$ (axis of doublet parallel to z axis)

H. Linearly varying line of doublets with strength proportional to distance from the origin

(1) $\mathbf{i}H[z \log (R+z)-R]$ (type $G(1)$ along z axis from $z=0$ to $z=-\infty$)
(2) $\mathbf{j}H[x \log (R+x)-R]$ (type $G(2)$ along x axis from $x=0$ to $x=-\infty$)
(3) $\mathbf{k}H[y \log (R+y)-R]$ (type $G(3)$ along y axis from $y=0$ to $y=-\infty$)
(4) $\mathbf{i}H[y \log (R+y)-R]$ (type $G(1)$ along y axis from $y=0$ to $y=-\infty$)
(5) $\mathbf{j}H[z \log (R+z)-R]$ (type $G(2)$ along z axis from $z=0$ to $z=-\infty$)
(6) $\mathbf{k}H[x \log (R+x)-R]$ (type $G(3)$ along x axis from $x=0$ to $x=-\infty$)

4. Force Normal to the Boundary of a Semi-Infinite Solid

The semi-infinite solid is considered to be bounded by the plane $z=0$, the positive z axis penetrating into the body. A force P is applied at point $(0, 0, +c)$ and acts in the positive z direction (Fig. 1).

Since the stress will be symmetrical about the z axis, we need consider only the $\mathbf{k}$ component of $\mathbf{F}$. Transforming to cylindrical coordinates r, θ, z and noting that Z is independent of θ, we find, from Eqs. (1) and (2), the following expressions for the displacements and stresses in symmetrical cylindrical coordinates:[10]

$$
\begin{aligned}
U &= -(1/2G)(\partial^2 Z/\partial r \partial z),\\
w &= (1/2G)[2(1-\mu)\Delta Z - \partial^2 Z/\partial z^2],\\
\sigma_r &= (\partial/\partial z)[\mu \Delta Z - \partial^2 Z/\partial r^2],\\
\sigma_\theta &= (\partial/\partial z)[\mu \Delta Z - (1/r)(\partial Z/\partial r)],\\
\sigma_z &= (\partial/\partial z)[(2-\mu)\Delta Z - \partial^2 Z/\partial z^2],\\
\tau_{rz} &= (\partial/\partial r)[(1-\mu)\Delta Z - \partial^2 Z/\partial z^2].
\end{aligned}
\tag{4}
$$

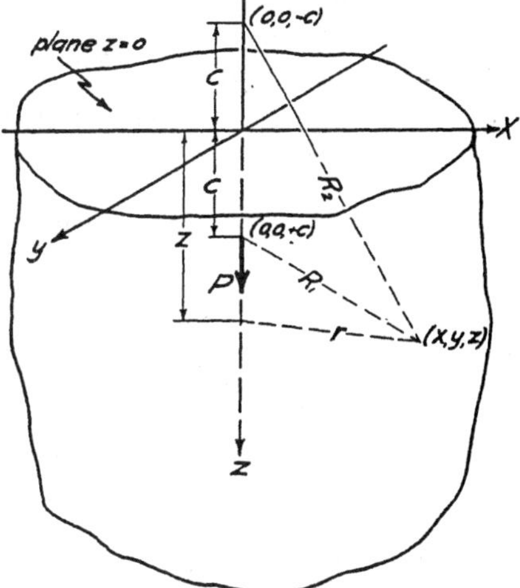

Fig. 1. Force normal to the boundary in the interior of a semi-infinite solid.

We first apply the Kelvin solution at point (0, 0, $+c$) by replacing R with $R_1=(r^2+(z-c)^2)^{\frac{1}{2}}$ in solution $A(3)$ of section 3. We note that stresses σ_z and τ_{rz} are produced on the plane $z=0$. The boundary conditions for $z=0$ are

$$[\sigma_z]_{z=0}=[\tau_{rz}]_{z=0}=0. \qquad (5)$$

It is therefore necessary to add further solutions to remove the boundary stresses and these solutions must introduce no new singularities in the region $z\geqq 0$. Nuclei of strain of types $A(3)$, $B(3)$, $E(3)$, $F(3)$ and $G(3)$, applied at (0, 0, $-c$), are found to satisfy these conditions if the proper values are assigned to the constants A, B, E, F and G.

To transfer nuclei of strain from the origin to (0, 0, $-c$), we replace z with $(z+c)$ and R with $R_2=(r^2+(z+c)^2)^{\frac{1}{2}}$ in the formulas of section 3.

The values of the constants are obtained from the boundary Eqs. (5) and the equilibrium condition

$$P=-\int_0^\infty 2\pi r\sigma_z dr, \quad (z>c). \qquad (6)$$

This procedure leads us to the required Galerkin vector:

$$\mathbf{F}=[P\mathbf{k}/8\pi(1-\mu)]\{R_1+(3-4\mu)R_2-2c(z+c)/R_2-4(1-2\mu)c\log(R_2+z+c) +4(1-\mu)(1-2\mu)[(z+c)\log(R_2+z+c)-R_2]+2c^2/R_2\}. \qquad (7)$$

The solution for a force in the interior of a semi-infinite solid and normal to the boundary is seen to be compounded of six nuclei of strain which, in an unlimited solid, represent: (1) a single force at (0, 0, $+c$); (2) a single force at (0, 0, $-c$); (3) a double force at (0, 0, $-c$); (4) a center of compression at (0, 0, $-c$); (5) a line of centers of compression extending from $z=-c$ to $z=-\infty$; (6) a doublet at (0, 0, $-c$).

Collecting terms in Eq. (7) we obtain:[11]

$$\mathbf{F}=[P\mathbf{k}/8\pi(1-\mu)]\{R_1+[8\mu(1-\mu)-1]R_2+4(1-2\mu)[(1-\mu)z-\mu c]\log(R_2+z+c)-2cz/R_2\}. \qquad (8)$$

Eqs. (8) and (4) yield the displacements and stresses:

$$U=\frac{Pr}{16\pi G(1-\mu)}\left[\frac{z-c}{R_1^3}+\frac{(3-4\mu)(z-c)}{R_2^3}-\frac{4(1-\mu)(1-2\mu)}{R_2(R_2+z+c)}+\frac{6cz(z+c)}{R_2^5}\right],$$

$$w=\frac{P}{16\pi G(1-\mu)}\left[\frac{3-4\mu}{R_1}+\frac{8(1-\mu)^2-(3-4\mu)}{R_2}+\frac{(z-c)^2}{R_1^3}+\frac{(3-4\mu)(z+c)^2-2cz}{R_2^3}+\frac{6cz(z+c)^2}{R_2^5}\right],$$

$$\sigma_r=\frac{P}{8\pi(1-\mu)}\left[\frac{(1-2\mu)(z-c)}{R_1^3}-\frac{(1-2\mu)(z+7c)}{R_2^3}+\frac{4(1-\mu)(1-2\mu)}{R_2(R_2+z+c)}-\frac{3r^2(z-c)}{R_1^5}+\frac{6c(1-2\mu)(z+c)^2-6c^2(z+c)-3(3-4\mu)r^2(z-c)}{R_2^5}-\frac{30cr^2z(z+c)}{R_2^7}\right], \qquad (9)$$

$$\sigma_\theta=\frac{P(1-2\mu)}{8\pi(1-\mu)}\left[\frac{(z-c)}{R_1^3}+\frac{(3-4\mu)(z+c)-6c}{R_2^3}-\frac{4(1-\mu)}{R_2(R_2+z+c)}+\frac{6c(z+c)^2}{R_2^5}-\frac{6c^2(z+c)}{(1-2\mu)R_2^5}\right],$$

$$\sigma_z=\frac{P}{8\pi(1-\mu)}\left[-\frac{(1-2\mu)(z-c)}{R_1^3}+\frac{(1-2\mu)(z-c)}{R_2^3}-\frac{3(z-c)^3}{R_1^5}-\frac{3(3-4\mu)z(z+c)^2-3c(z+c)(5z-c)}{R_2^5}-\frac{30cz(z+c)^3}{R_2^7}\right],$$

$$\tau_{rz}=\frac{Pr}{8\pi(1-\mu)}\left[-\frac{1-2\mu}{R_1^3}+\frac{1-2\mu}{R_2^3}-\frac{3(z-c)^2}{R_1^5}-\frac{3(3-4\mu)z(z+c)-3c(3z+c)}{R_2^5}-\frac{30cz(z+c)^2}{R_2^7}\right].$$

[11] R. D. Mindlin, Comptes rendus 201, 536 (1935).

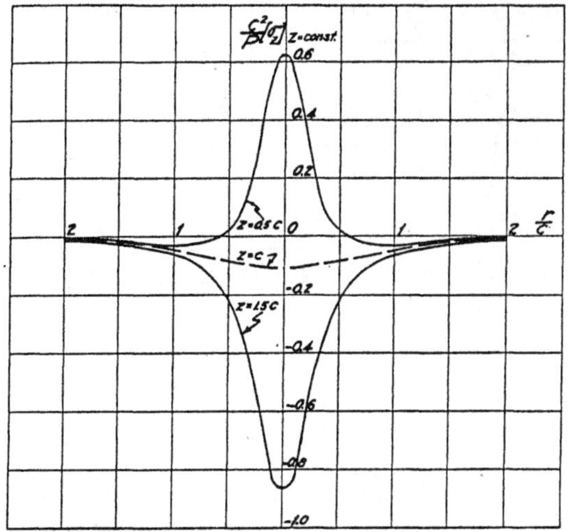

FIG. 2. Normal stress on planes parallel to boundary for the case of a force normal to the boundary ($\mu=0.3$).

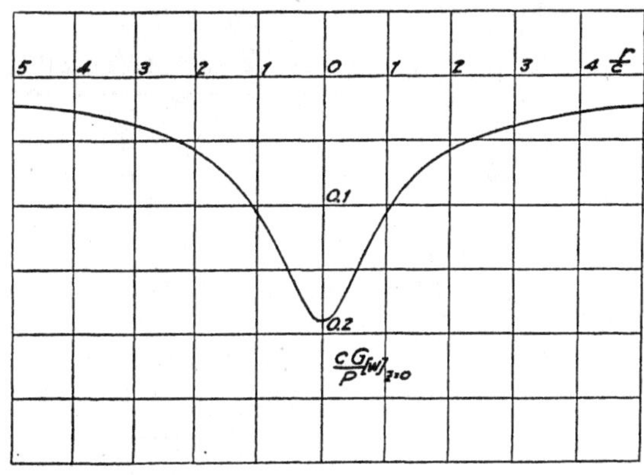

FIG. 3. Vertical deflection of the surface for the case of a force normal to the boundary ($\mu=0.3$).

Fig. 2 shows the distribution of normal stress on horizontal planes, and Fig. 3 shows the vertical deflection of the surface of the semi-infinite solid.

When $c\rightarrow\infty$ all terms in Eqs. (9) containing R_2 vanish and the solution becomes that for Kelvin's problem where the force is applied at $(0, 0, +c)$ in the positive z direction. The corresponding Galerkin vector is

$$\mathbf{F}=\mathbf{k}[PR_1/8\pi(1-\mu)]. \tag{10}$$

When $c\rightarrow 0$ Eqs. (9) give the stresses and displacements for Boussinesq's problem.[12] These stresses are represented by the Galerkin vector

$$\mathbf{F}=\mathbf{k}(P/2\pi)[2\mu R+(1-2\mu)z\log(R+z)]. \tag{11}$$

The solution for the problem of Boussinesq was given by Professor Westergaard[6] in the form of the Galerkin vector

$$\mathbf{F}=\mathbf{k}(PR/2\pi) \tag{12a}$$

combined with the potential function

$$\phi=-[(1-2\mu)/2\pi]P\log(R+z). \tag{12b}$$

The relation between ϕ and $\mathbf{F}$ is given by

$$\text{grad}\ \phi=2(1-\mu)\Delta\mathbf{F}-\text{grad div}\ \mathbf{F}. \tag{13}$$

As a further verification of the solution, we may compare the stresses with those obtained for the corresponding two-dimensional case by E. Melan.[13] While the approach to the solution in three dimensions was simplified through the use of cylindrical coordinates, the comparison with Melan's formulas is facilitated by referring the components of stress to rectangular coordinates. Eqs. (9) then become:

[12] The stresses and displacements for the problems of Boussinesq and Cerruti are given by F. Vogt, *Det Norske Videnskaps Akademi i Oslo*, Math.-Naturv. klasse, No. 2, pp. 5, 20 (1925).

[13] E. Melan, Zeits. f. angew. Math. und Mech. **12**, 343 (1932). This comparison was suggested by Professor S. Timoshenko.

$$\sigma_x=\frac{P}{8\pi(1-\mu)}\left[\frac{(1-2\mu)(z-c)}{R_1^3}-\frac{3x^2(z-c)}{R_1^5}+\frac{(1-2\mu)[3(z-c)-4\mu(z+c)]}{R_2^3}\right.$$
$$-\frac{3(3-4\mu)x^2(z-c)-6c(z+c)[(1-2\mu)z-2\mu c]}{R_2^5}-\frac{30cx^2z(z+c)}{R_2^7}$$
$$\left.-\frac{4(1-\mu)(1-2\mu)}{R_2(R_2+z+c)}\left(1-\frac{x^2}{R_2(R_2+z+c)}-\frac{x^2}{R_2^2}\right)\right],$$

$$\sigma_y=\frac{P}{8\pi(1-\mu)}\left[\frac{(1-2\mu)(z-c)}{R_1^3}-\frac{3y^2(z-c)}{R_1^5}+\frac{(1-2\mu)[3(z-c)-4\mu(z+c)]}{R_2^3}\right.$$
$$-\frac{3(3-4\mu)y^2(z-c)-6c(z+c)[(1-2\mu)z-2\mu c]}{R_2^5}-\frac{30cy^2z(z+c)}{R_2^7}$$
$$\left.-\frac{4(1-\mu)(1-2\mu)}{R_2(R_2+z+c)}\left(1-\frac{y^2}{R_2(R_2+z+c)}-\frac{y^2}{R_2^2}\right)\right], \quad (14)$$

$$\sigma_z=\frac{P}{8\pi(1-\mu)}\left[-\frac{(1-2\mu)(z-c)}{R_1^3}+\frac{(1-2\mu)(z-c)}{R_2^3}-\frac{3(z-c)^3}{R_1^5}\right.$$
$$\left.-\frac{3(3-4\mu)z(z+c)^2-3c(z+c)(5z-c)}{R_2^5}-\frac{30cz(z+c)^3}{R_2^7}\right],$$

$$\tau_{yz}=\frac{Py}{8\pi(1-\mu)}\left[-\frac{(1-2\mu)}{R_1^3}+\frac{(1-2\mu)}{R_2^3}-\frac{3(z-c)^2}{R_1^5}-\frac{3(3-4\mu)z(z+c)-3c(3z+c)}{R_2^5}-\frac{30cz(z+c)^2}{R_2^7}\right],$$

$$\tau_{zx}=\frac{Px}{8\pi(1-\mu)}\left[-\frac{(1-2\mu)}{R_1^3}+\frac{(1-2\mu)}{R_2^3}-\frac{3(z-c)^2}{R_1^5}-\frac{3(3-4\mu)z(z+c)-3c(3z+c)}{R_2^5}-\frac{30cz(z+c)^2}{R_2^7}\right],$$

$$\tau_{xy}=\frac{Pxy}{8\pi(1-\mu)}\left[-\frac{3(z-c)}{R_1^5}-\frac{3(3-4\mu)(z-c)}{R_2^5}+\frac{4(1-\mu)(1-2\mu)}{R_2^2(R_2+z+c)}\left(\frac{1}{R_2+z+c}+\frac{1}{R_2}\right)-\frac{30cz(z+c)}{R_2^7}\right].$$

We consider a uniform distribution of forces of magnitude p per unit of length along a line through $(0, 0, +c)$ parallel to the y axis. If db is a small element of this line at a distance b from the z axis, the stresses due to the uniform pressure p acting on db are found by substituting pdb for P and $(y-b)$ for y in Eqs. (14). If, now, we integrate these formulas with respect to b between the limits $-\infty$ and $+\infty$, we obtain the corresponding solution for plane strain. By taking account of the relations between the elastic constants for plane strain and plane stress and also noting an interchange of axes, this procedure leads directly to the stresses given by Melan.

5. Force Parallel to the Boundary of a Semi-Infinite Solid

In this case there is no axial symmetry and we employ rectangular coordinates (x, y, z), and Eqs. (1) and (2). The force P is applied at $(0, 0, +c)$ and acts in the positive x direction (Fig. 4).

We again begin by applying the Kelvin solution at $(0, 0, +c)$, this time using type $A(1)$. To remove the stresses on plane $z=0$, introduced by the single force, we employ nuclei of strain of types $A(1)$, $C(6)$, $D(6)$, $G(1)$ and $H(1)$, again substituting $(z+c)$ for z and $R_2=(x^2+y^2+(z+c)^2)^{\frac{1}{2}}$ for R. The constants associated with these solutions are determined from the boundary conditions $[\sigma_z]_{z=0}=[\tau_{yz}]_{z=0}=[\tau_{zx}]_{z=0}=0$ (15)

and the equilibrium condition

$$P=-4\int_0^\infty\int_0^\infty \tau_{zz}dydx, \quad (z>c). \qquad (16)$$

The result of these operations is the Galerkin vector[11]

$$\mathbf{F}=[P/8\pi(1-\mu)](\mathbf{i}\{R_1+R_2-2c^2/R_2+4(1-\mu)(1-2\mu)[(z+c)\log(R_2+z+c)-R_2]\}+\mathbf{k}\{2cx/R_2+2(1-2\mu)x\log(R_2+z+c)\}). \qquad (17)$$

Thus, for the case of a force in the interior of a semi-infinite solid, parallel to the boundary, the solution is obtained by superposition of six nuclei of strain for the unlimited solid: (1) a single force at $(0, 0, +c)$; (2) a single force at $(0, 0, -c)$; (3) a doublet at $(0, 0, -c)$; (4) a semi-infinite line of doublets extending from $z=-c$ to $z=-\infty$ with strength proportional to the distance from $z=-c$; (5) a double force with moment at $(0, 0, -c)$; (6) a semi-infinite line of double forces with moment extending from $z=-c$ to $z=-\infty$.

The displacements and stresses, obtained from Eqs. (1), (2) and (17), are found to be:

$$u=\frac{P}{16\pi G(1-\mu)}\left[\frac{3-4\mu}{R_1}+\frac{1}{R_2}+\frac{x^2}{R_1^3}+\frac{(3-4\mu)x^2}{R_2^3}+\frac{2cz}{R_2^3}\left(1-\frac{3x^2}{R_2^2}\right)+\frac{4(1-\mu)(1-2\mu)}{R_2+z+c}\left(1-\frac{x^2}{R_2(R_2+z+c)}\right)\right],$$

$$v=\frac{Pxy}{16\pi G(1-\mu)}\left[\frac{1}{R_1^3}+\frac{3-4\mu}{R_2^3}-\frac{6cz}{R_2^5}-\frac{4(1-\mu)(1-2\mu)}{R_2(R_2+z+c)^2}\right],$$

$$w=\frac{Px}{16\pi G(1-\mu)}\left[\frac{z-c}{R_1^3}+\frac{(3-4\mu)(z-c)}{R_2^3}-\frac{6cz(z+c)}{R_2^5}+\frac{4(1-\mu)(1-2\mu)}{R_2(R_2+z+c)}\right],$$

$$\sigma_x=\frac{Px}{8\pi(1-\mu)}\left[-\frac{(1-2\mu)}{R_1^3}+\frac{(1-2\mu)(5-4\mu)}{R_2^3}-\frac{3x^2}{R_1^5}-\frac{3(3-4\mu)x^2}{R_2^5}-\frac{4(1-\mu)(1-2\mu)}{R_2(R_2+z+c)^2}\left(3-\frac{x^2(3R_2+z+c)}{R_2^2(R_2+z+c)}\right)+\frac{6c}{R_2^5}\left(3c-(3-2\mu)(z+c)+\frac{5x^2z}{R_2^2}\right)\right],$$

$$\sigma_y=\frac{Px}{8\pi(1-\mu)}\left[\frac{(1-2\mu)}{R_1^3}+\frac{(1-2\mu)(3-4\mu)}{R_2^3}-\frac{3y^2}{R_1^5}-\frac{3(3-4\mu)y^2}{R_2^5}-\frac{4(1-\mu)(1-2\mu)}{R_2(R_2+z+c)^2}\left(1-\frac{y^2(3R_2+z+c)}{R_2^2(R_2+z+c)}\right)+\frac{6c}{R_2^5}\left(c-(1-2\mu)(z+c)+\frac{5y^2z}{R_2^2}\right)\right], \qquad (18)$$

$$\sigma_z=\frac{Px}{8\pi(1-\mu)}\left[\frac{(1-2\mu)}{R_1^3}-\frac{(1-2\mu)}{R_2^3}-\frac{3(z-c)^2}{R_1^5}-\frac{3(3-4\mu)(z+c)^2}{R_2^5}+\frac{6c}{R_2^5}\left(c+(1-2\mu)(z+c)+\frac{5z(z+c)^2}{R_2^2}\right)\right],$$

$$\tau_{yz}=\frac{Pxy}{8\pi(1-\mu)}\left[-\frac{3(z-c)}{R_1^5}-\frac{3(3-4\mu)(z+c)}{R_2^5}+\frac{6c}{R_2^5}\left(1-2\mu+\frac{5z(z+c)}{R_2^2}\right)\right],$$

$$\tau_{zx}=\frac{P}{8\pi(1-\mu)}\left[-\frac{(1-2\mu)(z-c)}{R_1^3}+\frac{(1-2\mu)(z-c)}{R_2^3}-\frac{3x^2(z-c)}{R_1^5}-\frac{3(3-4\mu)x^2(z+c)}{R_2^5}-\frac{6c}{R_2^5}\left(z(z+c)-(1-2\mu)x^2-\frac{5x^2z(z+c)}{R_2^2}\right)\right],$$

$$\tau_{xy}=\frac{Py}{8\pi(1-\mu)}\left[-\frac{(1-2\mu)}{R_1^3}+\frac{(1-2\mu)}{R_2^3}-\frac{3x^2}{R_1^5}-\frac{3(3-4\mu)x^2}{R_2^5}-\frac{4(1-\mu)(1-2\mu)}{R_2(R_2+z+c)^2}\left(1-\frac{x^2(3R_2+z+c)}{R_2^2(R_2+z+c)}\right)-\frac{6cz}{R_2^5}\left(1-\frac{5x^2}{R_2^2}\right)\right].$$

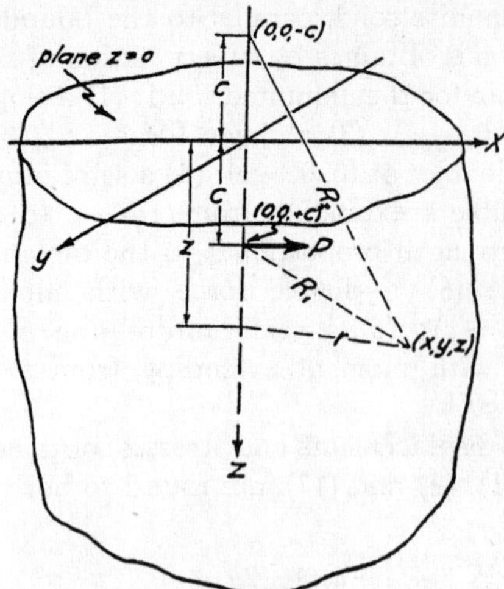

FIG. 4. Force parallel to the boundary in the interior of a semi-infinite solid.

Fig. 5 shows the distribution of normal stress on horizontal planes, and Fig. 6 shows the vertical deflection of the surface of the solid along the x axis.

When $c\rightarrow\infty$ all terms in Eq. (18) containing R_2 vanish and the solution becomes that for Kelvin's problem with the force applied at $(0, 0, +c)$ in the positive x direction. The corresponding Galerkin vector is

$$\mathbf{F}=\mathbf{i}[PR_1/8\pi(1-\mu)]. \tag{19}$$

When $c\rightarrow 0$ Eqs. (18) give the stresses and displacements for the problem of Cerruti.[12] These results are represented by the Galerkin vector

$$\begin{aligned}\mathbf{F}=[P/4\pi(1-\mu)](\mathbf{i}\{R+2(1-\mu)(1-2\mu)\\ \times[z\log(R+z)-R]\}\\ +\mathbf{k}\{(1-2\mu)x\log(R+z)\}).\end{aligned} \tag{20}$$

The solution for the problem of Cerruti was given by Professor Westergaard[6] in the form of the Galerkin vector

$$\begin{aligned}\mathbf{F}=[P/4\pi(1-\mu)][\mathbf{i}R\\ +\mathbf{k}(1-2\mu)x\log(R+z)]\end{aligned} \tag{21a}$$

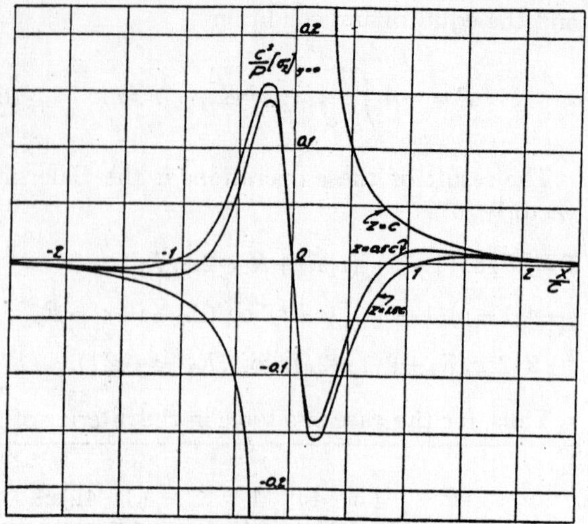

FIG. 5. Normal stress along lines parallel to the x axis on planes parallel to the boundary for the case of a force parallel to the boundary ($\mu=0.3$).

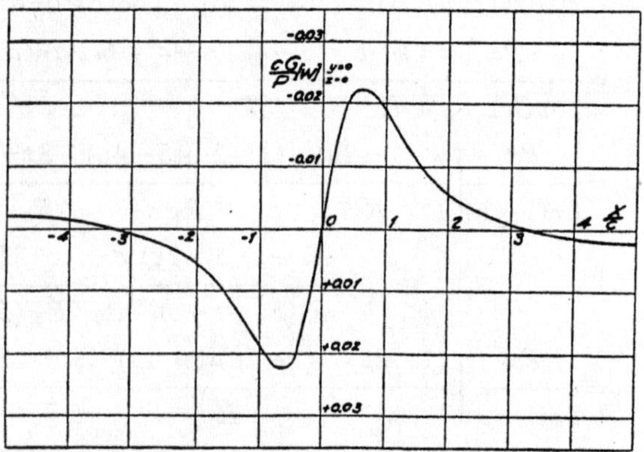

FIG. 6. Vertical deflection of the surface along the x axis for the case of a force parallel to the boundary ($\mu=0.3$).

combined with the potential function

$$\phi=[(1-2\mu)P/2\pi][x/(R+z)]. \tag{21b}$$

Following the same procedure as in the case of the normal force, the stresses given in Eqs. (18) may be reduced to Melan's formulas for the two-dimensional analog.[13]

The author is indebted to Professor H. M. Westergaard for valuable criticisms of the manuscript.

Originally published in
Journal of the Boston Society of Civil Engineers
Apr., 1936

A FUNDAMENTAL FALLACY IN EARTH PRESSURE COMPUTATIONS

BY DR. KARL TERZAGHI, MEMBER *

(To be presented at a meeting of the Boston Society of Civil Engineers on May 20, 1936)

FOR more than a century engineering textbooks have maintained that the pressure of sand on the back of a lateral support increases like a hydrostatic pressure in direct proportion to depth. This hydrostatic pressure distribution is supposed to exist regardless of whether the sand is retained by a masonry wall or by timbering. On the other hand, many engineers and contractors with broad experience in subway construction claim that the distribution of the lateral pressure on the timbering of cuts has no resemblance to the theoretical distribution. This attitude is based on the results of observations such as those which were described in the papers of Mr. J. C. Meem, "The Bracing of Trenches and Tunnels" (Trans. Am. Soc. C. E., Vol. LX, June, 1908), of Mr. H. G. Moulton, "Earth and Rock Pressures" (Trans. Am. Soc. of Min. and Metall. Eng., 1920), and in the discussions of these papers. In the following analysis the causes of these contradictions will be investigated.

RANKINE'S THEORY AND THE NATURAL STATE OF STRESS IN SANDS.

All the customary methods of earth pressure computation can be traced back either to Coulomb's or to Rankine's theory of earth pressure. Since Rankine's theory covers a broader field of application, it may be considered first.

According to Rankine's theory a negligible displacement should suffice to produce a failure by shear in sands. This means that every deposit of sand can be assumed to exist at the verge of the breakdown of its equilibrium. Taking this assumption for granted, an application of the accepted laws of applied mechanics to the state of stress in a mass of sand with an infinite horizontal surface leads to the following conclusions. In every point of such a mass the ratio between the horizontal and the vertical pressure is a constant, equal to—

$$K_r = tan^2\left(45 - \frac{\phi}{2}\right)$$

* Visiting Professor at the Graduate School of Engineering, Harvard University, Cambridge, Mass.

wherein ϕ is the angle of repose. The corresponding state of stress will briefly be called the *Rankine state of stress.* The value K_r represents the ratio between the lateral pressure in the sand and the lateral pressure in a liquid whose unit weight is equal to that of the sand. At the same time, K_r represents the smallest value which K can possibly assume in a mass of sand with an infinite horizontal surface. Therefore K_r is called the *minimum* hydrostatic pressure ratio. The index r indicates that it refers to the Rankine state of stress. Since the vertical pressure increases in simple proportion to the depth below the surface, and since, in addition, K_r is a constant, the pressure on any plane section through the sand increases like a hydrostatic pressure in direct proportion to

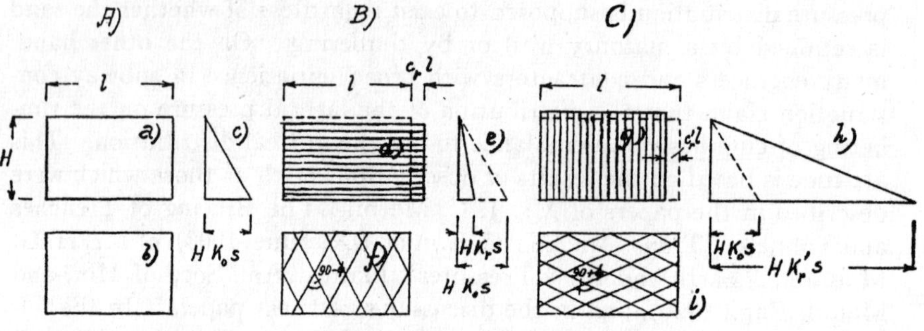

FIG. 1

Three states of stress in sand involving hydrostatic distribution of pressure along plane sections. A = natural state, B = Rankine state for minimum and C = Rankine state for maximum lateral pressure. Diagrams *d* and *g* show the deformation required to induce transition from state A into states B and C respectively; diagrams *e* and *h* represent the corresponding change of lateral pressure on vertical sections, and *f* and *i* show the two sets of planes of shear.

depth, as shown in Fig. 1*e*. A very small lateral displacement in the sand should suffice to induce failure by shear along two sets of inclined shearing planes rising at an angle of $45+\frac{\phi}{2}$ to the horizontal. These two sets are shown in Fig. 1*f*. Since the horizontal displacement is supposed to have no effect other than to induce the slip, the factor "strain" does not appear in the theory.

Again, according to Rankine's theory, a negligible lateral compression should suffice to increase the hydrostatic pressure ratio throughout the mass of sand from K_r to —

$$K'_r = tan^2\left(45+\frac{\phi}{2}\right)$$

Any further lateral compression induces sliding along two sets of planes rising at an angle of $45-\frac{\phi}{2}$ to the horizontal, while the hydrostatic pressure ratio K'_r remains unaltered.

In order to find out whether Rankine's assumptions are justified, we first determine the state of stress in a horizontal layer of sand after deposition, either by natural processes of sedimentation or otherwise. Then we investigate the means by which this state may be changed into either one of the two Rankine states of stress, indicated by the diagrams under the letters B and C in Fig. 1.

If a bed of sand is formed by sedimentation or by some artificial procedure the area covered by the individual layers remains constant throughout the process of formation, while the thickness of the lower layers decreases as the weight of the superimposed strata increases. The same stress conditions develop if we deposit the sand in a vessel with perfectly frictionless side walls such as that shown in Fig. 1*a*, provided that the side walls neither move nor deflect during the process of filling the vessel. The state of stress in the contents of this vessel will be called the *natural state of stress* because it is identical with the state of stress in a natural sand deposit before engineering operations of any kind have been started. Numerous experimental investigations concerning this natural state of stress have led to the following conclusions. This state differs from the two extreme Rankine states of stress inasmuch as the hydrostatic pressure has a value, K_o, intermediate between the two extreme Rankine values K_r and K'_r. The value K_o is called the natural hydrostatic pressure ratio. Since K_o is also a constant, the pressure on any plane section through the sand increases in simple proportion to the depth below the surface. The following values of K_o were obtained and verified by several independent experimental methods:

Dense sand	$K_o=0.40$ to 0.45
Loose sand	$K_o=0.45$ to 0.50

The validity of these figures is limited by the condition that the bed of sand to which they refer has never been subject to a temporary surcharge. The application and subsequent removal of a surcharge seems to increase the value of K_o and to eliminate the hydrostatic pressure distribution.

In order to change the natural state of stress shown in Fig. 1 under the letter *A* into the Rankine state of minimum lateral pressure shown under the letter *B*, the sand must be given an opportunity to expand laterally. This could be done by permitting one of the side walls of the vessel to yield parallel to its original position, as shown in Fig. 1*d*. However, the reduction of K_o to K_r produced by such a movement would be strictly limited to the immediate vicinity of the yielding side wall because at a greater distance from the wall the friction between the bottom of the vessel and the base of the fill prevents the expansion from spreading over the entire contents of the vessel. Hence, in order to produce the proposed modification of the state of stress, the friction between the bottom of the fill and the bottom of the vessel must be entirely eliminated. In practice no such elimination can be effectuated. Hence the process can only be performed in our imagination. On the other hand, the total amount of lateral expansion required to reduce the hydrostatic pressure ratio from K_o to K_r throughout the mass of sand can be reliably determined from what we know concerning the relation between stress and strain in sands. This expansion is equal to an empirical value, c_r, times the total width, l, of the deposit. The value c_r depends largely on the density of the sand. For a well-compacted sand the value of c_r is approximately 0.015, and for a loose sand it is considerably greater. The lateral expansion shown in Fig. 1*d* is always associated with a subsidence in a vertical direction.

On the other hand, the transition from the natural state of stress to the Rankine state of maximum lateral pressure requires a lateral compression of the sand through a distance $c'_r l$ such as shown in Fig. 1*g*. The value of c'_r seems to be almost identical with the value c_r. The lateral compression is associated with an expansion in a vertical direction.

As indicated in the above discussion, natural sand deposits exist exclusively in the natural state of stress. In order to change this state of stress into a Rankine state of stress for minimum lateral pressure over a width of not more than 100 feet of a bed of sand, the material must be given an opportunity for perfectly uniform lateral expansion through a distance of not less than 1.5 feet. Nowhere in nature is there a possibility for such an important expansion. Hence the states of stress shown in Fig. 1, *B* and *C*, exist exclusively in our imagination and in Rankine's earth pressure theory.

Eighty years ago, when Rankine's theory was published, nothing was known about the natural state of stress of sands and nothing was known regarding the importance of the lateral expansion required to produce the Rankine state of stress. Therefore the theory appeared to be acceptable. However, at the present state of our knowledge of the relations between strain and stress in sands, the theory can no longer be accepted.

Coulomb's Theory

In contrast to Rankine, Coulomb never attempted to investigate the state of stress within the backfill of the lateral support, but merely tried to interpret mathematically the phenomena observed when a lateral

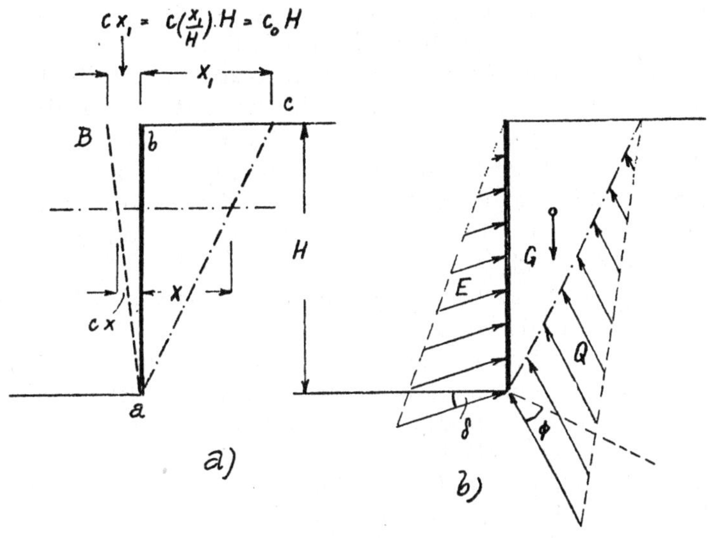

Fig. 2

Distribution of stresses along the boundaries of the sliding wedge according to Coulomb's theory. The abscissæ of aB represent the minimum lateral expansion of the fill required to establish the Coulomb condition.

support fails. The failure of the lateral support of a bank of sand always causes a slip along an inclined surface, *ac* in Fig. 2*a*, which passes through the lower edge of the back of the support. Coulomb's theory of this phenomenon can be divided into two parts which are based on two different assumptions. The first part, dealing with the total value of the lateral pressure, involves the assumptions that the surface of rupture, *ac* in Fig. 2*a*, is plane and that the shearing resistance along this plane is fully mobilized. The total lateral pressure, E, computed on the basis of these assumptions, will briefly be called the *Coulomb pressure.* It represents the smallest value the lateral pressure on the back of a retaining wall can possibly assume. In the computation of this pressure no assumption whatsoever is made regarding the state of stress within the wedge *abc* (Fig. 2*a*). As a consequence no information is obtained regarding the position of the force E with respect to the wall.

The second part of Coulomb's theory deals with the distribution of the lateral pressure over the back of the lateral support. In this connection Coulomb simply *assumes* that the lateral pressure increases like a hydrostatic pressure in simple proportion to depth. He does not even attempt to justify this assumption.

No serious objection can be raised against the first part of Coulomb's theory as applied to sands, because for practical purposes both assumptions involved in this part are justified. Experience has shown that the curvature of the plane of rupture is very small, and the lateral yield of retaining walls and of the strutting of cuts in sands is almost always sufficient to mobilize at least the major part of the shearing resistance along the boundary *ac* (Fig. 2*a*) of the wedge. In doubtful cases the validity of this second assumption should be examined. An example of such an analysis has been published by the author ("Engineering News-Record," May 17, 1934).

In order to pass an opinion on the conditions for the validity of the second part of Coulomb's theory concerning the distribution of the lateral pressure over the back of the lateral support, it is necessary to investigate the effect of the lateral yield of the support on the state of stress in the adjoining part of the backfill.

The state of stress in the backfill of a perfectly rigid wall can be assumed to be almost identical with the natural state of stress represented in Fig. 1, *A*. If the wall is allowed to yield in a horizontal direction, the sand adjoining the back of the wall expands laterally, with the result that the lateral pressure decreases. As soon as this expansion exceeds a certain critical value, depending on the elastic properties and the depth of the backfill, a slip occurs along an inclined surface *ac* (Fig. 2*a*). To the left of *ac* the outward movement of the wall produces a gradual transition from the natural state of stress into a state involving a minimum lateral pressure. To the right of *ac* (Fig. 2*a*) the lateral expansion produced by the outward movement of the wall is practically nil, because the lateral confinement of the sand located beneath the base of the fill prevents this part of the fill from participating in the expansion. Hence, in contrast to Rankine's assumptions, the lateral yield of the wall produces merely a *strictly localized* transition from the natural state into one of complete mobilization of the internal friction.

The lateral expansion required for reducing the hydrostatic pressure ratio to its minimum value throughout the wedge-shaped zone *abc*, in Fig. 2*a*, increases from the foot of the bank towards the crest of the wall, as indicated by the distance between the lines *ab* and *aB* (Fig. 2*a*). The expansion of the sand through the distance between the lines *ab* and *aB* in Fig. 2*a* can be produced by a rotation of the lateral support around its lower rim. After the expansion has occurred, the hydrostatic pressure ratio is equal to K_c throughout the entire wedge *abc*. The value K_c represents the minimum hydrostatic pressure ratio for the backfill of a rough lateral support. It is always smaller than the Rankine value K_r for the same backfill. If the back of the wall is perfectly frictionless, the value K_c becomes identical with K_r. A constant value of K throughout the wedge *abc* also involves a hydrostatic distribution of the lateral pressure over *aB* (Fig. 2*a*). This conclusion is identical with Coulomb's assumption regarding the distribution of the lateral pressure over the back of a retaining wall. Fig. 2*b* shows the distribution of the lateral pressure E over the back of the wall and of the reaction Q over the plane of rupture according to Coulomb's theory. The corresponding distribution of the lateral pressure over the back of the lateral support will be briefly called the *hydrostatic pressure distribution*.

Fig. 3*a* represents the ideal relation between the average yield of a wall, the location of the center of pressure and the hydrostatic pressure ratio K for a wall which yields by tilting around its lower edge. The back of the wall is supposed to be perfectly frictionless because the friction between the sand and the wall is likely to have at least some disturbing influence. The abscissæ of the diagrams represent the average distance of yield in fractions of the depth H of the fill. In the lower diagram the ordinates determine the vertical distance between the center of the pressure and the foot of the wall in fractions of the depth H of the fill. Since the tilting movement produces simultaneous mobilization of the internal friction in the sand adjoining the wall, the distribution of the pressure remains practically hydrostatic throughout the process, involving a location of the center of pressure at an elevation equal to one third of the depth of the fill. The ordinates of the curve C'_k in the upper diagram are equal to the hydrostatic pressure ratio K. The shape of the curve C'_k was estimated from the results of experimental investigations concerning the relation between stress and strain for sands. Its curvature is due to the fact that the strain in sands increases at a higher rate than the corresponding stresses.

The Non-Hydrostatic Pressure Distribution and the Arching Effect

The validity of the ideal diagram shown in Fig. 3*a* is limited to the ideal case in which the hydrostatic pressure ratio decreases in every

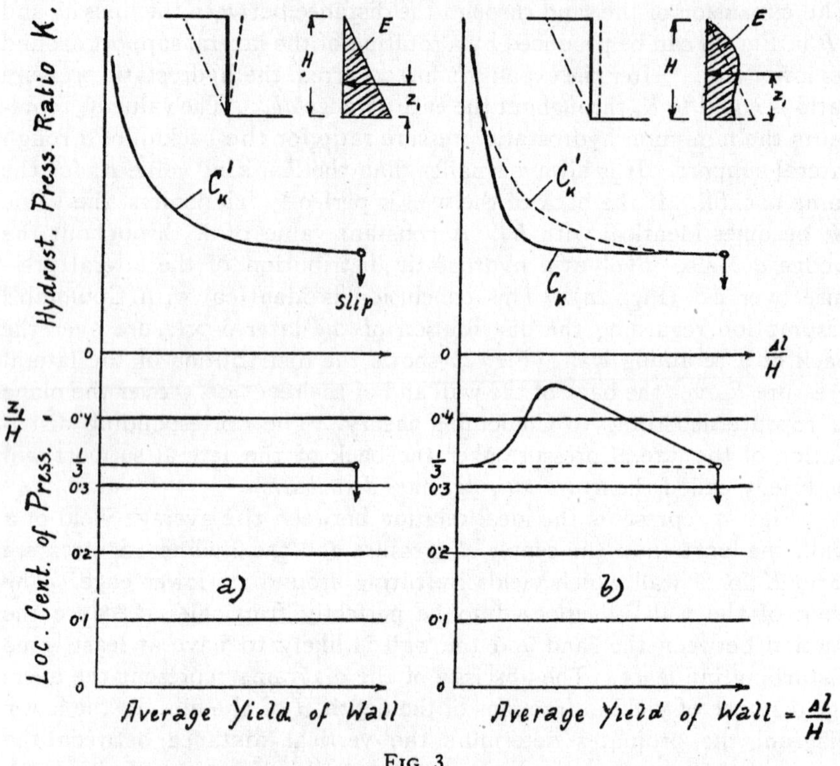

FIG. 3

Ideal relations between average yield of lateral support (abscissæ), hydrostatic pressure ratio K (ordinates upper diagram) and location of center of pressure (ordinates lower diagram) for a frictionless lateral support which yields by tilting (left hand diagram) or by advancing parallel to its original position (right hand diagram). Shaded areas represent the distribution of lateral pressure immediately after the total lateral pressure has become equal to Coulomb value.

point of the wedge at the same rate. To satisfy this condition the lateral support must yield by tilting around its lower rim. On the other hand, if a lateral support yields in any other fashion, involving an outward movement over the entire height, for instance as shown by the line a_1 b_1 in Fig. 4, the following events will occur. Owing to the relatively excessive lateral yield of the lower part, *ade*, of the wedge *abc* in Fig. 4, the hydrostatic pressure ratio for this lower part assumes its minimum value, K_c, at an early stage of the outward movement of the wall. At this same stage the upper part, *bcde*, of the wedge is still in an initial state of lateral expansion, involving a value $K > K_c$. The ener-

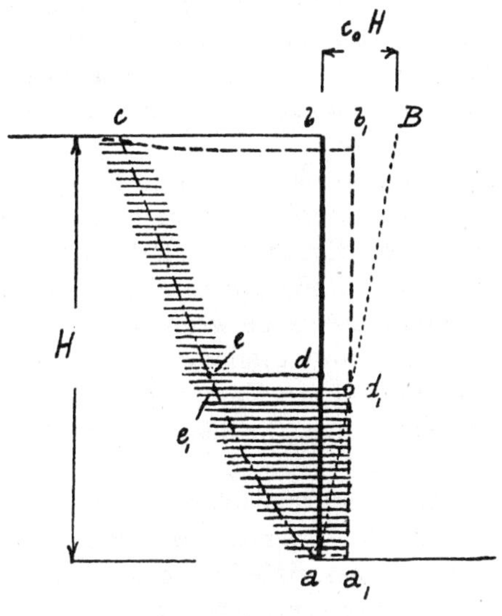

FIG. 4

Mechanical causes of non-hydrostatic pressure distribution over the back of lateral support after the support has yielded from *ab* to a_1b_1. abB = minimum yield required to produce hydrostatic pressure distribution, aa_1bb_1 = actual yield. Owing to the excessive lateral yield of the lower part, *ad*, of the lateral support, the lower, shaded part of the wedge *abc* subsides. The upper part must follow in spite of inadequate lateral expansion which, in turn, mobilizes the maximum shearing resistance along the balance *ec* of the surface of rupture.

getic horizontal expansion of the lower part of the wedge involves a subsidence of its upper boundary *de* into the position d_1e_1, which in turn causes the upper part, *bcde*, of the wedge to descend. This downward movement mobilizes the frictional resistance of the sand along the surface *ce* (Fig. 4). Along the section *ae* this resistance was already mobilized at an earlier stage. Once the frictional resistance along *ac* is fully mobilized, the total lateral pressure automatically assumes the theoretical Coulomb value, regardless of what the state of stress within the wedge *abc* may be. Since the upper part of the wedge is still in a state of incomplete lateral expansion, the friction along *ce* transfers part of the weight of this upper part on to the fill located at the left of *ce*, and this in turn reduces the vertical pressure on d_1e_1. The reduction of the vertical pressure on d_1e_1 (Fig. 4), due to friction along the lateral boundary, e_1c, of the superimposed strata $b_1cd_1e_1$ in a state of lateral restraint, is called the *arching effect.* Owing to the arching effect the lateral pressure exerted by the lower part of the backfill decreases. The total lateral pressure E remains unchanged. Therefore the lateral pressure of the upper part of the backfill is bound to increase, which in turn causes the center of the pressure to rise.

Any further yield of the lateral support beyond the position required to reduce the total lateral pressure to the theoretical Coulomb value E merely changes the distribution of the lateral pressure over the back of the lateral support while the total pressure remains constant. During this process the difference between the real and the hydrostatic pressure distribution gradually diminishes. Finally the hydrostatic pressure ratio K becomes equal to K_c throughout the wedge, whereupon a slip occurs. At the same time, the distribution of the lateral pressure becomes hydrostatic.

The non-hydrostatic pressure distribution involves a location of the center of pressure above the lower third. Since the last stage of the process prior to the slip is associated with an approach to the hydrostatic conditions, the relation between the average yield of the wall, the location of the center of pressure, and the average hydrostatic pressure ratio K should be as shown in Fig. 3*b*. The diagrams in the upper right corner of the figure show the position of the wall and the corresponding pressure distribution at the instant when the total lateral pressure assumes the Coulomb value. Any further outward movement of the wall causes a gradual transition of the distribution of the lateral pressure shown in Fig. 3*b* into that shown in Fig. 3*a*, while the total lateral pressure E represented by the shaded area remains unchanged.

It should be noticed that the curve C_k in Fig. 3*b* is shown to approach the horizontal tangent far more rapidly than does the curve C'_k in Fig. 3*a*. This assumption was suggested by the following reasoning. In the case represented by Fig. 3*a*, the wedge expands in a uniform fashion, because the wall is supposed to yield by tilting around the lower edge of its back. As a consequence the lateral pressure does not assume the Coulomb value until the hydrostatic pressure ratio becomes equal to the minimum value K_c throughout the wedge. On the other hand, if the wall yields parallel to its original position, as in the case represented by Fig. 3*b*, a very small horizontal displacement is sufficient to produce a general subsidence of the sand located between the lateral boundaries of the wedge. This subsidence mobilizes the frictional resistance within narrow zones as it is shown in Fig. 4, though the expansion of the upper part of the wedge is still incomplete. This phenomenon can be compared with the state of stress in a funnel-shaped bin. As soon as the bottom of such a bin is allowed to yield, the shearing stresses along the walls of the bin assume their maximum value, although there is no lateral expansion of the contents of the bin.

Empirical Data regarding the Non-Hydrostatic Pressure Distribution

According to Coulomb's theory the pressure of the sand on the back of any lateral support, the timbering of cuts included, should increase like a hydrostatic pressure in direct proportion to the depth. In contrast to this widespread conception many contractors maintain, on the strength of their personal experience, that this is never the case in connection with the vertical sides of a cut. This fact alone should suffice to produce a sceptical attitude towards the hydrostatic pressure theory.

Fig. 5*a* shows the results of direct measurements made by M. Miller on Flatbush Avenue, New York, in 1916, quoted by H. G. Moulton (reference given above). The lateral pressure was computed "from the deflection of rangers over a bank area measuring 22 feet in a horizontal direction and 55 feet in a vertical direction." Below a depth of 60 feet the cut was occupied by the completed structure. The width of the cut was 85 feet and the soil consisted of coarse sand mixed with 20 to 30 per cent of clay and some gravel. The dotted line *ab* in Fig. 5*a* shows what the distribution of the lateral pressure should be according to the hydrostatic theory. The difference between the hydrostatic and the observed pressure distribution is by far too important to be ignored as an item of no practical consequence.

Experimental data concerning the relation between the lateral yield of the wall, the location of the center of pressure, and the hydrostatic pressure ratio were obtained by the writer in connection with large-scale earth pressure tests at Massachusetts Institute of Technology in 1929. These tests were described in several articles in "Engineering News-Record," 1934. Fig. 5*d* shows the results of two series of pertinent tests, both of which were performed on a sand with an effective size of 0.54 mm. in a compacted state. In series I the wall yielded by tilting around a horizontal axis located at a depth of 1.67 feet below the base of the fill, as shown in Fig. 5*b*. In series II, the wall was moved out parallel to its original position, as shown in Fig. 5*c*. In the lower dia-

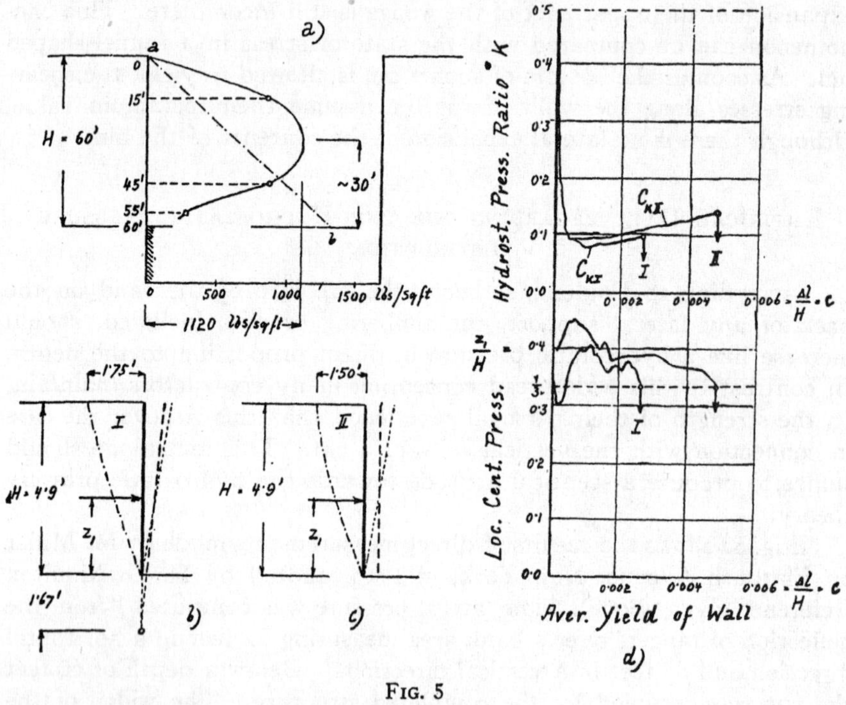

FIG. 5

Empirical evidence for non-hydrostatic pressure distribution. (*a*) Distribution of lateral pressure over timbering in Flatbush Avenue, New York. (*b*) and (*c*) Movement of model retaining wall in two different tests, I and II. (*d*) Relation between average yield of the wall (abscissæ), hydrostatic pressure ratio (ordinates upper diagram) and location of center of pressure (ordinates lower diagram) for tests I and II. The arrows I and II indicate slip in tests I and II.

gram of Fig. 5*d* the ordinates represent the location of the center of pressure, in the upper diagram the hydrostatic pressure ratio, and in both diagrams the abscissæ indicate the average yield of the wall. By comparing the theoretical diagram (Fig. 3*b*) with the empirical diagram (Fig. 5*d*) we recognize, first of all, that these two diagrams are practically identical as far as the essentials are concerned. This identity can be considered an experimental confirmation of the reasoning on which the theoretical diagram Fig. 3*b* was based. In addition, an analysis of the data shown in Fig. 5*d* leads to the following conclusions. The maximum arching effect disclosed by the highest position of the center of pressure develops at the instant when the total lateral pressure, expressed by the ratio K in Fig. 5*d*, assumes its smallest value. The smallest value of the total pressure of the sand on the lateral support is identical with the Coulomb value. Therefore the facts shown in Fig. 5*d* demonstrate that the Coulomb value is by no means necessarily associated with a hydrostatic distribution of the lateral pressure. A yield of the wall beyond the distance required to produce the maximum arching effect causes a slight but steady increase of the total lateral pressure. This fact suggests that the slip, which is indicated in Fig. 5*d* by arrows, is preceded by a decrease of the shearing resistance along the inclined boundary of the wedge. The highest position of the center of pressure depends on the type of the movement of the wall. The more this movement approaches simple tilting around the foot of the back of the wall, the closer the center of pressure remains to the lower third of the depth of the fill. At the same time, the average yield corresponding to the highest position of the center of pressure increases. The type of movement also has a marked influence on the curvature of the surface of rupture. In series II, involving a movement of the wall parallel to its original position, the width of the top of the wedge was 15 per cent smaller than it was in series I, though the minimum value of the total lateral pressure was in both series almost exactly the same. Finally it should be mentioned that the abscissæ of the arrows I and II in Fig. 5*d* correspond to almost equal yields of the upper rim of the wall in the two series of tests. In both cases the yield of the upper rim was approximately equal to 1.5 per cent of the top width of the wedge, though the average yield of the wall was different. Hence it seems that the slip which is indicated by the arrows does not occur until every part of the wedge, including its uppermost part, has expanded through a certain minimum distance regardless of the average yield of the wall. This minimum distance is c_oH in Fig. 4. In both tests this distance was approximately equal to 0.005 H. On the other hand, the hydrostatic pressure ratio K for the total lateral pressure seems to depend on the average yield only. In this connection attention should be called to the striking similarity between the curves C_{kI} and C_{kII} in Fig. 5*d*. Both curves closely resemble the curve C_k in Fig. 3*b*, but neither of them has any similarity to the curve C'_k in Fig. 3*a* corresponding to the imaginary case of a perfectly frictionless lateral support which yields by tilting around its lower edge. The rough model retaining wall to which case I in Fig. 5 refers also yielded by tilting, although the axis of rotation was located at some depth below the foot of the wall. Hence the striking difference between the curve C'_k in Fig. 3*b* and the curve C_{kI} in Fig. 5*d* suggests that the friction along the back of the model retaining wall (Fig. 5*b*) was sufficient to invalidate the reasoning on which Fig. 3*a* was based.

Limitations to the Validity of Coulomb's Earth Pressure Theory of the Lateral Pressure of Sand

Owing to the facts presented in the preceding paragraph, the validity of Coulomb's theory is limited to the case in which the lateral yield of the support exceeds certain minimum values. The yield required to reduce the total lateral pressure of the sand to the Coulomb value is very much smaller than the yield required to establish a hydrostatic distribution of the lateral pressure. Hence in the process of transition from the state of complete lateral confinement into the state described by Coulomb's theory, two well-defined stages can be distinguished. These two stages are illustrated by Fig. 6. During the first stage, involving an advance of the lateral support from its initial position *ab* into the position a_1b_1, the lateral pressure rapidly decreases from the initial value $K_o \cdot \frac{1}{2} H^2 s$ to the Coulomb value $K_c \cdot \frac{1}{2} H^2 s$. This transition requires an average yield of the wall through a certain distance which depends on the nature and the density of the sand and on the height of the support. For a dense sand this distance was found to be equal to 0.0005 H. In Fig. 6 this distance is represented by the abscissa of point A. Since point A is located at one half of the height of the wall, its abscissa represents the average yield regardless of the position of the lower rim, a, of the lateral support. However, at this stage the distribution of the lateral pressure over the lateral support has no similarity whatsoever with a hydrostatic pressure distribution. The importance of the difference between the real and the hydrostatic pressure distribution for the position a_1b_1 (Fig. 6) of the lateral support increases with the value of the ratio c_u/c_m. A difference in the distribution of the lateral pressure also involves a different elevation of the center of pressure above the foot of the support.

The second stage, consisting of an advance of the wall from the position a_1b_1 into the position a_2B in Fig. 6, involves a gradual transition from the non-hydrostatic into the hydrostatic distribution of the lateral pressure at a practically constant or slightly increasing value of the total lateral pressure. As soon as the upper rim of the wall reaches point B, the distribution of the lateral pressure becomes almost hydrostatic, and at the same time a visible and audible slip along an inclined surface of rupture, ac in Fig. 6, occurs.

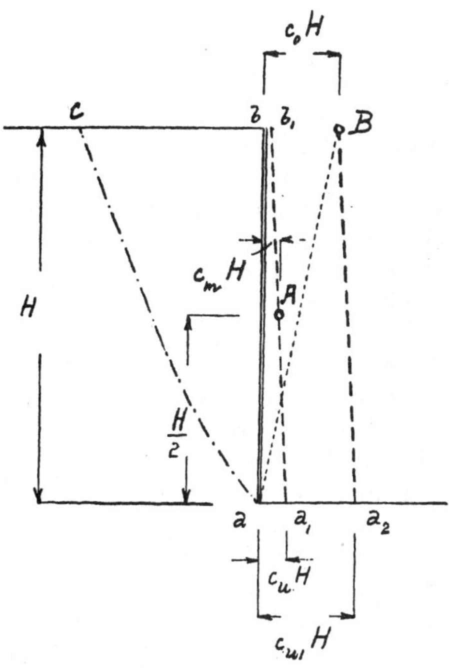

FIG. 6

Conditions for validity of Coulomb's theory. While the wall moves from ab into any position passing through point A, the total lateral pressure decreases from natural to Coulomb value. Yet at this stage the distribution of lateral pressure is non-hydrostatic. During further movement into any position passing through B, the distribution of lateral pressure gradually becomes hydrostatic while the total pressure remains practically unaltered. This second stage terminates with a slip.

Any further yield of the wall merely causes a slight increase of the lateral pressure due to progressive disintegration of the structure of the sand at an unaltered or slightly increased total lateral pressure. During all the transitions connected with an advance of the wall from the position ab (Fig. 6) to a position beyond a_2B the change of the state of stress in the sand is practically limited to the zone *abc*. To the left of the boundary *ac* of this zone the state of stress remains practically unchanged during the entire advance of the wall.

For any sand in a loose state the distribution of the lateral pressure seems to be almost hydrostatic from the very beginning of the first stage throughout the following ones. At the same time, the average yield $c_m H$ required to obtain the theoretical Coulomb pressure seems to be very much greater than 0.0005 H.

Any failure of a retaining wall by tilting or sliding involves the passage of the upper rim of the back of the wall through the position B in Fig. 6. This event is automatically associated with a transition from the non-hydrostatic into the hydrostatic distribution of the lateral pressure. Therefore the facts shown in Fig. 6 require no modification of the present method of the computation of the lateral pressure on retaining walls, provided that the backfill consists of dry sand, and that the upper rim of the retaining wall can yield at least through the distance $bB = c_o H$ shown in Fig. 6.

On the other hand, in connection with the lateral pressure of sand on a lateral support with a very limited capacity for yielding, such as the strutting of cuts, the non-hydrostatic pressure distribution associated with a position of the upper rim of the lateral support on the left side of point B in Fig. 6 becomes a factor of decisive practical importance. The assumption of a hydrostatic pressure distribution over supports of this type involves an error which may even be fatal.

Hence the fundamental misconception associated with the traditional earth pressure computations does not reside in the theories as such. It lies in the failure of the designers to consider the limitations on the validity of the theoretical results., Fig. 6 reveals these limitations at a single glance.

Owing to the important effect of apparently insignificant displacements of the lateral support on both the distribution and the intensity of the lateral pressure, no experimental data can be obtained concerning this important subject except by using extremely rigid model retaining walls capable of any type of lateral movement. Up to this time the only device which satisfies this fundamental requirement is at Massachusetts Institute of Technology in Cambridge, Massachusetts.

Lateral Pressure of Cohesive Soils.

The same type of reasoning which led to Fig. 6 applies without modification to cohesive soils. From general construction experience the following tentative conclusions were derived concerning the earth pressure exerted by cohesive soils. For sandy, cohesive soils the difference between the real and the hydrostatic pressure distribution may be still more important than it is for sands. In many cases the greatest lateral pressure develops in the vicinity of the upper rim of the cut. This fact is known to every experienced foreman. For plastic clays the value c_mH in Fig. 6 is likely to be far in excess of the greatest possible shortening of the struts under axial pressure. If such is the case the earth pressure theories fail to furnish any information on the lateral pressure which acts on the timbering. Further complications are likely to arise on account of volume changes produced by continuous rains. For this reason the cohesive soils were excluded from the preceding analysis.

Conclusions

(1) The fundamental assumptions of Rankine's earth pressure theory are incompatible with the known relation between stress and strain in soils, including sand. Therefore the use of this theory should be discontinued.

(2) The results obtained from the other classical theories can only be accepted for practical purposes if certain conditions concerning the lateral expansion of the supported sand are satisfied. These conditions are graphically shown in Fig. 6 of this paper.

(3) The total lateral pressure of the sand does not assume the theoretical value unless the average yield of the lateral support exceeds a certain minimum value, c_mH, wherein H is the depth of the fill and c_m is a coefficient whose value depends on the nature and the density of the sand. For a dense sand the value c_m is of the order of magnitude of 0.0005, and for a loose sand it is considerably greater. Under normal conditions this requirement is satisfied by both the retaining walls and the timbering in cuts.

(4) The distribution of the pressure of sand on the back of a lateral support does not agree with theory unless the yield of the support exceeds in every point the value determined by the abscissæ of $aB = c_oH$ in Fig. 6. For a dense sand this value c_o is of the order of magnitude of 0.005. In the upper part of timbered cuts this condition is very seldom satisfied. Therefore no agreement can be expected between the

real and the theoretical distribution of the pressure over the lateral support of the banks in cuts.

(5) For identical soil conditions and equal yield of the lateral support, the total lateral pressure on the support is the same. Yet the distribution of the pressure over this support varies enormously according to the distribution of the yield over the face of the bank. The more this distribution departs from a straight line through the foot of the bank, the more important the departure from the hydrostatic pressure distribution is likely to be. No pertinent information of practical value can be obtained except by direct measurements under field conditions similar to those which were made by M. Miller on Flatbush Avenue, in New York. However, the measurements should always be made not on the rangers but on the struts.

Originally published in
Journal of the New England Water Works Association.
June, 1937

New England Water Works Association

ORGANIZED 1882.

Vol. LI.	June, 1937.	No. 2

This Association, as a body, is not responsible for the statements or opinions of any of its members

SEEPAGE THROUGH DAMS.

BY ARTHUR CASAGRANDE.*

[Read February 28, 1935.]

CONTENTS.

*Assistant Professor of Civil Engineering, Harvard Graduate School of Engineering, Cambridge, Mass.

A. Introduction.

Until about ten years ago, the design of earth dams and dikes was based almost exclusively on empirical knowledge and consisted largely of adopting the cross-section of successful dams, with little regard to differences in character of soil and foundation conditions. At present, we are in a period in which the behavior of dams, particularly those which have failed, is analysed in the light of modern soil mechanics. The understanding and knowledge thus accumulating is being used as the basis for a more scientific approach to the design of such earth structures.

The most outstanding progress in this subject relates to the question of seepage beneath dams and dikes and to the effect of seepage on the stability of these structures. Foundation failures due to seepage, commonly known as "piping," were, for the first time, correctly explained by Terzaghi (1)* who developed what may be termed the "mechanics of piping." Later, Terzaghi (2, 3 and 4), called attention to the importance of the forces created within earth dams and concrete dams, due to the percolation of water. The practical application of this information has lagged behind our understanding of these forces partly because of theoretical difficulties of analyzing problems of seepage. It is only in recent years that substantial progress has been made in the solution of problems of seepage and ground water flow with free or open surface, of the flow through anisotropic materials, and of the conditions of flow through joint planes of different materials.

B. Darcy's Law for the Flow of Water through Soils.

The flow of water through soils, so far as it affects the question of seepage through dams, follows Darcy's empirical law, which states that *the amount of flow is directly proportional to the hydraulic gradient.* This law can be expressed either in the form:

$$\left.\begin{array}{ll} & v_o = k\,i \\ \text{or} & Q = k i A t \end{array}\right\} \qquad (1)$$

in which the symbols have the following meaning:

v_o = discharge velocity†
k = coefficient of permeability
i = hydraulic gradient
Q = quantity of water
A = area
t = time

In Figure 1 the meaning of Darcy's law is illustrated in simple form. A prismatic or cylindrical soil sample is exposed on the left side to a head

*Numerals refer to the bibliography at the end of this paper.

†This must not be confused with the seepage velocity $v_s = v_o \dfrac{1+e}{e}$, in which e = ratio of volume of voids to volume of solid matter. The average velocity through the soil is represented by the seepage velocity, while the discharge velocity determines the quantity of flow.

Darcy's Law
for Flow through Soils

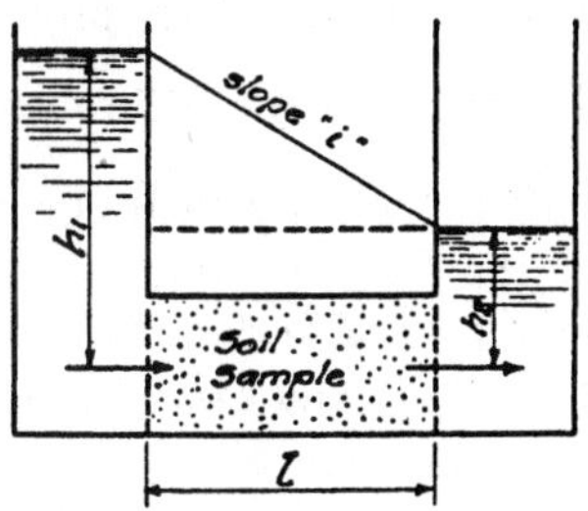

Quantity of Seepage	$Q = k \cdot i \cdot A \cdot t$
Discharge Velocity	$v = k \cdot i$
Hydraulic Gradient	$i = \dfrac{h_1 - h_2}{l}$
Area of Sample	A
Time	t
Coefficient of Permeability	k

FIG. 1.— DARCY'S LAW FOR FLOW THROUGH SOILS.

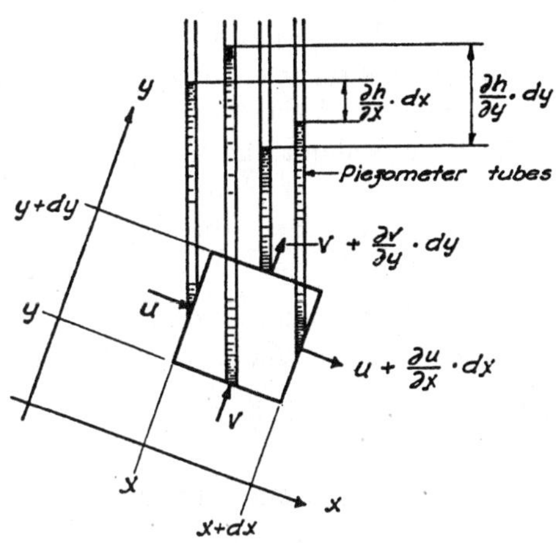

Condition of Continuity:

$$u\,dy + v\,dx = \left(u + \frac{\partial u}{\partial x}dx\right)dy + \left(v + \frac{\partial v}{\partial y}dy\right)dx$$

or

$$\frac{\partial u}{\partial x} + \frac{\partial v}{\partial y} = 0 \quad \cdots\cdots \quad (1)$$

Darcy's Law :

$$u = k \cdot \frac{\partial h}{\partial x} \quad \& \quad v = k \cdot \frac{\partial h}{\partial y}$$

(2) in (1)

$$\frac{\partial^2 h}{\partial x^2} + \frac{\partial^2 h}{\partial y^2} = 0$$

For three dimensions

$$\frac{\partial^2 h}{\partial x^2} + \frac{\partial^2 h}{\partial y^2} + \frac{\partial^2 h}{\partial z^2} = \nabla^2 h = 0$$

FIG. 2.— GENERAL DIFFERENTIAL EQUATION FOR THE FLOW OF WATER THROUGH HOMOGENEOUS ISOTROPIC SOILS.

of water h_1, and on the right side to a smaller head h_2. As a result, water will flow through the sample at a rate directly proportional to the hydraulic gradient, $i = \frac{h_1 - h_2}{l}$. If, for example, the difference in head $(h_1 - h_2)$ in Figure 1 is doubled, the quantity of seepage will also be doubled. This linear relationship suggests that the flow of water through the voids of most soils possesses the characteristics of laminar flow.

Darcy's law is frequently attacked as being incorrect. In general these attacks are based on misinterpretation of test results or improper technique of testing. In many cases they are due to loss of internal stability of the soil under the action of flowing water.

The reader may be assured that this law is valid for the study of seepage through dams.

C. General Differential Equation for the Flow of Water through Homogeneous Soil.

If water is percolating through a homogeneous mass of soil in such a manner that the voids of the soil are completely filled with water and no change in the size of the voids takes place, the quantity entering from one or several directions into a small element of volume of the soil (as shown in Figure 2) must be equal to the amount of water flowing out on the other faces of this element of volume during any given element of time. This condition, which is a statement of the fact that both water and soil are incompressible, can be expressed for the three-dimensional case by the following equation:

$$\frac{\partial u}{\partial x} + \frac{\partial v}{\partial y} + \frac{\partial w}{\partial z} = 0 \qquad (2)$$

This is known as the *equation of continuity*. (See Reference 5.) In this equation, u, v and w are the three components of the discharge velocity v_o. If $\frac{dh}{dl}$ represents the hydraulic gradient in the direction of flow and $\frac{dh}{dx}$, $\frac{dh}{dy}$ and $\frac{dh}{dz}$ are its three components, then Darcy's law:

$$v_o = k\frac{dh}{dl}$$

can also be expressed by the following equations:

$$\left.\begin{aligned} u &= k\frac{\partial h}{\partial x} \\ v &= k\frac{\partial h}{\partial y} \\ w &= k\frac{\partial h}{\partial z} \end{aligned}\right\} \qquad (3)$$

By substituting Equation (3) in Equation (2), one arrives at the general differential equation for the steady flow of water through isotropic soils. This has the form of a Laplace differential equation:

$$\frac{\partial^2 h}{\partial x^2} + \frac{\partial^2 h}{\partial y^2} + \frac{\partial^2 h}{\partial z^2} = 0 \qquad (4)$$

In our problem of seepage through dams we have to deal only with the two-dimensional case which is satisfied by the equation:

$$\frac{\partial^2 h}{\partial x^2} + \frac{\partial^2 h}{\partial y^2} = 0 \qquad (4a)$$

This equation represents *two families of curves intersecting at right angles.* (See Reference 5, p. 24 and 25.) In hydro-mechanics these curves are known respectively as the *flow lines* and the *equipotential lines* (or lines of equal head).

D. Forchheimer's Graphical Solution.

Although the general differential equation (4) has been solved only for few and simple cases of seepage, we can make use of certain geometric

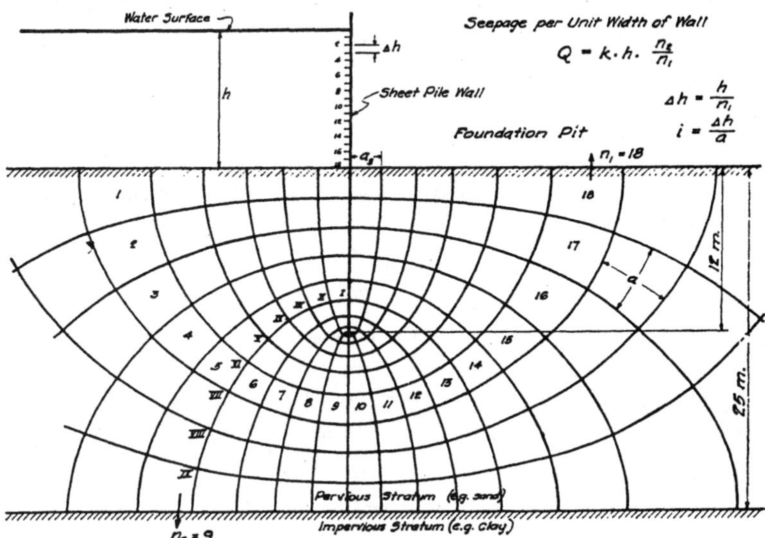

Fig. 3.— Flow Net beneath Sheet Pile Wall.

properties of flow lines and equipotential lines that permit graphical solutions for practically all two-dimensional seepage problems. This method was devised by Forchheimer (5), twenty-five years ago.

To explain this graphical method, the problem of determining the seepage beneath a sheet pile wall, shown in Figure 3 is chosen. The ground surface is a line of equal head or an equipotential line; the head being equal to the height of water standing above the ground surface, which is h on the left side, and zero on the right side of the wall. The bottom of the pervious soil stratum is a flow line; incidentally the longest flow line. The sides of the sheet pile wall and the short width at the bottom of the wall are the shortest flow line.

If, from the infinite number of flow lines possible within the given area, we choose only a few in such a manner that the same fraction Δq of the total seepage is passing between any pair of neighboring flow lines, and, similarly, if we choose from the infinite number of possible equipotential lines only a few in such a manner that the drop in head Δh between any pair of neighboring equipotential lines is equal to a constant fraction of the total loss in head h, then the resulting "flow net," Figure 3, possesses the property that the ratio of the sides of each rectangle, bordered by two flow lines and two equipotential lines, is constant. (See Reference 5, p. 82.) If all sides of one such rectangle are equal, then the entire flow net must consist of squares. Conversely, it can be proved that if one succeeds in plotting two sets of curves so that they intersect at right angles, forming squares and fulfilling the boundary conditions, then one has solved, graphically, equation (4a) for this problem. With experience, this method can be applied successfully to the most complicated problems of seepage and ground water flow in two dimensions, including seepage with a free surface.

After having plotted a flow net that fulfills satisfactorily these necessary conditions, one can derive therefrom, by simple computations, any desired information on quantity of seepage, seepage pressures, and hydrostatic uplift. For example, the total seepage per unit of length and per unit of time is determined from the following formula, which is simple to derive from Darcy's law:

$$q = kh\frac{n_2}{n_1} \tag{5}$$

in which n_1 is the number of squares between two neighboring flow lines, and n_2 the number of squares between two neighboring equipotential lines.

The maximum hydraulic gradient on the discharge surface, which influences the safety against "piping" or "blows," is equal to:

$$i_s = \frac{\Delta h}{a_s} \tag{6}$$

in which a_s is the length of the smallest square on the discharge surface, as indicated in Figure 3, and $\Delta h = \frac{h}{n_1}$, the drop in head between two adjacent equipotential lines.

To assist the beginner in learning the graphical method, the following suggestions are made:

1. Use every opportunity to study the appearance of well-constructed flow nets; when the picture is sufficiently absorbed in your mind, try to draw the same flow net

without looking at the available solution; repeat this until you are able to sketch this flow net in a satisfactory manner.

2. Four or five flow channels are usually sufficient for the first attempts; the use of too many flow channels may distract the attention from the essential features. (For examples see Figures 18b and c.)

3. Always watch the appearance of the entire flow net. Do not try to adjust details before the entire flow net is approximately correct.

4. Frequently there are portions of a flow net in which the flow lines should be approximately straight and parallel lines. The flow channels are then about of equal width, and the squares are therefore uniform in size. By starting to plot the flow net in such an area, assuming it to consist of straight lines, one can facilitate the solution.

5. The flow net in confined areas, limited by parallel boundaries, is frequently symmetrical, consisting of curves of elliptical shape. (For example see Figure 3.)

6. The beginner usually makes the mistake of drawing too sharp transitions between straight and curved sections of flow lines or equipotential lines. Keep in mind that all transitions are smooth, of elliptical or parabolic sharpe. The size of the squares in each channel will change gradually.

7. In general, the first assumption of flow channels will not result in a flow net consisting throughout of squares. The drop in head between neighbouring equipotential lines corresponding to the arbitrary number of flow channels, will usually not be an integer of the total drop in head. Thus, where the flow net is ended, a row of rectangles will remain. For usual purposes this has no disadvantages, and the last row is taken into consideration in computations by estimating the ratio of the sides of the rectangles. If, for the sake of appearance, it is desired to resolve the entire area into squares, then it becomes necessary to change the number of flow channels, either by interpolation or by a new start. One should not attempt to force the change into squares by adjustments in the neighbouring areas, unless the necessary correction is very small.

8. Boundary conditions may introduce singularities into the flow net, which are discussed more in detail in Appendix I, e.

9. A discharge face, in contact with air, is neither a flow line nor an equipotential line. Therefore, the squares along such a boundary are incomplete. However, such a boundary must fulfill the same condition as the line of seepage regarding equal drops in head between the points where the equipotential lines intersect.

10. When constructing a flow net containing a free surface one should start by assuming the discharge face and the discharge point and then work toward the upstream face until the correct relative positions of entrance point and discharge toe are attained. Hence, the scale to which a flow net with a free surface is plotted, will not be known until a large portion of the flow net is finished. For seepage problems with a free surface it is practically impossible to construct a flow net to a predetermined scale in a reasonable length of time.

E. Seepage through Dams; General Considerations.

In almost all problems concerning seepage *beneath* sheet pile walls or *through the foundation* of a dam, *all boundary conditions are known.* However, in the *seepage through* an earth dam or dike, the *upper boundary* or uppermost flow line is *not known*, but must first be found, thus introducing a complication. This upper boundary is a free water surface and will be referred to as the *line of seepage.*

Among the available theoretical solutions for seepage with a free surface there is one case which is of particular importance in connection with our problem. It is Kozeny's solution (6) of the flow along a horizontal impervious stratum that continues at a given point into a horizontal pervious stratum, thus representing an open horizontal discharge surface as shown in Figure 9d. In this case, all flow lines, including the line of seepage, and all equipotential lines are confocal parabolas with point *A* as the focus.

For the more common problem of seepage through cross-sections in which the discharge slope forms an angle with the horizontal between 0° and 180°, such as the open discharge on the downstream face of a dam or discharge into an overhanging slope of a very pervious toe, such as a rock

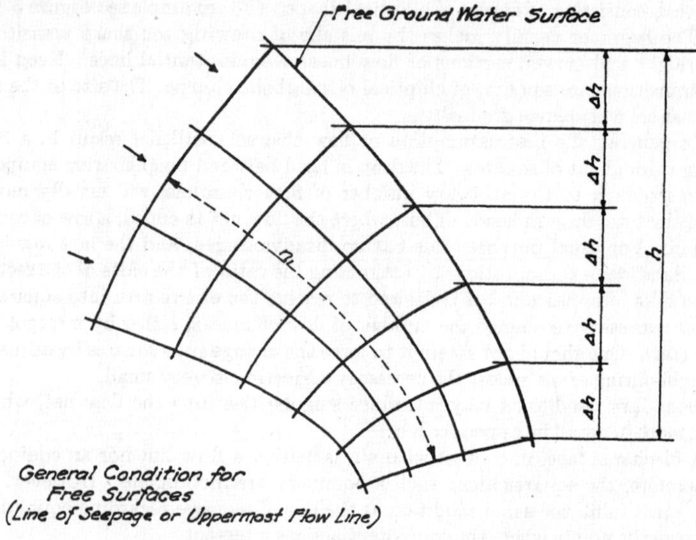

Fig. 4.— General Condition for Line of Seepage.

fill toe, one has to use either a graphical solution based on the construction of the flow net, or some approximate mathematical solution. In either case one must introduce certain conditions that the free water surface or line, of seepage must always fulfill.

The first condition is that the elevation of the point of intersection of any equipotential line with the line of seepage represents the head along this equipotential line. If we construct a flow net consisting of squares, then it follows that all intersections of equipotential lines with the line of seepage must be equidistant in the vertical direction. These distances, illustrated in Figure 4, represent the actual drop in head $\Delta h = \frac{h}{n_1}$ between any two neighboring equipotential lines.

The second condition refers to the slope of the line of seepage at the point of intersection with any boundary, as for example at the points of entrance and discharge and at the boundary line between different soils

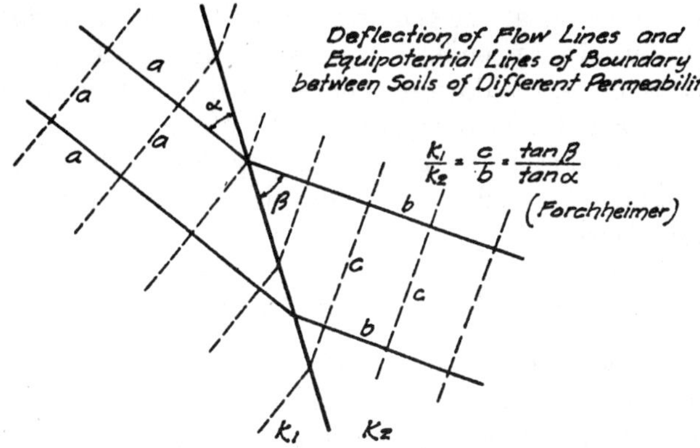

FIG. 5.— DEFLECTION OF FLOW NET AT BOUNDARY OF SOILS OF DIFFERENT PERMEABILITY.

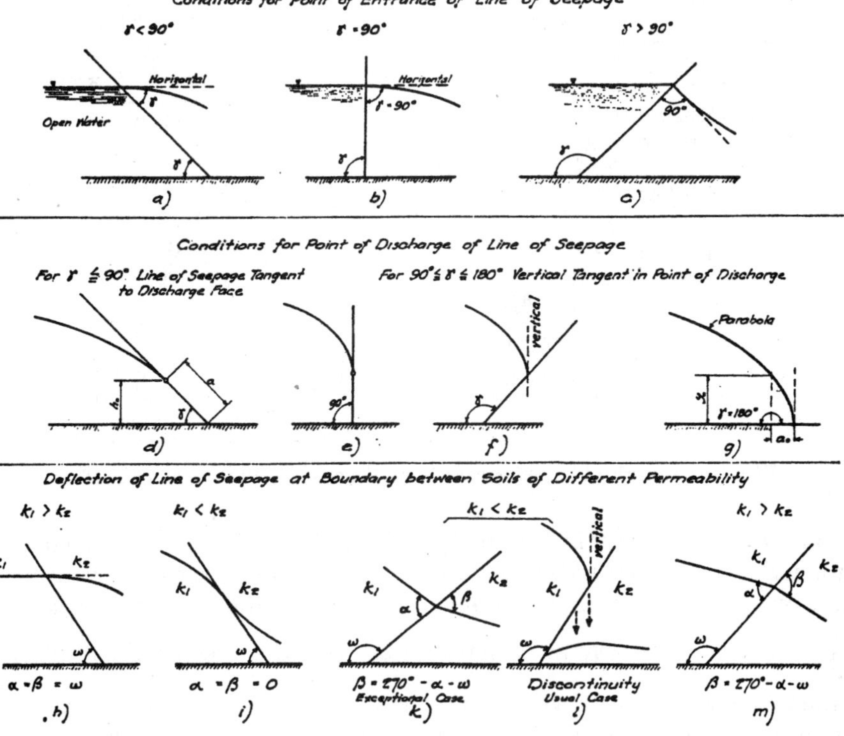

FIG. 6.— ENTRANCE, DISCHARGE AND TRANSFER CONDITIONS OF LINE OF SEEPAGE.

(See Figure 5). By considerations based on the general properties of a flow net, one can arrive at the conditions which must be fulfilled at such points of transfer. In Appendix I are assembled derivations for typical cases. If, for example, the downstream face is inclined less than or equal to 90°, one finds that the line of seepage must be tangent to that face at the discharge point. However, for all overhanging slopes, the tangent at the discharge point must be vertical. A summary of the possible combinations is assembled in Figure 6.

F. Seepage through Homogeneous Isotropic Earth Dams.

a. Approximate Solution for $\alpha < 30°$. The first approximate mathematical solution for determining the quantity of seepage and the line of seepage through a homogeneous earth section on an impervious base was

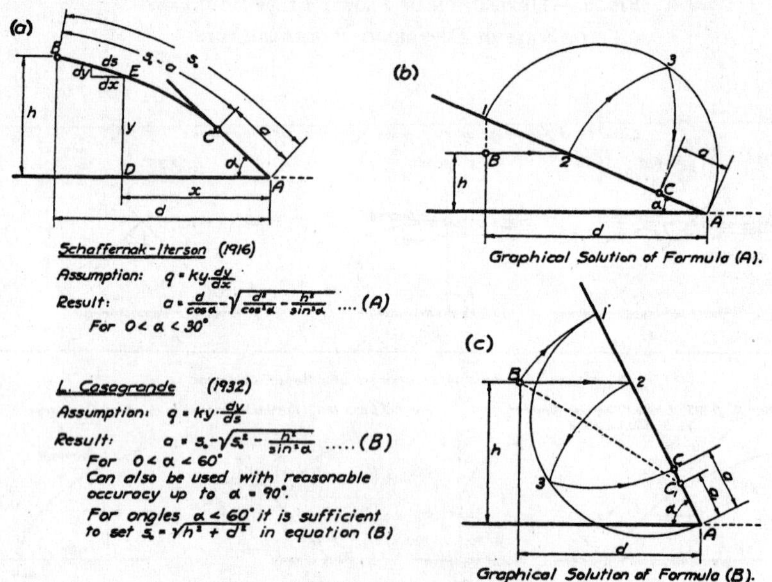

FIG. 7.—GRAPHICAL DETERMINATION OF DISCHARGE POINT FOR $\alpha < 60°$.

proposed independently in 1916 by Schaffernak (7) in Austria, and Iterson (8), in Holland. It is based on Dupuit's (9) assumption* that in every point of a vertical line the hydraulic gradient is constant and equal to the slope $\frac{dy}{dx}$ of the line of seepage at its intersection with that vertical line. This assumption represents a good approximation for the average hydraulic gradient in such a vertical line providing the slope of the line of seepage is relatively flat.

*On this same assumption are based the common methods of computing ground water flow toward wells.

With this assumption and the condition that the quantity of water flowing through any cross-section per unit of time must be constant, one can derive the differential equation for the line of seepage from Figure 7a:

$$q = ky\frac{dy}{dx} \tag{7}$$

The solution of equation (7) yields the equation of a parabola. Assuming that the quantities h, d, and α, in Figure 7, are known, and with the boundary conditions $y=h$ for $x=d$, and $dy/dx=tan\ \alpha$ for $x=a\ cos\ \alpha$, or $y=a\ sin\ \alpha$, integration leads to the following formula for the distance a which determines the discharge point C of the line of seepage on the downstream face of the dam:

$$a = \frac{d}{cos\ \alpha} - \sqrt{\frac{d^2}{cos^2\ \alpha} - \frac{h^2}{sin^2\ \alpha}} \tag{8}$$

$$q = k\ a\ sin\ \alpha\ tan\ \alpha \tag{9}$$

These equations differ from their original form in the use of the distance a, instead of its vertical projection, a change which provides a common basis for all theoretical developments in this paper. A further advantage

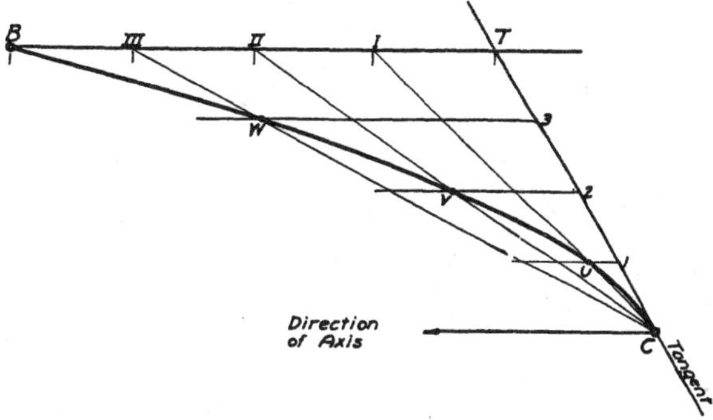

FIG. 8.— METHOD OF LOCATING POINTS ON A PARABOLA.

of this change resides in the possibility of determining graphically the distance a by means of a simple construction* which is shown in Figure 7b. The ordinate through the known point B of the line of seepage is extended to its intersection *1* with the discharge slope, and a semi-circle drawn through the points *1* and A, with its center on the discharge slope. Then a horizontal line through B is intersected with the discharge slope in point *2*, and the distance *2-A* projected onto the circle, yielding point *3*. The final step is to project the distance *1-3* onto the discharge slope. This yields the desired discharge point C. The proof for the validity of this method is readily found by comparing this construction with equation (8) and need not be discussed in detail.

*This construction is a simplification of another method which was proposed in References 10 and 11.

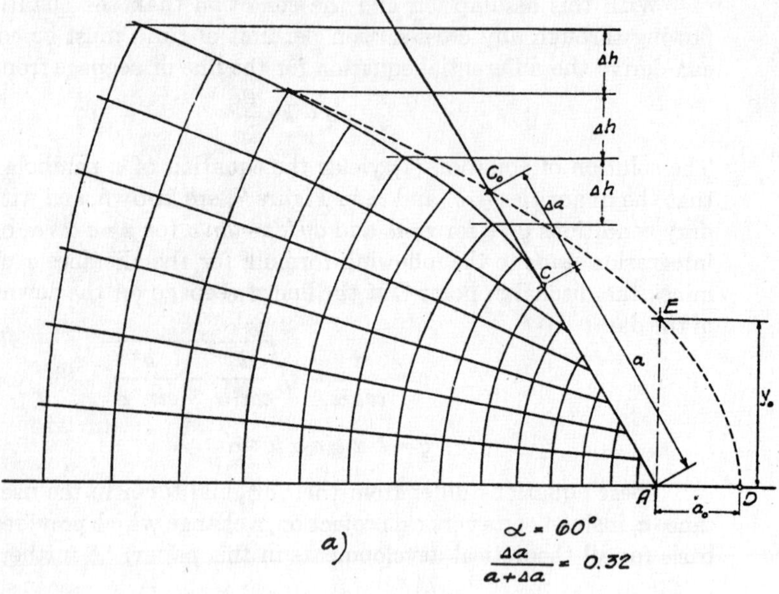

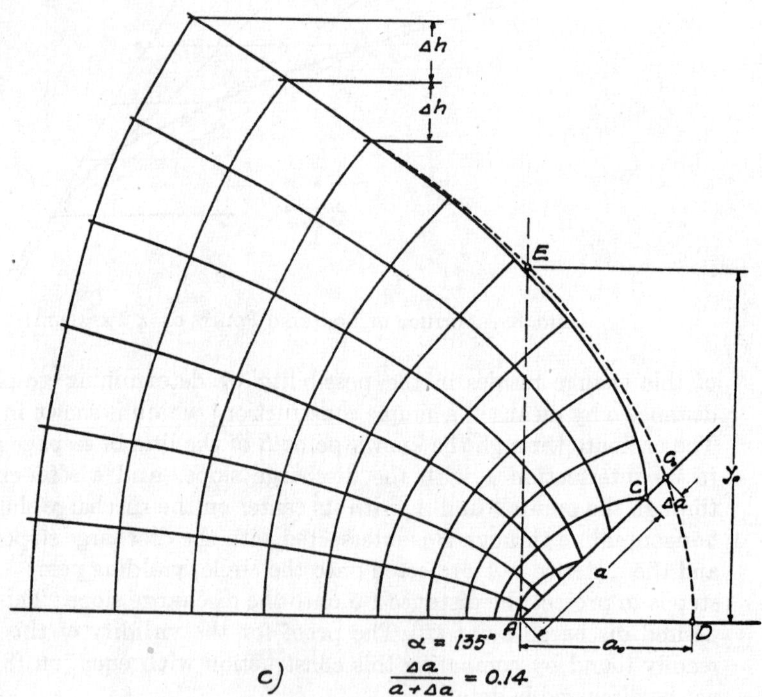

FIG. 9.— COMPARISON BETWEEN BASIC PARABOLA

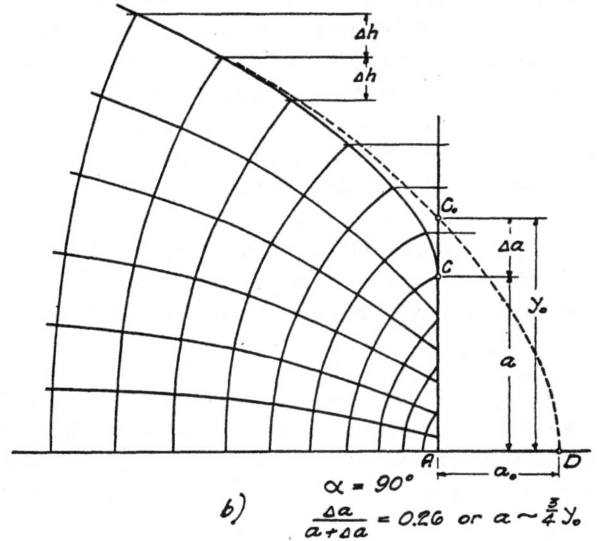

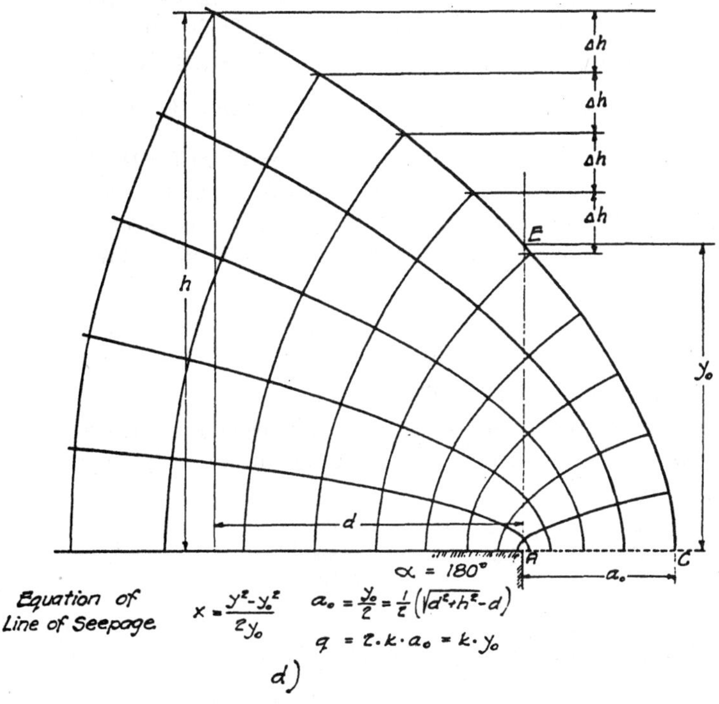

AND FLOW NETS FOR VARIOUS DISCHARGE SLOPES.

For the solution of many problems it is sufficient to know the discharge point of the line of seepage. If it is desired to draw the entire line of seepage, from the known point B to the discharge point C, one can make use of the graphical method, shown in Figure 8, for the rapid construction of any number of points on a parabola for which are known two points, the tangent to the parabola at one of these points, and the direction of the axis.

Through point B, Figure 8, one draws a line parallel to the axis and determines its intersection T with the tangent. Then one divides the distances B-T and C-T into an arbitrary number of equal parts, such as four parts. Points *I, II* and *III* are then connected with point C, and through points *1, 2* and *3* one draws lines parallel to the axis. The points where the lines through *1* and *I*, *2* and *II*, etc., intersect, are points of the parabola.

b. Approximate Solution for $\alpha > 30°$. The approximate solution by means of equation (8), or the corresponding graphical method shown in Figure 7b, gives satisfactory results for slopes of $\alpha < 30°$. For steeper slopes the deviation from the correct values increases rapidly beyond tolerable limits.

The causes for this deviation become apparent from a study of the flow net for a slope of $\alpha = 60°$, shown in Figure 9a. One can see that in the vicinity of the discharge point the size of the squares along the vertical line through the discharge point decreases only slightly towards the base. The average hydraulic gradient along this vertical line is larger than the hydraulic gradient $\frac{dy}{ds}$ along the line of seepage by less than 10 per cent. However, the sine of 60°, which is the true hydraulic gradient for the line of seepage at the discharge point, is only about one-half of the tangent of 60° used according to Dupuit's assumption. Hence the seepage can be analysed with a satisfactory degree of accuracy by means of the following equation:

$$q = ky\frac{dy}{ds} \tag{10}$$

This improvement was proposed by Leo Casagrande (10).

The difference between the use of the tangent and the sine of the slope of the line of seepage is best illustrated by the following numerical comparison for various angles:

Slope	*tan*	*sin*
30°	0.577	0.500
60°	1.732	0.866
90°	∞	1.000

Hence, for slopes <30° both methods may be used for practical purposes with equal advantage. For slopes > 30° the deviation by using $\frac{dy}{dx}$ becomes intolerably large, while the use of $\frac{dy}{ds}$ is very satisfactory for slopes up to 60°; and if deviations of 25 per cent. are permitted, it may even be used up to 90°, that is, for a vertical discharge face.

Gilboy (12) succeeded in finding an implicit solution of equation (10) which is recommended where greater accuracy is required than can be obtained by means of the graphical solution. The errors involved in the position of the discharge point, as obtained by one or the other method from equation (10), were investigated by G. P. Reyntjiens (13).

Using the symbols shown in Figure 7a, and assuming that in each vertical the hydraulic gradient is equal to $\frac{dy}{ds}$, equation (10) is the differential equation for the line of seepage. The solution of this equation cannot readily be expressed by rectangular coördinates x and y. (See References 12 and 13.) However, the use of s and s_o, measured along the line of seepage, does not represent any practical difficulty in the actual application of this method. The quantity a, which determines the discharge point for the line of seepage, is found by a simple integration:

$$qs = \frac{ky^2}{2} + \text{constant}$$

$$\begin{array}{l}\text{Boundary} \\ \text{Conditions}\end{array} \begin{cases} s = a, \; y = a \sin \alpha, \; q = ka \sin^2 \alpha \\ s = s_o, \; y = h \end{cases}$$

$$a = s_o - \sqrt{s_o^2 - \frac{h^2}{\sin^2 \alpha}} \tag{11}$$

$$q = k\,a \sin^2 \alpha \tag{12}$$

Again, the quantities employed in these equations differ from the original form as presented in References 10 and 11 to permit a simple graphical solution. This graphical solution of equation (11) is illustrated in Figure 7c, and can be easily verified. It requires first an assumption for the discharge point. The length $(s_o - a)$ is simply taken equal to the straight line from B to C_1, shown as a dotted line in Figure 7c. The slight error which is introduced when $(s_o - a)$ is replaced by a straight line has a negligible effect on the positions of the discharge point. In fact, for slopes $\alpha \leq 60°$, it is entirely tolerable to replace the length s_o by the straight distance from $\overline{AB} = \sqrt{h^2 + d^2}$, thus eliminating trial constructions. The construction is very similar to that shown in Figure 7b, except that point *1* is found by rotating distance $\overline{C_1B}$, or $\overline{AB}$, around point A.

If deviations up to 25 per cent. are permitted, the simplified value $s_o = \sqrt{h^2 + d^2} = \overline{AB}$ may be used also for slopes up to 90°. For a vertical slope the formula for a is reduced to the following simple form:

$$a = \sqrt{h^2 + d^2} - d \tag{13}$$

In other words, for a vertical discharge face the height of the discharge point for the line of seepage can be approximated by the difference between the distance $\overline{AB} = \sqrt{h^2+d^2}$ and its horizontal projection d.

c. Solution for a Horizontal Discharge Surface ($\alpha = 180°$). In 1931, Professor Kozeny (6) published a rigorous solution for the two-dimensional problem of ground water flow over a horizontal impervious surface which continues at a given point into a horizontal discharge face, as shown in Figure 9d. Kozeny's theoretical solution yields, for the flow lines and equi-potential lines, two families of confocal parabolas, with point A, where the impervious and pervious sections meet, as the focus.

The equation for the line of seepage can be conveniently expressed in the following form:

$$x = \frac{y^2 - y_o^2}{2y_o} \tag{14}$$

in which x and y are the coördinates with the focus as origin, and y_o the ordinate at the focus $x = 0$.

If the line of seepage is determined by the coördinates d and h of one known point, then the focal distance a_o and the ordinate y_o are computed from the following equation:

$$a_o = \frac{y_o}{2} = \frac{1}{2}\left(\sqrt{d^2+h^2} - d\right) \tag{15}$$

for which a graphical solution is recommended. (See Figure 11c.) The quantity y_o is simply equal to the difference between the distance $\sqrt{d^2+h^2}$ from the given point (d, h) to the focus of the parabola, and the abscissa d. The focal distance a_o is equal to one-half the ordinate y_o.

In addition to these simple relationships it is of advantage to remember that the tangent to the line of seepage at $x = 0$ and $y = y_o$ is inclined at 45°.

The quantity of seepage per unit of width is, according to Kozeny's solution:

$$q = 2ka_o = ky_o \tag{16}$$

It is indeed fortunate that the problem of seepage with a horizontal discharge face has such a simple solution, not only because of the fact that in modern earth dam and levee design horizontal drainage blankets in the downstream section are assuming considerable importance, but also because this solution permits fairly reliable and simple estimates for the position of the line of seepage for overhanging discharge slopes.

d. Approximate Solutions for Overhanging Discharge Surfaces ($90° < \alpha < 180°$). Although the determination of the line of seepage and its point of exit for an overhanging discharge face, such as a rock fill toe, is of importance in the design of earth dams, little attention has been paid to this problem. Experimental results were published by Leo Casagrande (10 and 11), which permit a reasonably accurate determination of the line of seepage. Later, in 1933, the author checked the results of these model tests by means of graphical solutions, of which a few typical examples are shown in Figure 9. These solutions check well with the experimental results just referred to, and are illustrated for one example in Figure 10. Such studies convinced the author that Forchheimer's graphical method for the determination of the flow net can be utilized for the solution of seepage problems with a free surface. The application of the graphical method to such problems requires considerable skill. This can be acquired only by extensive use of this method. Solutions such as those shown in Figure 9

FIG. 10.— MODEL TEST ON OTTAWA STANDARD SAND WITH OVERHANGING DISCHARGE SLOPE.

Note zone of capillary saturation above line of seepage — upper dye line.
After L. Casagrande (10).

required many hours of work; sometimes several days were spent on one case,— because of the complications that the unknown upper boundary introduces into such seepage problems.

After sufficient graphical solutions to permit a rapid determination of the line of seepage for any slopes $60° \leq \alpha \leq 180°$, had been accumulated, the author's attention was called to Kozeny's (6) theoretical solution for $\alpha = 180°$. This proved a splendid opportunity for checking the accuracy of a purely graphical solution of a seepage problem with a free water surface. Figure 9d represents the original graphical solution. The difference between this solution and the theoretical solution is not more than 3 per cent. for any point on the line of seepage. Therefore, no attempt was made to include the theoretical solution in Figure 9d. This remarkable accuracy of the graphical method should convince critics that the method is not a plaything but has great merit and that the time spent on acquiring sufficient skill in this method is well invested.

To simplify the application of the graphical solution for very steep and overhanging discharge slopes such as are shown in Figures 9, a, b and c, these flow nets were compared with Kozeny's theoretical solution for a horizontal discharge face. For the sake of simplicity, the line of seepage for $\alpha = 180°$, which is represented by equations (14) and (15), and Figure 9d, will be referred to as the "basic parabola."

In Figure 9, the basic parabola is plotted into every case illustrated. The basic parabola and the actual line of seepage approach each other very quickly and for practical purposes may be assumed to be identical for points whose ordinates h are less than their horizontal distances from the discharge point C. By comparing the actual line of seepage for a given discharge slope with the basic parabola, we find that the intersection of this parabola with the discharge face is a distance Δa above the discharge point of the line of seepage. The ratio $c = \frac{\Delta a}{a + \Delta a}$ (see Figure 9) gradually decreases with increasing angle α. The ratio C is equal to 0.32 for $\alpha = 60°$; for a vertical surface ($\alpha = 90°$) it is 0.26; and for $\alpha = 180°$ the ratio C is, of course, equal to zero.

In order to utilize these relationships for determining the line of seepage and the discharge point for steep vertical and overhanging slopes, there has been plotted in Figure 11 the relationships between the ratio $\frac{\Delta a}{a + \Delta a}$ and the angle α. The quantity $a + \Delta a$ is found by intersecting the basic parabola with the discharge slope, an operation that can be performed either graphically or mathematically. In both cases one computes or constructs first $y_o = \sqrt{d^2 + h^2} - d$, for the known or estimated starting point of the line of seepage. The graphical determination of the intersection C_o is usually preferred, since the basic parabola is needed for the determination of the line of seepage. The construction of the parabola is best performed in the manner illustrated in Figure 8. For tangent $\overline{CT}$, either the tangent at the vertex of the parabola, or the tangent under 45° at $x = 0$ and $y = y_o$ can be used.

The points on the curve representing the relation between α and $c = \frac{\Delta a}{a + \Delta a}$, in Figure 11b, are derived from the graphical solutions. Note how close to a smooth curve these points lie. This is another demonstration of the degree of accuracy that can be obtained by means of the graphical method.

The quantity c is not only a function of the angle α but it also varies somewhat with the relative position of points B_1, or B_2, and C_o (Figure 11d). The maximum variations in c, for the limits that would normally be encountered in earth dams, are about ±5 per cent. The curve in Figure 11b was determined for a relatively short distance from the entrance to the discharge point of the line of seepage, in consideration of the importance of stratification in earth dams which is discussed in the next chapter.

Having plotted the basic parabola and determined the discharge point by means of the $c-\alpha$ relation, Figure 11, and knowing the tangent to the seepage line at the discharge point, it is an easy matter to draw with a fair degree of approximation the entire line of seepage, as shown in the various cases in Figure 9.

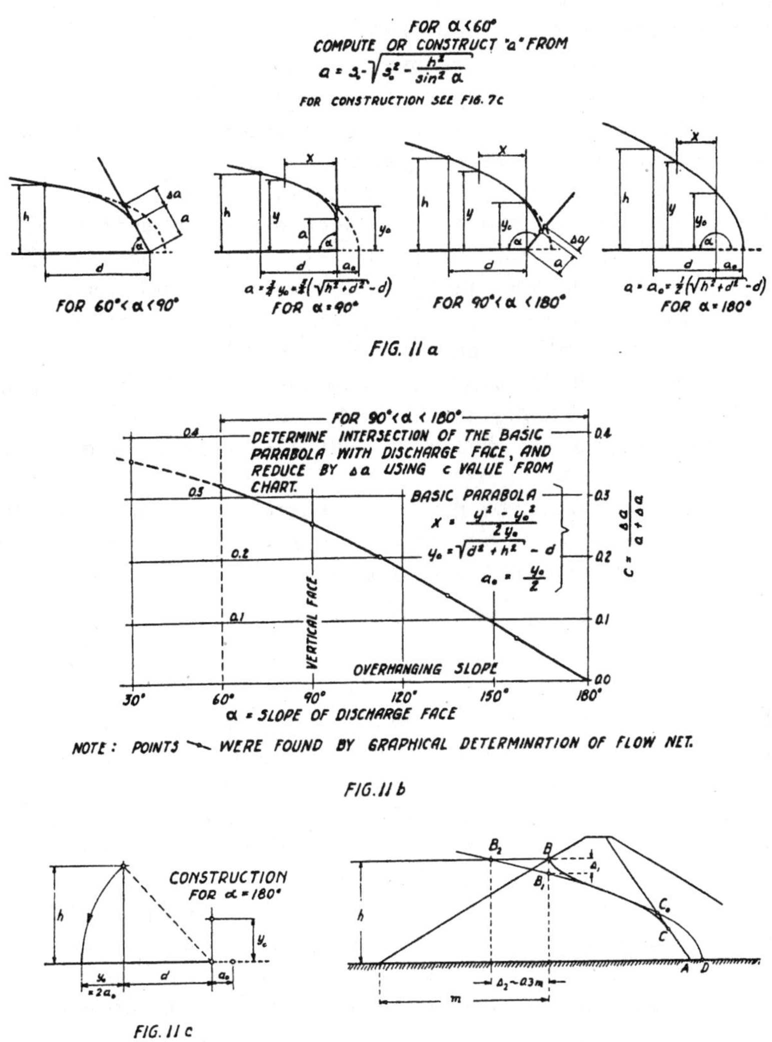

FIG. 11.— APPLICATION OF BASIC PARABOLA TO DETERMINATION OF DISCHARGE POINT OF LINE OF SEEPAGE.

e. Correction for Upstream Slope; Quantity of Seepage. Due to the entrance condition for the line of seepage and due to the fact that Dupuit's assumption is not valid for the upstream wedge of a dam, the line of seepage deviates from the parabolic shape. For the usual shape of a dam, there is an inflection point with a sharp curvature in the first section of the line of seepage, while for a vertical entrance face there is only an increase in curvature without reversal of direction.

For an accurate seepage analysis, these deviations should be taken into consideration. Referring to Figure 11d, it would be necessary to know in advance the position of one point of the parabolic curve in the vicinity of the entrance point B. L. Casagrande has chosen the intersection B_1 of the ordinate through the entrance point with the continuation of the parabolic line of seepage and has expressed the correction $\overline{BB_1} = \Delta_1$ as a function of d, h and the slope of the entrance face. A graphical presentation (11) facilitates the finding of the necessary correction Δ_1.

Somewhat simpler is the following approach. Instead of selecting point B_1 for the start of the theoretical line of seepage, we choose its intersection B_2 with the upstream water level. The corresponding correction Δ_2 is about ⅓ to ¼ of the horizontal projection m of the upstream slope, or for average conditions $\Delta_2 = 0.3m$. This is easy to remember and dispenses with the necessity for tables or graphs for the correction. The determination of the line of seepage is then carried out with point B_2 as the starting point. The actual shape of the first portion of the line of seepage, starting at point B, can easily be sketched in, so that it approaches gradually the parabolic curve, as shown in Figure 11d.

The quantity of seepage q per unit of length can be computed either from equations (12) or (16). If we substitute in these equations the known quantities, they appear in the following form:

$$q = k(\sqrt{h^2+d^2} - \sqrt{d^2 - h^2 \, cotan^2 \, \alpha}) \, sin^2 \alpha \tag{17}$$

and

$$q = k(\sqrt{d^2+h^2} - d) \tag{18}$$

For the great majority of cases encountered in earth dam design, both equations give practically the same result, so that the simpler equation (18) should be used for general purposes. In other words, the quantity of seepage is practically independent of the discharge slope, and is equal to the quantity that corresponds to the basic parabola. Only in those cases in which the starting point of the line of seepage is very near the discharge face will the difference between the two equations warrant the use of equation (17).

For the comparatively rare case in which the presence of tail water must be considered in the design, the determination of the line of seepage and of the quantity can be performed by dividing the dam horizontally at tail water level into an upper and lower section. The line of seepage is determined for the upper section in the same manner as if the dividing line were an impervious boundary. The seepage through the lower section is determined by means of Darcy's law, using the ratio of the difference in head over the average length of path of percolation as the hydraulic gradient. The total quantity of seepage is the sum of the quantities flowing through the upper section and the lower section. The results obtained by this rather crude approximation agree remarkably well with the values obtained from an accurate graphical solution.

Those readers who are interested in data showing how the results of seepage tests agree with the computed line of seepage and seepage quantity, using the methods described in this chapter, should consult References 10 and 11.

G. Seepage through Anisotropic Soils.

By a combination of the various methods of approach that have been outlined, and with proper consideration of boundary conditions as summarized in Figure 6 and discussed more in detail in Appendix I, one can arrive at a reliable determination of the line of seepage through even the most complicated cross-sections of earth dams. The cross-section may consist of portions with widely different permeabilities; however, each homogeneous section in itself is assumed to be isotropic, that is, possessing the same permeability in all directions. Unfortunately, this is practically never the case. Even a uniform clean sand, consisting of grains of the usual irregular shape, when placed in a glass flume for the purpose of building up a model dam section, does not produce an isotropic mass. The grains orientate themselves in such a manner that the coefficient of permeability is not uniform in all directions but larger in a more or less horizontal direction. As a consequence, the entire flow net is markedly influenced, resulting in considerable deviations from the theoretical flow net for an isotropic material.

Only by using a very uniform sand consisting of spherical grains, and by making tests on a sufficiently large scale, to reduce the capillary disturbance, can one arrive at test results that are in good agreement with theory. For this reason most of the tests described in References 10 and 11 were carried out on Ottawa standard sand.

Soils, in their natural, undisturbed condition, are always anisotropic in regard to permeability even if they convey to the eye the impression of being entirely uniform in character. If signs of stratification are visible, then the permeability in the direction of stratification may easily be ten times greater than that normal to stratification. For distinctly stratified soils this ratio can be very much larger than ten.

When soils are artificially deposited, as in the construction of a dam or dike, stratification develops to a greater or less degree. Such stratification has always been recognized by engineers as being undesirable, and for this reason special construction methods have been developed to disturb or destroy it. The hydraulic-fill core, during its construction, is frequently stirred with long rods in order to break up stratification as much as possible.

Sheepsfoot rollers are effective in compacting earth fills without creating distinct stratification. However, in spite of such precautionary measures, a certain amount of stratification remains. In addition, it is practically impossible to eliminate considerable variations in the general character of the material in the borrow pit; especially variations in permeability, which will result in substantial variations in the permeability of the dam from layer to layer. These cannot be eliminated by thorough rolling. Even the most carefully constructed rolled earth dams possess a considerably greater average permeability in a horizontal than in a vertical direction. Therefore, thorough investigation of variations in the character of the borrow pit materials forms an important part of preliminary studies. Taking into consideration the uncertainties that are always encountered in dealing with soil deposits and cannot be completely eliminated by the most elaborate investigations, it is essential that we should be conservative in the assumptions on which the design of an earth dam is based. This requires special attention to the possible degree of anisotropy in the dam.

The question of seepage through anisotropic soils was investigated for the first time and solved by Samsioe (14) in 1930. Fortunately the solution is simple and lends itself readily to practical application. The flow net of an anisotropic soil does not possess the usual characteristics of a flow net. However, it can be reduced, by the application of an appropriate geometric transformation, to an ordinary flow net. Designating the maximum and minimum coefficient of permeability for an anisotropic soil as k_{max} and k_{min}, it can be shown mathematically (see Reference 16) that by transforming the entire cross-section in such a manner that all dimensions in the direction of k_{max} are reduced by the factor $\sqrt{\frac{k_{min}}{k_{max}}}$, or that all dimensions in the direction of k_{min} are increased by the factor $\sqrt{\frac{k_{max}}{k_{min}}}$, the problem is again reduced to a solution of Laplace's equation. In other words, the flow net in the transformed section has the same characteristic flow lines and equipotential lines as previously discussed in this paper. Among others. Forchheimer's graphical method and all approximate methods suggested in this paper are applicable to the transformed section. After having found the line of seepage, or the entire flow net, in the transformed section, it is a simple matter to project this characteristic flow net back into the true section, in which flow lines and equipotential lines will not generally intersect at right angles. It should be noted that the hydraulic gradient at any point of the flow net and the magnitude of seepage pressures can only be determined in the true section, while the distribution of pore pressures and of hydrostatic uplift can be derived from either section.

The quantity of seepage can be computed from the transformed section on the basis of the coefficient of permeability $\bar{k} = \sqrt{k_{min.} \, k_{max.}}$ For proof see Reference 16.

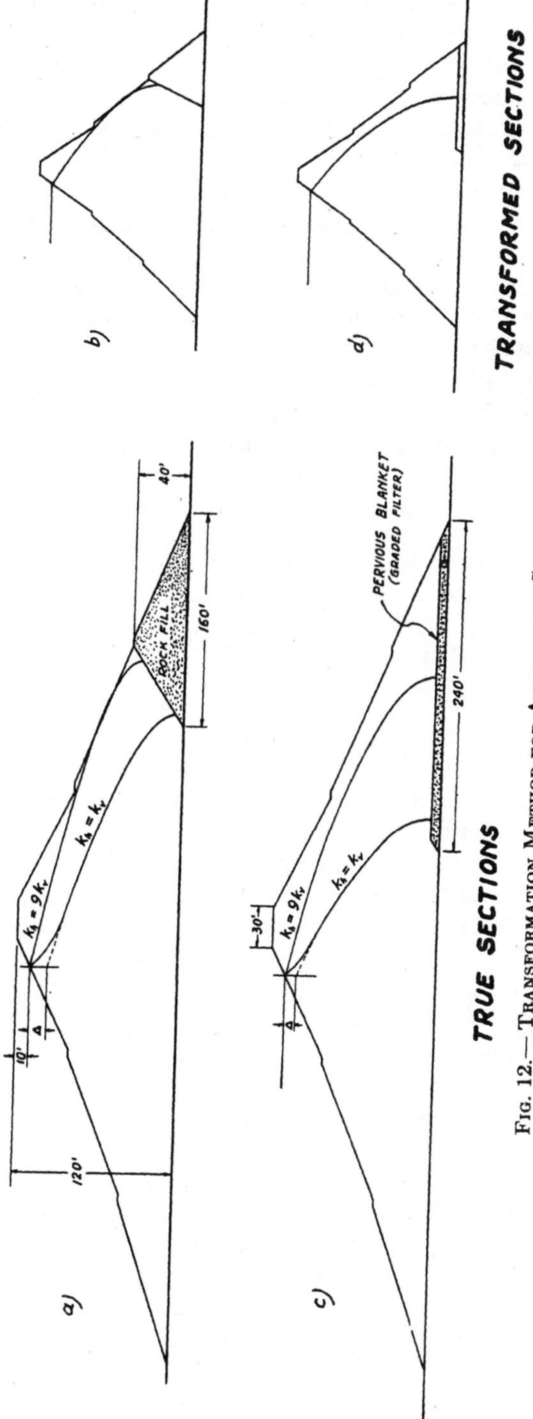

FIG. 12.—TRANSFORMATION METHOD FOR ANALYSIS OF STRATIFIED DAM SECTIONS.

Further information on seepage through stratified soils, the transformation theory, and examples may be found in References 15, 16 and 17.

The application of the transformation method is illustrated by the simple example of a rolled earth dam with a rock fill toe, shown in Figure 12. The dimensions and slopes of this dam are such that if a suitable soil is used, hardly any doubt would be raised regarding its stability. The rock fill toe seems to represent ample provision for safe discharge of seepage water. Indeed, the line of seepage, assuming isotropic soil, does fall well within the downstream face as shown in Figure 12a, (coefficient of permeability in horizontal direction k_h equal to coefficient in vertical direction k_v). However, if this dam is carelessly built of various types of soils with widely different permeabilities, a structure may well result that is many times more pervious in the horizontal direction than in the vertical direction. In the example shown in Figure 12, $k_h = 9k_v$ was chosen. On the right-hand side a new cross-section of the dam is plotted in which all horizontal dimensions are reduced by the factor $\sqrt{\frac{k_v}{k_h}} = \frac{1}{3}$. Then the line of seepage is determined in accordance with the methods outlined previously, and projected back into the true cross section. As can be seen in Figure 12a, the line of seepage for $k_h = 9k_v$ does intersect the downstream face, which is an undesirable condition that may in the course of time lead to a partial or complete failure of the structure.

H. Remarks on the Design of Earth Dams and Levees.

The question may arise of how to construct the downstream portion of a simple rolled earth dam, so that the line of seepage will remain a safe distance inside the downstream face of the structure, when only small quantities of coarse material are available. A simple solution is suggested in Figure 12c, in which a pervious blanket below the downstream portion of the dam is employed to control the position of the line of seepage to any desired extent. Such a blanket should be built up as a graded filter, carefully designed, to prevent erosion of any soil from the dam.

Whenever a dam or levee consists essentially of a uniform section of relatively impervious soil, e.g., possessing an average coefficient of permeability of less than 1 x 10^{-4} cm. per sec., the pervious blanket may well be extended as far as the centerline of the structure, as shown in Figure 13c. Such a design would add much more to the stability of the entire downstream portion, including the underlying foundation, than could be accomplished by a substantial flattening of the downstream slope. *A levee built in the conventional manner, with a downstream slope of 1 on 5, would possess less stability than a well-compacted levee in which the downstream slope is made as steep as 1 on 2, but which contains a filter blanket of the type shown in Figure 13c.* In the example illustrated in Figure 13c, it was assumed that a pervious foundation stratum lies beneath the levee and that the permea-

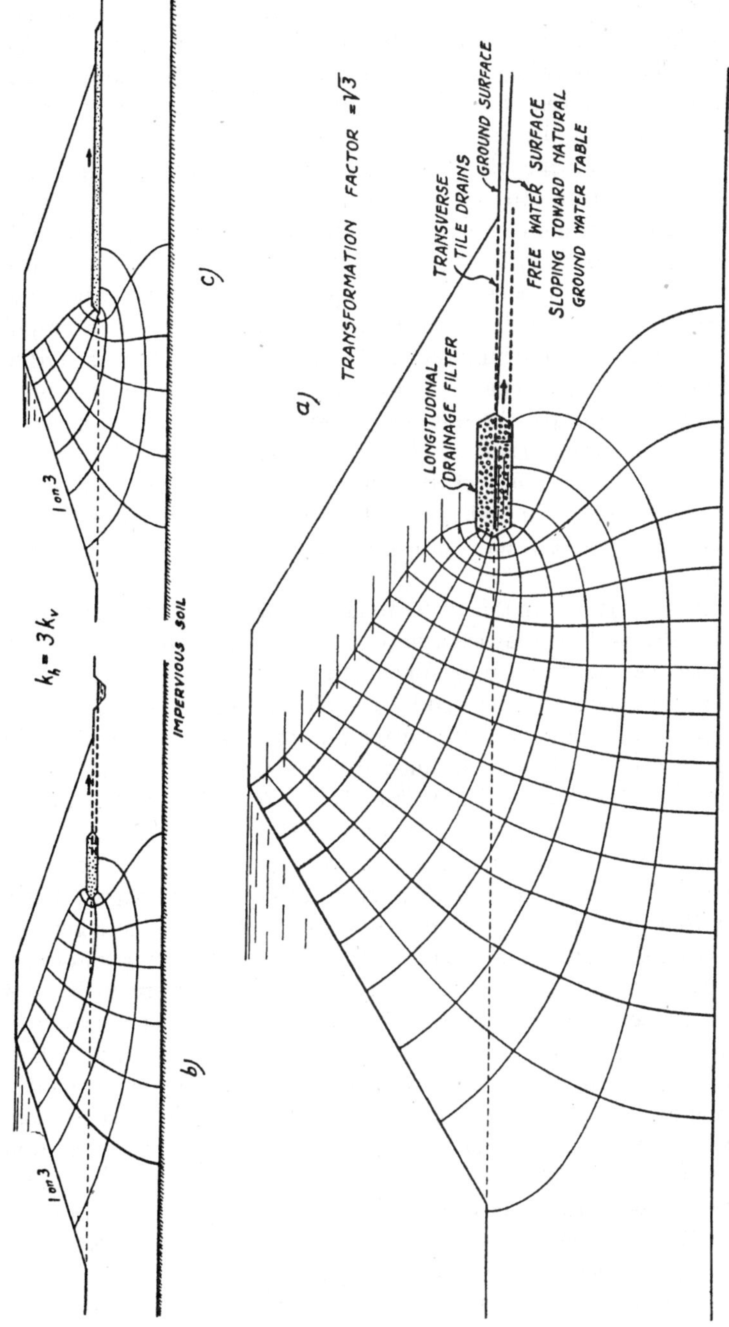

FIG. 13.— EFFECT OF INTERNAL DRAINAGE ON FLOW NET.

bilities of the levee material and the underlying foundation are the same, with a ratio of $k_h/k_v = 3$. The transformation which was required for obtaining the flow net, is illustrated by another example shown in Figures 13a and b. The assumptions are the same as in Figure 13c, except that the filter blanket is reduced to a longitudinal drainage strip with frequent transverse outlets. A tile drain may also be embedded within the core of the longitudinal drainage filter to increase the capacity of the drainage system if the structure consists of relatively pervious soils. This type of drainage would be employed where the quantities of suitable material for the drainage layer are very limited. To obtain the flow net, the true section was transformed into a steeper section, using as transformation factor $\sqrt{k_h/k_v} = \sqrt{3}$. The flow net was then obtained by Forchheimer's graphical method, by gradual approximation. Note the equi-distant horizontal lines intersecting the line of seepage. These were plotted before starting the flow net. In this example the line of seepage is not identical with the basic parabola, because the surface of the foundation on which the structure rests is not a flow line. However, it is convenient to use the basic parabola as a general guide for the first plot of the flow net. After a satisfactory solution is found, the flow net is projected back into the true cross section, Figure 13b.

No attempt is made in this paper to discuss in detail the important and interesting relationships that exist between the stability of earth dams or dikes and the seepage through and beneath them. However, it should be emphasized that the forces exerted by percolating water upon the soil can be very appreciable, and are often a maximum in critical points. These seepage forces are readily determined from a well-constructed flow net, and can then be combined with gravity forces for the stability analysis. Anyone who has made comparative studies of the seepage forces that may exist in dams and their foundations, must be impressed by the paramount importance of the design of those features that control seepage. It is not surprising that on the basis of empirical knowledge levees have been constructed with flat slopes. A levee built in the conventional manner of sandy soil, with slopes of 1 on 2 or 1 on 3, would be an unsafe structure. However, substantial flattening of the slopes is a very costly way of increasing its safety. Besides, even very flat slopes do not necessarily provide sufficient safety against undermining, particularly when a levee rests on a stratified, pervious foundation. In view of the large expenditures on levee construction which the next decade will bring, investment in research in this field would pay rich dividends if new designs for levees were developed that would not only be much safer than those built in the past, but considerably less expensive. *The widespread opinion among engineers that in earth dam and levee design "section makes for safety" needs to be revised.*

Many failures of levees are due to undermining caused by seepage through the foundation. Unless drainage provisions, as shown for example in Figure 13, are provided, the largest concentration of flow lines, both through and beneath the dam, occurs at the downstream toe where the soil is not confined. As stated before, flattening the slopes does not greatly improve this condition. In some cases, particularly when the foundation soils are porous and distinctly stratified, drainage provisions within a levee may not be sufficient protection. In such circumstances it may be beneficial to drill frequent holes into the foundation beneath the future longitudinal drain, and to fill these holes with coarse material. Such "drainage wells" have been employed already for another purpose, the relief of upward pressure on an overflow dam (see Reference 18). Properly designed and constructed drainage wells would effectively destroy any serious flow concentration at the toe of the structure. In other cases, particularly for a very pervious, but relatively thin foundation stratum, the use of a sheet-pile cut-off may be the ideal solution.

Improvements in levee design, as suggested here, are of a nature that will not produce interference with modern construction methods. Drainage wells, longitudinal and transverse drains or drainage blankets must all be built before construction of a levee is started. The building of the levee can then proceed in the same manner as if the drainage structures did not exist, thus permitting full use of large drag lines and tower machines.

I. Seepage through Composite Sections.

For the purpose of controlling seepage and utilizing available soils to best advantage, it is usually necessary to build dams of several sections with widely different coefficients of permeability. Since it is common to require that the ratio in permeability between neighboring sections should be at least one to ten, it is rarely necessary to determine the flow net for the entire dam if a careful study is made of the least pervious sections. However, in some cases it may be necessary to determine the position of the entire line of seepage. In Figure 14 is reproduced an example given by L. Casagrande (12), showing the line of seepage for a combination of two sections, with the downstream section built of soil which is five times more

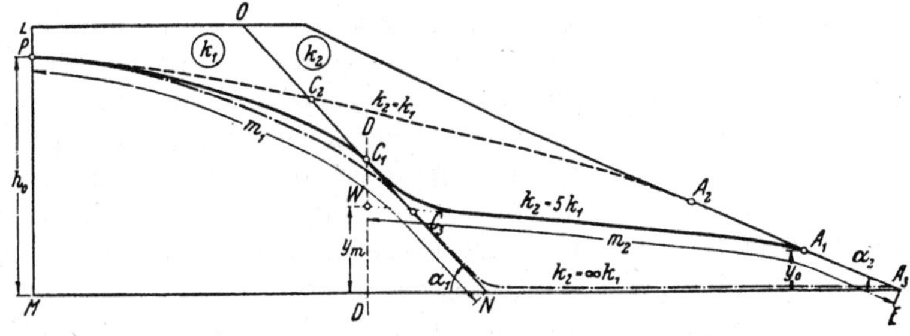

Fig. 14.— Line of Seepage for a Combined Section.
The computed result was verified by model experiments.
After L. Casagrande (11).

pervious than the central section. The line of seepage through such a composite section is found by changing the assumed position for the point of intersection of the line of seepage with the boundary until the quantities flowing through both sections are the same. After the correct position of the line of seepage has been determined, one can also develop any desired portions of the flow net.

J. Comparison between Forchheimer's Graphical Method, Hydraulic Model Tests and the Electric Analogy Method.

The purpose of model tests for seepage studies can be twofold (1) determination of the flow net for a given cross-section, assuming that the soil is isotropic; (2) determination of the flow net if the model is build up in such a manner that it resembles the prototype as to possible stratification, character of the soils, etc.

For the first purpose it is essential that the material used shall consist of grains as nearly spherical and as nearly of one size as possible. In addition, the models must be large enough so that the height of capillary rise will not distort the line of seepage, particularly in that portion of the flow net near the discharge point of the line of seepage. (See Figure 10.) Use of Ottawa Standard Sand has given good results. Since one should not go much below the size of Ottawa Standard Sand on account of the distorting effect of capillary rise, one is obliged to use artificial spheres, such as glass spheres, of appropriate sizes for models containing sections with different coefficients of permeability. For such coarse materials the validity of Darcy's law must be checked. The results of careful model tests conducted with such materials agree well with the solutions obtained by the graphical method, the electric analogy method, or rigorous theoretical solutions, so far as the latter are available.

For the second purpose, the testing of models similar to the prototype, one has to know first of all how the coefficient of permeability varies in the prototype, not only in its various sections, but particularly within each section due to anisotropy. Then one must build the model to imitate, on a small scale, these conditions. It is a waste of time and money to build a model using the same soil as in the prototype without attention to the anisotropic conditions in the prototype. Such a model does not represent the prototype; nor are the results comparable to the conditions for isotropic materials, because the inevitable irregularities and stratification due to the method of building the model are reflected in the resulting flow net to such an extent that the flow net looks very much like a beginner's attempt at employing the graphical method. The results of such model tests will lie somewhere between the conditions for an isotropic model and the actual conditions in the prototype and will tell practically nothing that can be of assistance in our problem; on the contrary, such results are often very confusing, particularly when the tests are made by men inexperienced in theoretical and graphical solutions and, therefore, are unable to interpret properly the test results.

The electric analogy method, when used by an experienced operator, is a useful and accurate method for *direct determination of flow nets between given boundaries.* Composite sections consisting of soils with different permeability can also be investigated by this method. Unfortunately, the method does not permit the *direct* determination of the line of seepage for those problems in which the upper surface is not a fixed boundary. Another disadvantage of this method is that it requires an accurate apparatus and the construction of a special testing model for every problem. If compared with graphical solutions, the electrical method is more expensive and requires more time. It should also be mentioned that a satisfactory presentation of the results obtained by means of the electrical method requires knowledge and application of the graphical method. In some instances the amount of work required to transform the test results into a good-looking flow net would have been enough to produce this flow net by the graphical method without assistance of the electrical apparatus. The graphical method will serve as an excellent check on the electrical method and should be used whenever accurate solutions are sought.

Another important advantage of the graphical method is that the process of finding the flow net for a proposed section almost inevitably suggests changes in the design which would improve the stability of the structure, and often its economy. With some experience in the use of the graphical method the effects of changes in one or the other detail of the design can quickly be appraised without the necessity of finding the complete flow net for a number of different cross-sections. Thus there can be explored in a short time many possibilities which would require months of work with any of the other methods. Such studies have already indicated desirable changes from the conventional design of earth dams and levees, some of which were briefly discussed in the preceding chapter.

Finally, there should be mentioned the pedagogical value of the graphical method. It gradually develops a feeling or instinct for streamline flow which not only improves, in turn, the speed and accuracy with which flow nets can be determined, but also develops a much better understanding of the hydromechanics of seepage and ground water movement. The investigator who is trained only in the use of "mechanical" methods for analysing seepage problems can check his tests only by performing additional tests. He is rarely able to detect inaccuracies by the appearance of the test results. In contrast to this, the author has been able to point out even minor inaccuracies in the results obtained from model tests, as a result of the sense for streamline flow developed by applying the graphical method for years.

In concluding this discussion, the author wishes to emphasize the almost obvious point, which nevertheless is frequently overlooked, that the investigator should consider carefully, before starting any model tests, what information he desires to obtain from these tests. In nine cases out of ten he will then come to the conclusion that he could obtain the results without tests. Particularly in those cases where he attempts to evaluate the effect of variations in the coefficient of permeability, he will arrive at a better conception of the probable limits within which the seepage conditions in the prototype may vary by making a careful study of the possible variations in the coefficient of permeability (e.g. from studies of the variations in the borrow pit material) and then applying these values in graphical solutions, utilizing the transformation method. A model test would yield only one result, the relation of which to the prototype is often unknown. Such a test would certainly not permit a conclusion in regard to the probable limits within which the actual flow conditions will vary.

The practical application of the graphical method would be promoted if, for all typical conditions encountered in dam design, carefully constructed flow nets were published. The beginner in the use of the graphical method in particular, would be greatly assisted and encouraged in his efforts to acquire skill in the use of this valuable tool.

Appendix I.

(a) Deflection of Flow Lines Due to Change in Permeability. Flow lines are deflected at the boundary between isotropic soils of different permeability in such a manner that the quantity Δq flowing between two neighbouring flow lines is the same on both sides of the boundary. Referring to Figure 5, in which the flow net is plotted on the basis of squares for the material on the left of the boundary, and designating by Δh the drop in head between any two neighbouring equipotential lines, the following relationship can be set up:

$$\Delta q = k_1 a \frac{\Delta h}{a} = k_2 c \frac{\Delta h}{b}$$

or

$$\frac{k_1}{k_2} = \frac{c}{b} \tag{19}$$

$$\frac{a}{\sin \alpha} = \frac{c}{\sin \beta} \text{ and } \frac{a}{\cos \alpha} = \frac{b}{\cos \beta}$$

By combining these relationships one arrives at:

$$\frac{c}{b} = \frac{\tan \beta}{\tan \alpha} = \frac{k_1}{k_2} \tag{20}$$

Expressed in words, the deflection of the flow lines occurs such that the tangent of the intersecting angles with the boundary is inversely proportional to the coefficients of permeability. Furthermore, the squares on one side of the boundary change on the other side into rectangles with the ratio of their sides equal to the ratio of the coefficients of permeability, such that the flow channels are wider in the material with the smaller coefficient of permeability.

It is probable that Forchheimer was the first one to use these relationships. However, he never took the trouble to publish them. In 1917, he communicated those relationships to Terzaghi, who made extensive use of them in his foundation investigations of dams, and also taught them in his course in Soil Mechanics at the Massachusetts Institute of Technology, during 1925–29.

(*b*) *Transfer Conditions for Line of Seepage at Boundaries; General Remarks.* L. Casagrande (10) made use of the general properties of a flow

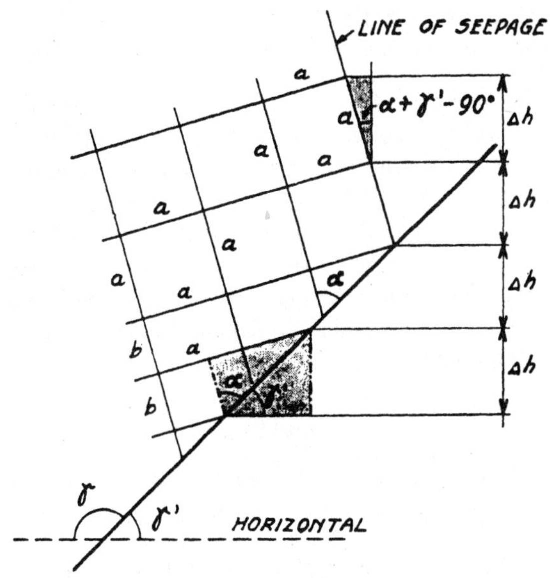

FIG. 15.— DERIVATION OF DISCHARGE CONDITION INTO OVERHANGING SLOPE.

net to analyze the condition at the entrance and discharge points of the line of seepage. Following the same general approach, the author determined the transfer conditions for other cases, including the transfer at the boundary between soils of different permeability. The results are assembled in Figure 6.

To acquaint the reader with the method used, it will be sufficient to present in the following the derivation for two typical cases.

(*c*) *Discharge into an Overhanging Slope.* In Figure 15 is shown the flow net in the immediate vicinity of the discharge point, sufficiently enlarged so that flow lines and equipotential lines appear straight. The slope of the line of seepage at the discharge point, the "discharge gradient," is assumed arbitrarily; then the flow net is plotted, starting with a series of equidistant horizontal lines which represent the head of consecutive equipotential lines. One can see immediately that the assumed discharge gradient in Figure 15 cannot be correct, because it is impossible to draw squares in the lower portion of the flow net. By setting up the condition that the sides a and b of the resulting rectangles must become equal, one can arrive at the necessary condition for the discharge gradient.

By projecting the sides a and b in the shaded triangles, Figure 15, one arrives at the following equation:

$$\frac{b}{\cos\alpha}\sin\gamma = a\cos(\alpha+\gamma-90°) = \Delta h$$

To fulfill the condition $a = b$, the only possible solution is $\alpha = 90-\gamma$; that means the line of seepage must have a vertical discharge slope.

(d) *Transfer Conditions for Line of Seepage at Boundary between Soils of Different Permeability.* To analyze the transfer conditions for the cases illustrated in Figures 6k and m, we start from the conditions that the hydraulic gradient at any point along the line of seepage is equal to the sine of the slope of the line of seepage of that point; and that the quantity flowing through a very thin flow channel along the line of seepage must be equal on both sides of the boundary. Referring to Figure 16a, we have the following velocities along the line of seepage, on both sides but in the immediate vicinity of the boundary:

$$v_1 = k_1 \sin(\alpha-\omega')$$
$$v_2 = k_2 \sin(\beta-\omega')$$

The quantity Δq flowing through the channel is:

$$\Delta q = a k_1 \sin(\alpha-\omega') = c k_2 \sin(\beta-\omega')$$

wherein the quantities a and c represent the widths of the flow channels, in accordance with Figure 16a. After replacing the quantities a and c by their projection onto the boundary, and substituting $k_1/k_2 = \tan\beta/\tan\alpha$, one arrives at the general condition:

$$\frac{\cos\alpha}{\cos\beta} = \frac{\sin(\beta-\omega')}{\sin(\alpha-\omega')}$$

or

$$\frac{\sin(90-\alpha)}{\sin(90-\beta)} = \frac{\sin(\beta-\omega')}{\sin(\alpha-\omega')}$$

Hence the only possi ble solution is:

$$\left.\begin{aligned} 90-\alpha &= \beta-\omega' \\ \text{or}\quad \beta &= 90+\omega'-\alpha \\ \text{or}\quad \beta &= 270°-\alpha-\omega \end{aligned}\right\} \qquad (21)$$

Since this condition does not contain the coefficients of permeability, it has to be fulfilled simultaneously with equation (20). Equations (20) and (21) determine, for a given slope ω of the boundary, the unknown angles α and β between the line of seepage and the boundary.

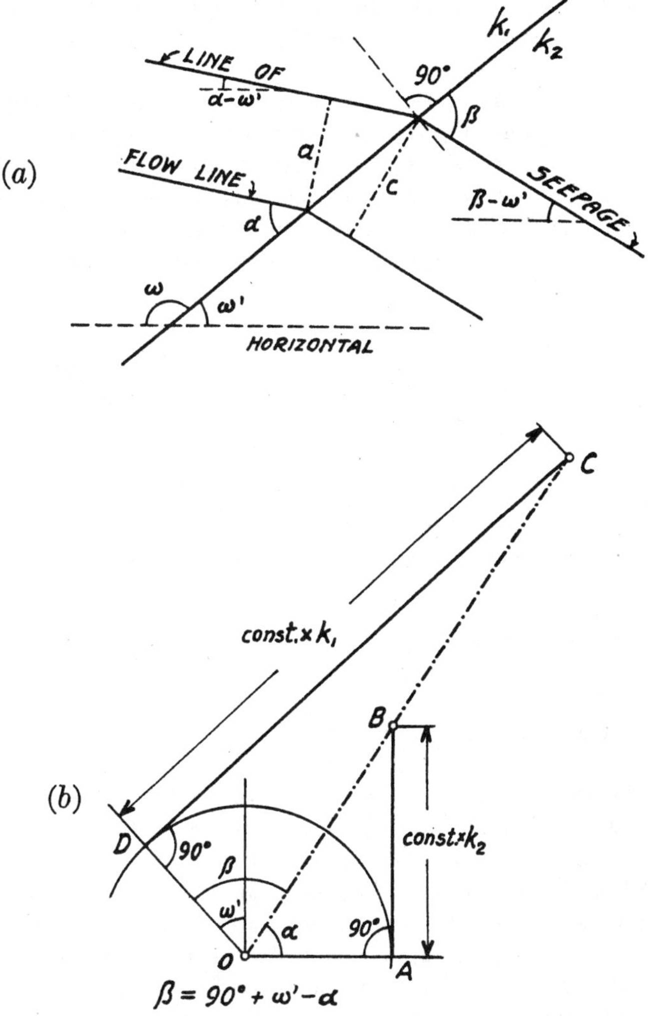

FIG. 16.— TRANSFER CONDITION OF LINE OF SEEPAGE AT OVERHANGING BOUNDARY.

The solution of these two equations can best be found graphically in the manner illustrated in Figure 16b. A circle is drawn with an arbitrary radius and the angle AOD is made equal to $(90+\omega')$. Then lines are drawn through points A and D perpendicular to the corresponding radii. The problem is to draw another line through the center O (shown as dot-dash line) which fulfills the condition that the ratio $\overline{AB}/\overline{CD} = k_2/k_1$. Such a line can be found quickly by trial. The unknown angles α and β are determined by the angles between the dot-dash line and lines OA and OD, respectively.

Depending on whether k_1 is larger or smaller than k_2, we arrive at solutions in which either point B or point C is nearer the center of the circle. The corresponding deflection of the line of seepage is illustrated in Figures 6k and m.

While this theoretical solution for $k_1 > k_2$ can easily be verified by model experiments, it is not generally true for $k_1 < k_2$. In this case, when the downstream section is more pervious, the boundary condition for the line of seepage is also influenced by all other dimensions of the dam, especially by the elevation of the discharge point and its distance from the boundary under consideration. Only in special cases, particularly for high tail water

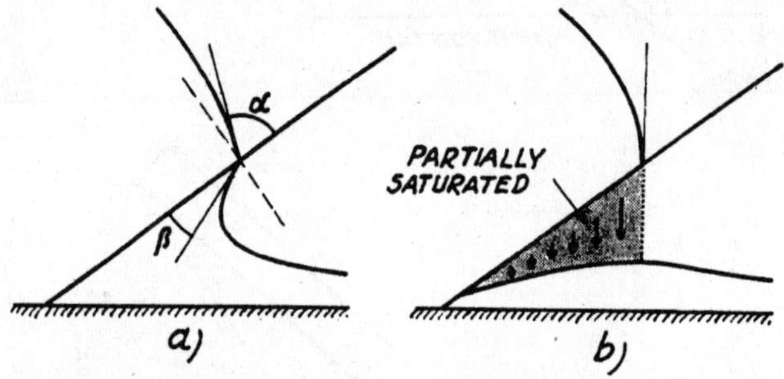

Fig. 17.— Transfer Condition of Line of Seepage at Overhanging Boundary.

level, and coefficients of permeability that do not differ greatly, does the line of seepage follow the theoretical solution. Whenever the theoretical solution has the appearance shown in Figure 17a, with the line of seepage deflected into an overhanging slope, it represents a condition that may be observed on a small scale in the laboratory but does not occur on a large scale. Instead of the continuous line of seepage of Figure 17a, a discontinuity develops, with the water seeping vertically into the more pervious soil, and only incompletely filling its voids. In other words, the quantity discharging vertically downward at the boundary is insufficient to fill the voids of the coarse material. Therefore, normal atmospheric pressure will act along that section of the boundary and the laws for open discharge are valid, forcing the line of seepage to assume a vertical discharge gradient at the boundary. That portion of the coarser soil which is only partially saturated, is illustrated in Figure 17b by the shaded area.

The graphical solution shown in Figure 16b also permits determination of the transfer conditions for the entrance of the line of seepage for the special case illustrated in Figure 6c. The open body of water on the upstream side may be considered a porous material with $k_1 \rightarrow \infty$, for which case Figure 16b yields $\beta = 90°$. The same conclusion may be reached from equation (21), remembering that for this case, $\alpha = \omega'$. This result means that the line of seepage enters perpendicularly to the upstream face of the dam, as shown in Figure 6c.

(*e*) *Singular Points in a Flow Net.* In trying to apply Forchheimer's graphical method, the beginner is frequently puzzled by the fact that some

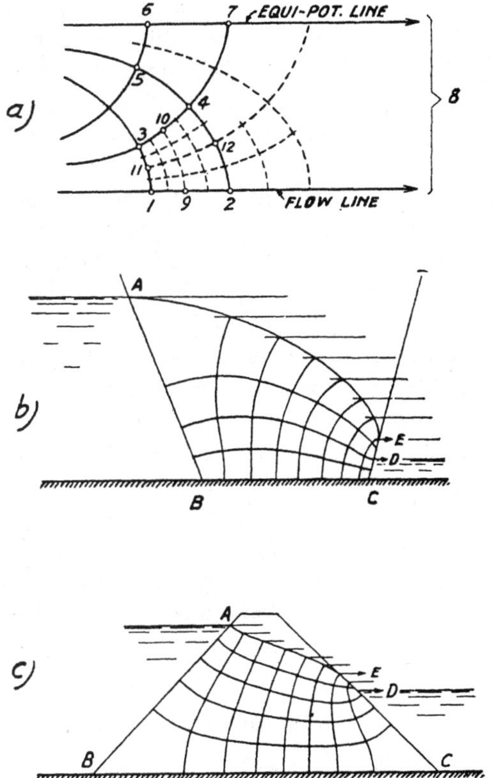

FIG. 18.— ILLUSTRATIONS OF SINGULAR POINTS OF FLOW NETS.

"squares" have no resemblance to real squares, and that, in some cases, flow lines and equipotential lines do not intersect at right angles. For example, in Figure 18a, the full-drawn areas 1, 2, 3, 4 and 4, 5, 6, 7 do not appear like "squares" to the inexperienced. However, by subdividing such areas by equal numbers of auxiliary flow lines and equipotential lines, one can easily check whether the original area is a "square" as defined for flow nets. By such sub-division one must arrive at areas which appear more and more like real squares. However, in most cases it is sufficient to compare the average distances between opposite sides, that is, e.g., the lengths 9–10 and 11–12, by means of a pair of dividers.

In Figure 18a, the entire area to the right of points 2, 4, 7 must also be considered a square, in spite of the facts that the fourth point lies at infinity and that the angle between the flow line and equipotential line at this point is zero instead of 90°. It is, indeed, possible to continue subdividing this area, as shown by the dotted lines, always leaving a semi-infinite strip as the "last square." By this process of subdivision the amount of water entering into the "last square" is continuously reduced and approaches zero. In this way it is possible to reconcile the irregularity of the fourth corner by the fact that there is no flow of water at that point.

Similar irregularities in the shape of squares appear wherever a given boundary of the soil, with water entering or discharging, and boundary flow lines (impervious base or line of seepage) intersect at a predetermined angle. If this angle is less than 90°, then the velocity of the water at the point of intersection is zero. Such points are the entrance point A of the line of seepage in Figure 18b, and points B and C in Figure 18c. On the other hand, if the intersecting angle is greater than 90°, then the theoretical velocity in that point is infinite. Such points are corner A in Figure 9d, corners B and C in Figure 18b, and point D in Figure 18c. The last, representing the concentration of flow lines at the elevation of tail water level, is the cause for the well-known erosion which is observed on the downstream slope of homogeneous dam sections at the line of wetting.

At points where the theoretical velocity is infinite, the actual velocity is influenced by the facts that for larger velocities Darcy's law loses its validity, and that changes in velocity head become so important that they cannot be neglected. Hence, in the vicinity of such points the general differential equation (4) is not valid, and the flow net will deviate from the theoretical shape. However, the areas affected are so small that these deviations may be disregarded.

Appendix II.

Additions to the Original Paper.

(*a*) *Graphical Procedure for Determining Intersection between Discharge Slope and Basic Parabola.* The intersection between the discharge face and the basic parabola, designated in Figure 9 by point C_o, can be determined by the following simple graphical procedure.

The ordinate h_1 of the intersection of the basic parabola $x = \dfrac{y^2 - y_o^2}{2y_o}$ with the discharge slope $y = \pm x \tan\alpha$ is found as the solution of these two equations in the following form:

$$h_1 = \pm \frac{y_o}{\tan\alpha} + \sqrt{\frac{y_o^2}{\tan^2\alpha} + y_o^2}$$

The first member, $\frac{y_o}{tan\ \alpha}$, is equal to the distance $\overline{EB}=f$, in Figure 19; the second member, under the square root, is equal to the distance $\overline{AB}=g$; the ordinate h_1 of the intersection is simply equal to the sum $(f+g)$ for angles $\alpha<90°$, and equal to the difference $(g-f)$ for angles $\alpha>90°$. These relationships are expressed by the construction shown in Figure 19, which needs no further explanation.

The discharge point of the line of seepage is then found as discussed in Section F-d, with the help of Figure 11.

(*b*) *Comparison between Hamel's Theoretical Solution and the Proposed Approximate Methods.* Hamel (19), has succeeded in arriving at a rigorous mathematical solution of the problem of seepage through a homogeneous

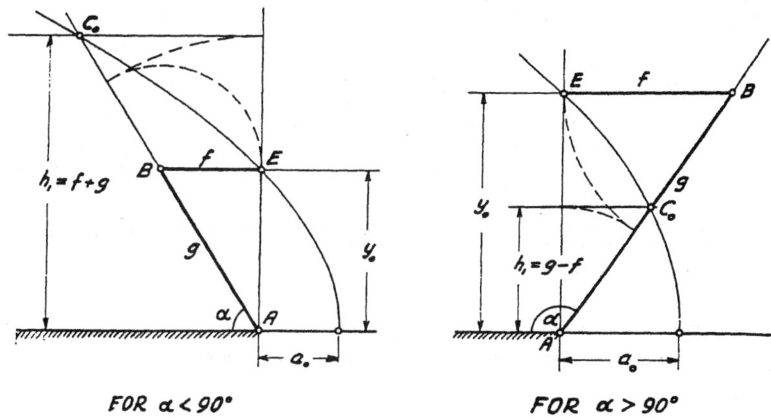

FIG. 19.—GRAPHICAL METHOD FOR DETERMINING INTERSECTION BETWEEN BASIC PARABOLA AND DISCHARGE FACE.

dam section. Unfortunately, the theory is so cumbersome that, in its present form, it is of little use to the engineering profession. It will be necessary to compute a sufficient number of typical cases, and publish the results in the form of tables or graphs, before engineers will be able to realize the advantages of this theoretical treatment. Recently, a few cases have been computed by Muskat (20) for coffer dam sections with vertical sides. These solutions presented an opportunity to investigate, at least for a few special cases, the accuracy of the approximate methods proposed in this paper. The results of this comparison were so encouraging that they are presented in the following paragraphs to permit the reader to formulate his own conclusions.

In Figure 20 are assembled three of the six cases which were published by Muskat (20). In each case the elevation of the discharge point, as computed from Hamel's theory, is designated by C_o, and its vertical distance from the tail water level, or from the impervious base in the absence of tail water, is designated a_H.

An approximate elevation of the discharge point was found by means of the graphical procedure shown in Figure 7c. To facilitate the comparison between these figures, all points in Figure 20 are marked to correspond to those in Figure 7c. The construction is shown with full lines. The resultant discharge point is marked C, and its elevation from the base, or the tail water level, is marked a.

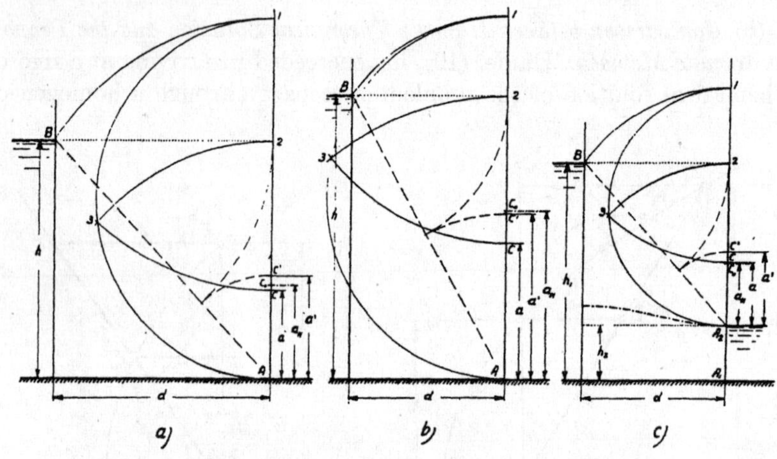

FIG. 20.— COMPARISON BETWEEN RIGOROUS AND APPROXIMATE DETERMINATIONS OF DISCHARGE POINT.

In addition to this construction, the simplified procedure was used in which $s_o = \sqrt{h^2+d^2}$. For a vertical discharge face the simplified formula for a becomes $a' = \sqrt{h^2+d^2}-d$ as proposed by Kozeny (6). The corresponding construction is shown in Figure 20 by dash lines and the resulting discharge point and elevation are designated by C' and a' respectively.

The case illustrated in Figure 20a, corresponding to Muskat's case No. 6, is identified by the ratio $d/h = 0.937$. Hamel's theory yields the quantities $a_H/h = 0.394$, and for the rate of seepage, $q_H = 0.539\ kh$.

As was shown by Muskat (20) and Dachler (23), the rate of seepage computed by means of Dupuit's formula $q = \dfrac{h_1^2 - h_2^2}{2d}k$, or without tail water $q = \dfrac{h^2}{2d}\ k$, represents an excellent approximation. For the case illustrated in Figure 20a, we have

$$q = \frac{h}{2d}\ kh = 0.539\ kh$$

The next case, Figure 20b, corresponds to Muskat's case No. 4 and is identified by the ratio $d/h = 0.556$. The theory by Hamel yields the quantities $a_H/h = 0.596$, and $q_H = 0.898\ kh$. The approximate rate of seepage, computed from Dupuit's formula, as described for the previous case, is $q = 0.900\ kh$.

The third case, shown in Figure 20c, is identical with Muskat's case No. 2. It differs from the other examples by the assumption of a definite tail water level, and is identified by the quantities $d/h_1 = 0.663$, and $d/h_2 = 2.81$. From the theory we get $a_H/h_1 = 0.301$, and $q_H = 0.717\ kh_1$. Dupuit's formula yields $q = 0.695\ kh_1$.

The comparison between the values for the elevation of the discharge point obtained by Hamel's rigorous solution and those by the approximate methods shows that, for engineering purposes, the approximate solutions are very satisfactory. It is interesting and of practical value to note that the approximate methods also give satisfactory results for ratios d/h considerably smaller than 1.0. Considering that the upstream portion of the flow net differs considerably from Dupuit's assumption of a constant hydraulic gradient in all verticals, this result is somewhat unexpected. For ratios of $d/h < 1.0$, it appears that Kozeny's formula (13) gives slightly better results than the formula by L. Casagrande.

The remarkable agreement between the theoretical rate of seepage and Dupuit's approximate solution deserves special emphasis.

(c) *Graphical Solution by Means of the Hodograph.* A new graphical method for determining the flow net was proposed by Weinig and Shields (30), in which the flow-net is determined graphically in the "hodograph plane" and then projected into the actual cross section.

The hodograph of a flow line is the curve which one obtains when plotting from one origin velocity vectors for all the points of the flow line. Therefore, the straight line connecting the origin with one point on the hodograph represents the magnitude and direction of the velocity for the corresponding point on the flow line.

Since the velocity along the free water surface is proportional to the sine of the slope, the hodograph for the line of seepage is a circle with diameter equal to the coefficient of permeability. The hodograph for a straight boundary is a straight line. Therefore, all boundaries of the hodograph that correspond to the flow net of a homogeneous isotropic dam section are known, and it is possible to set up equations that represent the solution of the problem in implicit form. That such a theoretical solution is rather complicated, even for the simplest dam section, has been mentioned before in the discussion of Hamel's theory (19). Therefore, Weinig and Shields follow the theoretical approach, using the hodograph, as far as mathematics permits conveniently; then they proceed to find the flow lines and equipotential lines in the hodograph by a graphical procedure which is essentially similar to Forchheimer's method.

One advantage of the method by Weinig and Shields is the possibility of determining numerically correct values for the velocity at certain points along the boundaries. In comparison with Forchheimer's graphical method the approach by Weinig and Shields is much more complicated and requires a thorough acquaintance with the hodograph, which very few engineers possess. Furthermore, this method is limited to simple cross-sections, while Forchheimer's method can be applied to complicated dam sections and foundation conditions.

Weinig and Shields (30) have solved a steep triangular dam section by means of the graphical solution of the hodograph. This cross-section is

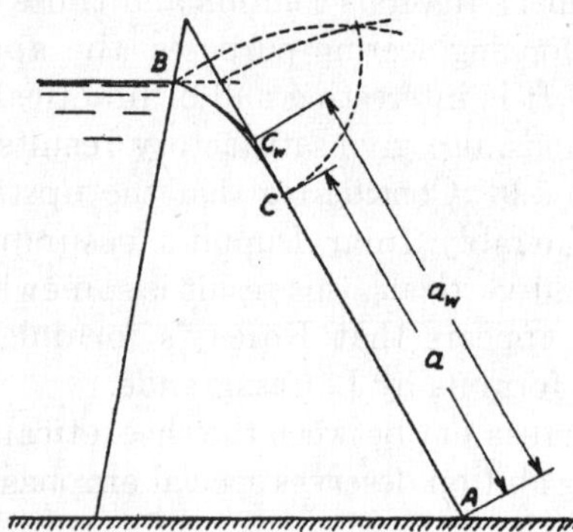

FIG. 21.—DISCHARGE POINTS OBTAINED BY GRAPHICAL SOLUTION OF HODOGRAPH AND METHOD ILLUSTRATED IN FIG. 7.

illustrated in Figure 21. Point C_w represents the discharge point as determined from the hodograph, and point C the discharge point using the method shown in Figure 7b. The elevation of point C is 15 per cent. lower than that of C_w. How much of this difference is due to inaccuracy in one or the other method is uncertain. Probably the hodograph solution is more accurate when the entrance point of the line of seepage is very close to the discharge face. In Figure 20, as well as in Figure 21, the discharge points obtained by the simple graphical procedure are situated lower than the other more accurate solutions. This would indicate the necessity for applying a correction in those cases where upstream and downstream face are very close.

BIBLIOGRAPHY.

(1) Terzaghi, K.v. Der Grundbruch an Staumauern und seine Verhuetung. Die Wasserkraft, 1922.

Terzaghi, K.v. Erdbaumechanik. Vienna 1925.

Terzaghi, K.v. Effect of Minor Geologic Details on the Safety of Dams. Am. Inst. of Min. and Metal. Engrs., Technical Publication No. 215, Feb. 1929.

(2) Terzaghi, K.v. Auftrieb und Kapillardruck an betonierten Talsperren. 1. Congres des Grands Barrages, Stockholm, 1933.

(3) Terzaghi, K.v. Beanspruchung von Gewichtsstaumauern durch das stroemende Sickerwasser. Die Bautechnik, 1934, No. 29.

(4) Terzaghi, K.v. Der Spannungszustand im Porenwasser trocknender Betonkoerper. Der Bauingenieur 1934, No. 29/30.

(5) Forchheimer, Philipp. Hydraulik, third edition 1930.

(6) Kozeny, J. Grundwasserbewegung bei freiem Spiegel, Fluss- und Kanalversickerung. Wasserkraft und Wasserwirtschaft, 1931, No. 3.

(7) Schaffernak, F. Ueber die Standsicherheit durchlaessiger geschuetteter Daemme. Allgemeine Bauzeitung, 1917.

(8) Iterson, F. K. Th. van. Eenige theoretische beschouwingen over kwel. De Ingenieur 1916 and 1919.

(9) Dupuit, J. Etudes théoretiques et pratiques sur le movement des eaux. Paris, 1863.

(10) Casagrande, Leo. Naeherungsmethoden zur Bestimmung von Art und Menge der Sickerung durch geschuettete Daemme. Thesis, Technische Hochschule, Vienna, July 1932.

This paper was translated into English for use in the U. S. Corps of Engineers by the staff of the U. S. Waterways Experiment Station, Vicksburg, Miss.

(11) Casagrande, Leo. Naeherungsverfahren zur Ermittlung der Sickerung in geschuetteten Daemmen auf undurchlaessiger Sohle. Die Bautechnik 1934, No. 15.

(12) Gilboy, Glennon. Hydraulic-Fill Dams. 1. Congres des Grands Barrages, Stockholm, 1933.

Gilboy, Glennon. Mechanics of Hydraulic-Fill Dams. Journal of the Boston Soc. of Civil Engrs., July, 1934.

(13) Reyntjiens, G. P. Model Experiments on the Flow of Water through Pervious Soils. Thesis, Mass. Inst. of Technology, May, 1933.

(14) Samsioe, A. Frey. Einfluss von Rohrbrunnen auf die Bewegung des Grund-Wassers. Zeitschrift fuer angewandte Mathematik und Mechanik, 1931, No. 2.

(15) Dachler, Robert. Ueber Sickerwasserstroemungen in geschichtetem Material. Die Wasserwirtschaft, 1933, No. 2.

(16) Schaffernak, Friedrich. Erforschung der physikalischen Gesetze, nach welchen die Durchsickerung des Wassers durch eine Talsperre oder durch den Untergrund stattfindet. Die Wasserwirtschaft, 1933, No. 30.

(17) Casagrande, Arthur. Discussion of E. W. Lane's paper on "Security from Underseepage," Proceedings Am. Soc. Civ. Engrs., March, 1935.

(18) Terzaghi, K.v. Discussion of L. F. Harza's paper on "Uplift and Seepage under Dams on Sand," Proceedings Am. Soc. Civ. Engrs., Jan., 1935.

(19) Hamel, G. Ueber Grundwasserstroemung, Zeitschrift f. angew. Math. u. Mech., Vol. 14, No. 3, 1934.

Hamel, G. and Günther, E. Numerische Durchrechnung zu der Abhandlung ueber Grundwasserstroemung. Zeitschrift f. angew. Math. u. Mech., Vol. 15, 1935.

Important Publications on the Question of Seepage and Its Effect on the Stability of Soil that have Appeared Since this Paper Was Written. (See Appendix II).

(20) Muskat, M. The Seepage of Water through Dams with Vertical Faces. Physics, Vol. 6, Dec. 1935.

(21) Wyckoff, R. D. and Reed, D. W. Electrical Conduction Models for the Solution of Water Seepage Problems. Physics, Vol. 6, Dec. 1935.

(22) Knappen, T. T. and Philippe, R. R. Practical Soil Mechanics at Muskingum. Eng'g. News-Record, April 9, 1936.

(23) Dachler, R. Grundwasserstroemung. Vienna, 1936.

(24) Terzaghi, K.v. Simple Tests to Determine Hydrostatic Uplift. Eng'g News-Record, June 18, 1936.

(25) Terzaghi, K.v. Critical Height and Factor of Safety of Slopes against Sliding. Proc. Int. Conf. on Soil Mech. and Found. Eng., Vol. I, No. G-6.

(26) Terzaghi, K.v. Distribution of the Lateral Pressure of Sand on the Timbering of Cuts. Proc. Int. Conf. on Soil Mech. and Found. Eng., Vol. I, No. J-3.

(27) Vreedenburgh, C. G. J. Electric Investigation of Underground Water Flow Nets. Proc. Int. Conf. on Soil Mech. and Found. Eng., Vol. I., No. K-1.

(28) Vreedenburgh, C. G. J. On the Steady Flow of Water Percolating through Soils with Homogeneous-Anisotropic Permeability. Proc. Int. Conf. Soil Mech. and Found. Eng., Vol. I., No. K-2.

(29) Brahtz, J. H. A. Pressures due to Percolating Water and Their Influence upon Stresses in Hydraulic Structures. Second Congress on Large Dams, Washington, D. C., 1936.

(30) Weinig, F., and Shields, A. Graphisches Verfahren zur Ermittlung der Sickerstroemung durch Staudaemme. Wasserkraft und Wasserwirtschaft, 1936, No. 18.

Originally published in
Journal of the Boston Society of Civil Engineers
July, 1937

JOURNAL OF THE

BOSTON SOCIETY OF CIVIL ENGINEERS

Volume XXIV **JULY, 1937** **Number 3**

STABILITY OF EARTH SLOPES

BY DONALD W. TAYLOR, MEMBER *

SYNOPSIS

THIS paper presents a review of several methods which have been proposed for analyzing the stability of earth slopes, with comparisons of the results they furnish. Two methods which appear to give consistent and reliable results have been studied in considerable detail, and complete mathematical solutions have been obtained. Although the equations of the mathematical solutions are somewhat involved, they have been evaluated, tabulated and plotted in such form that the solution to a practical problem may be found almost at a glance.

Numerous simplifying assumptions must be made in any approach to this problem, an important one being homogeneity of the soil. Consequently it must be recognized that in general no solution can be accepted as wholly reliable. However, as indications of the conditions which exist, and as aids to judgment, the results may often be of great value to the engineer, and they are presented herein with that end in view.

Measurement of the constants which describe the shearing strength of the soil is briefly discussed. As understanding of the phenomena of shear in soils becomes more advanced, the reliability of results of stability computations will increase and better methods of computation may be developed. At present, the accuracy with which shearing strength may be determined is decidedly the limiting feature, and it is

* Research Associate in Soil Mechanics, Massachusetts Institute of Technology, Cambridge, Mass.

probable that the methods recommended herein will prove adequate for some time to come.

Values for unit weight and their applicability are discussed. The effect of seeping water is analyzed, and a procedure is recommended which does not require determination of a flow net.

General Considerations and Assumptions

There are three distinct parts to an analysis of the stability of a slope.

1. *The Testing of Samples to determine the Cohesion and Angle of Internal Friction.* — If the analysis is for a natural slope it is essential that the samples be undisturbed. In such important respects as rate of shear application and state of initial consolidation, the conditions of testing must represent as closely as possible the most unfavorable conditions ever likely to occur in the actual slope.

2. *The Study of Items which are Known to Enter but which Cannot be Accounted for in the Computations.* — The most important of such items is the progressive cracking which will start at the top of the slope where the soil is in tension, and, aided by water pressure and perhaps frost action, may progress to considerable depth. In addition, there are the effects of the non-homogeneous nature of the typical soil and other variations from the ideal conditions which must be assumed. Decisions on these complicated points will tax the judgment of the most competent and experienced of engineers to the limit, and for important projects, where stability is essential, the best in consultation advice is to be recommended. For preliminary studies, however, conservative decisions on such points will lead to analyses which will at least give valuable indications of the condition of stability.

3. *The Computations Themselves.* — Parts 1 and 2 are both of fully as much importance as part 3. The principal aim of the main body of this paper, however, is to present a practical and simple method of handling part 3. Part 1 will be covered only in a general way, while phases of part 2 are discussed near the end of the paper.

If a slope such as that illustrated in Fig. 1 (*a*) is to fail, all shearing strength must be overcome along some surface, which then becomes a surface of rupture. The arc *AB* in this figure represents one of an infinite number of possible traces on which the failure might occur. For the simple case which is to be considered here the slope is assumed constant, the top surface level, and the soil homogeneous throughout.

Thus a simplified section is taken, but it may be noted that many actual slopes may be closely approximated by such an ideal section.

It is assumed that the problem is two-dimensional, which theoretically requires infinite running length of slope. However, if the cross section investigated holds for a running length of roughly two or more times the trace of the rupture, it is probable that the two-dimensional case holds within required accuracy.

Most methods of approach to the problem assume that the rupture surface passes through the toe of the slope, point *A* of Fig. 1 (*a*). It is known that the higher the slope the more likelihood of failure, thus, so long as the material is homogeneous to an unlimited depth, a rupture surface passing through a point part way up the slope is not to be expected. However, the possibility that the rupture surface may pass below the toe of the slope and break out at the surface some distance away must not be overlooked.

The determination of the shape and location of the surface of rupture is a very important consideration. If stability analyses could be made for each of the infinite number of possible surfaces, that one showing the most unfavorable condition of stability would be the rupture surface in case of failure. As such a procedure is at present impossible, the surface of rupture is determined in the various methods of solution (*a*) by pure assumption, (*b*) by mathematical analysis, or (*c*) by assumptions based on studies of actual slides. This point will be discussed later for the individual methods. In plastic soils, instead of a definite surface of rupture a zone of plastic flow may be expected. This need cause no concern here, however, since if a slope of such a material is just at the point of failure, the use of any surface composed of slip lines within the plastic zone as a rupture surface should show a limiting condition of equilibrium.

The shearing strength of the soil is assumed to follow the empirical relationship known as Coulomb's Law —

$$s = c + n \tan \phi$$

wherein s is the unit shearing strength, c the unit no-load shearing strength or unit cohesion, n the applied normal stress on the surface of rupture, and ϕ the effective angle of internal friction. It must be carefully noted that the values for cohesion and effective angle of internal friction are not necessarily constants for a given soil, but are intended to represent quantities which may be depended upon to hold for a definite set of conditions. Any attempt at a thorough discussion of the subject of shearing strength of cohesive soils is far beyond the scope of this paper. However, brief mention of a few of the most salient points may be desirable.

The linear relation expressed by Coulomb's Law must be looked upon as an approximation. If a series of laboratory tests is run, conforming to a given set of conditions, a plot of shearing strength against normal pressure will usually show a definite curvature. Generally, however, this curve may be approximated by a straight line without introducing serious error.

Foremost among the conditions in nature which must be reproduced in the laboratory are the rate of application of shearing strain and the state of initial consolidation. Terzaghi has discussed these items briefly (Fig. 9, reference 9*a*),* and they have been a subject of study by many investigators. In Sweden, where the principles of the method which is used herein originated, it has been found that for many cases actual conditions are best represented by an effective angle of internal friction of zero. In general, however, it would not be advisable to make this assumption without investigation.

Another item of great importance is the loss in shearing strength which many clays show when subjected to large shearing strain. The stress-strain curves for such clays show the stress rising with increasing strain to a maximum value, after which it decreases and approaches an ultimate value which may be much less than the maximum. Since a rupture surface tends to develop progressively rather than with all the points at the same state of strain, it is generally the ultimate value that should be used for the shearing strength rather than the maximum value.

In recent years much progress has been made in the development of a satisfactory understanding of the behavior of cohesive soils in shear, but many features are still only partially understood. It is probable that there are constant values for effective cohesion and effective angle of internal friction for Coulomb's Equation which approximately but adequately express the shearing strength for any given set of conditions. Undoubtedly the soil engineer of the future will be able to furnish better values for these constants, and quite possibly he may develop better methods for using them in stability analyses. However, the limiting factor of such analyses is the determination of the shearing strength which a soil possesses. Thus the methods presented herein appear to be adequate, and with the development of more accurate methods for determining values for shearing strength may be used in the future with increasing dependability.

* Numbers in parentheses refer to Bibliography at end of paper.

The subject of undisturbed sampling should not be passed without mention. If the structure of a soil is disturbed in sampling, the shearing strength is changed, and, regardless of how accurately other conditions are reproduced in the testing, this discrepancy is always present. When an analysis is made for a slope of a natural clay which has a structure such that disturbance causes radical change of properties, it is impossible to overemphasize the importance of requiring samples which are as nearly undisturbed as can be obtained.

The scope of the methods discussed herein is limited to clays or other cohesive materials which are not fissured. Considerations of stiff fissured clays, where the point in question is that of preventing disintegration rather than one of stability, have been discussed by Terzaghi (9*b*).

Summary of Methods

A number of methods for stability computations have been advanced. All assume homogeneous soil, constant angle of slope, level top surface and shearing strength as expressed by Coulomb's Equation. Unless otherwise noted, the assumptions are also made that cohesion is constant along the entire rupture surface and the rupture surface passes through the toe of the slope. The commonest methods may be summarized as follows:

1. The *Culmann Method* (1) assumes rupture will occur on a plane. It is of interest only as a classical solution, since actual failure surfaces are invariably curved. The general formula for this method is given later under Comparisons of Methods.

2. The *Résal-Frontard Method* (2, 3) assumes the soil mass to act as a slope of infinite extent, and, by use of conjugate stress relationships, results in an equation for the rupture surface. This method has been criticized because the results indicate that the mass above the rupture surface is not in static equilibrium when just at the point of failure with all shearing strength being utilized. However, the results are on the side of safety, and it is interesting to note that on one very important point they agree with actual conditions by indicating that the upper part of the mass is in tension. This method assumes that no cohesion may be depended upon within the depth to which tension occurs. Formulæ are given later under Comparisons of Methods.

3. The *Circular Arc Method* was first proposed by K. E. Petterson, based on his study of the failure of a quay wall in Goeteborg in 1915 or 1916. The justification that circular arcs are close approximations of actual rupture surfaces comes from field investigations of a large num-

ber of actual slides, especially in railroad cuts, by the Swedish Geotechnical Commission (4). An infinite number of arcs may be drawn for any given slope, the one to be used being that one which is least stable. That such arcs may either pass through the toe of the slope or may pass below this point is borne out by the field investigations; thus the method is not limited to arcs through the toe of the slope. This method has been widely accepted as satisfactory, and several methods of procedure based on the circular arc have been proposed, among which are the following:

3 (*a*). *Method of Slices* (4, 5, 6).— This method was advanced by the Swedish Geotechnical Commission and developed in quite some detail by Professor W. Fellenius. By dividing the mass above the rupture plane into vertical slices and assuming the forces on opposite sides of each slice are equal and opposite, a statically determinate problem is obtained, and semigraphical methods have been devised by which the stability of the mass may be analyzed for any given circle. The main objection to this or any other graphical method rests in the fact that the most dangerous one of an infinite number of possible circles must be found, and thus the graphical procedure must be repeated for a large number of circles. It is to be noted that some assumption must be introduced to make the problem statically determinate; the one mentioned above, which is but one of several which have been proposed, is generally recognized as the most suitable.

The work of Professor Fellenius includes a general solution which is the result of many series of graphical analyses based on this method, and which is presented (5) in the form of nests of curves. For the special case of $\phi = 0$ he also obtained a general mathematical solution. In addition he has given data on locations of critical circles which will be discussed later.

3 (*b*). *ϕ-Circle Method.*— This is one of the two methods upon which the general solution given in this paper is based. An assumption is introduced which may be more easily justified than that of method 3 (*a*), and the mathematical solution which has been derived eliminates all graphical work. A more complete description will be given later.

3 (*c*). *The Jáky Method* (9*c*) is a mathematical solution based on the equation of continuity of stress. The boundary conditions which locate the critical circle place too much dependence on the circular assumption and lead to results which are somewhat on the unsafe side, as discussed later.

4. *The Spiral Method* (7) assumes the rupture surface to be a logarithmic spiral. No further assumption is required to make the problem statically determinate, which constitutes the important advantage of this method. On the other hand, solutions based on spiral surfaces are not as easily handled by graphics or mathematics as are those based on circular arcs. As in the circular arc method, all possible logarithmic spirals, passing either through or below the toe of the slope, should be investigated to locate which is most dangerous. A mathematical solution for this method has been obtained by the writer and is covered in detail later.

The ϕ-Circle Method

The solution of the slope stability problem is developed below, using the ϕ-Circle Method. The basic assumption which characterizes this method is explained following the description of the forces which enter.

The ϕ-Circle Method was proposed some years ago by Professors Glennon Gilboy and Arthur Casagrande, its initial use being in the development of a completely graphical solution of the slope problem. For any circle that is analyzed, the result by this solution is a vector, the length of which represents the quantity $\frac{2c_1}{w}$, wherein w is the unit weight of the soil and c_1 is the unit cohesion required for equilibrium. It became evident, from the graphical layout for any given inclination of slope and any given angle of internal friction, that one plot could represent the case for various heights of slope merely by the use of different scales. Hence, if the length of the vector is divided by any characteristic linear dimension, such as H, the vertical height of the slope, the result is an abstract number $\frac{2c_1}{wH}$. This number, when determined for the most dangerous circle, describes the requirements for stability of any case involving the given slope and angle of internal friction. This relationship suggested the possibility of obtaining a general solution for all slopes and all angles of internal friction so that, once determined, there would be no further need of analyzing each individual case. Some years earlier Fellenius had originated the expression which in the notation of this paper is $\frac{4c_1}{wH}$, but in his paper (5) he did not emphasize the basic nature of this relationship.

In the mathematical solutions which have since been set up by the writer, the form used for this abstract number is $\frac{c}{FwH}$, wherein F is the factor of safety with respect to cohesion, and c is the actual unit cohesion for the soil in question. This basic dimensionless expression will be called the "Stability Number."

Two different factors of safety have been proposed for use in stability problems. Fellenius and most other investigators have used the ratio of actual shearing strength to critical shearing strength, which is in agreement with the usual conception of factor of safety, and will be called the true factor of safety. The other factor has been used by Jáky (9*c*) and Rendulic (7), and may be described as the ratio between actual cohesion and critical cohesion. Of the two items which comprise the shearing strength, namely, cohesion and friction, this latter factor has

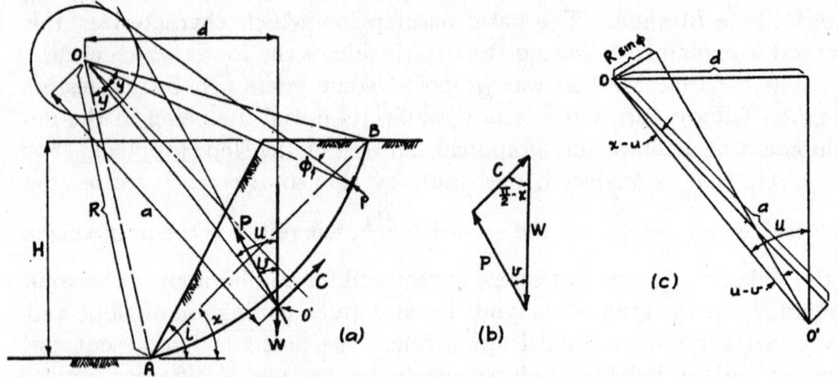

FIG. 1.—SKETCHES SHOWING ELEMENTS OF THE φ-CIRCLE METHOD FOR CIRCLES PASSING THROUGH TOE OF SLOPE

bearing on the cohesion only, and will be called herein the factor of safety with respect to cohesion. In setting up mathematical solutions, the use of this latter factor proves to be simpler, thus it is used in the derivations given herein. However, it will be shown later that the true factor of safety may easily be adapted in the results of these solutions.

In Fig. 1 (*a*), AB is any circular arc through A, the position of the center O being described by the two variable co-ordinate angles x and y. Arcs which pass below the toe of the slope will be treated later. The section considered extends a unit distance in the third dimension. The forces acting on the mass above the arc are shown in Fig. 1 (*a*), while Fig. 1 (*b*) shows the equilibrium polygon. The three forces entering are—

W, the Weight of the Mass. — The vector representing W will act vertically through the center of gravity of the mass, and its magnitude and line of action may be found easily by any one of several methods.

C, the Resultant Cohesion. — Its magnitude is $c_1\bar{L}$, where c_1 is the unit cohesion required for equilibrium and $\bar{L}$ is the length of the chord AB. Its line of action is parallel to the chord AB, and its moment arm, a, is described by —

$$c_1\bar{L}a = c_1\hat{L}R \quad \text{or} \quad a = \frac{R\hat{L}}{\bar{L}}$$

where $\hat{L}$ is the length of arc AB. Thus the line of action of C may be found, and its position is independent of the magnitude of the cohesion.

P, the Resultant Force transmitted from Grain to Grain of the Soil across the Arc AB. — The intersection of W and C is known. Since the three forces W, C and P must be concurrent, P must also pass through this intersection. With W known in magnitude and direction and the direction of C also known, the force polygon may be constructed if a second point can be obtained on P to determine its direction. This is accomplished by the basic assumption of the ϕ-Circle Method.

P is made up of small elementary forces, such as p, which must act at an obliquity of ϕ to the arc AB. If p is produced and a small circle is drawn with center at O tangent to the produced line, it will be seen that the p force for any other element of the arc AB will also be tangent to it. This small circle is called the ϕ-circle. Any two such p forces will intersect just outside of the ϕ-circle so their resultant would pass just outside this circle, and by the same reasoning P must pass outside. The assumption introduced is that P is tangent to the ϕ-circle. It is to be noted that this idea is not new, being only a new application of a scheme developed by Dr. H. Krey (8) and used by him in other types of analyses. Studies have been made to estimate the degree of inaccuracy involved by this assumption, and a discussion of these will be introduced later.

The mathematical solution is given, first for the case of circles passing through the toe of the slope, and followed by the case of circles passing below the toe of the slope.

From Fig. 1 (a):

$$C = c_1\bar{L} = \frac{c\bar{L}}{F} = \frac{2}{F}cR \sin y \qquad (1)$$

$$W = wR^2y - wR^2 \sin y \cos y + \frac{wH^2}{2} (\cot x - \cot i) \qquad (2)$$

$$R=\frac{H}{2}\csc x\csc y \qquad (3)$$

$$Wd=[wR^2y]\left[\frac{2R}{3}\frac{\sin y}{y}\sin x\right]-[wR^2\sin y\cos y]\left[\frac{2R}{3}\cos y\sin x\right]$$

$$+\left[\frac{wH^2}{2}\cot x\right]\left[\frac{H}{3}\cot x+R\sin(x-y)\right]$$

$$-\left[\frac{wH^2}{2}\cot i\right]\left[\frac{H}{3}\cot i+R\sin(x-y)\right]$$

which, reduced and combined with (3), gives—

$$Wd=\frac{wH^3}{12}[1-2\cot^2 i+3\cot i\cot x-3\cot i\cot y+3\cot x\cot y] \qquad (4)$$

From Fig. 1 (*c*):

$$OO'=d\csc u \qquad (5)$$

$$OO'=a\sec(x-u)=R\,y\csc y\sec(x-u) \qquad (6)$$

$$OO'=R\sin\phi\csc(u-v) \qquad (7)$$

From Fig. 1 (*b*):

$$\frac{W}{C}=\frac{\cos(x-v)}{\sin v}=\cos x\cot v+\sin x \qquad (8)$$

substituting (3) in (2) and the result in (4)

$$\frac{H}{2d}=\frac{\frac{1}{2}\csc^2 x\,(y\csc^2 y-\cot y)+\cot x-\cot i}{\frac{1}{3}(1-\cot^2 i)+\cot i\,(\cot x-\cot y)+\cot x\cot y} \qquad (9)$$

From (5), (6) and (3):

$$\cot u=\frac{H}{2d}\,y\sec x\csc x\csc^2 y-\tan x \qquad (10)$$

From (5), (7) and (3):

$$\sin(u-v)=\frac{H}{2d}\sin u\csc x\csc y\sin\phi \qquad (11)$$

placing (3) in (1) and (2), then setting (8) equal to the ratio of (2) and (1)

$$\frac{c}{FwH}=\frac{\frac{1}{2}\csc^2 x\,(y\csc^2 y-\cot y)+\cot x-\cot i}{2\cot x\cot v+2} \qquad (12)$$

The solution for the case where the rupture surface passes below the toe of the slope is almost as simple as that for the above case. The only important difference is that another variable enters which is designated by n, and is shown in Fig. 2 (a). However, there are still only two degrees of freedom, since a condition enters which must be satisfied.

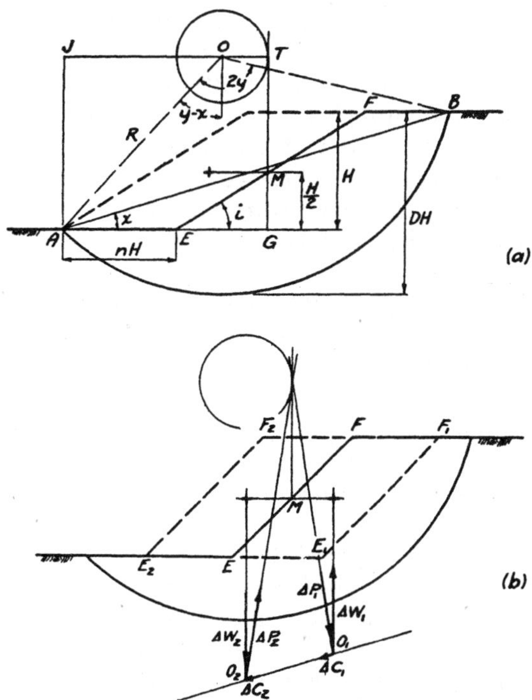

FIG. 2. — SKETCHES SHOWING ELEMENTS OF THE ϕ-CIRCLE METHOD FOR CIRCLES PASSING BELOW TOE OF SLOPE

Consider the case shown in Fig. 2 (b) where the right vertical tangent of the ϕ-circle passes through the mid-point of the slope M. If the slope EF is cut away parallel to itself to the line E_1F_1, the changes in forces W, C and P may be studied by a ϕ-circle analysis. Since W is decreased, its change ΔW_1 is an upward force which will act through the center of gravity of the area removed. The line of action of C must be the same before and after the change, thus ΔW_1 and ΔC_1 intersect at point O_1, and ΔP_1 must pass through this point and also be tangent to the ϕ-circle. ΔC_1 acts to the left, which indicates a decrease in the required cohesion, and it is seen that the removal of the mass EFF_1E_1 has given a more stable condition.

Similarly, if the slope EF is built up to E_2F_2, adding force ΔW_2, the analysis demonstrates that ΔC_2 also acts to the left. Therefore moving the mid-point of the slope to either side of the right vertical tangent of the ϕ-circle leads to greater stability, and the requirement that the mid-point must be on this tangent for the most dangerous circle must be introduced.

It may also be seen that there is little need of investigating cases of rupture arcs passing through the toe of the slope for which the mid-point of the slope is to the left of the right vertical ϕ-circle tangent, since a more dangerous case is obtained by moving the same circle to the left, allowing it to pass below the toe of the slope.

The condition may be obtained from Fig. 2 (a), as follows:

$$\overline{AE}+\overline{EG}=\overline{JO}+\overline{OT}$$

or

$$nH+\tfrac{1}{2}H\cot i=R\sin(y-x)+R\sin\phi$$

and inserting R from equation (3)

$$n=\tfrac{1}{2}(\cot x-\cot y-\cot i+\sin\phi\csc x\csc y) \qquad (13)$$

This equation could be used to eliminate n from the equations which follow. However, the requirement that n be positive must always be checked, and it is convenient to retain n as a dependent variable.

As compared to the case where the rupture arc passes through the toe of the slope, shown dotted in Fig. 2 (a), the change in W is —

$$-wnH^2=\frac{wH^2}{2}(-2n)$$

the change in Wd is —

$$(-wnH^2)(R\sin\phi-\tfrac{1}{2}nH)=\frac{wH^3}{4}(2n^2-2n\sin\phi\csc x\csc y)$$

If these changes are inserted in equations (9) to (12), inclusive, there results —

$$\frac{H}{2d}=\frac{\tfrac{1}{2}\csc^2 x\,(y\csc^2 y-\cot y)+\cot x-\cot i-2n}{\tfrac{1}{3}(1-2\cot^2 i)+\cot i\,(\cot x-\cot y)+\cot x\cot y+2n^2-2n\sin\phi\csc x\csc y} \qquad (14)$$

$$\cot u = \frac{H}{2d}\, y \sec x \csc x \csc^2 y - \tan x \tag{15}$$

$$\sin (u-v) = \frac{H}{2d} \sin u \csc x \csc y \sin \phi \tag{16}$$

$$\frac{c}{FwH} = \frac{\frac{1}{2} \csc^2 x\,(y \csc^2 y - \cot y) + \cot x - \cot i - 2n}{2 \cot x \cot v + 2} \tag{17}$$

To obtain the value of $\frac{c}{FwH}$ for any desired values of slope and friction angles and any chosen values of x and y, equation (13) may be tried first to determine whether or not the circle will pass through the toe of the slope. As a negative value of n is impossible, equation (13) must be discarded if it shows a negative result, and the circle will pass through the toe. For this case, evaluation of equations (9) to (12), inclusive, used in numerical order, give in succession $\frac{H}{2d}$, u, v and finally $\frac{c}{FwH}$. If equation (13) shows a positive value of n, the procedure is the same; equations (9) to (12) may be used for circles through the toe, while for circles below the toe equations (14) to (17) are to be used.

For the case of $\phi = 0$ the above equations may be simplified, and by differentiation with respect to the variable angles x and y the location of the critical circle may be obtained. The results for this special case will be discussed later. For the general case, where the friction angle is greater than zero, the very tedious method of trial must be resorted to, but the use of computing machines is of great assistance.

The procedure used in determining the tabulations was to choose values for x and y and compute $\frac{c}{FwH}$ as outlined. The center of the circle corresponding to these co-ordinates was plotted and at this point the value of $\frac{c}{FwH}$ recorded. Several such centers were tried, each helping to choose the following one until enough values were obtained to allow contouring and the determination of the maximum or critical value of $\frac{c}{FwH}$. After the procedure was standardized it was found that for each case of slope and friction angle from 10 to 20 trial centers were needed to locate the maximum. This procedure was carried out for a representative group of values of friction and slope angles, and Table I gives all pertinent data for all cases that were computed.

TABLE I

DATA ON CRITICAL CIRCLES BY ϕ-CIRCLE METHOD

i	ϕ	x	y	n	D	$\frac{c}{FwH}$	Corrected $\frac{c}{FwH^*}$
90 . . .	0	47.6	15.1	–	–	.261	.261
	5	50	14	–	–	.239	.239
	10	53	13.5	–	–	.218	.218
	15	56	13	–	–	.199	.199
	20	58	12	–	–	.182	.182
	25	60	11	–	–	.166	.166
75 . . .	0	41.8	25.9	–	–	.219	.219
	5	45	25	–	–	.195	.195
	10	47.5	23.5	–	–	.173	.173
	15	50	23	–	–	.153	.152
	20	53	22	–	–	.135	.134
	25	56	22	–	–	.118	.117
60 . . .	0	35.3	35.4	–	–	.191	.191
	5	38.5	34.5	–	–	.163	.162
	10	41	33	–	–	.139	.138
	15	44	31.5	–	–	.118	.116
	20	46.5	30.2	–	–	.098	.097
	25	50	30	–	–	.081	.079
45 . . .	0	(28.2)†	(44.7)	–	(1.062)	(.170)	(.170)
	5	31.2	42.1	–	1.026	.138	.136
	10	34	39.7	–	1.006	.110	.108
	15	36.1	37.2	–	1.001	.086	.083
	20	38	34.5	–	–	.065	.062
	25	40	31	–	–	.046	.044
30 . . .	0	(20)	(53.4)	–	(1.301)	(.156)	(.156)
	5 {	(23)	(48)	–	(1.161)	(.112)	(.110)
		20	53	0.29	1.332	.113	.110
	10	25	44	–	1.092	.078	.075
	15	27	39	–	1.038	.049	.046
	20	28	31	–	1.003	.027	.025
	25	29	25	–	–	.010	.009
15 . . .	0	(10.6)	(60.7)	–	(2.117)	(.145)	(.145)
	5 {	(12.5)	(47)	–	(1.549)	(.072)	(.068)
		11	47.5	0.55	1.697	.074	.070
	10 {	(14)	(34)	–	(1.222)	(.024)	(.023)
		14	34	0.04	1.222	.024	.023
All values‡	0	0	66.8	∞	∞	.181	.181

* With ϕ-circle correction applied.

† Figures in parentheses are values for most dangerous circle through the toe when a more dangerous circle exists which passes below the toe.

‡ A critical value at infinite depth. (See Fig. 8.)

As may be noted from this table, the critical circle through the toe of the slope was determined for each case. For cases of ϕ greater than zero in only three instances did the most dangerous circle pass below the toe, and for these double values appear in brackets. It is evident that the differences in these bracketed values are so small that circles passing below the toe might well have been neglected. An exception to this statement rests in the case of $\phi=0$ and $i<53°$, which is yet to be discussed. Where the circle passes below the toe the value of n is given in the table. When $y > x$ the lowest point on the rupture arc is at a lower elevation than the toe of the slope, and thus the existence of an underlying ledge or other surface of high shearing strength might have an effect on the Stability Number. The values of D (see key sketches of Fig. 8), which describe the depth to the lowest point of the rupture arc, are given in the table for all cases where $y > x$.

The final column of Table I gives the value of the Stability Number after the ϕ-circle correction has been applied. Returning now to the discussion of the ϕ-circle assumption, this correction will be described.

The distribution of the p forces (Fig. 1 (*a*)) is indeterminate, but investigations may be made for various assumed distributions. If k represents the ratio by which the moment arm of the force P exceeds the radius of the ϕ-circle, it may be shown that —

(*a*) Assuming a constant intensity of p forces —

$$k=\frac{y}{\sin y}-1$$

wherein y is half the central angle at the arc, as shown in Figs. 1 (*a*) and 2 (*a*).

(*b*) Assuming intensity of p force equal to zero at A and B and varying sinusoidally between —

$$k=\frac{1-\left(\frac{2y}{\pi}\right)^2}{\cos y}-1$$

For a given distribution assumption, k depends only on the central angle $2y$, and the relation between k and $2y$ is plotted in Fig. 2 for cases (*a*) and (*b*).

In all probability, assumption (*a*) is not even approximately correct, while assumption (*b*) is close enough to the actual to describe k with reasonable accuracy. By referring to the critical values of y as given in Table I, it may be seen that the highest value k will ever attain for cases where $\phi > 0$ is about 7 per cent.

It is of especial interest to note that the line of action of P as obtained by the use of the ϕ-circle assumption would be strictly correct for some slightly smaller value of ϕ; thus the assumption exactly corresponds to using a slightly conservative value of ϕ. Instead of the ϕ-circle, a circle of slightly larger radius should be used, and on this basis a correction which will be called the ϕ-circle correction may be

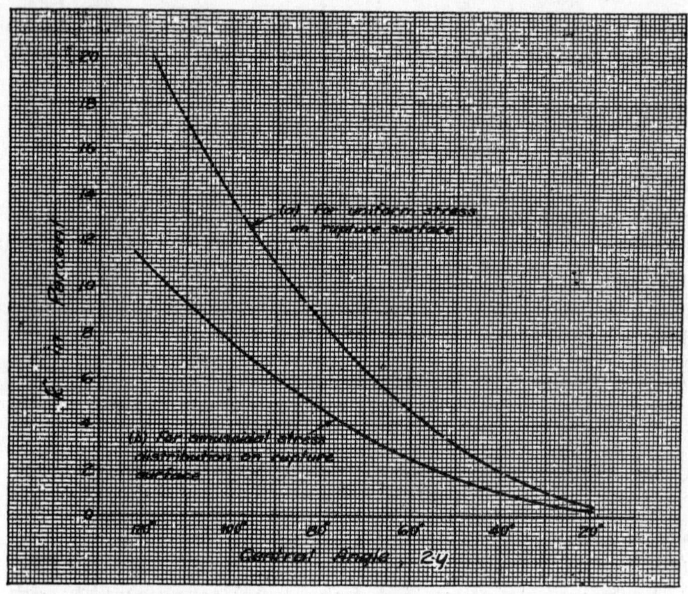

FIG. 3. — CHART FOR DETERMINATION OF FACTOR "K" IN THE ϕ-CIRCLE CORRECTION

made by the use of the k values of Fig. 3, curve (b). As an example, for $i = 30°$ and $\phi = 15°$, Table I gives $\frac{c}{FwH} = .049$ and $y = 39°$; hence from Fig. 3, $k = 4.5\%$. The radius of the ϕ-circle is $R \sin \phi$. If the corrected circle of slightly larger radius, $R \sin \phi'$, is introduced, the following is the relationship which must hold:

$$R \sin \phi' = R \sin \phi \ (1 + k)$$

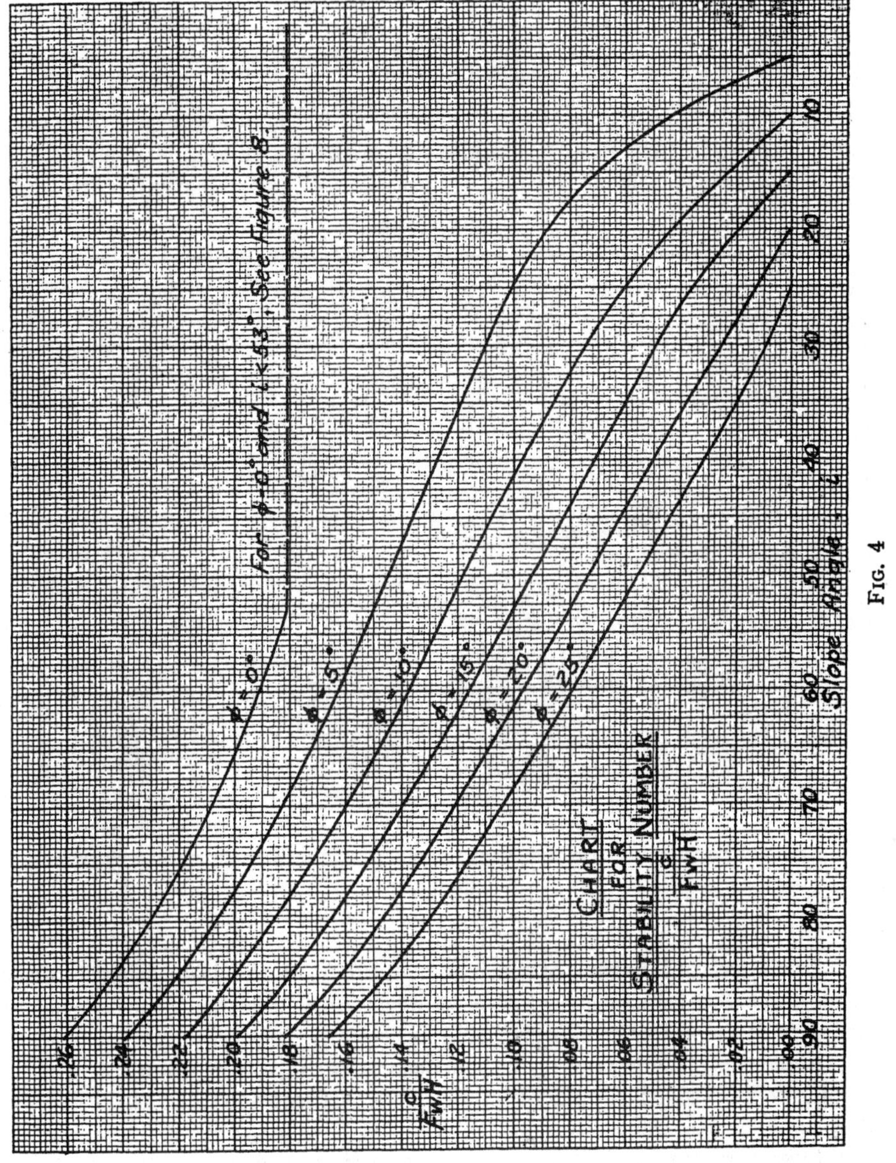
For φ=0° and i<53°, See Figure 8.
φ=0°
φ=5°
φ=10°
φ=15°
φ=20°
φ=25°
CHART
FOR
STABILITY NUMBER
c/FwH
c/FwH
.26
.24
.22
.20
.18
.16
.14
.12
.10
.08
.06
.04
.02
.00
90
80
70
60
50
40
30
20
10
Slope Angle, i

FIG. 4

By solving, $\phi' = 15° - 42'$, and by interpolation in Table I, for $i = 30°$ and $\phi = 15° - 42'$, $\frac{c}{FwH} = .046$, which is the corrected value. From a practical viewpoint, the original value of $\frac{c}{FwH} = .049$ is slightly conservative and sufficiently accurate. The correction is never larger than in this illustrative case, and it is evident from the above equation that the correction vanishes when $\phi = 0$.

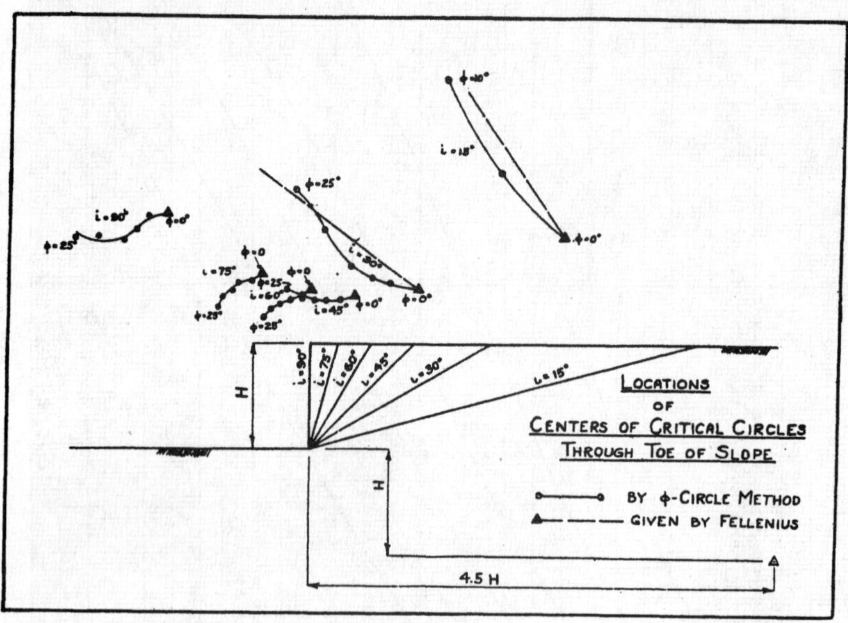

FIG. 5

The corrected values of Stability Number in the last column of Table I are accurate to three figures of decimals, which should always be sufficient for stability of slopes problems.

After the ϕ-circle correction has been applied, a rewording of the assumption involved should be made. The assumption that the force P is tangent to the ϕ-circle is replaced by the assumption that the distribution of the force P along the rupture is sinusoidal.

In Fig. 4 the results are presented as a nest of curves which allows easy interpolation. These curves give the critical values irrespective of the depth below the surface to which the critical circle may pass, and thus may be conservative if there is any limitation in this depth.

The co-ordinates describing the critical circles, as tabulated in Table I, are used for plotting Fig. 5, which shows the location of centers of most dangerous circles through the toe of the slope. As the majority of these centers were located only to the nearest degree in x and y, it may be noted that the plotting of the points is a bit irregular. This is, however, of no real importance. The data given by Fellenius for the locations of critical centers for circles through the toe consist of locations of centers for cases where $\phi=0$, together with the information that for flat slopes, when $\phi > 0$, the center will be approximately on a line through the point for $\phi=0$ and through a point at a depth H below and 4.5 H back from the bottom point of the slope. These points are plotted in Fig. 5 by triangles and the lines shown broken. As is to be expected, since Fellenius' work for $\phi=0$ was carried out mathematically, it checks exactly with the $\phi=0$ values of Table I. For slopes of 30 degrees and less the line given is also in fairly good agreement. For slopes of greater than 30 degrees, Fig. 5 would tend to show that this line begins to deviate appreciably from the most dangerous center, but it still may be used without serious error for slopes up to about 60 degrees as is indicated in Fig. 6. This figure shows the contouring of $\frac{c}{FwH}$ values with respect to the centers used, for $i=60$ and $\phi=20°$. Any point on the contour marked .090, if used as a center, gives $\frac{c}{FwH}=0.090$, and any point within this closed curve represents a value between .090 and the critical value .098. Thus any point within this rather large area, if used as a center, will give a value which will not differ from the critical value by more than 9 per cent. A line on the basis of that given by Fellenius is also shown on the figure, and, while it misses the critical point by an appreciable amount, it may be seen that the maximum along this line is .095, which is only 3 per cent below the critical value. On this same figure the traces obtained for the critical failure planes by the ϕ-Circle, Culmann and Jáky Methods are shown. The ϕ-circle correction has not been applied in Fig. 6. However, in all other figures applying to the ϕ-Circle Method, in which the Stability Number appears, this correction has been made.

When the effective friction angle is equal to zero, an unusual condition is met which requires special attention, as was pointed out by Fellenius as early as 1918. Mathematical analysis shows that there is a maximum value of the Stability Number which is independent of the slope angle and is equal to 0.181; which occurs when angle x is zero and the rupture surface has infinite radius and passes an infinite distance below the toe of the slope.

Zero friction angle and constant cohesion must mean constant shearing strength regardless of depth, a situation which cannot be pos-

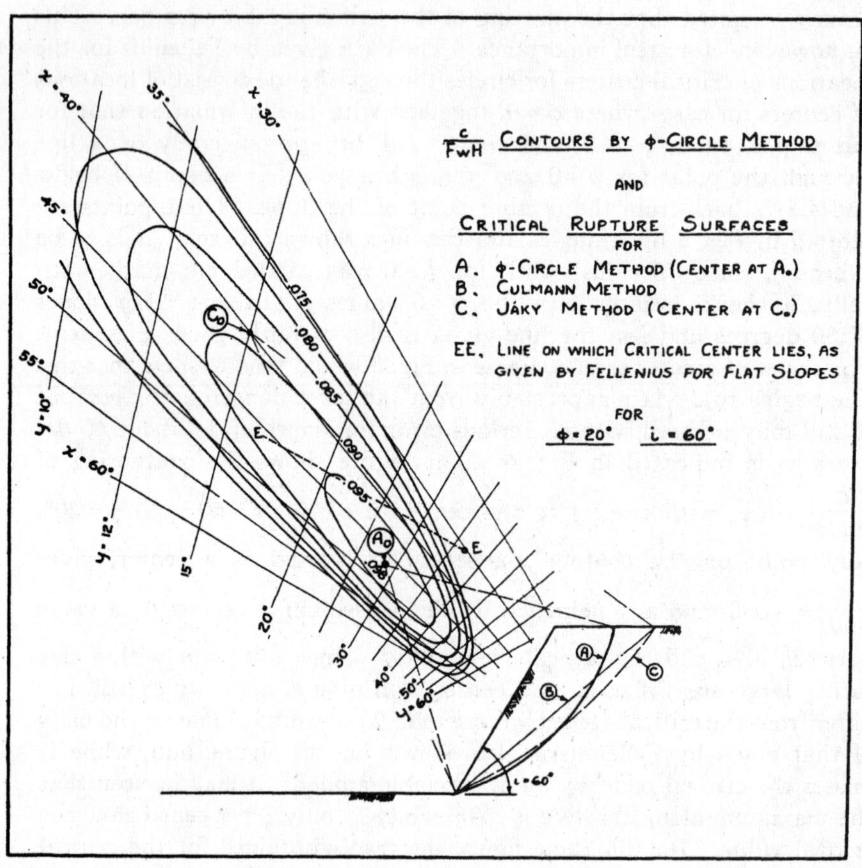

c/FwH Contours by ϕ-Circle Method
and
Critical Rupture Surfaces
for
A. ϕ-Circle Method (Center at A₀)
B. Culmann Method
C. Jáky Method (Center at C₀)
EE, Line on which Critical Center lies, as given by Fellenius for Flat Slopes
for
ϕ = 20° i = 60°

Fig. 6

sible to infinite depth. However, the shearing strength of embankments may often be essentially constant down to some definite strata of higher strength; for instance, to hardpan or ledge. Furthermore, the above maximum condition must mean that the Stability Number will increase with depth; thus to analyze such cases the relation between Stability Number and the available depth within which rupture may occur must be known.

Although this maximum value is independent of the slope, there is the maximum value for circles through the toe which occurs as shown in Fig. 5. This maximum through the toe gives Stability Numbers which are greater than 0.181 for slopes steeper than 53 degrees; thus the study of deep-seated rupture arcs is confined to slopes of less than 53 degrees. Fellenius called attention to the fact that for $\phi=0$ and slopes flatter than 53 degrees, the critical circle passes below the toe and has its center above the mid-point of the slope. He also gave data computed by mathematical solution for this case with a uniformly distributed surcharge at the top of the slope, but these data are limited to cases of vertical slope and values of depth factor D of 1, 1.5 and 2. In developing a complete solution of this case for the simplified conditions adopted in this paper, the following formulas were used. It should be noted that the ϕ-circle assumption does not enter when $\phi=0$.

When $\phi=0$, equation (12) reduces to —

$$\frac{c}{FwH}=\frac{\sin^2 x \sin^2 y}{2y}\left[\frac{1-2\cot^2 i}{3}+\cot x \cot y+\cot i\,(\cot x-\cot y)\right] \quad (18)$$

This simplified formula, in almost exactly the same form, was given by Fellenius.

Equation (13) becomes —

$$n=\frac{1}{2}\,(\cot x-\cot y-\cot i) \quad (19)$$

and for circles below the toe, equation (7) becomes —

$$\frac{c}{FwH}=\frac{\sin^2 x \sin^2 y}{2y}\left[\frac{1-2\cot^2 i}{3}+\cot x \cot y+\cot i\,(\cot x-\cot y)+2n^2\right] \quad (20)$$

or —

$$\frac{c}{FwH}=\frac{1}{4y}\left[\sin^2 x+\sin^2 y-\frac{2-\cot^2 i}{3}\sin^2 x \sin^2 y\right] \quad (20a)$$

From Fig. 2 (a):

$$D=\frac{1}{2}\,(\csc x \csc y-\cot x \cot y+1) \quad (21)$$

in which D may be designated as depth factor, and from which —

$$\csc x = (2D-1) \csc y + 2\sqrt{D^2 - D} \cot y \qquad (22)$$

For any chosen value of D, there is only one degree of freedom involved in the problem to be solved, since for any assigned value of y the value of x is available from equation (22). For critical circles through

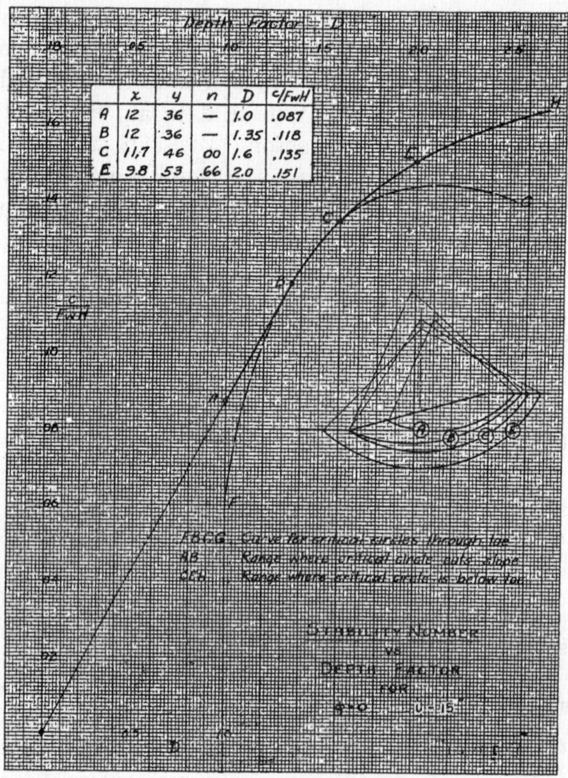

Fig. 7

the toe, by the use of equations (22) and (18), a curve of Stability Number against depth factor D may be obtained.

In Fig. 7, line $FBCG$ represents this condition for $i = 15°$. However, after equation (22) is evaluated, equation (19) should be tried, since positive values of n indicate a critical circle below the toe. If n is positive, as it will be for sufficiently high values of D, equation (20)

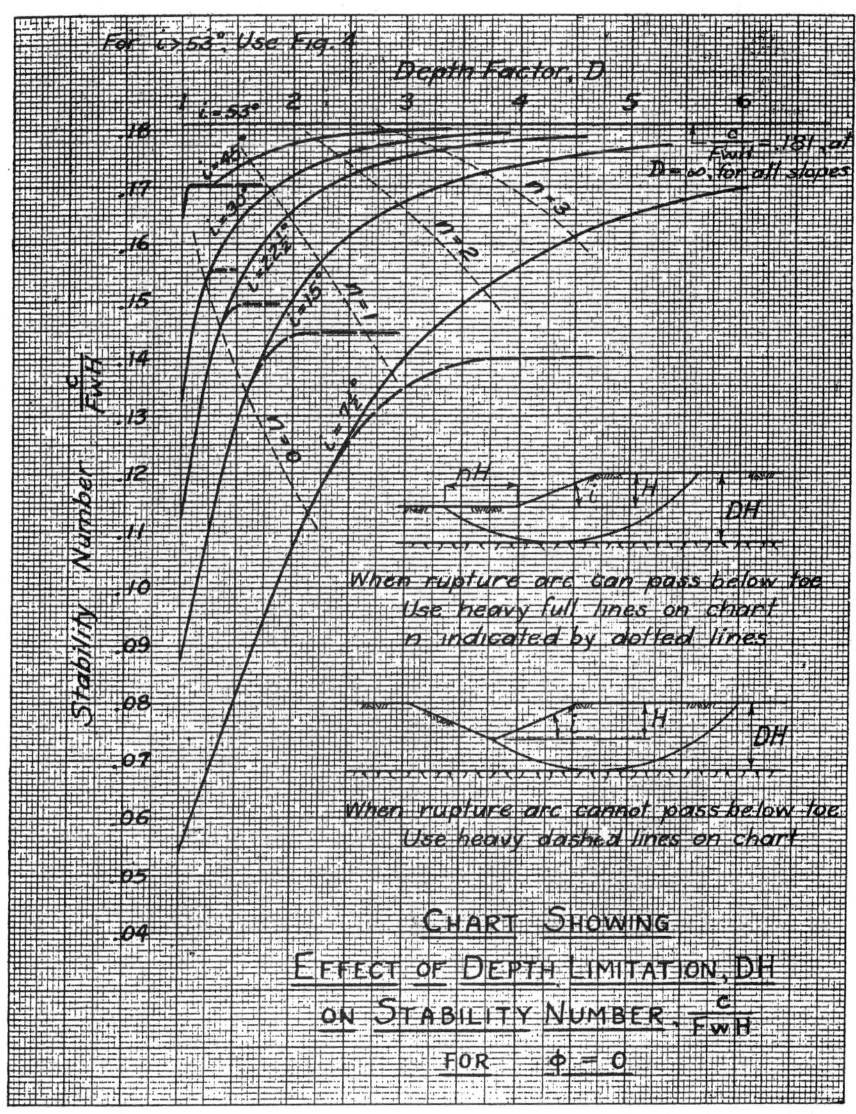
For i>53° Use Fig. 4
Depth Factor, D
Stability Number $\frac{c}{F_w H}$
$\frac{c}{F_w H}$ = .181 at D = ∞, for all slopes
When rupture arc can pass below toe
Use heavy full lines on chart
n indicated by dotted lines
When rupture arc cannot pass below toe
Use heavy dashed lines on chart
CHART SHOWING
EFFECT OF DEPTH LIMITATION, DH
ON STABILITY NUMBER, $\frac{c}{F_w H}$
FOR φ = 0

FIG. 8

or (20*a*) should be used instead of equation (18). Line *CEH* of the figure is the desired curve for the critical circle, in the range where this circle is below the toe.

If a rigid surface should exist at the level of the base of the slope, it is natural to expect a possible rupture arc passing above the toe of the slope. In such a case the ruling height is the depth to ledge, and

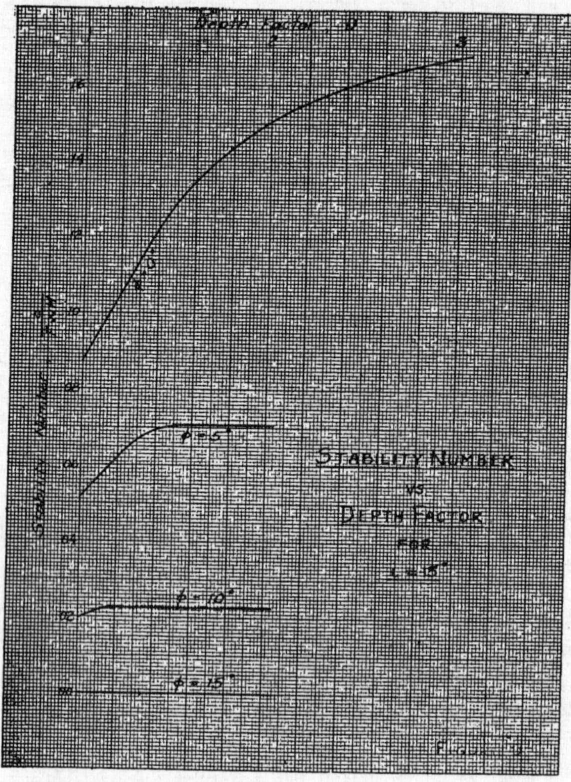

FIG. 9

the actual height of the slope is not important. In Fig. 7, if a line is drawn through the origin, tangent to the curve, this line, *AB*, is the required curve within the range where the most dangerous circle passes above the toe of the slope. For a mathematical approach within this range, it could be shown that x and y should be chosen so that the right-hand side of equation (18) divided by the right-hand side of equation (21) is a maximum. In the sketch of Fig. 7 the rupture arcs for points

A, B, C and E are drawn, while the data pertaining to them are given in the table. Fig. 8 presents a nest of curves of this type, covering the range of slopes from 7½ degrees to 53 degrees. The possibility of a failure through the toe of the slope may exist even though the critical circle is below the toe, if the material outside the toe is loaded in some manner. Such a case is indicated by the lower sketch of the figure, and the dashed lines provide values which may be used when such situations occur. Values of n may be obtained from the chart for cases where the critical arc is below the toe, and may be of some value in predicting the approximate outer edge of the mud wave which would be formed in case of failure. It must be noted, however, that if n is limited instead of d, an entirely different problem is presented, and this chart cannot be used to determine the Stability Number; an exception being that if n is restricted to a value of zero the dashed curves are applicable.

Reference to the depth factor column in Table I shows that the available depth may have some effect on the Stability Number when $\phi > 0$ and i is small. This effect is quite pronounced when $i = 15°$, and still exists when $i = 30°$. For $i = 15°$, Fig. 9 presents a solution of the effect of the depth factor.

The above derivations assume that an obliquity of ϕ is developed at all points along the rupture arc, or, in other words, that all available friction is utilized. If a part of this friction is not developed, there remains an excess of unused strength which represents a factor of safety with respect to friction. If at any point on an arc this factor with respect to friction is F_F, then the obliquity of stress, ϕ_D, which is developed across the arc at this point, is described by the following equation:

$$\tan \phi = F_F \tan \phi_D$$

and it may be noted that it will usually be sufficiently accurate to use the relation —

$$\phi_D = \frac{\phi}{F_F}$$

A constant value for ϕ_D would probably never occur simultaneously at all points of the arc because of the progressive nature of the development of the shear. Thus ϕ_D for any given arc must be described as the average developed obliquity, or the obliquity which, were it constant, would lead to the same total frictional shearing stress as actually is developed.

The friction angle ϕ, as used up to this point, was also the developed obliquity. If a true factor of safety, F_T, is desired, this factor to apply to both cohesion and friction, then the developed obliquity must be $\phi \div F_T$. If the value of ϕ in all the preceding derivations is superseded by the more general value of ϕ_D, then the factor of safety which was previously only relative to the cohesion becomes the true factor F_T. This replacement of ϕ by ϕ_D and F by F_T applies not only to all formulas but also to Figs. 4, 8 and 9. It could be demonstrated that any small change in the validity of the ϕ-circle assumption, because of ϕ_D being an average rather than a constant value, is probably favorable. Expressed mathematically, the above may be summarized as follows:

$$\frac{c}{FwH}=f\,(i,\phi) \quad \text{and} \quad \frac{c}{F_T wH}=f\,(i,\frac{\phi}{F_T})$$

Applications of both types of safety factors will be found in the illustrative problems. For the typical cohesive soil the effective cohesion can be determined with at least as good accuracy as the effective friction angle. Thus it will seldom be logical to use the factor of safety with respect to cohesion in preference to the true factor of safety.

The Logarithmic Spiral Method

In the Circular Arc Method it is necessary to introduce some assumption to make the problem statically determinate. To avoid this undesirable feature, Rendulic recommended the use of a logarithmic spiral as the surface of rupture.

The important property of the logarithmic spiral, expressed in polar co-ordinates by the equation —

$$r=r_1^{\,\theta \tan \phi} \qquad (1)$$

is that all radius vectors cut the curve at an angle of obliquity of ϕ. In Fig. 10 (*a*) such a curve is represented by the line ACB, r being equal to r_1 at OA where θ equals zero. As the forces on elements along the rupture surface must have an obliquity of ϕ, it may immediately be seen that with the use of this curve all elementary forces across the rupture surface, as well as their resultant force, must be directed toward the pole O. It should be noted that θ must increase counterclockwise in this figure, as the obliquity must be such as to resist a slide.

As in the circular scheme, there are two degrees of freedom in the logarithmic spiral. These will be taken as the central angle z and the chord slope t. A number of sets of values of these two variables must be tried until the critical location is obtained.

The principal steps in the mathematical solution are given below. As in the previous derivation, F is the factor of safety with respect to

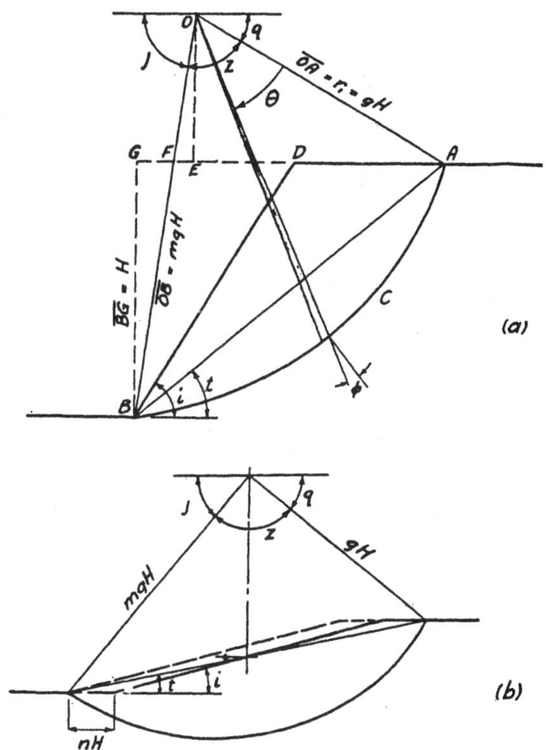

Fig. 10. — Sketches showing Elements of the Logarithmic Spiral Method

cohesion, but the solution is applicable for use with the true factor of safety as explained.

Four dependent variables are used to simplify the solution. They are as shown in Fig. 10, and are described by the following expressions, which hold for the curve either through or below the toe of the slope:

$$m = e^{z \tan \phi} \tag{2}$$

$$g = \frac{1}{\sin t \sqrt{1+m^2-2m \cos z}} = \frac{r_1}{H} \tag{3}$$

$$j = t + \sin^{-1}\left[\frac{\sin z}{\sqrt{1+m^2-2m \cos z}}\right] \tag{4}$$

$$q = \pi - z - j \tag{5}$$

Referring again to the three forces W, C and P, and taking moments about the pole, there results —

$$M_w + M_{c_1} = 0 \quad \text{or} \quad M_w + \frac{M_c}{F} = 0 \tag{6}$$

where

$$M_c = -\int_o^z c\, r^2\, d\,\theta = -\frac{c\, g^2\, H^2}{2 \tan \phi}\,(m^2-1) \tag{7}$$

and for the curve through the toe of the slope —

$$M_w = M_1 - M_2 - M_3 \tag{8}$$

in which M_1, M_2 and M_3 are moments about O for the masses represented by areas $OACB$, OAF and BDF, respectively, of Fig. 10 (a).

$$M_1 = w\int_o^z \frac{r^3}{3}\cos(\theta+q)\,d\theta = \frac{wg^3\,H^3}{3\,(9\tan^2\phi+1)}\left[(m^3\sin j-\sin q) - 3\tan\phi\,(m^3\cos j+\cos q)\right] \tag{9}$$

$$M_2 = \frac{1}{6} wg^3\, H^3 \sin^3 q\,(\cot^2 q - \cot^2 j) \tag{10}$$

$$M_3 = \frac{1}{6}\, wH^3 \left[\cot^2 i - \cot^2 j - 3mg\, \cos j\, (\cot i - \cot j)\right] \tag{11}$$

Substituting (7), (9), (10) and (11) in (8) and (6) gives —

$$\frac{c}{FwH} = \frac{\tan\phi}{3g^2\,(m^2-1)}\left[\frac{2g^3\left\{(m^3\sin j - \sin q) - 3\tan\phi\,(m^3\cos j + \cos q)\right\}}{9\tan^2\phi+1} + \right.$$

$$\left. + g^3 \sin^3 q\,(\cot^2 j - \cot^2 q) + 3mg\cos j\,(\cot i - \cot j) - \cot^2 i + \cot^2 j\right] \tag{12}$$

Equation (12) applies when the rupture curve passes through the toe of the slope. When it passes below the toe, the most dangerous condition is obtained when the mid-point of the slope is vertically below the origin, and from Fig. 10 (*b*) —

$$n=mg\cos j-\frac{1}{2}\cot i \qquad (13)$$

The increase in overturning moment as compared to the previous case is —

$$\frac{1}{2}wn^2H^3=\frac{wH^3}{2}\left(mg\cos j-\frac{1}{2}\cot i\right)^2$$

which, when included in equation (8), leads to the following alternate for equation (12):

$$\frac{c}{FwH}=\frac{\tan\phi}{3g^2(m^2-1)}\left[\frac{2g^3\left\{(m^3\sin j-\sin q)-3\tan\phi\,(m^3\cos j+\cos q)\right\}}{9\tan^2\phi+1}+\right.$$

$$\left.+g^3\sin^3 q\,(\cot^2 j-\cot^2 q)+3mg\cos^2 j\,(mg-\csc j)-\frac{1}{4}\cot^2 i+\cot^2 j\right] \qquad (14)$$

For any chosen values of z and t the dependent variables may be obtained from (2) to (5), inclusive, after which n may be evaluated by the use of (13). If n is negative, (12) must be used, while for positive values of n, (14) will give the greater value for the Stability Number.

It was found that a computation by this method requires about twice as long as one by the ϕ-Circle Method. For the critical values of the Stability Number, it developed that z and t, the central angle and chord slope for the Spiral Method, are approximately equal to the corresponding critical values, $2y$ and x of the ϕ-Circle Method. (See Table III.) This was of assistance in choosing points for trial computation by one method after the work of the other was completed. Because of the large amount of labor involved, the computations for the Spiral Method have been carried out for only enough representative cases to give a good comparison with results by the ϕ-Circle Method. These comparisons are given later.

Comparison of Methods

The general formula for the Culmann Method may be expressed as follows:

$$\frac{c}{FwH}=\frac{1-\cos(i-\phi)}{4\cos\phi\sin i}$$

By the Résal-Frontard Method, the depth to which tension exists, y_o, and the general formula expressed in the form of the Stability Number are —

$$y_o=\frac{2c}{w}\tan\left(45+\frac{\phi}{2}\right)$$

$$\frac{c}{FwH}=\frac{\sin(i-\phi)}{2\sin^2 i\cos\phi}\left[\frac{1}{\dfrac{\cos\phi}{\sin i\,(1-\sin\phi)}+\dfrac{\cos^{-1}\left\{\dfrac{\sin^2 i-\sin\phi}{\sin i\,(1-\sin\phi)}\right\}}{\sqrt{\sin(i-\phi)\sin(i+\phi)}}}\right]$$

The general mathematical expression as given by Jáky is too long to be reproduced here. However, his final results (9*c*) are submitted in the form of an expression $f(\beta)$ where —

$$\frac{c}{FwH}=\frac{1}{4}\,\frac{1}{f(\beta)}$$

Table II presents the results of computations for $\frac{c}{FwH}$ by the three above-mentioned methods, and also by the ϕ-Circle Method for rupture arcs through the toe of the slope. Where more dangerous circles exist passing below the toe, the values by the ϕ-Circle Method are given in parentheses. The ϕ-circle correction has not been applied in this table nor in the figure which follows. For the case $\phi=10°$ the results are plotted in Fig. 11, and the relationships between different methods as shown in this figure are very similar for any other value of ϕ. This figure illustrates the statement made previously, that the Culmann Method gives low values, while the Résal-Frontard Method gives results which are much too high. The values obtained by the ϕ-Circle Method and the Jáky Method show reasonably close agreement, and

TABLE II

VALUES OF $\frac{c}{FwH}$ BY VARIOUS METHODS

Slope Angle (i)	Friction Angle (φ)	Résal-Frontard (1)	Culmann (plane) (2)	Jáky $\frac{1}{4f(\beta)}$ (3)	Jáky Circle-φ-Circle Method (4a)	φ-Circle (4)
90 . . .	0	.500	.250	–	–	.261
	5	.458	.229	.229	.229	.239
	10	.420	.210	.210	.210	.218
	15	.384	.192	.192	.192	.199
	20	.350	.175	.175	.175	.182
	25	.319	.159	.159	.159	.166
75 . . .	0	.396	.192	–	–	.219
	5	.353	.171	.172	.183	.195
	10	.307	.152	.156	.163	.173
	15	.280	.134	.135	.144	.153
	20	.246	.117	.119	.126	.135
	25	.215	.102	.102	.110	.118
60 . . .	0	.328	.144	–	–	.191
	5	.284	.124	.135	.149	.163
	10	.244	.105	.116	.127	.139
	15	.208	.088	.098	.107	.118
	20	.174	.072	.081	.089	.098
	25	.143	.058	.063	.073	.081
45 . . .	0	.280	.104	–	–	(.170)
	5	.231	.083	.109	.121	.138
	10	.187	.065	.086	.097	.110
	15	.148	.049	.064	.075	.086
	20	.113	.035	.044	.057	.065
	25	.081	.023	.028	.041	.046
30 . . .	0	.244	.067	–	–	(.156)
	5	.183	.047	.082	.095	(.112)
	10	.130	.031	.051	.066	.078
	15	.085	.018	.027	.042	.049
	20	.047	.008	small	.022	.027
	25	.017	.002	0	.005	.010
15 . . .	0	.217	.033	–	–	(.144)
	5	.118	.015	.035	.059	(.072)
	10	.043	.004	0	.019	(.024)

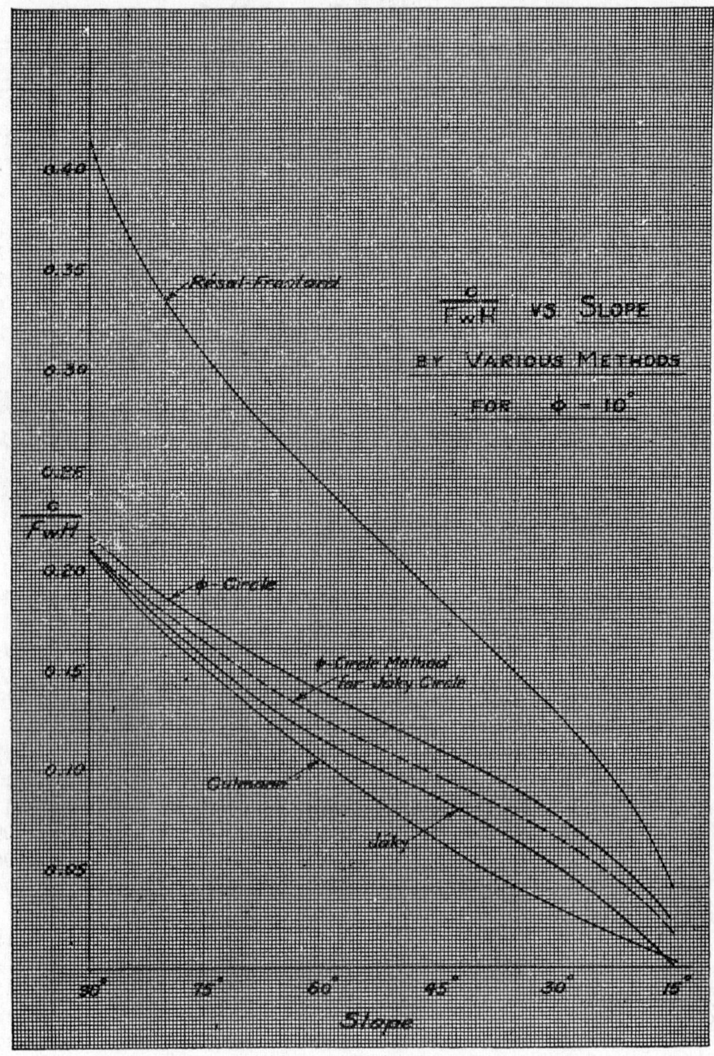

0.40
0.35
0.30
0.25
0.20
0.15
0.10
0.05
c/FwH
Résal-Frontard
c/FwH VS SLOPE
BY VARIOUS METHODS
FOR φ = 10°
φ-Circle
φ-Circle Method for Jáky Circle
Culmann
Jáky
90°
75°
60°
45°
30°
15°
Slope

Fig. 11

the difference is of little practical importance in computations of this type. The most questionable step in the Jáky Method is in the choice of critical circle, and if Jáky's circle is used in a computation by the ϕ-Circle Method the results given in Table II, column (4*a*), and the dotted curve of Fig. 11 are obtained. While it must be remembered that different assumptions may lead to different critical circles, there is at least an indication here that Jáky's choice of circle is questionable, and the results by his method are slightly low.

A comparison of results by the general mathematical solution based on the ϕ-Circle Method, and Fellenius' general solution by graphics using the Slices Method, is given in Table III. It may be seen that the

TABLE III

COMPARISON OF VALUES OF $\frac{c}{FwH}$ BY SLICES AND ϕ-CIRCLE METHODS.

i	$\phi=5$		$\phi=15$		$\phi=25$	
	Slices	ϕ-Circle	Slices	ϕ-Circle	Slices	ϕ-Circle
15	.072	.070	–	–	–	–
30	.114	.110	.048	.046	.012	.009
45	.141	.136	.085	.083	.048	.044
60	.165	.162	.120	.116	.082	.079
75	.196	.195	.154	.152	.118	.117
90	.239	.239	.199	.199	.165	.166

two solutions agree closely. The values by the Slices Method are slightly higher throughout, which checks the claim that the assumption used in the Slices Method is conservative.

Fellenius stated that for the steeper slopes the assumption of plane failure is not appreciably in error, thus there is little need to determine locations of critical circles for steep slopes. This is true only to a limited degree, as may be seen by comparison of columns (2) and (4) of Table II. As an example, it may be noted that on the basis of plane failure for $i=60$ and $\phi=20$ the Stability Number is about 25 per cent on the unsafe side as compared to the assumption of circular rupture arc.

For the cases for which computations have been made by both the ϕ-Circle and Spiral Methods, the comparisons given in Table IV are obtained. It is evident that for all practical purposes the results by the

TABLE IV

COMPARISON OF RESULTS BY THE ϕ-CIRCLE AND THE LOGARITHMIC SPIRAL METHODS

i	ϕ	ϕ-Circle Method			Logarithmic Spiral Method		
		Chord Slope x	Central Angle $2y$	$\frac{c}{FwH}$	Chord Slope t	Central Angle z	$\frac{c}{FwH}$
90	25	60	22	.1659	61	22	.1651
60	25	50	60	.0788	49	58	.0784
30	25	29	50	.0089	29	40	.0083
60	15	44	63	.1160	44	60	.1159
90	5	50	28	.2386	50	29	.2387
60	5	38.5	69	.1624	39	69	.1624
15	5	(12.5)	(94)	(.0682)	(13)	(88)	(.0681)
15*	5	11	95	.0695	11.7	92	.0696

* Indicates critical rupture surface passing below toe of slope: by ϕ-Circle Method n=.55; by Logarithmic Spiral Method n=.40.

two methods are identical. Fig. 12 illustrates the very close agreement for a typical case.

The conclusion may be drawn that because of its requiring no further assumption, the Spiral Method is the preferable of the two, at least from a theoretical viewpoint. However, the greater length of time required for computation more than overbalances this preference, and the final results given by this paper (Fig. 4) are figured by the ϕ-Circle Method. It may be noted that these two methods and the Slices Method as well become identical for the case of $\phi=0$.

EFFECTS OF SATURATION AND SEEPAGE

To study the various possible conditions to which a slope may be subjected, the following stages may be pictured:

A. Complete Submergence. — Free water level at top of slope, soil fully saturated.

B. Sudden Drawdown. — Case A, with free water suddenly removed.

C. Steady Seepage. — Continuous flow of water through the soil, requiring a constant supply of ground water or rainwater to maintain seepage.

D. Capillary Saturation. — No flow of water, no supply, no evaporation. Pores of soil filled with water held by capillarity, a condition common in typical cohesive soils.

Conditions other than the above, such as complete or partial drying, may also prevail. Here the problem is of an entirely different nature, involving prevention of disintegration rather than considera-

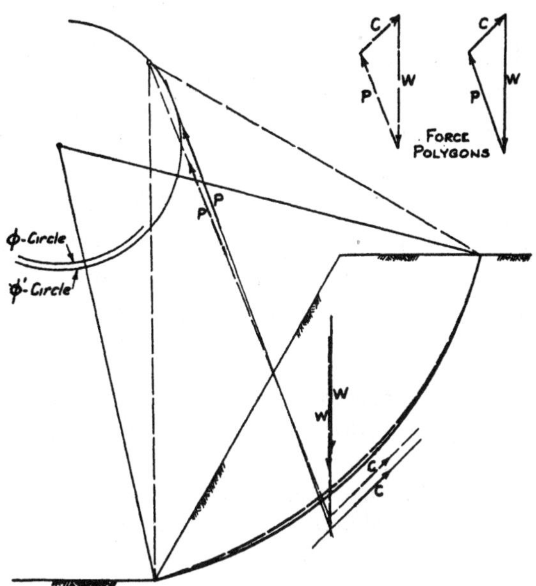

ϕ-Circle Method shown by full lines.
Logarithmic Spiral Method shown by dash lines.

Fig. 12. — Comparison of ϕ-Circle and Logarithmic Spiral Methods for $\phi = 15°$; $i = 60°$

tion of stability (9*b*). Also an indefinite number of intermediate states will occur in passing from one to another of those above outlined.

Cases A and D will be developed first, since they are important from the practical viewpoint and are easily handled.

A. Complete Submergence

The unit weight which is to be used in connection with the Stability Number for this case is the submerged unit weight, given by the expression $\frac{s-1}{1+e} w_o$, where s is the specific gravity of the solid particles of the soil, e is the void ratio or ratio between volume of voids and volume of solid matter, and w_o is the unit weight of water. For the general run of soils the submerged unit weight will be in the vicinity of 50 to 70 pounds per cubic foot. This low unit weight means that this case is the most favorable that can be obtained; therefore the submerged case cannot be assumed with safety for stability analyses unless permanent submergence is assured. With the use of the saturated unit weight, the equilibrium polygon of Fig. 1 (*b*) is applicable and no complications enter. It is to be noted that protection against such items as wave action are important but do not come within the scope of this paper.

D. Capillary Saturation

The unit weight which applies for this case is the total unit weight, that is, the weight of both solids and water, which is given by the expression $\frac{s+e}{1+e} w_o$. Its value will generally be in the neighborhood of 110 to 130 pounds per cubic foot. Compared to the saturated case, the unit weight for this case is approximately twice as large, thus the force W of Fig. 1 (*b*) would be about doubled, and the required shearing strength must also be about twice as great. However, the use of this case, where there is a possibility of future seepage within the slope, is not safe, as seepage introduces forces which lead to a still more unfavorable situation.

Where there is seepage of water through the embankment, or where there is the possibility of such seepage occurring at some season of the year, the development of a method for handling the situation is much more difficult. It has been suggested (9*a*) that flow nets may be sketched and from these a solution may be obtained. This involves the use of the tedious graphical trial method of solution plus the construction of the flow net. It can be demonstrated, however, that the shearing strength required for stability in case C, Steady Seepage, is greater than that in case D and less than that in case B; also that in many cases there is not a radical difference between the requirements of cases B and D. The percentages by which Stability Numbers for case D are less than those for case B are indicated in Table V, which shows that very large differences occur only for flat slopes with high friction angles. Thus the steady seepage case must represent a case somewhere within the rather narrow range between two cases for which computations may be made without difficulty. It can also be shown that, if submergence

TABLE V

PERCENTAGE BY WHICH STABILITY NUMBER FOR CAPILLARY SATURATION CASE IS LESS THAN STABILITY NUMBER FOR SUDDEN DRAWDOWN CASE FOR $\frac{s-1}{s+e}=\frac{1}{2}$

ϕ	Slope		
	90°	60°	30°
0°	0	0	0
10°	9	15	25
20°	17	30	65

or partial submergence ever occurs, subsequent drawdown may lead to a condition intermediate between cases B and C, and computations on the basis of case C would be unsafe.

These considerations lead to the recommendation that the requirements of case B, Sudden Drawdown, be used for slopes where there is seepage of water or any possibility of such seepage ever occurring. This is a procedure which is conservative, but which is probably sufficiently accurate for all cases of steep slopes and small friction angles. For cases where both flat slope and large friction angle occur, the bounding values as given by results for cases B and D may still be sufficient in many instances. It is doubtful if the accuracy which can be attached to stability computations is great enough to warrant the drawing of flow nets and the involved procedure that must follow, unless the difference between these bounding values is quite large.

A complete development of case C would require more space than can be allotted to it in this paper. An outline of the basic points involved would seem to be sufficient, and with this in mind points of theory behind such items as the principles of flow nets will be kept to the minimum. Later, a number of practical problems are given which will serve to illustrate various factors much better than any theoretical treatment could hope to do.

B. Sudden Drawdown

The force W of Fig. 1 (*b*) represents the effective weight of the sliding mass. For the submerged slope, Fig. 13 (*a*) shows the forces acting according to the ϕ-circle solution, wherein W_s, the effective weight for this case, equals the product of the area of the sliding mass and the submerged unit weight. The force polygon is shown in full lines in Fig. 13 (*b*). The corresponding diagram for the Capillary Saturation Case is shown in Fig. 13 (*c*), in which the effective weight W_T is based upon the total unit weight (soil plus water). The difference between the forces W_T and W_S will be designated by W_o, and it may be noted that this force is equal to the weight of a mass of water of the same total volume as the sliding mass. This weight W_o must be present in the submerged case, but it has no effect since it is just balanced by forces E_1 and E_2, the resultant water pressures across the slope and the rupture arc, respectively. Fig. 13 (*d*) shows the pressures which act along these surfaces; also the resultant forces E_1 and E_2 are shown in this figure, while the same forces appear dotted in the equilibrium polygon of Fig. 13 (*b*). E_2 must act normal to the slope at its lower third point, W_o must act vertically downward through the center of gravity of the sliding mass, and since the three forces, W_o, E_1 and E_2, are in equilibrium they must be concurrent. Since E_1 is made up of pressures which are everywhere normal to the rupture arc, and thus have no moment about the center of the circle, E_1 itself must have no moment about the center, and so must pass through O. Thus the lines of action of these three forces must be as shown in Fig. 13 (*e*). It may be noted that the moments of W_o and E_2 about O must just balance each other. It is correct to speak of W_o as an overturning force, but in this instance its overturning effect is just counterbalanced by the resisting effect of E_2.

The submerged case may be transformed into the sudden drawdown case by the sudden removal of the force E_2. Since the moment of E_2 just balances that of W_o, removal of E_2 introduces an additional overturning tendency equal to the moment of W_o. The weight W_o at the instant of sudden drawdown is carried by a temporary excess of pressure in the water, and intergranular stresses can replace this hydrostatic excess only as fast as the necessary strains in the mass can develop. Since W_o is not carried by the soil skeleton, no friction can be developed to help resist the shearing stresses it induces. Thus the overturning forces acting are W_o, which can be resisted only by cohesion, and W_S, which is resisted by cohesion and friction together. The force diagram for W_S alone is shown in Fig. 13 (*a*), while that for W_o, in which an effective friction angle of zero must be used, is given in Fig. 13 (*f*). Thus the cohesion required to overcome the combined overturning effects of W_S and W_o is the sum of C_S and C_o of Figs. 13 (*a*) and 13 (*f*).

A correct solution of this case would require trials on various rupture arcs for the determination of the critical value. The use of Fig. 4,

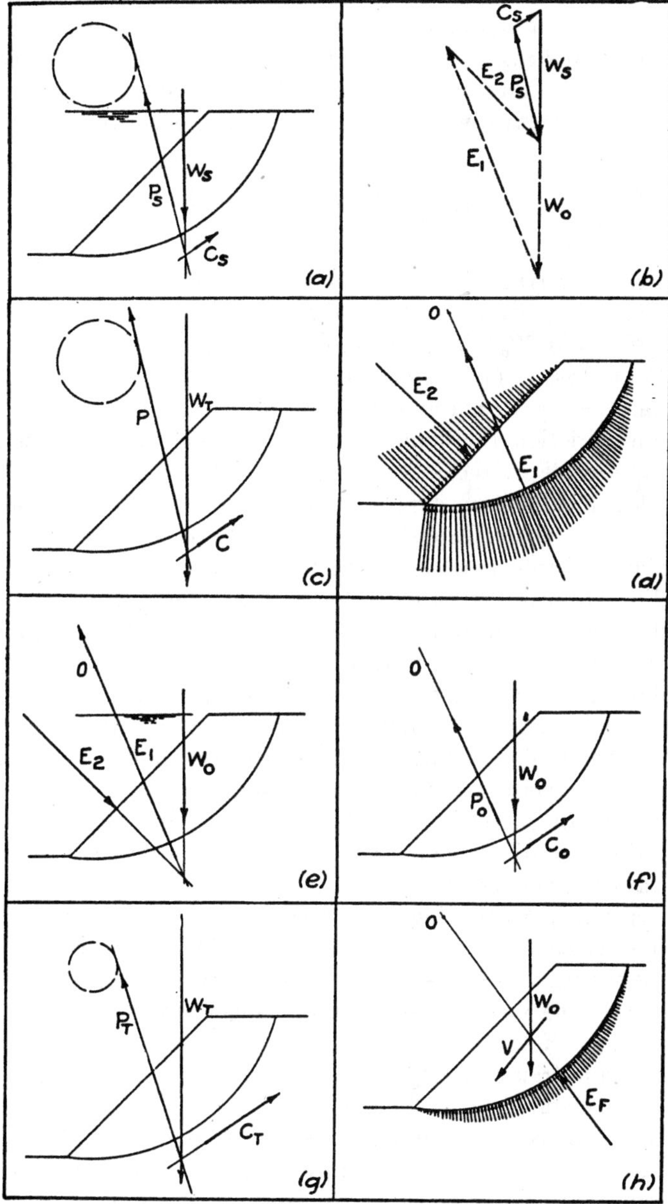
C_S
W_S
E_2
P_S
E_1
W_o
C_s
(a)
(b)
O
W_T
P
E_2
E_1
C
(c)
(d)
O
E_2
E_1
W_o
P_o
C_o
(e)
(f)
P_T
W_T
C_T
V
E_F
(g)
(h)

FIG. 13

with the two overturning forces handled separately, is not theoretically sound, as the two cases have different values of ϕ and thus will lead to different critical circles. A satisfactory method from a practical viewpoint is to use the total unit weight, with a weighted value for ϕ. The use of an arithmetically weighted value described by the following equation —

$$\phi_w = \frac{w_s}{w_s + w_o}\phi = \left[\frac{s-1}{s+e}\right]\phi$$

is not strictly correct, but the discrepancy involved will always be small. Fig. 13 (*g*) shows the use of a weighted value for ϕ, with W_T equal to W_S plus W_o. The required cohesion C_T equals the sum of C_S and C_o of Figs. 13 (*a*) and 13 (*f*).

It has been stated above that in the determination of cohesion and internal friction the conditions in nature must be reproduced as closely as possible in the laboratory tests. With this point in mind, for the sudden drawdown case it may be well to mention an alternate point of view to that given above. Let it be assumed that values for cohesion and friction have been determined for a case of a submerged slope. For the case of sudden drawdown a method of obtaining a weighted friction angle from these has been described. It may be possible that if another set of tests were to be run, using the conditions of consolidation, the time rate of shear, and other conditions, such as pressure on lateral planes which truly represent the situation, a value of ϕ might be obtained which would be a direct determination of that quantity which was obtained above by weighting. At present an accurate duplication of these natural conditions would be quite difficult, therefore the weighting scheme which offers a simple and conservative approximation is suggested.

The various possible stages have been discussed as if passing from one to another could occur without change of soil properties, but this may not be the case. The discussion of these stages aims to give a picture of the general situation. In addition to changes in this picture there will also be changes in the soil properties which must be covered by tests for cohesion and friction angle under conditions which as nearly as possible reflect the actual natural conditions.

When the force E_2 is suddenly removed, the relatively large unbalanced force P_o will produce an upward and outward flow of water, which may result in piping. At this instant the piping tendency is worse than later, when steady flow has been established. However, the condition is a transient one, and if investigations for piping are made for the steady state of seepage they may cover the situation. Further discussion of the possibility of piping will not be attempted in this paper.

The sudden drawdown case must be looked upon as a transient condition which is in effect only momentarily. The pressures within the sliding mass immediately begin to undergo adjustments, and after a sufficient lapse of time the condition of steady seepage is approached.

If water is seeping through a soil there must be a loss of potential in the direction of flow, this loss representing the energy required to force the water through the voids. This lost head is transmitted by viscous friction to the grain structure of the soil, thus producing forces in the direction of flow. It is the occurrence of these seepage forces which complicates the stability analysis for this case. It could be demonstrated that the resultant seepage force for the entire sliding mass would be a force such as V, Fig. 13 (h). While this force is developing, there will be an adjustment of pressures along the rupture arc starting from the initial state shown in Fig. 9 (d) and continuing to the ultimate as in Fig. 9 (h), where their resultant is the force E_F. The vector sum of forces E_F and W_o must equal force V. If no friction could be developed to resist these forces, the stability analysis would be exactly as for the sudden drawdown case, consisting of W_S as analyzed in Fig. 9 (a) and W_o as in Fig. 9 (e), with one other force for each case (E_1 and E_F) which produces no moment. Since friction will act to help resist the overturning effects, the actual situation must be somewhat safer than the above. Therefore the steady flow case is more stable than the sudden drawdown case.

There are an infinite number of flow nets which are possible for any given slope, and forces E_F and V will vary for different nets. For any given net, E_F may be determined for a given arc, but a rather complicated graphical procedure is involved.

If the supply of seepage water is completely shut off the magnitude of the pressures along the rupture arc decreases further, this time to zero if sufficient time is allowed, and meanwhile the force V disappears. This leads to the still more favorable case of capillary saturation which has already been discussed.

Surface Cracking

Foremost among the items which bring uncertainty into slope stability problems is the possibility of cracking at the top of the slope. As is indicated by the Résal-Frontard Method, the soil at this point is in tension. This tension may be counterbalanced, temporarily, at least, by the internal pressure of the soil, but the danger of cracks developing is always a disturbing possibility. It is evident that the cohesion which has been assumed constant over the rupture surface should not be depended upon within the zone of cracking.

If there were any rational way of figuring the depth of this zone, the item would not be of such great concern. The Résal-Frontard Method gives an expression for the depth to which tension is in effect, but it must be remembered that this derivation is based on the assumption of an infinite slope, and there is question as to its reliability when used adjacent to a discontinuity of the slope. Moreover, cracking is of a progressive nature, aided by water and often by ice pressure, and so is not limited to this depth in which tension occurs.

For an approximate analysis of this item some estimate must be made as to the depth to which cracks may progress. This estimate must be based largely on judgment and inspection of the soil and the conditions at the site, with perhaps some weight given to the value as determined by the Résal-Frontard Method. From this estimated depth the percentage of arc not affected by cracking may be determined, and this may then be multiplied by the unit cohesion. The average value for unit cohesion thus obtained does not lead to a resultant cohesion which is parallel to the chord, but it may be assumed that it does, noting that more nearly exact ways of handling this item could be introduced if the accuracy were warranted.

Terzaghi has suggested (9*a*) that within the zone of cracking the possibility of water pressure should be recognized and another overturning force introduced into the computations to cover it. Such a force is in reality a portion of a force of the nature of E_1 of Fig. 13 (*d*) which is fully covered when the sudden drawdown case is used.

As a means to obtaining a better understanding of this phase of the problem, it would be of value if data on a large number of slopes could be compiled, which should include estimates of depths of cracking and data on factors of safety as given by stability computations. The present scarcity of knowledge on this subject can be overcome only by the use of generous factors of safety.

Illustrative Examples

(1) A vertical cut of 15 ft. is to be made in a clay on which tests give $w=110$ lb. per cu. ft., $\phi=10°$, $c=750$ lb. per sq. ft. It may be assumed that there will be no seepage of water.

(*a*) What is the factor of safety with respect to cohesion?

(*b*) What cohesion would be required to give a true factor of safety of the same magnitude?

(*c*) Will the cut need sheeting?

(*a*) From Fig. 4, for $\phi=10°$ and $i=90°$, $\dfrac{c}{FwH}=.218$

$$\frac{c}{FwH}=.218=\frac{750}{F\times110\times15};\text{ whence } F=2.1$$

(*b*)
$$\phi_D=\frac{\phi}{F_T}=\frac{10}{2.1}=4.8°$$

For $\phi_D=4.8°$ and $i=90°$, Fig. 4 gives $\dfrac{c}{F_TwH}=.240$

$$\frac{c}{F_TwH}=\frac{c}{2.1\times110\times15}=.240;\text{ whence } c=\textit{830 lb. per sq. ft.}$$

(*c*) The small difference between given cohesion and the value in part (*b*) indicates the major portion of the shearing strength is cohesion. If the given value for cohesion is a dependable average value, the factor of safety is ample and no sheeting is needed.

(2) A 45° cut 20 ft. deep is to be made across flat land where the water table is just at ground surface. The soil has a specific gravity of 2.80 and a void ratio of 0.75. It is specified that the true factor of safety is to be 1.5. The effective angle of internal friction may be assumed to be constant and equal to 15°. What values of unit cohesion are required —

(*a*) With the cut submerged?

(*b*) With the cut unwatered and a steady state of seepage?

(*c*) After a drop in water table and with no seepage occurring?

(*a*) $w=\dfrac{s-1}{1+e}w_o=\dfrac{1.8}{1.75}\times62.5=64.3;\ \phi_D=\dfrac{\phi}{F_T}=\dfrac{15}{1.5}=10°$

For $\phi_D=10°$ and $i=45°$ from Fig. 4, $\dfrac{c}{F_TwH}=.108$

$$\frac{c}{F_TwH}=.108=\frac{c}{1.5\times64.3\times20};\text{ whence } c=\textit{210 lb. per sq. ft.}$$

(*b*) For the sudden drawdown case —

The weighted friction angle, $\phi_w = \frac{s-1}{s+e} \times \phi = \frac{1.8}{3.55} \times 15 = 7.6°$

$$\phi_D = \frac{\phi_w}{F_T} = \frac{7.6}{1.5} = 5.1°$$

From Fig. 4, for $\phi_D = 5.1°$ and $i = 45°$, $\frac{c}{F_T wH} = .136$

$$W_T = 64.3 + 62.5 = 126.8$$

$$\frac{c}{F_T wH} = .136 = \frac{c}{1.5 \times 126.8 \times 20}; \text{ whence } c = 520 \textit{ lb. per sq. ft.}$$

The cohesion would be slightly lower than this value for a steady state of seepage.

(*c*) Complete consolidation will be assumed.

As in part (*a*), $\phi_D = 10°$

$$\frac{c}{F_T wH} = .108 = \frac{c}{1.5 \times 126.8 \times 20}; \text{ whence } c = 410 \textit{ lb. per sq. ft.}$$

Previous to reaching complete consolidation the required value would be slightly higher.

(3) A railroad cut to a depth of 20 ft. must be made through a clay on which the value for cohesion corrected to allow for surface cracking is 400 lb./*SF*, $w = 115$ lb./*CF*, and $\phi = 10°$. A factor of safety of 1.75 is specified. It may be assumed that there will be no seepage. What is the allowable slope?

$$\phi_D = \frac{\phi}{F_T} = \frac{10}{1.75} = 5.7°$$

$$\frac{c}{F_T wH} = \frac{400}{1.75 \times 115 \times 20} = .099$$

From Fig. 4, when $\frac{c}{F_T wH} = .099$ and $\phi_D = 5.7°$, $i = 27°$

The allowable slope is thus about 2 to 1.

(4) A cut on a 1¼:1 slope is to be made in a clay on which tests show $c = 680$ lb. per sq. ft. and $\phi = 18°$. Unit weights are 64 lb. per cu. ft. submerged and 126 lb. per cu. ft. total. It is specified that a factor of safety with respect to cohesion of 2 is to be used. Seepage will occur during a part of each year. Surface cracking will occur, and it is estimated that this may invalidate the strength of 25 per cent of the rupture surface. What is the estimated allowable depth?

The available average cohesion becomes $.75 \times 680 = 510$ lb./*SF*

$$\cot i = 1.25,\ i = 39°;\ \phi_w = 18 \times \frac{64}{126} = 9.1°$$

For $i = 39°$ and $\phi = 9.1°$, from Fig. 4, $\frac{c}{FwH} = .102$

$\frac{c}{FwH} = .102 = \frac{510}{2 \times 126 \times H}$; whence H, the allowable depth, is *20 feet.*

(5) A 40° slope is 30 feet in height. Tests on the soil give the following values: $c = 350$ lb./*SF*, $w = 110$ lb./*CF*, $\phi = 20°$. The value for cohesion has been adjusted to allow for cracking at top of slope. It may be assumed that there will be no seepage. Determine —

(*a*) The factor of safety with respect to cohesion.

(*b*) The true factor of safety.

(*a*) For $\phi = 20°$; $i = 40°$; $\frac{c}{FwH} = .050$

$$\frac{c}{FwH} = .050 = \frac{350}{F \times 110 \times 30};\ \text{whence } F = 2.1$$

(*b*) A direct solution is not possible, but a solution by trial is not difficult.

Using the test values for the soil and assuming any true factor of safety, the allowable height for a 40° slope may be figured. A few trials will indicate what factor of safety corresponds to the actual height of 30 ft.

Assuming $F_T = 1.5$

$\phi_D = \frac{20}{1.5} = 13.3°$ and from Fig. 4, for $\phi_D = 13.3°$ and $i = 40°$

$$\frac{c}{F_T w H} = .080 = \frac{350}{1.5 \times 110 \times H}\text{; whence } H = \mathit{26.5\ ft.}$$

Similar computations for F_T equal to 1.4 and 1.3 lead to —

F_T	H
1.5	26.5
1.4	30.3
1.3	35.0

And it is seen that for the actual height of 30 ft. $F_T = \mathit{1.4}$.

(6) A rolled fill dam is to be constructed using a material which has an inappreciable amount of very fine particles. For this material, tests show that $\phi = 35°$ and cohesion is inappreciable. A factor of safety of 1.5 is desired. What side slopes should be used?

Assuming no freeboard and sudden drawdown —

$$\phi_D = \frac{\phi}{F_T} = \frac{35}{1.5} = 23.3°$$

$\phi_w = \dfrac{s-1}{s+e}\phi_D = \text{approx. } \dfrac{1}{2}\phi_D = \dfrac{1}{2} \times 23.3 = 11.7°$; whence $i = \mathit{11.7}°$, a slope of 1 on 5.

Assuming complete capillary saturation —

$$\phi_D = 23.3° \text{ as above}$$

and $i = 23.3°$, a slope of 1 on $2\frac{1}{3}$

This combination of high friction angle and flat slope leads to a large difference between the sudden drawdown and capillary saturation cases. The allowable slope depends on many factors which have not been mentioned, and the method of this paper can only bound it between the limits of 1 on $2\frac{1}{3}$ and 1 on 5. More detailed analyses which might be undertaken are a sudden drawdown analysis in which account is taken of the freeboard, or an analysis of the case of steady seepage using flow nets.

(7) A cut 30 ft. deep is to be made in a deposit of highly cohesive material which is 60 ft. deep and is underlaid by ledge. The shearing strength is essentially constant throughout the depth at 600 lb. per sq. ft. The unit weight is 120 lb. per cu. ft.

(*a*) For a factor of safety of 1.5, what side slopes are needed?

(*b*) What increased value of cohesion would be necessary for side slopes of 1.8 : 1?

(*a*) For $\phi=0$, $D=\frac{60}{30}=2$, $\frac{c}{FwH}=\frac{600}{1.5\times120\times30}=.111$

Fig. 8 gives $i=8°$, or a slope of 7 : 1

(*b*) $i=\cot^{-1} 1.8=29°$

For $D=2$ and $i=29°$, Fig. 8 gives $\frac{c}{FwH}=.171$

$$.171=\frac{c}{1.5\times120\times30};\quad \text{whence } c=\mathit{920\ lb./SF}$$

Fig. 8 also gives $n=0.8$, which means the rupture would break out at the surface 24 ft. from the toe. For a cut with a base width narrower than 24 ft., a somewhat lower cohesion would be required, but the absolute minimum would be for $n=0$, where

$$\frac{c}{FwH}=.154;\quad \text{whence } c=\mathit{830\ lb./SF}$$

Conclusion

By the use of the Stability Number, the solution of problems in stability of slopes becomes very simple. When the slope angle and the friction angle are known, this number may be obtained directly from Fig. 4, while for zero friction angle, if there is a limitation in the depth to which a rupture surface may extend, Fig. 8 will furnish the value. Nests of similar curves could be derived for more involved cross sections, such as those where the ground at the top of slope is not level, or where there is partial submergence, or for typical earth dam sections sloping in two directions. While such sections would apply to many actual problems, there is a question as to whether or not the large amount of work required for general solutions would be justified.

The dependability of the results of such solutions will depend entirely on the conditions. In a few ideal cases the results may be very dependable. More often there will be questionable items, such as the ever-present possibility of surface cracking at the top of the slope, which determine that the results can be accepted only as rough indications. If good judgment is used in estimating and attempting to evaluate the unknown factors, whatever the conditions may be, the results should be of value in arriving at logical conclusions.

The relative value of the different methods for solution of stability problems is largely one of academic interest. Accuracies far greater than those warranted have been adopted in this paper in an attempt to show that the two methods used to obtain the charts are at least as logical as any other methods available, and that they agree remarkably well with each other. The assumptions involved may be accepted as reasonable for any embankment which is essentially homogeneous and conforms approximately to the ideal cross section used, with a single exception. The assumption that cohesion is constant at all points may often vary considerably from the truth, and the use of the best average value that can be obtained is the only procedure available.

Although stability problems involve more questionable items than most engineering problems, it is seldom possible to make use of large factors of safety. Conditions are often such that about the largest factor of safety that may be chosen is 1.5, and sometimes a value this large may not be allowable.

Procedures for different conditions of saturation and seepage may be summarized as follows:

A. If the slope is continuously submerged, use the submerged unit weight. For partial submergence, the general solution is not applicable.

B. If the slope is submerged and values for cohesion and friction have been determined for this state, and if sudden drawdown is a possibility, use the total unit weight and the weighted friction angle.

C. For any embankment which is not submerged, through which there is seepage of water or any possibility of future seepage occurring, obtain the solutions for cases B and D. The solution for case C must fall between these two, and when they do not differ widely, as will often be the case, the results for case B may be used for C. Where B and D offer radically different requirements, no accurate answer is available by the method of this paper. However, it is to be hoped that future investigations of seepage cases by more involved methods will yield methods of approximate interpolation within the range between B and D.

D. For an intact cohesive soil, which is saturated or practically so, where the slope is dry and there is no possibility of future seepage of water, use the total unit weight and the friction angle as given by tests.

E. For a dried, fissured clay, this type of stability analysis does not apply.

Features which apply to all of the above items may be summarized as follows:

1. The depth to which surface cracks are likely to develop must be estimated or assumed. From this the percentage of arc which is not affected by cracking may be found. An average cohesion for use in computations may be obtained by multiplying this percentage by the unit cohesion as given by laboratory tests.

2. For slight variations of soil properties within the sliding mass, average values for cohesion, unit weight and friction angle may be used. For large variations the general solution is not strictly applicable, but in many cases, even where several distinctly different types of soil occur in a cross section, it has been found to give surprisingly good results. Thus it may at least be considered as a valuable check for complicated analyses.

The general type of solution which this paper presents may meet opposition by some engineers who feel that soil is too variable a material to be subjected to such analysis. However, such results in the hands of an experienced soil engineer, accompanied by data furnished by thorough investigations of the site and the soil involved, will give information which no responsible engineer can afford to overlook.

Acknowledgments

The writer wishes to acknowledge the advice and willing assistance of Dr. Glennon Gilboy in connection with the preparation of this paper. A large amount of valuable assistance was given by Mr. Harold A. Fidler of the Massachusetts Institute of Technology soil mechanics laboratory staff, who checked the mathematical derivations and performed a part of the computation work. To Professor Arthur Casagrande the writer wishes to express his thanks for reading the manuscript and for valuable suggestions.

Bibliography

(1) Culmann, K.: Die Graphische Statik. Zürich, 1866.
(2) Résal, J.: Poussee des Terres. Paris, 1910.
(3) Frontard, M.: Cycloides de Glissement des Terres, Comptes Rendues Hebdomadaires de l'Acadêmie des Sciences. Paris, 1922.
(4) Statens Järnvagärs Geoteknisca Commission Slutbetankande, 31 May, 1922.
(5) Fellenius, W.: Erdstatish Berechnungen mit Reibung und Kohäsion. Berlin, 1927. A summary of this material is given in English in a paper by the same author: Calculation of Stability of Earth Dams. Paper D–48, Second Congress on Large Dams, Washington, D. C., 1936.
(6) Terzaghi, Charles: The Mechanics of Shear Failures on Clay Slopes and the Creep of Retaining Walls, Public Roads, December, 1929.
(7) Rendulic, L.: Ein Beitrag zur Bestimmung der Gleitsicherheit. Der Bauingenieur, 1935, No. 19/20.
(8) Krey, H.: Erddruck, Erdwiderstand und Tragfähigkeit des Baugrundes.
(9) Proceedings of the First International Conference on Soil Mechanics and Foundation Engineering, Harvard School of Engineering, Cambridge, Mass., June, 1936.
   (*a*) von Terzaghi, Karl: Vol. I, Paper G–6. Critical Height and Factor of Safety of Slopes against Sliding.
   (*b*) von Terzaghi, Karl: Vol. I, Paper G–7. Stability of Slopes of Natural Clay.
   (*c*) Jáky, Joseph: Vol. II, Paper G–9. Stability of Earth Slopes.
   (*d*) Taylor, D. W.: Vol. III, Paper G–16. Notes on the Stability of Slopes.

Originally published in
Journal of the Boston Society of Civil Engineers
Jan., 1940

JOURNAL OF THE

BOSTON SOCIETY OF CIVIL ENGINEERS

Volume XXVII **JANUARY, 1940** **Number 1**

PLASTIC STATE OF STRESS AROUND A DEEP WELL

BY H. M. WESTERGAARD, MEMBER*

The analysis that follows is a result of conversations with Dr. Karl Terzaghi, who raised this question: What distributions of stress are possible in the soil around an unlined drill hole for a deep well? What distributions of stress make it possible for the hole not to collapse but remain stable for some time either with no lining or with a thin "stove-pipe" lining of small structural strength?

Notation. Let:

z = vertical coördinate, measured downward from the horizontal surface.

r = horizontal radial distance from the axis of z, which is the axis of the cylindrical drill hole.

a = radius of the hole.

b = value of r at the boundary between the regions of plasticity and elasticity; a function of z.

w = weight of the soil or rock with its content of water, per unit of volume, reduced below the water table by the weight of water per unit of volume; though w may be a function of z, the purposes of this analysis are served sufficiently well by assuming w to be a constant.

*Dean of the Graduate School of Engineering and Gordon McKay Professor of Civil Engineering, Harvard University, Cambridge, Mass.

$\sigma_r, \sigma_\theta, \sigma_z$ = horizontal radial pressure per unit of area, horizontal circumferential pressure per unit of area, and vertical pressure per unit of area, respectively, minus the hydrostatic pressure of the water in the hole.

τ_{rz} = shearing stress in the directions of r and z; positive when it acts upward on the parts having smaller values of r.

p = value of σ_r at the cylindrical surface of the hole, at $r = a$.

q = constant, appearing in equations (10) and (11) for the plastic state, measurable as a pressure in pounds per square inch.

k = ratio, appearing in equations (10) and (11) for the plastic state.

F, f, Z = stress functions; F and f are functions of both r and z; Z is a function of z only.

Equations of Equilibrium. Whether the material is in an elastic or plastic state, the stresses must obey the following two equations of equilibrium, which are derived by considering the forces on a small wedge-shaped block of the dimensions dr, $rd\theta$, dz*:

$$\frac{\partial(r\sigma_r)}{\partial r} - \sigma_\theta + \frac{\partial(r\tau_{r\theta})}{\partial z} = 0 \tag{1}$$

$$\frac{\partial(r\sigma_z)}{\partial z} + \frac{\partial(r\tau_{r\theta})}{\partial r} = wr \tag{2}$$

Stress Functions. It is possible to express the four stresses in equations (1) and (2) in terms of two stress functions F and f. No matter what functions F and f are chosen, if they permit the differentiations called for, the following stresses will be found to satisfy the equations of equilibrium, (1) and (2):

$$\sigma_r = wz + \frac{1}{r}\frac{\partial^2 F}{\partial z^2} + \frac{f}{r} \tag{3}$$

$$\sigma_\theta = wz + \frac{\partial f}{\partial r} \tag{4}$$

*Compare, for example, S. Timoshenko, "Theory of Elasticity," (McGraw-Hill Book Company, Inc.) 1934, p. 309.

$$\sigma_z = wz + \frac{1}{r}\frac{\partial^2 F}{\partial r^2} \tag{5}$$

$$\tau_{rz} = -\frac{1}{r}\frac{\partial^2 F}{\partial r \partial z} \tag{6}$$

Solution under Conditions of Elasticity. If the soil or rock could be counted on to obey Hooke's law for homogeneous, isotropic material, the following stress functions would be found applicable:

$$F = 0, f = -wza^2/r \tag{7}$$

The corresponding stresses,

$$\sigma_r = wz(1 - a^2/r^2), \sigma_\theta = wz(1 + a^2/r^2), \sigma_z = wz, \tau_{rz} = 0 \tag{8}$$

are of the type represented in Lamé's formulas for thick-walled cylinders; and they satisfy not only the equations of equilibrium, (1) and (2), but also the equations of compatibility of deformations.* The most severe combination of stresses defined at each depth by equations (8) occurs at $r = a$ and is

$$\sigma_r = 0, \sigma_\theta = 2wz, \sigma_z = wz, \tau_{rz} = 0 \tag{9}$$

Plasticity. The combination of pressures in equations (9) could not be maintained at a great depth. Before such a state could be reached, the material would either break or flow toward the hole. The flowing indicates a plastic state which relieves the stresses. It may be assumed then that within a radius $r = b$, dependent on z, the material is in a plastic state, but that outside this radius it remains in an elastic state so far as changes of stresses are concerned.

Let σ_1, σ_2, σ_3 denote the three principal pressures in the state of stress at a point, in the order of rising magnitude. It will be assumed for the purpose of this analysis that the law of plasticity can be stated in terms of these stresses by the formula†,

$$\sigma_3 - (k+1)\sigma_1 = q \tag{10}$$

with the values of k and q constant. The value of k may be as much as 3. Equation (10) is equivalent to the statement that the limiting curve for Mohr's circles is a straight line. It will be assumed further that an elastic state obeying Hooke's law, with unchanged values of

*S. Timoshenko, loc. cit., p. 56, p. 312.

†Compare, "A study of the Failure of Concrete under Combined Compressive Stresses," by Frank E. Richart, Anton Brandtzaeg, and Rex L. Brown, University of Illinois Engineering Experiment Station, Bulletin No. 185, 1928, 102 pp., especially pp. 78, 79, and 91.

the modulus of elasticity and Poisson's ratio, is possible when the left side of equation (10) is less than q. Finally it will be assumed that there are no sudden jumps in the stresses at the boundary $r = b$ between the regions of plasticity and elasticity.

Solution. The following assertions require verification, which is obtained after conclusions have been drawn from them: First, F will be such a function that at any depth that is not relatively small the term $\partial^2F/\partial z^2$ in equation (3) will be relatively so insignificant that it may be ignored. Secondly, F may be determined so that τ_{rz} in equation (6) will vanish at $r = a$ and be relatively insignificant everywhere. Thirdly, F may be determined so that in addition at each place σ_z will lie between σ_r and σ_θ.

When these assertions are accepted, it becomes possible to state that at each point σ_r, σ_z, σ_θ are the three principal pressures in the order of rising magnitude. Then equation (10) takes the form

$$\sigma_\theta - (k+1)\sigma_r = q \tag{11}$$

With the term $\partial^2F/\partial z^2$ omitted in equation (3), the pressures σ_r and σ_θ will depend on f only. The following form of f will be examined for the region of plasticity:

$$f = (p + q/k)r^{k+1}/a^k - (wz + q/k)r \tag{12}$$

Substitution in equations (3) and (4) gives

$$\sigma_r = (p + q/k)(r/a)^k - q/k \tag{13}$$

$$\sigma_\theta = (k+1)(p + q/k)(r/a)^k - q/k \tag{14}$$

Equation (13) gives the desired value $\sigma_r = p$ at $r = a$; and the two equations (13) and (14) satisfy the requirement of plasticity in equation (11). The pressure p may have to be assumed to be zero; but under some circumstances a small positive pressure p may perhaps be credited either to whatever strength the lining has, or to a "mud-head" added to the water pressure.

For the region of elasticity f may be taken as

$$f = -Za^2/r \tag{15}$$

with Z to be determined as a function of z. Substitution from equation (15) in equations (3) and (4) gives

$$\sigma_r = wz - Za^2/r^2, \sigma_\theta = wz + Za^2/r^2 \tag{16}$$

If dZ/dz varies only slowly with z, and if σ_z remains close to wz in this region, these pressures will conform sufficiently to the type represented in equation (8), and may therefore be accepted as stresses in the region of elasticity.

At the boundary $r = b$ between the regions of plasticity and elasticity the pressures should have the same values whether computed from equations (13) and (14) or from equations (16). By specifying coinciding values of $\sigma_\theta + \sigma_r$ one finds

$$\left(\frac{b}{a}\right)^k = \frac{2(kwz + q)}{(k+2)(kp + q)} \tag{17}$$

and similarly, by specifying coinciding values of $\sigma_\theta - \sigma_r$ one finds

$$Z = \tfrac{1}{2}(kp + q)\left(\frac{b}{a}\right)^{k+2} = \frac{kwz + q}{k+2}\left(\frac{b}{a}\right)^2 \tag{18}$$

At the boundary $r = b$ one finds now, either by equations (13), (14), and (17), or by equations (16) and (18):

$$\sigma_r = (2wz - q)/(k+2) \tag{19}$$
$$\sigma_\theta = [2(k+1)wz + q]/(k+2) \tag{20}$$

The circumferential pressure σ_θ in equation (20) is the greatest pressure at the depth z.

It remains to verify the assertions that were made concerning the stress function F. This may be done by drawing diagrams for σ_r and σ_θ and thereafter estimate diagrams for σ_z showing σ_z as an intermediate pressure. The values of σ_z determine $\partial F/\partial r$ through equation (5) except for an integration constant, which is defined by the condition $\tau_{rz} = 0$ at $r = a$. Results of such a procedure were found to support the assertions referred to.

Conclusion. The plastic action makes it possible for the great circumferential pressures that are necessary for stability to occur not at the cylindrical surface of the hole but at some distance behind the surface, where they may be combined with sufficiently great radial pressures. The formulas that have been derived serve to explain the circumstances under which the drill hole for a deep well may remain stable.

UNIVERSITY OF ILLINOIS
BULLETIN

Vol. 61 June, 1964 No. 92

ENGINEERING EXPERIMENT STATION
BULLETIN SERIES No. 338

INFLUENCE CHARTS FOR COMPUTATION OF STRESSES IN ELASTIC FOUNDATIONS

BY

NATHAN M. NEWMARK

PRICE: $1.00
PUBLISHED BY THE UNIVERSITY OF ILLINOIS
URBANA

Published nine times each month by the University of Illinois. Entered as second-class matter December 11, 1912, at the post office at Urbana, Illinois, under the Act of August 24, 1912. Office of Publication, 49 Administration Building (West), Urbana, Ill.

INFLUENCE CHARTS FOR COMPUTATION OF STRESSES IN ELASTIC FOUNDATIONS

I. Introduction

1. *Scope of Investigation and Summary.*—This bulletin describes a simple graphical procedure for computing stresses in the interior of an elastic, homogeneous, isotropic solid bounded by a plane horizontal surface and loaded by distributed vertical loads at the surface. The stresses are computed from charts given herein merely by counting on a chart the number of elements of area, or blocks, covered by a plan of the loaded area drawn to proper scale and laid upon the chart.

Influence charts of a size convenient for practical use are given in Plates 1, 2, and 3 for computing, respectively, vertical stress on horizontal planes, the sum of the principal stresses, and horizontal stress on vertical planes, the latter two charts being constructed for a value of Poisson's ratio of 0.5.

Influence charts are also given to a smaller scale for computing the components of shearing stress on horizontal and vertical planes, and for computing the corrections to the various stresses when Poisson's ratio is different from 0.5. Numerical data and instructions are given in the Appendix which permit drawing any of the charts to whatever scale is desired in particular applications. For all the charts, the influence value of the individual blocks is 0.001.

The calculation of maximum shearing stress at a point is discussed in Chapter IV. It is shown that the so-called octahedral shear, which in most materials is a more significant quantity than the maximum shear, and which lies between the limits of 81.6 and 94.3 per cent of the maximum shear, may be more easily computed than the maximum shear, from the six components of the stresses on horizontal and vertical planes.

The use of the influence charts is simple and rapid, and the accuracy of the calculations is sufficient for all practical purposes. After a few trials one can almost guess at accurate enough values for the stresses from a rough sketch of the loaded area.

This bulletin is not concerned with such questions as why stresses should be computed or what should be done with the stresses after the calculations are made. It is hoped that use of the procedure described herein will enable time to be saved in making calculations that are at present made by more laborious means.

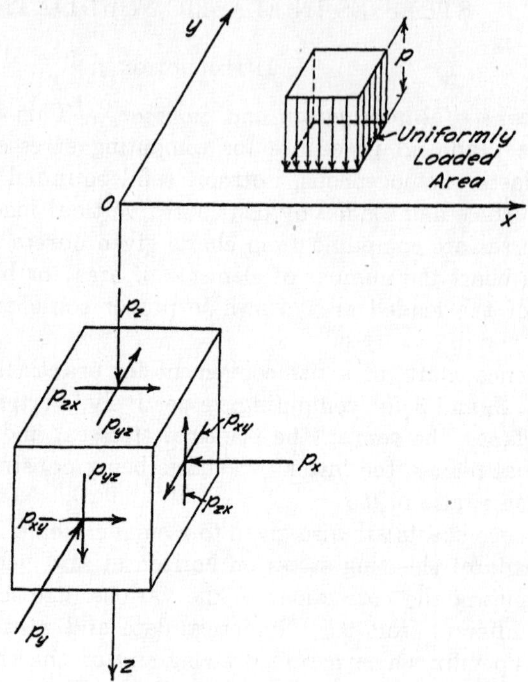

FIG. 1. SIGN CONVENTION AND NOTATION FOR AXES AND STRESSES

2. *Notation and Sign Convention.*—A homogeneous, elastic, isotropic solid of infinite extent, bounded by a horizontal plane surface, and loaded vertically at the surface, is considered. The x and y axes are in the plane of the surface and the z axis is positive downward.

The notation for the stresses acting on vertical and horizontal planes, parallel to the coördinate planes, is indicated in Fig. 1. It will be noted that the ordinary convention for positive stresses used in the theory of elasticity is reversed here in order that pressures or compressions may be positive.

The intensity of load, or the load per unit of area, is denoted by p. The stresses are denoted as follows:

p_z is the normal stress on horizontal planes positive for compression,

p_x and p_y are the normal stresses on vertical planes parallel to the yz and the zx planes, respectively, positive for compression.

p_{yz} and p_{zx} are horizontal shearing stresses on horizontal planes and also vertical shearing stresses on vertical planes, positive as shown in Fig. 1.

p_{xy} is the horizontal shearing stress on vertical planes, positive as shown in Fig. 1.

The following additional notation is used:

$p_{vol} = p_x + p_y + p_z$ = sum of principal stresses.

μ = Poisson's ratio of lateral contraction for the material.

r, α, β = quantities defined in Fig. 8.

$\left.\begin{matrix} a, b, c, d, e, f \\ A, B, C, D, E, F \end{matrix}\right\}$ = quantities defined in Equations (16) and (17).

3. *Acknowledgment.*—The investigation reported herein was conducted as part of the work of the Engineering Experiment Station of the University of Illinois, of which DEAN M. L. ENGER is the head. The calculations were performed by Mr. HAROLD CRATE, student in Civil Engineering.

A chart for computing vertical pressures, similar to that contained herein, was described by the writer in *Transactions*, Am. Soc. C. E., Vol. 103, 1938, p. 321-324, and in *Engineering News-Record,* Vol. 120, 1938, p. 23-24. A condensed description of the process of constructing such charts was also given by the writer in a paper entitled "Stress Distribution in Soils," Proceedings of Conference on Soil Mechanics and Its Applications, Purdue University, September, 1940, p. 295-303. However, the material contained in this bulletin is more complete, and the charts more convenient to use than those in the previous publications.

II. COMPUTATION OF VERTICAL STRESS ON HORIZONTAL PLANES

4. *Use of Influence Chart.*—In Fig. 2 is shown a rough chart for computing vertical stress. The chart represents a plan of the surface of the elastic body drawn to such a scale that the length marked OQ is the depth z at which the pressure is to be computed. The chart is constructed of arcs of concentric circle and radial lines drawn in such a way that each element of area bounded by two adjacent radii and two adjacent arcs contributes the same influence to the stress.

The influence of each element of area is 0.02 times the load on the element. Since there are 10 sectors in the diagram the influence values of the successive circles must be 0.20, 0.40, 0.60, and 0.80. The relative radii of the circles, in terms of the depth z from O to Q, are respectively, 0.401, 0.637, 0.918, and 1.387. The arc for an influence of 1.0 will have an infinite radius.

A more accurate chart in which the value of each elementary area is 0.001 is given in Plate 1. This chart is constructed in much the same

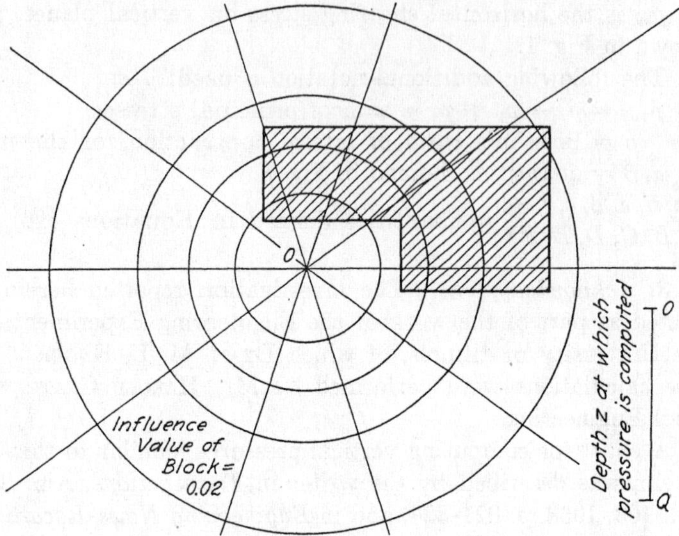

FIG. 2. SKETCH OF INFLUENCE CHART FOR VERTICAL STRESSES ON HORIZONTAL PLANES

manner as the chart in Fig. 2, but the areas are so chosen that they are of convenient shape with the x and y axes both axes of symmetry. The chart can be duplicated or drawn to other scales with the numerical data given in Tables 1 and 2 in the Appendix. The manner of using the chart is best illustrated by reference to Figs. 2 and 3.

Suppose the stress is desired at point Q' at a depth $z' = 80$ ft. under point O' in Fig. 3, where the area outlined is uniformly loaded with a load of 5 000 lb. per sq. ft. The plan of the loaded area is redrawn to such a scale that the distance $O'Q'$ in Fig. 3 becomes the same as the distance OQ in Fig. 2. Then the drawing is placed on Fig. 2 in such a way that point O' falls on point O. The number of blocks covered by the loaded area, multiplied by the magnitude of the load per unit of area and by the influence value of each block, gives the stress required. In the illustration, approximately 8 blocks are loaded, giving a stress of 0.16 times the intensity of load, or 800 lb. per sq. ft. Plate 1 is used in the same way, except that the influence value of each block is 0.001.

It may be noted that the stresses under similar loaded areas, at depths proportional to the sizes of the areas, are equal. For a given loading, different drawings of the plan of the loaded area are required to compute stresses at various depths.

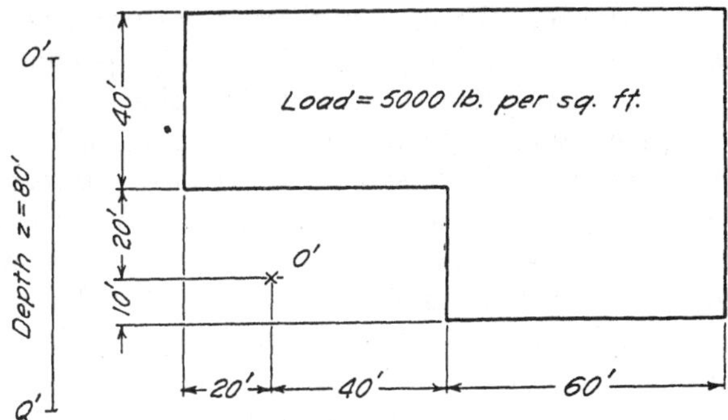

FIG. 3. PLAN OF LOADED AREA

The chart for vertical stress on horizontal planes is radially symmetrical. Therefore the loading plan may be rotated through any angle about a vertical axis through the point where the vertical stress is computed without changing the magnitude of the stress. Consequently the loading plan may be placed on the chart in the manner most convenient for the particular problem.

If the area is not uniformly loaded the chart may still be used, provided that each influence area is considered to be loaded by the average intensity of load on the particular block.

In using Plate 1, parts of blocks may be estimated when counting the blocks, with sufficient accuracy for all practical purposes. In counting the number of blocks when the number is large, the total amount of influence within a given region may be recorded on the chart where it can be taken into account conveniently. In general, the loaded area will be drawn on tracing paper and laid upon the chart so that blocks may be counted through the tracing.

III. COMPUTATION OF VARIOUS COMPONENTS OF STRESS AT A POINT

5. *General Considerations.*—The stresses on any plane at a particular point can be stated in terms of the six components of stress on an elementary cube, such as is indicated in Fig. 1. Of the six components of stress, three are normal stresses and three are shearing stresses. The normal and shearing stresses on any oblique plane can be determined from a consideration of the equilibrium of an infinitesimal pyramid of material. The largest and smallest normal stresses at a point occur on mutually perpendicular planes, called principal planes, on which there is no shearing stress, and these stresses are

called principal stresses. There is a third principal stress of intermediate magnitude on a third principal plane perpendicular to the other principal planes, on which there is no shearing stress either. The greatest shearing stress on any plane is found on the plane that makes angles of 45 degrees with the directions of greatest and least principal stresses, and that is perpendicular to the plane on which the intermediate principal stress acts. The magnitude of the maximum shearing stress is numerically equal to one-half the algebraic difference between the largest and the smallest principal stress.

The sum of three mutually perpendicular stresses at a point is a constant, and is therefore equal to the sum of the principal stresses at the point. In this bulletin this sum is denoted by the symbol p_{vol} since the sum of the principal stresses is associated with the change in volume of an elastic material.

6. *Sum of Principal Stresses.*—An influence chart for computing the sum of the principal stresses, p_{vol}, is given in Plate 2 for a value of Poisson's ratio of one-half. This chart is similar to that for the computation of vertical stress on horizontal planes, and is also radially symmetrical. Each elementary area or block on the chart has an influence value of 0.001. To compute p_{vol} for other values of Poisson's ratio than 0.5, one merely subtracts a correction equal to $\frac{1 - 2\mu}{3}$ times the value computed from Plate 2.

Since

$$p_{vol} = p_x + p_y + p_z, \tag{1}$$

and since p_z can be computed from Plate 1, one can determine the sum of p_x and p_y, the sum of horizontal stresses on two mutually perpendicular vertical planes, from the relation

$$p_x + p_y = p_{vol} - p_z. \tag{2}$$

In certain calculations this sum can be used without it being necessary to compute p_x and p_y individually. This would be the case if one wished to compute vertical strain at a particular point, taking into account the effect of Poisson's ratio. The vertical strain is given by the relation

$$\frac{1}{E_m}\left[p_z - \mu\,(p_x + p_y)\right] \tag{3}$$

or, in terms of p_z and p_{vol},

$$\frac{1}{E_m}[(1 + \mu)\,p_z - \mu p_{vol}] \tag{4}$$

where E_m is the modulus of elasticity of the material.

7. *Horizontal Stress on Vertical Planes.*—The horizontal stress p_x on the vertical plane containing the y axis can be computed for a value of Poisson's ratio of one-half from the influence chart given in Plate 3. It will be noted that the elementary influence areas are not of the same size in a given circle arc; hence, the position of the load must be determined relative to the x and y axes. Each block has an influence value of 0.001. Except for the fact that the chart is not radially symmetrical, Plate 3 is used in the same way as are Plates 1 and 2.

To determine p_x for a value of Poisson's ratio different from 0.5, one subtracts two corrections from the value of p_x for $\mu = 0.5$. One correction is $\frac{1 - 2\mu}{6}$ times the value of p_{vol} for $\mu = 0.5$ determined from Plate 2 and the other correction is $(1 - 2\mu)$ times the quantity obtained from a chart shown reduced in size in Fig. 4. For convenience in actual use such a chart should be drawn to the same scale as Plates 1, 2, and 3. This can be done by those who wish to do so with the numerical values of E and e given in Tables 1 and 2 of the Appendix.

In Fig. 4 each elementary block has an influence value numerically equal to 0.001, but the sign of the influence is positive for points closer to the x axis than to the y axis, and is negative for points closer to the y axis than to the x axis. Thus, for a quadrant of the surface bounded by the positive axes, the net amount of this correction is zero since there is as much negative influence as positive influence.

For certain loaded areas, extending to infinity in a sector, the correction becomes infinitely large. This means that p_x is infinite under certain conditions when μ is different from 0.5. This phenomenon has been noted by Love.* Of course, this means that the material will not remain elastic.

The same charts that are used to compute p_x can also be used to compute p_y by interchanging the x and y axes. More conveniently, p_y

*A. E. H. Love, "The Stress Produced in a Semi-Infinite Solid by Pressure on Part of the Boundary," Phil. Trans. Royal Society, London, Series A, Vol. 228, 1929, p. 377-420.

can be computed from the values of p_{vol}, p_z and p_x when these are known.

It should be noted that the vertical plane on which the stresses p_x act is part of the material and undergoes deformations. It is evident that when loads are applied symmetrically with respect to the y axis, there will be no deflection in the x direction of the vertical plane through the y axis. However there will be deformations in the vertical direction, and in the direction of y axis. Consequently, the stress acting on a perfectly smooth, or frictionless, rigid vertical plane can be computed for a given load by considering a fictitious additional load placed in a symmetrical position. However, this merely amounts to doubling the stress p_x due to the given load.

8. *Shearing Stresses on Horizontal and Vertical Planes.*—Influence charts for computing shearing stresses are shown reduced in size in Figs. 5, 6, and 7. These can be drawn to the same scale as Plates 1, 2, and 3 by those who will have opportunity to use them, with the data given in Tables 1 and 2 of the Appendix.

An influence chart for computing p_{zx} is shown in Fig. 5, constructed from values of B and b from Tables 1 and 2. Each elementary block has an influence value numerically equal to 0.001, but the sign of the influence is negative when x is positive, and positive when x is negative.

The value of p_{yz} can be computed from the same chart if the loaded area is rotated through an angle of 90 degrees clockwise about a vertical axis through O.

An influence chart for computing p_{xy} for a Poisson's ratio of 0.5 is shown in Fig. 6, constructed with values of F and f from Tables 1 and 2. An influence chart for computing the correction to be subtracted from the value of p_{xy} obtained from Fig. 6, when Poisson's ratio is different from 0.5, is shown in Fig. 7, constructed with values of E and f from Tables 1 and 2. The correction is $(1 - 2\,\mu)$ times the quantity obtained from Fig. 7.

In Figs. 6 and 7, each elementary block has an influence value numerically equal to 0.001, but the sign of the influence is positive when x and y have the same sign, and negative when x and y are of opposite sign.

It may be noted that Fig. 7 is identical with Fig. 4 rotated through an angle of 45 degrees counterclockwise.

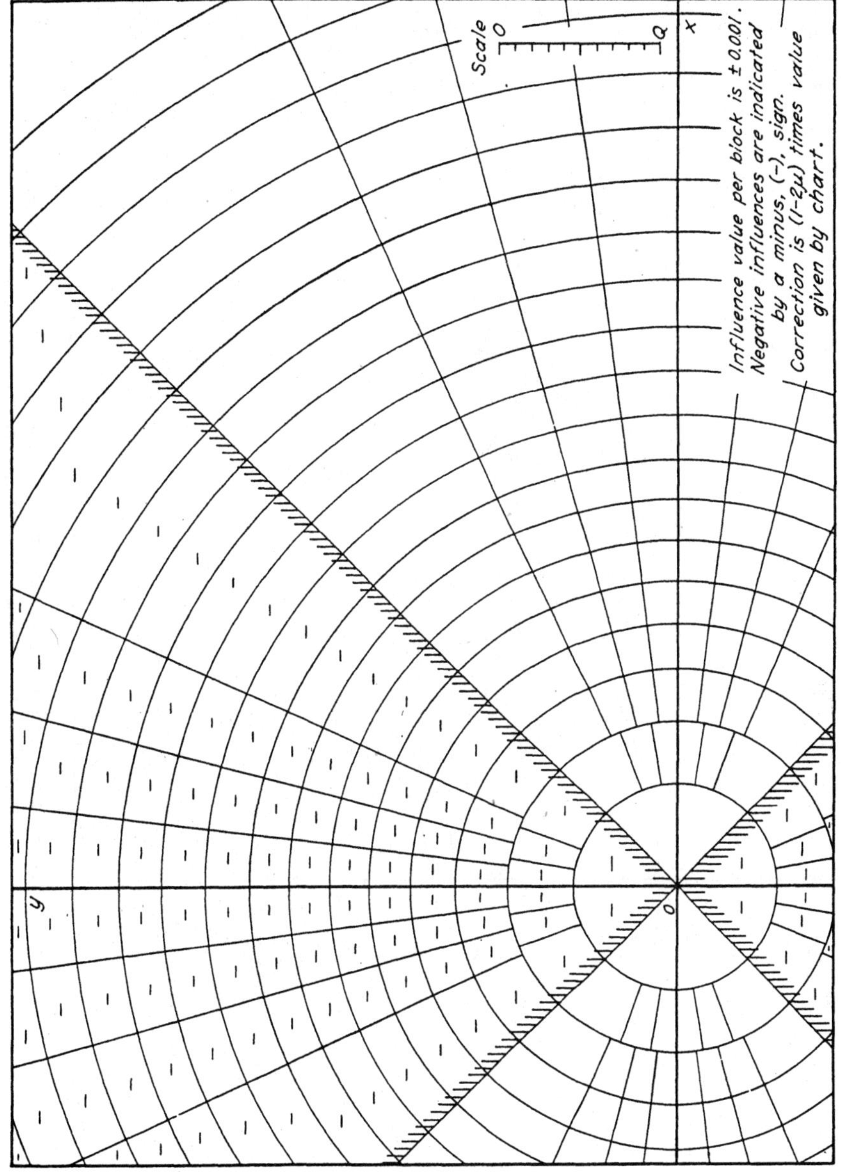
Scale
0
Q
x
y
0
Influence value per block is ±0.001.
Negative influences are indicated
by a minus, (–), sign.
Correction is (1–2μ) times value
given by chart.

FIG. 4. INFLUENCE CHART FOR PART OF CORRECTION TO p_z WHEN POISSON'S RATIO IS DIFFERENT FROM 0.5

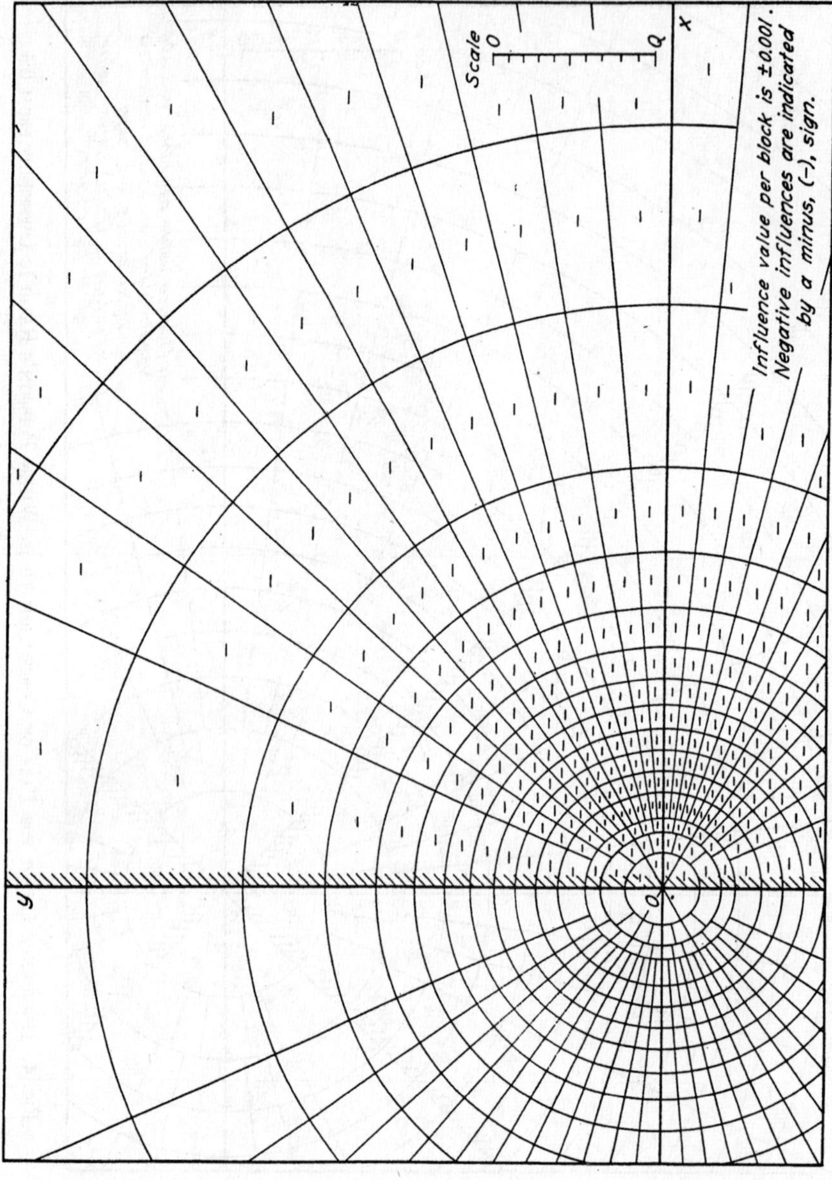

Fig. 5. Influence Chart for p_{zx}, Horizontal Shearing Stress on Horizontal Planes or Vertical Shearing Stress on Vertical Planes

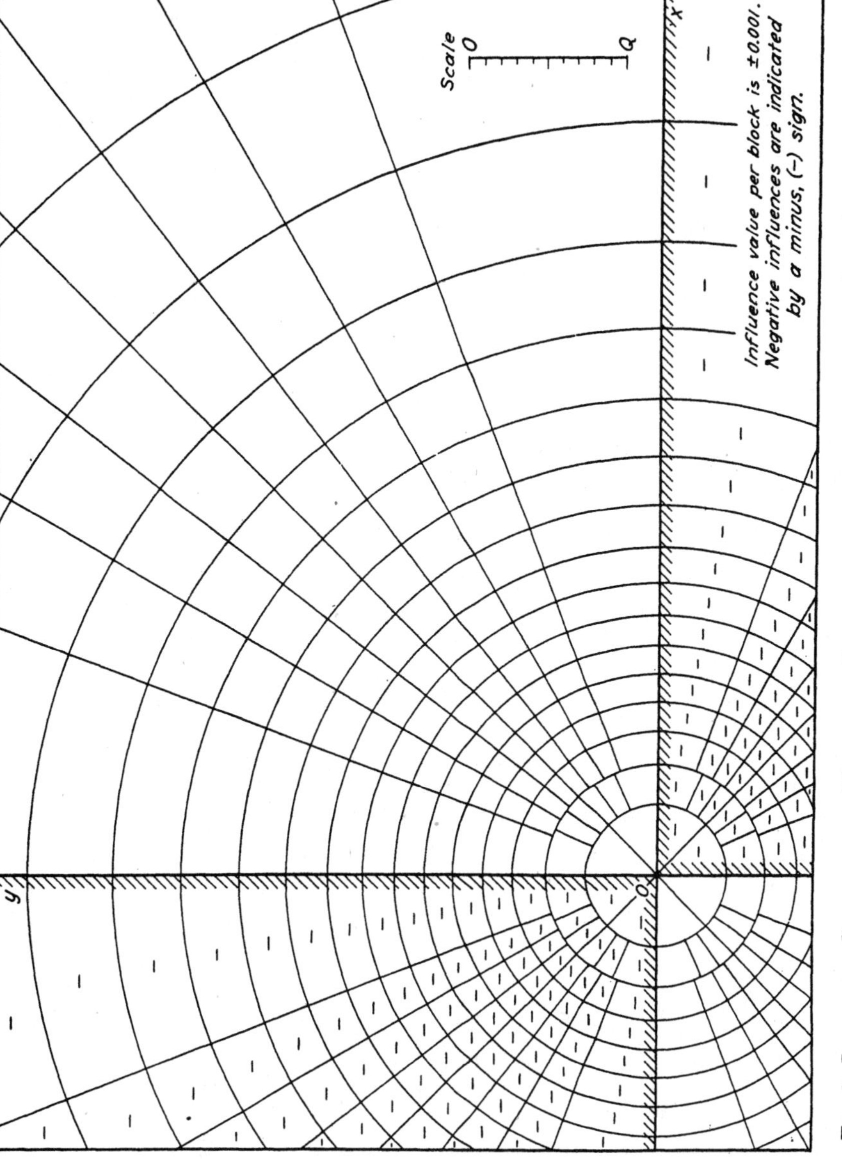

Fig. 6. Influence Chart for p_{xy}, Horizontal Shearing Stress on Vertical Planes, for Poisson's Ratio = 0.5

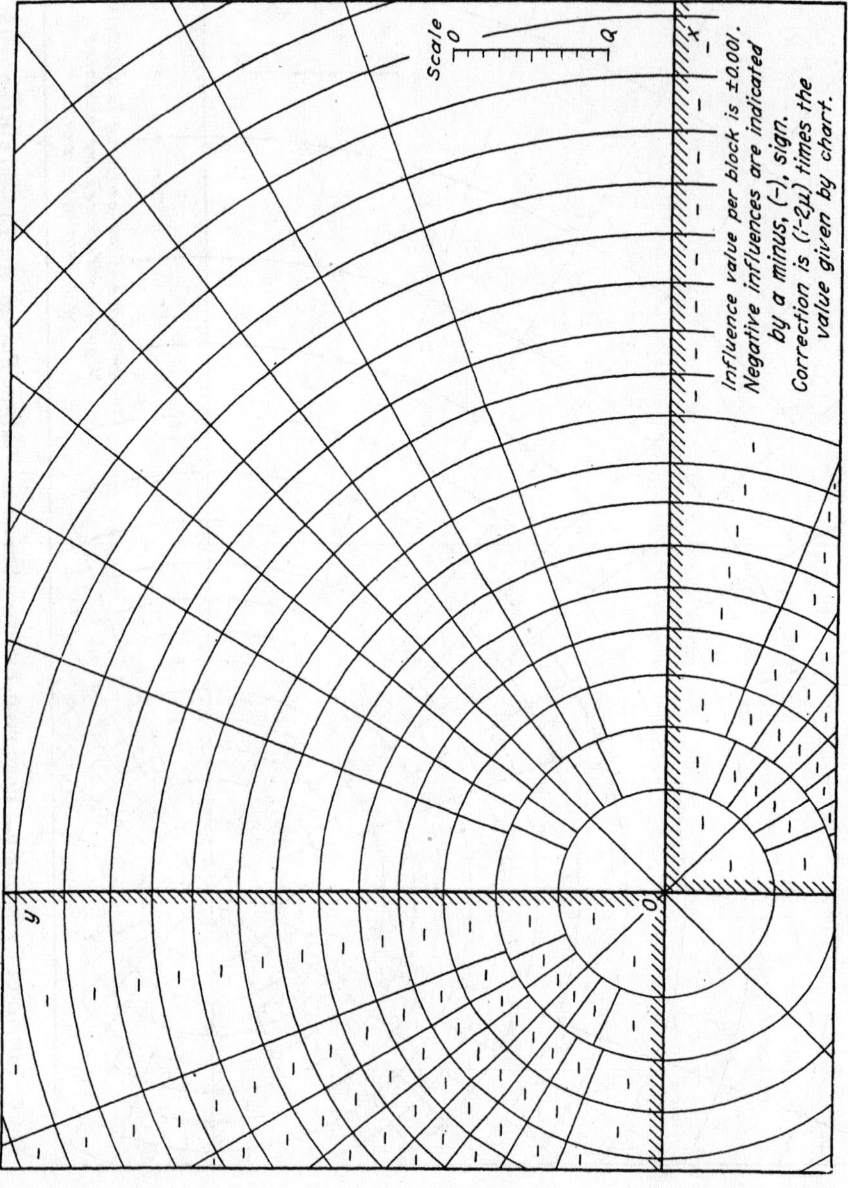
Scale
0
Q
x
y
O
Influence value per block is ±0.001.
Negative influences are indicated
by a minus, (-), sign.
Correction is (1-2μ) times the
value given by chart.

FIG. 7. INFLUENCE CHART FOR CORRECTION TO p_{xy} WHEN POISSON'S RATIO IS DIFFERENT FROM 0.5

IV. Determination of Maximum Shear

9. *Principal Stresses, Maximum Shear, and Octahedral Shear.*—With the components of stress p_x, p_y, p_z, p_{xy}, p_{yz}, and p_{zx}, known at a point the stresses on any plane through the point can be determined. Let l, m, and n be the direction cosines, with respect to the coordinate axes, x, y, and z, respectively, of the normal to the plane on which stresses are to be computed. Then the normal stress p_u on the plane is as follows:*

$$p_u = l^2p_x + m^2p_y + n^2p_z + 2lmp_{xy} + 2mnp_{yz} + 2nlp_{zx}. \quad (5)$$

The maximum shearing stress p_{uv} on the oblique plane is obtained from p_u and the resultant stress S_u on the oblique plane by the relation:

$$\left.\begin{aligned} p_u^2 + p_{uv}^2 = S_u^2 = {} & (lp_x + mp_{xy} + np_{zx})^2 \\ & + (lp_{xy} + mp_y + np_{yz})^2 \\ & + (lp_{zx} + mp_{yz} + np_z)^2 \end{aligned}\right\}. \quad (6)$$

In order that p_u be a principal stress S, p_{uv} must be zero. It can be shown that this condition requires the following relations:

$$\left.\begin{aligned} lS &= lp_x + mp_{xy} + np_{zx} \\ mS &= lp_{xy} + mp_y + np_{yz} \\ nS &= lp_{zx} + mp_{yz} + np_z \end{aligned}\right\}. \quad (7)$$

To find the principal planes these equations must be solved for l, m, and n, or rather for their ratios, since the equations are homogeneous. Then with the relation between the direction cosines of any line,

$$l^2 + m^2 + n^2 = 1, \quad (8)$$

one can obtain the direction cosines of the principal planes, when S is known.

In order that Equations (7) may have a solution for l, m, and n

*For a reference to derivation of the following equations, see, for example, S. Timoshenko, "Theory of Elasticity," McGraw-Hill, New York, 1934, p. 182-188.

different from zero the determinant of the coefficients must vanish. This condition leads to a cubic equation in S, which is:

$$S^3-(p_x+p_y+p_z)S^2+(p_xp_y+p_yp_z+p_zp_x-p^2_{xy}-p^2_{yz}-p^2_{zx})S$$
$$-(p_xp_yp_z+2p_{xy}p_{yz}p_{zx}-p_xp^2_{yz}-p_yp^2_{zx}-p_zp^2_{xy})=0 \quad (9)$$

where each of the terms in parentheses is an invariant with respect to directions of the axes at a point.

Equation (9) will always have three real roots which will be the three principal stresses. Let S_{max} and S_{min} denote the algebraically largest and smallest principal stresses. Then the maximum shear τ_{max} is numerically equal to the quantity

$$\tau_{max} = \frac{1}{2}(S_{max} - S_{min}). \quad (10)$$

It will be seen that the determination of the maximum shear requires the determination of the principal stresses, which is a tedious matter involving the solution of a cubic equation.

Part of the difficulty may be avoided by computing a quantity called the octahedral shear,* $\tau_{oct.}$, which differs only slightly from the maximum shear, and is generally a more significant quantity in determining the behavior of materials as they approach plastic action.

The octahedral shear is defined as the shearing stress on an oblique plane the normal to which makes equal angles with the three directions of principal stress at a point. The value of $\tau_{oct.}$ is proportional to the value of the so-called "shear-strain energy" or to the "energy of distortion" of a unit volume of the material. In terms of the principal stresses S_1, S_2, and S_3, the value of $\tau_{oct.}$ is as follows:

$$\tau_{oct.} = \frac{1}{3}\sqrt{(S_1 - S_2)^2 + (S_2 - S_3)^2 + (S_3 - S_1)^2}. \quad (11)$$

However, the octahedral shear can also be stated in terms of the 6 components of stress on planes parallel to the coordinate planes, as follows:

$$\tau_{oct.} = \frac{\sqrt{2}}{3}\sqrt{(S_1+S_2+S_3)^2-3(S_1S_2+S_2S_3+S_3S_1)} \quad (12)$$

*See A. Nadai, "Theories of Strength," Journal of Applied Mechanics, Vol. 1, No. 3, 1933, p. 111-129.

or

$$\tau_{oct.} = \frac{\sqrt{2}}{3} \sqrt{(p_x+p_y+p_z)^2 - 3(p_x p_y + p_y p_z + p_z p_x - p^2_{xy} - p^2_{yz} - p^2_{zx})}. \quad (13)$$

Therefore, the octahedral shear can be computed from the stresses obtained from the influence charts without first determining the principal stresses.

It is possible to show from Equation (11) that the value of $\tau_{oct.}$ is related to the maximum shear, τ_{max}, as follows:

$$0.8165\tau_{max} \leq \tau_{oct.} \leq 0.9428\tau_{max}. \quad (14)$$

That is, the octahedral shear always lies between the limits of 81.6 and 94.3 per cent of the maximum shear, and it is much simpler to compute.

Illustrative Problem

To illustrate the computation of shearing stress, consider the stresses at a depth of 20 ft. beneath the vertex of a uniformly loaded sector of a circle of radius 100 ft., where the central angle of the sector is 45 degrees. One side of the sector lies along the x axis, and the other side along the line $x = y$. Let the load on the area be 1000 lb. per sq. ft. The value of Poisson's ratio is 0.5.

The following stresses are obtained from the formulas in the Appendix, and may be checked by use of the influence charts.

$p_z =$ 124 lb. per sq. ft.

$p_{vol} =$ 301 " " " "

$p_x =$ 145 " " " "

$p_y =$ 32 " " " "

$p_{zx} =$ −106 " " " "

$p_{yz} =$ −44 " " " "

$p_{xy} =$ 56 " " " "

The principal stresses are determined from the following equation obtained by substituting the foregoing numerical values in Equation (9).

$$S^3 - 301\,S^2 + 10\,280\,S - 68\,592 = 0.$$

The magnitude of the principal stresses can be obtained by solving this equation by trial, or by any of the methods available for the solution of a cubic equation. The results are, to the nearest unit,

$$S_1 = 263 \text{ lb. per sq. ft.}$$
$$S_2 = 29 \text{ " " " "}$$
$$S_3 = 9 \text{ " " " "}$$

Therefore the maximum shear is

$$\tau_{max} = \tfrac{1}{2}(263 - 9) = 127 \text{ lb. per sq. ft.}$$

The octahedral shear is determined from Equation (13) and is as follows:

$$\tau_{oct.} = 115 \text{ lb. per sq. ft.}$$

Appendix

Construction of Influence Charts

1. *Stresses Under Vertex of Uniformly Loaded Sector of a Circle.*—Formulas have been given by Boussinesq from which one can obtain the stresses at a point in the interior of a semi-infinite solid bounded by a plane surface, due to a concentrated load applied normal to the surface.* From Boussinesq's formulas one can readily obtain by integration the stresses at a depth z under the vertex of a sector of a circle located as in Fig. 8, uniformly loaded with an intensity of load p.

With an intensity of loading of unity, the following stresses are obtained at the point Q in Fig. 8:

$$\left.\begin{aligned} p_z &= aA \\ p_{xz} &= -bB \\ p_{yz} &= -cB \\ p_{vol} &= p_x + p_y + p_z = aC - \frac{1-2\mu}{3}aC \\ p_x &= dD - \frac{1-2\mu}{6}aC - (1-2\mu)\,eE \\ p_{xy} &= fF - (1-2\mu)\,fE \end{aligned}\right\} \qquad (15)$$

*See, for example, N. M. Newmark, "Stress Distribution in Soils," Proceedings of Conference on Soil Mechanics and its Applications; Purdue University, September 1940, p. 295-303, especially Equation (1).

where

$$\left.\begin{aligned} a &= \frac{\beta}{2\pi} \\ b &= \sin\beta \\ c &= 1 - \cos\beta \\ d &= \frac{2\beta + \sin 2\beta}{4\pi} \\ e &= \frac{\sin 2\beta}{2} \\ f &= \frac{1 - \cos 2\beta}{2} \end{aligned}\right\} \quad (16)$$

$$\left.\begin{aligned} A &= 1 - \cos^3 \alpha \\ B &= \frac{1}{2\pi}\sin^3 \alpha \\ C &= 3\,(1 - \cos\alpha) \\ D &= 1 - \frac{3}{2}\cos\alpha + \frac{1}{2}\cos^3\alpha \\ E &= \frac{1}{4\pi}\left(\log_e \frac{1 + \cos\alpha}{2\cos\alpha} + \cos\alpha - 1\right) \\ F &= \frac{1}{4\pi}\,(2 - 3\cos\alpha + \cos^3\alpha) \end{aligned}\right\}. \quad (17)$$

It will be noted that for $\mu = \frac{1}{2}$, the value of $(1 - 2\mu)$ is zero. For any value of μ one has the results:

$$\left.\begin{aligned} p_{\text{vol}} &= [p_{\text{vol}}]_{\mu=1/2} - \frac{1 - 2\mu}{3}\,[p_{\text{vol}}]_{\mu=1/2} \\ p_x &= [p_x]_{\mu=1/2} - \frac{1 - 2\mu}{6}\,[p_{\text{vol}}]_{\mu=1/2} - (1 - 2\mu)\,eE \\ p_{xy} &= [p_{xy}]_{\mu=1/2} - (1 - 2\mu)\,fE \end{aligned}\right\}. \quad (18)$$

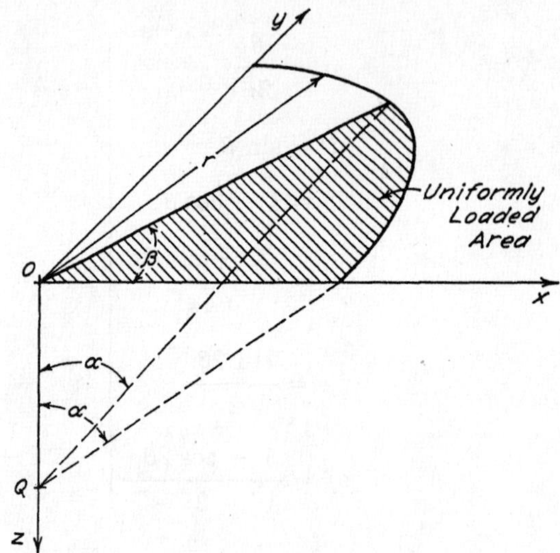

FIG. 8. UNIFORMLY LOADED SECTOR OF A CIRCLE

2. *Numerical Data for Construction of Charts.*—The formulas in the preceding section give stresses at a depth z under the vertex of a uniformly loaded sector of a circle of radius r and central angle β. By combining numerical values of stresses for different sectors one can obtain the stress at Q due to uniform load on an element of area bounded by two radial lines and by two concentric circle arcs. The stress at Q due to such a loading may be interpreted as an influence coefficient. Different areas will in general produce different influences, but values of β and r can be so chosen that elements of area are defined which produce the same influence.

By proper choice of the values of $a, b, \ldots A, B, \ldots F$ one can find the values of β and α, or preferably, of β and $\tan \alpha = r/z$, which give, when combined, influences for stresses in steps of uniform amount. In Tables 1 and 2 values of β and r/z are reported, which, when combined, correspond to elementary influence areas of value 0.001. The numerical values are so chosen that the values of β may be laid out with the x and y axes always axes of symmetry for the influence diagrams. By referring to the charts in this bulletin, the manner of using the values of β and r/z to construct influence charts can be seen.

The chart for p_z involves a and A and is independent of μ. Since the values of a correspond to equal divisions of a circle the chart will

TABLE 1

VALUES OF r/z CORRESPONDING TO GIVEN VALUES OF A, B, C, D, E, AND F IN FORMULAS FOR STRESSES, EQUATION (15)

A	Increment in A	r/z	B	Increment in B	r/z	C	Increment in C	r/z	D	Increment in D	r/z	E	Increment in E	r/z	F	Increment in F	r/z
0	0	0	0	0	0	0	0	0	0	0	0	0	0	0	0	0	0
0.008	0.008	0.073	0.002	0.002	0.239	0.008	0.008	0.073	0.008	0.008	0.408	0.002	0.002	0.673	0.002	0.002	0.464
0.024	0.016	0.128	0.006	0.004	0.379	0.024	0.016	0.127	0.024	0.016	0.565	0.008	0.006	1.070	0.008	0.006	0.722
0.048	0.024	0.183	0.012	0.006	0.466	0.048	0.024	0.181	0.048	0.024	0.710	0.016	0.008	1.414	0.016	0.008	0.940
0.072	0.024	0.226	0.024	0.012	0.629	0.080	0.032	0.236	0.096	0.048	0.922	0.024	0.008	1.706	0.024	0.008	1.128
0.096	0.024	0.264	0.036	0.012	0.768	0.112	0.032	0.281	0.144	0.048	1.103	0.032	0.008	1.980	0.032	0.008	1.308
0.144	0.048	0.330	0.048	0.012	0.904	0.144	0.032	0.322	0.192	0.048	1.276	0.040	0.008	2.247	0.040	0.008	1.493
0.192	0.048	0.391	0.060	0.012	1.045	0.192	0.048	0.376	0.240	0.048	1.450	0.048	0.008	2.516	0.048	0.008	1.688
0.240	0.048	0.448	0.072	0.012	1.198	0.240	0.048	0.426	0.288	0.048	1.634	0.056	0.008	2.785	0.056	0.008	1.900
0.288	0.048	0.504	0.084	0.012	1.372	0.288	0.048	0.473	0.336	0.048	1.831	0.064	0.008	3.062	0.064	0.008	2.136
0.336	0.048	0.560	0.096	0.012	1.580	0.336	0.048	0.521	0.384	0.048	2.048	0.072	0.008	3.347	0.072	0.008	2.405
0.384	0.048	0.617	0.108	0.012	1.841	0.384	0.048	0.561	0.432	0.048	2.291	0.080	0.008	3.640	0.080	0.008	2.718
0.432	0.048	0.677	0.120	0.012	2.197	0.480	0.096	0.646	0.480	0.048	2.576	0.088	0.008	3.945	0.088	0.008	3.091
0.480	0.048	0.739	0.132	0.012	2.744	0.576	0.096	0.729	0.528	0.048	2.898	0.096	0.008	4.261	0.096	0.008	3.549
0.528	0.048	0.806	0.144	0.012	3.808	0.672	0.096	0.813	0.576	0.048	3.289	0.104	0.008	4.591	0.104	0.008	4.129
0.576	0.048	0.879	0.150	0.006	4.983	0.768	0.096	0.898	0.624	0.048	3.772	0.112	0.008	4.935	0.112	0.008	4.894
0.624	0.048	0.959	0.156	0.006	8.630	0.864	0.096	0.986	0.672	0.048	4.385	0.120	0.008	5.294	0.120	0.008	5.957
0.672	0.048	1.050	0.159	0.003+	inf.	0.960	0.096	1.078	0.720	0.048	5.198	0.128	0.008	5.679	0.128	0.008	7.553
0.720	0.048	1.156	($\frac{1}{2\pi}$)			1.056	0.096	1.175	0.768	0.048	6.335	0.136	0.008	6.062	0.136	0.008	10.230
0.768	0.048	1.284				1.152	0.096	1.279	0.816	0.048	8.049	0.144	0.008	6.474	0.144	0.008	15.700
0.816	0.048	1.446				1.248	0.096	1.390	0.864	0.048	10.954	0.152	0.008	6.905	0.152	0.008	33.344
0.864	0.048	1.668				1.344	0.096	1.511	0.912	0.048	16.996	0.160	0.008	7.357	0.159	0.007+	inf.
0.912	0.048	2.014				1.440	0.096	1.643	0.960	0.048	37.397	0.168	0.008	7.831	($\frac{1}{2\pi}$)		
0.944	0.032	2.415				1.536	0.096	1.789	1.000	0.040	inf.	0.176	0.008	8.328			
0.976	0.032	3.315				1.632	0.096	1.952				0.184	0.008	8.850			
0.992	0.016	4.899				1.728	0.096	2.136				0.192	0.008	9.397			
1.000	0.008	inf.				1.824	0.096	2.347				0.200	0.008	9.972			
						1.920	0.096	2.592				inf.	inf.	inf.			
						2.016	0.096	2.880									
						2.112	0.096	3.227									
						2.208	0.096	3.653									
						2.304	0.096	4.193									
						2.352	0.048	4.520									
						2.400	0.048	4.899									
						2.448	0.048	5.342									
						2.496	0.048	5.868									
						2.592	0.096	7.285									
						2.688	0.096	9.563									
						2.784	0.096	13.853									
						2.880	0.096	24.971									
						2.976	0.096	125.012									
						3.000	0.024	inf.									

TABLE 2

VALUES OF β FOR GIVEN INCREMENTS IN VALUES OF a, b, c, d, e, AND f IN FORMULAS FOR STRESSES, EQUATION (15)

Increment in β deg.	Increment in a	β deg.	Increment in b	β deg.	Increment in c*	β deg.	Increment in d	β deg.	Increment in e	β deg.	Increment in e	β deg.	Increment in f†	β deg.	Increment in f†
3.75	1/96	0		0		0		0		0		0		0	
7.5	1/48	±4.78	1/12	23.56	1/12	±3.76	1/48	±7.24	1/8	±9.74	1/6	20.70	1/8	29.04	1/6
45.0	1/8	±9.59	1/12	33.56	1/12	±7.54	1/48	±15.00	1/8	±20.91	1/6	30.00	1/8	35.26	1/6
etc.,		±14.48	1/12	41.41	1/12	±11.40	1/48	±24.30	1/8	±45.00	1/6	37.76	1/8	45.00	1/6
uniform		±19.47	1/12	48.19	1/12	±15.36	1/48	±45.00	1/8	±69.09	−1/6	45.00	1/8	54.74	1/6
division		±24.62	1/12	54.32	1/12	±19.48	1/48	±65.70	−1/8	±80.26	−1/6	52.24	1/8	65.91	1/6
		±30.00	1/12	60.00	1/12	±23.83	1/48	±75.00	−1/8	±90.00	−1/6	60.00	1/8	90.00	1/6
		±35.68	1/12	65.38	1/12	±28.48	1/48	±82.76	−1/8	±99.74	−1/6	69.30	1/8	114.09	−1/6
		±41.81	1/12	70.53	1/12	±33.59	1/48	±90.00	−1/8	±110.91	−1/6	90.00	1/8	125.26	−1/6
		±48.59	1/12	75.52	1/12	±39.40	1/48	±97.24	−1/8	±135.00	−1/6	110.70	−1/8	135.00	−1/6
		±56.44	1/12	80.41	1/12	±46.39	1/48	±105.00	−1/8	±159.09	1/6	120.00	−1/8	144.74	−1/6
		±66.44	1/12	85.22	1/12	±55.90	1/48	±114.30	−1/8	±170.26	1/6	127.76	−1/8	155.91	−1/6
		±90.00	1/12	90.00	1/12	±90.00	1/48	±135.00	−1/8	±180.00	1/6	135.00	−1/8	180.00	−1/6
		±113.56	−1/12	94.78	1/12	±124.10	1/48	±155.70	1/8			142.24	−1/8		
		±123.56	−1/12	99.59	1/12	±133.61	1/48	±165.00	1/8			150.00	−1/8		
		±131.41	−1/12	104.48	1/12	±140.60	1/48	±172.76	1/8			159.30	−1/8		
		±138.19	−1/12	109.47	1/12	±146.41	1/48	±180.00	1/8			180.00	−1/8		
		±144.32	−1/12	114.62	1/12	±151.52	1/48								
		±150.00	−1/12	120.00	1/12	±156.17	1/48								
		±155.38	−1/12	125.68	1/12	±160.52	1/48								
		±160.53	−1/12	131.81	1/12	±164.64	1/48								
		±165.52	−1/12	138.59	1/12	±168.60	1/48								
		±170.41	−1/12	146.44	1/12	±172.46	1/48								
		±175.22	−1/12	156.44	1/12	±176.24	1/48								
		±180.00	−1/12	180.00	1/12	±180.00	1/48								

*For corresponding values of β between 0 and −180, the increment in c is negative in sign.
†For corresponding values of β between 0 and −180, the increment in f is opposite in sign to that given.

be made up of circles and uniformly spaced radial lines. For convenience in counting it is desirable to make the elementary areas as nearly square as possible. For this reason different numbers of radial divisions are used in different parts of the diagram.

The same general comments apply also to p_{vol} which involves a and C. The value of p_{vol} for $\mu = ½$ may be used also to obtain p_{vol} for other values of μ merely by subtracting a correction as indicated in the previous section.

The chart for p_{zx} involves b and B, and is independent of μ. The elementary influence areas are negative where x is positive and positive where x is negative. It may be observed that the same chart can be used for p_{zx} and p_{yz} provided that the x and y axes are interchanged; or also, to prevent redrawing of the loading diagram, if the negative y axis is taken as a new x axis and the x axis is taken as a new y axis.

Since p_x involves μ it is convenient to construct a chart for p_x when $\mu = ½$, which involves d and D, and correction charts for the effect of μ. One correction is proportional to p_{vol} as indicated by Equation (18), and the other requires a separate chart involving e and E. In this chart the elementary influence areas are positive when y is numerically less than x, and negative otherwise.

The same charts can be used for p_x and p_y with proper change of axes, but since p_z and p_{vol} will usually be easier to determine, p_y can be obtained from the relation

$$p_y = p_{vol} - p_z - p_x.$$

The value of p_{xy} also depends on μ, and it is expedient to construct a chart for p_{xy} when $\mu = ½$, which involves f and F, and a correction chart for the effect of μ. The correction involves f and E. For both charts the elementary influence areas are positive when x and y have the same sign, and are negative otherwise.

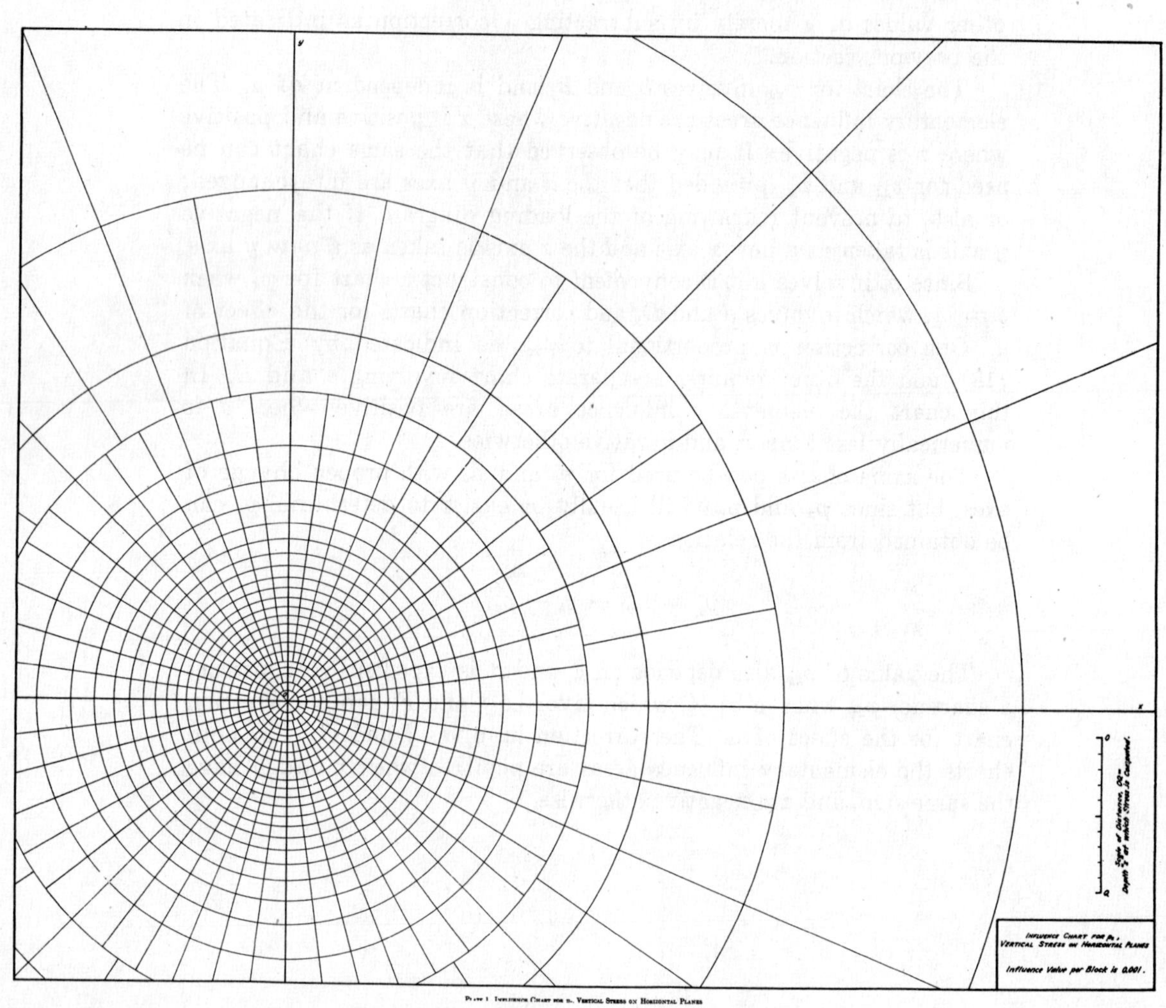

Plate 1. Influence Chart for p_z, Vertical Stress on Horizontal Planes

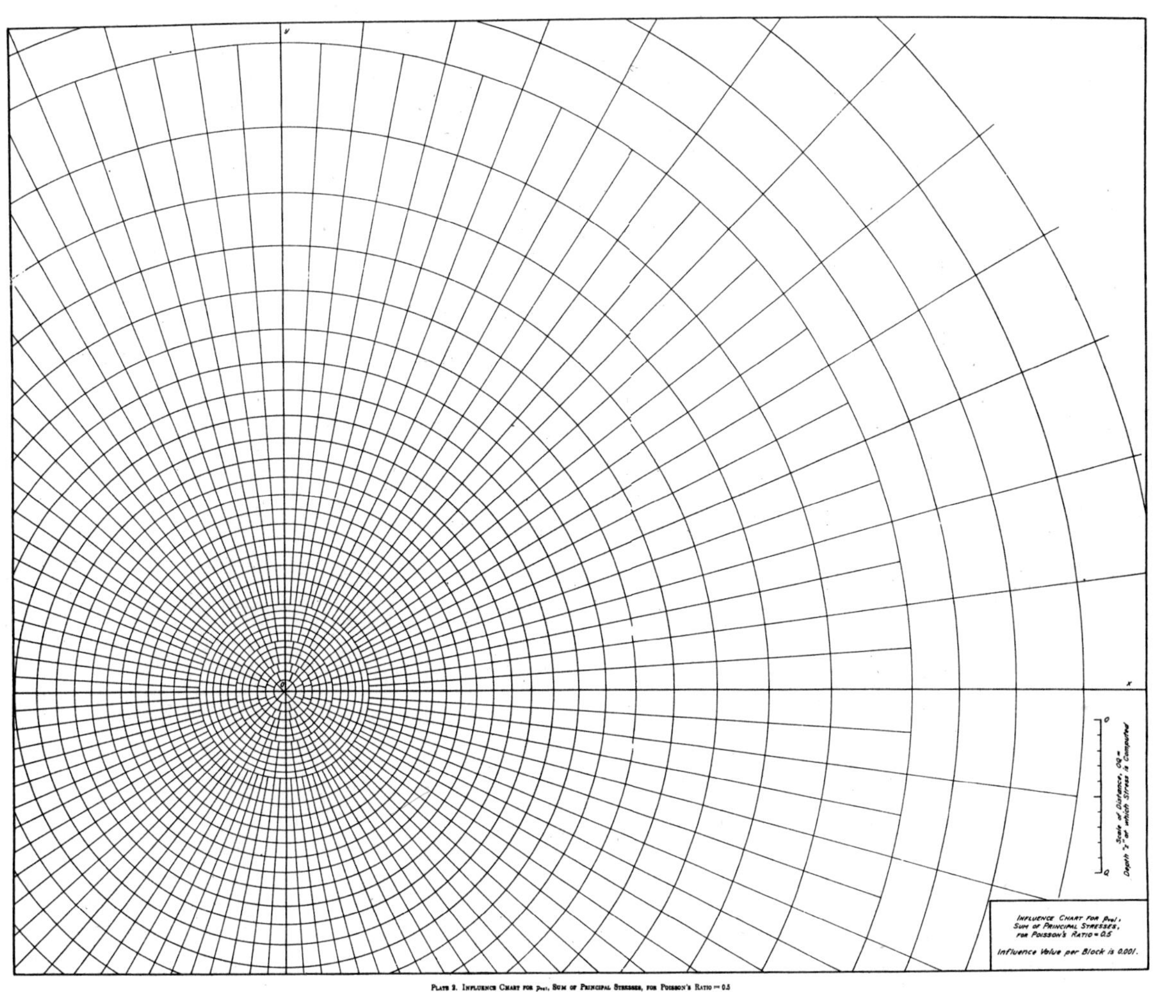

PLATE 2. INFLUENCE CHART FOR p_{vol}, SUM OF PRINCIPAL STRESSES, FOR POISSON'S RATIO = 0.5

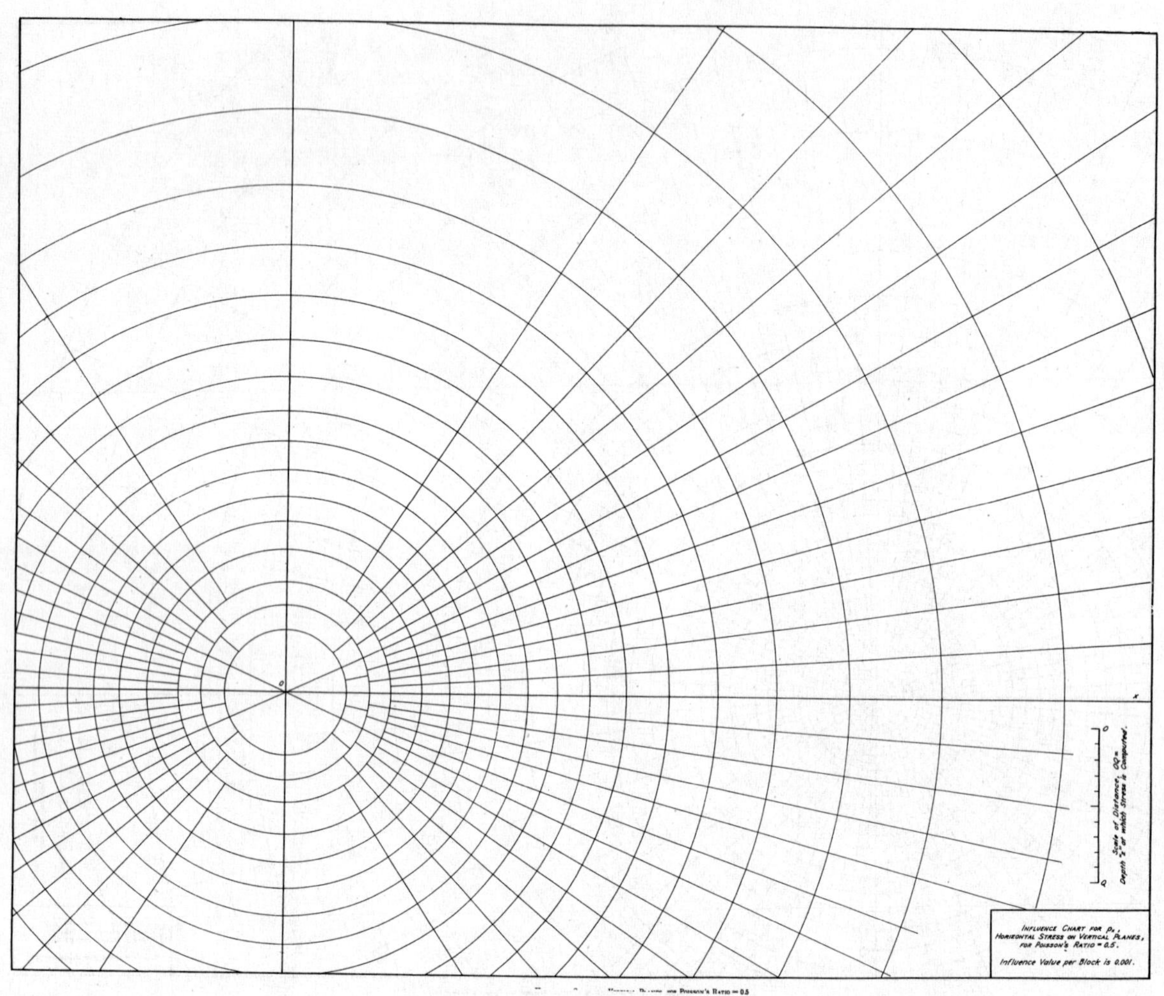

Poisson's Ratio = 0.5

EARTH-PRESSURE MEASUREMENTS IN OPEN CUTS, CHICAGO (ILL.) SUBWAY

By Ralph B. Peck,[7] Jun. Am. Soc. C. E.

Synopsis

Systematic field observations on open-cut portions of the Chicago subway provided a constant check on the loads carried by the bracing as well as the soil movements associated with the excavations. Results of the measurements are given in this paper, together with a comparison of the measured earth pressures with generally accepted theories. It was found that the magnitude of the total lateral pressure was in satisfactory agreement with either the plane or general wedge theories for purely cohesive soils, having no effective internal friction, but that the distribution of the pressure was non-hydrostatic. Measured movements of the sheeting were found to be in accordance with those theoretically necessary for non-hydrostatic distribution. Simple rules are given which are believed to be applicable to the design of bracing for similar cuts in plastic clay deposits.

Part I.—Field Measurements

Although the Chicago subway was constructed largely by tunneling methods, a number of special structures were built in open cut. The excavations, frequently of considerable magnitude, were in congested districts near the downtown area in a clay deposit of variable stiffness. It was considered of vital importance to obtain definite information concerning the safety of the bracing of the temporary structures, and to determine the nature and magnitude of the movements of the soil mass and ground surface associated with the excavation. Furthermore, it was considered desirable to obtain quantitative data which could be used as the basis for economical design of the bracing for future excavations. As a result, a systematic program of measurements was undertaken to determine the necessary information.

The open cuts varied in type of bracing, in dimensions, in soil properties, and in method of construction. All these factors were included in the information obtained. The proposed excavations were always outlined by steel piling before any material was removed. Both continuous steel sheeting and H-piles spaced at suitable distances for the insertion of timber lagging were used. All of the excavations were braced with horizontal wales and struts, some of steel and others of wood. The excavations varied in width from 15 ft to 55 ft, in depth from 35 ft to 46 ft, and in length from 100 ft to 1,300 ft. Although the

[7] Research Asst. Prof., Soil Mechanics, Univ. of Illinois, Urbana, Ill.

construction procedures were all standard practice, each cut was made by a different contractor, with inevitable variations in detail.

The measurements included systematic observations of the settlement of

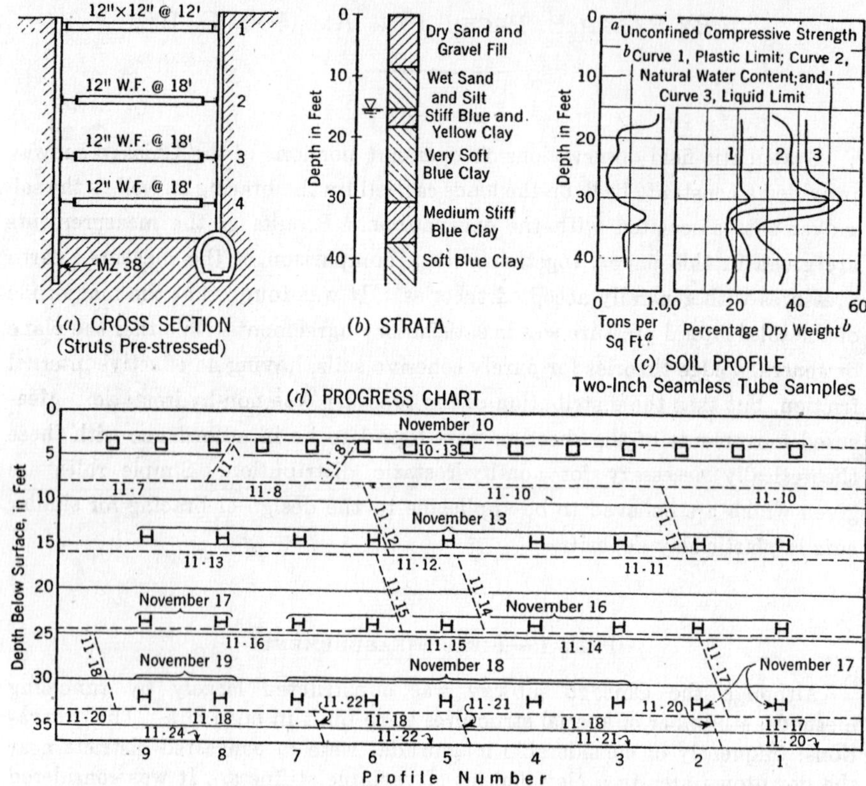

(e) SUMMARY OF STRUT LOADS, IN KIPS (ESTIMATED ERROR ±15%)

Depth (ft)	Date (Nov. 1939)	Profile (See Fig. (d)) 9	8	7	6	5	4	3	2	1
4.0	20–27	27	....	33	44	30	....	44	....	....
14.5	20	322	257	270	254	292	220	262	242	226
	27	....	252	249	188	238	....	....	....	....
23.5	20	147	185	161	....	230	164	188	....	140
	27	185	222	199	....	269	218	210	....	157
31.5	20	58	49	40	77	98	81	145	112	150
	27	182	....	252	295	355	248	323	266	262

FIG. 19.—STRUT LOADS IN OPEN CUT, CONTRACT S-3

the adjacent ground surface, of the lateral movements of the sheeting during the excavation process, of the rise of the soil at the bottom of the cuts, and of the settlements of the sheeting itself. In each cut, at least one vertical profile of bracing was selected for determination of the loads carried by the struts.

In all, more than fifty profiles were subjected to measurement, some repeatedly. For steel struts the loads were usually determined by strain gage, and for wooden struts a hydraulic jacking arrangement was used. Hydraulic jacking, when used, was so arranged that errors due to jack friction or to over-jacking

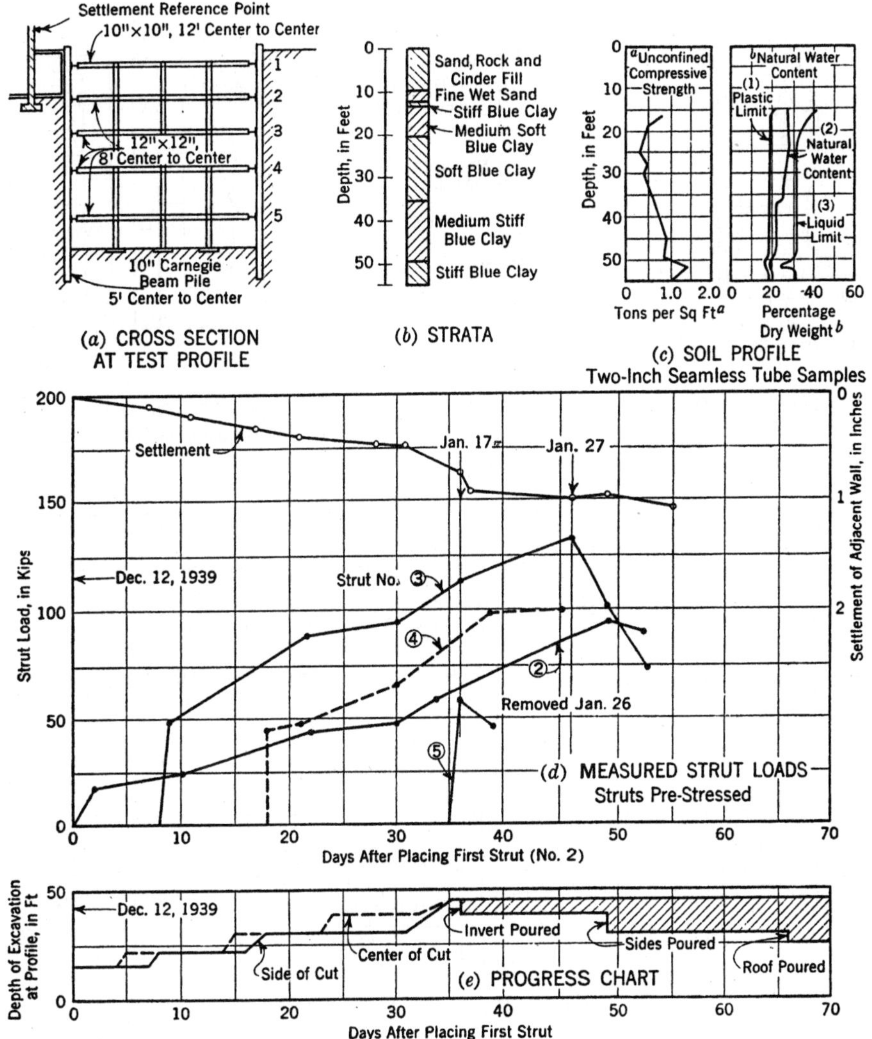

Fig. 20.—Strut Loads in Open Cut, Contract S-4B

were eliminated. A detailed discussion of the techniques of measurement and the errors involved was presented elsewhere by the writer.[8] The soil properties adjacent to each cut were determined from the results of laboratory tests on

[8] "The Measurement of Earth Pressures on the Chicago Subway," by Ralph B. Peck, *Bulletin*, A. S. T. M., August, 1941, p. 25.

2-in. diameter samples obtained in seamless tubes from borings taken at 300-ft intervals along the subway right of way.[9] For the comparison of results of the open-cut measurements, the most important physical property of the clay was the unconfined compressive strength, determined on samples 2 in. in diameter and 3.5 in. long at 2.5-ft intervals in the vertical direction. The samples ob-

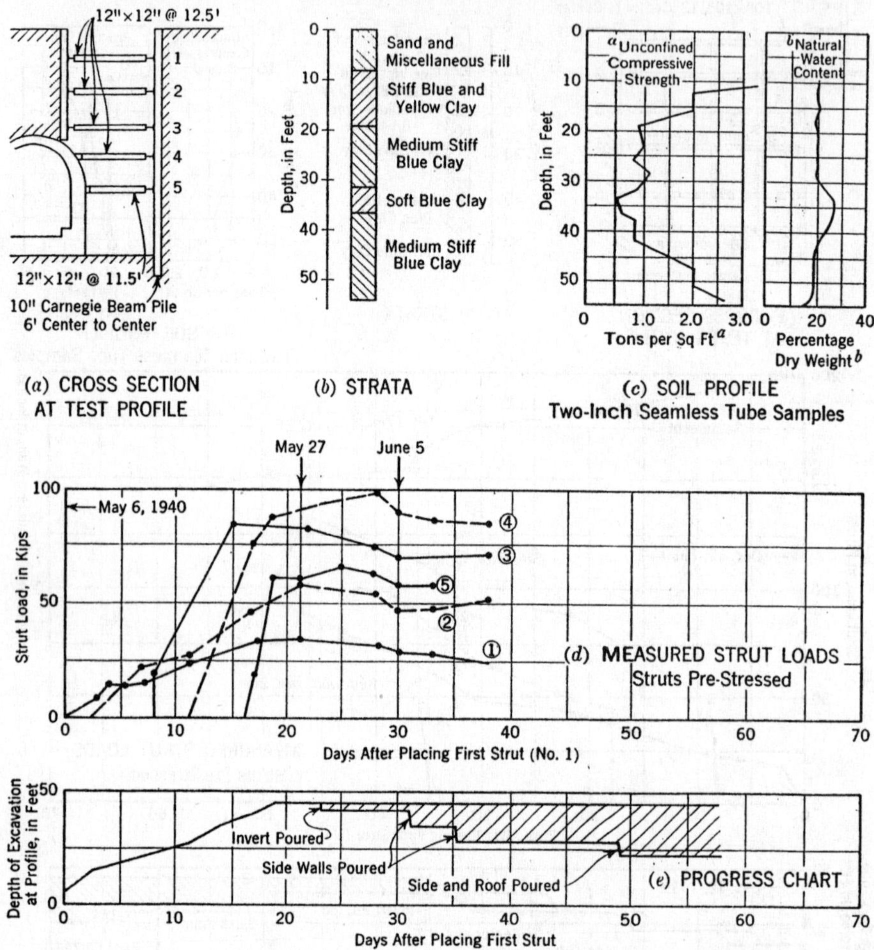

FIG. 21.—STRUT LOADS IN OPEN CUT, CONTRACT S-9C

tained from the borings were disturbed to a certain extent because of the small diameter of the tubes, and an extensive series of tests was conducted to determine the loss in strength during the sampling process. As a result, it was determined that the average unconfined compressive strength of large relatively undisturbed blocks of clay was approximately 1.35 times the value ob-

[9] For details of the soil exploration program, see "Soil Tests Check Chicago Subway Work," by Ralph B. Peck, *Engineering News-Record*, December 7, 1939, p. 87.

tained from the seamless tube samples on the same soil. It was also considered important to determine the order of magnitude of the error in the average unconfined compressive strength for a single boring, introduced because samples were tested only at 2.5-ft intervals. On certain borings, two samples were tested in each 2.5-ft interval and the average value of compressive strength determined from all of the tests. This value was compared with the average

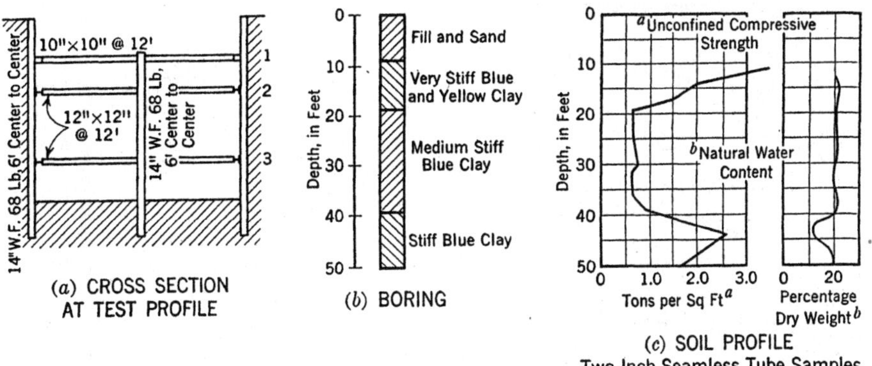

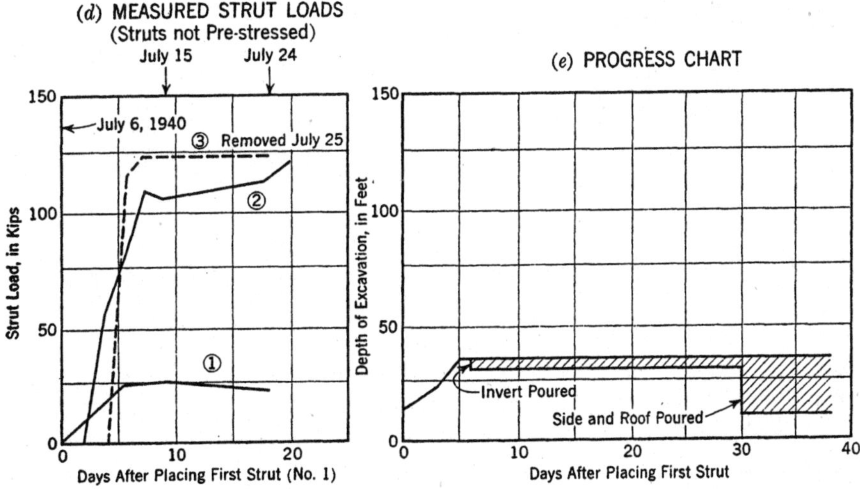

Fig. 22.—Strut Loads in Open Cut, Contract S-8A

for alternate samples, representing the usual 2.5-ft spacing. It was found that the difference was about 6%, and that the average compressive strength of the entire boring was probably determined within 10% by the use of the 2.5-ft spacing.

The pertinent information and test results for each open cut are shown in Figs. 19 to 26. In Fig. 19 the progress of the excavation is indicated by the dates along the broken lines. The remaining dates indicate when the struts were placed. In Fig. 19(*c*) the left-hand curve gives the unconfined compressive

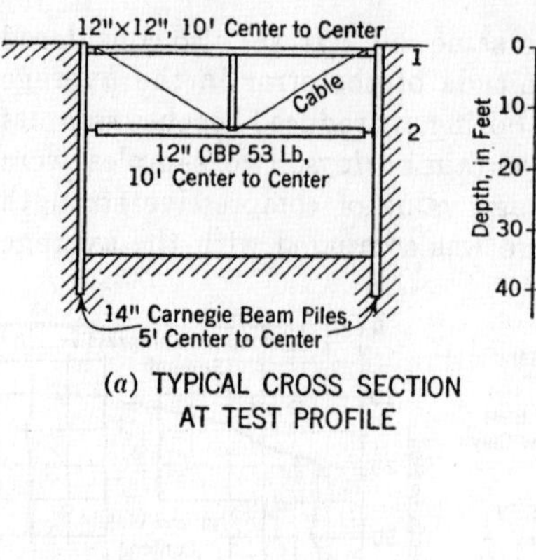

(*a*) TYPICAL CROSS SECTION AT TEST PROFILE

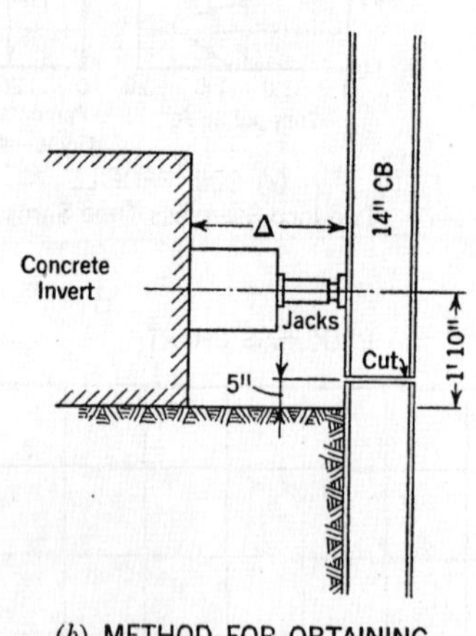

(*b*) METHOD FOR OBTAINING SOIL REACTION, STA. 213+85

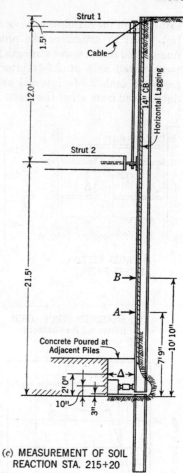

(*c*) MEASUREMENT OF SOIL REACTION STA. 215+20

(*d*) SOIL PROFILE

Two-Inch Seamless Tube Samples

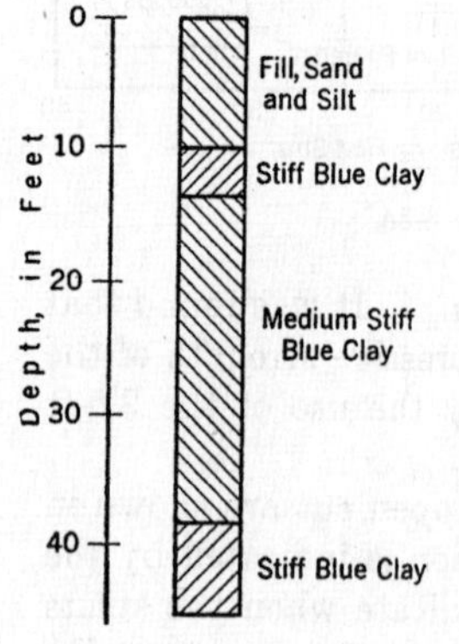

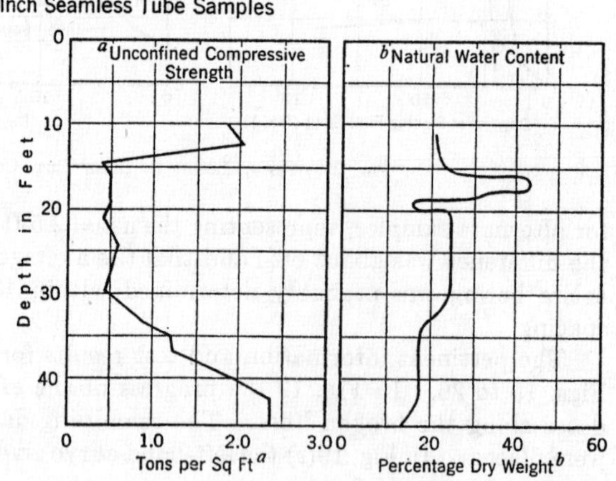

Fig. 23.—continued on page 1014.

(*e*) SUMMARY OF TEST RESULTS (*Continued from page 1013*)

No.	Description	STATIONS:	
		213+85	215+20
1	Load in strut 1 (kips)	9±5	1±2
2	Load in strut 2 (kips)	167±10	137±5
3	Twice the reaction (in kips) at the embedded part of the pile	83±4	82±10
4	Distance of the pile cutoff above the invert (inches)	5	3
5	Measured load in the hydraulic jacks (kips)	41±2	41±5
6	Distance Δ held unchanged at both stations	√	√
7	Subsequent test reduction of load (kips)	8	0
	Result of load reduction (Item 7):		
8	Decrease in Δ (inches)	0.2	0.21
9	Inward movement of point A, Fig. (*c*), as measured on the cut pile (inches)	....	0.4
10	Inward movement of point B, Fig. (*c*), as measured on the adjacent pile (inches)	....	0.0013
11	Excavation completed (1940)	Oct. 8	Oct. 24
12	Invert poured (1940)	....	Nov. 12
13	Measurements made	Oct. 14–18	Nov. 13

FIG. 23.—LATERAL EARTH PRESSURE MEASUREMENTS, CONTRACT D-6E

strength in tons per square feet; and curve 1 is the plastic limit, curve 2 is the natural water content, and curve 3 is the liquid limit. In Figs. 20 to 22, the estimated error in strut loads is approximately 10 kips. Cross sections of each cut are shown in Figs. 19 to 25, together with a description of the soil profile and the results of unconfined compressive strength and natural water-content determinations. Diagrams showing the progress of excavation are shown in Figs. 19 to 22, as well as the measured strut loads corresponding to various excavation stages for contracts S-3, S-4B, S-9C, S-8A, and D-6E (for location, see Fig. 1). Figs. 25 and 26 show the excavation progress, strut-load determinations and soil movements at one profile on contract S-1A, and Table 1 contains the measured strut loads for 21 profiles in contract S-1A, measured soon after excavation reached grade at each profile.

TABLE 1.—STRUT LOADS, CONTRACT S-1A; DEPTH 36.5 FT

Station	Profile	LOAD (IN KIPS) IN THE STRUT, AT DEPTHS (IN FT) OF:				Total load, Q	Center of pressure,[d] N
		4	13	23	31		
61+05	3	...[b]	342	...[c]	...[b]	....	....
60+90	4	...[b]	335	...[c]	...[b]	....	....
60+45	7	...[b]	466	...[c]	...[b]	....	....
60+30	8	8.5	...[b]	...[c]	...[b]	....	....
60+15	9	...[b]	...[b]	...[c]	237	....	....
59+85	11	...[b]	338	82	232	652	0.43
59+70	12	3.8	223	316	89	628	0.43
59+25	15	...[b]	286	284	82	652	0.45
59+10	16	...[b]	258	196	88	542	0.46
58+95	17	...[b]	300	254	130	684	0.45
58+80	18	...[b]	271	199	142	612	0.44
58+65	19	...[b]	294	185	214	693	0.42
58+50	20	...[b]	228	254	123	605	0.43
58+35	21	...[b]	244	249	146	639	0.42
58+20	22	...[b]	196	202	147	545	0.41
58+05	23	1.0	249	193	105	547	0.45
57+90	24	...[b]	261	224	83	568	0.51
57+75	25	...[b]	278	146	124	548	0.46
57+60	26[a]	...[b]	170	147	109	426	0.42
57+45	27	...[b]	259	187	148	594	0.43
57+30	28	...[b]	219	166	116	501	0.44
57+15	29	...[b]	189	173	131	493	0.42

[a] At a stair well; wales not blocked to the sheeting. [b] Strut load not measured. [c] No strut at this level. [d] Center of pressure N is the ratio of the height of the center of gravity of the measured strut loads above the bottom of the cut, to the total depth of the cut.

The closeness of superimposed loads to the excavation is not indicated in Figs. 19 to 24. Although the lateral pressure produced by the weight of adjacent structures was insignifi-

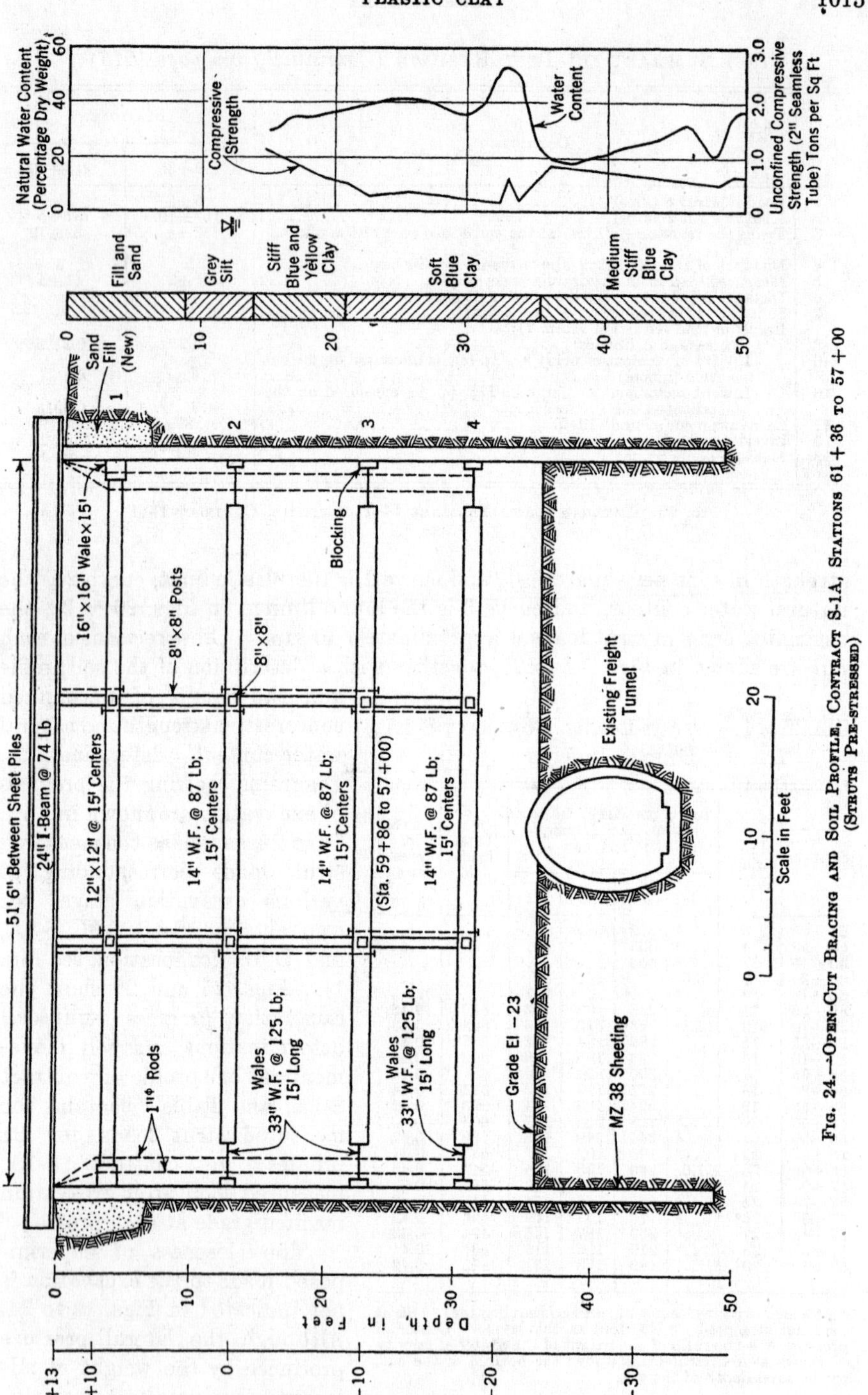

Fig. 24.—Open-Cut Bracing and Soil Profile, Contract S-1A, Stations 61 + 36 to 57 + 00 (Struts Pre-stressed)

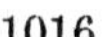

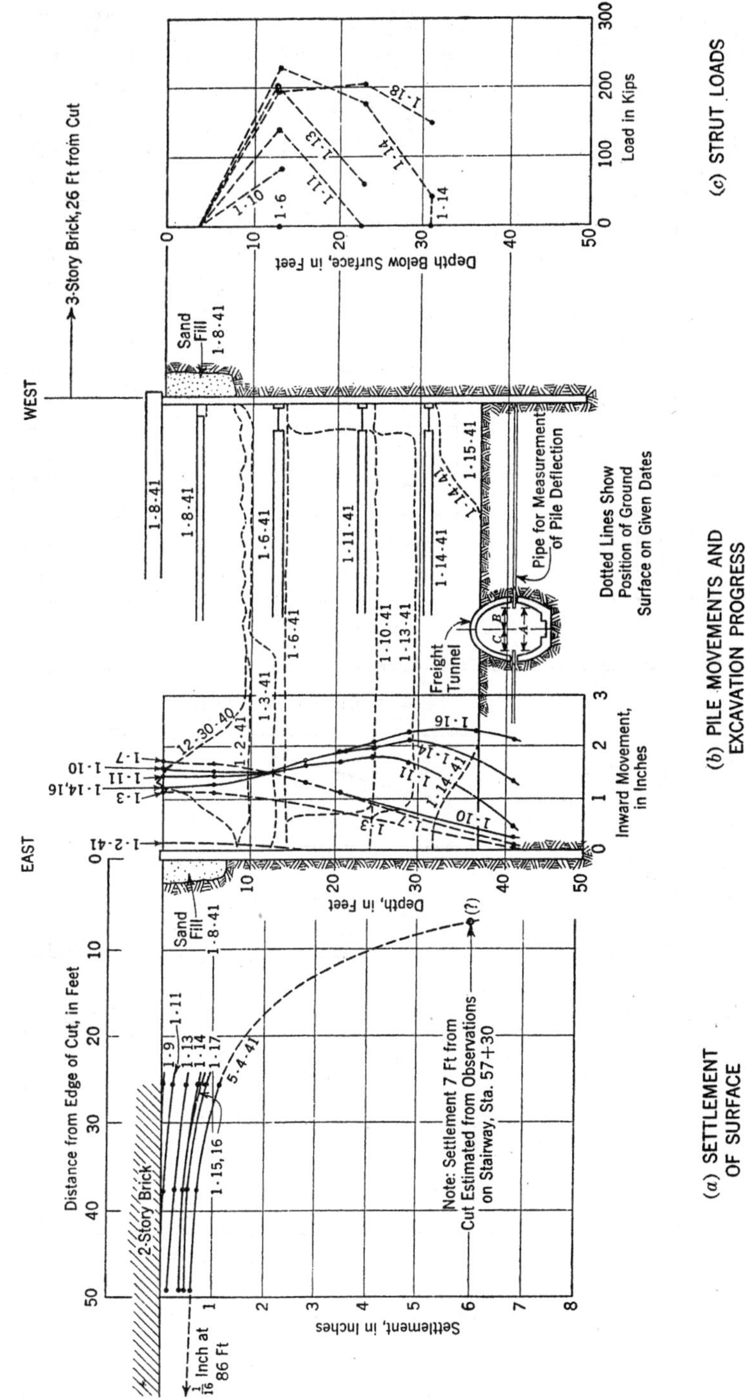

FIG. 25.—MEASUREMENTS ON OPEN-CUT PROFILE 22, STATION 58+20, CONTRACT S-1A

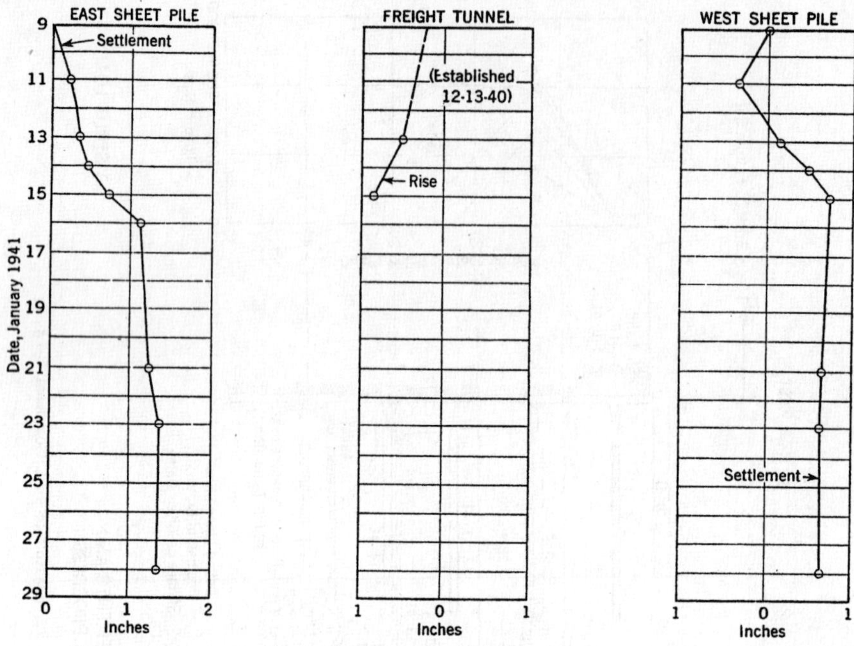

FIG. 26.—SETTLEMENT OF SHEET PILE WALLS, AND RISE OF FREIGHT TUNNEL (See Fig. 25(b))

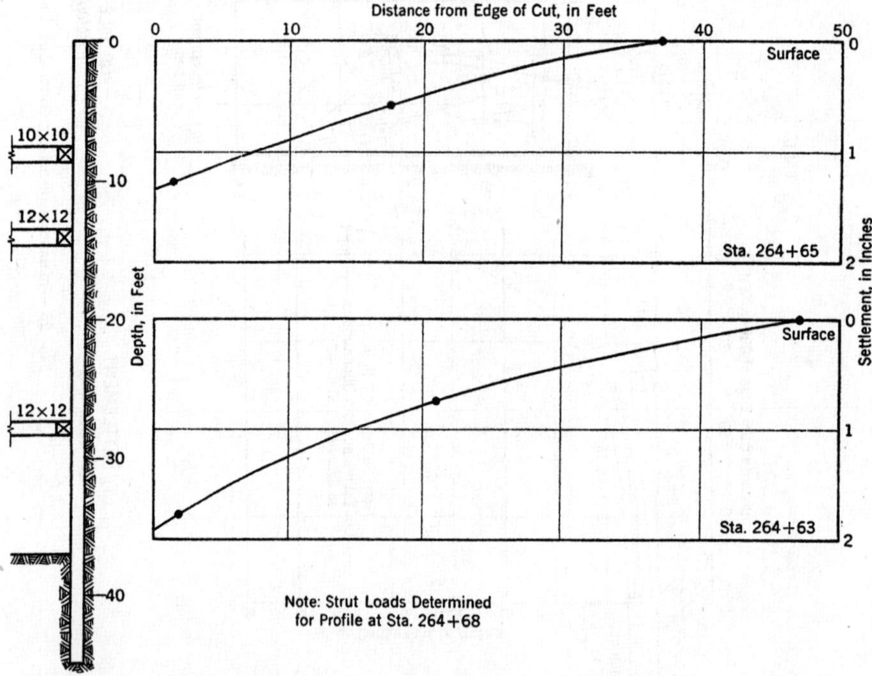

FIG. 27.—SETTLEMENT ADJACENT TO OPEN CUT, CONTRACT S-8A, FOUR MONTHS AFTER COMPLETION

cant compared to the total lateral pressure, the distance from each open cut to the nearest structure is recorded in Table 2.

The presence of the two-story frame building immediately adjacent to the cut on contract S-8A permitted accurate settlement observations very close to the excavation. The results are shown in Fig. 27.

TABLE 2.—DISTANCE FROM EACH OPEN CUT TO THE NEAREST STRUCTURE

Cut (see Fig. 1)	Distance (ft)	Structure	
		Stories	Type
S-3	50	5	Brick[a]
S-4B	9	3	Brick[b]
S-9C	50	3	Brick
S-8A	0	2	Wood frame
D-6E	40	2	Brick
S-1A	26	2	Brick

[a] On 60-ft piles. [b] 7,000 lb per foot of wall.

PART II.—THEORY OF TOTAL LOADS

In order to generalize the results of the earth-pressure measurements, it is desirable to determine a theoretical basis upon which the results can be compared. The so-called classical earth-pressure theories have been developed for many years to include homogeneous materials whose shearing resistance is made up of both cohesion and internal friction. Although the Chicago soils in which the cuts were made were predominantly plastic clay, the presence of an overlying sand layer required modification of the existing theories. Furthermore, the influence of progressive failure in reducing the average strength of cohesive deposits, and the real angle of internal friction exhibited by large masses of clay in the field, represented unknown factors which could not be determined by laboratory tests. In this section, the theoretical expressions for total lateral earth pressure are modified to include the effect of the sand layer, and are written in such a form as to permit the determination, from the measured pressures and soil properties, of the effect of progressive failure, and of the angle of internal friction which was actually effective.

1. Rankine's Earth-Pressure Theory.—In a normally consolidated clay deposit with a horizontal surface, the intensity of vertical stress on a horizontal section is given by the expression

$$\sigma_v = \gamma H_z \qquad (5a)$$

in which γ = density of the overlying soil mass; and H_z = vertical distance from the ground surface to the horizontal section. In an undisturbed soil mass, the intensity of horizontal pressure on a vertical plane is given by the expression

$$\sigma_h = K_0 \gamma H_z \qquad (5b)$$

in which K_0 = coefficient of natural earth pressure, or earth pressure at rest. The value of K_0 for clays in their natural state is not yet known.

If the bed of clay is given an opportunity for lateral expansion to a very great depth until a state of plastic equilibrium is reached, the lateral intensity of pressure decreases to the smallest value compatible with equilibrium. Its value, known as the active earth pressure, is

$$\sigma_A = \gamma H_z \left[\tan^2 \left(45° - \frac{\phi}{2} \right) - \frac{2 c}{\gamma H_z} \tan \left(45° - \frac{\phi}{2} \right) \right] \ldots\ldots (6)$$

in which ϕ = effective angle of internal friction of the clay; and c = cohesion of the clay. The total lateral pressure from the surface to depth $H_z = H$ is equal to

$$P_A = \int_0^H \sigma_A \, dH_z = \tfrac{1}{2} \gamma H^2 \tan^2 \left(45° - \frac{\phi}{2} \right) - 2 c H \tan \left(45° - \frac{\phi}{2} \right) \ldots (7)$$

Eqs. 6 and 7 were probably first derived by Résal (1910) as an extension of Rankine's theory. A derivation in English was given by the late William Cain, M. Am. Soc. C. E., in 1916.[10] Eq. 7 may be expressed in another useful form: The unconfined compressive strength of a clay sample, q, may be expressed in terms of the cohesion and effective angle of internal friction by the equation

$$q = \frac{2 c}{\tan \left(45° - \frac{\phi}{2} \right)} \ldots\ldots (8)$$

Substituting Eq. 8 in Eq. 7, the expression

$$P_A = \tfrac{1}{2} \gamma H^2 \left(1 - \frac{2 q}{\gamma H} \right) \tan^2 \left(45° - \frac{\phi}{2} \right) \ldots\ldots (9)$$

is obtained.

The resultant earth pressure given by Eqs. 7 and 9 is horizontal, and its point of application is below the lower third-point. For $H = \frac{2 q}{\gamma}$, it is at infinite depth below the bottom of the cut. The magnitude of the resultant earth pressure is often compared with that of a fluid of equal unit weight, which exerts a total lateral pressure

$$P_F = \tfrac{1}{2} \gamma H^2 \ldots\ldots (10)$$

The ratio of P_A to P_F is known as the hydrostatic pressure ratio for active earth pressure, and is denoted by K_A. Its value is given by either of the equations

$$K_A = \tan^2 \left(45° - \frac{\phi}{2} \right) - \frac{4 c}{\gamma H} \tan \left(45° - \frac{\phi}{2} \right) \ldots\ldots (11a)$$

or

$$K_A = \left(1 - \frac{2 q}{\gamma H} \right) \tan^2 \left(45° - \frac{\phi}{2} \right) \ldots\ldots (11b)$$

In actual experience, opportunity for lateral expansion sufficient to develop a state of plastic equilibrium throughout the soil mass is never realized. Such a movement would result in the full development of shearing resistance on two

[10] "Earth Pressure, Retaining Walls and Bins," by William Cain, New York, N. Y., 1916, p. 186.

intersecting sets of sliding planes, each of which makes an angle of $45° - \frac{\phi}{2}$ with the vertical as indicated in Fig. 28. Actually the expansion is restricted to a wedge-shaped zone extending behind the wall supporting the earth mass, and having a depth limited approximately to the depth H of the wall. If sufficient movement of the wall is permitted to develop the shearing resistance along the plane which intersects the wall at depth H, the validity of Eqs. 11 is not impaired, provided that two other conditions are also fulfilled. A bond

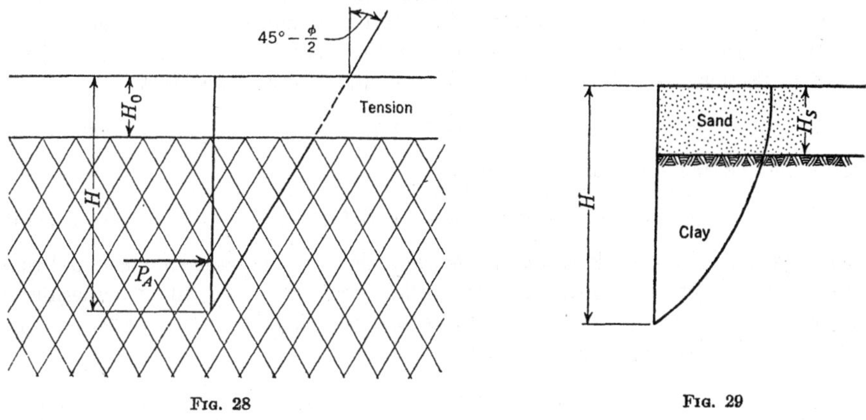

Fig. 28

Fig. 29

must exist between the lateral support and the clay, capable of resisting tension to a depth, according to Eq. 6, of

$$H_0 = \frac{q}{\gamma} \qquad (12)$$

and the deformation conditions must be such that the point of application of the resultant pressure is identical with the theoretical location. In an actual open cut, these latter two conditions are never satisfied, because bracing is always inserted near the ground surface, and lateral expansion is greatly restricted. The zone of tension in the upper part of the deposit no longer can exist, the center of pressure rises, and Eq. 6 is no longer applicable. The sliding surface becomes curved, and the resultant earth pressure is not horizontal. Ignorance of these factors can be expected to introduce errors into the computed values of the total pressure. The theory is adequate to serve, nevertheless, as the starting point in the analysis of the experiment results.

Eqs. 11 refer strictly to a homogeneous clay mass possessing everywhere an unconfined compressive strength q and an angle of internal friction ϕ. Natural clay deposits are not homogeneous, and the unconfined compressive strength is variable. To use equations containing the quantity q, it is then necessary to substitute in place of it the arithmetic average of the compressive strength of the clay, denoted by q_a. In the case of the Chicago clays, another complication is introduced by the presence of a deposit of cohesionless material in the upper portion of the soil mass. For computation, it is convenient to transform the cohesionless layer into an equivalent layer of clay possessing the same shearing resistance along the potential sliding surface as the existing sand, but having an angle of internal friction equal to that of the actual clay deposit.

The shearing resistance of the sand along the portion of the sliding wedge located above the clay is equal to

$$S_s = \sigma \tan \phi_1 \qquad (13)$$

in which: σ = the average normal stress on the sand portion of the sliding surface; and ϕ_1 = angle of internal friction of the sand. The average normal stress (neglecting the inclination of the shearing surface which actually is almost vertical, as indicated in Fig. 29) is equal approximately to

$$\sigma = K_s \times \tfrac{1}{2} \gamma_1 H_s \qquad (14a)$$

and the average shearing resistance

$$S_s = K_s \times \tfrac{1}{2} \gamma_1 H_s \tan \phi_1 \qquad (14b)$$

In Eqs. 14: K_s = hydrostatic pressure ratio for the cohesionless material; γ_1 = density of the cohesionless material; and H_s = depth of cohesionless layer. The shearing resistance of the substitute clay layer must also be equal to S_s and is given by the expression

$$S_s = c_s + \sigma \tan \phi \qquad (15)$$

in which: c_s = cohesion of the substitute clay layer; and ϕ = angle of internal friction of the clay deposit. Substituting Eq. 14*b* for S_s in Eq. 15 and solving for the cohesion, it is found that

$$c_s = \tfrac{1}{2} K_s \gamma_1 H_s (\tan \phi_1 - \tan \phi) \qquad (16a)$$

The unconfined compressive strength of the substitute clay layer would then be, from Eq. 8,

$$q_s = \frac{2c}{\tan\left(45° - \frac{\phi}{2}\right)} = \frac{1}{\tan\left(45° - \frac{\phi}{2}\right)} \times K_s \gamma_1 H_s (\tan \phi_1 - \tan \phi) \qquad (16b)$$

The average unconfined compressive strength of the entire deposit from the surface to depth H would become

$$q_a = \frac{1}{H}\left[\frac{K_s \gamma_1 H_s^2}{\tan\left(45° - \frac{\phi}{2}\right)} (\tan \phi_1 - \tan \phi) + (H - H_s)\, q_c\right] \qquad (17)$$

Eqs. 16*a* and 17 contain the quantity K_s, which can only be estimated, by a procedure to be described subsequently. They afford only estimates of the influence of the sand layer, rather than exact evaluations.

One further soil property must be considered in the adaptation of the theory to an actual soil mass. The value of the average unconfined compressive strength used in the computations is based upon tests of individual samples as representative of the clay deposit as possible. For each sample, the maximum strength is recorded as the unconfined compressive strength. For most natural clays, the maximum strength is attained at relatively small strain, and the compressive strength decreases rapidly to a constant value which persists throughout the additional strain, as shown in Fig. 30. Since some of the layers in a natural deposit are stiffer than others, at a given deformation the clay at certain points along the sliding surface may have attained its maximum resistance and experienced the decrease in strength, whereas at other points the maximum may not have been reached. As a result, the effective shearing resistance along the surface is always less than the average computed from the maximum strengths of individual samples. When sufficient deformation has taken place to reach the constant portion of the stress-strain curve for each component of the soil, it is apparent that the average resistance is only a fraction of that determined from the tests. The ratio between the shearing resistance actually developed along the sliding surface, and that determined by averaging the separate shearing resistances of the individual samples is denoted by n. Eq. 17 can be modified to take account of this phenomenon (known as progressive failure) by replacing q_c by $n\,q'_c$, in which q'_c is the average unconfined compressive strength of practically undisturbed clay samples as determined in the laboratory. No reduction is experienced in the sand layer because the stress-strain curve for loose sands, as exist over the Chicago clay deposits, does not exhibit a peak before attaining a constant shearing resistance as deformation proceeds. The final form of Eq. 17 is therefore

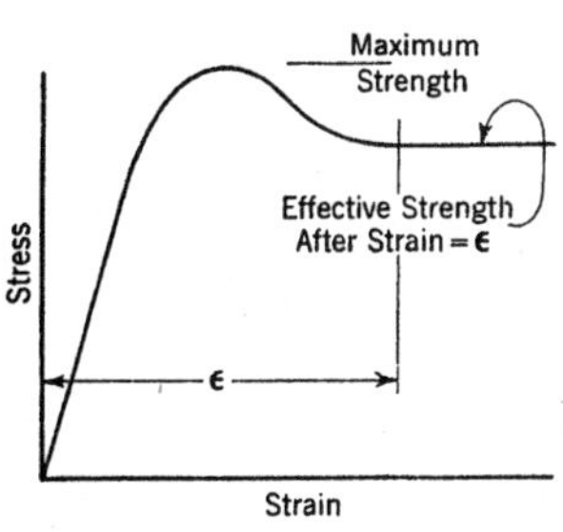

Fig. 30.—Stress-Strain Curve

$$q_a = \frac{1}{H}\left[\frac{K_s\,\gamma_1\,H^2_s}{\tan\left(45° - \frac{\phi}{2}\right)}(\tan\phi_1 - \tan\phi) + (H - H_s)\,n\,q'_c\right] \quad\text{..(18a)}$$

and the theoretical relation Eq. 11*b* becomes

$$K_A = \left(1 - \frac{2\,q_a}{\gamma_a\,H}\right)\tan^2\left(45° - \frac{\phi}{2}\right) \quad\text{............(18b)}$$

in which the average density, γ_a, is determined by the equation

$$\gamma_a = \frac{1}{H}[H_s\,\gamma_1 + (H - H_s)\,\gamma_c] \quad\text{................(19)}$$

γ_c being the average density of the clay.

It is important to note that Eq. 18*b* represents a linear relationship between K_A and $\frac{2\,q_a}{\gamma_a\,H}$, as shown in Fig. 31. Furthermore, the intercept of the straight line on the K_A-axis must be equal to $\tan^2\left(45° - \frac{\phi}{2}\right)$, and the intercept on the

$\frac{2\,q_a}{\gamma_a\,H}$-axis must be equal to unity. These facts are of the greatest importance, because they enable the values of ϕ and n for the clay deposit to be determined from the experimental values of earth pressure. The procedure is described in Part III.

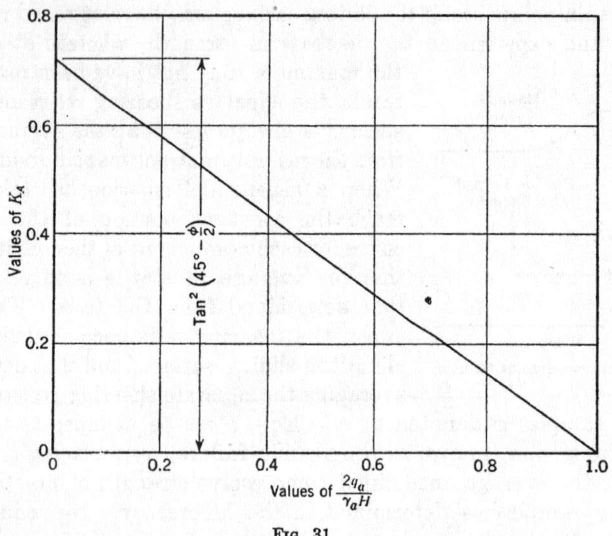

FIG. 31

2. General Wedge Theory for Earth Pressures.—When an open cut is braced with struts, the deformation at the top of the cut is greatly restricted and the conditions for tensile stresses in the upper part of the soil mass are no longer satisfied. Pressure exists everywhere within the wedge, and the center of pressure is correspondingly raised. The sliding surface is no longer plane, but takes the shape of a curve which intersects the surface at a right angle. The resultant earth pressure is not horizontal, in general, but is inclined to the horizontal at an angle θ.

The total pressure against the sheeting under these circumstances may be computed by consideration of the equilibrium of the failure wedge, under the assumption that the shape of the curved boundary is that of a logarithmic spiral. This curve was chosen for its close approach to observed conditions and its simplicity, because the resultant of the normal and shearing forces on the curved surface passes through the center of the spiral and may be eliminated from computations if this point is taken as the center of moments. For the special case when $\phi = 0$, the logarithmic spiral becomes a circle with its center on a horizontal line at the elevation of the ground surface. The forces considered are shown in Fig. 32. They include W, the weight of the wedge; P_A, the resultant earth pressure; S, the total shearing resistance along the curved sliding surface; and F, the resultant normal force on the curved surface. Denoting the lever arms of these forces with respect to the center of rotation

O by the symbol l, the value of P_A is found to be

$$P_A = \frac{W\, l_w - S\, l_c}{l_r} \quad \text{. . . . . (20)}$$

The total shearing resistance S is equal to the product of one half the average unconfined compressive strength and the length L of the curved surface. The weight W acts at the center of gravity of the wedge. Several positions are

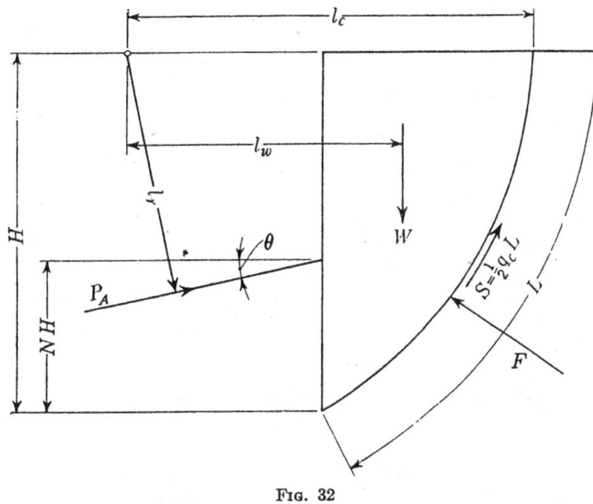

FIG. 32

assumed for the center of rotation along the horizontal line, and the maximum value of P_A is taken as the active earth pressure. The value of the hydrostatic pressure ratio is defined as the ratio of the horizontal component of P_A to the lateral pressure exerted by the fluid of equal density. Thus

$$K_A = \frac{P_A \cos \theta}{\frac{1}{2}\, \gamma_a\, H^2} \quad \text{. . . . . (21)}$$

Values of K_A have been computed for different values of the inclination of the resultant earth pressure, for different positions of the center of pressure, and for different values of cohesion. The results are shown in graphical form in Fig. 33. Each of these charts pertains to a particular value of the inclination of the resultant pressure. Values of K_A may be determined for any location of the center of pressure, and for any value of cohesion, expressed in terms of the quantity $\frac{H\,\gamma}{2\,q}$. It should be recalled that Fig. 33 is constructed on the assumption that $\phi = 0°$. The values of K_A determined by Rankine's theory are also shown in Fig. 33. They appear as horizontal lines, inasmuch as the values are independent of the location of the center of pressure. For either of the values of θ it is observed that a particular value of the position of the center of pressure gives results by the general wedge theory which are practically identical to those given by Rankine's theory. For values of $\theta = 0°$, and 20°, the respective centers of pressure for this condition are approximately at 0.37 H and 0.45 H. Similar charts have also been prepared for other values of ϕ, in which case the cohesion of the clay is determined by Eq. 8. However, the

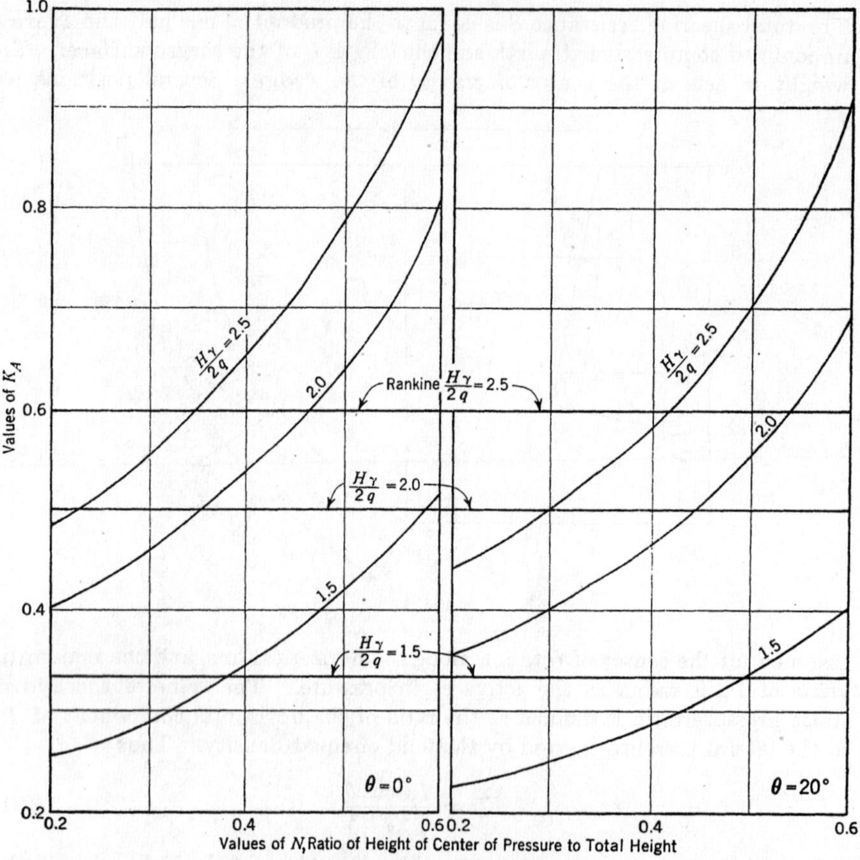

FIG. 33.—RESULTS OF EARTH PRESSURE COMPUTATIONS; GENERAL WEDGE THEORY ($\phi = 0°$)

agreement with the field data is more satisfactory in the case of Fig. 33, and the other charts are not shown.

PART III.—COMPARISON OF MEASURED AND THEORETICAL TOTAL LOADS

The essential data regarding the open cuts are shown in Figs. 19 to 27 and Table 1. The results are summarized in tabular form in Table 3. In Table 3(*a*), the symbol H indicates the total depth of the cut; H_s the thickness of the overlying sand layer, q'_c the average unconfined compressive strength of practically undisturbed clay samples between the clay surface and the bottom of the cut; K_s the hydrostatic pressure ratio for the sand; P_l the lateral pressure of a fluid having a density equal to γ_a; and L_p the horizontal distance between vertical profiles of struts. In Table 3(*b*), the symbols Q_1, Q_2 . . . indicate the

TABLE 3.—COMPARISON OF MEASURED AND

Cut (see Fig. 1)	Station	(a) TOTAL LATERAL PRESSURE												
		H	H_s	$H - H_s$	q'_c	K_s	γ_c	γ_a	P_l	Q	L_p	P_A	K_A	$\frac{2\,q_a}{\gamma_a H}$
		(1)[a]	(2)[a]	(3)[a]	(4)[b]	(5)	(6)[c]	(7)[c]	(8)[d]	(9)[d]	(10)[a]	(11)[e]	(12)	(13)
S9C	....	44	9	35	3,280	1.0	130	126	124	331	12.5	26.5	0.214	0.748
S8A	....	37	9	28	2,950	1.0	130	125	86	258	12	21.5	0.250	0.783
D6E	213+85	35	8	27	2,270	1.0	126	123	75	259	10	25.9	0.346	0.664
D6E	215+20	35	8	27	2,270	1.0	126	123	75	220	10	22.0	0.293	0.664
S4B	....	23	12	11	2,050	0.5	122	116	31	73	8	9.13	0.295	0.701
S4B	....	30	12	18	1,730	0.5	124	118	53	181	8	22.7	0.428	0.526
S3	Average	33	16	17	1,170	1.0	114	112	61	558	18	31.0	0.508	0.511
S3	Average	39	16	23	1,310	1.0	118	115	87	750	18	41.7	0.478	0.443
S1A	Average	36.5	15	21.5	1,320	1.0	118	115	76	626	15	41.7	0.547	0.464
S1A	Average	36.5	15	21.5	1,320	1.0	118	115	76	586	15	39.1	0.515	0.464
S1A	58+21	25	15	10	1,470	1.0	114	112	35	140	15	9.33	0.267	0.722
S1A	58+21	27	15	12	1,420	1.0	114	112	41	264	15	17.6	0.429	0.661
S1A	58+21	34.5	15	19.5	1,270	1.0	117	114	68	448	15	29.9	0.433	0.484
S1A	58+21	36.5	15	21.5	1,320	1.0	118	115	76	550	15	36.6	0.482	0.464

Units of Table (a) are: [a] feet; [b] pounds per square foot; [c] pounds per cubic foot; [d] kips = 1,000 lb; [e] kips per foot.

measured load in individual struts, and H_1, H_2 . . . the vertical distance between the surface and the strut.

Table 3 also contains the total measured lateral pressure Q per lineal foot of the cut, and the ratio N defined as the distance from the point of application of the lateral pressure to the bottom of the cut divided by the total depth of the cut. To determine these quantities, it is necessary to know the bottom reaction, or the shear in the vertical members of the sheeting at depth H. Whether or not the value of this quantity is important enough to warrant inclusion in the total load Q depends upon the nature of the different cuts. In contract S-1A (see Fig. 1) the greatest deflection of the sheeting occurred at the bottom of the excavation. This type of deflection corresponds to a pressure distribution having small values at the bottom, and the bottom reaction is likely to be small.[11, 12, 13] In contract S-3, the sheeting on one side rested on top of an existing freight tunnel of unreinforced concrete which was capable of supporting an inconsequential lateral load after exposure. Similarly in contract S-4B the results of the strut loads indicate plainly that the intensity of lateral pressure decreased from the midheight to the bottom of the cut and that the major part of this pressure has been taken by the struts. On the contrary, the vertical H-piles on contract D-6E penetrated a stiff layer below the bottom of the cut, and the large clearance between the lower strut and the bottom of the cut induced a deflection curve with a maximum value at some distance above the bottom of the cut. As a result, the bottom reaction was considered likely to be of considerable importance, and it was measured by the

[11] "Distribution of the Lateral Pressure of Sand on the Timbering of Cuts," by Karl Terzaghi, *Proceedings*, International Conference on Soil Mechanics, Vol. I, Cambridge, Mass., 1936.

[12] "A Fundamental Fallacy in Earth Pressure Computations," by Karl Terzaghi, *Journal*, Boston Soc. of Civ. Engrs., April, 1936, p. 78.

[13] "General Wedge Theory of Earth Pressure," by Karl Terzaghi, *Transactions*, Am. Soc. C. E., Vol. 106, 1941, pp. 68, 94.

THEORETICAL TOTAL LOADS IN OPEN CUTS

(b) INDIVIDUAL STRUT LOADS

H	Strut 1		Strut 2		Strut 3		Strut 4		Strut 5		N	Q
	H_1	Q_1	H_2	Q_2	H_3	Q_3	H_4	Q_4	H_5	Q_5		
(14)	(15)	(16)	(17)	(18)	(19)	(20)	(21)	(22)	(23)	(24)	(25)	(26)
44	5.4	35	12.6	58	19.5	82	25.8	91	33.0	65	0.52	331
37	7.8	25	12.2	109	28.1	124	....	....	....	....	0.48	258
35	5.5	9	13.6	167	33.2	(83)[f]	....	....	....	....	0.44	259
35	1.5	1	13.5	137	34	(82)[f]	....	....	....	....	0.40	220
23	0	0	11	25	18.6	48	....	....	....	....	0.30	73
30	0	0	11	43	18.6	88	27.4	50	....	....	0.36	181
33	4	36	14.5	260	23.5	173	31.5	89	....	....	0.42	558
39	4	36	14.5	232	23.5	209	31.5	273	....	....	0.42	750
36.5	4	8	13	381	....	....	31	237	....	....	0.46	626
36.5	4	2	13	251	23	203	31	130	....	....	0.44	586
25	4	0	13	140	....	....	....	....	....	....	0.48	140
27	4	0	13	205	23	59	....	....	....	....	0.45	264
34.5	4	0	13	230	23	178	31	40	....	....	0.46	448
36.5	4	0	13	196	23	204	31	150	....	....	0.40	550

[f] Values in parentheses obtained by burning H-pile free at bottom of cut.

method indicated in Fig. 23. For cuts S-8A and S-9C, it was concluded that the bottom reaction was not important, although it is possible that in S-9C a small reaction might have been required to produce the state of stress measured in the bracing. As a result, the cut on contract D-6E was the only one in which the bottom reaction was taken into consideration in the computations.

To compute the real value of K_A, the average density of the soil throughout the height H must be determined by Eq. 19. For this computation, the density of the sand above the clay surface was estimated at 110 lb per cu ft. The density of the clay was computed from the average natural water content and a specific gravity of the solid matter assumed equal to 2.74. The values of γ_c and γ_a are given in Table 3(*a*).

The theoretical equation (Eq. 18*a*) for the average equivalent unconfined compressive strength of the deposit contains two values, K_s and tan ϕ_1, which can only be estimated. Since the sand above the clay is usually loose and somewhat silty, it is likely that ϕ_1 varies between the limits of 30° and 35°. It is most likely closer to 30°, but both values have been used in subsequent calculations. The value of K_s depends upon the lateral strain that develops in the sand during excavation. It must be selected for each cut individually, because of its dependence upon the method of excavation and bracing. As an example of the method of evaluation, cut S-9C is used. In the vicinity of the cut, the depth of sand was 9 ft. For $K_s = 1$, the earth pressure exerted by the sand is $\frac{1}{2} \times 110 \times 9^2 = 4{,}450$ lb per ft = 4.45 kips per ft. The top row of struts was located at a depth of 5.4 ft, and the next at 12.6 ft below the surface. Thus the top row received almost the full lateral pressure of the sand. The greatest load in this strut was 35 kips, or 2.80 kips per ft, corresponding to a hydrostatic pressure ratio $K_s = 0.63$. However, while excavation proceeds, the vertical H-beams pivot about the second row of struts, and by virtue of the stiffness of the H-beams the strut load is relieved. As a result, the pressure actually exerted by the sand is greater than that indicated by the strut load, and a value of $K_s = 1.0$ was estimated. Conditions were similar for all other cuts except contract S-4B, where the top strut was inoperative and the pressure of the sand was entirely carried by bending in the upper part of the H-beams. Therefore, much greater opportunity for lateral expansion of the sand existed in S-4B than in the other cuts, and a value of $K_s = 0.5$ was estimated correspondingly. An error of 25% in the estimated values of K_s would result in an error in the computed shearing resistance of less than 2% on contracts S-9C, S-8A, and S-4B, less than 10% on D-6E, and approximately 15% on S-1A and S-3.

In Table 3, the results of measurements on contract S-4B after the cut exceeded a depth of 30 ft are not included because measurements were not made on all profiles simultaneously, and certain strut loads would have to be determined by interpolation. This operation is not considered reliable enough in comparison with the other data. On cuts subsequently measured, care was taken to obtain readings on all struts on the same day when possible.

Table 3(*a*), therefore, contains the experimental values of K_A, and all of the quantities necessary to compute the value of q_a in Eq. 18(*a*), except the unknown properties of the clay, ϕ and n. The most probable values of ϕ and n are those which produce the most satisfactory agreement with the theoretical straight-line relationship shown in Fig. 31, provided a straight-line relationship is found to exist. In Figs. 34(*a*), 34(*b*), and 34(*c*), the experimental values of K_A are plotted against values of $\frac{2\,q_a}{\gamma_a\,H}$, in which q_a has been computed by Eq. 18*a* on the assumption that $\phi = 17°$. Values of n equal to 0.50, 0.75, and 1.0 were also assumed. A straight-line relationship is seen to exist, indicating that the reasoning leading to Eq. 18*b* is satisfactory. However, in none of these cases do the plotted test results indicate a satisfactory agreement with the theoretical relationship for $\phi = 17°$, shown by the heavy line. It appears evident that the value of ϕ must be appreciably less than 17°. In Figs. 34(*d*), 34(*e*), and 34(*f*), a similar series of plots has been made, in which q_a was determined on the assumption that $\phi = 0°$. In this case, it is observed that the experimental data are in excellent agreement with the theory when $n = 0.75$. It was found by similar computations that no other combination of ϕ and n besides $\phi = 0°$ and $n = 0.75$ gave a reasonable fit to the theoretical relationship. In the charts of Fig. 34, the value of ϕ_1 was taken as 30°. Repetition of the computations with a value of $\phi_1 = 35°$ led to almost identical results. The solution of Eq. 18*b*, therefore, results in the conclusions that, for all practical purposes, in the Chicago subway cuts, the value of ϕ for the clay is equal to 0°, and the factor $n = 0.75$.

The correctness of the values for ϕ and n is influenced by the fact that the upper end of the line is largely determined by points representing contracts S-3 and S-1A, in which an error in the value of K_s has an important effect. Nevertheless, if the position of the line is determined from only those profiles in which a 25% error in K_s involves a negligible error in the result, the value of ϕ cannot possibly exceed 3°, and the value of n is not significantly altered. It

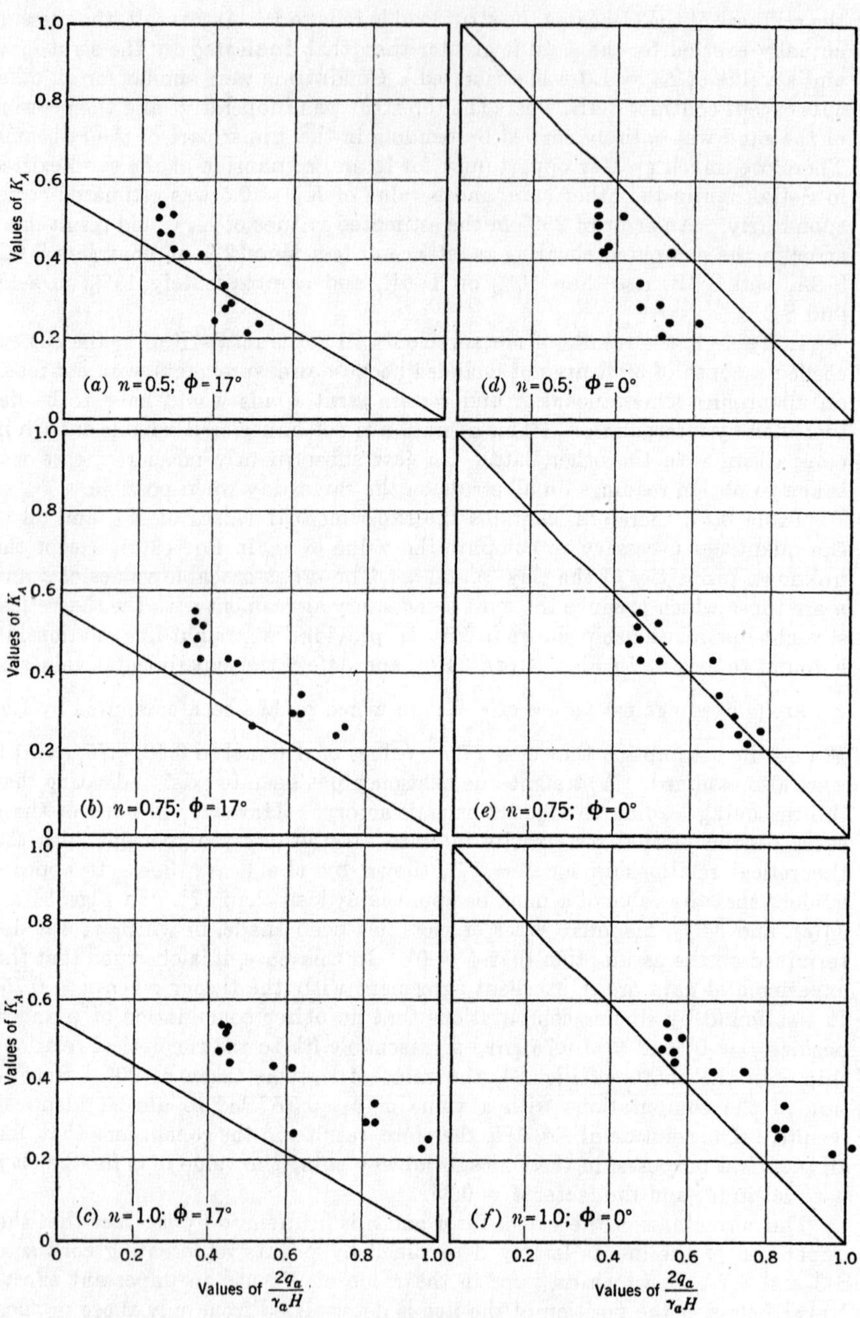
Values of K_A
(a) $n=0.5$; $\phi=17°$
(b) $n=0.75$; $\phi=17°$
(c) $n=1.0$; $\phi=17°$
(d) $n=0.5$; $\phi=0°$
(e) $n=0.75$; $\phi=0°$
(f) $n=1.0$; $\phi=0°$
Values of $\frac{2q_a}{\gamma_a H}$
0 0.2 0.4 0.6 0.8 1.0

Fig. 34

is apparently safe to conclude, therefore, that the mass of clay involved in the deformation accompanying the construction of the Chicago cuts behaved as a purely cohesive mass, without internal friction.

The incomplete development of the shearing resistance, as evidenced by the value $n = 0.75$, might be considered due to the fact that insufficient lateral expansion had taken place to develop the full resistance, and that therefore the measured earth pressure was not the smallest possible value (active earth pressure), but some value between active earth pressure and earth pressure at rest. This interpretation of n, instead of progressive failure, is considered unlikely. The cuts upon which measurements were made were built by different contractors using different systems of lateral support. As long as the shearing resistance of the clay is incompletely developed, a slight change in the amount of lateral yield should have a marked effect on the value of the earth pressure. Nevertheless, the points in Fig. 34 which represent the hydrostatic pressure ratio are all located close to a straight line, indicating that the ratio between active and maximum shearing resistance was consistently equal to 0.75. This fact is in sharp disagreement with the insufficient expansion hypothesis. It appears therefore that the amount of inevitable deformation associated with the construction of such structures is sufficient to reduce the earth pressure to its smallest value, and therefore every effort should be made to keep the lateral yield, which results in settlement, as low as possible.

A study of Table 1, for profiles 15–29 in contract S-1A, indicates the variation from the mean value of the total lateral pressure. These fifteen consecutive profiles were identical except for the ordinary variations in construction procedure. The greatest deviation from the mean was approximately $\pm$ 20%. The physical causes of this scattering are probably numerous, but in any case, provision in design should be made to take it into account.

It is of interest and importance to compare the measured results of the total lateral pressure with the values computed by the general wedge theory, which takes into account the curvature of the sliding surface, the location of the center of pressure, and the inclination of the resultant earth pressure. Inasmuch as no direct measurements were made of the vertical load carried by the sheeting, the value of θ in the experiments in unknown, and only a general comparison is possible. It is observed that the average center of pressure for all the measurements was at 0.43 H. This value, according to the curves of Fig. 33, gives results by the general wedge theory when $\theta = 20°$ which are on the average nearly equal to those given by Rankine's theory. It should be emphasized, however, that the center of pressure in Rankine's theory can never be as high as the lower third point, and for small values of $\frac{H\gamma}{2q}$ may even be below the bottom of the cut. The fact that the total lateral pressure is approximately equal to the Rankine value, even when the distribution of the pressure is entirely different, is not altogether surprising after a study of Fig. 33, from which it can be ascertained that the effect of the curvature of the sliding surface is to increase the total lateral pressure, whereas the effect of the vertical component of the pressure is to reduce it. Actually, the observed total pressures were found to fit the general wedge theory for $\theta = 20°$ slightly better than the plane wedge theory, represented by Eq. 18*b*.

Part IV.—Distribution of Pressure

It has been mentioned in Part II that if the shearing resistance is fully developed along the sliding surface forming one of the boundaries of the failure wedge, the total earth pressure can be computed by considering the equilibrium of the wedge. This procedure, however, does not produce any information concerning the distribution of the lateral pressure. The position of the center of pressure depends upon the deformation conditions within the wedge itself. Professor Terzaghi[14] has stated the conditions necessary for various types of pressure distributions. He has shown that if the supporting wall does not yield in a manner approaching rotation about the base of the cut, through a sufficiently large angle to reduce all of the material in the wedge to a state of plastic equilibrium, then the center of pressure will be higher than that predicted by the theory of plasticity. In particular, if yielding is restricted at the top of the cut by the presence of struts, the center of pressure may be expected to rise.

Theoretical considerations do not permit the calculation of the location of the center of pressure, or of the loads carried by individual struts, because the deformation conditions are a function of the manner in which the excavation and bracing are performed. Inspection of Fig. 25, showing the measured deformations accompanying the excavation of one Chicago subway cut, demonstrates that the type of yielding is of the nature to produce a high center of pressure, but quantitative information about the distribution can only be obtained from measurements in the field. An important fact, however, is disclosed by Fig. 25. Almost all of the deformation at a given elevation occurs before excavation reaches that depth. This yielding obviously cannot be prevented entirely by the use of struts. This observation leads to the conclusion that minor variations in construction procedure cannot greatly influence the nature of the deformation, and thus may not have a pronounced effect on the location of the center of pressure. It also leads to the conclusion that most of the deformation occurs before the placement of wales and struts, and that it is unlikely that important arching in the horizontal direction will develop.

Inspection of Table 3(*b*) shows that the measured centers of pressure for the various open cuts were quite consistent, and were in general slightly below midheight of the cuts. In the cut S-4B the highest strut that was acted upon by earth pressure was 11 ft below the surface, and the next one at a depth of 19 ft. Hence the center of pressure while the cut was at shallow depths was necessarily low.

Table 1 gives the individual strut loads measured on contract S-1A. The deviation of the center of pressure from its average value is less than 15%. However, the variation in individual strut loads is very much greater. That this is due to minor details of construction was demonstrated in many instances such as the following:

[14] "A Fundamental Fallacy in Earth Pressure Computations," by K. Terzaghi, *Journal*, Boston Soc. of Civ. Engrs., April, 1936, p. 78.

On contract S-1A, excavation reached a depth of 24 ft in the bay between profiles 11 and 12 (see Table 1), and a strut at the 23-ft level was installed in each profile on the same day. In profile 11, excavation on the next day proceeded to a depth of 33 ft and a strut was placed immediately at the 31-ft level. In profile 12, excavation did not proceed from the 24 to the 33-ft level until the third day, at which time the entire bay to profile 13 was mined to 33 ft. The lower strut in profile 12 was not placed until 8 hr after the 33-ft depth was reached. As a result, the lower strut in profile 11 attained a final load of 232 kips, whereas the corresponding strut in profile 12 never exceeded 89 kips. Nevertheless, the center of pressure was at 0.43 H in both profiles.

It is apparent, therefore, that it is useless to attempt to compute the real distribution of lateral pressure over the sheeting. Of far greater practical importance is the statistical investigation of the variation in strut loads actually measured, in order to determine the maximum loads that may be expected under ordinary construction procedures. A useful method of investigation is by means of an equivalent beam, as demonstrated by Professor Terzaghi[15] in analyzing the results of measurements on the Berlin subway. The method is shown in Fig. 35. The vertical members of the sheeting are

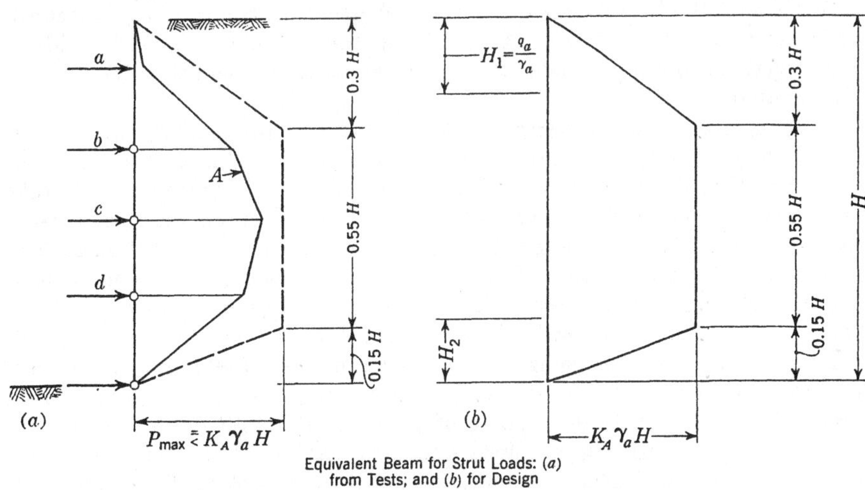

Equivalent Beam for Strut Loads: (a) from Tests; and (b) for Design

Fig. 35

assumed to be hinged at each strut except the uppermost one, and a hinge is assumed to exist at the bottom of the cut. The abscissas of the pressure diagram "A" represent the intensity of horizontal pressure required to produce the measured strut loads. A study of such diagrams for all of the measured profiles disclosed that the maximum abscissa never exceeded the value $K_A \gamma_a H$. Every measured set of strut loads resulted in a different pressure diagram "A," all of which were found to lie within the boundaries of the trapezoid indicated by the dotted lines. Thus, if strut loads are computed on the basis

[15] "General Wedge Theory of Earth Pressure," by Karl Terzaghi, *Transactions*, Am. Soc. C. E., Vol. 106 (1941), p. 93.

of this trapezoid, they will most probably be on the safe side. The area of the trapezoid is approximately 50% in excess of the computed total lateral pressure, which is equal to

$$P_A = K_A \times \tfrac{1}{2} \gamma_a H^2 \ldots\ldots\ldots\ldots\ldots\ldots\ldots\ldots(22)$$

per unit of length of the cut. The excess takes care of the statistical deviation of individual strut loads from the average.

PART V.—SETTLEMENT CONSIDERATIONS

The settlement of the ground surface is a direct result of the lateral squeeze of the clay toward the cut. Although construction conditions made it impossible to determine accurately the complete record of soil movements at the sides and bottom of the cut at the same profiles where complete settlement cross sections were available, a study of the partly incomplete results at numerous profiles established the approximate equivalence of the settlement volume and the volume of lateral squeeze. The most complete information at any of the measured profiles of movements of the sides and bottom of a cut is assembled in Figs. 25 and 26. In this particular cut, the presence of an existing freight tunnel, located as shown in Fig. 25, permitted measurements to be made of the lateral movement of the sheeting at the level of the tunnel while excavation had not yet reached the bottom of the cut. It is evident that the lateral squeeze extended for a considerable distance below the bottom of excavation.

The movements encountered at this profile are particularly interesting because they provide information from which can be deduced the maximum excavation that should be permitted before placement of the uppermost strut. On January 3, 1941, the earth pressure on the sheet piles was practically zero, because the trench, dug to permit driving the sheet piles, was not yet backfilled. In spite of this fact, the tops of the piling moved toward the cut more than 1 in. because of the unbalanced weight of the earth outside of the excavation. Most of this movement occurred while the depth of excavation increased from 10 to 13 ft, corresponding to an increase of the unbalanced part of the load from 1,100 to 1,440 lb per sq ft. The average unconfined compressive strength of the clay between a depth of 13 ft and the bottom of the cut was 1,340 lb per sq ft. Hence, the important deformation of the sheeting began when the unbalanced load became approximately equal to the unconfined compressive strength of the underlying clay. The disturbance of the clay was deep-seated, as evidenced by the appearance of horizontal cracks in the side walls of the freight tunnel when excavation reached a depth of approximately 8 ft. It seems possible that the abrupt and extensive yielding represents the spontaneous change in the clay from its relatively rigid undisturbed state to a plastic state. Installation of the vertical strut before reaching a depth H_1 equal to

$$H_1 = \frac{q_c}{\gamma_a} \ldots\ldots\ldots\ldots\ldots\ldots\ldots\ldots(23)$$

would probably have resulted in the prevention of this sudden change and effected a reduction in the corresponding settlement. The struts actually placed at a depth of 4 ft after excavation had reached 13 ft never received an appreciable load. On the other hand, in contract S-3, which was adjacent to contract S-1A with practically identical soil conditions, the top struts were placed when excavation reached a depth of only 8 ft. These struts experienced a substantial load, and no such sudden increase in lateral deflection of the sheeting, or of settlement, occurred. As a matter of fact, in every other open cut observed, the top strut was placed before a depth equal to H_1 was reached, and in each of these cases, the spontaneous settlement was absent. On contract D-6E, the soil was excavated 13 ft before struts were placed, exactly as in S-1A, but the value of H_1 for this case was 18 ft. The maximum movement experienced by the piling for the complete excavation of this cut was 0.5 in.

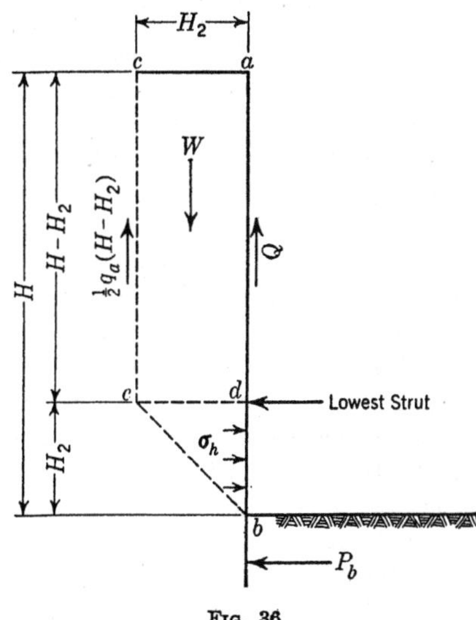

Fig. 36

The maximum permissible spacing between the bottom strut and the bottom of the excavation is less simple to formulate. No satisfactory analytical formulation has as yet been devised, which gives results in close accordance with the observations. The factors involved may be recognized by study of the simplified diagram in Fig. 36.

The total vertical force acting on a horizontal section (e d) at the level of the lowest strut is equal to the weight W of the prism a d e c minus the shearing resistance along c e, which is equal approximately to $\frac{1}{2} q_a (H - H_2)$, and minus the vertical load Q carried by the sheeting. The width of the prism is taken as H_2, equal to the distance between the lowest strut and the bottom of the cut, because the failure surface e b is likely to be inclined at 45° when $\phi = 0$. The intensity of vertical pressure at midheight of the space between b and d is, therefore,

$$\sigma_v = (H - H_2)\,\gamma_a - \frac{1}{2} q_a \frac{(H - H_2)}{H_2} - \frac{Q}{H_2} + \frac{1}{2}\gamma_c H_2 \quad \ldots\ldots (24)$$

in which: γ_a = average density of material above lowest strut; γ_c = average density of material between b and d; and q_a = average effective unconfined compressive strength of material above lowest strut. The lowest value of the normal stress σ_h on the vertical face b d which is compatible with the average vertical stress σ_v is

$$\sigma_h = \sigma_v - q_c = (H - H_2)\,\gamma_a - \frac{1}{2} q_a \frac{(H - H_2)}{H_2} - \frac{Q}{H_2} + \frac{1}{2}\gamma_c H_2 - q_c \ldots (25)$$

in which q_c = unconfined compressive strength of material between b and d.

For the particular case when the entire vertical load is carried by shearing stresses on the wall and on the plane c e, the value of σ_h becomes equal to zero (the most desirable condition) when

$$H_2 = \frac{2\,q_c}{\gamma_c} \qquad (26)$$

or, in other words, when the distance below the bottom strut equals the critical height of the material. (The critical height is the theoretical height at which a vertical bank of cohesive material will stand without support.)[16] If the vertical load is greater than zero, or if it is desired to increase H_2 beyond the critical height of the material, it is necessary that the vertical sheeting be stiff enough to carry the horizontal load σ_h, and that the clay beneath the bottom of the cut be adequate to carry the reaction P_b with sufficient factor of safety. Calculation for the Chicago open cuts indicates that the method of Eq. 23 is likely to be ultraconservative, but it is seen that until measurements are available for evaluation of the shearing forces on the sheeting, a theoretical solution, even of more refined nature, would not be of great practical value.

Repeated calculations of the distances H_2 as excavation and successive struts advanced would determine the maximum vertical spacing between struts in order to prevent excessive lateral squeeze and settlement. This distance was empirically checked for the open cuts on the Chicago subway by systematic observations of inward movement of the sheeting. Excessive movements were readily determined and corrected by relocating struts. An instructive example was again furnished by contract S-1A. Between a depth of 24 and 34 ft, the critical height of the clay ranged between 10 and 15 ft. In the first part of the cut, the vertical distance between the second and third row of struts, at depths of 13 and 31 ft, was equal to 18 ft, considerably greater than the critical height. The corresponding lateral movement and settlement were very important. As a consequence, an intermediate row of struts was established which reduced the spacing at the softest part of the clay to 8 ft. Thereafter the lateral squeeze dropped from $3\frac{1}{4}$ to $2\frac{1}{4}$ in. at this elevation, and the settlement on adjacent buildings was reduced approximately 35%. The total pressure on all the struts of one vertical row remained practically identical.

On one early open cut, 46 ft deep, upon which measurements were not made, a lateral yield of 10 in. developed, and evidence of considerable distress in some of the bracing appeared. From this observation, together with those already described, it may be concluded that for the Chicago clays, a lateral yield of slightly less than 0.25% of the height of the cut is all that is necessary to reduce the earth pressure to the active value, and that the magnitude of the pressure remains substantially unchanged for a yield up to 1%. Beyond this value, the lateral pressure again increases. Furthermore, it is apparent that nature takes care that a yield of at least 0.25% will occur, regardless of the type of bracing.

Part VI.—Rules for Design

For open cuts constructed in a normally consolidated clay soil in accordance with standard construction procedure, the following rules for design seem justified on the basis of the Chicago subway experiments:

[16] See "Poussée des Terres," Pt. II, by J. Résal, Paris, 1910, p. 77.

1. From test borings, secure samples for unconfined compression tests and other soil data needed to establish the value q in the equation (compare Eq. 18*a*)

$$q_a = \frac{1}{H} [\gamma_1 H^2_s K_s \tan \phi_1 + (H - H_s) n q_c] \ldots\ldots\ldots (27a)$$

and for clays with no overlying sand layers,

$$q = n q_c \ldots\ldots\ldots (27b)$$

For Chicago, $n = 0.75$. For other locations, the value of n must be determined on the basis of a few representative field tests.

2. Compute the total lateral pressure by use of the formula

$$K_A = 1.2 \left(1 - \frac{2 q_a}{\gamma_a H}\right) \ldots\ldots\ldots (28)$$

The factor 1.2 in Eq. 28 merely takes account of the scattering from the mean value. It should not be taken to indicate that K_A can be greater then 1.0 for a perfect fluid.

3. Construct the pressure diagram Fig. 35, with maximum abscissa, $K_A \gamma_a H$, and compute the maximum probable strut loads. The struts may be designed for these loads with a conservative factor of safety, inasmuch as provision for all normal variations has been made in the equations. Wales should be designed with no reduction for horizontal arching.

4. Determine allowable depth of excavation prior to placing the first row of struts, from Eq. 23—$H_1 = \frac{q_c}{\gamma_a}$.

Conclusive information concerning the maximum permissible vertical spacing between remaining struts has not been obtained. The proper depth of penetration for the piling depends to a large extent on the stiffness of the clay, and the change in stiffness below the bottom of the cut. Therefore, no general rules can be established for determining this depth.

It should be cautioned that although the results of the Chicago tests are likely to be applicable to other deposits of normally consolidated clays, they almost surely will not apply to clays which have been overconsolidated by heavy pressures in their previous geological history, nor to clays possessing cracks and fissures. They are not likely to apply to shafts or to other excavations in which the deformation conditions of the sheeting are likely to be restricted or significantly different from those in the excavations described. Although the general applicability of Eqs. 11 for total pressure in the Chicago clays has been strongly indicated, the final test in any other locality resides in field measurements.

AMERICAN SOCIETY OF CIVIL ENGINEERS
Founded November 5, 1852

TRANSACTIONS

Paper No. 2346

CONSOLIDATION OF FINE-GRAINED SOILS BY DRAIN WELLS

By Reginald A. Barron,[1] Jun. ASCE

With Discussion by Messrs. Kenneth S. Lane, Philip Keene, Walter Kjellman, and Reginald A. Barron

Synopsis

Drain wells have been used to accelerate the consolidation of fine-grained compressible soils subjected to new loads. However, mathematical analyses of such wells have become available only recently, and are confined to wells having infinitely pervious well backfill and no peripheral smear. Complete formulas for consolidation by vertical and radial flow to wells, for cases with or without peripheral smear and drain well resistance, are presented in this paper; necessary derivations are given in Appendix II. Because soils are not homogeneous, and because of incomplete knowledge of stress-strain consolidation characteristics of soils, these solutions should be regarded as approximate when applied to practical problems.

Introduction

When a soil mass is subjected to a new load, compressive stresses cause a reduction of pore space and, if saturated, the excess water is expelled. For slightly pervious soil, time is required to accomplish this, and, initially, before any reduction of pore space has occurred, the new compressive stresses are carried by the water (called hydrostatic excess or excess pore-water pressure). As the excess water drains, the new compressive stresses are transferred to the soil grains increasing the intergranular pressures. This process is known, in soil mechanics terminology, as consolidation.

New loads applied to foundation soils create shearing stresses; to maintain stability the soil must not be overstressed. For slightly pervious compressible soils, the factors affecting the shearing strength are not fully understood at present; however, it is known that the shearing strength increases with a reduction of the soil pore water.

Most compressible soils are alluvial deposits and are more pervious in the direction of the bedding plane than in a perpendicular direction. This is very

Note.—Published in June, 1947, *Proceedings*. Positions and titles are those in effect when the paper or discussion was received for publication.

[1] Engr., Soil Mechanics, Embankment and Foundation Branch, U. S. Waterways Experiment Station, Vicksburg, Miss.

pronounced in varved deposits[2] laid down in lakes during the wasting away of the continental glaciers. When such soils are loaded, horizontal flow accelerates the consolidation of the soil mass as compared with strictly vertical flow. Because of the horizontal migration of water, however, this process imposes a dangerous condition on the stability of the foundation due to increased pore-water pressure[3] reducing the intergranular stresses at the toes of fills.

The limitation that the shearing strength shall not be exceeded may be met by widening the loading area (thus reducing the shearing stresses in the foundation), or by stage construction, thereby permitting a gain in shearing strength from the consolidation process. These methods are costly. Sometimes a more economical method may be to accelerate the consolidation process, and thus hasten a gain in shearing strength, by the use of vertical sand-filled holes known as drain wells. Drain wells are especially efficient in stratified soils because of the greater perviousness parallel to the bedding. Furthermore, drain wells permit control over the water migrating in a horizontal direction, thus reducing the excess pore-water pressures that might be built up in the toe area of an earth fill.

Drain wells have been used in this country mainly by the California State Highway Department,[4,5] and also at an eastern shipyard.[6] They have also been used by the Providence (R. I.) District, United States Engineer Department, as a repair measure to a dike failure. Further use is planned by the latter organization to permit construction of a moderately high dam on a foundation containing a large deposit of soft varved silt.

The theory of consolidation by drain wells was developed by the author during the design of the two Providence District projects previously noted. The theory for ideal wells having no resistance to flow up the well bore and having no smear at the well periphery has been presented elsewhere,[7,8,9,10,11] but consolidation theory which includes the effect of well smear and well resistance is presented here publicly for the first time.

Notation.—The letter symbols in this paper are defined where they first appear, in the text or by diagrams, and are assembled for convenience in Appendix I. In so far as possible the symbols conform with those given in "Soil Mechanics Nomenclature," *Manual of Engineering Practice No. 22.*

---

[2] "The Recession of the Last Ice Sheet in New England," by Ernst Antevs, *Research Series No. 11.* Am. Geographical Soc., Washington, D. C., 1922.

[3] "Stability of Fills Above Horizontal Clay Strata," by Karl Terzaghi, *Proceedings*, 6th Texas Conference on Soil Mechanics and Foundation Eng., Univ. of Texas, Austin, Tex., 1943.

[4] "Studies of Fill Construction Over Mud Flats Including a Description of Experimental Construction Using Vertical Sand Drains to Hasten Stabilization," by O. J. Porter, *Proceedings*, International Conference on Soil Mechanics and Foundation Eng., Harvard Univ., Cambridge, Mass., Vol. I, 1936, pp. 229–235.

[5] "Studies of Fill Construction Over Mud Flats Including a Description of Experimental Construction Using Vertical Sand Drains to Hasten Stabilization," by O. J. Porter, *Proceedings*, Highway Research Board, Vol. 18, Pt. VI, pp. 129–141.

[6] "Submerged Shipways with Steel Sheeting Walls," by Adolph J. Ackerman and C. B. Jansen, *Civil Engineering*, July, 1943, p. 309.

[7] "The Influence of Drain Wells on the Consolidation of Fine-Grained Soils," by R. A. Barron, Providence (R. I.) Dist., U. S. Engr. Office, 1944.

[8] "Der Hydrodynamische Spannungsausgleich in Zentral Entwasserten Tonzylindern," by L. Rendulic, *Wasserwirtsch, u, Technik*, Vol. 2, Jahigang, 1935, pp. 250–253 and pp. 269–273.

[9] "Consolidation of a Soil Stratum Drained by Wells," by N. Carrillo, Harvard Univ., Cambridge, Mass., 1941 (mimeographed).

[10] "Simple Two and Three Dimensional Cases in the Theory of Consolidation of Soils," by N. Carrillo, *Journal of Mathematics and Physics*, March, 1942.

[11] "Theoretical Soil Mechanics," by Karl Terzaghi, John Wiley & Sons, Inc., New York, N. Y., 1943, pp. 290–296.

BASIC CONSIDERATIONS

The basic theory of consolidation has been developed by Karl Terzaghi, M. ASCE.[12] For the purpose of this paper the following basic assumptions are made:

(1) All vertical loads are initially carried by excess pore-water pressure u.

(2) All compressive strains within the soil mass occur in a vertical direction. Thus, the basic partial differential equation for consolidation by three-directional flow in stratified soils is:

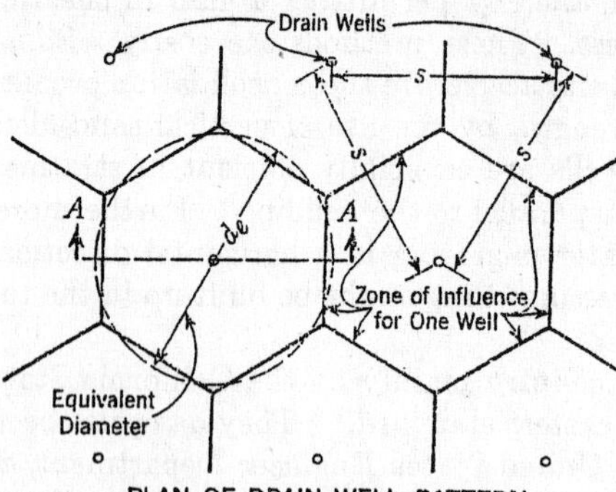

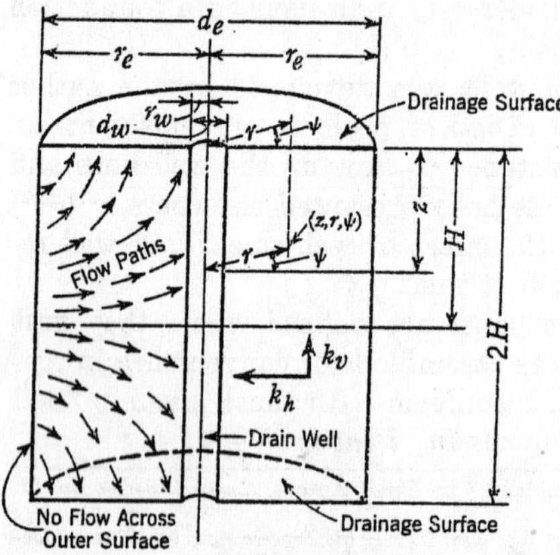

FIG. 1.—PLAN OF DRAIN WELL PATTERN AND FUNDAMENTAL CONCEPTS OF FLOW WITHIN ZONE OF INFLUENCE OF EACH WELL

$$\frac{k_h}{\gamma_w}\left(\frac{\partial^2 u}{\partial x^2} + \frac{\partial^2 u}{\partial y^2}\right) + \frac{k_v}{\gamma_w}\left(\frac{\partial^2 u}{\partial z^2}\right) = \frac{a_v}{1+e}\left(\frac{\partial u}{\partial t}\right) \quad \text{(1a)}$$

in which k_h is the coefficient of permeability of the soil in a horizontal direction; k_v is the coefficient of permeability in a vertical direction, γ_w is the unit weight of water; x, y, and z are rectangular coordinates; e is the void ratio of the soil; t is time; and a_v, the coefficient of compressibility of the soil, equals $-\frac{de}{dp}$. For symmetrical flow to a central drain well Eq. 1a becomes:

$$\frac{k_h}{\gamma_w}\left(\frac{1}{r}\frac{\partial u}{\partial r} + \frac{\partial^2 u}{\partial r^2}\right) + \frac{k_v}{\gamma_w}\left(\frac{\partial^2 u}{\partial z^2}\right) = \frac{a_v}{1+e}\left(\frac{\partial u}{\partial t}\right) \quad \text{(1b)}$$

in which r and z are cylindrical coordinates as defined by Fig. 1.

(3) The most economical pattern of drain wells is that shown in Fig. 1. An exact analysis should include the load distribution and the effect of each well on the rate of consolidation at any point in the foundation. However, it is believed that such a solution would be extremely cumbersome and rather academic when consideration is given to the heterogeneity of the foundation, and the present incomplete knowledge of the three-dimensional stress-strain consolidation properties of fine-grained soils.

(4) The zone of influence of each well is a circle (see Fig. 1).

(5) Load distribution is uniform over this area.

[12] "Erdbaumechanik auf Bodenphysikalischer Grundlage," by Karl Terzaghi, F. Deutiche, Vienna, 1925.

Ideal Drain Wells

Wherever drain wells are installed construction operations generally cause some remolding or smear of the fine-grain soil adjacent to the well periphery, especially in varved soils. Also, the backfilling of the well, even with pervious material, results in some resistance to flow of the expelled water up the well bore. For ideal drain wells these effects are ignored, although solutions for these effects are discussed subsequently.

Case of Free Strain.—The initial solution for ideal drain wells is made on the assumptions that the load is uniform over the circular zone of influence for each well and that differential settlements occurring over such a zone, as consolidation progresses, have no effect on redistribution of stresses by arching of the fill. Furthermore, it is assumed that shear strains developed in the

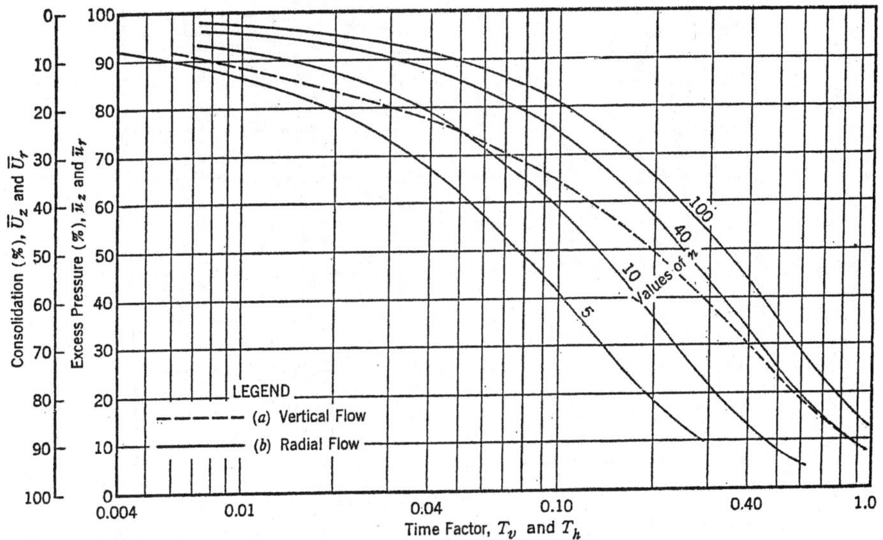

Fig. 2.—Average Consolidation Rates
(a) For Vertical Flow in a Clay Stratum of Thickness 2 H Drained on Both Upper and Lower Surfaces
(b) For Radial Flow to Axial Drain Wells in Clay Cylinders Having Various Values of n

foundation by differential settlements will have no effect on the consolidation process. The boundary conditions are:

(1) Initial excess pore-water pressure, u_0, is uniform throughout the soil mass when $t = 0$.

(2) Excess pore-water pressure at the drain well surface (r_w) is zero when $t > 0$.

(3) The external radius, r_e, is impervious or, because of symmetry, no flow occurs across this boundary; that is, $\frac{\partial u}{\partial r} = 0$ when $r = r_e$.

(4) Excess pore-water pressure at the upper horizontal boundary of the soil mass ($z = 0$) is zero when $t > 0$.

(5) The lower horizontal boundary of the soil mass ($z = H$) is impervious or, because of symmetry, no flow occurs across this boundary; that is, $\frac{\partial u}{\partial z} = 0$ at $z = H$.

Subject to the foregoing conditions a solution of Eq. 1b for excess pore-water pressure, $u_{r,z}$, at any point due to consolidation by both radial and vertical flow is:

$$u_{r,z} = \frac{u_r \, u_z}{u_0} \qquad (2)$$

in which u_r and u_z are, respectively, the excess pore-water pressures due to radial flow only and vertical flow only. Similarly, the average excess pore-water

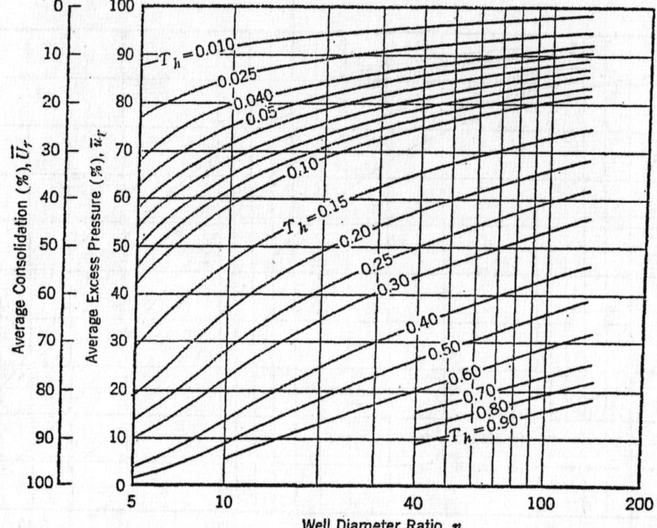

FIG. 3.—AVERAGE CONSOLIDATION RATES FOR RADIAL DRAINAGE ONLY, IN TERMS OF WELL DIAMETER RATIO, n

pressure throughout the entire soil mass is:

$$\bar{u}_{r,z} = \frac{\bar{u}_r \, \bar{u}_z}{u_0} \qquad (3)$$

in which $\bar{u}_r$ and $\bar{u}_z$ are the average excess pore-water pressures throughout the entire soil mass due to radial flow and vertical flow, respectively. Equations for u_r and $\bar{u}_r$ were developed by R. E. Glover;[13] those for u_z and $\bar{u}_z$ were developed by Professor Terzaghi.[11,12] A more general solution may be obtained when the initial excess pore-water pressure is a function of z, provided no expansion of the soil occurs. Equations for u_z and $\bar{u}_z$ for this case were developed by G. Gilboy,[14] Assoc. M. ASCE.

[13] *Technical Memorandum No. 155*, by R. E. Glover, Bureau of Reclamation, Denver, Colo., 1930.

[14] "Soil Mechanics," by G. Gilboy, Mass. Inst. Tech., Cambridge, Mass., 1930 (mimeographed notes).

Curves of average excess pore-water pressures $\bar{u}_r$ due to purely radial flow, versus time factor $T_h = \frac{k_h (1 + e) t}{\gamma_0 a_v d^2_e}$ for various values of n are shown in Fig. 2, in which n is the ratio of the zone of well influence to well diameter. Also given in Fig. 2 is a curve of average excess pore-water pressure $\bar{u}_z$ due to vertical flow only, versus time factor $T_v = \frac{k_v (1 + e) t}{\gamma_0 a_v H^2}$. Curves of $\bar{u}_r$ versus well diameter n for various time factors T_h are shown in Fig. 3. These curves permit the construction of average pore-water pressures versus time factor curves similar to those shown in Fig. 2 for any value of n between 5 and 100. Variations of excess pore-water pressures u_r at various points, due to radial flow, for a well diameter ratio $n = 40$, are shown in Fig. 4, indicating a much

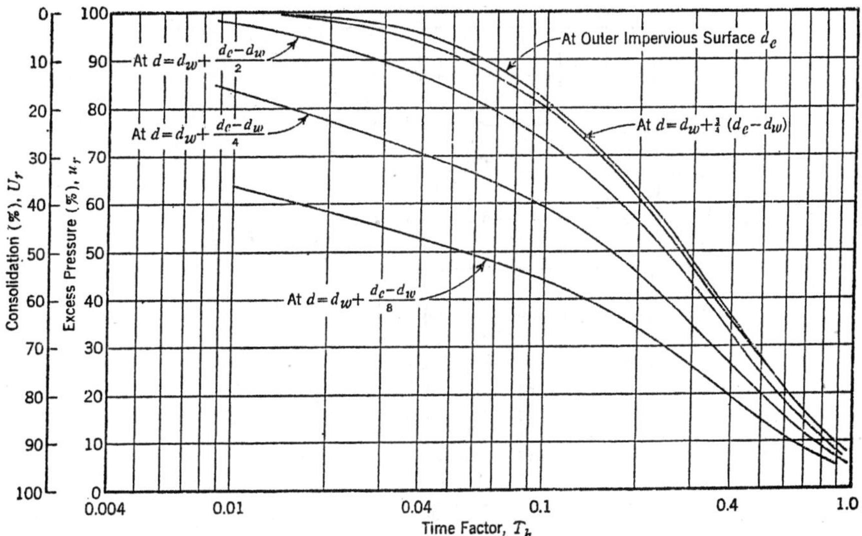

FIG. 4.—RADIAL CONSOLIDATION RATES AT VARIOUS RADIAL SURFACES FOR CLAY CYLINDER HAVING WELL DIAMETER RATIO, $n = 40$

faster rate of consolidation nearer to the well as compared to that nearer the outer zone of well influence. Shown in Fig. 5 is excess pore-water pressure, u_z, due to vertical flow only, versus relative depth, $Z\%$, of clay layer for various time factors T_v. The relative depth, $Z\%$, is defined as the ratio of the depth z to the point to the total thickness H of the clay deposit. Use of the curves shown in Fig. 2 and Fig. 5 is explained in the following example.

Example.—A bed of varved clay, 40 ft thick, resting on impervious rock, and having free drainage at its upper surface is quickly loaded with a very wide fill to 1 ton per sq ft. Tests on undisturbed soil samples furnish the following average data: $a_v = 13 \times 10^{-5}$ cm² per g; $e = 1.500$; $k_v = 100 \times 10^{-8}$ cm per sec; $k_h = 500 \times 10^{-8}$ cm per sec (assumed); and $\Delta e = 0.125$. Settlement under the 1-ton loading will be:

$$\Delta H = \frac{H\,\Delta e}{1 + e} = \frac{(40\text{ ft})\,(0.125)}{1 + 1.500} = 2.0\text{ ft}$$

Fig. 6(*a*) shows a time-average consolidation curve for vertical flow only. (The effect of horizontal flow has been neglected.) For time in months,

$$t_v = \frac{T_v\,\gamma_w\,a_v\,H^2}{k_v\,(1 + e)} = \frac{T_v\,(1)\,(13 \times 10^{-5})\,(40 \times 30.5)^2}{(100 \times 10^{-8})\,(1 + 1.500)\,(60 \times 60 \times 24 \times 30)} = 30\,T_v$$

Values for T_v at various stages of consolidation are obtained from time factor consolidation curve for vertical flow only, as given in Fig. 2. After the average

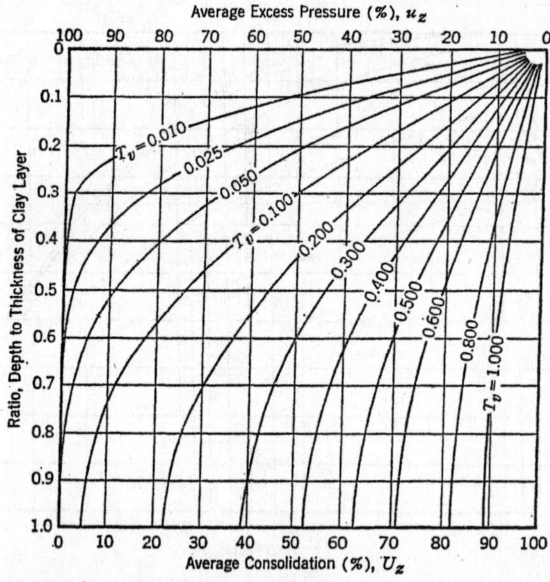

FIG. 5.—VARIATION IN CONSOLIDATION BY VERTICAL FLOW FOR VARIOUS TIME FACTORS AND DEPTH RATIOS

consolidation has reached 50%, sand-filled drain wells 1 ft in diameter are installed 40 ft apart in an equilateral pattern. For time in months,

$$t_h = \frac{T_h\,\gamma_w\,a_v\,d^2_e}{k_h\,(1 + e)} = \frac{T_h\,(1)\,(13 \times 10^{-5})\,(40 \times 30.5)^2}{(500 \times 10^{-8})\,(1 + 1.500)\,(60 \times 60 \times 24 \times 30)} = 6\,T_h$$

Values for T_h are obtained from the time factor consolidation curve for $n = 40$ in Fig. 2. The time for 50% consolidation by vertical flow only is

$$t_v = 30 \times 0.195 = 5.85\text{ months}$$

which is, therefore, the origin for the radial-average consolidation curve. At any time, t, after installations of the wells, $\bar{u}_{r,z}$ is obtained by use of Eq. 3.

For example, at nine months

$$\bar{u}_{r,z} = \frac{\bar{u}_r \times \bar{u}_z}{u_0} = \frac{24\% \times 39\%}{100\%} = 9.4\%$$

The average consolidation is 100 − 9.4 = 90.6%.

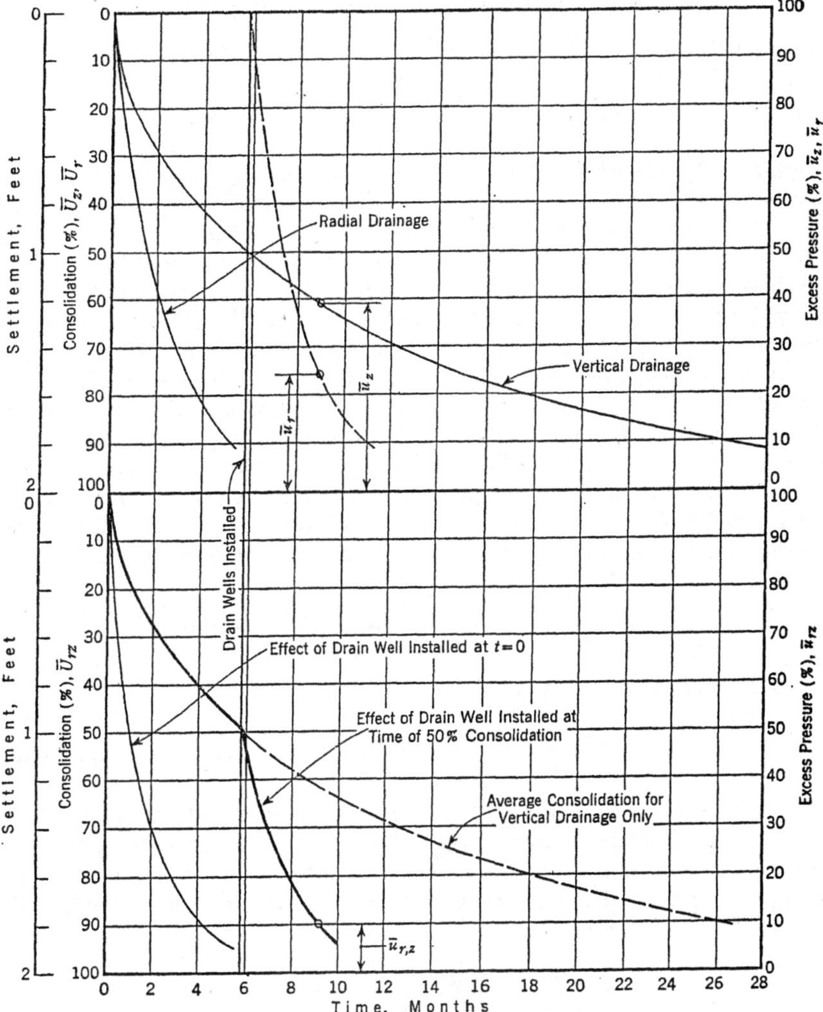

FIG. 6.—AVERAGE CONSOLIDATION RATES FOR RADIAL AND VERTICAL DRAINAGE IN CLAY CYLINDER IN WHICH $D = H$, $n = 40$, AND $k_h = 5\,k_v$, SHOWING EFFECT OF DRAIN WELLS INSTALLED WHEN LOAD IS APPLIED AND WHEN 50% CONSOLIDATION HAS TAKEN PLACE

For wells installed before the load has been added, the solid curve for $\bar{u}_r$ shown in Fig. 6 may be used. The effect of drain wells installed when 50% consolidation had been obtained and before application of load is shown in Fig. 6. Where $k_h \geqq 5\,k_v$, the effect of wells is obvious.

The determination of excess pressure at a point on the bottom of the clay layer and at the outer edge of the influence zone of a well nine months after load is applied is as follows: $Z\% = 40/40 = 1.0$, and $T_v = 9/30 = 0.3$. From Fig. 5, u_z for a relative depth of $Z\% = 1$ and $T_v = 0.3$ is 61%. The value of u_r at the outer surface of the zone of influence (for $n = 40$) is obtained from Fig. 4 at $T_h = \frac{9 - 5.85}{6} = 0.525$, and is 26%. Therefore the excess pressure at this point, which, incidently, consolidates at the slowest rate of all, by Eq. 2, is

$$u_{r,z} = \frac{u_r \times u_z}{u_0} = \frac{26\% \times 61\%}{100\%} = 16\%$$

Case of Equal Vertical Strain.—For consolidation by radial flow with free strains permitted, the soil adjacent to the well consolidates and compresses faster than soil farther away from the drain. This difference in rate of consolidation develops differential settlement of the upper surface of the soil mass and shear strains within the mass. For free strains, it was assumed that these effects did not influence the redistribution of load to the soil nor the rate of consolidation. It is obvious, however, that these effects will redistribute the load to some extent, depending on the amount of arching developed in the material above the compressible soil. The extreme case would be where the arching process redistributes the load to the consolidating soil so that all vertical strains are equal and no differential settlement develops. This condition is rather severe. It can be obtained in the laboratory by use of a rigid loading platform, and probably is approached in the field if the ratio of H to d_e is large.

Using Eq. 1*b*, the rate at which water is expelled from a differential soil cube is:

$$\frac{\partial Q}{\partial t} = -\left[\frac{k_h}{\gamma_w}\left(\frac{1}{r}\frac{\partial u}{\partial r} + \frac{\partial^2 u}{\partial r^2}\right) + \frac{k_v}{\gamma_w}\frac{\partial^2 u}{\partial z^2}\right] 2\,\pi\,r\,dr\,dz \ldots\ldots\ldots (4)$$

The volume lost at any point is:

$$-\frac{\partial V}{\partial t} = \frac{-\partial e}{1 + e}\left(\frac{2\,\pi\,r\,dr\,dz}{\partial t}\right) \ldots\ldots\ldots\ldots\ldots\ldots (5a)$$

which is the same at any depth for any given time. At any depth, z, the relation between the average hydrostatic excess pressure, $\bar{u}$, and the average intergranular pressure, $\bar{p}$, is $\bar{u} + \bar{p} = p_f$, in which p_f is the final intergranular pressure upon completion of consolidation. Thus, $-\partial\bar{u} = \partial\bar{p}$, $\partial e = a_v\,\partial\bar{u}$, and Eq. 5*a* may be rewritten as:

$$\frac{-\partial V}{\partial t} = \frac{-a_v\,\partial\bar{u}}{(1 + e)\,\partial t}\,2\,\pi\,r\,dr\,dz \ldots\ldots\ldots\ldots\ldots\ldots (5b)$$

Equating Eqs. 4 and 5*b* and simplifying,

$$c_h\left(\frac{1}{r}\frac{\partial u}{\partial r} + \frac{\partial^2 u}{\partial r^2}\right) + c_v\frac{\partial^2 u}{\partial z^2} = \frac{\partial\bar{u}}{\partial t} \ldots\ldots\ldots\ldots\ldots (6a)$$

in which c_h is a coefficient of consolidation for horizontal flow and is equal to

$\frac{k_h (1 + e)}{a_v \gamma_w}$. For radial flow only, this becomes:

$$c_h \left(\frac{1}{r} \frac{\partial u}{\partial r} + \frac{\partial^2 u}{\partial r^2} \right) = \frac{\partial \bar{u}}{\partial t} \qquad (6b)$$

A solution for this expression is—

$$u_r = \frac{4 \bar{u}}{d^2_e F(n)} \left[r^2_e \log_e \left(\frac{r}{r_w} \right) - \frac{r^2 - r^2_w}{2} \right] \qquad (7)$$

—in which

$$\bar{u} = u_0 \epsilon^{\lambda} \qquad (8)$$

ϵ is the base of natural logarithms,

$$\lambda = \frac{-8 T_h}{F(n)} \qquad (9)$$

and

$$F(n) = \frac{n^2}{n^2 - 1} \log_e (n) - \frac{3 n^2 - 1}{4 n^2} \qquad (10)$$

The initial distribution of the hydrostatic excess pressure is not uniform (see

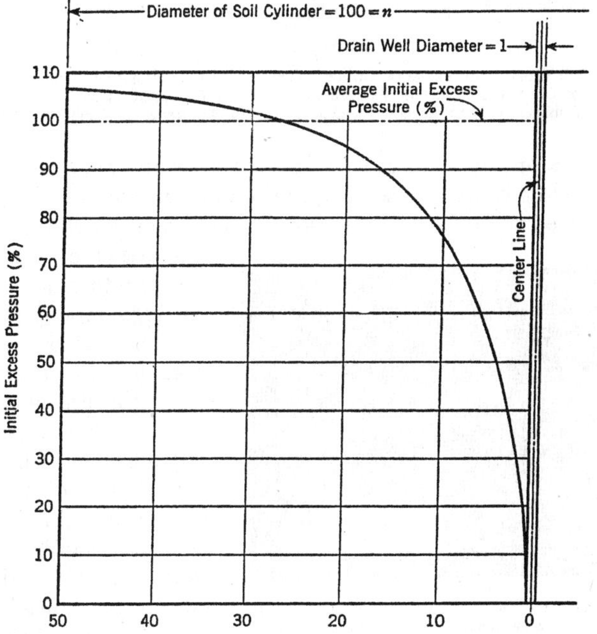

FIG. 7.—DISTRIBUTION OF INITIAL EXCESS PORE-WATER PRESSURE IN SOIL CYLINDER WITH $n = 100$

Fig. 7); but with passage of time the distribution approaches that for the free strain case (see Fig. 8).

Effect of Peripheral Smear

If drain wells are installed by driving cased holes and then backfilling as the casing is withdrawn, the driving and pulling of the casing will distort and

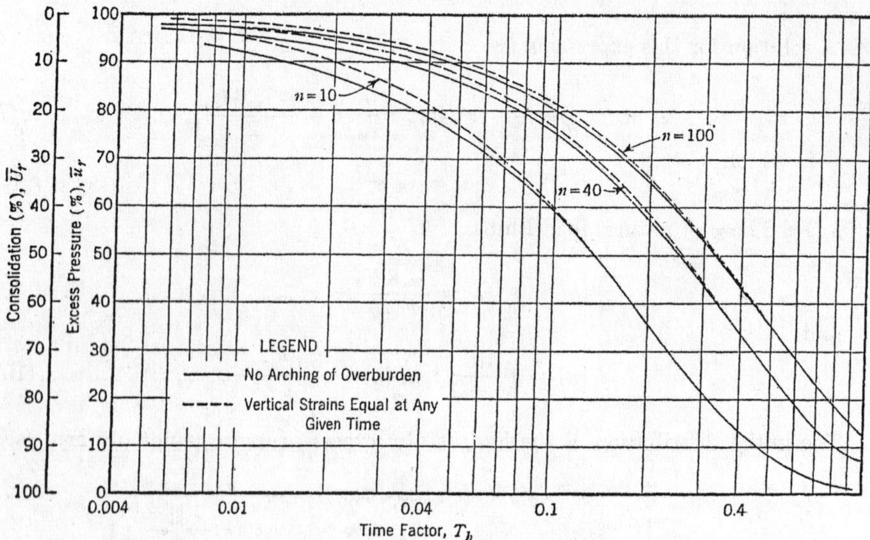

Fig. 8.—Comparison of Average Consolidation Rates in Clay Cylinders by Radial Drainage Only for Various Values of n Under Conditions of Equal Vertical Strains at Any Given Time and of No Arching of Overburden

remold the adjacent soil. In varved soils the finer and more impervious layers will be dragged down and smeared over the more pervious layers, resulting in a zone of reduced permeability in the soil adjacent to the well periphery. This distortion of the soil is very similar to that which occurs during sampling operations (see Fig. 9). It is not difficult to visualize a far greater smear from the use of a heavy casing or hollow mandrel with a bottom plug which displaces entirely the soil at the location of a drain well.

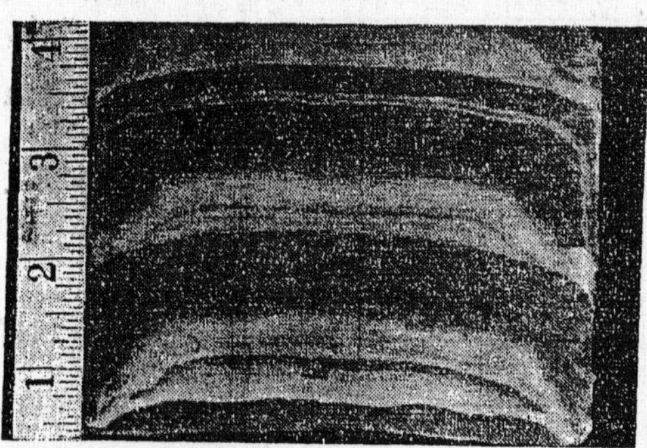

Fig. 9.—Distorted or Smeared Varved Clay Sample Taken with Early Massachusetts Institute of Technology Type 4½-In.-Diameter Sampler

The remolded or smeared zone creates additional resistance which must be overcome by the excess water being expelled. This, in turn, will retard the rate of consolidation. The smeared zone will not be of constant thickness nor will it be homogeneous as far as soil properties are concerned. For this analysis, however, to obtain solutions for smear effect it is assumed that the thickness of the smeared zone is constant and homogeneous. It is further assumed that, because the remolded zone is adjacent to the drain well and will consolidate very fast, its consolidation can be ignored and the zone treated as an incompressible material.

In the smeared zone Eq. 1*b* then becomes:

$$\frac{-k_s}{\gamma_w}\left(\frac{1}{r}\frac{\partial u'}{\partial r} + \frac{\partial^2 u'}{\partial r^2}\right) = 0 \quad \text{.................. (11)}$$

in which k_s is the coefficient of permeability of the smeared soil.

It follows that variation of excess water pressure between r_w and r_s, the outer limit of the smeared zone, is:

$$u'_r = u'_{rs}\left[\frac{\log_e\left(\frac{r}{r_w}\right)}{\log_e(s)}\right] \quad \text{.................... (12)}$$

in which u'_{rs} is the excess pressure at r_s, and $s = r_s/r_w$. Added boundary conditions required because of the presence of the smeared zone are: First, that the excess water pressure at the boundary of the smeared zone be the same in the undisturbed zone as in the smeared zone ($u_{rs} = u'_{rs}$); and, second, that the rate of flow out of the undisturbed zone be equal to that into the smeared zone, so that at r_s:

$$k_h\frac{\partial u}{\partial r} = k_s\frac{\partial u'}{\partial r} \quad \text{...................... (13)}$$

Case of Free Strain.—In consolidation by radial flow to a central drain well with a smear zone at its periphery, Eq. 1*b* reduces to—

$$c_h\left(\frac{1}{r}\frac{\partial u}{\partial r} + \frac{\partial^2 u}{\partial r^2}\right) = \frac{\partial u}{\partial t} \quad \text{.................... (14)}$$

—from which a solution for u_r, for uniform initial excess pore-water pressure distribution, subject to the previously noted boundary conditions, is:

$$u_r = u_0 \sum_{\alpha_1, \alpha_2, \alpha_3 \cdots}^{\alpha = \infty} \frac{-\frac{2}{s\,\alpha} U_1(\alpha\, s)\, U_0\left(\frac{\alpha\, r}{r_w}\right)\epsilon^{\mu}}{\frac{4}{\pi^2\alpha^2 s^2} - U^2_0(\alpha\, s) - U^2_1(\alpha\, s)} \quad \text{......... (15)}$$

in which

$$U_0(\alpha\, s) = J_0(\alpha\, s)\, Y_1(\alpha\, n) - J_1(\alpha\, n)\, Y_0(\alpha\, s) \quad \text{.......... (16a)}$$

$$U_1(\alpha\, s) = J_1(\alpha\, s)\, Y_1(\alpha\, n) - J_1(\alpha\, n)\, Y_1(\alpha\, s) \quad \text{.......... (16b)}$$

$$U_0\left(\frac{\alpha\, r}{r_w}\right) = J_0\left(\frac{\alpha\, r}{r_w}\right) Y_1(\alpha\, n) - J_1(\alpha\, n)\, Y_0\left(\frac{\alpha\, r}{r_w}\right) \quad \text{...... (16c)}$$

The functions $J_0(\ )$ and $J_1(\ )$ are Bessel functions of the first kind of zero and first order, respectively, and $Y_0(\ )$ and $Y_1(\ )$ are Bessel functions of the second kind of zero and first order, respectively, and α, α_1, α_2, α_3, $\cdots$, are the roots of

$$\frac{k_s\, U_0\, (\alpha\, s)}{k_h\, \alpha\, s\, \log_e\, (s)} + U_1\, (\alpha\, s) = 0 \ldots\ldots\ldots\ldots\ldots\ldots (17)$$

and

$$\mu = -\, 4\, n^2\, \alpha^2\, T_h \ldots\ldots\ldots\ldots\ldots\ldots\ldots\ldots (18)$$

The average excess pore-water pressure between r_s and r_e is:

$$\bar{u}_r = u_0 \sum_{\alpha_1, \alpha_2, \alpha_3 \ldots}^{\alpha = \infty} \frac{U^2_1\, (\alpha\, s)\, \epsilon^{\mu}}{\alpha^2\, (n^2 - s^2) \left[\dfrac{4}{\pi^2\, \alpha^2\, s^2} - U^2_0\, (\alpha\, s) - U^2_1\, (\alpha\, s) \right]} \ldots (19)$$

If consolidation occurs by both radial and vertical flow, an approximate solution for the excess water pressure is:

$$u_{r,z} = \frac{u_r\, u_z}{u_0} \ldots\ldots\ldots\ldots\ldots\ldots\ldots\ldots (20)$$

in which u_r is found by Eq. 15; the average excess pressure for the entire body between r_s and r_e is:

$$\bar{u}_{r,z} = \frac{\bar{u}_r\, \bar{u}_z}{u_0} \ldots\ldots\ldots\ldots\ldots\ldots\ldots\ldots (21)$$

in which $\bar{u}_r$ is obtained from Eq. 19, and values of u_z and $\bar{u}_z$ are obtained from Figs. 2 and 5. Eqs. 20 and 21 are approximate because the effect of vertical flow in the smeared zone has been neglected. However, for reasonable values of k_h and k_v this effect is believed to be negligible.

Case of Equal Vertical Strain.—Solution of Eq. 6b for consolidation by radial flow to a central drain well with a smeared zone at its periphery is—

$$u_r = \bar{u}_r \frac{\left[\log_e \left(\dfrac{r}{r_s} \right) - \dfrac{r^2 - r^2_s}{2\, r^2_e} + \dfrac{k_h}{k_s} \left(\dfrac{n^2 - s^2}{n^2} \right) \log_e\, (s) \right]}{\nu} \ldots\ldots (22)$$

in which

$$\nu = F\, (n, s, k_h, k_s) = \left[\frac{n^2}{n^2 - s^2} \log_e \left(\frac{n}{s} \right) - \frac{3}{4} + \frac{s^2}{4\, n^2} + \frac{k_h}{k_s} \left(\frac{n^2 - s^2}{n^2} \right) \log_e\, (s) \right] \ldots (23)$$

and $\bar{u}_r$ can be expressed as—

$$\bar{u}_r = u_0\, \epsilon^{\xi} \ldots\ldots\ldots\ldots\ldots\ldots\ldots\ldots (24)$$

—in which

$$\xi = \frac{-\, 8\, T_h}{\nu} \ldots\ldots\ldots\ldots\ldots\ldots\ldots\ldots (25)$$

For combined radial and vertical flow, approximate solutions may be obtained using Eqs. 22 and 24 with u_z and $\bar{u}_z$ obtained as noted previously.

For purposes of studying the effect of smear in retarding the average rate of consolidation, Eq. 24 is preferred to Eq. 19 because of its simplicity. The effect of smear is indicated in Fig. 10. For a greater ratio of k_h/k_v the difference is still larger.

Effect of Well Resistance

The foregoing developments are for ideal drain wells of infinite permeability offering no resistance to flow up the well. Actually, head losses will occur due to resistance of well backfill material to flow. If the flow is large, or if the well area is small, then the back pressure due to well resistance will be high.

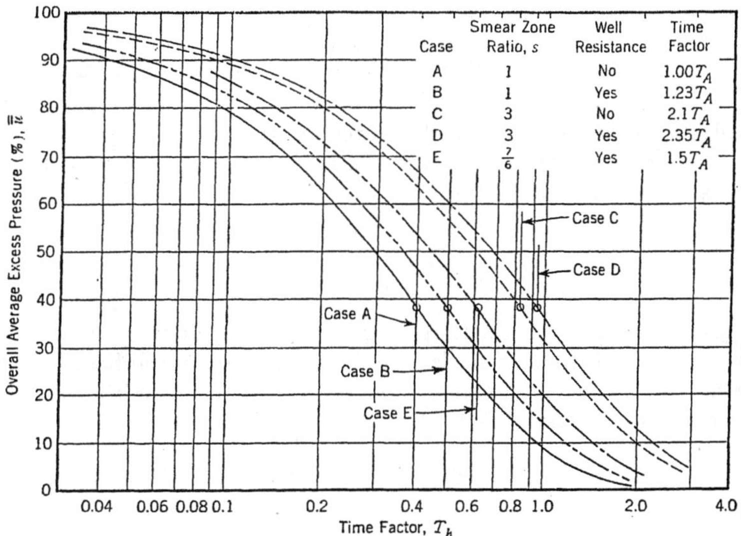

Fig. 10.—Effect of Smear and Well Resistance on Equal Strain Consolidation by Radial Flow to Drain Wells in Which $n = 60$, $k_w/k_h = 20{,}000$, $d_s/H = \frac{60}{80}$, and $k_h/k_s = 10$

On the other hand, if the flow is small, as for shallow deposits and tight deposits, or, if the well area is large, then the resistance of the well to flow will be small. A very pervious filling would be ideal from the standpoint of minimizing well resistance but filter action is needed with sufficiently small voids in the backfill to prevent "inwash" of the surrounding soil from clogging the well. This need is especially required for cohesionless silts and for varved soils in which the more pervious part of the varves is fine sand or coarse silt. For clays the filter requirements are less certain.

A solution for the effect of well resistance on the rate of consolidation for free strain has not been developed. A solution has been developed for equal vertical strain, with or without smear, where k_v of the consolidating soil is equal to zero; that is, $\frac{\partial u}{\partial z} \neq 0$, but because $k_v = 0$ no vertical flow exists.

A solution of Eq. 6*b* for $u_{r,z}$ at any point where smear and well resistance are involved is—

$$u_{r,z} = \bar{\bar{u}}_z \left\{ \frac{f(z)}{\nu} \left[\log_e \frac{r}{r_s} - \frac{r^2 - r_s^2}{2\, r_e^2} + \frac{k_h}{k_s} \left(\frac{n^2 - s^2}{n^2} \right) \log_e (s) \right] + 1 - f(z) \right\} \quad (26)$$

—in which $\bar{\bar{u}}_z$, the average excess pressure between r_e and r_s at depth z, is:

$$\bar{\bar{u}}_z = u_0\, \epsilon^{\xi f(z)} \quad (27)$$

$$f(z) = \frac{\epsilon^{\beta(z-2H)} + \epsilon^{-\beta z}}{1 + \epsilon^{-2\beta H}} \quad (28)$$

and

$$\beta = \left[\frac{2\, k_h\, (n^2 - s^2)}{k_w\, r_e^2\, \nu} \right]^{\frac{1}{2}} \quad (29)$$

in which k_w is the permeability of the well backfill.

The over-all average excess pressure over the entire soil mass between r and r_s and between 0 and $z = H$ is:

$$\bar{\bar{u}} = \int_0^H \bar{\bar{u}}_z\, dz \quad (30)$$

This may be solved approximately by graphical integration or by use of Simpson's rule. In this solution horizontal flow in the drain well backfill has been ignored. Where there is no smear $F(n, s, k_h, k_s)$ reduces to $F(n)$. An example of the effect of well resistance is given in Fig. 10.

Acknowledgments

Mathematical solutions presented in this paper, except as otherwise noted, were developed by the writer for use in special design studies made by the Providence District, U. S. Engineer Department. The writer wishes to express his appreciation to Prof. W. Prager, Brown University, Providence, who checked the derivation of Eqs. 2 and 3, and to Kenneth S. Lane, Assoc. M. ASCE, formerly head, Providence District Soils Laboratory, and now head, Garrison (N. Dak.) District Soils Laboratory, United States Engineer Department, for his valuable aid and encouragement.

APPENDIX I. NOTATION

The following letter symbols, adopted for use in this paper and for the guidance of discussers, conform essentially to "Soil Mechanics Nomenclature," ASCE *Manual of Engineering Practice No. 22*, prepared by a Committee of the Soil Mechanics and Foundations Division, and adopted by the Society in April, 1941. In addition, the symbols of linear concepts are defined by Fig. 1.

A = constant of integration;

a_v = coefficient of compressibility $= -\dfrac{de}{dp}$;

B = constant of integration;

C = constant of integration;

c = coefficient of consolidation:

c_h = for horizontal flow $= \dfrac{k_h (1 + e)}{a_v \gamma_w}$;

c_v = for vertical flow $= \dfrac{k_v (1 + e)}{a_v \gamma_w}$;

d = diameter:

d_e = of zone of influence;

d_s = of remolded zone;

d_w = of drain well;

e = void ratio of soil;

H = depth of clay mass from a free-draining horizontal surface to an impervious one;

i = hydraulic gradient;

k = coefficient of permeability:

k_h = in horizontal direction;

k_s = of remolded zone;

k_v = in vertical direction;

k_w = of well backfill;

n = a ratio $= r_e/r_w = d_e/d_r$;

p = intergranular pressure:

p_f = final;

$\bar{p}$ = average;

$\bar{p}_z$ = average, at depth z;

Q = rate of flow of water;

r = radius, a coordinate in the cylindrical system:

r_e = of zone of influence;

r_s = of remolded zone;

r_w = of drain well;

S = spacing of drain wells;

s = a ratio $= r_s/r_w$;

T = time factor:

T_h = for radial flow $= \dfrac{c_h t}{d^2_e}$;

T_v = for vertical flow $= \dfrac{c_v t}{H^2}$;

t = time of consolidation:

t_h = due to horizontal flow;

t_v = due to vertical flow;

u = excess pore-water pressure:

u_0 = initial uniform;

u_r = due to radial flow;

u_{rs} = due to radial flow at outer limit of remolded zone;

$u_{rs,z}$ = due to radial flow at outer limit of remolded zone and depth z;
u_{rz} = due to radial flow only at depth z;
$u_{r,z}$ = due to combined radial and vertical flow;
$u_{w,z}$ = in drain well at depth z;
u_z = due to vertical flow;
u'_r = due to radial flow in the smeared zone at point r;
u'_{rs} = due to radial flow on remolded side of remolded zone boundary;

$\bar{u}$ = average excess pore-water pressure:
$\bar{u}_r$ = due to radial flow;
$\bar{u}_{r,z}$ = throughout entire mass, due to combined radial and vertical flow;
$\bar{u}_z$ = due to vertical flow;
$\bar{\bar{u}}_z$ = in undisturbed zone at depth z, due to radial flow with well resistance;
$\bar{\bar{u}}$ = in entire undisturbed soil mass, due to radial flow with well resistance;

V = volume of soil mass;
x = a coordinate in the rectangular system;
y = a coordinate in the rectangular system;
$Z\%$ = relative depth $= \dfrac{z}{H} = \dfrac{\text{depth } z}{\text{depth of clay mass}}$;
z = a coordinate in both the rectangular and the cylindrical systems;
α = root of Eq. 17;
β = a substitution factor defined by Eq. 29;
γ_w = unit weight of water;
ϵ = base of natural logarithms;
θ = a substitution factor defined by Eq. 39;
λ = a substitution factor defined by Eq. 9;
μ = a substitution factor defined by Eq. 18;
ν = a substitution factor defined by Eq. 23;
ξ = a substitution factor defined by Eq. 25;
$\bar{\sigma}_z$ = total average stress at depth z; and
ψ = a coordinate in the cylindrical system.

APPENDIX II. DEVELOPMENT OF SOLUTIONS

Case of Free Strain—No Well Resistance Nor Peripheral Smear

Derivations of Eqs. 2 and 3 are given elsewhere,[11,12,13,14] as is also proof of their validity.[10] Therefore, these derivations and proofs are not presented here.

Case of Free Strain with Peripheral Smear

Assume a cylindrical body of fine-grained, saturated compressible soil with an outer impervious boundary and a central drain having no resistance to flow

($k_w = \infty$). Also, assume that the horizontal permeability of the soil mass is k_h while that in a vertical direction, k_v, is zero. Further assume that the central drain is surrounded by a zone of remolded soil whose permeability is k_s, and which is bounded by the well surface, r_w, and by the outer radius of the remolded zone, r_s. The outer radius of the cylindrical body of undisturbed soil is r_e. It is further assumed that the cylindrical body is loaded so that an excess pore-water pressure $u_r = f(r)$ is induced at time zero and that during the process of consolidation vertical strains do not redistribute the load in the soil mass. The remolded soil is assumed to undergo no consolidation.

The boundary conditions are:

(1) When $t = 0$, $u_r = f(r)$ for $r_e > r > r_s$.

(2) No flow occurs accross the impervious surface at the outer boundary r_e of the soil cylinder $\left(\frac{\partial u_r}{\partial r} = 0\right)$.

(3) No vertical flow takes place because $k_v = 0$.

(4) The flow of water through the remolded zone surrounding the drain, for $r_s > r > r_w$, is given by Eq. 12.

(5) The seepage to the interface between the undisturbed and remolded soils at r_s is equal to the seepage away (Eq. 13).

The basic partial differential equation in the undisturbed zone ($r_e > r > r_s$) is expressed by Eq. 6*b*, a solution for which is

$$u_r = \left[A\, J_0\left(\frac{\alpha\, r}{r_w}\right) + B\, Y_0\left(\frac{\alpha\, r}{r_w}\right)\right] \epsilon^{\mu} \quad \ldots\ldots\ldots\ldots (31)$$

—in which A and B are constants of integration. At r_e the second boundary condition is operative and

$$\frac{\partial u_r}{\partial r} = \frac{-\alpha}{r_w}\left[A\, J_1(\alpha\, n) + B\, Y_1(\alpha\, n)\right] \epsilon^{\mu} = 0 \quad \ldots\ldots\ldots\ldots (32)$$

Therefore,

$$A\, J_1(\alpha\, n) + B\, Y_1(\alpha\, n) = 0 \quad \ldots\ldots\ldots\ldots (33)$$

and

$$B = -A\left[\frac{J_1(\alpha\, n)}{Y_1(\alpha\, n)}\right] \quad \ldots\ldots\ldots\ldots (34)$$

Eq. 31 may be written, therefore, as

$$u_r = A\left[J_0\left(\frac{\alpha\, r}{r_w}\right) - \frac{J_1(\alpha\, n)}{Y_1(\alpha\, n)}\, Y_0\left(\frac{\alpha\, r}{r_w}\right)\right] \epsilon^{\mu} = A\, V_0\left(\frac{\alpha\, r}{r_w}\right) \epsilon^{\mu} \ldots (35)$$

The excess pressure in the remolded zone ($r_s > r > r_w$) may be expressed as

$$u'_r = \frac{\log_e\left(\frac{r}{r_w}\right)}{\log_e(s)}\, V_0(\alpha\, s)\, A\, \epsilon^{\mu} \quad \ldots\ldots\ldots\ldots (36)$$

Also, at r_s,

$$k_s \frac{\partial u'_{r_s}}{\partial r} = \frac{k_s}{r_s \log_e (s)} V_0 (\alpha s) A \epsilon^{\mu} \ldots\ldots\ldots\ldots\ldots (37)$$

and

$$k_h \frac{\partial u_{r,s}}{\partial r} = - k_h \frac{\alpha}{r_w} \left[J_1 (\alpha s) - \frac{J_1 (\alpha n)}{Y_1 (\alpha n)} Y_1 (\alpha s) \right] A \epsilon^{\mu}$$

$$= - k_h \frac{\alpha}{r_w} V_1 (\alpha s) A \epsilon^{\mu} \ldots (38)$$

Equating Eqs. 37 and 38, simplifying, and substituting,

$$\frac{k_s}{\alpha k_h s \log_e s} = \theta \ldots\ldots\ldots\ldots\ldots\ldots\ldots (39)$$

$$\theta V_0 (\alpha s) + V_1 (\alpha s) = 0 \ldots\ldots\ldots\ldots\ldots\ldots (40)$$

If α is a root of this equation, the fifth boundary condition is satisfied and the roots are infinite. Therefore,

$$u_r = \sum_{\alpha_1, \alpha_2, \alpha_3, \ldots}^{\alpha=\infty} A V_0 \left(\frac{\alpha r}{r_w} \right) \epsilon^{\mu} \ldots\ldots\ldots\ldots\ldots (41)$$

is also a solution of Eq. 6*b*, provided A can be so determined that, when $t = 0$,

$$f(r) = \sum_{\alpha_1, \alpha_2, \alpha_3, \ldots}^{\alpha=\infty} A V_0 \left(\frac{\alpha r}{r_w} \right) \ldots\ldots\ldots\ldots\ldots\ldots (42)$$

to satisfy the first boundary condition.

To solve for A, multiply both sides of Eq. 42 by $r V_0 \left(\frac{\alpha r}{r_w} \right) dr$ and integrate from r_s to r_e. Noting that $V_1 (\alpha n) = 0$, and if $\alpha \neq \alpha'$,

$$\int_{r}^{r_e} A V_0 \left(\frac{\alpha r}{r_w} \right) V_0 \left(\frac{\alpha' r}{r_w} \right) r \, dr$$

$$= \frac{+ A r^2_w}{\alpha^2 - (\alpha')^2} \left[- \frac{\alpha' r}{r_w} V_0 \left(\frac{\alpha r}{r_w} \right) V_1 \left(\frac{\alpha' r}{r_w} \right) + \frac{\alpha r}{r_w} V_1 \left(\frac{\alpha r}{r_w} \right) V_0 \left(\frac{\alpha' r}{r_w} \right) \right]_{r_s}^{r_e}$$

$$= \frac{+ A r^2_w}{\alpha^2 - (\alpha')^2} [\alpha' s V_0 (\alpha s) V_1 (\alpha' s) - \alpha s V_1 (\alpha s) V_0 (\alpha' s)]$$

$$= \frac{- A r^2_w s}{\alpha^2 - (\alpha')^2} [\alpha' \theta' V_0 (\alpha s) V_0 (\alpha' s) - \alpha \theta V_0 (\alpha s) V_0 (\alpha' s)] = 0 \ldots (43)$$

Therefore, all the right-hand terms of Eq. 42 when multiplied and integrated as previously noted vanish except where $\alpha = \alpha'$, and

$$\int_{r_s}^{r_e} f(r) V_0 \left(\frac{\alpha r}{r_w} \right) r \, dr = \int_{r_s}^{r_e} A V^2_0 \left(\frac{\alpha r}{r_w} \right) r \, dr \ldots\ldots\ldots (44)$$

Thus,

$$A = \frac{\int_{r_s}^{r_e} f(r)\, V_0\left(\frac{\alpha\, r}{r_w}\right) r\, dr}{\int_{r_s}^{r_e} V^2{}_0\left(\frac{\alpha\, r}{r_w}\right) r\, dr} \quad \text{.................(45)}$$

If $f(r) = u_0$,

$$u_0 \int_{r_s}^{r_e} V_0\left(\frac{\alpha\, r}{r_w}\right) r\, dr = u_0 \left[\frac{r\, r_w}{\alpha} V_1\left(\frac{\alpha\, r}{r_w}\right)\right]_{r_s}^{r_e} = \frac{-\, u_0\, r_s\, r_w\, V_1\,(\alpha\, s)}{\alpha} \quad \text{..(46)}$$

and

$$\int_{r_s}^{r_e} V^2{}_0\left(\frac{\alpha\, r}{r_w}\right) r\, dr = \left\{\frac{r^2}{2}\left[V^2{}_0\left(\frac{\alpha\, r}{r_w}\right) + V^2{}_1\left(\frac{\alpha\, r}{r_w}\right)\right]\right\}_{r_w}^{r_e}$$

$$= \frac{r^2{}_e}{2} V^2{}_0\,(\alpha\, n) - \frac{r^2{}_s}{2} V_0\,(\alpha\, s) - \frac{r^2{}_s}{2} V^2{}_1\,(\alpha\, s) \quad \text{..........(47)}$$

Therefore,

$$A = \frac{-\frac{2\, u_0}{s\, \alpha} U_1\,(\alpha\, s)\; Y_1\,(\alpha\, n)}{\frac{4}{\pi^2\, \alpha^2\, s^2} - U^2{}_0\,(\alpha\, s) - U^2{}_1\,(\alpha\, s)} \quad \text{...............(48)}$$

in which

$$U_0\,(\alpha\, n) = \frac{-\,2}{\pi\, \alpha\, n} \quad \text{.........................(49a)}$$

$$U_0\left(\frac{\alpha\, r}{r_w}\right) = V_0\left(\frac{\alpha\, r}{r_w}\right) Y_1\,(\alpha\, n) \quad \text{..............(49b)}$$

$$U_1\left(\frac{\alpha\, r}{r_w}\right) = V_1\left(\frac{\alpha\, r}{r_w}\right) Y_1\,(\alpha\, n) \quad \text{..............(49c)}$$

Substituting Eq. 48 in Eq. 41 for the initial uniform excess pressure, $u_0 = f(r)$:

$$u_r = u_0 \sum_{\alpha_1,\, \alpha_2,\, \alpha_3 \cdots}^{\alpha=\infty} \frac{-\frac{2}{s\, \alpha} U_1\,(\alpha\, s)\; U_0\left(\frac{\alpha\, r}{r_w}\right) \epsilon^{\mu}}{\frac{4}{\pi^2\, \alpha^2\, s^2} - U^2{}_0\,(\alpha\, s) - U^2{}_1\,(\alpha\, s)} \quad \text{........(50)}$$

which is Eq. 15.

The average excess pressure in the cylinder of compressible soil between r_e and r_s is found by integrating Eq. 50 and dividing by the area. Thus,

$$\bar{u}_r = \int_{r_s}^{r_e} \frac{2\, \pi\, r\, u_r\, dr}{\pi\,(r^2{}_e - r^2{}_s)} \quad \text{....................(51)}$$

Substitution of Eq. 50 in Eq. 51, followed by integration and simplification produces the following, which is Eq. 19 of the main paper:

$$\bar{u}_r = u_0 \sum_{\alpha_1, \alpha_2, \alpha_3 \cdots}^{\alpha = \infty} \cdot \frac{U^2_1(\alpha\, s)\, \epsilon^{\mu}}{\alpha^2 (n^2 - s^2) \left[\dfrac{4}{\pi^2 \alpha^2 s^2} - U^2_0(\alpha\, s) - U^2_1(\alpha\, s) \right]} \ldots (52)$$

Case of Equal Vertical Strain with or Without Well Resistance or Peripheral Smear

Assume a cylinder of soil with conditions similar to those of the previous case except that the well backfill has a finite permeability, k_w, which furnishes resistance to flow of expelled water up the well bore. Further assume that all vertical strains at any depth z are equal. This means that there are no shearing strains and consequently no differential settlement of the upper surface. Because of this last condition the same amount of water is expelled from all differential soil cubes at a given depth for any given time. All flow in the well bore is assumed to be vertical.

The boundary conditions are then as follows:

(1) Initial uniform load placed on upper surface of cylinder.

(2) No flow across the impervious surface at outer boundary of soil cylinder $\left(\dfrac{\partial u}{\partial r} = 0 \right)$

(3) No vertical flow in soil cylinder because $k_v = 0$.

(4) Flow of water through the remolded zone surrounding the drains is, for $r_s > r > r_w$,

$$\frac{1}{r}\frac{\partial u'}{\partial r} + \frac{\partial^2 u'}{\partial r^2} = 0 \ldots\ldots\ldots\ldots (53)$$

for conditions of no consolidation in this zone.

(5) Seepage to the interface between the remolded soil and undisturbed soil is equal to the seepage away $\left(k_h \dfrac{\partial u}{\partial r} = k_s \dfrac{\partial u'}{\partial r} \right)$

(6) At r_s the excess pressures on both sides of the interface are equal; that is, $u_{rs,z} = u'_{rs,z}$.

(7) Excess pressure in the drain well is zero at $z = 0$.

(8) At base of drain well vertical flow ceases; that is, $\partial u_{w,z}/\partial z = 0$ at $z = H$.

(9) The radial flow from the soil into the well at depth, z, is equal to the increase in flow up the well:

$$2\,\pi\, r_w\, k_s \frac{\partial u'_{rw,z}}{\partial r} dz = -\,\pi\, r^2_w\, k_w \frac{\partial^2 u_{w,z}}{\partial z^2} dz \ldots\ldots\ldots\ldots (54)$$

Because the vertical strains are all equal at any given depth, z, the amount of water expelled from any small soil mass is the same throughout the entire mass at any given time. Therefore the amount of water passing any radial surface in the zone $r_e > r > r_s$ is proportional to the volume of the cylinder between that radial surface and the external boundary. Thus,

$$\frac{\partial Q}{\partial t} = k\,i\,A' = \frac{-k_h}{\gamma_w}\frac{\partial u}{\partial r}\,2\,\pi\,r\,dz = C_1\,(r^2_e - r^2)\,\pi\,\frac{dz}{\partial t} \quad\text{.......(55)}$$

in which i is the hydraulic gradient; A' is the area of flow; and C, with appropriate subscripts, represents integration constants. Therefore,

$$\frac{\partial u}{\partial r} = \frac{C_1\,\gamma_w}{2\,k_h}\frac{r^2_e - r^2}{\partial t\quad r} = C_2\left(\frac{r^2_e - r^2}{r}\right) \quad\text{.............(56)}$$

and the pressure distribution at any depth, z, at time t is:

$$u_{r,z} = C_2\int(r^2_e - r^2)\,\frac{dr}{r} = C_2\left[r^2_e\,\log_e(r) - \frac{r^2}{2}\right] + C_3 \quad\text{......(57)}$$

The excess pressure, $u'_{r,z}$, in the remolded zone by condition (4) is:

$$u'_{r,z} = (u'_{rs,z} - u_{w,z})\,\frac{\log_e\left(\frac{r}{r_w}\right)}{\log_e(s)} + u_{w,z} \quad\text{..............(58)}$$

in which $u_{rs,z}$ is the excess pressure at r_s and depth z and $u_{w,z}$ is the excess pressure in the drain well at depth z and condition (6) requires that $u_{rs,z} = u'_{rs,z}$ at the interface. Equating Eq. 57 with Eq. 58, at r_s,

$$C_2\left[r^2_e\,\log_e(r_s) - \frac{r^2_s}{2}\right] + C_3 = u'_{rs,z} \quad\text{.............(59)}$$

In addition, condition (5) requires that, at r_s, $k_h\,\dfrac{\partial u}{\partial r} = k_s\,\dfrac{\partial u'}{\partial r}$. Therefore,

$$k_h\,C_2\left(\frac{r^2_e - r^2_s}{r_s}\right) = k_s\,\frac{(u'_{rs,z} - u_{w,z})}{r_s\,\log_e(s)} \quad\text{..............(60)}$$

so that

$$u'_{rs,z} = u_{rs,z} = \frac{k_h}{k_s}\,C_2\,(n^2 - s^2)\,r^2_w\,\log_e(s) + u_{w,z}$$

$$= C_2\left[r^2_e\,\log_e(r_s) - \frac{r^2_s}{2}\right] + C_3 \quad\text{..(61)}$$

This makes

$$C_3 = C_2\left[\frac{k_h}{k_s}\,(n^2 - s^2)\,r^2_w\,\log_e(s) - r^2_e\,\log_e(r_s) + \frac{r^2_s}{2}\right] + u_{w,z} \quad\text{....(62)}$$

so that substitution in Eq. 61 gives:

$$u_{r,z} = C_2\left[r^2_e\,\log_e\left(\frac{r}{r_s}\right) - \left(\frac{r^2 - r^2_s}{2}\right) + \frac{k_h}{k_s}\,(n^2 - s^2)\,r^2_w\,\log_e(s)\right] + u_{w,z} \quad\text{..(63)}$$

The average excess pressure between r_e and r_s is:

$$\bar{\bar{u}}_z = \frac{1}{\pi (r^2_e - r^2_s)} \int_{r_s}^{r_e} u_{r,z}\, 2\,\pi\, r\, dr$$

$$= C_2\, r^2_e \left[\frac{n^2}{n^2 - s^2} \log_e \left(\frac{n}{s}\right) - \frac{3}{4} + \frac{s^2}{4\, n^2} + \frac{k_h}{k_s}\left(\frac{n^2 - s^2}{n^2}\right) \log_e (s) \right] + u_{w,z}$$

$$= C_2 \frac{d^2_e}{4} F(n, s, k_h, k_s) + u_{w,z} \quad \ldots\ldots (64)$$

in which $F(n, s, k_h, k_s) = \nu$, which is defined by Eq. 23.

In accordance with condition (9) and Eq. 60, at r_w:

$$\frac{2\, k_s}{r_w\, k_w} \frac{\partial u'_{r,z}}{\partial r} = - \frac{\partial^2 u_{w,z}}{\partial z^2} = \frac{2\, k_s}{r^2_w\, k_w} \frac{(u'_{rs,z} - u_{w,z})}{\log_e (s)} = 2\, C_2 \frac{k_h}{k_w} (n^2 - s^2) \ldots (65)$$

However, from Eq. 64:

$$C_2 = \frac{4\,(\bar{\bar{u}}_z - u_{w,z})}{d^2_e\, F(n, s, k_h, k_s)} \ldots\ldots\ldots\ldots\ldots\ldots\ldots (66)$$

and Eq. 65 may be written as,

$$\frac{-\,\partial^2 u_{w,z}}{\partial z^2} = \frac{8\, k_h\, (n^2 - s^2)\, (\bar{\bar{u}}_z - u_{w,z})}{k_w\, d^2_e\, F(n, s, k_h, k_s)} = \beta^2\, (\bar{\bar{u}}_z - u_{w,z}) \ldots\ldots (67)$$

When $t = 0$, $\bar{\bar{u}}_z = u_0$ and Eq. 67 becomes—

$$\frac{-\,\partial^2 u_{w,z}}{\partial z^2} = \beta^2\, (u_0 - u_{w,z}) \ldots\ldots\ldots\ldots\ldots\ldots\ldots (68)$$

—a solution for which is:

$$u_0 - u_{w,z} = C_4\, \epsilon^{\beta z} + C_5\, \epsilon^{-\beta z} \ldots\ldots\ldots\ldots\ldots\ldots (69)$$

At $z = 0$, $u_{w,z} = 0$ (condition (7)) so that $u_0 = C_4 + C_5$; and, at $z = H$, $\frac{\partial u_{w,z}}{\partial z} = 0$ (condition (8)), so that:

$$\frac{\partial u_{w,z}}{\partial z} = + \beta\, C_4\, \epsilon^{\beta H} - \beta\, C_5\, \epsilon^{-\beta H} = 0 \ldots\ldots\ldots\ldots (70)$$

Substituting these in Eq. 69 gives:

$$u_0 - u_{w,z} = u_0 \frac{\epsilon^{\beta(z - 2H)} + \epsilon^{-\beta z}}{1 + \epsilon^{-2\beta H}} = u_0\, f(z) \ldots\ldots\ldots\ldots (71)$$

Thus, at $t = 0$,

$$u_{w,z} = u_0\, [1 - f(z)] \ldots\ldots\ldots\ldots\ldots\ldots\ldots\ldots (72)$$

Also, at $t = 0$,

$$\bar{\bar{u}}_z = u_0 = \frac{C_2 d^2_e}{4} F(n, s, k_h, k_s) \qquad (73)$$

therefore,

$$C_2 = \frac{4 u_0 f(z)}{d^2_e F(n, s, k_h, k_s)} \qquad (74)$$

The relationship between average excess pressure $\bar{\bar{u}}_z$ and average soil pressure $\bar{p}_z$ is:

$$\bar{\bar{u}}_z + \bar{p}_z = \bar{\sigma}_z \qquad (75)$$

in which $\bar{\sigma}_z$, the total average stress, is a constant for any fixed loading. Therefore,

$$\partial\bar{\bar{u}}_z = -\partial\bar{p}_z \qquad (76)$$

and, because all strains are equal,

$$\partial e = -a_v \partial\bar{p}_z = \partial e_s = a_v \partial\bar{\bar{u}}_z \qquad (77)$$

The volume change then at any depth z between r and r_e is:

$$-\frac{\Delta\partial V}{\partial t} = -\frac{\partial e}{1+e}(r^2_e - r)\frac{\pi\, dz}{\partial t} = -\frac{a_v \partial\bar{\bar{u}}_z}{1+e}(r^2_e - r^2)\frac{\pi\, dz}{\partial t} \qquad (78)$$

and the total rate of volume change at depth z for time t is:

$$\frac{-\partial V}{\partial t} = \frac{-a_v \partial\bar{\bar{u}}_z}{1+e}(r^2_e - r^2_s)\,\pi\frac{dz}{dt} \qquad (79)$$

From Eqs. 55 and 56, the total water flowing into the well at depth z is:

$$\frac{\partial Q}{\partial t} = \frac{k_h}{\gamma_w} C_2 \frac{r^2_e - r^2_s}{r_s} 2\pi r_s\, dz \qquad (80)$$

which is equal to Eq. 79; and, thus, when $t = 0$,

$$C_2 = \frac{-a_v \partial\bar{\bar{u}}_z \gamma_w}{2(1+e) k_h \partial t} = \frac{4 u_0 f(z)}{d^2_e F(n, s, k_h, k_s)} \qquad (81)$$

and

$$\frac{\partial\bar{\bar{u}}_z}{\partial t} = -\frac{u_0 8 f(z)}{F(n, s, k_h, k_s)}\left[\frac{(1+e) k_h}{a_v \gamma_w d^2_e}\right] = -\frac{u_0 8 f(z) T_h}{F(n, s, k_h, k_s)\, t} = \frac{\xi u_0 f(z)}{t} \qquad (82)$$

If it is assumed that

$$\bar{\bar{u}}_z = u_0\, \epsilon^{\xi f(z)} \qquad (83)$$

then,

$$\frac{\partial \bar{\bar{u}}_z}{\partial t} = \frac{\xi\, u_0\, f(z)}{t}\, \epsilon^{\xi f(z)} \qquad (84)$$

and, at $t = 0$,

$$\frac{\partial \bar{\bar{u}}_z}{\partial t} = -\frac{u_0\, 8\, f(z)}{F(n, s, k_h, k_s)} \left[\frac{(1 + e)\, k_h}{a_v\, \gamma_w\, d^2_e}\right] \qquad (85)$$

which will be recognized as Eq. 82. From Eq. 64 excess pressure in the drain well is:

$$u_{w,z} = \bar{\bar{u}} - \frac{C_2\, d^2_e\, \nu}{4} = \bar{\bar{u}}_z + \frac{a_v\, \gamma_w}{2\,(1 + e)\, k_h}\, \frac{d^2_e\, \nu}{4}\, \frac{\partial \bar{\bar{u}}_z}{\partial t}$$

$$= \bar{\bar{u}}_z - \frac{a_v\, \gamma_w}{(1 + e)\, k_h}\, \frac{d^2_e\, \nu}{8\, \nu}\, \frac{u_0\, 8\, f(z)\, (1 + e)\, k_h}{a_v\, \gamma_w\, d^2_e}\, \epsilon^{\xi f(z)}$$

$$= \bar{\bar{u}}_z - u_0\, \epsilon^{\xi f(z)} = \bar{\bar{u}}_z\, [1 - f(z)] \qquad (86)$$

From Eqs. 37 and 38 the excess pressure at any point within the undisturbed soil is:

$$u_{r,z} = \bar{\bar{u}}_z \left\{ \frac{f(z)}{F(n, s, k_h, k_s)} \times \left[\log_e\left(\frac{r}{r_s}\right) - \frac{r^2 - r^2_s}{2\, r^2_e} + \frac{k_h}{k_s}\left(\frac{n^2 - s^2}{n^2}\right) \log_e (s)\right] + 1 - f(z) \right\} \qquad (87)$$

The average excess pressure, $\bar{\bar{\bar{u}}}$, throughout the entire undisturbed soil mass is:

$$\bar{\bar{\bar{u}}} = \frac{1}{H} \int^{H} \bar{\bar{u}}_z\, dz \qquad (88)$$

which may be solved by such approximate method as Simpson's rule. For no smear but with well resistance, $F(n, s, k_h, k_s)$ reduces to $F(n)$ as given in Eq. 10. For smear but no well resistance, $f(z)$ becomes equal to 1.

DISCUSSION

KENNETH S. LANE,[15] ASSOC. M. ASCE.—As the drain well promises to become a very useful tool in foundation engineering, the author's development of the theory of consolidation with drain wells is particularly timely. The process of consolidation involves the expulsion of water from the soil voids, thus permitting the grains to attain closer contact under the influence of applied loads. In a homogeneous material the time required for consolidation varies as the square of the distance the water must travel. Insertion of sand-filled drain wells in a foundation shortens the water escape path and thus greatly accelerates the rate of consolidation. With water escaping radially to the wells, in addition to escaping in the normal vertical direction, the time of consolidation can be reduced to a fraction of that for vertical flow only (see Fig. 6) —governed mainly by the choice of well spacing. This benefit from drain wells is far greater in a horizontally stratified material (as the author has noted) where the horizontal or radial permeability is generally much larger than that in a vertical direction.

Benefits of Drain Wells.—There are three major benefits from using drain wells in a weak compressible foundation: (1) Acceleration in the rate of settlement; (2) acceleration in the rate of gain of shear strength; and (3) reduction in the lateral transmission of excess pressure.

1. *Acceleration in Rate of Settlement.*—The percentage of ultimate settlement at a given time is the same as the percentage of consolidation.

2. *Acceleration in Rate of Gain of Shear Strength.*—Let σ_t be the total unit stress on an element of soil; σ_o be the initial stress; $\Delta\sigma$ be the increment of stress added to σ_o; $\bar{\sigma}$ be the intergranular pressure; and σ_f be the final intergranular pressure. In the notation of the paper $z\,\gamma_w$ is the hydrostatic pressure and u is the pore water pressure in excess of hydrostatic pressure. The total stress on an element of soil consists of intergranular pressure plus water pressure. With the addition of a new load to a soil which has reached full consolidation under its former loads, this results in:

$$\sigma_t = \sigma_o + \Delta\sigma + z\,\gamma_w \qquad (89a)$$

$$\sigma_t = \bar{\sigma}_o + u + z\,\gamma_w \qquad (89b)$$

$$\sigma_t = \bar{\sigma}_f + z\,\gamma_w \qquad (89c)$$

Under a common hypothesis for the shear strength of clays, only the intergranular pressure, $\bar{\sigma}$, is effective in creating shear strength from internal friction. The process of consolidation involves the transfer of added stress, $\Delta\sigma$, initially carried in the pore water as excess pressure, u, to intergranular stress with resulting gain in shear strength. Eqs. 89b and 89c represent the beginning and end of consolidation, the hydrostatic portion of the pore-water pressure, $z\,\gamma_w$, remaining unchanged.

[15] Chf., Soils and Geology Branch, Garrison Dist., Corps of Engrs., Bismarck, N. Dak.

As the author has indicated, the factors governing the shear strength of clays are not well understood. The other common approach is based on empirical discoveries that shear strength varies inversely with water content. Under either approach shear strength increases as consolidation progresses by decreasing both pore pressure and water content.

3. *Reduction in Lateral Transmission of Excess Pressure, u, in the More Pervious Layers of Stratified Soil.*—Beneath the center of an embankment the added stress, $\Delta\sigma_c$, and the resulting excess pressure, u_c, reach their maximum values. If the horizontal permeability is much greater than the vertical permeability, as in the varved clay pictured in Fig. 9, a portion of the center line excess pressure, u_c, will be transmitted laterally to the zone beneath the embankment toe. For such conditions at the toe, Eqs. 89 become:

$$\sigma_t = \sigma_{ot} + \Delta\sigma_t + z\,\gamma_w \qquad (90a)$$

$$\sigma_t = \underbrace{\overbrace{\bar{\sigma}_{ot} + f(u_c)}^{\downarrow} + u_t}_{\downarrow} + z\,\gamma_w \qquad (90b)$$

$$\sigma_t = \bar{\sigma}_{ft} + z\,\gamma_w \qquad (90c)$$

in which $f(u_c)$ is the part of u_c transmitted to beneath the toe. The effective intergranular pressure is reduced below its initial value to

$$\bar{\sigma}_{ot} = \sigma_{ot} - f(u_c) \qquad (91)$$

The shear strength is correspondingly reduced, and may be dangerously so, since an embankment induces high shear stresses in the zone beneath its toe.

By intercepting such laterally transmitted excess pressure, drain wells serve to minimize reduction in shear strength beneath an embankment toe. Although the danger from this lateral transmission of excess pressure becomes greater as the ratio of horizontal to vertical permeability increases, fortunately the same condition operates to improve the efficacy of the drain wells in attracting radial drainage.

Usage of Drain Wells.—Sand piles have been used to strengthen soft foundations for many years in the Low Countries of Europe and such sand columns undoubtedly accelerated consolidation when loads were added; although it is questionable if constructors of the sand piles contemplated this added benefit or relied on it. To the writer's knowledge the first intended application as true drain wells occurred on the "SVIR 3" power dam constructed from 1930 to 1935 in Russia.[16] Here 12-in. sand-filled wells were installed 40 ft deep in a bed of interstratified clay and sand of Devonian age and appear to have served several purposes. Although they were initially installed for water control during excavation, they seem to have been used later for seepage control as relief wells for reducing seepage pressure in the more pervious buried strata. They also served to accelerate settlement of this concrete dam, as well as rate of gain of shear strength and corresponding resistance against sliding.

In American practice it appears that O. J. Porter, M. ASCE, and his associates at that time in the California Highway Department were the first to design

[16] "Some Features in Connection with the Foundation of SVIR 3 Hydro-electric Power Development," by H. Graftio, *Proceedings*, International Conference on Soil Mechanics and Foundation Eng., Harvard Univ., Cambridge, Mass., Vol. 1, 1936, pp. 284–290.

and install sand-filled drain wells for the expressed purpose of accelerating consolidation—an experimental installation in 1934 followed by several highway embankments over marshes, peats, and harbor silts.[4,5] Early work was characterized by relatively large wells, up to about 30 in. in diameter. Subsequently, drain wells have been used considerably on the west coast of the United States on several important war construction and other projects. Recent large installations are on the Terminal Island Freeway near Los Angeles, Calif.,[17] and on the Bayshore Freeway near San Francisco, Calif.,[18] employing drain wells from 18 in. to 22 in. in diameter and from 30 ft to 60 ft deep. The objective of these installations has been largely for accelerating settlement and, to a lesser degree, for increasing shear strength to reduce toe displacements.

For similar purposes, the Connecticut Highway Department has employed drain wells of 12 in. in diameter and 25 ft in depth on 15-ft centers in crossing a marsh near Old Lyme, Conn.[19] An extensive installation is contemplated on a very wide or "dual-dual" highway near Elizabeth, N. J., using wells 18 in. in diameter on 10-ft centers up to 24 ft deep across a marsh.[20] It is particularly interesting that alternate bids on this New Jersey project showed a saving of more than 30% in favor of drain wells over excavation of the marsh deposit. An application almost solely for accelerating settlement is the experimental installation at LaGuardia Field, New York, N. Y.[21]—17-in. wells about 80 ft deep in silt with spacing varied from 8 ft to 14 ft.

Drain wells for the major purpose of increasing rate of gain in shear strength were employed during the construction of cellular cofferdam walls for submerged shipways.[6] Here 12-in. wells were placed inside the cofferdam cells to stiffen the filling of clay-like material formed from dredged-in marl. As an interesting innovation to speed further the consolidation of the cell fill, the central wells were pumped.

For the prime purpose of reducing lateral transmission of excess pressure, the first installation to the writer's knowledge was employed on the reconstruction of a slide on Riverfront Dike in Hartford, Conn.[22] As a matter of interest, this particular slide is one of the few that was analyzed and predicted in advance. A highway of dredged fill was being added landward soon after substantial completion of the dike, and the highway engineers decided to take the risk in the face of warnings on the probability of a slide, which occurred substantially as predicted. In designing the earthwork phases of the reconstruction for the Providence (R. I.) District of the Corps of Engineers, the writer and the author concentrated three lines of drain wells just inside the toe of the rebuilt embankment to insure against loss of shear strength at this critical zone.

Some ninety drain wells were constructed, in February and March, 1942, through 40 ft of sand and 40 ft of soft, varved clay (see Fig. 9). Wells were

---

[17] "Vertical Sand Drains," *Roads and Streets*, August, 1946, p. 72.

[18] "Vertical Sand Drains Speed Consolidation of Soft Water-Bearing Subsoil," *Engineering News-Record*, March 6, 1947, p. 88.

[19] "Sand Drainage Wells Provide Possibilities for Economy," Highway Research Abstracts, Highway Research Board, October, 1947, p. 6.

[20] "Plan Use of Vertical Sand Drains Under New Jersey Superhighway," *Engineering News-Record*, October 23, 1947, p. 7.

[21] "Sand Drains for LaGuardia Field," *ibid.*, August 28, 1947, p. 7.

[22] "Foundation Failure Causes Slump in Big Dike at Hartford, Conn.," *ibid.*, July 31, 1941, p. 142.

spaced in a triangular pattern on 20-ft centers and, to facilitate construction by hired labor equipment, were 6 in. in diameter. A 6-in. temporary casing was used through the sand, advancing the hole with a churn drill. To minimize smear, the clay was excavated by water jets of a special clay auger designed by F. E. Fahlquist, M. ASCE. Construction cost was about $1.00 per ft of well.

A group of closed-system piezometers was in place, with which it was hoped to measure the effect of these drain wells. However, the effect of daily temperature variations from darkness to daylight made the closed-system piezometers very unsatisfactory and, by removing the Bourdon gages, they were converted to the open-system type (open pipe with filter at the tip). After this change the piezometers also failed to give dependable readings because of the effect of dampening on changing pressures, from the flow into an open-system piezometer from a highly impervious soil. Accordingly, no dependable measurements were obtained on the performance of these drain wells except that, from general observations, it appeared satisfactory.

One question that arose in the design of these Riverfront Dike drain wells concerned a possible tendency of the well to act as a sand pile or hard point in the foundation, attracting load and somewhat relieving stresses in the surrounding compressible deposit. Although this tendency would be less for wells of small diameter than for larger wells with a smaller slenderness ratio, it was decided to place two cushion layers of cinders on the Riverfront Dike wells, with the thought that this material would be more compressible than the sand backfill, and substantially as pervious. However, the cinders obtainable proved to be a poor filter against intrusion of the silt strata in the surrounding varved clay and were difficult to install, tending to sink through water slowly and to mix with the sand. Accordingly, use of the cinder cushion layers was abandoned and most of the wells were backfilled entirely with sand, poured in from the surface.

If a drain well should act as a stiff point in attracting load in spite of the loose state of its sand backfill, the effect on arching of load would apparently be the opposite of the author's solution for equal vertical strain. The effect on consolidation rate would probably not be large, judging from the small difference in rates for free strain and for equal vertical strain shown in Fig. 8. Of more concern is the action of the sand column in deforming under a load concentration as the surrounding, and probably more compressible, soil settles. This deformation may be accomplished either by a bulging of the well column of sand with a small increase in density, or by a local shearing with possibly a relative displacement of the sand column. Bulging is not objectionable and is believed to be the more probable action in soft foundation soils where most drain wells have been installed to date, except on the SVIR 3 dam.[16] For example, to accommodate about 4 ft of settlement, an 18-in. drain well 80 ft long would have to increase uniformly in diameter only about 0.4 in., and considerably less than this by allowing for the densification of the sand backfill from its initially loose state. A definite shearing of the well column could be very undesirable because, if continuity of the drainage path were broken, the break might even nullify the effect on the consolidation contributed by the region of the well below the blocked zone. Such shearing of the sand column would seem more likely to occur in brittle, but still quite compressible, clays and, pending further knowledge, is probably best handled by conservatism in selecting well spacing and size.

In what is believed to be the first application of drain wells to the design of an earth dam, from 1943 to 1945, the writer and the author designed a system of about 240 drain wells to be placed in a triangular pattern on 60-ft centers over substantially the entire valley bottom of Claremont Dam, near Claremont, N. H. Although the design of this dam was essentially completed in 1945 by the Providence District, Corps of Engineers, its construction has been deferred pending further negotiation with local interests. The dam is intended to be about 120 ft high over a foundation of approximately 15 ft of sand and gravel lying over from 40 ft to 80 ft of soft, varved silt (silt, clay, and fine sand interstratified, as shown in Fig. 9, with silt predominant). The weak silt is underlain by impervious glacial till which is likely to create a case of single drainage where excess pore water can escape only in an upward direction. Drain wells 18 in. in diameter, extended to the bottom of the silt with methods to minimize smear, were included in the design because flat slopes to reduce shear stresses and an extended construction period of gradual load application were not in themselves considered sufficient for stability with this very weak silt foundation. The wells are intended both to reduce lateral transmission of excess pressure and to accelerate rate of gain of shear strength.

Conclusions.—For the design of a drain well installation, the author's solutions are excellent examples of the case where a mathematical solution is invaluable in indicating the relative importance of the different variables—ratio of horizontal to vertical permeability, well spacing, well resistance, peripheral smear, and well size. In general, these can be evaluated from the curves which the author has laboriously computed and presented in this paper and in his earlier paper.[7] If the author has other similar curves available, it is felt they would be a worthwhile inclusion in the closing discussion, particularly where they would aid in showing the effect of single variables and comparing the relative importance of the different variables.

The author has properly indicated that certain solutions were developed by L. Rendulic[8] in 1934 and by N. Carillo[9,10] in 1940.[23] For the record, it is worthy of mention that the author also developed substantially the same solutions independently from 1940 to 1942 before learning of the solutions of Messrs. Rendulic and Carillo.

Philip Keene,[24] Assoc. M. ASCE.—A valuable contribution to the subject of drainage wells for hastening consolidation of fine-grained soils, a subject which is receiving increased attention in the field of soil mechanics, has been made by the author. After a study of the paper, the graphs become useful tools in analyzing actual problems and in designing installations, as mentioned by the author. Mr. Barron is also to be congratulated for his solutions showing the effects of smear and well resistance which were not presented before. When the wells are installed without casing or mandrel, the smear zone is probably

[23] "Drainage of Clay Strata by Filter Wells," by K. Terzaghi, *Civil Engineering*, October, 1945, p. 463.
[24] Associate Highway Engr. (Soils and Foundations), State Highway Dept., Hartford, Conn.

less than ½ in. thick and can be ignored for wells larger than 6 in. in diameter. For other cases, the author's solution indicates that the smear zone materially affects the time factor and rate of settlement.

During 1947 the writer designed and supervised the installation of about 400 sand drainage wells on a highway project in Old Lyme, Conn.[25] A fill from 15 ft to 20 ft high was placed for a four-lane divided highway on a foundation of approximately 30 ft of soft organic clayey silt. Final settlements will average about 5.4 ft; 3.5 ft (or 65%) consolidation occurred during the six weeks necessary to place the fill. Assuming the average date of load application to be three weeks after commencement of placing the fill, 4.0 ft (or 75%) consolidation occurred five and one-half weeks after this average loading date. This is about 6% of the time estimated for 75% consolidation if no wells had been installed. Calculations made with the aid of Fig. 3 indicate that the actual time-settlement curve for that project will roughly approximate a theoretical time-settlement curve, if it is assumed that horizontal permeability is ten times as great as vertical permeability. Well resistance was not included in these calculations, although it was actually present in approximately the same value used in Fig. 10.

WALTER KJELLMAN,[26] ESQ.—The consolidation of a fine-grained soil subjected to a load can be accelerated by vertical drains inserted in the soil. In the United States such drains consist of circular sand-filled wells, having normally about a 20-in. diameter and a spacing of from 10 ft to 15 ft. In Sweden, band shaped cardboard wicks of ⅛-in. by 4-in. cross section (Fig. 11) are used. These are furnished with inner longitudinal channels and spaced about 4 ft apart. It may perhaps interest readers to learn how the problems, so clearly dealt with in this paper, are looked upon in Sweden.

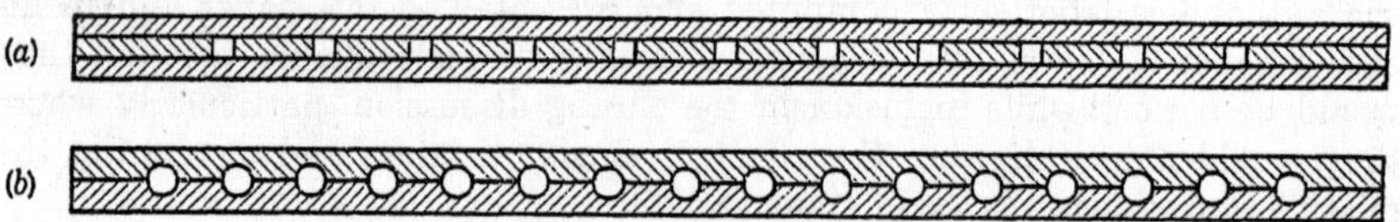

FIG. 11.—CROSS SECTION OF BAND SHAPED CARDBOARD DRAIN
(a) Present Type
(b) Prospective Type

Swedish investigations concerning "deep drainage" of soils were started in 1936. Owing to arching of ground and overburden in all practical cases, the momentary strains were deemed to be nearly equal everywhere in the ground. Therefore, all Swedish calculations refer to this very simple case of equal strain. In the United States, on the contrary, all calculations of this kind prior to Mr. Barron's paper seem to refer to the very complicated case of free strain. Fig. 8 shows that both cases give nearly the same result. Consequently, there is no reason for anyone to become further involved with free strain.

When determining the spacing of the drains in the United States, great reliance is placed on the horizontal draining effectuated by coarse-grained layers.

[25] "Sand Wells Speed Marsh Crossing," by Philip Keene, *Better Roads*, September, 1947.

[26] Head, Statens Geotekniska Institut (The Royal Swedish Geotechnical Institute), Narvavägen 25· Stockholm, Sweden.

In Sweden such layers exist in the lower part of the glacial clay deposits, but they are of little use since the spacing must be determined with regard to the upper part of the deposits (the upper part of the glacial clay and, on top of it, the postglacial clay), which as a rule contains no coarse-grained layers. Furthermore, the fine-grained soils in Sweden seem on the whole to be considerably less pervious than those in America. For these reasons, and because the equivalent radius of the wicks used in Sweden is much smaller than the radius of the American wells, the spacing is considerably smaller in Sweden.

By laboratory tests it has been found that the permeability of a clay without coarse-grained layers is not appreciably reduced by remolding. The unfavorable effect of peripheral smear, discussed by the author, must therefore refer to coarse-grained layers only. No reliance is placed on such layers in Sweden, so that peripheral smear need not be considered.

In Sweden the spacing of the drains is rather small. Thus, the vertical water flow is unimportant when compared to the radial flow and is disregarded. This approximation, which simplifies the calculation and leads to safe results, is justified by the uncertainty of the calculation.

In principle the triangular drain pattern is the most economical one, as stated by Mr. Barron. However, certain considerations have shown that in this respect the difference between the triangular and square patterns is quite unimportant. Therefore, in order to attain certain small advantages when driving the wicks, no fixed pattern is used, other than to make the spacing of the wicks in each row equal to the distance between rows.

The longitudinal channels in the cardboard wick easily can be made numerous enough and wide enough to let through (without appreciable resistance) any water flow that the one drain may be called on to carry. Furthermore, the permeability of the cardboard is very great when compared to that of the fine-grained soils in question. Therefore, there is no need to complicate the calculations by taking into account the flow resistance in the drain. Because the cardboard serves as a perfect filter, there is, of course, no risk of the channels becoming clogged as may occur in sand-filled wells. Laboratory tests have shown that the channels do not collapse even under a very high clay pressure on the wick. This is probably due to arching in the clay, causing the pressure to be much lower on the covers of the channels than between them.

The question has been raised as to whether a well (like a pile) attracts a great part of the load, and whether it can be damaged when the soil settles. A cardboard wick cannot, of course, affect the load distribution. Tests have shown that it may crease, if the vertical compression is great, but that it is not damaged.

In Swedish practice (as follows from the foregoing statements), the time-settlement curve of a deep-drained soil is calculated in the simplest possible manner—assuming equal strain, no coarse-grained layers, radial flow only, no peripheral smear, and no drain resistance. The calculation, made in 1937 but never published, is contained in principle in Mr. Barron's paper. As a result of the calculation the following procedure is used for determining the spacing of the drains (square pattern):

(a) Having assumed a time, t_h, in months, in which a certain percentage of the final settlement is to occur, Fig. 12 is used to find the value of g, which is a quantity of calculation with no physical meaning.

(b) Knowing the value of g, and also the coefficient of consolidation of the soil, c_h (in square centimeters per second), the value of m, which is a quantity of calculation with no physical meaning, is computed from

$$m = g\, c_h \quad \text{(92)}$$

(c) Knowing the value of m, Fig. 13 is used to find a convenient combination of drain radius r_w and spacing S.

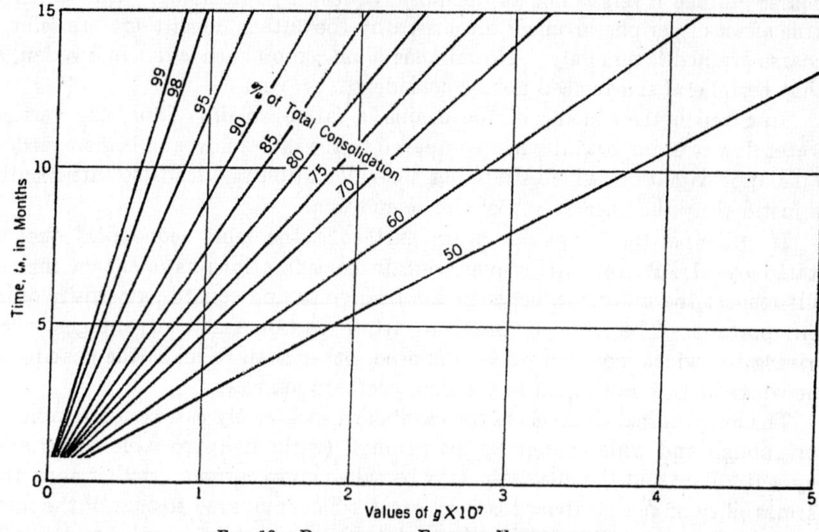

Fig. 12.—Diagram for Finding Values of g

This procedure may be used in cases not too divergent from Swedish conditions.

Obviously, the draining effect of a drain depends to a great extent upon the circumference of its cross section, but very little upon its cross-sectional area. Therefore, the circular cross section has been abandoned in Sweden in favor of the band shaped drain. Certain considerations show that the cardboard wick is as effective as a circular drain with a 1-in. radius. (The circumference of the latter is slightly greater than that of the former.) Thus, the curve in Fig. 13 for a 1-in. radius can be used for the wicks. The curve for a 10-in. radius is valid for the sand wells with a 10-in. radius frequently used in the United States. Comparing these two curves, it appears that wells with 10-ft spacing are equivalent to wicks spaced 6.2 ft apart. Thus, in this case, one well is equivalent to 2.5 wicks.

The wick has several advantages over the well. First, very little material is consumed in its manufacture, and, second, as both its weight and its volume per unit of length are small and as it can be rolled on a drum, it is easy to handle. This means that production can be concentrated in a factory, whence the wicks can be transported inexpensively to the different sites and driven into the ground by a machine working at high speed in about the same manner as a sewing machine.

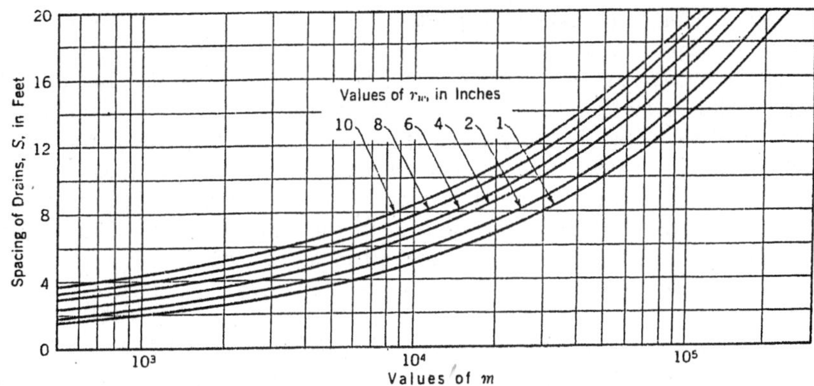

FIG. 13.—DIAGRAM FOR FINDING THE SPACING OF DRAINS

At the Halmsjön Airport outside Stockholm (Sweden) some 3,000,000 ft of cardboard wicks are to be driven in 1947–1949. The cost is between 10¢ per ft and 15¢ per ft. Further information about the wick method will appear in the *Proceedings* of the Second International Conference on Soil Mechanics to be held in Rotterdam, Holland, in 1948.

REGINALD A. BARRON,[27] Assoc. M. ASCE.—The benefits of drain wells have been brought out by Mr. Lane's discussion. Although, probably, they are not quantitatively correct, Eqs. 89, 90, and 91 are of great use qualitatively. The writer believes that Mr. Lane is correct in his statement that column action by the well backfill would redistribute the soil stresses and would result in a slower rate of consolidation. However, Fig. 8 indicates that the difference in consolidation rates is very minor and of no practical importance. In fact, such column action may be much less than the possible compressive effect of the horizontal seepage gradient. This possibility was not considered by the writer but its effect would be to compress the soil radially in a horizontal plane adjacent to the well where the seepage gradients are high. Because of this radial compression, the soil farther out will probably cause vertical settlement as it deforms radially in extension.

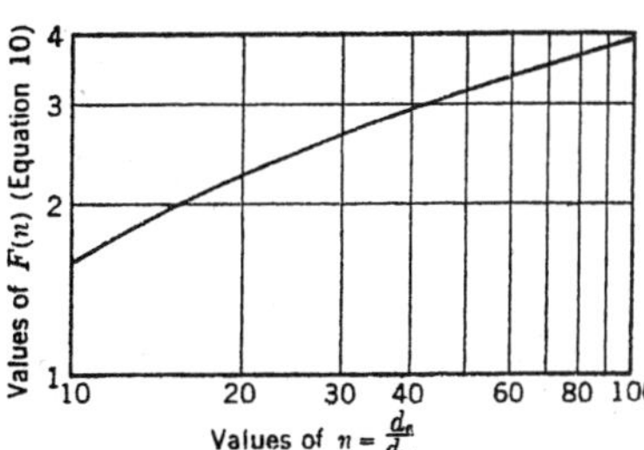

FIG. 14.—CURVE FOR THE DETERMINATION OF $F(n)$

Mr. Keene has discussed the use of consolidation drain wells on a highway project. The use of drain wells to accelerate the consolidation of foundations for highway fills is valuable, especially considering the effect of foundation consolidation on the finished quality of the pavement. On the other hand, such measures are not always necessary for earth dams, and it may be more economical to obtain stability by using flatter embankment slopes.

[27] Head, Special Studies Section, Embankment and Foundation Branch, Soils Div., U. S. Waterways Experiment Station, Vicksburg, Miss.

Mr. Kjellman discusses a very interesting development of the consolidation drain well method. It is possible that, should wick material and installation machines become available in the United States, sand wells may be outmoded. Mr. Kjellman states that "* * * great reliance is placed on the horizontal draining effectuated by coarse-grained layers" and "* * * the fine-grained soils in Sweden seem on the whole to be considerably less pervious than those in America." The writer is not in a position to state exactly what degree of reliance is placed on horizontal drainage in the United States. Furthermore, considering the variable geological history of this country, it is very possible that some of the clay deposits in the United States are quite similar to those of

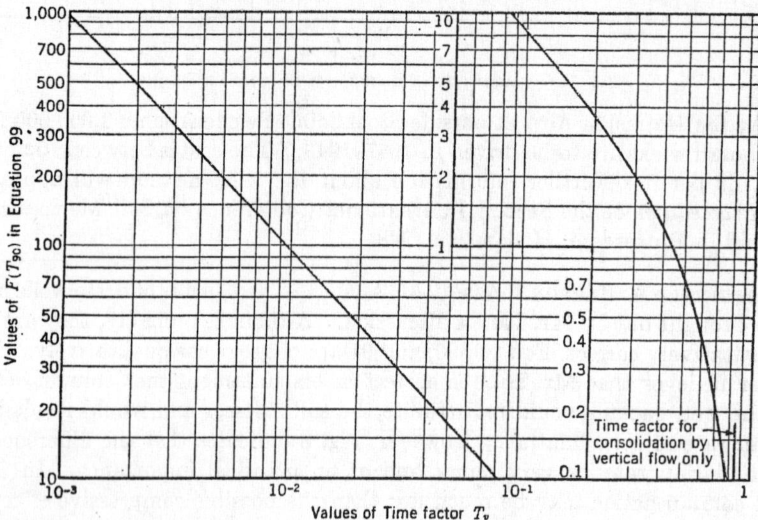

Fig. 15.—Curve for the Determination of Time Factor of 90% Consolidation Due to Combined Radial and Vertical Flow

Sweden. The writer considered the effect of horizontal permeability in the theoretical developments because such conditions existed at sites for which the studies were made. For other cases this difference in permeability may not be so marked.

Mr. Lane requests additional curves indicating the effect of variables. Fig. 14 shows the variation of $F(n)$ as defined in Eq. 10 versus n which is used in the following development for instantaneously applied loads using Eq. 21 in which $\bar{u}_r$ is obtained from Eq. 8. For 90% average consolidation the average hydrostatic excess pressure is 10%. Thus:

$$\bar{u}_{r,z} = \bar{u}_z \, \epsilon^{\lambda} = 0.1 \qquad (93)$$

Expressions for the time factors are

$$T_h = \frac{k_h (1 + e) t}{\gamma_o \, a_v \, d^2_e} \qquad (94)$$

and

$$T_v = \frac{k_v (1 + e) t}{\gamma_o a_v H^2} \quad \ldots\ldots (95)$$

and, therefore,

$$T_h = T_v \frac{H^2}{d^2_e} \frac{k_h}{k_v} \quad \ldots\ldots (96)$$

and

$$\lambda = \frac{-8 T_h}{F(n)} = \frac{-8 T_v H^2 k_n}{k_v d^2_e F(n)} \quad \ldots\ldots (97)$$

Consequently,

$$2.3026 (1 + \log_{10} \bar{u}_z) = \frac{8 T_v H^2 k_n}{k_v d^2_e F(n)} \quad \ldots\ldots (98)$$

or

$$\frac{(1 + \log_{10} \bar{u}_z)}{T_v} = \frac{8 k_h H^2}{2.3026 k_v d^2_e F(n)} = F(T_{90}) \quad \ldots\ldots (99)$$

A curve of $F(T_{90})$ versus T_v is given in Fig. 15.

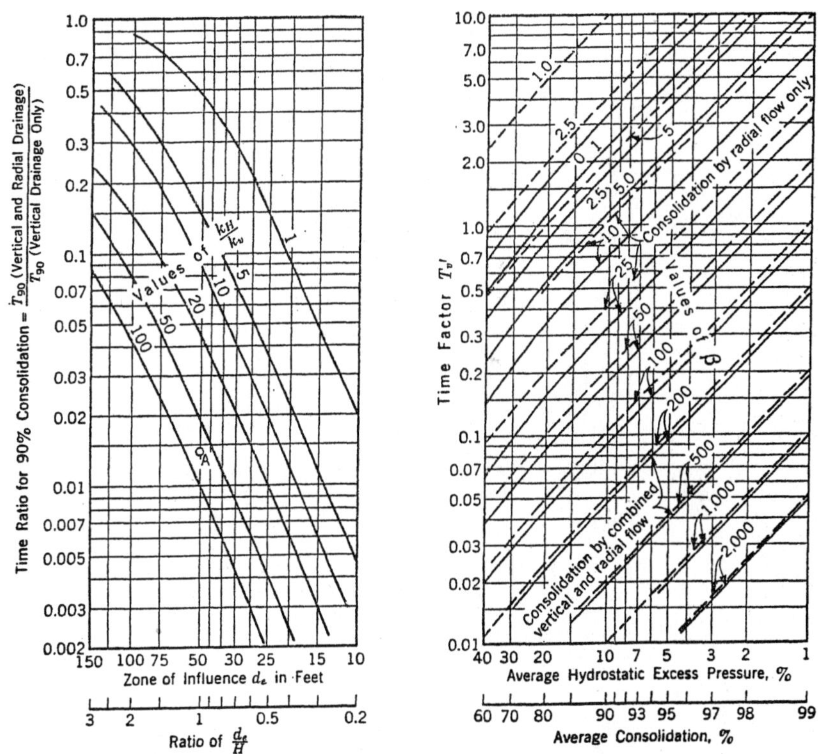

FIG. 16.—EFFECT OF WELL SPACING AND PERMEABILITY RATIO ON TIME REQUIRED FOR 90% CONSOLIDATION

FIG. 17.—AVERAGE PERCENTAGE OF HYDROSTATIC EXCESS PRESSURE AND CONSOLIDATION AT END OF UNIFORM RATE OF LOADING PERIOD

Using the curve in Fig. 15, the curves in Fig. 16 for various permeability ratios were determined for the conditions $d_w = 1$ ft and $H = 50$ ft, with no smear or well resistance. The influence of permeability ratio k_h/k_v and well spacing is evident.

Point A in Fig. 16 was determined for the case of no smear but with well resistance for the condition that $k_w/k_v = 2{,}000{,}000$ and $k_h/k_v = 100$. Well resistance thus requires about 40% more time to reach 90% consolidation; for small ratios of k_h/k_v the effect would be less.

The curves in Fig. 17 for several values of

$$\beta = \frac{8\ k_h\ H^2}{k_v\ d^2_e\ F(n)} \qquad (100)$$

indicate the effects of k_h, k_v, H, d_e, and n on the average percentage of consolidation and average hydrostatic excess pore water pressure at the end of uniform rate of load application.

AMERICAN SOCIETY OF CIVIL ENGINEERS

Founded November 5, 1852

TRANSACTIONS

Paper No. 2351

CLASSIFICATION AND IDENTIFICATION OF SOILS

BY ARTHUR CASAGRANDE,[1] M. ASCE

WITH DISCUSSION BY MESSRS. RALPH E. FADUM; JAMES H. STRATTON; DONALD J. BELCHER; J. A. HAINE AND J. W. HILF; JACOB FELD; KENNETH S. LANE; GEORGE F. SOWERS; RENÉ S. PULIDO Y MORALES; RAYMOND F. DAWSON; D. F. GLYNN; L. F. COOLING; A. W. SKEMPTON, AND R. GLOSSOP; MILTON VARGAS; DONALD M. BURMISTER; M. G. SPANGLER; E. W. LANE; B. K. HOUGH, JR.; AND ARTHUR CASAGRANDE.

SYNOPSIS

The purpose of this paper is threefold: (1) To review and compare existing soil classifications used in civil engineering, with emphasis on their limitations; (2) to present a new soil grouping tentatively adopted by the United States Engineer Department in 1942 for use on airfield projects; and (3) to outline field identification procedures, requiring no apparatus. In abbreviated form, this paper was presented at the Annual Meeting of the Society in New York, N. Y., in January, 1943.

1. INTRODUCTION

The two most controversial chapters in the study of soils for civil engineering purposes are: (*a*) Soil classification, the most confused chapter; and (*b*) the shear strength of soils, the most difficult chapter. If the art of soil classification is far from satisfactory, the confusion is often made worse in that users are unaware of its limitations and apply it for purposes other than that originally intended.

Thorough knowledge of not only one but all classification systems important in civil engineering is the best protection against pitfalls and is the best assurance of an intelligent application of an individual classification. In addition, it provides tools with which the engineer can fashion, if necessary, a new classification to fit his needs in applying soil mechanics to a particular problem. It also frees him from the danger of becoming narrow-minded in this field. Those who really understand soils can, and often do, apply soil mechanics without any formally accepted soil classification.

NOTE.—Published in June, 1947, *Proceedings*. Positions and titles given are those in effect when the paper or discussion was received for publication.

[1] Prof. of Soil Mechanics and Foundation Eng., Graduate School of Eng., Harvard Univ., Cambridge, Mass.

Limitations of Soil Classification Systems.—It is not possible to classify all soils into a relatively small number of groups such that the relation of each soil to the many divergent problems of applied soil mechanics will be adequately presented. Some systems require detailed tests to permit proper grouping of soils; others are specifically designed so that classification is possible on the basis of field identification without quantitative test results. Some classifications were created principally to facilitate making soil surveys, whereas others are chiefly to aid in digesting and correlating empirical knowledge in a specific engineering problem.

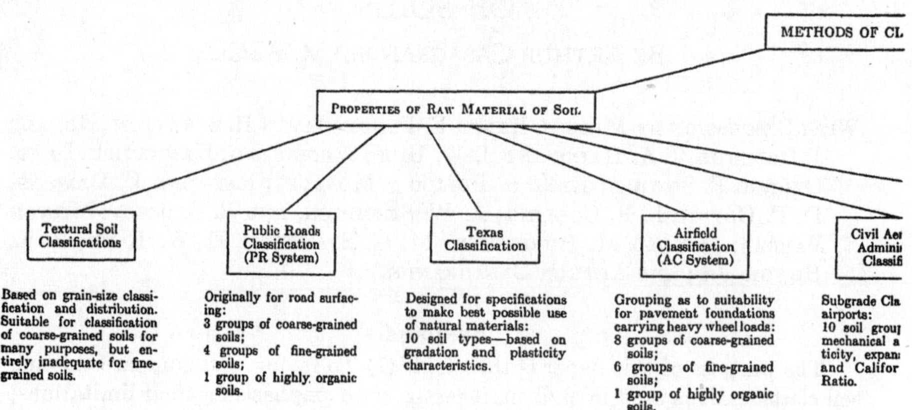

FIG. 1.—METHODS OF

Development of Systems.—To some extent, soil classifications used in civil engineering reflect in their chronological order the development of soil mechanics. In its beginnings this leaned heavily on the soil testing methods developed by agricultural soil science. Since this science originally stressed classification by grain size, the first soil classifications used by civil engineers were textural classifications. Most of these were based on the relative contents of three grain-size fractions called sand, silt, and clay. Later classifications used the entire grain-size curve.

Probably the first departure from the grain-size analysis as the exclusive basis for classification was proposed in 1908 by A. Atterberg in Sweden for agricultural purposes. Later, the Swedish Geotechnical Commission devised and used a similar classification in its extensive research. Still later, the United States Bureau of Public Roads classification was created, with identification and classification of fine-grained soils based almost exclusively on tests other than the grain-size analysis. Finally, new soil classifications have been developed to fill the needs in design and construction of airfields for heavy airplanes.

The essential elements of all classifications described in this paper are summarized in Fig. 1.

2. GRAIN-SIZE SCALES

The division of grain sizes into fractions with arbitrarily standardized boundaries is usually called grain-size classification. It is not really a soil classification but rather a grain-size scale, and will be so termed in this paper, to prevent confusion. Its purpose is to form the basis of grain-size classifications of soils, usually termed textural soil classifications.

The most important grain-size scales, summarized in Fig. 2, will be discussed.

United States Bureau of Soils Grain-Size Scale.—This scale is the one most commonly used in the United States. For practical purposes, and particularly for the textural soil classification as developed by the Bureau, the fraction between 1 mm and 2 mm has always been grouped together with the sand fraction (1).[2]

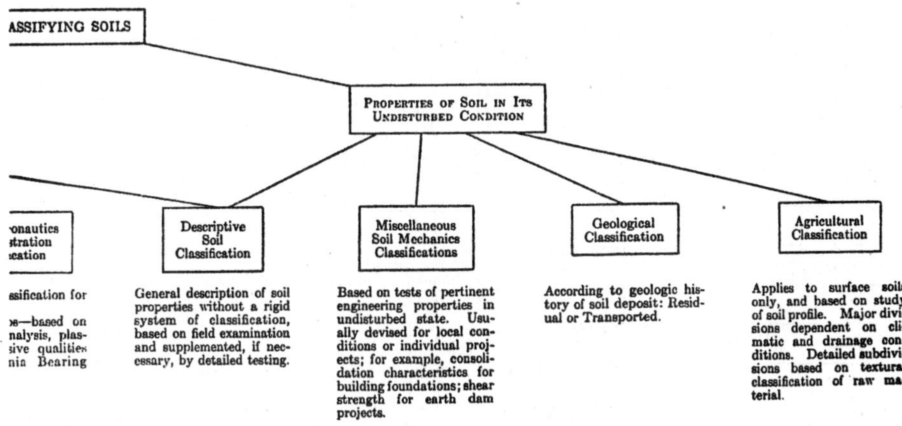

CLASSIFYING SOILS

International Grain-Size Scale.—A. Atterberg's (2) scale was proposed as international standard in 1913 when the International Society of Soil Science (3) stated:

> "It was decided to accept the group division of soil grains recommended by Dr. Atterberg. For the time being the following nomenclature shall be used in the German language [English terms added by the author]:
>
> | | | |
> |---|---|---|
> | Grains larger than 20 mm........... | Stein und Geröll | [stones] |
> | Grains from 20 mm to 2 mm........ | Kies | [gravel] |
> | Grains from 2 mm to 0.2 mm....... | Grobsand | [coarse sand] |
> | Grains from 0.2 mm to 0.02 mm..... | Feinsand | [fine sand] |
> | Grains from 0.02 mm to 0.002 mm... | Schluff | [silt] |
> | Grains finer than 0.002 mm......... | Kolloidale Teilchen oder Rohton | [colloidal particles or raw clay]" |

A noteworthy exception to this proposal was taken in 1914 by J. Kopecky (4), who recommended a modification based on the following divisions: 0.2, 0.06, 0.02, 0.006, and 0.002 mm. It is of interest that the scale adopted by the Massachusetts Institute of Technology (M.I.T.), Cambridge, Mass. (Fig. 2), has essentially the same numerical values.

M.I.T. Grain-Size Scale.—The advantages of the M.I.T. scale are (*a*) that the 0.002-mm limit for the clay fraction has been adopted to correspond to the

[2] Numerals in parentheses, thus: (1), refer to corresponding items in the Bibliography (see Appendix).

international grain-size scale; and (*b*) that it, like that of the Bureau of Soils, possesses the simplicity of division into three major fractions, and permits the plotting of results on a triangular chart, as discussed in Section 3, textural soil classifications.

United States Department of Agriculture (U.S.D.A.) Grain-Size Scale.—The U.S.D.A. modified the original Bureau of Soils scale by (1) lowering the upper boundary for the clay fraction from 0.005 mm to 0.002 mm, to bring this scale in line with the international standard and (2) designating the entire range from 2.0 mm to 0.05 mm as sand.

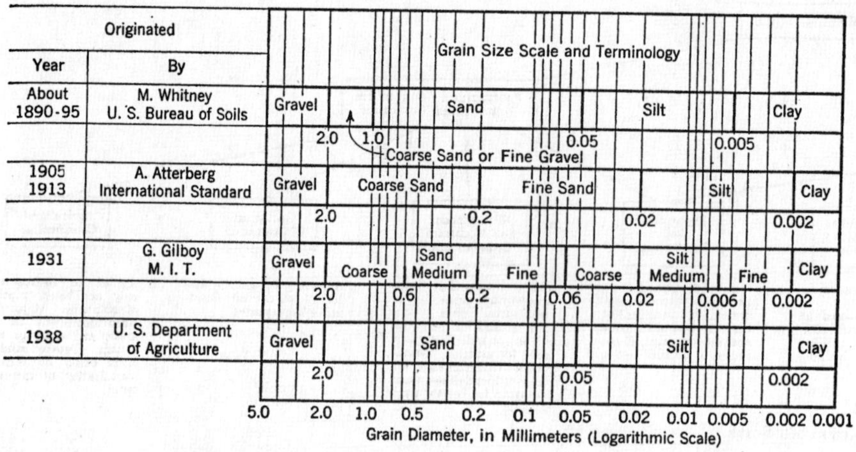

Fig. 2.—Comparison of Principal Grain-Size Scales

Comparison of the M.I.T. and the new U.S.D.A. scales (Fig. 2) shows that their major divisions are practically identical. The various subdivisions are of no consequence since, in fact, they are never used in practical applications.

Limitations of Grain-Size Scales.—Below the range of the sieve analysis, all preceding grain-size scales represent merely two points of the grain-size curve. This is a very incomplete presentation as compared with the complete curve, which can be obtained readily by modern methods of wet-mechanical analysis.

The use of the terms "silt" and "clay" for arbitrary grain-size fractions is misleading. Such terms should be reserved exclusively for the description of soils having the corresponding physical characteristics. To designate the size fractions, one should either use the terms "sand sizes," "silt sizes," "clay sizes," together with the name of the classification used, or state the size without a name. For example, "30% clay sizes according to the Bureau of Soils Classification" would mean "30% < 0.005 mm." Obviously, designation of the grain size itself is much more brief and leaves less doubt as to its meaning. If a publication merely refers to 12% clay, the reader must usually guess what particular grain-size fraction is meant, or whether perhaps 12% of the total material consists of a plastic soil.

Because of the limited usefulness of the grain-size scales, the lack of a uniform standard is less serious than would be expected. Since the new U.S.D.A. scale is almost identical with the M.I.T. scale, either would probably cause less confusion if adopted as a United States standard. In the absence of such a uniform standard, it is desirable to adhere to the designation of the numerical limits of the grain-size fraction, omitting any names for the individual fractions.

In 1945, R. Glossop and A. W. Skempton (5) proposed the M.I.T. grain-size scale as a standard in their valuable contribution on this subject.

3. Textural Soil Classifications

Basis and Types.—The principle of classifying soils purely on the basis of their grain-size distribution dates back to the time when it was not yet realized that the physical properties of fine-grained soils, having the same grain-size distribution, can be widely different. The simplicity of textural soil classifications, and the fact that they can be applied with little experience, are the main reasons why they are still widely used.

In the United States the original textural classifications are based on the relative percentages of the sand, silt, and clay size fractions. They are facilitated by plotting these three fractions on a triangular chart. Therefore, they will be referred to as triangle textural classifications.

For individual engineering projects, textural classifications have been devised which are based on a comparison of the entire grain-size curves of soils with a set of master curves. These classifications will be called master curve textural classifications.

Triangle Textural Classifications.—The most widely used triangle textural classification is the U. S. Bureau of Soils classification (6) shown in Fig. 3. Soils are divided into the ten groups whose names appear inside the triangle. This classification is applied only to that part of a soil smaller than 2 mm in size, with the sand fraction extending from 2.0 mm to 0.05 mm.

The principle of the triangular plot is that any point represents a number on each of the three scales, and these three numbers always add up to 100. Therefore, to classify a soil, it is only necessary to determine the sand and silt fractions in per cent, and to find the corresponding point in the triangle. For example, if the fraction of a soil smaller than 2 mm in size contains 26% of sand sizes, and 32% of silt sizes (automatically making the contents of clay sizes 42%), then the position of this soil on the chart is found by the intersection of the 26% line on the sand scale with the 32% line on the silt scale. It can be seen that this point P (Fig. 3(*a*)) is to be classified as a "clay." If particles larger than 2 mm in size were present in quantities sufficient to be especially noticeable, this would be indicated by adding the term "gravelly" to the name of the soil.

Engineers who have used this classification have often noticed serious discrepancies between the names as derived from the triangle and the actual physical characteristics of the soils. Therefore, they have introduced other subdivisions and, in some cases, other names. For example, Fig. 3(*b*), used by the Lower Mississippi Valley Division of the U. S. Engineer Department, represents an attempt to adjust this method of classification to the soils of the lower Mississippi Valley. It should be noted that "loam," which is really an agricultural term, is not used in Fig. 3(*b*).

The difficulty of all triangle classifications lies in the fact that those physical characteristics of fine-grained soils which are of interest to the civil engineer often are not reflected by the three grain-size fractions. Point P in Fig. 3(*a*) may represent a highly plastic soil in one part of the United States and a slightly plastic silt in another locality.

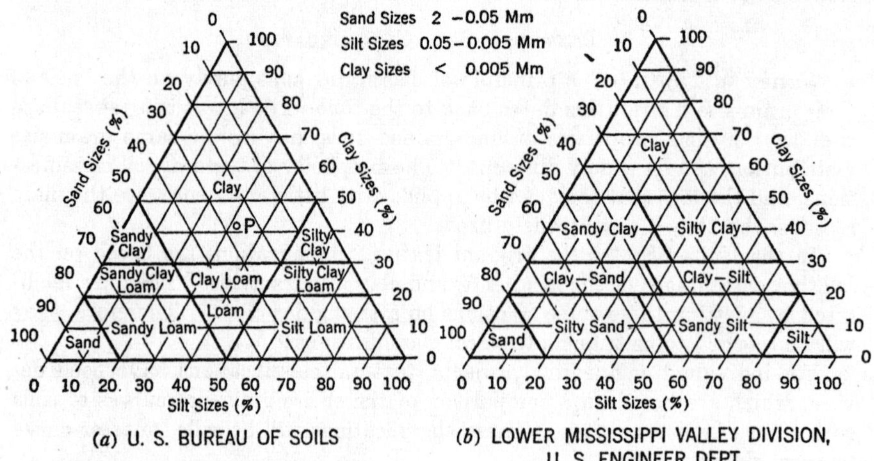

FIG. 3.—COMPARISON OF SOILS CLASSIFICATION TRIANGLES (SIZES IN MILLIMETERS: SAND FRACTION, 2 TO 0.05; SILT FRACTION, 0.05 TO 0.005; AND CLAY FRACTION, LESS THAN 0.005)

The more successful use of the classification in Fig. 3(*a*) in agricultural soil studies may be because in well-developed surface soils certain fine-grained materials such as uniform, nonplastic, rock flour do not exist. It would seem as if the weathering due to the mechanical, chemical, and biological forces, which are responsible for the development of the surface soils, tends to reduce the range of physical properties which a soil with a given grain-size distribution may possess; or, in other words, the grain-size distribution is more likely to reflect the physical properties of a weathered surface soil than of the underlying unweathered materials.

Master Curve Textural Classifications.—The development of modern methods of wet-mechanical analysis which yield the entire grain-size curve as readily as the three fractions used in the triangle classification has led to utilizing the entire grain-size curve for soil classification. This can be accomplished by plotting a set of typical grain-size curves, called master curves, covering the desired range of soils.

Examples of master curve textural classifications are the classifications developed by the Boston Metropolitan District Water Supply Commission (7), by the U. S. Engineer Office in Providence, R. I. (8), and by Donald M. Burmister, Assoc. M. ASCE (8*a*). In the Providence classification, the necessity for further differentiation of fine-grained soils into nonplastic and plastic soils is also recognized.

Textural soil classifications by means of master curves have been reasonably successful for noncohesive soils of the same general geologic origin. However, for most fine-grained soils the same objections apply as for the triangle classification.

TABLE 1.—Comparison of Grain-Size Distribution and Physical Characteristics of Some Fine-Grained Soils

No.	Location (1)	Material (2)
1	Winnipeg, Canada	Very tough, highly plastic clay
2	Massena, N. Y.	Soft, medium plastic, Laurentian clay
3	Massena, N. Y.	Soft, somewhat organic, silty Laurentian clay of low plasticity
4	Vera, Wash.	Kaolin-type clay, with typical lack of toughness
5	California	Diatomaceous earth with characteristics of clayey rock flour
6	Wayland, Mass.	Diatomaceous earth with characteristics of nonplastic rock flour
7	Missouri	Lean, silty clay
8	Hartwick, N. Y.	Typical rock flour

TABLE 1.—(*Continued*)

No.	Percentage by Weight Smaller Than: (Diameters in Millimeters)				Liquid limit	Plastic limit	Plasticity index	Dry strength	AC group symbols[a]
	0.05	0.005	0.002	0.001					
	(3)	(4)	(5)	(6)	(7)	(8)	(9)	(10)	(11)
1	90	75	54	35	96	29	67	Very high	CH
2	96	77	62	50	55	26	29	Medium	CH
3	92	70	50	37	37	24	13	Medium	CL-OL
4	96	76	50	34	55	31	24	Low	MH-CH[b]
5	96	70	47	34	122	80	42	Very low	MH
6	92	52	27	15		Nonplastic		Almost none	MH
7	90	26	16	10	35	22	13	Medium	CL
8	88	20	10	6	24	21	3	Very low	ML

[a] Airfield Classification System symbols; see Table 4. [b] The new KH-group.

Limitations of Textural Soil Classifications.—Table 1 shows the large range of soils covered by almost identical grain-size curves. Soils Nos. 1 to 5 range from a highly plastic, very tough clay having high dry strength, through kaolin-type clays lacking toughness and having only low dry strength, to diatomaceous soils with very low plasticity and only slight dry strength.

It is also instructive to compare the grain-size distribution of soils Nos. 6 to 8 (Table 1) with their other characteristics. The highest percentage of grains smaller than 0.005 mm is shown by the diatomaceous earth which is practically noncohesive and nonplastic. The clay (No. 7) with only 26% material smaller than 0.005 mm actually shows more plasticity and greater dry strength than soil No. 5 containing three times as much material smaller than 0.005 mm.

Many similar discrepancies between grain-size distribution and physical characteristics of fine-grained soils could be cited. Nevertheless, the grain-size curves of fine-grained soils are still extensively used as the exclusive basis for soil classification. In contrast to the misleading impressions which these curves may convey, the pertinent physical properties of such soils can easily be distinguished on the basis of simple field identification procedures.

4. Agricultural Soil Classification

A classification of surface soils for agricultural purposes, called pedological classification, is a complicated system in which the classification of the physical properties of an individual soil layer represents merely one element. Since surface soils have "grown" under the influence of climate, vegetation, type of underlying parent material, topographic features, and age, all these factors, and particularly their influence on the chemistry of the surface soils, play a major role in a pedological soil classification.

Of main interest to the civil engineer is that part of such a classification which describes the physical properties and composition of the surface soils and of the underlying parent materials. In the United States, such description is confined to a textural classification.

Because in a textural classification soils with different physical characteristics would be classified the same, A. Atterberg in Sweden suggested the classification (3) of the physical properties of soils for agricultural purposes shown in Table 2. Further subdivisions are made to distinguish between soils rich in

TABLE 2.—Atterberg Soil Classification

Major divisions	Secondary divisions	Description
A.—Clays (Plastic Soils)		
	I.—Sticky Clays (Highly Plastic)	This group contains only the heaviest clays
	II.—Loamy Clays (Not Sticky)	Subdivided into medium heavy and fairly heavy clays
B.—Loams (Nonplastic, More or Less Cohesive Soils)		
	I.—Fairly Heavy Loams	Clayey loams
	II.—Light Loams	Sandy loams and loess soils
C.—Sand, *Mo*, and Silt Soils (Noncohesive Soils)		
	I.—Capillarity Greater Than 34 Cm	Fine-grained sandy soils; dust loess; subdivided by mechanica analysis
	II.—Capillarity Less Than 34 Cm	Coarse, dry sandy soils, useful only for forestry

humus or lime, and soils containing gravel, stones, and diatoms. To assist in identification, Atterberg proposed the use of his plasticity tests as well as a dry strength number based on the dry strength of the soil. Thus, his classification emphasizes pertinent physical characteristics whose differentiation is not possible by using grain-size analysis alone.

Use of Agricultural Soil Maps in Civil Engineering.—The surface soils of large areas of the United States and other countries have been surveyed and mapped from an agricultural standpoint. Often it is possible to correlate, by means of field observations and tests, the classifications used on such maps with the pertinent engineering properties of these soils so as to save much time in soil surveys for roads and airfields, and in finding sources of satisfactory material for fill construction. Also, with sufficient experience, or advice from agricultural soil experts, the engineer can derive information on the subdrainage characteristics of a given area.

A systematic correlation between agricultural soil classifications, as found on the soil maps, and the engineering properties, has been published for the State of Indiana (9).

5. Geologic Soil Classifications

Every comprehensive soil investigation should also include a geologic classification. In most cases differentiation between the two major subdivisions, that is, residual and transported soils, will be possible on the basis of a general field examination. Further subdivisions of the transported soils, such as glacial, fluvial, lacustrine, marine, aeolian, and volcanic, are a useful and sometimes necessary part of a comprehensive soil classification which, however, may require the assistance of a geologist. When geologic maps are available, they should be utilized in connection with foundation investigations for important civil engineering projects.

6. United States Bureau of Public Roads Classification System

Development and Limitations.—The Public Roads (PR) Classification (10) was developed in the years 1927 to 1929 from extensive research that the U. S. Bureau of Public Roads had directed particularly toward the use of soils in secondary road construction. The original definition of its groups is based on the stability characteristics of soils when directly acted upon by wheel loads. This classification, therefore, is essentially a grouping applicable to the use of soil in a road surface or as a base beneath a thin bituminous wearing surface. Unfortunately, it has been termed "subgrade classification." Attempts to use it for judging the characteristics of subgrades beneath a substantial thickness of pavement and base have frequently led to disqualification of excellent subgrade and subbase materials. The susceptibility of soils to frost action is also not properly recognized in this classification because at the time it was developed, knowledge of frost action was still limited.

A simplified presentation of the PR classification system is shown in Table 3. It differs from the usual presentation (10) to place more emphasis on those characteristics which aid in field identification. To understand the use of the letter A in the group designations, it is noted that the original PR classification also provided for three additional groups designated B-1, B-2, and B-3. These soon fell into disuse and need not be considered here.

Classification Procedure.—The original publication (10) contains recommendations for classifying soils based on grain-size distribution, plasticity, and shrinkage characteristics, as well as on other tests which are not so commonly used. The numerical limits of the test results, which determine the group classification of a particular soil, have undergone many revisions, the latest (11) being dated February, 1942.

Modifications of PR System.—Some of the shortcomings of the PR system were recognized by soils engineers in various highway departments, who tried either to adapt it to their local soil conditions and problems, or to develop new classifications. In New Hampshire, where frost action is the most destructive agent, engineers have eliminated the A–1 group entirely, since soils in this group (such as some well-graded glacial tills) were found to be just as subject to frost heaving as soils in the A–2 group. Therefore, all sandy and gravelly soils of the A–1 and A–2 groups were classified as A–2 materials, and the new A–2 group was subdivided into three subgroups, depending on their frost-heaving characteristics. The definition of the A–3 group was not changed; all soils in this group are considered better subgrades than the soils in the combined A–1 and A–2 groups (12).

TABLE 3.—SIMPLIFIED CHART OF PUBLIC ROADS CLASSIFICATION SYSTEM

Major divisions		Group	General description	Examples
Coarse-Grained Soils		A–1	Coarse with fines—well graded; considerable dry strength and practically no swelling and shrinkage.	Sand-clays in Southern States; some types of glacial till.
		A–2	Coarse with fines—compared with A–1 either deficient in dry strength or subject to volume change.	Many types of glacial deposits and of residual soil.
		A–3	Coarse-grained soils containing less than 10% passing No. 200 mesh and having no dry strength.	Clean, well graded glacial gravel; beach and dune sands.
Fine-Grained Soils	Silts	A–4	Silts and clays of low plasticity	Rock flour; many sandy and silty clays; loess.
		A–5	Elastic silts	Highly micaceous silt; diatomaceous earth.
	Clays	A–6	Highly colloidal types of inorganic clays; medium to high plasticity.	Some inorganic clays from southern and tropical regions.
		A–7	The great majority of inorganic clays and some types of organic clays; medium to high plasticity.	Glacial clays; some types of gumbo; kaolin-type clays; some organic clays.
Peat and Muck		A–8	Materials having very high compressibility.	Peat and organic clays.

Realizing that the PR system is seriously deficient for classifying subgrades, highway engineers have given consideration to a radical revision of this system without changing the designations A–1, A–2, etc. Under a tentative proposal, all gravelly and coarse sandy soils of the A–3 group, which are excellent subgrades, would be classified as A–1 materials, together with those originally defined as A–1. Equally radical changes are proposed for the fine-grained soils. If this is adopted, it would not be possible to gain a clear idea of the type of soil from the PR designation, except by noting the "edition" of the classification. Such confusion could easily be avoided by giving this radical modification an entirely new designation not mistakable for the original, as by using a prefix letter other than A.

7. TEXAS HIGHWAY DEPARTMENT SOIL CLASSIFICATION SYSTEM

About 1939, the Texas Highway Department developed a classification of Texas soils for writing specifications such that natural materials would be used to best advantage in road construction (13). The ten soil types are based essentially on gradation and plasticity characteristics. Type 1, which is typical, is defined as follows:

> Type 1 soil is a material which will retain high stability under the direct impact of traffic loads. Soils of this type are the only materials permitted in the upper six inches of the finished roadway. This type of material can also be used for the roadbed below six inches, if economically available. Type 1 soil includes caliche, rock, shell, iron ore, or gravel with good grading, and excellent soil binder. The detailed requirements vary depending upon whether the aggregate is angular or rounded by nature.

The other soil types are classified in the order of decreasing value in road construction. Types 7 to 9 represent clays of increasing plasticity and type 10 represents a highly compressible material, usually of organic character, having a liquid limit greater than 90.

8. Civil Aeronautics Administration (CAA) Soil Classification System

The CAA system, established in 1944 for classifying soils in the order of their value as subgrade of airport pavements, is based on the mechanical analysis, plasticity characteristics, expansive qualities, and California Bearing Ratio (CBR).

Soils are divided into groups designated by the symbols E–1 to E–10. Their characteristics are briefly described by A. H. Hadfield (14):

> "* * * there are two groups: the E–1 to E–4 'granular soils,' containing 50% or more of sand and the E–5 to E–10 'non-granular soils,' containing less than 55% sand (the E–5 has 65% or less sand).
>
> "The granular soils are further divided into non-frostheave soils, groups E–1 and E–2, and soils subject to frost heave, groups E–3 and E–4. The E–1 soil is a free draining, non-plastic sand corresponding to the PRA classification A–3. The E–2 soil is a sand containing slightly more silt and clay than the E–1. The E–3 corresponds to a non-plastic and moderately plastic PRA A–2 type, and the E–4 is equivalent to the PRA A–2 plastic type.
>
> "In the non-granular group the E–5 soil corresponds to the non-plastic or moderately plastic A–4 silt, and the E–6 to the more plastic A–4 silts, the A–4 and A–6 silty clays, and the A–6 or A–7 clays of low plasticity. The E–7, E–8 and E–9 groups include clay soils of average plasticity, high plasticity and very high plasticity, respectively, covered by PRA's A–6 and A–7 soils groups. The E–10 is the highly elastic soil classed as A–5 by PR."

9. Airfield Classification System for U. S. Engineer Department

Development.—As a result of difficulties in the use of other classifications for the design and construction of military airfields, the Office of the Chief of Engineers tentatively adopted in 1942 a new soil classification which will be referred to as the Airfield Classification (AC) System. It was originally developed by the writer for instruction in Army courses on "Control of Soils in Military Construction" offered at the Graduate School of Engineering, Harvard University, Cambridge, Mass., from 1942 to 1944.

TABLE 4.—AIRFIELD CLASSIFICATION SYSTEM—PHYSICAL CHARACTERISTICS OF SOIL GROUPS AS AFFECTING CONSTRUCTION OF FOUNDATIONS FOR AIRFIELD PAVEMENTS[a]

1		2	3	4		5	6
Major divisions		Soil groups and typical names	Group symbol	GENERAL IDENTIFICATION (ON DISTURBED SAMPLES)		Observations and tests relating to material in place	Principal classification tests (on disturbed samples)
				Dry strength[b]	Other pertinent examinations		
Coarse-grained soils	Gravel and gravelly soils	Well-graded gravel and gravel-sand mixtures, little or no fines	GW	None	Gradation Grain shape Examination of binder wet and dry Durability of grains	Dry unit weight or void ratio Degree of compaction Cementation Stratification and drainage characteristics Ground water conditions Traffic tests Large-scale load tests California bearing ratio tests	Mechanical analysis
		Well-graded gravel-sand mixtures with excellent clay binder	GC	Medium			Mechanical analysis, liquid and plastic limits on binder
		Poorly graded gravel and gravel-sand mixtures, little or no fines	GP	None			Mechanical analysis
		Gravel with fines, silty gravel, clayey gravel, poorly graded gravel-sand-clay mixtures	GF	Very slight to high			Mechanical analysis, liquid and plastic limits on binder if applicable
	Sands and sandy soils	Well-graded sands and gravelly sands, little or no fines	SW	None			Mechanical analysis
		Well-graded sand with excellent clay binder	SC	Medium to high			Mechanical analysis, liquid and plastic limits on binder
		Poorly graded sands, little or no fines	SP	None			Mechanical analysis
		Sand with fines, silty sands, clayey sands, poorly graded sand-clay mixtures	SF	Very slight to high			Mechanical analysis, liquid and plastic limits on binder if applicable

TABLE 4.—(*Continued*)

1		2	3	4		5	6
Major divisions		Soil groups and typical names	Group symbol	General Identification (On Disturbed Samples)		Observations and tests relating to material in place	Principal classification tests (on disturbed samples)
				Dry strength[b]	Other pertinent examinations		
Fine-grained soils containing little or no coarse grained material	Fine-grained soils having low to medium compressibility; liquid limit < 50	Silts (inorganic) and very fine sands, *mo*, rock flour, silty or clayey fine sands with slight plasticity	ML	Very slight to medium	Shaking test and plasticity	Dry unit weight, water content and void ratio	Mechanical analysis, liquid and plastic limits if applicable
		Clays (inorganic) of low to medium plasticity sandy clays, silty clays, lean clays	CL	Medium to high	Examination in plastic range	Consistency, undisturbed and remolded	Liquid and plastic limits
		Organic silts and organic silt-clays of low plasticity	OL	Slight to medium	Examination in plastic range, odor, color	Stratification, root holes and fissures Drainage and ground-water conditions	Liquid and plastic limits from natural condition and after oven drying
	Fine-grained soils having high compressibility; liquid limit > 50	Micaceous or diatomaceous fine sandy and silty soils, elastic silts	MH	Very slight to medium	Shaking test and plasticity	Traffic tests Large-scale load tests	Mechanical analysis, liquid and plastic limits if applicable
		Clays (inorganic) of high plasticity, fat clays	CH	High to very high	Examination in plastic range	California bearing ratio tests	Liquid and plastic limits
		Organic clays of medium to high plasticity	OH	Medium to high	Examination in plastic range, odor, color	Compression tests	Liquid and plastic limits from natural condition and after oven drying
Fibrous organic soils with very high compressibility		Peat and other highly organic swamp soils	Pt	Readily identified		Consistency, texture and natural water content	

TABLE 4.—(*Continued*)

3	7	8		9	10	11	12	13[d]	14	15
Group symbols	Value as foundation when not subject to frost action	VALUE AS WEARING SURFACE FOR STAGE OR EMERGENCY CONSTRUCTION		Potential frost action	Compressibility and expansion	Drainage characteristics[c]	Field compaction characteristics and equipment	Solids at optimum compaction (lb per cu ft) and void ratio, e	California bearing ratio for compacted and soaked specimen	Comparable groups in PR classification
		With dust palliative	With bituminous surface treatment							
GW	Excellent	Fair to poor	Excellent	None to very slight	Almost none	Excellent	Excellent; crawler tractor, rubber tired equipment	>125 e <0.35	>50	A-3
GC	Excellent	Excellent	Excellent	Medium	Very slight	Practically impervious	Excellent; tamping roller, rubber tired equipment	>130 e <0.30	>40	A-1
GP	Excellent	Poor	Poor to fair	None to very slight	Almost none	Excellent	Good to excellent; crawler tractor, rubber tired equipment	>115 e <0.45	25-60	A-3
GF	Good to excellent	Poor to good	Fair to good	Slight to medium	Almost none to slight	Fair to practically impervious	Good to excellent; crawler tractor, rubber tired equipment, tamping roller	>120 e <0.40	>20	A-2
SW	Excellent	Poor	Good	None to very slight	Almost none	Excellent	Excellent; crawler tractor, rubber tired equipment	>120 e <0.40	20-60	A-3
SC	Excellent	Excellent	Excellent	Medium	Very slight	Practically impervious	Excellent; tamping roller, rubber tired equipment	>125 e <0.35	20-60	A-1
SP	Good	Poor	Poor	None to very slight	Almost none	Excellent	Good to excellent; crawler tractor, rubber tired equipment	>100 e <0.70	10-30	A-3
SF	Fair to good	Poor to good	Poor to good	Slight to high	Almost none to medium	Fair to practically impervious	Good to excellent; crawler tractor, rubber tired equipment, tamping roller	>105 e <0.60	8-30	A-2

TABLE 4.—(*Continued*)

3	7	8		9	10	11	12	13[d]	14	15
Group symbols	Value as foundation when not subject to frost action	VALUE AS WEARING SURFACE FOR STAGE OR EMERGENCY CONSTRUCTION		Potential frost action	Compressibility and expansion	Drainage characteristics[c]	Field compaction characteristics and equipment	Solids at optimum compaction (lb per cu ft) and void ratio, e	California bearing ratio for compacted and soaked specimen	Comparable groups in PR classification
		With dust palliative	With bituminous surface treatment							
ML	Fair to poor	Poor		Medium to very high	Slight to medium	Fair to poor	Good to poor; close control essential; rubber tired roller	>100 e <0.70	6–25	A–4
CL	Fair to poor	Poor		Medium to high	Medium	Practically impervious	Fair to good; tamping roller	>100 e <0.70	4–15	A–4 A–6 A–7
OL	Poor	Very poor		Medium to high	Medium to high	Poor	Fair to poor; tamping roller	>90 e <0.90	3–8	A–4 A–7
MH	Poor to very poor	Very poor		Medium to very high	High	Fair to poor	Poor to very poor	>100 e <0.70	<7	A–5
CH	Poor to very poor	Very poor		Medium	High	Practically impervious	Fair to poor; tamping roller	>90 e <0.90	<6	A–6 A–7
OH	Very poor	Useless		Medium	High	Practically impervious	Poor to very poor	<100 e >0.70	<4	A–7 A–8
Pt	Extremely poor	Useless		Slight	Very high	Fair to poor	Compaction not practical			A–8

[a] Table 4 represents the third revision since it was first prepared by the author in 1942. [b] For binder = fraction passing No. 40 U. S. standard mesh. [c] These characteristics do not apply to undisturbed materials having fissures and root holes, such as most surface soils. [d] These weights apply only to soils having specific gravities from 2.65 to 2.75.

In the AC system, it is recognized that a clean gravelly soil, which would be classified A–3 in the PR system, is a better foundation material for pavements than a gravelly soil of the A–1 or A–2 groups, because it is not subject to frost action and is free draining. The AC system also recognizes that a gravelly soil is generally a better foundation material for heavy wheel loads than a sand.

The fifteen groups of the AC system represent a considerably greater number than are used in the other systems. However, the grouping is specifically so devised that soils can readily be classified by an experienced soils engineer on the basis of visual and manual examination.

10. Soil Groups and Group Symbols of AC System

Table 4, representing the third revision by the writer, summarizes the essential information on the AC system including the group symbols. Each symbol consists of two letters which may be considered as initials of the name of the most typical soil of that group. Their meaning will be understood from the following outline:

Coarse-grained soils are subdivided into:

(1) Gravel and gravelly soils; symbol G.
(2) Sands and sandy soils; symbol S.

The gravels and the sands are each divided into four groups:

(*a*) Well-graded, fairly clean material; symbol W, in combinations GW and SW.
(*b*) Well-graded material with excellent clay binder (corresponding to PR A–1 group); symbol C, in combinations GC and SC.
(*c*) Poorly graded, fairly clean material; symbol P, in combinations GP and SP.
(*d*) Coarse materials containing fines, not covered by preceding groups; symbol F, in combinations GF and SF.

Fine-grained soils are subdivided into three types:

(1) The inorganic silty and very fine sandy soils; symbol M, which stands for the Swedish terms *mo* and *mjala* (flour), used for fine-grained nonplastic, or slightly plastic, soils and, in the international grain-size scale, designating the fractions from 0.2 mm to 0.02 mm and from 0.02 mm to 0.002 mm, respectively.
(2) The inorganic clays; symbol C.
(3) The organic silts and clays; symbol O.

Each of these three types of fine-grained soils is grouped according to its liquid limit into:

(*a*) Fine-grained soils having liquid limits less than 50; that is, of low to medium compressibility; symbol L, in combinations ML, CL, and OL.
(*b*) Fine-grained soils having liquid limits greater than 50; that is, of high compressibility; symbol H, in combinations MH, CH, and OH.

Important Note.—In classifying fine-grained soils the term compressibility refers to the slope of the virgin pressure-void ratio curve and not to the actual condition of the undisturbed soil which may be partly dried or otherwise preconsolidated. Experience has proved that the liquid limit is a good measure of relative compressibility.

Highly organic soils, usually fibrous, such as peat and swamp soils having very high compressibility, are not subdivided and are placed in one group; symbol Pt.

Use of Group Symbols.—The sequence of the group symbols need not be memorized, but the meaning of the symbols and the sequence of the major divisions, that is, G–S–L–H should be learned.

When a material does not clearly fall into one group, boundary classifications such as GW–SW or ML–CL should be used.

A study of the soil names in Col. 2, Table 4, will show that these represent no new soil terminology, but are those that experienced soils engineers use when describing a soil on the basis of visual and manual examination.

Emphatically the group symbols are not intended to be substituted for detailed descriptions. However, the use of these symbols is proper when comparing physical characteristics of different soil groups, as in Table 4, references (15)(16).

11. Coarse-Grained Soil Groups of AC System

GW–SW Groups.—These soils are well graded. They comprise gravelly and sandy soils having little or no fines. The presence of the fines must not noticeably change the strength characteristics of the coarse-grained fraction and must not interfere with its free-draining characteristics. In areas subject to frost action, the material should not contain more than about 3% of soil grains smaller than 0.02 mm in diameter. Where frost action is not a factor, contents of the order of 10% passing No. 200 mesh may not be objectionable. The GW and SW soils would be classified in the PR system as A–3 materials.

GC–SC Groups.—These soils also are well graded. They comprise gravel-sand mixtures and sands with an excellent clay binder, or other natural cementing agent, so proportioned that the mixture shows negligible swelling and shrinkage. Thus, dry strength of GC and SC materials is usually provided by a small amount of highly colloidal clay, and in some cases by the cementation of calcareous material (as in coral), or iron oxides (as in iron ore gravels). The letter C in these group symbols conveys the meaning of both "clay" and "cementation."

Materials in the GC and SC groups correspond to the A–1 group in the PR system. They may be used as road surfacing material because they are stable under wheel loads both in wet and dry condition. However, they are usually subject to frost action.

GP–SP Groups.—These soils are poorly graded. They comprise gravels and sands, containing little or no fines, as stated for the GW and SW groups. Materials in the GP and SP groups can be subdivided into:

(*a*) Uniform gravels, such as the clean, round-grained gravel deposited in gravel banks along the middle courses of rivers;

(*b*) Uniform sands, such as beach sand and dune sand; and

(c) Nonuniform mixtures of very coarse material with fine sand, with intermediate sizes lacking, as often results during excavation from mixing of gravel and sand layers.

These groups would be classified A-3 in the PR system.

GF–SF Groups.—Any coarse-grained material which contains detrimental fines (either silty or clayey) and which does not qualify for the GC–SC groups is classified in the GF–SF groups. Silty gravels, silty sands, poorly graded gravel-sand-clay mixtures, and sand-clay mixtures are typical examples. In the PR system they would be classified as A-2 materials.

Identification of Coarse-Grained Soils.—In field identification of coarse-grained materials, the gradation, grain size, shape, and the mineral composition are examined by spreading a dry sample on a flat surface.

The division between gravelly sands and sandy gravels has not been standardized. Many highway engineers classify gravelly sands containing only 20% to 30% of gravel sizes (retained on No. 10 mesh) as gravelly soils. An exact point of division is unimportant, except for specifications, and boundary cases may be classified as GW–SW, and so forth.

Considerable experience is required to differentiate, on the basis of a visual examination, between some well-graded soils and poorly graded soils. When in doubt, a mechanical analysis should be made and the grain-size curve plotted, from which the gradation characteristics can be judged.

Examination of fines is confined to the fraction passing No. 40 mesh (United States Standard), which fraction is usually called the "binder." In the absence of a sieve, this separation can be performed tolerably well by hand. The binder is mixed with water, and its plasticity characteristics and dry strength are examined. The dry strengths are listed in Col. 4, Table 4. For more accurate identification, quantitative determination of the liquid and plastic limit of the binder, and the grain-size curve of the entire material are required.

Identification of active cementing agents other than clay is usually not possible by visual and manual examination, since such agents require a curing period of days, and even weeks. It is on the basis of the actual use of such soils that they are classified into the GC–SC groups. In the absence of such experience, they may be classified tentatively into other groups, neglecting any possible development of strength due to cementation.

Depending on the use to which a coarse-grained soil is put, the durability of its grains may require careful examination. Pebbles and sand grains consisting of sound, igneous rock are easily identified. Weathered material is recognized from its discolorations and the relative ease with which the grains can be crushed. Pebbles consisting of weathered granitic rocks and quartzite are not necessarily objectionable when used for the foundations of pavements. In the process of compaction, some round-grained quartzite gravels, found in southeastern United States, will break up into much smaller, but angular and quite durable, quartz fragments which will form a well-graded stable base. Disintegrated granite in California is extensively used in highway construction. On the other hand, coarse-grained soils containing fragments of shaley rock may be dangerous because alternate wetting and drying may result in partial or complete disintegration. This tendency can be identified by a slaking test. The particles are first thoroughly oven-dried, or sun-dried, then submerged in water for at least 24 hours; finally, their strength is examined and compared with their original strength. Some types of shales will completely disintegrate when subjected to such a slaking test.

The properties of coarse-grained soils in the undisturbed state (natural deposite or man-made fill) require examination if they are to be used in that condition. Depending on the problem, several of the observations and tests listed in Col. 5, Table 4, will be required.

12. Fine-Grained Soil Groups of AC System

In the transition between coarse-grained and fine-grained soils, usually those soils containing more than 50% of material smaller than 0.1 mm in size (passing No. 150 mesh) are classified as fine-grained soils (ML or CL); and soils containing less than 50% smaller than 0.1 mm in size are classified as coarse grained (SF or SC). A non-plastic, silty, fine sand with 40% to 60% passing a No. 150 mesh would be classified as an SF–ML soil.

Of basic importance in the classification of fine-grained soils is the plasticity as measured by the Atterberg liquid and plastic limit tests, or estimated by manual examination (17)(18)(19)(20)(21).

The Plasticity Chart.—The writer's method of using the liquid and plastic limit values for soil identification is illustrated in the plasticity chart, Fig. 4.

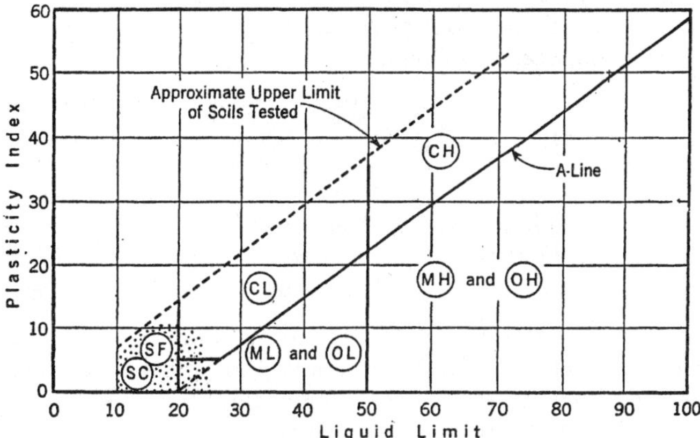

Fig. 4.—Plasticity Chart for AC System

The inclined A-line represents an important empirical boundary. It goes through the point on the base line having a liquid limit of 20, and through the point with a liquid limit of 50 and a plasticity index of 22.0. Therefore, its equation is: Plasticity index = 0.73 × (liquid limit − 20).

The A-line represents the empirical boundary between typical inorganic clays (CL–CH groups) which are generally above the A-line, and plastic soils containing organic colloids (OL–OH groups) which are below it. Also located below the A-line are typical inorganic silts and silty clays (ML and MH groups), except for liquid limits less than 30, for which values inorganic silts may range slightly above the A-line.

For soils having liquid limits less than 25, there is considerable overlapping of several soil groups as shown by the dotted area in Fig. 4. For example, a soil with a liquid limit of 20 and a plasticity index of 8 could be classified according to its grain-size composition and physical characteristics either as a sandy clay (CL), or a very clayey sand (SF), or it may be such a well-graded mixture of sand and clay as to fall into the SC group. Frequently, such soils will be classified as boundary cases CL–SC or CL–SF; or SF–SC. A soil with a liquid limit of 20, and a plasticity index of 0 to 2, may be either a typical rock flour (ML), or a very silty sand with a trace of plasticity (SF), or a boundary case designated SF–ML. In spite of this overlapping, an experienced soils engineer will be able to differentiate between these soil groups purely on the basis of field identification.

Relation Between Position on Plasticity Chart and the Physical Properties.—Experience has shown that the compressibility, when compared at equal preconsolidation load, is approximately proportional to the liquid limit, and that materials having the same liquid limit possess approximately equal compressibility. In comparing the physical characteristics of soils having the same liquid limit, it will be found that, with increasing plasticity index, the cohesive characteristics (toughness and dry strength) increase, whereas permeability decreases.

The behavior of soils at equal liquid limit, with plasticity index increasing, as compared with their behavior at equal plasticity indexes, with liquid limit increasing, is summarized as follows:

Characteristic	Comparing soils at equal liquid limit with plasticity index increasing	Comparing soils at equal plasticity index with liquid limit increasing
Compressibility	About the same	Increases
Permeability	Decreases	Increases
Rate of volume change	Decreases	....
Toughness near plastic limit	Increases	Decreases
Dry strength	Increases	Decreases

After plotting the results of limit tests on a number of samples from the same fine-grained deposit (Fig. 5), generally the points lie on a straight line approximately parallel to the A-line. For example, for the thick stratum of the so-called "Soft Boston Blue Clay," which underlies large areas of Greater Boston, the liquid limit ranges from 30 to more than 70 along a line corresponding approximately to the lower boundary of the area marked "Glacial Clays." Of hundreds of limit determinations on this clay, the average liquid limit is 47 and the average plastic limit is 23.

Position of Typical Inorganic Clays.—The belt of glacial clays in Fig. 5 seems to cover most inorganic clays from Boston, Detroit, Chicago, and northeastern Canada. Much tougher clays are represented by the line showing test results on clays from one locality in Venezuela and by the belt which covers gumbo clays from several southern states.

Kaolin-Type Clays.—Interesting types of fine-grained soils are the kaolin-type clays, usually derived from feldspar in granitic rocks. On the plasticity chart, they plot below the A-line. Although they are generally referred to as clays, they possess some of the characteristics of inorganic silts. They have relatively low dry strength; and in the wet state show some reaction to the shaking test, which is characteristic of the ML–MH groups. However, they are extremely fine grained and smooth, and often contain as great a clay-size fraction as many tough clays located well above the A-line, as shown by soil No. 4 in Table 1. In Fig. 5 the kaolin-type clays in a belt slightly below the A-line are from widely separate deposits. They may be classified as boundary cases, ML–CL, and MH–CH. Test results on a different type of kaolin (from Mica, Wash.) are located along a line which lies in the MH area well below the A-line.

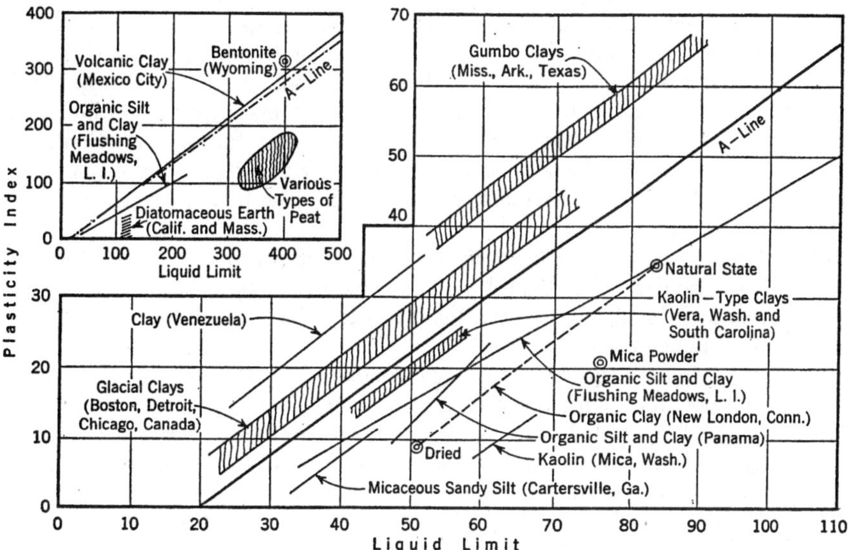

Fig. 5.—Relation Between Liquid Limit and Plasticity Index for Typical Soils

Inorganic Silts, Rock Flour, Micaceous, and Diatomaceous Soils.—The common types of inorganic silts and rock flour have liquid limits less than 30 and are located in the ML area, Fig. 4. A typical rock flour, such as is found in glaciated areas, is usually nonplastic.

Eolian deposits of the loess-type have liquid limits ranging usually between 25 and 35 and vary from nonplastic (that is, the base line) to the A-line, and in exceptional cases also into the CL group slightly above the A-line.

Micaceous soils have a substantially greater liquid limit than a similar soil without the admixture of mica, as illustrated in Fig. 5 by the micaceous, sandy silt from Cartersville, Ga.

Fairly pure diatomaceous earth is generally nonplastic, with liquid limits above 100. With other fine-grained admixtures, diatomaceous soils were found to range throughout the ML and MH groups. The locations on the plasticity chart of two types of fairly pure diatomaceous earth are shown by the small shaded area in the inset of Fig. 5.

Organic Soils.—Even a very small admixture of organic colloidal matter will cause an inorganic clay to increase its liquid limit, without an appreciable change in the plasticity index, thus causing a relative shift to a position below the A-line. In Fig. 5 are shown the positions of several organic silts and clays. Most of the organic silts, silt clays, and clays deposited by the rivers along their lower reaches near the Atlantic seaboard (for example, Hudson River silt, and Potomac River silt) range in liquid limit between 40 and more than 100 and are located on the plasticity chart from 5 units to 15 units below the A-line.

An important property of all fine-grained, organic soils is the radical drop in plasticity due to oven-drying. This is due to the fact that organic colloids are very sensitive and undergo irreversible changes upon drying. In Fig. 5 this drying effect is illustrated by the two points, connected by a dashed line, for an organic clay from New London, Conn. The liquid limit dropped from 84 to 51, and the plastic limit from about 50 to 42, with the corresponding drop in plasticity index from 34 to 9. (Note: Oven-drying also affects the limits of inorganic soils, but to a much more limited extent. Furthermore, the limits may be either raised or lowered by drying, depending on the soil. For this reason, it is not advisable to dry soil samples before subjecting them to the limit tests.)

The limit tests can also be performed on most peaty soils after they are thoroughly remolded. The liquid limit is generally several hundred per cent, and the position of such materials on the plasticity chart is well below the A-line, as illustrated in the inset of Fig. 5.

Volcanic Clays.—Liquid limits much greater than 100 are uncommon for typical, sedimentary, inorganic clays. However, clays which are formed by chemical breakdown of volcanic ash, such as bentonite and the volcanic clays underlying Mexico City, Mexico, have been found to possess liquid limits as high as 500. The inset in Fig. 5 shows the position for a commercially available bentonite, as well as the tremendous range over which the Mexico City clays extend.

13. Field Identification of Fine-Grained Soils

Basis of Field Identification.—It has been shown how the limit tests, supplemented if necessary by other tests, are used to classify fine-grained soils. However, the AC system is so devised that it is possible, with experience, to classify most soils correctly on the basis of field identification procedures alone. The easiest way of learning field identification is under the guidance of an experienced man. To learn without such assistance, one should compare systematically the numerical test results for typical soils in each group with the "feel" of the material while performing the field identification procedures.

The distinguishing characteristics between coarse-grained soils and fine-grained soils have already been described. The principal procedures for field identification of fine-grained soils are (*a*) the shaking test, (*b*) the examination of plasticity characteristics, and (*c*) the examination of dry strength. In addition, observation of color and odor is of value, particularly for organic soils.

Shaking Test.—In this method a wet pat of soil is alternately shaken in the palm of the hand and then squeezed between the fingers. A nonplastic, fine-grained soil will become "livery" and show free water on the surface while being shaken. Squeezing will cause the water to disappear from the surface and the sample to stiffen and finally crumble under increasing finger pressure, like a brittle material. If the water content is just right, shaking the broken pieces will cause them to liquefy again and flow together. The speed with which the pat changes its consistency and the water on the surface appears or disappears, is distinguished as (1) rapid, (2) sluggish or slow, and (3) no reaction. Rapid reaction to the shaking test is typical for nonplastic, uniform, fine sand (SP, SF) and inorganic silts (ML), particularly of the rock flour type; also for diatomaceous earths (MH). With decreasing uniformity the reaction becomes more sluggish. Even a slight content of colloidal clay will impart some plasticity and slow up materially the reaction to the shaking test; this is true, for example, of slightly plastic, inorganic, and organic silts (ML, OL); very silty clays (CL–ML); and many kaolin-type clays (ML, ML–CL, MH, and MH–CH). Extremely slow or no reaction to the shaking test is typical for all clays above the A-line (CL, CH), as well as for highly plastic organic clays (OH).

Plasticity Characteristics.—Examining of the plasticity characteristics should start with a small sample having a soft, but not too sticky, consistency. This sample is alternately rolled into a thread and folded into a lump, and the increase in stiffness of the thread is observed as the plastic limit is approached. After the water content has dropped below the plastic limit, the pieces of the threads should be lumped together, and a slight kneading action on the lump continued until it crumbles.

The higher the position of a soil above the A-line on the plasticity chart (CL, CH), the stiffer are the threads near the plastic limit, and the tougher are the lumps below the plastic limit. Examples for such tough clays are the gumbo clays (CH) shown in Fig. 5. For soils slightly above the A-line (as the glacial clays (CL, CH)), the threads at the plastic limit are medium stiff and the lumps soon start crumbling when the water content drops. Soils below the A-line (ML, MH, OL, and OH) show, with very few exceptions, weak threads near the plastic limit. For organic soils and micaceous inorganic soils, located considerably below the A-line, the threads are very weak and spongy. All soils below the A-line, except OH soils near the A-line, lack cohesiveness when the water content drops below the plastic limit, and the lumps will readily crumble.

When working in a locality where the air humidity is fairly constant, the length of time to reach the plastic limit furnishes a rough indication of the magnitude of the plasticity index. For example, for a gumbo clay with a liquid limit of 70 and a plasticity index of 50 (CH), or a highly plastic organic clay with a liquid limit of 100 and a plasticity index of 50 (OH), a very much longer time will elapse before the plastic limit is reached, as compared with a glacial clay of the CL group. For a slightly plastic silt (ML) the plastic limit will be reached very quickly. To prevent a misleading conclusion it is important that the examination of the plasticity characteristics be started for all soils at about the same consistency.

Dry Strength.—The resistance of a piece of dried soil to crushing by finger pressure is an indication of the character of the colloidal fraction of a soil.

Almost no dry strength characterizes nonplastic ML and MH soils. The dried lumps will fall apart under the slightest finger pressure. Typical rock flour and fairly pure diatomaceous earth are examples.

Low dry strength is representative of all soils of low plasticity, located below the A-line, and of very silty, inorganic clays in the CL group, slightly above the A-line.

Medium dry strength is typical of most clays of the CL group, and of those clays of the CH, MH (kaolin-type clays), and OH groups which are located near the A-line.

High dry strength is indicative of most CH clays, also of those CL clays located high above the A-line, as well as some organic clays of the OH group having very high liquid limits and located near the A-line.

Very high dry strength is developed only by those clays of the CH group located high above the A-line, as the gumbo clays shown in Fig. 5.

Color.—In field soil surveys color is often helpful in distinguishing between various soil strata, and, with sufficient preliminary experience with the local soils, also in identifying the individual soils. To the experienced eye certain dark or drab shades of gray and brown, including almost black colors, are indicative of fine-grained soils containing organic-colloidal matter (OL, OH). In contrast, clean and bright-looking colors, including medium and light gray, olive green, brown, red, yellow, and white, are generally associated with inorganic soils.

Odor.—Organic soils of the OL and OH groups usually have a distinctive odor which, with experience, can be used as an aid in identification. This odor is especially apparent in fresh samples. It is gradually reduced by exposure to air, but can again be brought out by heating a wet sample.

Observations and Tests Relating to Material in Place.—For foundations, the properties of soils in the undisturbed state require special examination. Most important are the denseness of coarse-grained soils and the consistency of fine-grained soils. Also, the consistency of the same soil remolded with its natural water content should be observed. For example, a clay may be described as "stiff and brittle in undisturbed state and very soft and sticky in remolded state." Where the drainage characteristics are of importance, one should observe stratification, root holes, and fissures. Depending on the problem, other tests, such as those listed in Col. 5, Table 4, may be required.

14. Possible Expansion of AC System

General.—Numerous suggestions have been received for the expansion of the airfield classification system. Some have recommended the use of subscripts or letters and numbers added in parentheses, to supply more detailed information. Such additional symbols are useful abbreviations but have the disadvantage that they can be read only by those who have taken the trouble to acquire a working knowledge of such a "shorthand" system. In general, readers of reports and publications should not be burdened with more than a minimum number of symbols; full description in words is preferred.

For soil specialists it may be desirable to arrive at an agreement on the use of possible symbols for describing soils more in detail or on the addition of soil groups. For this purpose the most practical of the suggestions received are summarized below, including the writer's ideas, based on experience accumulated since the new system was formulated.

Coarse-Grained Soils.—Simple field identification methods permit differentiation within the GP and SP groups between uniform soils (such as beach sand) and those materials covering a wide range of sizes but of poor gradation (as a mixture of gravel and fine sand which is deficient in intermediate sizes). By adding the new groups GU and SU for uniform gravels and sands, respectively, and restricting the use of the GP and SP symbols to those materials which have a wide but poor gradation, it is easy to provide for this differentiation.

To distinguish between those soils in the GF and SF groups which possess a silt binder and those having a plastic (clay) binder, the subscripts m and c, respectively, can be used, for example, GF_c, SF_m. The gradation of the coarse fraction could also be described by subscripts, such as, S_uF_m, S_pF_c, and S_wF_m.

Fine-Grained Soils.—The entire range of plasticity is at present divided into the two main groups designated with the letters L and H. Experience has indicated that the L group covers too large a range and that with practice one can distinguish by field identification between soils having a liquid limit ranging well below 35 and those from 35 to 50. The soils in such an intermediate group could be identified with the letter I. Thus, CL would designate a very lean clay having a liquid limit below 35, and located in the plasticity chart above the A-line; CI would be a clay with a liquid limit between 35 and 50, above the A-line; and CH would be a fat clay with a liquid limit larger than 50, above the A-line.

In localities where kaolin-type clays require attention, it would be desirable to add such clays as separate groups for which the symbols KL, KI, and KH suggest themselves. This illustrates the flexibility of this classification, which permits adding new groups without affecting the main framework. It should be kept in mind, however, that any expansion should not be beyond the possibilities of field identification. From that viewpoint there is no objection to introducing the kaolin-type clays as a separate group, because anyone who has handled such clays and learned their "feel" will be able to recognize them readily.

A comparison of the original groups of fine-grained soils with this possible expansion is illustrated in Fig. 6. It will be noted that the order has also been changed to conform better to the order of usefulness of soils in foundation and earthwork engineering.

15. Descriptive Soil Classification

Describing a soil in detail on the basis of a careful visual and manual inspection, utilizing more or less generally accepted nomenclature, classifies this soil, although no rigid system is used. This approach will be designated as a "descriptive soil classification." As such, it may be considered the oldest and most widely used classification. Not only field engineers and boring foremen, but also soil mechanics experts use this extensively, often in preference to a more rigid system.

The disadvantage of the descriptive classification lies in a lack of uniformity of soil terminology. The abuse of classifications by giving only the group number of a soil without any further information, has perhaps been the greatest obstacle to the development of a more uniform soil nomenclature. Therefore, the application of any system of classification as an excuse to reduce the description to a minimum or to eliminate it entirely must be strongly condemned. Instead, the main emphasis should always be on a complete verbal description of every pertinent characteristic of a soil, in its undisturbed state as well as after it is reworked.

Original Groups	Suggested Expansion
ML CL OL	CL KL ML OL —Liquid Limit = 30
	CI KI MI OI —Liquid Limit = 50
MH CH OH	CH KH MH OH

Fig. 6.—Suggested Expansion of AC Groups for Fine-Grained Soils

For example, a soil might be described as "olive green, homogeneous, medium-plastic, silty clay, medium stiff and brittle in the undisturbed state, very soft when remolded." This descriptive classification will convey to the reader, familiar with this terminology, a fairly clear picture of this soil. In contrast, if it is designated as an A–4 or A–7 soil, the information is entirely inadequate for any engineering use.

If a soil surveyor is forced to describe each soil in detail, he will develop the habit of examining it with greater thoroughness than if he is asked merely to give the group number or symbol. He can, for convenience, use his own abbreviations for frequently recurring characteristics thereby creating a more detailed classification than is afforded by any existing rigid classification. Such abbreviations may be in the form of numerous subdivisions of certain classes within an existing system.

For some purposes the descriptive classification should emphasize specific soil characteristics. Regarding compaction, it was often found useful to group soils into (*a*) noncohesive, coarse-grained soils (that is, soils having no dry strength); (*b*) typical cohesive soils (soils having considerable dry strength); and (*c*) those fine-grained soils having little or almost no dry strength.

Descriptive Adjectives.—The most common adjectives used in describing soils are:

Grain Size

(1) Very coarse
(2) Coarse
(3) Medium coarse
(4) Medium fine
(5) Fine
(6) Very fine

Grain Shape

(1) Angular
(2) Subangular
(3) Subrounded
(4) Rounded
(5) Well rounded

Gradation

(1) Very uniform
(2) Uniform
(3) Poorly graded
(4) Fairly well graded
(5) Well graded
(6) Very well graded

Compactness

(1) Very loose
(2) Loose
(3) Medium dense
(4) Dense, firm, compact
(5) Very dense, well compacted

Consistency

a. Consistency Grading—
(1) Very soft
(2) Soft
(3) Medium stiff
(4) Stiff
(5) Hard

b. Other Consistency Characteristics—
(1) Brittle
(2) Friable
(3) Elastic
(4) Spongy
(5) Sticky

Plasticity

a. Plasticity Grading—
(1) Nonplastic
(2) Trace of plasticity
(3) Medium plastic
(4) Highly plastic

b. Cohesiveness Near Plastic Limit—
(1) Very weak
(2) Weak
(3) Firm
(4) Medium tough
(5) Tough
(6) Very tough

Dry Strength

(1) Very slight
(2) Slight
(3) Medium
(4) High
(5) Very high

Examples of Descriptive Soil Classification.—Examples of descriptive soil classification will indicate how the adjectives can be combined into a detailed description:

(1) Uniform, fine, clean sand with rounded grains. Equivalent to A–3; SP (or SU in expanded AC system).

(2) Well-graded, gravelly silty sand; angular gravel particles, $\frac{1}{2}$-in. maximum size; rounded and subangular sand grains; a very silty binder of low dry strength with no plasticity; well compacted and moist in undisturbed state. Equivalent to A–2–3; GW–GF.

(3) Organic silt; medium stiff and brittle in undisturbed state and very soft and sticky in remolded state; medium dry strength; weak and spongy at plastic limit; crumbles readily below plastic limit. Equivalent to A–7, OL–OH.

(4) Highly plastic (fat) clay; stiff at the plastic limit; forms a tough lump when drying is continued considerably below the plastic limit; medium stiff in undisturbed state, soft and sticky when remolded. Equivalent to A–6–7; CH.

Acknowledgment.—The writer wishes to express his appreciation to S. D. Wilson for stimulating assistance.

APPENDIX. BIBLIOGRAPHY

(1) **"Mechanical Analysis of Soils,"** *Bulletin No. 4*, U.S.D.A., 1896.

(2) **"Die Rationelle Klassifikation der Sande und Kiese,"** by A. Atterberg, *Chemiker-Zeitung, No. 15*, 1905.

(3) *Internationale Mitteilungen für Bodenkunde* (rept., meeting of the International Comm. for Mech. and Physical Soil Investigations, October 31, 1913), Vol. IV, No. 1, 1914.

(4) **"Ein Beitrag zur Frage der neuen Einteilung der Körnungsgruppen,"** by J. Kopecky, *Internationale Mitteilungen für Bodenkunde*, Vol. IV, No. 2/3, 1914.

(5) **"Particle-Size in Silts and Sands,"** by R. Glossop and A. W. Skempton, *Journal*, Inst. C. E. (Great Britain), December, 1945, p. 81. (*a*) p. 102, Table V.

(6) **"Grouping of Soils on the Basis of Mechanical Analysis,"** by R. O. E. Davis and H. H. Bennett, *Circular No. 419*, U.S.D.A., July, 1927.

(7) **"Permeability Determination, Quabbin Dam,"** by Stanley M. Dore, *Transactions*, ASCE, Vol. 102, 1937, p. 682.

(8) Discussion of the paper, **"Graphical Representation of the Mechanical Analyses of Soils,"** by Frank B. Campbell, *ibid.*, Vol. 104, 1939, p. 169. (*a*) p. 156.

(9) **"The Formation, Distribution and Engineering Characteristics of Soils,"** by D. J. Belcher, L. E. Gregg, and K. B. Woods, *Research Series No. 87*, Purdue Univ., Lafayette, Ind., January, 1943.

(10) **"Interrelationship of Load, Road and Subgrade,"** by C. A. Hogentogler and Charles Terzaghi, *Public Roads*, May, 1929, p. 1.

(11) **"Classification of Soils and Control Procedures Used in Construction of Embankments,"** by H. Allen, *ibid.*, February, 1942, p. 263.

(12) **"The Application of Soil Mechanics to Highway Foundation Engineering,"** by J. O. Morton, *Proceedings*, International Conference on Soil Mechanics and Foundation Eng., Cambridge, Mass., Vol. III, 1936, p. 244.

(13) **"Materials and Tests,"** *Manual of Instructions*, Texas Highway Dept., Vol. 3, Pt. 4, 1939.

(14) **"Airport Pavement Design,"** by A. H. Hadfield, *Journal*, Boston Soc. of Civ. Engrs., July, 1944, p. 157.

(15) *Engineering Manual*, Chapter XX, Office of the Chf. of Engrs., War Dept., March, 1943.

(16) **"Soil Tests for Design of Runway Pavements,"** by T. A. Middlebrooks and G. E. Bertram, *Proceedings*, 22d Annual Meeting, Highway Research Board, National Research Council, December, 1942, p. 144.

(17) **"Über die physikalische Bodenuntersuchurg und über die Plastizität der Tone,"** by A. Atterberg, *Internationale Mitteilungen für Bodenkunde*, Vol. 1, 1911.

(18) Statens Järnvägars Geotekniska Kommission 1914–1922, Statens Järnvägar, Geotekniska Meddelanden, Stockholm, 1922.

(19) **"Erdbaumechanik auf Bodenphysikalischer Grundlage,"** by K. Terzaghi, Vienna, 1925.

(20) **"Adaptation of Atterberg Plasticity Tests for Subgrade Soils,"** by A. M. Wintermyer, *Public Roads*, Vol. 7, August, 1926, p. 119.

(21) **"Research on the Atterberg Limits of Soils,"** by A. Casagrande, *ibid.*, Vol. 13, October, 1932, p. 121.

(22) **"Preparation of Soil Samples for Mechanical Analysis and Determination of Subgrade Constants,"** Designation D 421–39, *A.S.T.M. Standards*, 1944, pp. 614–615.

(23) **"Exploration of Soil Conditions and Sampling Operations,"** by H. A. Mohr, *Publication No. 376*, Graduate School of Eng., Harvard Univ., Cambridge, Mass., 3d Ed., November, 1943, pp. 9–12.

(24) **"Definitions of Soil Components,"** Rept. of Committee VII on Foundations and Soil Mechanics, *Civil Engineering Bulletin*, Civ. Eng. Div., Am. Soc. for Eng. Education, March, 1947.

(25) "The Preparation of Subgrade," by O. J. Porter, *Proceedings*, Highway Research Board, National Research Council, Washington, D. C., Vol. 18, 1938.

(26) **"Soil Survey Practice in the United States,"** by Levi Muir and William F. Hughes, *ibid.*, Vol. 19, 1939, p. 467.

(27) **"Engineering Significance of National Bureau of Standards Soil Corrosion Data,"** by K. H. Logan, *Paper No. RP 1171*, National Bureau of Standards, Washington, D. C., January, 1939.

(28) **"Soil-Cement Mixtures, Laboratory Handbook,"** Portland Cement Assn., Chicago, Ill., 1939–1942.

(29) **"State-Wide Highway Planning Survey Soil Study,"** *Bulletin No. 6*, Nebraska Dept. of Roads and Irrig., Lincoln, Nebr., 1939.

(30) **"Special Papers on the Pumping Action of Concrete Pavements,"** *Research Reports No. 1–D*, Highway Research Board, National Research Council, Washington, D. C., Vol. 25, 1945.

(31) **"Characteristics and Uses of Loess in Highway Construction,"** by R. E. Bollen, *American Journal of Science*, May, 1945, p. 283.

(32) **"Special Papers on the Pumping Action of Concrete Pavements,"** *ibid.*, 1946.

(33) **"Flexible Pavement Test Section For 300,000 lb. Airplanes, Stockton, California,"** by R. A. Freeman and O. J. Porter, *Proceedings*, Highway Research Board, National Research Council, Washington, D. C., Vol. 25, 1945, p. 23.

(34) **"Longitudinal Cracking of Concrete Pavements on State Highway 13 in Clark and Taylor Counties, Wisconsin,"** by H. F. Janda, *ibid.*, Vol. 15, 1935, p. 157.

(35) **"Experimental Soil-Cement Stabilization at Cheboygan, Michigan,"** by W. S. Housel, *ibid.*, Pt. II, Vol. 17, 1937, p. 46.

(36) **"Soil-Cement Stabilization in Missouri,"** by F. V. Reagel, *ibid.*, p. 66.

(37) **"Sampling, Soil Classification and Cement Requirement—North Carolina,"** by L. D. Hicks, *ibid.*, Vol. 19, 1939, p. 521.

(38) **"Experimental Sand-Clay Base Course in North Carolina,"** by James A. Kelley, *Public Roads*, April–May–June, 1947, p. 299.

(39) *Transactions*, ASCE, Vol. 110, 1945, p. 324.

(40) **"Application of Geologic and Soils Principles to Highway Research,"** by F. R. Olmstead, paper presented at the seminar on engineering geology, Section on Eng. Geology, U. S. G. S., Denver, Colo., November, 1946.

(41) **"Aviation Engineers,"** U. S. War Dept., *Technical Manual TM 5-255*, U. S. Govt. Printing Office, Washington, D. C., April 15, 1944.
(42) **"Terra, a Philosophical Essay of Earth,"** by John Evelyn, London, 1678.
(43) **"Traite Experimental et Analytique de la Poussee des Terres,"** by M. Mayniel, D. Colas, Paris, 1808, p. 102.
(44) *Ibid.*, Book IV, p. 166.
(45) **"Treatise on Soils and Manures,"** by a Practical Agriculturist, London, 1818.
(46) **"Classification System for Composite Soils,"** by D. M. Burmister, *Engineering News-Record*, July 31, 1941, p. 61.
(47) **"Practical Methods for the Classification of Soils,"** by D. M. Burmister, *Proceedings*, Purdue Conference on Soil Mechanics, Purdue Univ., Lafayette, Ind., July, 1940, p. 129.
(48) **"Soil Investigations,"** Eng. Manual for Civil Works, Exhibit 4, Pt. VI, Office of the Chf. of Engrs., Dept. of the Army, Washington, D. C., September, 1945.
(49) **"Engineering For Dams,"** by W. P. Creager, J. D. Justin, and J. Hinds, John Wiley & Sons, Inc., New York, N. Y., Vol. III, 1945, p. 625.
(50) **"L'Analyse Méchanique,"** by H. Geesner, Dunod, Paris, 1936, p. 329.
(51) **"Classification of Highway Subgrade Materials,"** *Proceedings*, Highway Research Board, National Research Council, Washington, D. C., Vol. 25, 1945, pp. 376–392.
(52) **"Mechanical Composition of Soil in Relation to Field Descriptions of Texture,"** by T. J. Marshall, *Bulletin No. 224*, Council for Scientific and Industrial Research, Commonwealth of Australia, Melbourne, 1947.
(53) **"Micropedology,"** by Walter L. Kubiena, Iowa State College Press, Inc. (formerly Collegiate Press, Inc.), Ames, Iowa, 1938.
(54) **"Soil-Cement Mixtures, Laboratory Handbook,"** Portland Cement Assn., Chicago, Ill., 1946.
(55) **"Report of Subcommittee on Sediment Terminology,"** *Transactions*, Am. Geophysical Union, December, 1947, pp. 936–938.
(56) **"A Scale of Grade and Class Terms for Clastic Sediments,"** by C. K. Wentworth, *Journal of Geology*, Vol. 30, 1922, pp. 377–392.
(57) **"Mechanical Composition of Clastic Sediments,"** by J. A. Udden, *Bulletin*, Geological Soc. of America, Vol. 25, 1914, pp. 655–744.
(58) **"Sealing the Lagoon Lining at Treasure Island with Silt,"** by Charles H. Lee, *Transactions*, ASCE, Vol. 106, 1941, p. 577.
(59) **"Earth Materials Investigation Manual,"** Bureau of Reclamation, U. S. Dept. of the Interior, Denver, Colo., November, 1947.

DISCUSSION

RALPH E. FADUM,[3] ASSOC. M. ASCE.—The critical review of the soil classification systems used in civil engineering, as presented by the author, is a worthy contribution to the literature of soil mechanics. The author is to be commended for directing attention to the inadequacies of the much used textural classification systems according to which all soils are distinguished on the basis of grain-size composition. He has submitted ample evidence to support his thesis that the grain-size composition does not, by itself, reflect the significant physical properties of fine-grained soils.

In the proposed Airfield Classification (AC) System the author distinguishes between two primary soil groups—the so-called coarse-grained group and the fine-grained group. Sands and gravels comprise the first group and silts and clays the latter group. Soils of the coarse-grained group are distinguished primarily on the basis of grain-size composition. In general, a visual examination will suffice to differentiate the eight soil types included in this group. The six soil types included in the fine-grained group, on the other hand, are distinguished primarily by the degree of plasticity that they exhibit as measured by the simple, manual tests described by the author. It is noteworthy that in accordance with this system fourteen soil types, which have significantly different physical properties, can, with experience, be recognized by simple visual and manual tests that can be made without the aid of laboratory facilities.

The proposed system is flexible and, accordingly, suitable for classifying soils with respect to performance characteristics under a wide variety of conditions. The author has referred to the AC system as the Airfield Classification System. This name connotes a restrictive meaning that does not do justice to its general applicability.

The author recommends (see heading, "12. Fine-Grained Soil Groups of AC System") that 0.1 mm be used as the boundary size between the coarse-grained and fine-grained groups—

> "* * * usually those soils containing more than 50% of material smaller than 0.1 mm in size (passing No. 150 mesh) are classified as fine-grained soils (ML or CL); and soils containing less than 50% smaller than 0.1 mm in size are classified as coarse grained (SF or SC)."

This arbitrary limit provides a convenient method of separating a soil into its component fractions for detailed laboratory analysis. If a soil is dispersed in water, the particles larger than approximately 0.1 mm will settle out very quickly; those smaller than this size will remain in suspension. Thus, the coarse-grained and fine-grained fractions can be separated conveniently by a process of successive washing and decanting. After drying, the coarse fraction s suitably prepared for a dry sieve analysis. The wash water containing the fines can be used for hydrometer analysis (if the gradation of this fraction is considered significant), or the fines can be recovered from the wash water

[3] Prof. of Soil Mechanics, School of Civ. Eng. and Eng. Mechanics, Purdue Univ., Lafayette, Ind.

and subjected to other tests that may be prescribed for the so-called binder fraction.

The binder fraction is usually defined (22) as that fraction of a soil the constituents of which pass a standard No. 40 sieve. The writer is of the opinion that the results of the physical tests such as the Atterberg limit tests as determined from the binder fraction, consisting of the fraction of a soil smaller in size than 0.1 mm, would be more significant than the results obtained from the same tests performed on a binder fraction consisting of all material passing the No. 40 sieve.

In his discussion of the various methods of classifying soils the author includes one that he refers to as descriptive soil classification (see heading, "15. Descriptive Soil Classification"). This method, which is used extensively, consists in "Describing a soil in detail on the basis of a careful visual and manual inspection, utilizing more or less generally accepted nomenclature * * *." The author states (see heading, "15. Descripture Soil Classification") that: "The disadvantage of the descriptive classification lies in a lack of uniformity of soil terminology." The following remarks are directed particularly toward the soil terminology used in a descriptive soil classification system.

Confusion in terminology has arisen from the practice of differentiating soil types quantitatively on the basis of grain size composition, from the use of provincial names such as Boston Blue Clay, Chicago Blue Clay, etc., and from the use of terms such as marl and loam that do not convey sufficiently specific meanings. To be acceptable and useful, terminology used in a descriptive soil classification system should:

1. Consist of terms with which a boring foreman, an earthwork engineer, and a soil mechanician are equally conversant;
2. Describe the significant characteristics that can be determined quickly without the aid of laboratory facilities;
3. Be discriminating; and
4. Convey a well-defined meaning.

For purposes of soil identification and descriptive classification it is suggested that the following three size fractions be recognized as significant in describing the grain-size composition of a soil—the gravel fraction, the sand fraction, and the fine-grained fraction (composed of particles smaller than 0.1 mm in size). The composition in terms of these fractions can be expressed conveniently by a number system, as follows: 50–10–40 designates a soil composed 50% by weight of gravel, 10% by weight of sand, and 40% by weight of fines; and 0–20–80 designates 20% by weight of sand and 80% by weight of fines. If graphical representation is preferred, a triangular chart similar to those shown in Fig. 3 could be used. Reasonable estimates of composition according to these fractions suitable for identification purposes can be made without the aid of laboratory facilities.

As stated, sands and gravels are distinguished solely on the basis of the size of their constituent particles. Factors other than grain size, however, must be considered in distinguishing between silts and clays. The author has directed attention to the fact that the grain-size composition of two fine-grained soils may be identical; yet, the physical properties can be significantly different. The effect of mineralogical composition, which is reflected by the shape of the soil particles, dominates the effect of grain size on the properties of a fine-grained soil. Soils composed essentially of "clay-size" minerals, as defined by the various classification scales in current use, may exhibit either silt-like or clay-like properties. According to engineering concepts a fine-grained soil or fraction is identified on the basis of the degree of plasticity and dry strength that it exhibits. A clay-like soil can be made to exhibit plastic properties by an adjustment of its water content and it has considerable strength when air-dried. (The term plastic is used, herein, to describe that property of a material which enables it to be deformed quickly without exhibiting a noticeable elastic rebound or volume change and without cracking or crumbling.) A silt-like soil cannot be made plastic and it exhibits little or no strength when air-dried. There are no simple, direct methods suitable for routine identification and classification purposes by which the clay-like constituents of the fine-grained fraction can be separated quantitatively from the silt-like constituents. It is suggested, therefore, that the terms silt and clay be used in a qualitative rather than in a quantitative sense.

In light of the foregoing remarks and in the interest of promoting agreement on soil terminology to be used in a descriptive soil classification system, the following summarized recommendations are proposed:

1. The terms "gravel," "sand," "silt," and "clay," used singly or in combination, shall form the basic elements of a soil name. The name shall be qualified by the use of carefully selected adjectives so as to make the description of a soil complete and discriminating.
2. Distinction between silts and clays shall not be based upon the size of the constituent particles.
3. The terms "silt" and "clay" shall be used in a qualitative rather than in a quantitative sense.
4. The gravel fraction, the sand fraction, and the fine-grained fraction shall be recognized as the significant fractions in describing the grain-size composition of a soil.
5. The fine-grained fraction (or binder fraction) shall be defined as that fraction of a soil the constituents of which are smaller than 0.1 mm in size.
6. The fine-grained fraction shall be designated "silt" if it cannot be made plastic and if it exhibits little or no strength when air-dried. (Careful distinction must be made between inorganic and organic silts (23).)
7. The fine-grained fraction shall be designated "clay" if it can be made plastic by an adjustment of its water content and if it exhibits considerable strength when air-dried.

The foregoing recommendations were submitted by the writer to the American Society for Engineering Education for use in formulating a set of standard definitions (24).

James H. Stratton,[4] M. ASCE.—The very nature of military operations in active theaters of war precludes other than "rule of thumb" methods in the

[4] Col., U. S. Army, Superv. Engr., The Panama Canal, Diablo Heights, Canal Zone.

design of airfields and roads, and the fluidity of the military situation generally requires that their construction and improvement be undertaken by stages. Improvements in bearing capacities must generally be accomplished in the face of enemy activity and increasing severity of use by the engineer's own forces.

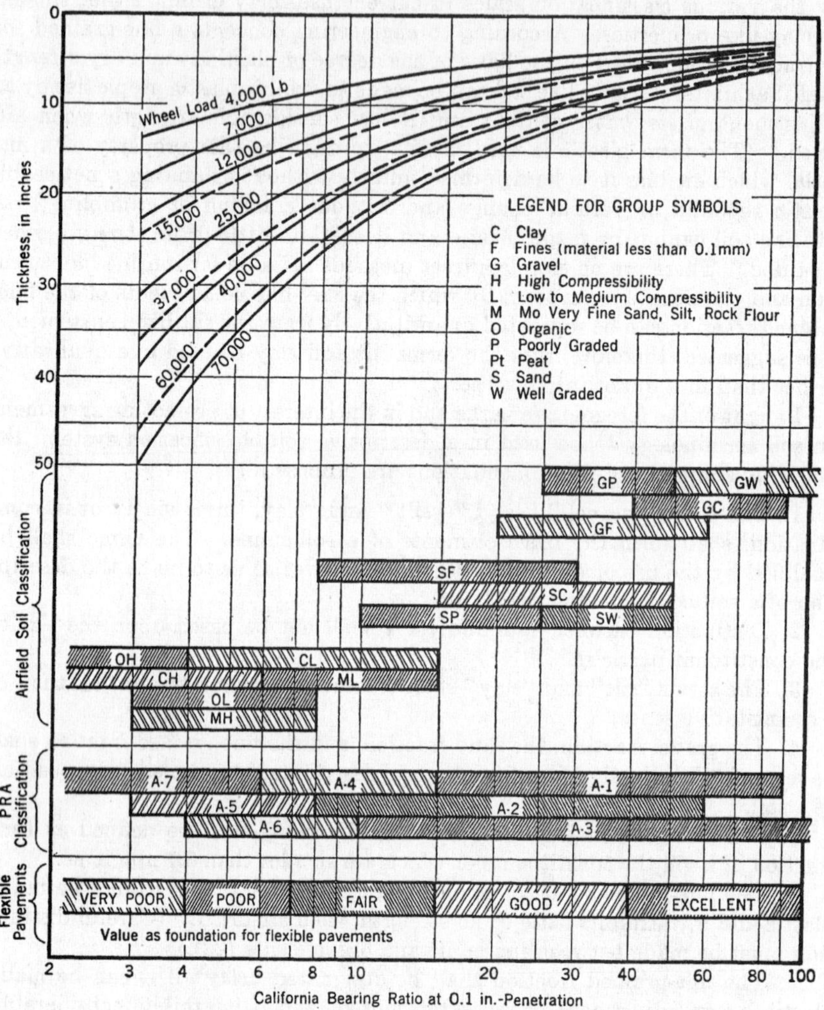

FIG. 7.—TENTATIVE RECOMMENDATIONS FOR COMBINED THICKNESS OF PAVEMENT AND BASE HAVING SATISFACTORY BEARING RATIO

The military engineer must know, at all times, the capacities of his roads and airfields in terms of both traffic volume and loading intensities. This calls for rapid interpretations or analyses, and therefore some standardization of the "rule of thumb" design method is required to insure against faulty designs or interpretations of the strengths of roads and airfields which would adversely affect the plans of the military commander.

In the spring of 1942, the Corps of Engineers, United States Army, adopted the California highway method of evaluating the bearing values of road subgrades and extended the design curves developed by O. J. Porter, M. ASCE (25),[4a] to the higher loadings later required for airfields (see Fig. 7). An attempt was made to correlate the Public Roads Administration soils classification with the bearing-value curves in order to provide a design method for military road and airfield construction in the zone of operations through simple soil classification tests. As a result of the inability of the Corps of Engineers satisfactorily to correlate the Public Roads Administration soils classification with design procedures, Professor Casagrande was requested to study methods of soils classification. The material for the paper was taken from the resulting study.

Prior to adopting the Casagrande classification, a canvass was made of all district and division engineers with the interesting result that, of the forty-four offices canvassed, thirty nine were in favor of the adoption and only five raised objections. The objections were largely concerned with the confusion that would result in the adoption of revised soils classification when construction was at a peak. However, in the light of the favoring majority opinion, it appeared that a change at that time would actually facilitate the transition by giving a wide opportunity for correlation with the Public Roads Administration system of classification that would aid in later revisions and improvements of the Casagrande classification. Accordingly, the Casagrande classification was adopted as standard for the Corps of Engineers.

Objections to the soil classification method were interposed by some of the most experienced soil technicians, who condemn all soils classification methods on the ground that their use encourages casual design conclusions whereas exhaustive investigations and study are required. In weighing this objection it must be noted that "symbols" identified with soils classification are a form of "engineering shorthand" which will always be popular in the profession despite the concern of those who recognize the complexity of soils and the difficulties in predicting their behavior. The Corps of Engineers does not propose to substitute a soil classification method for needed investigations and tests and orderly design procedures except where circumstances preclude other than rough or rudimentary methods of design.

It was the conclusion of the Corps of Engineers that the correlation of the soil classification methods proposed by Professor Casagrande with the bearing values of subgrades and base courses would be helpful in developing a "rule of thumb" method of design that could be employed where extensive investigations, tests, and deliberate design methods were infeasible.

The virtues ascribed by districts and divisions of the United States Engineer Department (U.S.E.D.) to the Casagrande method of classification may be summarized by stating that the improved classification is in consonance with general progress in soils engineering. One division of the U.S.E.D. commented favorably on the method because of its mnemonic characteristics.

The Lower Mississippi Valley Division, U.S.E.D., noted that a soil derivative of loess, frequently encountered in the Mississippi Valley, consisted of from 40% to 60% of silt and partakes of the nature of both an A-4 and an A-5 soil, and thus is not classifiable by the Public Roads Administration method. This division found that this soil also was not classifiable by the Casagrande method and that further study with a view to cataloging this soil is required. The same division also pointed out that soils classification correlation with the California Bearing Ratio (CBR) values must be predicated on standard methods of soil compaction and testing for the bearing values. The United States Waterways Experiment Station at Vicksburg, Miss., investigated a wide variety of soils and reported that the bearing values correlated very well with its interpretation of the Casagrande classification with one exception, which fell 3% outside the estimated range of CBR values as specified by Professor Casagrande for the particular soil type.

One district suggested that the percentage of soil binder and its plasticity index be indicated by numerical values in the classification of gravels and sands; for example, GW-3,1 would indicate a well-graded gravel with approximately 30% soil binder and a plasticity index of approximately 10%; likewise, SP-0 would indicate a poorly graded sand with no fines. If this plan were adopted, the GC, GF, SC, and SF groups would be eliminated. For the fine-graded soils, this district proposed correlating the plasticity index and moisture condition with the classification letters. For example, C5-3 would represent a "fat" clay with a plasticity index of 50% and a moisture content of 30%. There are certain virtues in this suggestion which may be realized indirectly by adopting the suggestion of the Upper Mississippi Valley Division, U.S.E.D., that the CBR number be added after the letter symbols; thus, for example, SF-25 would indicate a soil to be a sand with an excess of fines and would give a picture of its internal structure. This suggestion is of value since the CBR is indicative of behavior under load. It is probable that this form of identification will be used when bearing tests have been made.

The North Atlantic Division, U.S.E.D., stated that the suggested soil groups are satisfactory with the exception of the gravels and sands in the fine groups—that is, GF and SF, respectively. These could be subdivided advantageously into additional groups reducing the wide latitude now covered by them. It was suggested by this division that an additional gravel and sand group, designated GL and SL, respectively, be added to the Casagrande classification, to include gravels and sands with excess of fines.

The Casagrande classification is not expected to supplant local names entirely for special soils, such as loess, buckshot, gumbo, etc. These names have special meaning in the localities of occurrence and it is expected that they will be used to supplement the Casagrande classification.

The foregoing discussion was prepared for delivery at the Annual Meeting of the Society in January, 1943. Subsequently, the discusser's duties deprived him of opportunity to participate in engineering activities where he might have noted the results of use of the Casagrande soils classification system. It would be of interest to learn if the method of classification found favor and if it assisted military engineers in the construction of roads and airfields in the theater of operations.

DONALD J. BELCHER,[5] ASSOC., M. ASCE.—Professor Casagrande's review of the present status of soil classification and identification provides a well-organized framework that is necessary to a perspective of the current situation in this field. The trial and error process, plus native ingenuity, has produced a "system" for each large organization dealing with soil in the engineering sense. This review organizes soil classifications and identification and places them in juxtaposition so that they may be evaluated and their relationships noted. All too often various methods are treated with an intolerance that indicates a lack of scientific maturity or perspective. Much of this intolerance appears when something "fails"—that is, the system is wrong. The same people throw golf clubs and tennis rackets because their game is poor.

Perhaps there have been many instances where each system has been applied too broadly—used beyond the limitations originally perscribed. Professor Casagrande points this out with respect to the Public Roads (PR) system. It is true of other systems as well and cannot be considered as a fundamental fault. In this respect a highly desirable development of Fig. 1 would be to introduce a vertical distribution based on grain size. The author indicates this particularly under the heading, "3. Textural Soil Classifications," where it is to be noted that such a method is inadequate for fine-grained soils. Similar limitations can be placed on several of the others, particularly those in Fig. 1 under the division dealing with "Properties of Soil in its Undisturbed Condition."

In the detailed discussions of the various classifications, especially those under the heading, "4. Agricultural Soil Classification," and "5. Geologic Soil Classification," some additional attention will add to the perspective that can be gained from the over-all treatment of the subject. Primarily, it is erroneous to describe the first method as an "Agricultural Soil Classification." Engineers are in fact unintentionally perpetuating a careless description of a science; this is the proper place to correct this concept in the minds of civil engineers. To tell an engineer that this is an agricultural method (as has been done for so long) is to send him elsewhere looking for an engineering method. Unfortunately, to describe it as a pedological soil classification is a dubious improvement. It is, more simply, the science of soil formation; the classification is based on all the properties of the soils. Therein lies an advantage not found in other methods.

First, and for the record as it stands in the original paper, it would be more complete to state that it is a classification for the purpose of describing a soil and the conditions under which that soil exists. On that basis there is no fundamental leaning toward any one field of application, but any field of interest may interpret the data for its own particular use. If there is a technical predilection toward agriculture on the part of some pedologists, it is attributable to the fact that agriculture supports their programs by the allotment of funds.

It is to be regretted that an influential paper such as this should label the pedological system as a complicated one. On the contrary, its principles are simple and logical, particularly when the method is interpreted for engineering use. It also possesses simplicity of another order: With this system a soil is classified in three dimensions, length, width, and depth. With one well chosen auger boring a soil area can be classified, whereas in other methods the area

[5] Associate Prof., Civ. Eng., Cornell Univ., Ithaca, N. Y.

shrinks to the size of the auger hole and a sample has little significance beyond the hole from which it was obtained.

It has been commonplace to find the pedological classification described in the literature as a classification of surface soils for agricultural purposes. Although Professor Casagrande described it in this way, he also states that the underlying parent materials are included. The term "surface soils" is a relative one. When it is used in conjunction with "agricultural purposes," "surface soil" is immediately interpreted as meaning 6, 8, or at most 10 in. of depth below the surface. For the sake of clarification this point should be examined in some detail and emphasis added to Professor Casagrande's inclusion of the parent material, which may extend to considerable depths. Actually, there is no arbitrary boundary or basement beyond which lies the realm of the geologist, and there is no particular need for such a limitation.

No one will advocate this method for material existing at depths common to deep foundations. Therefore, at depths of from 80 ft to 100 ft it will never be of value. Depths of from 60 ft to 80 ft are still much too great for any general use of this method, but sand dunes of that depth are found in some areas that are or can be classified on the pedological basis. Also, there are extensive areas of deep wind-blown silt that reach these and even greater depths (from 100 ft to 200 ft). These may be called Aeolian (geologic classification) or loess deposits but they also coincide with pedologic units. Along the Lower Mississippi Valley these are classified as the Memphis soil; to the north, especially in Iowa, Illinois, and Missouri, they are described as Clinton soils; and in the State of Washington as Palouse.

As the surface is approached, in the 40-ft to 60-ft depth range, the application of pedology becomes somewhat greater although only geologic deposits that form the parent materials are as yet under consideration. In this range are included much rock and drift in the glaciated areas. In those soils weathered from rock there are valid relationships in many areas between the soil mantle and the character of the rock at such depths. In the drift areas the character of the mass of of material below the surface is sufficiently well described, so that general compaction and several other requirements (26) can be anticipated on the basis of the series name of a soil (a unit of pedological classification).

Each 10-ft increment, as the surface is approached, emphasizes two trends: First, that the application of the pedological principles increases to its greatest value in the 10-ft zone immediately below the surface; and, secondly, that the dollar volume of highway and airport construction increases to its greatest volume within the same general zone.

This somewhat brief illustration is intended to show that, on a national basis, a sliding scale of applicability is necessary. In smaller units, such as within a state or district, the scale becomes more fixed. It will be found that the significance of the pedological method, as it applies to soils in the depth, is a function of the type of deposit or formation. In areas where bare rock is exposed, or, where very shallow soils exist on rough terrain, the value of the method vanishes.

Climate is also important. Within the United States climatic differences are sufficiently great that they strongly influence soil development. In the humid regions much water percolates downward through the soil, constantly leaching the soil and rock, and in general building up accumulations of clay at depths of several feet below the surface. In this process soluble materials are removed. In the west, however, the situation is often reversed. There, rainfall and melting snows in the mountains are a contrast to the desert climate prevailing in the intervening valleys a few thousand feet below. Much of the water proceeds as ground water to the valley area, where it is largely dissipated by evaporation from the surface. This reversed process precipitates salts near the surface to form the alkali and caliche soils. These are also areas where the clays may easily become saturated with exchangeable ions and thus take on characteristics of abnormally high volume change and plasticity. Such soils occur as far east as western Minnesota, where small areas of alkaline soil are associated with the Fargo (clay) series.

At present relatively little correlation has been accomplished in this way, but it is known that pavement performance, base thickness requirements, and many construction details are related to the individual soil series. The alkali attack on concrete slabs, head walls, culverts, and steel pipe is a severe action in many of these areas (27).

If this method is adaptable to engineering use, then who is using it, where is it being used, and how? An examination of this field reveals the following:

The Portland Cement Association in its work on soil-cement stabilization uses and recommends the use of the pedological method as a basis for obtaining soil data (28).

> "Missouri [(26)] uses the pedological system as used by the U. S. Bureau of Chemistry and Soils, in making soil surveys for highway purposes. Missouri takes an attitude similar to that of Michigan, in that, of the three best known methods of classifying soils, the pedological system seems best to suit their needs. The other two methods seem to be limited in their scope because the classification by texture, as in the triaxial chart takes into account no other soil properties and the classification as developed by the Public Roads Administration is more an index of the behavior of disturbed soils and so does not take into account field conditions. The pedological system is essentially a reflection of drainage, topography, parent material, weathering agencies, as well as textural class. Missouri, however, does require that samples of each horizon of each soil type be taken and submitted to the testing laboratory for analysis. Field density tests are also made of each horizon in order that the theoretical soil shrinkage can be calculated for each horizon.

> "The Michigan State Highway Department [(26)] has found that the pedological scheme of soil classification provides an excellent means of studying soils for general highway purposes, and that it serves as a foundation for all other soil studies in the department. The soil type, in addition to serving as a unit in mapping, also serves as a unit to which laboratory test results, road behavior observations, construction problems, and maintenance experiences are related. In other words, the soil type serves as a means for classifying and coordinating the accumulated soil experience of the entire highway department personnel."

The Nebraska Department of Roads and Irrigation (29) states:

> "The pedological system of soil classification which has international recognition, was selected.
>
> "By adopting this method of classifying and indentifying soils, it was possible to make use of a wealth of available scientific information on soils,

including the published county soil maps and reports, and to supplement such knowledge with the test data and information obtained by observation on this survey as well as that which may be obtained through future investigations."

Tennessee and Nebraska have established (30)(31) a correlation between the soil series name and the texture of the material, as well as a further correlation between these and pavement performance.

The writer has observed a distinct positive correlation between the pumping of rigid pavements and the occurrence of the Superior clay series near Superior, Wis.; the Sharkey clay series near New Orleans, La.; and the Susquehanna clay series in Texas (partially reported in (32)).

There is a fair but incomplete correlation between the California Bearing Ratio (CBR) value and the soil series. This is indicated in the results of United States Engineer Department tests on the Clermont and Brookston soil series in Indiana; and by inference in the reported CBR experiments at Stockton, Calif. (33).

The North Carolina Department of Highways reports (32) "Pumping was not encountered on soils of the Norfolk or Ruston series. Pumping occurred on the Portsmouth and Plummer series * * *" It occurred also on the Lufkin, Elkton, and Coxville series.

In 1935, H. F. Janda (34) reported that excessive longitudinal cracks in a particular concrete pavement were confined to an area of Colby silt loam. W. S. Housel, M. ASCE (35), and F. V. Reagel (36) have related soil-cement construction to the soil series and L. D. Hicks (37) has established the relationship between the principal horizons of several soil series and the cement required for proper stabilization. Public Roads Administration (PRA) studies (38) show results of typical soils of the coastal plain (Norfolk, Portsmouth, and Ruston) and their relationship to base course performance in North Carolina.

Professor Casagrande's description of "5. Geologic Soil Classifications" is significant. It would be of considerable value if the method could be evaluated in greater detail, or perhaps less subtly; for truly "it is damned by faint praise."

The weakness of the method seems to be lodged in the broad basis of the classification and in the recognition of age and general origin as the most important factors. For those reasons it can seldom be used in engineering without considerable modification. An evaluation of this situation was concisely expressed by Karl Terzaghi, M. ASCE (39), in summing up the results of an investigation in this field:

> "There are no definite relations [between geologic classifications and engineering properties], because for a given geologic origin and petrographic character of a stratum the strength of the material, rock, or soil depends on conditions that are not, and cannot be, recognized by a purely geological investigation. For almost every geologically well-defined stratum the possible scattering of its physical properties from the statistical average is so important, that one cannot even make an acceptable guess based on precedent."

Essentially the pedological classification includes a refined geological classification and the parent material of a soil is directly related to materials that are mapped and described by geologists. F. R. Olmstead (40) and others have shown this in various sections of the United States.

J. A. Haine,[6] Esq., and J. W. Hilf,[7] Esq.—A development of soils classification for engineering purposes is presented to the profession by the author's excellent Airfield Classification (AC) System, which one of the writers studied under him at Harvard University in Cambridge, Mass., in 1943. From personal observation in the Southwest Pacific theater of operations, the AC system proved definitely valuable to United States Army engineer officers who were required to design airdromes rapidly in the field. The AC system was found to have two advantages over other current classification systems:

(1) The symbols used by Professor Casagrande are descriptive and easy to associate with actual soils, and experience in teaching this system to military personnel, who had no previous soils training, showed that it could be learned more quickly than systems depending solely on memory.

(2) The ability to assign definite engineering properties and even numerical values of design criteria, within reasonable limits, to the soil groups (such as the California Bearing Ratio) greatly facilitated the design of foundations for flexible pavements.

Although the AC system was devised for military roads and airdromes, it can be used profitably in foundation explorations and in prospecting for materials for rolled earth dams. In adapting this system to earth dams, it becomes necessary to expand two of the soil groups, and to specify certain limits of particle size and percentages that the author has not definitely stated in the paper.

Two of the fifteen soil groups, GF and SF, should be subdivided in order to distinguish the character of the fines in these soils which may be critical in high embankments because of the phenomenon of pore fluid pressures. This modification is easily accomplished without affecting the definitions of the original soil groups, by substituting GF (silty), GF (clayey), SF (silty), and SF (clayey), for GF and SF, respectively—mentioned by the author as a possible expansion of the AC system.

Specifications for rolled earth dams usually limit the maximum size of gravel in the rolled fill to 5 in. for a 6-in. compacted layer. In the investigation of sources of borrow materials for such structures, it follows that sizes larger than 5 in. should be excluded from the soil classification, although the importance of reporting oversize rock in exploration logs is recognized. With a range of coarse grains, from 0.1 mm to 5 in., the lower limit of gravel size (defined as 1 mm or 2 mm by the various grain-size classifications) should be raised to $\frac{1}{4}$ in. in order to simplify field identification, and to make the sand and gravel ranges more nearly equal on the grain-size curve. This definition more nearly corresponds with the coarse aggregate of concrete (No. 4 sieve), and was used by the Army (41).

A question often asked by those studying the AC system is, "How much gravel must be present in order to classify a coarse-grained soil as gravel, rather than as sand?" Professor Casagrande does not list quantitative limits for the soil groups in Table 4. Undoubtedly, such limits could be provided, but

[6] Civ. Engr., Bureau of Reclamation, U. S. Dept. of the Interior, Denver, Colo.

[7] Civ. Engr., Bureau of Reclamation, U. S. Dept. of the Interior, Denver, Colo.

TABLE 5.—SOIL CLASSIFICATION FOR ROLLED EARTH DAMS (ADAPTED FROM AIRFIELD CLASSIFICATION SYSTEM)

1				2			3	4	5	6			
Major divisions				Field identification (Excluding material more than 5 in. in diameter)			Group symbol	Soil groups and typical names	Permeability	RELATIVE DESIRABILITY FOR EMBANKMENT USE			
										Homogeneous	Zone 1[c]	Zone 2[c]	Zone 3[c]
Coarse-grained soils	Gravel and gravelly soils	More than 50% of sample is coarse grains (>0.1 mm[a])	More than 50% of coarse grains >¼ in.	Good gradation	Substantial amounts of all coarse sizes present (up to maximum size in sample)	Just enough clayey fines to bind coarse grains together	GC	Well-graded gravel-sand mixtures with excellent clay binder	Impervious	1	1	—	—
						"Clean" material (absence of appreciable amount of fines)	GW	Well-graded gravels and gravel-sand mixtures; little or no fines	Pervious	—	—	1	1
				Poor gradation	Predominantly more of one or more coarse sizes		GP	Poorly graded gravels and gravel-sand mixtures; little or no fines	Very pervious	—	—	—	2
					"Dirty" material (excess of fines)	Fines are silty (nonplastic)	GF[b] Silty	Very silty gravel, poorly graded gravel-sand-silt mixtures	Semipervious to impervious	2	2	2	—
						Fines are clayey (plastic)	GF[b] Clayey	Clayey gravel, poorly graded gravel-sand-clay mixtures	Impervious	3	3	—	—
	Sand and sandy soils		More than 50% of coarse grains <¼ in.	Good gradation	Substantial amounts of all coarse sizes present (up to maximum size in sample)	Just enough clayey fines to bind coarse grains together	SC	Well-graded sand with excellent clay binder	Impervious	4	4	—	—
						"Clean" material (absence of appreciable amount of fines)	SW	Well-graded sands and gravelly sands; little or no fines	Impervious	—	—	3	3[d]
				Poor gradation	Predominantly more of one or more coarse sizes		SP	Poorly graded sands and gravelly sands; little or no fines	Pervious	—	—	4	4[d]
					"Dirty" material (excess of fines)	Fines are silty (nonplastic)	SF[b] Silty	Very silty sands, poorly graded sand-silt mixtures	Semipervious to impervious	6	7	5	—
						Fines are clayey (plastic)	SF[b] Clayey	Clayey sands, poorly graded sand-clay mixtures	Impervious	5	5	—	—

TABLE 5.—(*Continued*)

1		2					3	4	5	6			
Major divisions		Field Identification (Excluding material more than 5 in. in diameter)					Group symbol	Soil groups and typical names	Permeability	Relative Desirability for Embankment Use			
		Visual	Tests on fraction < $\frac{1}{16}$ in. in diameter			Other means of identification							
			Shaking	Breaking	Thread					Homogeneous	Zone 1[c]	Zone 2[c]	Zone 3[c]
Fine-grained soils	Fine-grained soils of low to medium compressibility	More than 50% of sample is fine grains (<0.1 mm[a])	Moderate to fast reaction	Very slight to medium dry strength	Soft weak thread or none	Will form cracks when kneaded while moist	ML	Inorganic silts and very fine sands; rock flour; silty or clayey fine sands with slight plasticity	Semipervious to impervious	8	8	6	—
			Extremely slow reaction or none at all	Medium to high dry strength	Medium to tough thread	Can be kneaded while moist without cracking	CL	Inorganic clays of low to medium plasticity; sandy clays; silty clays; lean clays	Impervious	7	6	—	—
			Slow reaction	Slight to medium dry strength	Soft weak thread	Odor of decay which can be intensified by heating	OL	Organic silts and organic silt-clays of low plasticity	Semipervious to impervious	9	9	—	—
	Fine-grained soils of high compressibility		Slow to moderate reaction	Very slight to medium dry strength	Soft weak thread	Small shiny particles of mica or limestone can be seen	MH	Micaceous or diatomaceous fine sandy or silty soils; elastic silts	Semipervious to impervious	11	11	—	—
			No reaction	High dry strength	Very tough thread	In moist state, can be kneaded for a considerable period without cracking	CH	Inorganic clays of high plasticity; fat clays	Impervious	10	10	—	—
			No reaction	Medium to high dry strength	Soft weak to medium thread	Odor of decay which can be intensified by heating	OH	Organic clays of medium to high plasticity	Impervious	12	12	—	—
Fibrous organic soils of high compressibility		Fibrous, decayed, organic vegetable matter; noticeably compressible					PT	Peat and other highly organic swamp soils	Entirely unsuitable				

[a] 0.1 mm is the smallest visual size. [b] Subdivisions of the GF-groups and SF-groups of the AC system. [c] Zone 1 is inner, impervious zone; Zone 2 is intermediate, semipervious zone; Zone 3 is outer, free-draining zone. [d] If gravelly.

the percentages by weight would be different for each soil group, making it necessary to memorize additional information. Also, these percentages would be inappropriate in field classification, because of the difficulty of estimating weights of materials in various states of compactness by visual inspection. To avoid these objections and still provide a basis for dividing coarse-grained soils into sands and gravels, the writers have rearranged Table 4 for field use. As shown in Table 5, identification of coarse-grained soils (ten of the seventeen groups) may be accomplished by a step by step procedure without requiring the estimation of any fraction other than one half.

Col. 1, Table 4, lists fine-grained soils as "containing little or no coarse-grained material," but (under the heading, "12. Fine-Grained Soil Groups of AC System") the author states:

> "In the transition between coarse-grained and fine-grained soils, usually those soils containing more than 50% of material smaller than 0.1 mm in size (passing No. 150 mesh) are classified as fine-grained soils (ML or CL); and soils containing less than 50% smaller than 0.1 mm in size are classified as coarse grained (SF or SC)."

The writers agree with the latter division between the major soil groups and have used it in Table 5. The same reasoning is used in dividing coarse-grained soils into the gravel (G) and sand (S) groups. Regardless of the amount of fine-grained material present in coarse-grained soil, the latter is classified as gravel or sand, depending on the size of the coarse grains (whether more than 50% of them are larger or smaller than ¼ in.).

Persons who have not had the benefit of considerable laboratory experience in sieve analysis often find it difficult to estimate even the simple fraction, one half. They are inclined to judge quantity by volume, rather than by weight; for example, since the percentage by volume of particles larger than ¼ in. in a soil mass depends on whether the soil is in a loose state or a dense state, accuracy cannot be expected. However, the AC system boundary classifications, suggested by the author, such as GW-SW, or SF (silty)-ML, can be used in all cases where there is any doubt about the percentages.

Table 5 was devised to enable a relatively inexperienced investigator, or one unfamiliar with the AC system, to classify a soil—first, into the coarse-grained division or the fine-grained division, and then, if coarse-grained, into the gravel or sand groups when the soil is typical. For a boundary case in the coarse-grained division, the procedure is to assume that the soil is a gravel and then continue in the chart until the final soil group, say, GC, is reached, using the indicated criteria of gradation, and the amount and character of the fines. Since it could have been assumed that the doubtful coarse-grained soil was a sand, the correct field classification is GC-SC, which follows from the duplicate criteria for the gravel and sand subgroups. Similarly, if there is doubt as to whether a soil should be classified as coarse-grained or fine-grained, the assumption of first one type, and then the other, will result in the proper boundary classification. For fine-grained soils, Professor Casagrande's simple field tests, plus other identification aids, are used to determine the proper soil group. Here, the fraction to be tested is stated in Table 5 as material smaller than $\frac{1}{16}$ in., rather than No. 40 mesh. This is merely a field expedient for separation by hand, which will not affect the resulting classification in the large majority of cases, it is believed.

Together with instructions for performing the shaking, breaking, and thread tests, Table 5 should be sufficient to enable a proper classification in the field without any laboratory equipment. If it is desired, a representative sample can be classified in the laboratory, using the criteria in Table 5 and the Atterberg limits criteria proposed by Professor Casagrande; but it must be recognized that only a very small percentage of the soils examined in the field ever reach a laboratory, and that proper field classification is the primary objective in soils surveys. The writers heartily agree with the author that a complete description of the material is necessary in reporting a soil in a log. However, the facility in identification and classification of soils that must be developed by the use of such a chart as Table 5, is a necessary first step toward that goal.

After a soil is properly classified, as Professor Casagrande has shown, it is possible to indicate the engineering properties typical of the various soil groups and their use in engineering structures. For rolled earth dams, the permeability of a material is of outstanding importance and this property is listed for each group in Table 5. Also, the prospector for embankment materials often desires to know how the soil he has classified compares with other kinds of borrow material that may be available for a homogeneous dam, or for various zones of a zoned dam. To aid in evaluating possible sources of material, the writers have attempted to rate the various soil groups according to their experience, for use in rolled earth dams, considering the permeability, workability, shear strength, and resistance to piping of typical materials. In Table 5, Zone 1 represents the impervious zone, Zone 2, the semipervious zone adjacent to Zone 1 in the dam, and Zone 3, the free-draining outer zone of a rolled earth dam. It is recognized that the numerical ratings are not rigid, but in a qualitative way they will enable the investigator to judge the suitability of a material for rolled earth dam construction.

For final design of a dam, a soil classification alone is entirely insufficient. Laboratory tests must be made on representative samples of the foundation, on the required excavation, and on the proposed borrow areas to determine, among other properties, the permeability, cohesion, angle of internal friction, and compressibility characteristics of the materials. Nevertheless, when adequate soil classifications and descriptions are available in the logs of exploration, the designer has at his disposal timely information which he can use in making preliminary estimates, in determining the extent of additional field investigations needed for final design, and in planning an economical laboratory testing program.

It would be desirable if there were a sufficiently large number of experienced soil engineers to personally perform all the identifications and classifications of soils which are required for earth dam projects, but the fact is that often the investigation is performed by personnel relatively inexperienced in soil mechanics. Under these conditions, the simplest possible descriptive classification system, consistent with adequate utility, appears to be desirable.

The AC system without undue additional refinements is admirably suited for this purpose.

JACOB FELD,[8] M. ASCE.—This authoritative and complete program of soil classification is most timely. An extrapolation of reported experience to engineering problems is required to make the subject of soil mechanics usable in actual practice. To permit such procedure, the soil type must be known, when such experience is a report of someone else's work. A universal adoption of the author's system of classification and, at the same time, a recording of the soil type encountered in each problem reported in the technical literature will make a true wealth of data available in contrast to the present volume of contributions in soil mechanics; most contributions are useless because the authors employ secret codes or untranslatable language in describing the soils encountered and controlled.

The author clearly develops the thesis that soil classification must proceed on the basis of soil action and reaction, and not on the basis of component materials. The former method corresponds to the classification of human beings by their expected reactions, rather than by their height or weight. Textural or geometric soil classifications of granular or noncohesive soils have been reasonably successful, because the interaction of grains, in most cases, is not affected by the gas or liquid occlusions on the grain surfaces. In cohesive soils, there is a great variety of possible soil reactions, with identical textural make-up, depending on the mechanical and colloidal extent of the surface coatings. In addition, textural or sieving classifications do not distinguish between grains of unlike shape.

By magnifying the problem of soil classification, the difficulty of true classification by geometric methods can be easily understood. If a grouping of humans by size (let us say, equivalent volume) is made at a shore resort, and an identical size grouping is made at a winter resort, the two groups are geometrically identical; but they cannot be expected to react alike, because the factor of their clothing has been disregarded. Similarly, as the author described, the kaolin-type clays derived from a single type of feldspar may have particle sizes equivalent to many tough clays, but they react more nearly like a silt than a clay. The particles may be of the same size and, therefore, of equal gravitational adhesion; yet the chemical nature of the grain coverings and the effect on wetness (caused by the kind of electrostatic charge on the grain surfaces) are more important factors in determining the total adhesion and the tightness of possible grain packing, or the cohesion and possible consolidation, respectively.

In soil classification, it is unfortunate that the division between the sand and gravel designations does not agree with the accepted sieve separation in concrete and road base technology. The general agreement, that sand contains all particles passing a No. 4 sieve in all cement and bituminous concrete specifications, seems to require a restudy of this item for soil classification. To make the Airfield Classification (AC) System complete, definitions of the terms used, such as "compressibility" and "dry strength" should be included. It is

[8] Cons. Engr., New York, N. Y.

important to distinguish between compressibility or dilatation as an elastic phenomenon, and changes in shape without change in volume, resulting from plastic flow or viscous deformation.

The author's recommendation of field identification tests brings to mind a very early treatise on soil classification, presented by John Evelyn before the Royal Society of London almost three hundred years ago (42). Although he was chiefly interested in the agricultural possibilities in artificially prepared soils (as well as natural soils), the recommendations for classifications and field identification are quite similar to recent writings, and are summarized herein for the reader who does not have access to this rare book.

In the "Introduction," Evelyn cites "De Arte Combinatoria" by Athanasius Kircher in which it is stated that there are "* * * no fewer than 179,100,060 different sorts of earths, but only eight or nine are useful to our purpose." Corresponding to modern soil horizons, three strata are identified as: (1) The top layer of "underturf earth," a foot or so deep; (2) the next strata may be "loam, clay, plastic, figuline or smectic"; and (3) the lowest strata may be "chalk, marle, fullers-earth, sandy, gravelly, stony, rock, shelly, coal or mineral." Incidentally, Evelyn states, "The Ancients called them: Creta, Argilla, Smectica, Tophacea, Pulla, Alba, Rufa, Columbina, Macra, Cariosa, Rubrica." He further subdivides the top layer into:

"I. A virgin-earth, black, fat, porous, light and sufficiently tenacious, without any admixture of sand or gravel, which in the lower series may be darkest-gray or tawney, becoming veined with yellow or sometimes red.
II. An obscure color mould, with some loam and sand.
III. A mixture of the former two types with small flints and pebbles.
IV. A totally sandy with a bottom of gravel, rock and 'not seldom' clay."

Then, a general description of soil types follows, which is quoted in full because of its picturesque language:

"Pure sand is white, black, bluish, red, yellow, harsher and milder and some mere dust in appearance, none of them to be desired alone. But the grey black and ash coloured, or of the travelling kind, volatile and exceedingly light, is the most insipid and worst of all.

"Clays there are of many Kinds—some more pinguid, some more slippery, all of them tenacious of water on the surface where it stagnates without penetration, most of them pernicious and untractable. The blue, white and red clay are all unkind, the stony and looser sort is sometimes tolerable, but the light brick-earth does very well with most fruit trees.

"Loams and brick earth are of several sorts, some approaching to clay, others nearer marle.

"Marle is usually at greater depths, of many colours, all of them unctuous, slippery nature, slackens upon drying after a shower and crumbles into dust."

Evelyn than advised that general textural description of soils may be of less value than classification of properties by the use of the senses. Four tests are described:

(1) By odor or smell—since "Lord Verulam affirms that vegetable odours exist in the soil."

(2) By taste—since the water percolation through soil shows the kind of chemicals therein, "although some say that there is no taste or odour in the best earths."

(3) By touch—fatty and slippery, or gritty, porous and friable, if it sticks to the fingers or dissolves on the tongue. The "best earth is blackish, cuts like butter, sticks not obstinately, is sweet, and becoming mortar when wet, without crusting when dry."

(4) By sight—"grey is pre-eminent, next russett, clear tawney is worse, light and dark coloured are good for nothing, yellowish red is worst of all."

Having analyzed and shown how various basic properties could be identified, Evelyn did not entirely discredit the laboratory technician of his day; he advised the following procedure for the synthesis of desired properties in soils:

> "All these are fit to be known, as contributing to noble and useful experiments, upon due and accurate comparisons, and inquiries from the several particles of their Constitutions, Figures and Modes, as far at least, as we can discover them by the best auxiliaries of microscopes, lotions, strainers, calcinations, triturations and grindings; upon such discovery to judge of their qualities, and by essaying variety of mixtures, and imitating all sorts of mold, foreign or indigen, to compound earths as near as may be resembling the natural, for any special or curious use, and be thereby enabled to alter the genius of Grounds as we see occasion."

He then demonstrated how artificial soils had been made by mixing samples of earths from various sources with types of chemicals and fertilizers, and concluded:

> "Therefore let no man be over-confident, that because some earths are soft, fat and slippery, they may not possibly consist of sands, (of which there are so many Kinds), since 'tis evident, that even all fossile bodies which can be reduced and brought to sands, may by contrition of the particles be rendered so minute, as to emulate the finest earths we have enumerated."

The early French experimenters and theorists in soil problems distinguished only between "loose soils" and "firm soils." However, when M. Chauvelot (43) presented a paper before the French Academy of Sciences in 1783, showing how the lateral earth pressure could be derived from a knowledge of the weight, the internal friction, and the cohesion of a soil without any other description or classification, that learned body announced that the problem of the determination of lateral earth pressure was now susceptible to a rigorous solution. Later, contributors went back to a textural soil classification and M. Mayniel (44), in 1808, reports on his extensive experiments and tables to cover all kinds of soil, namely: Soil, sand, gravel, rubbish or ash, clay and debris fills. Incidentally, Mayniel's division between large and small gravel is denoted by the size of a pigeon's egg.

An anonymous author (45) quotes from the works of Sir Humphrey Davy, classifying the four earths found in soils as: "* * * aluminous (clay), siliceous (flint), calcareous (limestone) and magnesian (not to be mistaken for limestone)." However, a soil is to be classified on how it acts and not on what it contains.

To quote that author:

> "In framing a system of definitions, a soil is to take a particular designation from a particular Kind of earth, not exactly in proportion as that earth may preponderate, or not over others in forming the basis of the soil, but rather in proportion to the influence which a particular Kind of earth, forming part of the sample, has on tillage and vegetation."

That might also be said for the influence on engineering uses of soils.

The need for a universal soil classification is an old one and much credit must be given to the author of this paper in presenting a comprehensive answer to it, in the form of the AC system, which permits and provides a unique name for every type of soil, using expected action and reaction as criteria.

KENNETH S. LANE,[9] Assoc. M. ASCE.—Of the eight classification systems presented in the paper, the writer has had occasion, since 1937, to work with four—the triangle textural classification, the master curve textural classification, the Public Roads Classification, and the Casagrande system. From this experience the writer can certainly agree with the author's designation of soil classification as the most confused chapter in soil mechanics. Although the author modestly terms the Casagrande system as the Airfield Classification System, the writer prefers to designate it as the Casagrande classification, since it has been widely known as such from its initial publication in 1942. This classification is easily the best of the aforementioned four systems (in their forms as described in the paper) for the airport-highway type of work for which it was originally designed. In its application to dams and levees, it has certain limitations which can be best considered from their relations to the primary requirements of a soil classification system intended for wide usage.

Requirements of a Classification System.—The requirements particularly considered are those for a large and widespread organization handling many projects of all sizes; for less intensive operations, or for work concentrated in a particular region, simpler requirements might suffice. In this type of work, boring logs generally include a classification of the soil into a group or type, plus a word description similar to the descriptive classification covered in the paper (under the heading, "15. Descriptive Soil Classification").

On a large airport, levee, dam, or canal, involving a hundred or more foundation explorations, it is generally not feasible to consult the original boring logs whenever questions arise, so these are usually filed as reference material. Rather, the usefulness of a soil classification system comes from its application to graphic abstracts of boring logs, wherein the most pertinent information is abstracted and placed on boring drawings. Hence, for a large project, this graphic abstract is the form of boring information most used by engineers in designing, by bidders in estimating, and by contractors and field inspectors in planning construction. For this purpose, it is essential that the classification system include some type of shorthand notation to convey maximum information at a glance. All engineering classifications described by the author are so

[9] Chf., Soils and Geology Branch, Garrison Dist., Corps of Engrs., Dept of the Army, Fort Lincoln, Bismarck, N. Dak.

designed, except for one variation of the textural system where the words "some, little, and trace" indicate decreasing percentages of size fractions, and require slow reading for understanding.

For this type of work, the writer feels that a soil classification system should satisfy the following four requirements:

Requirement A.—It should describe the soil in well-understood terms, which convey an idea of its type and behavior.

1. In addition to soil specialists, the terms used should convey the intended meaning to designing engineers, inspectors, geologists, contractors, and courts of law, which last are the recipients of altogether too many disputes, involving boring information versus foundation conditions encountered.

2. The system should make allowance for regional use of certain terms, particularly meaningful in that region, such as glacial till, buckshot clay, caliche, and others.

Requirement B.—The system should furnish an indication of soil properties and performance.

1. It should permit a crude estimate of permeability, compressibility, and shear strength—the three primary properties of a soil. When combined with other boring information (compactness and water table location), the classification should permit a rough estimate of soil performance, such as frost susceptibility, drainage, settlement, slope stability, wheel load capacity, and compactibility (probable range of compacted density and normally suitable compaction procedure).

2. It should be capable of subdivision. Soil is so complex that no classification system can replace tests. Nevertheless, subdivision can often reduce testing by narrowing the range of properties for soils, at least within a particular region. Approximate correlations of soil properties and performance experiences with the indicators of a classification system often can be facilitated by subdivision, particularly when the testing program extends over a length of time sufficient to accumulate experience from past tests. Furthermore, such correlations, even though rough, are often a "must" for rapid handling of a large number of projects whereas necessity frequently requires that testing be limited, or omitted entirely, for smaller projects.

Requirement C.—It should be applicable from visual examination—both in a simplified form and, with experience, in a more refined form.

1. The simplified form should be suitable for determination by boring inspectors from visual examination.

2. The refined form should be such that: (*a*) It can be assigned by a well-trained classifier or chief boring inspector from visual examination, supplemented by occasional check tests; or (*b*) it can be determined from tests in the case of a less experienced classifier. For the latter case, and also so that different laboratories will arrive at the same result, reasonably definite numerical limits are necessary.

Requirement D.—It should employ a simple system of notation for graphic abstracts of boring logs on drawings recording boring information.

1. The notation should register rapidly in the mind while the reader is scanning and comparing plots of several different borings.

2. The drafting effort should be minimized in the preparation of boring drawings, including those suitable for reduction. These are generally of two types: (*a*) Contract drawings to serve bidders, contractors, and construction engineers, and which also usually form one group of exhibits in any court case; and (*b*) study drawings (usually foundation profiles) to serve soils engineers, geologists, and designing engineers in developing the design treatment and in illustrating its adequacy in reports. Unlike study drawings, contract drawings should be made to avoid interpretation, from the legal standpoint that information furnished to bidders should be basic data—not interpretations.

Requirement A, for naming a soil to cover its type and behavior, has been complicated by the practice of using the soil names, "gravel, sand, silt, and clay," to represent, also, grain-size fractions as arbitrarily delineated by one of the grain-size scales in Fig. 2. The terms, "gravel, sand, silt, and clay," describe common soil types, and like all natural soils, each is usually a composite mixture of from two to four of the size fractions designated as "gravel, sand, silt, and clay sizes." Good definitions of these four soil types have been given by H. A. Mohr, M. ASCE (23), and also in dictionaries; and the terms convey about the same meaning to engineers, geologists, contractors, and courts. The general behavior of each of these four soil types was fairly uniformly understood until soil technicians began to confuse themselves and others by using the same words for size-fraction names. This practice has degenerated to the extent that some laboratories assign a name to a soil from the percentages of different size fractions, according to certain charts or classification triangles like those in Fig. 3; sometimes this is done as a routine matter from a grain-size curve by a classifier who never even sees the sample.

If the size-fraction percentages were a reliable measure of soil behavior, this practice might be condoned; but such is far from the case, as the author has indicated, mentioning the common illustration of plastic lean clay and fine cohesionless silt or rock flour of the same gradation. As one example, samples of Manhattan silt having 20% in the clay-size fraction would be designated in some classifications as silty or lean clay. Since this material becomes quick on the slightest provocation, it would indeed be unfortunate for any contractor (and later for the owner in court) who relied on the cohesion of a true lean clay in attempting to excavate below ground water.

Although frequently of the same gradation, classification of loess as silt ignores the fact (and the consequence) that loess often undergoes a serious rapid settlement from collapse of structure under the influence of seepage, whereas silt is seldom so affected. Because of the angularity of its coarse particles and its greater compactness, glacial till generally has greater shear strength, lower permeability, and higher resistance to excavation than a rounded particle gravel of equal gradation. These and other misleading results obtained with textural classifications have convinced the writer of the necessity for requirement A in naming the soil to describe its type and behavior.

Casagrande Classification.—In most usage to date, the Casagrande classification has generally been applied in the form of the group symbols GW to Pt, as shown in Table 4, Col. 3. When so applied, it is actually a plasticity classification for fine-grained soils, and a textural classification for coarse-grained soils. Since it is based on the position and shape of the entire gradation curve, the textural classification phase is superior to the triangle classification, which is based on only three points of the gradation curve. The shape and position of the entire gradation curve have considerable bearing on important soil properties. This fact is widely recognized and is illustrated by many past attempts at better definition—such as the effective size and uniformity coefficient presented by Allen Hazen; the mean diameter and grade line introduced by Frank B. Campbell, M. ASCE (8); and the mean size, fineness dispersion, and curve shape proposed by D. M. Burmister, Assoc. M. ASCE (46)(47).

Hence, although this textural phase of the Casagrande classification is one of the best of the systems described in the paper, it is subject to the general objections to a textural classification when used alone. For example, such unlike materials as gravel, cohesionless till, talus, and caliche would be classed in group GW, and silt and loess in group ML. From the closing paragraphs of the paper, it is apparent that the author is aware of this deficiency (or possible misapplication) of the system and intends that the group symbol should be supplemented with a soil name. The writer is in agreement with this form of the system, as it then meets requirement A.

For furnishing indicators of soil properties (requirement B), the writer considers the Casagrande classification the best of any yet devised in regard to compressibility and shear strength of fine-grained soils. However, its performance has been relatively poor in indicating the permeability of coarse-grained soils. The writer has experienced ranges in permeability in ratios of from 1 to 10,000 and even from 1 to 100,000 which are considered undesirably wide for a single-class group. (Although many past attempts have shown uncertain relationships between permeability and results of simpler classification tests, the lack of success in deriving these relationships is probably explained by the difficulties encountered in permeability test procedure. With more refined testing techniques, considerable data have been accumulated since 1942 (unpublished) which, it is believed, indicate that such relationships do exist; and, of the simpler classification systems available at present, the data correlate best with a master curve system.)

If subdivisions were feasible, they would probably narrow this permeability range, as well as the range in compacted density, which has also been experienced as relatively widespread in certain groups. After several unsatisfactory attempts at subdivision, the writer has joined in the frequently-voiced comment that the Casagrande classification is not easily adapted to subdivision. In discussing possible expansion of the system in Section 14 of the paper, the author offers a scheme of subdivision, partly by the addition of new groups

(GU, SU, KL, KI, and KH) and partly by refinement using boundary groups (GW–SW, GC–SC, and so on) and by use of subscripts m, c, u, w, and p. When subdivision is practiced, it is preferably accomplished by refinement, rather than by addition, because the latter makes less feasible any later attempts at correlating data gathered prior to expanding a classification system by addition. The proposed expansion of the Casagrande classification to represent a soil as $GF_m - S_wF_c$ (to give, possibly, an extreme case with the binder near the boundary between plastic and nonplastic) would seem to indicate (by its complexity) the difficulties of subdivision.

In respect to requirement C, for determination by visual examination, the Casagrande classification is one of the best, and the author's methods for field identification of fine-grained soils in Section 13 of the paper are worthy of widespread application. However, for checking between different laboratories, it is desirable to have numerical limits for the results of usual classification tests, without the necessity of transmitting a sample for examination. Such limits are also of considerable value for checking classification of a soil by tests, especially in the case of an inexperienced classifier. For fine-grained soils, the Casagrande classification provides such limits, but only to a minor degree for coarse-grained soils.

A good feature of the Casagrande classification is the use of a simple system of notation for boring drawings (requirement D), unless it is complicated by the aforementioned scheme of subdivision. Possibly the best of the simpler systems yet devised for registering quickly in the mind is the system of rectangular and diagonal hatching usually applied with the triangle classification in Fig. 3(*b*), in which heavy symbols indicate the predominant fractions, and lighter symbols denote the two minor fractions of the three main grain-size fractions (48). However, the drafting cost required for suitable reduction is so much greater for this system of hatching (generally requiring inked tracings) that it more than offsets the advantage of quick registering.

QPC Classification.—These difficulties experienced in applying the Casagrande classification, and the greater limitations of the other systems discussed by the author, have led the writer to seek a combination of the better features of several systems. Since 1943, the result has gradually evolved into a combination of soil names from the descriptive classification; a gradation identification from the master curve system started on Quabbin Dam in Massachusetts by Karl R. Kennison and Stanley M. Dore, Members, ASCE (7), and expanded by Frank E. Fahlquist, M. ASCE, and Waldo I. Kenerson into the Providence classification (8); and the plasticity characteristics of the Casagrande classification. The combination, in these three parts, has been tentatively termed the QPC classification after its principal sources (Quabbin, Providence, and Casagrande).

Part 1. Name of Soil Type.—This consists of a soil name, generally with one modifying adjective, selected primarily to describe soil type and behavior rather than gradation. It is chosen from visual examination (preferably on a moist sample) to agree with local semitechnical usage by engineers, geologists, contractors, and courts.

Examples of Soil Names.—The following are typical examples of names, used as applicable to local conditions, with additions as necessary.

Material	Description
Gravel	Variably graded combination of pebbles and sand, sometimes including some silt and clay. Coarser particles, from 6 in. to ¼ in., at least partly rounded from fluvial action.
Sand	Uniformly graded, bulk of material smaller than ⅛ in., individual particles readily visible to the naked eye.
Silt	Fine grained, lacking plasticity, little or no cohesion when dried, majority of individual grains not visible to the naked eye. Shaking test shows quick or "livery" condition.
Clay	Fine grained, plastic, and cohesive, strong when dried.
Till	Variably graded, unsorted glacial deposit, characterized by angularity of coarser fragments, generally very compact. Indicate whether cohesive or cohesionless (dependent on character of fines).
Loess	Unstratified aeolian deposit, homogeneous powdery soil of silt sizes, structure loose and inclined toward vertical cleavage.
Varved Clay	Alternating thin beds of clay and silt, interstratified, clay-like behavior.
Gumbo	Highly plastic clay, extremely sticky when wet, often a product of weathered shale (localized term).
Talus	Variably sized angular rock fragments, fallen to base of slope, voids seldom all filled.
Decomposed Rock	Structure preserved in place, but chemically disintegrated to soil sizes (with bonding largely destroyed). Also include name of rock.
Coral Sand	Ground-up coral.
Blow Sand (Dune Sand)	Wind-borne, free-moving sand, particles at least partly rounded.
Buckshot Clay	Compact clay, persistently breaking into uniformly small-sized fragments (localized term).
Diatomaceous Earth	Siliceous soil composed chiefly of skeletons of minute organisms.
Topsoil	Weathered surface soil containing organic matter, usually supporting vegetation. Termed loam in some regions.

Examples of Stratum Names Where Grain-Size Number Not Applicable.—

Material	Description
Peat	Partly carbonized vegetable matter, odoriferous, and fibrous. Swamp deposit.
Fill	A man-made deposit. Describe constituents, as cinder fill, rubbish fill, sawdust, ashes, cans, bricks, broken concrete, and so on. Grain-size number used only when soil a major constituent, or when knowledge of gradation desirable, as with slag, cinders, or earth fill.

Examples of Modifying Adjectives.—As frequently used on graphic abstracts of boring logs:

Coarse	Gravelly	Organic	Kaolinitic
Medium	Sandy	Micaceous	Bentonitic
Fine	Silty	Cemented	Carbonaceous
	Clayey		Calcareous

Fat, Lean } Applied to Clays

Generally not more than one adjective is used to supplement the soil name on graphic abstract logs and it is omitted when not particularly significant.

The system is flexible in allowing for the addition of terms of particular regional significance and certain geologic terms well characterizing soil behavior. Devices for obtaining a soil name, such as a classification triangle, have been purposely omitted in order to avoid classification by rote, and its resultant misleading information. For selection of the soil name, guidance of geologic training is helpful; and, for an inexperienced classifier, it might be advisable to add a few remarks to the foregoing definitions to cover behavior similar to those given by H. A. Mohr (23).

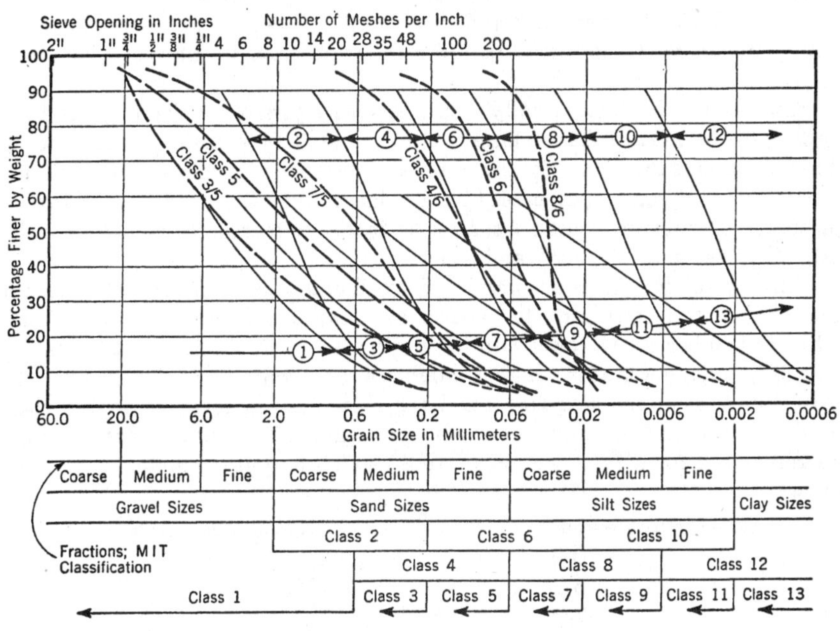

FIG. 8.—LIMITS OF GRAIN-SIZE NUMBER CLASSES

Part 2A. Grain-Size Number.—Part 2A consists of a grain-size number representing position and shape of the entire gradation curve. Thirteen basic classes are recognized as defined by the master curve bands in Fig. 8. Even numbers indicate predominantly uniform gradation; odd numbers indicate predominantly variable gradation.

When a word description is advisable for those unaccustomed to thinking in terms of gradation curves, the thirteen basic classes are defined as follows:

Class	Description of Gradation
1	Graded from gravel to coarse sand sizes with little medium sand.
2	Coarse to medium sand sizes with little gravel and fine sand.
3	Graded from gravel to medium sand sizes with little fine sand.
4	Medium to fine sand sizes with little coarse sand and coarse silt.
5	Graded from gravel to fine-sand sizes with little coarse silt.
6	Fine sand to coarse silt sizes with little medium sand and medium silt.
7	Graded from gravel to coarse silt sizes with little medium silt.
8	Coarse to medium silt sizes with little fine sand and fine silt.
9	Graded from gravel or coarse sand to medium silt sizes with little fine silt.
10	Medium to fine silt sizes with little coarse silt and clay.
11	Graded from gravel or coarse sand to fine silt sizes with little clay.
12	Fine silt to clay sizes with little medium silt and fine clay.
13	Graded from coarse sand to clay sizes with little fine clay.

In these descriptions the terms designate size fractions of the M.I.T. classification. A much better system would be the eventual acceptance of other terms for the size fractions which would not conflict with "gravel," "sand," "silt," and "clay," as soil types. Names from a dead language such as Latin, or the use of Greek letters might be acceptable, or possibly terms such as "coarse," "fines," "superfines," and "colloida." However, any possible terms would need careful consideration for correlation with languages other than English.

Samples are classified visually, using reference jars of materials graded according to the thirteen basic classes, supplemented by occasional check tests. When determined from a test, the grain-size number is obtained with a transparent celluloid reproduction of Fig. 8, placed over the gradation curve in question. For distinguishing visual classifications, these are followed by the letter "v," as "4 v," "7/9 v," and so on.

Primary subdivision is accomplished by double class numbers to represent soils grading from one band to another as illustrated in Fig. 8, thus: "8/6." A higher number over a lower number indicates a more uniform gradation (as class 8/6) and a lower over a higher number a more variable gradation (as class 4/6) than that of the standard band (as class 6). In forming double class numbers, even and odd numbers are never combined. Double class numbers are recognized for a spread of two bands, which is quite common, and is as follows:

Basically variable gradation		Basically uniform gradation	
1/3	3/1	2/4	4/2
3/5	5/3	4/6	6/4
5/7	7/5	6/8	8/6
7/9	9/7	8/10	10/8
9/11	11/9	10/12	12/10
11/13	13/11		

With this primary subdivision, the thirteen basic classes are expanded to thirty-five grain-size number classes which have been found to cover most needs of the several hundred thousand samples to which this system has been applied.

Rare cases of extremely variable soils grading over a wide range can be handled by recognizing a spread of three bands from lower to higher odd numbers: 3/7, 5/9, 7/11, and so on. However, it is not advisable to reverse this by recognizing a spread in odd numbers of three bands from higher to lower numbers, since the result is one of the uniform basic classes. Thus gradations would not be designated as 7/3 or 11/7; rather these would be noted as classes 2 and 4, respectively.

If further refinement is desired, it may be accomplished by a secondary subdivision to recognize border classes, although the writer has seldom considered this necessary. A border gradation may be defined as one lying near the boundary of two of the adjacent master bands in Fig. 8. Such a gradation is represented as class 4–6 to indicate gradation near the boundary between classes 4 and 6. Other than for attempting close refinement, border classes allow for cases of uncertainty in visual classification, as class 4–6 v.

TABLE 6.—PLASTICITY-CHARACTERISTICS SYMBOLS FOR FINE-GRAINED SOILS WITH LITTLE OR NO COARSE AGGREGATE

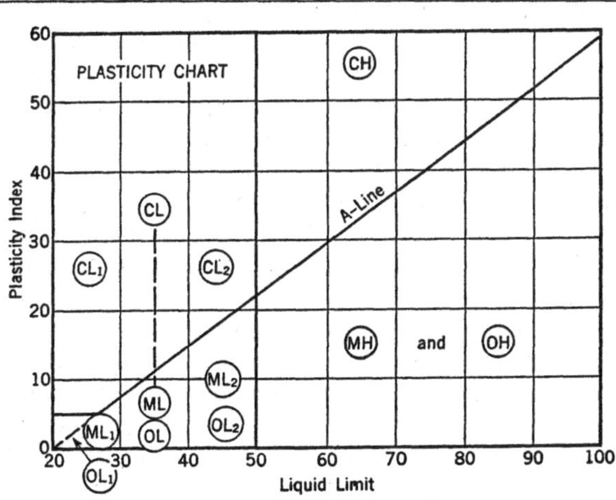

Low plasticity When subdivided: L_1 = slight plasticity L_2 = moderate plasticity	L	M C O	Behavior characteristics of silt Behavior characteristics of lean clay Organic content causing plasticity
High plasticity	H	M C O	Behavior characteristics of quite plastic silts, such as micaceous, diatomaceous, or elastic silts Behavior characteristics of fat clay Organic content causing high plasticity, such as organic clays

In such cases of uncertainty, the last number, as "6 v," is chosen as the more probable one. Then, for simplicity, combined class numbers (whether 4/6, 8/6, or 4–6) can always be considered from the standpoint that the last number (in this case, class 6) approximates the gradation—and conservatively so, because it indicates the range of finer sizes. Hence, the thirteen basic classes are the ones to be fixed in the mind. Once these are understood, combined class numbers are easily visualized as variations.

Part 2B. Plasticity Characteristics.—This phase consists of plasticity characteristics symbols taken from the Casagrande system and, likewise, applicable only to fine-grained soils with little or no coarse aggregate. The proper symbol is determined visually as described by the author; or, where Atterberg limit tests are conducted, it is determined from the plasticity chart in Fig. 4. In Table 6 this chart has been redrawn to follow the author's suggestion of dividing the L-group at a liquid limit of 35. However, the writer prefers to retain the original L-group as the simple form (requirement C) and to use the two subdivisions as the refined form designated by subscripts as shown in Table 6.

In application of the QPC system, classification is accomplished visually by examining moist samples—for fine-grained soils, employing the shaking test and examination of plasticity characteristics as the author describes (under the heading, "13. Field Identification of Fine-Grained Soils"). Laboratory tests are employed only when the classifier feels uncertain, or for purposes of checking, or for training an inexperienced classifier.

Boring logs include soil name, grain-size number, and, for a fine-grained soil, the plasticity-characteristics symbol. Also included is a complete description covering color, particle angularity, consistency, plasticity, and dry strength. Such description is similar to that in Section 15 of the paper, except that, considering the legal viewpoint, the writer prefers to avoid interpretation by using adjectives describing compactness. Instead, it is preferred to give basic data without interpretation wherever such basic data are available, as blows per foot required to drive the sampling spoon (sizes of spoon and hammer, and height of fall being indicated).

On contract drawings, graphic abstracts of logs include the symbols aforementioned under Part 2, plus a brief name of the soil from Part 1 for the benefit of bidders. On study drawings, the name is generally included only when it would add some significant information to that already indicated by the grain-size number and plasticity-characteristics symbol. The simpler form (requirement C) includes soil names plus the thirteen basic grain-size number classes with only limited use of border classes. The refined form includes soil names plus grain-size numbers from the thirteen basic classes and from the primary subdivision if applicable, plus plasticity-characteristics symbols.

Summary.—In addition to reasonably satisfying requirements A to D, the features of the QPC system have been demonstrated to be sound from considerable experience with many types of work—airports, highways, dams, levees, and foundations. The use of soil names in Part 1 of this classification follows the oldest system—the descriptive classification. The grain-size numbers (Part 2A) have been used since 1937 with soils from New England to the Gulf Coast, and have been recommended by Joel D. Justin, M. ASCE (49), as well as by others who have given this system a serious trial. As a part of the Casagrande system, the plasticity-characteristics symbols (Part 2B) have been used intensively in airport construction since 1942, and are based on data collected by the author over a much longer period. Taken together, these three phases of the QPC system reasonably cover the results of classification tests for gradation, Atterberg limits, and organic content. For judging the major soil properties, these three classification tests are considered the best currently available when used in conjunction with information on the grain shape and mineral composition. Although additional symbols could be added to cover the last two factors, the writer feels that this can be handled adequately by proper choice of the soil name.

The author is to be congratulated for the excellence of his critical review of the various classification systems, and for devising a classification which is a decided improvement over prior systems. If, from his efforts and those of others, the profession eventually succeeds in evolving a truly adequate system of soil classification, that would, indeed, be a forward step.

George F. Sowers,[10] Jun. ASCE.—The purpose of a classification is to make it possible for materials which have similar characteristics to be considered and treated as a group. Those characteristics of immediate interest determine the grouping and the boundaries of the groups. The Airfield Classification (AC) System classifies soils by solid ingredients. Soils whose grains are coarser than 0.1 mm are classified by grain size and grain-size distribution; soils finer than 0.1 mm are classified by the liquid limit and plastic index. These properties serve as guides to the capabilities of the soil as an airfield subgrade, but they may be misleading in other applications, as implied by the author in the "Introduction."

Two properties closely linked with the performance of the soil as a foundation material are not considered in the AC system. These are the water content and the consistency. A soft clay of low "plasticity" may be a worse foundation material than a hard, dried-out clay of high "plasticity." A highly plastic clay is not highly compressible unless it has a high water content. Although the author stresses this fact when he states that compressibility must be compared at the same preconsolidation loads, further qualification should be made. Compressibility is dependent on water content; to ignore this in classification may lead to trouble. The meaning of the terms plasticity and consistency are often confused, and the inclusion of the one factor and the exclusion of the other in the AC system might make the situation worse. To a soils engineer, plasticity refers to the range in water content through which the soil is in a plastic state. However, many engineers and especially contractors and drillers use "plastic clay" to mean just what the name implies—clay in a soft or plastic state. To the man who has to dig the soil, or to build on it, the important point is the consistency of the soil, or its water content with regard to its plastic range. Ignoring these points may be dangerous, especially in the light of the confused terminology.

The greatest difficulty with all classifications is the fact that the soils are grouped according to laboratory tests. The nature of the soil deposit is usually given only secondary consideration, although it is often extremely important. For example, a varved clay consisting of alternate layers of silt and clay probably would be tested in the laboratory as a mixture of both. The classification

[10] Asst. Prof., Civ. Eng., Georgia School of Technology, Atlanta, Ga.

tests might reveal a sandy or silty clay (perhaps CL) whereas the actual deposit may exhibit entirely different characteristics. Glacial deposits with erratic inclusions of clay, silt, and gravel defy simple classification; their variations in structure are far more important that the characteristics of each individual material.

Loess soils, which cover large areas in the Mississippi Valley, are as hard as concrete in their dry state, but are soft when saturated. These cannot adequately fit into any system of classification that depends on the index tests of soil solids in a disturbed condition. The same holds true for cemented soils, such as the red sands of the south and southwest. The density of natural sand deposits may be more important than the size or size distribution of the grains, and the presence of water only complicates the possibilities.

The discussion of the uses and of the limitations of soil classification systems by the author in the "Introduction" is a sensible approach to this problem. In order to reduce the soil groups to a finite number, many important characteristics must be ignored. Special classifications, such as the AC system applied to airfields, serve an important purpose, but their limitations should be clearly understood. Too often a classification such as this, coupled with such tabulations as Table 4, give some technicians a false sense of security.

The descriptive system in Section 15 has none of these defects because the descriptive terms can be modified to suit conditions at hand. The greatest difficulty with such a classification is a lack of standard terminology. This obstacle need not be condemnatory, for it has been encountered in other fields, such as meteorology and mineralogy. In Section 15, various degrees of each characteristic in the list of descriptive adjectives (as applied to soils, covering such points as consistency, plasticity, and gradation) could be indicated by numbers, similar to Moh's scale of mineral hardness, or to Beaufort's scale of winds. The degrees should be few so as to be distinguishable through the field tests and through the hand identification methods described in the paper, and they should be standardized by appropriate laboratory tests.

The author's plasticity chart (Fig. 4) would be well suited to standardizing the plasticity and the potential compressibility of the fine-grained soils. Consistency could be defined by the unconfined compressive strength as suggested by K. Terzaghi, M. ASCE. On the other hand, engineers are at such great odds as to the meaning of "silt" and "clay" that it probably will be better to use these terms only in a qualitative sense.

In the application of the descriptive method, both the trained soils specialist and the design engineer will have a better picture of the important characteristics of the soil, unhampered by insertion into limited pigeonholes. As Professor Casagrande points out in the "Introduction," "Those who really understand soils can, and often do, apply soil mechanics without any formally accepted classification."

René S. Pulido y Morales,[11] Esq.—The subject of soil classification, as stated by the author, is the most confused chapter in the study of soil mechanics. Most people believe erroneously that any soils classification can be

[11] Chf. Engr., Soil Laboratory, Public Works Ministry, Havana, Cuba.

used indiscriminately. In order to be successful, the classification must be in accord with the physical properties so that, by using the identification tests, the two most important requisites of compaction and plasticity can be found. It is necessary to know the specific gravity, apparent specific gravity, water content, mechanical analysis, Atterberg limits, and even the elementary chemical tests, so as to establish a relation between the compaction of the soil and the load applied to it.

A knowledge of these properties is important, but one must also study the performance of each soil within a given profile for similar work in other places (highways, foundations, dams, and others) so that, once the relations between the tests and their performance are established, the soils can be classified as types.

The research required to correlate the tests with the performance of the finished works can be obtained only after long periods of time. After the completion of a project the tests can be correlated with experience only by extensive research. The first and most important characteristic to be determined is cohesion, which is a function of the resistance of a soil to consolidation. Therefore, it is necessary, and most important, to include the consolidation test in the research program. The granulometric test is also important, and it must be taken into account as well as a measure of the structure as an index type.

The disintegration of igneous rocks will be more or less in an advanced state if the percentage of fine particles is more or less high. Hence, the result of the disintegration state may be determined by the granulometric curve, which is concave upward if the state is advanced, and downward if it proceeds gently. Intermediate grades are represented by a straight line.

The Public Roads (PR) classification has been used to classify for soils for highways. One of the uncertainties of this system is in group A-4 (Table 3), in which, frequently, one finds silty clays with high capillarity; but this is not the general case since silty clays and silty loams, with little cohesion and low capillarity, are also found. Another uncertainty occurs in the elimination of the material retained in the 40-mesh sieve for the determination of the physical properties and also in the exclusion of the physical determinations of the state of compaction of the soil.

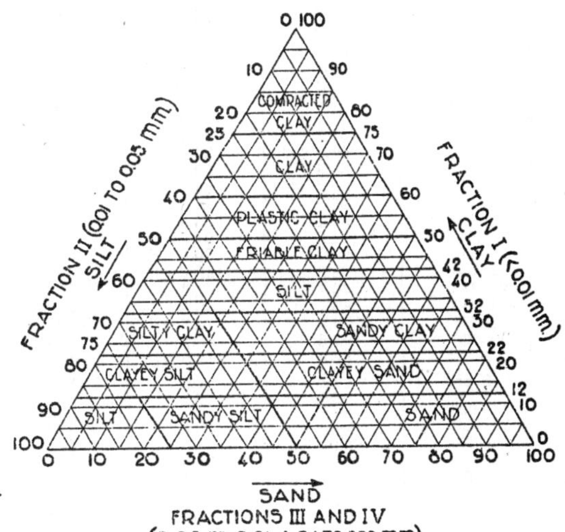

FIG. 9.—CLASSIFICATION BY SWISS ENGINEERS

The author reviews many systems of soils classification with great success. The trilinear graphs in Fig. 3 find an interesting comparison in the textural classification of Swiss engineers (50) as illustrated by Fig. 9.

The latest modification of the PR classification, presented for the Highway Research Board to its Committee on Classification of Materials for Subgrades and Granular Type Road (51), is probably the best tool yet devised for the highway engineers. This system eliminates overlapping, subdivides several groups, and introduces the "group index" which is very useful for evaluating the soil within a given group.

In this new classification the silty clay (group A-4 in the previous PR system) has been made A-6, limiting group A-4 to silty materials. The indeterminancy of the previous system has been eliminated by inserting other subgroups.

TABLE 7.—COMPARISON OF PR AND MODIFIED PR CLASSIFICATIONS

Laboratory No.	Percentage passing mesh No. 200	Liquid limit	Plasticity index	PR group	Modified PR
S-108-1662	4.9	24.7	1.9	A-1	A-1-a (0.2)
M-56-1676	37.6	21.8	0.0	A-1-2	A-4 (0.9)
C-102-560	54.4	25.0	8.9	A-2	A-4 (4.3)
H-22-668	34.5	33.1	3.4	A-2	A-2-4 (0.0)
H-22-908	36.6	46.6	8.7	A-2	A-6 (0.0)
H-22-919	36.7	0.0	0.0	A-2	A-4 (0.0)
H-41-1236	29.2	28.0	7.4	A-2	A-2-4 (0.4)
V-85-1319	2.6	30.1	4.6	A-2	A-1-a (0.0
P-14-1381	40.0	32.1	8.5	A-2	A-4 (1.0)
V-70-1401	38.0	20.4	2.6	A-2	A-4 (0.8)
H-25-1448	27.4	34.4	1.3	A-2	A-2-6 (0.6)
P-14-1465	41.8	32.7	11.8	A-2	A-6 (2.0)
M-56-1654	46.8	23.6	3.2	A-2	A-4 (0.3)
H-41-1223	59.4	33.5	7.9	A-2-4	A-4 (5.3)
H-25-1451	26.9	31.6	8.4	A-2-4	A-2-4 (0.3)
M-43-927	33.3	0.0	0.0	A-3	A-3 (0.0)
P-4-806	36.4	26.3	2.7	A-4	A-4 (0.0)
S-105-489	54.6	36.7	5.3	A-4	A-4 (4.5)
S-105-491	75.8	32.5	1.2	A-4	A-4 (8.6)
S-105-492	37.8	25.7	4.6	A-4	A-4 (0.95)
S-105-494	41.4	31.5	7.1	A-4	A-4 (1.4)
S-105-504	81.1	40.1	12.0	A-4	A-6 (9.0)
S-105-506	68.7	39.4	14.8	A-4	A-6 (8.3)
V-72-550	85.3	22.1	17.7	A-4	A-6 (11.4)
V-80-598	15.7	34.6	11.1	A-4	A-2-6 (0.2)
M-59-949	35.8	44.3	9.5	A-5	A-5 (0.0)
S-108-1051	60.1	48.4	7.56	A-5	A-5 (6.4)
H-22-634	94.8	92.5	46.6	A-5	A-7-6 (20.0)
H-22-635	84.9	89.0	43.3	A-5	A-2-5 (0.0)
H-22-636	23.8	49.6	7.8	A-5	A-2-5 (0.0)
H-22-748	70.7	51.1	14.9	A-5	A-7-5 (11.0)
H-22-749	69.4	60.0	23.8	A-5	A-7-5 (17.2)
H-22-751	73.2	74.7	30.8	A-5	A-7-6 (19.8)
H-22-778	63.6	48.2	7.3	A-6	A-5 (7.1)
C-102-844	82.2	114.0	60.3	A-6	A-7-5 (20.0)
C-102-845	8.6	64.8	33.3	A-6	A-2-7 (0.5)
C-109-891	58.2	35.0	10.5	A-6	A-4 (5)
H-41-1222	48.4	39.7	12.9	A-6	A-6 (3.66)
H-22-633	84.0	54.7	21.8	A-6	A-7-5 (15.2)
V-80-593	42.2	84.5	41.9	A-6	A-7-5 (7.1)
H-22-632	82.1	62.0	27.9	A-7	A-7-6 (18.8)
H-22-755	40.8	44.6	8.9	A-7	A-5 (1.2)
H-22-757	39.2	76.8	44.1	A-7	A-7-5 (4.8)
V-80-587	37.6	54.5	22.0	A-7	A-7-5 (15.7)

It is difficult to discuss soil classification for highway engineering purposes without mentioning the California Bearing Ratio (Porter method). This system provides the best way to determine the soil properties for a definite project and, with the help of the modified PR classification, is the more logical method. Table 7 is an example of some soils tests, made by the writer at the Soil Laboratory of the Public Works Department in Havana, Cuba, which have been classified under the PR and the modified PR classification (51) for comparison and also to show the "group index."

In conclusion the writer wishes to congratulate Professor Casagrande on his clear presentation of this soil classification, a problem that is sorely in need of international standardization.

Raymond F. Dawson,[12] M. ASCE.—A stimulating review of the various soil classification systems which have been used is presented in this paper. A new method of classification is also described which will be a real help to engineers working with soils; but, equally important, Professor Casagrande emphasizes the limitations of any soil classification system. Regardless of these cautions many engineers will adopt or continue to use a single classification system for all types of earthwork, and sooner or later this practice will cause difficulty.

A review of the soil investigations published since 1917 reveals an amazing number of classification triangles that have been proposed. It appears that every investigator wished to insure his name for posterity by attaching it to a classification triangle. Actually, these men were attempting to determine all the soils characteristics from the classification triangle, and this was impossible. Instead of seeking new information they continued to juggle old charts, looking for a combination that would be the panacea.

Not until the Atterberg limits were included in the criteria was any real progress made in the classification of soils for engineering purposes. One of the first schemes using the Atterberg limits was the Public Road (PR) classification system, of which the author correctly states (in Section 6): "This classification, therefore, is essentially a grouping applicable to the use of soil in a road surface or as a base beneath a thin bituminous wearing surface." Regardless of this limitation, highway engineers continually use the PR classification system for all types of earthwork construction. Sometime ago, an engineer called the writer's attention to a paper describing a construction method using an A-7 soil in a northern state. He said, "They cannot use that procedure with an A-7 soil; I know, because we tried it." Thereupon he visited the project and found that they were using the method as described. The soil in the northern state had a liquid limit of 42 and a plasticity index of 25, whereas the Texas soil had a liquid limit of 86 and a plasticity index of 50. Both soils were classified as A-7 by the PR system; however, they were two entirely different materials for the construction involved.

The author's Airfield Classification (AC) System is a forward step in the classification of soils for use as subgrades of pavements in airports and similar structures. The AC system will undoubtedly be the standard of classification for subgrade soils in airfield construction.

Another subgrade classification system (51) uses the group index which utilizes both the grading and the Atterberg limits of the soil. To the writer, the group index appears to offer a valuable classification system, although he is of the opinion that modifications of the limit ranges may be necessary. It will be interesting to have the author's discussion of the group index system.

[12] Associate Prof., Civ. Eng.; Associate Director, Bureau of Eng. Research, The Univ. of Texas, Austin, Tex.

Professor Casagrande comments (in Section 10) that "Experience has proved that the liquid limit is a good measure of relative compressibility." It is of considerable importance, and further discussion of this statement is desirable.

Engineers must be cautious, or they may be misled by local terminology in describing soil formations. In the Texas gulf coast area, old-time drilling operators call the first clay stratum beneath a sand stratum a "shale" regardless of how soft it may be. Under such conditions the engineer who is unfamiliar with these colloquialisms and who does not examine the cores may not be aware of a serious condition that exists.

D. F. Glynn,[13] Jun. ASCE.—Since a soils classification provides a uniform standard of nomenclature for field identification and for condensing engineering experience with soils, the confusion reviewed by Professor Casagrande's paper is most disturbing. Thus, immediate considerations arising from it concern: "What order could be produced in the existing position?" and "which of the systems described are suitable for incorporation in a coordinated classification?"

Any comprehensive standard classification system would have to meet the needs of agricultural scientists and geologists in addition to civil engineers. Any single system, sufficiently comprehensive to cover all these different interests, would probably be too complex for convenient use. An alternative would be the adoption of a multistage system, the first stage being a simplified field identification and the second a series of detailed functional classifications for each profession. If these were kept as simple logical systems they would form the uniform standards and allow sections with special interests to formulate additional stages to cover their own local interests. It would appear from Professor Casagrande's description that agricultural scientists and geologists already have functional classifications that meet their requirements. Thus the problem is reduced to providing a uniform field identification for general use and to formulating a standard engineering classification.

For a uniform system of field identification, a modified descriptive soil classification would seem very attractive. As a result of wartime cooperation by engineers, geologists, and agricultural scientists in the Southwest Pacific Theater, such a system has already been widely adopted in Australia. This system has sufficiently standardized soil terminology to avoid the disadvantage referred to by Professor Casagrande. It is essentially a field identification system, however, and the ultimate test of the description of any soil is the majority verdict of the experienced field surveyors. In order to keep such opinions up to date and uniform, periodic conferences are held, to which all interested persons are invited. Thus, the formal descriptive standards for adjectives and soil textures are subordinate to field usage and are periodically amended where necessary (52).

An alternative to this simple descriptive standard would be the field identification system based on the Airfield Classification (AC) System. It is difficult for anyone who has not served an apprenticeship with this system to assess its

[13] Superv. Engr., Soils and Hydr. Testing, Melbourne and Met. Board of Works, Melbourne, Australia.

value in practice. Nevertheless, it would seem more complex and less flexible than the simpler descriptive type of classification. As such, it would probably be much harder to reach an international agreement on the basis of this system than with the more widely accepted descriptive classification.

As a functional engineering classification, the Texas and Civil Aeronautics Administration (CAA) systems would appear too limited in their acceptance. The Public Roads (PR) system is widely used by highway engineers, but in addition to Professor Casagrande's criticism as to the difficulty in keeping abreast of the latest "edition," its basic structure is too closely tied to highway requirements to be a good general system.

From its first appearance in Australia in 1943 the AC system has been regarded by many soils engineers as the most suitable general system. After use by the Corps of Engineers, United States Army, it is probably the most widely adopted of all systems. For these reasons the most recent revision and the discussion on the soils of the various groups are extremely welcome. It impresses one as being simple and easily understood and as following logically from the simpler field description. The principal objection offered in Australia has been due to the lack of definition in the group boundaries. It is suggested, however, that this may be a distinct advantage in that it prevents indiscriminate use by people who have no knowledge of the significance of soil test results.

If the AC system could be accepted as the basic classification to be used for samples, selected from the results of simple descriptive field mapping, it is suggested that the proposed further subdivision of classes should be limited to special applications. The general system, summarized in Fig. 4, would appear sufficient for general use and more detailed subdivision might lead to unnecessary confusion.

Another feature referred to by Professor Casagrande, the grain-size scale, might be also amended to achieve greater uniformity. In Australia an increasing number of laboratories have reverted to the international standard. At a conference of regional officers of the Division of Soils, Council for Scientific and Industrial Research, at Adelaide, Australia, in September, 1946, the sand-silt size boundary of 0.02 mm was discussed (52), in the light of the proposal by R. Glossop and A. W. Skempton (5). Although no change was recommended, it is believed that change to 0.05 mm or 0.06 mm would be readily accepted if sufficiently valid reasons could be offered. On other hand, it is understood that the United States Department of Agriculture is by no means irrevocably committed to 0.05 mm. The time would thus seem ripe for discussion to achieve uniformity on at least this one point.

In view of the proposed international soil conference in Holland, Professor Casagrande's able summary of the present soils classifications is most timely. It is to be hoped that this favorable opportunity for using Professor Casagrande's wide experience in soils mechanics, soils classification, and international cooperation with engineers will not be allowed to pass.

In conclusion, it is desired to thank Professor Casagrande for the comprehensive review of the present state of soils classification and particularly for his discussion on the AC system.

L. F. Cooling,[14] Esq., A. W. Skempton,[15] Esq., and R. Glossop,[16] Esq.—One of the most important, present needs in applied soil mechanics is for a universal acceptance of at least some broad general classification of soils, and in his paper Professor Casagrande has done much to clarify the position in a field where there is admittedly much confusion. The writers met the difficulties involved in this problem a few years ago when they were engaged in preparing drafts for a British code of practice on site investigations. It was soon realized, as stated by the author, that it is not possible to classify all soils into a relatively small number of groups in such a way that the system would adequately meet the requirements of the many divergent problems of applied soil mechanics. However, it was felt that it would be desirable first to adopt a broad classification which would serve as a basis from which, by expansion and amplification within the original framework, the more detailed and specific classification systems required for various practical problems could be built up. For this reason, the approach was to formulate a general basis for the field identification and classification of soils using a system similar to that presented by the author in Section 15. It was considered sufficient for this purpose to take into account two essential characteristics—the size and nature of the particles composing the soil and the strength and structural features of the soil as it exists in the ground. The type of general arrangement suggested is indicated in Table 8.

Six principal types are recognized on the basis of the size and nature of the particles (Col. 1, Table 8), and the simple tests by which these types are distinguished in the field (Col. 2, Table 8) are similar to those described by the author. In this connection the term "dilatancy" corresponds to the property described by the author under "shaking test."

The subdivision of the uniform, well-sorted sands into coarse, medium, and fine is perhaps an elaboration, but it can be readily made and, since the subdivisions can often be associated with important changes in engineering properties, has much to commend it. The division of clays into "lean" and "fat" can be made on the basis of the extent to which the soil exhibits the typical colloidal properties of plasticity, cohesion, and shrinkage. Col. 3, Table 8, gives the more important types of composite soils. These are primarily natural mixtures of two or more of the principal soil types, but mineralogical variations which have an important influence on the soil properties (mica, organic matter, and calcareous material) are also recognized.

Cols. 4 and 5, Table 8, are concerned with the "in-situ" strength and structural characteristics. Since, in most engineering problems, the strength or density of the soil in its natural condition is of controlling importance and since, similarly, the structural features of the deposit (such as laminations and fissures) can be very significant, the writers would stress the importance of their inclusion in any basic classification.

Finally, the color of the soil should be stated since this often has local significance and is helpful for identification. From the terms defined in Table 8,

[14] In Chg. of Soil Mechanics Div., Bldg. Research Station, Watford, England.
[15] Reader in Soil Mechanics, Univ. of London, London, England.
[16] Engr., Messrs. John Mowlem and Co., London, England.

therefore, it is possible to give a fairly accurate picture of a particular soil in a manner that could be generally recognized. The author's suggestion, that with fine-grained soils a note on the effect of remolding should also be included, is a valuable one.

The writers attach particular significance to the strength classification of clays in all problems except those concerned with near-surface soils, and it would be appreciated if the author would give his opinion on this question. The grouping suggested in Table 8 is based largely on the Boston Building Code with an additional class at each end of the scale. These descriptive terms are implicitly connected with shear strength and, within the writers' experience, they are roughly related in the following manner:

Class	Unconfined compression strength (lb per sq in.)
Very soft	<5
Soft	5 to 10
Firm	10 to 20
Stiff	20 to 40
Hard	>40

From this relation it is possible to develop more specialized classifications for such purposes as foundation bearing capacities and earth pressures.

In cases of near-surface soils and soils used for constructional purposes when the "in-situ" properties are likely to be drastically changed, it is agreed that a

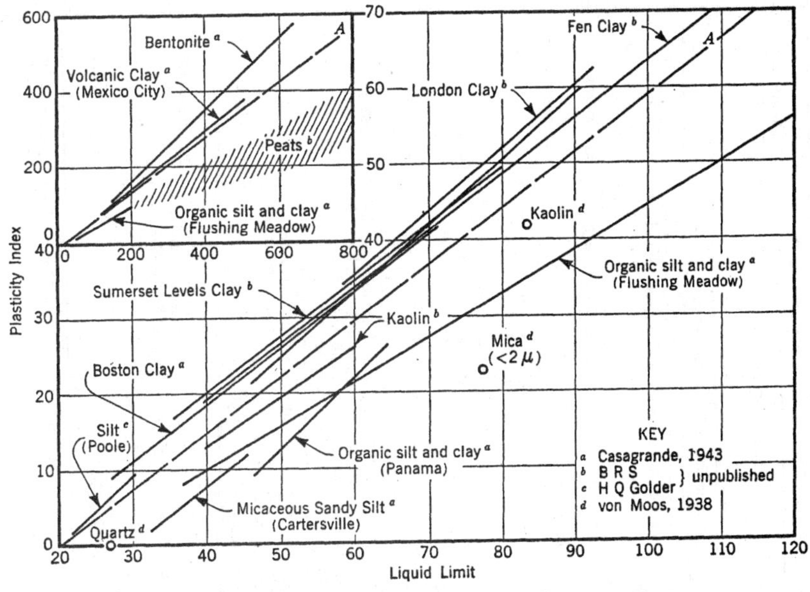

Fig. 10.—Relation Between Liquid Limit and Plasticity Index

classification system leaning more heavily on the composition of the soil is required. Consequently, for site investigations in relation to roads and airfields, the Airfield Classification (AC) System, adopted in 1942 by the United States Engineer Department, was recommended in the aforementioned code of practice.

TABLE 8.—General Basis for Field Identification and Classification of Soils

Principal Soil Types; Size and Nature of Particles			Composite types	Strength Characteristics		Structural Characteristics	
Divisions (1)	Types (2)	Field identification (3)	(4)	Term (5)	Field test (6)	Term (7)	Field identification (8)
Coarse grained, noncohesive	Stones	Boulders; larger than 8 in. in diameter. Cobbles; mostly between 8 in. and 3 in.	Boulder gravels	Loose	Can be excavated with spade. 2-in. wooden peg can be easily driven.	Homogeneous	Deposit consisting essentially of one type.
	Gravels	Mostly between 3 in. and No. 7 British Standard sieve.	Hoggin[a] sandy gravels				
	Sands	Particles mostly between Nos. 7 and 200 British Standard sieves. Particles visible to naked eye. No cohesion when dry. { Uniform / Graded } { Coarse—between Nos. 7 and 25 British Standard sieves. Medium—between Nos. 25 and 72 British Standard sieves. Fine—between Nos. 72 and 200 British Standard sieves. }	Silty sands Micaceous sands Lateritic sands Clayey sands	Compact Slightly cemented	Requires pick for excavation. 2-in. wooden peg hard to drive more than few inches. Visual examination. Pick removes soil in lumps which can be abraded with thumb.	Stratified	Alternating layers of varying types.
Fine grained, cohesive	Silts	Particles mostly passing No. 200 British Standard sieve. Particles mostly invisible or barely visible to the naked eye. Some plasticity and exhibits marked dilatancy. Dries moderately quickly and can be dusted off the fingers. Dry lumps possess cohesion but can be powdered easily in the fingers.	Loams (silt, sand, clay) Organic silts Micaceous silts	Soft Firm	Easily molded in fingers. Can be molded by strong pressure in the fingers.	Homogeneous Stratified	Deposit consisting essentially of one type. Alternating layers of varying types.
	Clays	Smooth touch and plastic, no dilatancy. Sticks to the fingers and dries slowly. Shrinks appreciably on drying, usually showing cracks. Dry lumps can be broken but not powdered. They also disintegrate under water { Lean clays show these properties to low degree. Fat clays show these properties to high degree. }	Sandy clays Silty clays Marls[b] Organic clays Boulder clays Lateritic clays	Very soft Soft Firm Stiff Hard	Exudes between fingers when squeezed in fist. Easily molded in fingers. Can be molded by strong pressure in the fingers. Cannot be molded in fingers Brittle or very tough.	Fissured Intact Homogeneous Stratified Weathered	Breaks into polyhedral fragments along fissure planes. No fissures. Deposits consisting essentially of one type. Alternating layers of varying types. If layers are thin the soil may be described as laminated. Usually exhibits crumb or columnar structure.
Organic	Peats	Fibrous organic material, usually brown or black in color.	Sandy, silty or clayey peats.	Firm Spongy	Fibers compressed together. Very compressible and open structure.	....	....

[a] Hoggin: Sandy gravel with small admixture of clay. [b] Marl: Calcareous clay.

However, it was considered that this classification system could be legitimately regarded as a development and expansion of the basic descriptive soil classification and that it was in the same category as others designed for special engineering purposes, such as foundation bearing capacities, limits of the applicability of various geotechnical processes (5*a*), and so on.

The author's discussion on the properties of fine-grained soils and their field identification is a valuable contribution to the subject. His method of using the liquid and plastic limit values for soil identification by use of the plasticity chart has been followed with interest in Great Britain, and Fig. 10 summarizes results on typical British soils.

A further relationship for fine-grained soils, which seems to the writers to be significant, is that between the liquid limit and the percentage clay fraction—that is, the proportion of particles of a size less than 2 μ. For samples from a given deposit the liquid limit increases with percentage clay fraction and the relation for some typical British soils is given in Fig. 11. Most clays fall within the zone marked "normal clays." However, the kaolins and boulder clays show quite low liquid limits even for fairly high clay contents. This is probably explained by the "inactive" nature of kaolinite compared with the minerals in most clays and by the fact that the particles of the boulder clay have been produced by mechanical comminution and not by weathering. On the other hand, some of the highly weathered clays and organic muds show a high liquid limit even with a comparatively small percentage clay fraction. This may be indicative of a more than usually "active" clay mineral or the presence of unusually fine particles in the clay fraction or, more probably, of a combination of these two effects. In this connection it is interesting to note that the alluvial clay of the River Lea in England has been derived largely from the weathering of the London clay, itself a fairly fat clay, being the product of the weathering and transport of the soil mantle in Eocene times. The indication of the "quality" of the clay fraction which this relationship reveals is considered helpful in assessing the probable characteristics of the soil. As a matter of interest the relationship for seven of the soils given by the author in Table 1, Section 3 are plotted as spot points in Fig. 11. The results agree with the characteristics described with the exception of the diatomaceous earth (No. 5) which, the author states, shows little plasticity although its plasticity index is 42. It would be of interest to know if the diatomaceous earth behaves in an anomalous manner in other respects.

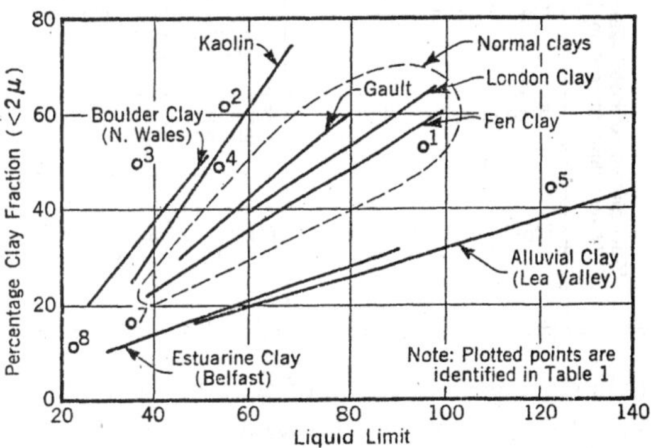

FIG. 11.—RELATION BETWEEN LIQUID LIMIT AND CLAY CONTENT

Milton Vargas,[17] Jun. ASCE.—In Section 1 (under the heading, "Development of Systems") Professor Casagrande states that "* * * soil classifications used in civil engineering reflect in their chronological order the development of soil mechanics." This is likewise true of other sciences. Classification systems only reach perfection when the corresponding sciences reach a certain final stage of equilibrium. The existence and success of the Airfield Classification (AC) System shows that this final stage, in the case of soil mechanics, is already in sight.

However, there are some points where explanations are needed. First, in order to be considered perfect, a classification system must be able to divide the things to be classified according to the fundamental characteristics that are significant for the entire class. The AC system begins classifying the soils into two major divisions, according to grain size—the first division referring to coarse-grained soils and the second to fine-grained soils. However, the author was compelled to add a third major division for fibrous organic soils which cannot be classified according to grain-size characteristics. This shows that grain size is a significant characteristic only for a certain class of soils and not generally for all soils. The existence of the third major division

TABLE 9.—Suggested Subdivision of Soil Groups According to Mineralogical Composition.

1		2		3
Major divisions		Soils groups and subgroups and typical names		Group symbols
Fine-grained soils containing little or no coarse-grained material	Low to medium compressibility	Silts and very fine sands, silty or clayey fine sands, silty clays, slight to low plasticity	Inorganic (except kaolin type)	ML
			Organic	OL
	Liquid limit < 50	Clays of low to medium plasticity, sandy clays, silty clays, lean clays	Inorganic (except kaolin type)	CL
			Kaolin type	KL
	High compressibility	Fine sandy and silty soils, elastic silts	Micaceous Diatomaceous Others	MH
			Organic	OH
	Liquid limit > 50	Clays of medium to high plasticity, fat clays	Inorganic (except kaolin type)	CH
			Kaolin type	KH

without a clear criterion of classification allows the engineer to expect that, in the future, it may be necessary to add new major divisions, more or less arbitrarily, as soils become better known. What then should be the most general characteristic of soils for the purpose of classification? That is an open question.

Following the same line of thought, the adopted criterion for separating the major divisions into subdivisions (according to grain-size composition, in the case of coarse-grained soils, and compressibility, or value of liquid limit, in

[17] Chf., Soils and Foundations Div., Inst. for Technological Research, Sao Paulo, Brazil.

the case of fine-grained composition) appears to be excellent, and the inclusion of intermediate subdivisions among them would do no harm to the system. The same could be said of the groups that divide the major divisions of the coarse-grained soils. However, the groups of the two subdivisions for the fine-grained soils are separated according to two different criteria—grain size and organic matter content. If a new group for the kaolin-type clays is added, as suggested by the author, certainly a third criterion will be introduced —that is, the mineralogical composition.

It would be much simpler, in any future expansion, if the two major divisions of the fine-grained soils were both subdivided into only two groups, that is, the silt group and the clay group, according to grain-size composition. These groups could then be divided into subgroups according to the mineralogical composition and organic matter content, as suggested in Table 9. The systematic framework of the AC system would thus be improved with practically no change. Thus, the way is open for future additions of new groups without the possibility of confusion.

Another requirement for a perfect classification system is that it be suitable for general use under different conditions. A good soil classification must be as useful for the design of airfields as for foundations or tunnels. The author appears to limit his classification system in referring to it as an Airfield Classification System. Its applicability seems to be very much wider.

In his paper the author does not explain whether the group symbols he proposes are intended to be simple initials of the group names, in order that the users may easily remember them, or whether his intentions were to propose symbols of world-wide significance for soils, in the way that chemical symbols represent the chemical elements. In the first case it would be logical that the group symbols should be changed in each language; on the other hand, in order to maintain the universal significance of the symbols they should not be translated. This would be really advantageous for the proper understanding of engineers all over the world—at least as far as soil properties are concerned. The fact that the group symbols are initials of English words would be no impediment when one realizes that the English language is as widely recognized today as Latin was in the days when botanic nomenclature or chemical symbols were established.

Donald M. Burmister,[18] Assoc. M. ASCE.—A worthwhile contribution to the engineering profession has been made in this critical review of the soil classification systems now in use and in the presentation of the Airfield Classification (AC) System, which served such a useful purpose in the work of the Corps of Engineers, United States Army, during World War II. It is significant that the AC system has been developed on the basis of the important behavior characteristics of soils, and therefore has a broader application than to airfield construction alone.

The greatest practical use of the AC system in the postwar period is in the preliminary exploration, design, and construction stages of projects involving earthwork and foundation problems, where the information obtained is put

[18] Associate Prof. of Civ. Eng., Columbia Univ., New York, N. Y.

to immediate and practical use. Those engaged in the soil engineering work on a given project become intimately familiar with the class designations and symbols of the different classes of soils encountered, and they are able to infer accurately the general qualities and behavior characteristics. The group symbols become a convenient tool and expedite the work wherever a large number of samples must be examined and identified. The author has properly warned against the abuse of classification symbols wherein only the group symbol is given without more specific information. Use of symbols alone short-circuits the process and assumes the triple role of identification, description, and classification. After a lapse of some time following the completion of a project such a classification of the soils encountered on a project inevitably becomes more and more vague and less significant as to the actual character and behavior of the soils, because symbols or class designations can never convey a sufficiently precise meaning.

For all permanent records of an organization engaged in soil engineering work, as well as for presentation of information in papers and articles in the technical journals, the more precise and significant descriptive classification, such as that discussed by the author, should be used so as to convey a definite meaning to those who have not had the benefit of examining and working with the soils firsthand. The greatest field of progress in soil engineering lies in the accumulation of detailed, accurate information on the physical character and properties of soils and on their behavior under the varied conditions encountered in practice. Thus, a body of authoritative knowledge on soil behavior, which would be of great practical value and use, can be built up, not only in a given organization but also for the profession as a whole. Therefore, in the interest of greatest progress in soil engineering, and of maximum usefulness of data on soil behavior, greater emphasis should be placed on the development and continued use of thoroughness in examining, identifying, and describing soils. These considerations far outweigh any consideration of convenience.

In order to satisfy these essential requirements and needs in the postwar period the base of the AC system should be broadened to make the system more fundamental, and at the same time more precise and significant for all types of foundation and earthwork problems. There are two basic steps in such a process.

The first step is to come to some agreement on the definitions of soil components, which may find general acceptance and which may form a common basis for work in soil engineering. Recommendations have been presented by the Foundations and Soil Mechanics Committee, Civil Engineering Division, American Society for Engineering Education (A.S.E.E.), for the consideration of those interested in soil engineering work (24). These definitions are intended to define terms to be used in identifying and describing soils, and are not a textural classification. The following is quoted from the report of this committee:

> "The definitions of the principal soil components: GRAVEL, SAND, SILT, and CLAY should be based on significant criteria. From an engi-

neering point of view the primary difference between sand and gravel is the size of the constituent grains, which can be recognized visually. The primary differences between sand and silt are that the constituent particles of silt cannot be readily distinguished by the unaided eye and that silt exhibits considerable capillarity. The significant and distinctive difference between silt and clay is that clay exhibits plastic properties and silt does not. In the case of fine-grained soils containing clay the influence of grain size is dominated by the influence of mineralogical and chemical composition. Thus, gravel and sand should be defined on the basis of grain size; sand and silt should be defined on the basis of grain sizes and of capillarity; while silt and clay should be defined on the basis of plasticity. These universally comprehended terms form an adequate and satisfactory basis for the definitions of soil components."

In order to have a workable system of definitions for practical use, it was considered desirable by this committee to define the size limits of the principal soil components and of the coarse, medium, and fine fractions directly in terms of sieve sizes, which convey a definite idea of particle size. The term "CLAY-SOIL" is used instead of "CLAY" in order to be more precise in terminology, because the silt admixture cannot be separated out to determine the proportion of the true clay fraction. The A.S.E.E. recommended definitions are given in Table 10. [Capitalized terms, as "CLAY-SOIL," and soil names are given in quotation marks throughout this discussion to indicate that they are proposed special terminology for describing and naming soils.—Ed.]

TABLE 10.—AMERICAN SOCIETY FOR ENGINEERING EDUCATION RECOMMENDATIONS FOR SOIL COMPONENTS AND FRACTIONS

Principal components	Description	Sieve limit	SIEVE SIZES FOR SUBCOMPONENTS		
			Coarse	Medium	Fine
(1)	(2)	(3)	(4)	(5)	(6)
BOULDERS and ROCK[a]	Retained on 3-in. sieve	Lower	....	....	3 in.
GRAVEL and STONE[a]	Passes 3-in. sieve; retained on No. 10 sieve	Upper Lower	3 in. 1 in.	1 in. ⅜ in.	⅜ in. No. 10
SAND	Passes No. 10 sieve; retained on No. 200 sieve	Upper Lower	No. 10 No. 30	No. 30 No. 60	No. 60 No. 200
SILT	Passes No. 200 sieve; nonplastic, little or no strength when air-dried	Upper	No. 200	....	....
CLAY-SOIL	Passes No. 200 sieve; exhibits plastic properties and clay qualities within a certain range of moisture content; considerable strength when air-dried	Upper	No. 200	....	....

[a] BOULDERS and GRAVEL refer to waterworn material; ROCK and STONE refer to angular fragments.

The second step is to establish a common language for description of soils so that the soils examined and identified can be properly named and described in simple engineering and technical terms. The author points out that at present there is no uniformity in the terms used to describe soils. The descriptive terms, however, should be sufficiently precise and accurate, and at the same time sufficiently flexible in nature, so that the entire range of natural soils can be readily described. The soil name should have a clear, well-defined meaning to enable people who have not had the benefit of examining the soils firsthand to readily understand and to interpret them. These descriptive terms should be suitable, not only for rapid field identification by visual and manual means, but also for the more detailed laboratory identification based on soil tests.

The descriptive terms shown in Table 11, which have been used and found to be satisfactory by the writer, are suggested. Such terms have been used to a considerable extent by field engineers and boring foremen, but without having the precise, well-defined meanings given in Table 11. Although natural soils are composite materials composed of various proportions of the principal soil components in almost infinite variety, recognizable proportions can be defined for practical use. Descriptions of typical soils are given at the end of this discussion to illustrate the use of this terminology in naming soils.

The proportions given in Table 11 cover recognizable ranges, sufficiently obvious for practical purposes. In borderline cases between two proportions either designation is acceptable. The descriptive terms for these recognizable ranges of proportions are used instead of percentages directly because they

TABLE 11.—Descriptive Terms for Cohesionless Soils to be Used in Forming the Soil Name

Soil component (1)	As written in the soil name (2)	Descriptive or qualifying terms as written (3)	Range of proportions (4)
Principal	GRAVEL, SAND, SILT	.... and	50% or more[a] 35% to 50%[a]
Others	Gravel, Sand, Silt	some little trace	20% to 35%[a] 10% to 20%[a] 1% to 10%[a]
Subcomponents		coarse to fine coarse to medium medium to fine coarse medium fine	all sizes <10% fine <10% coarse <10% medium and fine <10% coarse and fine <10% coarse and medium

Additional descriptive terms:
(1) Color, grain shape, etc.
(2) Degree of compactness, degree of plasticity.
(3) Inorganic constituents (mica, shells, and foreign matter).
(4) Organic matter (roots, humus, peat, and muck).
(5) Geological origin (alluvial, glacial, wind, beach, swamp, etc., also horizon).

[a] Finer than, or coarser than, the principal soil component.

form a more suitable soil name, which is sufficiently precise for all practical purposes, considering the variation to be expected from sample to sample in the same deposit. The proportions of the gravel and sand components can be readily determined visually; the proportions of the sand and silt components can be determined by feel and by simple manual tests. In the laboratory or in the field office a standard set of bottle samples should be made up for reference. Thus, a person who is not continuously working with soils and identifying them can refer to these standards and check himself, until he has again acquired the sense of identification and of proportions, which really involves seeing and comparing volumes, at the same time giving due consideration to weight and size of grains, and other characteristics.

Such a set of standard bottle samples should be composed of at least fourteen combinations covering the limits of the proportions corresponding to the descriptive terms in Table 11: (*a*) Seven of dry gravel-sand mixtures in the percentage proportions 90–10, 80–20, 65–35, 50–50, 35–65, 20–80, and 10–90; (*b*) three of dry sand of coarse, medium, and fine sizes, respectively; and (*c*) four of moist sand-silt mixtures in the percentage proportions 90–10, 80–20, 65–35, and 50–50. All samples should be at least one pint in volume, and preferably larger.

In examining and identifying soils one should start with the coarsest component and work by stages to the finest component. The proportion of the

TABLE 12.—DESCRIPTIVE NAMES OF CLAY-SOILS BASED ON THE DEGREE OF PLASTICITY[a]

Degree of plasticity (1)	Plastic index (2)	Descriptive name as written (3)	Qualities (4)
Nonplastic	0–1	SILT	Friable
Slight plasticity	1–5	trace Clay	Desirable
Low plasticity	5–10	little Clay	Cohesiveness
Medium plasticity	10–20	CLAY and SILT	Increasingly objectionable plastic displacements and compressibility
High plasticity	20–35	Silty CLAY	
Very high plasticity	>35	CLAY	

[a] Over-all plasticity of sand-clay-soil fraction.

gravel component and its character as to coarse, medium, and fine fractions should be determined on a fairly large sample. The sand component and finer fractions are then examined to determine the proportion and character of the sand component and of the silt or clay-soil component. Visual means are satisfactory for sands containing not more than a trace of silt. The fine-grained soils or fractions of a soil are more difficult to identify; it is also more important to identify them accurately. Simple manual tests, such as the shaking test for silty soils and the plasticity test (thread rolling test) for clay-soils, can be used satisfactorily for proper identification by adjusting the moisture content to some particular condition. Plasticity and certain clay qualities are the most distinctive and characteristic properties of clay-soils. Therefore, they form a satisfactory and significant basis for identifying and naming these soils. Table 12 lists suggested descriptive terms based on the degree of plasticity of the sand-clay-soil fraction to be used in forming the soil name. These terms have also been used and found to be satisfactory by the writer for a number of years.

For a field and laboratory identification the over-all plasticity of the sand-silt-clay proportion of the soil as determined by the simple plasticity test is the most significant property, because it has a more direct relation to the behavior characteristics of the soil, and because a separation is not conveniently possible. The gravel sizes can be picked out by hand for the plasticity test.

The range of the degree of over-all plasticity expressed in terms of the plasticity index for each clay-soil type in Table 12 is broadened and adjusted to be more suitable and significant for identifying the soils. The proportion of sand can be determined by the feel of the soil.

The author has given a few examples of descriptive soil classification to illustrate how the AC system can be expanded. The descriptions do not, however, convey a sufficiently definite meaning to a person working in another field of soil engineering, who has not had the benefit of examining the soil firsthand, but who wishes to use the information on the behavior of such soils for the practical purposes of sizing up and judging his own particular problem and for reaching a satisfactory solution. Examples 1, 2, and 3, hereinafter, illustrate some advantages and the general usefulness for all purposes of the descriptive names of soils, based on the terminology in Tables 10, 11, and 12. It is intended that the soil name be put together in a certain definite fashion to emphasize the significant characteristics of the soils examined. The capitalized term in the soil name stands out as the principal component of the soil, so that it can be spotted at a glance. The significant detail then follows, expressed by the qualifying terms giving the proportions of the other soil components, usually in order of their importance. As is frequently necessary one can glance down a boring record, soil profile, or a list of soils from an area, and pick out immediately the types which have certain significant characteristics bearing directly on a given problem. The descriptive names of the soils can be abbreviated somewhat for a large group of similar soils, provided that the upper and lower limits of the group are given in detail.

Example 1.—

(*a*) "Light tan coarse to fine GRAVEL, and coarse to fine SAND, trace Silt." (Maximum size 1½ in.; plus No. 4 sieve, waterworn.)

(*b*) "Light brown coarse to fine GRAVEL, some coarse to fine Sand, trace Silt." (Maximum size 1½ in.; plus No. 4 sieve, waterworn.)

Certain important conclusions can be drawn from the descriptions of these soils based on a general knowledge of the behavior characteristics of different classes of soils. First, the soils will be free draining because there is not more than a trace of silt. Second, the soils ranging from "GRAVEL and SAND" to "GRAVEL, some Sand" (as an abbreviated descriptive classification) represent about the limits of drainable soils that can be considered well-graded. Third, the large gravel content and the large sizes indicate that high supporting capacity can be expected in the compacted condition. The quality of being well-graded is significant only for soils containing a large proportion of gravel, not only because it connotes compactibility to high densities (exceeding 135 lb per cu ft), but also because the close packing and contacts, together with the interlocking and proper bedding of the large particles, provide a high supporting capacity.

Example 2.—

(*a*) "Light brown coarse to fine SAND, trace Silt"; medium compact.

(*b*) "Light brown medium to fine SAND, little Silt"; medium compact.

(*c*) "Light brown medium to fine SAND, some Silt"; loose.

From these descriptions certain significant characteristics are immediately apparent. These soils are nonplastic in character. Soil (*a*) is free draining, has only slight capillarity, and is not susceptible to frost heaving. Soil (*b*) is not as free draining, but can be drained by well points, has some capillarity, and is slightly susceptible to frost heaving with high ground-water levels. Soil (*c*) is more difficult to drain with well points, has rather objectionable capillarity, and has frost heaving characteristics. On the basis of natural compactness, soil (*a*) would be much superior to soil (*c*) in the natural state as to supporting capacity. On the other hand, soils (*b*) and (*c*) would show better compactibility because the silt fills the void spaces and provides a bedding for the sand.

Example 3.—

(*a*) "Brown medium to fine SAND, some Silt, trace Clay"; slight plasticity.

(*b*) "Gray-brown SILT, some medium to fine Sand, little Clay"; low plasticity.

(*c*) "Gray CLAY and SILT, little medium to fine Sand"; medium plasticity.

All these soils are relatively impervious. Soil (*a*) would have fairly good compactibility. The supporting capacity would be only slightly affected by capillary saturation in the compacted state. Soil (*b*) is on the borderline of objectionable plastic displacement characteristics and softening by capillary saturation in the compacted state, particularly with a high ground-water level. Soil (*c*) would compact to a low density and would definitely have objectionable characteristics. However, this soil might have a fairly good supporting capacity in the natural state, if the natural consistency was stiff or better.

The amount of extra effort and time required to identify and describe soils in the more thorough fashion described herein is small when compared to the greater usefulness and value of the information obtained for all purposes, especially to people who wish to use such information to aid the solution of their own particular problems. Such descriptions and soil names, if generally used in boring records and in soil profiles and soil sections, would give far more significant and valuable information to the designing engineer, and, as a part of the contract drawings, to the contractor and construction engineer, than is obtainable at present from the vague and often wholly inaccurate terms generally used. Such information can be interpreted, as noted in the foregoing examples, on the basis of a person's particular experience, into terms of significant behavior characteristics, which would have a more direct bearing on the solution of the practical problems of foundation design and of the practical problems of foundation construction. If a typical soil near the upper limit and another near the lower limit of a group of soils, which have in common similar characteristics, are tested in detail in the laboratory to determine all the pertinent physical properties and strength characteristics, and the results are reported with the descriptive classifications of the whole group, then the range of behavior characteristics to be expected for the entire group can be fairly reliably known. When the actual behavior and performance of the different groups of soils have been observed in sufficient detail in the field and have been correlated with significant soil characteristics, the information would become an important part of a body of authoritative knowledge, which would be extremely useful and valuable to the engineering profession.

M. G. Spangler,[19] M. ASCE.—The fundamental purpose of a soil classification system is to provide a language by means of which one person's knowledge of the general characteristics of a particular soil can be conveyed to another person or group in a brief and concise manner, without the necessity of entering into lengthy descriptions and detailed analyses. Since soils are so widely heterogeneous in character, it is not to be expected that one kind of classification system can cover all the possible features of a soil which it may be desirable for one person to convey to another—hence, the desirability of several bases for classification. In engineering work, it is frequently appropriate to use at least several of, or all, four classifications for describing soils, such as: (1) The geological classification, which provides a language for stating the geological history and background of a soil; (2) the textural classification which provides names for describing the relative proportions of sand, silt, and clay size particles in a soil; (3) the pedological system of classification which reveals the nature of the soil profile as affected by the climatic and other environmental conditions under which the profile developed; and (4) an engineering classification that groups soils on the basis of their characteristics which influence engineering performance.

Great care should be exercised not to read into these various classifications more information than they are intended to convey. Particularly, caution is urged against leaning too heavily upon soil classification as a basis for design, since many of the factors that influence design cannot be incorporated into an engineering soil classification. For example, the susceptibility of a road or runway to frost damage is a question that involves not only the properties and characteristics of the soil itself, but climatic and geophysical factors as well.

With reference to the grain-size scales used in various branches of applied soil science, it would be more convenient for civil engineering usage if the arbitrary division between silt size and sand size were established at 0.074 mm (the size of opening of a standard No. 200 sieve) instead of sizes 0.02 mm, 0.05 mm, or 0.06 mm, shown in the various scales quoted by the author in Fig. 2. This is for the purely practical reason that a No. 200 sieve is about the smallest size that can be used in routine soil testing. Sieves with much smaller openings are manufactured, as is well known, but they are much more delicate and expensive and will not wear under rigorous usage as well as the No. 200 sieve. At present, many engineering organizations distinguish between coarse-grained soils and fine-grained soils on the basis of the quantity of material passing the No. 200 sieve, and it would be convenient if this size also marked the boundary indicated by the terms sand size and silt size.

The author mentions the difficulty involved in the fact that textural classifications based on the percentages of sand, silt, and clay size in a soil do not reflect all the physical properties of interest to the civil engineer. The term "texture," as originally used in the United States (53) was not intended to convey any information concerning a soil beyond the relative proportions of the size groups of the material less than 2 mm in diameter. It was not intended to indicate such properties as plasticity, shrinkage, swell, and others. Rather, those additional properties should be expressed by other kinds of

[19] Research Prof., Civ. Eng. Dept., Iowa State College, Ames, Iowa.

classification or by specific test results. The term "texture" is used in some European countries (German *textur*) to embrace properties of soil other than relative amounts of the various size groups, such as soil structure and the form and relative size of the void spaces, but in the United States, so far as this writer is aware, "texture" has always referred only to grain-size relationships. It is the writer's belief that less confusion will ensue if this meaning of "texture" is preserved and emphasized.

No discussion of triangle textural classification is complete without pointing out that all the information given on an equilateral classification triangle can be shown on a right triangle (54) and that the process of determining the textural class of a soil is much simplified by using the right triangle chart. Since the sum of the percentages of sand, silt, and clay size in the portion of a soil smaller than 2 mm is 100%, the values of only two of these percentages are required to establish a point in an equilateral triangle such as shown in Fig. 3. By virtue of this fact, a right triangle can be constructed, as shown in Fig. 12, on which the percentages of any two of the three size groups can be scaled along the legs of the triangle. The point of intersection of the percentage lines representing the two groups for which the triangle is constructed can be more quickly spotted on such a diagram because of the orthogonal arrangement of the coordinates.

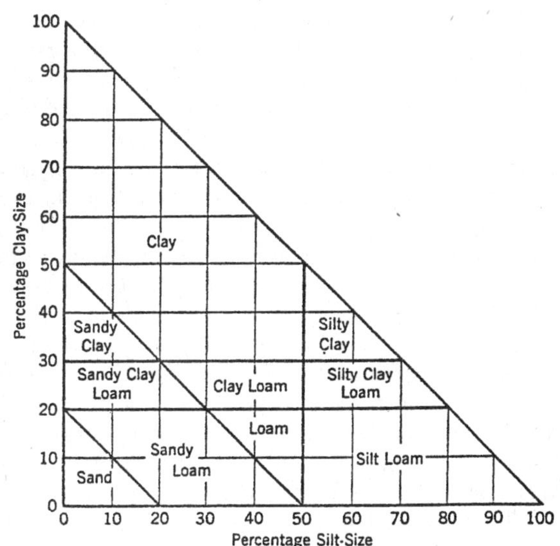

Fig. 12.—Right Triangle Texture Chart

Soils that contain sufficient material larger than 2 mm to be especially noticeable (say, about 8% or 10%) should be qualified by prefixing the term "stony" or "gravelly" to the textural class indicated by the triangle chart, as stated by the author. The principal textural class name of such a soil is determined on the basis of that part of the material smaller than 2 mm. Obviously if a soil contains an appreciable quantity of +2 mm material, the percentage that each of the three smaller sizes bears to the total soil will not be the same as the percentage of these groups in relation to the amount of the sand, silt, and clay only. It is this latter percentage which should be used in the triangle chart classification. The percentages relative to the whole soil can be quickly converted to the percentages of the sand, silt, and clay by multiplying each of the original percentages by the ratio, $\frac{100}{100 - \text{gravel } (\%)}$.

Thus, if the mechanical analysis of a soil shows it to consist of 17% gravel, 26% sand, 36% silt, and 21% clay, the ratio by which the percentages of the three smaller sizes should be increased is 1.2. The new values for these sizes then become 31% sand, 43% silt, and 25% clay. Entering the triangle chart (Fig. 3(*a*)) with these revised figures, the point of intersection falls within the clay loam area and the textural classification of this soil is "gravelly clay loam." If the material larger than 2 mm consists of broken stone fragments instead of rounded gravel particles, the soil would be classed as a "stony clay loam."

If one attempts to use the original percentages of sand, silt, and clay (that is, the percentages referred to the whole soil) three different points of intersection can be obtained on the equilateral triangle chart, depending on which pair of the three sizes is used. In the foregoing example 26% sand and 36% silt gives a point within the area labeled clay; 26% sand and 21% clay gives a point in the silty clay loam area; and 36% silt and 21% clay yields a point in the clay loam area. The necessity for reducing the percentages to the weight of material smaller than 2 mm as a base is apparent.

This writer has no particular criticism relative to the Airfield Classification (AC) System proposed by the author and tentatively adopted by the United States Engineer Department. It is a good system of classification. He does feel, however, that it is unfortunate for the civil engineering profession as a whole that the Office of the Chief of Engineers deemed it necessary to adopt any entirely new system for classifying soils for engineering purposes. Here is a new "language" added to the existing Babel of tongues which must be learned, and a new source of confusion to be surmounted, if an engineer is to avoid pitfalls in the matter of soil classification, as suggested by the author in the opening paragraphs of his paper. There would be no objection to the necessity for learning this new language if the need for an additional classification system were clear cut and positive. However, careful study of the AC system fails to reveal any substantial superiority over the revised Public Roads (PR) System of soil classification. They are tweedledum and tweedledee in many major respects. Viewed objectively it seems somewhat specious to argue that a soil needs to be classified by a different system just because the wheel loads to which it will be subjected are those of an airplane instead of a truck; and, in view of the fact that the Civil Aeronautics Administration also has its own classification system, one might conclude that the needs of the situation are different if the airplane wheels are those of a military plane rather than those of a civil aircraft.

There is need here for "engineering statesmanship" to bring these three federal agencies closer together in the matter of engineering soil classification. Without doubt, the problems of the three agencies are different and must be solved in different ways, but all have a common interest in engineering properties of soil. A great deal of "compartmentation" and confusion could be eliminated if all three would use the same language when discussing those properties.

The revised PR system of soil classification provides an excellent medium for expressing the general engineering characteristics of soils with sufficient clarity for practically all engineering applications and with a degree of precision commensurate with that which can be expected of any classification system.

It has been used by engineers since about 1928, and on an ever widening scale, until nearly all who deal with the soil have a general understanding of the meaning of the symbols A-1, A-4, and others. It is true that local engineering organizations often have had to modify the system to fit their local situations but, by and large, such modifications were readily made within the framework of the basic system, so that the language was essentially preserved.

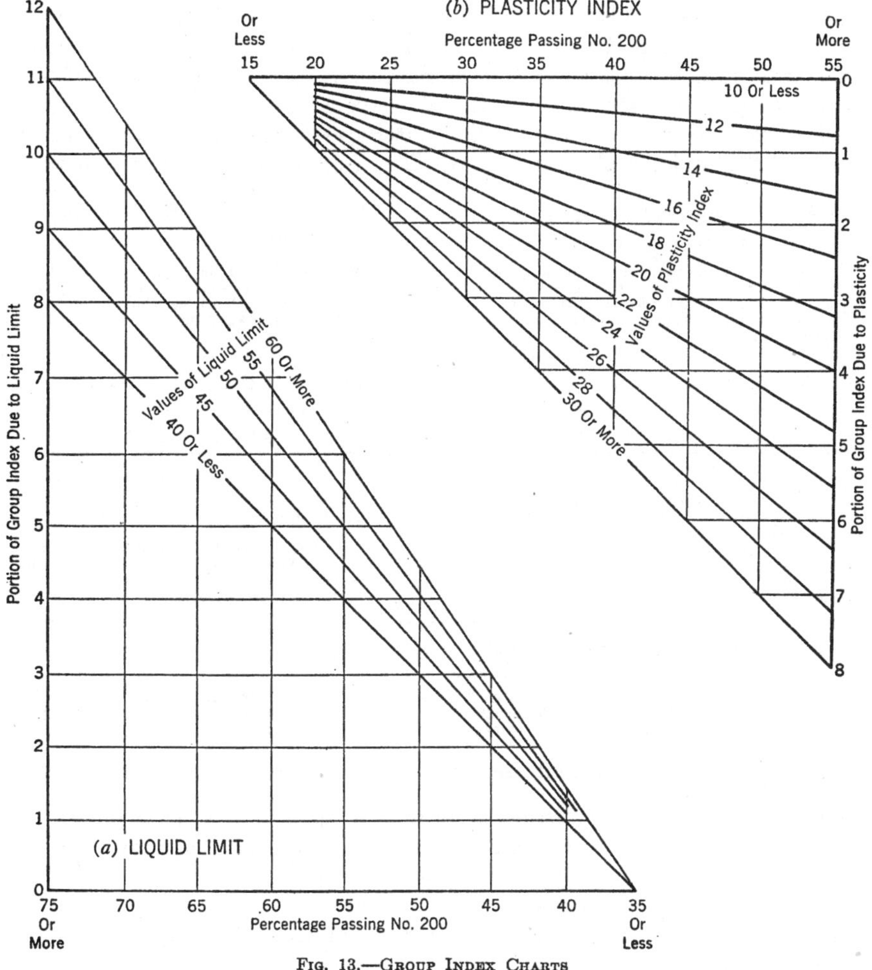

Fig. 13.—Group Index Charts

It is also true that the PR system has undergone a number of revisions since its introduction, which is to be expected in the case of any new tool or device as use demonstrates the need for modification. The author states that the AC system has been revised three times since its inception in 1942. Undoubtedly both systems will undergo further revision as time passes.

The most recent and the most extensive revision of the PR system is that which has been made by a committee of experienced engineers designated by the Highway Research Board, under the chairmanship of Harold Allen, M. ASCE, of the United States Public Roads Administration. The findings of a highway subcommittee of this committee (51) constitute a major revision in the PR system which brings it up to date in the light of many years of experience. This revision retains the basic framework of the system, with changes in detail here and there which are clearly set forth in the report. In other words, the classification language has not been changed, but the meaning and connotation of some of the terms has been modified as dictated by experience.

A major accomplishment by this committee has been the simplification of the procedure by which a soil may be classified. It is now possible to classify a soil, quickly and definitely, into one of twelve major groups and subgroups on the basis of three properties of the material—the mechanical analysis, the liquid limit, and the plasticity index. The committee did not minimize the

TABLE 13.—GROUP CLASSIFICATION OF HIGHWAY SUBGRADE MATERIALS

Procedure: With the required test data available, proceed from left to right in the following chart and the correct group will be found by the process of elimination. The first group from the left into which the test data will fit is the correct classification. All limiting test values are shown as whole numbers. If fractional numbers appear on test reports, convert them to the nearest whole number for purposes of classification.

GENERAL CLASSIFICATION	GRANULAR MATERIALS[a]			SILT-CLAY MATERIALS[b]			
Group Classification	A-1	A-3	A-2	A-4	A-5	A-6	A-7
Sieve Analysis[c]							
Sieve No. 10..........	....	....	....	....	....	....	....
Sieve No. 40..........	50 max.	51 min.	....	....	....	....	....
Sieve No. 200..........	25 max.	10 max.	35 max.	36 min.	36 min.	36 min.	36 min.
Characteristics of:[d]							
Liquid limit..........	....	....	....	40 max.	41 min.	40 max.	41 min.
Plasticity index........	6 max.	nonplastic	....	10 max.	10 max.	11 min.	11 min.
Group index................	....	....	4 max.	8 max.	12 max.	16 max.	20 max.
General rating as a subgrade..	Excellent to good			Fair to poor			

[a]35% or less passing sieve No. 200. In the "left to right" elimination process it is necessary to place group A-3 before group A-2; but this does not signify that group A-3 is superior to group A-2. [b] More than 35% passing sieve No. 200. [c] Percentage by weight passing sieves No. 10, No. 40, and No. 200. [d] Fraction passing sieve No. 40.

value of other test constants such as shrinkage factors and centrifuge moisture equivalent in connection with design procedures, but felt that they were not essential in a simplified classification procedure.

Another major achievement by the committee was the introduction of a "group index" into the classification, by means of which a soil in any one group or subgroup of fine-grained materials may be rated within its own group on the basis of variations in percentages of coarse material, liquid limit, and plasticity index. This group index or "in-group" rating is determined from an empirical formula, which is derived from experience with the original PR classification. It is based on the relationships between the amount of fine-grained material (material passing a No. 200 sieve) in a soil and the liquid limit and plasticity index within the significant ranges of variation of these properties. It can be quickly determined for a given soil by two simple charts which provide a

TABLE 14.—Subgroups Classification of Highway Subgrade Materials

(Letter M Designates Maximum Values; Letter m Designates Minimum Values)

Procedure: With the required test data available, proceed from left to right in the following chart and the correct group will be found by the process of elimination. The first group from the left into which the test data will fit is the correct classification.

General Classification	Granular Materials[a]							Silt-Clay Materials[b]			
Groups	A-1		A-3	A-2				A-4	A-5	A-6	A-7
Subgroups	A-1-a	A-1-b	....	A-2-4	A-2-5	A-2-6	A-2-7	....	....	....	A-7-5 A-7-6
Sieve Analysis:[c]											
Sieve No. 10	50 M	....	....	....	....	....	....	....	....	....	....
Sieve No. 40	30 M	50 M	51 m	....	....	....	....	....	....	....	....
Sieve No. 200	15 M	25 M	10 M	35 M	35 M	35 M	35 M	36 m	36 m	36 m	36 m
Characteristics of:[d]											
Liquid limit	....		....	40 M	41 m	40 M	41 m	40 M	41 m	40 M	41 m
Plasticity index	6 M		nonplastic	10 M	10 M	11 m	11 m	10 M	10 M	11 m	11 m[e]
Group index[f]	0		0	0		4 M		8 M	12 M	16 M	20 M
Usual types of significant constituent materials	Stone fragments, gravel, and sand		Fine sand	Silty or clayey gravel and sand				Silty soils		Clayey soils	
General rating as a subgrade	Excellent to good					Fair to poor					

[a] 35% or less passing sieve No. 200. As in Table 13, group A-3 precedes group A-2; but this does not signify that group A-3 is superior to group A-2. [b] More than 35% passing sieve No. 200. [c] Percentage by weight passing sieves No. 10, No. 40, and No. 200. [d] Fraction passing sieve No. 40. [e] The plasticity index of subgroup A-7-5 is equal to, or less than, the liquid limit −30. The plasticity index of subgroup A-7-6 is greater than the liquid limit −30. [f] The group index is computed by reference to Fig. 13 and is shown in parentheses after the group symbol; thus: A-2-6(5).

Granular Materials—Containing 35% or Less Passing a No. 200 Sieve.—

Group A-1.—The typical material of this group is a well-graded mixture of stone fragments or gravel, coarse sand, fine sand, and a nonplastic or feebly plastic soil binder. However, this group includes also stone fragments, gravel, coarse sand, volcanic cinders, etc., without soil binder.

Subgroup A-1-a includes materials consisting predominantly of stone fragments or gravel, either with or without a well-graded binder of fine material.

Subgroup A-1-b includes those materials consisting predominantly of coarse sand, either with or without a well-graded soil binder.

Group A-3.—The typical material of this group is fine beach sand or fine desert blow sand, without silty or clay fines or with a very small amount of nonplastic silt. The group includes also stream deposited mixtures of poorly graded fine sand and limited amounts of coarse sand and gravel.

Group A-2.—This group includes a wide variety of "granular" materials which are border line between the materials falling in groups A-1 and A-3 and the silt-clay materials of groups A-4, A-5, A-6, and A-7. It includes all materials containing 35% or less passing a No. 200 sieve, which cannot be classified as A-1 or A-3 due to fines content or plasticity, or both, in excess of the limitations for those groups.

Subgroups A-2-4 and A-2-5 include various granular materials containing 35% or less passing a No. 200 sieve and with a minus No. 40 portion having the characteristics of the A-4 and A-5 groups. They include such materials as gravel and coarse sand with silt content or plasticity index in excess of the limitations of group A-1, and fine sand with nonplastic silt content in excess of the limitations of group A-3.

Subgroups A-2-6 and A-2-7 include materials similar to those described under subgroups A-2-4 and A-2-5, except that the fine portion contains plastic clay having the characteristics of the A-6 or A-7 group. The approximate combined effects of plasticity indexes in excess of 10 and percentages passing a No. 200 sieve in excess of 15 are reflected by group index values of from 0 to 4.

Silt-Clay Materials—Containing More Than 35% Passing a No. 200 Sieve.—

Group A-4.—The typical material of this group is a nonplastic or moderately plastic silty soil usually having 75% or more passing a No. 200 sieve. The group includes also mixtures of fine, silty soil and up to 64% of sand and gravel retained on a No. 200 sieve. The group index values range from 1 to 8, with increasing percentages of coarse material being reflected by decreasing group index values.

Group A-5.—The typical material of this group is similar to that described under group A-4, except that it is usually of diatomaceous or micaceous character and may be highly elastic as indicated by the high liquid limit. The group index values range from 1 to 12, with increasing values indicating the combined effect of increasing liquid limits and decreasing percentages of coarse material.

Group A-6.—The typical material of this group is a plastic clay soil usually having 75% or more passing a No. 200 sieve. The group includes also mixtures of fine clayey soil and up to 64% of sand and gravel retained on a No. 200 sieve. Materials of this group usually have a high volume change between wet and dry states. The group index values range from 1 to 16, with increasing values indicating the combined effect of increasing plasticity indexes and decreasing percentages of coarse material.

Group A-7.—The typical material of this group is similar to that described under group A-6, except that it has the high liquid limits characteristics of the A-5 group and may be elastic as well as subject to high volume change. The range of group index values is from 1 to 20, with increasing values indicating the combined effect of increasing liquid limits and plasticity indexes and decreasing percentages of coarse material.

Subgroup A-7-5 includes those materials with moderate plasticity indexes in relation to liquid limit and which may be highly elastic as well as subject to considerable volume change.

Subgroup A-7-6 includes those materials with high plasticity indexes in relation to liquid limit and which are subject to extremely high volume change.

graphical solution for the group index formula. The group index of a soil is reported as a whole number in parentheses after the group or subgroup symbol, as, A-4 (5) or A-6 (12). This method of reporting automatically indicates the "edition" of the PR system which was used to classify a soil.

The classification system embodied in this committee report has been variously called the Highway Research Board system and the group index system. The writer prefers to call it the "Revised PR" system, since that is actually what it is. It is strongly recommended as a complete and comprehensive language for expressing the engineering characteristics of soil material—a system whose terms are widely understood and recognized, which is based on long usage and experience, and which is easy to apply in day to day practice.

The following description of classification groups is extracted from the report, as are Tables 13 and 14 and Fig. 13.

E. W. Lane,[20] M. ASCE.—It is believed that engineers working in soils would be glad to know of the size classification recently proposed by a Sub-Committee on Sediment Terminology of the American Geophysical Union

TABLE 15.—Scale of Large Particle Sizes, in Metric and English Units

Class name	Metric unit (mm)	English unit (in.)
(1)	(2)	(3)
Very large boulders	4,096–2,048	160–80
Large boulders....	2,048–1,024	80–40
Medium boulders..	1,024–512	40–20
Small boulders....	512–256	20–10
Large cobbles.....	256–128	10–5
Small cobbles.....	128–64	5–2.5
Very coarse gravel.	64–32	2.5–1.3
Coarse gravel.....	32–16	1.3–0.6
Medium gravel....	16–8	0.6–0.3
Fine gravel.......	8–4	0.3–0.16
Very fine gravel...	4–2	0.16–0.08

TABLE 16.—Scale of Small Particle Sizes

Class name	Metric Units	
	(Mm)	(Microns)
(1)	(2)	(3)
Very coarse sand..	2.000 -1.000	2,000–1,000
Coarse sand......	1.000 -0.500	1,000–500
Medium sand.....	0.500 -0.250	500–250
Fine sand........	0.250 -0.125	250–125
Very fine sand....	0.125 -0.062	125–62
Coarse silt.......	0.062 -0.031	62–31
Medium silt......	0.031 -0.016	31–16
Fine silt.........	0.016 -0.008	16–8
Very fine silt.....	0.008 -0.004	8–4
Coarse clay size...	0.004 -0.0020	4–2
Medium clay size.	0.0020–0.0010	2–1
Fine clay size.....	0.0010–0.0005	1–0.5
Very fine clay size.	0.0005–0.00024	0.5–0.24

(55) for use in work in sediments as given in Tables 15 and 16. The addition of a discussion of this classification to Professor Casagrande's excellent paper, therefore, seems justified.

[20] Hydr. Consultant on Chf. Engr.'s Staff, Bureau of Reclamation, Denver, Colo.

The classification is practically a modification of the classification (56) of C. K. Wentworth substituting the term gravel for his terms granule and pebble, subdividing his silt and clay sizes according to the widely used classifications proposed by J. A. Udden (57), and adding subdivisions of Mr. Wentworth's boulder classification. Most of the Wentworth-Udden classification has been widely used among geologists for many years and has been adopted by engineers in many sediment studies, but the granule and pebble classifications were not generally accepted.

The purpose of the Sub-Committee was to obtain agreement among engineers and geologists on a large number of terms in the sediment field, on which there was considerable difference of usage, including that of particle size. The Sub-Committee was composed of engineers and geologists, the engineers being G. H. Matthes, Hon. M. ASCE, L. G. Straub and the late G. C. Dobson, Members, ASCE, C. B. Brown, Assoc. M. ASCE, and the writer. The other members were equally well-known geologists.

The principal advantages of the proposed classification are believed to be as follows: (1) It gives exact subdivisions over the full range of particle sizes of sedimentary materials; (2) the terms used are common ones, and are applied to the range of sizes in common usage; and (3) the lower limit of each size is exactly one half of the upper limit, thus making the division points easy to remember and causing them to plot equally spaced on the semilogarithmic diagrams so widely used to indicate particle size distribution.

This equal size of classes is also a great advantage in the use of statistical methods in treating size ranges. Since the sizes of most sediment samples tend to form straight lines when plotted on logarithmic probability paper, the standard deviation of the logarithm of the sizes indicates the spread of the particle sizes. The composition of a sample can thus be closely represented by two numbers, the median (or geometric mean) and the standard deviation of the logarithm of the size. With these numbers, the composition of a large number of samples can be represented more conveniently than with a large number of size composition curves. It is sometimes possible to analyze a large mass of data by this method in illuminating ways which would be impossible by using the curves. This method of analysis was largely introduced into sedimentary science over a decade ago by W. C. Krumbein and has been widely used. It has more recently been independently introduced into soil mechanics by Donald M. Burmister, Assoc. M. ASCE.

B. K. Hough, Jr.,[21] Assoc. M. ASCE.—One of the many controversial aspects of this significant paper is the distinction between, and the relative importance of, the subject processes of soil classification and soil identification. It is noted that the author gives most attention to classification of soils, said classification being based on soil characteristics which are of various but generally limited interest, and which generally do not include identification. The question is raised herein as to whether identification may not be of major importance in all applications and the basis for more reliable classification and indication of probable soil behavior under service conditions.

[21] Associate Prof. of Soil Mechanics, Cornell Univ., Ithaca, N. Y.

It is suggested that such terms as clay and silt, for example, are inadequate for true identification, although they may be sufficient in some applications as classifications. To designate a soil as clay, for instance, is scarcely more satisfactory than to designate one of a number of unidentified construction materials as wood. The differences between clay minerals are as basic and, in some respects, may be as important as the differences between oak and balsa, or pine and cedar. Not only mineralogical, but also chemical distinctions of great importance, notably base exchange capacity (BEC) and type of exchangeable ion, are in some cases of major interest.

Many engineers would agree that identification of soils (especially fine-grained soils) on the basis of their fundamental properties, mineral type, and composition in particular, is sound and desirable in principle but so difficult in practical application as to be of very limited utility, especially in the field. This rather general view may be influenced by misconception of the difficulty of such identification and insufficient knowledge of its practical value. It may now be the case that the practical value can be demonstrated and that identification methods have been simplified to such an extent that in the near future there will be a trend in engineering toward soil identification rather than toward the development of more complex empirical classification systems.

Mineral identification for many years was based on optical methods, which required highly specialized training and experience, usually in the field of geology or physics. For engineering applications this type of analysis has never seemed to be necessary and this may still be true although the increasing evidence of the value of such analysis might justify it. Other methods are available, however, such as the differential thermal analysis method, which can be simplified in application to almost any reasonable degree. Equipment for field use was developed during World War II by the United States Department of Agriculture and is now available for engineering purposes. This equipment can be operated quite satisfactorily by engineers in the junior professional categories. More elaborate equipment for analysis by the same method may be obtained and utilized in soils engineering laboratories without employment of specialists in this field.

The chemical analysis of a soil can also be organized in any reasonably well-equipped laboratory so that it may be conducted by technicians with minimum technical training, and in fact it is now on this basis in many agricultural laboratories. Equipment for field use is also available for certain types of chemical determinations.

In most, if not all soils engineering laboratories, the mechanical analysis is now a standard test and the data from such tests are considered essential in many systems of soil classification. It may well be, however, that in fine-grained soils specific surface is a more significant measurement than gradation or particle diameter, particularly when the latter are determined by such methods as the analysis of sedimentation by the specific gravity hydrometer. This observation is made as an extension of the foregoing comments on chemical analysis, inasmuch as specific surface, although in itself a physical characteristic, is usually indicative of chemical activity. In this respect it is fundamentally preferable to grain-size measurements as the basis for estimating such soil characteristics as permeability, capillarity, plasticity, stability, and many other features which are of engineering interest. Data from hydrometer tests are of little use in such determinations and, in fact, are often accorded no more attention by engineers than a cursory examination despite the considerable cost of performing such tests.

The practical value of the identification of the clay minerals has been more clearly established in recent years. The swelling propensites of the montmorillonite group are generally recognized, although it is still not common practice to make an analysis of mineral types in cases where swelling is of particular importance. Many other correllations between mineral type and engineering behavior may be possible as studies in this area are undertaken. During recent soil tests under the writer's supervision, for example, standard Proctor compaction tests on an unknown clayey soil indicated an optimum moisture content of approximately 30% (dry weight basis). This value considerably exceeded that for any other soil tested by the writer in the course of many years of practical testing experience. Mineralogical analysis revealed the predominant clay mineral to be illite, which with its distinctive potassium bonding may well have characteristics that differ appreciably from those of the more common kaolinite group.

With reference to determination of the type of exchangeable ion, an almost classic example of the practical value of this soil characteristic is the treatment of the Treasure Island (Calif.) Lagoon lining (58) with ocean water to reduce its permeability. None of the soil classification systems described by the author would have indicated the feasibility of this eminently practical and effective treatment. Other examples of the practical importance of chemical reactions in soil in the field of highway engineering are almost too numerous to cite.

In conclusion, it is recommended that efforts be made by engineers to identify soils on the basis of fundamental properties—specifically, mineral type, exchangeable ions, BEC, and specific surface—and that thereafter soils be classified for various purposes. The basic identification will then be valid, no matter what the application or use of the soil may be, and might well serve as the common denominator for the many diversified interests now engaged in soil study and research.

ARTHUR CASAGRANDE,[22] M. ASCE.—The discussions have rounded out the contents of the paper by the addition of valuable material and of many helpful suggestions, for which the writer is grateful.

There is one misunderstanding noticeable in several of the discussions which is caused by the brevity of paper—namely, that the purpose of the Airfield Classification (AC) System is merely to classify the raw material of which the soil is composed, without regard to the engineering properties of the material in its undisturbed state. This interpretation is definitely not the writer's intent. All the data listed in Cols. 1 to 6, Table 4, must be considered as an integral part of the complete classification. Cols. 1, 2, 3, 4, and 6 classify the material in the "disturbed state"—that is, the raw material. In Col. 5 are listed characteristics which refer to the engineering properties of the material in the condition in which

[22] Prof. of Soil Mechanics and Foundation Eng., Graduate School of Eng., Harvard Univ., Cambridge, Mass.

it will be used by the engineer. Since Table 4 is prepared specifically for the use of the AC system in airfield projects, the observations and tests listed in Col. 5 are selected for this purpose. When dealing with other types of engineering projects, Col. 5 will obviously require revision. No formal grouping has been proposed for the data which are to be included in Col. 5. However, it is conceivable that, for specific engineering projects, the observations and tests relating to the material in place could be subjected to further classification.

Several of the discussions have pointed out that in many cases a geologic classification of the soils should be included in the general classification, and that use should be made of local soil names. The writer agrees with these views and suggests that another column should be added in Table 4 with the heading, "Geologic Description and Local Name." In this connection it should be noted that there is also some danger in the use of local soil names because in different parts of the United States they may refer to materials differing greatly in their engineering characteristics. For example, the term "hardpan" has led to numerous lawsuits in which the lawyers had no difficulty in finding in the literature definitions which suited their individual purposes. It seemed to the writer that the older the quotation the more weight it carried in the Court of Law.

Professor Fadum offers several practical suggestions for promoting agreement on soil terminology. The writer agrees with Professor Fadum that it would be more significant if the Atterberg limits were performed on material smaller than 0.1 mm in diameter, rather than on the fraction passing the No. 40 sieve. Since separation of the fraction smaller than 0.1 mm is cumbersome and time consuming, general acceptance of such a suggestion is not probable until a much faster procedure is developed for the separation of this fraction.

Colonel Stratton reviews the circumstances leading to the development of the AC system, and demonstrates in Fig. 7 how the system was combined with the design procedure for flexible pavements, which was developed under his direction by the Office of the Chief of Engineers, United States Army. Then he reviews the comments made in 1942 by various divisions and districts of the United States Engineer Department—that is, soon after the AC system was adopted as a tentative standard by the Corps of Engineers for military construction. More recently, a modification was developed by the Office of the Chief of Engineers for civil works purposes, based on the experience gained in the use of the AC system since its inception.

Professor Belcher presents a competent discussion on the advantages of the pedological classification. It is true that the thoroughness of exploration on which a pedological classification of a given locality is based results in information which is of value to the engineer, and that much time and effort can be saved by utilizing such available information. However, the writer does not believe that the pedological classification is suitable in general for engineering purposes.

Messrs. Haine and Hilf describe an adaptation of the AC system which is used by the Bureau of Reclamation for use with earth dam projects (59). They emphasize that proper field identification is the primary objective in soil surveys, and found it necessary to subdivide the GF and SF groups according to the type of fines. This step is logical, particularly in connection with the use of soils for the construction of rolled earth dams. In addition, Messrs. Haine and Hilf have rated the various groups of materials according to their usefulness in the design of earth dams—a helpful supplement when using this classification for earth dam projects.

Mr. Feld has brought a welcome diversion into this rather prosaic subject, by reviewing some of the history of soil classification.

Kenneth S. Lane first presents an analysis of the requirements which a classification system should fulfil, and then proposes a new system referred to as QPC (Quabbin, Providence, and Casagrande) classification, in which he combines the advantages of several systems by using a descriptive name, a grain-size number, and the group symbol of the AC system. The writer is in full accord with Mr. Lane's discussion of typical soil names, with emphasis on the geology of the soils. The use of a grain-size number is a desirable supplement in classifying many coarse-grained soils of a great variety, such as are encountered in glacial deposits. In such instances, a man could be trained to estimate the grain-size class by visual inspection. However, when working in areas where coarse-grained soils are rather limited in extent, the necessary skill for a fair differentiation between thirty-five grain-size classes will usually not be available. If high accuracy in classifying the grain-size distribution is needed, one will have to depend on laboratory analyses. Because of Mr. Lane's extensive experience, his recommendations deserve serious consideration by those who wish to use a comprehensive method for classifying soils.

Professor Sowers points out the difficulty resulting from the fact that the term "plastic" is used by builders and drillers to designate the consistency of a soil in its natural state. He also emphasizes the importance of a detailed description of the properties of the undisturbed material and suggests the use of numerical scales to replace the uncertainty of descriptive adjectives. The writer believes that in visual and manual examination only a small number of grades can be distinguished, which are better described by generally understood adjectives. However, there is no question that a quantitative basis should be established for the purpose of standardizing the use of the adjectives.

Mr. Pulido y Morales expresses preference for the modified Public Roads (PR) Classification. He compares (Table 7) the group symbols for forty-four different soils according to the original and the modified PR classification, which to the writer demonstrates the confusion that must result if both classifications, which differ so greatly, make use of the same group symbols.

Because of space limitation the writer must forego a detailed discussion of the group index in the modified PR classification, and of the relationship between the liquid limit and compressibility, which Professor Dawson desires.

Mr. Glynn points out that in Australia a descriptive soil classification has been widely used as the basis of field identification. He suggests that, to prevent confusion, further subdivisions in the AC system should be limited to special applications. The writer is inclined to agree with this suggestion.

Many of the features which are insufficiently presented in the AC system are very well covered in the classification proposed by Messrs. Cooling, Skempton, and Glossop in Table 8. Although specifically adapted to the soil conditions in England, Table 8 may well serve as a general basis for establishing similar classifications in other parts of the world. In Fig. 11, the relationship between the liquid limit and the clay fraction (percentage smaller than 2 microns) is plotted, to illustrate that, for two clays having the same liquid limit, the clay with the more potent colloidal content has the smaller clay fraction. Whether such a relationship will yield information in addition to what can be derived from the plasticity chart, is questioned. Besides, difficulties arise when one also plots the results of materials with high liquid limit which lie well below the A-line, such as organic clays, or micaceous and diatomaceous soils.

Mr. Vargas suggests that the organic and inorganic silts may be grouped together into a major division. The writer questions the value of such a combination, since organic silts and inorganic silts are really quite different types of soils.

Mr. Vargas raises the question as to whether the group symbols, which are abbreviations of English names, should in other languages be changed to the initials of the corresponding names. He answers his question by suggesting that the original symbols should also be retained in other languages. This would certainly prevent confusion. The writer admits that in devising the symbols he has not thought about the possibility that this classification might spread to other countries. The purpose of the group symbols was to give the user something that would not tax his memory. Thus, it would be logical to create new symbols in each language. On the other hand, if this classification should become widely used, it would be desirable to have only one set of symbols which are internationally accepted. There is already one symbol in these groups which is not derived from the English language, namely, the letter, M. Considering that the majority of important references in soil mechanics are, and probably in the future will be, in the English language, and that English is the only language used for the International Soil Mechanics Conferences, the adoption of the present symbols as an international standard would not be too great a hardship. It would mean that, in teaching the classification, the student would have to learn the English names of about one dozen soils. After he has mastered these words, the use of these symbols would be just as understandable to him as if the symbols referred to the names of these soils in his own language.

Professor Burmister proposes to broaden the AC system by a better definition of the principal soil components, and by establishing a common language in describing soils. His discussion contains many valuable suggestions, which merit attention in any future improvements of existing soil classifications.

Professor Spangler has presented a competent defense of the new PR system. By including sufficiently detailed information on this new system, he has contributed a necessary supplement to the paper. Professor Spangler expresses regrets that other organizations deemed it necessary to adopt an entirely new system, when the PR system would have been satisfactory. However, that the original PR system was not satisfactory was admitted tacitly by those who created the new PR system, after the AC system was already in existence. In fact, it seems to the writer that, except for the symbols, the new PR system is closer to the AC system than to the 1929 PR system. Therefore, the writer reiterates that it would have been less confusing and more straightforward if the new PR system had received a new and more logical set of symbols, instead of retaining the original symbols.

Professor Spangler misunderstood the writer's remark regarding the revisions of Table 4 to mean that the AC system has been revised three times since its inception in 1942. Some minor revisions were made in Cols. 12 to 15, Table 4, which are not an inherent part of the AC system.

If none of the existing methods for identifying and classifying soils for engineering purposes makes use of their mineralogical and chemical composition and surface area, the reason is simply that the very complex relationships between these characteristics and the engineering properties of soils have not been established to any useful and practical extent. The writer, however, agrees with Professor Hough that more research along these lines is desirable.

E. W. Lane discusses a new grain-size classification recently proposed by the American Geophysical Union. In soil mechanics there has not developed, so far, any need for such detailed grain-size classifications, and the writer doubts that they will ever be needed.

The writer should like to point out that the soil classifying procedures actually used by the Bureau of Reclamation and by the Corps of Engineers, as well as by the Bureau of Yards and Docks, United States Navy (the latter according to written information received from L. A. Palmer), differ only in minor points. It would seem to the writer feasible and eminently desirable for these three large users of soil mechanics to initiate a cooperative effort, which should eventually lead to a uniform soil classification. If such an effort succeeds, it would constitute the most important step toward resolving order from the chaos of the past.

TIME LAG AND SOIL PERMEABILITY IN GROUND-WATER OBSERVATIONS

BULLETIN NO. 36

WATERWAYS EXPERIMENT STATION

CORPS OF ENGINEERS, U. S. ARMY

VICKSBURG, MISSISSIPPI

PREFACE

With the advance of soil mechanics and its applications in the design and construction of foundation and earth structures, the influence of ground-water levels and pore-water pressures is being considered to a much greater extent than a decade or two ago. Rapid and reliable determination of such levels and pressures is assuming increasing importance, and sources of error which may influence the measurements must be eliminated or taken into account.

A review of irregularities in ground-water conditions and the principal sources of error in ground-water observations is presented in the first part of this paper. Many of these sources of error can be eliminated by proper design, installation, and operation of observation wells, piezometers, or hydrostatic pressure cells. However, other sources of error will always be present and will influence the observations to a greater or lesser degree, depending on the type of installation and the soil and ground-water conditions. Conspicuous among the latter sources of error is the time lag or the time required for practical elimination of differences between hydrostatic pressures in the ground water and within the pressure measuring device.

Theoretical and experimental methods for determination of the time lag and its influence on the results of ground-water observations are proposed in the second part of the paper. Simplifications are obtained by introducing a term called the basic time lag, and solutions are presented for both static, uniformly changing, and fluctuating ground-water conditions. The influence of a secondary or stress adjustment time lag, caused by changes in void ratio or water content of the soil during the observations, is discussed.

The third part of the paper contains data which will assist in the practical application of the proposed methods. Formulas for determination of the flow of water through various types of intakes or well points are summarized and expanded to include conditions where the coefficients of the vertical and horizontal permeability of the soil are different. Examples of computations and a table facilitate preliminary estimates of the basic time lag for the principal types of installations and soils, and determination of the actual time lag is illustrated by several examples of field observations and their evaluation.

Determination of the coefficients of vertical and horizontal permeability for the soil in situ by means of time lag observations is theoretically possible and is discussed briefly in the closing section of the paper. Such field determinations of permeability have many potential advantages, but further research is needed in order to eliminate or determine the influence of various sources of error.

An abstract of the paper was presented in January 1949 at the Annual Meeting of the American Society of Civil Engineers, and a limited number of copies of the first draft were distributed. In this final version of the paper the individual sections have been rearranged and amplified to some extent, and some new sections have been added.

NOTATION

x Distance from ref. level to piezometer level for steady state, cm.
y Distance from ref. level to piezometer level, transient state, cm.
z Distance from reference level to the outside piezometric level, cm.

x_o, y_o, z_o Values of x, y, and z for $t = 0$, cm.

x_a Amplitude of fluctuating piezometer levels for steady state, cm.
z_a Amplitude of fluctuating outside piezometric levels, cm.

h Increment change in active head, cm.
H Active head, $H = z - y$, cm.
H_o Active head for $t = 0$, cm.
H_c Constant piezometric head, cm.

h' Increment change in transient differential head, cm.
H' Transient differential head, $H' = y - x$, cm.
H'_o Transient differential head for $t = 0$, cm.

A Area of casing, piezometer, manometer, or pressure cell, cm^2.
d Diameter of piezometer, manometer, or pressure cell, cm.
D Diameter of effective intake, boring, or well point, cm.
e Base of natural logarithms, no dimension.
E Equalization ratio, $E = (H_o - H)/H_o$, no dimension.
F Intake shape factor, from $q = FkH$, cm.
L Length of effective intake or well point, cm.

k Coefficient of permeability, cm/sec.
k_h Coefficient of horizontal permeability, undisturbed soil, cm/sec.
k_m Mean coefficient of permeability, $k_m = \sqrt{k_h \cdot k_v}$, cm/sec.
k_v Coefficient of vertical permeability, undisturbed soil, cm/sec.
k'_v Coefficient of vertical permeability, soil in casing, cm/sec.
m Transformation ratio, $m = \sqrt{k_h/k_v}$, no dimension.

q Rate of flow at time t and head H, cm^3/sec.
q_o Rate of flow at time $t = 0$ and head H_o, cm^3/sec.
t Time, seconds unless otherwise indicated.

- s Seconds)
- m Minutes)
- h Hours) Used only in Figs. 14, 16, 17.
- d Days)

t_s Phase shift of sinusoidal wave, seconds unless otherwise indicated.
T Basic time lag, $T = V/q$, seconds unless otherwise indicated.
T_w Period of sinusoidal wave, seconds unless otherwise indicated.
V Total volume of flow required for pressure equalization, cm^3.

α Rate of linear change in pressure, cm/sec.
γ Unit weight, g/cm^3.
ε Deflection of diaphragm in pressure cell, cm.

TIME LAG AND SOIL PERMEABILITY IN GROUND-WATER OBSERVATIONS

by

M. Juul Hvorslev*

INTRODUCTION

Accurate determination of ground-water levels and pressures is required, not only in surveys of ground-water supplies and movements, but also for proper design and construction of most major foundation and earth structures. The depth to the free ground-water level is often a deciding factor in the choice of types of foundations, and it governs the feasibility of and the methods used in deep excavations. A recent fall or rise in ground-water levels may be the cause of consolidation or swelling of the soil with consequent settlement or heaving of the ground surface and foundations. The existence of artesian or excess pore-water pressures greatly influences the stability of the soil; determination of pore-water pressures permits an estimate of the state or progress of consolidation, and it is often essential for checking the safety of slopes, embankments, and foundation structures. In general, determination of both free ground-water levels and pore-water pressures at various depths is usually a necessary part of detailed subsurface explorations, and the observations are often continued during and for some period after completion of foundation and earth structures.

Ground-water levels and pore-water pressures are determined by means of borings, observation wells, or various types of piezometers and hydrostatic pressure cells. During the advance of a bore hole or immediately after installation of a pressure measuring device, the hydrostatic pressure within the hole or device is seldom equal to the original pore-water pressure. A flow of water to or from the boring or pressure measuring device then takes place until pressure differences are eliminated, and the time required for practical equalization of the pressures is the time lag. Such a flow with a corresponding time lag also occurs when the pore-water pressures change after initial equalization. It is not always convenient or possible to continue the observations for the required length of time, and adequate equalization cannot always be attained when the pore-water pressures change continually during the period of observations. In such cases there may be considerable difference between the actual and observed pressures, and the latter should then be corrected for influence of the time lag.

---

* Consultant, Soils Division, Waterways Experiment Station.

The magnitude of the time lag depends on the type and dimensions of the pressure measuring installation, and it is inversely proportional to the permeability of the soil. A preliminary estimate of the time lag is necessary for the design or selection of the proper type of installation for given conditions. The actual time lag should be determined by field experiments so that subsequent observations may be corrected for its influence, when conditions are such that corrections are required or desirable.

Theoretical and experimental methods for determination of the time lag and its influence on the results of pressure measurements are presented in this paper. These methods are based on the assumptions usually made in the theories on flow of fluids through homogeneous soils, and the results are subject to corresponding limitations. In addition to the time lag, ground-water observations may be influenced by several other sources of error and by irregular and changing ground-water conditions. Therefore, an initial review of ground-water conditions in general and of the principal sources of error in determination of ground-water levels and pressures is desirable in order to clarify the assumptions on which the proposed methods are based, and to delimit the field of application of these methods.

PART I: GROUND-WATER CONDITIONS AND OBSERVATIONS

Irregularities and Variations

Several sources of error in determination of ground-water levels and pressures occur primarily when irregular and/or rapidly changing ground-water conditions are encountered. Regular conditions, with the piezometric pressure level equal to the free ground-water level at any depth below the latter, are the exception rather than the rule. Irregular conditions or changes in piezometric pressure level with increasing depth may be caused by: (a) perched ground-water tables or bodies of ground water isolated by impermeable soil strata; (b) downward seepage to more permeable and/or better drained strata; (c) upward seepage from strata under artesian pressure or by evaporation and transpiration; and (d) incomplete processes of consolidation or swelling caused by changes in loads and stresses. For a more detailed description of these conditions reference is made to MEINZER (20)* and TOLMAN (30); a general discussion of ground-water observations is found in a recent report by the writer (16).

Ground-water levels and pressures are seldom constant over considerable periods of time but are subject to changes by: (a) precipitation, infiltration, evaporation, and drainage; (b) load and stress changes and/or seepage due to seasonal or diurnal variations in water levels of nearby rivers, lakes, estuaries, and the sea; (c) construction operations involving increase or decrease in surface loads and removal or displacement of soil; (d) pumping and discharge of water; (e) variations in temperature and especially freezing and thawing of the upper soil strata; and (f) variations in atmospheric pressure and humidity. The last mentioned variations may cause appreciable and rapid changes in ground-water levels, but the interrelationship between atmospheric and ground-water conditions is not yet fully explored and understood; see HUIZINGA (13), MEINZER (20), and TOLMAN (30). The possibility that minor but rapid changes in ground-water levels and pressures may occur should be realized, since such changes may be misinterpreted and treated as errors, and since they may affect the determination of corrections for actual errors.

Sources of Error in Measurements

The principal sources of error in determination of ground-water levels and

---

* Numbers in parentheses refer to references at end of paper.

pressures are summarized in Fig. 1, and some further details are presented in the following paragraphs.

Hydrostatic time lag

When the water content of the soil in the vicinity of the bottom of a bore hole or intake for a pressure measuring device remains constant, and when other sources of error are negligible, the total flow or volume of water required for equalization of differences in hydrostatic pressure in the soil and in the pressure measuring device depends primarily on the permeability of the soil, type and dimensions of the device, and on the hydrostatic pressure difference. The time required for water to flow to or from the device until a desired degree of pressure equalization is attained, may be called the *hydrostatic time lag*. In order to reduce the time lag and increase the sensitivity of the installation to rapid pressure changes, the volume of flow required for pressure equalization should be reduced to a minimum, and the intake area should be as large as possible.

Stress adjustment time lag

The soil structure is often disturbed and the stress conditions are changed by advancing a bore hole, driving a well point or installing and sealing a pressure measuring device, and by a flow of water to or from the device. A permanent and/or transient change in void ratio and water content of the affected soil mass will then take place, and the time required for the corresponding volume of water to flow to or from the soil may be called the *stress adjustment time lag*. The apparent stress adjustment time lag will be increased greatly by the presence of air or gas bubbles in the pressure measuring system or in the soil; see Items 6 to 8, Fig. 1. This time lag and its influence on the results of observations are discussed in greater detail in Part II, pages 21-29.

General instrument errors

Several sources of error may be found in the design, construction, and method of operation of the pressure measuring installation. Among such sources of error may be mentioned: (a) inaccurate determination of the depth to the water surface in wells and piezometers; (b) faulty calibration of pressure gages and cells; (c) leakage through joints in pipes and pressure gage connections; (d) evaporation of water or condensation of water vapors; (e) poor electrical connections and damage to or deterioration of the insulation; (f) insufficient insulation against extreme temperature variations or differences, especially inactivation or damage by frost. The effect of leakage through joints and connections is similar to that of seepage along the outside of conduits, discussed below.

Seepage along conduits

Seepage along the casing, piezometer tubing, or other conduits may take

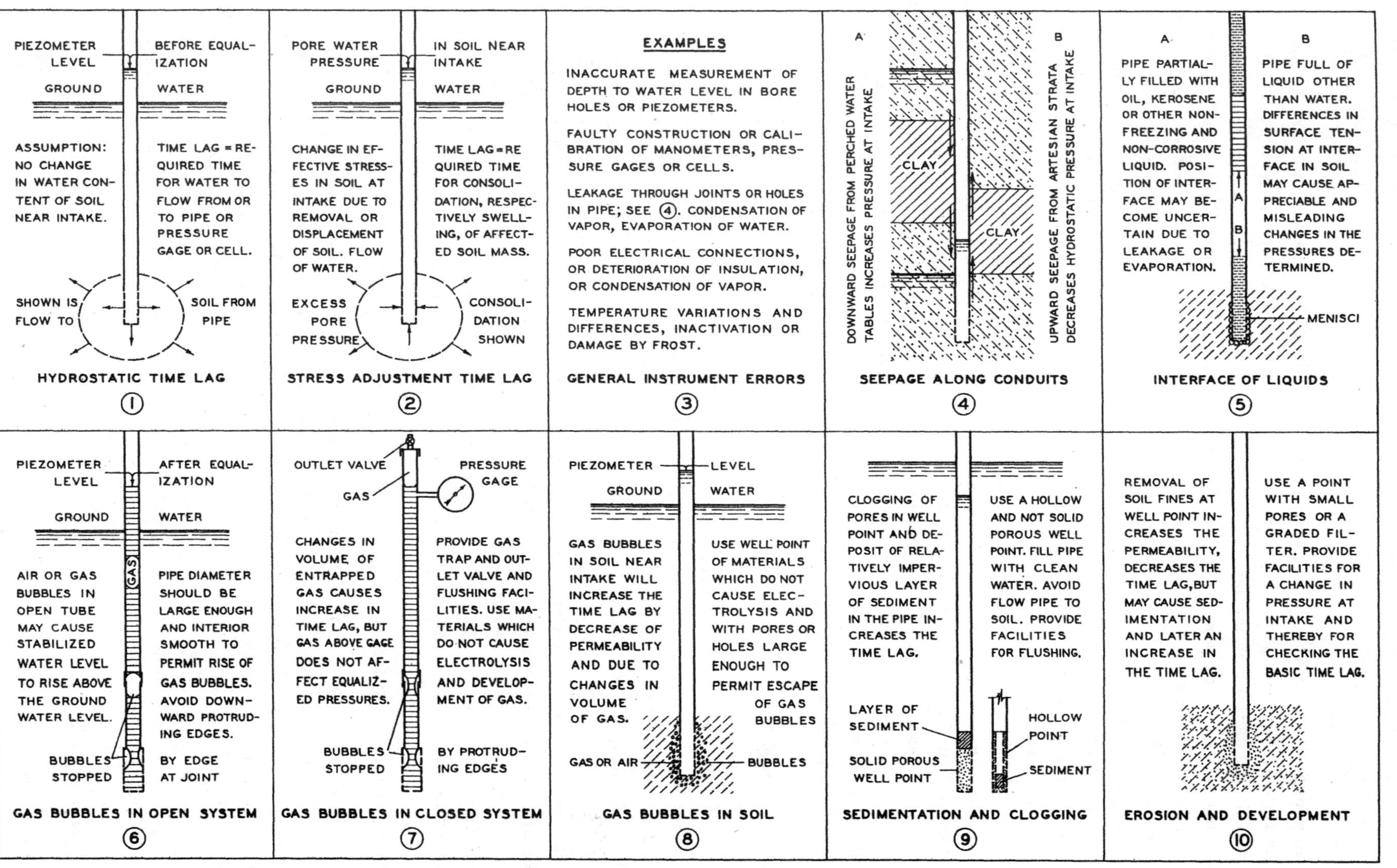

Fig. 1. Sources of error in determination of ground-water pressures

place, especially when irregular ground-water conditions are encountered. As shown in the figure, such seepage may increase or decrease the pore-water pressure in the soil at the bottom of the hole or at the intake for a pressure measuring device. Even under regular ground-water conditions seepage may occur in closed systems with attached manometer or pressure gage, and it will always affect experimental determination of the time lag of the system and of the permeability of the soil. To avoid seepage, the entire piezometer or the well point is often driven into the soil; but this method causes increased disturbance of the soil, and in many cases it is also desirable to surround the well point with a graded sand filter. When the well point is installed in an oversized bore hole, the space between the standpipe and the wall of the hole must be sealed above the well point, preferably in a fairly impermeable stratum. Puddled clay, bentonite mixtures, and cement grout have been used for sealing, but it is not always easy to obtain a tight seal and at the same time avoid stress changes in the surrounding soil because of swelling of the sealing material. A seal consisting of alternate layers of sand and clay balls, compacted by means of an annular tamping tool, has been developed and used successfully by A. CASAGRANDE (2) and (3).

Interface of liquids

To avoid corrosion or inactivation and damage by frost, manometer and pressure gages and the upper part of piezometers may be filled with kerosene or other oils. The difference in specific gravity of water and the liquid used, as well as the position of the interface, must be taken into consideration in determining the pore-water pressure. However, when observations are extended over long periods of time, the position of the interface may change because of evaporation and/or leakage and be difficult to determine. If the interface is in the wall of a well point with very fine pores, or in fine-grained soil outside the well point, additional and considerable errors may be caused by the menisci formed in the pores and by the difference in surface tension of water and the liquid in the pipe and well point.

Gas bubbles in open systems

Air or gas bubbles in an open observation well or piezometer may influence the time lag and cause the stabilized level in the pipe to rise above the ground-water or the piezometric pressure level for the soil. Therefore, the interior of the pipe should be smooth, downward protruding edges or joints should be avoided, and the diameter of the pipe should be large enough to cause the bubbles filling the cross section to rise to the surface. These requirements are fulfilled by use of seamless and jointless plastic tubing, CASAGRANDE (2) and (3), and when the inside diameter of such tubing is 3/8 in. or more.

Gas bubbles in closed systems

Air or gas bubbles in a closed pipe connected to a manometer or pressure gage will increase the time lag, but gas above the connection to the pressure gage, and small gas bubbles adhering to the walls of the pipe, will not affect the stabilized pressure indicated by the gage. Gas bubbles below the gage connection and filling the entire cross section of the pipe will influence the indicated stabilized pressure. The pipe should be provided with an air trap and outlet valve at top, and should be smooth, without protruding joints, and of a diameter large enough to permit free rise of gas bubbles. At least, facilities for occasional flushing should be provided and the entire installation should be composed of materials which do not cause development of gases through electrolysis.

Gas entrapped in the water-filled space below the diaphragm of a hydrostatic pressure cell of the type shown in Case 9, Fig. 13 -- or in the perforated cover plate or porous stone -- will not influence the ultimate pressure indicated but will greatly increase the time lag of the pressure cell. It is conceivable that a material accumulation of gas below the diaphragm may cause the time lag of a hydrostatic pressure cell to be considerably greater than that of a closed piezometer with attached manometer or Bourdon pressure gage.

Gas bubbles in soil

Air and other gases are often entrapped in the pores of the soil, even below the ground-water level, or dissolved in the water. When the gas bubbles migrate to and cluster around the well point or are released there from solution in the water, the time lag will be increased on account of volume changes of the gas and because the gas bubbles decrease the permeability of the soil. The well point should consist of materials which do not cause development of gases through electrolysis. It is also advisable to avoid an excessive decrease of the hydrostatic pressure inside the well point and a consequent decrease of the pore-water pressure in the surrounding soil, since a decrease in hydrostatic pressure may cause release of gases dissolved in the water.

Sedimentation and clogging

Sediment in the water of the standpipe or piezometer will ultimately settle at the bottom of the pipe. When a solid porous well point is used, the sediment may form a relatively impervious layer on its top and thereby increase the time lag. Therefore, a hollow well point should be used, the pipe should be filled with clean water, and facilities for occasional cleaning and flushing are desirable. An outward flow of water from the pipe and well point may carry sediment in the pipe into the pores of the walls of the point or of the surrounding soil and may thereby cause clogging and a further increase in time lag. Therefore and insofar as possible, a strong outward flow of water from well point should be avoided.

Erosion and development

A strong inward flow of water may carry fine particles from the soil into the pipe, thereby increasing the permeability of the soil in the vicinity of the well point and decreasing the time lag of the installation. An initial strong inward flow of water and "development" of the well point may in some cases be desirable in order to decrease the time lag, provided the well point and pipe thereafter are cleaned out and filled with clean water. Uncontrolled erosion or development is undesirable on account of consequent unknown changes in the time lag characteristics of the installation, and because the soil grains may cause clogging of the well point, or the soil grains may be carried into the pipe, settle at the bottom, and ultimately increase the time lag. The porosity of, or openings in, the well point should be selected in accordance with the composition and character of the soil, or the well point should be surrounded with a properly graded sand or gravel filter.

Summary comments

It should be noted that several of the above mentioned sources of error require conflicting remedial measures, and for each installation it must be determined which one of these sources of error is most serious. Those listed under Items 3, 4, 5, and 6 in Fig. 1 will affect the results of the observations, even when these are made after practical equalization of the inside and outside pressures is attained. Those described under Items 7, 8, 9, and 10 primarily influence the time lag, but they may also affect the final results when the direct field observations are corrected for influence of the time lag. It is possible that these sources of error may develop or may disappear and that their influence on the observations may vary within wide limits during the life of a particular installation. Therefore, it is desirable that facilities be provided for controlled changes of the hydrostatic pressure inside the well point, so that the time lag characteristics may be verified or determined by methods to be described in the following sections of the paper.

The time lag characteristics of a hydrostatic pressure cell may be determined by laboratory experiments, but it should be realized that these characteristics may be radically altered and the time lag greatly increased by an accumulation of gases below the diaphragm after the pressure cell has been installed. When a hydrostatic pressure cell is to be left in the ground for prolonged periods, it would be desirable but also very difficult to provide means for releasing such gas accumulations and for verifying the basic time lag of the pressure cell in place.

PART II: THEORY OF TIME LAG

The Basic Hydrostatic Time Lag

In this and the following sections concerning the hydrostatic time lag, it is assumed that this time lag is the only source of error or that the influence of the stress adjustment time lag and other sources of error, summarized in Fig. 1, is negligible. Derivation of the basic differential equation for determination of the hydrostatic time lag, Fig. 2, is similar to that of the equations for a falling-head permeameter and is based on the assumption that Darcy's Law is valid and that water and soil are incompressible. It is also assumed that artesian conditions prevail or that the flow required for pressure equalization does not cause any perceptible draw-down of the ground-water level. The active head, H, at the time t is $H = z - y$, where z may be a constant or a function of t. The corresponding flow, q, may then be expressed by the following simplified equation,

$$q = F k H = F k (z - y) \qquad (1)$$

where F is a factor which depends on the shape and dimensions of the intake or well point and k is the coefficient of permeability. This equation is valid also for conditions of anisotropic permeability provided modified or equivalent values $\overline{F}$ and $\overline{k}$ are used; see pages 32-35. It is assumed that the friction losses in the pipe are negligible for the small rates of flow occurring during pressure observations. Considering the volume of flow during the time dt, the following equation is obtained,

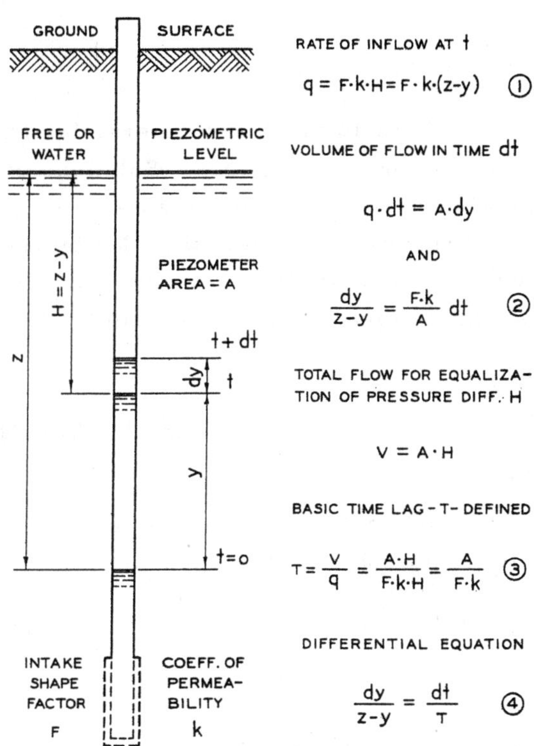

Fig. 2. Basic definitions and equations

$$q \, dt = A \, dy$$

where A is the cross-sectional area of the standpipe or an equivalent area expressing the relationship between volume and pressure changes in a pressure gage or cell. By introducing q from equation (1), the differential equation can be written as,

$$\frac{dy}{z - y} = \frac{F k}{A} dt \qquad (2)$$

The total volume of flow required for equalization of the pressure difference, H, is $V = AH$. The *basic time lag*, T, is now defined as the time required for equalization of this pressure difference when the original rate of flow, $q = FkH$, is maintained; that is,

$$T = \frac{V}{q} = \frac{AH}{FkH} = \frac{A}{Fk} \tag{3}$$

and equation (2) can then be written,

$$\frac{dy}{z - y} = \frac{dt}{T} \tag{4}$$

This is the basic differential equation for determination of the hydrostatic time lag and its influence. Solutions of this equation for both constant and variable ground-water pressures are derived in the following sections, and methods for determination of the basic time lag by field observations are discussed. Examples of theoretical shape factors, F, and preliminary estimates of the basic time lag by means of equation (3) are presented in Part III, pages 30-37.

Applications for Constant Ground-Water Pressure

When the ground-water level or piezometric pressure is constant and $z = H_o$, Fig. 3, equation (4) becomes

$$\frac{dy}{H_o - y} = \frac{dt}{T}$$

and with $y = 0$ for $t = 0$, the solution is,

$$\frac{t}{T} = \ln \frac{H_o}{H_o - y} = \ln \frac{H_o}{H} \tag{5}$$

The ratio t/T may be called the time lag ratio. The head ratio, H/H_o, is determined by the equation

$$\frac{H}{H_o} = e^{-\frac{t}{T}} \tag{6}$$

and the equalization ratio, E, by

$$E = \frac{y}{H_o} = 1 - \frac{H}{H_o} = 1 - e^{-\frac{t}{T}} \tag{7}$$

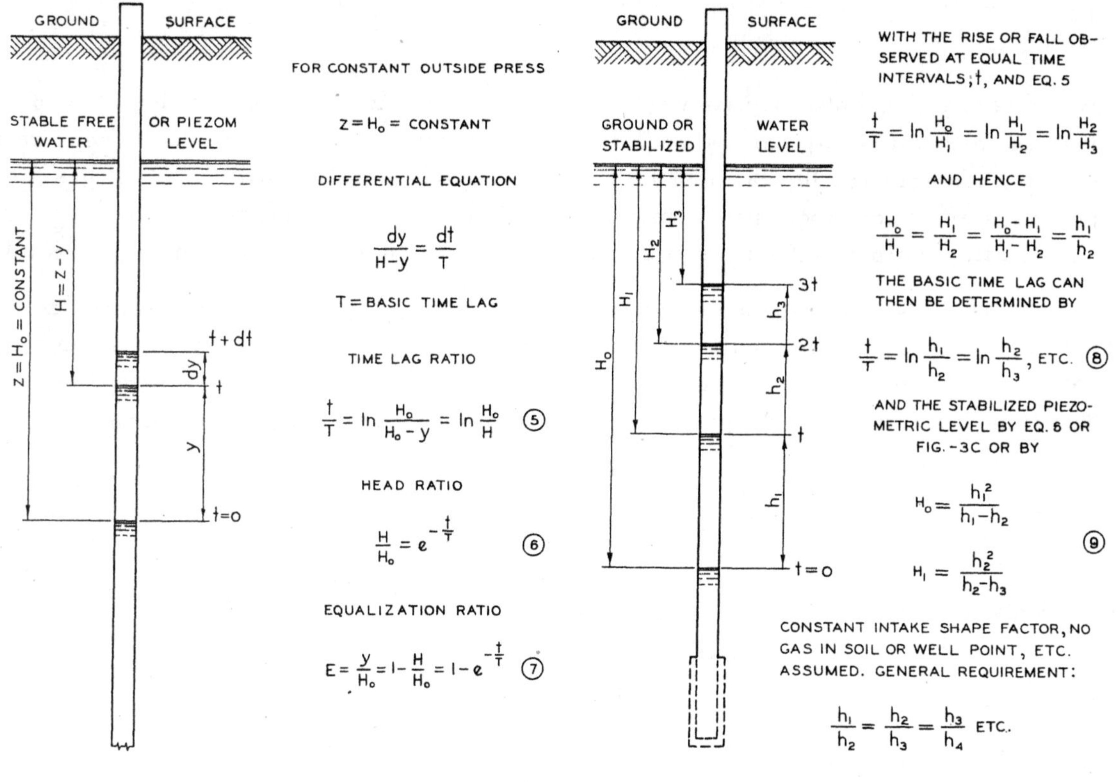

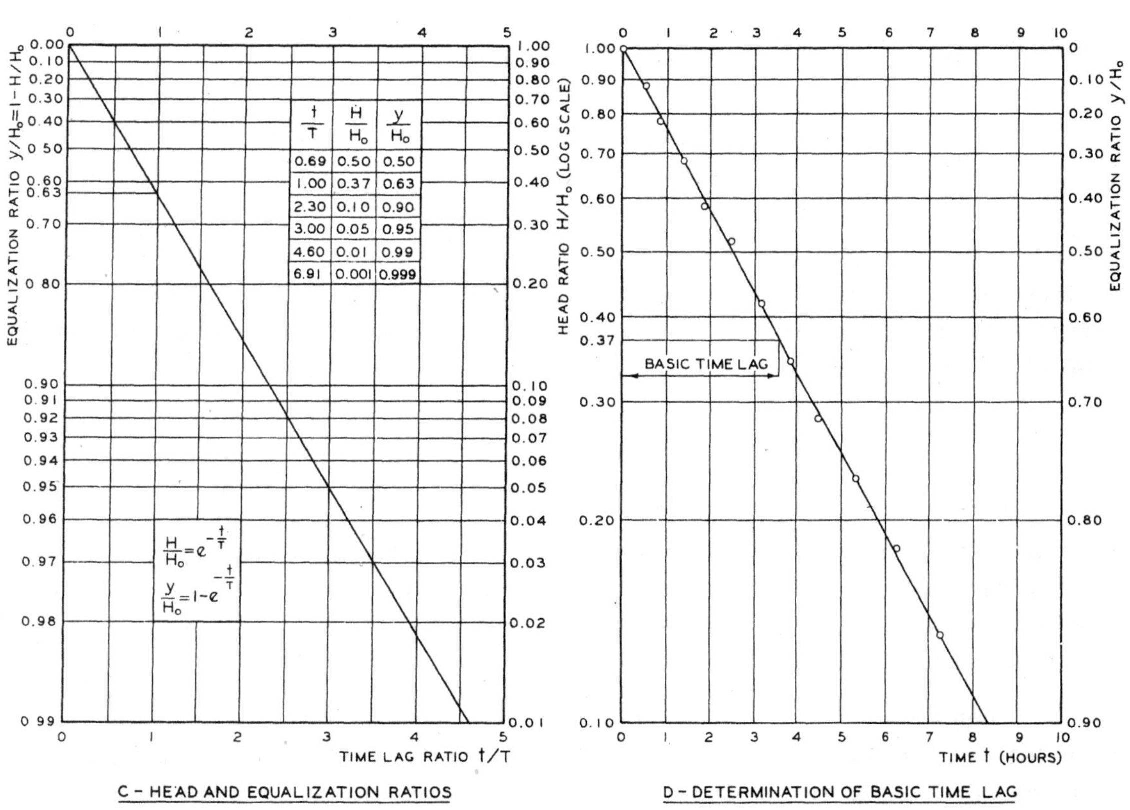

t/T	H/H_o	y/H_o
0.69	0.50	0.50
1.00	0.37	0.63
2.30	0.10	0.90
3.00	0.05	0.95
4.60	0.01	0.99
6.91	0.001	0.999

Fig. 3. Constant ground-water levels and pressures

A diagram representing equations (6) and (7) is shown in Fig. 3-C. It should be noted that the basic time lag corresponds to an equalization ratio of 0.63 and a head ratio of 0.37. An equalization ratio of 0.90 may be considered adequate for many practical purposes and corresponds to a time lag equal to 2.3 times the basic time lag. An equalization ratio of 0.99 requires twice as long time as 90 per cent equalization.

When the stabilized pressure level, or initial pressure difference, is not known, it may be determined in advance of full stabilization by observing successive changes in piezometer level, h_1, h_2, h_3, etc., for equal time intervals; see Fig. 3-B. The time lag ratio is then equal for all intervals, or according to equation (5),

$$\frac{t}{T} = \ln \frac{H_o}{H_1} = \ln \frac{H_1}{H_2} = \ln \frac{H_2}{H_3}, \text{ etc.}$$

and hence,

$$\frac{H_o}{H_1} = \frac{H_1}{H_2} = \frac{H_o - H_1}{H_1 - H_2} = \frac{h_1}{h_2}$$

or,

$$\frac{t}{T} = \ln \frac{h_1}{h_2} = \ln \frac{h_2}{h_3}, \text{ etc.} \tag{8}$$

and since $H_1 = H_o - h_1$, $H_2 = H_1 - h_2$, etc.,

$$H_o = \frac{h_1^2}{h_1 - h_2} \quad \text{or} \quad H_1 = \frac{h_2^2}{h_2 - h_3}, \text{ etc.} \tag{9}$$

It is emphasized that these equations can be used only when the influence of the stress adjustment time lag, air or gas in soil or piezometer system, clogging of the intake, etc., is negligible, or when

$$\frac{h_1}{h_2} = \frac{h_2}{h_3} = \frac{h_3}{h_4}, \text{ etc.}$$

Equations (9) form a convenient means of estimating the stabilized pressure level. In actual practice it is advisable to fill or empty the piezometer to the computed level and to continue the observations for a period sufficient to verify or determine the actual stabilized level.

When the head or equalization ratios, or the ratios between successive pressure changes for equal time intervals, have been determined, the basic time lag may be found by means of equations (5), (7), or (8). However, due to observational errors, there may be considerable scattering in results, especially when the pressure changes are small. In general, it is advisable to prepare an equalization diagram or a semi-logarithmic plot of head ratios and time, as shown in Fig. 3-D. When the assumptions on which the theory is based are fulfilled, the plotted points should lie on a straight line through the origin of the diagram. The basic time lag is then determined as the time corresponding to a head ratio of 0.37. Examples of both straight and curved diagrams of the above mentioned type are discussed in Part III, pages 38-43.

Applications for Linearly Changing Pressures

When the ground-water or piezometric pressure level, as shown in Fig. 4, is rising at a uniform rate, $+\alpha$, or falling at the rate $-\alpha$, then

$$z = H_o + \alpha t \tag{10}$$

and equation (4) may be written,

$$\frac{dy}{H_o + \alpha t - y} = \frac{dt}{T} \tag{11}$$

With $y = 0$ for $t = 0$, the solution of equation (11) is,

$$\frac{y - \alpha t}{H_o - \alpha T} = 1 - e^{-\frac{t}{T}} \tag{12}$$

which corresponds to equation (7) for constant ground-water pressure. Theoretically α, T, and H_o may be determined, as shown in Fig. 4-B, by observing three successive changes in piezometer level at equal time intervals, t, and expressing the results by three equations similar to equation (12). By successively eliminating $(H_o - \alpha T)$ and $e^{-\frac{t}{T}}$ from these equations, the following solutions are obtained,

$$\alpha t = \frac{h_1 h_3 - h_2^2}{h_1 + h_3 - 2h_2} \tag{13}$$

$$\frac{t}{T} = \ln \frac{h_1 - \alpha t}{h_2 - \alpha t} \tag{14}$$

$$H_o = \alpha T + \frac{(h_1 - \alpha t)^2}{h_1 - h_2} \tag{15}$$

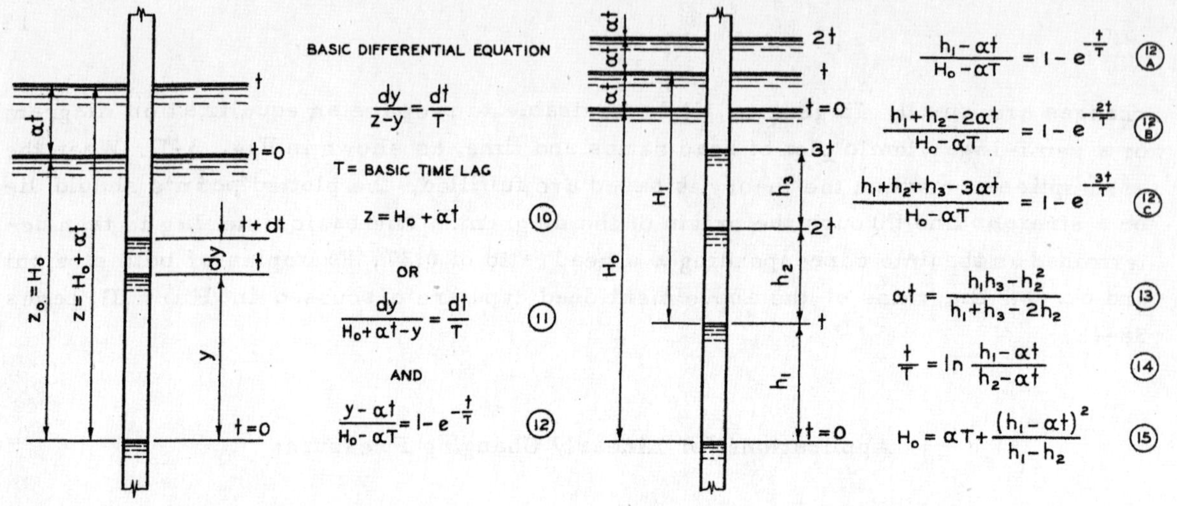

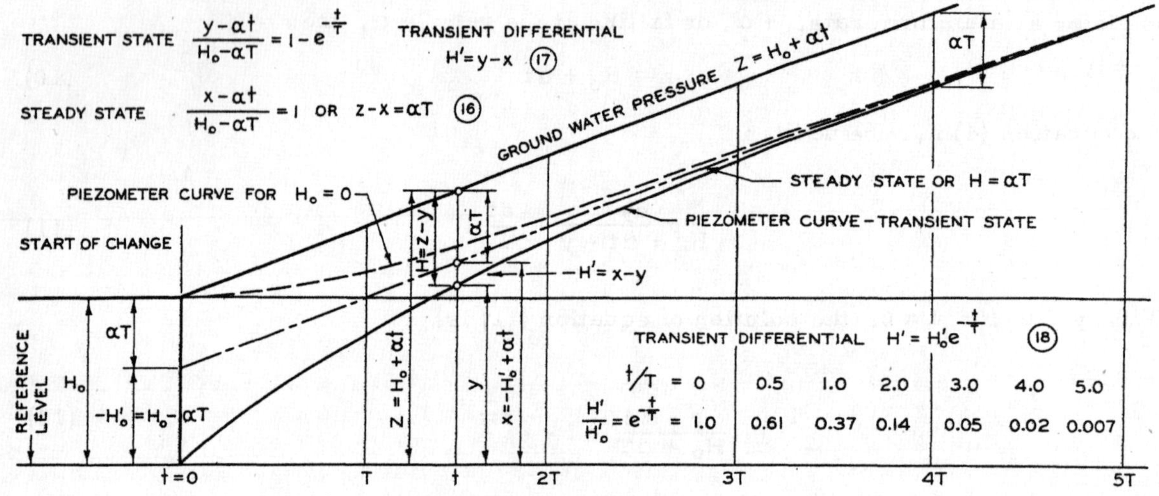

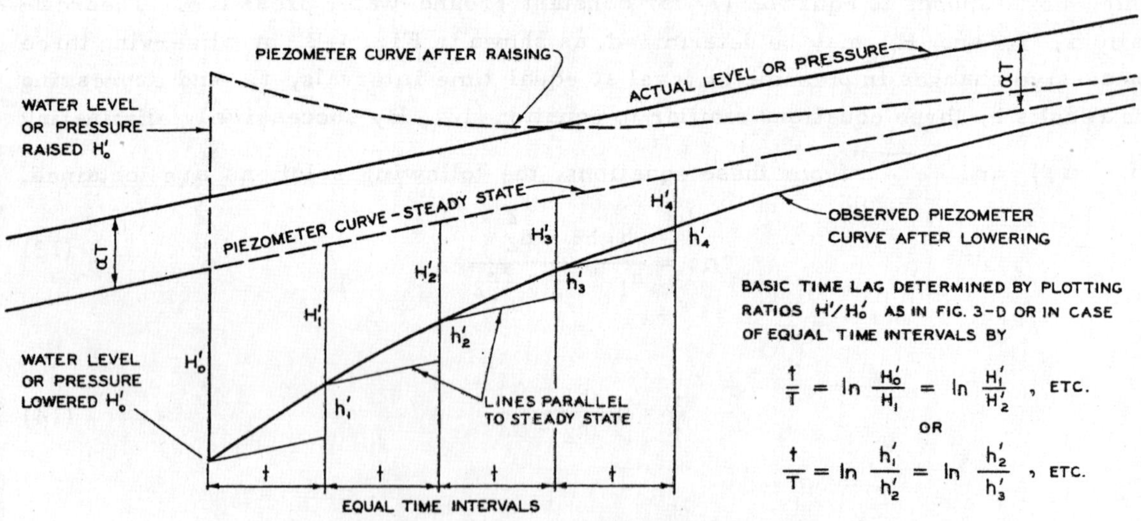

Fig. 4. Linearly changing ground-water pressures

These equations correspond to equations (8) and (9) for constant ground-water pressure. However, the form of equation (13) is such that a small error in determination of the increment pressure changes may cause a very large error in the computed value of αt. In general, it is better to determine the basic time lag and the actual ground-water pressures after the steady state, discussed below, is attained.

Referring to Fig. 4-C, equation (12) represents the transient state of the piezometer curve. With increasing values of t, the right side of this equation approaches unity and the curve the steady state. Designating the ordinates of the steady state of the piezometer curve by x, this curve is represented by,

$$\frac{x - \alpha t}{H_o - \alpha T} = 1$$

or by means of equation (10),

$$z - x = \alpha T = \text{constant} \tag{16}$$

That is, the difference between the actual ground-water pressure and that indicated by the piezometer is constant and equal to αT during the steady state. The difference between the pressures corresponding to the transient and steady states of the piezometer curve

$$H' = y - x \tag{17}$$

may be called the transient pressure differential. For the conditions shown in Fig. 4-C, this differential is negative. With

$$x = H_o - \alpha T + \alpha t \quad \text{and} \quad H'_o = \alpha T - H_o$$

equation (17) can be written,

$$H' = (y - \alpha t) + H'_o$$

and by means of equation (12)

$$H' = H'_o \, e^{-\frac{t}{T}} \tag{18}$$

This equation is identical with equation (6) for constant ground-water pressure; that is, the transient pressure differential can be determined as if the line representing the steady state were a constant piezometric pressure level. As will be seen in Fig. 4-C and also the diagram in Fig. 3-C, the steady state may for practical purposes be considered attained at a time after a change in piezometer level, or start of a change in the rate α, equal to three to four times the basic time lag.

When the piezometer level increases or decreases linearly with time, it may be concluded that the steady state is attained and that the rate of change, α, is equal to that for the ground-water pressure. If the piezometer level now is raised or lowered by the amount H'_o, and the transient pressure differentials are observed, then the basic time lag may be determined by means of a semi-logarithmic plot of the ratios H'/H'_o and the time, t, as in Fig. 3-D; that is, the basic time lag is the time corresponding to $H'/H'_o = 0.37$. To complete the analogy with constant ground-water pressures, the transient pressure differential may be observed at equal time intervals, t, and the basic time lag determined by,

$$\frac{t}{T} = \ln \frac{H'_o}{H'_1} = \ln \frac{H'_1}{H'_2}, \text{ etc.}$$

or by

$$\frac{t}{T} = \ln \frac{h'_1}{h'_2} = \ln \frac{h'_2}{h'_3}, \text{ etc.}$$

where h'_1, h'_2, h'_3 are the increment pressure differentials. However, it is generally advisable to use the ratios H'/H'_o and a diagram of the type shown in Fig. 3-D. Having thus determined the basic time lag, the difference between the piezometer and ground-water levels, αT, can be computed.

Applications for Sinusoidal Fluctuating Pressures

Periodic fluctuations of the ground-water pressure, in form approaching a sinusoidal wave, may be produced by tidal variations of the water level of nearby open waters, Fig. 5-A. Such fluctuations of the ground-water pressure may be represented by the equation

$$z = z_a \sin \frac{2\pi t}{T_w} \tag{19}$$

where z_a is the amplitude and T_w the period of the wave. By means of the basic differential equation (4) the following equation for the fluctuations of the piezometer level is obtained,

$$\frac{dy}{dt} = \frac{1}{T} \left(z_a \sin \frac{2\pi t}{T_w} - y\right) \tag{20}$$

Through the temporary substitution of a new variable v and $y = v e^{-\frac{t}{T}}$, setting $\frac{2\pi T}{T_w} = \tan \frac{2\pi t_s}{T_w}$, and with $y = y_o$ for $t = 0$, the following solution of the equation is obtained,

$$y = z_a \cos \frac{2\pi t_s}{T_w} \sin \frac{2\pi}{T_w} (t - t_s) + \left[y_o + z_a \cos \frac{2\pi t_s}{T_w} \sin \frac{2\pi t_s}{T_w} \right] e^{-\frac{t}{T}}$$

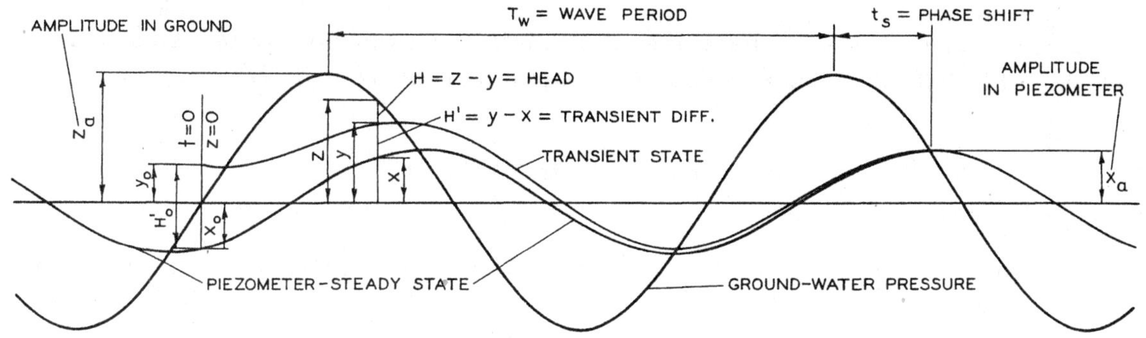

GROUND-WATER PRESSURES

T_w = PERIOD Z_a = AMPLITUDE

$$z = z_a \sin\frac{2\pi}{T_w} t \qquad (19)$$

PRESSURES INDICATED BY PIEZOMETER

T = BASIC TIME LAG

$$\frac{dy}{dt} = \frac{1}{T}\left(z_a \sin\frac{2\pi t}{T_w} - y\right) \qquad (20)$$

STEADY STATE

t_s = PHASE SHIFT x_a = AMPLITUDE

$$x = x_a \sin\frac{2\pi}{T_w}(t - t_s) \qquad (23)$$

$$\tan\frac{2\pi}{T_w}t_s = \frac{2\pi}{T_w}T \quad (21) \qquad \frac{x_a}{z_a} = \cos\frac{2\pi}{T_w}t_s \quad (22)$$

TRANSIENT STATE

$$y = x_a \sin\frac{2\pi}{T_w}(t - t_s) + \left(y_o + x_a \sin\frac{2\pi}{T_w}t_s\right)e^{-\frac{t}{T}} \qquad (24)$$

TRANSIENT DIFFERENTIAL H' = y − x

$$H' = \left(y_o + x_a \sin\frac{2\pi}{T_w}t_s\right)e^{-\frac{t}{T}} = H'_o e^{-\frac{t}{T}} \qquad (25)$$

A - TRANSIENT AND STEADY STATES

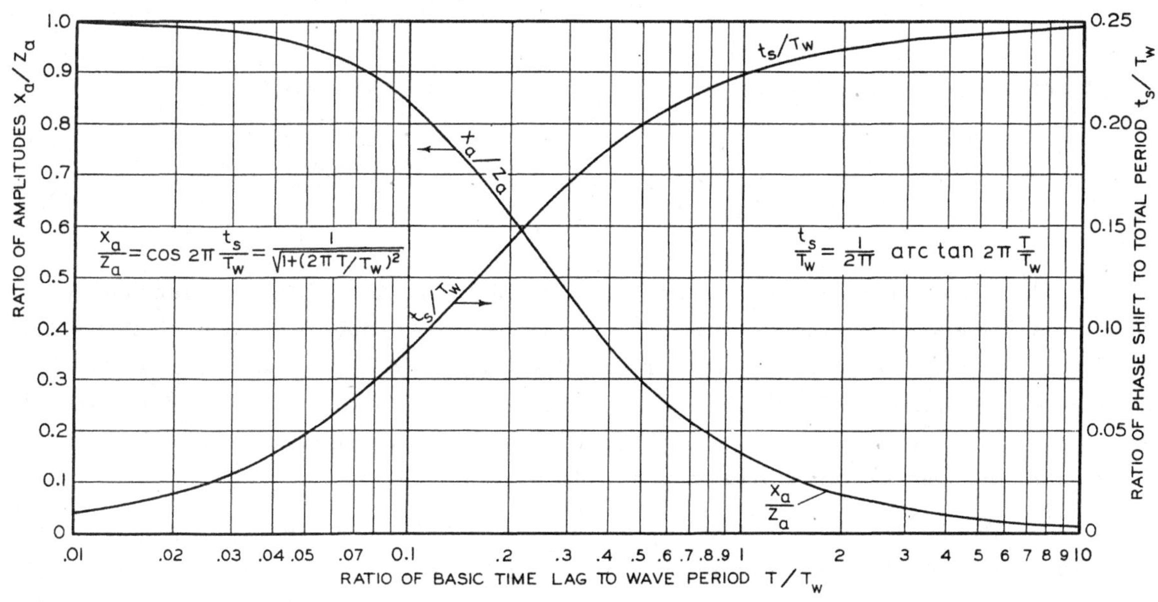

B - CHANGE IN AMPLITUDE AND PHASE OF STEADY STATE WITH BASIC TIME LAG

Fig. 5. Sinusoidal fluctuating ground-water pressure

For large values of t, $e^{-\frac{t}{T}}$ becomes very small and is zero for the steady state, for which the following equation applies, substituting x for y,

$$x = z_a \cos \frac{2\pi t_s}{T_w} \sin \frac{2\pi}{T_w} (t - t_s)$$

This equation represents a sinusoidal wave with the phase shift t_s, determined by,

$$\tan \frac{2\pi t_s}{T_w} = \frac{2\pi T}{T_w} \tag{21}$$

and the amplitude

$$x_a = z_a \cos \frac{2\pi t_s}{T_w} = \frac{z_a}{\sqrt{1 + (2\pi T/T_w)^2}} \tag{22}$$

The equation for the steady state can then be written,

$$x = x_a \sin \frac{2\pi}{T_w} (t - t_s) \tag{23}$$

and the equation for the transient state,

$$y = x_a \sin \frac{2\pi}{T_w} (t - t_s) + (y_o + x_a \sin \frac{2\pi t_s}{T_w}) e^{-\frac{t}{T}} \tag{24}$$

The transient pressure differential, $H' = y - x$, is determined by

$$H' = (y_o + x_a \sin \frac{2\pi t_s}{T_w}) e^{-\frac{t}{T}} = H'_o e^{-\frac{t}{T}} \tag{25}$$

where H'_o is the transient differential for $t = 0$. Equation (25) is identical with equations (6) and (18), and the transient pressure differential can also in this case be computed as if the steady state were a constant pressure level. H' may be determined as a function of H'_o by means of the diagram shown in Fig. 3-C, and it will be seen that for practical purposes the steady state is reached after elapse of a time equal to three to four times the basic time lag.

Equations (22) and (23) are represented by the diagram in Fig. 5-B, by means of which the phase shift and the decrease of amplitude in the piezometer can easily be determined. If the fluctuations of the piezometer level have reached the steady state and the wave period, T_w, and the phase shift, t_s, can be observed in the field, it is theoretically possible to determine the basic time lag by means of the diagram in Fig. 5-B. However, it is difficult to determine the phase shift by direct observation, since it cannot be assumed that the pressure fluctuations in the ground water are in phase with those of the surface waters. When the fluctuations in the ground-water pressure are caused by load and stress changes without material seepage and volume changes of the soil, it is possible that the phase shift in pore-water pressures, with respect to the surface water, may be insignificant even though a material decrease in amplitude occurs. On the other hand, when pressure changes in the pore water in part are caused by infiltration or are accompanied by changes in water content of the soil, then it is possible that there also will be a material shift in phase of the pressure fluctuations. The basic time lag may be determined during the steady state by raising or lowering the piezometer pressure, observing the transient pressure differentials, and plotting the ratios H'/H'_o and the elapsed time in a diagram similar to that shown in Fig. 3-D.

Corrections for Influence of the Hydrostatic Time Lag

The characteristics of an installation for determination of ground-water levels and pressures may change with time because of sedimentation, clogging, and accumulation of gases in the system or in the soil near the intake. When observations of such levels and pressures are to be corrected for influence of the hydrostatic time lag, the first task is to determine the basic time lag and verify that the assumptions, on which the general theory is based, are satisfied. This is best accomplished during periods when the ground-water pressure is constant, but as shown in the foregoing sections, the verification may also be performed during the steady state of linear and sinusoidal variations in the ground-water and piezometer levels.

Verification by means of transient pressure differentials can be used irrespective of the form of the curve representing the steady state of pressure variations. The pressure variations may be represented by the following general equations, $z = F(t)$ for the ground-water pressure; $x = f(t)$ for the steady state of the piezometer pressure; and $y = g(t)$ for the transient state or after the piezometer pressure has been raised or lowered by an arbitrary amount H'_o. The transient pressure differential is the $H' = y - x$, and according to equation (4), which applies to all conditions,

$$\frac{dy}{z - y} = \frac{dt}{T} = \frac{dx}{z - x} = \frac{dy - dx}{x - y} = \frac{dH'}{H'}$$

or

$$\ln H' = -\frac{t}{T} + C$$

and with $H' = H'_o$ for $t = 0$

$$\frac{t}{T} = \ln \frac{H'_o}{H'}$$

which is identical with equation (5). Therefore, when the piezometer pressure varies in such a manner that the pressures can be predicted with sufficient accuracy for a future period of reasonable length, the basic time lag may be determined by raising or lowering the piezometer pressure by an arbitrary amount, H'_o, observing the transient pressure differentials, H', and plotting the ratios H'/H'_o as a function of time as shown in Fig. 3-D. Application of the basic equation (4) requires that the points in the semi-logarithmic plot fall on a straight line through the origin of the diagram.

Having determined the basic time lag and verified that the assumptions are satisfied, corrections for influence of the time lag in case of linear or sinusoidal variations may be determined as shown in Figs. 4 and 5. In case of irregular fluctuations, it should first be noted that when the piezometer curve passes through a maximum or minimum, the pressure indicated by the piezometer must be equal to that of the ground water. In this connection it is again emphasized that the fluctuations of the ground-water pressure are not necessarily in phase with those of the water level of nearby surface waters. The maxima or minima of the piezometer variations may be used as starting points for the corrections, which may be determined by assuming either an equivalent constant value or, alternatively, an equivalent constant rate of change of the ground-water pressure during each time interval.

The first of these methods is shown in Fig. 6-A. The difference, H_c, between the equivalent constant ground-water pressure and the piezometer pressure at the start of the time interval may be determined by equation (7) and substituting H_c for H_o and h for y; that is,

$$H_c = \frac{h}{E} \tag{26}$$

where h is the change in piezometer pressure and E is the equalization ratio for the time interval, t, or time lag ratio t/T; see Fig. 3-C. It is now assumed that the actual ground-water pressure in the middle of the time interval is equal to the equivalent constant pressure during the interval.

In applying the second method of correction, Fig. 6-B, it is assumed that the pressure difference at the beginning of the time interval, H_o, has been determined, for example by starting the operations at a maximum or minimum of the piezometer curve. Designating the equivalent uniform rate of change in ground-water pressure by α, the total change during the time interval, $H_t = \alpha t$, can be computed by means of equation (12), or when solving for αt and introducing the equalization ratio E,

$$H_t = \frac{h - E H_o}{1 - E \frac{T}{t}} \tag{27}$$

This method will usually give more accurate results than the method of equivalent constant pressure, but the latter method is easier to apply. The results obtained by the two methods are compared in Fig. 6-C, and it will be seen from the equations and the diagram that the difference in results is only a few per cent when the initial pressure difference is large and the time interval is small, in which case the easier method of equivalent constant pressures may be used. On the other hand, there is considerable difference in results and the method of equivalent constant rate of change should be used when the initial pressure difference is small and the time lag ratio is large.

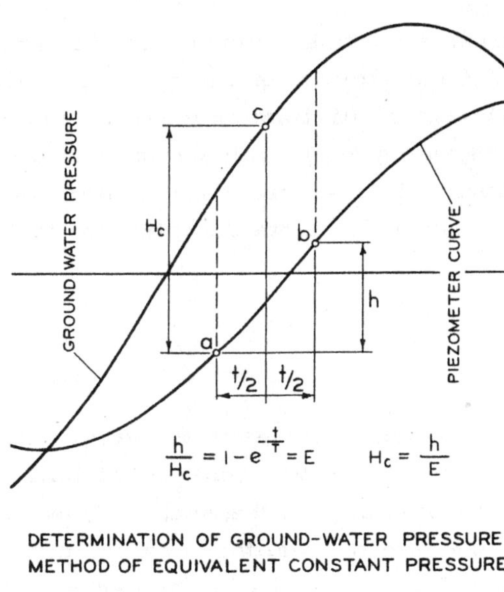

DETERMINATION OF GROUND-WATER PRESSURE
METHOD OF EQUIVALENT CONSTANT PRESSURE

A

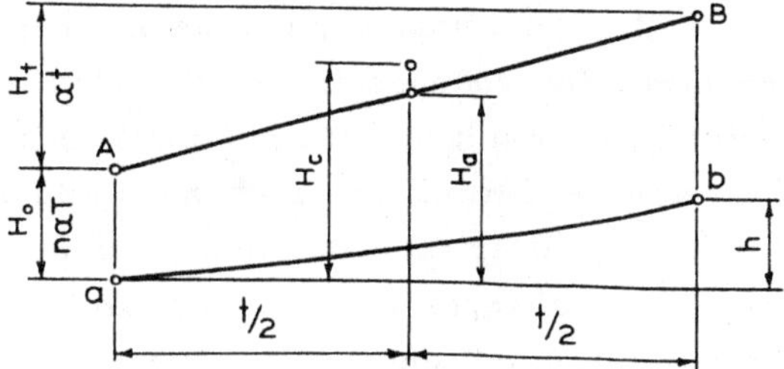

ASSUME LINEAR CHANGE A TO B AND $H_t = \alpha t$, $H_o = n\alpha T$

$$\frac{h - \alpha t}{H_o - \alpha T} = 1 - e^{-\frac{t}{T}} = E, \quad h = \alpha t + E(n-1)\alpha T, \quad n = \frac{H_o}{H_t} \cdot \frac{t}{T}$$

$$H_c = \frac{h}{E} = \frac{\alpha t}{E} + (n-1)\alpha T \qquad H_a = H_o + \frac{1}{2}\alpha t = n\alpha T + \frac{1}{2}\alpha t$$

$$\frac{H_c}{H_a} = \frac{n + \frac{1}{E}\frac{t}{T} - 1}{n + \frac{1}{2}\frac{t}{T}}$$

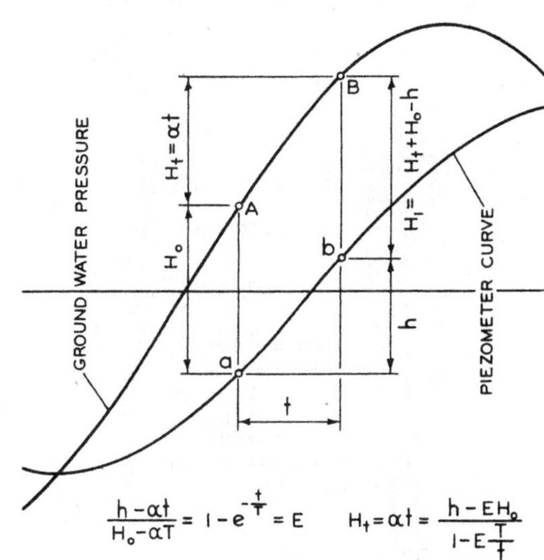

DETERMINATION OF GROUND-WATER PRESSURE
METHOD OF LINEAR CHANGE IN PRESSURE

B

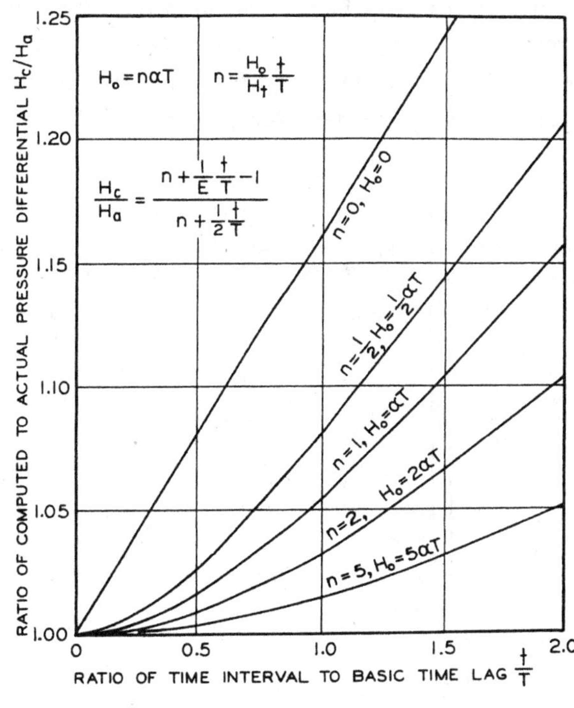

C - RELATIVE ACCURACY OF METHODS

Fig. 6. Corrections for influence of hydrostatic time lag

Influence of the Stress Adjustment Time Lag

In absence of detailed theoretical and experimental investigations of the stress adjustment time lag and its influence on pressure observations, the following discussion is tentative in character, and its principal object is to call attention to the problems encountered.

As mentioned in discussing Fig. 1, the stress adjustment time lag is the time required for changes in water content of the soil in the vicinity of the intake or well point as a result of changes in the stress conditions. A distinction must be made between the initial stress changes and adjustments, which occur only during and immediately after installation of a pressure measuring device, and the transient but repetitive changes which occur each time water flows to or from the intake or well point during subsequent pressure observations.

Initial disturbance and stress changes

When a boring is advanced by removal of soil, the stresses in the vicinity of its bottom or section below the casing will be decreased with a consequent initial decrease in pore-water pressure and tendency to swelling of the soil. A flow of water from the boring to the soil will increase the rate of swelling, and the combined initial hydrostatic and stress adjustment time lags will probably be decreased when the initial hydrostatic pressure inside the boring or well point is slightly above the normal ground-water pressure, Fig. 7-A.

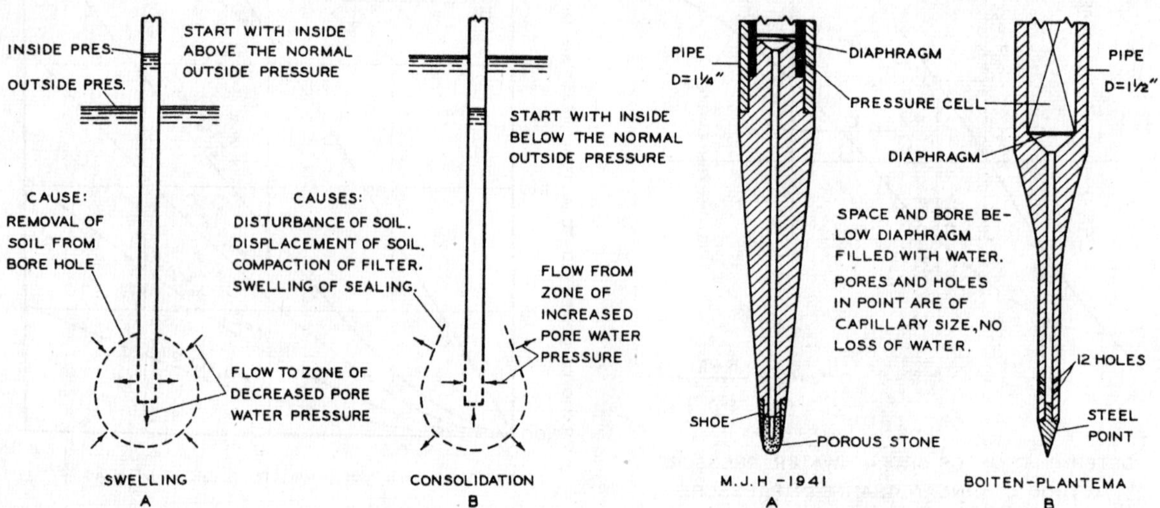

Fig. 7. Initial disturbance and stress changes

Fig. 8. Points for pressure sounding rod

A zone with increased pore-water pressures and a tendency to consolidation of the soil may be caused by disturbance and displacement of soil during the driving of a well point and by compaction of a sand filter or a seal above a well point or pressure cell installed in an oversize bore hole, Fig. 7-B. Subsequent swelling of the sealing material may also cause consolidation of the surrounding soil, but its effect on the pore-water pressures in the vicinity of the well point is uncertain. A flow of water from the soil to the well point will increase the rate of consolidation, and when the basic time lag of the installation is large, the combined initial hydrostatic and stress adjustment time lags will probably be decreased when the initial hydrostatic pressure inside the well point is below the normal ground-water pressure.

The initial stress adjustment time lag depends on the dimensions of the zone of stress changes and on the permeability, sensitivity to disturbance, and consolidation characteristics of the soil. The initial stress adjustment time lag will be small compared to the hydrostatic time lag when the total volume change of the soil is small compared to the required increase or decrease of the volume of water in the pressure measuring device, as in case of a boring or observation well in coarse-grained soils. On the other hand, the stress adjustment time lag may be very large compared to the hydrostatic time lag for a pressure cell installed in fine-grained and highly compressible soils.

The initial stress adjustment time lag can be reduced by decreasing the dimensions of the well point and/or filter, but this will increase the hydrostatic time lag. When the ground-water observations are to be extended over a considerable period of time, the hydrostatic time lag is usually governing and the well point should be large. On the other hand, when it is desired to make only a single or a few measurements at each location and depth, and when a sensitive pressure measuring device is used, then the well point should be small in order to reduce the zone of disturbance and the initial stress adjustment time lag. Even then there is an optimum size, and when the dimensions of the well point are made smaller than that size, the consequent decrease in the initial stress adjustment time lag may be more than offset by an increase in the hydrostatic time lag.

Examples of points for pressure measuring devices, similar to sounding rods and intended for reconnaissance exploration of ground-water conditions in soft or loose soils, are shown in Fig. 8. The one to the left, designed by the writer (14, 15), has a larger intake area than the one shown to the right and designed by BOITEN and PLANTEMA (1), but the latter is sturdier and will probably cause less disturbance of the soil in the immediate vicinity of the point.

Transient consolidation or swelling of soil

When water is flowing to or from a pressure measuring device, the pore-water pressures, the effective stresses in, and the void ratio of the soil in the vicinity of the well point or intake will be subject to changes. As a consequence, the rate of flow of water to or from the intake will be increased or decreased, and this will influence the shape of the equalization diagrams. The above mentioned changes are more or less transient, and with decreasing difference between the piezometer and ground-water pressures, the stress conditions and void ratios will approach those corresponding to the pore-water pressures in the soil mass as a whole. The probable sequence of consolidation and swelling of the soil around a rigid well point when the piezometer level is lowered or raised is shown in Fig. 9.

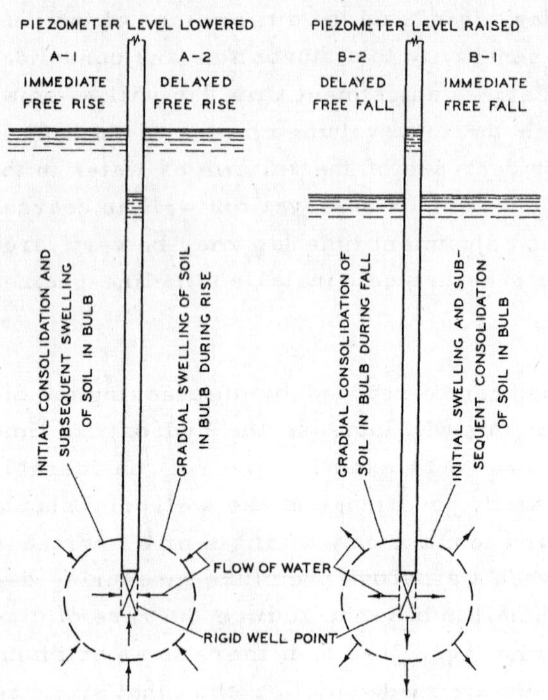

Fig. 9. Transient changes in void ratio

It is difficult by theory or experiment to determine the changes in void ratio and water content around a well point, but similar changes occur during soil permeability tests with a rising or falling head permeameter, and observations made immediately after the head is applied in such a permeameter usually furnish too high values for the coefficient of permeability and are discarded as unreliable. Although the stress conditions around a rigid well point are more complicated than in a soil test specimen in a permeameter, the results of permeability tests, which are extended until practical equalization of the water levels is attained, will furnish an indication of the magnitude of the transient consolidation and swelling and on the resulting shape of equalization diagrams for a rigid well point*. A series of such tests were performed with Atlantic muck, a soft organic clay, and the testing arrangement and some test results are shown in Fig. 10. The volume changes during these permeability tests were very small since the test specimens were overconsolidated in order to obtain nearly equal consolidation and swelling characteristics.

When the water level in the standpipe, Fig. 10, is raised and immediately thereafter allowed to fall -- corresponding to Case B-1 in Fig. 9 -- an initial swelling of the soil takes place, since the total vertical stresses remain constant whereas the pore-water pressure has been increased and the effective stresses tend to decrease. As a consequence, the rate of flow from the standpipe to the soil sample is increased and the initial slope of the equalization diagram becomes steeper. As the swelling progresses and the water level in the standpipe falls, the rate of excess flow decreases; the equalization diagram acquires a concave curvature, and a condition will be reached where the void ratio of the soil corresponds to the pore-water

* The relatively simple conditions shown in Fig. 9, and a comparison with the conditions in a permeameter, may not apply in case of an open bore hole, when the well point or intake is not rigid, and when the pressure in Case B is so great that the soil is deflected and a clearance is created between the well point and the soil.

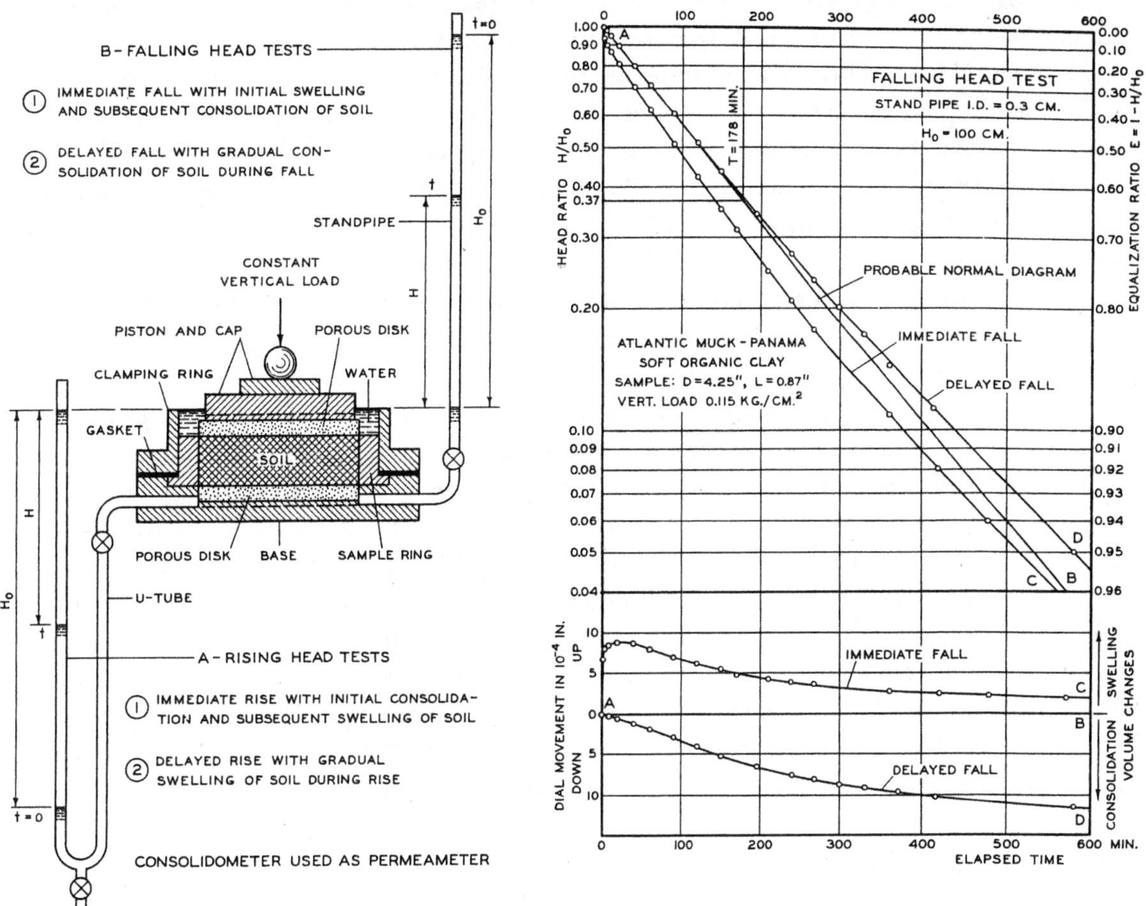

Fig. 10. Volume changes during laboratory permeability tests

pressure indicated by the standpipe level. With further fall in this level and decrease in pore-water pressures, a reconsolidation of the soil takes place with a consequent deficiency in rate of flow from the standpipe. The curvature of the equalization diagram decreases; the diagram becomes fairly straight and may even acquire a slight convex curvature as it approaches the normal diagram, obtained when there is no change in void ratio of the soil. However, the ultimate shape and slope of the diagram could not be determined from the results of tests so far performed, since these results were influenced by very small temperature changes in the laboratory.

When the water level in the standpipe is raised and maintained in its upper position until the initial swelling of the soil sample is completed and then allowed to fall -- Cases B-2 in Figs. 9 and 10 -- a gradual re-consolidation of the soil takes place during the actual test, and an equalization diagram which lies above the normal diagram is obtained, but its lower part is more or less parallel to the lower part of the diagram for immediate fall.

Similar diagrams were obtained by rising head tests. When the water level in the U-tube is lowered and immediately thereafter allowed to rise, Case A-1 in Figs. 9 and 10, the soil will be subjected to an initial consolidation with a consequent increase in rate of flow to the U-tube, but this volume decrease of the soil will later be eliminated by a swelling and a corresponding deficiency in rate of flow to the U-tube. The resulting equalization diagram has a concave curvature and lies below the normal diagram. When the water level in the U-tube is maintained in its lower position until the initial consolidation is completed and then allowed to rise, a gradual swelling of the soil takes place; the rate of flow to the U-tube is decreased, and the equalization diagram lies above the normal diagram.

All the above mentioned tests were repeated several times with both undisturbed and remolded soil, and the results obtained were all similar to those shown in Fig. 10. A slight sudden drop in head ratio in case of immediate fall -- or rise -- is probably due to a small amount of air in the system. As already indicated, the shape of the lower part of the diagrams was influenced by small amounts of leakage and evaporation and by temperature changes. The temperature in the laboratory did not vary more than 1.5^{o} F from the mean temperature, but even such small variations are sufficient to cause conspicuous irregularities in the test results when the active head is small. However, it is believed that the results are adequate for demonstration of the consolidation and swelling of the soil during permeability tests and of the resulting general shape of the equalization diagrams.

Volume changes of gas in soil

The influence of gas bubbles in an open or closed pressure measuring system is summarized in Fig. 1 and discussed briefly on pages 6 and 7. Whereas such gas bubbles may cause a change in both the ultimate indicated pressure and the time lag or slope of the equalization diagram, they will not materially influence the shape of the latter, since changes in pressure and volume of the gas bubbles occur nearly simultaneously with the changes in hydrostatic pressure within the system. On the other hand, when the gas bubbles are in the soil surrounding the well point and their volume and the water content of the soil are changed, there will be a time lag between changes in hydrostatic pressure in the system and corresponding changes in pressure and volume of the gas bubbles, and this time lag will cause a change in both slope and shape of the equalization diagrams. The general effect of the gas bubbles is an increase in the apparent compressibility of the soil, and the equalization diagrams should be similar to those shown in Fig. 10.

The change in volume of the gas bubbles, when the piezometer level is lowered or raised, and probable resulting equalization diagrams are shown in Fig. 11. This figure and the following discussion are essentially a tentative interpretation of the results of the laboratory permeability tests and the field observations shown in Figs. 10 and 17.

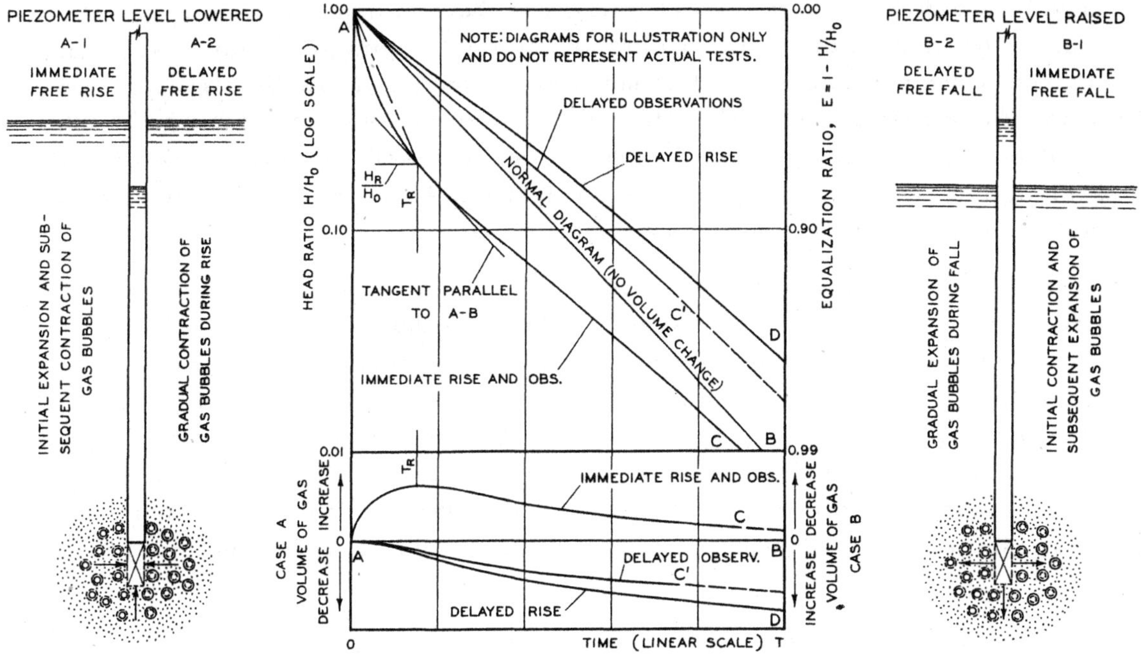

Fig. 11. Influence of volume changes of gas in soil

When the piezometer level suddenly is lowered and immediately thereafter allowed to rise, Case A-1, the pressure in the pore water is decreased, and the gas bubbles tend to expand and force an excess amount of water into the well point; that is, the initial rate of rise of the piezometer level will be increased and the equalization diagram, A-C, will have a steeper slope than the normal diagram, A-B, and a concave curvature. It is emphasized that the normal diagram, A-B, corresponds to the condition of no volume change of the gas bubbles and not to complete absence of gas bubbles in the soil. Even when the volume of the gas bubbles does not change, the presence of these bubbles will decrease the effective permeability of the soil and increase the time lag of the piezometer. As the piezometer level rises, the difference between the pressures in the gas bubbles and the surrounding pore water decreases. At the time T_r these pressures are equalized, and the rate of excess inflow ceases; that is, the tangent to the equalization diagram, A-C, at the time T_r should be parallel to the normal diagram, A-B. With a further rise in piezometer level, the pore-water pressure around the well point increases; the volume of the gas bubbles decreases, and there will be a deficiency in inflow of water. The curvature of the equalization diagram decreases and may eventually become zero or, perhaps, even change to a slight convex curvature as the volume of the gas bubbles approaches its original value.

If the observations were started at the time of reversal of the volume changes, T_r, the volume of the gas bubbles would decrease throughout the observations; there would be a deficiency in the rate of inflow, and the equalization diagram, A-C', would be above the normal diagram. A similar but higher-lying diagram, A-D, would be obtained if the piezometer level is not allowed to rise immediately after lowering but is maintained in its lower position until the initial swelling of the gas bubbles is completed, Case A-2. The two diagrams A-C and A-D should ultimately become parallel, and the normal diagram is a straight line between these limiting diagrams and is tangent at "A" to diagrams A-C' and A-D.

When the piezometer level suddenly is raised and immediately thereafter is allowed to fall, Case B-1, the volume of the gas bubbles at first decreases with a consequent excess outflow of water from the piezometer. Later on the gas bubbles expand until their original volume is attained, and during this period there will be a corresponding deficiency in rate of outflow. The resulting equalization diagram is similar in form to A-C for Case A-1. When the piezometer level is maintained in its upper position until the initial contraction of the gas bubbles is completed and then is allowed to fall, an equalization diagram similar to A-D is obtained.

Normal operating conditions

The discussions in the foregoing sections concern mainly time lag tests during which the piezometer level suddenly is changed whereas the general ground-water level or pore-water pressure remains constant. In normal operation the ground-water pressure changes first, and the piezometer level follows these changes with a certain pressure difference or time lag. When the ground-water level or pore-water pressure changes, the void ratio of the soil and the volume of gas bubbles below the ground-water level also tend to change, but the rate of such changes generally decreases in the immediate vicinity of a well point or intake for a pressure measuring installation on account of the pressure difference and time lag. However, all changes progress in the same direction and there is no initial increase in void ratio and water content followed by a decrease -- or vice versa -- as in the case of time lag tests.

In general, normal operating conditions resemble in most cases those of delayed fall or rise, or rather delayed observations, shown in Figs. 10 and 11. *It is probable that the time lag during normal operating conditions corresponds to an equalization diagram which, for practical purposes, may be represented by a straight line through the origin of the diagram and parallel to the lower portions of the diagrams obtained in time lag tests.* However, sufficient experimental data for verification of the suggested approximation -- especially comparative tests during rapidly changing ground-water pressures and with several pressure measuring installations having widely different basic time lags -- are not yet available.

As indicated by permeability tests of the type shown in Fig. 10, it is probable that the influence of swelling or consolidation of the soil is very small or negligible when observation wells or open piezometers are used in ground-water observations, but it is also possible that such changes in void ratio may cause appreciable distortion of the equalization diagrams and increase in actual time lag when pressure gages or cells with a small basic time lag are used and the soil is relatively compressible. On the other hand, gas bubbles in the soil around a well point may cause considerable distortion of the equalization diagrams and increase in actual time lag even for open piezometers; see Fig. 17. Accumulation of gas in the pressure measuring system causes no curvature of the equalization diagram but materially decreases its slope and increases the effective time lag under normal operating conditions.

PART III -- DATA FOR PRACTICAL DETERMINATION AND USE OF TIME LAG

Flow through Intakes and Well Points

For the purpose of designing or selecting the proper type of pressure measuring installation for specific soil and ground-water conditions, the basic time lag may be computed by means of equation (3). In order to facilitate such computations, formulas for flow through various types or shapes of intakes or well points are assembled in Fig. 12. These formulas are all derived on the assumption that the soil stratum in which the well point is placed is of infinite thickness and that artesian conditions prevail, or that the inflow or outflow is so small that it does not cause any appreciable change in the ground-water level or pressure. Except when otherwise noted by subscripts, as in k_v and k_h, it is also assumed that the permeability of the soil, k, is uniform throughout the stratum and equal in all directions.

The formula for Case 1 is that for a point source, and by reasons of symmetry the flow in Case 2 is half as great, but the formula for this case has also been derived directly by DACHLER (6). Derivation of the formula for Case 3 is given in the books by FORCHHEIMER (9) and DACHLER (6). A simple formal mathematical solution for Case 4 is not known to the writer, and the formula shown in Fig. 12 is empirical and based on experiments by HARZA (12) and a graphical solution through radial flow nets by TAYLOR (28). The formulas for Cases 5 and 6 are derived by addition of the losses in piezometric pressure head outside the casing -- Cases 3 and 4 -- and in the soil inside the casing. The formulas are only approximately correct since it is assumed that the velocity of flow is uniformly distributed over the length and cross section of the soil plug. It is taken into consideration that for soil within the casing the vertical permeability is governing and may be different from that of the soil below the casing on account of soil disturbance and sedimentation.

The formula given for Case 7 is derived by DACHLER (6) on basis of flow from a line source for which the equipotential surfaces are semi-ellipsoids. Therefore, and as emphasized by DACHLER, the formula can provide only approximate results when it is applied to a cylindrical intake or well point. In Case 8 it is assumed that the flow lines are symmetrical with respect to a horizontal plane through the center of the intake, and the formula for Case 7 is then applied to the upper and lower halves of the intake. The accuracy of these formulas probably decreases with decreasing values of L/R and L/D. When these ratios are equal to unity, Cases 7 and 8 correspond to Cases 2 and 1, respectively, but furnish 13.4 per cent greater values for the flow. For large values of L/R and L/D the following simplified formulas may be used,

$$\text{CASE 7.} \qquad q = \frac{2\pi L k H}{\ln (2L/R)}$$

$$\text{CASE 8.} \qquad q = \frac{2\pi L k H}{\ln (2L/D)}$$

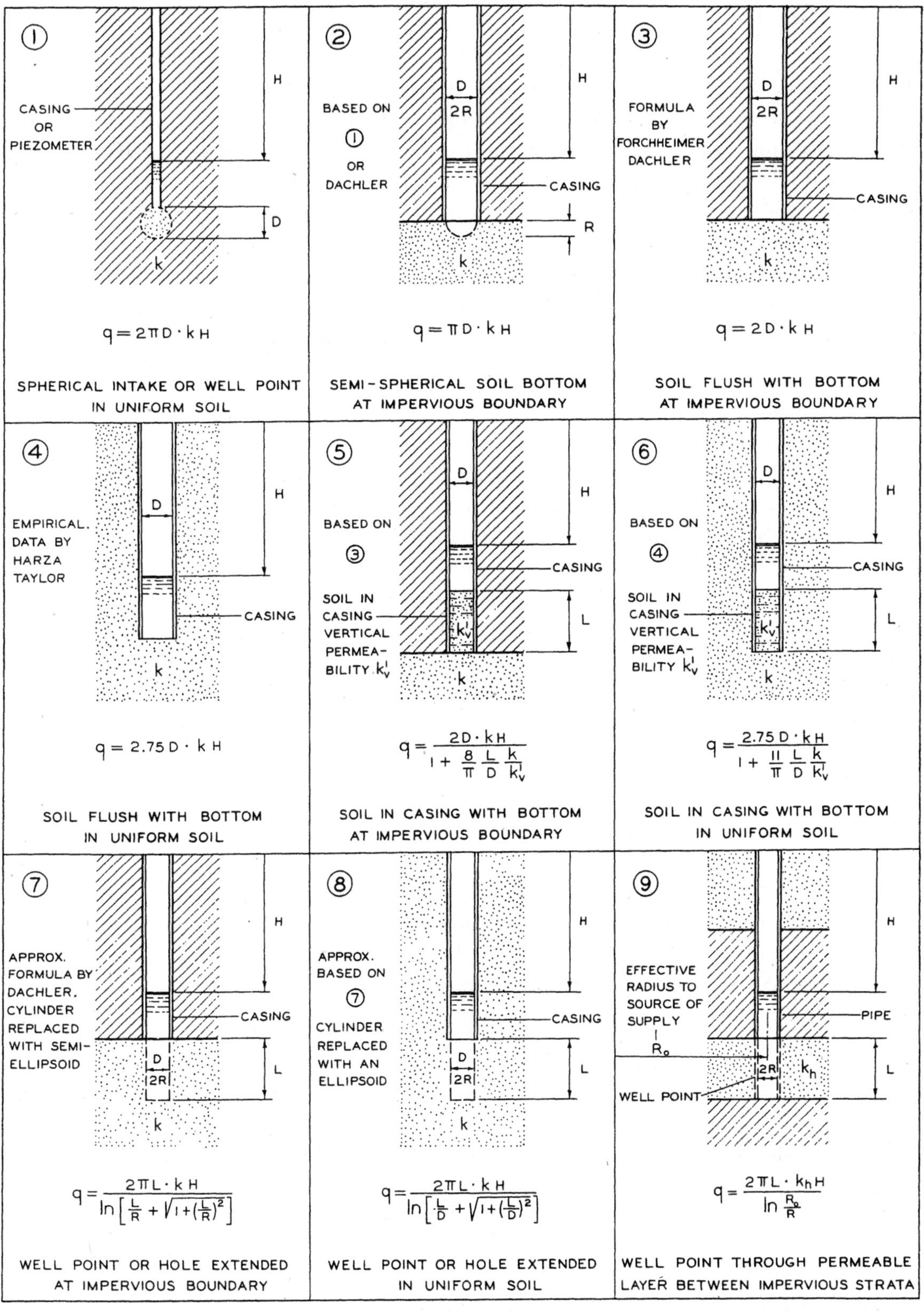

q = RATE OF FLOW IN cm^3/SEC, H = HEAD IN cm, k = COEF. OF PERMEABILITY IN cm/SEC, $\ln = \log_e$, DIMENSIONS IN cm.

CASES I TO 8: UNIFORM PERMEABILITY AND INFINITE DEPTH OF PERVIOUS STRATUM ASSUMED

FORMULAS FOR ANISOTROPIC PERMEABILITY GIVEN IN TEXT

Fig. 12. Inflow and shape factors

In this form the formulas were derived earlier by SAMSIOE (26). When L/R or L/D is greater than four, the error resulting from use of the simplified formulas is less than one per cent. In Case 9 the flow lines are horizontal and the coefficient of horizontal permeability, k_h, is governing. The effective radius, R_o, depends on the distance to the source of supply and to some extent on the compressibility of the soil, MUSKAT (22) and JACOB (17, 18). It may be noted that the simplified formula for Case 7 is identical with the formula for Case 9 when $R_o = 2L$. For flow through wells with only partial penetration of the pervious stratum, reference is made to MUSKAT (22) and the paper by MIDDLEBROOKS and JERVIS (21).

The assumptions, on which the derivation of the formulas in Fig. 12 are based, are seldom fully satisfied under practical conditions. It is especially to be noted that the horizontal permeability of soil strata generally is much larger than the vertical permeability. Correction of the formulas for the effect of anisotropic permeability is discussed in the following section. Even when such corrections are made, the formulas should be expected only to yield approximate results, since the soil strata are not infinite in extent and are rarely uniform in character. However and taking into consideration that the permeability characteristics of the soil strata seldom are accurately known in advance, the formulas are generally adequate for the purpose of preliminary design or selection of the proper type of pressure measuring installation, but the basic time lag obtained by the formulas should always be verified and corrected by means of field experiments.

Influence of Anisotropic Permeability

As first demonstrated by SAMSIOE (26) and later by DACHLER (6) for two-dimensional or plane problems of flow through soils, the influence of a difference between the coefficients of vertical and horizontal permeability of the soil, k_v and k_h, may be taken into consideration by multiplying all horizontal dimensions by the factor $\sqrt{k_v/k_h}$ and using the mean permeability $k_m = \sqrt{k_v \cdot k_h}$, whereafter formulas or flow nets for isotropic conditions may be used.

A general solution for three-dimensional problems and different but constant coefficients of permeability k_x, k_y, and k_z in direction of the coordinate axes is given by VREEDENBURG (31) and MUSKAT (22). With k_o an arbitrarily selected coefficient, the following transformation is made,

$$x' = x\sqrt{k_o/k_x} \qquad y' = y\sqrt{k_o/k_y} \qquad z' = z\sqrt{k_o/k_z} \tag{28}$$

and when an equivalent coefficient of permeability

$$k_e = k_o\sqrt{\frac{k_x}{k_o} \cdot \frac{k_y}{k_o} \cdot \frac{k_z}{k_o}} \tag{29}$$

is used, then the problem may be treated as if the conditions were isotropic. In applying these transformations to problems of flow through intakes or well points in soil with horizontal isotropic permeability, k_h, and vertical permeability k_v, it is convenient to use the following substitutions,

$$k_o = k_z = k_v \qquad k_x = k_y = k_h \qquad \text{and} \qquad m = \sqrt{k_h/k_v} \tag{30}$$

whereby the transformations assume the following form,

$$x' = x/m \qquad y' = y/m \qquad \text{or} \qquad r' = r/m \qquad \text{and} \qquad z' = z \tag{31}$$

$$k_e = k_v \sqrt{m^2 \cdot m^2} = k_v \cdot m^2 = k_h \tag{32}$$

That is, the problems can be treated as for isotropic conditions when the horizontal dimensions are divided by the square root of the ratio between the horizontal and vertical coefficients of permeability and the flow through the transformed well points is computed for a coefficient of permeability equal to k_h. When these transformations are applied to Cases 1 and 2 in Fig. 12, the sphere and semi-sphere become an ellipsoid, respectively a semi-ellipsoid, and formulas corresponding to those for Cases 7 and 8 should then be used. In Cases 5 and 6 the transformations should be applied only to flow through soil below the casing and not to soil within the casing. With introduction of the mean coefficient of permeability,

$$k_m = \sqrt{k_v \cdot k_h} = m \cdot k_v = k_h/m \tag{33}$$

the flow through the intakes and well points shown in Fig. 12 can be expressed as follows:

CASE 1. $$q = \frac{2\pi D k_h H}{\ln (m + \sqrt{1 + m^2})}$$

CASE 2. $$q = \frac{\pi D k_h H}{\ln (m + \sqrt{1 + m^2})}$$

CASE 3. $$q = 2 D k_m H$$

CASE 4. $$q = 2.75 D k_m H$$

CASE 5. $$q = \frac{2 D k_m H}{1 + \frac{8}{\pi}\frac{L}{D}\frac{k_m}{k_v'}}$$

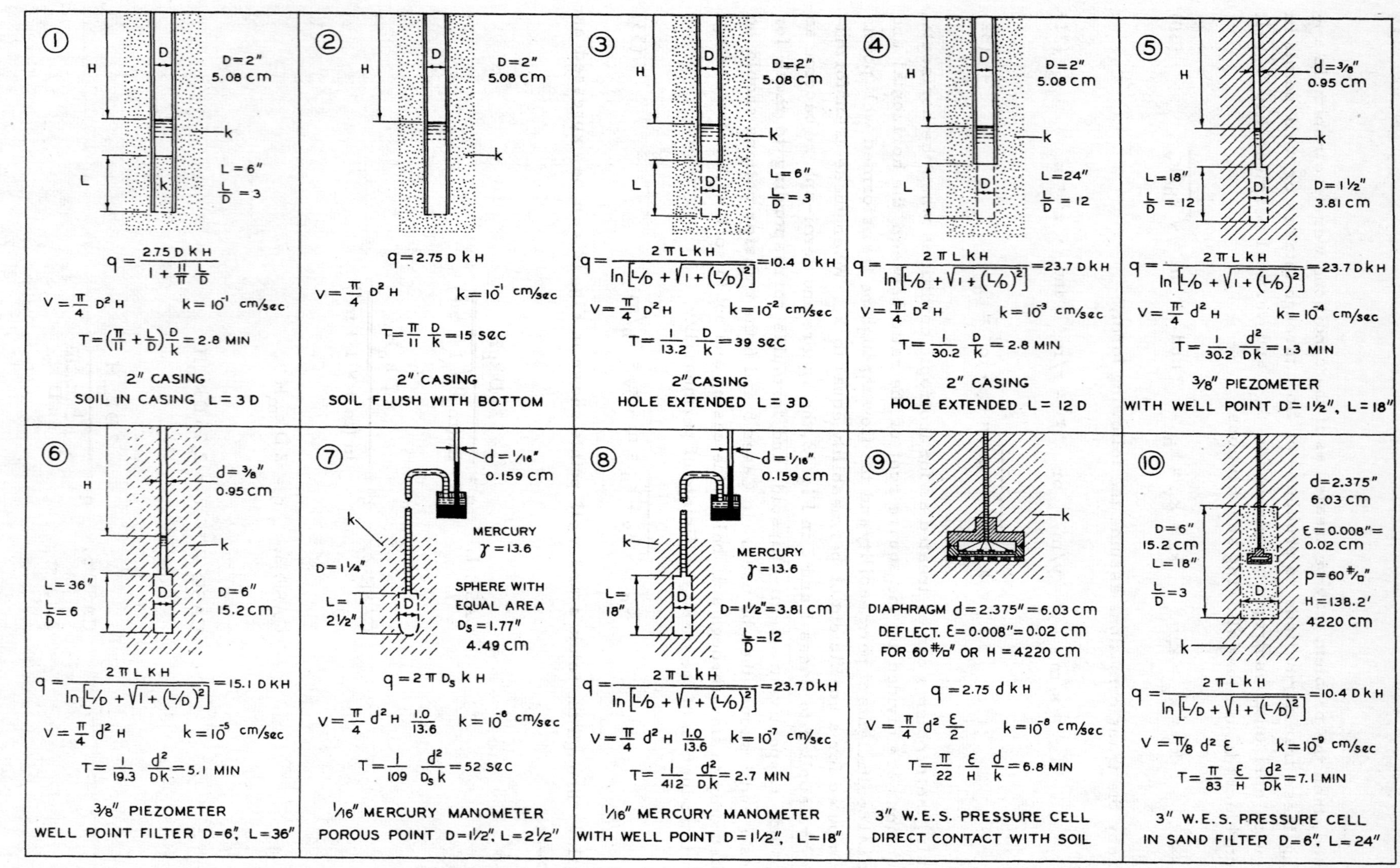

q = RATE OF FLOW FOR HEAD H, V = TOTAL VOLUME OF FLOW TO EQUALIZE PRESSURES, T = V/q = BASIC TIME LAG, ISOTROPIC SOIL CONDITIONS ASSUMED

Fig. 13. Examples of computation of basic time lag

CASE 6. $$q = \frac{2.75\, D\, k_m H}{1 + \frac{11}{\pi} \frac{L}{D} \frac{k_m}{k_v'}}$$

CASE 7. $$q = \frac{2\pi L\, k_h H}{\ln\left(mL/R + \sqrt{1 + (mL/R)^2}\right)}$$

CASE 8. $$q = \frac{2\pi L\, k_h H}{\ln\left(mL/D + \sqrt{1 + (mL/D)^2}\right)}$$

The formula for Case 9 in Fig. 12 is already expressed in terms of the horizontal permeability and is not affected by the transformation. The modified formulas for Cases 1 and 2 should be considered as being only approximately correct, and for isotropic conditions or $m = 1$ they yield 13.4 per cent greater values of flow than obtained by the basic formulas in Fig. 12. In Cases 7 and 8 and for large values of mL/R or mL/D the denominators may be replaced with $\ln(2mL/R)$, respectively $\ln(2mL/D)$.

Computation of Time Lag for Design Purposes

Examples of computation of the basic time lag, using the flow formulas in Fig. 12, are shown in Fig. 13. In all cases it is assumed that the soil is uniform and the permeability equal in all directions; this applies also to soil in the casing as shown in Case 1. The porous cup point in Case 7 is replaced with a sphere of equal surface area and the flow computed as through a spherical well point. This transformation furnishes a time lag which is slightly too small, since flow through a spherical well point is greater than through a point of any other shape and equal surface area. The pressure cell shown in Cases 9 and 10 is similar to the one described in a report by the WATERWAYS EXPERIMENT STATION (33). It may be noted that hydrostatic pressure cells with a diaphragm diameter of only 3/4 in. have been built and used successfully by the Waterways Experiment Station, and that a pressure cell with a diaphragm diameter of about one inch is described in a paper by BOITEN and PLANTEMA (1); see also Fig. 8-B. It is emphasized that the basic time lags for Cases 9 and 10 are computed on the assumption that there is no accumulation of gases below the diaphragm or in the sand filter; see discussion on pages 7 and 8.

A few general rules may be deduced from the examples shown in Fig. 13. In all cases the basic time lag is inversely proportional to the coefficient of permeability. When the ratio between the effective length and the diameter of the intake,

TIME LAGS		FOR 90 PERCENT EQUALIZATION = T_{90}										BASIC TIME LAG T
APPROXIMATE SOIL TYPE		SAND			SILT			CLAY				
COEFFICIENT OF PERMEABILITY IN cm/sec		10^{-1}	10^{-2}	10^{-3}	10^{-4}	10^{-5}	10^{-6}	10^{-7}	10^{-8}	10^{-9}	10^{-10}	10^{-6}
1	2" CASING - SOIL IN CASING, L = 3D = 6"	6^m	1^h	10^h	4.2^d							193^d
2	2" CASING - SOIL FLUSH BOTTOM CASING	0.6^m	6^m	1^h	10^h	4.2^d						17^d
3	2" CASING - HOLE EXTENDED, L = 3D = 6"		1.5^m	15^m	2.5^h	25^h	10^d					4.5^d
4	2" CASING - HOLE EXTENDED, L = 12D = 24"			6^m	1^h	10^h	4.2^d	42^d				47^h
5	3/8" PIEZOMETER WITH WELL POINT DIAMETER 1 1/2", LENGTH 18"				3^m	30^m	5^h	50^h	21^d			130^m
6	3/8" PIEZOMETER WITH WELL POINT AND SAND FILTER, D = 6", L = 36"					12^m	2^h	20^h	8.3^d	83^d		51^m
7	1/16" MERCURY MANOMETER, SINGLE TUBE WITH POROUS CUP POINT, D = 1 1/4", L = 2 1/2"		ONE-HALF OF VALUES FOR 1/16" MERCURY U-TUBE MANOMETER OR 4 1/2" BOURDON GAGE.				2^m	20^m	3.3^h	33^h	14^d	52^s
8	1/16" MERCURY MANOMETER, SINGLE TUBE WITH WELL POINT, D = 1 1/2", L = 18"							6^m	1^h	10^h	4.2^d	16^s
9	3" W. E. S. HYDROSTATIC PRESSURE CELL IN DIRECT CONTACT WITH SOIL								16^m	2.6^h	26^h	4^s
10	3" W. E. S. HYDROSTATIC PRESSURE CELL IN SAND FILTER, D = 6", L = 18"									16^m	2.6^h	0.4^s

SYMBOLS: s = SECONDS, m = MINUTES, h = HOURS, d = DAYS – ASSUMPTIONS: CONSTANT GROUND-WATER PRESSURE AND INTAKE SHAPE FACTOR, ISOTROPIC SOIL, NO GAS, STRESS ADJUSTMENT TIME LAG NEGLIGIBLE. THE COMPUTED TIME LAGS HAVE BEEN ROUNDED OFF TO CONVENIENT VALUES

Fig. 14. Approximate hydrostatic time lags

L/D, remains constant, the basic time lag is inversely proportional to the diameter of the intake and directly proportional to the cross-sectional area or the square of the diameter of the piezometer or manometer tube. When furthermore the diameters of the intake and piezometer are equal, Cases 1 to 4, the basic time lag is directly proportional to the diameter.

The results of the examples in Fig. 13 are summarized in a slightly different form in the last column in Fig. 14. The basic time lags are here given for a coefficient of permeability $k = 10^{-6}$ cm/sec., and these time lags may be used as a rating of the response to pressure changes for the various types of installations. For the examples shown in Figs. 13 and 14 this rating time lag varies from 193 days for a 2-in. boring with 6 in. of soil in the casing to 0.4 seconds for a 3-in. pressure cell placed in a 6-in. by 18-in. sand filter.

In the central part of Fig. 14 the basic time lags for various coefficients of permeability have been multiplied by 2.3 and indicate the time lags for 90 per cent equalization of the original pressure difference, which approximately is the time lag to be considered in practical operations. As mentioned on page 12, the time lag for 99 per cent equalization is twice as great as for 90 per cent equalization. According to data furnished the writer by Dr. A. WARLAM, the volume change of a 4-1/2-in. Bourdon pressure gage is 0.5 to 1.0 cm^3 for 1.0 kg/cm^2 change in pressure, or approximately half of that for a 1/16-in., single-tube, mercury manometer. Therefore, when the standpipe in Cases 7 and 8 is connected to a 4-1/2-in. Bourdon gage or to a double-tube mercury manometer with 1/16-in. inside diameter, the time lags will be about one-half those shown for a 1/16-in., single-tube mercury manometer. It is possible that the above mentioned volume change for a Bourdon pressure gage includes deformations of pliable rubber or plastic tube connections used in the experiments, and that the volume changes and corresponding time lags are smaller when rigid connections are used.

In all cases the computed time lags should be considered as being only approximate values, and they have been rounded off to convenient figures. The actual time lags may be influenced by several factors not taken into consideration in the above mentioned computations, such as stress adjustment and volume changes of soil and gases in the soil or pressure measuring system, sedimentation or clogging of the well point, filter, or surrounding soil, etc. The actual time lags may therefore be considerably greater or smaller than those indicated in Figs. 13 and 14, and special attention is called to the fact that the horizontal permeability of the soil, because of stratifications, often is many times greater than the vertical permeability as generally determined by laboratory tests and often used as a measure of the permeability of the soil stratum as a whole. Nevertheless, the examples shown in Figs. 13 and 14 will furnish some indication of the relative responsiveness of the various types of installations and permit a preliminary selection of the type suited for specific conditions and purposes.

Examples of Field Observations and Their Evaluation

Logan International Airport, Boston

Observations of pore-water pressures in the foundation soil of Logan International Airport at Boston are described in papers by CASAGRANDE (3) and GOULD (10). Most of the piezometers used were of the Casagrande type, shown diagrammatically in Fig. 15-A. The results of a series of time lag tests for piezometer C are

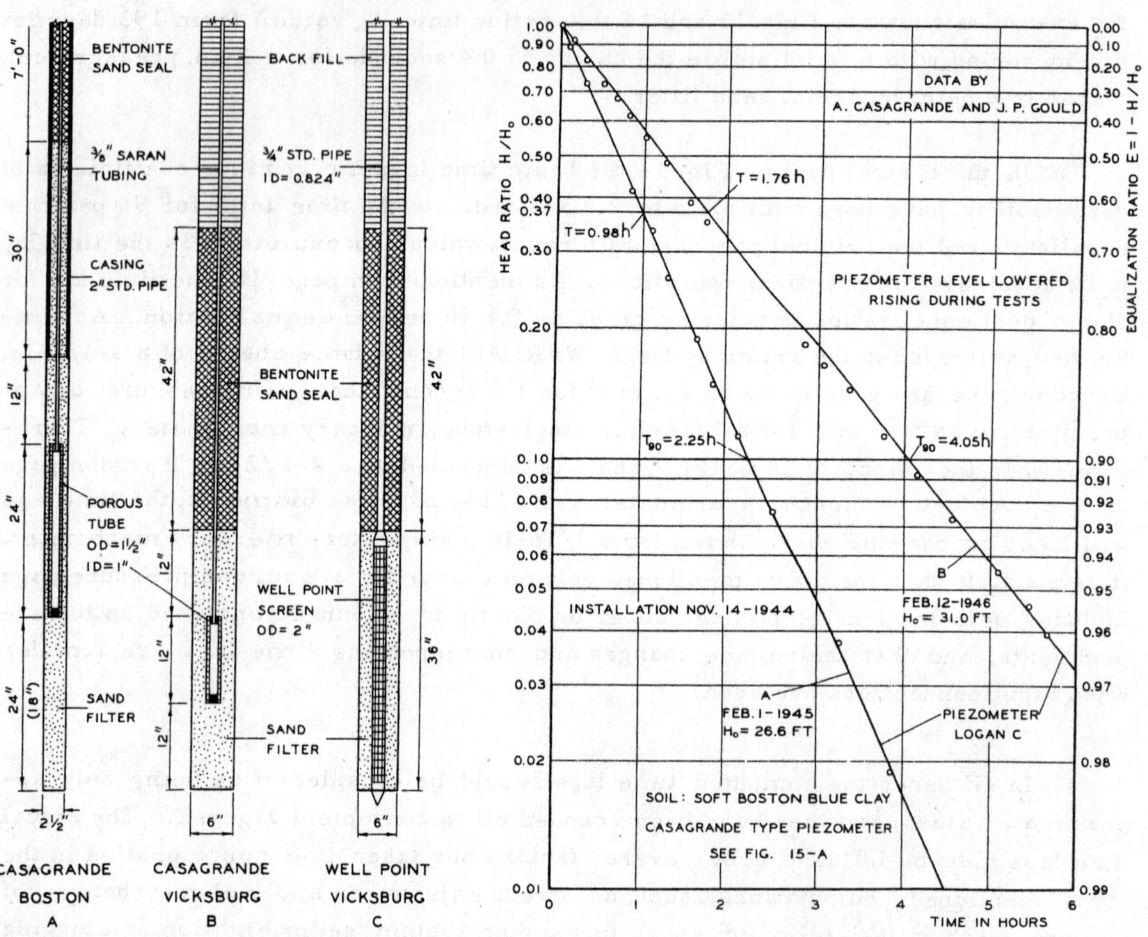

Fig. 15. Piezometers used in tests

Fig. 16. Time lag tests at Logan Airport, Boston

summarized in the paper by GOULD and further details were placed at the writer's disposal by CASAGRANDE. The filter or intake for this piezometer is installed in soft Boston Blue clay at a depth of 47 ft below the finished grade of fill.

The equalization diagrams obtained in two of the above mentioned tests, performed a year apart, are shown in Fig. 16. The first of these diagrams is straight, thereby indicating that the influence of transient stress adjustments or volume changes of the soil and gas in the voids is negligible; the basic time lag determined by this diagram is 0.98 hours. The equalization diagram obtained a year later shows a slight curvature and a basic time lag of 1.76 hours. Since the curvature is very small, the increase in time lag is probably caused by clogging of the porous tube or point and the filter. Estimates of the coefficients of permeability of the soil were obtained by means of new methods of settlement analysis, GOULD (10), and it was found that k_v varies between (28 and 35) x 10^{-9} cm/sec and k_h between (940 and 1410) x 10^{-9} cm/sec. Using the average values k_v = 31.5 x 10^{-9} cm/sec and k_h = 1175 x 10^{-9} cm/sec, the transformation ratio, m, is then

$$m = \sqrt{k_h/k_v} = \sqrt{37.5} = 6.1$$

The dimensions of the installation as given in the paper by GOULD are: diameter of filter D = 2.5 in. = 6.35 cm; length of filter L = 54 in. = 137.2 cm; inside diameter of piezometer d = 0.375 in. = 0.95 cm. The rate of flow for the active head H is obtained by the simplified formula for Case 8 on page 35

$$q = \frac{2\pi L k_h H}{\ln (2mL/D)}$$

and the total volume of flow required for equalization is,

$$V = \frac{\pi}{4} d^2 H$$

The basic time lag as determined by equation 3 is then,

$$T = \frac{V}{q} = \frac{d^2 \ln (2mL/D)}{8 L k_h} = \frac{0.95^2 \ln (263.6)}{8 \cdot 137.2 \cdot 1175} 10^9 = 3910 \text{ sec} = 1.09 \text{ hours} \qquad (34)$$

which agrees closely with the actual time lag, T = 0.98 hours.

Vicinity of Vicksburg, Mississippi

A preliminary series of comparative tests with various types of observation wells and piezometers has been performed by the WATERWAYS EXPERIMENT STATION (34). The wells and piezometers were installed behind the Mississippi River levees at two locations, Willow Point and Reid Bedford Bend. Time lag tests were made one to eight months after installation, and some of the equalization diagrams obtained in these tests are shown in Fig. 17. All the diagrams show a distinct initial curvature, and the period of observations was often too short, covering only the first and curved part of the diagrams. It was observed that gas emerged from some of the piezometers, and it is probable that the initial curvature of the equalization diagrams is caused by transient volume changes of gas bubbles accumulated in the

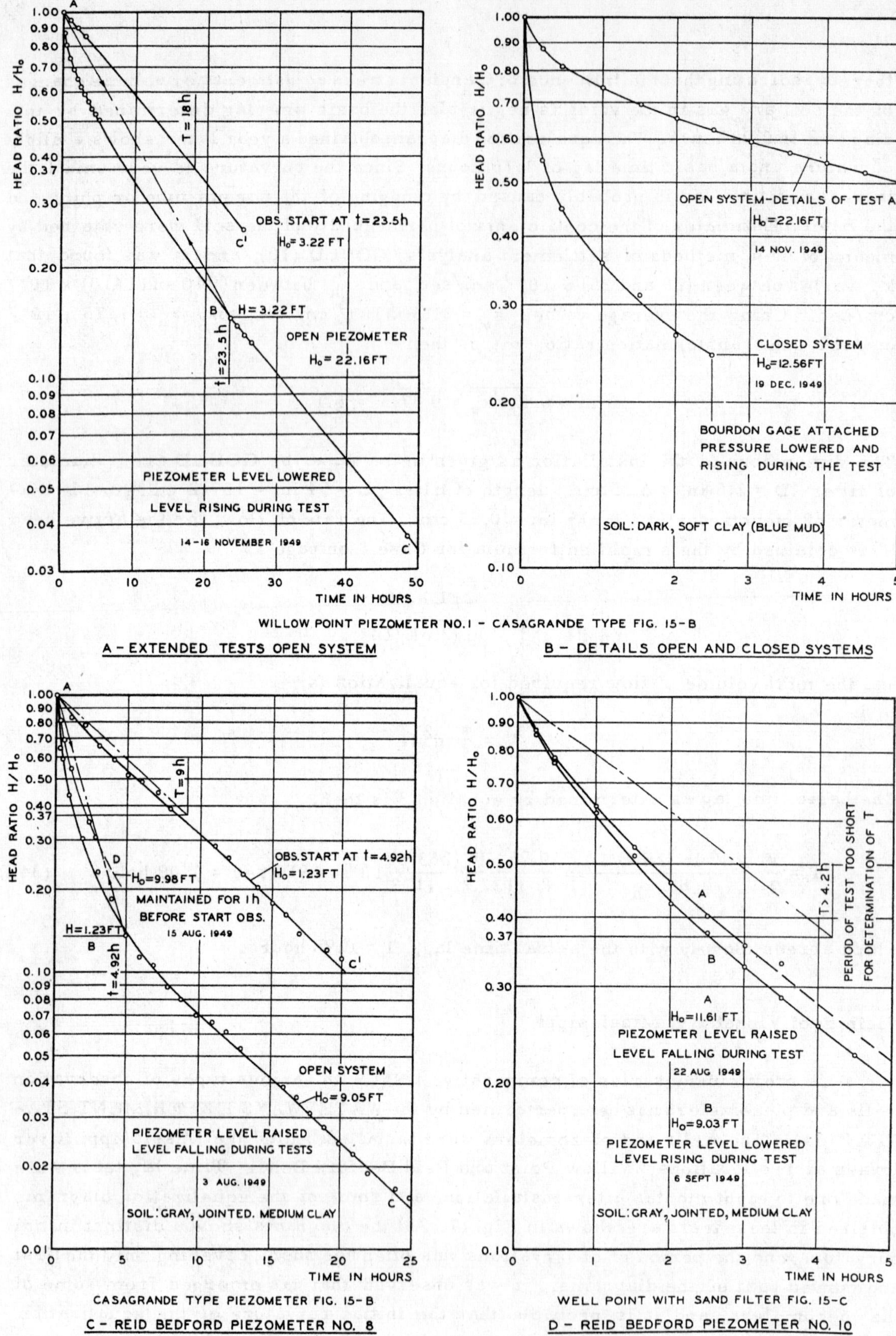

Fig. 17. Time lag tests by the Waterways Experiment Station, Vicksburg

soil near the well points or filters. The individual piezometers in the two groups are only 15 ft apart, and it is possible that time lag tests on a piezometer to a minor extent were influenced by flow to or from neighboring piezometers.

Laboratory tests on soil samples from the vicinity of the intakes for these installations indicate that the coefficients of vertical permeability vary between (10 and 150) x 10^{-9} cm/sec. Data on the coefficients of horizontal permeability are not available, and the soils at Reid Bedford Bend were jointed. Therefore, reliable estimates of the theoretical basic time lags cannot be made, but the basic time lags obtained by means of the equalization diagrams fall between those computed on basis of isotropic conditions and coefficients of permeability equal to the above mentioned upper and lower limits of the coefficients of vertical permeability.

Piezometer No. 1 at Willow Point is of the modified Casagrande type, Fig. 15-B, and is installed 92.5 ft below ground surface in a soft dark clay, locally known as "blue mud." The first part of the equalization diagram, Fig. 17-A, is curved but the lower part is fairly straight, possibly with a slight reverse curvature. If the observations are started 23.5 hours after the piezometer level was lowered, the diagram A-C' would be obtained; this diagram is parallel to the lower part, B-C, of the main diagram. As indicated on page 28, it is probable that the effective equalization diagram for the piezometer under normal operating conditions may be represented by a straight line through the origin and parallel to the lower and fairly straight part of the diagram obtained in a time lag test. By drawing such a line in Fig. 17-A, an effective basic time lag $T = 18$ hours is obtained.

In a second time lag test a Bourdon pressure gage was attached to the piezometer so that a closed system was formed. The pressure in the system was lowered by bleeding off a small amount of water, but the piezometric pressure level was above the gage level throughout the test. The equalization diagram obtained by observing the subsequent rise in pressure, Fig. 17-B, is lower and has considerably greater curvature than the one for an open system, which can be explained by the fact that the total amount of flow required for pressure equalization in the closed system is materially decreased, and the influence of volume changes of the gas bubbles and the soil consequently is greater.

Piezometer No. 8 at Reid Bedford is also of the modified Casagrande type and is installed 30 ft below ground surface in a gray, jointed, medium clay. The irregular, closely spaced joints in this clay are probably caused by previous drying, and the surfaces of some of the joints are covered with a thin layer of silt, but the joints at the depth of the piezometer intake are probably closed. The equalization diagram, A-B-C in Fig. 17-C, shows a pronounced initial curvature, but the lower part of the diagram is fairly straight. A straight line through the origin and parallel to the lower part of the diagram indicates an effective basic time lag $T = 9$ hours. In a second test the head -- $H_o = 9.98$ ft -- was maintained for one hour before the piezometer level was allowed to fall and the observations were started. The resulting equalization diagram, A-D, is above the first diagram and not so strongly curved. If the full head had been maintained for at least 24 hours, it is probable that a diagram similar to A-C' or the lower portion, B-C, of the main diagram would have been obtained.

Piezometer No. 10 at Reid Bedford is installed 15 ft from piezometer No. 8 and at the same depth. The sand filter has the same dimensions as for No. 8, but the porous tube is replaced with a well point screen extending through the whole length of the filter, and the piezometer proper is a 3/4-in. standard pipe; Fig. 15-C. Equalization diagrams were obtained for both falling and rising piezometer levels and are shown in Fig. 17-D. The periods of observation are too short for definite determination of the effective basic time lag, which is greater than 4.2 hours but probably smaller than the 9 hours obtained for piezometer No. 8. The initial curvature of the diagrams is considerably less than that of the diagrams for piezometer No. 8, which may be explained by the fact that the cross-sectional area of the piezometer pipe is $(0.824/0.375)^2 = 4.8$ times as great and that the influence of volume changes of soil and gas bubbles consequently is smaller. However, the basic time lag should then also be 4.8 times as great, since the dimensions of the sand filters for piezometers 8 and 10 are identical, but the equalization diagrams indicate a smaller time lag. This inconsistency may be due to local joints and other irregularities in soil conditions, but it is also probable that the well point screen is less subject to clogging than a porous tube, and that gases can escape more easily since the screen extends to the top of the sand filter.

Piezometer No. 11 at Reid Bedford consists of a 3/4-in. standard pipe with its lower end in the center of a sand filter at the same depth and with the same dimensions as the filters for piezometers 8 and 10. The time lag observations for piezometer No. 11 are incomplete but indicate that the effective basic time lag is at least 25 hours. It is probable that this increase in time lag, in comparison with piezometers 8 and 10, is caused by clogging of the sand in the immediate vicinity of the end of the pipe and of sand which may have entered the lower part of the pipe. Cleaning of the pipe and subsequent careful surging would undoubtedly decrease the time lag, but it is probable that clogging would re-occur in time.

Piezometer No. 15 at Reid Bedford is a 3/4-in. standard pipe with a solid drive point and a 4-in.-long, perforated section above the point. The pipe was driven to the same depth as the other piezometers and then withdrawn one foot. In a time lag test the piezometer level was raised 7.48 ft, and in 22.7 hours it fell only 0.12 ft. The lower part of the equalization diagram, during which the piezometer level fell from 7.45 ft to 7.36 ft in 17 hours, is fairly straight. For such a small drop in piezometer level it is better to compute the effective basic time lag by means of equation (5) than to determine it graphically; that is,

$$T = \frac{t}{\ln (H_o/H)} = \frac{17}{\ln (7.45/7.36)} = 1730 \text{ hours} = 72 \text{ days} \tag{35}$$

Because of the solid drive point, it is doubtful that withdrawal of the pipe for one foot materially affects flow to or from the perforated section, and the effective length of the latter would then be less than 4 in., even when the perforations remain open. However, it is possible that the perforations have been filled with molded soil during the driving, that a smear layer of remolded soil is formed around the pipe, and that this layer has covered the joints in the clay and decreased its effective permeability.

Determination of Permeability of Soil in Situ

Basic formulas

When the dimensions or shape factor, F, of a pressure measuring installation are known, it is theoretically possible to determine the coefficients of permeability of the soil in situ by field observations.

For constant head, H_c, and rate of flow, q, equation (1) yields,

$$k = \frac{q}{F H_c} \tag{36}$$

For variable head but constant ground-water level or pressure, the heads H_1 and H_2 corresponding to the times t_1 and t_2, and $A = \frac{\pi}{4} d^2$ the cross-sectional area of the standpipe, the following expression is obtained by means of equation (5),

$$t_2 - t_1 = T \left(\ln \frac{H_o}{H_2} - \ln \frac{H_o}{H_1}\right) = \frac{A}{F k} \ln \frac{H_1}{H_2}$$

$$k = \frac{A}{F (t_2 - t_1)} \ln \frac{H_1}{H_2} \tag{37}$$

This is also the formula commonly used for determination of coefficients of permeability in the laboratory by means of a variable head permeameter.

The simplest expression for the coefficient of permeability is obtained by determination of the basic time lag, T, of the installation and use of equation (3); that is,

$$k = \frac{A}{F T} \tag{38}$$

The shape factors, F, for various types of observation wells and piezometers may be obtained from the formulas in Fig. 12 and on pages 33 and 35 by eliminating the factors $(k H)$, respectively $(k_m H)$ or $(k_h H)$, from the right side of the

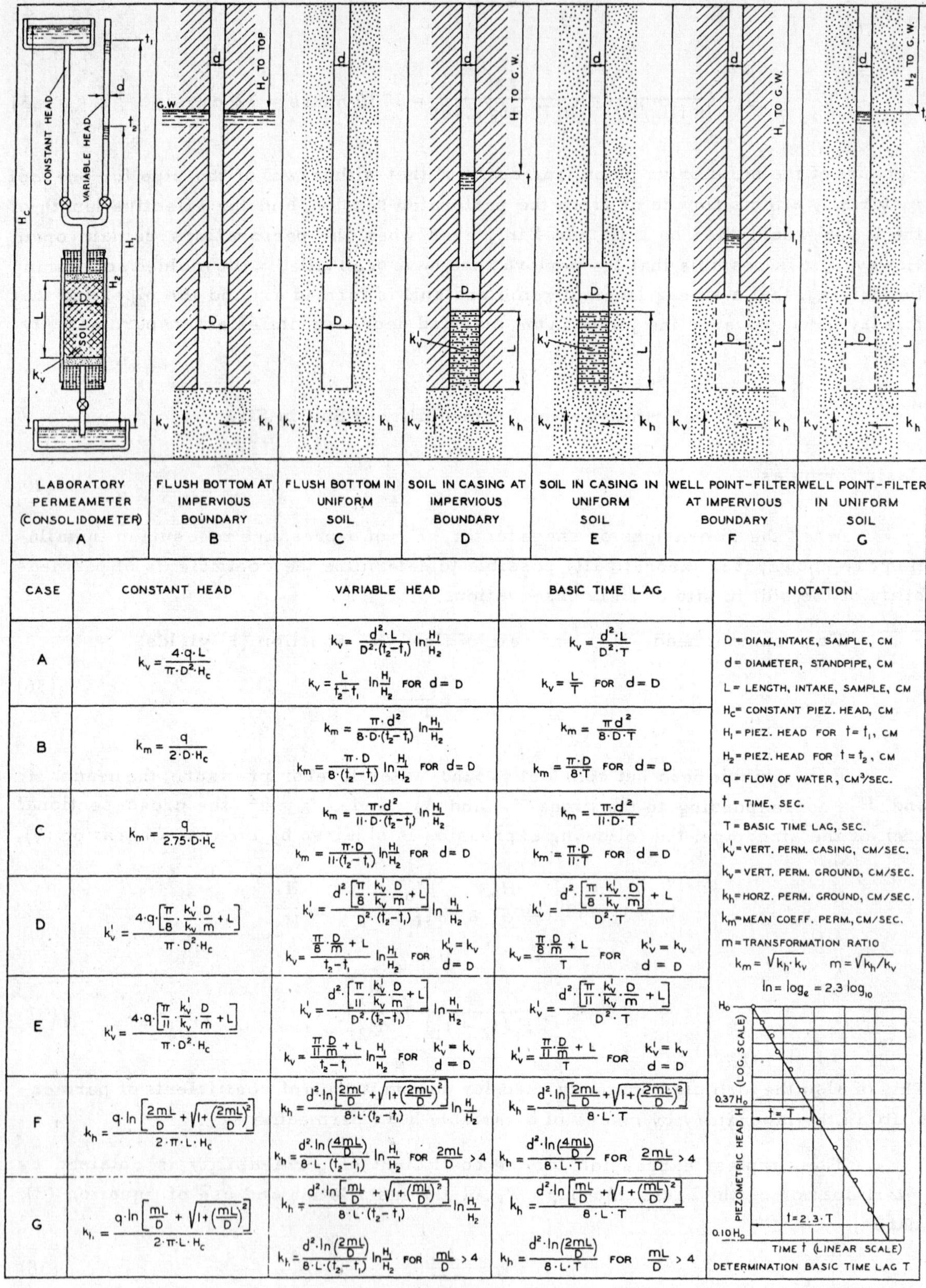

CASE	CONSTANT HEAD	VARIABLE HEAD	BASIC TIME LAG
A	$k_v = \frac{4 \cdot q \cdot L}{\pi \cdot D^2 \cdot H_c}$	$k_v = \frac{d^2 \cdot L}{D^2 \cdot (t_2 - t_1)} \ln \frac{H_1}{H_2}$ $k_v = \frac{L}{t_2 - t_1} \ln \frac{H_1}{H_2}$ FOR $d = D$	$k_v = \frac{d^2 \cdot L}{D^2 \cdot T}$ $k_v = \frac{L}{T}$ FOR $d = D$
B	$k_m = \frac{q}{2 \cdot D \cdot H_c}$	$k_m = \frac{\pi \cdot d^2}{8 \cdot D \cdot (t_2 - t_1)} \ln \frac{H_1}{H_2}$ $k_m = \frac{\pi \cdot D}{8 \cdot (t_2 - t_1)} \ln \frac{H_1}{H_2}$ FOR $d = D$	$k_m = \frac{\pi d^2}{8 \cdot D \cdot T}$ $k_m = \frac{\pi \cdot D}{8 \cdot T}$ FOR $d = D$
C	$k_m = \frac{q}{2.75 \cdot D \cdot H_c}$	$k_m = \frac{\pi \cdot d^2}{11 \cdot D \cdot (t_2 - t_1)} \ln \frac{H_1}{H_2}$ $k_m = \frac{\pi \cdot D}{11 \cdot (t_2 - t_1)} \ln \frac{H_1}{H_2}$ FOR $d = D$	$k_m = \frac{\pi \cdot d^2}{11 \cdot D \cdot T}$ $k_m = \frac{\pi \cdot D}{11 \cdot T}$ FOR $d = D$
D	$k_v' = \frac{4 \cdot q \cdot \left[\frac{\pi}{8} \cdot \frac{k_v'}{k_v} \cdot \frac{D}{m} + L\right]}{\pi \cdot D^2 \cdot H_c}$	$k_v' = \frac{d^2 \cdot \left[\frac{\pi}{8} \cdot \frac{k_v'}{k_v} \cdot \frac{D}{m} + L\right]}{D^2 \cdot (t_2 - t_1)} \ln \frac{H_1}{H_2}$ $k_v = \frac{\frac{\pi}{8} \cdot \frac{D}{m} + L}{t_2 - t_1} \ln \frac{H_1}{H_2}$ FOR $k_v' = k_v$, $d = D$	$k_v' = \frac{d^2 \cdot \left[\frac{\pi}{8} \cdot \frac{k_v'}{k_v} \cdot \frac{D}{m}\right] + L}{D^2 \cdot T}$ $k_v = \frac{\frac{\pi}{8} \cdot \frac{D}{m} + L}{T}$ FOR $k_v' = k_v$, $d = D$
E	$k_v' = \frac{4 \cdot q \cdot \left[\frac{\pi}{11} \cdot \frac{k_v'}{k_v} \cdot \frac{D}{m} + L\right]}{\pi \cdot D^2 \cdot H_c}$	$k_v' = \frac{d^2 \cdot \left[\frac{\pi}{11} \cdot \frac{k_v'}{k_v} \cdot \frac{D}{m} + L\right]}{D^2 \cdot (t_2 - t_1)} \ln \frac{H_1}{H_2}$ $k_v = \frac{\frac{\pi}{11} \cdot \frac{D}{m} + L}{t_2 - t_1} \ln \frac{H_1}{H_2}$ FOR $k_v' = k_v$, $d = D$	$k_v' = \frac{d^2 \cdot \left[\frac{\pi}{11} \cdot \frac{k_v'}{k_v} \cdot \frac{D}{m} + L\right]}{D^2 \cdot T}$ $k_v' = \frac{\frac{\pi}{11} \cdot \frac{D}{m} + L}{T}$ FOR $k_v' = k_v$, $d = D$
F	$k_h = \frac{q \cdot \ln \left[\frac{2mL}{D} + \sqrt{1 + \left(\frac{2mL}{D}\right)^2}\right]}{2 \cdot \pi \cdot L \cdot H_c}$	$k_h = \frac{d^2 \cdot \ln \left[\frac{2mL}{D} + \sqrt{1 + \left(\frac{2mL}{D}\right)^2}\right]}{8 \cdot L \cdot (t_2 - t_1)} \ln \frac{H_1}{H_2}$ $k_h = \frac{d^2 \cdot \ln \left(\frac{4mL}{D}\right)}{8 \cdot L \cdot (t_2 - t_1)} \ln \frac{H_1}{H_2}$ FOR $\frac{2mL}{D} > 4$	$k_h = \frac{d^2 \cdot \ln \left[\frac{2mL}{D} + \sqrt{1 + \left(\frac{2mL}{D}\right)^2}\right]}{8 \cdot L \cdot T}$ $k_h = \frac{d^2 \cdot \ln \left(\frac{4mL}{D}\right)}{8 \cdot L \cdot T}$ FOR $\frac{2mL}{D} > 4$
G	$k_h = \frac{q \cdot \ln \left[\frac{mL}{D} + \sqrt{1 + \left(\frac{mL}{D}\right)^2}\right]}{2 \cdot \pi \cdot L \cdot H_c}$	$k_h = \frac{d^2 \cdot \ln \left[\frac{mL}{D} + \sqrt{1 + \left(\frac{mL}{D}\right)^2}\right]}{8 \cdot L \cdot (t_2 - t_1)} \ln \frac{H_1}{H_2}$ $k_h = \frac{d^2 \cdot \ln \left(\frac{2mL}{D}\right)}{8 \cdot L \cdot (t_2 - t_1)} \ln \frac{H_1}{H_2}$ FOR $\frac{mL}{D} > 4$	$k_h = \frac{d^2 \cdot \ln \left[\frac{mL}{D} + \sqrt{1 + \left(\frac{mL}{D}\right)^2}\right]}{8 \cdot L \cdot T}$ $k_h = \frac{d^2 \cdot \ln \left(\frac{2mL}{D}\right)}{8 \cdot L \cdot T}$ FOR $\frac{mL}{D} > 4$

NOTATION

D = DIAM., INTAKE, SAMPLE, CM
d = DIAMETER, STANDPIPE, CM
L = LENGTH, INTAKE, SAMPLE, CM
H_c = CONSTANT PIEZ. HEAD, CM
H_1 = PIEZ. HEAD FOR $t = t_1$, CM
H_2 = PIEZ. HEAD FOR $t = t_2$, CM
q = FLOW OF WATER, CM^3/SEC.
t = TIME, SEC.
T = BASIC TIME LAG, SEC.
k_v' = VERT. PERM. CASING, CM/SEC.
k_v = VERT. PERM. GROUND, CM/SEC.
k_h = HORZ. PERM. GROUND, CM/SEC.
k_m = MEAN COEFF. PERM., CM/SEC.
m = TRANSFORMATION RATIO

$k_m = \sqrt{k_h \cdot k_v}$ $\quad m = \sqrt{k_h / k_v}$

$\ln = \log_e = 2.3 \log_{10}$

ASSUMPTIONS

SOIL AT INTAKE, INFINITE DEPTH AND DIRECTIONAL ISOTROPY (k_v AND k_h CONSTANT) - NO DISTURBANCE, SEGREGATION, SWELLING OR CONSOLIDATION OF SOIL - NO SEDIMENTATION OR LEAKAGE - NO AIR OR GAS IN SOIL, WELL POINT, OR PIPE - HYDRAULIC LOSSES IN PIPES, WELL POINT OR FILTER NEGLIGIBLE

Fig. 18. Formulas for determination of permeability

equations. Explicit formulas for determination of coefficients of permeability by constant head, variable head, and basic time lag tests with permeameters and various types of borings and piezometers are summarized in Fig. 18. For a permeameter, Case A, the rate of flow for the head H is $q = \frac{\pi}{4} D^2 kH/L$, or $F = \frac{\pi}{4} D^2/L$. In cases D and E the coefficient of vertical permeability of soil in the casing is usually governing, and the equations have been solved for this coefficient and appear in a form slightly different from that corresponding to Cases (5) and (6) in Fig. 12 and on pages 33 and 35. Simplified formulas for $d = D$, $k'_v = k_v$, and the ratio (mL/D) greater than 2 or 4, are given below the main formulas in each case.

The basic time lag is easily determined by means of an equalization diagram -- or a semilogarithmic plot of time versus head -- as the time T corresponding to $H = 0.37H_o$; i.e., $\ln (H_o/H) = 1$. The work involved in plotting the diagram is offset by simpler formulas for computing the coefficient of permeability, compared to the formulas for variable head, and the diagram has the great advantage that it reveals irregularities caused by volume changes or stress adjustment time lag and permits easy advance adjustment of the results of the tests. It is emphasized that the above mentioned methods and formulas are applicable only when the basic assumptions for the theory of time lag, page 9, are substantially correct.

Examples of applications

The following dimensions apply to the permeability tests on Atlantic muck, Fig. 10: D = 4.25 in. = 10.8 cm; L = 0.87 in. = 2.21 cm; d = 0.30 cm. The basic time lag obtained from the probable normal diagram in Fig. 10 is T = 178 minutes, and hence

$$k_v = \frac{d^2 L}{D^2 T} = \frac{0.30^2 \cdot 2.21}{10.8^2 \cdot 178 \cdot 60} = 159 \times 10^{-9} \text{ cm/sec.}$$

The slope of the lower parts of the equalization diagrams corresponds to a basic time lag T = 210 min and $k_v = 135 \times 10^{-9}$ cm/sec. Larger basic time lags and correspondingly smaller values of the coefficients of permeability were obtained in similar tests with other undisturbed samples of Atlantic muck.

The first test with piezometer C at Logan International Airport, Fig. 16, gave a basic time lag T = 0.98 hours = 3530 seconds. With $k_v = 31.5 \times 10^{-9}$ cm/sec and the dimensions given on page 39, the coefficient of horizontal permeability of Boston Blue clay may be determined as follows:

$$k_h = \frac{d^2 \ln (2mL/D)}{8 \cdot L \cdot T} = \frac{0.95^2 \ln (m \cdot 43.2)}{8 \cdot 137.2 \cdot 3530} = 233.5 \cdot 10^{-9} \cdot \ln (m \cdot 43.2)$$

This equation may be solved by estimating the value of $m = \sqrt{k_h/k_v}$ and successive corrections, which yield

$$k_h = 1310 \times 10^{-9} \text{ cm/sec} \quad \text{and} \quad k_h/k_v = 1310/31.5 = 41.6$$

These values lie within the limits obtained by other methods, GOULD (10), and discussed on page 39.

The second time lag test with piezometer C gave T = 1.76 hours and indicated thereby that clogging of the porous tube had taken place. Therefore, reliable values of the coefficient of permeability can no longer be obtained by means of this installation. This applies also to the installations at Willow Point and Reid Bedford, Fig. 17, since the strong initial curvature of the equalization diagrams indicates large transient volume changes and probably accumulation of gas bubbles in the sand filters and surrounding soil with a consequent decrease in permeability of this soil and increase in time lag.

Advantages and limitations

Observation of the basic time lag for borings and piezometers provides theoretically a very simple method for determination of the permeability of soil in situ, even for anisotropic conditions. However, many difficulties are encountered in the practical execution of such permeability tests and evaluation of the results obtained, since the latter are subject to the same sources of error as those of pressure observations discussed in Part I, and since methods of correction for the influence of some of these sources of error have not yet been devised.

The shape factor of the installation must be computed, but some of the formulas in Figs. 12 and 18 are empirical or only approximately correct, and they are all based on the assumption of infinite thickness of the soil layer in which the well point or intake is installed. When sand filters are used, the dimensions must be determined with greater accuracy than is required for pressure observations. The greatest part of the hydraulic friction losses occur near the intake, and the results of a test consequently indicate the permeability of the soil in the immediate vicinity of the intake. Misleading results are obtained when the permeability of this soil is changed by disturbance of the soil during advance of a bore hole or installation of filters or well points. Leakage, clogging of the intake or removal of fine-grained particles from the surrounding soil, and accumulation of gases near the intake or within the pressure measuring system may render the installation wholly unreliable as a means of determining the permeability of the undisturbed soil. Gas bubbles in the soil near the intake will decrease the permeability, cause curvature of the equalization diagram, and increase the effective basic time lag. Gas bubbles in a coarse-grained filter or within the pressure measuring system will not cause any appreciable curvature of the equalization diagram but will materially decrease the slope of the diagram and increase the basic time lag so that too small values of the coefficients of permeability are obtained.

Many of the above mentioned sources of error are avoided in the commonly used pumping tests, during which the shape of the draw-down curve is determined for a given rate of flow, but such tests are expensive and time consuming. Determination of the permeability of soil in situ by means of the time lag of observation wells and piezometers has so many potential advantages that it is to be hoped that systematic research will be undertaken in an effort to develop reliable methods of calibration or experimental determination of shape factors, and also of methods for detection, correction, or elimination of the various sources of error in the observations. Until such research is successfully completed, it is advisable to exert great caution in the practical application of the results obtained by the method.

REFERENCES

1. BOITEN, R. G. and PLANTEMA, G., An Electrically Operating Pore Water Pressure Cell. Proc. Second Int. Conf. Soil Mech., Rotterdam 1948, v. 1, p. 306-309.

2. CASAGRANDE, A., Piezometers for Pore Pressure Measurement in Clay. Memorandum. Graduate School of Engineering, Harvard University, Cambridge, Mass., July 1946.

3. CASAGRANDE, A., Soil Mechanics in the Design and Construction of the Logan International Airport. Journal Boston Society of Civil Engineers, April 1949, v. 36, p. 192-221.

4. CHRISTIANSEN, J. E., Effect of Trapped Air on Soil Permeability. Soil Science, 1944, v. 58, p. 355-365.

5. CUPERUS, J. L. A., Permeability of Peat by Water. Proc. Second Int. Conf. Soil Mech., Rotterdam 1948, v. 1, p. 258-264.

6. DACHLER, R., Grundwasserströmung (Flow of Ground Water). 141 p. Julius Springer, Wien 1936.

7. DE BEER, E. and RAEDSCHELDERS, H., Some Results of Waterpressure Measurements in Clay Layers. Proc. Second Int. Conf. Soil Mech., Rotterdam 1948, v. 1, p. 294-299.

8. FIELDS, K. E. and WELLS, W. L., Pendleton Levee Failure. Paper contains drawing and description of piezometer and pressure gage, p. 1402. A second pore-water pressure gage is described in a discussion by R. B. Peck, p. 1415. Trans. Am. Soc. Civ. Eng., 1944, v. 109, p. 1400-1413. Discussions p. 1414-1429.

9. FORCHHEIMER, PH., Hydraulik (Hydraulics), 3 ed., 596 p. B. G. Teubner, Leipzig und Berlin, 1930.

10. GOULD, J. P., Analysis of Pore Pressure and Settlement Observations at Logan International Airport. Harvard Soil Mechanics Series No. 34, Department of Engineering, Harvard University, Cambridge, Mass., December 1949.

11. HARDING, H. J. B., Some Recorded Observations of Ground Water Levels Related to Adjacent Tidal Movements. Proc. Second Int. Conf. Soil Mech., Rotterdam 1948, v. 4, p. 317-323.

12. HARZA, L. F., Uplift and Seepage under Dams. Paper contains results of model tests to determine inflow into wells. Trans. Am. Soc. Civ. Eng., 1935, v. 100, p. 1352-1385. Discussions p. 1386-1406.

13. HUIZINGA, T. K., Measurement of Pore Water Pressure. Proc. Second Int. Conf. Soil Mech., Rotterdam 1948, v. 1, p. 303-306.

14. HVORSLEV, M. J., Discussion of paper by George L. Freeman, The Application of Soil Mechanics in the Building of the New York Worlds Fair. Discussion deals with the need of and methods for determination of pore-water pressures. Presented at the ASCE Meeting, New York, Jan. 1940.

15. HVORSLEV, M. J., Survey of Methods for Field and Laboratory Investigation of Soils. Report to the Waterways Experiment Station, Corps of Engineers, Vicksburg, Miss., June 1941.

16. HVORSLEV, M. J., Subsurface Exploration and Sampling of Soils for Civil Engineering Purposes. Report on a research project of the American Society of Civil Engineers, sponsored by the Engineering Foundation, Harvard University, Waterways Experiment Station. Final printing 521 p., November 1949. Obtainable through the Engineering Foundation, New York.

17. JACOB, C. E., On the Flow of Water in an Elastic Artesian Aquifer. Trans. American Geophysical Union, 1940, v. 21, p. 574-586.

18. JACOB, C. E., Drawdown Test to Determine Effective Radius of Artesian Well. Trans. Am. Soc. Civ. Eng., 1947, v. 112, p. 1047-1064. Discussions p. 1065-1070.

19. KIRKHAM, D., Proposed Method for Field Measurement of Permeability of Soil below the Water Table. Proc. Soil Science Soc. Amer., 1947, v. 12, p. 54-60.

20. MEINZER, O. E., Hydrology. Volume IX of Physics of the Earth. 712 p. McGraw-Hill Book Co., New York and London, 1942.

21. MIDDLEBROOKS, T. A. and JERVIS, W. H., Relief Wells for Dams and Levees. Trans. Am. Soc. Civ. Eng., 1947, v. 112, p. 1321-1338. Discussions p. 1339-1402.

22. MUSKAT, M., The Flow of Homogeneous Fluids through Porous Media. 737 p. McGraw-Hill Book Co., New York and London, 1937.

23. RICHARDS, L. A. and GARDNER, W., Tensiometer for Measuring the Capillary Tension of Soil Water. Jour. Amer. Soc. Agron., 1936, v. 28, p. 352-358.

24. RICHARDS, L. A., Soil Moisture Tensiometer Materials and Construction. Soil Science, 1942, v. 53, p. 241-248.

25. ROGERS, W. S. A., A Soil Moisture Meter. Jour. Agr. Science, 1935, v. 25, p. 326-343.

26. SAMSIOE, A. F., Einfluss von Rohrbrunnen auf die Bewegung des Grundwassers (Influence of Wells on the Movement of Ground Water). Zeitschrift für angewandte Mathematik und Mechanik, April 1931, v. 11, p. 124-135.

27. SPEEDIE, M. G., Experience Gained in the Measurement of Pore Pressures in a Dam and its Foundation. Proc. Second Int. Conf. Soil Mech., Rotterdam 1948, v. 1, p. 287-294.

28. TAYLOR, D. W., Fundamentals of Soil Mechanics. 700 p. John Wiley & Sons, New York, 1948.

29. TERZAGHI, K., Measurement of Pore Water Pressure in Silt and Clay. Civil Engineering, 1943, v. 13, p. 33-36.

30. TOLMAN, C. F., Ground Water. 593 p. McGraw-Hill Book Co., New York and London, 1937.

31. VREEDENBURG, G. G. J., On the Steady Flow of Water Percolating through Soils with Homogeneous-Anisotropic Permeability. Proc. First Int. Conf. Soil Mech., Harvard 1936, v. 1, p. 222-225.

32. WALKER, F. C. and DAEHN, W. W., Ten Years of Pore Pressure Measurements. Proc. Second Int. Conf. Soil Mech., Rotterdam 1948, v. 3, p. 245-250.

33. WATERWAYS EXPERIMENT STATION, CORPS OF ENGINEERS, Soil Pressure Cell Investigation. Interim Report. The report also contains drawings and descriptions of hydrostatic pressure cells. Technical Memorandum No. 210-1, Vicksburg, Miss., 1944.

34. WATERWAYS EXPERIMENT STATION, CORPS OF ENGINEERS, Investigation of Pore Pressure Measuring Devices. Inter-office report in preparation; deals principally with various types of observation wells and piezometers.

THE GEOLOGICAL SOCIETY OF AMERICA
ENGINEERING GEOLOGY (BERKEY) VOLUME NOVEMBER 1950

MECHANISM OF LANDSLIDES

BY KARL TERZAGHI
Harvard University, Cambridge, Mass.

CONTENTS

ILLUSTRATIONS

VARIETIES OF SLOPE MOVEMENTS

DEFINITIONS

The term *landslide* refers to a rapid displacement of a mass of rock, residual soil, or sediments adjoining a slope, in which the center of gravity of the moving mass advances in a downward and outward direction. A similar movement proceeding at an imperceptible rate is called *creep*. The velocity of the masses involved in a typical landslide increases more or less rapidly from almost zero to at least 1 foot per hour. Then it again decreases to a small value. By contrast, typical creep is a continuous movement which proceeds at an average rate of less than 1 foot per decade. Higher rates of creep movements are rather uncommon.

Slides on the slopes of man-made cuts are sometimes referred to as *slope failures.* In this paper, for the sake of convenience, the term landslide will be retained for failure on both natural slopes and slopes of cuts.

DIFFERENCES BETWEEN LANDSLIDES AND CREEP

A landslide is an event which takes place within a short period of time as soon as the stress conditions for the failure of the ground located beneath the slope are satisfied. By contrast, creep is a more or less continuous process. A landslide represents the movement of a relatively small body of material with well-defined boundaries, whereas creep may involve the ground located beneath all the slopes in a whole region and no sharp boundary exists between stationary and moving material. Most landslides are produced only by the force of gravity, whereas creep movements can also be due to the combined action of the force of gravity and various other agents.

Within the zone of seasonal changes of moisture and temperature, at least part of the horizontal component of the ground movement is produced by thermal expansion and contraction, by swelling and shrinking, freezing and thawing, and other seasonal processes (Sharpe and Dosch, 1942, p. 46–48). These processes result in a downhill movement of a sheet of earth, with depth equal to or smaller than the depth of seasonal variations in the condition of the ground. Below this depth, creep can be produced only by the force of gravity, unaided by other agents. Since the force of gravity does not change with the seasons, the rate of the resulting gravity creep is fairly constant. This type of creep will be referred to as *continuous creep* in contrast to the *seasonal creep* which can occur in the top layer of the ground only. In the following comparison between landslides and creep the seasonal creep is not considered.

If the difference between landslides and continuous creep resided only in the velocity of the movement, it would hardly be justifiable to consider landslides and creep as different types. However, experience has disclosed another much more significant distinguishing feature. It consists in the difference between the pattern of the deformations produced by these two processes. Patterns of equally different character can be obtained in the laboratory by loading tests on blocks of a nonhomogeneous mass composed of materials such as asphalt which have the properties of a very viscous liquid. A heavy load on such a block causes an almost sudden failure by separation along one or more surfaces of rupture, which cut across the boundaries between the strong and weak portions of the block. By contrast, the instantaneous effect of a very small load on a similar block is imperceptible; but if the load acts on the block for many years the block undergoes very important and very intricate deformations which reflect all the details of the internal structure of the block.

The striking difference between the resulting deformation patterns is due to the fact that the laws which determine the deformations are as different as those of hydraulics and of the mechanics of elastic solids. If a system composed of strata with very different elastic properties is acted upon for a long time by shearing stresses which are smaller than the average shearing strength of the system, the most rigid members only will behave like solids, whereas the balance will be deformed like a very viscous liquid. The deformation of the system will be like that of a sheet of asphalt containing layers of a brittle material. As the shearing stresses increase, a higher and higher percentage of the members of the system will perform like solids, and, if the stresses are rapidly increased to the point of failure, the entire system will behave like a solid.

The load per unit of area under which a block fails by shear or splitting is commonly known as the *compressive strength* of the material of which the block is composed. The load at which creep begins is very much smaller. It is called the *fundamental strength* (Griggs, 1936, p. 364). As long as the shearing stresses in the material beneath a slope are smaller than the "fundamental" shearing resistance of the material, the slope is at rest. If they exceed this value the slope creeps, and if they become equal to the stress required to produce a shear failure a landslide occurs.

The characteristics of typical creep deformations are illustrated by Figure 1 and Plate 1. Figure 1 of Plate 1 is an oblique view of two of the vertical sides of a prismatic block that was cut out of a layer of soft Devonian clay at the bottom of a test pit at the site of the hydro-electric power development Swir III east of Leningrad. The layer had a thickness of about 2 feet, and it was located between two very much thicker layers of stiff clay. It was distinctly stratified and contained several thin seams of brittle clay.

The clay strata are almost horizontal, and on the vertical walls of the test pits it could be seen that the thick layers of stiff clay were almost intact. By contrast, the soft clay strata exhibited shear deformations strikingly reminiscent of alpine tectonics. All the thin layers of brittle clay were intensely folded and broken up into small fragments, displaced with reference to each other by sliding along reverse faults, but all the faults died out in the plastically deformed material within a short distance of the sheared-off competent layers. Figure 1 of Plate 1 shows one of the brittle layers which was folded and broke. Another one, visible at the upper edge of the photograph, was relatively intact because it was located close to the boundary between soft and stiff clay.

The intact condition of the stiff clay strata indicated that the force which produced the intense deformation of the soft ones has acted in a horizontal direction. It is also apparent that the shearing stresses, due to the external force, were considerably smaller than the shearing resistance of the soft clay. Otherwise the force would have caused a slip along a smooth surface of sliding, parallel to the bedding planes, and not an intricate deformation. When shear tests were performed *in situ*, on the bottom of the test pits, the surfaces of sliding were almost plane, and the deformations of the clay on both sides of the surface were imperceptible, although the displacements along the plane exceeded 1 inch.

The nature of the force which deformed the soft clay is a matter of speculation. In this connection it is possibly significant that the axes of the folds in the clay are approximately at right angles to the direction in which the Pleistocene ice sheet, coming from the Scandinavian shield, advanced over the top surface of the clay deposit. This fact suggests that the folds may have been produced by the movement of the ice.

Another example of the effect of large-scale creep has been disclosed by quarrying operations in the Northampton ironstone field in central England and during the excavation of dam trenches in this district (Hollingworth *et al.*, 1944). The stresses which caused the creep were produced by the formation of shallow erosion valleys. The local removal of load resulting from the erosion induced a slow flow of stiff Lias clay toward the bottom of the valleys. Beneath the bottom, the clay was squeezed up into "valley bulges," and the strata above the bulging clay were intensely folded, broken, and sheared, although there is no evidence that the shearing stresses in the deformed material ever exceeded a small fraction of its average shearing strength. Figure 1 shows the deformation of thin, hard layers, separating thick beds of Lias clay beneath the bottom of one of the valleys.

Figure 2 of Plate 1 is a view of an exposure on the left bank of the Sulak River in Daghestan, close to the point where the river emerges from a gorge cut across a brachy-anticline. The anticline is composed of Upper Cretaceous limestones interbedded with shales. The Tertiary strata (limestone and phyllite), located north of the anticline, appeared to be almost horizontal. However, they had advanced toward the anticline in a southerly direction. According to the Soviet geologists who worked in this region, the horizontal displacement east of the anticline exceeded 5 miles. Since the brachy-anticline was located in the path of the movement, the advancing Tertiary strata were rolled under at the foot of the anticline, as shown in the photograph. No rapid displacement could possibly produce such an effect.

The known manifestations of creep suggest that creep is nothing but a small-scale and superficial replica of what takes place at depth under the influence of tectonic forces. Both processes go on continously over vast areas, and the mechanism of both is essentially the same although the driving forces are different.

From time to time, at geographically widely separated points, the intensity of the shearing stresses, within a zone of creep, becomes equal to the shearing resistance of the material, or the shearing resistance of the material decreases until it becomes equal to the shearing stresses. Under such circumstances, a landslide takes place.

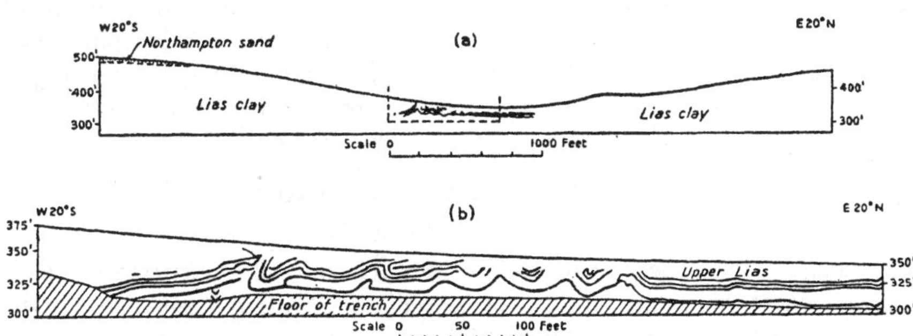

FIGURE 1.—*Hollowell valley near Northampton, England*

(a) Geologic cross section showing creep deformation of Lias clay beneath valley bottom. Dashed line indicates boundaries of section which was exposed in trench for foundation of Hollowell dam. (b) Large-scale view of exposure in dam trench. (By permission of Mr. E. Sandeman, M. Inst. C. E.)

Creep, like tectonic movements, may lead to intricate deformations, revealing and accentuating the resistance pattern of the masses subject to deformation. In contrast to creep deformations, landslides are characterized by sliding movements along well-defined surfaces, which cut across the boundaries between competent and incompetent strata like the shear planes produced by a shear test of short duration. In spite of these radical differences, no sharp boundaries between these two groups of slope movements can be drawn. On any one slope located above unconsolidated material such as residual soil or sedimentary clay, creep may develop into a slide, and the slide may be followed by creep in the material which has moved out of the slope (Sharpe and Dosch, 1942).

Both creep and landslides require the attention of the engineer. Creep deformation of strata, located beneath the bottom of erosion valleys like those shown in Figure 1, may have a profound effect on the foundation conditions at the site of proposed storage dams (Lapworth, 1911). Lugeon (1922) has published a brief account of repeated failures of a water main near Lausanne due to the rapid creep of weathered Flysch shales underlying the slope which supported the piers of the pipe line. Haefeli (1944) described serious damage to a railroad viaduct caused by creep. It required expensive underpinning operations. The foundations for the piers of the bridge rest on talus. Similar creep phenomena, leading to bridge defects in Switzerland were thoroughly investigated and described by Mohr *et al.* (1947). However, the implications of creep are beyond the scope of this paper, which deals exclusively with typical landslides.

VARIETIES OF LANDSLIDES

Landslides may involve materials of any kind, ranging between hard rock and soft clay, or any combination of materials. A similar variety prevails among the processes which may lead to landslides. They include undercutting by river erosion, man-made excavation, change in the ground water regime, and progressive structural changes in the material adjoining the slopes.

A phenomenon involving such a multitude of combinations between materials and disturbing agents opens unlimited vistas for the classification enthusiast. The result of the classification depends quite obviously on the classifier's opinion regarding the relative importance of the many different aspects of the classified phenomenon.

In this paper no new classification will be added to the numerous existing ones. In exchange, an attempt will be made to discriminate between the processes which may conceivably lead to landslides, and to analyze each one of them.

PROCESSES RESPONSIBLE FOR LANDSLIDES

CAUSES OF LANDSLIDES

The causes of landslides can be divided into external and internal ones. External causes are those which produce an increase of the shearing stresses at unaltered shearing resistance of the material adjoining the slope. They include a steepening or heightening of the slope by river erosion or man-made excavation. They also include the deposition of material along the upper edge of slopes and earthquake shocks. If an external cause leads to a landslide, we can conclude that it increased the shearing stresses along the potential surface of sliding to the point of failure.

Internal causes are those which lead to a slide without any change in surface conditions and without the assistance of an earthquake shock. Unaltered surface conditions involve unaltered shearing stresses in the slope material. If a slope fails in spite of the absence of an external cause, we must assume that the shearing resistance of the material has decreased. The most common causes of such a decrease are an increase of the pore-water pressure, and progressive decrease of the cohesion[1] of the material adjoining the slope. Intermediate between the landslides due to external and internal causes are those due to rapid drawdown, to subsurface erosion, and to spontaneous liquefaction.

EXTERNAL CHANGE OF STABILITY CONDITIONS

One of the most common and most obvious causes of landslides consists in the undercutting of the foot of a slope or the deposition of earth or other materials along the upper edge of the slope. Both operations produced an increase of the shearing stresses in the ground beneath the slope. If and as soon as the average shearing stress on the potential surface of sliding[2] becomes equal to the average shearing resistance, a landslide occurs.

A slope failure on a man-made slope may occur during or at any time after construction. If the slope fails several weeks after construction or later, the slide can be ascribed only to an internal cause which reduced the shearing resistance of the slope material after the completion of the construction operations. Delayed slides occur most commonly during heavy rainstorms.

[1] The term cohesion indicates the resistance of a material, rock, or sediment against shear along a surface which is under no pressure.

[2] The term *potential surface of sliding* indicates that surface located beneath a slope for which the ratio between average shearing stress and average shearing resistance is a maximum. If the material beneath the slope is fairly homogeneous, the cross section of the surface of sliding resembles a cycloid. Stability computations are commonly based on the simplifying assumption that the profile has the shape of an arc of a circle; the error due to this assumption is unimportant.

EARTHQUAKE SHOCKS

Earthquake shocks are considered external causes of landslides because they increase the shearing stresses along the potential surface of sliding, whereas the shearing resistance remains unchanged. The conventional method for evaluating the effect of an earthquake shock on the stability of a slope is illustrated by Figure 2, represent-

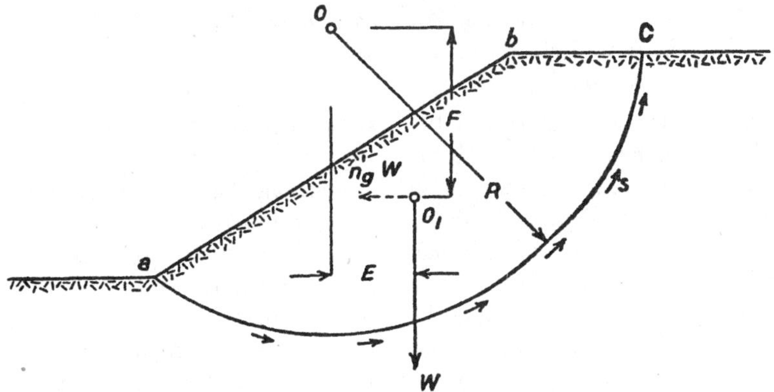

FIGURE 2.—*Diagram illustrating conventional method for computing effect of earthquake on stability of a slope*

The earthquake produces a horizontal acceleration n_g-times the acceleration of gravity, which increases the moment tending to rotate the wedge *abc* clockwise by n_gFW.

ing a cross section of a slope. The section through the potential surface of sliding is assumed to be an arc of a circle, *ac*, with the center *O*. Let

W = weight of the earth (water and solid combined) located above the surface of sliding, per unit of length of the slope measured at a right angle to the plane of the section shown in Figure 2,
l = length of the arc *aC*,
s = average shearing resistance per unit of area of the surface of sliding,
g = acceleration due to gravity,
n_g = ratio between the greatest horizontal acceleration produced by the earthquake and the acceleration g due to gravity, and
O_1 = center of gravity of the slice *abC*.

The weight W, acting at the lever arm E, tends to produce a rotation of the slice *abc* about the axis O and the rotation is resisted by the shearing resistance, sl, acting at a lever arm R. Hence, prior to the earthquake the safety factor G_s of the slope with respect to sliding is

$$G_s = \frac{\text{resisting moment}}{\text{driving moment}} = \frac{slR}{EW} \tag{1}$$

An earthquake with an acceleration equivalent n_g produces a mass force acting in a horizontal direction of intensity n_g per unit of weight of the earth. (*See*, for instance, Terzaghi, 1943b, p. 473–479.) The resultant of this mass force, n_gW, passes, like the weight W, through the center of gravity O_1 of the slice *abc*. It acts at a lever arm with length F and increases the moment which tends to produce a rotation of the slice *abc* about the axis O by n_gFW. Hence the earthquake reduces the factor of safety of the slope with respect to sliding from G_s, equation (1), to

$$G'_s = \frac{slR}{EW + n_g FW} \tag{2}$$

The numerical value of n_g depends on the intensity of the earthquake. Independent estimates (Freeman, 1932) have led to the following approximate values:

Severe earthquakes, Rossi-Forel scale IX, $n_g = 0.1$
Violent, destructive, Rossi-Forel scale X, $n_g = 0.25$
Catastrophic $n_g = 0.5$

The earthquake of San Francisco in 1906 was violent and destructive (Rossi-Forel scale X), corresponding to $n_g = 0.25$.

Equation (2) is based on the simplifying assumptions that the horizontal acceleration $n_g g$ acts permanently on the slope material and in one direction only. Therefore the conception it conveys of earthquake effects on slopes is very inaccurate, to say the least. Theoretically a value of $G'_s = 1$ would mean a slide, but in reality a slope may remain stable in spite of G'_s being smaller than unity and it may fail at a value of $G'_s > 1$, depending on the character of the slope-forming material.

The most stable materials are clays with a low degree of sensitivity, in a plastic state (Terzaghi and Peck, 1948, p. 31), dense sand either above or below the water table, and loose sand above the water table. The most sensitive materials are slightly cemented grain aggregates such as loess and submerged or partly submerged loose sand.

If a violent earthquake shock strikes a slope on plastic clay with low sensitivity, it will hardly have any effect beyond the formation of tension cracks along the upper edge associated with a slight bulging of the slope, because the viscosity of the clay interferes with more extensive displacements under impact. The slopes of earth dams or dikes, consisting of sand, may bulge slightly, and the crest of the fills may settle, but the slopes will not fail, provided the fills rest on a rough and stable base. After the earthquake, the fills will be more stable than before, because the earthquake vibrations tend to compact the material. Toward the end of the last century two earth dams have been built across the San Andreas fault in California—the San Andreas Dam, 95 feet high, and the Upper Crystal Springs Dam, 85 feet high. During the earthquake of San Francisco in 1906 the horizontal displacement along the fault at the site of the dams amounted to more than 10 feet. Yet the slopes of the dams remained intact except at the point where they were warped by shear (Eng. News-Rec., 1932).

If a mass of stable material, such as dense sand, rests on a slippery base, like the surface of a layer of soft clay, a slope failure may occur by sliding of the stable material on its base.

The destructive effect of earthquakes on slightly cemented grain aggregates, such as loess, and on submerged loose sand, seems to be chiefly due to the rapid vibratory movement of the particles with reference to each other and not to the quasi-static effects described by equation (2). In loess these movements are likely to break the connection between the grains, whereupon the material assumes the character of cohesionless sand. In December 1920 a catastrophic earthquake occurred in the heart of the loess district of the province of Kansu in China. "In each case the earth which came down bore the appearance of having shaken loose clod from clod and grain from grain, and then cascaded like water, forming vortices, swirls, and all the convolutions into which a torrent might shape itself" (Close and McCormick, 1922, p. 463). During the earthquake of New Madrid in 1811, many loess slopes failed (Fuller, 1912).

Submerged masses of loose sand may, under the impact of an earthquake shock, temporarily assume the character of a suspension which flows like a viscous liquid. (*See* the subheading "Spontaneous Liquefaction".)

LUBRICATING EFFECT OF WATER

If a slide takes place during a rainstorm at unaltered external stability conditions, most geologists and many engineers are inclined to ascribe it to a decrease of the shearing resistance of the ground due to the "lubricating action" of the water which seeped into the ground. This explanation is unacceptable for two reasons.

First of all, water in contact with many common minerals, such as quartz, acts as an anti-lubricant and not as a lubricant. Thus, for instance, the coefficient of static friction between smooth, dry quartz surfaces is 0.17 to 0.20 against 0.36 to 0.41 for wet ones (Terzaghi, 1925, p. 42–64).

Second, only an extremely thin film of any lubricant is required to produce the full static lubricating effect characteristic of the lubricant. Any further amount of lubricant has no additional effect on the coefficient of static friction between them (Hardy, 1919). In humid regions such as the eastern United States and within less than 1-2 feet from the sloping surface every sediment—sand included—permanently contains far more than the quantity of water needed for "lubricating" the surfaces of the grains (Terzaghi, 1942, Fig. 15, p. 356). Yet in humid regions rainstorms start landslides as often as they do in arid ones. In other words, since practically all the sediments located beneath slopes are permanently "lubricated" with water, a rainstorm cannot possibly start a slide by lubricating the soil or boundaries between soil strata.

However, the rain water which seeps into a slope affects the stability of the slope in various other ways. If the voids of the ground are partly filled with air, the water eliminates the surface tension which imparts to fine-grained, cohesionless soils a considerable amount of apparent cohesion (Terzaghi and Peck, 1948, p. 114–128). The water which enters the voids also increases the unit weight of the soil though, as a rule, this increase is commonly unimportant.

Some soils, such as typical loess, owe their cohesion to a soluble binder. If a slope on such a soil is submerged for the first time, or if the soil becomes saturated by seepage from a newly created, artificial source of water, the binder is removed by solution and the soil loses its cohesion.

Last—but not least—water which enters the ground beneath a slope always causes a rise of the piezometric surface[3], which, in turn, involves an increase of the pore-water pressure and a decrease of the shearing resistance of the soil. Since water can affect the stability of slopes in several radically different ways, its actions will be discussed under different subheadings.

[3] The piezometric surface is the locus of the points to which the water would rise in piezometric tubes. If the permeability of a soil, such as a soft clay, is too low to permit locating the position of the piezometric surface by means of observation wells, pressure gages must be used.

RISE OF PIEZOMETRIC SURFACE

Throughout a saturated mass of jointed rock, soil, or sediment, the water which occupies the voids is under pressure. Let

p = pressure per unit of area at a given point P of a potential surface of sliding, due to the weight of the solids and the water located above the surface,
h = the piezometric head at that point,
w = the unit weight of the water, and
ϕ = the angle of sliding friction for the surface of sliding.

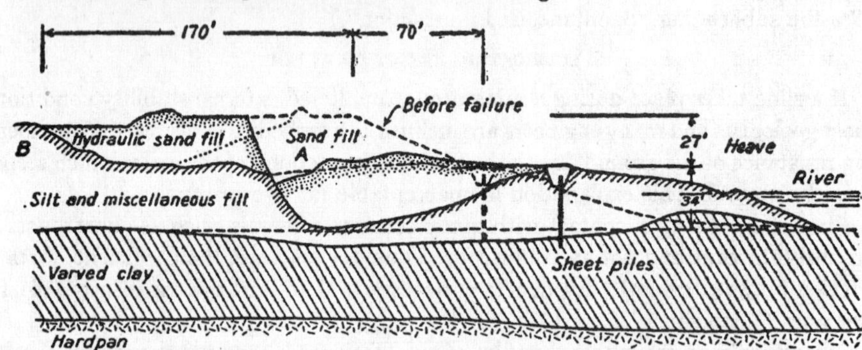

FIGURE 3.—*Section through a slide which was caused by an excess hydrostatic pressure in the silt layers of a stratum of varved clay*

The row of sheetpiles advanced in a few minutes over a maximum distance of up to 60 feet toward the river.

Regarding the relation between these four quantities, soil mechanics has led to the following conclusions (Terzaghi and Peck, 1948, p. 51–55). If the potential surface of sliding is located in a layer of sand or silt, the shearing resistance s per unit of area at the observation point is equal to

$$s = (p - hw) \tan \phi \tag{3}$$

Hence, if the piezometric surface rises, h increases, and the shearing resistance s decreases. It can even become equal to zero. The action of the water pressure hw can be compared to that of a hydraulic jack. The greater hw, the greater is the part of the total weight of the overburden which is carried by the water, and as soon as hw becomes equal to p the overburden "floats." If a material has cohesion, c per unit of area, its shearing resistance is equal to the sum of s, equation (3), and the cohesion value c, whence

$$s = c + (p - hw) \tan \phi \tag{4}$$

The effect of a decrease of the shearing resistance s on the stability of slopes on stratified sediments is illustrated by Figure 3. The figure shows a vertical section through sand dikes which were constructed along a river for flood-protection purposes. The dikes rest on a layer of soft silt and miscellaneous fill which covers the surface of a horizontal stratum of varved clay with a thickness of about 50 feet. The dash line shows the position of the dike and of the boundaries between the underlying strata prior to the slide.

Some time after the dike A was completed by depositing and compacting moist sand in layers, the space between the landward slope of this dike and the outer slope of an older dike, B, was filled with sand. The sand was excavated by means of a hydraulic dredge and deposited in a semiliquid condition. The sluicing operations raised the piezometric level in the pore water of the silt layers in the varved clay to a considerable height h above the original water table. As a consequence the resistance s, equation (3), against sliding decreased.

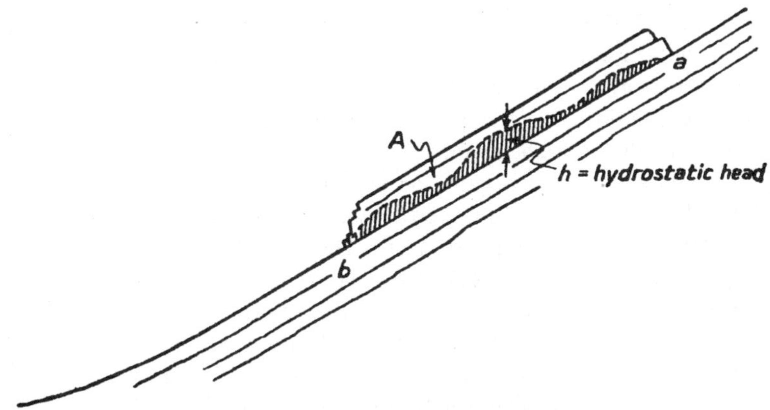

FIGURE 4.—*Diagrammatic section through site of rock slide of Goldau (1806) prior to slide*

Slab A was separated from its base by a thin layer of weathered rock. The dashed line represents the piezometric surface in this layer during a heavy rainstorm.

During the construction of the hydraulic fill the dike A suddenly subsided, and the row of sheet piles, together with the foreland, moved over a distance up to 60 feet and over a length of about 1200 feet toward the river. The row of sheet piles remained perfectly intact. This fact showed that the failure had occurred by sliding along one or more horizontal surfaces of sliding located in the varved clay. If the shearing resistance along these surfaces had not been extremely low, the wedge-shaped body of silt, located on the river side of the sheet piles, could not possibly have advanced over a distance up to 60 feet without undergoing intense compression and shortening in the direction of the movement.

After a slide has occurred the excess pore-water pressure in the zone of shear always decreases, on account of progressive consoiidation, and approaches a value zero. In order to get information on the rate of decrease of the pressure which prevailed in the varved clay, a great number of pressure gages were installed. The first readings were made more than 3 months after the slide occurred. Yet, even beneath the banks of the river, the piezometric elevation h, equation (3), still amounted to more than 10 feet, with reference to the river level.

The relation expressed by equation (3) also applies to stratified or jointed rocks. To illustrate its bearings on rockslides, the classical slide of Goldau in Switzerland will be discussed. This slide has always been ascribed to the "lubricating action" of the rain- and meltwater. Figure 4 is a diagrammatic section through the slide area. It shows a slope oriented parallel to the bedding planes of a stratified mass of Tertiary Nagelflue (conglomerate with calcareous binder) which rises at an angle of 30° to the horizontal. On this slope rested a slab of Nagelfiue 5000 feet long, 1000 feet wide, and about 100 feet thick. It was separated from its base by a porous layer of weathered rock.

The fact that the slab had occupied its position since prehistoric times indicates that the shearing force, which tended to displace the slab, never exceeded the shearing strength, in spite of the effects of whatever hydrostatic pressures, hw, in equation (3), may have temporarily acted on the base of the slab in the course of its existence.

On September 2, 1806, during heavy rainstroms, the slab moved down the slope, wiped out a village located in its path, and killed 457 people (Heim, 1882). This catastrophe can be explained in at least three ways. One explanation is that the angle of inclination of the slope had gradually increased on account of tectonic movements, until the driving force which acted on the slab became equal to the resistance against sliding. A second explanation is based on the assumption that the resistance of the slab against sliding was due not only to friction, but also to a cohesive bond between the mineral constituents of the contact layer. The total shearing resistance due to the bond was gradually reduced by progressive weathering, or by the gradual removal of cementing material, either in solution or by the erosive action of water veins. The third explanation is that h in equation (3) or (4) assumed an unprecedented value during the rainstorm, whereas the cohesion c, in equation (4), remained unchanged, provided cohesion existed. In Figure 4 the value h is equal to the vertical distance be-between the potential surface of sliding, ab, and the dash line interconnecting ab which represents the piezometric line. During dry spells h is equal to zero. In other words, the piezometric surface is located at the slope. During rainstorms the rain water enters the porous layer located between slab and slope at a and leaves it at b. Since the permeability of this layer is variable, the piezometric line descends from a to b in steps, and the value h in equations (3) and (4) is equal to the average vertical distance between the piezometric line and the slope.

The maximum value of h changes from year to year, and if the exits of the water veins at b are temporarily closed by ice formation while rain- or melt-water enters at a, h assumes exceptionally high values. However, the seasonal variations of h, the corresponding variations of s, equations (3) and (4), and the occasional obstruction of the exits at b have occurred in rhythmic sequence for thousands of years, without catastrophic effects. It is very unlikely that h assumed a record value in 1806, in spite of unaltered external conditions. Therefore it is more plausible to assume that the slide was caused by a process which worked only in one direction, such as a gradual increase of the slope angle or the gradual decrease of the strength of the bond between slab and base. In no event can the slide be explained by the "lubricating effect" of the rain water. One might as well ascribe a theft to some mysterious effects of the presence of the thief in the house instead of inquiring about his physical actions.

PROGRESSIVE STRUCTURAL CHANGES IN THE SLOPE-FORMING MATERIALS

Every rainstorm causes an increase of the value h in equations (3) and (4) and, as a consequence, a decrease of the shearing resistance along potential surfaces of sliding. Therefore the factor of safety G_s, equation (1), of every slope with respect to sliding is subject to cyclic changes. The minor variations have a period of a few weeks or months, and the major ones of many years. (*See* section on "Periodicity of Landslides".). These variations are part of the routine of the slopes. Hence the probability that an old slope should be exposed in our lifetime to unfavorable conditions without

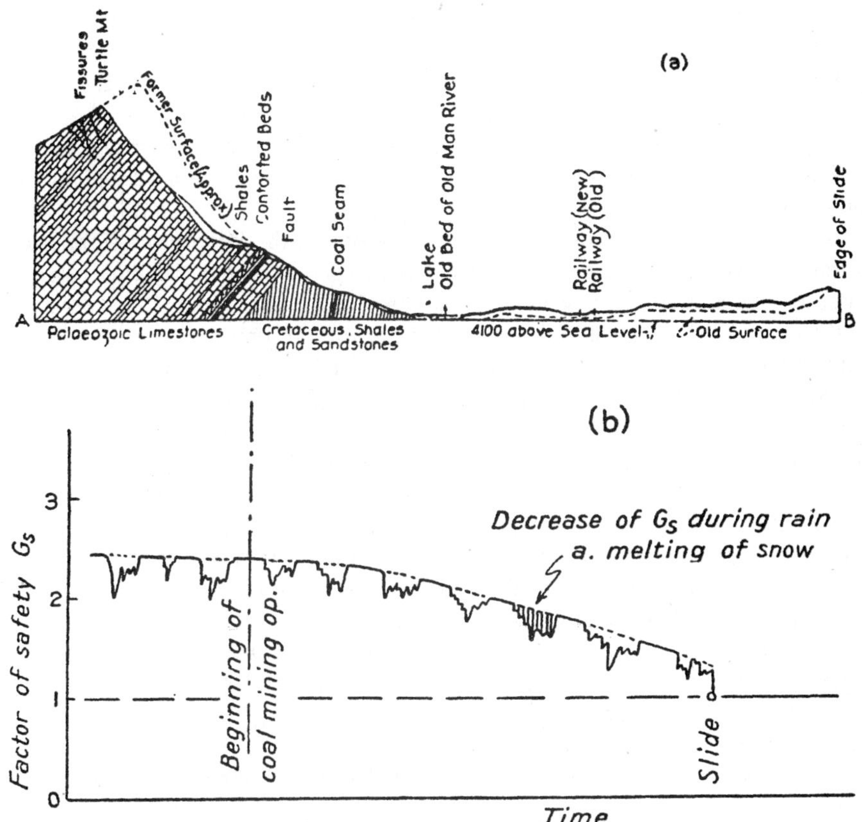

FIGURE 5.—*Turtle Mountain after the Great Frank Alberta Slide in 1903*

After McConnell and Brock (1904). (a) Cross section (b) Diagram illustrating the writer's concept of the changes of the safety factor of the slope prior to the slide.

any precedent is almost nil. If such a slope fails without external provocation it is much more probable that it failed on account of a gradual decrease of the cohesion of the slope-forming materials.

In hard, jointed rocks, resting on softer rocks, a decrease of the cohesion of the rock adjoining a slab may occur on account of creep of the softer rocks forming their base. The great Turtle Mountain slide of 1903 near Frank, Alberta (Fig. 5a), seems to belong to this category. Percolating waters and frost action have contributed to the breakdown (Sharpe, 1938, p. 79). They always do, but they have done it for many thousands of years. Percolating waters cannot move blocks located between joints at great depth, and the frost action is only skin deep. Hence neither water nor frost could have altered the stability conditions in the rock adjoining the slope beyond a distance of a few feet from the slope. However, the limestones, forming the bulk of the peak, rested on weaker strata which certainly "crept" under the influence of the unbalanced pressure produced by the weight of the limestone, and the rate of creep was accelerated by coal-mining operations in the weaker strata.

The total cohesion along the potential surface of sliding in a jointed rock is equal to the combined shearing strength of all those blocks of rock which interfere, like dowels, with the sliding movement. The yield of the base of the limestone caused an increase of the shearing stresses; the increases of the stresses caused one dowel after another to "snap," and the slope failed when it was ripe for failure, at a time when the factor of safety assumed one of its periodic minimum values. Figure 5 b is a graph illustrating the process which led to the slide.

Another incident of a similar kind was the collapse of the Pulverhörndl, an isolated limestone tower in the northern Alps, which rested on a bed of shale. The tower was a favorite training ground for mountaineers. It had a volume of about 260,000 cu. yds. In the fall of 1920 the tower suddenly collapsed, without any provocation. The fragments struck the shale, whereupon the shale assumed the character of a mudstream, and about ten million cubic yards of shale advanced on a gentle slope toward the mouth of a valley (Lehmann, 1926). This catastrophe, too, was probably caused by the gradual yield of the base of the tower, whereas all the other circumstances attending the slide were only a repetition of what had happened many times before.

The writer had an opportunity to investigate a flow slide which was indirectly caused by the swelling of a homogeneous clay stratum. The slide occurred on the side of an open excavation for one of the locks adjoining the powerhouse of the hydroelectric power development, Swir III east of Leningrad.

The slope rose at an angle of about 35° and intersected three horizontal strata. The uppermost stratum consisted of a well-compacted and slightly cohesive glacial till, the middle one of a stiff, greenish Devonian clay, which could be cut with a knife, and the lower one of a very stiff, reddish, sandy clay. The slope made a perfectly sound impression, and nobody, the writer included, felt the slightest concern about its stability.

In the summer of 1930, during a heavy rainstorm, the till turned into a mudflow and descended into the lock excavation. The only explanation which appeared to be acceptable was that the breakdown of the till was due to a horizontal expansion of its base. Since the till is too rigid to participate in a lateral expansion of its base it is torn into fragments by tension. The rain water accumulates in the open spaces between the fragments. The fragments disintegrate, and the supersaturated mixture of water and till flows into the cut.

In order to find out whether this explanation is correct, a niche was carved out of the green clay, a few hours after the slide had started. Figure 1 of Plate 2 is a photographic view of the niche 1 hour after the excavation was completed. One could already see that the distance between the sides of the niche, above the base of the green clay, had decreased and that the rear face had advanced with reference to the underlying red clay. A few hours later the relative displacements were very conspicuous, and the overhanging parts of the expanding green clay started to crumble (Pl. 2, fig. 2).

FIGURE 1. OBLIQUE VIEW OF PRISMATIC BLOCK OF SOFT DEVONIAN CLAY
Showing deformation by creep under low stress. Clay encountered between two layers of stiff clay in test shaft for foundation of storage dam Swir III, U. S. S. R.

FIGURE 2. EXPOSURE ON LEFT BANK OF SULAK RIVER (DAGHESTAN, U. S. S. R.) AT NORTHERN EDGE OF CAUCASUS
Showing south edge of an overthrust coming from the north. Brachy-anticline interfered with further advance of strata whereupon they curled up at foot of obstacle. East of the anticline, south edge of thrust was located about 5 miles farther south.

DEFORMATION OF STRATA

FIGURE 1. VISIBLE EFFECTS OF HORIZONTAL EXPANSION OF HORIZONTAL LAYER OF STIFF DEVONIAN CLAY
Exposed on slope of excavation for navigation lock, Swir III, U. S. S. R. Protrusion of expanding clay stratum with reference to underlying, relatively rigid stratum of sandy clay, half an hour after niche was excavated.

FIGURE 2. SAME AS FIGURE 1; PROTRUSION 6 HOURS LATER

FIGURE 3. SEVEN OAKS SLIDE ON SOUTHERN RAILWAY SOUTH OF LONDON
Slide occurred in stiff, fissured Weald clay, 70 years after cut was excavated.

SLIDES

Figure 1. Head of a "Bossoroca" in Province of Sao Paulo, Brazil
Slope at head of valley is being undermined by subsurface erosion of spring which emerges at foot of slope; slope fails, slide material is removed by erosion, and the process of undermining starts again (By permission of Mr. E. Pichler, Sao Paulo).

Figure 2. View of Southeastern Part of Kenogami Slide Area after Ground Movements Came to Standstill
Courtesy of Mr. G. F. Layne, Price Brothers and Co., Quebec.

SLIDES

To obtain supplementary information on the movements of the base of the till a test shaft was excavated about 20 feet beyond the original position of the upper edge of the cut. The test shaft went through the till and the green clay into the reddish one. As soon as the shaft was completed that part of the shaft located above the boundary between the greenish and the reddish clay advanced with reference to the lower one, and within a few days the shaft lining was sheared off.

This slide, like the Turtle Mountain slide and the collapse of the Pulverhörndl, was due to a loss of cohesion caused by a slow plastic deformation of the base of the slope-forming material. Far more common are slides due to a decrease of cohesion produced by direct action of the weakening agents on the ground located beneath the slope. The slides on the slopes of cuts in stiff fissured clays are typical examples.

The excavation of an open cut always produces a stress relaxation in the ground adjoining the slopes. In stiff fissured clay this process causes the joints to open; rain water invades the joints; the fragments between joints swell, break up, the average shearing resistance decreases, and as soon as it becomes equal to the average shearing stress on the potential surface of sliding the slope fails. It is by no means uncommon that the shearing resistance of a stiff fissured clay decreases from an original value of 1 or 2 tons per square foot to 0.3 or less (Terzaghi, 1936).

On account of the low permeability even of jointed clays, the joints are likely to be permanently filled with water. Therefore the factor of safety of the slopes on fissured clay decreases steadily, and not intermittently, as indicated in Figure 5b. Slides on such slopes may fail at any time of the year. Figure 3 of Plate 2 shows an example of a clay slide which occurred in fair weather. It took place in 1939 in a railway cut between London and Folkestone, about 70 years after the cut had been excavated. The height of the slope was 60 feet, and the rise 1:2.5. The eyewitnesses were surprised about the "dry" appearance of the slide material. No springs or other indications of seepage could be observed.

Skempton (1948) analyzed the available records of slope failures on London clay, which is a highly colloidal, stiff, fissured clay of Eocene age, and he arrived at the following conclusions. In this clay a vertical slope with a height of 20 feet may stand up for several weeks. A 1:2 slope with the same height fails after 10 to 20 years. If the rise of the slope is 1:3 the life of the slope is likely to exceed 50 years. Yet the rise of the steepest slopes, which are encountered on the side of erosion valleys in London clay, rarely exceeds 1:6. Based on these findings Skempton constructed a set of curves (Fig. 6) which represent approximately the effect of time on the shearing resistance of London clay forming the slopes of open cuts in different localities.

Exceptionally the cohesion of the material forming the slopes of an open cut may also be reduced by the formation of shrinkage cracks during a long, dry spell. A slide due to this cause was described by Ladd (1934). The slide occurred in a railway cut and wrecked two trains.

RAPID DRAWDOWN

The term *rapid drawdown* refers to the lowering of the water level in a storage reservoir or to the descent of the water level in a river after a flood at a rate of at least several feet per day. The effect of this process on the stability of the slopes forming

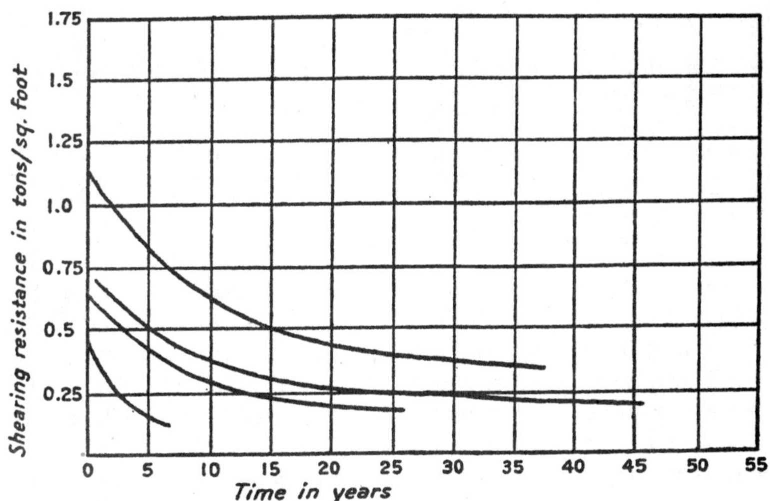

FIGURE 6.—*Diagram showing gradual decrease of shearing resistance of stiff, fissured London clay*

The curves are based on the results of a statistical study of slope failures in the London area. Each curve represents a different locality. (After A. W. Skempton 1948.)

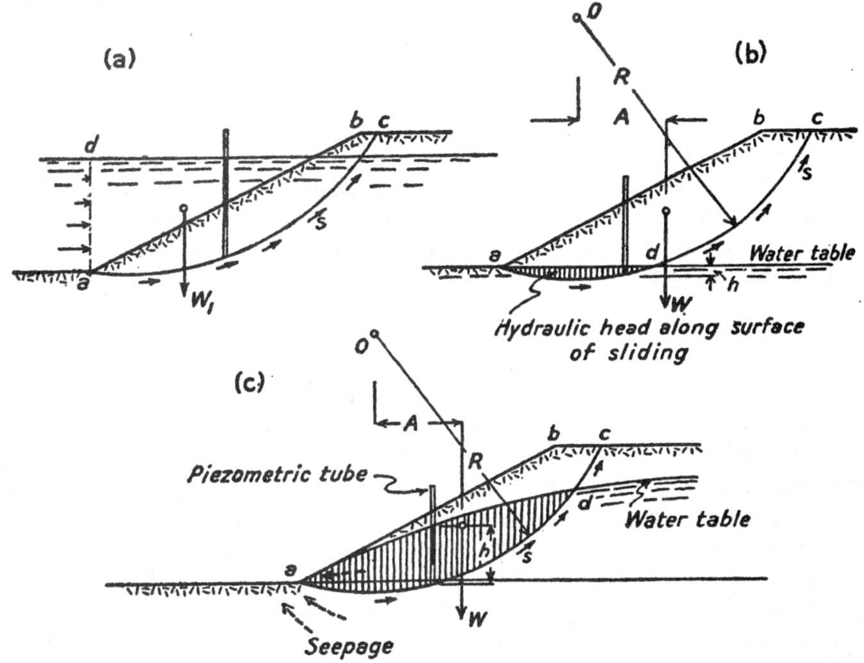

FIGURE 7.—*Diagrams illustrating effect of speed of drawdown on stability of temporarily submerged slope*

(a) Section through the slope prior to drawdown. (b) Forces acting on ground above potential surface of sliding after very slow drawdown, and (c) after rapid drawdown. Dashed arrows in (c) indicate directions of seepage toward foot of slope.

the sides of the reservoir or the river is illustrated by Figure 7. Figure 7a shows a vertical section through a partly submerged slope *ab*. The potential surface of sliding is indicated by the arc *ac* of a circle. Let

W = Weight of slice *abc*, solid and water combined, per unit of length of the slope,
l = length of the arc *ac*,
c = cohesion of the slope-forming material,
ϕ = angle of internal friction of this material,
p = average unit pressure on the surface of sliding *ac* due to the weight W of the slice *abc*,
h = piezometric head for any point of the potential surface of sliding, at any time,
h_1 = average of the piezometric heads h for the surface of sliding after a very slow drawdown, and
h_2 = as before, after a rapid drawdown.

The following analysis is based on the assumption that the voids of the soil are completely filled with water both below and above the piezometric surface. The error due to this assumption is very small unless the slope material consists of very course-grained sediments such as coarse sand or gravel without an admixture of finer fractions. The effect of the capillary forces on the stability of the slope will be disregarded. These forces increase the stability of the slope under any circumstances.

If the level of the body of water adjoining the slope goes down very slowly, the water table remains horizontal and descends at the same rate as the water level of the reservoir. After the drawdown is complete, the piezometric surface is a horizontal surface passing through the foot of the slope (Fig. 7b). The average shearing resistance s of the material adjoining the surface of sliding is determined by equation (4), and the factor of saftey of the slope with respect to sliding is

$$G_s = \frac{Rl\,[c + (p - h_1 w)\tan\phi]}{AW} \qquad (5)$$

On the other hand, if a drawdown takes place very rapidly the descent of the piezometric surface lags behind the descent of the free water level, and at the end of the drawdown the piezometric surface rises from the foot of the slope as indicated in Figure 7c and intersects the potential surface of sliding at a point d which is located high above d in Figure 7b. The corresponding factor of safety with respect to sliding is

$$G_s = \frac{Rl\,[c + (p - h_2 w)\tan\phi]}{AW} \qquad (6)$$

In Figures 7b and 7c the total water pressure on the surface of sliding *ac* is indicated by shaded areas. Since the total water pressure on *ac* in Figure 7b (slow drawdown) is very much smaller than that on *ac* in Figure 7c (rapid drawdown), h_1 is very much smaller than h_2, and, as a consequence, G_s', equation (6), is smaller than G_s, equation (5). Hence, even if a slope has survived a great number of slow drawdowns it may fail after a rapid drawdown, because G_s' is smaller than G_s.

Landslides caused by rapid drawdown are very common, and many records of such slides have been published. (*See*, for instance, Pollack, 1912). The sediments which are most seriously affected by a rapid drawdown are those intermediate between sand and clay.

As long as the piezometric surface in the ground beneath the slope has a gradient, the water percolates through the ground toward the surfaces adjoining the foot a of the slope, as indicated in Figure 7c by dashed arrows. On account of its viscosity the percolating water exerts on the soil particles a pressure known as *seepage pressure.*

This pressure acts in the direction of the flow, and its intensity increases in simple proportion to the seepage velocity (Terzaghi and Peck, 1948, p. 54). At the foot of the slope the seepage velocity and the corresponding seepage pressure are much greater than higher up, and the seepage pressure tends to move the soil particles along the flow lines which are directed toward the foot of the slope. As a consequence, at the foot of the slope, the point of failure is reached much earlier than at higher elevations, and once the lower part of the slope has failed, the upper part follows because it has lost its support. The mechanics of this process and the means of preventing its detrimental effects have been investigated by Reinius (1948).

SPONTANEOUS LIQUEFACTION

The arrangement of the grains of fine sand or coarse silt can be so unstable that a slight disturbance of the equilibrium of the grains may cause a rearrangement of the grains, whereby the grains settle into more stable positions, and the porosity of the sediment decreases.

If this process takes place above the water table it has no noticeable effect other than a settlement of the ground surface. By contrast, if it occurs below the water table its consequences can be catastrophic, because the viscosity of the water, which occupies the voids of the sand, prevents a rapid decrease of the porosity. During the time between the collapse of the structure and the reconsolidation under the new conditions of equilibrium, the sediment has the properties of a thick viscous liquid which spreads laterally, until its surface becomes almost horizontal. The transformation into the liquid state is known as *spontaneous liquefaction*. (*See*, for instance, Terzaghi and Peck, 1948, p. 100–105.)

The liquefaction of an unstable sediment can be caused by vibrations such as those produced by pile driving or quarry blasts. It can also be produced by the rapid rise or fall of the water table. The best-known slides, due to spontaneous liquefaction, are those which occur from time to time at the coast of the province of Zeeland (Holland), on sand slopes whose rise may be as gentle as 1:4. Between 1881 and 1946 no less than 229 slides have been reported. The quantity of sand which moves out during a slide ranges between about 80 cubic yards and 3 million cubic yards.

The slides are probably preceded by erosion caused by shore currents, associated with an increase of the average slope angle. The slides commonly occur after exceptionally high tides, particularly if the tide coincides with a heavy gale. The moving sand spreads out like a fan on the bottom of the sea, and after the slide the slope angle of the surface of the slide material may be as small as 3° to 4° (Koppejan *et al.*, 1948).

SEEPAGE FROM ARTIFICIAL SOURCES OF WATER

Seepage from artificial sources of water, such as storage reservoirs or unlined canals, may compromise the stability of existing slopes in at least four ways, depending on the character of the slope-forming material and on the conditions of stratification. It may reduce the shearing resistance of the ground by increasing the item hw in equations (3) and (4); it may eliminate apparent cohesion produced by the surface tension in drained soils; it may eliminate real cohesion by removing cementing materials in solution; it may also cause a slope failure by retrogressive underground erosion by water veins emerging at the foot of the slope. The term "artificial source of water" implies that the source is of recent origin. Otherwise the slope failure caused by the action of the seepage derived from the source would have occurred long ago.

Figure 3 illustrates a slope failure caused by an increase of the pore-water pressure, hw. The mechanics of this slide are explained in the figure caption. The seepage water came from a freshly deposited hydraulic fill located between the dikes A and B.

Moist, fine, silty sand can form permanent vertical slopes with a height of several tens of feet. This can be seen in any sand pit. The cohesion required for maintaining the equilibrium of such slopes is due to the friction produced by the surface tension of the contact moisture (water particles, surrounding the points of contact of the grains; *see* Terzaghi and Peck, 1948, p. 127). Hence the stability of such slopes requires the existence, within the slope-forming material, of a large area of contact between air and soil moisture.

Experience shows that the water, which seeps toward steep slopes during rainstorms, does not displace enough air to destroy the apparent cohesion of sand or silt. However, if water percolates through the ground toward the slope in large quantities and without any intermissions, the air is almost completely expelled, the apparent cohesion is eliminated, and the slope fails. A similar failure would occur if a steep slope on fine sand or silty sand is submerged for the first time in its history, for instance by the creation of a storage reservoir.

Slope failures due to the removal of a binder by solution are a common phenomenon in loess regions. Loess owes its cohesion to a soluble binder which consists chiefly of calcium carbonate. Since typical loess contains numerous vertical root holes, it commonly forms vertical cliffs, and the cliffs remain stable for years or decades, provided the water table is permanently located below the level of the base of the cliffs. Furthermore many artificial caves in loess are known to have existed for centuries without their unsupported roofs showing any signs of deterioration. These facts lead to the conclusion that the water, which percolates through the voids of loess during rainstorms, does not perceptibly weaken the bond between the loess particles. However, if loess is submerged or if a permanent flow of seepage through loess is established, the bond between the loess particles perishes within a few weeks or months and the loess assumes the character of supersaturated rock flour which flows like molasses. After a loess flow has come to rest, the excess water gradually drains out, and the final product of the process of drainage has the density and the properties of a very fine, loose sand.

The intensity of the effects of saturation on the physical properties of loess increases with increasing initial porosity, which ranges between about 40 and 60 per cent. The results of a comprehensive survey of the physical properties of loess and their engineering implications were presented by Scheidig (1934).

The effects of saturation on loess were impressively demonstrated by a large-scale test which was performed some 15 years ago in southeastern Turkestan in connection with an irrigation project. In order to investigate the reaction of loess to seepage through the bottom of a proposed irrigation canal, a bowl-shaped excavation 13 feet deep was made. The bottom of the excavation was rectangular, and it covered an area of 70 by 35 feet. The rise of the slopes was 1:1.5. The base of the loess stratum was located at a depth of about 80 feet below the bottom of the excavation. After the excavation was finished it was filled with water, and the water level was maintained at a height of about 10 feet above the bottom, by pumping.

A short time after the bottom was flooded it started to subside, and the slopes began to slough. The subsidence continued, first at an accelerated and later at a decreasing rate. At the end of 6 weeks the bottom of the excavation was located at a depth of about 2½ feet below its original position. The sloughing had spread about 20 feet beyond the original upper edge of the slope, and within the zone of sloughing the loess was so soft that it was not possible to walk on its surface. In its original state the loess, in the vicinity of the site of the test, is so coherent that vertical cliffs, 60 feet high, remain permanently stable.

Seepage coming from a storage reservoir or an unlined canal may also decrease the stability of a slope by subsurface erosion, proceeding from the exit of water veins toward the source of water supply. As the length of the underground conduit increases, the quantity of water seeping into the conduit and the cross section of the conduit also increase. Finally the width of the conduit becomes so great that the roof collapses, whereupon the mass of sediment above the roof breaks up and a slide ensues.

Under natural conditions this process may produce, in the course of time, deep gullies with a considerable length. The *Bossorocas* in the State of Sao Paulo (Brazil) are an example. A spring undermines the foot of the slope forming the rear wall of the niche which surrounds the spring. The slope fails, the slide mass is removed by erosion, and the process starts again. Some of the resulting gullies have a depth of about 160 feet. The upper part of the formation exposed on the walls of the gullies, with a thickness of about 50 feet, consists of a loesslike sediment and the lower part of a fairly coarse sand with an argillaceous binder. The mechanics of the formation of the Bossorocas has been investigated by E. Pichler (results not yet published). Figure 1 of Plate 3 shows the rear wall of one of the gullies. The water which performs the subsurface erosion leaves the ground through a cave at the foot of the rear wall (not visible in the photograph).

Subsurface erosion is also the most common cause of the dreaded failure of storage dams by piping. The erosion works back from springs emerging close to the foot of the downstream slope of the dam (Terzaghi and Peck, 1948, p. 507–510). If part of the enclosure of a storage reservoir consists of a natural ridge composed of unconsolidated sediments or decomposed rock, subsurface erosion may cause a failure of the downstream slope of the ridge, followed by a discharge of the contents of the reservoir through the gap. The failure of the Cedar Reservoir, Washington, in December 1918, is a typical example. A quantity of morainal material, estimated between 800,000 and 2,000,000 cubic yards, moved out of the slope. The slide initiated a flood which wrecked the tracks of the Milwaukee Railroad and destroyed the town of Edgewick and several industrial plants (Mackin, 1941).

PERIODICITY OF LANDSLIDES

The periodicity in the occurrence of certain types of landslides came to the writer's attention for the first time in connection with the landslides in the Folkestone Warren, at the north coast of the English Channel, between Folkestone and Dover. The Warren can be described as a giant niche, about 10,000 feet long and 1000 feet wide, located between the channel coast and a steep cliff with a height of about 400 feet. The upper, vertical part of the cliff consists of chalk. The bottom of the niche consists of Gault clay buried beneath a chaotic accumulation of large fragments of chalk (Fig. 8a). The Southern Railway enters the niche from the West through the Martello Tunnel and leaves it at the east end through a tunnel leading toward Dover.

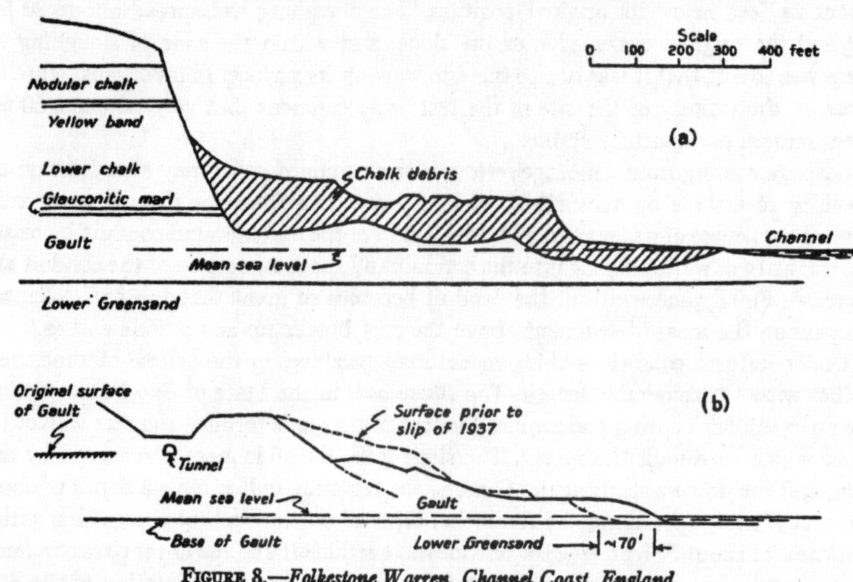

FIGURE 8.—*Folkestone Warren, Channel Coast, England*

(a) Diagrammatic cross section. (b) Displacement which occurred during the slide of 1937. The outer part of the slice of Gault clay involved in the slide advanced over a distance of about 70 feet toward the Channel by sliding along the boundary between Gault clay and Lower Greensand. Yet the surface topography of this part of the slice remained almost unchanged.

The railway was constructed in the middle of the nineteenth century. Prior to the construction of the railway, huge slides occurred at different points of the Warren in 1765, 1800, and 1839. After the railroad was built the periodic recurrence of slides continued. The first recorded slide affecting the railroad took place in 1877. In 1896 large movements occurred in the western part of the Warren. In December 1915 a slip affected nearly the whole of the Warren and effectively blocked the railway until the end of the first World War. It involved the movement of several millions of cubic yards. In the spring of 1937, slides took place over an area of 35 acres in the western part of the Warren and caused the formation of a crack across the lining of the Martello Tunnel.

Figure 8b is a diagrammatic section through the area affected by the slide of 1937. The major part of the sliding mass advanced along the almost horizontal boundary between the Gault, which is a very stiff clay, and the Lower Greensand, which is a soft sandstone. A detailed account of the physical characteristics of the materials involved in the slide was published by Toms (1946). In Figure 8b the original position of the ground surface is indicated by a dash line, and the final one by a plain line. It should be noted that the front part of the sliding mass moved bodily, without undergoing more than a slight deformation, over a distance of about 70 feet toward the channel. Such a movement would not be conceivable unless the resistance against sliding along the base of the moving section of ground was very low. Furthermore the slide was not preceded by a change of the external conditions for the equilibrium of the slope. Hence we are compelled to assume that the resistance against sliding has decreased, which can be accounted for only by an increase of the hydrostatic pressure [*hw* in equation (3)] on the base of the sliding body.

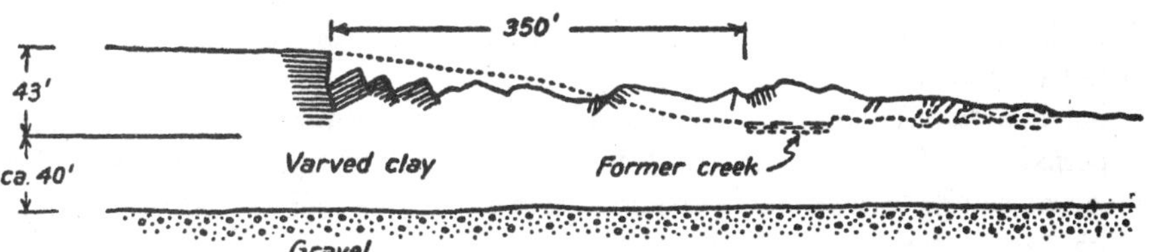

FIGURE 9.—*Section through the Hudson slide of 1915*

After Newland (1916). The slide was probably caused by exceptionally high hydrostatic pressure in the pore water of the gravel. According to the writer's hypothesis the varved clay moved along the boundary between the gravel and the clay.

The borings, made in 1939, showed that the piezometric surface for the Lower Greensand was located at elevations up to 27 feet above sea level. These elevations are subject to seasonal variations and to variations within a longer period, caused by variations in the average rainfall. However, in the Warren, the rainwater has no opportunity to get into the Lower Greensand from above, because the Greensand is covered with a blanket of Gault clay having a very low permeability. The variations of the piezometric heads can be caused only by similar variations in the elevations of the water table in that region where the Greensand emerges at the surface. This region is located many miles north of the Warren.

The fact that the major slides in the Warren occurred once every 19 or 20 years suggests that the movements were due to corresponding maxima in the amount of rainfall in those regions where the aquifer located beneath the slide area reaches the surface. In agreement with this assumption the Warren slide of 1937 was preceded by abnormally heavy rainfalls, "between 15 and 16 inches of rain falling during the first three months" (Seaton, 1938, discussion by Ellson on p. 438).

Considering the mechanics of the Warren slide one might expect a similar periodicity in connection with landslides on all those slopes whose factor of safety, with respect to sliding, varies with the elevation of the water table in distant aquifers. The slide of Hudson, N. Y., belongs in this category. It occurred on August 2, 1915, and wrecked the powerhouse of the Knickerbocker Cement Factory. Figure 9 is a section through the slide area. The slope was located on varved clays which rested at a depth of about 40 or 50 feet below the bottom of the valley on morainal gravels (Newland, 1916). The varved clay corresponds to the Gault at the Warren site (Fig. 8), and the morainal gravels to the Lower Greensand. Like the Warren slide, the Hudson slide was not preceded by any change in external conditions of equilibrium, and it can be explained only by an increase of the pore-water pressure in the morainal gravels or in the silt seams located close to the base of the varved clay. The Hudson slide of 1915, like the Warren slide which took place the same year, was preceded by "unusually heavy rainfall, the July precipitation, reported by the Albany weather station, amounting to 5.05 inches." Yet the permeability of the varved clay, like that of the Gault clay in the Warren, is very small in vertical directions. Therefore the increase of the pore-water pressure in the gravel below the varved clay can only be due to the rise of the water table in the region where the gravel, or an aquifer communicating with the gravel, comes to the surface. Considering the striking similarity between the hydrologic conditions attending the Hudson and Warren slides, it is probably more than a mere coincidence that both slides took place with a period of less than 3 months.

Landslide statistics are still too incomplete to permit definite conclusions regarding the periodicity of major earth movements of the Warren type. Yet the subject is important enough to deserve the attention of geologists as well as that of hydrologists and civil engineers.

REVIEW OF LANDSLIDE-PRODUCING PROCESSES

Table 1 contains a review of the processes which may cause landslides. It is intended as an aid to memory and as a guide in landslide investigation.

In connection with landslide investigations the following fact should be remembered. Most slope failures take place during periods of exceptionally heavy rainfall or in spring, when the snow melts. However, exposure to rain or melting snow belongs to the normal existence of a slope. Hence, if a slope is old, heavy rainstorms or rapidly melting snow can hardly be the sole cause of a slope failure, because it is most unlikely that they are without any precedent in the history of the slope. They can only be considered contributing factors.

The circumstances attending the Hudson slide of 1915 (Fig. 9) are an example. There can be no doubt that the slope failed on account of exceptionally high pore-water pressures in the proximity of the boundary between clay and gravel. Yet the slope was very old. Therefore the slide must have been preceded by an unprecedented change in the conditions of the existence of the slope. One of them consisted in the accumulation of stockpiles of crushed rock, with a total weight of about 25,000 tons, along the upper edge of the slope (Newland, 1916, p. 104). Furthermore, it is conceivable that the deforestation of the outcrops of the gravel or of an adjacent aquifer has produced an unprecedented increase of the highest elevation of the water table, associated with unprecedented pore-water pressures at the base of the varved clay. Either one of the two changes may account for the catastrophe. Therefore the slide can be ascribed to action 1 (column D, Table 1), action 14, or to a combination of both. In connection with the rock slide of Goldau (Fig. 4), it was also impossible to decide whether the slide was caused by tectonic movements (action 2), by progressive weathering (action 8), or by a combination of both.

In order to facilitate discrimination between causes and contributing factors, Figure 10 has been prepared. It shows the factor of safety G_s with respect to sliding as a function of time. Each curve represents an individual slope. Curve A, (Fig. 10*a*) refers to the failure of an old slope due to progressive weathering of the slope-forming material or to tectonic movements. Curve B illustrates the failure of an old slope

TABLE 1.—*Processes leading to landslides*

A	B	C	D	E	F
Name of agent	Event or process which brings agent into action	Mode of action of agent	Slope materials most sensitive to action	Physical nature of significant actions of agent	Effects on equilibrium conditions of slope
Transporting agent	Construction operations or erosion	1. Increase of height or rise of slope	Every material	Changes state of stress in slope-forming material	Increases shearing stresses
			Stiff, fissured clay, shale	Changes state of stress and causes opening of joints	Increases shearing stresses and initiates process 8
Tectonic stresses	Tectonic movements	2. Large-scale deformations of earth crust	Every material	Increases slope angle	Increase of shearing stresses
Tectonic stresses or explosives	Earthquakes or blasting	3. High-frequency vibrations	Every material	Produces transitory change of stress	
			Loess, slightly cemented sand, and gravel	Damages intergranular bonds	Decrease of cohesion and increase of shearing stresses
			Medium or fine loose sand in saturated state	Initiates rearrangement of grains	Spontaneous liquefaction
Weight of slope-forming material	Process which created the slope	4. Creep on slope	Stiff, fissured clay, shale, remnants of old slides	Opens up closed joints, produces new ones	Reduces cohesion, accelerates process 8
		5. Creep in weak stratum below foot of slope	Rigid materials resting on plastic ones		

Water	Rains or melting snow	6. Displacement of air in voids	Moist sand	Increases pore-water pressure	Decrease of frictional resistance
		7. Displacement of air in open joints	Jointed rock, shale		
		8. Reduction of capillary pressure associated with swelling	Stiff, fissured clay and some shales	Causes swelling	
		9. Chemical weathering	Rock of any kind	Weakens intergranular bonds (chemical weathering)	Decrease of cohesion
	Frost	10. Expansion of water due to freezing	Jointed rock	Widens existing joints, produces new ones	
		11. Formation and subsequent melting of ice layers	Silt and silty sand	Increases water content of soil in frozen top-layer	Decrease of frictional resistance
	Dry spell	12. Shrinkage	Clay	Produces shrinkage cracks	Decrease of cohesion
	Rapid drawdown	13. Produces seepage toward foot of slope	Fine sand, silt, previously drained	Produces excess pore-water pressure	Decrease of frictional resistance
	Rapid change of elevation of water table	14. Initiates rearrangement of grains	Medium or fine loose sand in saturated state	Spontaneous increase of pore-water pressure	Spontaneous liquefaction
	Rise of water table in distant aquifer	15. Causes a rise of piezometric surface in slope-forming material	Silt or sand layers between or below clay layers	Increases pore-water pressure	Decrease of frictional resistance

TABLE 1.—*Continued*

A	B	C	D	E	F
Name of agent	Event or process which brings agent into action	Mode of action of agent	Slope materials most sensitive to action	Physical nature of significant actions of agent	Effects on equilibrium conditions of slope
Water—(*Cont.*)	Seepage from artificial source of water (reservoir or canal)	16. Seepage toward slope	Saturated silt	Increases pore-water pressure	Decrease of frictional resistance
		17. Displaces air in the voids	Moist, fine sand	Eliminates surface tension	Decrease of cohesion
		18. Removes soluble binder	Loess	Destroys intergranular bond	
		19. Subsurface erosion	Fine sand or silt	Undermines the slope	Increase of shearing stress

which occurred during a heavy rainstorm several years after the foot of the slope was undercut. The curves C to G in Figure 10b refer to slopes of recent origin, such as the sides of open cuts, and to old slopes which were exposed to unprecedented conditions (curves D and G). Slope C failed on account of spontaneous liquefaction

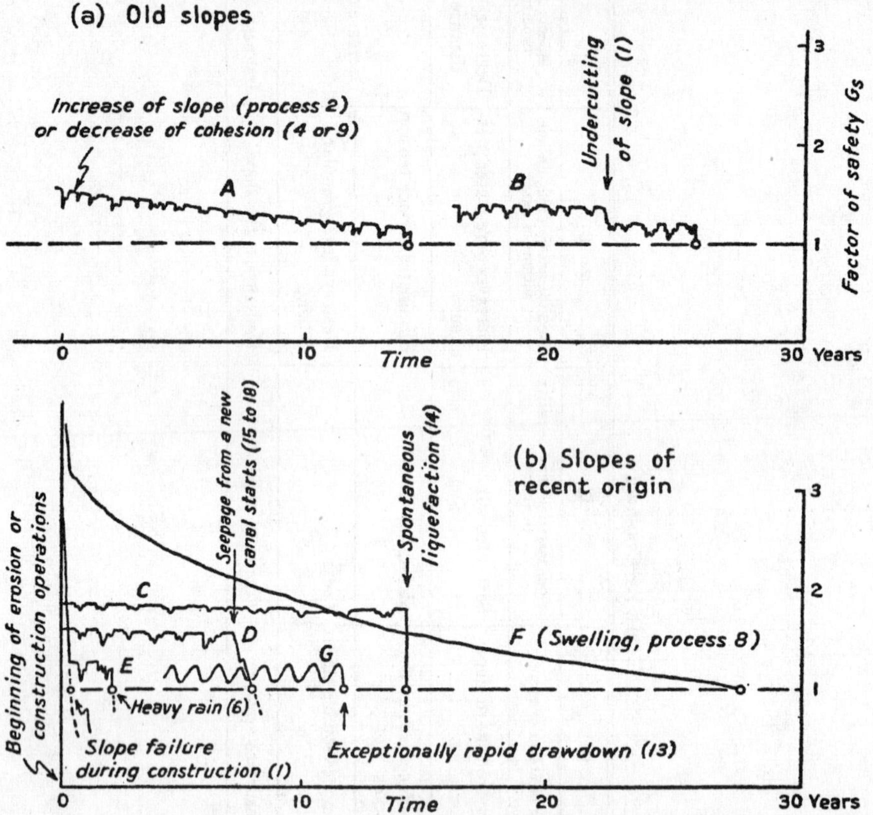

FIGURE 10.—*Diagram illustrating the variations of the factor of safety of different slopes prior to a landslide*

The ups and downs of the lines *A* to *E* are due to the changes in the pore-water pressure in the slope-forming material, associated with the alternation of dry and wet spells. Those of line *G* represent the effects of filling and emptying of the reservoirs adjoining the slope. The numerals in parentheses indicate the slide-producing processes listed in column C of Table 1.

caused by a near-by blast; D on account of seepage through the bottom of a recently constructed unlined canal located beyond the upper edge of the slope; E as a result of the heaviest rainfall since the time when the slope was formed; F on account of the gradual softening of stiff, fissured clay on which the slope was located, and G on account of an exceptionally rapid drawdown produced by a partial failure of the storage dam or the failure of an outlet valve.

The ups and downs of the curves A to E reflect the sequence of dry and wet spells, and those of curve G the normal variations of the free water level adjoining the slope of a storage reservoir. Special attention is called to curve B in Figure 10*a*, which represents a slope failure caused by a slight undercutting of the foot of the slope. The effects of this operation were not important enough to cause immediate failure. Yet, without it, the heavy rains that fell on the slope several years later would not have produced a slide. Hence the undercutting was the real cause of the failure, but the rainstorm was equally essential. This is one of many instances in which a slope failure can be accounted for only by a combination of two of the processes listed in column D of Table 1.

The degree of stability of an existing or a proposed slope cannot be reliably estimated, unless the process or processes, which may conceivably lead to a failure, are clearly understood and quantitative information regarding the controlling factors is available. The means for securing this information will be discussed in the last part of this paper.

DYNAMICS OF A LANDSLIDE

SURFACE MOVEMENTS PRECEDING A SLIDE

It has often been stated that certain slides occurred without warning. Yet no slide can take place unless the ratio between the average shearing resistance of the ground and the average shearing stresses on the potential surface of sliding has previously decreased from an initial value greater than one to unity at the instant of the slide. The only landslides which are preceded by an almost instantaneous decrease of this ratio are those due to earthquakes (Table 1, column D, action 3) and to spontaneous liquefaction (action 13). All the others are preceded by a gradual decrease of the ratio which, in turn, involves a progressive deformation of the slice of material located above the potential surface of sliding and a downward movement of all points located on the surface of the slice. Hence if a landslide comes as a surprise to the eyewitnesses, it would be more accurate to say that the observers failed to detect the phenomena which preceded the slide. The slide of Goldau, (Fig. 4) took the villagers by surprise, but the horses and cattle became restless several hours before the slide, and the bees deserted their hives (Heim, 1882). The clay slides which occurred in the spring of 1935, during the construction of the German superhighway from Munich to Saltzburg, came as a surprise to the supervising engineers, but one week before the movements started the laborers claimed that "the slope becomes alive."

In Figure 11 the ordinates of the curve located above the horizontal axis represent the factor of safety G_s of the slope with respect to sliding, and those of the curve below the axis the downhill movement of a point P in the upper reaches of the slide area. The abscissas represent the time which elapsed since the instant when the internal or external conditions for the equilibrium of the slope became worse than they ever had been. The line Oa shows the movement which preceded the slide. The distance OD_1 through which the point moved was covered in a time t_1.

The distance OD_1 (Fig. 11) depends primarily on the thickness of the zone within which the state of stress approaches the state of failure. If this zone is very thin, as in the slide of Goldau (Fig. 4), the distance OD_1 (Fig. 11) will hardly exceed an inch. Yet its effects were conspicuous enough to attract the attention of animals. On the other hand, if the potential surface of sliding is located in a fairly homogeneous mass of clay, or of residual soil which does not contain exceptionally weak seams or layers, the distance OD_1 in Figure 11 may amount to several feet, and the displacement is likely to be associated with the formation of tension cracks along the upper boundary of the slide area.

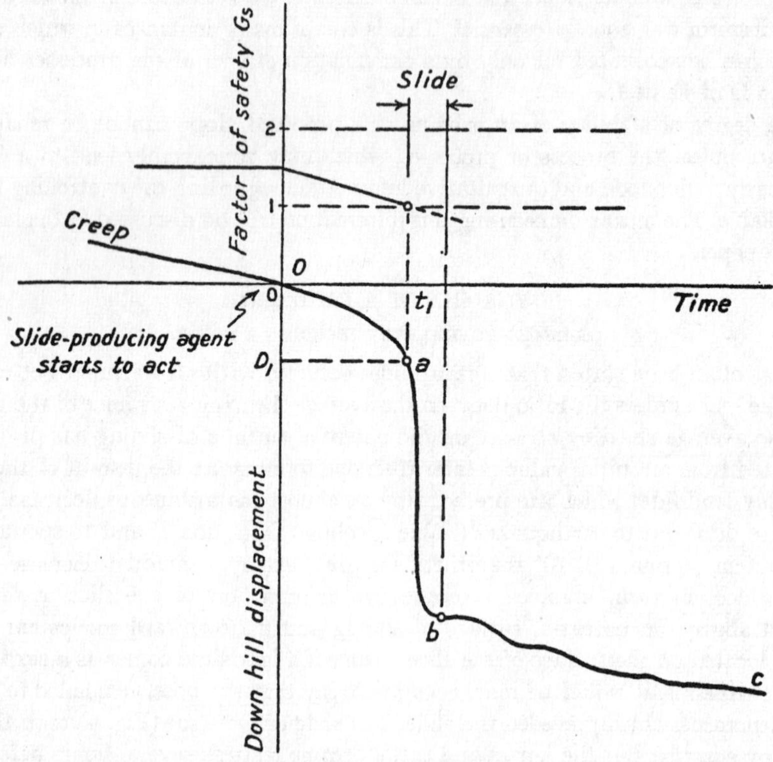

FIGURE 11.—*Diagram illustrating the ground movements which precede a landslide*

MOVEMENTS DURING THE SLIDE

Shear failures, along a surface of sliding through any material, are associated with a decrease of the shearing resistance. (*See* for instance Terzaghi and Peck, 1948, p. 81, 88.) Therefore, during the first phase of the slide, the sliding masses advance at an accelerated rate as shown by the upper part of the curve *ab* in Figure 11. However, as the slide proceeds, the force which tends to maintain the sliding movement decreases, because the mass comes into more and more stable positions. Therefore the accelerated movement changes into a retarded one, and finally it stops or assumes the character of creep.

The maximum velocity of the movement depends on the average slope angle of the surface of sliding, the importance of the effect of the slip on the resistance against sliding, and the nature of the stratification. The steepest surfaces of sliding develop in fairly homogeneous materials, such as irregularly jointed rocks (Fig. 5), cemented sands, and loess, which combine cohesion with high internal friction. Therefore slides in such materials are likely to be sudden.

The decrease of the shearing resistance produced by the slip ranges between about 20 per cent for fairly loose sand and clays with low sensitivity and probably 90 per cent for very loose, saturated sand or silt and soft extrasensitive clays. Slides in very sensitive clays and loose, saturated sand occur so rapidly that passing trains are likely to be wrecked. The slide of Vita Sikudden in Sweden is an example. It occurred in 1918 and claimed not less than 41 victims (Jarnvagrstyrelsen, 1922).

If a slide occurs on account of an excess hydrostatic pressure at the horizontal boundary between sand or silt and clay, the maximum velocity of the movement is likely to be very high even if the surface of sliding is horizontal. This is probably due to the fact that the deformation of the clay stratum remains very small until the instant when the slip removes the restraint on the base of the clay. The slide shown in Figure 3 and the Hudson slide (Fig. 9) are examples. The row of sheet piles shown in Figure 3 advanced in a few minutes over a length of about 1200 feet to a distance up to 60 feet from its original position. The Hudson slide was associated with a heave of the bottom of the valley, and the course of the creek which flowed in the valley was diverted by the formation of a ridge (Fig. 9). "So quickly were the waters discharged that fish were stranded on the bottom" (Newland, 1916, p. 103).

By contrast, slides due to an increase of the shearing stresses on a potential surface of sliding in more or less homogeneous masses of residual soil or of clay with low sensitivity seldom attain a velocity of more than a foot per minute, and the velocity may be as low as a foot per hour.

MOVEMENTS AFTER THE SLIDE

After the descent of the sliding mass has eliminated the differences between the driving and the resisting force (point b, Fig. 11), the movement passes into a slow creep, unless the slide has radically altered the physical properties of the sliding mass. The change can be due either to mechanical mixture of the slide material with water or to a destruction of the intergranular bonds at unaltered water content of the material.

The slide at Swir (Pl. 2) is an example of a process leading to a mechanical mixture of slide material with water. Before the slide occurred, the glacial till overlying the Devonian clay was firm and stable, and its porosity hardly exceeded 25 per cent. The expansion of the underlying clay broke the glacial till into large fragments. Rain water accumulated in the crevices between fragments. It caused disintegration and collapse of the fragments; and the mixture of water and till fragments flowed into the cut. During this process the porosity of the till must have increased from 25 to at least 40 per cent. Otherwise the slide material could not possibly have flowed like molasses.

The other extreme possibility, involving a great reduction of the average shearing resistance of the sliding material at almost constant water content, was strikingly illustrated by the slide which took place in 1924 in Kenogami, south of the Saguenay River in the Province of Quebec. The area is located at the head of a shallow gully which descends from the site of a pulp and paper mill toward the Au Sable River, a tributary of the Saguenay River The gully was carved out of clay.

Upstream from the head of the gully the Au Sable River was damned to form a pond, as shown in the sketch (Fig. 12). Prior to the slide the bottom of the gully

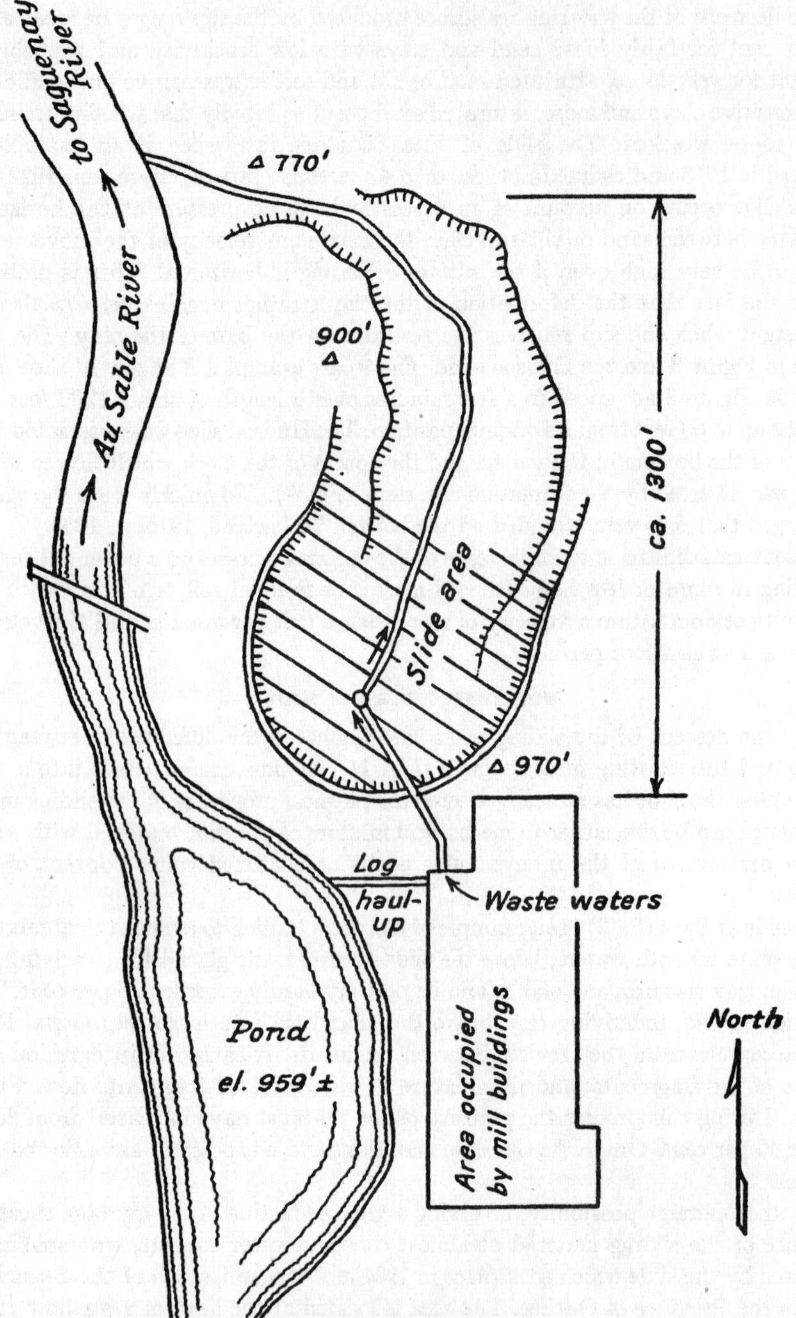

Figure 12.—*Sketch map of the site of the Kenogami slide south of the Saguenay River, province of Quebec, Canada*

descended from the plateau occupied by the mill toward the Au Sable River with an average gradient of about 1:10, and the side slopes of the gully were gentle. Most of the water which flowed through the gully came out of the discharge conduits of the mill.

In 1923 the capacity of the mill was considerably increased. As a consequence the waste-water discharge also increased, and the waste-water streamlet began to erode, starting at the head of the gully. During the spring and summer of 1924 the canyon, carved out by the streamlet, attained a maximum depth of about 70 feet, and its sides rose at an angle of about 65°. At that stage the walls of the canyon started to collapse. Slice after slice broke down, leaving vertical cliffs, probably representing the rear walls of tension cracks.[4] The collapsing slices crumbled, and the fragments formed a mud flow which descended on the bottom of the canyon with a velocity of 8 to 10 miles per hour into the valley of the Au Sable River and further on to the Saguenay River. The erosive action of the mud flow further increased the depth of the canyon to a maximum of about 100 feet and the gradient of the mud flow decreased to about 6°. The quantity of material which came out of the gully amounted to about 2.5 million cubic yards. Figure 2 of Plate 3 shows the cliffs at the head of the canyon a short time after the walls became stable. An eyewitness, who had the misfortune to stand on top of one of the slices while it was collapsing, described the mudflow (Personal communication, Mr. H. L. Munro) in the following words:

> "... after reaching the bottom I was thrown about in such a manner that at one time I found myself facing upstream toward what had been the top of the gully ... The appearance of the stream was that of a huge, rapidly tumbling, and moving mass of moist clayey earth .... At no time was it smooth looking, evenly flowing or very liquid. Although I rode in and on the mass for some time my clothes afterwards did not show any serious signs of moisture or mudstains.... as I was carried further down the gully away from the immediate effect of the rapid succession of collapsing slices near its head ... it became possible to make short scrambling dashes across its surface toward the solid ground at the side without sinking much over the ankles.[5]"

The topographic and hydrologic conditions attending the Kenogami slide strictly excluded the possibility that the water content of the sediment increased during the brief interval between the collapse of a slice and the descent of its fragments by flow on a very gentle slope. In other words, the same sediment which was stiff enough to form vertical cliffs with a height of 50 feet (Pl. 3, fig. 2) flowed a few seconds later on a gentle slope, at unaltered water content. Such a rapid transformation is conceivable only if the sliding sediment consists of an extrasensitive clay. In its natural state such a clay has the character of a brittle material with a considerable compressive strength. However, if it is remolded by kneading in the hand, it becomes so soft that it can easily be extruded between the fingers.

The preceding account of the eyewitness of the Kenogami slide indicates that the clay fragments themselves remained intact. However, at the surfaces of contact between the fragments the clay became remolded. At these surfaces it assumed the consistency of a viscous liquid, which acted as a lubricant and transformed the accumulation of relatively stiff chunks into a mud flow.

Extrasensitive clays are commonly very thixotropic. (*See*, for instance, Terzaghi

[4] The collapse of the slices was probably preceded by settlement which in turn produced the steep fissures. If this assumption is correct the peculiar character of the slides was due to process 5, Table 1 (Creep in weak stratum below the level of the foot of the cliff).

[5] When the writer of this paper stepped onto the mass of disintegrated boulder clay which overflowed the upper edge of the clay stratum at Swir III (Pl. 2), he sank rapidly more than kneedeep and he could not extricate himself without speedy asistance. This fact illustrates the great difference between the mud flow at Swir III and the Kenogami flow.

and Peck, 1948, p. 16.) Hence the strength of such clays is likely to increase considerably, at unaltered water content, after the clay has moved out of the slope. In Finland, where extrasensitive clays with high water content are rather common, the following observation has been made. A few days after a slide in such a clay has occurred, the clay, which flowed out of the slope, is likely to be stiff enough to support the weight of light construction equipment; yet, immediately after the slide, one can hardly walk on the slide material without sinking into it.

If a slide causes the disintegration and breakdown of a mass of nonthixotropic sediments, such as glacial till or loess, during a period of heavy rainfalls, the slide material is likely to creep very actively for many years after the slide, and every new rainstorm accelerates the rate of creep. This process is indicated by the steps in the trend of the line *bc* in Figure 11. Examples are the soil movements which followed the Drynock slide in Canada (Eng. Rec., 1909) and the failure of the base of a railroad fill in Austria (Raschka, 1912). The seasonal acceleration may be important enough to damage lines of communication crossing the slide material, and remedial measures are indicated.

LANDSLIDE PROBLEMS

PREVENTIVE MEASURES

The practical importance of a thorough investigation of the degree of stability of existing and of proposed slopes is illustrated by the following statistics published by Ladd (1934):

> "Within the last three years landslides have resulted in more than 3,000 deaths and very great material losses. Since the spring of 1931 landslides have led to at least thirteen railroad disasters, four of which were in foreign countries and nine in the United States. By these, 227 people were killed and 31 were injured."

The foremost requirements for slide prevention are reliable information on the geologic structure of the ground adjoining the slope under consideration, to be obtained by test borings combined with a geologic survey and a clear conception of the processes which may conceivably lead to a failure of the slope. The processes that produce slides are listed in Table 1 (Nos. 1 to 19).

The first step toward slide prevention is to take all the measures required to make the slide-producing processes as ineffective as conditions permit. The rise of the piezometric surface behind the slope, associated with a displacement of air during heavy rainstorms (process 7), can be reduced by covering the slope and a broad strip of the area beyond the crest of the slope with a layer or lining having a low permeability. The formation of ice layers (process 11) and subsequent sloughing, known as solifluction, can be counteracted by drainage and various other means. (*See*, for instance, Terzaghi and Peck, 1948, p. 131–134.) The formation of deep shrinkage cracks (process 12) can be avoided by covering the slope with sod or a thick layer of sand. The danger of spontaneous liquefaction (processes 3 and 14) can be eliminated by compaction. The available technical means for accomplishing it was described by Terzaghi and Peck (1948, p. 379–381). The risk of slope failures, due to the concentration of flow lines at the foot of a slope after rapid drawdown (process 13), can be avoided by covering the lower part of the slope with an inverted filter of a weight sufficient to counteract the seepage pressure exerted by the percolating water (Reinius, 1948). Seepage from an artificial source of water, such as a canal or a storage reservoir (process 16–18), can be prevented by a water-tight lining of the bottom of the body of open water, a cut off, deep drains, or a combination of these. The danger of subsurface erosion (process 19) can be removed by covering the exit area of the water veins with an inverted filter or by adequate drainage (Terzaghi and Peck, 1948, p. 502–514). If the soil located beneath the slope contains layers or pockets of relatively permeable material such as sand or silt which may conceivably communicate with a distant aquifer (process 15), drainage is imperative. If the process which led to the Hudson slide of 1915 (Fig. 14) had been recognized, the catastrophe could have been avoided at a moderate expense, by "bleeding" the gravel beneath the varved clay.

By means of these and similar provisions it is commonly possible to prevent a decrease of the degree of stability of the slope. There are, as a matter of course, exceptions to this rule. Slides on slopes on stiff, fissured clay, due to the gradual softening of the slope-forming material (process 8), are an example. However, the effect of time on the stability of such slopes may be evaluated on the basis of experience. Figure 6 is a graphic representation of what experience can teach a competent observer.

The next step is to estimate the degree of stability of the existing or proposed slope under the conditions prevailing at the time of the investigation. If a slope is located on clay, which does not contain layers or pockets of relatively permeable materials, a fairly accurate computation of the factor of safety of the slope, with respect to sliding, can be made on the basis of the results of laboratory tests on undisturbed samples. This investigation belongs in the realm of soil mechanics.

The stability of slopes on clay containing layers of soil or sand and tha of slopes on relatively permeable materials, such as silt or coarser sediments, depends not only on the physical properties of the slope-forming material but also to a large extent on the pore-water pressure in the most permeable members of the geologic system. In such instances it is necessary to secure reliable information concerning the pore-water pressure.

Standard observation wells can be relied upon only if their lower ends are located in relatively permeable material, not finer than fine sand. Otherwise it is necessary to use pore-pressure gages (Terzaghi, 1943a). The relative merits of the existing types of gages are being investigated by the Subcommittee on the Measurement of Pore Water Pressures in Earth Dams and their Foundations (chairman Prof. R. B. Peck, Univ. of Illinois), Committee on Earth Dams, Soil Mechanics & Foundations Division, A.S.C.E.

Figure 13 is an example of the surprising results which may be obtained by gage observations. Figure 13*a* is a section through the upper part of a stratum of stiff, glacial clay consisting of three layers. The uppermost layer, labeled C_1, is remarkably homogeneous. In order to get information on the gradual consolidation of the clay, under the influence of the weight of an ore pile, pressure gages were installed in the stratum C_1, at different depths below the surface. To the surprise of everybody concerned, the gage readings showed that the pore water was under artesian pressure before the ore was piled up. The results of the gage readings are shown in Figure 13*b*.

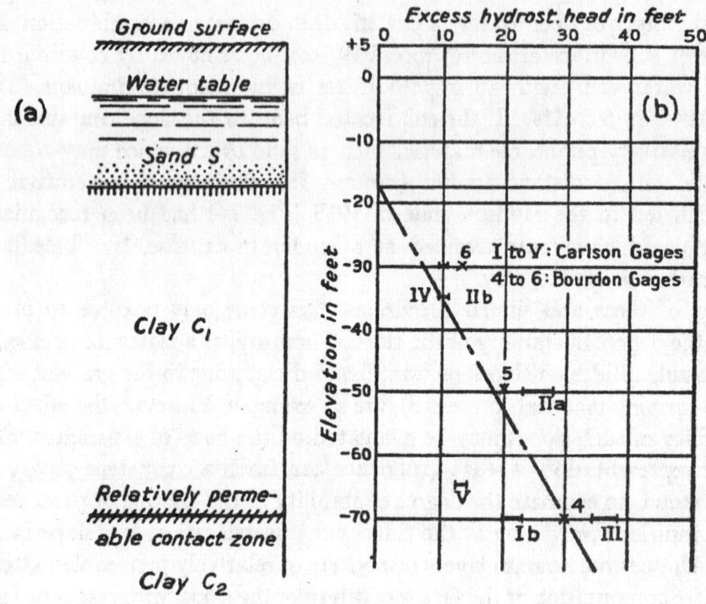

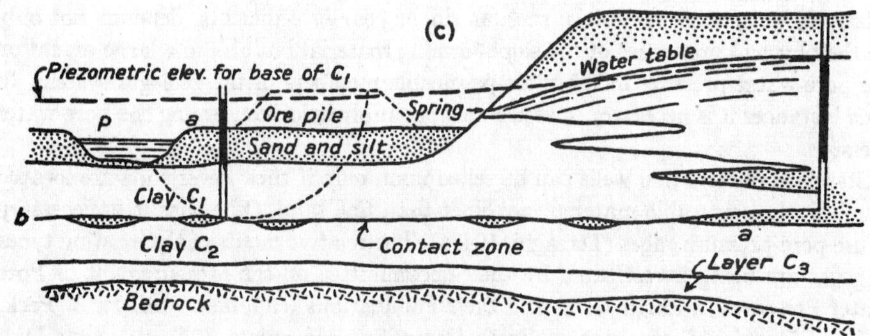

FIGURE 13.—*Artesian conditions*

(a) Section through upper part of stratum of glacial clay. (b) Results of pore-pressure gage readings which disclosed the existence of artesian conditions in top layer C_1 prior to application of ore load. (c) Diagram illustrating the writer's hypothesis concerning the origin of the artesian pressure.

The probable origin of the artesian pressure is illustrated by Figure 13*c*. South of the valley floor, occupied by the ore yard, the clay stratum C_1 interfingers with the sand and gravel stratum located beneath a terrace formerly occupied by a city park. It was learned from well records that the water table beneath the plateau is between

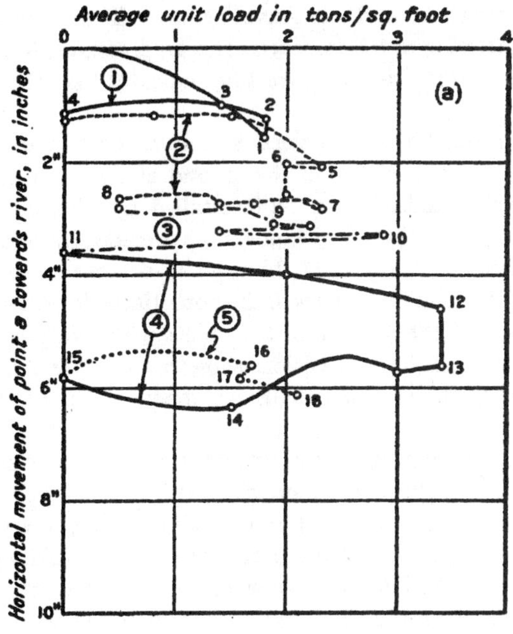

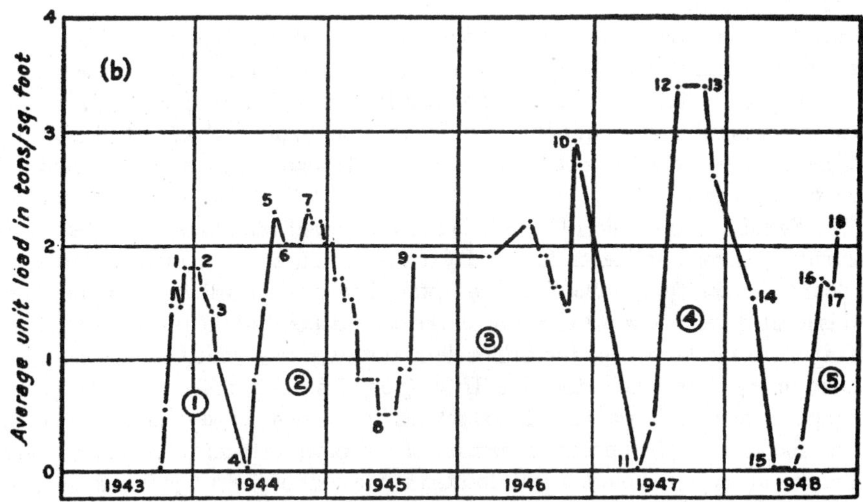

FIGURE 14.—*Ore loading*

(a) Relation between unit load on base of ore pile shown in Figure 13b and the horizontal outward movement of point a, Fig. 13b. (b) Seasonal variations of ore load. The numerals 1 to 18 in (a) indicate the stages of loading 1 to 18 in diagram (b). The numerals surrounded by circles represent the numbers of the cycles of loading.

50 and 90 feet above the valley floor. The test borings have shown that the clay beneath the ore yard is locally stratified. Hence the artesian pressure seems to be due to the existence of relatively permeable connections between the sand and silt seams

in the clay and the water-bearing mass of sand and gravel beneath the plateau south of the ore yard. However, the scattering of the recorded artesian pressures, from the average represented by the dash-dotted line in Figure 13*a*, indicates that the communications are imperfect.

If the geologic structure of the ground beneath a slope is complex, or if an accurate determination of the average shearing resistance of the ground along the potential surface of sliding is impracticable, a landslide can be forestalled by adequate field observations. These are based on the fact that every landslide is preceded by creeplike movements represented by the line *Oa* in Figure 11. In order to detect these movements, reference points are established on those parts of the potential slide area which are likely to move most actively, and the position of the points with reference to a stable base is periodically determined. If the observations show that the rate of movement increases without any change in external conditions, the rise of the slope is reduced by excavation.

Figure 14 shows the results of the periodic survey of reference points. The observations were made for determining the height to which ore can be piled up on the valley floor shown in Figure 13*c*, without the risk of a failure by sliding toward the river. The location of the ore pile and the position of the potential surface of sliding are indicated in Figure 13*c* by dotted lines. For the sake of convenience the reference points were established at the waterfront, *a*, in Figure 13*c*. Figure 14*b* shows the seasonal variations of the load on the ore yard, and Figure 14*a* represents the relation between the unit load on the ore yard and the corresponding outward movement of point *a* (Fig. 13*c*).

Based on the knowledge of the physical properties of the clay obtained by laboratory tests on undisturbed samples of clay C_1, and on the results of the observations shown in Figure 14*a*, it was concluded that it is safe to increase the load on the yard until the outward movement of point *a* (Fig. 13*b*) amounts to 2 feet. If the soil tests and field observations had not been made, the safe load on the ore yard could only have been guessed.

The Swedish Geotechnical Commission has proposed the following procedure for preventing railroad accidents due to clay slides in railway cuts. In cuts with slopes of doubtful stability, flexible pipes are installed in vertical drill holes, whose lower end is located well below the elevation of the deepest point of the potential surface of sliding. Into the bottom of the drill hole is driven a rod, which, at the outset, is located in the center line of the pipe. If the ground starts to move the upper part of the pipe moves with it, whereas the rod retains its position. As soon as the upper part of the rod touches the uppermost section of the pipe, an electric circuit is closed; automatically operated block signals located at the ends of the cut go into action, and traffic through the cut is discontinued until the danger is eliminated by trimming the slopes or other appropriate means (Jarnvagrastyrelsen, 1922).

LANDSLIDE CORRECTION

If a slope has started to move, the means for stopping the movement must be adapted to the processes which started the slide (Table 1, column *C*). It is hardly an exaggeration to say that most slides are due to an abnormal increase of the pore-water pressure in the slope-forming material or in a part of its base (processes 6, 7, 13, 15, 16). In such instances radical drainage is indicated.

The extraordinary efficacy of drainage has recently been demonstrated by the following observation. During a tropical cloudburst, involving a precipitation of 9 inches in 24 hours, a slide occurred on a slope rising at an average angle of 30°.

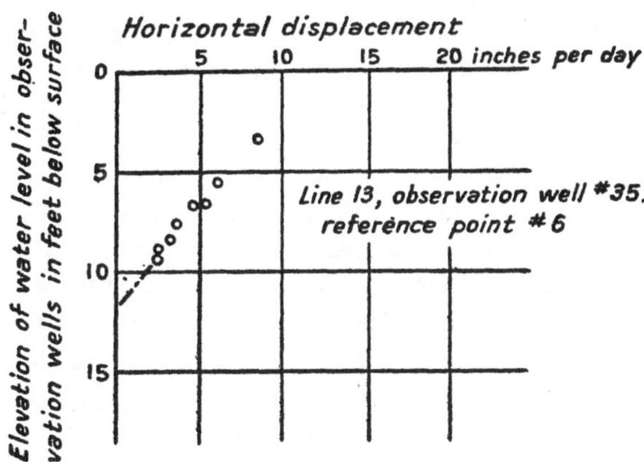

FIGURE 15.—*Diagram showing the relation between the position of the water table with reference to a slope after failure (ordinates) and the horizontal component of the corresponding downhill movement of the surface of the slope*

The slope is located on deeply weathered metamorphic rocks, and the deepest part of the surface of sliding was about 130 feet below the surface. The slide area was about 500 feet wide, 1000 feet long, and the quantity of material involved in the slide exceeded half a million cubic yards.

Since the slide occurred in the close proximity of a hydro-electric power station, immediate action was indicated. In order to get quantitative information concerning the ground movements and the factors which determine the rate of movement, reference points were established on several horizontal lines across the slide area, and observation wells were drilled in the proximity of the reference points. By plotting the vertical distance between slope and the water level in the wells as ordinates, and the corresponding rate of movement of the adjoining reference points as abscissas, diagrams like Figure 15 were obtained. Although the moving mass had a depth up to 130 feet, the diagrams showed that the lowering of the water table by not more than about 15 feet would suffice to stop the movement.

Drainage was accomplished by means of toe trenches, drainage galleries, and horizontal drill holes extending from the headings into water-bearing zones of the jointed rock. The movements ceased while the drainage was still in an initial state. The following rainy season brought record rainfalls; yet the ground movements in the slide area remained imperceptible.

Drainage can also be used to advantage if the water seeps through open joints between chunks of relatively impermeable material such as shale (Forbes, 1947).

If the permeability of the slide material is too low to permit drainage by pumping from wells or bleeding through galleries, the resistance against sliding can be increased and the ground movements can be stopped by means of the vacuum method or the electro-osmotic method (Terzaghi and Peck, 1948, p. 337–340). Both procedures create a reduction of the pore-water pressure associated with a permanent decrease of the water content of the drained strata which, in turn, increases the cohesion and shearing resistance of the ground. Similar effects can be obtained by circulating hot, dry air through galleries located within the unstable material (Hill, 1934).

If drainage is difficult or if its success is doubtful, the ground movements can be stopped either by reducing the slope angle or by constructing artificial barriers, such as heavy retaining walls or rows of piles across the path of the moving material. A list of the current procedures and comments on their efficacy have been published by Ladd (1935). (*See also* Seaton, 1938.)

Every landslide or slope failure is the large-scale experiment which enables competent investigators to draw reliable conclusions regarding the shearing resistance of the materials involved in the slide. Once a slide has occurred on a construction job, the data derived from the failure may permit reliable computation of the factor of safety of proposed slopes on the same job and modification of the design in accordance with the findings. This procedure was successfully used during the construction of several of the German superhighways on treacherous ground (Gottstein, 1936).

CO-OPERATION BETWEEN GEOLOGIST AND ENGINEER ON LANDSLIDE PROBLEMS

If a geologist is called upon to report on the degree of stability of an existing or a proposed slope, he is likely to furnish an adequate account of the geology of the site and of the hydrologic conditions. However, his understanding of the physical processes, which may impair the stability of the slope, is commonly deficient because he has not been trained to think in terms of exact physical concepts. This is demonstrated by the indiscriminate use of the term "lubrication" and other misnomers. Very few geologists have a clear conception of the difference between total and effective pressure, of the effect of the pore-water pressure and of surface tension on the shearing resistance of sediments, and of the relation between stress, strain, and time for cohesive soils. Yet, an opinion concerning the means for increasing the stability of a slope is merely guesswork unless it is based on a knowledge of fundamental physical relationships, and the guess may be wrong.

A civil engineer, trained in soil mechanics, may have a better grasp of the physical processes leading to slides. However, he may have a very inadequate conception of the geologic structure of the ground beneath the slopes, and he may not even suspect that the stability of the slope may depend on the hydrologic conditions in a region at a distance of more than a mile from the slope.

On account of the wide range of specialized knowledge and experience required for judging the stability of slopes, important landslide problems call for co-operation between geologist and engineer. To get satisfactory results the geologist should be familiar with the fundamental principles of soil mechanics, and the engineer should know at least the elements of physical geology.

REFERENCES

Close, U., and McCormick, E. (1922) *Where the mountains walked*, Nat. Geog. Mag., vol. 41, p. 445–464.

Eng. News-Rec. (1932) *Three dams on San Andreas fault have resisted earthquakes*, vol. 109, p. 218–219.

Eng. Rec. (1909), *Landslides*, vol. 59, p. 737–740.

Forbes, H. (1947) *Landslide investigations and correction*, Am. Soc. Civil Eng., Tr., vol. 112, Paper 2303, p. 377–442.

Freeman, John R. (1932), *Earthquake damage and earthquake insurance*, New York, McGraw-Hill Book Co., Inc.

Fuller, M. L. (1912) *The New Madrid earthquake*, U. S. Geol. Survey, Bull. 494, 119 pages.

Gottstein, E. v. (1936) *Two examples concerning underground sliding caused by construction of embankments*, 1st Inter. Conf. Soil Mech., Pr. Cambridge, Mass., vol. III, p. 122–128.

Griggs, David T. (1936) *Deformation of rocks under high confining pressures*, Jour. Geol., vol. 44, p. 541–577.

Haefeli, R. (1944) *Zur Erd- und Kriechdruck Theorie*, Schweiz. Bauztg., vol. 124.

Hardy, W. B. and T. V. (1919) *Note on static friction and on the lubricating properties of certain chemical substances*, Philos. Mag. London, vol. 39, no. 223, p. 32–35.

Heim, A. (1882) *Über Bergstürze*, Naturf. Gesell. Zürich, Neujahrsblatt 84.

Hill, R. A. (1934) *Clay stratum dried out to prevent landslips*, Civil Eng., vol. 4, p. 403–407.

Hollingworth, S. E., Taylor, J. H., and Kellaway, G. A. (1944) *Large-scale superficial structures in the Northampton Ironstone Field*, Geol. Soc. London, Quart. Jour., vol. C, p. 1–44.

Jarnvagrastyrelsen (1922), *Statens Jarnvagars Geotekniska Commission, 1914–1922*, Final Rep. to Royal Bd. State Railroads, May 31.

Koppejan, A. W., van Wamelen, B. M., and Weinberg, L. J. H. (1948) *Coastal flow slides in the Dutch Province of Zeeland*, 2d Inter. Conf. Soil Mech. Found. Eng., Pr., Rotterdam, Holland, vol. V, p. 89–96.

Ladd, G. E. (1934), *Bank slide in deep cut caused by draught*, Eng. News-Record, vol. 112, p. 324–326.

——— (1935) *Landslides, subsidences and rockfalls*, Am. RR. Eng. Assoc., Pr., vol. 36, p. 1091–1162.

Lapworth, H. (1911) *The geology of dam trenches*, Inst. Water Eng., Tr., vol. 16, p. 25.

Lehmann, O. (1926) *Die Verheerungen in der Sandlinggruppe*, Denkschrift Ak. Wiss. Wien, Math. natw. Klasse, vol. 100, p. 263–299.

Lugeon, M., and Oulianoff, N. (1922) *Sur le balancement des couches*, Univ. de Lausanne, Lab. Géol.. Géog. Bull., no. 32.

Mackin, J. H. (1941) *A geologic interpretation of the failure of the Cedar Reservoir, Washington*. Eng. Exper. Sta., Bull. 107. Univ. Washington.

Marmer, H. A. (1930) *Chart datums*, U. S. Coast Geod. Survey, Spec. Pub. 170.

McConnell, R. G., and Brock, R. W. (1904) *Report on the great landslide at Frank, Alberta*, Canada Dept. Interior, Ann. Rept. 1902–1908, pt. 8, 17 pages.

Mohr, C., Haefeli, R., Meisser, L., Waltz, F. and Schaad, W. (1947) *Umbau der Landquartbrücke der Rhätischen Bahn in Klosters*, Schweiz. Bauztg., vol. 65, p. 5–8, 20–24, 32–37.

Newland, D. H. (1916) *Landslides in unconsolidated sediments*, N. Y. State Mus., Bull. 187, p. 79–105.

Pollack, V. (1913) *Über Seeuferbewegungen*, Österr. Wochenschr. für öffentl. Baudienst, vol. 19, Heft 35, p. 595–603.

Raschka, H. (1912) *Die Rutschungen im Abschnitt Ziersdorf-Eggenburg der Kaiser-Franz-Josefs-Bahn*, Ztschr. Österr. Ing. Arch. Ver., p. 561.

Reinius, E. (1948) *On the stability of the upstream slope of earth dams*, Doctor's Thesis, Kungl. Tekniska Högskolan, Stockholm, Sweden.

Scheidig, A. (1934) *Der Löss und seine geotechnischen Eigenschaften*, Theodor Steinkopf, Dresden and Leipzig, 233 pages.

Seaton, T. H. (1938) *Engineering problems associated with clay, with special reference to clay slips*, Inst. Civil Eng. Jour. (London), Paper 5170, p. 457–498.

Sharpe, C. F. S. (1938) *Landslides and related phenomena*, Columbia Univ. Press, New York, 136 pages.

Sharpe, C. F. S., and Dosch, E. F. (1942) *Relation of soil-creep to earth-flow in the Appalachian Plateaus*, Jour. Geomorph., vol. 5, p. 312–324.

Skempton, A. W. (1948) *The rate of softening of stiff, fissured clays*, 2d Inter. Conf. Soil Mech. Found. Eng. Pr., Rotterdam, vol. II, p. 50–53.

Terzaghi, K. (1925) *Erdbaumechanik*, Franz Deuticke, Wien, 399 pages.

——— (1936) *Stability of slopes on natural clay*, 1st Inter. Conf. Soil Mech. Found. Eng., Pr., Cambridge, Mass., vol. I, p. 161–165.

——— (1942) *Soil moisture and capillary phenomena in soils*, Physics of the Earth, vol. IX (Hydrology), McGraw-Hill Co., New York, p. 331–363.

——— (1943a) *Measurements of pore-water pressure in silt and clay*, Civil Eng., vol. 13, p. 33–36.

——— (1943b) *Theoretical soil mechanics*, John Wiley and Sons, New York, 510 pages.

——— and Peck, R. B. (1948) *Soil mechanics in engineering practice*, John Wiley and Sons, New York, 566 pages.

Toms, A. H. (1946) *Folkestone Warren landslips: Research carried out in 1939 by the Southern Railway Company*, Inst. Civ. Eng., RR. Eng. Div., Railway Paper no. 19.

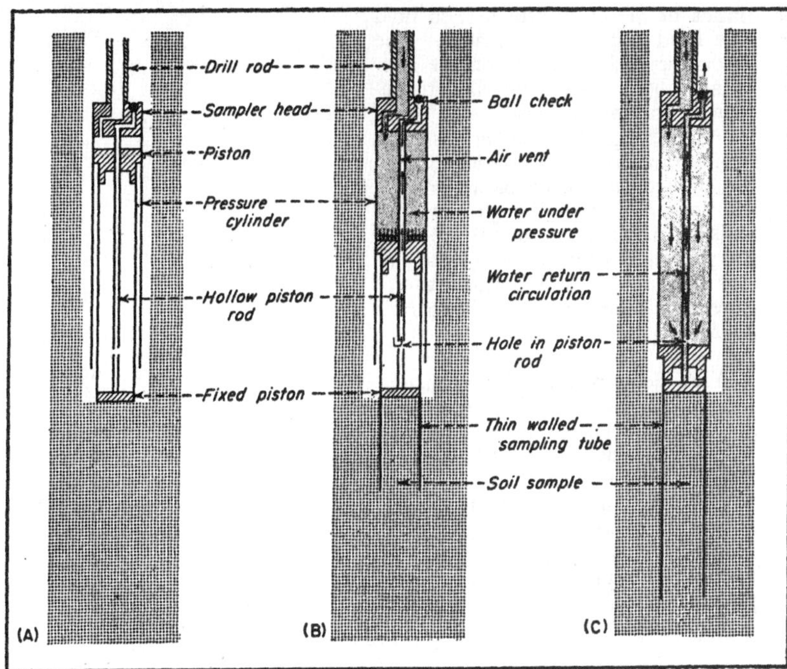

SUCCESSFUL METHOD: (A) Sampler is set in drilled hole; (B) sampling tube is propelled hydraulically into soil; (C) pressure is released through hole in piston rod.

ADVANTAGE: only one string of rods is needed.

New piston-type soil sampler

J. O. Osterberg
Associate Professor of Civil Engineering
Northwestern University
Evanston, Ill.

A SIMPLE, hydraulically-operated piston sampler—recently developed by the author—has been found to be unusually successful in obtaining full recovery on 5-in. dia tube samples 4 ft long in both sands and clays. A major advantage of this sampler over the usual fixed piston type is that only one string of rods is needed to operate it. In addition, the pushing is done hydraulically, and it is impossible to overdrive.

• How it operates—The sampler is lowered into a previously drilled and cleaned-out hole (A). When water pressure is applied through the drill rod, a piston to which a thin walled sampling tube is attached is forced out of the pressure cylinder (B). A second piston—the fixed piston inside the sampling tube—is connected to the sampler head by a hollow rod. As the piston is forced down in the pressure cylinder, air in the sampler tube escapes through the hollow rod and ball check. The reaction for the pushing is obtained by clamping the drill rod either to the casing or to the drill rig.

When the piston has reached its full stroke and the sampling tube penetrated its full depth in the soil, water pressure is automatically relieved by allowing circulation through a hole in the hollow piston rod and through the ball check (C). The sampler is then turned 1½ revolutions (the inside tube being held to the outside tube by means of a friction clutch) to break off the soil at the bottom of the tube. The sampler is then ready for removal from the drill hole.

• Developed for Milwaukee job—The sampler was developed in connection with a comprehensive foundation investigation for the Milwaukee Sewage Commission at its sewage disposal plant on Jones Island. Since there was no hydraulically operated rig available, and since it would be impractical to push 5-in. dia tubes into the bore hole with the cable tool rig on the site, the need was apparent for the development of an independently operated sampler.

The sampler constructed for this job has a 4-ft stroke and uses 5-in. OD tubing for the samples. Unfortunately, 14-ga seamless tubing was not available so ⅛-in. wall steel tubing was used. Since both sands and clays were expected, no inside clearance was left on the cutting edge. Unusually good recovery was obtained, however, in both sands and clays in spite of the thick tube wall and no inside edge clearance.

• Tests went 81 ft deep—The samples were taken in both cased and uncased holes, but drilling mud was used in all holes. The holes were advanced and cleaned by a fish-tail bit and mud circulation. Continuous samples were taken between 24-ft and 81-ft in depth. Pressure was applied to the sampler with clean water through a hand operated pump capable of exerting 250 psi pressure. This corresponded to a 4,000-lb thrust on the sampling tube.

At first it was thought that a hand operated pump would not be satisfactory because of the pulsations in pressure when stroking. However, constant pressure was maintained easily. Moreover, the excellent recovery and lack of disturbance in the samples was proof of the adequacy of a simple hand pump.

Of the three holes drilled from which 5-in. samples were taken, a total of 38 tube samples were taken in material varying from fine silt to coarse sand with some gravel, and from soft uniform clay to a hard varved clay. In addition, considerable amounts of organic clayey silts with fibrous material and numerous shells were encountered.

• High rate of recovery—Of the 38 tubes, 26 had full recovery (99-100%), 7 had between 96 and 99% recovery, and 3 were empty. About 1-ft of sample was missing out of each of the other two tubes. In the case of sands, when 4 to 5 in. had dropped out of the bottom of the tube and the top of the soil was flush with the piston, full recovery was assumed. Full recovery was obtained on all samples taken entirely in clay. The soil samples were left in the tubes until ready for laboratory analysis.

• **Fixed-piston vs open-drive**—Basically, fixed-piston samplers offer distinct advantages over open-drive samplers: During driving, the fixed piston prevents entrance of shavings and disturbed soil. During sampling, the fixed piston aids recovery by pulling hard soils into the tube when wall-friction tends to prevent such entrance, and in soft soils prevents excess soil from entering.

Another advantage: During withdrawal, the fixed piston eliminates the pressure over the sample, preventing it from dropping out of the tube.

• **Saves time and labor**—The hydraulically-operated piston sampler has additional advantages: The time-consuming job of stringing a set of rods inside of the outer drill rods to hold the fixed piston is eliminated. The hydraulic pushing action on the sampling tube is easily controlled to obtain a continuous push.

Another great advantage is that while ejecting, the sampler cannot be twisted, struck by blows or over driven. And even though the tube strikes a boulder or stratum too hard to penetrate—thus preventing a full 4-ft stroke—it is still possible to secure 100% recovery for the partial stroke.

• **Who developed it**—The hydraulic sampler was developed in the Northwestern University machine shops, with helpful suggestions of P. C. Rutledge, chairman, Department of Civil Engineering; and John Schmertmann, former graduate student. Klug & Smith Company, Engineers and Contractors, Milwaukee, directed all sampling at Jones Island and suggested modifications of the sampler.

• **Who makes it**—Arrangements have been made with Soil Testing Services, Inc., 4520 West North Ave, Chicago, for manufacturing the hydraulic sampler in 3-in. and 5-in. sizes.

JOURNAL OF THE

BOSTON SOCIETY OF CIVIL ENGINEERS

Volume xxxx JANUARY, 1953 Number 1

CONTROL OF FOUNDATION SETTLEMENTS BY PRELOADING

BY STANLEY D. WILSON,* Member

(Presented at a meeting of the Structural Section of the Boston Society of Civil Engineers, held on October 8, 1952.)

INTRODUCTION

THE foundation engineer has a costly problem on his hands when the site of a proposed structure is underlain by compressible soils. The engineer, after analyzing the many possible solutions, must choose one which is both effective and economical. If an economical solution is not found, the foundation costs may be so excessive that the owner is discouraged from building. In arriving at a decision as to foundation type, the method of preloading to control settlements is quite frequently overlooked. Provided that foundation conditions are suitable and provided also that the foundation consultants are given an opportunity to study the problem well in advance of start of construction operations, the method of preloading may not only be the most effective, but also the most economical.

Preloading consists of applying a dead load or surcharge, equal to or greater than the weight of the proposed structure, over the site, thus developing the settlements prior to construction. The surcharge load is then removed and replaced by the building itself.

One may well ask why this method is not used more frequently if it is so simple. One reason is that modern construction practice does not permit the long period of waiting often required for the settlements to occur. In the case of a building founded on a thick bed of clay, this required time might range from a few years up to several decades. However, this is not always the case, and in a few moments I will describe four recent projects on which the method of preloading was used successfully, including one case in which the settlement rate was accelerated by "overloading".

Load-Settlement Relationships. In order to explain why preloading is so effective, it is necessary to review briefly the load-settlement relationships of compressible soils.

*Assistant Professor of Soil Mechanics and Foundation Engineering, Harvard University, Cambridge, Mass.

During the process of sedimentation of a fine-grained soil, water is squeezed out under the weight of the overlying sediment and the structure becomes more dense. This process is called consolidation, and each additional overburden load produces additional consolidation which reduces the thickness of the compressible strata. The relationship between void ratio, or thickness, and effective overburden pressure is shown by the line V-V in Fig. 1. This line is called the

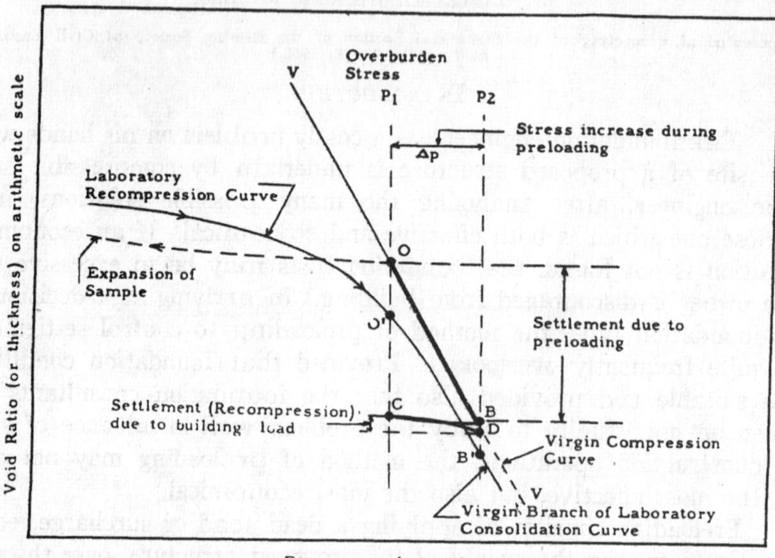

Fig. 1. Comparison of settlements during preloading with settlements during recompression.

virgin compression curve, and for most soils becomes a straight line when the pressure is plotted on a log scale.

If the clay has never been subjected to a stress greater than the present overburden stress, it is said to be normally consolidated, and the relationship between thickness and overburden stress may be represented by point 0 (Fig. 1) which lies along the virgin compression curve. In order to determine the slope of this line, it is necessary to test undisturbed samples of the material in the laboratory. Even the best samples that can be obtained, however, are subjected to some disturbance and stress changes such that when reconsolidated in the laboratory to the original overburden stress, point 0′ falls below point 0. As the stress is increased to larger values, a straight-line virgin branch of the laboratory compression curve develops which is parallel to the virgin compression curve V-V. Drawing a line through point 0 with this slope establishes the virgin compression curve.

If a preload (ΔP) equal to the weight of the proposed structure is placed on the ground surface, the stratum will consolidate along curve OB, producing settlements as indicated in Fig. 1. If the preload is now removed, reducing the stress to its original value p_1, the clay will swell slightly along the rebound curve BC. When the building or other structure has been erected, and the stress increased again to p_2 the settlement resulting from recompression along the curve CD is seen to be only a small fraction of that which developed under the preload.

It should be noted that the recompression branch CD is much flatter than the laboratory recompression branch, and the change in void ratio BD much less than 00′. The reason for this is that the partial stress relief which occurs when the preload is removed causes little disturbance to the structure of the clay in contrast to the relatively severe sampling operations.

The above discussion does not take into consideration the long-time effects of secondary consolidation. For most inorganic clays these effects are relatively minor and may be disregarded. For organic soils, however, the secondary time effects may be of considerable importance and preloading with a load greater than the weight of the structure may be required.

Examples of Preloading

The following four examples illustrate specific applications of the method of preloading to control settlements. Each example represents a different type of problem.

Example No. 1

The first example I have chosen is that of the new American Stores Warehouse in Philadelphia,* a one story structure approximately 370 ft. wide and 1130 ft. long, nearly ten acres under one roof. Interior columns at 21 ft. centers support the roof, and, in

*Designed by the Boston firm of Ganteaume & McMullen. The method of preloading was recommended by A. Casagrande. The field observations were planned and supervised by S. B. Avery, Jr. and the author.

addition, carry a portion of the live load through cantilevered supports. The live floor load is 400 lb. per sq. ft. The walls are supported on a five-foot high exterior concrete wall with spread footing.

Site Conditions. The structure is underlain by an old valley which cuts diagonally across the site as shown in Fig. 2. The plan

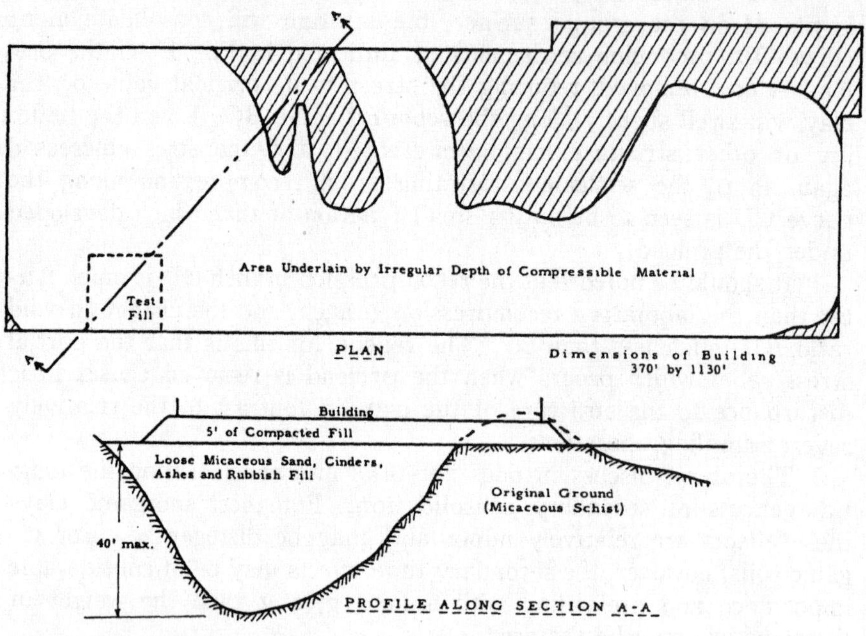

Fig. 2. Foundation conditions at site of American Stores Warehouse

shows that about one-third of the warehouse rests on original ground (micaceous schist), and the remainder on an existing fill of loose micaceous sand, cinders, ashes and miscellaneous rubbish. The depth of this irregular and erratic fill varies from 0 to 40 ft. as shown by the profile.

A bakery, which adjoins one end of the warehouse, is founded directly on original ground and its foundation design presented no particular problem despite high intensity of loading. Under the warehouse, however, it was anticipated that settlements of as much as one foot could occur, and that differential settlements would be abrupt and erratic.

Full-Scale Load Test. The foundation conditions were so variable that it could not be predicted from the results of laboratory tests how long the surcharge load would have to remain in place. It was also not known whether or not the soil would show appreciable rebound after the surcharge was removed. Therefore it was necessary to make a full-scale load test at the actual site. It was decided to preload with six feet of sand which would load the subsoils somewhat in excess of the weight of the completed structure. Since about five feet of fill was required to bring the existing ground up to floor grade, a test fill approximately 100 ft. by 100 ft. in plan and 11 ft. high was constructed at the location marked in Fig. 2. Steel plates with one-inch pipes extending to above the top of the fill were installed at various depths below the surface of the original ground, and settlements determined by level observations of the tops of the rods.

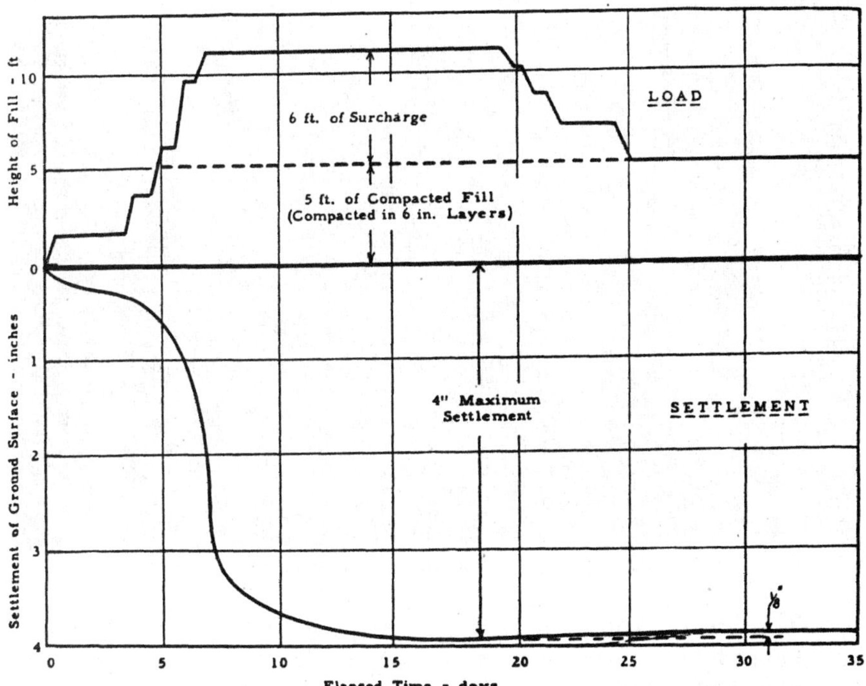

Fig. 3. Load and settlement curves for test fill at site of American Stores Warehouse.

A typical load-settlement curve under the test fill is shown in Fig. 3. It can be seen that most of the settlement occurred as fast as the load was applied, and that all settlement ceased about eight days after the final increment of load was applied. The maximum settlement at the site of the test fill was about 4 inches, and the total rebound following removal of the surcharge did not exceed 0.12 in. The architect-engineers and the owners were convinced by the full-scale load test that preloading would be effective in controlling settlements.

Construction Procedure. The entire site was first brought up to floor grade with about three feet of well-graded gravelly sand. This was spread in six-inch layers, each layer being compacted with three complete coverages of the treads of a heavy crawler type tractor as shown in Fig. 4.

FIG. 4.—HEAVY TRACTORS COMPACTING FILL.

The preload consisted also of the gravelly sand, but this was end-dumped to a total thickness of six feet without compaction, as shown in Fig. 5. Since the settlements occurred rapidly, it was not necessary to preload the entire site at one time. Instead, the area

FIG. 5.—PRELOAD IN PLACE AT SITE OF AMERICAN STORES WAREHOUSE.

was divided into three zones, and each zone successively preloaded for a period of at least two weeks. The sand surcharge was then shifted from one zone to the next, with a 20 ft. overlap between zones to assure complete prestressing of the underlying strata. The surcharge fill also extended outside the boundary of the structure to preload the foundation soils under the exterior wall footings. At the conclusion of the preloading operations all surcharge material was removed and utilized in surfacing adjacent parking lots and driveways.

Settlement During Preloading. After the compacted fill had been placed, and prior to preloading, stakes were driven to exact grade at a number of locations. The elevations of these grade stakes were obtained immediately after the surcharge had been removed. The settlements, as expected, were quite erratic and varied from zero to a maximum of about ten inches, with differential settlements of more than six inches in 60 ft. being recorded. A second set of readings, taken more than two months after the first set, showed that some rebound had occurred, but in no instance did the rebound exceed one-half inch.

Interior Column Footings. Interior columns were on spread

footings at a depth of one foot ten inches below the bottom of the floor slab. In order to verify that the design footing load of 2000 lb./sq. ft. was conservative, a load test was made on a two foot x two foot concrete slab. The 24-hour settlement at 150% of design load was approximately 0.10 in., of which 50% was recovered elastically when the load was removed.

Exterior Wall Footings. The exterior wall was on spread footings designed for 1000 lb./sq. ft. at a depth of two feet below original ground surface. After preloading, a load test made on a four foot by four foot concrete slab showed a 24-hour settlement at 150% of design load of 0.08 in., again with 50% elastic recovery.

The more conservative design loads on the exterior wall footings were chosen because the type of wall construction was more sensitive to differential settlements than the interior steel structure.

The general appearance of the structure after completion is outstanding. A careful examination of the concrete floor slab and of the exterior wall has failed to disclose any evidence of cracking or unevenness several months after the building has been occupied and fully loaded. The owners also are pleased with the great savings as compared with the originally contemplated pile foundation with self-supporting floor.

*Example No. 2**

Example No. 2 is a one-story steel frame warehouse in the Boston area, approximately 180 ft. x 330 ft. in plan, where the subsoil conditions, as shown in Fig. 6, were about 8 ft. of miscellaneous fill, below which was five feet of peat, then the usual sequence of hard, firm and soft clay. The original design called for the columns and foundation wall to be supported on concrete piers, with the concrete floor slab resting on a sand and gravel fill six feet thick above the ground surface.

After the piers and a short section of the foundation wall were installed it was realized that the live load would cause serious settlements of the concrete slab because of the high compressibility of the peat, and the owners arranged for an investigation of the foundation design. The logical change would have been to put in more caissons, or to drive piles between the caissons, and carry the floor loads through the peat to the hard clay, and of course omit the gravel fill.

*This material was made available to the writer by A. Casagrande.

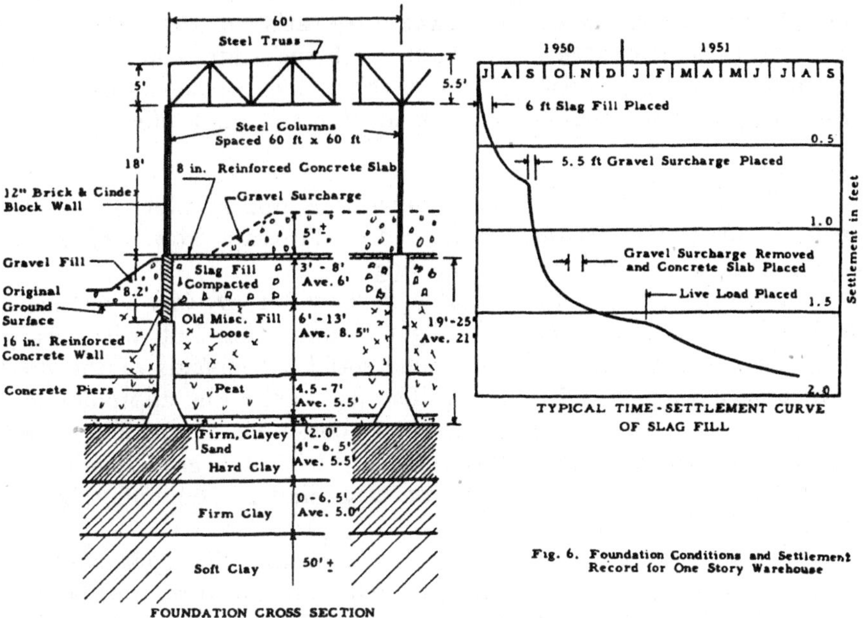

Fig. 6. Foundation Conditions and Settlement Record for One Story Warehouse

However, the owners objected to the additional cost, and while they could not tolerate the 18 to 24 inches of settlement, which were estimated would develop with the original design, they were willing to accept floor settlements of the order of six inches. Therefore it was proposed to use below the floor slab a lightweight slag and to preload the area by a surcharge of 5.5 ft. of sand and gravel which would later be used for filling certain areas outside the building.

The settlement curve on the right of Fig. 6 shows that during the short period during which the slag fill was in place, about nine inches of settlement developed. Under the gravel surcharge which was in place about two months, the settlements increased to 1.5 ft. Then, after this surcharge was removed, the concrete slab placed and live loads applied, some additional settlements developed as indicated by the flat curve which seems to approach the two ft. total settlement. However, of the total, less than six inches represent settlements after the slab was placed. Special provisions were made to permit this movement between the slab and caissons.

The building has now been in use for a year, and because the floor settlements have not resulted in any unsightly cracks in the floor, or in any operating difficulties, the owners are satisfied with this solution and feel that it would not have been worth the additional cost to put in a self-supporting floor. If the problem had been studied ahead of time, preloading could have been made much more effective.

*Example No. 3**

The third example I wish to describe is one in which, as it developed, preloading was ***not*** used. It does illustrate, however, several advantages of the method of preloading.

The proposed structure, located in New Jersey, was a one-story industrial building of relatively light construction but with heavy floor loads. It was essential that differential settlements of the floor slab be avoided if at all possible. The floor area was approximately 260 x 300 ft. in plan. The surface layer consisted of about five feet of miscellaneous fill, including cinders, sand and gravel, with an appreciable amount of a kaolin-type clay binder, and underlain by firm sand and gravel.

It was necessary to raise the grade of the site approximately four feet, and this fill was to be obtained from an excavation at one end of the structure for the furnace room. The architects were concerned seriously about the possibility of settlements in both the upper portion of the existing ground, and in the new fill. The soils consultant was able to convince them that proper moisture control of the fill material combined with adequate compaction by crawler type tractors would assure a stable fill. Neither the owner nor the architects could be convinced, however, that detrimental settlements would not develop in the upper five feet of the existing ground. It was therefore recommended that preloading be adopted, as this was definitely much cheaper than excavating and replacing the layer.

After the loose top soil was removed, the original ground was compacted with six coverages by the treads of a heavy crawler type tractor, operating at top speed. The fill was then compacted in six inch layers up to grade. Settlement plates were placed on top of the compacted fill, and a four foot layer of uncompacted earth was placed over about one-quarter of the area. This was approximately the weight of the proposed floor loading. Careful level surveys were made during and after placing of the surcharge, and they disclosed no measurable settlement. The owner was so impressed by these results

*The foundation consultant for this project was A. Casagrande, and field observations were planned by the writer.

that of his own initiative he proposed that preloading of the remainder of the site be dispensed with, to which the consultant readily agreed.

In this example, the preloading actually constituted a full-scale load test. The estimated savings in construction cost, which originally led to the choice of preloading over other methods, were substantially increased by proving to the owner's satisfaction that even preloading was not needed. The flexibility of the method of preloading, as illustrated by this example, is one of its greatest advantages.

At this point I wish to emphasize the importance of settlement observations during the preloading operations. In this instance they were responsible for substantial savings in construction cost. In other cases they might disclose the unknown presence of areas underlain by peat or other compressible materials and permit corrective measures to be undertaken prior to construction. Such observations are imperative when the time the surcharge must remain in place cannot be estimated in advance.

Example No. 4

In the introduction to this paper it was stated that one of the limitations of this method was often the length of time required to develop full consolidation under the preload. This limitation can sometimes be overcome by increasing the weight of the surcharge. The last example to be described is the one in which this procedure was used.

This project was the extension of a runway at the Naval Auxiliary Landing Field, Mayport, Florida, constructed by the Bureau of Yards and Docks, Department of the Navy, with foundation investigations being handled by the Soil Mechanics Section of the Bureau of Yards and Docks. The data and illustrations are based on a description of the project published by the U. S. Navy Civil Engineer Corps.*

The runway extension was approximately 4000 ft. long and 700 ft. wide, and the only practicable location was in a salt marsh covered with reeds and underlain to a depth of from 10 ft. to 20 ft. with a relatively homogeneous organic clay (OH) having high compressibility, an average in-place unit weight of 83 lbs. per cu. ft. and an average moisture content of 142%. Below this soft organic clay was a good bearing stratum of silty sand and shell.

*"Sand Over Muck", by CDR Richard L. Mann, CEC, USN, Civil Engineer Corps Bulletin, July, 1952.

About 10 ft. of fill were required to bring the site up to grade. A typical time-settlement curve for this depth of fill, as computed from laboratory consolidation tests, is shown as curve A in Fig. 7. The computed settlement was 5.5 inches at five months.

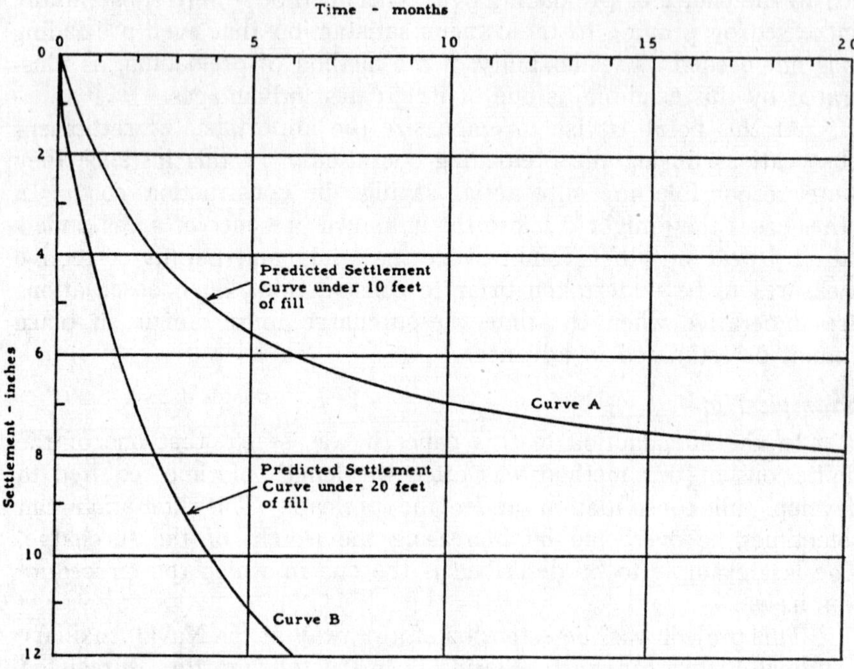

Fig. 7. Comparison of normal and accelerated time - settlement curves.

The following schemes were considered by the Bureau for either reducing the total settlement or accelerating the rate of settlement:

1. Excavate about half the depth of muck, place the fill to final grade and allow to consolidate for two to three months. Since the thickness is reduced by one half the rate of consolidation would have been increased four-fold. The total time prior to paving was estimated at nine months, and the cost about $600,000.
2. Remove all the muck and replace with sand. Estimated time was seven months and cost about $750,000.
3. Sand drains were considered, but the method did not appear economical because of the relatively thin stratum and would

have involved a considerable delay because of the time required to prepare a new contract and mobilize the necessary specialized equipment. Estimated cost of sand drains was about $550,000 and time for completion about 12 months.

4. Accelerate the consolidation of the muck layer by temporarily doubling the height of fill to be placed upon it. The estimated cost was $427,000 and the time about nine months.

These studies indicated that consolidation by surcharge would not only be the most economical, but would utilize the available materials to the best advantage and not consume too much time. The predicted rate of settlement under a 10 ft. fill plus a 10 ft. surcharge fill is shown as curve B in Fig. 7. It can be seen that the 5.5 inches of settlement which was originally predicted in five months for the 10 ft. fill, would now develop in approximately one month.

Actual time-settlement curves under the combined fill and surcharge are shown in Fig. 8. The surcharge was completed on September 30, 1951, but settlement surveys were not started until late in October. By the end of February an additional eight to twelve inches of settlement had developed. It was felt at this time that settlements had developed which were larger than ultimate settlements under the

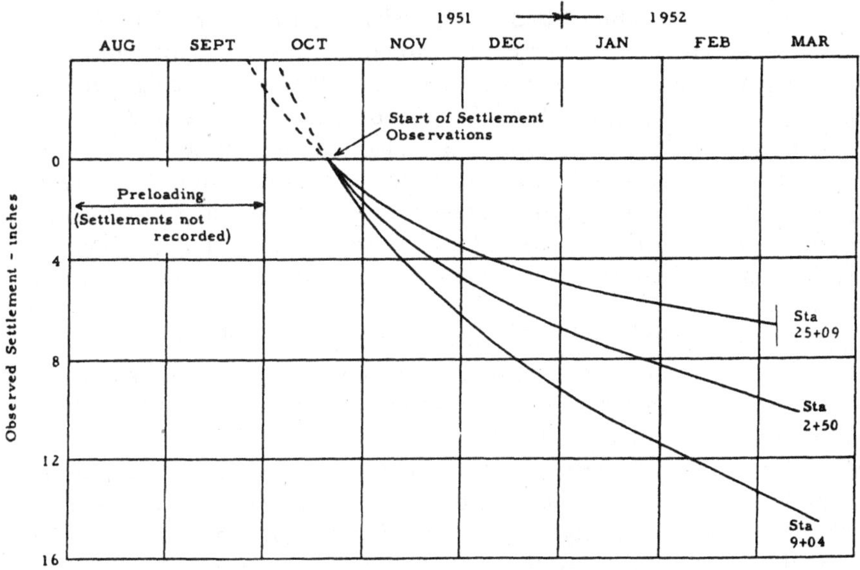

Fig. 8. Observed time-settlement curves.

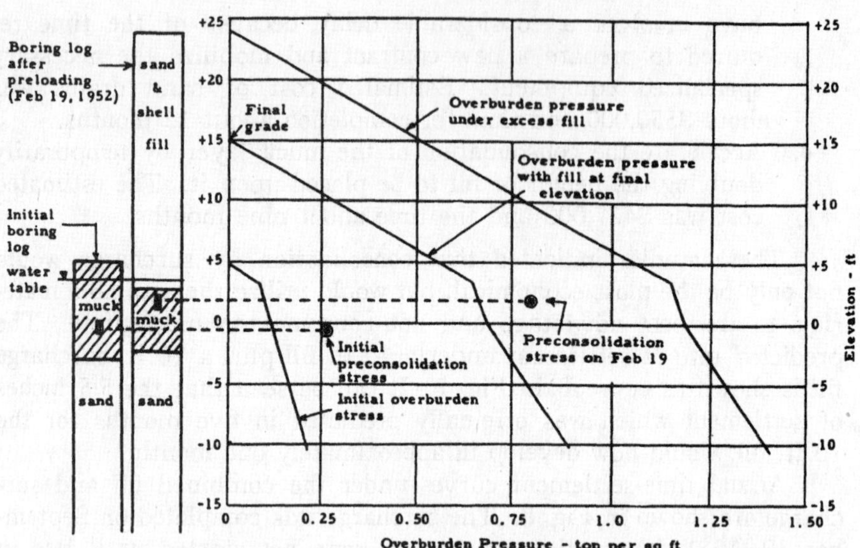

Fig. 9. Comparison of boring logs and stress conditions before and after preloading.

normal load. Additional borings made at this time disclosed that the organic clay had been compressed to half its original thickness, and the moisture content had decreased from 142% to 67%.

Fig. 9 shows a comparison of the borings made before and after surcharging. On the right hand side are plotted overburden pressure vs. depth at a typical station, for the original state, the state after surcharging and the state after removal of the excess load.

On the basis of this and other evidence, the Bureau concluded that accelerated consolidation under the surcharge had taken place as predicted, and that it was safe to proceed with final grading and paving without danger of settlement and consequent pavement failure.

Miscellaneous Comments on Preloading

The two most serious objections to the method of preloading are (1) the time required to squeeze out the excess water, and (2) the large quantity of material required.

The actual time required to squeeze out the excess water is always less than that predicted from laboratory tests on undisturbed samples. The most important reason for this is that in an actual case the consolidation is three-dimensional, i.e., the water escapes outwards in all directions from the zone of stress, whereas the laboratory tests are based upon one-dimensional flow. The time rate is further accelerated by the fact that natural soil deposits are generally much more pervious in horizontal direction than in vertical.

There are at least two possibilities for further accelerating the consolidation. The first is to place additional fill as described in Example 4 and the second is to install sand drains. The successful application of sand drains has been spectacularly illustrated on the New Jersey Turnpike and has been described in many recent papers. In connection with sand drains, however, one must not increase the vertical load too rapidly because the horizontal displacements may shear off the drains and make them inoperative.

If the quantity of material required for preloading is excessive, or if sufficient local materials are not available, there still remain several possibilities. The preferred method is to preload in successive stages, as was described in Examples 1 and 3. This procedure can be used only where the subsoils consolidate rapidly, or when sufficient time is available.

An interesting possibility for preloading was proposed in a recent paper by Kjellman.* This procedure consists of first placing a layer of pervious sand or gravel over the site, then stretching over the gravel an impervious membrane which is sealed at the edges. If the air is now evacuated from the gravel by means of a vacuum pump, an effective vertical stress of up to one ton per square foot is applied which is the equivalent of nearly 20 ft. of fill. Although serious technical problems are posed by such a procedure, it is probable that these difficulties may be overcome.

Another possibility would be to lower the ground water table, thus increasing the effective overburden stress. Thorough subsurface exploration would be required in advance, and a theoretical analysis might prove difficult, but since foundation excavation often requires ground water lowering, the installation of well points at a somewhat earlier date should not involve excessive costs. If such a method were adopted, however, one would have to be concerned about the settlements of nearby structures.

In addition to the four projects described in this paper, other recent projects also show that if foundation conditions are suitable and if the project is planned intelligently, the method of preloading will not only control the settlements, but also result in substantial savings.

*"Consolidation of Clay Soil by Means of Atmospheric Pressure", W. J. Kjellman, presented at MIT Conference on Soil Stabilization, June 20, 1952.

ANCHORED BULKHEADS

By Karl Terzaghi,[1] Hon. M. ASCE

With Discussion by Messrs. T. H. Wu; W. D. Bigler; J. Brinch Hansen; Gregory P. Tschebotarioff; Paul Baumann; Rudolf Briske; Norman D. Lea; S. Packshaw; J. Verdeyen and V. Roisin; and Karl Terzaghi

September 1953, Proceedings–Separate 262
Transactions, Vol. 119, 1954, p. 1243
1955 Norman Medal

Synopsis

In Part I of this paper the fundamental assumptions made in existing methods of bulkhead design are compared with observational data published in the past few decades. The investigation shows that the discrepancies are too important to be ignored in the design.

Part II deals with the evaluation of the forces acting on anchored bulkheads and with the safety requirements. The importance of the errors involved in the estimates of the bending moments and the soil reactions depend, to a large extent, on the type of soils involved in the design problem, on the degree of complexity of the structure of the soil strata, on the degree of uniformity of the backfill material, and on the time and labor invested in the subsoil exploration. Therefore, it would be economically unjustified to select the allowable stresses and safety factors in the construction materials without taking into consideration the degree of reliability of the available data. The selection of these values requires experience and good judgment.

If the designer has only the results of a conventional subsoil exploration available, the problem must be solved on the basis of empirical values such as those contained in this paper. In connection with the design of bulkheads, involving unusual features not treated in this paper, the designer must improvise the procedure; and, in doing so, it will be found advantageous to consult the results of the experimental investigations summarized in Part I.

Notation

The letter symbols in this paper are defined where they first appear, in the text or by illustration, and are assembled alphabetically, for convenience of reference, in the Appendix.

[1] Prof. of the Practice of Civ. Eng., Harvard Univ., Cambridge, Mass.

Note.—Published, essentially as printed here, in September, 1953, as *Proceedings-Separate No. 262*. Positions and titles given are those in effect when the paper or discussion was received for publication.

I. THEORETICAL AND REAL PRESSURES ON ANCHORED BULKHEADS

Types of Bulkheads

Bulkheads are the retaining walls of the waterfront. Depending on the depth of penetration of the sheet piles, they are commonly divided into bulkheads with free earth support (Fig. 1 (*a*)) and bulkheads with fixed earth support (Fig. 1 (*b*)). The sheet piles of the second type are driven so deep into the ground that the bulkhead can fail only by bending or because of inadequate anchorage. If the sheet piles have been driven their full length (or almost full length) into natural ground, and the natural ground has then been removed by excavation, the bulkhead is referred to as a "dredge bulkhead." On the other hand, if the surface of the natural ground is approximately at the level of the dredge line, and the material between the natural ground surface and the level of the upper edge of the sheet piles was deposited after the sheet piles were driven, the bulkhead is called a "fill bulkhead."

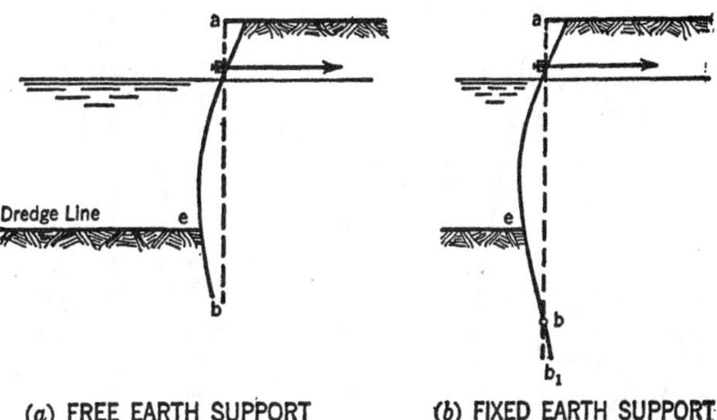

Fig. 1.—The Two Extreme Types of Earth Support

Classical Design Assumptions

Although anchored bulkheads were probably in use in pre-Roman times, it appears that no attempt was made to design them on the basis of earth pressure computations until about 1910, when H. Krey, in Berlin, Germany, began to investigate the problem. Somewhat later he published an analytical procedure for bulkhead design. During the subsequent decades the Krey method was supplemented by various refinements including the "elastic line" and the "equivalent beam" method. In the United States, methods for the design of bulkheads on the basis of the classical earth pressure theories were developed and introduced into the practice of structural design by A. P. Pennoyer.[2] The fundamental assumptions on which all these procedures are based are illustrated in Fig. 2.

Fig. 2 (*a*) represents a bulkhead with free earth support. The inner face of the bulkhead is assumed to be acted upon by the active Coulomb pressure

[2] "Design of Sheet-Piling Bulkheads," by Raymond A. Pennoyer, *Civil Engineering*, November 1933, p. 615.

(pressure triangle, abc). The force that resists the outward movement of the buried part, eb, is represented by the shaded triangle, bef.

Fig. 2(*b*) shows a bulkhead with fixed earth support. The triangle, abc, represents the active earth pressure on the inner face. The assumed resistance against the outward movement of the buried part of the bulkhead is indicated by the pressure area, edb. Since the elastic line of the sheet piles forming an anchored bulkhead with fixed earth support is assumed to pass below a certain point b onto the right-hand, or inner, side of the original position of the bulkhead, it is also assumed that the earth pressure on the inner face changes at the

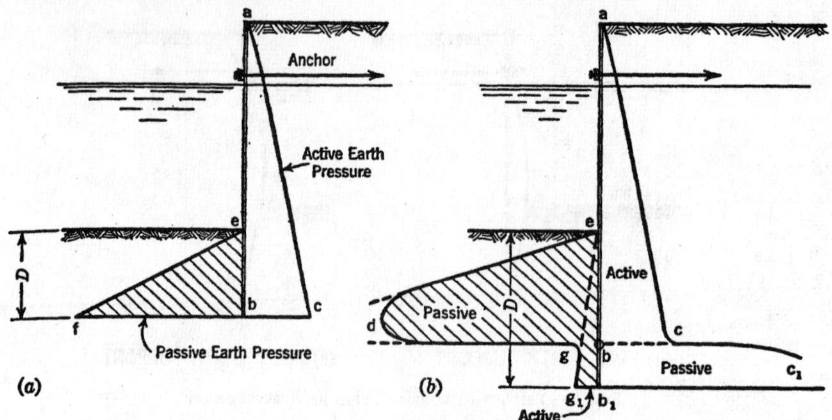

FIG. 2.—CLASSICAL CONCEPTIONS CONCERNING DISTRIBUTION OF EARTH PRESSURE ON BULKHEADS WITH (*a*) FREE AND (*b*) FIXED EARTH SUPPORT

elevation of point b from active to passive (line cc_1, Fig. 2(*b*)), whereas the pressure on the outer face changes at the same point from passive to active (line gg_1). The depth of penetration of the sheet piles (D) is computed in such a manner that the elastic line of the sheet piles satisfies the condition of fixed earth support. The computation is performed either by trial and error or on the basis of supplementary simplifying assumptions.

At the time when the assumptions illustrated by Fig. 2 were originated, the physical conditions for the validity of these assumptions were still unknown. Since that time, many observational data have been accumulated which are incompatible with the original assumptions. Nevertheless, the methods of bulkhead design remained practically unchanged. The following sections of Part I contain a summary of those observational data which are at variance with the original assumptions. These data will then be used in Part II as a basis for a revision of the methods of design.

EFFECT OF WALL MOVEMENTS ON EARTH PRESSURE

The design methods illustrated in Fig. 2 involve the assumption that an infinitesimal yield of a lateral support is sufficient to reduce the intensity of the earth pressure to its minimum value and that a further yield has no influence on this pressure. This assumption is incompatible with the results of large-scale earth pressure tests[3] which furnished accurate information concerning relations among (1) the yield ratio $d = Y/H$ of a vertical wall (Fig. 3(*b*)) with height $H = 5$ ft, backfilled with coarse, clean sand in a loose or in a compacted state; (2) the coefficient of active earth pressure K_A; (3) the mobilized part ϕ' of the angle of internal friction ϕ; and (4) the angle of wall friction δ. The tests led to the following conclusions:

The assumed value K_A for the dense sand was its minimum value corresponding to $d = 0.0005$. This value was retained until the yield ratio became equal to 0.002. Further yield was associated with an increase of K_A toward the minimum value of K_A for loose sand as shown in Fig. 3. At $d = 0.0046$ an

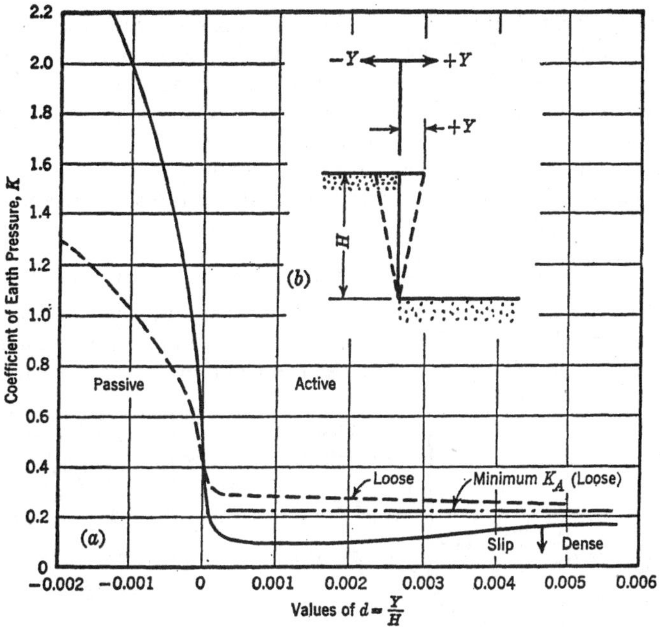

FIG. 3.—INFLUENCE OF RELATIVE DENSITY OF SAND AND EFFECTIVE YIELD Y ON COEFFICIENT OF EARTH PRESSURE

audible slip occurred in the backfill. Along the line of intersection between the surface of sliding and the surface of the backfill a low fault scarp appeared.

The value of K_A for loose sand decreased from 0.4 to 0.30 while the yield ratio d increased from zero to 0.0003. Further yield was associated with a less important decrease of K_A (Fig. 3(*a*)). At a yield ratio of $d = 0.007$, corresponding to the maximum distance through which the model retaining wall could be advanced, the value of K_A was still considerably greater than the minimum value for the dense sand (0.23), and no slip had yet occurred.

The angle of wall friction δ assumed its full value before the internal friction was completely active. On the other hand, when the wall was advanced

[3] "Large Retaining-Wall Tests. I. Pressure of Dry Sands," by Karl Terzaghi, *Engineering News-Record*, Vol. 112, 1934, pp. 136–140.

toward the backfill over a distance of 0.002 H, the wall friction was still much smaller than its maximum value.

At any stage of the tests, as soon as the wall ceased to yield, both the angle of internal friction ϕ' and the angle of wall friction δ decreased slightly at a decreasing rate. Part of the decrease was probably caused by the fact that the backfill of the model retaining wall was subject to intermittent vibrations caused by passing trains.

The backfill of fill bulkheads is almost never compacted by artificial methods, and the average yield of the bulkhead hardly exceeds a fraction of 1% the height of the bulkhead. Therefore, it is unlikely that the lateral pressure of a sand backfill on an anchored bulkhead is as low as the active earth pressure of the fill material.

In the tests illustrated in Fig.3, the wall was not allowed to yield until the entire backfill had been placed. In practice, backfilling operations and yield take place simultaneously. In this case distinction must be made between total and effective yield. The term "effective yield" refers to that part of the total horizontal movement of a point on the back of a lateral support which occurs after the point has been buried. The value of the coefficient of earth pressure depends on the average effective yield of the support and not on the average total yield.

EFFECT OF WALL FRICTION ON PASSIVE EARTH PRESSURE

At the time when the analytical methods for bulkhead design came into existence, it was generally believed that the Coulomb method for computing passive earth pressure was as reliable as the Coulomb procedure for computing

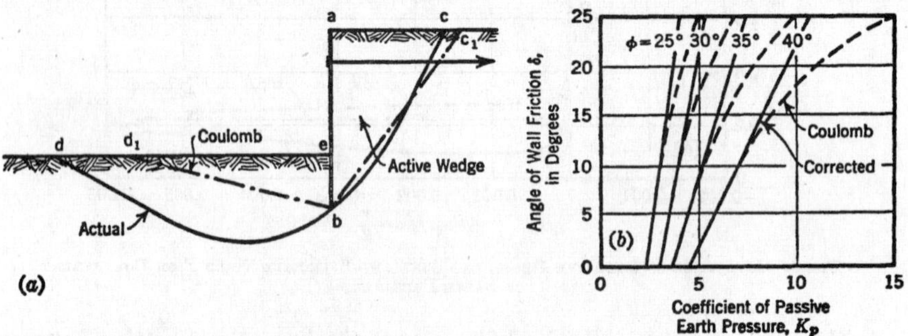

FIG. 4.—DIAGRAMS ILLUSTRATING ERRORS INVOLVED IN COULOMB'S ASSUMPTION THAT THE PASSIVE FAILURE TAKES PLACE ALONG A PLANE SURFACE OF SLIDING

active earth pressure. Both are based on the assumption that the surface of sliding is plane. In Fig. 4(*a*) the surfaces of sliding are indicated by lines bc_1 (active wedge) and bd_1 (passive wedge).

If the Coulomb assumption represented by line bd_1 were justified, the coefficient of passive earth pressure K_P, of a sand with an angle of internal friction ϕ, would increase with increasing angle of wall friction δ, as shown in Fig. 4(*b*) by dashed lines. Subsequent theoretical investigations[4,5] have shown

[4] "Theoretical Soil Mechanics," by Karl Terzaghi, John Wiley & Sons, Inc., New York, N. Y., 1943, pp. 42–53.

[5] *Ibid.*, pp. 100–105.

that the base of the active wedge (line bc, Fig. 4(*a*)) really is almost a plane. Therefore, the Coulomb minimum value for the active earth pressure is almost correct. However, in connection with the passive earth pressure, both the theory of plasticity and various experimental investigations led to the conclusion that the base of the passive wedge is not even approximately a plane. It consists of one curved section and one plane section, as indicated in Fig. 4(*a*) by line bd. On the basis of the knowledge of the actual shape of the surface of sliding, it was found by computation that the coefficient K_P of the passive earth pressure on a vertical lateral support (ratio between horizontal and vertical pressure at any depth below the surface) increases in accordance with the solid lines, and not the dashed lines, in Fig. 4(*b*).

Unbalanced Water Pressure

If an anchored bulkhead is located at the seashore, the earth pressure on the inner face of the sheet piles is a maximum at low tide. At the same time, the inner face is acted upon by an unbalanced water pressure because the water table behind the bulkhead lags behind the receding tide. Unbalanced

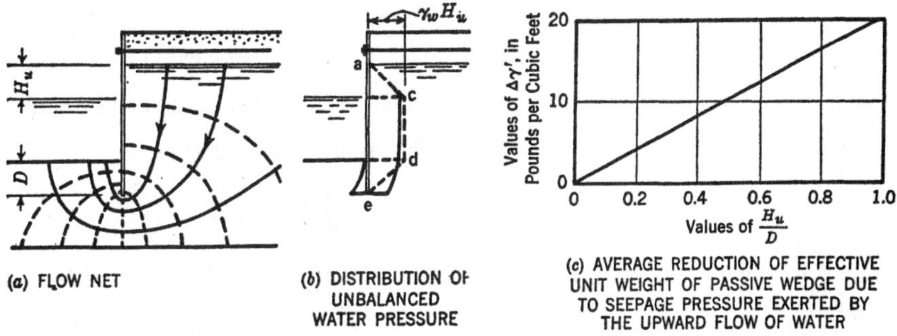

Fig. 5.—Unbalanced Water Pressure

water pressures may also develop at bulkheads located at the shores of rivers or lakes, during a rapidly receding high water or during heavy rainstorms.

If the coefficients of permeability of the strata in contact with the bulkhead are known, the distribution of the unbalanced water pressure on the two faces of the bulkhead corresponding to a hydraulic head H_u can be determined by means of the flow net method, as shown in Fig. 5, for a dredge bulkhead with sheet piles driven into a homogeneous mass of fine, uniform sand.[6] Fig. 5(*a*) represents the flow net and Fig. 5(*b*), the corresponding distribution of the unbalanced part of the water pressure over the two faces of the bulkhead. Accurate flow nets for anchored bulkheads in contact with homogeneous sand have been published by J. McNamee.[7]

If the permeability of all the soil strata in contact with the bulkhead is practically the same, it can be assumed, without serious error, that the inner face of the bulkhead at any depth between the dredge line and the outside

[6] "Theoretical Soil Mechanics," by Karl Terzaghi, John Wiley & Sons, Inc., New York, N. Y., 1943, p. 252.

[7] "Seepage into a Sheeted Excavation," by J. McNamee, *Geotechnique*, Vol. I, 1949, pp. 229–241.

water level is acted upon by an unbalanced water pressure,

$$P_u = \gamma_w H_u \qquad (1)$$

in which γ_w is the unit weight of water. Below the dredge line, P_u decreases from $\gamma_w H_u$ to zero at the lower edge of the sheet piles, as indicated in Fig. 5(*b*) by the straight line, de.

If the permeability varies in vertical directions between wide limits, the determination of the distribution of the unbalanced water pressure may require the construction of a flow net.

When the water table in the backfill is above the free water level, the water percolates through the backfill in a downward direction, flows around the lower edge of the sheet piles, and rises beyond the outer face, as indicated in Fig. 5(*a*). The seepage pressure exerted by the rising ground water reduces the effective unit weight of the soil in contact with the outer face of the bulkhead and, as a consequence, it reduces the passive earth pressure. If i is the average hydraulic gradient in the soil adjoining the outer face, the corresponding reduction of the submerged unit weight γ' of the soil is[8]

$$\Delta\gamma' = i\,\gamma_w \qquad (2)$$

Under the conditions illustrated in Fig. 5(*a*), the average value of i is somewhat smaller than $\frac{H_u}{3\,D}$. Hence, the effective unit weight of the soil in contact with the outer face of the bulkhead will be slightly greater than

$$\gamma' - \Delta\gamma' = \gamma' - \gamma_u \frac{H_u}{3\,D} \qquad (3)$$

The relationship between $\Delta\gamma'$ and H_u/D is shown in Fig. 5(*c*). Within the backfill, the water percolates in a downward direction and it produces an increase of the effective unit weight of the fill by $\Delta\gamma''$. However $\Delta\gamma''$ is very much smaller than $\Delta\gamma'$ and does not require consideration.

Lateral Pressure Resulting from Line Loads

The first attempts to compute the lateral earth pressure p' per unit of length of a wall due to line loads, q', per unit of length, were made in the nineteenth century, before experimental data were available. Based on the Coulomb earth pressure theory, they led to the conclusion[9] that the intensity and the position of the center of the pressure caused by the line load depend on the angle of internal friction ϕ and the angle of wall friction δ. These conclusions are illustrated in Figs. 6(*a*) and 6(*b*). In Fig. 6(*b*) the abscissa of point s_1 represents the ratio m_1 between the width of the top surface of the "sliding wedge" (line ac, Fig. 4(*a*)) and the height H of a wall with a smooth back ($\delta = 0$) acted upon by the lateral pressure of very loose sand ($\phi = 30°$). The abscissa of point s_2 represents the corresponding m_1-value for a wall with a rough back ($\delta = 25°$), backfilled with dense sand ($\phi = 40°$). The ordinates

[8] "Soil Mechanics in Engineering Practice," by Karl Terzaghi and R. B. Peck, John Wiley & Sons, Inc., New York, N. Y., 1948, p. 54.

[9] *Ibid.*, p. 160.

of the dash-double-dot curves that pass through these points represent the ratio between the lateral pressure p' and the intensity q' of the line load computed by means of the Coulomb theory.

At any given value of ϕ and δ the width $m_1 H$ of the top of the sliding wedge decreases slightly with increasing values of q'. If the line load is applied at a distance of less than $m_1 H$ from the crest of the wall, the value of p'/q' is independent of the position of the line load with reference to the crest of the wall. Therefore, on the left-hand side of points s_1 and s_2 the lines representing the (p'/q')-values are horizontal. If the line load moves to the right-hand side of points s_1 (loose sand) or s_2 (dense sand), the lateral pressure decreases rapidly and finally equals zero. The Coulomb theory also leads to the conclusion that the lateral pressure caused by a line load acts only on a narrow, horizontal strip. The elevation of this strip above the foot of the wall depends on the values ϕ and δ, and on the value m, which determines the position of the line load q' with reference to the crest of the wall (Fig. 6(*c*)). In Fig. 6(*a*) the position of the strips acted upon by the lateral force p' is indicated by horizontal arrows. To each arrow are added the values of ϕ and δ on which the computation of its position was based. In each of the four diagrams in Fig. 6(*a*) the two arrows indicate the extreme positions between which the theoretical center of the pressure caused by line loads on sand fills with different density may be located.

The conclusions based on the Coulomb theory concerning the lateral pressure resulting from line loads remained practically unchallenged until both the intensity and distribution of the lateral pressure were determined experimentally by E. Gerber[10] and M. G. Spangler,[11] M. ASCE. The backfill used by Mr. Gerber consisted of clean, uniform river sand with a grain size between 0.2 mm and 1.5 mm. The lateral support was practically rigid. It consisted of the concrete side wall of a rectangular pit, with a depth of 31 in. The lateral pressures caused by the line load were measured by means of pressure cells arranged in vertical rows. Mr. Spangler used as a backfill material pit run gravel with 13% particles passing the 200-mesh sieve. The lateral support consisted of a reinforced concrete cantilever wall, 84 in. high and 6 in. thick, which was free to tilt about the outer edge of the base of the base plate. In spite of the differences in the test conditions in these two sets of tests, there are no essential differences between the test results.

The solid curves in Fig. 6(*a*) represent the results of one of Mr. Gerber's series of tests performed on backfills with a height $H = 32$ in. (Fig. 6(*c*)). The surcharge, 0.4 ton per sq ft, covered a strip 6.3 in. wide and 25 in. long. The center line of the loaded strip was established, successively, at a distance $x =$ 3.1 in., 9.4 in., 15.5 in., and 22 in. from the upper edge of the wall, corresponding to values of m $(= x/H)$ of 0.1, 0.3, 0.5, and 0.7. The abscissas of the solid curves represent the values of the ratio $p \frac{H}{q'}$, p being the horizontal pressure per unit area on the back of the wall, and q' being the surcharge per unit of length

[10] "Untersuchungen über die Druckverteilung im örtlich belasteten Sand," by E. Gerber, Zürich, Switzerland, 1929.

[11] "Horizontal Pressures on Retaining Walls Due to Concentrated Surface Loads," by M. G. Spangler, *Bulletin No. 140*, Iowa Eng. Experiment Station, Iowa City, Iowa.

of the loaded strip. In Fig. 6 (*b*) the ordinates of the solid curve represent the ratio p'/q' between the measured lateral pressure p' per unit length of the wall and the line load q'.

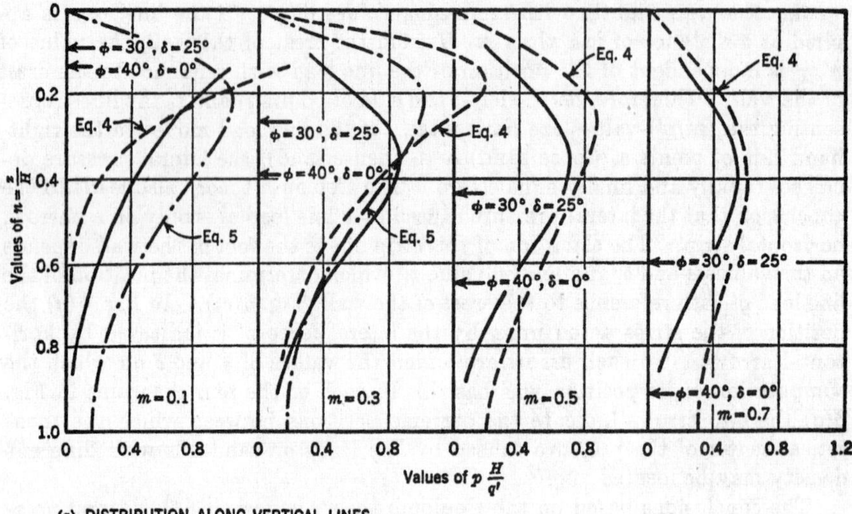

(*a*) DISTRIBUTION ALONG VERTICAL LINES

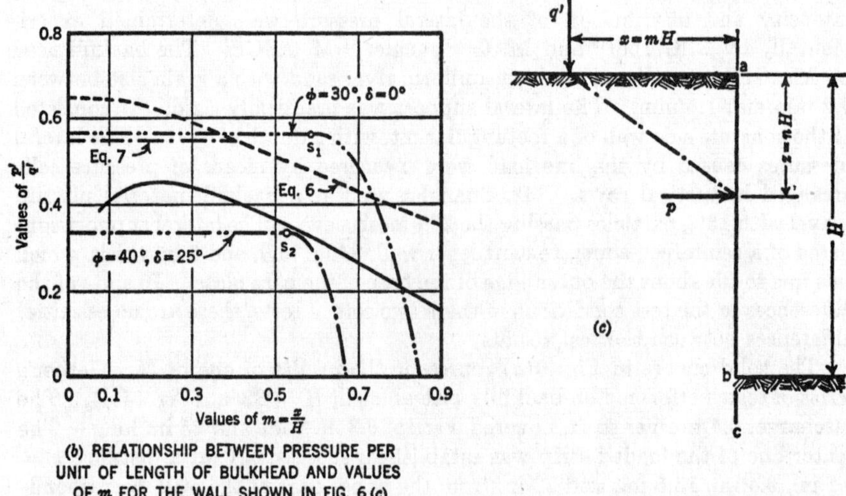

(*b*) RELATIONSHIP BETWEEN PRESSURE PER UNIT OF LENGTH OF BULKHEAD AND VALUES OF *m* FOR THE WALL SHOWN IN FIG. 6 (*c*)

Fig. 6.—Measured (Plain Curves) and Computed (Dash and Dash-Dot Curves) Lateral Pressure Due to Line Loads

According to the data shown in Fig. 6 the information furnished by the Coulomb theory concerning the intensity and distribution of the lateral pressure resulting from line loads is incompatible with the experimental data. More satisfactory is the agreement between the measured pressures and the theory of Boussinesq.[12] According to this theory the horizontal unit pressure σ_x on a vertical section, ac, in Fig. 6 (*c*) through a semi-infinite elastic medium, at a depth $n\,H$ below the surface, caused by a surface load q' per unit of length, acting on a line at a distance $m\,H$ from the vertical section is equal to $\sigma_x = \frac{2}{\pi}\frac{q'}{H}\frac{m^2\,n}{(m^2+n^2)^2}$. However, the application of the line load tends to produce a lateral deflection of the vertical section, and the flexural rigidity of the bulkhead interferes with that deflection. In order to obtain the lateral pressure on a relatively rigid diaphragm, ab, in Fig. 6 (*c*), located at the site of the vertical section,[13] a second and equal line load q' must be applied at a distance $m\,H$ on the right-hand side of point a. This second line load doubles the unit pressure; therefore, the unit pressure on the wall at depth $n\,H$ below the surface is $p = 2\,\sigma_x = \frac{4\,q'}{\pi\,H}\frac{m^2\,n}{(m^2+n^2)^2}$; and

$$p\frac{H}{q'} = \frac{4}{\pi}\frac{m^2\,n}{(m^2+n^2)^2} \qquad (4)$$

Eq. 4 is represented in Fig. 6 (*a*) by dash-line curves. For values of m greater than about 0.4 the agreement between theory and observation is fair. However, for values smaller than 0.4, the discrepancy between observed and computed values increases with decreasing values of m (as shown by dash-line curves in Fig. 6 (*a*) for $m = 0.1$ and $m = 0.3$). For such values of m (less than 0.4) it was found, by trial and error, that the observed pressure distribution has greater similarity to the computed distribution for $m = 0.4$ which is determined by the equation:

$$p\frac{H}{q'} = \frac{0.203\,n}{(0.16+n^2)^2} \qquad (5)$$

Eq. 5 is represented in Fig. 6 (*a*) by dash-dot curves.

For values of m greater than 0.4 the lateral pressure p' per unit of length of the wall is $p' = \int_{n=0}^{n=1} p\,H\,dn = \frac{2}{\pi}\frac{q'}{m^2+1}$; from which

$$\frac{p'}{q'} = \frac{2}{\pi\,(m^2+1)} \qquad (6)$$

For values of m smaller than 0.4 Eq. 6 must be replaced, in accordance with Eq. 5, by

$$\frac{p'}{q'} = \frac{2}{\pi\,(0.16+1)} = 0.55 \qquad (7)$$

In Fig. 6 (*b*), Eq. 6 is represented by a dash line and Eq. 7 is represented by a dash-dot line.

The diagrams in Figs. 6 (*a*) and 6 (*b*) show that the values obtained by use of Eqs. 4 to 7 are consistently greater than the measured values. Part

[12] "Theoretical Soil Mechanics," by Karl Terzaghi, John Wiley & Sons, Inc., New York, N. Y., 1943, p. 376.

[13] "Pressure Distribution on Retaining Walls," *Proceedings*, First International Conference on Soil Mechanics and Foundation Eng., Discussion by R. D. Mindlin, Cambridge, Mass., 1936, Vol. 3, pp. 155–156.

of the difference is due to the fact that the Gerber tests[10] were made with line loads having a length of not more than 0.8 H, whereas the computed values refer to line loads with infinite length. The remainder of the difference results from the fact that the Boussinesq theory is strictly applicable only to perfectly elastic materials. Since the computed values are too high, the differences partly compensate for the fact that the distribution of the (p'/q')-values under field conditions, may deviate to some extent from the observational curves which have been constructed on the basis of the test results reported by Mr. Gerber. The deviations are chiefly caused by the complex stress-strain relations for sands. Because of these relations the pressure distribution depends not only on the values of m and n contained in Eqs. 4 and 5, but also, to some extent, on the value of the ratio q'/H. Mr. Gerber's data were obtained only for one value of this ratio.

The observational data shown in Fig. 6 eliminated the Coulomb theory as a source of information concerning the lateral pressure produced by line loads, and they made it possible to establish empirical equations for estimating upper limiting values of the pressures produced by such loads.

Lateral Pressure Resulting from Point Loads

Intensity and distribution of the lateral pressure resulting from point loads were investigated by Messrs. Gerber[10] and Spangler.[11] The test results were practically identical. The point load in Mr. Gerber's tests consisted of a

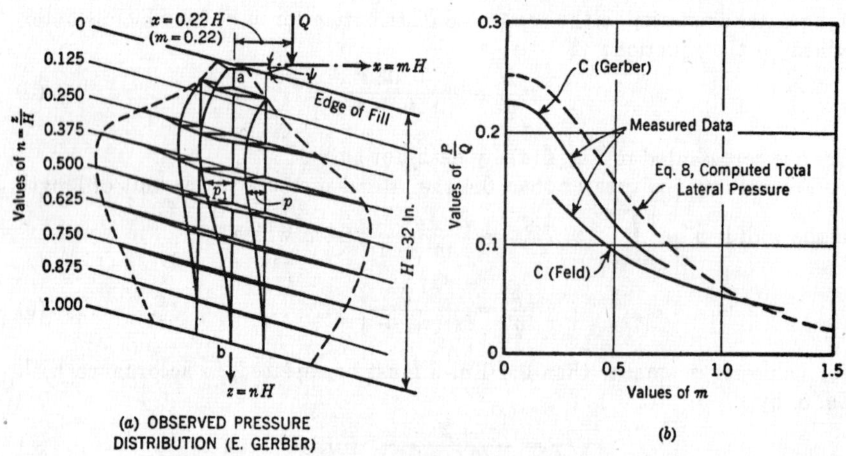

Fig. 7.—Lateral Pressure Due to Point Loads

loaded circular slab with a diameter of 14 in. The slab rested upon the horizontal surface of a layer of clean, coarse sand with a depth of 32 in., at varying distances from the crest of the wall. Fig. 7 shows the distribution of the lateral pressure over the back of the wall. The pressure is greatest along the line of intersection, ab, between the wall and a vertical plane through the center of the load at right angles to the wall. Along this line, the unit pressure p_1 first increases with increasing depth, assumes a maximum value at a depth which is somewhat greater than the distance between load and wall, and then decreases again. At any depth the pressure decreases in horizontal directions with increasing distance from line ab.

Fig. 7(*b*) shows the relationship between $m = x/H$ and the total intensity P of the lateral pressure caused by the point load Q. Curve C(Gerber) is based on test results[10] and curve C(Feld) on test data published by Jacob Feld,[14] M. ASCE. The lack of perfect agreement between observational curves obtained by different investigators is chiefly a result of the fact that the values of P/Q depend not only on m but also on various other factors such as the deflection ratio d for the lateral support and the ratio between Q and the ultimate bearing capacity of the backfill. Upper limiting values for P can be obtained by use of the empirical equation:

$$\frac{P}{Q} = \frac{0.25}{(1 + m^2)^2} \qquad (8)$$

which is based on Mr. Gerber's test results. In Fig. 7(*b*) Eq. 8 is represented by a dashed curve.

None of the existing theories (1953) account satisfactorily for the distribution over the inner face of a wall of the lateral pressure produced by a point load Q. For values of m greater than 0.4 the unit pressures p_1 along line ab (Fig. 7(*a*)) can be estimated roughly by use of the empirical equation:

$$p_1 \frac{H^2}{Q} = 1.77 \frac{m^2 n^2}{(m^2 + n^2)^3} \qquad (9)$$

For values of m less than 0.4, a better approximation is obtained by assigning to m in Eq. 9 a constant value $m = 0.4$, thus—

$$p_1 \frac{H^2}{Q} = \frac{0.28\, n^2}{(0.16 + n^2)^3} \qquad (10)$$

In Fig. 8(*a*) the solid curves represent the Gerber test results, and the dashed curves represent Eqs. 9 and 10.

The intensity of the lateral pressure on the back of the wall on both sides of the line ab in Fig. 7(*a*) is a complicated function of the depth below the crest of the wall and the horizontal distance from the line ab. If the point load Q is located at a distance $m\,H$ from the wall, an upper limiting value for the unit pressure p at a depth $n\,H$ below the surface of the fill and at a horizontal distance $m\,H \tan \psi$ (Fig. 8(*b*)) from the vertical line through point a can be obtained by use of the empirical equation:

$$p = p_1 \cos^2 (1.1\, \psi) \qquad (11)$$

In Eq. 11, p_1 is the unit pressure on the wall at depth $n\,H$ for $\psi = 0$. The value of p_1 is determined by Eq. 9 for m greater than 0.4 and by Eq. 10 for values of m smaller than 0.4. Eq. 11 is an empirical equation based on Mr. Gerber's test results.

[14] "Lateral Earth Pressure: The Accurate Experimental Determination of the Lateral Earth Pressure, Together with a Resume of Previous Experiments," by Jacob Feld, *Transactions*, ASCE, Vol. LXXXVI, 1923, pp. 1448–1505.

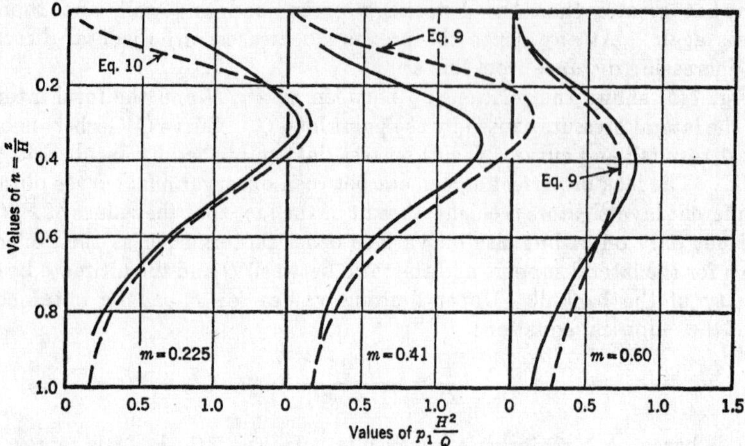

(a) PRESSURE PER UNIT OF AREA ALONG LINE a b IN FIG. 7 (a)

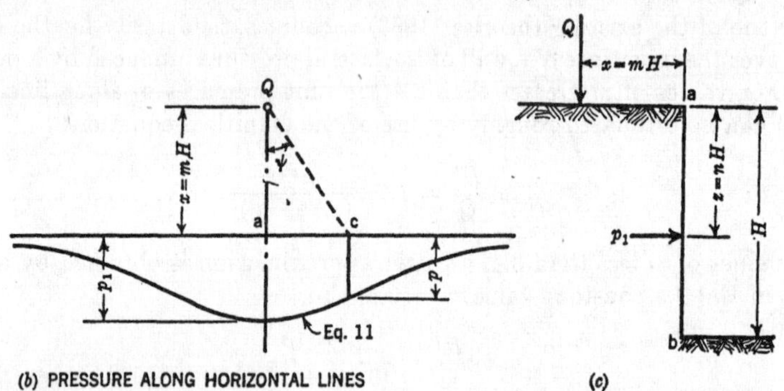

(b) PRESSURE ALONG HORIZONTAL LINES (c)

Fig. 8.—Lateral Pressure Produced by Point Loads on Lateral Support

Distribution of Earth Pressure

Theory and observation have shown that the distribution of the pressure on a lateral support is by no means necessarily in accordance with the Coulomb theory because it depends largely on the type of yield.[15] This fact is illustrated by Fig. 9 which represents the distribution of the lateral pressure on the back of a lateral support for three different types of yield. The effective yield at any depth below the surface is indicated by the width of the shaded area at that depth.

In connection with anchored bulkheads, the validity of the Coulomb theory was questioned for the first time in 1906 by Danish engineers on purely empirical grounds. It was argued that the lateral pressure on bulkheads is a minimum midway between the dredge line and the anchor line, as shown in Fig. 9(c). This conception received experimental support by tests performed by J.P.R.N. Stroyer[16] and received considerable attention be-

[15] "General Wedge Theory of Earth Pressure," by Karl Terzaghi, *Transactions*, ASCE, Vol. 106, 1941, pp. 68-80.

[16] "Earth Pressure on Flexible Walls," by J. P. R. N. Stroyer, *Journal*, Inst. of C. E., London, England, Vol. I, 1935, p. 94.

cause a lateral pressure with the distribution shown in Fig. 9(*c*) produces very much smaller bending moments in the sheet piling than a pressure with Coulomb distribution (Fig. 9(*a*)) and equal total intensity. In 1938

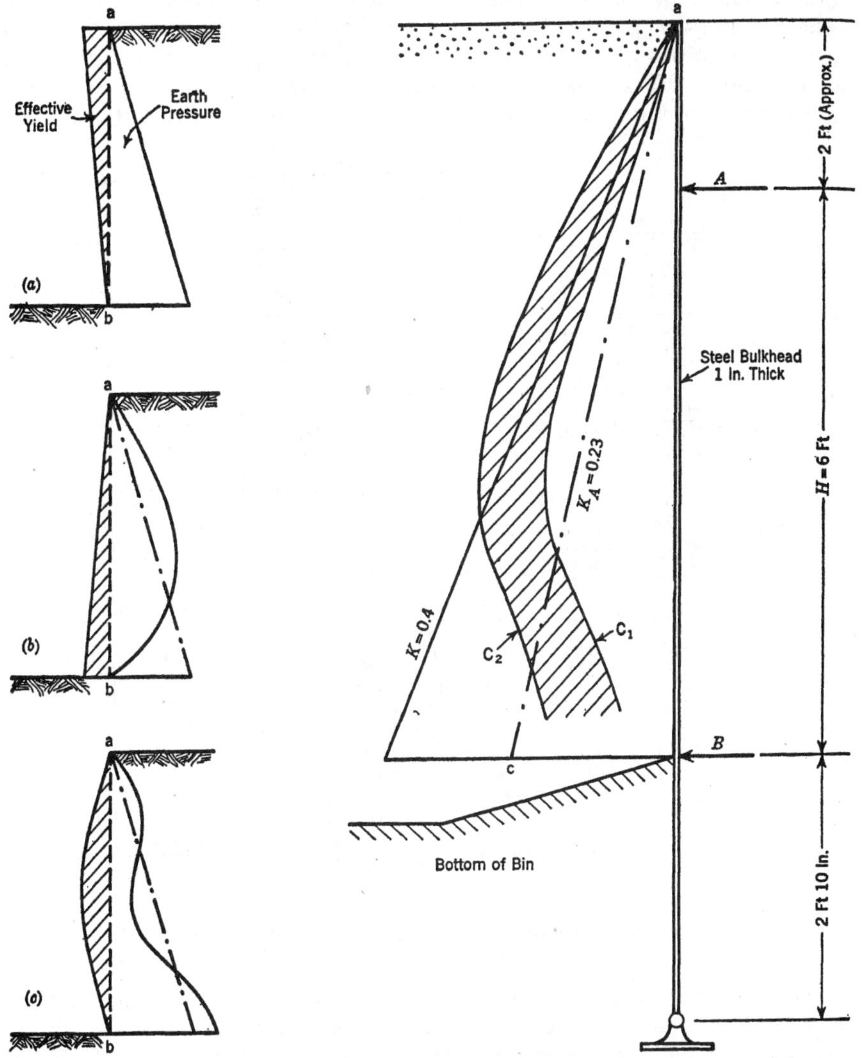

Fig. 9.—Influence of Movement Type on Pressure Distribution

Fig. 10.—Intensity and Distribution of Active Earth Pressure on Inner Face of Relatively Stiff Model Bulkhead

J. Ohde computed the distribution of the earth pressure on flexible walls with fixed upper and lower edges and he also found that the pressure distribution should have the characteristics shown in Fig. 9(*c*). His theoretical conclusions were confirmed by large-scale tests performed by H. Press in 1948,[17] on a

[17] "Über die Druckverteilung im Boden hinter Wänden verschiedener Art," by H. Press, *Bautechnik Archiv.*, W. Ernst and Son, Berlin, Germany, 1948, pp. 36–54.

flexible wall 27 ft high. The inner face of the wall was paved with pressure cells.

Between the years 1944 and 1948 G. P. Tschebotarioff, M. ASCE, performed large-scale tests on bulkhead models.[18] He measured the lateral deflection and the extreme fiber stresses in the bulkheads at different elevations above their lower edge and he computed the pressure distribution on the basis of these data. He obtained the pressure distribution illustrated by Fig. 9(*c*) for bulkheads of the dredge type only. The distribution of the earth pressure

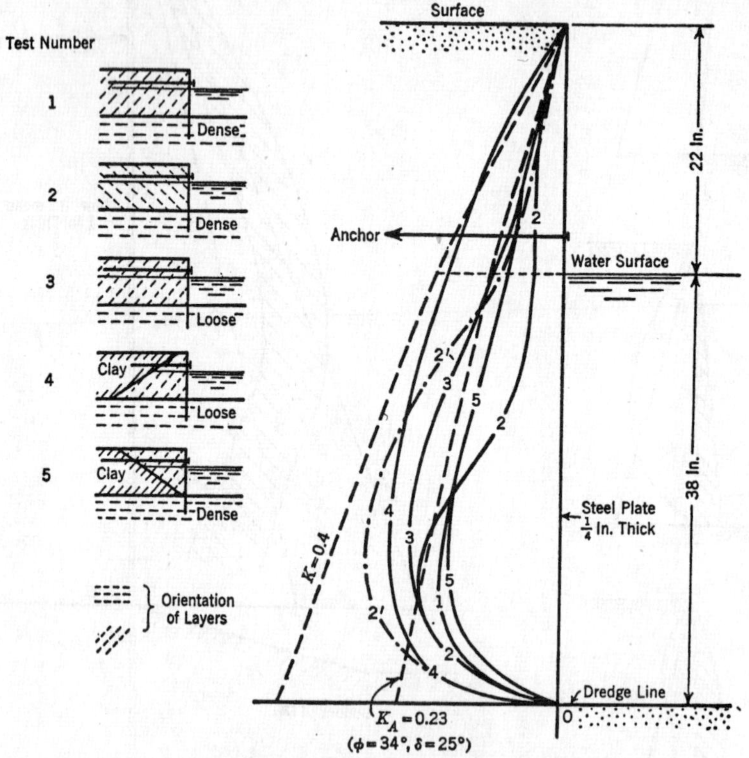

FIG. 11.—INTENSITY AND DISTRIBUTION OF ACTIVE EARTH PRESSURE ON THE INNER FACE OF RELATIVELY FLEXIBLE MODEL BULKHEADS

on the inner face of the models of fill bulkheads was found to be similar to that shown in Figs. 10 and 11. The tests represented by Fig. 10 were made with a relatively stiff bulkhead with a maximum deflection equal to about 0.1% of the vertical distance H between the anchor line and the dredge line. The corresponding K_A-value was nearly 0.4, which is approximately equal to the coefficient of earth pressure at rest. The intensity and the distribution of the active earth pressure on more flexible bulkheads, having a deflection ratio of about 0.5%, is shown in Fig. 11. The corresponding K_A-value approximated

[18] "Final Report. Large Scale Earth Pressure Tests with Model Flexible Bulkheads," by G. P. Tschebotarioff, Princeton Univ., Princeton, N. J., 1949.

that of the active earth pressure. Curve 2′ represents Test No. 2 after compaction of the fill by vibration. The line marked $K_A = 0.23$ represents the Coulomb pressure computed on the assumption that $\phi = 34°$ and $\delta = 25°$. The line marked $K = 0.4$ represents the earth pressure at rest. All the test results showed that the lateral earth pressure was a maximum at some elevation above the dredge line. They also showed that the real pressure distribution depends on factors, such as the method of placing the fill, which do not receive any consideration in earth pressure theory. Hence, the agreement between the real pressure distribution and the Coulomb pressure distribution is by no means perfect because the intensity of the lateral earth pressure is a maximum at some elevation above the dredge line (Fig. 10 and Fig. 11) and not at the dredge line. The computation of the bending moments in the sheet piles on the basis of a Coulomb pressure with equal total intensity involves an error on the unsafe side.

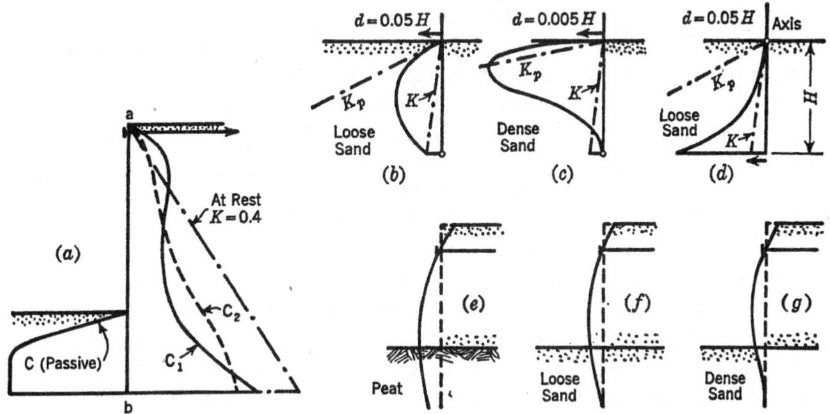

FIG. 12.—DIAGRAMS ILLUSTRATING THE EFFECT OF SUBSOIL CONDITIONS ON DISTRIBUTION OF PASSIVE EARTH PRESSURE AND ON TYPE OF BULKHEAD DEFLECTION

Mr. Tschebotarioff also experimented with composite backfills, Fig. 11, Tests No. 4 and No. 5. The lateral pressure exerted by sand fills backed by clay fills in Test No. 4 was found to be slightly greater than the lateral pressure of a continuous sand backfill with identical properties. The lateral pressure exerted by sand fills located between the bulkhead and a clay slope as in Test No. 5 was slightly smaller than that exerted by the fill in Test No. 4. P. W. Rowe determined the distribution of the earth pressure on a flexible wall directly by means of pressure cells in 1952.[19] In agreement with Mr. Tschebotarioff's findings, he obtained the pressure distribution for dredge bulkheads as shown in Fig. 9(*c*). However, he found that an anchor yield of 0.1% of the height of the bulkhead is sufficient to change the pressure distribution into one that agrees fairly closely with the Coulomb theory. The test results are illustrated in Fig. 12(*a*). In Fig. 12, curve C_1 represents the distribution of the earth pressure on the inner face of the model bulkhead prior to yielding of the

[19] "Anchored Sheet-Pile Walls," by P. W. Rowe, *Proceedings*, Inst. of C. E., London, England, Vol. 1, Pt. 1, 1952, pp. 27–70.

anchorage and C_2 represents distribution after the bulkhead had yielded. The total intensity of the pressure remained almost unchanged.

The anchorage of Mr. Rowe's model bulkhead was allowed to yield after the sand in contact with the upper part of the outer face of the bulkhead had been removed by excavation, whereas in practice the anchorage yields gradually during the process of excavation. This difference may involve a considerable difference in the type of pressure distribution. However, the yield of the anchorage may exceed considerably the limiting value of 0.001 H, and the pounding of waves or traffic vibrations may contribute further to a modification of the pressure distribution. Hence, even in cases of dredge bulkheads it does not seem justified to depend on the benefits to be derived from a difference between the real pressure distribution and the distribution computed on the basis of the Coulomb theory.

Curve C(Passive), Fig. 12(*a*), shows the results of the measurement of the passive earth pressure that acted upon the buried part of the model bulkhead. In order to obtain supplementary information concerning the effect of the type of wall movement on the distribution of the passive earth pressure, Mr. Rowe experimented with a $\frac{3}{8}$-in. steel plate that was buried to a depth of 2 ft in clean sand. The plate could be advanced toward the sand by rotation about a horizontal axis. The passive earth pressure on the wall was measured by use of seven pressure cells, spaced 3 in. on centers along the vertical axis of the area acted upon by the passive earth pressure. The test results are shown in Figs. 12(*b*), 12(*c*), and 12(*d*).

The pressure distribution represented by curve C(passive) in Fig. 12(*a*) is intermediate between those shown in Figs. 12(*c*) and 12(*d*), but none of them involves an increase of the passive earth pressure in simple proportion to the depth below the dredge line. With increasing flexibility of the buried part of the bulkhead, the movement of this part changes from a displacement almost parallel to the original position of the buried part into a movement by rotation about the lower edge of the bulkhead so that the distribution of the passive earth pressure becomes increasingly similar to that shown in Fig. 12(*b*), or Fig. 12(*c*).

Influence of Flexural Rigidity on Bending Moment

According to the theories of bulkhead analysis explained under the heading, "Classical Design Assumptions," and illustrated in Fig. 2, the conditions of end support and (as a consequence) the maximum bending moment in the sheet piles are independent of the flexural rigidity of the sheet piles. According to the same theories, the maximum bending moment decreases with increasing depth of sheet-pile penetration, whatever the flexural rigidity may be. These postulates are incompatible with what has been learned concerning the relation between horizontal displacement and horizontal soil reaction. The fallacies involved in the postulates have already been emphasized by Paul Baumann, M. ASCE, in connection with an analysis of the causes of the failure of Pier B in the Outer Harbor of the City of Long Beach, Calif.[20] In fact, if the sheet piles were perfectly rigid, the maximum bending

[20] "Analysis of Sheet-Pile Bulkheads," by Paul Baumann, *Transactions*, ASCE, Vol. 100, 1935, pp. 707–797.

moment would increase with increasing depth of pile penetration. However, no observational data were available concerning these important relationships until the results of Mr. Rowe's experimental investigations were published.[19]

Mr. Rowe's model bulkheads consisted of metal plates of height H equal to from 20 in. to 36 in., and with different thicknesses. The dredge line (Fig. 13) was at a variable depth αH below the surface of the fill and the anchor line was at a depth βH. The vertical strains on the two surfaces of the metal plates were measured by strain gages spaced 2.5 in. vertically. The tests were performed using four different materials—coarse, clean sand; crushed rock (chips); pea gravel; and ashes. In one series of tests the materials were placed in a loose state and in a second series they were placed in a dense state. In each series of tests the extreme fiber stresses in the plate were measured by use of strain gages for different values of the free-height ratio α, anchor-level ratio β, surcharge ratio $\frac{q}{\gamma H}$, modulus of elasticity E of the wall, and the moment of inertia I of the cross section of the wall. The strain-gage readings furnished the data for computing the maximum bending moment in the plate for each set of test conditions.

According to the results of Mr. Rowe's analytical studies, which preceded the tests, the condition for similitude between the bulkhead model and the prototype is satisfied if the values α, β, $\frac{q}{\gamma H}$, and ρ are the same. The symbol ρ indicates the flexibility number of the sheet piles—

$$\rho = \frac{H^4}{E I} \qquad (12)$$

in which I represents the rectangular moment of inertia of the piles. This conclusion is based on the tacit assumption that the modulus of elasticity of the sand increases in simple proportion to the depth below the dredge line. For loose sand, this assumption is at least approximately correct. For dense sand, the modulus of elasticity seems to increase more nearly with the square root of depth. Hence, if the sheet piles of a bulkhead are driven into dense sand, the conditions of end support will be less favorable than those of the model sheet piles embedded in sand with the same relative density.

The investigations led to the following conclusion, illustrated in Fig. 13. For very stiff bulkheads, the maximum bending moment M in the sheet piles is practically independent of the flexibility number and is equal to the value M(max) computed on the assumption of free earth support, Fig. 2(*a*). However, if ρ exceeds a certain value, the maximum bending moment M decreases with increasing values of ρ and finally approaches a value approximately equal to one third of M(max). The critical flexibility ρ_c at which the maximum bending moment starts to drop below the value of M(max) increases with decreasing relative density of the sand, as shown in Fig. 13. The value ρ_c is almost independent of the values of α, β, and $\frac{q}{\gamma H}$ in the range over which these quantities are likely to vary under actual conditions.

The fact illustrated by Fig. 13—that the bending moment in sheet piles decreases with increasing flexibility of the piles—is chiefly the result of the interdependence between the type of deflection of the buried part of the sheet piles and the corresponding distribution of the passive earth pressure. If the sheet piles, with a depth of penetration D, were perfectly rigid and the anchorage unyielding, the buried part of the sheet piles would rotate about the anchor line. The corresponding distribution of the earth pressure would be similar to that shown in Fig. 12(d), and the center of the pressure would be located at an elevation of less than $D/3$ above the lower edge of the piles.

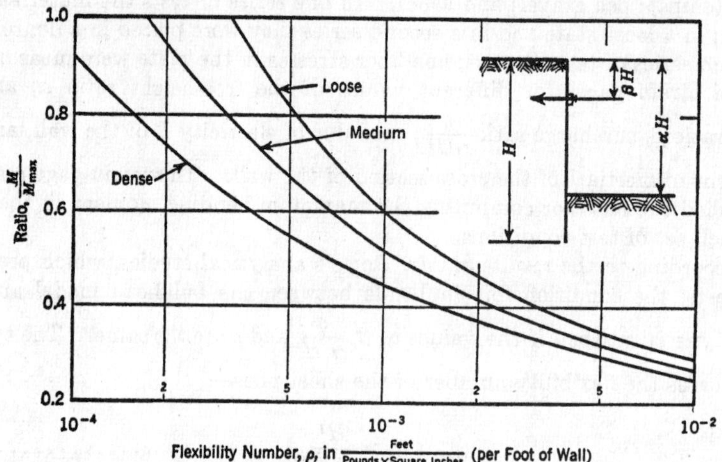

FIG. 13.—RELATION BETWEEN THE FLEXIBILITY NUMBER, ρ, OF SHEET PILES, AND BENDING-MOMENT RATIO, $\frac{M}{M\,(\max)}$ (LOGARITHMIC SCALE)

This condition corresponds to the ideal "free earth support." As the flexibility increases, the outward movement of the lower edge of the piles becomes smaller and smaller. The yield assumes the character of a yield by rotation about the lower edge, involving a pressure distribution as shown in Figs. 12(b) and 12(c). The elevation of the center of the passive pressure increases to more than $D/2$, whereby the "free span"—equal to the distance between anchor line and center of the passive pressure—decreases, and the maximum bending moment decreases with the third power of the span. Finally, if the piles are extremely flexible, the lowest part of the sheet piles will neither advance nor rotate. In other words, the lower ends of the sheet piles will be "fixed" as shown in Fig. 2(b).

The critical value ρ_c of the flexibility number increases with increasing compressibility of the soil because the resistance of the soil against tilt and outward movement of the buried part of the sheet piles decreases. This interdependence is illustrated in Figs. 12(e) to 12(g). If sheet piles are driven into peat (Fig. 12(e)), they receive "free earth support" even if they are made of a flexible material such as wood.

If sheet piles are driven into silt or clay, the initial end restraint may be important enough to produce "fixed earth support." However, as time goes on, the soil yields under the lateral pressure because of progressive consolidation. The final lateral displacement of the buried part of the sheet piles can be even greater than that of sheet piles driven into loose sand. The yield is associated with a transition from fixed earth support to free earth support, whereby the maximum bending moment in the sheet piles increases. A permanent fixed earth support for piles driven into a cohesive soil can hardly be expected, unless the soil has been heavily precompressed by overburden pressures that have subsequently been removed by natural processes such as erosion.

Approximate information concerning the influence of the flexibility of sheet piles on the maximum bending moment can be obtained by computation, using the theory of subgrade reaction. For the first time, this theory was applied to anchored bulkheads by Mr. Baumann in 1934.[20] Further contributions to this subject were made by H. Blum.[21] The sources of error involved in the procedure are rather important. They are common to all the computations based on the concept of subgrade reaction, such as the computations of the bending moments in nonuniformly loaded beams resting on an elastic subgrade.[22] Reliable methods for determining the coefficient of horizontal subgrade reactions are not available. In many instances the value of this coefficient has been seriously misjudged and the results of the computations were therefore misleading. The computations require considerable time and labor.

Anchor Pull

The decrease of the bending moments associated with an increase of the flexibility number (Fig. 13) results from a transition from the condition of free earth support (Fig. 12(*e*)) to the fixed end condition (Fig. 12(*g*)). If the lowest part of the sheet piles is fixed, the fixed ends are acted upon by moments that carry part of the lateral pressure on the inner face, and as a consequence the anchor pull is reduced.

Mr. Rowe's test results concerning the influence of the flexibility number on the anchor pull are shown in Fig. 14 for a value of $\beta = 0.2$. In Fig. 14, the ordinates represent the measured anchor pull in percentage of the pull corresponding to free earth support. According to the test results, the relieving effect of the fixed end condition depends not only on the relative density of the backfill, as does the bending moment (Fig. 13), but also on the values of the anchor-level ratio β and the free-height ratio α. The investigation also showed that the anchor pull decreases to some extent with increasing yield of the anchorage.

Since Mr. Rowe's data (illustrated in Fig. 13) have been published, there is no longer any justification for assuming fixed earth support without considering flexibility of the sheet piles.

The Shearing Resistance of Soils and the Angle of Repose

Whatever the design assumptions may be, the computation of the numerical values of the forces acting upon the bulkhead requires adequate knowledge of the shearing resistance of all the soils involved in the problem.

[21] "Beiträge zur Berechnung von Bohlwerken," by H. Blum, W. Ernst and Son, Berlin, Germany, 1951.

[22] "Theoretical Soil Mechanics," by Karl Terzaghi, John Wiley & Sons, Inc., New York, N. Y., 1943, pp. 345–346.

In the early days of bulkhead design, it was generally believed that the shearing resistance of a soil was equal to the normal pressure on the potential surface of sliding multiplied by the tangent of the slope angle at which the soil came to rest after it was dumped. This angle was called the angle of repose. For more than twenty years it has been known that there is no relationship between the shearing resistance and the tangent of the angle of repose except for clean, loose, and dust-dry sand. The angle of repose of any

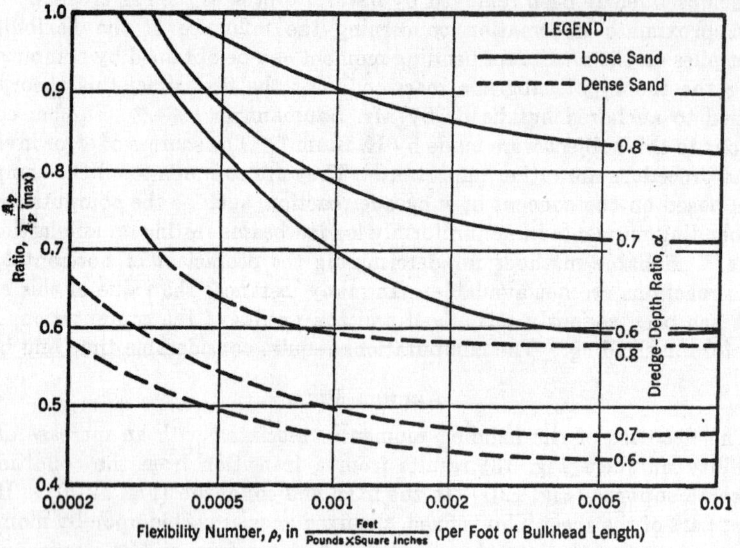

FIG. 14.—RELATION BETWEEN THE FLEXIBILITY NUMBER, ρ OF SHEET PILES, AND ANCHOR-PULL RATIO, $\frac{A_p}{A_p\,(\text{max})}$ FOR A β-VALUE OF 0.2

other material depends on both the characteristics of the material and the height of the slope. Therefore, it cannot be used as a basis for estimating the shearing resistance of the soil.

Whatever the characteristics of a soil may be, its shearing resistance can be divided into two components. One of them, commonly known as cohesion, is independent of the pressure, whereas the other one increases with increasing effective pressure $\bar{p}$ on the surface of sliding. The effective pressure $\bar{p}$ is equal to the difference between the total unit pressure p and the pore water pressure p_w. This fundamental relationship can be expressed approximately by the empirical equation:

$$s = c + (p - p_w) \tan \phi \qquad (13)$$

The limits of the validity of this equation are discussed in most references on soil mechanics.

For clean sand, $c = 0$ and $s = (p - p_w) \tan \phi$, in which s is the shearing resistance per unit of area. Cohesive soils, such as silt and clay, are commonly much more compressible than sand, and their permeability is very low. If the load on saturated, cohesive soil is increased—for example, by the deposition of a backfill—the application of the load, p, per unit of area, is associated with an equally important increase of the pore water pressure p_w because the excess water drains out of the loaded soil stratum very slowly. The corresponding increase of the second term on the right-hand side of Eq. 13 becomes equal to zero and, as a consequence, the loaded stratum performs during and immediately after the application of the load as if its angle of internal friction were zero. Experience shows that the initial shearing strength of such soils is approximately equal to one half of their unconfined compressive strength q_u, and $s = q_u/2$.

If γ' is the submerged weight of a cohesive soil stratum in contact with a bulkhead and $\bar{p}$ is the effective unit load on the top surface of the stratum caused by the submerged weight of the immersed part of the soil located above the stratum and the total weight of the materials located above the water table, the active earth pressure at any depth z below the surface of the cohesive stratum is

$$p_A = \bar{p} + \gamma' z - q_u \qquad (14)$$

per unit of area, and the passive earth pressure is

$$p_P = \bar{p} + \gamma' z + q_u \qquad (15)$$

Eqs. 14 and 15 were derived on the assumption that the surfaces of contact between soil and bulkhead are frictionless.[23] Because of the adhesion between the soil and the bulkhead, the values of p_A are somewhat smaller and those of p_P somewhat greater than those determined by Eqs. 14 and 15, respectively. Information concerning the values of q_u for clays with different consistencies is given in Table 1.

TABLE 1.—UNCONFINED COMPRESSIVE STRENGTH OF SILT AND CLAY

Consistency (the drillman's designation)	COMPRESSIVE STRENGTH, q_u, IN TONS PER SQUARE FOOT	
	Minimum	Maximum
Very soft	0	0.25
Soft	0.25	0.50
Medium	0.50	1.00
Stiff	1.00	2.00
Very stiff	2.00	4.00
Hard	4.00	∞

In the course of time, the shearing resistance of cohesive soils, such as silt or clay, increases because of progressive consolidation. Nevertheless, this process is harmful because the consolidation is associated with an increase of the maximum bending moment in the sheet piles, as previously explained under the heading, "Influence of Flexural Rigidity on Bending Moment."

If a cohesive soil, such as fine silt or clay, is deposited under water, consolidation hardly starts during construction. At any depth below the surface,

[23] "Soil Mechanics in Engineering Practice," by Karl Terzaghi and R. B. Peck, John Wiley & Sons, Inc., New York, N. Y., 1948, pp. 147–150.

the pore water pressure p_w almost equals the overburden pressure p at that depth, so that $(p - p_w)$ in Eq. 15 equals zero. Since the cohesion c of such backfills is also very small, the lateral earth pressure exerted by such a backfill and its resistance against lateral displacement are at any point equal to the vertical pressure in the fill at that point; hence,

$$K_A = K_P = 1.00 \qquad (16)$$

Because of the absence of any definite relationship between the angle of repose and the shearing resistance of soils, reliable information concerning the active and passive earth pressure of the different soils in contact with an anchored bulkhead can be obtained only on the basis of: (1) The results of laboratory tests simulating the conditions under which the soils will be subject to shear in the field, or (2) empirical values derived from the results of tests such as those which will be proposed in Part II of this paper.

II. DESIGN OF ANCHORED BULKHEADS

Uncertainties Involved in the Design of Anchored Bulkheads

The experimental investigations described in Part I have made it possible to eliminate the most serious misconceptions associated with the customary methods of bulkhead design. On the basis of the findings, one can reliably estimate the forces exerted on anchored bulkheads by homogeneous layers of soil with known physical properties. Hence, the uncertainties involved in the design of bulkheads no longer result from inadequate knowledge of the fundamental principles involved. They are caused only by the fact that the structure of natural soil deposits is usually complex, whereas the theories of bulkhead design inevitably presuppose homogeneous materials. Not even the backfills composed of excavated and transported soils can be considered homogeneous. Because of local variations of the soil properties within the borrowpit area and segregation according to grain size during the process of underwater deposition, the characteristics of the backfill may change from place to place and its properties cannot be reliably determined by tests in advance of construction.

Because of these conditions, the most economical and expedient procedure consists in estimating the constants and coefficients—such as the unit weights and the coefficients of earth pressure—on the basis of the results of exploratory borings and of routine tests performed on representative samples, and compensating for the uncertainties involved in this procedure by an adequate margin of safety. More elaborate investigations are justified only in exceptional cases.

General Design Procedure

The design of anchored bulkheads requires several successive operations: (*a*) Evaluation of the forces that act on the inner face of the bulkhead, (*b*) determination of the depth of sheet-pile penetration, (*c*) computation of the maximum bending moments in the sheet piles, (*d*) evaluation of the anchor pull, and (*e*) selection of allowable stresses in the construction materials in accordance with the uncertainties involved in the evaluation of the acting forces.

Each operation is described herein under a separate subheading. The essential data required for performing the operations are obtained by exploratory borings at the site of the bulkhead and in the borrowpit area and by certain routine tests as specified subsequently.

At the site of the bulkhead, the borings must be supplemented by the standard penetration test[24] or an equivalent procedure to obtain information concerning the relative density of the strata of sand or of silty sand that will be in contact with the sheet piles. Drive-samples of the sand are secured, and a mechanical analysis is performed. The results define the general nature of the cohesionless strata. If layers of silt or clay are encountered, fairly undisturbed samples should be recovered and the natural water content, Atterberg limits, and the unconfined compressive strength q_u of these samples should be determined. If possible, the boring operations should be supplemented by field tests with the vane borer. This method of testing was developed in Sweden and it has been extensively used in the United States and in Canada.[25] The value of the natural water content is required for computing the full and the submerged unit weight, γ and γ', respectively, of the materials, and the value of the unconfined compressive stress q_u determines, according to Eqs. 14 and 15, the active and passive earth pressure of clay and silt.

The soils encountered at the bulkhead site and the backfill materials are classified on the basis of the following system: Clean sand (dense, medium,or loose); silty sand (dense, medium, or loose); and silt or clay (hard to very soft, as in Table 1). A sand should be classified as silty, if more than 5% passes the 200-mesh sieve.

Forces Acting on Inner Face of Bulkhead

Varieties of Forces.—Fig. 15 shows diagrammatic sections through bulkheads, ab. The sheet piles are driven into sand (Fig. 15(*a*)) and clay (Fig. 15(*b*)), respectively. Both bulkheads are backfilled with clean sand which has been sluiced into place.

The intensity of the forces acting on the inner face of each bulkhead is represented by the width of the pressure areas, numbered from I to IV on the left-hand side of ab. These forces include the following:

I. Active earth pressure produced by the weight of the backfill;
II. Active earth pressure produced by the uniformly distributed surcharge, q;
III. Unbalanced water pressure (see also Fig. 5); and
IV. Lateral pressure caused by line load q'.

The computation of the intensity of these forces must be preceded by an evaluation of the full and submerged unit weights of the various soils in contact with the inner face and of their coefficients of active earth pressure. In each figure the symbol $\overline{p}$ with subscripts indicates the effective vertical unit pressure on the different horizontal sections through the soil.

The active earth pressure exerted by the backfill depends on the coefficient of active earth pressure K_A of the fill material and on its effective unit weight.

[24] "Soil Mechanics in Engineering Practice," by Karl Terzaghi and R. B. Peck, John Wiley & Sons, Inc., 1948, p. 265.

[25] "The Vane Borer," by L. Cadling and S. Odenstad, *Proceedings*, Royal Swedish Geotechnical Inst., No. 2, 1950; Stockholm, Sweden.

It may be increased temporarily or permanently by a uniformly distributed surcharge, by line loads, or by mobile or stationary concentrated loads.

Unit Weights.—The effective unit weight depends on the position of the soil with reference to the water table. Above the water table, it is equal to the sum of the dry weight of the soil and the weight of the water contained in

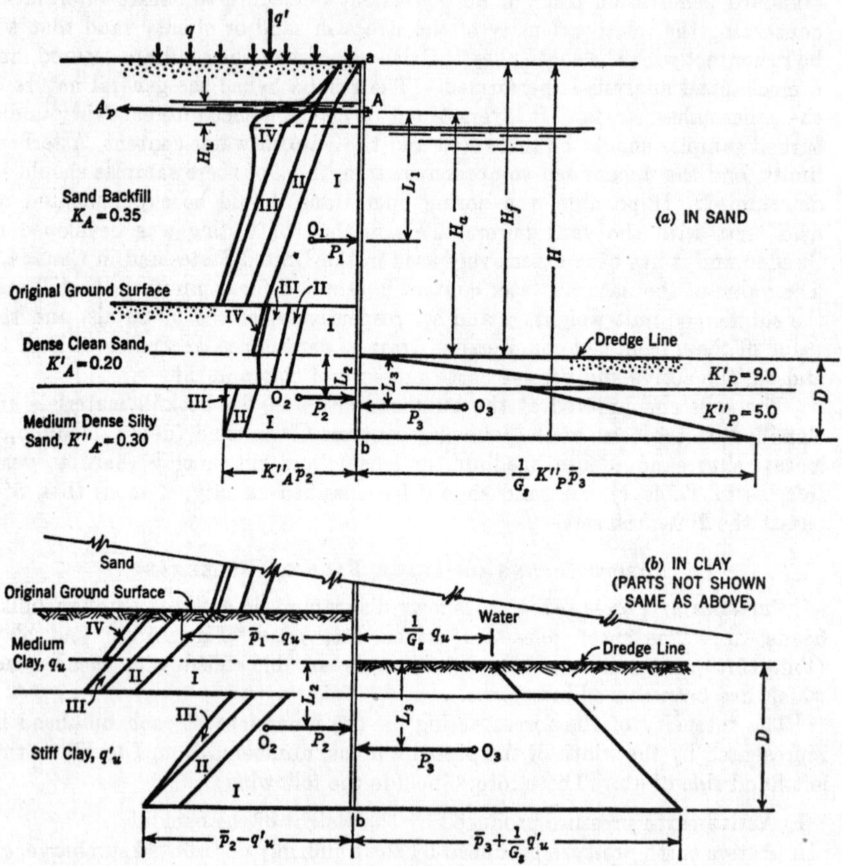

FIG. 15.—FORCES ACTING ON BULKHEADS CONSISTING OF SHEET PILES DRIVEN INTO SAND AND INTO CLAY

the voids (moist unit weight). Below the water table it is equal to the submerged weight of the soil particles. Limiting values for the moist and the submerged unit weights for cohesionless soils are given in Table 2. In the computation of the active earth pressure, the upper limiting values should be used unless the real unit weights have been estimated on the basis of test results. The unit weight of cohesive soils is determined by their natural water content w, and it can be computed by use of the equations given in Table 2.

Coefficients of Active Earth Pressure.—Design values for the coefficient of active earth pressure are in Table 2. The backfills are usually deposited under-

TABLE 2.—UNIT WEIGHTS OF SOILS, AND COEFFICIENTS OF EARTH PRESSURE

Type of soil	Unit Weight[a] of Moist Soil, γ		Unit Weight[a] of Submerged Soil, γ'		Coefficient of Active Earth Pressure, K_A				Coefficient of Passive Earth Pressure, K_P		
	Minimum	Maximum	Minimum	Maximum	For backfill	For soils in place	Friction Angles[b] ϕ	Friction Angles[b] δ	For soils in place	Friction Angles[b] ϕ	Friction Angles[b] δ
(1)	(2)	(3)	(4)	(5)	(7)	(8)	(9)	(10)	(11)	(12)	(13)
Clean Sand:											
Dense......	110	140	65	78		0.20	38	20	9.0	38	25
Medium....	110	130	60	68		0.25	34	17	7.0	34	23
Loose......	90	125	56	63	0.35	0.30	30	15	5.0	30	20
Silty Sand:											
Dense......	110	150	70	88		0.25			7.0		
Medium....	95	130	60	68		0.30			5.0		
Loose......	80	125	50	63	0.50	0.35			3.0		
Silt and clay[c].......	$\frac{165(1+w)}{1+2.65w}$		$\frac{103}{1+2.65w}$		1.00	$1-\frac{q_u}{\bar{p}+\gamma\bar{z}}$			$1+\frac{q_u}{\bar{p}+\gamma\bar{z}}$		

[a] In pounds per cubic foot. [b] These angles, expressed in degrees, are ϕ, the angle of internal friction, and δ, the angle of wall friction, and are used in estimating the coefficients under which they are listed. [c] The symbol γ represents γ or γ', whichever is applicable; $\bar{p}$ is the effective unit pressure on the top surface of the stratum; q_u is the unconfined compressive stress; w is the natural water content, in percentage of dry weight; and z is the depth below the top surface of the stratum.

water and their structure is likely to be loose. The deflection of the bulkhead is not sufficient to mobilize the full value of the shearing resistance of the backfill (as shown in Fig. 3). Therefore, the values of K_A assigned to the backfill materials are higher than the corresponding values for the same materials in place.

The values of K_A for cohesionless soils in place were estimated on the basis of the ϕ-values and δ-values in Table 2. In selecting values of δ, the angle of wall friction, it was reasoned that, since the active earth pressure is equal to the sum of the anchor pull and passive earth pressure (as shown in Fig. 2(*a*)), the total value of the active earth pressure is considerably greater than that of the passive earth pressure. Hence, for any specified value of the angle of wall friction, the resultant of the friction forces would tend to pull the sheet piles down. If the lower edge of the sheet piles were rigidly supported, the resultant would be carried by this rigid support. In reality, the resistance against a downward movement of the lower edge of the sheet piles is small. Therefore, the sheet piles will settle so that the total wall friction on the inner face will decrease until it becomes nearly equal to the friction force on the outer face. Because of this condition, the angle of wall friction for the active pressure was assigned smaller values than that for the passive earth pressure.

The K_A-value for dense sand can be much lower than the value of 0.2 given in Table 2, but this is only a possibility because the value of K_A depends, to a large extent, on the uniformity of the sand and the shape of the grains.[3] Therefore, the use of a lower value would be justified only if it has been computed on the basis of test results on representative samples.

Silty sand has been assigned greater K_A-values than clean sand with equal relative density because its angle of internal friction is likely to be smaller and its compressibility higher than that of clean sand with equal relative density. Furthermore, the evaluation of the relative density of silty sands by use of the standard penetration test appears to be much less reliable than that of clean sands. The K_A-values for silt and clay are represented in Table 2 by an equation that gives results which are on the safe side because the pressure reducing effect of the adhesion between soil and sheet piles has been disregarded.

Uniformly Distributed Surcharge.—The lateral unit pressure resulting from a uniformly distributed surcharge, q, per unit of area, is at any depth equal to q times the value of K_A for that depth (area II in Fig. 15).

Unbalanced Water Pressure.—The first step in studying the unbalanced water pressure is to estimate the greatest hydraulic head, H_u, in Fig. 5(*a*), which can be anticipated at the site of the bulkhead. This can be done on the basis of regional hydrographic data such as local tidal curves or flood records. If the coefficient of permeability of all the soils in contact with the bulkhead is of the same order of magnitude, the unbalanced water pressure associated with a head H_u can be estimated on the basis of the assumption represented by the broken line cde in Fig. 5(*b*). Area III in Fig. 15(*a*) has been plotted on this assumption. Otherwise, the effects of stratification on the distribution of the unbalanced water pressure must be considered. In any event, precautions must be taken to prevent the unbalanced water pressure from exceeding the estimated value even under exceptional conditions such as an outgoing spring tide combined with a heavy rainstorm. This may, for example, be accomplished by adequate surface drainage.

If the bulkhead is backfilled with a material—such as silt or soft clay—which does not begin to consolidate during construction, the initial water table in the backfill is located at the surface of the backfill. The coefficient of active earth pressure of such materials is equal to unity (Table 2). Therefore, the total lateral pressure of the backfill, earth and water pressure combined, is equal to the fluid pressure of a liquid the unit weight of which is equal to that of the saturated backfill material.

Line and Point Loads.—Line loads and point loads can be established only on the surface of backfills with adequate bearing capacity. Otherwise the loads must be carried by piles, and pile-supported loads have no influence on the lateral pressure.

Line loads are stationary, and the lateral pressure resulting from the loads is determined by Eqs. 4 and 5. Point loads are either stationary or mobile. The wheel loads of cranes that can be displaced on rails along the waterfront constitute mobile point loads. The unit pressures produced by a point load along the line of intersection ab (Fig. 7(*a*)) between the bulkhead and a vertical plane through the load at right angles to the bulkhead are determined by Eqs. 9 and 10. A mobile point load may occupy any position on the line along which it travels. Hence, every sheet pile may be acted upon by these pressures. However, if the point load is stationary, the unit pressures produced by the load decrease on both sides of the line ab, as shown in Fig. 7(*a*). Upper limiting values for these pressures are obtained by use of Eq. 11 (Fig. 8(*b*)).

Figs. 6(*a*) and 8(*a*) show that the computed unit pressures on the lower part of the lateral supports are consistently greater than the real pressures. As a partial compensation for this difference, it should be assumed that the lateral pressure produced by line loads and point loads decreases, at the dredge line, to zero. Hence, in Eqs. 4, 5, 9, and 10, the value H should be made equal to the vertical distance H_f between the dredge line and the surface of the backfill. In Fig. 15 the lateral pressure caused by point loads or line loads is represented by area IV.

Depth of Sheet-Pile Penetration

Factors Determining Penetration Depth.—The investigations of Mr. Rowe have shown that no tangible benefits can be obtained by driving the sheet piles deeper than the depth required to assure an adequate margin of safety with respect to a failure resulting from an outward movement of the buried part of the sheet piles, and a sufficiently small horizontal displacement of the lower edge of the sheet piles.[19]

The resistance of a cohesionless soil against an outward movement of the buried part of the sheet pile depends on its effective unit weight and on the coefficient of passive earth pressure. The resistance of a cohesive soil such as silt or clay depends only on its unconfined compressive strength.

Unit Weights.—Because the passive earth pressure of a cohesionless soil decreases with decreasing unit weight, the computations should be made on the basis of the lower limiting values in Table 2, unless more accurate values have been obtained by tests on representative samples. Should the water table in the backfill be temporarily located above the free water level, as in Fig. 5(*a*), the reduction of the effective unit weight resulting from the seepage pressure associated with upward percolation through the soil in contact with the outer face of the bulkhead must be considered. If the coefficient of permeability of all the soils in contact with the bulkhead is of the same order of magnitude, the effective unit weight of the soil acted upon by the seepage pressure can be estimated by use of Eq. 3. Otherwise, an appropriate value for the hydraulic gradient i in equation Eq. 2 must be selected on the basis of the soil profile. In many instances the gradient i is too small to require any consideration.

Coefficients of Passive Earth Pressure.—Lower limiting values for the coefficients of passive earth pressure are in Col. 11, Table 2. The values of ϕ and δ used in the estimates of K_P-values for clean sand (Fig. 4) are in Cols. 12 and 13. The K_P-value for a well-graded dense sand or a mixture of sand and gravel can be almost twice as high as the tabulated value for dense sand. This, however, is only a possibility. Hence, higher values should be tolerated only if they are based on test results.

In the rare event that the buried part of an anchored bulkhead derives its lateral support from a sand fill, the sand can be assigned a K_P-value of 3.

Silty sand has been assigned smaller K_P-values than clean sand with equal relative density. This was done because of the conditions previously explained (under the heading, "Forces Acting on Inner Face of Bulkhead: Coefficients of Active Earth Pressure"). Very loose, silty sand cannot be expected to provide adequate lateral support for the buried part of sheet piles because of its high compressibility.

If the surface of the soil constituting the lateral support of the buried part of the bulkhead is not horizontal, the passive earth pressure must be determined by a graphical procedure on the basis of the ϕ-values and δ-values in Table 2. Since the assumption of a plane surface of sliding may involve important errors on the unsafe side (Fig. 4(*b*)), the logarithmic spiral method should be used.[26]

Computation of Depth of Sheet-Pile Penetration

The general principles in the computation of penetration depth are illustrated in Fig. 15. The area on the left-hand side of the bulkhead, ab, represents the active pressures on the inner face, and the area on the right-hand side represents the passive earth pressures, plotted on the assumption of free earth support.

The actual distribution of the passive earth pressure under conditions of free earth support (Fig. 12(*a*)) is approximately trapezoidal. However, the assumption of a trapezoidal distribution would complicate the computations considerably except in the rare event of a bulkhead with sheet piles driven into homogeneous ground. Therefore, the Coulomb concept is retained. In accordance with this concept, the passive earth pressure of homogeneous material increases in a manner similar to hydrostatic pressure (in simple proportion to the depth below the surface). The error resulting from this simplifying assumption is unimportant and on the safe side.

In order to provide for an adequate margin of safety with respect to a failure of the lateral earth support of the buried part of the sheet piles, the pressure areas on the right-hand side of line ab in Fig. 15(*a*) are constructed on the assumption that the coefficient of passive earth pressure is equal to the K_P-values in Table 2, divided by a safety factor G_s. If the sheet piles are driven into silt or clay, as in Fig. 15(*b*), the safety requirements are satisfied by assigning to the soil in contact with the outer face of the bulkhead a reduced unconfined compressive strength, $\frac{1}{G_s} q_u$. The conditions requiring consideration in the choice of G_s will be discussed, subsequently, under the heading, "Safety Requirements."

In Fig. 15, points O_1,O_2, and O_3 represent the centers of gravity of the pressure areas above the dredge line (O_1), and below the dredge line on the active side (O_2) and passive side (O_3) of the bulkhead. The total pressures represented by these areas are P_1, P_2, and P_3. The position of their lines of action are defined by the distances L_1, L_2, and L_3. The values of L_1 and P_1 are independent of the depth D of sheet-pile penetration. The depth D must satisfy the condition that the sum of all the moments about the anchor point A is equal to zero,

$$P_1 L_1 + P_2(H_a + L_2) = P_3(H_a + L_3) \quad \text{.............}(17)$$

[26] "Theoretical Soil Mechanics," by Karl Terzaghi, John Wiley & Sons, Inc., New York, N. Y., 1943, pp. 100–113.

If the values of P_2, P_3, L_2, and L_3 are expressed in terms of the unknown quantity D, an equation of the third degree is obtained which can be solved for D.

Computation of Maximum Bending Moments

Free Earth Support.—If the profile of the soil into which the sheet piles will be driven is erratic or if no reliable data concerning the details of the soil profile are available, the maximum bending moment in the sheet piles should be computed on the assumption of free earth support. The acting forces are shown in Fig. 15(*a*) for bulkheads with lateral sand support and in Fig. 15(*b*) for bulkheads with clay support. The maximum bending moment can be determined by analytical or graphical methods in the usual manner.

Mr. Tschebotarioff has suggested computation of the maximum bending moment in the sheet piles on the assumption of fixed earth support irrespective of flexibility and of the relative density of the sand into which the sheet piles are driven.[18] Mr. Rowe showed that the errors involved in such a procedure can be very important and they are on the unsafe side.[19] Moment reduction caused by partly or completely fixed earth support can only be expected under the conditions to be described subsequently.

Moment Reduction Resulting from Flexibility Effects.—If the sheet piles are to be driven into a fairly homogeneous stratum of clean sand with known relative density, the maximum bending moment for free earth support can be reduced on the basis of the results of Mr. Rowe's investigations that are represented in Fig. 13. As an example, the reduction of the bending moments in sheet piles driven into sand with medium density will be considered.

The first step consists of computing the maximum bending moments for free earth support, M(max), and of the cross section of the piles required to withstand M(max). The conditions that must be considered in choosing the allowable fiber stresses f will be discussed subsequently (under the heading, "Safety Requirements"). The flexibility number ρ_1 of these piles is determined by Eq. 12.

At a specified value of M(max) the moment of inertia I and the corresponding flexibility number ρ_1 depend on the value assigned to the allowable fiber stresses and on the construction material. Hence, at a specified value of M(max), very different values of ρ_1 may be obtained, such as ρ_{1t} for timber piles, ρ_{1s} for steel piles, and ρ_{1c} for reinforced concrete piles.

The next step consists in plotting the moment-reduction curve, which is shown as the heavy curve in Fig. 16. This curve is obtained by multiplying the ordinates of the curve for sand with medium density in the Rowe diagram (Fig. 13) by M(max). The abscissa ρ_c of point C at which the curve starts to descend represents the critical flexibility number. If the flexibility number of the pile required by M(max) is smaller than ρ_c, the maximum bending moment in this pile is determined by the condition of free earth support and is equal to M(max).

In Fig. 16 the sheet piles with flexibility numbers ρ_1, such as ρ_{1t}, ρ_{1s}, and ρ_{1c}, are represented by points S. At a given value of M(max), the flexibility number of timber sheet piles with the required moment of inertia has the greatest value—designated as point S(timber)—and that of reinforced concrete piles the smallest value—designated as point S (concrete). Point S (steel), representing steel sheet piles, occupies an intermediate position.

If point S is located on the left-hand side of C, no moment reduction can be tolerated. A position of point S on the right-hand side of C indicates that the pile represented by the point is stronger than necessary because the maximum bending moment in the pile M_1 will be less than M(max). In order to select a more economical profile for the sheet piles, the allowable bending moments

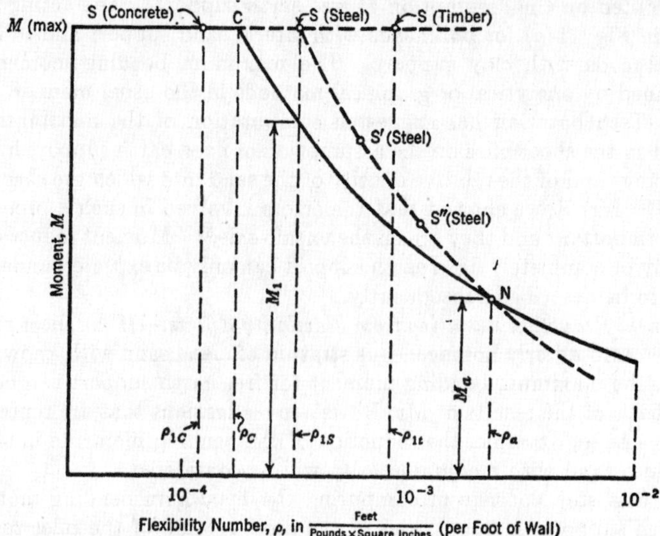

FIG. 16.—GRAPHIC PROCEDURE FOR DETERMINING TOLERABLE MOMENT REDUCTION

M', M'', ⋯ and the corresponding flexibility numbers ρ', ρ'', ⋯ for various weaker profiles are computed. In Fig. 16 these weaker sheet piles are represented by points such as S′ (steel) with the ordinate M' and the abscissa log ρ'. All these points are located in proximity of a curve that intersects the moment-reduction curve at N, with abscissa log ρ_a and ordinate M_a. The corresponding moment reduction is $M(\text{max}) - M_a$. However, because of the rudimentary state of present (1953) knowledge of the performance of bulkheads under field conditions, the computed bending moment M(max) should be reduced by not more than $\frac{1}{2}(M(\text{max}) - M_a)$.

If the sheet piles are to be driven into a homogeneous stratum of dense or medium silty sand, Mr. Rowe's moment-reduction curves for medium and loose sand should be used instead of those for dense and medium sand. Sheet piles to be driven into loose, silty sand should be dimensioned for free earth support because the compressibility of such sands may be very high.

Sheet Piles Driven into Silt or Clay.—In Part I it was shown that the initial earth support for sheet piles driven into silt or clay is likely to be fixed. However, as time elapses, the end restraint decreases because of progressive consolidation of the soil resisting the lateral pressure of the buried part of the bulkhead. As the present (1953) state of knowledge concerning the effect of the progressive yield on the end restraint is limited, no moment reduction should be tolerated. The fundamental principles of the procedure have been explained in Part I.

Anchor Pull

Anchor Pull for Free Earth Support.—The anchor pull A_p for free earth support is determined by the condition that the sum of all the horizontal forces that act on the bulkhead (Fig. 15) is equal to zero:

$$A_p = (P_1 + P_2 - P_3)\,l \qquad (18)$$

in which l is the spacing between anchor rods.

The force P_1 also includes the lateral pressure caused by mobile or stationary point loads. In Fig. 15 this pressure is represented by area IV, which is equal to the lateral pressure per unit of length of the bulkhead along section ab in Fig. 7(*a*). According to this figure, the lateral pressure produced by the point load decreases from ab in both horizontal directions. Hence, if the backfill carries a point load, computation of A_p by use of Eq. 18 involves an error on the safe side. The importance of the error can be estimated and corrected on the basis of the experimental data shown in Figs. 7 and 8. Usually the error is too small to require consideration.

Anchor-Pull Reduction.—As indicated in Fig. 14, the anchor pull A_p decreases with increasing flexibility number, but the decrease is not as important as the corresponding decrease of the maximum bending moment M_1 in Fig. 13. The anchor pull depends on several factors other than the properties of the backfill material and the flexibility number. Therefore, the anchor pull should be computed on the assumption of free earth support.

Safety Requirements

General Considerations.—The importance of the error involved in estimating the forces which act on the inner face of a bulkhead depends to a large extent on the degree of uniformity of the fill material, the complexity of the soil profile, and the amount of information which can be, or has been, secured concerning the significant soil properties. Therefore, it would be uneconomical to establish a rigid code concerning the safety factors and the allowable stresses. Designers who are not thoroughly familiar with the principles and techniques of soil mechanics are advised to consider only free earth support and to use large safety factors.

Passive Earth Pressure.—The safety factor G_s (Fig. 15(*a*)), with respect to a failure of the lateral earth support of piles driven into clean or silty sand should be assigned a value between 2 and 3, depending on the accuracy with which the active earth pressure on the inner face of the bulkhead can be estimated.

If the earth support is derived from silt or clay, the limiting values for G_s (as shown in Fig. 15(*b*)) can be reduced to 1.5 and 2.0 because the K_P-value defined by Eq. 15 is on the safe side.

Whatever the subsoil conditions may be, the computed depth of sheet-pile penetration should always be increased by 20% as insurance against the effect of unintentional excess dredging, unanticipated local scour, and the presence of pockets of exceptionally weak material in the zone of passive earth pressure which have not been revealed in the borings. The corresponding increase of the cost of the bulkhead is small as compared to the increase of the margin of safety for all parts of the bulkhead associated with the excess penetration. The bending moments in the sheet piles and the anchor pull should be determined on the basis of the computed depth—and not of the increased depth—of sheet-pile penetration.

Allowable Fiber Stresses in Sheet Piles.—The maximum bending stresses f in the sheet piles of anchored steel bulkheads backfilled with clean sand can be made equal to at least two thirds of the yield point stress f_y. This recommendation also applies to dredge bulkheads supporting clean sand in place. If the backfill consists of hydraulically excavated, silty sand or a mixture of clean and silty sand, the real value of the coefficient of earth pressure can locally be considerably higher than the values given in Table 2 for silty sand because of the inevitable segregation of fine and coarse material during the process of deposition. Therefore, the fiber stress in the sheet piles of steel bulkheads backfilled with such materials should be assigned a value of not more than two thirds of the yield point. However, if a bulkhead is to be backfilled with hydraulically excavated clay or silt to which a coefficient of active earth pressure equal to unity is assigned, the sheet piles can safely be dimensioned on the basis of $f = f_y$ because the real lateral earth pressure cannot be greater than that which is computed.

The extreme fiber stresses in sheet piles having locks on the neutral axis depend largely on lock friction. Little information is available (1953) concerning the values that can safely be assigned to the lock friction in straight rows of sheet piles. Hence, it is advisable that such sheet piles be dimensioned on the assumption that the locks are lubricated. The flexibility number should be computed on the same basis. At a given value of M(max), the flexibility number for sheet piles dimensioned on the assumption of lubricated locks is considerably greater than the corresponding number for piles with "frozen locks." Therefore, a decrease of the friction in the interlocks can be associated with a decrease of the maximum bending moment in the sheet piles.

The allowable stresses for construction materials other than steel should be selected on the basis of the considerations that have been set forth for steel.

Allowable Stress in Anchor Rods.—Eq. 18, which is used for computing the anchor pull, involves the assumption that the distribution of the active earth pressure on the inner face is in accordance with the Coulomb theory. The real distribution may be somewhat different—as shown in Figs. 11 and 12—and the corresponding anchor pull may be greater than the computed anchor pull. Furthermore, if the uppermost part of the sheet piles is in contact with a soil having low compressibility and the lower part moves out (for example, because of progressive consolidation of the soil that provides the lateral earth support), the anchor pull increases. The anchor pull may also increase because of repeated application and removal of heavy surcharges. Finally, an unequal yield of adjacent anchorages produces an increase in the pull in some anchor rods, associated with a decrease in others. Because of these possibilities, the anchor rods should be dimensioned on the basis of more conservative allowable stresses than are applied to the design of sheet piles.

Supplementary Safety Provisions.—The first and foremost requirement for the success of a bulkhead project is avoiding, during construction, loading conditions which the designer has not anticipated. This vital aspect of the bulkhead problem has been investigated by J. R. Ayers, M.ASCE, and R. C. Stokes.[27] In addition to providing for an adequate margin of safety in the design and for protection of the construction materials against deterioration, it is also necessary to protect the bulkhead against the consequences of processes and events beyond the scope of theory.

If a bulkhead is at the seashore or at the bank of a swiftly flowing river, consideration should be given to the possibility that the scour along the foot of the exposed part of the outer face of the sheet piles may increase considerably the free height of the bulkhead, H_f (Fig. 15). The upper limiting value for the increase of H_f should be estimated on the basis of experience with local scouring action.

After the sheet piles for a fill bulkhead with submerged outer face are driven, a careful inspection should be made of the condition of the locks or joints. If gaps between adjacent sheet-piles are found, they should be plugged by divers. A case of important loss of soil caused by a flow of sand through gaps between sheet piles was reported by J. C. Gebhard.[28]

If the backfill material is very compressible or if the subsoil of the backfill contains layers of soft silt or clay, it is necessary to encase the anchor rods in large conduits with a circular or rectangular cross section that can follow the downward movement of the backfill without transmitting vertical forces onto the anchor rods. As an alternative, the rods could be supported at several points by piles, but then allowance must be made for the stresses in the tie rods caused by the earth load which acts on the rods between the points of support. On Pier C at Long Beach, it was found that the anchor rods embedded in the backfill carried a large part of the weight of the backfill lying above the level of the anchor line. This load increased considerably the tension in the rods.[29]

Bulkhead Failures.—All the bulkhead failures that have come to the writer's attention can be attributed to one of two causes. The designer has estimated earth pressure and earth resistance on the basis of the "angle-of-repose" concept, or else he has failed to notice a source of weakness in the ground below the level of the lower edge of the sheet piles.

It is known that there is no relationship between the angle of repose of a soil and its angle of internal friction except in the case of loose and dust-dry sand. Nevertheless, several bulkhead computations have been made in which soft clay was assigned an angle of internal friction of 11°, on the strength of the observation that the front part of a sheet of hydraulic-fill clay commonly has a slope approximating one on five. The bulkheads to which these computations referred have failed.

---

[27] "The Design of Flexible Bulkheads," by J. R. Ayers and R. C. Stokes, *Transactions*, ASCE, Vol. 119, 1954, p. 373.

[28] "Cave-Ins of Sandy Backfills," by J. C. Gebhard, *ibid.*, Vol. 114, 1949, pp. 490–498.

[29] "Field Study of a Sheet-Pile Bulkhead," by C. Martin Duke, *ibid.*, Vol. 118, 1953, p. 1131.

Some bulkheads with sheet piles driven into sand have failed because of an outward movement of both bulkhead and fill on a soft clay stratum beneath the sand. The design of the bulkheads and of the anchorage was satisfactory, but the designers failed to take heed of the subsoil of the stratum that provided the lateral support for the buried part of the sheet piles.

If the submerged natural earth in front of a bulkhead slopes downward in an offshore direction, the slope may fail along a surface of sliding located below the anchor and the lower edge of the sheet piles. Several bulkhead failures of this kind have occurred on the Whangpoo River at Shanghai, China. The subsoil consisted of river silt, the offshore slope was about one on two, and most of the failures occurred within two hours after low tide following spring tide.

At the present state of knowledge of the physical properties of soils (1953), all the aforementioned bulkhead failures could have been avoided by proper consideration of the results of exploratory borings and of a few identification tests on representative samples of the earth materials involved.

Conclusions

1. Since the time (1910 to 1930) when the current procedures (1953) for bulkhead computations originated, knowledge of the physical properties of earth materials and of the mechanics of earth pressure has increased vastly; yet the procedures remained practically unaltered. This paper contains suggestions for revisions on the basis of the experimental and observational data which have been secured during the last few decades.

2. Many bulkheads have been designed without adequate information concerning the relative density of the sand layers and the shearing strength of clay strata in contact with the bulkheads. Bulkheads designed in such a manner can be badly overdimensioned or they may be unsafe, depending on factors unknown to the designers. One of the most common causes of faulty design is the erroneous assumption that the shearing resistance of cohesive soils can be evaluated on the basis of an angle of repose.

3. The importance of the errors involved in the estimate of the soil constants appearing in the equations depends to a large extent on the complexity of the structure of the strata into which the sheet piles are driven, on the degree of uniformity of the material in the borrowpit area, and on the quality of the subsoil exploration. Therefore, it would be unwarranted to establish rigid rules for the factors of safety that should be used. One of the responsibilities of the designer is to evaluate the prevailing uncertainties and to choose the factors in accordance with his findings. Limiting values are proposed in this paper.

4. No bulkhead theory can possibly anticipate all the varieties of subsoil and hydraulic conditions that may be encountered in practice, and every case requires a certain amount of independent judgment. Hence, if the subsoil conditions do not conform to a standard pattern, designers who are not thoroughly familiar with the basic principles and techniques of soil mecahnics are advised to assume free earth support and to use conservative factors of safety. When solving unusual problems, the designer should consult the observational data (summarized in Part I) on which the design procedures are based.

ACKNOWLEDGMENTS

The writer expresses his thanks to A. E. Cummings, M. ASCE, S.D. Wilson, A.M. ASCE, and S. B. Avery, Jr., for valuable comments and suggestions. He is also indebted to Mr. Rowe who kindly placed at his disposal copies of the original records of various tests.

APPENDIX. NOTATION

The following letter symbols conform essentially with ASCE *Manual of Engineering Practice No. 22* ("Soil Mechanics Nomenclature") and American Standard Letter Symbols for Structural Analysis (ASA Z10.8—1949) prepared by a Committee of the American Standards Association, with ASCE participation, and approved by the Association in 1949:

A_p = anchor pull;
D = depth of sheet-pile penetration;
d = the ratio of deflection to the height of a wall;
E = the modulus of elasticity;
f = the allowable stress in bending of the piles;
f_y = the yield point;
G_s = the factor of safety;
H = the height of a lateral support, total length of sheet piles;
H_a = the vertical distance from the anchor to the dredge line;
H_f = the vertical distance between the dredge line and the surface of the backfill;
H_w = the vertical distance between the free water level and the water table in the backfill;
I = the moment of inertia of the cross section of a sheet pile;
i = the hydraulic gradient;
K_0 = the pressure coefficient for earth at rest;
K_A = the coefficient of active earth pressure (the ratio between the normal component of the earth pressure on the lateral support and the corresponding fluid pressure);
K_P = the coefficient of passive earth pressure;
L_1, L_2, and L_3 = the vertical distances from centers of pressure;
l = the anchor spacing;
M = the maximum bending moment in a sheet pile:
- $M(\max)$ = the maximum bending moment in a sheet pile computed on the assumption of free earth support;
- M' = the allowable bending moment for a sheet pile with a given flexibility number;
- M_a = the maximum bending moment in a sheet pile with flexibility number ρ_a;
- M_1 = the maximum bending moment in a sheet pile with a given flexibility number;

m = the ratio between the horizontal distance from the wall to the height of the wall;
n = the ratio between the depth below the surface of the backfill and the height of the wall;
P = the total normal pressure on a lateral support produced by a point load;
P_u = the unbalanced water pressure;
p = the unit pressure:
p' = the horizontal pressure produced by a line load per unit of wall length;
$\overline{p}$ = the effective unit pressure;
p_A = intensity of active earth pressure;
p_P = intensity of passive earth pressure;
p_w = pore water pressure;
p_1 = the horizontal unit pressure produced by a point load along the intersection between the inner face of a wall and a vertical section through the point load, perpendicular to the wall;
Q = a point load;
q = a uniformly distributed surcharge, per unit of area:
q' = a unit line load;
q_u = the unconfined compressive strength of cohesive soil;
s = the shearing resistance, per unit of area;
w = the natural water content, in percentage of the dry weight;
$x, y, z,$ = the variable distances in the directions X, Y, and Z;
α = the ratio between the depth of the dredge line below the surface of the backfill and the length of the sheet piles;
β = the ratio between the depth of the anchor line below the surface of the backfill and the length of the sheet piles;
γ = the unit weight of a soil including the weight of water contained in its voids:
γ' = the submerged unit weight;
$\overline{\gamma}$ = the effective unit weight of silt or clay (the saturated unit weight above the water table and the submerged unit weight below the water table);
$\Delta\gamma'$ = the reduction in the submerged unit weight resulting from seepage pressure exerted by rising ground water;
$\Delta\gamma''$ = the increase in the effective unit weight;
γ_w = the unit weight of water;
δ = the angle of wall friction;
μ = the Poisson ratio;

ρ = the flexibility number;
ρ_a = the flexibility number corresponding to the point of intersection N in Fig. 16.
ρ_c = the critical flexibility number at which M becomes smaller than M(max);
σ = the normal stress;
σ_x = the horizontal unit pressure;
ϕ = the angle of internal friction;
ϕ' = the angle of partly mobilized internal friction; and
ψ = an angle defined by Fig. 8(*b*).

INFLUENCE OF GEOLOGICAL FACTORS ON THE ENGINEERING PROPERTIES OF SEDIMENTS

KARL TERZAGHI

CONTENTS

ABSTRACT

The paper contains a condensed review of the influence of geological factors on all those physical properties of sediments that have a significant influence on the performance of sediments in open cuts, beneath building foundations and storage dams, and on those properties which determine their suitability for use as construction materials. The geological factors are divided into three large groups: petrographic factors, modes of deposition, and changes after deposition. The paper also contains brief explanations of those terms and physical constants that are needed to describe quantitatively the significant properties of the sediments.

557

INTRODUCTION

THE term "engineering properties of sediments" refers to all those properties of sediments that have a significant influence on the performance of sediments during and after the completion of construction operations.

The influence of geological factors on these properties has been known in a general way for more than a century. Yet the benefits derived from the knowledge of the existence of this influence were very slight until quite recently because engineers were not yet able to define by numerical values the significant properties of sediments and the influence of geological factors on these properties.

This unfortunate condition was still reflected twenty years ago by the building codes in many cities. According to these codes the *"allowable unit load"* for the design of footings on fine sand was assigned a very much smaller value than that for coarse sand because it was not yet known that the settlement of a given footing on sand depends on the relative density of the sand and not on its grain-size characteristics. Allowable bearing values were specified for "soft," "medium" and "stiff" clay, but the codes did not provide the engineer with any means for distinguishing between these three categories of clay. As a consequence the terms "soft" and "stiff" left a wide margin for interpretation. In this connection the writer had the following experience.

In 1929 he received two boring records, one from southern Texas and one from the coast of Washington. According to the Texas record representing the site for a 25-story office building the subsoil consisted of "soft clay" to a depth of at least 40 feet. It was intended to establish the structure on a pile foundation. Subsequent laboratory tests showed that the clay was stiff enough to sustain the weight of the structure without the assistance of piles. The building was erected on a shallow foundation and the settlement turned out to be less than two inches.

The boring record from the state of Washington indicated the presence of a layer of "stiff" clay, 30 feet thick, at a depth of about 100 feet beneath a relatively light factory building. At the time the record was received the building had already settled up to 8 inches. On the basis of the results of soil tests the writer found that the settlement would increase within ten years to a value of about 30 inches and the forecast was confirmed by the subsequent settlement observations.

Conditions did not change until the mechanical properties of sediments became a subject of systematic investigations. Such investigations were started about 1915 to establish procedures for expressing the significant mechanical properties of sediments by numerical values to be derived from laboratory tests. It soon became evident that the performance of this task required a large-scale cooperative effort. It was also realized that a radical modification of the techniques of subsoil exploration was needed because the change of the structure of sediments produced by the action of conventional drilling tools is associated with an equally radical change of the mechanical properties of the sampled material. The mechanical properties of samples furnished by an earth auger have no resemblance to those of the material *in situ.* To fill this need, procedures for the recovery of undisturbed samples were developed. The relative merits of the newly developed techniques were investigated and appraised by Hvorslev (41).

During the last four decades the study of the engineering properties of sediments developed into an independent branch of engineering science, known as Soil Mechanics. Research in soil mechanics created a rational basis for several important branches of civil engineering such as foundation engineering and earth dam design. At the same time it provided the geologist with new and efficient tools for unraveling the origin and history of sedimentary deposits.

Before soil mechanics came into existence, many of the significant properties of sediments, such as the activity and sensitivity of clays and the critical density of sands, were unknown and the relation between stress and strain for sediments received no attention. Hence, the discoveries in the field of soil mechanics made necessary the establishment of a rather elaborate terminology. Without this terminology the engineering properties of sediments cannot adequately be described. Explanations of the newly created terms can be found in any up-to-date textbook on soil mechanics, but the terms have not yet made their appearance in textbooks on physical geology. Therefore, the writer has expanded the attached list of symbols by adding to each symbol a definition of the property indicated by it.

The geological factors that influence the engineering properties of sediments can be divided into three large categories: (A) Petrographic factors, (B) Modes of transportation and deposition, and (C) Changes after deposition. Our knowledge of the influence of these factors on the mechanical properties of sediments is still very incomplete. The following sections contain a condensed review of present concepts.

PETROGRAPHIC FACTORS

Varieties of Petrographic Factors.—The petrographic factors that have or can have a significant influence on the engineering properties of sediments include the grain-size distribution, shape, and mineralogical composition of the solid particles, the water and gases contained in its voids, and the chemical composition of the adsorbed layers.

Practically all of the sediments encountered in nature consist of particles with different size, shape, and mineralogical composition. If a sample of a sediment is divided by mechanical analysis into grain-size fractions, the mineralogical composition of the different fractions is commonly found to be very different. The meaning of the terms that are used to designate the grain-size fractions is illustrated by Figure 1. In this paper the MIT classification will be used.

The sand- and silt-size fraction is dominated by fresh or moderately altered fragments of rock-forming minerals such as quartz, feldspar, calcite and mica. The ratio between the percentage of platy and equidimensional grains increases with decreasing grain size. The clay-size fraction is dominated by clay minerals such as kaolinite, illite, and montmorillonite, produced by chemical weathering of rock-forming minerals.

The relationship between grain size and mineralogical composition is commonly represented by frequency distribution curves. Grim (30) has published frequency diagrams of this kind for a great number of clay sediments with very different mechanical properties. These diagrams demonstrate clearly the existence of the aforementioned relationships.

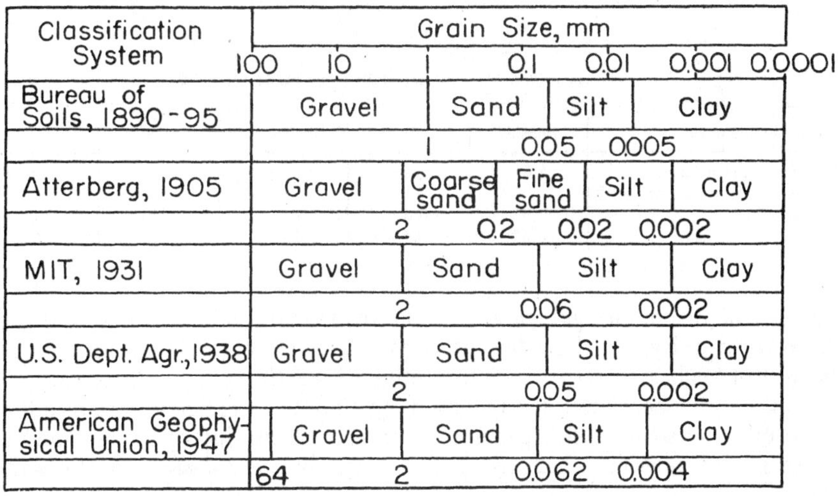

Classification System	Grain Size, mm (100 10 1 0.1 0.01 0.001 0.0001)
Bureau of Soils, 1890-95	Gravel | Sand | Silt | Clay
	1 0.05 0.005
Atterberg, 1905	Gravel | Coarse sand | Fine sand | Silt | Clay
	2 0.2 0.02 0.002
MIT, 1931	Gravel | Sand | Silt | Clay
	2 0.06 0.002
U.S. Dept. Agr., 1938	Gravel | Sand | Silt | Clay
	2 0.05 0.002
American Geophysical Union, 1947	Gravel | Sand | Silt | Clay
	64 2 0.062 0.004

FIG. 1. Comparison of several common textural classification systems. (After Peck et al.)

Below the water table the voids of sediments are filled either entirely with water or with water and gas bubbles. Even if gas bubbles are present the *degree of saturation* S_r of the sediment *in situ* is commonly close to 100 percent. When a sample of the sediment is taken out of the ground, however, additional gas is released from the water and as a consequence the sample may expand rapidly and noticeably.

Within a certain distance from the surface of the solid particles the physical properties of the water contained in a sediment are radically different from those of free water, owing to molecular interaction between water and solid. That part of the interstitial water that is significantly affected by the molecular interaction with the solids constitutes the *adsorbed* water. In a general way the ratio between the quantity of adsorbed and free water contained in a sediment increases with increasing specific surface A_s. At a given specific surface, however, it depends to a large extent on the mineralogic composition of the solid particles and on the adsorption complex.

The properties of sediments affected by the petrographic factors include the void ratio immediately after sedimentation, the permeability, plasticity, sensitivity, activity, compressibility, shearing resistance, and rate of compression.

Void Ratio.—In soil mechanics terminology, the term *void ratio, e,* indicates the ratio between the total volume of voids and the volume occupied by the solid constituents of a grain-aggregate.

The void ratio of a well sorted and perfectly cohesionless aggregate of equidimensional grains, such as the particles obtained by crushing a quartz specimen, can range between the extreme values of about 0.35 and 1.00. Yet if a sediment is formed from a suspension of quartz powder the void ratio e of the sediment is greater than unity and it increases with decreasing grain size as shown in Table I.

TABLE I

VOID RATIO e OF CRUSHED QUARTZ. (Terzaghi, 1925)

Grain size in microns............	700–250	100–20	20–6	6–2	2
e_0 (after sedimentation)	1.0	1.21	2.23	2.57	2.66
e_1 (after sedimentation and vibration)	0.67	0.80	1.10	1.50	2.16
Ratio e_1/e_0	0.67	0.66	0.49	0.58	0.81

The structure of an aggregate of equidimensional grains, with a void ratio of more than unity, will be called a *collapsible* or *metastable* structure. The existence of such a structure and the influence of the grain size on the void ratio can only be accounted for by assuming that the particles start to adhere to each other as soon as they come into contact. This intergranular bond results in a true cohesion which has been termed *initial cohesion* by the writer (82).

With decreasing grain size, the number of intergranular bonds per unit of weight of solid constituents increases. As a consequence, the ratio between the weight of an individual particle and the force that resists its tendency to move into a more stable position decreases, so that the void ratio of the sediment increases. For the same reason the reduction of the void ratio produced by vibrating the sediment decreases with decreasing grain size.

If the quartz powder is mixed with powdered mica the void ratio at a given grain size increases rapidly with increasing mica content and the effect of vibrations on the density of the sediment decreases.

If the void ratio of a sediment is changed, all its other properties are affected. Therefore, the engineering properties of a sediment *in situ* cannot be evaluated unless the void ratio *in situ* or some other equally significant property, such as its relative density D_r or the liquidity index L_l, is known.

Permeability.—The permeability affects the engineering properties of sediments because it has a decisive influence (a) on the rate at which the void ratio decreases after an increase of the load on the sediment and (b) on the relationship between shearing resistance and rate of application of the shearing force.

The coefficient of permeability k of a sand composed of more or less equidimensional grains depends almost entirely on the effective grain size D_{10} and the void ratio e. Therefore the value k of such sand can roughly be estimated by means of semi-empirical equations, for instance

$$k = 200\, D^2_{10}\, e^2 \tag{1}$$

If the sand contains more than a few percent of mica flakes this equation loses its validity and no substitute is available. Yet even for sand-mica mixtures the k-value depends only on particle size, particle shape, and void ratio. On the other hand the k-value of sediments containing a clay-size fraction is also influenced to a large extent by the adsorption complex. If, for instance, the Na-ions in a bentonite are replaced by Ca-ions, the k-value becomes many times greater than its original value, at the same void ratio.

On account of the great variety of factors that determine its value, the k-value of sediments other than clean sand and the relation between this value and the void ratio can only be determined by experiment. The k-value of clean, fine sand such as dune sand is of the order of magnitude of 10^{-2} cm/sec, that of silt 10^{-4} cm/sec, and of clay 10^{-7} cm/sec, with a wide scattering from the average.

Plasticity.—Experience has shown that a sediment cannot be expected to be very compressible or to acquire a high degree of cohesion unless it possesses, within a certain range of water content, the properties of a plastic material. Since experience has also disclosed the existence of statistical relationships between the degree of plasticity of sediments and all their other physical properties, the plasticity of sediments has been the subject of extensive investigations.

In connection with the classification of sediments for engineering purposes it was found convenient to express the plastic characteristics of sediments by the *plastic limit* P_w, the *liquid light* L_w, and the *plasticity index* I_p. This procedure was proposed originally by Atterberg (1) for agricultural soil classification. Subsequently, it was adapted with notable success to the needs of soil mechanics by the author (82).

The plastic limit P_w is the lowest water content, in percent of the dry weight, at which the sediment can still be rolled out into threads with a diameter of about one eighth of an inch. If the water content of the sediment is increased beyond P_w, the material becomes softer and at the liquid limit L_w it acquires the consistency of a paste with low strength. The difference

$$I_p = L_w - P_w$$

represents the *plasticity index* or the plastic range. The values P_w and L_w are known as *Atterberg limits.* If w is the water content of a sediment ***in situ,*** the ratio

$$L_i = \frac{w - P_w}{L_w - P_w}$$

is called the *liquidity index* of the sediment.

Since there is no sharp boundary between solid and liquid state, the liquid limit is inevitably an arbitrary one. The experimental procedure that has been adopted for determining the liquid limit is such that the limit corresponds to a shearing resistance of several tens of grams per square centimeter. In other words, if the water content of a sediment is at the liquid limit, its shearing resistance in a remolded state is equal to a few tens of grams per square centimeter. However, the strength of the sediment *in situ* at a water content equal to the liquid limit can amount to several hundred grams per square centimeter. The difference is due to thixotropic hardening and other physico-chemical processes that take place after deposition.

Experience has shown that the liquidity index of the top layer of a normally consolidated clay deposit is commonly close to one (see, for instance, Hogentogler (35), p. 178, and Skempton (72)). Yet, the liquidity index of a sediment formed on the bottom of a vessel filled with a clay suspension is always greater than two. The physical cause of the gradual decrease of the void ratio of the top layer from a high initial value to a void ratio close to that corresponding to the liquid limit is still unknown. The decrease may be due to a process similar to the syneresis in gels.

At a given grain size, the values of P_w and L_w depend on the mineralogical composition of the grains and at a given mineralogical composition both values increase with decreasing grain size, as shown by the data assembled in Table II.

TABLE II
INFLUENCE OF GRAIN SIZE OF CRUSHED MINERALS ON ATTERBERG LIMITS
(After Atterberg 1)

Material	Grain size in microns											
	100–20			20–6			6–2			<2		
	L_w	P_w	I_p	L_w	P_w	I_p	L_w	P_w	I_p	L_w	P_w	I_p
Quartz	34	34	0	34	34	0	34	34	0	35	35	0
Feldspar	37	37	0	38	38	0	38	38	0	39	39	0
Biotite	—	—	—	—	—	—	46	46	0	53	45	8
Serpentine	—	—	—	—	—	—	41	41	0	76	59	8
Muscovite	—	—	—	—	—	—	49	49	0	91	77	14
Haematite	—	—	—	—	—	—	—	—	—	36	20	16
Chlorite	33	33	0				44	44	0	72	47	25
Talcum	33	33	0				48	48	0	76	48	28

At given mineralogical composition and grain-size characteristics, P_w and L_w also depend to a large extent on the adsorption complex. For instance, if the Na-ions in a montmorillonite are replaced by Ca-ions, the liquid limit decreases from 700 to 124 and the plastic limit from 93 to 72 (29).

If the particles of a fine-grained sediment would touch each other, the plastic limit should be practically independent of particle size. Therefore, the general relationships described in the preceding paragraphs indicate that even at the plastic limit the solid particles are separated from each other by adsorbed layers and that the seat of strength and cohesion is located in these layers. As the grain size decreases the ratio between the quantities of adsorbed and free water contained in the sediment increases and as a consequence the Atterberg limits increase.

The influence of grain size and mineralogical composition on the Atterberg limits is clearly brought out by the plasticity chart, Figure 2 (10). The abscissas represent the liquid limit L_w, the ordinates the plasticity index I_p and the A-line the average relationship between L_w and I_p. If a mineral is ground up and the Atterberg limits of the grain-size fractions are determined, the points representing the test results in Figure 2 are located on a straight line roughly parallel to the A-line. The line may be located either above or below the A-line, depending on the mineralogical composition of the grains. Furthermore, the points representing different samples from a geologically well-defined sedimentary deposit are also located on such a line, because the mineralogical composition of their clay-size fraction is likely to be similar.

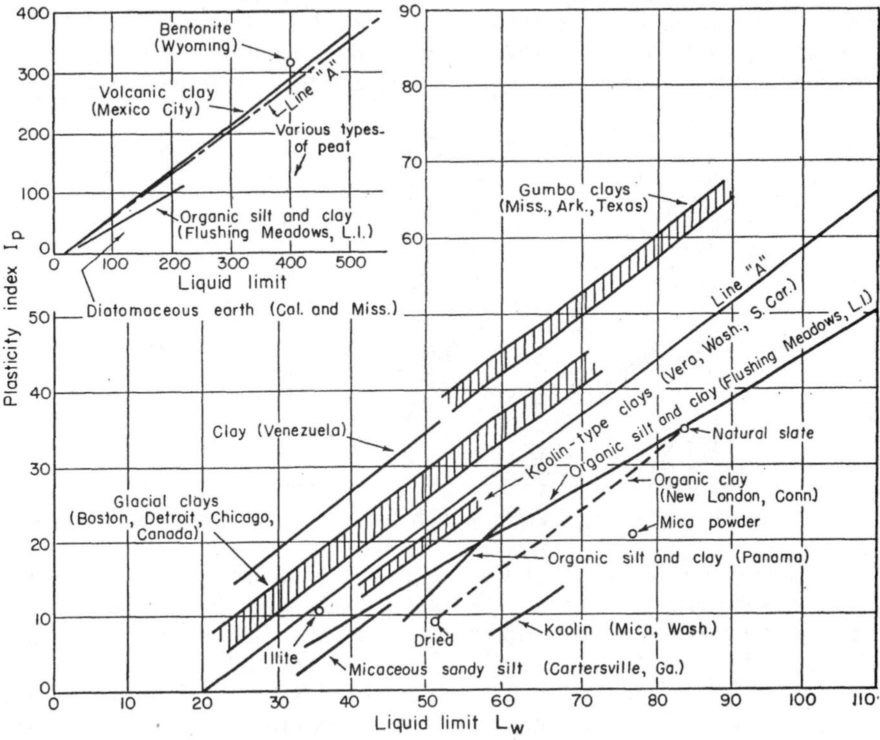

Fig 2. Plasticity chart.

The determination of Atterberg's limits is a simple routine procedure and the position of a point in the plasticity chart, combined with the results of a few supplementary routine tests, permits fairly reliable identification of sedimentary deposits with others that have been encountered at previous occasions. Therefore, the plasticity chart represents a satisfactory basis for the classification of cohesive soils for engineering purposes. A classification on the basis of mineralogical composition and exchangeable bases is still impracticable; indeed, it is even doubtful whether it is desirable. The plasticity chart also provides the geologist with an expedient and reliable means for the classification, identification, and correlation of the individual members of composite sedimentary deposits. If the points representing two members in the plasticity chart are located on very different lines it is almost certain that the sediments were derived from two different sources.

Sensitivity.—The sensitivity s_t of a cohesive sediment such as a clay is equal to the ratio between the unconfined compressive strength of the material in an undisturbed and in a remolded state. Its value may range between 1 (*insensitive clays*) and more than 16 (*quick clays*).

The sensitivity of clays with a liquidity index of less than unity is commonly low to medium ($s_t = 1.5$ to 4). The major portion of the loss of strength can be accounted for by the fact that the process of remolding eliminates the strength that was previously acquired by thixotropic hardening. The balance is due to the disturbance of the structure of the sediment. If the remolded specimen is allowed to rest under unaltered external conditions it

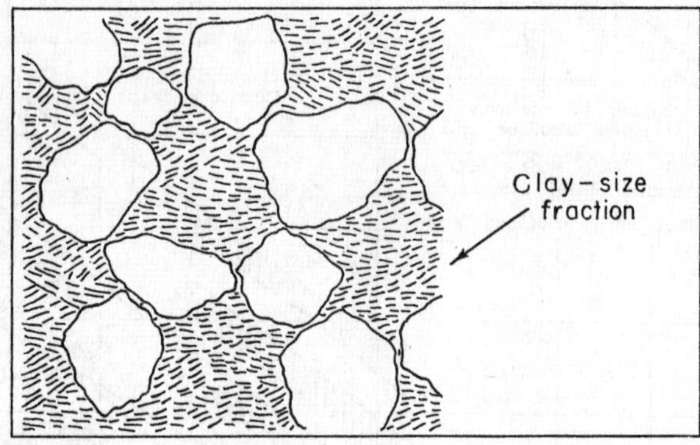

FIG. 3. Metastable arrangement of equidimensional grains in sensitive clays.

regains a large portion of its original strength. The strength that can be acquired by a sediment as a result of thixotropic hardening depends on the mineralogical composition of the prevalent clay mineral. It is smallest for kaolinite, intermediate for illite, and greatest for montmorillonite (79).

Sensitive ($s_t = 4$ to 8), extra-sensitive ($s_t = 8$ to 16), and quick clays (s_t greater than 16) have, with very few exceptions, a liquidity index far above unity. Most of them belong to one of two categories, (a) glacial lake clays, and (b) marine clays which were subsequently lifted above sea level and lost their salt content by leaching.

Glacial lake clays commonly contain a high percentage of rock flour with equi-dimensional particles. Both the high liquidity index and the high sensitivity of these clays appear to be due to a metastable arrangement of the equidimensional particles as shown in Figure 3 (8, 93). The strength of the undisturbed clay is chiefly due to the strength of the framework formed by these particles and the bond between particles at their points of contact. If the framework is destroyed by remolding, the clay loses most of its strength and any subsequent gain in strength from thixotropic hardening does not exceed a small fraction of its original value (79).

Marine clays, in their original state, commonly have a liquidity index close to unity and their sensitivity is low to medium. If such clays, however, are subsequently lifted above sea level, they can, under certain conditions, turn into extrasensitive clays with a high liquidity index. A clay of this kind has been encountered in the proximity of St. Thuribe in Quebec. It has a liquidity index of 1.9 and a sensitivity of more than 150 (58, 66). The Norwegian quick-clays belong to the same category.

The high sensitivity of these clays is attributable to leaching whereby the Na in the adsorption complex is replaced by Ca. This base exchange reduces the thickness of the adsorbed layers surrounding the clay particles and therefore lowers the liquid limit of the clay. The water content, however, of the clay remains unaltered, because the framework formed by the equidimensional particles prevents consolidation under the influence of the overburden pressure. The resulting structure is practically identical with that of extra-sensitive freshwater clays that acquired their sensitivity during the process of deposition.

The degree of sensitivity has a decisive influence on the performance of a clay in the event of a slope failure. The failure of a slope on clay with a low or medium sensitivity commonly starts with the formation of tension cracks along the upper boundary of the slide area. During the slide, the slide material moves rather slowly. The sliding segment breaks up but it retains more or less its original shape and it comes to rest at a short distance from the foot of the slope. By contrast, the slides in very sensitive or quick clays are not preceded by the formation of tension cracks at the upper boundary of the slide area. Moreover, the failure spreads with great speed from the foot of the slope in an uphill direction to a distance which is many times greater than the width of the belt affected by the initial slope failure (96).

As soon as a slice of clay breaks down, the clay assumes the character of a viscous liquid and it flows even on horizontal surfaces to a distance far in excess of the length of the slide area. During a slide at Vaerdalen in Norway, about 100 million cubic yards of slide material flowed on the almost horizontal valley floor at a rate of more than five miles per hour to a distance of seven miles from the slide area. Yet, at the upper boundary of the slide area, the clay stood in vertical cliffs with a height of several tens of feet (37).

In the Vaerdalen slide, the slide of St. Thuribe, and many others, the slide material emerged from the slide area through a narrow gap between steep cliffs, comparable to a bottleneck.

Activity.—As indicated in Table II, the plasticity of cohesive sediments is due almost exclusively to the interaction between water and the particles of the clay-size ($< 2\,\mu$) fraction. At a given mineralogical composition of the clay-size fraction, the plasticity index of the sediment increases approximately in simple proportion to the clay-size content in percent of the total weight. However, at a given clay-size percentage, the plasticity index I_p can have very different values depending on the mineralogical composition of the clay-size fraction. The ratio between the plasticity index I_p and the clay-size fraction in percent of the total weight represents the *activity* a_c of the sediment (76). *"Inactive"* sediments have an a_c value of less than 0.75, *"normal"* ones $a_c =$ 0.75 to 1.25, and *"active"* ones a_c greater than 1.25. Skempton (76) obtained for the most common constituents of the clay-size fraction the following activity values.

Quartz	Calcite	Mica (muscovite)	Kaolinite	Illite
0.0	0.18	0.23	0.33, 0.46	0.90

Ca-Montmorillonite	Na-Montmorillonite
1.50	7.2

The activity values for montmorillonite show that these values depend not only on the mineralogical composition of the solid particles but also to a large extent on the adsorbtion complex. It is obvious that the activity also depends on the average grain size of the most active portion of the clay-size fraction, everything else being equal (Table II). Since the clay minerals are not subject to attrition and to sorting, however, their average particle size is not necessarily as variable as that of the coarser fractions.

Compressibility.—The term *compressibility* refers to the relationship between the increase of the unit load on a laterally confined specimen of a sediment and the corresponding decrease of its void ratio. Application and subsequent removal of a load p on a sediment permanently reduces its compressibility under loads that are smaller than p. Therefore distinction is made between *normally consolidated* and *pre-compressed* or *pre-consolidated* sediments.

The term *normally consolidated* refers to sediments that have never been subject to a load reduction, such as the removal of superimposed sediments by erosion. The following paragraphs refer only to normally consolidated sediments. The effects of pre-compression will be discussed under the heading *Pre-Consolidation under Load.*

The relationship between unit load and void ratio for a sediment can conveniently be represented by plotting the void ratio e against the logarithm of the unit load p as shown in Figure 4. In the laboratory, the relationship between e and p is investigated by successively increasing the unit load on a disk-shaped specimen confined within a ring and measuring the corresponding decrease of its thickness. Tests of this kind are called *confined compression* or *consolidation tests.*

For every sediment, the e-log p curve starts with a horizontal tangent and at a certain unit load joins an inclined straight line with the equation

$$e = -C_c \log_{10} p + C \qquad (2)$$

wherein C_c represents the *compression index* and C is a constant. The value C_c is equal to the decrease of the void ratio produced by an increase of the unit load from p to $10\,p$.

For sand composed of equidimensional particles, C_c is of the order of magnitude of 0.01. If mica flakes are added to the sand, the value of C_c increases rapidly with increasing percentage of mica. The C_c-value of a mixture of 50 percent sand and 50 percent mica is greater than 1.0. In other words, if the unit load on such a mixture is increased from p to $10\,p$ the corresponding decrease of the volume of the mixture is greater than the volume occupied by its solid constituents.

Curve s in Figure 4 represents the e-log p curve for a remolded specimen of clay with an initial water content equal to the liquid limit. The left-hand end of the straight, inclined section of the e-log p curve corresponds for any clay to a p-value of about 1 kg/cm². According to Skempton (73) the value C_c, which is equal to the slope of the straight inclined section, increases with increasing liquid limit L_w approximately in accordance with the empirical equation

$$C_c = 0.007\ (L_w - 10\%) \qquad (3)$$

The value of C_c ranges between about 0.15 for lean, sandy clays and more than 1.0 for highly colloidal, bentonitic clays.

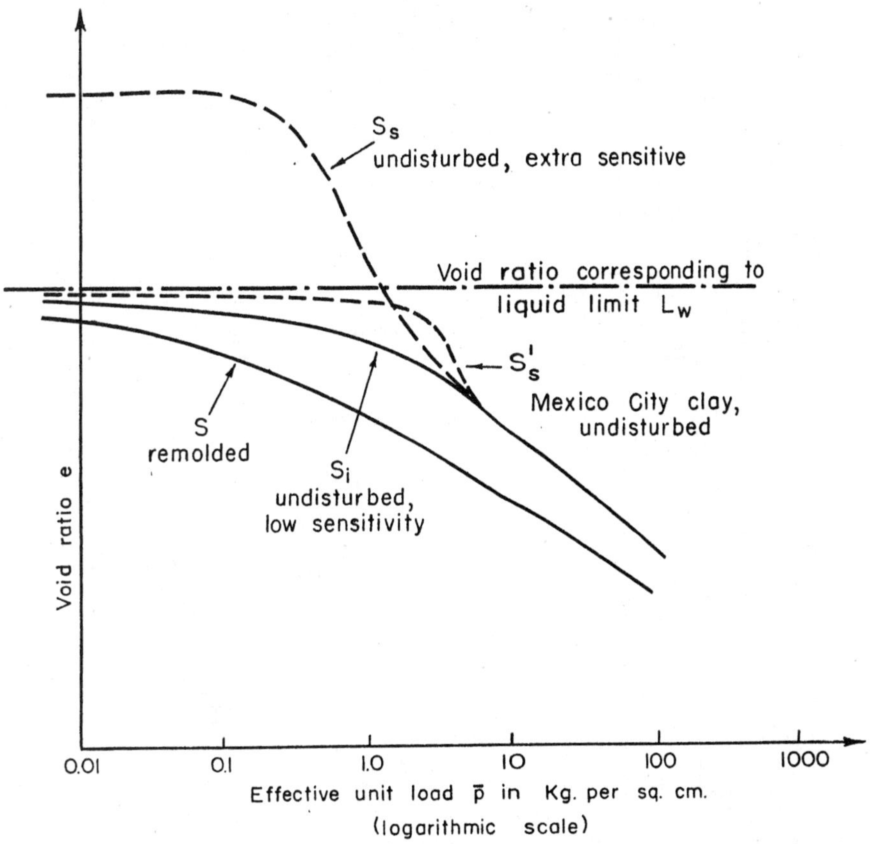

FIG. 4. Principal types of relationships between unit load and void ratio for cohesive sediments.

If the consolidation test is performed on an undisturbed specimen of a clay with low sensitivity, an e-log p curve S_i, Figure 4, is obtained which is very similar to the e-log p curve S for the same clay in a remolded state, except that the straight inclined part of the e-log p curve is steeper. The C_c-value of the undisturbed specimen exceeds that of the specimen in a remolded state by amounts up to about 30 percent. Since the water content of the uppermost layer of a normally consolidated deposit of clay with a low sensitivity is commonly close to the liquid limit, the horizontal tangents of the curves S and S_l in Figure 4 are almost identical.

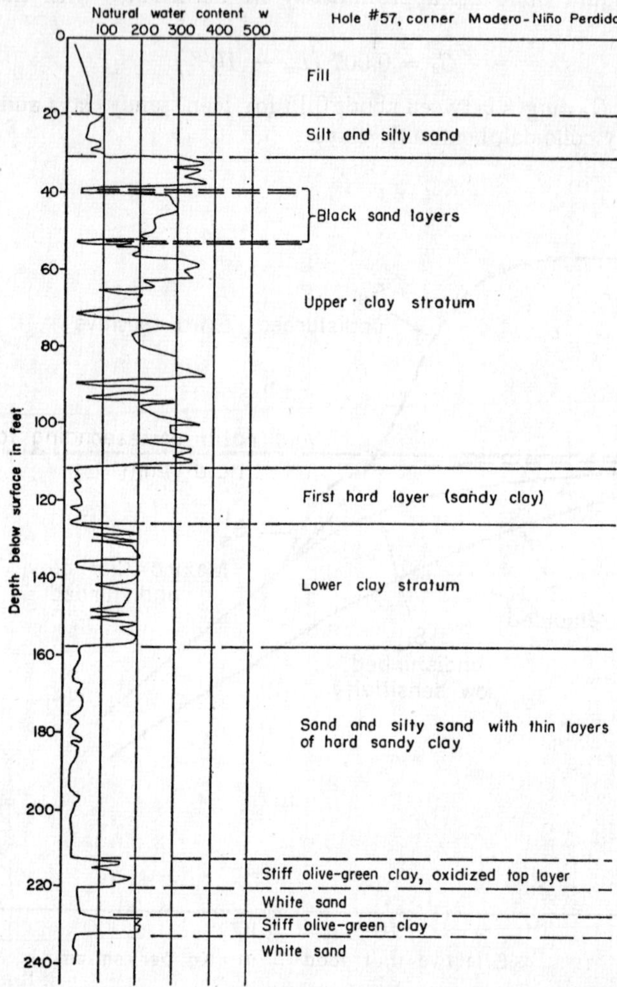

FIG. 5. Typical soil profile, subsoil of Mexico City. (After L. Zeevaert.)

If the consolidation test is performed on an undisturbed specimen of a highly sensitive clay, the e-log p curve has the characteristics of the dashed curve S_s in Figure 4. Since the liquidity index L_l of such clays is commonly very much greater than unity, the horizontal tangent to curve S_s is located high above the horizontal tangent to the e-log p curve S for the sediment in a remolded state. The unit load at which the upper part of curve S_s descends abruptly towards the lower inclined part represents the load at which the metastable structure shown in Figure 2 collapses.

In exceptional cases, e-log p curves similar to S_s in Figure 4 have also been obtained by tests on undisturbed samples of clay with an initial water content close to the liquid limit. The clay underlying Mexico City is an example. The uppermost two hundred feet of the subsoil of the city contain two thick layers of a highly colloidal, laminated lacustrine clay with thin seams of fine sand and silt of pyroclastic origin. Figure 5 is a typical boring record. Figures 6*a* and 6*b* represent histograms of the liquid limit (*a*) and the natural water content (*b*) of the clay. The mean value of the natural water content,

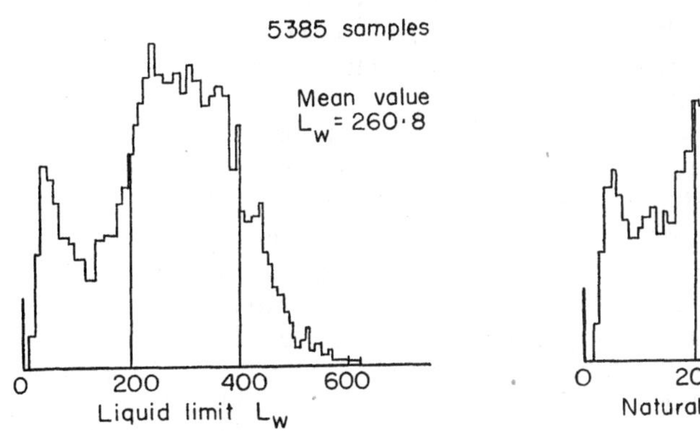

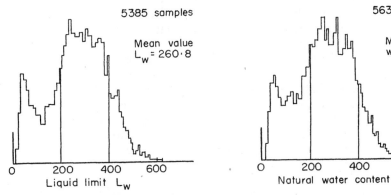

FIG. 6. Histograms for Mexico City clay. (After Marsal and Mazari Marcos, 1952.)

which is 254.1, is close to that of the liquid limit (260.8). Nevertheless the e-log p curve for this clay, curve S'_s in Figure 4, has the shape of curve S_s and not that of curve S_i (52).

The sensitivity of the clay ranges between 6 and 16, which is high. Sensitivities of this magnitude are commonly associated with low activity. Yet the activity of the Mexico City clay is of the order of magnitude of 5 which is also very high. These unusual properties of the clay appear to be due to a high montmorillonite content combined with a high organic content. Nevertheless, its permeability is many times greater than that of normal clays with a very much smaller clay content. This discrepancy suggests that the Mexico City clay consists of firm clusters composed of clay-size particles. The high sensitivity of the clay indicates that the structure formed by the clusters is metastable.

It has been mentioned before that the e-log p curve of remolded specimens of clay of any kind and that of undisturbed specimens of clay with low sensitivity have the same shape as that of sand-mica mixtures. The ratio between the weight of the flaky particles and the dry weight of the sediment will be designated as *grain-shape factor*, F_s. With an increasing value of F_s, both the initial void ratio and the compression index C_c of the sand-mica mixture increases rapidly. This is due to the shape and the flexibility of the mica particles. If the grain-shape factor F_s is zero (clean sand composed of equidimensional grains), the compression of the sand under load is almost exclusively caused by a slight displacement of the sand grains with reference to each other, and the compression index is very low. As the grain-shape factor increases, more and more of the compression of the sand is due to the bending of mica flakes. Since the flexural rigidity of the mica flakes is low, the compressibility of the sand-mica mixture is inevitably high.

Sediments with a high clay content also consist of a mixture of equidimensional particles and of flaky constituents. Therefore, the compression index C_c of the clay must also be a function of the grain-shape factor which depends exclusively on the shape of the grains and not on their chemical composition. Yet it depends also on physico-chemical factors. According to Salas and Serratosa (67), the replacement of sodium by calcium in a bentonite reduced the compression index from 7.72 to 2.04 in spite of the fact that the grain-shape factor remained unaltered. Hence it is evident that the compressibility of clay is determined by two independent factors: the grain-shape factor, and the forces of molecular attraction and repulsion within the adsorbed layer.

Yet, even if the grain-shape factor were entirely eliminated, the *e*-log *p* curve of the clay would still have the shape of that for a sand-mica mixture. This is demonstrated by the fact that there is very little difference between the *e*-log *p* curve of such a mixture and that of pure gelatine and other organic gels that do not contain any constituents with a flaky habit. As a consequence it is impossible to find out which part of the compressibility of a clay is due to the flexibility of its flaky constituents and which part to the interplay of molecular attraction and repulsion within the adsorbed layers. Fortunately, from a practical point of view, the answer to this question is irrelevant.

In a natural clay deposit, the load on the sediment increases with increasing depth below the surface. Therefore, the relation between depth and void ratio should be determined by the same laws as those representd by the *e*-log *p* curves in Figure 4. This conclusion is confirmed by field data collected by Skempton (73). These data were obtained by correlating the void ratio of sediments at different depths below the surface with the effective overburden pressure prevailing at that depth. Curves of this kind are referred to as *sedimentation-compression curves* (89). Some of them have the general characteristics of curve S_t in Figure 4 and others those of curves S_s and S'_s, depending on the liquidity index L_i and the sensitivity s_t of the clays which they represent.

In order to bring out the influence of L_i on the shape of the sedimentation-consolidation curves, Skempton and Northey (79) plotted the liquidity index against the logarithm of the effective overburden pressure, as shown in Figure 7. In this diagram, the thin dashed lines represent the results of consolidation tests on clay slurries. The average slope of these curves and their position in the diagram are practically the same for all clays. The plain curves were obtained from field data, by plotting the liquidity index of samples from different depths against the effective overburden pressure at these depths.

The values of the liquid limit L_w are shown in parentheses and those of the sensitivity s_t in brackets.

In Figure 7 it can be seen that the vertical distance between the field and the slurry curves increases in a general way with increasing sensitivity. The

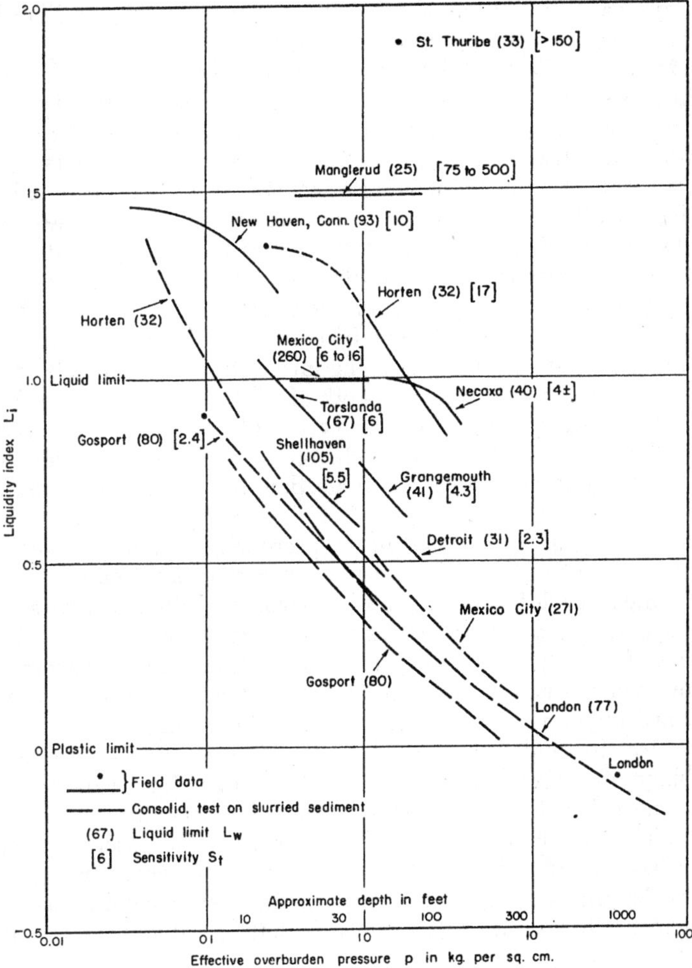

FIG. 7. Relation between effective overburden pressure and liquidity index for normally consolidated clays *in situ*. (Data collected by Skempton, Bjerrum, and Terzaghi.)

curves for extra-sensitive ($s_t = 8$ to 16) and for quick clays ($s_t > 16$) are located high above the slurry curves, and their slope is still very slight in a range of pressure within which the slope of the curves for the less sensitive clays is already equal to that of the slurry curves. The line representing the clay from Mexico City occupies an exceptional position in the diagram, inasmuch as its L_t-log p curve is almost horizontal within a range of $p = 0.4$ to 1.1 kg/cm² in spite of the fact that its L_t-value is close to unity. The Necaxa clay (core of a hydraulic fill dam) exhibits similar properties. The Mexico City clay is partly derived from pyroclastic material and Necaxa clay from deeply weathered basalt. The probable causes of the exceptional properties of these clays have been discussed before. The clays of St. Thuribe (58) and of Manglerud (3) are typical quick clays.

As the thickness of a sedimentary deposit increases both pressure and temperature increase at any given point in simple proportion to the depth of overburden, whereupon chemical changes may take place which invalidate equation 2. According to Dawson (21), chemically treated mud starts to solidify rapidly as a result of chemical changes at a temperature of about 260°F. In the Wilmington oilfield in California, the compressibility of the clay strata located between the oil sand layers decreases rather abruptly at a depth of about 4,000 feet corresponding to an effective overburden pressure of about 150 kg per sq cm and a temperature of about 180° F. Yet the void ratio of the sediments decreases to a depth of 6,000 feet in accordance with equation 2. This fact suggests that the changes in the lower strata responsible for the decrease of the compressibility below a depth of 4,000 feet took place during the time that elapsed since the end of the period of sedimentation (end of the Pliocene). The investigation of the relation between temperature, pressure, and compressibility is of interest in connection with subsidence problems in oil fields.

According to the data presented in the preceding paragraphs the compressibility of normally consolidated sediments may range between extremely wide limits. A deposit of dense sand is almost incompressible, whereas a deposit containing a high percentage of active clay minerals may be as compressible as a mixture of equal parts of sand and mica flakes, except inasmuch as the compression lags behind the increase of the load as explained below. This fact is of equal importance to the engineer and the geologist.

The engineer is compelled to adapt the foundations of his structures to the compressibility of the underlying strata. Therefore, he needs adequate information concerning the subsoil of the sites of his structures. In the realm of geology high compressibility of sedimentary strata is associated with progressive surface subsidence, with the development of secondary structure in sedimentary deposits resting on a steeply inclined surface, and with various other phenomena that cannot be explained without taking compressibility into consideration.

Time-Rate of Compression.—The sudden application of a load on a layer of cohesionless sediment composed of sound, equidimensional mineral particles produces an instantaneous compression followed by a slight additional compression at a decreasing rate. At low pressure, both instantaneous and gradual compression are almost exclusively due to slippage at contact points. As the load increases an increasing percentage of the compression is due to the crushing of grains.

Croce (20) described the results of consolidation tests on a volcanic sediment with an effective grain size $D_{10} = 0.01$ mm and a uniformity coefficient

$U = 6$, composed chiefly of fragments of volcanic glass, many of which were porous. The decrease Δe of the void ratio e took place approximately in accordance with the equation

$$\Delta e = a + b \log (t - t_0) \tag{4}$$

wherein a is the instantaneous decrease of e, t is the time and t_0 and b are constants. After five days Δe was already equal to $2a$. In this case the major part of the compression was due to the crushing of grains.

If the load on a saturated layer of clay is increased, the corresponding compression takes place gradually, and at a decreasing rate. The compression is associated with a decrease of the water content of the clay and the rate at which the excess water drains out of clay is very slow.

At the instant in which the load p on the layer is suddenly increased by Δp, the thickness of the layer remains unchanged. As a consequence, the application of the load Δp produces an equal increase Δu of the hydrostatic pressure in the pore water of the clay,

$$\Delta u = \Delta p$$

As time goes on the excess pore pressure gradually decreases and finally becomes zero. At the same time the grain-to-grain pressure increases from an initial value p to $p + \Delta p$.

At any time t after the load Δp was applied, the decrease Δe of the void ratio is equal to the decrease that would be produced by the sudden application of a load

$$\Delta \bar{p} = \Delta p_1 - \Delta u \tag{5}$$

if no time lag were involved. The value $\Delta \bar{p}$ represents the increase of the effective or grain-to-grain pressure on the clay at time t. The ratio U_c between the decrease of the void ratio Δe at time t and the ultimate decrease Δe_1 (decrease after time infinity) represents the degree of consolidation at time t,

$$U_c = 100 \frac{\Delta e}{\Delta e_1} \tag{6}$$

For time infinity,

$$\Delta \bar{p} = \Delta p_1, \qquad \Delta u = 0, \qquad \Delta e = \Delta e_1, \qquad \text{and} \qquad U_c = 100\%$$

At a given thickness H of the clay layer the degree of consolidation at time t depends exclusively on the coefficient of consolidation c_v,

$$c_v(\text{cm}^2/\text{sec}) = \frac{k(\text{cm/sec})}{\gamma_w(\text{gm/cm}^3)m_v(\text{cm}^2/\text{gm})} \tag{7}$$

wherein k is the coefficient of permeability of the clay for the load interval p to $p + \Delta p$, m_v the coefficient of compressibility for the same load interval (decrease of void ratio per gram of increase of the unit load), and γ_w the unit weight of water. With increasing values of p, both k and m_v decrease. Therefore c_v is fairly independent of p. It decreases for normally consolidated clays from about 10^{-2} cm²/sec for very lean clays to about 10^{-6} cm²/sec for highly colloidal clays. At a given value c_v, the time at which a given degree of consolidation U_c is reached increases in simple proportion to the square of the thickness H of the layer.

If c_v is known, U_c can be computed for any given time t by means of the theory of consolidation (82, 91). According to this theory, the value U_c should increase with time t as shown in Figure 8. The theory is based on the assumption that the time-lag between increase in unit load and the correspond-

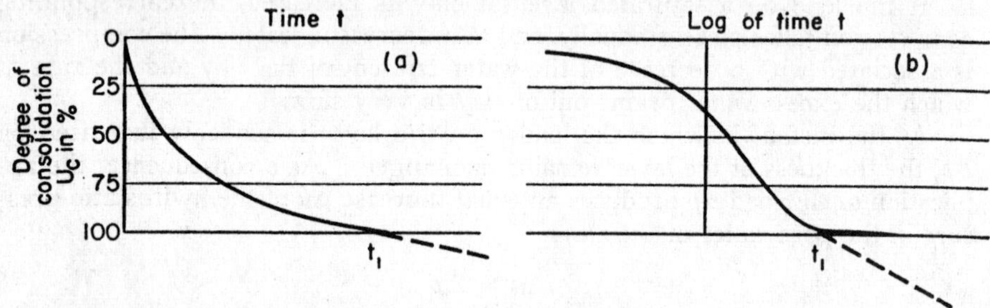

FIG. 8. Relation between time and degree of consolidation. (a) arithmetic and (b) logarithmic time scale.

ing decrease of void ratio is exclusively due to the low permeability of the clay. According to the theory, the slope of the curve representing the relation between U_c and the logarithm of time (Fig. 8 *b*) should steadily increase until the time t becomes approximately equal to

$$t_1 = \frac{H^2}{c_v}$$

when the corresponding degree of consolidation is about 95%. Then the slope should rapidly become flatter and approach the value zero.

In reality the U-log t curve continues to descend after time t, as indicated in Figure 8 by a dashed line. In the semi-log plot, Figure 8 *b*, this line can either be straight or slightly convex downward. Its average slope is very different for different clays. For organic clays its initial slope can be almost equal to that of the adjacent section of the plain curve. The consolidation represented by the plain curve for time $t < t_1$ is commonly referred to as *primary consolidation* and the consolidation represented by the dashed curve is the *secondary consolidation.*

If water contained in the voids of a clay is replaced by an organic liquid such as ethyl alcohol, the primary consolidation becomes negligible compared to the secondary one. The influence of the organic liquid on the relative importance of the secondary consolidation decreases with decreasing polar moment of the molecules of the liquid whereas the compressibility of the clay increases (67). The results of these experiments may account for the exceptionally conspicuous secondary time effects observed by Buisman (5) in connection with loading tests on organic soils in Holland.

The secondary time effect is probably due to the fact that the compression of a layer of clay is associated with slippage between grains. Since the bond between grains consists of layers of adsorbed water with a very high viscosity, the resistence of these layers to deformation by shear would delay the compression even in the event that the time lag resulting from the low permeability of the clay were negligible.

The rate of settlement of structures owing to secondary time effects ranges between almost zero and about one inch per year. Although the secondary time effects can be observed and measured during the performance of consolidation tests, the results of various attempts to predict the settlement of full-sized structures owing to secondary time effects on the basis of the results of laboratory tests (for instance Koppejan (45)) are not yet satisfactory.

If clay has been deposited on the bottom of a lake, at the mouth of a river, or in the core pool of a hydraulic fill dam, the deposition is followed by a subsidence of the surface as a result of progressive consolidation of the deposit. At a given initial thickness of the deposit, the time-rate of the subsidence after deposition depends on the time-rate at which the deposit was formed, the compressibility of the sediment, and the coefficient of consolidation c_v. If these quantities are known, the surface subsidence can be computed for any time t between the end of the period of sedimentation and the time t_1 (Fig. 3) at which the secondary time effect starts (55).

The surface subsidence is associated with a decrease of the excess hydrostatic pressures in the pore water of the sediment. The decrease of these pressures can be ascertained by observations on piezometric tubes. The agreement between measured and computed pore water pressures is satisfactory, provided that the stratum subject to consolidation is fairly homogeneous. (See for instance A, Casagrande (12).)

The preceding discussions referred exclusively to clay sediments, the voids of which are completely filled with water. If the sediment contains gas bubbles, a sudden increase of the load on the sediment produces an equally sudden decrease of its void ratio, owing to the compression of the gas bubbles which takes place without any time lag. The subsequent process of consolidation is associated with an expansion of the gas bubbles, which calls for a modification of the equations determining the rate of consolidation of completely saturated sediments (91).

Removal of the load on a clay sediment containing gas bubbles is associated with an expansion of the gas bubbles. Since it also produces a temporary decrease, $-\Delta u$, of the pore water pressure the quantity of gas contained in the bubbles increases, because the decrease of the pressure in the liquid is followed by the release of dissolved gas. The consequences of this process have repeatedly been observed.

Moran (54) reported that samples of organic silt recovered from drillholes at the site of piers for the Oakland Bay Bridge in San Francisco continued to swell for weeks after they had been received in the laboratory and the swelling pressure was high enough to break the connection between the cylindrical container and the caps. Since the water content of the samples remained unchanged, the volume increase was exclusively due to the release and expansion of gas. Samples of organic silt from drillholes in Flushing Meadows on Long Island, New York, exhibited similar properties. In both instances the gas contained in the samples was inflammable.

The mechanics of the relationship between pressure and void ratio of sediments with a high gas content have not yet been adequately investigated and the influence of chemical factors on these relationship is still unknown.

Shearing Resistance.—Let

p = unit load on a potential surface of sliding,
u = hydrostatic pressure in the interstitial liquid prior to the application of the shearing force, and
s = shearing resistance per unit of area.

For cohesionless sediments, such as clean sand, the relation between p and s can be approximately expressed by the equation

$$s = (p - u) \tan \phi \tag{8}$$

wherein ϕ is the angle of internal friction. The value of ϕ depends on the shape of grains, the degree of uniformity, and on the relative density of the sand. It ranges between 30° for loose uniform sand with rounded particles and 45° for dense well-graded sand with angular grains.

If the void ratio of the sand is above a certain value known as the ***critical void ratio*** e_c, the shear failure is associated with a volume decrease. Otherwise it produces volume expansion (dilatancy). The presence of water in the voids of a sand has practically no influence on the value of ϕ. However, if the sand is fine and the voids are completely filled with water, the volume change associated with the application of the shearing force is delayed by the time required for the corresponding change of the water content of the sand. Therefore the shearing resistance changes in spite of unaltered value of the angle ϕ.

If the void ratio of the sand is smaller than the critical value e_c the sand tends to expand. Since the expansion involves the flow of water into the voids of the expanding material, the application of the shearing force produces a temporary drop of the pore water pressure by Δu. Hence Δu is negative. On the other hand, if the void ratio is greater than e_c, the water content decreases and the temporary change Δu of the pore water pressure, required for the expulsion of the excess water, is positive. In any case the shearing resistance of a fine, saturated sand at unaltered water content is determined by the equation

$$s_{cq} = (p - u - \Delta u) \tan \phi = (p - u) \tan \phi_{cq} \tag{9}$$

In this equation ϕ_{cq} is a fictitious value which depends on the relative density of the sand. If the void ratio of the sand is below the critical value e_c, Δu is negative and ϕ_{cq} is greather than ϕ. Otherwise Δu is positive and ϕ_{cq} is smaller than ϕ.

If a sand has a metastable structure (void ratio e close to unity), a slight disturbance, such as the vibrations produced by blasting in the proximity or a rapid descent of the water table, produces a collapse of the structure. At the instant of the breakdown the entire submerged weight is transferred onto the interstitial liquid,

$$p - u = \Delta u, \qquad \text{and} \qquad \phi_{cq} = 0 \tag{10}$$

As a consequence, the sand temporarily assumes the character of a viscous liquid. Surcharges resting on the sand surface sink into the sand and if the sand forms a slope the sand starts to flow. On the coast of Zeeland, Holland, failures of this type have repeatedly occurred on slopes rising at an angle of

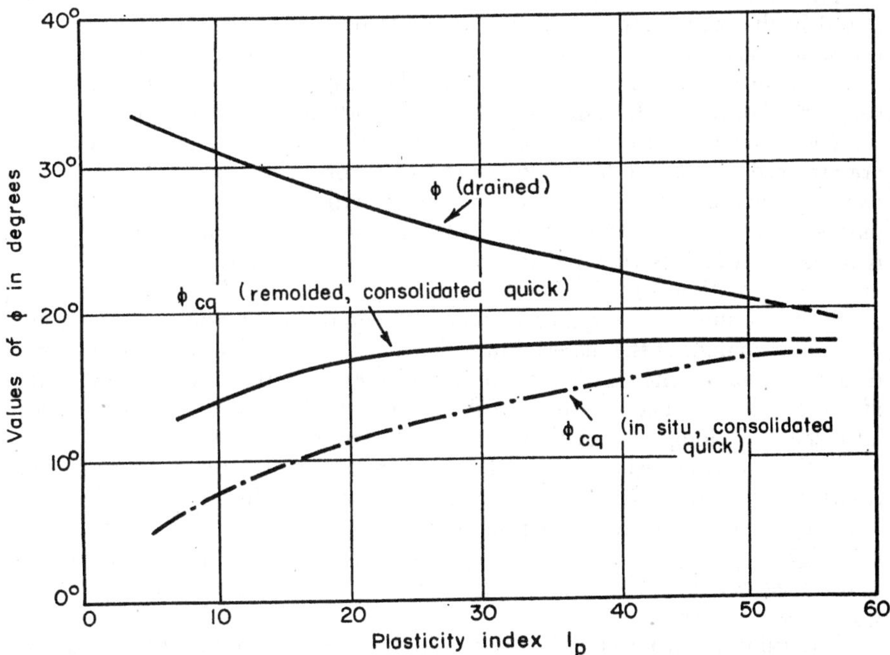

FIG. 9. Statistical relations between plasticity index and angle of shearing resistance (ϕ = drained test, ϕ_{cq} = consolidated quick for remolded sediment, and ϕ_{cq} [*in situ*] = shear failure *in situ* at unaltered water content).

not more than 15°, on no provocation other than the recession of spring tides (46). This process is known as *spontaneous liquefaction.* The conditions for spontaneous liquefaction have been investigated by Geuze (28).

For clays in a remolded and normally consolidated state, the relation between unit pressure and shearing resistance is also approximately determined by equations 8 and 9, provided that the water content of the clay, prior to the application of the load p, is at or above the liquid limit. On account of the low permeability and high compressibility of clay, however, the value Δu in equation 9 is not negligible unless the shearing force is very slowly applied. For normally consolidated clays, furthermore, Δu is always positive and as a consequence ϕ_{cq} in equation 9 is always smaller than ϕ in equation 8. The values ϕ

decrease in a general way with increasing plasticity index I_p, whereas the values of ϕ_{cq} probably have a maximum for values of I_p of the order of magnitude of 50. In Figure 9 this general relationship is indicated by the plain curves marked ϕ (drained) and ϕ_{cq} (consolidated-quick). The average scattering of the real values from those represented by the curves is not yet known. The figure shows that the difference between the values ϕ and ϕ_{cq} for the remolded state decrease in a general way with increasing values of the plasticity index. This fact indicates that the volume decrease associated with the shear deformation decreases with increasing plasticity index.

The values of ϕ and ϕ_{cq} for sand and for remolded clay can only be determined in the laboratory by direct shear tests or, preferably, by triaxial compression tests. The value ϕ_{cq} for normally consolidated clay in an undisturbed state can also be computed from the results of vane tests performed below the bottom of drillholes (Cadling and Odenstad (6)).

The vanes consist of four narrow rectangular pieces of sheet steel which are attached lengthwise to the lower end of a rod; they are oriented at right angles to each other. The end of the rod to which the vanes are attached is pushed into the sediment below the bottom of a drillhole and then the torque is measured which is required to start a rotation of the rod about its centerline. The rotation is resisted by the shearing resistance of the sediment along a cylindrical surface the radius of which is equal to the width of the vanes. The water content of the material in which the surface is located remains practically constant. The test is repeated at different depths D below the surface of the sediment. These tests furnish the unit resistance s_{cq} of the sediment against shear along vertical surfaces at different depths D.

If the shearing resistance s_{cq} of a homogeneous deposit of normally consolidated clay with a submerged surface is plotted against the effective overburden pressure at different depths, the points thus obtained are located on a straight line, Figure 10, which intersects the horizontal axis at a distance s_o from the origin. The value s_o represents the s_{cq}-value for the top layer of the deposit for which $p = 0$.

The value s_o for clays with low sensitivity is inevitably greater than zero, because the water content of the top layer of deposits of such clay is commonly close to the liquid limit L_w. According to definition, L_w is the water content at which the shearing resistance of the clay in a remolded state has a value of several tens of grams per sq cm. As time goes on, the shearing resistance further increases on account of thixotropic hardening.

The water content of the top layer of extra-sensitive and of quick clays is very much higher than the liquid limit. Yet, according to Jakobson (42), the value of s_o for a very sensitive, postglacial, normally consolidated, and submerged lake clay in southern Sweden with a liquidity index of about 1.3 ranged between the high values of 0.06 and 0.08 kg/cm². In any event the value of s_o is different for different sediments.

If it is assumed that the effective unit pressure p_v on a vertical plane at any depth D below the surface is equal to the effective overburden pressure p at the same depth, then

$$\tan \phi_{cq} = \frac{s_{cq} - s_0}{p} \tag{11}$$

Since p_v represents the earth pressure at rest at depth D, its value is likely to be smaller than the overburden pressure p at the same depth. The ratio p_v/p is always smaller than unity. Therefore the values of ϕ_{cq} computed by means of equation 11 represent lower limiting values. On the other hand, the values of ϕ_{cq} determined by means of consolidated-quick triaxial tests are greater than the *in-situ* values because at a given confining pressure p the water content of the specimen is lower than that of the sediment *in situ* at a depth at which the effective overburden pressure is equal to p. According to Skempton and Henkel (78) the difference between the two sets of values increases

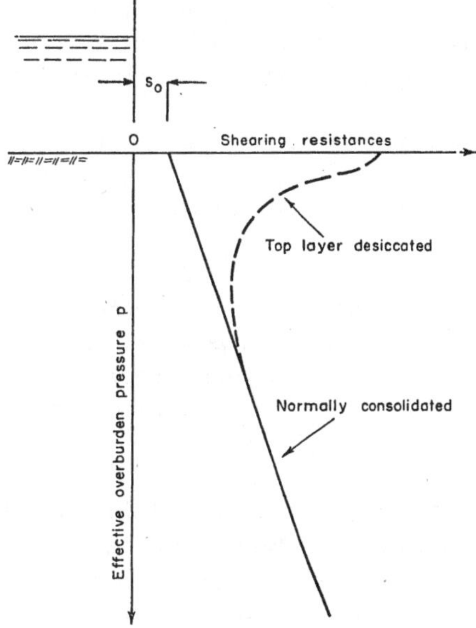

FIG. 10. Increase of shearing resistance at unaltered water content with depth below surface in normally consolidated submerged clay deposit (solid line) and in dried crust on normally consolidated deposit (dash line).

with decreasing values of the plasticity index I_p, from about zero for $I_p = 80$ to about 30 percent for $I_p = 55$. For clays with an I_p-value of less than 55 the difference may be still more important.

Skempton and Henkel (78) have also observed that the ϕ_{cq}-value for clays in an undisturbed state decreases with decreasing plasticity index. In Figure 9 all the points representing ϕ_{cq} for clays in an undisturbed state are located in the proximity of the dash-dot curve marked ϕ_{cq} (*in situ*). The scattering of the measured values from the average represented by the curve appears to be less than $\pm 3°$. Subsequent investigations performed by Bjerrum (3) on Norwegian quick clays with a low liquid limit confirmed Skempton's findings.

The decrease of the plasticity index is associated statistically not only with a decrease of the value ϕ_{cq} for the clay *in situ* (dash-dot curve in Fig. 9) but also with an increase of the liquidity index L_i, of the sensitivity, and of the percentage of equidimensional grains contained in the clay. There can hardly be any doubt that these properties are interdependent.

It has been mentioned under the heading *Void Ratio* that fine-grained sediments composed of equidimensional particles have a metastable structure, because the particles adhere to each other as soon as they touch. If the process of sedimentation takes place in a clay slurry and not in clear water, the tendency of the coarser equidimensional particles to assemble into a metastable framework is further increased, because the capacity of the particles to settle into more stable positions under the influence of their own weight decreases with increasing density of the surrounding medium. However as the plasticity index increases, the percentage of equidimensional particles becomes so small that skeletons of the type shown in Figure 3 can no longer be formed. As a consequence, liquidity index and sensitivity decrease and the ϕ_{cq}-value increases.

In the preceding discussions nothing has been said about the seat of the shearing resistance of clay sediments. Hvorslev (40) has shown that the shearing resistance of remolded, normally consolidated clays, equation 8,

$$s = (p - u) \tan \phi$$

consists of two parts. One part, $f(w)$, is a function of the water content, w, of the clay and has the same value on every section through a given point in the clay subject to shear. The other part is equal to the effective unit pressure $(p - u)$ on the section times a constant, so that

$$s = f(w) + (p - u) \tan \phi_t \tag{12}$$

The value ϕ_t represents the "true" angle of internal friction of the clay. Skempton (76) has shown that the value of the ratio

$$m = \frac{f(w)}{(p - u) \tan \phi_t}$$

increases from almost zero for inactive clays to almost unity for highly colloidal clays with an activity a_c of more than two.

All the data concerning the relationship between unit pressure p and the friction angles ϕ and ϕ_{cq} were derived from tests during which the specimen was subject to shear for a relatively short period. Therefore the shearing resistance computed on the basis of the test results and defined by equations 8 to 12 represents what is sometimes referred to as the "technical" shearing resistance of the sediments. For cohesionless sediments such as sand, the duration of the shearing stress appears to be irrelevant. In other words, if the shearing stress in a sand is smaller than the shearing resistance s in equation 8, the corresponding shear deformation takes place almost instantaneously. On the other hand if the shearing stress in a clay is increased to a value of more than about one half of the technical shearing resistance s in equation 8, the clay is likely to "creep" at constant shearing stress and the creep-displacement takes place in accordance with the time-displacement curves shown in Figure 11. The inclined portions of these curves are almost straight and their slope increases with increasing intensity of the shearing stress (99). If the shearing stress is increased very slowly, failure may occur at a stress which is smaller than s in equation 8 (14).

The creep phenomena illustrated by Figure 11 may be responsible for part of the secondary settlement of structures located above thick clay deposits. They are definitely responsible for the formation of the cambers, gulls, and valley bulges in the Northampton area in England, described by Hollingworth et al. (36) and they may account for similar secondary structures elsewhere. So far, however, the creep phenomena illustrated by Figure 11 have not yet been adequately investigated.

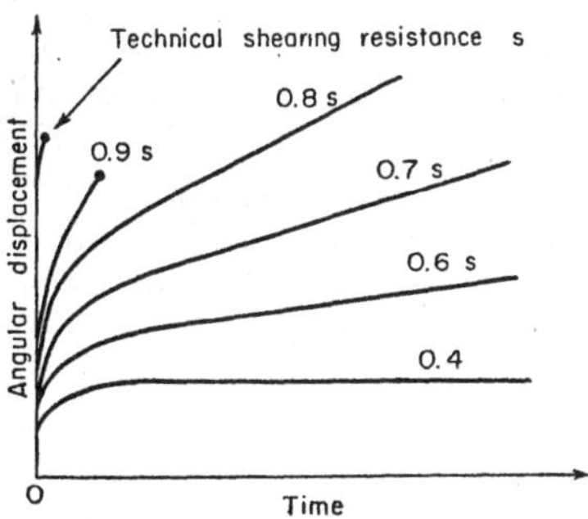

FIG. 11. Increase of shear deformation of clay with time at constant shearing stress.

Under exceptional conditions the rate of creep attributable to gravity stresses beneath gentle slopes can be high enough to cause detrimental displacement of bridge piers, resting on the slope. According to Haefeli et al (32), one of the abutments of the Landquart Bridge advanced at such a rate towards the other one that it became necessary to insert a strut between the two abutments. The ground movements responsible for the displacement of the abutment occur in silty landslide material containing dolomite boulders and extend to a depth of more than 100 feet below the ground surface. They involve a large area with a gently sloping surface and proceed at a maximum rate of about 1½ inches per year. Similar mass-movements have also damaged the Casteler Viaduct on the Chur-Arosa railroad in Switzerland to such an extent that a radical reconstruction became necessary, 27 years after the bridge was built (31).

MODE OF TRANSPORTATION AND DEPOSITION

Influences of Geological Origin on Engineering Properties.—The geological origin of a deposit determines both its pattern of stratification and the physical properties of its constituents. The pattern of stratification is of outstanding engineering importance because it determines the degree of accuracy with which the performance of the subsoil under given conditions of loading can be predicted on the basis of boring records and test data (98). On the other hand, the performance itself depends primarily on the physical properties of the deposit in its present state, such as sorting, relative density, consistency, and others. Some of these properties such as the grain-size characteristics are likely to remain unchanged after deposition is completed, whereas others may undergo more or less radical modifications on account of a change in environment.

The knowledge of the processes responsible for the physical properties that the sediments acquire during the period of deposition is still very fragmentary. Yet every significant advance of this knowledge is of considerable practical importance because it reduces the margin of error involved in the interpretation of boring records and test results. Under the following subheadings some of the more recent findings will be summarized.

Sorting of Stream Channel Deposits.—Stream channel deposits are commonly well graded, but the grain-size characteristics vary erratically from point to point. This is due to unequal distribution of the velocities over the stream bed at any given time and to seasonal variations of the discharge that produce alternate periods of prevalent erosion and prevalent deposition.

The influence of unequal distribution of the velocities over the streambed on the grain size of the bottom sediments is illustrated by the results of observations made by Straub (81) concerning the relation between bottom velocities and grain-size characteristics of the sediments on the bottom of a crossing in the Missouri River at Baltimore Bend, 300 miles upstream from the mouth of the river. Straub's paper contains maps showing the distribution of the bottom velocity, the depth of water, and the grain size of the bed sediments. The coarsest sediments were encountered at and immediately downstream from the crossing, whereas the finest sediments occupied areas at the downstream end of longitudinal sand bars. The location of crossings changes in the course of time and the position of the channel line and the sand bars varies from year to year. Therefore cross bedding and erratic variations in grain size are inevitable.

In order to find out whether there is any relation between grain size and sorting, Shockley and Garber (70) represented the results of the mechanical analysis of a great number of samples of sand from the banks of the lower Mississippi in frequency diagrams on a double-logarithmic scale, as was previously done by Bagnold (2) in connection with dune sand investigations. All of the diagrams are roughly triangular, as shown in Figure 12. Therefore, the grain-size characteristics can adequately be described by three numerical values, the *peak* or *modal* value D_p, the slope S_c of the side of the triangle on the coarse side and the slope S_f on the fine side of the grain-size scale. The value D_p of the investigated sands ranged between 0.1 and 0.7 mm. The slope S_c was found to be the same for all the samples. The slope S_f was steepest for $D_p = 0.2$ mm and decreased with both decreasing and increasing values of D_p.

In any given drainage area the average grain size of the stream channel deposits commonly increases with increasing depth below the flood plain. More rarely it remains constant or it decreases. An increase in grain size with increasing depth creates serious problems in connection with underseepage on flood-protection dikes and dams (102).

Some stream deposits contain lenses or pockets of well-sorted gravel known as *open-work gravel.* The existence of this type of stream deposits was recognized, probably for the first time, by W. M. Davis in 1890. Yet it did not receive special attention until Cary (7) published the results of his observations concerning the characteristics of the open-work gravel deposits that he encountered in the Pacific Northwest. He divided these deposits into two categories. Those of the first category consist of lenses or pockets with a thickness up to 6 feet and a length of several tens of feet, exceptionally more than 100 feet. The formation of these bodies of open-work gravel is ascribed to the turbulence at the downstream end of gravel bars in swift flowing streams or to similar processes on the fore-set beds in deltas. Inclusions with a similar shape, composed of open-work detritus, were encountered by the writer in alluvial fans in the western Alps.

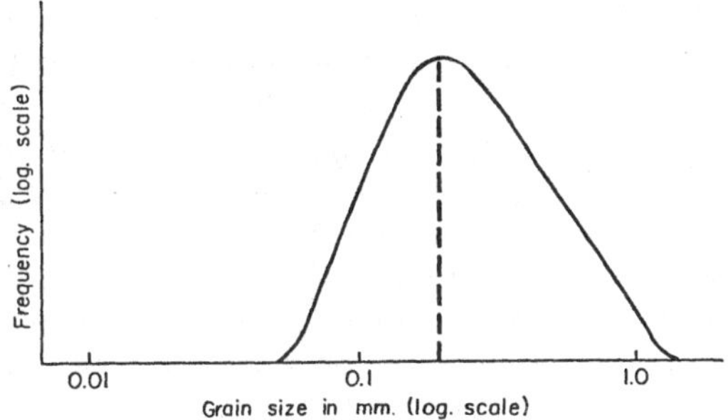

FIG. 12. Grain-size frequency diagram, double-logarithmic plot.

The second category consists of thin, horizontal layers of well-sorted gravel. No explanation was offered for the origin of these deposits.

The discussers of Cary's paper furnished the description of numerous open-work gravel deposits that were encountered in the eastern United States and in Europe. As a rule the distances between pockets of open-work gravel is so great that the presence of the pockets has no important influence on the average permeability of the sediments in which they are located. One exception to this rule was experienced at the site of the Sautet dam in eastern France. The reservoir formed by this dam communicates with an adjacent valley through a narrow, buried valley with a length of about 1½ miles, occupied by fluvioglacial sediments with inclusions of open-work gravel. Although the average permeability of the sediments is low, the quantity of water that escapes through the chain of pockets of open-work gravel amounts to about 30 cu ft per second. Attempts to stop the leaks by grouting have not yet been successful.

If water percolates through a deposit of well-graded sediment such as glacial outwash, containing pockets of open-work gravel, the finer constituents of the well-graded sediment may gradually be washed into the interstices of the gravel. If the pocket is located close to the bottom of a reservoir, this process leads to the formation of sinkholes above the pockets. Numerous sinkholes of this kind can be seen on the bottom of the Cedar Reservoir in the State of Washington, which is located on glacial outwash. Water escapes from the reservoir through a buried valley toward a slope located downstream from the reservoir.

All of the bodies of open-work gravel that have been described so far form lenses or pockets in strata composed of well-graded sediments. An exception to this rule was encountered in Plains, Montana, where an area of about five square miles is covered by a stratum, 300 feet thick, which is composed entirely of open-work gravel. The voids of the uppermost fifty feet are filled with finer sediments, whereas those of the balance are open. The downstream portion of the deposit consists entirely of boulders measuring 1 to 4 feet across (7). Deposits of this kind can be formed only by catastrophic floods.

The most important practical consequences of the occurrence of open-work gravel are the uncertainties involved in the estimate of the loss of seepage through sedimentary deposits containing pockets of open-work gravel and the difficulties which arise if open-work gravel is encountered in tunneling operations.

If bodies of open-work gravel are located in a buried valley leading from the site of a reservoir into an adjacent valley, it is impracticable to estimate the quantity of water that will be lost by seepage. It is probable that the losses will not exceed the quantity computed on the assumption that the effect of the inclusion of the open-work gravel on the permeability of the deposit is negligible, but it is possible that the losses will be many times greater.

If open-work gravel is encountered in a tunnel above the water table, it assumes the character of "running" ground, which is bad enough. An encounter with open-work gravel below the water table can be catastrophic, on account of the excessive permeability of this material.

Relative Density of Cohesionless, Waterlaid Sediments.—The mechanical properties of cohesionless sediments depend almost entirely on their relative density, D_r. Therefore the description of such a sediment is entirely inadequate unless it contains reliable information concerning its relative density.

The most expedient method for determining the relative density involves static or dynamic penetration tests in the field. In the static tests the resistance against slow penetration of a cone is measured below the bottom of a drillhole, at different depths below the surface; in the dynamic tests the number of blows is counted per foot of penetration of a rail or of a rod with a drive point, also at different depths (98). If the sediment is loose, the resistance against penetration is low and practically independent of depth, whereas the resistance of the same sediment in a dense state increases approximately in simple proportion to depth.

Construction experience and penetration tests have shown that some waterlaid, cohesionless sediments are predominantly dense. In some of them, the relative density varies erratically from point to point between loose and dense; in other localities, trough- or channel-shaped depressions of the surface of dense sediments are occupied by loose material, or the bulk of the entire deposit may be loose. If the sediment is dense it is almost incompressible. Piles driven into the sediment meet refusal at a depth of a few feet below the surface and submerged slopes on dense sand are permanently stable even if they rise at the angle of repose of the material. Deposits of this kind, however, are rather rare.

If the relative density of a sand deposit varies erratically from point to point, the differential settlement of buildings on footings that rest on the deposit may be almost as great as the maximum settlement. This type of settlement can be more detrimental than the settlement that would ensue if all of the footings rested on loose sand. Piles driven into the deposit meet refusal at very different depths. On Vancouver Island, piles were driven through soft silt into fluvio-glacial sand. In some clusters the depth at which the piles met refusal varied by as much as 25 feet (98).

Bodies occupying trough- or channel-shaped depressions were encountered, for instance, near Rotterdam in Holland and in the vicinity of Houston, Texas. At the waterfront of Rotterdam, piles were driven through about 70 feet of silt and peat into a sand stratum, to serve as a foundation for a quay wall. Everywhere the piles met refusal at a depth of a few feet below the surface of the sand stratum, except in a section with a length of about 200 feet where the piles met very little resistance to penetration within the depth of about 20 feet below the sand surface. Subsequent borings showed that there was no noticeable difference between the sand samples recovered from this section and those obtained from the drillholes in the dense sand (86).

During the excavation of the Houston (Texas) ship canal in relatively dense sand to a depth of about 30 feet a deposit of loose sand was encountered. Within the body of loose sand, the sand moved into the cut almost as rapidly as it could be removed by the dredge. A quay wall located at a distance of about 300 feet from the centerline of the cut subsided 10 feet over a length of about 300 feet (64).

If the bulk of a sand stratum located above the water table is loose, the sand is relatively compressible but otherwise stable. However, if the same stratum is partly or wholly submerged, the possibility of spontaneous liquefaction must be taken into consideration. This phenomenon has been discussed under the heading *Shearing Resistance.* It is a problem that arises in connection with the foundation of dams and with the stability of partly submerged slopes.

A large body of fluvio-glacial sand in a loose state was encountered at the site for the Franklin Falls Dam in New Hampshire (4). If the sand had been left in its original state, the dam might have failed in the event of an earthquake as a result of spontaneous liquefaction of its base. According to Lyman (50) the danger was eliminated by detonating charges of dynamite on the bottom of drill holes at a depth of about thirty feet. As a result of the vibrations produced by the blasts the ground surface subsided by amounts up to 2 feet.

The existence of the danger of flow slides of submerged slopes on loose sand is demonstrated by the catastrophic sand slides that occur from time to time at the coast of Zeeland in Holland. Reference was made to these slides under the heading *Shearing Resistance.* Figure 13 shows two typical depth-resistance diagrams for the sand, which were obtained by the static penetration procedure (46). In both diagrams the abscissas of the dash-dot line represent the resistance that the sand would have offered to the penetration of the drive point of the penetrometer, if its void ratio had been equal to the critical value e_c (borderline between dense and loose).

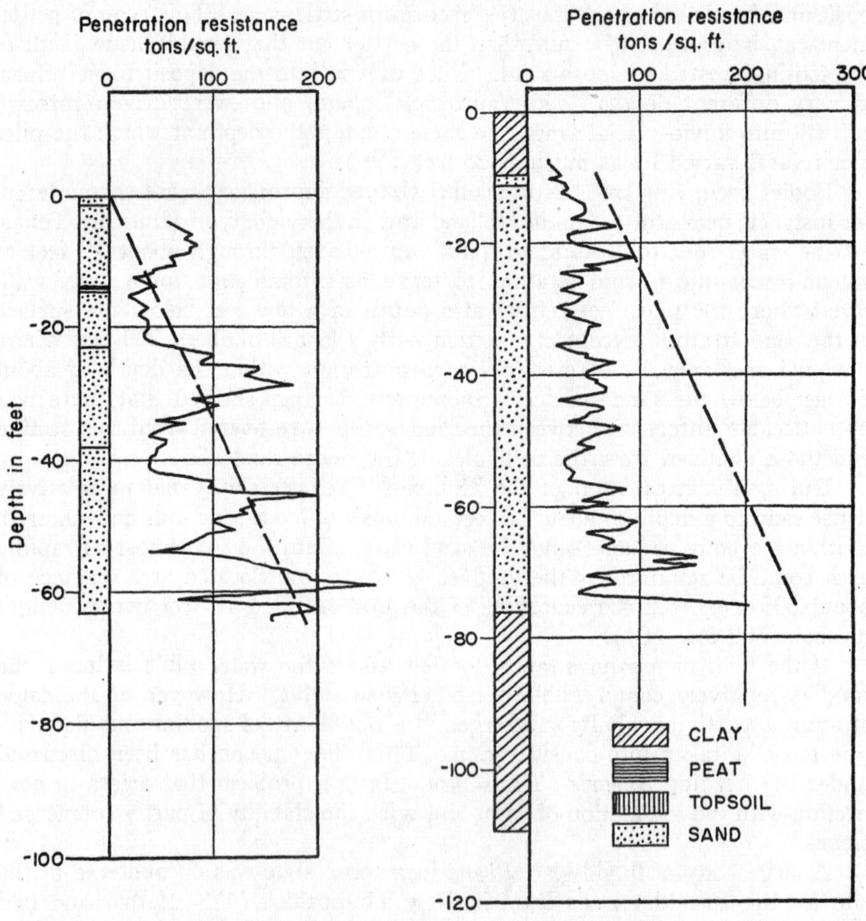

FIG. 13. Relation between static penetration resistance and depth below surface of a stratum of loose sand at the coast of Zeeland, Holland. (After Koppejan et al., 1948.)

According to the diagram the relative density of the Zeeland sand varied at the site of test (a) between loose and dense, and at site (b) the sand can be regarded as loose over the full depth of penetration. It is not yet known whether the presence of layers of dense sand in a deposit of loose sand, like those shown in Figure 13 *a*, is sufficient to prevent the occurrence of flow slides; but it is rather unlikely.

In spite of the outstanding practical importance of the relative density of waterlaid cohesionless sediments, the factors that determine it are not yet reliably known. According to the data given in Table I one should expect that D_r would decrease in a general way with decreasing grain size. This conclusion is corroborated by various observations. Shockley and Garber (70) found that the relative density of sand strata deposited by the lower Mississippi decreased from 88 percent for a peak value D_p (Fig. 12) of 0.7 mm to 59 percent for $D_p = 0.10$ mm. However, there is also ample evidence that the relative density of waterlaid sands with identical grain-size characteristics can vary between wide limits, as shown by the following observations.

No relationships could be detected between D_r and the grain-size characteristics of the fluvio-glacial sands encountered on Vancouver Island, B. C. and in downtown New York (98). A very fine and well-sorted sand near the shore at Lynn, Mass. is among the densest, and a coarse well-graded fluvio-glacial sand on the Muskingum River in Michigan is one of the loosest sands ever encountered by the writer. In the sand deposits beneath the flood plain of the Godavari River in Madras, at the point where the river emerges from the Eastern Ghatt, the relative density of the sand changed almost abruptly from loose to dense at approximately the depth of the seasonal scour, with values up to sixty feet, whereas the grain-size characteristics remained unchanged. Hence, it appears that the grain size is only one, and by no means the most important, of several factors that determine the relative density of sands deposited by flowing water. The rate of deposition may be another one.

In connection with beach sands, similar conditions prevail. Peck and Reed (59) constructed profiles showing variations of D_r for the sand deposits along the west shore of Lake Michigan over a length of several miles and found that D_r decreased with increasing silt content of the sand. This observation shows that the relative density of beach sands depends to some extent on grain size. On the other hand Krumbein (47) reported that the relative density of beach sands at the shore of Lake Michigan was a maximum within about 10 feet from the shore line. From there to a distance of about 50 feet the density decreased. Beyond this zone of transition it remained low. These variations have nothing to do with grain size. Since the lake level and, as a consequence, the position of the shore line changes, it is not surprising that the resulting density pattern of the shore deposit is rather complex.

Engineering Properties of Till Sheets.—Till sheets are primarily subglacial deposits that were formed either on the ground surface or below the base of floating ice sheets.

Till deposited on the ground surface commonly occurs in the form of a non-stratified veneer of well-graded material with a thickness ranging between a few inches and many tens of feet. Some till sheets are composed of a very dense lower layer (*basal till*) and a less dense upper layer (*superglacial till*). The basal till is sometimes referred to in boring records as *hardpan.* It owes its strength to compaction under a heavy ice load and it may have the bearing capacity of a soft rock. In thick till sheets, inclusions consisting of pockets or lenses of well-sorted and stratified material are not uncommon.

Experience has shown that till sheets of equal age have similar grain-size characteristics over large areas. For instance, the tills in New England are well graded from very coarse to fine, but their clay content is very low. Therefore, they furnish excellent dam construction materials that combine great stability with adequately low permeability. On account of their properties they can be placed in dams without special precautions. By contrast the till sheets in Central British Columbia are much less well graded and their clay content is much higher than that of the New England tills. Therefore the use of these materials in earth dams requires rigorous water-content control and intense compaction.

The differences that can exist between the grain-size characteristics of till sheets of different age in the same region are illustrated by the observations of Shepps (69) concerning the land-laid till sheets of northeastern Ohio. He found that the sheets of equal age had practically identical grain-size characteristics. With greater age the percentage of clay-size particles increased from 16 percent in the Tazewell to 42 percent in the late Cary tills.

Tills formed by deposition beneath the base of floating ice sheets have so far been identified and investigated only in the Chicago area on Lake Michigan and the Cleveland area on Lake Erie.

A description of the till sheets of the Chicago area was published by Peck and Reed (59). In the Loop District the till formation consists of five sheets with different age. The second sheet from the top consists of a soft to very soft clay and it is remarkably homogenous. All the others are heterogeneous and exhibit an erratic pattern of stratification. Each layer is stiffer than the one located above it, except for the uppermost layer which has a stiff crust produced by desiccation during temporary exposure to the air.

The physical properties acquired by the Chicago clays during the process of deposition had a decisive influence on the methods and cost of construction of the subway of Chicago. The bulk of the subway tunnels are located in the Deerfield till. North of the Chicago River the tunnel invert rests on the underlying Parkridge till, which is stiff enough to provide reliable support for arch-rib footings. Therefore the tunnels were handmined (92). South of the Chicago River the invert is located well above the top of the Parkridge till and as a consequence the more expensive shield method had to be used (90). The influence of the "dry crust" on the shield tunnel operations will be discussed under the heading *Desiccation.*

In the Cleveland area the floodplain of the Cuyahoga River is located above three water-laid till sheets that are separated from each other by layers of glacial lake deposits. After the last retreat of the ice all of these deposits were buried beneath a stratum of fluvio-glacial sand and gravel with a thickness of about 100 feet; the present river valley was formed by erosion in the fluvio-glacial deposits. The recent flood plain deposits form a veneer on the surface of the uppermost till sheet. On account of these developments, all of the glacial clay strata were precompressed under a load of about four tons per square foot. The effect thereof on the engineering properties of the clay strata will be discussed under the heading *Preconsolidation Under Load.*

Glacial Lake Deposits.—Glacial lake deposits are commonly composed of light-colored layers of silt with a low clay content alternating with darker layers containing a higher percentage of clay. The thickness of each layer remains constant over long distances. Clay deposits with such a pattern of stratification are called *varved clays.* It is generally agreed that each double layer was formed in one year. The segregation of the sediment into a light-colored summer and a dark-colored winter layer is commonly ascribed to the following process. Every summer a large quantity of meltwater carrying silt and clay enters the lake. The silt grains are deposited within a short time after they enter the lake. The clay particles remain in suspension and settle out during the winter. Because no meltwater enters the lake in the winter, the winter layers contain most of the clay fraction of the suspended material which was washed into the lake during the summer. According to Leggett and Bartley (48) chemical processes are also likely to participate in and to influence the results of particle segregation.

The thickness of the double-layers varies from year to year. It may range between a fraction of an inch and several inches in a single deposit. Exceptional double-layers with a thickness of several feet have been encountered.

The segregation of glacial lake clays into silty layers and layers with a higher clay content has had a twofold influence on the engineering properties of these sediments. First, the silty layers are likely to be extra-sensitive and second, on account of the continuity of the individual layers and the relatively high permeability of the summer layers, the permeability of deposits of varved clay in horizontal directions is very much greater than in the vertical direction.

It was shown earlier that the disturbance of the structure of an extra-sensitive sediment at unaltered water content reduces the shearing resistance to a small fraction of the original value and it increases the compressibility. If piles are driven into a normally consolidated sediment of this kind, the sediment located between the piles assumes the consistency of a thick slurry. Hence pile-supported structures on varved clay may settle excessively unless the piles are driven to refusal into a firm stratum located beneath the clay stratum.

The high sensitivity of normally consolidated varved clays also determines their performance in the event of a land slide. If a slope on such a clay fails, the slide may assume the character of a typical quick-clay slide. A slide of this kind occurred in April 1952 in varved clay on the west shore of the northern part of Lake Roosevelt, formed by the Grand Coulee Dam. The slide started at the shore line, but in a short time the slide scar assumed the shape of a crater surrounded by almost vertical walls with a height of more than 100 feet. The slide material had the consistency of a thick slurry and flowed from the crater into the lake through a narrow gap in the walls of the crater. The diameter of the crater steadily increased owing to the collapse of vertical slices. When the writer saw the slide, about two months after it had started, the vertical rear face of the slide scar was already located at a distance of about 600 feet from the gap through which the slide material flowed out of the crater.

On Steep Rock Lake in Canada, a deep excavation was made in varved clay for the purpose of uncovering the surface of an ore deposit located at a depth of about 150 feet below the bottom of the lake. Before excavation was started the area was drained and it was kept in a drained state while excavation proceeded. The slides that occurred on the slopes of the excavation pit had most of the characteristics of quick-clay slides (48).

Subsequent laboratory investigations showed that the light layers of the Steep Rock varved clay had a water content w of 20 to 40, an activity a_c of 0.25, and a sensitivity s_t of more than 100; whereas for the dark layers $w =$ 60 to 100, $a_c = 0.55$, and $s_t = 12$. The geometric mean of the grain size of the light-colored layers was 4 μ against 0.5 μ for the dark-colored ones. The light-colored layers contained calcium carbonate, the dark ones none. The high sensitivity of the light-colored layers appears to account for the semi-liquid condition of the slide material.

The second important characteristic of the glacial lake deposits is the low ratio between the permeability in vertical and horizontal directions. On account of this characteristic, the rise of the water table in a sand stratum communicating with the clay deposit along its margin increases the value of the hydrostatic pressure u in the varved clay to great distances from the margin. The possible consequences of such an increase will be discussed under the heading *Changes in Groundwater Conditions.*

Windlaid Deposits.—Windlaid deposits can be divided into two categories, dune sand and loess. Each of them has remarkably well-defined grain-size characteristics.

Dune sand is composed either of the coarser grain-size fraction of the products of weathering under arid conditions or of those constituents of graded sediments that can be transported by the wind without rising to an elevation of more than a few feet above the ground surface. Therefore sand dunes are formed only within a short distance beyond the outer boundaries of the windswept area.

Dune sand consists chiefly of well-rounded quartz grains. The peak value (Fig. 12) ranges between 0.15 and 0.3 mm and the uniformity coefficient U rarely exceeds 3. According to Bagnold (2) the relative density of the bulk of sand dunes is low. Patches of very dense sand, comparable to patches of wind-packed snow, are formed on the windward slope by accretion and on the crest. Patches of exceptionally loose sand, referred to as *pools,* are also quite common. The transition from loose to dense sand is likely to be abrupt.

The silt-size particles picked up by the wind while sweeping over barren areas are carried by turbulent air currents to high altitudes and deposited in thin layers at a long distance from the source. Those that come to rest on grass-covered surfaces accumulate in the course of centuries to form loess deposits with a thickness ranging between a few inches and many hundred feet. Exceptionally, loess deposits are also formed in reed-studded lakes (swamp loess) or on thin forest. Since loess deposits cover large areas on every continent their physical properties are of outstanding importance in connection with engineering operations.

The petrographic and grain-size characteristics of loess and its physical properties in a remolded state are remarkably uniform. The material is well sorted.

About fifty percent of the particles have a size intermediate between 10 and 50 μ and the uniformity coefficient U is of the order of magnitude of 10. According to Smith (49) the mean particle size of loess in Illinois decreases with increasing distance from bluffs in accordance with a simple logarithmic law.

About 50 to 75 percent of the loess particles consist of quartz, 10 to 25 percent of feldspar, and 10 to 30 percent of calcite. The percentage of clay minerals contained in unweathered loess ranges between a few percent and a maximum of about 15 percent. The liquid limit L_w is 25 to 30, and exceptionally up to 45; the plasticity index I_p is 4 to 9 and ϕ in equation 8 is between 30° and 34°.

Loess *in situ* owes its outstanding characteristics to the properties that it acquired during the process of deposition, and not to those of the constituents. These characteristics include metastable structure combined with great cohesion and the presence of closely spaced, more or less vertical rootholes lined with a film of calcium carbonate.

Although the loess is a subaerial deposit, its metastable structure originates in the same manner as that of waterlaid sediments with silt-size particles, as in Table I. As soon as a descending grain touches a grain on the surface of the deposit it adheres to it and the adhesion is strong enough to prevent it from settling into a more stable position. However, in a waterlaid sediment the cohesive bond remains unchanged and it is so weak that a slight overburden pressure is sufficient to change the metastable arrangement of the silt particles into a stable though very loose one. In loess the initial bond is also very weak, but it becomes progressively stronger, probably on account of successive precipitation of cementing material from evaporating films of moisture. As a consequence the bond between particles becomes so strong that the particles retain their metastable arrangement while the depth of overburden increases, even if this depth assumes a value of many hundred feet.

According to the United States Bureau of Reclamation (103), the binder between loess particles consists chiefly of clay minerals. Yet the unconfined compressive strength q_u of unweathered loess located above the water table may amount to several kilograms per sq cm and vertical cliffs with a height of more than one hundred feet may remain stable.

Since the binder between loess particles consists chiefly of clay minerals, the strength of the top layer of loess deposits varies with the seasons. If loess is permanently submerged its metastable structure breaks down, whereupon the loess temporarily assumes the consistency of a thick slurry. As time goes on the void ratio decreases and the surface of the deposit subsides. The influence of saturation on the relation between unit load and void ratio of loess is shown in Figure 14. The line AB represents this relation for loess in its natural condition above the water table. After the unit load was increased to 7.2 kg/sq cm water was admitted to the sample whereupon the void ratio decreased at constant load by the amount $\overline{BC}$. Point C is located close to the curve *AD* which was obtained by saturating the specimen before the consolidation test was started.

The importance of the decrease of the void ratio, $\overline{BC}$ in Figure 14, produced by saturation depends on the initial void ratio e_o. For $e_o = 0.8$ the decrease is imperceptible but it becomes more apparent with increasing values of e_o. The value of e_o depends primarily on the climatic conditions which pre-

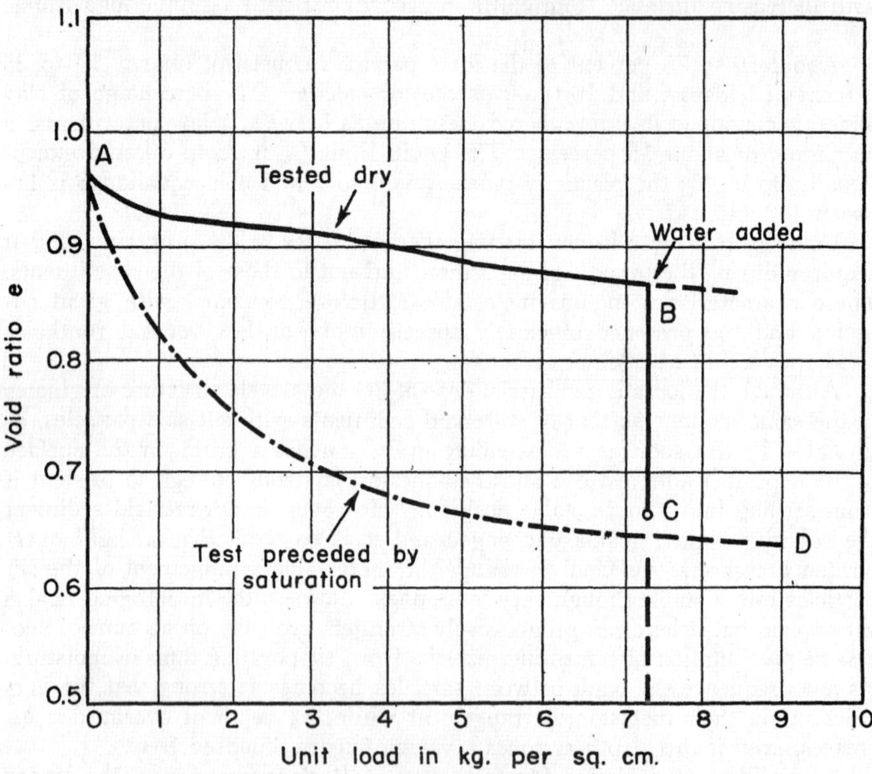

FIG. 14. Relation between unit load and void ratio for loess. (After Holtz and Gibbs, 1951.)

vailed during the deposition of the loess and on the changes in these conditions since deposition. Furthermore, the void ratio of loess derived from glacial outwash, such as the Mississippi loess, is commonly lower than that of desert loess. According to Scheidig (68) the void ratio of the loess deposits in Russia increases from less than 1.0 north of the Black Sea to values up to 1.4 in Central Asia.

On account of the loss of strength and surface subsidence associated with saturation, leakage from defective water mains or tanks into the subsoil of foundations on loess may cause detrimental differential settlement. Incidents of this kind that have occurred in Utah were reported by Peck and Peck (57). According to Holtz and Gibbs (38) irrigated areas in Nebraska and Montana settled and warped after water was admitted into the ditches. A concrete-lined flume connecting the Glendive Pumping Station with an unlined main irrigation canal deteriorated progressively, starting at the downstream end. The pile-supported abutments of highway bridges across the canal settled by amounts up to 6 inches, while the loess into which the piles were driven became saturated.

In order to reduce the settlement of canals and of storage dams on loess subject to saturation after construction, Johnson (44) reduced the void ratio of the loess prior to construction by the injection of a silt slurry under pressure. The procedure has been extensively used on irrigation projects in Nebraska.

In spite of the undesirable properties of loess *in situ,* Turnbull (101) reports that loess has successfully been used as a construction material for earth dams.

Water-table observations in loess districts in Germany have shown that local heavy rainfalls may cause an almost simultaneous rise of the water table by several feet and the subsequent descent of the water table takes place very slowly. This observation indicates that the coefficient of permeability even of saturated loess in a vertical direction is much higher than in horizontal ones. Hence it appears that saturation does not necessarily involve complete deterioration of the system of vertical rootholes, which is characteristic of typical loess.

Also on account of the vertical rootholes, loess exhibits vertical cleavage. If exposed on an inclined slope it is subject to intense and rapid soil erosion whereas vertical slopes remain intact and stable for long periods. If a vertical slope fails, the failure assumes the character of the collapse of a vertical slice, exposing again a vertical face.

Violent earthquakes or explosions may break the bond between the loess particles above the water table whereupon the loess assumes the character of a suspension of silt-size particles in air, which can flow with great speed to long distances from the site of the breakdown. According to Close and McCormick (17) a major catastrophe of this kind occurred in December 1920 in the Kansu Province at the headwaters of the Hwang-Ho River in China.

Landslide Deposits.—If a slope on quick-clay fails, the adjacent valley floor may be covered over a length of many miles from the seat of the slide with a thick layer of semi-liquid slide material. Many slides of this kind have been described by Holmsen (37). On account of thixotropic hardening the slurry "freezes" within a few days, but the compressive strength of the solidified slide material does not exceed a small fraction of that of the parent material, as explained under the heading "Sensitivity." The thixotropic hardening is followed by consolidation under the influence of gravity associated with a decrease of the void ratio, but this process goes on very slowly. At the same time a stiff crust is likely to be formed by superficial desiccation.

Extensive landslide deposits can also be formed by major slides on slopes above deeply weathered rock. Such a slide occurred about 1850 on a vertical cliff composed of weathered basalt close to the upper end of Rubble Creek west of Mt. Garibaldi in British Columbia, at an elevation of about 3,000 feet above the forest-covered valley floor of the Cheakamus River. The slide material flooded the floor over a length of several miles. It formed a deposit with a thickness up to 40 feet and an uneven, but more or less horizontal, surface. It has the appearance of a weak and lean concrete, containing boulders with a diameter up to several feet. The deposit is well graded and completely void of stratification. Although about thirty-five percent of the constituents have a size of more than 3 inches and the clay content is almost nil, the average permeability is of the order of magnitude of 10^{-4} cm per sec. Most of the trees that grew on the valley floor prior to the slide retained their vertical position except for those that grew along the margins of the flooded area. The base of the landslide deposit is separated from the underlying river sediments by a porous layer composed of topsoil and brush.

According to Crandell and Waldron (19) a post-Pleistocene mud flow extends from the northeast side of Mt. Rainier down the White River Valley to the west front of the Cascade Range and spreads out in the form of a lobe 20 miles long and 3 to 10 miles wide. The thickness of the deposit ranges from a few feet to 350 feet. It is unsorted, unstratified, and highly plastic. It contains abundant andesite lava fragments from Mt. Rainier.

Landslide deposits composed of weathered rock fragments and detritus with a high clay content may retain their mobility for decades. In the valley of the Reventazon River in Costa Rica the railroad from Port Lemon to San José crosses several deposits of this kind. In every rainy season the movements became important enough to require readjustment of the tracks until the landslide material has been stabilized by artificial means.

CHANGES AFTER DEPOSITION

Varieties of Changes after Deposition.—The changes that may occur after deposition are summarized below and described in more detail in the following pages. These changes may be considered as: (a) chemical changes, (b) jointing, (c) pre-consolidation, (d) desiccation, and (e) changes in temperature and moisture.

If the surface of a water or ice-laid deposit is exposed to the air, the top layer becomes subject to chemical changes from weathering and leaching. In many cohesive sediments with a high clay content, changes take place even if the surface of the deposit remains submerged. These changes lead to the formation of a system of joints.

If the load on a sediment is reduced, for instance by the melting of an ice sheet covering its surface or by the removal of superimposed sediments by erosion, the sediment assumes the characteristics of a pre-consolidated or pre-compressed sediment. These characteristics are very different from those of the same sediment in a normally consolidated state.

If the top layer of a cohesive sediment is exposed to the air, additional and permanent consolidation of its top layers takes place owing to the volume change associated with desiccation. Seasonal or permanent changes in the rate of surface evaporation are accompanied by volume changes in the top layer of the sediment that may be important enough to damage structures resting on the top layer.

The least conspicuous but in some instances most important or even catastrophic change after deposition is an increase of the pore-water pressure in a fine-grained sediment resulting from a rise of the water table in a more permeable formation located below the area occupied by the sediments or beyond its boundaries.

Chemical Changes.—After the surface of a sediment is exposed to the air the top layer becomes subject to weathering. Weathering commonly involves an increase of the clay content owing to a breakdown of chemically unstable constituents. As a consequence, sediments with an initially low clay content may acquire a certain degree of plasticity and their permeability may decrease. These changes have manifold practical consequences.

On loess deposits, for instance, weathering produces a top layer of "loess loam" with relatively low permeability. This layer assists in maintaining the integrity of the binder between loess particles in the lower portions of the deposit because it reduces the rate at which rain water percolates into the loess. If the top layer is removed the compressibility of the underlying loess increases and surface subsidence may ensue.

On cohesionless sediments, weathering produces a top layer that contains at least a moderate amount of silt- and clay-size particles. Hence in regions where clay deposits are absent the top layer can be used as construction material for the "impervious" section of earth dams, as reported by Esmiol (25).

The weathering of the top layer of sediments is associated with leaching whereby soluble constituents are removed and deposited in a lower horizon. This process leads to a subdivision of the uppermost portion of the sediment into an *A*-and *B*-horizon. In some instances the leaching proceeds to a considerable depth below the surface and is associated with a marked increase of the porosity of the leached zone. In some regions, such as that of Sao Paulo in Brazil, this process has a profound influence on foundation conditions.

Over large areas in the vicinity of Sao Paulo the uppermost stratum consists of a thick layer of stiff, red, Tertiary clay. To a depth of about ten feet, the clay has a honeycombed structure owing to leaching that increased the void ratio from its initial value of less than one to about 1.5. Below the leached layer the clay contains large and numerous concretions of limonite (61, 104).

The increase of the void ratio to 1.5 was associated with a very important increase of the compressibility of the leached portion of the clay although the water content remained low, and excessive settlements ensue under moderate loads. Therefore foundations cannot be established on this layer, unless the void ratio of the layer is reduced by some process of artificial compaction. This process was used for preparing the foundation for a municipal water reservoir. The clay was excavated to the base of the honeycombed stratum and the excavation was immediately refilled with the excavated material in a compacted state. During the process of compaction the volume of the clay decreased by about 25 percent.

A radical change in environment can also be associated with a significant change of the physical properties of a sediment. Forbes (27) described changes of this kind which occurred in the western United States in earth-dams composed of materials that have not yet been subject to weathering and leaching.

Another chemical alteration resulting from a change in environment appears to have occurred in the subsoil of Cairo, Egypt. According to Hanna (33), the subsoil of this city contains two clay strata both of which are flood-plain deposits of the Nile. The upper stratum has a brown color and is relatively stiff and insensitive ($L_w = 60$ to 70, $P_w = 25$ to 30, and $w = 30$ to 40). The lower stratum has a darker color and it is softer and more sensitive than the upper one ($L_w = 50$ to 60, $P_w = 25$ to 35, and $w = 40$ to 50). Grim (30) found the following differences between the two clay strata: the upper (brown) stratum contains montmorillonite with calcium and sodium in equal proportions in the adsorption complex, whereas the dark stratum contains only calcium-montmorillonite. It will be shown below that the replacement of sodium by calcium is associated with a decrease of the liquid limit and an increase of the sensitivity. These relationships may account for some of the differences between the properties of the brown and the dark-colored clay strata. They corroborate the opinion expressed by Hanna (33) that the dark clay represents the product of an alteration of a layer of brown clay which took place after this layer became permanently submerged.

Profound changes resulting solely from base exchange have occurred in deposits of marine, glacial clays that subsequently rose above sea level. While clay is still located below sea level the voids are occupied by sea water with a salt content of about 35 gm per liter and the adsorption complex is dominated by sodium. By correlating the physical properties of Norwegian clays with their salt content Bjerrum (3) found that the effect of a decrease of the salt content from 35 to about 15 gms per liter is rather unimportant. However a decrease from 15 to zero is associated with a decrease of the activity from 0.6 ($\pm$) to 0.25 ($\pm$), an increase of the sensitivity from about 3 to several hundred, a decrease of both the liquid and the plastic limit, and an increase of the liquidity index L_i from a value close to one to values up to 3. According to Figure 7 an important increase of the liquidity index is associated with a radical change of the compression characteristics of the clay, and according to Figure 9 it involves an important decrease of the *in situ* value of the angle of shearing resistance ϕ_{cq}. On account of these changes, marine glacial clays that are perfectly normal sediments turn into dangerous quick-clays. The characteristics of slides on quick-clays have been described under the heading *"Sensitivity."*

The transformation of salt-water clays with low sensitivity into extra-sensitive clays has also been performed by a process of leaching in the laboratory (79, 66).

Jointing.—Some normally consolidated clays, practically all flood plain clays, and many pre-compressed clays are weakened by a network of joints.

The formation of joints in normally consolidated clays is not yet satisfactorily explained. It appears to be due to a process comparable to syneresis and it may be identical with the process that gradually reduces the void ratio of the top layer of many submerged sediments to a value corresponding to the liquid limit. Yet, even some of the extra-sensitive clays in Canada, with a liquidity index of more than 1.5, contain joints. Skempton and Northey (79) have observed that remolded samples of London and Shellhaven clay, stored under water for a period of one year, developed a system of joint simliar to that which exists in the same clays in a natural state. The water content of the clay next to the walls of the joints was slightly above the average. According to L. Casagrande (15) systems of joints are also produced by electro-osmotic processes. These are probably shrinkage cracks.

The jointing in flood-plain clays can be accounted for by seasonal variations in water content associated with alternating expansion and contraction. Joints in pre-consolidated clays were probably produced by unequal expansion upon removal of load or/and deformation by tectonic movements.

The variety in the origin of joints is matched by an equal variety in the spacing of joints and the appearance of the walls of the joints. The spacing of the joints may range between a fraction of an inch and several feet. The walls of the joints may be dull or slickensided, fairly even or intricately warped. If a specimen of jointed clay is submerged, all of the joints may open up simultaneously or the specimen may gradually break up into smaller and smaller fragments. An adequate classification of joints in cohesive sediments is not yet available.

In normally consolidated clays the existence of joints has no significant practical consequences whereas in pre-consolidated clays the joints have a decisive influence on the stability of slopes.

If the joints are closely spaced a slide may occur during the excavation of an open cut with sloping sides. Slides of this kind occurred during and after the excavation of a railway cut near Frankfurt am Main (Rosengarten cut) in 1911, in an intensely jointed, highly colloidal, and heavily pre-compressed clay. The spacing between joints did not exceed a small fraction of an inch and the walls of the joints were slickensided. The descent of the slide material into the cut resembled the movement of glaciers (62).

In most instances the slides of slopes in cuts in stiff jointed clay occur several years or decades after the cut is excavated. This is due to the fact that the excavation of the cut is followed by a gradual decrease of the shearing resistance of the material on both sides of the cut. This decrease proceeds in the following manner.

The removal of the lateral support of the material located on both sides of the excavation produces in this material shearing stresses associated with a slight opening-up of the joints. The joints are invaded by water, the shearing resistance of the clay located next to the joints decreases, and the slide occurs as soon as the average shearing resistance of the clay becomes equal to the average shearing stress on the potential surface of sliding (87).

Experience has shown that the average shearing resistance of stiff fissured clay at the instant of sliding commonly ranges between 0.15 and 0.30 kg per sq cm, whereas the initial shearing resistance of such clays ranges between 1 and 3 kg per sq cm. The slide material always contains many chunks of clay which have retained most of their original strength, embedded in a softer matrix.

In connection with the delayed occurrence of slides, Skempton (74) has correlated the time that elapsed between the excavation operations in London clay and subsequent shear failures in the clay with the shearing resistance at the instant of failure. Some of the results of his investigations are shown in Figure 15. On account of the gradual decrease of the shearing resistance of fissured clay, slides on such clays may occur without provocation and without conspicuous signs of warning. However it can be considered certain that the slide is preceded by a gentle bulging of the slope proceding at an accelerated rate.

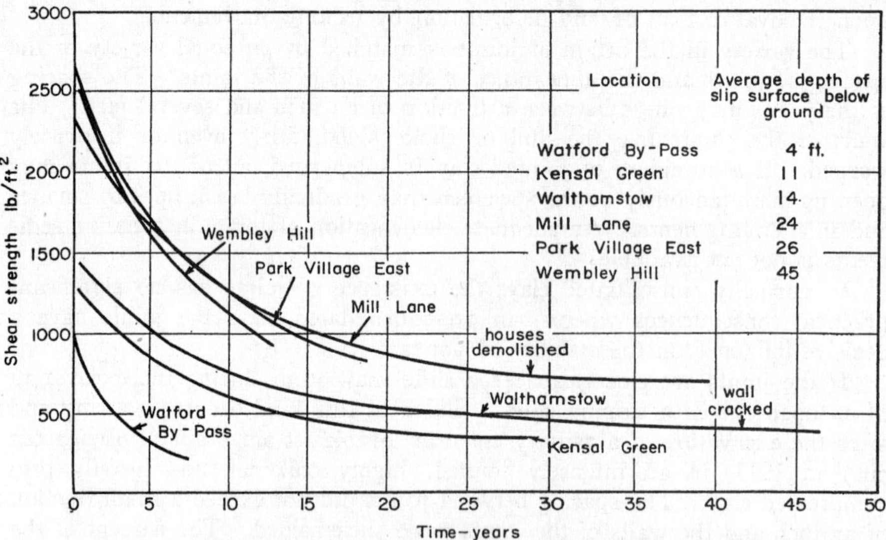

FIG. 15. Tentative relation between average shear strength along slip surface and time, for cuttings and retaining walls in London clay. (After Skempton, 1948*a*.)

A description of slides in cuts of the Great Western Railway of England through hard, fissured Mesozoic clays was published by Cassel (16). The shearing resistance of these clays at the instant of sliding ranged between 4 percent and 20 percent of that of the intact material, whereas the values obtained by Skempton for the relatively plastic Tertiary London clay range between 20 and 30 percent.

In tunnels through jointed sediments, the sediment exposed on roof and walls assumes the character of raveling ground. The rate at which the deterioration proceeds depends on the physical properties of the sediment, the spacing between joints and, in some instances, on the relative humidity of the air in the tunnel. The time that elapses between exposure and the beginning of the deterioration ranges from a few minutes to several days (95).

Pre-Consolidation Under Load.—If the load on a sediment is reduced, for instance by the melting of an ice sheet or the removal of superimposed sediments by erosion, its void ratio increases. The relation between the decrease of the void ratio resulting from load application and its increase owing to subsequent load reduction is shown by curves *AB* and *BC* in Figure 16, in which both the unit load and the void ratio have been plotted on an arithmetic scale.

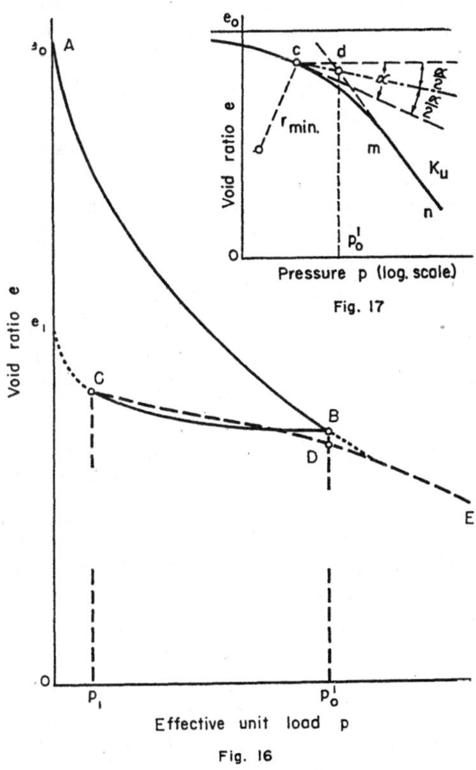

FIG. 16. Influence of reduction of load on relation between unit load and void ratio of a sediment.

FIG. 17. Graphical construction for estimating value of pre-consolidation pressure. (After A. Casagrande, 1936.)

If the load reduction from p'_0 to p_1 is again followed by a load increase, the void ratio decreases as indicated by the dash curve *CDE*. Within the range of pressure p_1 to p'_0 the average slope of the "re-compression" curve *CD* is much smaller than that of the "virgin" branch *AB*, whereas beyond p'_0 these two curves become identical. The load p'_0 is commonly referred to as *pre-consolidation* or *pre-compression load.*

If the pre-consolidation load for a cohesive sediment in the field is unknown, its intensity can be estimated on the basis of the results of consolidation tests on undisturbed samples by means of the simple construction shown in Figure 17 (9). Through point c at which the radius of curvature of the e-log p curve K_u is a minimum, a horizontal line is drawn. The bisector of the angle α between this line and the tangent to curve K_u at c intersects the upward continuation of the straight, inclined lower part of K_u at point d. The abscissa of d is approximately equal to the preconsolidation load p'_o.

The properties of pre-compressed clays illustrated by Figure 16 have four important practical consequences. (1) In cohesive sediments the reduction of the void ratio from e_o to values of less than e_l is associated with a permanent increase of the cohesion. Therefore the shearing resistance of a preconsolidated clay is always greater than that of the same clay under the same load, in a normally consolidated state, even if the clay has been weakened by a network of joints. This condition has a decisive influence on the stability of slopes. (2) Under loads smaller than the pre-consolidation pressure p'_o the compressibility of any sediment is smaller than that of the same sediment under identical loading conditions. (3) The heave of the bottom of deep excavations in heavily pre-compressed clays can be very important and troublesome, because the volume increase associated with the removal of the last load increments increases with increasing values of p'_o. (4) In some sediments subject to load reduction the horizontal pressure decreases less than the vertical pressure. In such instances the horizontal pressure on tunnel supports can be greater than the vertical pressure. In the following paragraphs these conclusions will be illustrated by the results of field observations.

Under the heading *Engineering Properties of Till Sheets* it was mentioned that each of the till sheets underlying the district of Chicago had been consolidated under a heavier ice load than its successor and that this condition had a decisive influence on the cost and the method of construction of the subway of Chicago. In the Cleveland area, all of the sediments underlying the flood plain of the Cuyahoga River have been precompressed under the weight of fluvio-glacial sediments which have been removed subsequently to a depth of about 100 feet by erosion. As a consequence it was possible to establish a heavy structure on one of the glacial lake deposits without supporting the foundation on piles, because the unit load on the base of the building was considerably smaller than the preconsolidation load p'_o. If the clay had been in a normally consolidated state, the settlement of the structure would have been excessive.

Local pre-consolidation of a clay deposit, for instance by the weight of an isolated and subsequently demolished heavy structure, may give rise to unequal surface settlement under the influence of renewed application of load over a larger area. Such a development took place in Mexico City where a regional subsidence is occurring as a result of the withdrawal of water from aquifers located between the soft clay layers shown in Figure 5. At the place where a street crosses an area formerly occupied by a heavy Aztec temple, the street which was originally horizontal acquired conspicuous dips away from the pre-loaded area.

The heave of the bottom of excavations in heavily precompressed clays has recently received careful attention in connection with irrigation and flood-control projects in the valleys of the Missouri and Saskatchewan rivers, because unequal heave may cause an undesirable tilt of turbine shafts and detrimental movements of the concrete lining of discharge canals with reference to adjacent structures. Inasmuch as the expansion of the clay is associated with an increase of its water content and since, in addition, the permeability of the clay is very low, the expansion takes place very slowly, at a decreasing rate. It is governed by laws similar to those that determine the rate of consolidation (see Part *A*).

In Missouri as well as in Saskatchewan the pre-compressed clays are referred to as shales. However, the fact that they can be completely dispersed indicates that they have not even started to acquire the properties of a real shale.

The physical properties of the pre-compressed clays of Missouri (Fort Union shale) were described by Smith and Redlinger (80). This early Tertiary sediment covers an area of about 120,000 sq mi on both sides of the Missouri River. It contains thin beds of lignite, sandstone, and limestone and it was preconsolidated under a load ranging between 80 and 100 tons per sq ft by an overburden consisting of similar sediments which were subsequently removed. Further temporary increases of the overburden pressure were produced by repeated advance and retreat of the Pleistocene ice sheets. The lignite beds are jointed and cracked and normally waterbearing. They are commonly associated with layers of highly colloidal and slickensided clay.

The Fort Union shale has a liquid limit up to 140. The clay fraction consists of 85 to 90 percent illite and 10 to 15 percent montmorillonite. On the basis of test results the authors computed that the lignite beds located beneath the floor of the river valley should have risen, on account of heave owing to removal of the overburden, by an amount of 4.2 feet with reference to their position beneath the slopes of the valley. In reality the differential heave ranges between 5.6 and 7.4 feet.

Peterson (60) published the results of extensive field and laboratory investigations performed on a pre-compressed clay in Saskatchewan (Bearpaw shale) at the site for the proposed Saskatchewan Dam across the Saskatchewan River south of Saskatoon in Canada. The Bearpaw shale is an upper Cretaceous clay that has been pre-consolidated under a load ranging between 100 and 150 tons per sq ft, owing to the weight of sediments that were subsequently removed and, at a later date, to the weight of ice during the advances and retreats of the Pleistocene ice sheets.

The clay has a liquid limit of more than 100. It is intensely jointed and slickensided, but the spacing between joints decreases with increasing depth below the surface. The secondary time effects, represented by the slope of the dash lines in Figure 8, are unusually important. Two hundred days after the load was applied on a specimen, $\frac{3}{4}$ in. thick, the secondary compression was already fifty percent greater than the primary one.

A. Casagrande (11) has shown that the water content of the Bearpaw shale decreases with increasing depth much more rapidly than the laboratory swelling curve, *BC* in Figure 16, indicates. The difference appears to be due to the importance of the secondary time effects combined with the fact that the shales *in situ* continued to swell during a period of many millions of years, whereas the swelling tests in the laboratory were performed within less than a year. Secondary time effects may also account for the aforementioned difference between the computed and the observed heave of the lignite beds in the Fort Union shale beneath the floor of the valley of the Missouri River.

After the sedimentary surcharge on the Bearpaw shale was removed by erosion the Pleistocene ice sheets advanced over the area surrounding the dam site and the last one left in its wake a layer with a thickness of about 100 feet composed of till, fluvio-glacial sand and gravel, and glacial lake clay. Finally the present valley of the Saskatchewan River was formed by erosion. The downward cutting through the shale to a maximum depth of about 300 feet below the original ground surface was associated with an energetic lateral expansion of the shale on both sides of the valley towards the trough formed by erosion. It appears that the lateral movement assumed the character of a gravity creep, whereby the gradient of the slopes was reduced to about 1 vertical on 15 horizontal. On account of their rigidity the glacial sediments were torn along lines parallel to the thalweg and transformed into series of broad, parallel ridges separated from each other by elongated troughs. As a consequence, typical landslide topography developed on both sides of the river, over a length of many miles.

It is rather unlikely that the rate of movement ever exceeded that of a slow gravity creep. Yet the results of observations in an experimental drift indicate that the upper portions of the shale are severely stressed and minor disturbances of the present state of equilibrium are sufficient to produce deformations measured in inches.

Both the Fort Union and the Bearpaw shales are completely saturated. Incompletely saturated, heavily pre-compressed clays were encountered by Florentin and l'Heriteau (26) in various parts of France. In such clays an increase of the load produces instantaneous compression followed by a progressive compression. In each of the deposits investigated, the degree of saturation S_r decreased with increasing depth to values ranging between 90 and 70 percent. Information concerning the performance of these clays in connection with construction operations is not yet available.

Data regarding the value of the ratio between horizontal and vertical pressure in heavily pre-compressed clays are also still very fragmentary. Cooling and Ward (18) have found that the average pressure on the lining of tunnels with a circular cross section in London clay is roughly equal to the full overburden pressure. On the other hand, according to K. Langer (88) the failure of timber supports in the drifts in a clay mine in Provins northeast of Paris, at depths ranging between 100 and 130 feet below the ground surface, can only be accounted for by assuming that the horizontal pressure exerted by the surrounding material, a highly colloidal and heavily pre-compressed clay ("Argile plastique"), is considerably greater than the overburden pressure.

Desiccation.—As soon as the surface of a sedimentary deposit is exposed to the air the layer located immediately below the surface starts to dry out; thereupon water is drawn by capillary action from the deeper layers towards the surface of evaporation. The capillary rise is associated with a decrease of the pressure in the pore water in the layer located beneath the surface of evaporation and an increase $\Delta\bar{p}$ of the effective (grain-to-grain) pressure of equal intensity. This supplementary pressure is known as *capillary pressure.* In sediments composed of silt-size particles, the maximum value which the capillary pressure can assume depends exclusively on the surface tension of the water and the size of the voids (34). In clays it also depends on various physico-chemical factors and it can assume values up to several tens of kgs per sq cm (83). If water is admitted to a specimen acted upon by capillary pressure without permitting a volume increase to take place, the pressure on the walls of the confining vessel increases from zero to the full value of the capillary pressure.

The capillary pressure has the same mechanical effects as a heavy surcharge, but its effects are limited to the layer subject to desiccation. If the sediment is very compressible, surface evaporation produces a conspicuous decrease of the void ratio in the layer subject to desiccation whereby this layer assumes the character of a *dried crust.*

The initial rate of evaporation at the surface of the sediment depends primarily on the relative humidity of the air and the conditions of exposure. As soon as the water content of the uppermost layer of the sediment reaches a certain value known as the *shrinkage limit* S_w, which is at least somewhat lower than the plastic limit P_w, air invades the voids of the top layer, the color of the top layer changes from dark to light, the surface of evaporation recedes to increasing depths below the surface, and the rate of evaporation decreases. If the maximum value p_c of the capillary pressure, the coefficient of compressibility m_v of the sediment, and the coefficient of consolidation c_v are known, the rate at which desiccation progresses in a downward direction can be computed (82).

The process comes to a standstill as soon as the thickness of the dry crust becomes approximately equal to the depth to which desiccation proceeds within one or two exceptionally dry seasons. Therefore the thickness of the dry crust depends in part on the climatic conditions. Under humid conditions it may vary between 5 and 15 feet and in arid regions it can be considerably greater. In the uppermost portion of the dry crust the voids of the sediment are partly or wholly filled with air, the structure of the sediment becomes crumbly, and if the sediment has a plasticity index of more than 15 or 20, the seasonal variations in the moisture content of the top layer are associated with a noticeable seasonal heave and subsidence of the ground surface, which will be discussed under the following subheadings.

The decrease of the void ratio produced by the desiccation of a cohesive sediment such as clay is associated with an increase of the shearing strength, because it involves an increase of the effective pressure $\bar{p}$ in the layer. As a consequence the shearing resistance of the sediment changes with depth as indicated by the dash line in Figure 10. If a dry crust is again permanently submerged, it retains the character of a heavily pre-consolidated stratum with a shearing resistance which is very much greater than that of the same layer in a normally consolidated state.

The geographic distribution of dry crusts and their location within stratified, sedimentary formations depends on the changes that the position of the free water level underwent, in the course of time, with reference to the elevation of the adjacent sedimentary deposits. In many coastal regions these changes represent the combined result of the eustatic rise of the sea level after the end of the Pleistocene and the simultaneous isostatic rise of the land. The influence of these changes on the location and elevation of dry crusts on marine glacial and post-glacial clays on both sides of the North Atlantic has been described by Skempton (75). On the continents, within the areas formerly occupied by the Pleistocene ice sheet, the number and elevation of the dry crusts are determined chiefly by the fluctuations of lake levels associated with the advances and retreats of the ice front. For instance, in the area formerly occupied by Lake Agassiz northwest of the Great Lakes region, Rominger and Rutledge (65) located several dry crusts, at different depths below the ground surface, each of which corresponds to one of several successive stages in the history of this lake.

Wherever a dry crust is located at or not far below the ground surface it has a decisive influence on the foundation conditions in this region because it serves the function of a raft, resting on the surface of softer material, which is capable of supporting the weight of fairly heavy structures. Such a dry crust is located beneath a large portion of downtown Boston on the soft "Boston blue clay," at a depth of about 20 feet below the present ground surface (13) and beneath the Loop District of Chicago, on the water-laid Blodgett moraine, at a depth of about 15 feet below street level (59). Also in Chicago, the presence of the dry crust on the Blodgett till sheet reduced very substantially the heave produced by advancing the shield through the soft Deerfield till during the construction of the Chicago subway tunnels (90).

The thickness of dry crusts is likely to vary erratically. Hence, in connection with foundation operations, it is necessary to investigate the variations of the thickness of the crust by closely spaced drill holes over the entire area to be occupied by the structure.

If the clay underlying a surface of evaporation has previously been preconsolidated under a heavy load there is no conspicuous difference between the top layer and the balance of the deposit, except inasmuch as the top layer may be weathered or more intensely jointed. In arid and semi-arid regions there may be no noticeable difference at all. On account of surface evaporation, however, the top layer is permanently acted upon by capillary pressure. Local flooding of the top layer, for instance by leakage from a conduit, eliminates the capillary pressure whereupon the void ratio of the wetted portion increases. If the plasticity index I_p of the sediment is greater than about 15 or 20, the expansion of the ground can be so important that structures resting on the swelling ground may be severely damaged. Numerous incidents of this kind were investigated by Holtz and Gibbs (38). The volume increase produced by the saturation of air-dry samples of the clays subject to investigation had values up to forty percent. Clays with a high swelling capacity are referred to by Holtz as *expansive clays*. The swelling is due exclusively to the elimination of the capillary pressure and the process of swelling is represented by curve BC in Figure 16. At any stage of the process the pressure required to prevent further swelling is determined by the abscissa of the corresponding point on the swelling curve BC.

Flood-plain deposits are subject to successive desiccation in thin layers, within a few months after each layer is formed. Therefore, such deposits have all the characteristics of heavily pre-consolidated sediments. In wet years, however, the consolidation is less intense than in dry ones. Therefore the intensity of the pre-consolidation load varies with depth. The permeability of the uppermost 10 or 15 feet of clay deposits of this kind can be very high on account of the presence of closely spaced and interconnected root holes.

In contrast to the flood-plain clays, which become precompressed as rapidly as they are formed, clays that were deposited in oxbow lakes remain in a normally consolidated state and their liquidity index is likely to be close to unity. Within a few years the surface layer is transformed into a dry crust and the crust becomes buried beneath normal flood plain deposits, whereas the bulk consolidates only under the influence of gravity. Oxbow deposits, therefore, constitute crescent-shaped bodies of highly compressible sediments embedded in stiff silt and clay with low compressibility. If the presence of such a lens beneath the site of a proposed structure remains undetected, severe differential settlements may ensue. In some cases such settlements have damaged the structure beyond repair. If the flood plain is still in its natural condition the existence of oxbow deposits at a shallow depth is disclosed by gentle, crescent-shaped depressions, produced by progressive consolidation of the oxbow sediments.

In arid regions, silty sediments washed down by sheet floods onto bahadas are likely to acquire and to retain a metastable structure, because deposition is almost immediately followed by desiccation and the deposition of cementing material comparable to the binder in loess. Subsequent saturation, such as by leakage from canals or reservoirs, causes a collapse of the structure within the zone subject to wetting. In one instance, the surface of the zone has settled by amounts up to several feet.

If a heavy rain descends on the desiccated top layer of a playa sediment, water is pulled by capillary action into the sediment causing the air in the voids between the water table and saturated top layer to become compressed. The compressed air may cause a failure of the top layer by tension (99) or it may blow out along lines of least resistance. In silty sediments of some of the seasonal lakes of Algeria, the formation of a system of tubes, similar to root holes with a diameter up to several millimeters, was ascribed to such processes by Drouhin et al. (24).

Variations in Moisture and Temperature Conditions.—Above the water table the moisture content of the top layer of sediments varies with the seasons and in regions with a moderate and cold climate it is also subject to alternate freezing and thawing. In coarse-grained sediments such as clean sand, the mechanical effects of these processes are inconsequential because the maximum value that the capillary pressure can assume is very small and the water freezes and thaws *in situ*. On the other hand, in fine-grained cohesive soils, the maximum value of the capillary pressure and, as a consequence, the volume changes associated with alternating wetting and drying increase rapidly with increasing plasticity index I_p. During the process of freezing, water is drawn from lower strata into the top layer, where it accumulates and forms layers or pockets of clear ice. The seasonal or progressive volume changes associated with these processes can be important enough to break up hard-surface roads or to damage existing structures. As a consequence, they are receiving increasing attention.

Wherever the surface of a clay stratum is covered with vegetation, the capillary pressure increases during each growing season and the corresponding moisture content decreases. As a result, the ground surface subsides temporarily and rises again in the fall or spring. Ward (105) measured this seasonal movement of the grass-covered surface of London Clay at various points in southeastern England and obtained values up to 2.4 inches.

The evaporation from trees reduces the moisture content of the body of sediment invaded by its root system and produces a bowl-shaped subsidence of the ground surface surrounding the tree. Skempton (77) described the gradual breaking-up of the walls of a theater building in Samford Hills, with footings on London Clay, as a result of the draining action of poplar trees located about 30 feet from the nearest part of the structure. At the time of the investigation the trees had reached a height of about 45 feet. According to Skempton's observations, subsidences associated with the growing of trees are noticeable to distances up to 1½ times the height of the tree. Within a distance equal to the height of the tree they may be harmful. In close proximity of the tree the desiccation may extend to a depth of 8 to 10 feet and the corresponding settlement may assume values up to four inches.

On areas covered by buildings, the rate of evaporation is almost equal to zero. If a building has been established on a heavily pre-compressed clay with a high plasticity index, in a region with a long dry season, the water content of the clay located below the central part of the building gradually increases owing to the capillary rise from the water table, which is not compensated by evaporation. As a consequence the footings of the inner columns or partitions rise with reference to those of the outside walls. This phenomenon was observed, probably for the first time, by Simpson (72) in the region of San Antonio, Texas, on buildings with footings on a heavily pre-compressed Cretaceous clay.

On account of alternate drying and wetting, the San Antonio clays have acquired a crumbly structure to a depth of about 20 feet below the ground surface. Most of the expansion takes place within the uppermost 3 to 6 feet. The base of the central portion of one of the buildings examined by Simpson rose about 14 inches with reference to the foundations of the outer walls within four years. Owing to the rise the partitions were wrecked. On some structures Simpson prevented further increase of the heave by the construction of small ventilating tunnels beneath the basement floor.

The observations in Texas were followed by similar ones in Burma, northern Argentina, South Africa, Morocco (23), Palestine (63), India (56) and other regions.

Woolterton (106) reported that the most conspicuous heaves in Burma occur on alkaline clays, $L_w = 56$ to 81, $I_p = 38$ to 57 and activity $a_c = 1 \pm$. Swelling pressures up to 2 tons per sq ft have been measured. The heaving clays of the region of Austin, Texas, investigated by R. F. Dawson (22) are described as very hard, fissured and slickensided, $L_w = 37$ to 104, $I_p = 22$ to 81. Extensive investigations of the heaving clays of South Africa have been carried out during the last six years under the direction of J. E. Jennings.

According to Jennings (43) the prerequisites for the heaving of South African clays are high activity, low liquidity index, low position of the water table, and evaporation in excess of rainfall. These conclusions do not have general validity, because the activity of the heaving Burma clays is normal. Moreover, Tschebotarioff (100) described damage to buildings as a result of heave in eastern Cuba, where the fainfall exceeds the rate of evaporation by about fifty percent. However, all of the investigated heave phenomena have the following features in common. The water table is located at a depth of about 15 feet or more below the ground surface. The clay has an average plasticity index of more than 30, and a liquidity index close to or below zero; also it is intensely jointed. Under arid conditions such clays would perform as "expansive" clays and in deep excavations or tunnels as "swelling" clays.

In contrast to moisture heave, which occurs only in heavily pre-consolidated clays with a high plasticity index, the phenomena of frost heave are most conspicuous in normally consolidated sediments with a low plasticity index. The expansion associated with seasonal freezing is strictly limited to the depth of seasonal frost penetration, which increases from a few inches at the outer boundary of the subtropical belts to a maximum of 10 to 12 feet in regions of degrading permafrost. Beyond these regions the ground is permanently frozen to a depth that increases in a general way towards the poles and the zone of seasonal freezing is replaced by a zone of seasonal thawing.

In coarse-grained sediments the water freezes *in situ* and its volume increases by about ten percent. The corresponding rise of the ground surface is inconsequential. In cohesive soils, on the other hand, layers of clear ice are formed. The water contained in these lenses is drawn by capillarity from greater depth towards the centers of freezing. The total quantity of water contained in the top layer increases and the surface rises perceptibly. This rise is referred to as *frost heave*. The worst frost heaves occur on sediments dominated by the silt-size fraction, because these sediments combine great height of capillary rise with relatively high permeability. On account of the high permeability, large quantities of water can reach the seats of ice-layer formation in a short time.

When the ice melts in the spring the stratum containing the ice layers breaks up and turns into a thick slurry because its water content is higher than it was before the beginning of the frost season. In areas where the vertical distance between the ground surface and the water table is smaller than the height of capillary rise of the sediments located within this distance, the frost heave can amount to as much as six inches and the subsequent melting of the ice layers leads to the formation of patches or layers of very soft material. Owing to the detrimental effects of these processes on hard-surface roads frost heave has been the subject of extensive field and laboratory investigations for the past three decades (34).

North of a line located in the proximity of the 0° C isotherm of the mean annual temperature, the ground located below the depth of seasonal thawing remains permanently frozen. This condition can adequately be described as a duplicate of the seasonal frost phenomena on an exaggerated space and time scale. The only conspicuous difference between these two phenomena resides in the shape of the bodies of clear ice that are formed during the process of freezing. In the zones of seasonal frost, these bodies consist of thin layers of ice and the frozen ground assumes the appearance of a layered sediment; whereas in the permafrost zone the ice forms huge, irregular concretions, not uncommonly as large as a cabin. The difference is probably due to the difference in the rate of ice formation.

The shifting of the 0° C isotherm is associated with a shifting of the outer boundary of the permafrost zone, but the shifting takes place with a considerable time lag. The thawing takes place on the upper surface of the permafrost layer much more rapidly than at its base. Therefore the surface of a degrading permafrost layer assumes an uneven, hummocky appearance, whereas the base remains more or less even. Finally, the body of permafrost breaks up into individual slabs of frozen ground located at depths up to twenty feet or more below the ground surface. The relationships between and increase of the mean annual temperature, thermal conductivity, and the shrinkage of permafrost layers were described by Terzaghi (97). Since the frozen ground is impermeable, the formation and degradation of permafrost layers have a profound influence on the groundwater conditions.

If a frozen body of silty sediments thaws, deep kettleholes are formed above the spaces formerly occupied by pockets of clear ice. The ground surface assumes the character of a limestone terrain and therefore has been called *thermokarst.* The original structure of the sediments located beneath the kettleholes is completely distorted.

The degradation of permafrost also takes place beneath heated structures. If such a structure is located above a silt stratum containing ice pockets the structure may be damaged beyond repair. Therefore, the design of the foundations must be preceded by an investigation of the permafrost conditions beneath the site of the building.

On slopes located above permanently frozen ground the seasonal thawing produces various flow phenomena known as *solifluction.* These represent the arctic equivalent of moisture creep in regions with a moderate climate and they add considerably to the cost of maintenance of highways and railroads in such regions (71).

Changes in Groundwater Conditions.—If the piezometric level of the water contained in the voids of a sediment underlying a slope rises the value u in equation 8 rises in every point of the sediments located below the piezometric surface; the corresponding value of the shearing resistance of these sediments s decreases and a slide may occur. This is one of the most common causes of the failure of slopes on fine-grained sediments (96).

The rise of the piezometric level can be caused by the filling of a storage reservoir, leakage from an irrigation canal, rise of the water table in a jointed rock formation located below the base of the sediments, and various other incidents. Before the consequences of the rise become visible, the change in the pore water pressures can be detected only by observations on piezometric tubes or pore-pressure gages. If no such observations were made, the ground movements produced by the change convey the impression that they occurred without provocation. The following examples illustrate the variety of the consequences of a rise of the piezometric level.

Figure 18 is a vertical section through the subsoil of a powerhouse located at the shore of a lake in the Coast Range. According to the results of borings performed prior to construction the site is located on a deposit *D* composed of silty and sandy detritus that rests on a layer of stiff, glacial, slightly varved clay *C*.

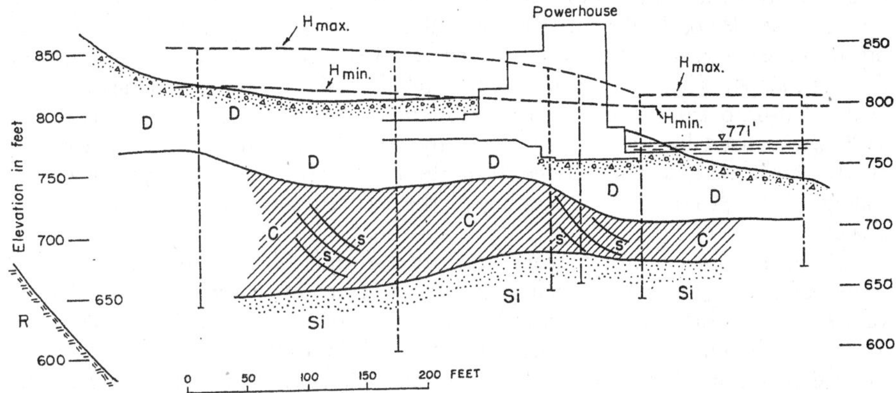

FIG. 18. Excess hydrostatic pressure in silt stratum located below layer of stiff clay with shear zones, east of the Coast Range in British Columbia.

Soon after construction was completed it was noticed that the powerhouse advanced towards the lake at a rate of about ½ inch per year and tilted away from the lake. In the following spring tension cracks with a width of several inches opened up at a distance of about 200 feet from the lake shore, extending from the site of the powerhouse to a distance of about 1,000 feet in opposite directions.

On account of these developments supplementary borings were made and furnished the following information, as illustrated by Figure 18. The stiff clay rests at a depth of about 100 feet below lake level on a very thick stratum of silt and silty sand. The strength of the clay stratum is impaired by several shear zones, which indicated that a slide or slides have occurred at some time in the past at the site of the power house. During the boring operations it was noticed that the water rose overnight in the casings and finally overtopped their upper ends. Therefore Bourdon gages were installed on the casings and the gage readings showed that the piezometric levels at the time of installation were close to the line marked H_{min} in Figure 18. During the following spring the piezometric elevations rose rapidly and when they reached the values represented by the ordinates H_{max} a new set of tension cracks were formed parallel to the lake shore and close to the preceding ones.

These observations explained the mechanism of the ground movements. The resistance against sliding along the base of the clay stratum is determined by equation 8. As the water pressure u at the base of the clay stratum increased, the factor $(p - u)$ in equation 8 decreased. When it approached the value zero the ground beneath the margin of the lake was deformed by "spreading" on the surface of contact between silt and clay.

In August 1952 a similar process led to the partial destruction of a power house on the west shore of Whatshan Lake in British Columbia. The power house is located on the lake shore, at the edge of a low terrace consisting of silt and silty sand. The terrace material rests on the steeply inclined slope of the bedrock, a jointed granite containing shear zones. The surface of the terrace rises towards the rock slope and at the foot of the rock slope it is buried beneath granite boulders and scree. The slide occurred at the foot of the rock slope, where the rock was blanketed with silt and the slide material descended onto the power house. Subsequent investigation showed that the slide was caused by an increase of the hydrostatic pressure, u in equation 8, on the surface of contact between granite and terrace material, produced by leakage through cracks in the lining of the pressure tunnels into the jointed granite.

Slides also occur almost every year, at the time of the snow melt, on the slopes of silt deposits that occupy the lower part of many of the *U*-shaped valleys in southern British Columbia, such as the valleys of the Frazer, Big Thompson, Columbia and Okanogan rivers. According to Matthews (53) these deposits were formed at the end of the Pleistocene by sedimentation in ice-dammed lakes. The present river valleys were carved out of the silt deposits by erosion to a depth of several hundred feet and their sides rise in ter races to the level of the original silt surface.

In many sections of these valleys the rock is intensely jointed and its permeability is much higher than that of the silt. In these sections the water table rises during the period of melting snow to a level high above that of the present river bottom. The pressure in the pore water of the silt increases and slides ensue wherever the shearing resistance along potential surfaces of sliding drops below the value of the shearing stresses. In the valley of the Big Thompson River these conditions were responsible for the slides that occurred from time to time along the lines of the Canadian Pacific and the Canadian National railway.

A fatal rise of the pressure in the pore water can also be caused by the filling of a storage reservoir or by raising its level. This process was the cause of the catastrophic slide which occurred in December 1918 at the east end of the Cedar Reservoir in the State of Washington. The geological conditions prevailing at the site of the reservoir and of the slide were described by Mackin (51).

If a storage reservoir formed by a dam resting on sediments is filled for the first time, springs are likely to come out of the ground near the foot of the slopes downstream from the area occupied by the dam. Subsurface erosion starting at one or more of these springs may produce tunnel-shaped conduits following the water veins in an upstream direction. If the upstream end of one of these tunnels approaches the bottom of the reservoir, water rushes into the tunnel, the tunnel is rapidly enlarged, its roof caves in and the dam fails. These dreaded accidents are commonly referred to as failures due to *piping*. They may occur without any warning signs of danger many years after a reservoir has been filled for the first time.

Terzaghi (84) has shown that the factor of safety of a dam resting on sediments, with respect to piping, may range between less than unity and more than ten, depending on details of stratification that cannot be ascertained by any practicable means, prior to the filling of the reservoir. However, the danger of piping can be reliably eliminated, irrespective of the details of stratification, by covering the exit area of the seepage water by an inverted, adequately graded filter, after the location of the danger points has been determined by observations.

The conditions leading to piping can also be established by natural processes. An incident of this kind has occurred at the outskirts of the city of Memphis, Tenn. The city is located at the edge of a terrace, at an elevation of about 100 feet above the low-water level of the Mississippi, on the surface of a layer of stiff, yellow clay with a thickness of about 60 feet. The clay rests on a stratum of clean to silty, waterbearing sand with a thickness of about 40 feet. On the river side, the outcrop of the sand stratum is blanketed by a practically impervious layer of slide material. As a consequence the piezometric level of the water in the sand is located high above the low water of the Mississippi.

On July 25, 1927, during the recession of a flood in the Mississippi River a strip of ground with a length of about 700 feet and a width of about 100 feet, located next to the upper edge of the slope descending towards the Mississippi, started to subside at a rate of about one foot per hour and continued at that rate for about thirty hours. At the end of that period the subsided area was surrounded by almost vertical clay cliffs. Subsequent borings showed that the thickness of the sand stratum had decreased beneath the area of subsidence by the full amount of subsidence.

According to Terzaghi (85) the subsidence was caused by piping; the sand flowed towards an opening in the clay blanket that was formed by erosion in the blanket during the preceding highwater stage of the Mississippi. The opening was located at a distance of not less than 1000 feet downstream from the south end of the area of subsidence.

CONCLUSIONS

In this paper it has been shown that the performance of sediments in connection with engineering operations can be predicted with reasonable assurance only if the significant physical properties of the sediments involved in the operations are known. Experience has shown that the classification of sediments on the basis of visual inspection and even of petrographic investigation of samples furnished by conventional exploratory boring is inadequate and often misleading. The significant properties of the materials encountered in boring operations can be recognized and adequately described only on the basis of the results of laboratory tests on undisturbed samples or of field tests as explained in the paper. Yet, even when the required investigations have been carried out, the available information is still more or less fragmentary, because the properties of natural sedimentary deposits change from point to point and the construction of the soil profiles requires a considerable amount of extrapolation.

The extrapolation consists in establishing the pattern of stratification on the basis of the information obtained along vertical lines spaced no less than about fifty feet. The pattern of stratification is determined by the geological history of the deposits which, in turn, is reflected by the physical properties of the sediments, as explained in this paper. Therefore, the knowledge of the relation between physical properties and geological history is of outstanding practical importance. On the other hand the results of the detailed subsoil investigations performed in connection with engineering operations provide the geologist with a new source of significant information in the realm of physical geology.

HARVARD UNIVERSITY,
CAMBRIDGE, MASS.

APPENDIX. LIST OF SYMBOLS

A_s = Specific surface = ratio between total surface area of soil particles per unit of volume and that of a sphere with a volume equal to unity.
a_c = Activity = ratio between the liquid limit L_w of the sediment and the weight of its clay-size fraction in percent of total dry weight.
C_c = Compression index = decrease of void ratio e of a sediment due to increase of effective unit load from $\bar{p}$ to 10 $\bar{p}$.
c_v (cm²/sec) = coefficient of consolidation = constant, defined by equation 7, which determines the rate of consolidation of a sedimentary stratum with a given thickness H at a given increase of the unit load p.
D (cm) = Depth.
D_p (cm) = Peak or modal value of the grain size, corresponding to the abscissa of the apex of the curve in the double-logarithmic grain size frequency diagram, Figure 12.
D_r = Relative density of a cohesionless sediment = $\frac{e_{max} - e}{e_{max} - e_{min}}$, wherein e is the void ratio of the sediment *in situ*, e_{max} that of the sediment in the loosest state and e_{min} its void ratio after thorough compaction.
D_{10} (cm) = Effective grain size of a sediment, corresponding to 10% in the cumulative grain size diagram (10% of the particles by weight are smaller and 90% larger then D_{10}).
D_{60} (cm) = Grain size corresponding to 60 percent in the grain size diagram.
e = Void ratio = ratio between volume occupied by voids and by the solid constituents respectively.
e_c = Critical void ratio = void ratio at which the volume change associated with a shear deformation of the sediment is zero. If the void ratio is greater than e_c the sediment contracts and if it is smaller, it expands.
F_s = Grain-shape factor = ratio between weight of particles with flaky habit and total dry weight of the sediment.
H (cm) = Height, thickness.

I_p = Plasticity index = difference between liquid limit L_w and plastic limit P_w.
i = Hydraulic gradient = ratio between loss of head and length of path of percolation.
K_0 = Coefficient of earth pressure at rest = ratio between unit pressure on vertical sections and on a horizontal section through any point in a sediment with a horizontal surface.
k (cm/sec) = Coefficient of permeability = quantity of percolation in cm^3 at a hydraulic gradient $i = 1$, through a section at right angles to the direction of the flow, per cm^2 of this section.
L (cm) = Length.
L_i = Liquidity index = $\frac{w - P_w}{L_w - P_w}$.
L_w = Liquid limit or lower limit of liquid state = water content at which the sediment in a remolded condition is at the boundary between plastic and liquid state.
m_v (cm^2/gm) = Coefficient of volume compressibility = compression per unit of volume produced by an increase of the unit load by 1 gm/cm^2. For every sediment the value of m_v decreases with increasing unit load.
P_w = Plastic limit or lower limit of plastic state = lowest water content at which the sediment can still be rolled out into threads.
p (gm/cm^2) = Unit load.
$\bar{p}$ (gm/cm^2) = Effective unit load or grain-to-grain pressure = difference between unit load and hydrostatic pressure in the pore water.
p_c (gm/cm^2) = Capillary pressure = pressure exerted on the sediment by the surface tension of the water on a surface exposed to the air. As the pressure p_c on the solid increases due to evaporation the pressure in the interstitial liquid decreases by an equal amount.
p_0 (gm/cm^2) = Pre-consolidation pressure = greatest unit load which has acted on a sediment in the course of its history.
p_u (gm/cm^2) = Confining pressure = pressure per unit of area, acting on the sediment in every direction with equal intensity.
p_v (gm/cm^2) = Unit pressure on a vertical section through a given point in the sediment.
S_a = Degree of aeration = $100 - S_r$.
S_c = Slope of the triangular log-log grain size frequency diagram, Figure 12, on the coarse side.
S_f = As before, on the fine side.
S_r = Degree of saturation = ratio between pore space occupied by water, in percent of total pore space.
S_w = Shrinkage limit = void ratio at which further shrinkage ceases during the process of desiccation and air starts to invade the voids of the sediment.
s (gm/cm^2) = Shearing resistance per unit of area, if the state of failure is reached at unaltered porewater pressure.
s_{cq} (gm/cm^2) = Shearing resistance per unit of area, if the state of failure is reached at unaltered water content (consolidated-quick value of the shearing resistance).
s_0 = Value of s_{cq} for the top layer of a normally consolidated, waterlaid sediment, the surface of which was never exposed to the air.
s_t = Sensitivity = ratio between the unconfined compressive strength of the sediment in an undisturbed and in a remolded state.
t (sec) = Time.
t_1 (sec) = Time corresponding to the point at which the empirical time-consolidation curve in Figure 8 deviates from the theoretical curve.
U = Uniformity coefficient = D_{60}/D_{10}.
U_c = Degree of consolidation = compression of a sediment under constant load at time t, in percent of the computed ultimate compression.
u (gm/cm^2) = Hydrostatic pressure in the pore water of a sediment.

w = Water content of a sediment in percent of its dry weight.
γ (gm/cm^3) = Unit weight of the solid constituents of a sediment.
γ_w (gm/cm^3) = Unit weight of water.
Δ = Increment.
ϕ = Angle of shearing resistance; tan ϕ = ratio between increase of unit pressure and corresponding increase of shearing resistance, if pressure in pore water remains unchanged.
ϕ_{cq} = Consolidated-quick value of angle of shearing resistance; tan ϕ_{cq} = ratio between increase of shearing resistance and increase in unit pressure, if the shearing stress is increased at constant water content.
ϕ_t = True angle of shearing resistance = angle ϕ_t in Hvorslev's equation 12.

REFERENCES

1. Atterberg, A., 1913, Die Konsistenz und Bindigkeit der Böden: Int. Mitt. Bod., vol. II.
2. Bagnold, R. A., 1941, The Physics of Blown Sand and Desert Dunes: London, 1941.
3. Bjerrum, L., 1954, Geotechnical properties of Norwegian marine clays: Geotechnique, vol. IV, p. 49–69.
4. Brown, F. S., 1940, Foundation investigation for the Franklin Falls Dam: Jour. Boston Soc. C. E., vol. 28, p. 126–143.
5. Buisman, A. S. K., 1936, Results of long duration settlement tests: Proc. First. Int. Conf. on Soil Mech. and Found. Eng., Cambridge, Mass., vol. I, p. 103–106.
6. Cadling, L., and Odenstad, S., 1950, The vane borer: Proc. Royal Swedish Geotechnical Institute, No. 2, Stockholm.
7. Cary, A. S., 1951, Origin and significance of openwork gravel: Trans. Am. Soc. C. E., vol. 116, p. 1296–1318.
8. Casagrande, A., 1932, The Structure of Clay and its Importance in Foundation Engineering: Jour. Boston Soc. C. E., vol. 19.
9. ——, 1936, The determination of the pre-consolidation load and its practical significance: Proc. First Int. Conf. on Soil Mechanics, vol. III, p. 60–64.
10. ——, 1948, Classification and identification of soils: Trans. Am. Soc. C. E., vol. 113, p. 901–930.
11. ——, 1949 *a*, Notes on Swelling Characteristics of Clay-Shales: Harvard University, Pierce Hall, Cambridge, Mass.
12. ——, 1949 *b*, Soil mechanics in the design and construction of the Logan airport: Jour. Boston Soc. C. E., vol. 36.
13. Casagrande, A., and Fadum, R. E., 1942, Application of soil mechanics in designing building foundations: Proc. Am. Soc. C. E., vol. 68, no. 9, p. 1487–1520.
14. Casagrande, A., and Wilson, S. D., 1951, Effect of rate of loading on the strength of clays and shales at constant water content: Geotechnique, vol. 2, p. 251–263.
15. Casagrande, L., 1948, Structures produced in clays by electric potentials and their relation to natural structures: Nature, vol. 160, p. 470.
16. Cassel, F. L., 1948, Slips in fissured clay: Proc. Sec. Intern. Conf. Soil Mech., Rotterdam, vol. II, p. 46–50.
17. Close, V., and McCormick, E., 1922, Where the mountains walk: Nat. Geogr. Mag., vol. XLI, p. 442–464.
18. Cooling, L. F., and Ward, W. H., 1953, Measurement of loads and strains in earth supporting structures: Proc. Third Int. Conf. Soil Mech. and Found. Eng., Zürich, vol. II, p. 162–167.
19. Crandell, D. R., and Waldron, H. H., 1954, Abstract of papers submitted for the April meeting of the Geol. Soc. of Amer., Seattle, Wash.
20. Croce, A., 1948, Secondary time effect in the compression of unconsolidated sediments of volcanic origin: Proc. Sec. Int. Conf. Soil Mech., Rotterdam, vol. I, p. 166–169.
21. Dawson, D. D., Jr., 1954, Mud problems in drilling deep wells: The Petroleum Engineer, p. B-25–34.
22. Dawson, R. F., 1953, Movement of small houses erected on expansive clay soil: Proc. Third Int. Conf. Soil Mech., Zürich, vol. I, p. 346–350.
23. Delarue, J., Mariotti, M. V., and Berthier, R. L., 1953, Caractéristiques mécaniques des argiles préconsolidés Marocaines: Proc. Third Int. Conf. Soil Mech., Zürich, vol. I, p. 351–357.
24. Drouhin, G., Cervieux, F., and Gautier, M., 1953, Existence et formation de canalicules dans certains sols argileux: Proc. Third Int. Conf. Soil Mech., Zürich, vol. I, p. 225–231.

25. Esmiol, E. E., 1953, The diversity of impervious soils used in Bureau of Reclamation earth dams: Proc. Third Int. Conf. Soil Mech., vol. II, p. 224–229.
26. Florentin, G., and L'Heriteau, G., 1948, Remarques sur quelques Marnes Fortement Préconsolidées: Géotechnique, vol. 1, p. 59–65.
27. Forbes, H., 1951, The geochemistry of earthwork: Trans. Am. Soc. C. E., vol. 116, p. 637–670.
28. Geuze, E. C. W. A., 1948, Critical density of some Dutch sands: Proc. Sec. Int. Conf. Soil Mech. and Found. Eng., Rotterdam, vol. III, p. 125–130.
29. Grim, R. E., 1948, Some fundamental factors influencing the properties of soil materials: Proc. Sec. Int. Conf. Soil Mech., Rotterdam, vol. III, p. 8–12.
30. ——, 1949, Mineralogical composition in relation to the properties of certain soils: Géotechnique, vol. 1, p. 139–147.
31. Haefeli, R., 1944, Zur Erd- und Kriechdruck Theorie: Schweiz. Bauztg, vol. 124.
32. Haefeli, R., Schaerer, Ch., and Amberg, G., 1953, The behavior under the influence of soil creep pressure of the concrete bridge built at Klosters by the Rhaetian Railway Com pany, Switzerland: Proc. Third. Int. Conf. Soil Mech., Zürich, vol. II, p. 175–179.
33. Hanna, W. S., 1948, Soil mechanics and soil conditions in Egypt: Geotechnique, vol. 1, p. 90–97.
34. Highway Research Board, 1953, Soil temperature and ground freezing: Nat. Acad. Sc., Nat. Res. Council Bull. 71, Washington, D. C.
35. Hogentogler, C. A., 1937, Engineering Properties of Soil: McGraw-Hill, New York.
36. Hollingworth, S. E., Taylor, J. H., and Kellaway, G. A., 1944, Large-scale superficial structures in the Northampton Ironstone Field: Quart. Jour. Geol. Soc. London, vol. C, p. 1–44.
37. Holmsen, P., 1953, Landslips in Norwegian quick-clays: Géotechnique, vol. 3, p. 187–194.
38. Holtz, W. G., and Gibbs, H. J., 1951; Consolidation and related properties of loess soils: Am. Soc. Test Mat. Special Technical Publication No. 126, "Symposium on Consolidation Testing of Soils."
39. ————, 1952, Engineering properties of expansive clays: Am. Soc. C. E., Convention Preprint 57 (Centennial Convocation).
40. Hvorslev, J., 1936, Conditions of failure for remolded cohesion soils: Proc. First Int. Conf. on Soil Mech. and Found. Eng., Cambridge, Mass., vol. III, p. 51–53.
41. ——, 1948, Subsurface exploration and sampling of soil for civil engineering purposes: Waterways Experiment Station, Vicksburg, Miss., Nov. 1948.
42. Jakobson, B., 1953, Origin cohesion of clay: Proc. Third Int. Conf. Soil Mech., Zürich, vol. 1, p. 35–37.
43. Jennings, J. E., 1953, The heaving of buildings on desiccated clay: Proc. Third Int. Conf. Soil Mech., Zürich, vol. I, p. 390–396.
44. Johnson, G. E., 1953, Stabilization of soil by silt injection method: Proc. Am. Soc. C. E., vol. 79, Sep. No. 323.
45. Koppejan, A. W., 1948, A formula combining the Terzaghi load-compression relationship and the Buisman secular time effect: Proc. Sec. Int. Conf. on Soil Mech. and Found. Eng., Rotterdam, vol. III, p. 32–37.
46. Koppejan, A. W., Van Wamelen, B. M., and Weinberg, L. J. H., 1948, Coastal flow slides in the Dutch province of Zeeland: Proc. Sec. Int. Conf. Soil Mech., Rotterdam, vol. V, p. 89–96.
47. Krumbein, W. C., 1950, Geological aspects of beach engineering: Geol. Soc. Am., Application of Geology to Engineering Practice (Berkey Volume), p. 195–223.
48. Legget, R. F., and Bartley, M. W., 1953, An engineering study of glacial deposits at Steep Rock Lake, Ontario: Econ. Geol., vol. 48, p. 513–540.
49. Leighton, M. M., and Willman, H. B., 1950, Loess formations of the Mississippi Valley: Jour. Geology, vol. 58, p. 599–623.
50. Lyman, A. K. B., 1942, Compaction of cohesionless foundation soils by explosives: Trans. Am. Soc. C. E., Paper no. 2160, vol. 68, p. 1330–1348.
51. Mackin, J. H., 1941, A geologic interpretation of the failure of the Cedar Reservoir, Washington: University of Washington, Engin. Exper. Sta., Bull. No. 107.
52. Marsal, R. J., and Mazari Marcos, M., 1952, Arcillas del Valle de Mexico: Ingen. Civ. Assoc. Mexico D. F., Ingeniera Experimental, series B, no. 12, p. 1–6.
53. Matthews, W. H., 1944, Glacial lakes and ice retreat in South Central British Columbia: Trans. Roy. Soc. Canada, 3rd series, vol. 38, sect. IV, p. 39–57.
54. Moran, D. E., 1936, Discussion: Section B, Proc. First Int. Conf. Soil Mech., Cambridge, Mass., vol. III, p. 24–25.
55. Olsson, R. G., 1953, Approximate solution of the progress of consolidation in a sediment: Proc. Third Int. Conf. on Soil Mech. and Found. Eng., Zürich, vol. I, p. 38–42.

56. Palit, R. M., 1953, Determination of swelling pressure of black cotton soil: Proc. Third Int. Conf. Soil Mech., Zürich, vol. I, p. 170–172.
57. Peck, O. K., and Peck, R. B., 1948. Settlement of foundation due to saturation of loess subsoil: Proc. Sec. Int. Conf. Soil Mech. and Found. Eng., vol. IV, p. 4–5.
58. Peck, R. B., Ireland, H. O., and Fry, T. S., 1951, Studies of soil characteristics—the earth flows of St. Thuribe, Quebec: Dept. Civ. Eng. Univ. of Illinois, Soil Mech. Series no. 1.
59. Peck, R. B., and Reed, W. C., 1954, Engineering properties of Chicago subsoils: Univ. of Illinois. Engin. Exper. Sta., Bull. no. 423.
60. Peterson, R., 1952. Studies—Bearpaw shale at damsite in Saskatchewan: Am. Soc. C. E., Centennial Convocation, Chicago, Ill., Preprint no. 52.
61. Pichler, E., 1948, Regional study of the soils from São Paulo, Brazil: Proc. Soc. Int. Conf. Soil Mech., Rotterdam, vol. III, p. 222–223.
62. Pollack, V., 1917, Uber Rutschungen im Glazialen und die Notwendigkeit einer Klassifikation loser Massen: Jb. b. Geol. Reichsanst., bd. 67, h. 3 and 4.
63. Polonsky, M., 1948, A problem of foundation on loamy soil in subtropical countries: Proc. Sec. Int. Conf. Soil Mech., Rotterdam, vol. IV, p. 78–81.
64. Redlich, K., Terzaghi, K., and Kampe, R., 1929, Ingenieurgeologie: Julius Springer, Wien.
65. Rominger, J. F., and Rutledge, P. C., 1952, Use of soil mechanics data in correlation and interpretation of Lake Agassiz sediments: Jour. Geology, vol. 60, p. 160–180.
66. Rosenqvist, I. Th., 1953, Considerations on the sensitivity of Norwegian quick-clays: Géotechnique, vol. 3, p. 195–200.
67. Salas, J. A. J., and Serratosa, J. M., 1953, Compressibility of clays: Proc. Third Int. Conf. Soil Mech., Zürich, vol. I, p. 192–198.
68. Scheidig, A., 1934, Der Löss: Theodor Steinkopf, Dresden and Leipzig.
69. Shepps, V. C., 1953, Correlation of the tills of northeastern Ohio by size analysis: Sedimentary Petrology, vol. 23, p. 34–48.
70. Shockley, W. G., and Garber, P. K., 1953, Correlation of some physical properties of sand: Proc. Third Int. Conf. Soil Mech., Zürich, vol. I, p. 203–206.
71. Sigafoos, R. S., and Hopkins, D. M., 1952, Soil instability on slopes in regions of perennially-frozen ground; *also* Soil instability in relation to highway construction in tundra regions in Alaska: *in* Highway Research Board, Special Report No. 2, Frost Action in Soils, a Symposium, Nat. Res. Counc. Publ. no. 213.
72. Simpson, W. E., 1934, Foundation experiences with clay in Texas: Civil Engin., vol. 4, p. 581–584.
73. Skempton, A. W., 1944, Notes on the compressibility of clays: Quart. Jour. Geol. Soc. London, vol. C, p. 119–135 (July 28, 1944).
74. ——, 1948 *a*, The rate of softening in stiff fissured clays, with special reference to London clay: Sec. Int. Conf. Soil Mech., Rotterdam, vol. 2, p. 50–53.
75. ——, 1948 *b*, A study of the geo-technical properties of some post-glacial clays: Geotechnique, vol. 1, p. 7–22.
76. ——, 1953, The colloidal activity of clays: Proc. Third Int. Conf. Soil Mech., Zürich, vol. I, p. 57–61.
77. ——, 1954, A foundation failure due to clay shrinkage caused by poplar trees: Proc. Inst. C. E. (London), Paper no. 5959, p. 66–86.
78. Skempton, A. W., and Henkel, D. J., 1953, The post-glacial clays of the Thames estuary at Tilbury and Shellhaven: Proc. Third Int. Conf. Soil Mech., Zürich, vol. I, p. 302–308.
79. Skempton, A. W., and Northey, R. D., 1952, The sensitivity of clays: Geotechnique, vol. III, p. 30–53.
80. Smith, K. S., and Redlinger, J. F., 1953, Soil properties of Fort Union clay shale: Proc. Third Int. Conf. Soil Mech., Zürich, vol. I, p. 62–66.
81. Straub, G., 1935, Some observations of sorting of river sediments: Trans. Am. Geophys. Union, 16th Meeting.
82. Terzaghi, K., 1925, Erdbaumechanik auf Bodenphysikalischer Grundlage: Franz Deuticke, Wien.
83. ——, 1926, The mechanics of adsorption and of the swelling of gels: Fourth Colloid Symposium Monograph, Chem. Catal. Co., Inc., New York.
84. ——, 1929, Effect of minor geologic details on the safety of dams: Am. Inst. Min. Met. Eng., Technical Publ. no. 215, p. 31–44.
85. ——, 1931, Underground erosion and the Corpus Christi dam failure: Eng. News-Record, vol. 107, p. 90–92.
86. ——, 1935, The actual factor of safety in foundations: The Structural Engineer, vol. 13.
87. ——, 1936*a*, Stability of slopes on natural clay: Proc. First Int. Conf. Soil Mech. and Found. Eng., Cambridge, Mass., vol. I, p. 161–165.

88. ——, 1936*b*, Discussion: Proc. First Int. Conf. of Soil Mech. and Found. Eng., Cambridge, Mass., vol. III, p. 152–155.
89. ——, 1941, Undisturbed clay samples and undisturbed clays: Journ. Boston Soc. C. E., vol. 28, p. 211–231.
90. ——, 1942, Shield tunnels of the Chicago subway: Journ. Boston Soc. C. E., vol. 29, p. 163–210.
91. ——, 1943*a*, Theoretical Soil Mechanics: John Wiley and Sons, New York.
92. ——, 1943*b*, Liner plate tunnels on the Chicago (Ill.) subway: Paper No. 2200, Trans. Am. Soc. C. E., vol. 69, p. 970–1007.
93. ——, 1947, Shear characteristics of quicksands and soft clay: Proc. Second Texas Conf. Soil Mech.
94. ——, 1949, Soil Moisture and Capillary Phenomena in Soils: Physics of the Earth. IX. Hydrology. Edit. by O. Meinzer, p. 331–363. First Edition. Dover Publications, Inc., New York.
95. ——, 1950*a*, Geologic Aspects of Soft Ground Tunneling. Applied Sedimentation: Edited by P. D. Trask, John Wiley and Sons, New York.
96. ——, 1950*b*. Mechanism of landslides: Geol. Soc. Am., Application of Geology to Engineering Practice (Berkey Volume), p. 83–123.
97. ——, 1952, Permafrost: Jour. Boston Soc. C. E., vol. 39, p. 1–50.
98. ——, 1953, Fifty years of Subsoil exploration: Proc. Third Int. Conf. Soil Mech., Zürich, vol. III, p. 227–237.
99. Terzaghi, K., and Peck, R. B., 1948, Soil Mechanics in Engineering Practice: John Wiley and Sons, New York.
100. Tschebotarioff, G. P., 1953, A case of structural damages sustained by one-storey high houses founded on swelling clays: Proc. Third Int. Conf. Soil Mech., Zürich, vol. I, p. 473–476.
101. Turnbull, W. J., 1948, Utility of loess as a construction material: Proc. Sec. Int. Conf. Soil Mech., Rotterdam, vol. 5, p. 97–103.
102. U. S. Army, Corps of Engineers, 1944, Conference on Control of Underseepage, Cincinnati, Ohio: U. S. Waterways Exper. Sta., Vicksburg, Miss.
103. U. S. Bureau of Reclamation, 1949, Petrographic Characteristics of Loess—Trenton Dam —Frenchman—Cambridge Division—Missouri River Basin Project: Petr. Lab. Proj. No. Pet. -93, Denver, July 19, 1949.
104. Vargas, M., 1953, Some engineering properties of residual clay soils occurring in southern Brazil: Proc. Third Int. Conf. Soil Mech., Zürich, vol. I, p. 67–71.
105. Ward, W. H., 1953, Soil movement and weather: Proc. Third Int. Conf. Soil Mech., Zürich, vol. I, p. 477–482.
106. Woolterton, F. L. D., 1951, Movements in the desiccated alkaline soils of Burma: Trans. Am. Soc. C. E., vol. 116, p. 433–479.

UNDERSEEPAGE, MISSISSIPPI RIVER LEVEES, ST. LOUIS DISTRICT

By Charles I. Mansur,[1] M. ASCE, and Robert I. Kaufman,[2] J. M. ASCE

January 1956, Journal of the Soil Mechanics
and Foundations Division
Transactions, Vol. 122, 1957, p. 985
1958 Thomas Fitch Rowland Prize

Synopsis

Contained herein is a description of the underseepage problem along the Mississippi River levees between Alton (Ill.) and Gale (Ill.) and of the procedures used to design seepage-control measures. The paper includes: (1) A description of the investigations conducted; (2) the methods used to control the seepage; (3) examples of the design of the control measures; and (4) a description of the construction procedures used to install the control measures.

Introduction

Soil conditions in the alluvial valley of the Mississippi River are such that seepage beneath the levees along the river has always posed a serious problem during high waters. In fact, some crevasses in the levees along the Mississippi River have occurred as a result of sand boils and subsurface piping.

Prior to 1938, many of the levees along the Mississippi River in the St. Louis (Mo.) District, Corps of Engineers, United States Department of the Army, were below project-flood grade, and only relatively low heads could be developed against them. Although some underseepage had been experienced along these levees, only minor problems had occurred, mainly because of the low height of the levees. By authority of the Flood Control Acts of 1936 and 1938, the St. Louis District initiated a program to complete the levee systems and to raise the grade of existing levees so that they could withstand the project flood. Approximately 90% of this program was accomplished by 1955. A plan of the present levee systems along the Mississippi River from Alton (Ill.) to Gale (Ill.) is shown in Fig. 1. The increased height of the levees, however, has

Note.—Published, essentially as printed here, in January, 1956, as *Proceedings Paper 864*. Positions and titles given are those in effect when the paper was approved for publication in *Transactions*.

[1] Asst. Chief, Embankment and Foundation Branch, Soils Div., Waterways Experiment Station, Vicksburg, Miss.

[2] Chief, Design and Analytical Section, Embankment and Foundation Branch, Soils Div., Waterways Experiment Station, Vicksburg, Miss.

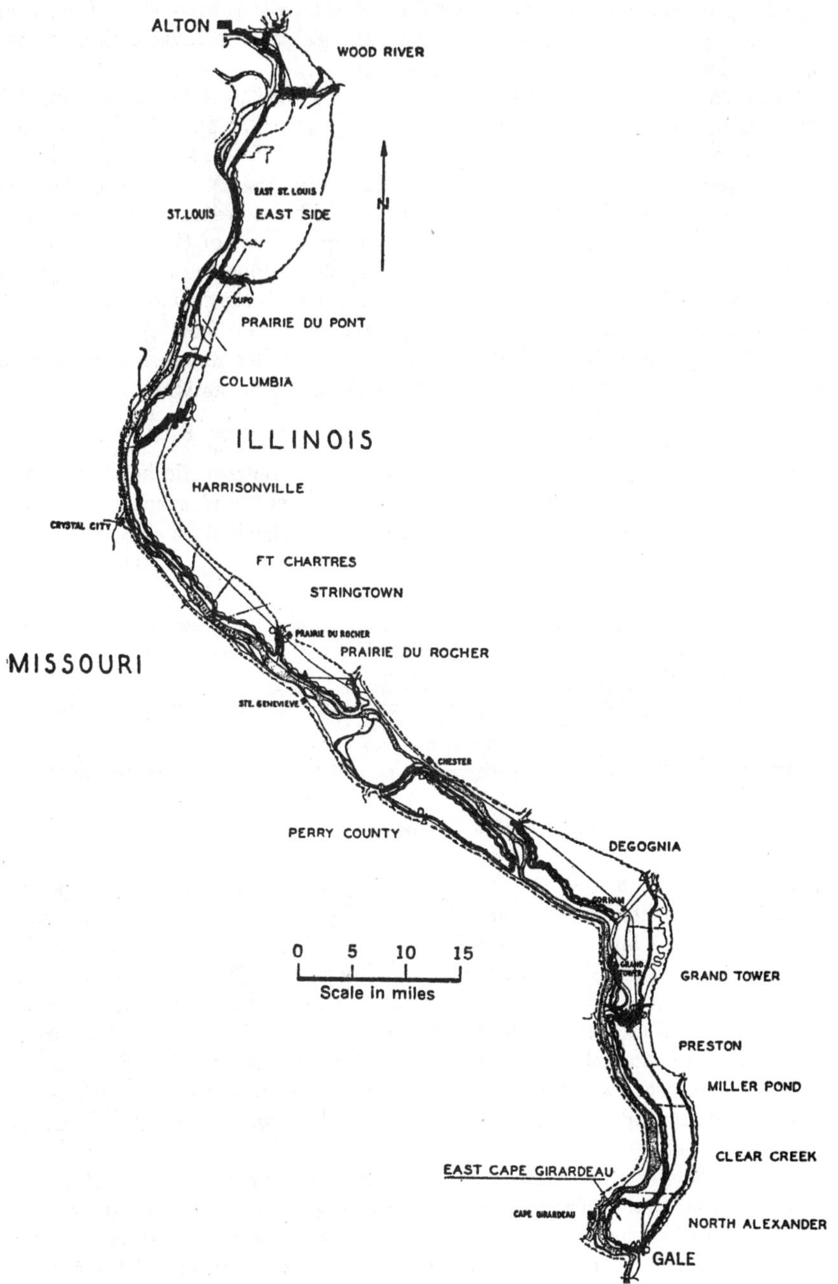

Fig. 1.—Plan of Levees Along the Mississippi River—Alton (Ill.) to Gale (Ill.)

created a problem of underseepage which did not exist previously. Inasmuch as considerable head can now be developed against the levees, their safety against underseepage has become questionable.

Although it was realized that underseepage would be a problem along the new levee system, the control of underseepage was not specifically provided for in the design of the new levees. Following the 1951 and 1952 floods in the Omaha (Nebr.) and Kansas City (Mo.) Districts, the Office of the Chief of Engineers and the St. Louis District requested that the Waterways Experiment Station, Corps of Engineers, United States Department of the Army, make a comprehensive investigation of the seepage problem along the new levee system and design the necessary control measures. This investigation was initiated in October, 1952, and was essentially completed by June, 1955. Approximately 70% of the control measures indicated by these studies have been completed (as of 1955), and considerable additional work is under contract.

Development of Underseepage and Sand Boils

Whenever a levee is subjected to a differential hydrostatic head of water, as a result of river stages being higher than the adjacent land, seepage enters the pervious substratum beneath the levee through the bed of the river, and

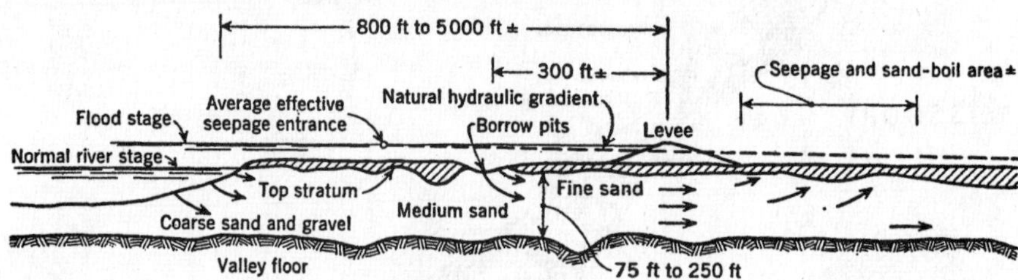

Fig. 2.—Cross Section of Levee and Alluvial Valley

through riverside borrow pits and/or through the riverside top stratum. This seepage creates an artesian head and hydraulic gradient in the sand stratum under the levee. The gradient, in turn, causes a flow of seepage beneath and landward of the levee (Fig. 2). The seepage from the previous substratum which emerges at or landward of the levee toe is generally termed underseepage. Concentrated underseepage which carries sand up to the surface through an open channel in the top stratum is called a sand boil. Piping is the active erosion of sand or other soil from below the ground surface which occurs as a result of substratum pressure and the concentration of seepage in localized channels.

If the hydrostatic pressure in the pervious substratum landward of a levee becomes greater than the submerged weight of the top stratum, the excess pressure will cause heaving of the top blanket, or will cause it to rupture at one or more weak spots with a resulting concentration of seepage flow in the form of sand boils. Where the foundation and top strata are heterogeneous in character, as is usually the case, seepage tends to appear at localized spots instead of causing the entire top stratum to heave or become "quick."

The hydraulic gradient required to cause heaving is called "the critical hydraulic gradient." This gradient is the ratio of the submerged unit weight of the soil to the unit weight of water and is usually approximately 0.85. Any tendency for the hydraulic gradient to increase above the critical gradient causes additional sand boils or increased percolation. In nature, high exit gradients and concentrations of seepage are usually found along the landside toe of the levee, at thin or weak spots in the top stratum, and adjacent to clay-filled swales or channels.

Where seepage is concentrated to the extent that it creates turbulent flow, the flow causes erosion in the top stratum and causes the development of a channel in the underlying silts and fine sands which frequently exist immediately below the base of the top stratum. As the channel increases in size or length, or in both, the concentration of flow into the channel becomes progressively greater, as does the tendency for erosion to occur beneath the levee. This is especially true if the top stratum is cohesive and the underlying soils are susceptible to erosion. Shrinkage cracks, root holes, and holes dug by man or by animals form channels in the top stratum susceptible to the development of localized flows and attendant piping. Not only is underseepage a hazard from the standpoint of underground erosion, but it also may saturate the landside slope of a levee and reduce its stability to such an extent that the slope may slough with attendant dangers to the levee.

Geology of Alluvial Valley

Detailed geological studies of several sites along the levees in the Lower Mississippi Valley where underseepage had been a problem during previous high waters show a definite correlation between the distribution of alluvial deposits of sand, silt, and clay and the location and occurrence of underseepage and sand boils. Thus, before it was possible to make an intelligent investigation of underseepage and to design control measures, an understanding of the alluvial fill in the Upper Mississippi Valley from Alton to Gale was necessary. The geological studies made for this project included field reconnaissances, studies of aerial mosaics, and a review of boring logs.

A few miles south of St. Louis, the Mississippi River cuts across an eastward extension of the Ozark Plateaus and, for a distance of approximately 100 miles, it flows in a narrow 3-miles-to-10-miles-wide, rock-walled, alluviated valley incised in ancient bedrock composed predominately of limestone. The river emerges from the plateaus through a narrow gorge called Thebes Gap into the broad Lower Mississippi Valley. The depth of the alluvial deposits in the St. Louis District is variable, ranging approximately from 75 ft to 200 ft with an average depth of approximately 125 ft.

In general, the river is adjacent to the bluffs on the Missouri side, with the floodplain east of the river in the State of Illinois. Tributary streams originate in the bluffs on the east, flow through the floodplain, and empty into the Mississippi River at frequent intervals between Alton and Gale. The levee system on the left bank in this area consists of levees which originate at the bluff, run parallel to the tributary streams toward the Mississippi River, and then along the Mississippi River to the next tributary stream (Fig. 1). Thus, the alluvial valley is subdivided into small areas, each with its own encircling levees.

Melting of the glaciers during late or postglacial times caused the entrenched valley to become filled with a series of sandy gravels, sands, silts, and clays, which can be grouped into two broad units: (1) A sand and gravel substratum and (2) a fine-grained top stratum.

The Pervious Substratum.—During the gradual rise of sea level which accompanied the final retreat of glacial ice, the Mississippi River became an overloaded, braided stream in which large quantities of gravel-bearing sands were deposited. As the sea level continued to rise and the deposits on the entrenched valley floor continued to thicken, stream slopes were progressively reduced and both the quantity and grain size of the material transported by the river decreased. The gravel-bearing sands were succeeded by coarse sands grading upward into progressively finer materials and terminating in deposits of very fine sand. This upward gradation from coarse to fine materials reflects the gradual adjustment between the transporting capacity of the river and the available load. Fine to very fine sands were not deposited until the sea level had essentially reached its present stand and the river had begun to change from a braided to a meandering stream. The substratum sands are quite pervious and have a high seepage-carrying capacity, as will be subsequently described.

The Top Stratum.—The predominantly fine-grained, top-stratum material was deposited by the Mississippi River as it meandered across the floodplain. Meander-belt deposits are characterized by marked lateral and vertical discontinuities and by wide variations in grain size and permeability. They are, therefore, of primary importance in the localization of underseepage and the design of control measures. Meander-belt deposits may be grouped as: (1) Point bar, (2) channel fill, (3) natural levee, and (4) backswamp. Such deposits can usually be identified in the field and on aerial mosaics.

Point-Bar Deposits.—As the radii of meander loops increase by erosion, deposition occurs on the inside bank where low sandy ridges are built up with intervening, elongated depressions which usually become filled with fine-grained deposits. Ridges and swales formed in this manner are known as point bars. Long sandy ridges, separated by clay-filled and silt-filled depressions which approximately parallel the margins of former river courses, are also characteristic of shifting river valleys and are especially common in the St. Louis District. These features appear to have been formed in migrating channels rather than on points and might well be referred to as channel bars and slough fillings. The upper part of these ridge and swale deposits is usually covered with clayey silts and silty clays which are deposited during gradual migration of the river channel from its former banks. Normally the slough fillings are approximately from 10 ft to 20 ft deep.

Channel-Fill Deposits.—When the river abandons its former course as a result of a cutoff, the central and lower parts of the old cutoff channel usually become filled with silts and clays which are relatively impermeable. Such deposits in the area studied may be from 40 ft to 80 ft deep.

Natural Levees.—When the river overflows its banks, the water spreads out, its velocity is reduced, and some of its sedimentary load is deposited. In this manner, long ridges known as natural levees are formed on the outside of meander loops and along both banks in straight reaches. Natural levee deposits found in the Upper Mississippi Valley appear to be predominantly sandy, approximately 5 ft high, and seldom more than 200 yd wide. Clearly distinct natural levees are relatively rare in the St. Louis District.

Backswamp Deposits.—Low-lying areas on the landside of natural levees are known as backswamps. These areas receive only quiet floodwaters and, as a result, the sediments deposited consist of fairly uniform silts and clays. Backswamp deposits create an almost impervious block to the emergence of subsurface seepage. There do not appear to be many, if any, true backswamp deposits in the St. Louis District; however, some thick clay top-stratum deposits resembling backswamp deposits were found adjacent to the valley walls.

Effects of Alluvial Deposits.—The emergence of seepage landward of a levee is influenced by: (1) Configuration of geological features, such as swale fillings

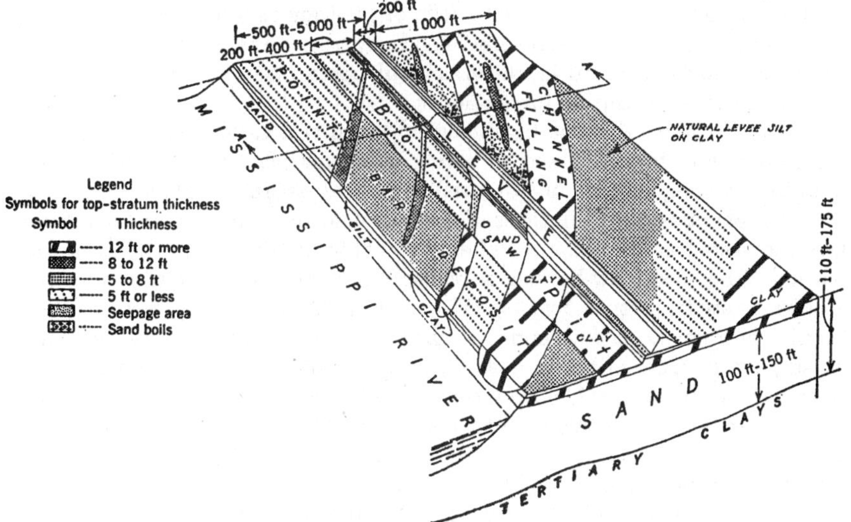

FIG. 3.—SEEPAGE THROUGH POINT–BAR DEPOSITS

and channel fillings, and their alinement relative to the levee; (2) characteristics and thickness of the top stratum; (3) cracks or fissures formed by drying and other natural causes; (4) borrow pits, post holes, seismic shot holes, ditches, and other works of man; and (5) decay of roots, uprooting of trees, animal burrows, crayfish holes, and other organic agencies. The severity of underseepage along a levee is frequently dependent on the configuration of geological features in that area. The greatest concentration of seepage always occurs along the edges of swales and the landside levee toe (Fig. 3).

FIELD INVESTIGATIONS

An extensive field investigation was made to ascertain the need for underseepage control measures and to obtain data on which to base their design. The field investigations consisted of a reconnaissance of all areas adjacent to the levees, the study of aerial mosaics, an intensive boring program, a geophysical investigation to determine the depth of the alluvial valley, and field pumping tests.

Field Reconnaissance.—Field reconnaissances were made along the levees for the purpose of noting features appearing on the ground surface that might affect the underseepage problem. The presence of geological features such as swales and sloughs and their orientation with respect to the levee were noted. The locations and extent of man-made features such as drainage ditches, stock ponds, and borrow pits were ascertained. The layout of new borings, made in the office on the basis of available soils data, levee heights, and a study of aerial mosaics, was checked in the field to insure adequate exploration of those features which would affect the underseepage problem and the design of control measures. The condition and depths of riverside borrow pits also were noted.

Aerial Mosaics and Maps.—Because there were no complete maps of the new levee system available to which the new borings, piezometers, and control measures could be referenced, a new folio of photomaps was prepared, showing all the Mississippi River levees and flood-control works in the area. These new maps will be used during flood-control operations and to record underseepage observations during future floods.

Boring Program.—At the beginning of the investigation the existing soils information was reviewed and a boring program planned to delineate soil conditions more precisely in potentially critical areas. Shallow borings to define the character and thickness of the top stratum along the landside toe of the levees were located on approximately 200-ft to 250-ft centers and at critical locations. Some shallow borings were also made farther landward of the levee in order to study pertinent geological features. A few borings were made in riverside borrow pits to determine the character and thickness of the remaining top stratum in these pits. In general, blanket borings were made with hand augers whenever water-table conditions would permit. Deeper borings (60 ft to 80 ft) were made on 1,000-ft to 1,500-ft centers to determine the characteristics of the subsurface sands for both the design and installation of relief wells in reaches where it was thought that control measures might be necessary. Some borings were extended to rock to determine the thickness of the pervious aquifer and to check geophysical determinations of the depth to rock.

Most of the deep borings were made using a 2-in., split-spoon sampler and using drilling mud to keep the hole open. Samples were generally taken at approximately 5-ft intervals and at points at which there were major changes in soil strata. A few undisturbed samples of the sand foundation were obtained with a 3-in. Shelby tube and drilling mud.[3] The reason for obtaining the undisturbed samples was to compare the grain size of the sands, as determined from such samples, with that of samples obtained with split-spoon and bailer samplers. Another reason for taking a few undisturbed sand samples was to study the stratification of the sand. A typical soil profile developed from field investigations is shown for a short reach of levee in the Harrisonville (Ill.) Levee District (Fig. 4).

---

[3] "Undisturbed Sand Sampling below the Water Table," *Bulletin No. 35*, Waterways Experiment tation, Vicksburg, Miss., June, 1950, pp. 1–19.

Geophysical Investigations.—Because of the difficulty and cost of ascertaining the depth to rock by means of borings, it was decided to obtain this information by means of geophysical explorations. At the beginning of the investigation the depths to rock were determined by the electrical resistivity method. In analyzing the data obtained, it became apparent that the depth to rock determined by the resistivity method was extremely sensitive to the method of interpretation and did not check very well the depths determined by borings. As a result, this method was abandoned in favor of the seismic method which involved making initial checks with a seismograph at locations where the depth to rock had been determined from deep borings and from excavation for the Chain of Rocks Canal Lock on the Mississippi River near St. Louis. Because

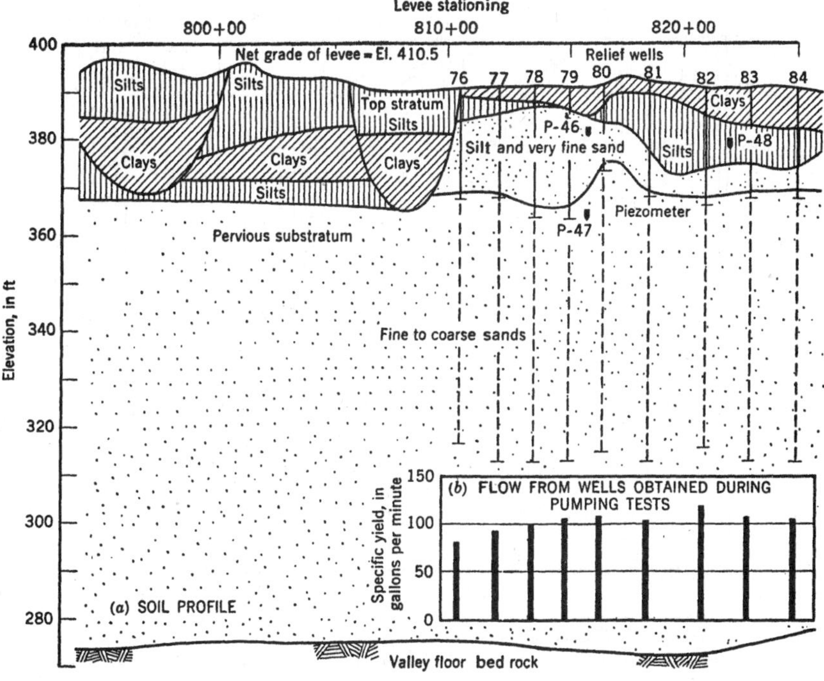

Fig. 4.—Soil Profile, and Relief-Well and Piezometer-Installation Data

good agreement was obtained, the seismic method was used for the remainder of the investigation.

Field Pumping Tests.—The design of relief wells along the levees in the St. Louis District was based on the assumption that the screens for the wells would penetrate the pervious aquifer underlying the levees to a depth which would result in the performance of a well system comparable to the performance obtained from a 50% penetration of a homogeneous, isotropic aquifer. Because the pervious strata underlying the levees along the Mississippi River generally become more pervious with depth, special pumping tests were performed on wells which fully penetrated the pervious aquifer. This was done for the purpose of determining the depth required for the well screens in order to achieve an effective penetration of 50%. The over-all permeability of the sand foundation at the sites of the test wells was also determined from these tests. The horizontal permeability of the various sand strata encountered at the test wells was determined by measuring the flow into the well screen from each sand stratum as previously delineated by a boring.

The field pumping tests consisted of pumping the wells at three different rates of flow, measuring the drawdown in the wells, and observing the drawdown in adjacent piezometers and wells. The flow in the well screen was measured at changes in strata by means of a calibrated, well-flow meter.[4] Head losses

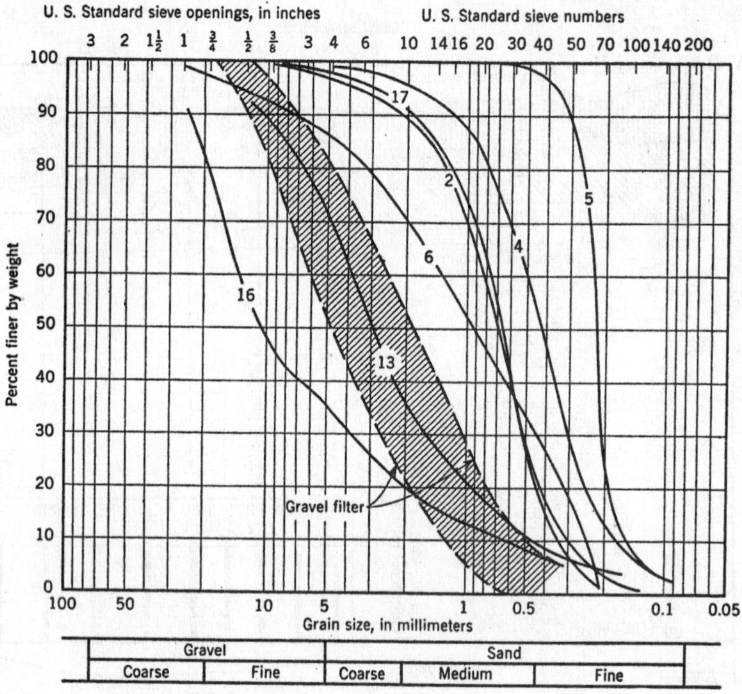

Fig. 5.—Foundation Sands and Filter Gravel–Well FC-105

through the filter and the well screen and inside the well were measured by piezometers installed at the periphery of the gravel filter around the well screen. The over-all horizontal permeability, k_H, of the sand aquifer and of the individual sand strata was computed from the formula for artesian radial flow to a single well.

Grain-size curves of typical sands taken from a boring at one of the test wells (FC-105) together with the gradation of the filter gravel used around the well screen are shown in Fig. 5. The permeabilities of the different sand strata

[4] "Waterways Experiment Station Relief Well Flow Meter," *Miscellaneous Paper 5-83*, Waterways Experiment Station, Vicksburg, Miss., April, 1954.

encountered at well FC-105, as determined from laboratory tests and field pumping tests, are shown in Fig. 6.

The flow of the test wells was essentially artesian because of the existence of upper and lower impervious strata bounding the principal aquifer and because the water in the well was not drawn down below the top of the main sand stratum. As would be expected with artesian flow, drawdown in the test wells

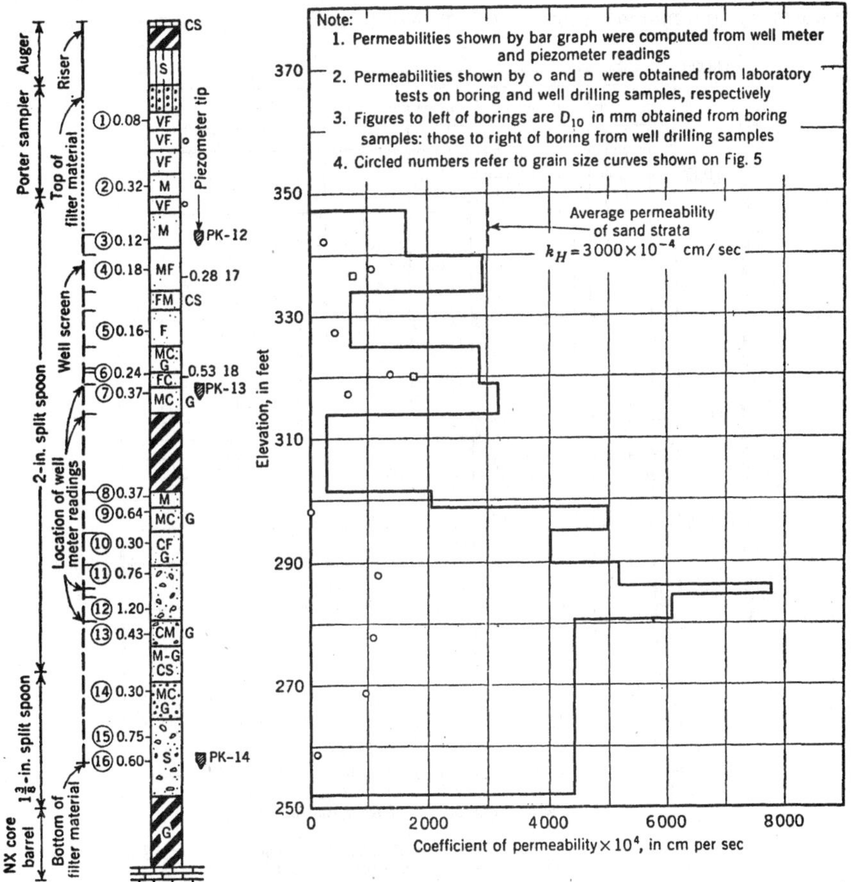

Fig. 6.—Coefficient of Permeability and Effective Grain Size of Individual Sand Strata (Well FC-105, Located at Levee Station 301 + 25, Ft. Chartres Levee District, Approximately 45 River Miles South of St. Louis)

stabilized rapidly at a constant rate of pumping. Approximately 80% stabilization was achieved within 15 min to 30 min of pumping; full stabilization was achieved after 2 hr of pumping. Drawdown curves to the test wells, as plotted on semilog graph paper, resulted in straight lines, another indication of artesian flow to the test wells. The radius of influence of the different test wells ranged approximately from 500 ft to 1,000 ft. The specific yield from the test wells ranged approximately from 150 to 375 gal per min per ft of drawdown at the well. Head loss through the filter and wooden well screen was quite small, amounting to about 0.10 ft to 0.25 ft for a flow through the well screen of 10 gal per min per ft of screen. In analyzing the well test data, it was necessary to consider the hydraulic head losses in the well screen and riser pipe.

The results of the field pumping tests may be summarized as follows:

1. Little agreement was found between permeabilities determined in the laboratory on remolded samples and those obtained from the field pumping tests.

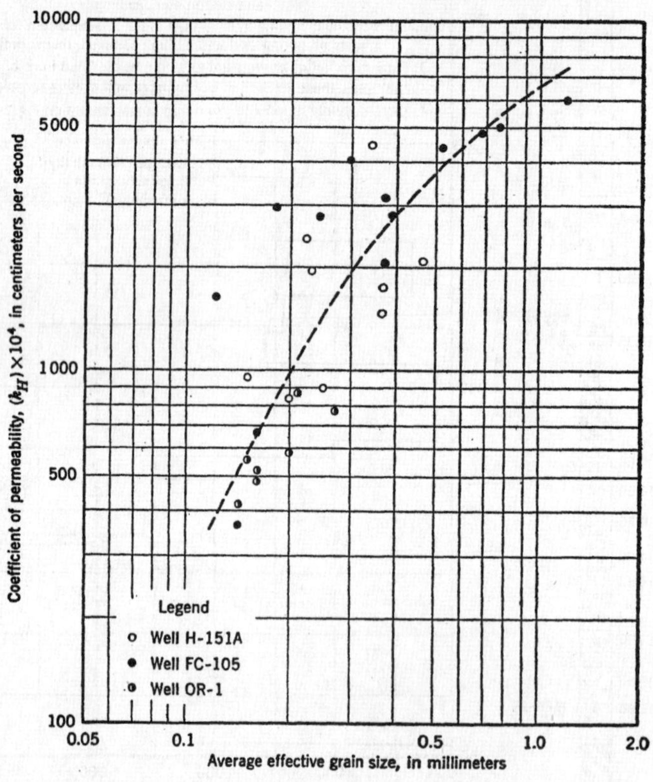

FIG. 7.—COEFFICIENT OF PERMEABILITY (k_H) VERSUS EFFECTIVE GRAIN SIZE (D_{10}) BASED ON FIELD PUMPING TESTS

There is no reason why the permeabilities should have agreed considering that the aquifer was stratified and that lenses of coarse sand and fine gravel existed. Generally, the field permeabilities for any given strata exceeded by from 2 to 4 times the permeability as normally determined in the laboratory.

2. A relationship between the effective grain size, D_{10}, and the coefficient of permeability, k_H, as determined from flow from individual sand strata is shown in Fig. 7. The scatter of points can probably be attributed to variation in uniformity of the grain-size curves and to the fact that the values of D_{10} used in the plot are based on the average of relatively few values of D_{10} for

each sand stratum tested. In the absence of pumping-test data, the curve shown in Fig. 7 may be of use to engineers concerned with seepage, water supply, or irrigation problems in the Mississippi River Valley.

3. The permeability of the principal seepage-carrying stratum at the test well sites ranged from 1,500 $\times 10^{-4}$ cm per sec to 3,000 $\times 10^{-4}$ cm per sec. The permeabilities of the individual sand strata varied from 200 $\times 10^{-4}$ cm per sec to 6,000 $\times 10^{-4}$ cm per sec. Pumping tests on relief wells subsequently installed throughout the St. Louis District indicated that the permeability of the principal seepage-carrying sand stratum generally ranges from 1,000 $\times 10^{-4}$ cm per sec to 2,000 $\times 10^{-4}$ cm per sec.

4. In order to achieve an effective penetration of 50% of the principal water-carrying stratum, the well screen should penetrate approximately 60% of the depth of the principal sand aquifer.

Laboratory Tests

Laboratory studies included visual classification of all samples, grain-size determinations and permeability tests on numerous representative samples, and filter tests on the gravel filter used around the well screens.

Classification.—Samples were classified both visually and, when in doubt, by means of mechanical analyses. The classification used was based on the Unified Soil Classification System.[5]

Grain-Size Determinations.—Mechanical analyses were made of any top-stratum samples of doubtful classification and also on numerous sand samples obtained from the foundation for the purpose of determining the D_{10}-size and D_{85}-size. These values were used in delineating the principal seepage-carrying stratum (sands with values of $D_{10} > 0.15$ mm) and in determining those strata in which well screens should not be set because of the possibility of the foundation sands migrating into and through the filter gravel Although tests showed that the filter would satisfactorily drain sands with values of $D_{85} \geqq 0.15$ mm, well screens, in general, were not set in sand with values of $D_{85} < 0.25$ mm.

Permeability Tests.—These tests were conducted on a number of remolded samples of sand. The coefficients of permeability as determined in the laboratory were adjusted to a temperature of 20° C and to the estimated natural void ratio of the sands. The natural void ratio of the sands was estimated from the D_{50}-value of the sand.[6] Coefficients of permeability, as determined from laboratory tests on split-spoon samples of the alluvial sands, ranged from about 100 $\times 10^{-4}$ cm per sec to 2,000 $\times 10^{-4}$ cm per sec.

In connection with the permeability tests made on samples obtained with split-spoon samplers in mudded holes, it is emphasized that such samples are contaminated frequently by drilling mud. Therefore, it was necessary to wash all such samples thoroughly before performing any permeability tests. Some contamination of the samples taken with a Shelby tube in a mudded hole was also noted.

[5] "Unified Soil Classification System," *Technical Memorandum No. 3-357*, Waterways Experiment Station, Vicksburg, Miss., March, 1953.

[6] "Summary Report of Soil Studies," *Potamology Investigation Report 12-2*, Waterways Experiment Station, Vicksburg, Miss., October, 1952.

Filter Tests.—Several laboratory tests were also made to check the ability of the filter gravel specified for relief wells to protect foundation sands against piping into the well. The filter sample tested was on the coarse side of the gradation band specified. The thickness of filter tested was 6 in., the minimum required in the field.

Three relatively uniform sands, typical of the finer foundation sands encountered in the St. Louis District and having D_{85}-values of 0.20 mm, 0.15 mm, and 0.12 mm, respectively, were used in the tests. The tests proved that the filter would safely drain all the samples tested at gradients of less than 8. The maximum gradient existing at the interface between the sand foundation and filter during either pumping tests or flood periods when wells are in operation will probably not exceed 1. Therefore, it is concluded that if the filter material is satisfactorily placed and has a minimum thickness of 6 in. it should protect any sand with a D_{85}-value coarser than 0.15 mm.

Methods of Underseepage Control

There are several methods which may be used for controlling seepage beneath levees. The choice depends on a number of factors, including the character of the foundation, cost, permanency, availability of right of way, maintenance, and disposal of seepage water. Devices that may be used to control underseepage are (1) landside berms, (2) relief wells, (3) impervious riverside blankets, (4) sublevees, (5) cutoffs, (6) drainage blankets, and (7) drainage trenches. Only the first two foregoing methods were considered practicable for the control of underseepage along the levees in the St. Louis District.

Landside Berms.—Landside berms have been used extensively for the control of underseepage. A berm serves to strengthen the top stratum, increase the base width of the levee, reinforce the landward slope of the levee against sloughing during high water, and move landward the emerging seepage or sand boils and piping from the levee proper. Landside berms were used along the levees in the St. Louis project wherever they were the most economical and where the top stratum landward of a levee was very thin or nonexistent and relief wells would have had to be installed excessively close together. Landside berms were also used where a levee was relatively low, but it appeared that certain precautions against underseepage and possible sloughing of the landside slope were desirable.

Relief Wells.—One method of controlling underseepage is to tap the underlying pervious strata with a series of relief wells. These wells will provide pressure relief and controlled seepage outlets which offer little resistance to flow and, at the same time, will prevent erosion of the foundation soils. This method has been used successfully by the Corps at a number of dams and along numerous levees underlain by a pervious foundation. Relief wells offer an advantage as compared to gravel toes, pervious blankets, and other surface measures where the foundation consists of stratified deposits of pervious materials in that they penetrate the more pervious strata in which pressure relief is necessary. With proper spacing and penetration, relief wells will reduce hydrostatic pressures landward of levees underlain by impervious foundations for a wide range of seepage entrances, foundation stratification, and landward top strata. In addition, seepage which formerly emerged landward of a levee in an uncontrolled manner without relief wells will be materially reduced, although the total of well flow plus seepage with wells will be increased.[7,8]

Impervious Riverside Blanket.—The natural top stratum riverside of levees frequently has been removed as a result of borrow operations in the construction of the levee, thereby exposing the pervious substratum and providing a ready source of seepage entry. In the St. Louis District no artificial, impervious blankets were placed riverward of the levees as a means of seepage control. However, where the river had scoured deep holes or channels in riverside borrow pits, these holes or channels were filled either by dredged or by hauled material, and abatis dikes were constructed to promote silting and the growth of willows in the pits.

Sublevees.—Areas known to be susceptible to dangerous underseepage may be encircled by one or more sublevees in which water can be impounded during high water to reduce the net effective head acting in this critical area. Construction of sublevees requires considerable right of way landward of a levee, skilful operation during high water to insure proper balancing of heads so as to prevent dangerous sand boils or blowouts, adequate spillways, and adequate section to insure that the sublevee itself will not fail during maximum flood stages because of a head of water in the sublevee basin. For these reasons, sublevees were not considered practicable as permanent underseepage control measures in the St. Louis District.

Cutoffs.—Where practicable, the most positive method of underseepage control is to cut off all pervious strata beneath a levee by means of an impervious barrier which will eliminate both excess substratum pressures and the problem of seepage water landward of a levee. However, completely cutting off pervious strata from 80 ft to 200 ft deep along extensive reaches of levees is not economically feasible. The installation of partly penetrating cutoffs will not reduce seepage and excess pressures significantly unless the cutoff penetrates 95% or more of the pervious aquifer. Because of this fact, which has been demonstrated by both mathematical and model studies,[9] no deep cutoffs were attempted along the levees in the St. Louis District. However, a few shallow cutoffs along the riverside toe of the levee were constructed to cut off relatively thin layers of either natural levee sands or crevasse sands which lie immediately under the base of the levee and which, in turn, are underlain by more impervious strata.

Drainage Blankets.—These blankets may be used for controlling underseepage where the levee is built on exposed sands and gravels of fairly homogeneous character. However, they are not effective for controlling seepage in deep substrata where the drain is underlain by impervious top strata or where stratified fine sands exist between the drain and the deeper, more pervious sand. No drainage blankets were used to control deep underseepage in the St. Louis District.

---

[7] "Relief Well Systems for Dams and Levees," by W. J. Turnbull and C. I. Mansur, *Transactions*, ASCE, Vol. 119, 1954, pp. 842–878.

[8] "Control of Underseepage by Relief Wells, Trotters, Miss.," *Technical Memorandum No. 3-341*, Waterways Experiment Station, Vicksburg, Miss., April, 1952, p. 61.

[9] "Efficacy of Partial Cutoffs for Controlling Underseepage Beneath Levees," *Technical Memorandum No. 3-267*, Waterways Experiment Station, Vicksburg, Miss., January, 1949, p. 63.

Drainage Trenches.—Drainage trenches may be used to control underseepage where the pervious foundation is of limited depth so that the trench substantially penetrates the formation. Where the pervious foundation is deep, a drainage trench of any practicable depth would attract only a small part of underseepage. The effect of this trench would be local, and detrimental underseepage would bypass the trench. Because of the depth of the pervious substratum along the Mississippi River levees, drainage trenches were not considered feasible for this project.

Design of Seepage Control Measures

Seepage control measures were considered necessary wherever the estimated head beneath the top stratum (at the landside toe of the levee or existing berm) at flood stages equal to the net grade of the levee would create an excessively high upward gradient through the top stratum. In agriculatural areas where it appeared that the upward gradient, i, would become equal to or exceed 0.85, seepage control measures were considered necessary. Where the gradient was between 0.67 and 0.85, piezometers were installed to measure the substratum pressure during future high-water periods; from these data, decisions will be made later as to the need for control measures. Where the gradient was less than 0.67, no control measures or piezometers were considered necessary. In industrial areas, control measures were considered necessary wherever i was expected to exceed 0.67 and were designed so that the gradient at the levee toe would not exceed 0.50.

The head, h_o, beneath the top stratum at the landside toe of existing levees was computed from the following:

$$h_o = \frac{H x_3}{s + x_3} \qquad (1)$$

in which H is the design net head on the levee; s denotes the distance from the landside toe of the levee to the effective seepage source; and x_3 is the distance from the landside toe to the effective seepage exit.

The design net head for riverfront levees was taken as the difference in elevation between the net grade of the levee and the average ground surface landward of the levee. Along the flank levees the design flood stage was taken as the elevation of the net grade of the river front levee at the intersection of the flank and riverfront levees. The distance to the effective source of seepage, s, was estimated from field reconnaissances of borrow pits, and the river, boring data, and piezometric data were obtained at similar sites along levees in the Lower Mississippi River Valley. For riverfront levees, values of s ranging from 600 ft to 1,000 ft were used for analytical and design purposes. Along flank levees where the seepage entrance is restricted to the bed of the tributary stream, values of s ranging from 800 ft to 1,500 ft were used. The distance from the landside-levee toe to the effective seepage exit, x_3 ,was computed from the following formula by Preston T. Bennett,[10] M. ASCE, which assumes that the top

[10] "The Effect of Blankets on Seepage Through Pervious Foundations," by Preston T. Bennett, *Transactions*, ASCE, Vol. 111, 1946, pp. 215–252.

stratum is level and infinite in landward extent:

$$x_3 = \frac{1}{c} = \sqrt{\frac{k_f}{k_b} z\, d} \qquad (2)$$

in which c denotes a constant; k_f/k_b is the ratio of horizontal permeability of the foundation to vertical permeability of the top stratum; z represents the top-stratum thickness in feet; and d equals the effective thickness of the aquifer in feet.

To simplify the preparation of design charts, it was assumed that $d = 80$ ft (variations in d between 50 ft and 120 ft do not greatly affect the values of h_o or x_3). Evaluation of the top-stratum thickness and of the permeability ratio was predicted largely on the thickness of the clayey part of the top stratum. The value of z was obtained by transforming the thickness of each layer of the top stratum into an equivalent thickness of material of constant vertical per-

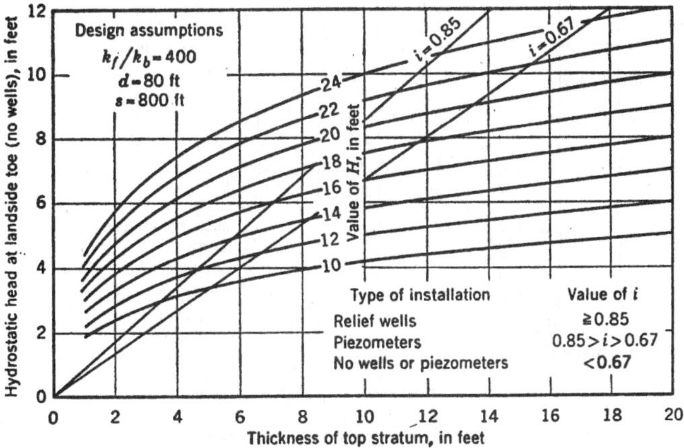

FIG. 8.—COMPUTED HYDROSTATIC HEADS AT LANDSIDE TOE OF LEVEES—NO RELIEF WELLS

meability. Where the top stratum was predominantly silty with less than 5 ft of clay, k_f/k_b was assumed equal to 400. When the top stratum consisted of 5 ft or more of fat clay, k_f/k_b was assumed equal to 1,000. The foregoing values of k_f/k_b were based on analyses of piezometric data from similar sites in the Lower Mississippi River Valley. By substituting x_3 (as given by Eq. 2) in Eq. 1, the head, h_o, can be expressed in terms of H and z for given values of s, k_f/k_b, and d. Design curves used for the computation of hydrostatic head at the landside toe of the levee without relief wells were developed from Eq. 1 and Eq. 2. A typical set of curves for $s = 800$ ft is shown in Fig. 8.

Relief Wells.—Relief well systems were designed so that the gradient midway between wells would not exceed 0.67 in agricultural areas and 0.50 in industrial areas. The effective well radius, r_w, was taken as 0.9 ft, which is the distance to the periphery of the gravel filter. Design curves were developed from the equations of Reginald A. Barron,[11] A.M. ASCE, for an infinite line of

[11] "The Effect of a Slightly Pervious Top Blanket on the Performance of Relief Wells," by Reginald A Barron, *Proceedings*, 2d International Conference on Soil Mechanics, Rotterdam, Vol. IV, 1948, p. 324.

fully penetrating wells with a semipervious top stratum. The curves were developed by computing the head midway between wells and then multiplying this head by the ratio of the head computed for 50% and for fully penetrating wells as obtained from formulas by T. A. Middlebrooks and William H. Jervis,[12] M. ASCE, and Morris Muskat,[13] respectively, for an impervious top stratum. Because the computed head midway between wells does not include hydraulic head losses in the well, these head losses were computed from the estimated flow from the well system and from hydraulic formulas, well-tank tests, and observations made on relief wells at Trotters (Miss.)[8] and added to the computed head midway between wells. The well spacing, a_c, was then adjusted so that the final head between wells, including hydraulic head losses, would not exceed the design values stated in the foregoing. Design curves were developed for $k_f/k_b = 400$ and for various distances to the source of seepage. Where

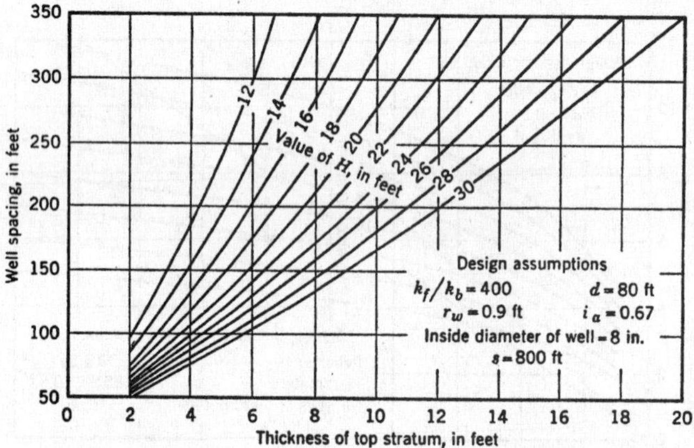

FIG. 9.—COMPUTED WELL SPACING ADJUSTED FOR HEAD LOSS IN WELL

$k_f/k_b = 1{,}000$, $a_c \cong 0.9\ a_c$ for $k_f/k_b = 400$. A typical set of design curves showing the relationship between the computed well spacing, a_c, adjusted for the head losses in the wells, and the top-stratum thickness for various heads on a levee is shown in Fig. 9. For the curves in Fig. 9 the penetration of the pervious aquifer by the well screen was 50%, and the head loss in the well was computed on the basis of a 25-ft riser and a 55-ft screen, assuming that $k_f = 1{,}000 \times 10^{-4}$ cm per sec and the top of the well at 0.33 ft above the ground surface.

The flow from relief well systems was obtained from Mr. Barron's formulas for fully penetrating wells, multiplied by the ratio of flow from 50% penetrating wells to that for 100% penetrating wells for an impervious top stratum as determined by Messrs. Middlebrooks and Jervis, and Mr. Muskat, respectively. The relationship between well flow, Q_w/H, for various top-stratum thicknesses and well spacings is shown in Fig. 10, in which the penetration of the pervious

[12] "Relief Wells for Dams and Levees," by T. A. Middlebrooks and William H. Jervis, *Transactions*, ASCE, Vol. 112, 1947, p. 1321.

[13] "The Flow of Homogeneous Fluids Through Porous Media," by Morris Muskat, McGraw-Hill Book Co., Inc., New York, N. Y., 1937.

aquifer by the well screen is 50%. From the results of pumping tests conducted to date it now appears that the average k_f is approximately from $1{,}200 \times 10^{-4}$ cm per sec to $1{,}500 \times 10^{-4}$ cm per sec for the pervious substratum in the Upper Mississippi River Valley.

In a short, finite line of wells, the heads midway between the wells exceed those obtained for an infinite line of relief wells, both at the center and near the ends of the well system. Because numerous well systems were fairly short (less than 1,500 ft) it was necessary to reduce the computed well spacing for an infinite line of wells so that heads midway between wells in a short line of wells would not create an i-value in excess of the design value. The ratio of the head, midway between wells at the center of finite well systems, to the head between wells in an infinite line of wells is shown in Fig. 11 for various well spacings and exit lengths. In developing Fig. 10, it was assumed that (1) the penetration of the pervious aquifer by the well screen was 100%, (2) the semipervious top stratum was infinite in landward extent, and (3) the effective well radius was

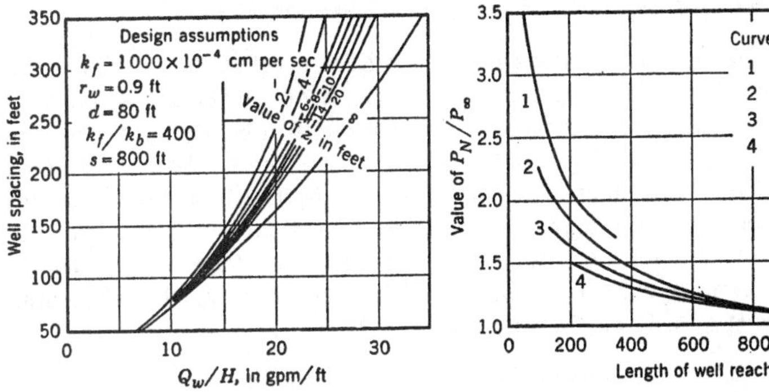

Fig. 10.—Computed Well Flow

Fig. 11.—Ratio of Head Midway Between Wells, at Center of a Finite Well System, to Head Midway Between Wells in an Infinite Line of Wells

1.0 ft. In Fig. 11, P_∞ is the head midway between wells in an infinite well line, computed from equations developed by Mr. Barron, and P_N denotes the head midway between wells at the center of a finite well line, computed from unpublished equations developed by W. E. Miner. Where a short line of wells was required, the spacing required was computed for an infinite line of wells, and this spacing then was divided by the ratio indicated in Fig. 11. In any finite line of wells of constant penetration and spacing, the head midway between wells near the ends of the system exceeds that at the center of the system. Thus, at the ends of both short and long well systems, the wells were generally made deeper to provide additional penetration of the pervious substratum so as to obtain the same head reduction as in the central part of the well line.

The average well spacing for a given reach of levee was determined from design curves similar to those shown in Figs. 8 through 11 taking cognizance of the effect of geological features on seepage conditions. However, the location of each well was checked in the field and adjusted where necessary so that the wells would fit natural topographic features.

Seepage Berms.—These berms were designed by an empirical method developed by the Kansas City District that is based on electrical-analogy model studies in which it was assumed that the permeability of the berm would be the same as that of the top stratum. Adding a berm on the landside of a levee

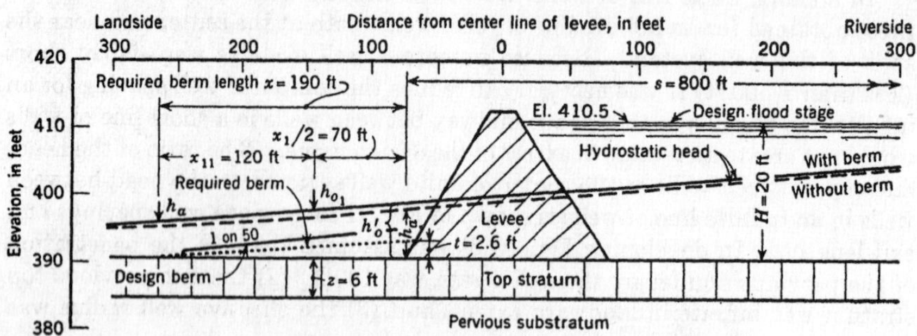

FIG. 12.—NOMENCLATURE FOR DESIGN OF SEEPAGE BERMS

restricts the natural relief of pressure as a result of natural seepage through the top stratum and, thus, increases the hydrostatic head at the levee toe with respect to the original ground surface. This increase in head depends on the length, thickness, and permeability of the berm. The electrical-analogy model studies indicated that the effect of the berm on landside substratum pressures would be the same as if the impervious base width of the levee had been increased by 50% of the length of the berm. The nomenclature used in the design of seepage berms is shown in Fig. 12. The procedure used in designing seepage berms in the St. Louis District was as follows:

1. The head, h_o, at the landside toe of the levee without a berm was determined from Eq. 1. The allowable value of i_o with a berm was taken as 0.60 in agricultural levee districts and as 0.50 in industrial districts.
2. The allowable head at the berm toe, h_a, was taken as 0.85 z in agricultural levee districts and as 0.67 z in industrial districts.
3. The first trial length of berm x_1 was computed as that length of berm where the head, h_x, at the berm toe would be equal to h_a; this length was determined by means of Mr. Bennett's[10] blanket formulas.
4. The value of $x_1/2$ was then added to s, and the head, h_{o_1}, was computed from Eq. 1 at a point $x/2$ distant from the landside toe of the levee. The following equation results:

$$h_{o_1} = \frac{H\, x_3}{s + \dfrac{x_1}{2} + x_3} \qquad (3)$$

5. A new distance, x_{11}, from the assumed impermeable section ($x_1/2$ distant from the landside toe) was computed by means of the blanket formulas so that the head, h_x, at distance x_{11} would then be equal to the allowable head, h_a.
6. The required berm length, X, is then $x_1/2 + x_{11}$. To simplify the foregoing computations, it was found expedient to represent the blanket

formulas graphically as shown in Fig. 13, which shows the relationship between the ratio of head h_x (at a distance, x, landward of the levee) to the head, h_o, (at the toe) and the distance, x, divided by the exit length, x_3. By selecting h_a and substituting this value for h_x and knowing h_o and x_3, the distance, x, at which $h_x = h_a$ can readily be found.

To compute the required thickness of the berm, it was necessary to estimate the head, h^1_o, at the actual levee toe with the berm in place. This was accomplished by

$$h^1_o = \frac{H\left(\frac{x_1}{2} + x_3\right)}{s + \frac{x_1}{2} + x_3} \qquad (4)$$

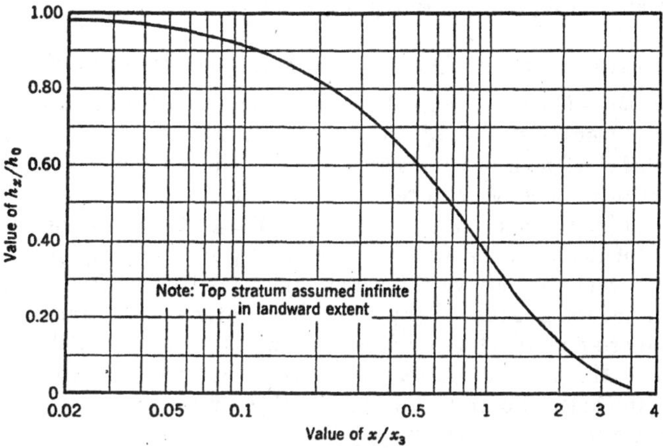

FIG. 13.—RATIO OF THE HEAD LANDWARD OF LEVEE TO THE HEAD AT LANDSIDE TOE OF LEVEE

The average upward gradient through the top stratum and the berm was expressed by

$$i = \frac{h_x - t}{z + t} \qquad (5)$$

in which t is the berm thickness. From Eq. 5, the required berm thickness, t at the levee toe can be found as

$$t = \frac{h^1_o - i_o\, z}{1 + i_o} \qquad (6)$$

in which i_o is the allowable upward gradient at the levee toe.

From Mr. Bennett's blanket formulas and from unpublished equations by Mr. Bennett for the design of seepage berms, an equation for determining the required length, X, of a berm was developed. This equation is as follows:

$$X = \frac{-A \pm \sqrt{A^2 - 24\,(2 + r)\left(1 + s\,c - \frac{H}{h_a}\right)}}{2\,c\,(2 + r)} \qquad (7)$$

in which A is equal to $6 + 3\,s\,c\,(r + 1)$; r denotes i_o/i_1; i_o is the allowable upward gradient at the landside toe of the levee; and i_1 is the allowable upward gradient at the landside toe of the berm. It should be noted that two values of X are obtained from Eq. 7. As a result of the sign convention adopted in developing the formula, the positive value of X gives the required berm length. The head, h^1_o, at the toe of the levee with the berm in place can be computed from the following:

$$h^1_o = h_a \left[1 + c\,X + \left(\frac{2 + r}{6}\right)(c\,X)^2\right] \ldots\ldots\ldots\ldots\ldots (8)$$

The required thickness of the berm at the levee toe is given by Eq. 6. The foregoing equations indicate the following required dimensions for the seepage berm illustrated in Fig. 12: $X = 175$ ft and $t = 2.5$ ft.

Installation and Construction of Seepage Control Measures

Relief Wells.—The relief wells along the levees in the St. Louis District have been and are being installed by the reverse-rotary method. This procedure for drilling holes for wells in sand is basically a suction-dredging method in which the material in the hole is removed by a suction pipe. The walls of the hole are supported by seepage forces acting against a thin film of fine-grained soil on the walls of the hole; the forces are created by maintaining a head of water in the hole several feet above the water table. A film of fine-grained soil is deposited on the walls of the hole by the drilling water as water and soil from the suction pipe are circulated through a sump pit. The sand settles out in the sump pit, and the water containing the fine-grained particles flows back from the pit into the hole. Successful drilling by the reverse-rotary method requires that the water table be approximately 7 ft or more below the ground surface and that there be an adequate supply of drilling water.

A typical reverse-rotary drilling unit is normally equipped with a large-capacity centrifugal pump and a 6-in.-diameter drill pipe, and is a bit similar in appearance to the cutterhead of a dredge. Where cobbles larger than approximately 3¾ in. are encountered, it is necessary to provide this type of equipment with a rock trap. Holes for relief wells also may be advanced by the reverse-rotary method using compressed air or a jet educator system, neither of which requires the use of rock traps.

A typical relief well consists of a screen section, riser pipe, gravel filter, sand backfill from the top of gravel filter to an elevation 10 ft below finished ground surface, and concrete backfill from the top of the sand backfill to ground surface. The riser and screen are of 8-in.-inside-diameter wood pipe. The screen is perforated with slots $\frac{3}{16}$ in. wide. The bottom of the well screen is closed with a wood plug. Tops of relief wells discharging on the ground surface are protected by metal well guards. (A schematic drawing of a typical relief well is shown in Fig. 14.)

As the boring for a relief well is advanced, samples of the foundation soil are obtained at 2-ft intervals by catching samples of the effluent from the drill rig in a bucket. The purpose of this sampling is to determine the depth at which the screen section of the well can be started and also to locate strata of silts, silty sands, and very fine sands through which unslotted sections of pipe are used instead of slotted screen. During or immediately after drilling operations, the well screen and riser pipe are assembled and guides are attached to the pipe to keep it centered in the hole as the filter gravel is placed. The hole for each well is overdrilled at by least 4 ft so as to provide a space at the bottom for filter gravel which may become segregated when the tremie pipe is first filled. The

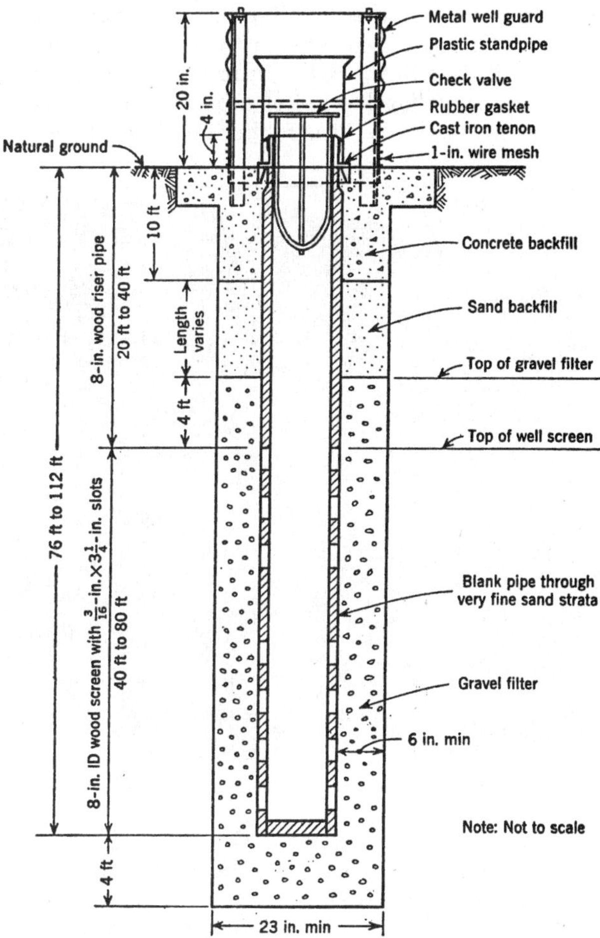

Fig. 14.—Relief Well and Appurtenances

filter gravel is placed only after the screen and riser have been lowered to the correct depth and the top of the riser piper set about 4 in. above the natural ground surface. The filter gravel is placed by first lowering a perforated tremie to the bottom of the hole and filling it with filter gravel. The tremie then is slowly raised to allow the filter gravel to run out of the bottom as it is fed into the top. The tremie pipe is kept filled at all times with the filter gravel being kept above the water surface so as to prevent any segregation of the gravel as it is placed. The filter gravel is placed to at least 4 ft above the top of the screen section so as to insure the existence of filter above the top of the screen after the well has been surged and developed.

After the filter is placed, material which has entered the well during the placing operations is removed, and the well is surged and pumped to remove drilling mud or other fines from the filter gravel. Surging and development pumping are started within 2 hr after the filter has been placed and are continued until the amount of material pulled through the screen between surging operations is less than 0.2 ft deep in the well. Surging is usually accomplished by lowering and raising a surge block made of a heavy rubber disk between two steel disks mounted on a steel rod. The rubber disk has a diameter of approximately 1 in. less than the inside diameter of the well, and the steel disks are 1 in. smaller than the rubber disk. The surge block is raised and lowered at a rate of travel of approximately 5 ft per sec.

Fig. 15.—Completed Relief-Well System

Each well is subjected to a pumping test after the completion of surging so as to achieve a drawdown of 5 ft in the well or a flow of at least 500 gal per min. The inflow of sand into a well during the pumping test is carefully checked for each well to prove its stability. This is accomplished by pumping the flow into a carefully baffled 1,000-gal sedimentation tank for 15 min. During this interval the sand in the tank is collected and measured, and the change in the amount of sand in the bottom of the well is determined. The pumping test is continued until the rate of inflow of sand into the well is less than 1 pt per hr. Fig. 4 illustrates a completed reach of a well system, showing the soil profile, spacing of wells, head on levee, and location of well screen and filter gravel; it also includes a bar graph of the specific yields of each well. A photograph of a completed well system is shown in Fig. 15.

The top of each well is protected against the backflow of muddy surface water by means of a specially designed rubber gasket and simple check valve as shown in Fig. 14. Accelerated full-scale laboratory tests have indicated this check valve to be very effective in preventing backflow of muddy water into a well under adverse conditions.

During periods of relatively low stages on the levees, relief wells will flow more than natural seepage. In agricultural areas each well is provided with a plastic sleeve which will raise the discharge elevation of the well to either 1 ft or 1.5 ft above the natural ground surface. These low sleeves or standpipes will prevent well flow at low river heads on the levee when no pressure relief is necessary. As soon as artesian pressure develops to the extent that water begins to spill over the top of the standpipes, the pipes will be removed and the well system allowed to operate as originally designed. The flow from relief wells in highly developed areas will be pumped over the levee during high water.

Piezometers.—Piezometers consisting of a 2-ft section of brass wellpoint (1¼ in. in diameter), plastic coupling, and 1¼-in. galvanized iron riser pipe have been installed in the pervious substratum immediately beneath the top stratum in almost all well reaches so as to check the efficacy of the well systems during high water. Generally these piezometers are installed between the wells near the end and in the middle of any sizeable reach of wells. At certain typical locations in each levee district, a few piezometers have been installed on a line perpendicular to the levee and beneath it. The purpose of these piezometers is to estimate the effective source of seepage entry at those locations and to check on design assumptions and performance of the well systems. Piezometers were also installed in areas where the estimated upward gradient at the design flood was between 0.67 and 0.85. The screens for the piezometers have either 40-mesh brass screen or No. 18 slots. The tops of the piezometers are protected with a column of concrete 6 in. in diameter which extends from just below the piezometer cap to 30 in. below the ground surface. After each piezometer is installed, it is pumped and tested to check its performance.

Seepage Berms.—These berms may be constructed by either hydraulic-fill methods or by hauling. Berms constructed of sand are preferable; however, the berms constructed to date (1955) in the St. Louis District have been generally built of random soil. In no event should a berm be constructed of soil that, when placed, is more impervious than the natural top stratum.

THE RELEVANCE OF THE TRIAXIAL TEST TO THE SOLUTION OF STABILITY PROBLEMS

By Alan W. Bishop[1] and Laurits Bjerrum[2]

1. INTRODUCTION

The purpose of the present paper is to show how the actual properties of cohesive soils measured in the standard undrained, consolidated-undrained and drained triaxial test are applied to the solution of the more important classes of stability problem encountered by the practicing engineer. The failure criteria chosen and the shear parameters by which they are expressed are those found most convenient and most appropriate to the methods of stability analysis used. The relation of the practical shear parameters to the more basic shear parameters proposed, for example, by Hvorslev (1937)[3] is outside the scope of the present paper and is discussed elsewhere (Skempton and Bishop, 1954; Bjerrum, 1954 b).

The practical shear parameters serve to take full account of the principal differences between cohesive soils and other structural materials, such as the dependence of strength on the state of stress and on the conditions of drainage.

Although the purpose of the paper is to present the logical relationship of the various standard tests to the different classes of stability problem, attention must also be drawn to the various limitations of the apparatus in which the triaxial test is usually performed. These include non-uniformity of stress and strain particularly at large deformations, and the inability of the apparatus to simulate the changes in direction of the principal stresses which occur in many practical problems. To enable the quantitative importance of these limitations to be seen in perspective, emphasis is laid on the direct correlation between laboratory tests and field observations of stability (or instability) wherever case records are available.

It may seem to the practicing engineer that many of his problems are too small in scale or are in soils too lacking in homogeneity to apply detailed quantitative methods of stability analysis. However, even for the application of semi-empirical rules it is important to determine into which class the stability problem falls.

2. THE PRINCIPLE OF EFFECTIVE STRESS

One of the main reasons for the late development of Soil Mechanics as a systematic branch of Civil Engineering has been the difficulty in recognising

1. Reader in Soil Mech., Imperial College of Science and Tech., Univ. of London, England.
2. Dir., Norwegian Geotech. Inst., Oslo, Norway.
3. Items indicated thus, Hvorslev (1937), refer to corresponding entries listed alphabetically in the Appendix Bibliography.

that the difference between the shear characteristics of sand and clay lies not so much in the difference between the frictional properties of the component particles as in the very wide difference—about one million times—in permeability. The all-round component of a stress change applied to a saturated clay is thus not effective in producing any change in the frictional component of strength until a sufficient time has elapsed for water to leave (or enter), so that the appropriate volume change can take place.

The clarification of this situation did not begin until the discovery of the principle of effective stress by Terzaghi (1923 and 1932) and its experimental investigation by Rendulic (1937). An examination of current design methods might suggest that the impact of Terzaghi's discovery had yet to be fully felt.

For soil having a single fluid, either water or air, in the pore space, the principle of effective stress may be expressed in relation both to volume change and to shear strength:

(a) The change in volume of an element of soil depends, not on the change in total normal stress applied, but on the difference between the change in total normal stress and the change in pore pressure. For an equal all-round change in stress this is expressed quantitatively by the expression:

$$\Delta V/V = -C_c(\Delta\sigma - \Delta u) \tag{1}$$

where $\Delta V/V$ denotes the change in volume per unit volume of soil,
$\Delta\sigma$ denotes the change in total normal stress,
Δu denotes the change in pore pressure,
and C_c denotes the compressibility of the soil skeleton for the particular stress range considered.

It may be noted in passing that this equation shows that a decrease in pore pressure at constant total stress is as effective in producing a volume change as an increase in total stress at constant pore pressure, a fact which is confirmed by field experience.

(b) The maximum resistance to shear on any plane in the soil is a function, not of the total normal stress acting on the plane, but of the difference between the total normal stress and the pore pressure. This may be expressed quantitatively by the expression:

$$\tau_f = c' + (\sigma - u)\tan\Phi' \tag{2}$$

where τ_f denotes the shear stress on the plane at failure,
c' denotes the apparent cohesion, } in terms of effective
Φ' denotes the angle of shearing resistance, } stress.
σ denotes the total stress on the plane considered,[a]
and u denotes the pore pressure.

In both cases the effective normal stress is thus the stress difference σ - u, usually denoted by the symbol σ'.

The validity of the principle of effective stress has been amply confirmed, for saturated soils, by the experimental work of Rendulic (1937), Taylor (1944), Bishop and Eldin (1950),[b] and Laughton (1955)[b]; and indirectly by the

a. Stresses and pressures are here considered as measured with respect to atmospheric pressure as zero (i.e. gauge pressures). The actual datum does not of course affect the value of the effective stress.

b. A treatment of the influence of contact area is given in these papers.

field records referred to in later sections of this paper. For partly saturated soils, however, a more general form of expression must be used, since the pore space contains both air and water which may be in equilibrium at widely different pressures, due to surface tension. A tentative expression has been suggested for the effective stress under these conditions (Bishop, 1959[b]; 1960), of the form:

$$\sigma' = \sigma - u_1 + x\,(u_1 - u_2) \tag{3}$$

where u_1 denotes the pressure in the air in the pore space,
u_2 denotes the pressure in the water in the pore space,
and x is a parameter closely related to the degree of saturation S and varying from unity in saturated soils to zero in dry soils.

The parameter x and its values under various soil conditions are discussed in more detail elsewhere (Bishop, 1960; Bishop, Alpan, Blight, and Donald, 1960). It may be noted in passing that for a given soil condition, the value of x measured in relation to shear strength may differ from its value measured in relation to volume change. However, the large positive pore pressures likely to lead to instability in rolled fills will in general only occur if the degree of saturation is high, where x may be equated to unity with little error. The additional complication of observing or predicting pore air pressure may therefore hardly be justified in such cases.

In most stability problems the magnitude of the body forces and of the applied loads is known quite accurately. It is in the magnitude of the shear strength that the main uncertainty lies and it is therefore useful to examine the variables controlling the value of τ_f in equation (2).

The magnitude of the total normal stress σ on a potential slip surface may be estimated with reasonable accuracy from considerations of statics. The shear parameters c' and Φ' are properties which depend primarily on the soil type and to a limited extent on stress history (see Table I in section 6). Provided representative samples are taken and tested in the appropriate stress range, little error need arise in evaluating c' and Φ'. This aspect of any investigation does, however, call for sound judgment and a knowledge of geology.

It is in the prediction of the value of the pore pressure u that in many problems the greatest uncertainty lies. The development of cheap and reliable field devices for measuring pore pressure in soils of low permeability[c] has, however, transformed the situation as far as the practicing engineer is concerned by enabling predictions to be checked and a control to be kept on stability during construction work.

Much of the uncertainty about the pore pressure prediction has arisen from a failure to distinguish between the two main classes of problem[d]:

(a) Problems where pore pressure is an independent variable and is controlled either by ground water level or by the flow pattern of impounded or underground water, for example, and

(b) Problems in which the magnitude of the pore pressure depends on the magnitude of the stresses tending to lead to instability, as in rapid construction or excavation in soils of low permeability.

---

c. See, for example, Casagrande, 1949; U.S.B.R., 1951; Penman, 1956; Sevaldson, 1956; Kallstenius and Wallgren, 1956; Bishop, Kennard, and Penman, 1960.

d. This distinction is discussed in detail by Bishop (1952).

In problems which initially fall into class (b) the pore pressure distribution will change with time and at any point the pore pressure will either decrease or increase to adjust itself to the ultimate condition of equilibrium with the prevailing conditions of ground water level or seepage. The rate at which this adjustment occurs depends on the permeability of the soil (as reflected in its coefficient of consolidation) and on the excess pore pressure gradients, which depend both on the stress gradients and on the distance to drainage surfaces.

The least favourable distribution of pore pressure may occur either in the initial stage or at the ultimate condition or, in special cases, at an intermediate time depending, for example, on whether load is applied or removed, and on other specific details of the problem, as discussed in section 6.

3. PORE PRESSURE PARAMETERS

In slope stability problems the influence of pore pressure on the factor of safety is most conveniently expressed in terms of the ratio of the pore pressure to the weight of material overlying the potential slip surface. This ratio was used by Daehn and Hilf (1951) in the form of an overall ratio of the sum of resolved components of the pore pressure and of the weight of soil, to express the results of the stability analysis of four earth dams, based on the field measurement of pore pressure.

Bishop (1952 and 1954 b) showed that, for a slope in which the ratio of the pore pressure u to the vertical head[e] of soil γh above the element considered was a constant, the value of the factor of safety F decreased almost linearly with increase in pore pressure ratio $u/\gamma h$. Subsequent work by Bishop and Morgenstern (in course of preparation for publication) has shown that, both for pore pressures obtained from flow patterns (class (a) problems) and for those obtained as a function of stress (class (b) problems), the average value of the pore pressure ratio $u/\gamma h$ is the most convenient dimensionless parameter by which to express the influence of pore pressure stability (Fig. 1). The ratio is denoted r_u.

Where the pore pressure is independent of stress, its value is obtained directly from the ground water conditions or flow net and expressed as the average[f] value of r_u. In all other cases (class (b) problems) the ratio must either be obtained from field measurements or predicted from the observed relationships between pore pressure and stress change under undrained conditions and from the theory of consolidation. This in turn necessitates an estimate of the stress distribution within the soil mass.

For the inclusion of the laboratory results in the stability calculation it is convenient to express them in terms of pore pressure parameters. The development of these parameters[g] and their application to practical problems is described in detail elsewhere (Skempton, 1948 b; Bishop, 1952; Skempton, 1954; Bishop, 1954 a; Bishop and Henkel, 1957; Bishop and Morgenstern, 1960).

---

e. γ is the average bulk density of the soil and h the vertical distance of the surface above the element.

f. Details of the averaging method are given by Bishop and Morgenstern, 1960.

g. Attention was drawn to the possibility of pore pressure changes in clays under the action of a deviator stress before it proved practicable to measure them (Terzaghi, 1925: Casagrande, 1934).

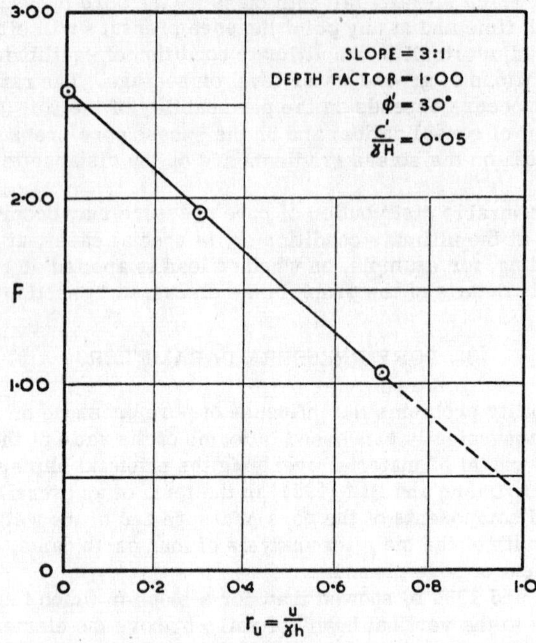

Fig. 1.—The linear relationship between factor of safety and pore pressure ratio for a slope or cut in cohesive soil.

For a change in stress under undrained conditions the change in pore pressure may be expressed as Δu, where

$$\Delta u = B\,[\Delta\sigma_3 + A\,(\Delta\sigma_1 - \Delta\sigma_3)] \tag{4}$$

where $\Delta\sigma_1$ denotes the change in major principal stress,
$\Delta\sigma_3$ denotes the change in minor principal stress,
(in both cases total stresses are considered)
and A and B denote the pore pressure parameters (Skempton, 1954).

Triaxial compression tests show that for fully saturated soils B = 1 to within practical limits of accuracy, and that the value of A depends on stress history and on the proportion of the failure stress applied. This is illustrated in Fig. 2. where the A values for normally and overconsolidated clay are given. The values of A at failure are seen to be very dependent on the over-consolidation ratio (defined as the ratio of the maximum consolidation pressure to which the soil has been subjected to the consolidation pressure immediately before the undrained test is performed).

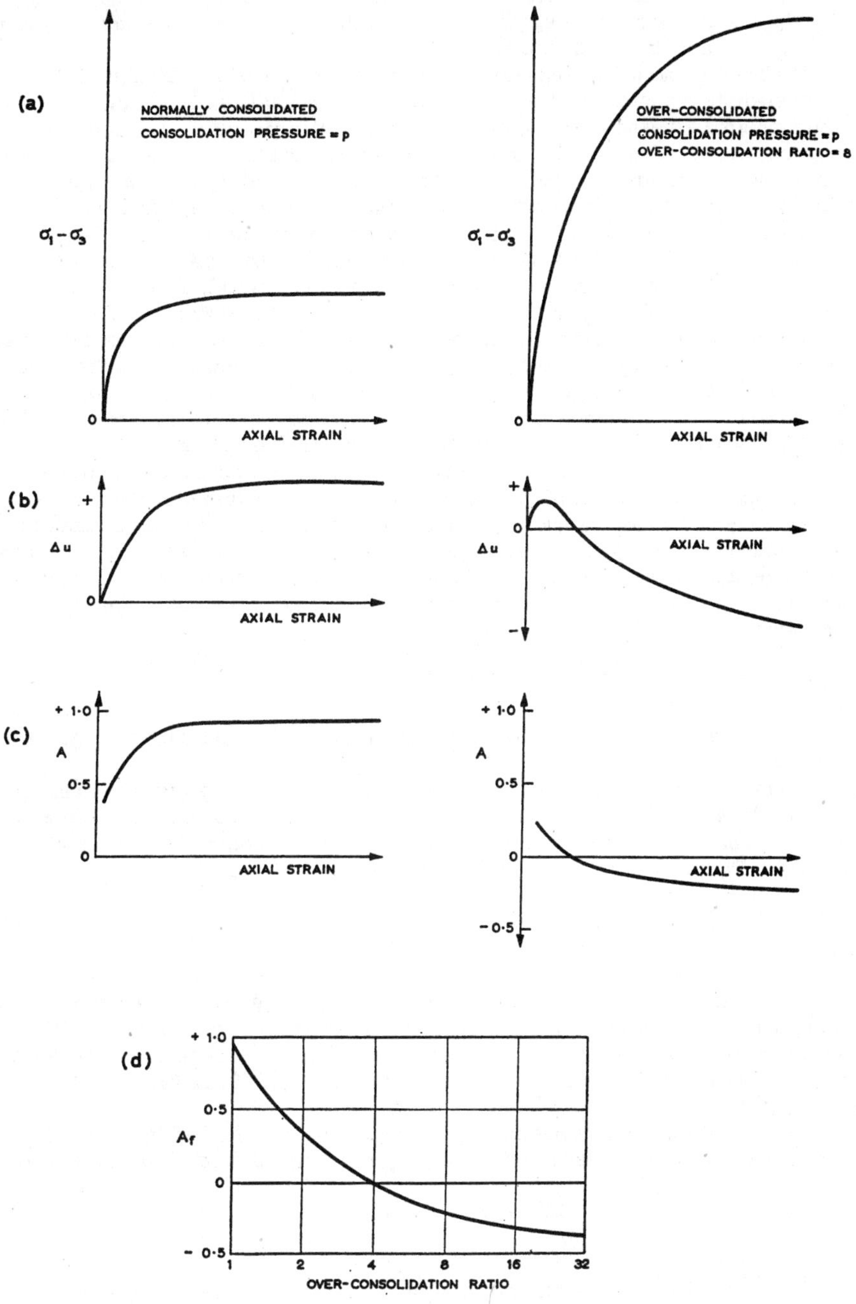

Fig. 2.—The dependence of the pore pressure parameter A on stress history.

For partly saturated soils the value of B lies between 0 and 1 depending on the degree of saturation and the compressibility of the soil skeleton. Typical values of A and B are used in section 6[h].

It should be noted that equation 4 takes no account of the change in the intermediate principal stress $\Delta\sigma_2$, or of possible changes in the directions of the principal stresses. In the majority of stability problems the conditions approximate to plane strain, in which the intermediate principal stress does not equal the minor principal stress as in the standard cylindrical compression test. Theoretical studies (Skempton, 1948 a and b; Hansen and Gibson, 1949; Bishop and Henkel, 1957) indicate that the form of the equation remains the same, but that the triaxial test underestimates the value of A, and the limited amount of test data so far available (Wood, 1958; Cornforth, 1960; Henkel, 1960) supports this view. Little is yet known about the influence of the rotation of the principal stresses on the value of A. The importance of these limitations in practice can at present only be assessed from the overall check with observed pore pressures in the field.

It should also be noted that the principle of superposition can be applied to pore pressure changes in soil only in a very restricted sense. Where the purpose of the test is the accurate prediction of pore pressure at states of stress other than failure, a more accurate result is obtained if the stress increments occurring in practice are closely followed in the test by making simultaneous changes in the values of both σ_1 and σ_3. The test result is then conveniently expressed in terms of the relationship between pore pressure and major principal stress, for the specified stress ratio, using the expression:

$$\Delta u = \bar{B} \, . \, \Delta\sigma_1 \qquad (5)$$

The influence of stress ratio on this parameter is illustrated in Fig. 3 for a compacted earth fill.

It should be noted that the value of the parameter $\bar{B}$ only gives the change in pore pressure due to stress change under undrained conditions. The actual pore pressure depends also on the initial value u_0 before the stress change is made (Fig. 4), and is given by the expression:

$$u = u_0 + \Delta u \qquad (6)$$

$$\text{i.e.} \quad u = u_0 + \bar{B} \, . \, \Delta\sigma_1 \qquad (7)$$

In natural strata u_0 is determined from the initial ground water conditions, being positive below ground water level and negative above. In rolled fill the initial value is usually negative, reaching quite high values in cohesive soils placed at or below the optimum water content (Hilf, 1956; Bishop, 1960; Bishop, Alpan, Blight, and Donald, 1960).

In cases where no dissipation of pore pressure is assumed to occur, the pore pressure ratio r_u used in the stability analysis is directly related to $\bar{B}$:

$$r_u = u/\gamma h = 1/\gamma h \, . \, (u_0 + \bar{B}\Delta\sigma_1) \qquad (8)$$

---

h. The value of B is generally given with respect to changes in pore water pressure. A slightly different value relates the change in air pressure to the change in total stress.

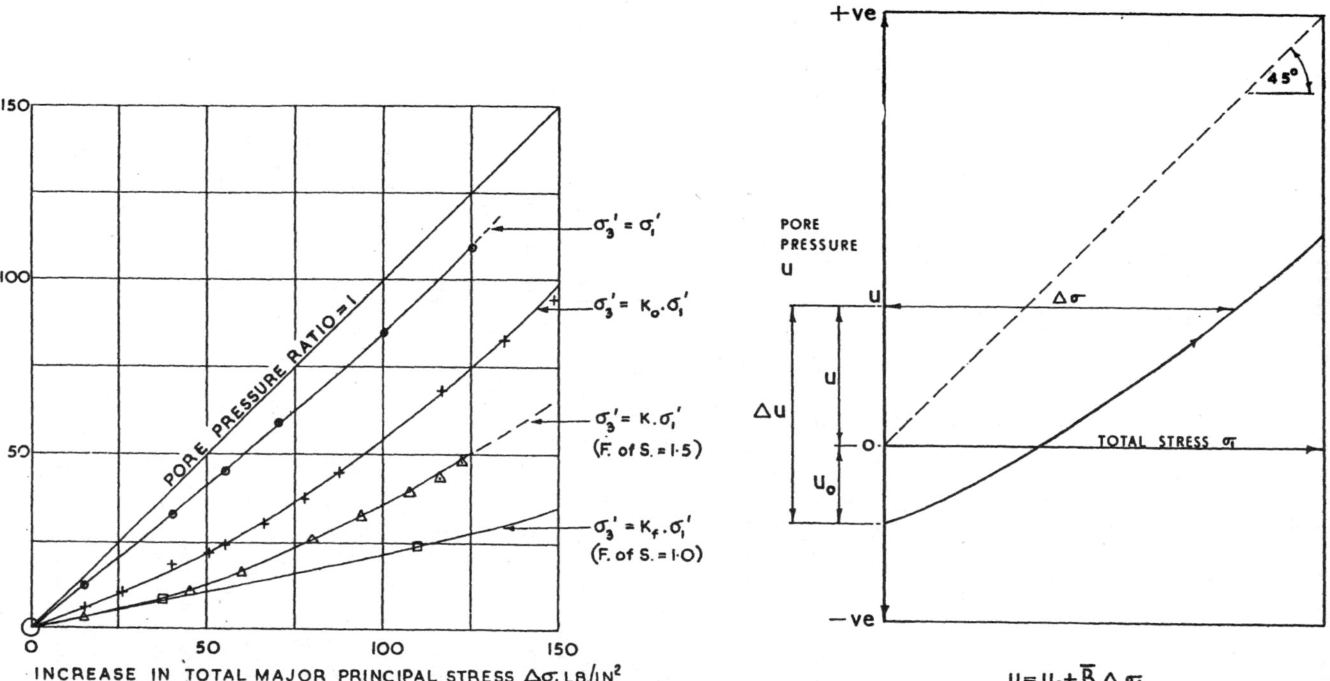

Fig. 3.—The influence of principal stress ratio on the pore pressure parameter $\bar{B}$: boulder clay compacted at optimum + 1%.

Fig. 4.—Pore pressure change under undrained test conditions.

In the special case of the construction of earth fill embankments the average value of $\Delta\sigma_1$ along a potential slip surface approximates to γh (Bishop, 1952), and equation (8) becomes:

$$r_u = \bar{B} + u_0/\gamma h \tag{9}$$

For earth fills of low plasticity placed wet of the optimum the term $u_0/\gamma h$ is small, and a further approximation is sometimes used in preliminary design:

$$r_u = \bar{B} \tag{10}$$

Some typical examples of the use of pore pressure parameters are given in section 6.

4. STANDARD TYPES OF TRIAXIAL TEST

The type of triaxial test most commonly used in research work and in routine testing is the cylindrical compression test. A diagrammatic layout of the apparatus is given in Fig. 5.

The cylindrical specimen is sealed in a thin rubber membrane and subjected to fluid pressure. A load applied axially, through a ram acting on the top cap, is used to control the deviator stress. In a compression test the axial stress is thus the major principal stress σ_1; the intermediate and minor principal stresses (σ_2 and σ_3 respectively) are both equal to the cell pressure.

Connections to the ends of the sample permit either drainage of water or air from the voids of the soil or, alternatively, the measurement of pore pressure under conditions of no drainage.

In most standard tests the application of the allround pressure and of the deviator stress form two separate stages of the test; and tests are therefore classified according to the conditions of drainage obtaining during each stage:

(1) Undrained tests[i].—No drainage, and hence no dissipation of pore pressure, is permitted during the application of the all-round stress. No drainage is permitted during the application of the deviator stress (σ_1 - σ_3).

(2) Consolidated-undrained tests.—Drainage is permitted after the application of the all-round stress, so that the sample is fully consolidated under this stress. No drainage is permitted during the application of the deviator stress.

(3) Drained tests.—Drainage is permitted throughout the test, so that full consolidation occurs under the all-round stress and no excess pore pressure is set up during the application of the deviator stress.

In order to illustrate the inter-relation between results of the different types of test, saturated and partly saturated soils will be considered separately.

---

i. Alternative nomenclatures have been used both in Europe and the U.S.A. The present terms are considered to be the most descriptive of the test conditions.

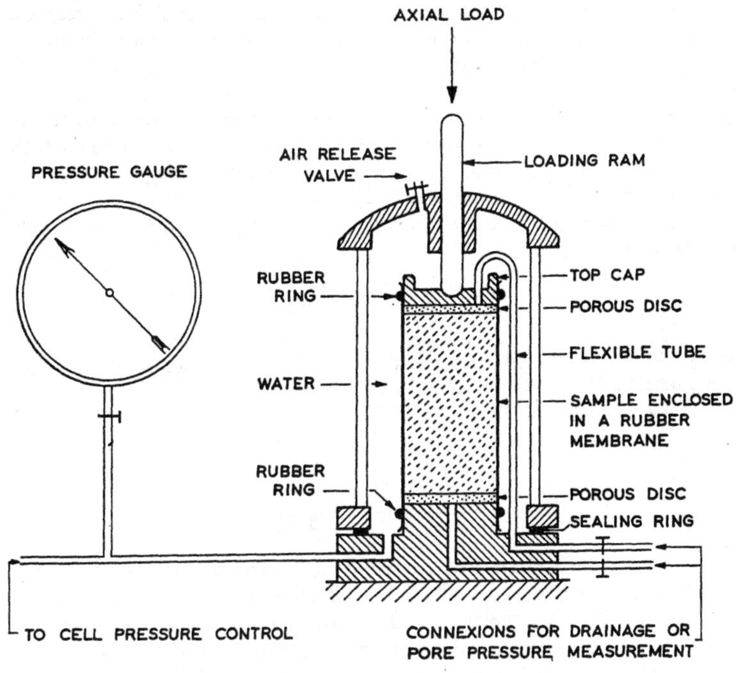

Fig. 5.—Diagrammatic layout of the triaxial cell.

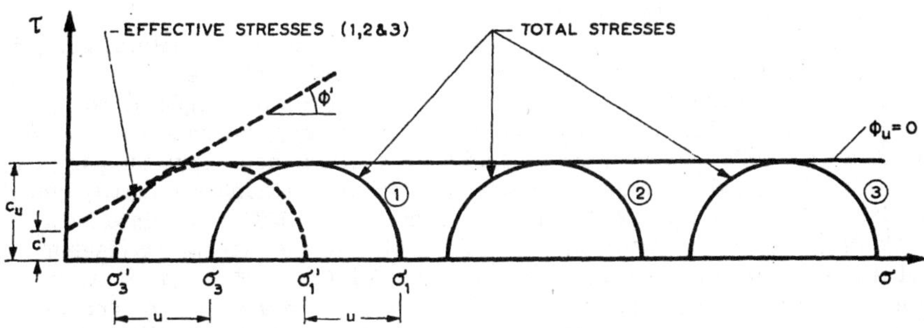

Fig. 6.—Undrained tests on saturated soil: total and effective stress circles.

(a) Undrained Tests on Saturated Cohesive Soils.—

These tests are carried out on undisturbed samples of clay, silt and peat as a measure of the existing strength of natural strata, and on remoulded samples when measuring sensitivity or carrying out model tests in the laboratory.

The compression strength (i.e. the deviator stress at failure) is found to be independent of the cell pressure, with the exception of fissured clays (discussed in section 6) and compact silts at low cell pressures. The corresponding Mohr stress circles are shown in Fig. 6.

If the shear strength is expressed as a function of total normal stress by Coulomb's empirical law:

$$\tau_f = c_u + \sigma \tan \Phi_u \qquad (11)$$

where c_u denotes apparent cohesion, Φ_u denotes angle of shearing resistance; in terms of total stress, it follows that, in this particular case,

$$\left. \begin{aligned} \Phi_u &= 0 \\ c_u &= \frac{1}{2}(\sigma_1 - \sigma_3)_f \end{aligned} \right\} \qquad (12)$$

The shear strength of the soil, expressed as the apparent cohesion, is used in a stability analysis carried out in terms of total stress, which, for this type of soil, is known as the $\Phi_u = 0$ analysis (Skempton, 1948 a and b). Since the value of c_u may be obtained directly from the unconfined compression test (where $\sigma_3 = 0$), and from the vane test in the field, it is a simple and economical test, but is often used without regard to the class of stability problem under consideration.

For fully saturated soils the increase in cell pressure is reflected in an equal increase in pore pressure and the effective stresses at failure remain unchanged. If pore pressure measurements are made during the test only one effective stress circle is obtained (Fig. 6), and tests at other water contents must be carried out to obtain the failure envelope in terms of effective stress.

In Fig. 7(a) an example is given of the changes in pore pressure during shear in an unconfined compression test and in Fig. 7(b) the Mohr circles are given in terms of total and effective stresses.

The A value measured in the undrained test on a sample of natural ground is very different from the value in situ under a similar change in shear stress. This results from the stress history given to the sample by changes in pore pressure which occur during sampling and preparation due to the removal of the insitu stresses, quite apart from disturbance due to the sampler itself. The release of the deviator stress existing in samples normally consolidated with no lateral yield is a major factor contributing to this effect.

Tests on samples anisotropically consolidated in the laboratory (Bishop and Henkel, 1953) and on undisturbed samples (Bishop, 1960) show that the effective stress in the sample when under an all-round pressure or unconfined can be less than half the effective overburden pressure in situ. Yet when the shear stress is increased to bring the sample to failure, the undrained strength closely corresponds to the in situ strength deduced from stability analysis or from vane tests. This is consistent with the experimental observation that, for a limited range of soil types and stress paths, strength

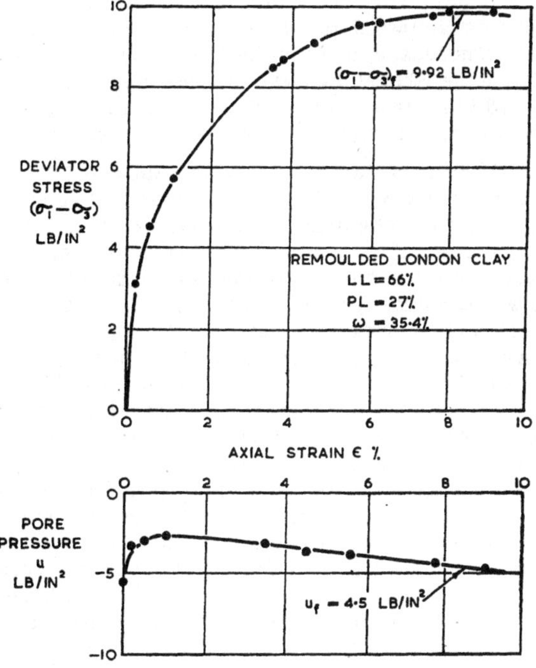

Fig. 7a.—Pore pressure change during shear in an unconfined compression test.

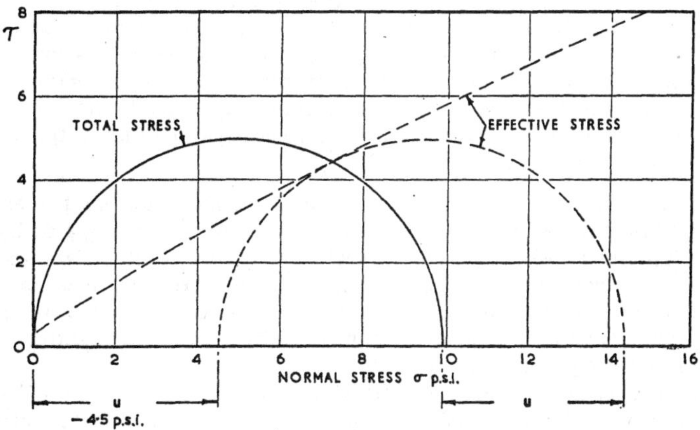

Fig. 7b.—Total and effective stress circles for the unconfined compression test.

and water content are uniquely related (Waterways Experiment Station, 1947; Henkel, 1959).

If it is indeed this fact which provides the empirical justification for the use of undrained compression tests in the $\Phi_u = 0$ analysis, then to reconsolidate the samples in the laboratory under the existing overburden pressure will inevitably lead to an overestimate of the in situ strength of the soil, since reconsolidation is almost always accompanied by a decrease in water content.

(b) Consolidated-Undrained Tests on Saturated Soils.—

These tests are carried out on both undisturbed and remoulded samples of cohesive soils, primarily to determine the values of c' and Φ', but also to determine the values of A and to study the effect of stress history.

In the standard test the sample is allowed to consolidate under a cell pressure of known magnitude (p), the three principal stresses thus being equal. The sample is then sheared under undrained conditions by applying an axial load. As in the case of the undrained test in the previous section, the cell pressure at which the sample is sheared does not influence the strength (except in dilatant silts at low pressures) as illustrated in Fig. 8e. The test result, in terms of total stresses, may thus be expressed by plotting the value of c_u against consolidation pressure p, Fig. 8b.

For normally consolidated soils the ratio c_u/p is found to be constant, its value depending on soil type. However, strengths measured in undrained triaxial tests and vane tests on strata existing in nature in a normally consolidated state, when plotted against the effective overburden pressure, lead to a lower estimate of c_u/p than is found with samples consolidated under equal all-round stress in the laboratory. The difference increases as the plasticity index decreases and appears to be due to two causes:

(i) A naturally deposited sediment is consolidated under conditions of no lateral displacement, and hence with a lateral effective stress considerably less than the vertical stress. The ratio of the effective stresses, termed the coefficient of earth pressure at rest, is generally found from laboratory tests to lie in the range 0.7-0.35, the lower values occurring in soils with a low plasticity index (Terzaghi, 1925; Bishop, 1958 a; Simons, 1958). This cause alone can account in soils of low plasticity for a difference of 50% in the value of c_u/p (for example, Bishop and Henkel, 1953; Bishop and Eldin, 1953).

(ii) Reconsolidation in the laboratory after the stress release associated with even the most careful sampling technique leads to a lower void ratio than would occur in the natural stratum under the same stress. The value of the pore pressure parameter A in particular is sensitive to the resulting modification in soil structure and this in turn leads to a higher undrained strength.

For these reasons the use of the results of consolidated-undrained tests, expressed in terms of total stress either by the parameter c_u/p or by the value of Φ_{cu} (see appendix 1), can be justified in few practical applications. However, if the pore pressure is measured during the undrained stage of the test, the results can be expressed in terms of the effective parameters c' and Φ'. Experience has shown that these parameters can be applied to a wider range of practical problems.

The relationships between the total stress, pore pressure and effective stress characteristics obtained in a typical series of consolidated-undrained

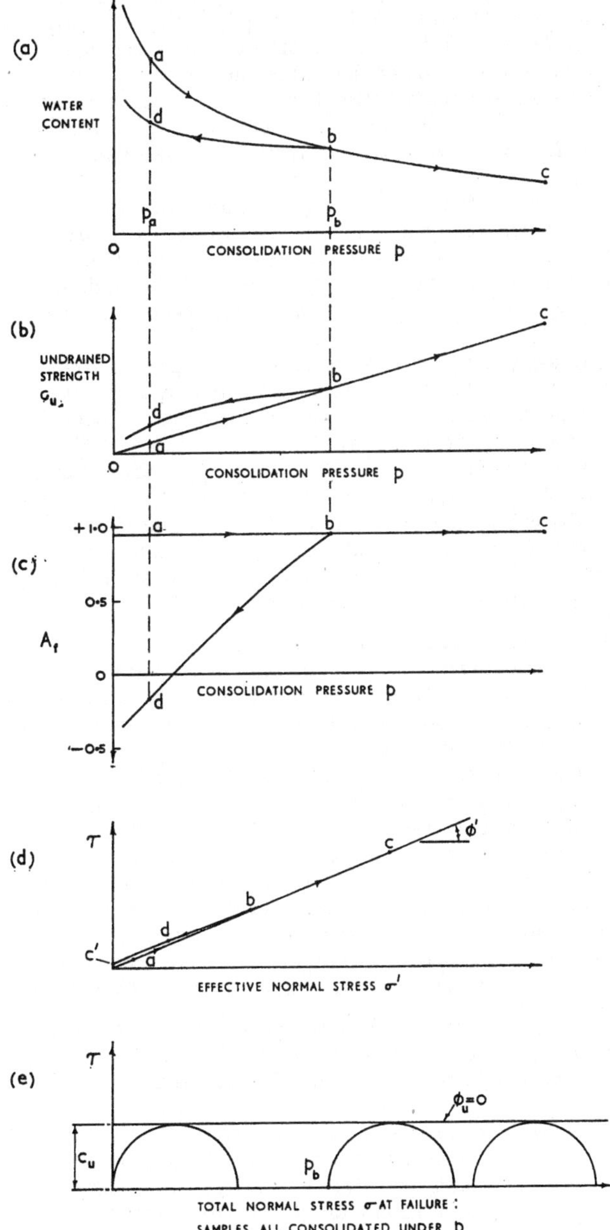

Fig. 8.—The relationships between the total stress, pore pressure and effective stress characteristics for a series of consolidated-undrained triaxial tests on saturated cohesive soil.

triaxial tests are illustrated in Fig. 8. The points, a, b, and c represent normally consolidated samples; the point d represents an over-consolidated sample, the overconsolidation ratio being p_b/p_d, Fig. 8a. For normally consolidated samples the effective stress envelope is a straight line with c' equal to zero (Fig. 8d), the value of Φ' depending on soil type. Over-consolidation results in an envelope lying a little above this straight line; the section of this envelope relevant to any particular practical problem can generally be represented with sufficient accuracy by a slightly modified value of Φ' and a cohesion intercept c'.

The most marked effect of over-consolidation is, however, on the value of A, which, with increasing over-consolidation ratio, drops from a value typically about 1 at failure to values in the negative range (Figs. 2 and 8c). These low A-values are, in turn, largely responsible for the high undrained strength values resulting from over-consolidation (compare point d with point a in Fig. 8b).

Values of c' and Φ' are usually based on the effective stress circles corresponding to maximum deviator stress. However, in some over-consolidated clays in which large decreases in pore pressure during shear are associated with very large failure strains, a slightly larger value of Φ' is obtained by plotting the state of stress at a smaller strain approximating to the point at which the ratio of the principal effective stresses σ_1'/σ_3' reached its maximum value. The difference in the value of Φ' is generally not important from a practical point of view, but in making comparisons between the values of Φ' obtained from consolidated-undrained and drained tests it is necessary to specify which definition of Φ' is being used[j].

(c) Drained Tests on Saturated Soils.—

Drained tests are carried out on both undisturbed and remoulded samples of cohesive soils to obtain directly the shear strength parameters relevant to the condition of long term stability, when the pore pressures have decreased (or increased) to their equilibrium values.

In the standard test the sample is allowed to consolidate under a cell pressure of magnitude p and is then sheared by increasing the axial load at a sufficiently slow rate to prevent any build-up of excess pore pressure. The effective minor principal stress σ_3' at failure is thus equal to p, the consolidation pressure; the major effective principal stress σ_1' is the axial stress. The test results lead directly to the effective stress shear parameters c' and Φ', which for drained tests are often denoted c_d and Φ_d.

The drained tests also provides data on the volume changes which occur during the application of the equal all-round stress and the deviator stress.

(d) Inter-Relationship between the three Types of Test on Saturated Soil.—

Two aspects of this inter-relationship are of practical interest to the engineer concerned with stability problems: (1) The degree of reliability with

j. Whether this difference reflects an actual characteristic of frictional materials or merely the increasing nonuniformity of stress in the cylindrical compression tests at large strains is still open to question. It is, however, clear from tests on sand published by the Waterways Experiment Station (1950) and other unpublished tests at the Norwegian Geotechnical Institute and at Imperial College that for very loose soil structures the maximum deviator stress may occur at smaller strains than the maximum stress ratio, and here the difference in Φ' (of up to 15^0 or so) undoubtedly represents a physical property of the soil.

which the effective stress envelope defined by the parameters c' and Φ' can be assumed to be the same for undrained, consolidated-undrained and drained tests; and (2) the extent to which volume changes in drained tests are an indication of the magnitude of pore pressure changes in consolidated-undrained tests.

In Fig. 9d are compared the results of undrained, consolidated-undrained, and drained tests on clay from the foundation of the Chew Stoke Dam (described by Skempton and Bishop, 1955). The close agreement between the effective stress failure envelopes may be noted. It is also of interest to note from the low values of A_f that an undisturbed sample reconsolidated in the laboratory behaves as though it were 'over-consolidated' even at cell pressures greatly in excess of the in situ pre-consolidation pressure. The intercept c' will in general not be zero for this part of the failure envelope.

That there should be close agreement between the effective stress envelopes for consolidated-undrained and drained tests on normally consolidated samples has been shown theoretically by Skempton and Bishop (1954) using the concept of true cohesion and friction due to Hvorslev (1937). Since Φ' is to some extent time-dependent, it is necessary to use similar rates of testing in making an experimental comparison, and to ensure adequate time for pore pressure measurement in the consolidated-undrained test and for drainage in the drained test. The predicted values of Φ' from the consolidated-undrained test are the higher, but only by 0-1° in typical cases[k].

However, for heavily over-consolidated clays the position is generally reversed[l] and the drained test is usually found to give the higher value, due to the work done by the increase in volume during shear in the drained test, and to the smaller strain at failure.

The volume changes in drained tests have for some time been known to correlate qualitatively with the pore pressure changes in undrained tests. Experimental data on two remoulded clays have recently been presented by Henkel (1959 and 1960) who has described a simple graphical procedure from which the quantitative relationship may be obtained.

(e) Undrained Tests on Partly Saturated Cohesive Soils.—

These tests are most commonly carried out on samples of earth-fill material compacted in the laboratory under specified conditions of water content and density. They are also applied to undisturbed samples of strata which are not fully saturated, and to samples cut from existing rolled fills and trial sections.

The compression strength is found to increase with cell pressure (Fig. 10a), as the compression of the air in the voids permits the effective stresses to increase. However, the increase in strength becomes progressively smaller as the air is compressed and passes into solution, and ceases when the stresses are large enough to cause full saturation, Φ_u the approximating to zero. The failure envelope expressed in terms of total stress is thus non-linear, and values of c_u and Φ_u can be quoted only for specific ranges of normal stress.

---

k. The sign and magnitude of this difference may change if the failure strains are very dissimilar, as in long term tests reported by Bjerrum, Simons, and Torblaa (1958).

l. If the failure envelopes corresponding to maximum deviator stress are compared.

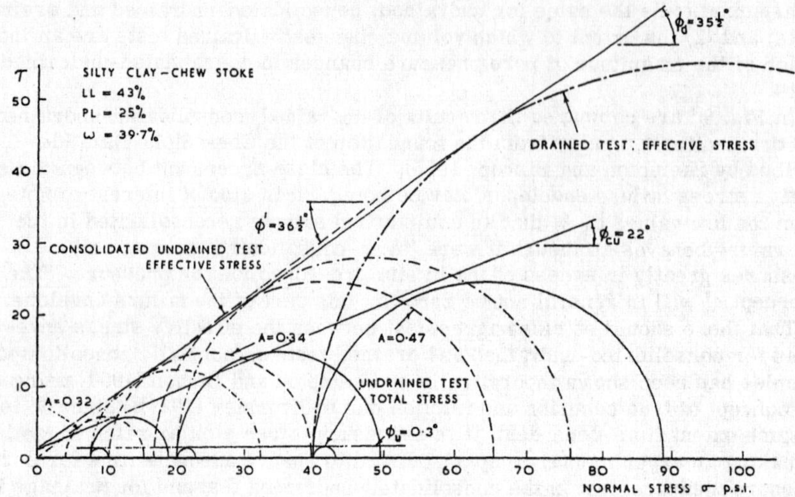

Fig. 9.—Undrained, consolidated-undrained and drained tests on undisturbed samples of Chew Stoke silty clay: maximum deviator stress.

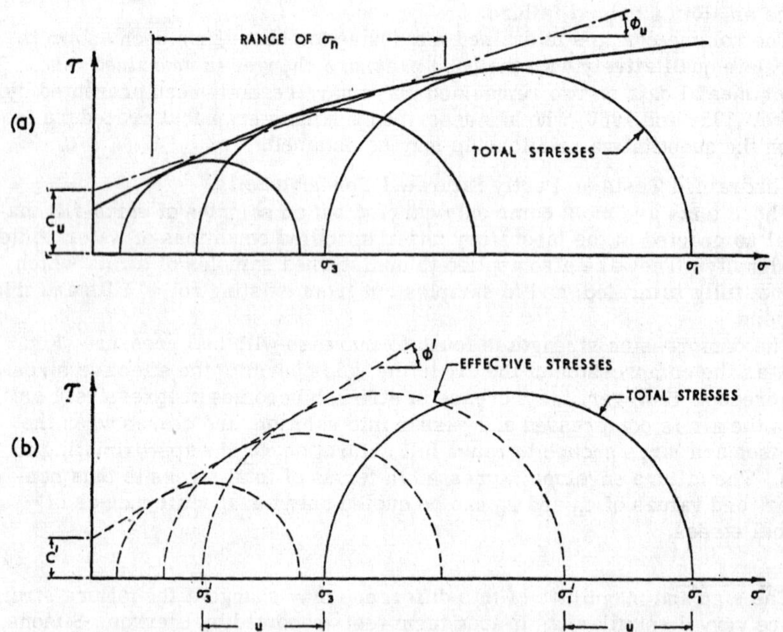

Fig. 10.—Undrained tests on partly saturated cohesive soil (a) in terms of total stress, (b) in terms of effective stress.

If the pore pressure is measured during the test, as is usual where field pore pressure measurements are to be used to check the stability during construction, then the failure envelope can be expressed in terms of effective stress, Fig. 10b. The effective stress envelope is found to approximate very closely to a straight line over a wide range of stress.

However, rather more difficulty arises in defining accurately the effective stress envelope for a partly saturated soil than at first apparent. The first difficulty lies in testing technique, to which attention was drawn by Hilf (1956). This problem is discussed in detail by Bishop (1960) and Bishop, Alpan, Blight, and Donald (1960), where it is concluded that accurate pore water pressure measurements can be made in the triaxial apparatus in partly saturated cohesive soils provided a porous element of very high air entry value is used and provided a considerably reduced rate of testing is accepted.

The second difficulty lies in the form of the expression for effective stress (equation 3), which includes a term for pore-air pressure as well as pore-water pressure for values of the factor x other than unity. The use of the simple expression for effective stress of total stress minus porewater pressure leads to an over-estimate of effective stress of $(1 - x)(u_1 - u_2)$ where $(u_1 - u_2)$ is the difference between pore-air pressure and pore water pressure. Since values of $(u_1 - u_2)$ of up to 40 lb. per sq. inch have already been measured on rolled fill in the triaxial test, and the value of x approximates to the degree of saturation, significant errors in effective stress result from the use of the simpler expression. This is particularly marked near the origin of the Mohr diagram and may lead to the apparent anomaly of a negative 'cohesion' intercept (Bishop, Alpan, Blight, and Donald, 1960). However, pore pressures set up under construction conditions are only critical if the water content of the fill and the magnitude of the stresses lead to almost full saturation, and in this case the error is small enough to be ignored in many practical problems.

(f) Consolidated-Undrained Tests on Partly Saturated Cohesive Soils.—

These tests are carried out on samples of compacted earth-fill material and on undisturbed samples. They may be necessary to determine c' and Φ' when the degree of saturation of the samples is not low enough to result in a sufficient range of strengths in the undrained tests to define a satisfactory failure envelope.

Consolidated-undrained tests in which a backpressure is applied to the pore space to ensure full saturation before shearing are carried out to examine the effect on the values of c' and Φ' of the submergence of fill or foundation strata. Back-pressures of up to 100 lb. per sq. inch are often required to give full saturation on a short term basis.

(g) Drained Tests on Partly Saturated Cohesive Soils.—

Drained tests are carried out on both compacted and undisturbed samples to obtain directly the values of c' and Φ' for the condition of long term stability. Generally a backpressure is applied to ensure full saturation of the sample before the application of the deviator stress, during which the backpressure is held constant.

(h) Inter-Relationship between the Three Types of Test on Partly Saturated Soil.—

Here again two aspects of this inter-relationship are of practical interest to the engineer concerned with stability problems: (1) The comparison of the values of c' and Φ' obtained from the different types of test; and (2) the prediction of pore pressure changes from volume changes.

Tests carried out at Imperial College have generally shown that the difference between the values of Φ' measured in the different types of test are not very significant from a practical point of view. The value of c', however, tends to correlate with water-content at failure. Where all the samples defining a failure envelope show a marked increase in water content in the consolidated-undrained or drained test with a back-pressure, c' is generally reduced. With the lower values of c' obtained by using the improved pore pressure technique described elsewhere (Bishop, 1960; Bishop, Alpan, Blight, and Donald, 1960), the difference in c' obtained in the different tests are less marked, and in some soils are not of practical significance[m] (Fig. 11). The range of soil types so far tested using this technique is, however, rather limited.

It is generally easier to make accurate measurements of pore water pressure under undrained conditions than to make the necessarily very accurate measurements of volume change and degree of saturation on which pore pressure predictions depend. Studies at the Bureau of Reclamation by Bruggeman et al. (1939), Hamilton (1939), Hilf (1948 and 1956) have shown that the change in pore-air pressure can be related to observed volume changes by the use of Boyle's law and Henry's law. However, the magnitude of the difference between pore-air and porewater pressure still has to be found experimentally. For practical purposes, where the pore water pressure is the more significant factor, it is therefore more convenient to measure it directly, particularly if the effect of stress ratio on pore pressure is also to be studied[n].

(i) Advantages and Limitations of the Triaxial Test.—

The advantages and limitations of the triaxial test have been discussed in some detail elsewhere (Bishop and Henkel, 1957), and will be referred to only briefly here.

The principal advantages of the triaxial test as performed on cylindrical specimens are that it combines control of the drainage conditions and the possibility of the measurement of pore pressure with relative simplicity in operation.

The principal limitations are that the intermediate principal stress cannot be varied to simulate plane strain conditions, that the directions of the principal stresses cannot be progressively changed, and that end restraint may modify the various relationships between stress, strain, volume change, and pore pressure.

For most practical purposes the advantages outweigh the limitations, and it will be apparent from section 6 that a very satisfactory correlation does in fact exist between laboratory tests and field observations of stability in many important engineering problems.

---

m. Some difference will in general arise from such factors as the different strains at which 'failure' is taken to occur, the different rates of volume change at failure, and, in soils having true cohesion in the Hvorslev sense, the different water contents of the samples defining the failure envelope.

n. The effect of stress ratio is discussed by Bishop (1952 and 1954 a) and Fraser (1957).

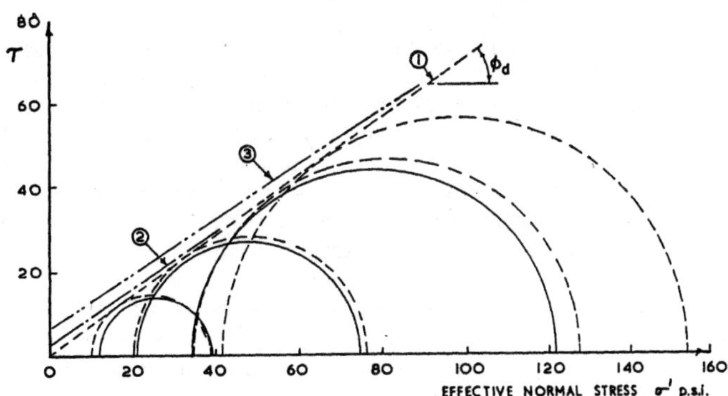

STRESS CIRCLES FOR DRAINED TESTS WITH FULL SATURATION SHOWN BY BROKEN LINE
STRESS CIRCLES FOR CONSOLIDATED UNDRAINED TESTS WITH FULL SATURATION, PLOTTED IN TERMS OF EFFECTIVE STRESS AT MAX. DEVIATOR STRESS, SHOWN BY SOLID LINE

ENVELOPE (1) REPRESENTS DRAINED TESTS
ENVELOPE (2) REPRESENTS UNDRAINED TESTS IN TERMS OF σ−u (CIRCLES NOT SHOWN)
ENVELOPE (3) REPRESENTS UNDRAINED TESTS IN TERMS OF EQN (3) WITH ASSUMED X-VALUES

STRAIN RATE 0·38% PER HOUR IN ALL TESTS

Fig. 11.—Undrained, consolidated-undrained and drained tests on boulder clay compacted at an initial water content 2% dry of optimum: clay fraction 4%.

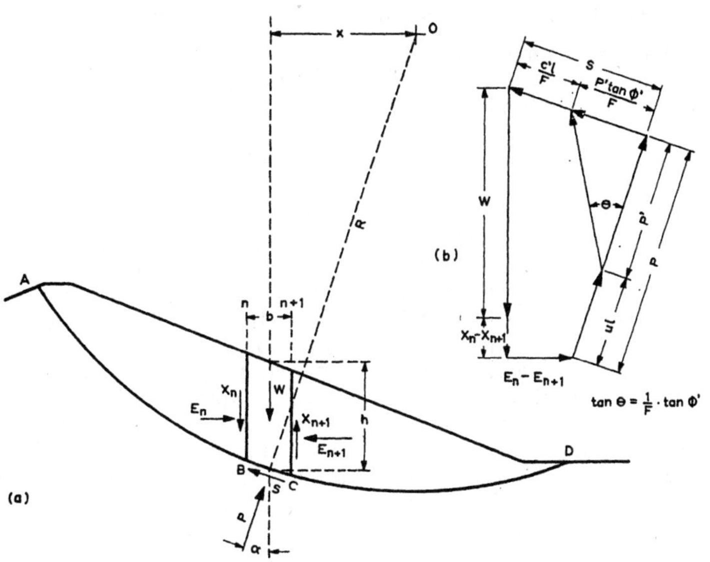

Fig. 12.—Forces in the slices method of stability analysis.

5. METHODS OF STABILITY ANALYSIS

The stability of soil masses against failure under their own weight, or under the action of applied loads, can be examined either by methods based on elastic theory or by methods based on the principle of limit design.

In the first case the stress distribution is calculated and the maximum stresses are then compared with the strength of the soil. As a practical method it is, however, open to several serious objections. Firstly, it is difficult to assess the error resulting from the assumption that the soil mass is a homogeneous elastic material having elastic constants which are independent of the magnitude of the stresses. Secondly, it has been shown that, even if these assumptions were true, local overstress would occur in a typical earth dam section when its factor of safety (by a slip circle method) lay below a value of about 1.8 (Bishop, 1952). The same applies in principle to earth slopes and foundations.

In consequence elastic methods are not applicable to the calculation of the factor of safety when studying observed failures or for design work on embankments and cuts where 1.5 is accepted as a working value for factor of safety. Elastic methods are, however, useful in giving an estimate of the stress distribution studies and for pore pressure prediction.

In most practical stability problems, therefore, the engineer is concerned with the factor of safety against complete failure, rather than against local overstress. The most general definition of factor of safety against complete failure, which can be applied irrespective of the shape of the failure surface, is expressed in terms of the proportion of the measured shear strength that must be mobilized to just maintain limiting equilibrium. The shear strength parameters to which the factor of safety is applied in setting up the equations expressing the condition of limiting equilibrium depend on whether the analysis is carried out in terms of effective stress (c', Φ' analysis) or total stress ($\Phi_u = 0$ analysis). The two cases will be treated separately.

(a) Effective Stress Analysis.—

In the effective stress analysis the proportion of the shear strength mobilized for limiting equilibrium is expressed:

$$\tau = (c'/F) + (\sigma - u)(\tan \Phi'/F) \tag{13}$$

The value of the factor of safety F is obtained by assuming limiting equilibrium along a trial slip surface (usually the arc of a circle in cross-section), balancing the forces and solving for F. The value of σ is determined from the equilibrium of the soil mass above the failure surface by an appropriate graphical or numerical method. The method of determining the value of u will depend on the class of stability problem.

(I) In class (a) problems, where the pore pressure is an independent variable, the value of u will be obtained from ground water level if there is no flow, or from a flow net if a state of steady seepage exists. The flow net can either be calculated or based on field measurements of pore pressure.

(II) In class (b) problems, where the magnitude of the pore pressure depends on the stress changes tending to lead to instability, the most practical method of approach is that adopted in earth dam design. Here a prediction is made of the actual pore pressure likely to obtain in the stable dam, which should thus check with the field pore pressure measurements usually made during construction. This prediction is based on an approximate stress distribution within the dam, the undrained pore pressure parameter $\bar{B}$ and a calculated allowance for pore pressure dissipation, the value of $\bar{B}$ being re-adjusted if necessary to match the calculated factor of safety.

Where field measurements of pore pressure are available they are of course substituted directly in the analysis.

While any method of stability analysis can be used which correctly represents the statics of the problem, the more complex soil profiles or dam sections involving a number of zones of c' and Φ' and irregular distributions of pore pressure can be handled most readily by a numerical form of the method of slices (Bishop, 1954 b).

As applied to the slip circle analysis (Fig. 12), the method leads to an expression for the factor of safety:

$$F = \frac{1}{\Sigma W \sin \alpha} \cdot \sum \left[\{c'b + \tan \Phi' . [W(1 - r_u) + (X_n - X_{n+1})]\} \frac{\sec \alpha}{1 + \frac{\tan \Phi' . \tan \alpha}{F}} \right] \tag{14}$$

This expression takes full account of both horizontal and vertical forces between the slices. The vertical shear force term, which cannot be eliminated mathematically, can however be put equal to zero with little loss in accuracy. The method agrees to within about 1% with the modified friction circle method described by Taylor (1948) in two cases which have been checked.

The programming of the digital electronic computer 'DEUCE' for the numerical method by Little and Price (1958) has given it an additional and overwhelming advantage, since any specified pattern of slip circles can be analysed at a rate of about 5 seconds per circle using about 30-50 slices. This leaves the engineer free to investigate the effect of varying his assumptions about soil properties and pore pressure, and to modify his design, without the heavy burden of computation previously involved.

The extension of the slices method to noncircular surfaces has been undertaken by Janbu (1954 and 1957) and Kenney (1956) and it is at present being programmed for the computer.

(b) Total Stress Analysis.—

In the total stress analysis the proportion of the shear strength mobilized is expressed, for the $\Phi_u = 0$ condition, as:

$$\tau = c_u / F \tag{15}$$

In the notation of Fig. 12, the expression for the factor of safety using the slip circle analysis becomes:

$$F = \Sigma c_u l / \Sigma W \sin \alpha \tag{16}$$

When $\Phi_u = 0$ the inter-slice forces enter into the calculation only if a non-circular slip surface is used.

For saturated soils the apparent cohesion c_u is equal to one half of the undrained compression strength (equ. 12) and its value is obtained from undrained tests on undisturbed samples or from vane tests. The value of c_u usually varies with depth and appropriate values must be used around the trial failure surface.

It should be noted that the use of this method is correct only where the field conditions correspond to the laboratory tests conditions, i.e. where the shear stress tending to cause failure is applied under undrained conditions[o]. It cannot in general be applied using undisturbed samples from slopes, for example, where the water content has had time to adjust itself to the stress changes set up by the formation of the slope.

The validity of the $\Phi_u = 0$ method is in fact restricted to saturated soils[p] and to problems in which insufficient time has elapsed after the stress change considered for an increase or decrease in water content to occur. It is therefore an 'end of construction method'. Whether the factor of safety subsequent to construction will have a lower value depends on the sign and magnitude of the stress changes. The particular cases are discussed in Section 6.

The use of total stress methods in which Φ_u is not zero, or in which the angle of consolidated undrained shearing resistance Φ_{cu} is used, is, in the opinion of the authors, to be avoided except in special cases, owing to the difficulty of determining the physical significance of the factor of safety thus obtained.

(c) Relationship between Total and Effective Stress Methods of Stability Analysis.—

Since the failure criterion and the associated method of stability are only convenient means of linking the stability problem with the appropriate laboratory test, a soil mass in limiting equilibrium should be found to have a factor of safety of 1 by whichever method the analysis is performed. As total stress methods can only be applied under undrained conditions, it is convenient to demonstrate this point by a simplified analysis of a vertical cut in saturated clay immediately after construction (Fig. 13).

To simplify the mathematics of the problem it is assumed that the undrained strength c_u does not vary with depth, and that the effective stress failure envelope is represented by $c' = 0$ and a constant value of Φ'. The failure surface is assumed to approximate to a plane without tension cracks.

The critical height H under these conditions is known to be equal to $4c_u/\gamma$ where γ is the density of the soil. The factor of safety of the soil adjacent to a vertical cut of depth H can be calculated in terms of either total or effective stress:

(I) Total Stress.—From equation 16:

$$\begin{aligned} F &= (\Sigma\, c_u\, .\, l)/(\Sigma\, W \sin\alpha) \\ &= (c_u\, .\, H \operatorname{cosec}\alpha)/(1/2\, .\, \gamma H^2\, .\, \cot\alpha \sin\alpha) \\ &= (2c_u)/(\gamma\, .\, H \cos\alpha \sin\alpha) \end{aligned} \tag{17}$$

Putting $dF/d\alpha = 0$ to obtain the value of α giving the lowest value of F we obtain $\alpha = 45^o$.

Substituting in equ. 17 gives:

$$F = 4c_u/\gamma H \tag{18}$$

Substituting $4c_u/\gamma$ for H we obtain F = 1.

---

o. The error introduced by the fact that the principal stress directions in most practical problems differ from those in the laboratory is discussed by Hansen and Gibson (1949).

p. Stiff fissured clays under a reduction of normal stress are an exception.

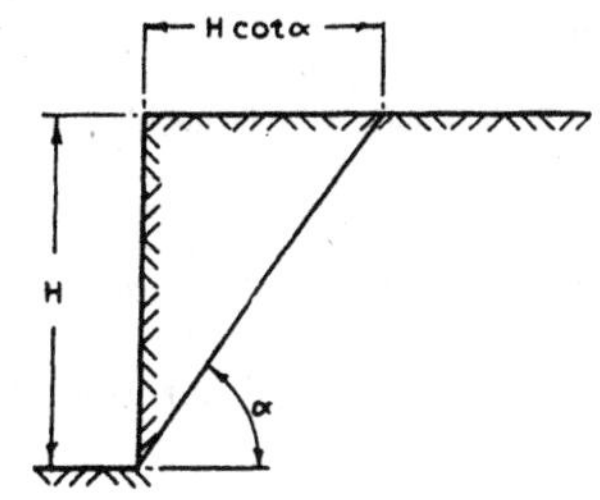

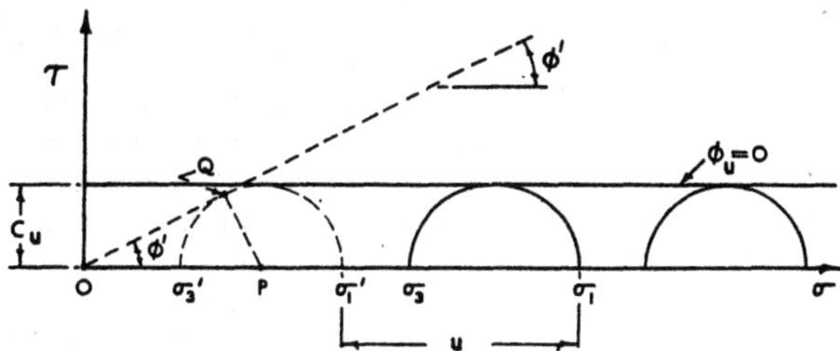

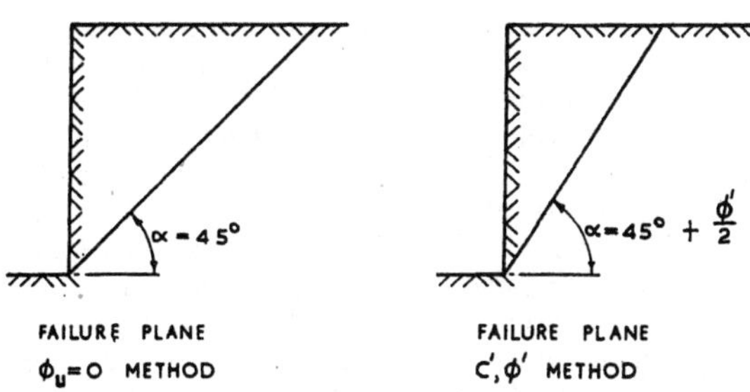

Fig. 13.—Simplified analysis of the stability of a vertical cut in saturated cohesive soil immediately after excavation, using both total and effective stress methods.

(II) Effective Stress.—From the Mohr diagram in Fig. 13 we can obtain the pore pressure in an element of soil at failure in terms of the major principal stress. It follows from the geometry of the triangle OPQ that:

$$(\sigma_1 - \sigma_3)/2 = [(\sigma_1 - u) - \{(\sigma_1 - \sigma_3)/2\}] \sin \Phi' \qquad (19)$$

Putting $(\sigma_1 - \sigma_3)/2 = c_u$ and rearranging we obtain:

$$u = \sigma_1 - c_u \, . \, [(1 + \sin \Phi')/(\sin \Phi')] \qquad (20)$$

For a plane slip surface, and with $c' = 0$ the expression for F given in equ. 15 simplifies to the form:

$$F = [1/(\Sigma W \sin \alpha)] . \Sigma[\tan \Phi'(W \cos \alpha - ul)] \quad (21)$$

For failure on a plane, the state of stress corresponds in this case to the Rankine active state, and thus the major principal stress σ_1 is equal to γh, the vertical head of soil above the element.

Substituting in equ. 21 the value of u given by equ. 20, and putting $c_u = \gamma H/4$, we obtain the expression for F:

$$F = \tan \Phi' (\cot \alpha - [1/\sin 2\alpha] . [(\sin \Phi - 1)/(\sin \Phi)]) \quad (22)$$

Putting $dF/d\alpha = 0$ we now find that the minimum value of F is given by the inclination $\alpha = 45^0 + \Phi'/2$. Substituting this value in equation 22 and expressing the angles in terms of $\Phi'/2$, we find that the expression again reduces to $F = 1$.

This comparison illustrates two important conclusions. Firstly, both total and effective stress methods of stability analysis will agree in giving a factor of safety of 1 for a soil mass brought into limiting equilibrium by a change in stress under undrained conditions. Secondly, although the values of factor of safety are the same, the position of the rupture surface is found to depend on the value of Φ used in the analysis. The closer this value approximates to the true angle of internal friction, the more realistic is the position of the failure surface[q], and this is confirmed by the analysis of the Lodalen slide in terms of c' and Φ' (Sevaldson, 1956 and Section 6).

The choice of method in short term stability problems in saturated soil is thus a matter of practical convenience and the $\Phi_u = 0$ method is generally used because of its simplicity, unless field measurements of pore pressure are to be used as a control. It should be noted, however, that for factors of safety other than 1 the two methods will not in general give numerically equal values of F. In the effective stress method the pore pressure is predicted for the stresses in the soil, under the actual loading conditions, and the value of F expresses the proportion of c' and $\tan \Phi'$ then necessary for equilibrium. The total stress method on the other hand implicitly uses a value of pore pressure related to the pore pressure at failure in the undrained test. The high factor of safety shown for example in the $\Phi_u = 0$ analysis of a slope of over-consolidated clay in which the pore pressure shows a marked drop during the latter stages of shear will therefore not be reflected in the effective stress analysis in such a marked way.

It cannot be too strongly emphasized that a comparison between effective stress and total stress methods can only be logically made when the shear stress tending to cause instability has been applied under undrained conditions. The use of the $\Phi_u = 0$ method under other conditions cannot be justified theoretically and in practice often leads to very unrealistic results (see Table V).

q. This point is discussed more fully by Terzaghi, 1936 b; Skempton, 1948 a; and Bishop, 1952.

6. THE APPLICATION OF STABILITY ANALYSIS TO PRACTICAL PROBLEMS

In this section the stability analysis of a number of typical engineering problems will be examined. The purpose of the examination is in the first place to obtain a clear qualitative picture of what happens to the variables controlling stability during and after the construction operation or load change under consideration. The second purpose of the examination is to indicate the most dangerous stage from the stability point of view and to select the appropriate shear parameters and method of stability analysis.

It is not possible to generalise about the solution of practical problems without considering the principal properties of the soil in each case. It will have been apparent from section 2 that the permeability of the soil has an important bearing on the way in which the stability problem is treated. In the more permeable soils (e.g. sands and gravels) the pore pressure will be influenced by the magnitude of the stresses tending to lead to instability only under conditions of transient loading. Both end of construction and long term problems will fall into class (a) in which pore pressure is an independent variable. Only in the less permeable soils do the relative merits of alternative methods of analysis have to be considered in most practical cases.

In Table I are listed representative values of the shear strength parameters of some typical soils arranged in order of decreasing permeability. The wide range of permeability values will be noted, and it will be apparent that it is here that the largest quantitative difference between the soil types lies.

Table I.—Permeability and Shear Strength Parameters of Typical Soils. (* Signifies Undisturbed Samples).

Material	Plasticity Index P I %	Permeability K cm/sec. (Approx.)	c' lb./sq.ft.	Φ' Degrees
Rock fill: tunnel spoil	–	5	0	45
Alluvial gravel: Thames Valley	–	5×10^{-2}	0	43
Medium sand: Brasted	–		0	33
Fine sand	–	1×10^{-4}	0	20–35
Silt: Braehead	–	3×10^{-5}	0	32
Normally consolidated clay of low plasticity – Chew Stoke*	20	1.5×10^{-8}	0	32
Normally consolidated clay of high plasticity – Shellhaven*	87	1×10^{-8}	0	23
Over-consolidated clay of low plasticity – Selset boulder clay*	13	1×10^{-8}	170	$32^1/_2$
Over-consolidated clay of high plasticity – London clay*	50	5×10^{-9}	250	20
Quick clay*	5	1×10^{-8}	0	10–20

The more important problems are:

(a) Bearing Capacity of a Clay Foundation.—

This problem may be illustrated most simply in terms of the construction of a low embankment on a saturated soft clay stratum with a horizontal surface. In Fig. 14a is shown diagrammatically the variation with time of the factors which govern stability, i.e. the average shear stress along a potential sliding surface and the average pore pressure ratio.

The excess pore pressure set up in an element of clay beneath the embankment is given by the expression:

$$\Delta u = B[\Delta\sigma_3 + A(\Delta\sigma_1 - \Delta\sigma_3)] \quad (4)$$

For points beneath the embankment Δu will in general be positive and have its greatest value at the end of construction, since B = 1 and A is positive for normally or lightly overconsolidated clay. Unless construction is slow or the clay contains permeable layers, little dissipation of pore pressure will occur during the construction period. After construction is completed the average value of r_u will decrease as redistribution and dissipation of the excess pore pressures occur, until finally the pore pressures correspond to ground water level.

The factor of safety given by the effective stress analysis will thus show a minimum value at or near the end of construction, after which it will rise to the long term equilibrium value[r]. For the long term stability calculation it is obviously appropriate to take the values of c' and Φ' from drained tests. For the end of construction case the same values may also be used, for, though it is more logical to take the value from undrained and consolidated-undrained tests expressed in terms of effective stress, the error is on the conservative side and is likely to be small.

The use of the effective stress method for the end of construction case means, however, that the pore pressures must be predicted or measured in the field. Typical field measurements of pore pressure under an oil storage tank are illustrated in Fig. 14b (after Gibson and Marsland, 1960). However, field measurements are usually limited to the more important structures and to earth dams, and the application of this method to the end of construction case will in other instances have to depend on estimated pore pressure values. As this estimate involves an assumption about the stress distribution (which is influenced by how nearly limiting equilibrium is approached) and the determination of the value of A, it is usually avoided by going directly to the $\Phi_u = 0$ analysis which is applicable to the end of construction case with zero drainage.

The undrained shear strength to be used in the $\Phi_u = 0$ analysis is obtained from undrained triaxial tests (or unconfined compression tests) on undisturbed samples, or from vane tests in the field. In the majority of problems involving foundations on soft clay, where it is quite clear that the long term factor of safety is higher than the value at the end of construction, there is then no need for the more elaborate testing and analysis required by the effective stress method. However, if appreciable dissipation of pore pressure is likely to occur during construction, it is uneconomical not to take advantage of it in calculating the factor of safety and the effective stress method is then required.

The failure of a bauxite dump at Newport (reported by Skempton and Golder, 1948), may be taken as an example of the use of the $\Phi_u = 0$ analysis for end of construction conditions (Fig. 15). After relatively rapid tipping,

r. The position of the most critical slip surface will of course change as the pressure pattern alters.

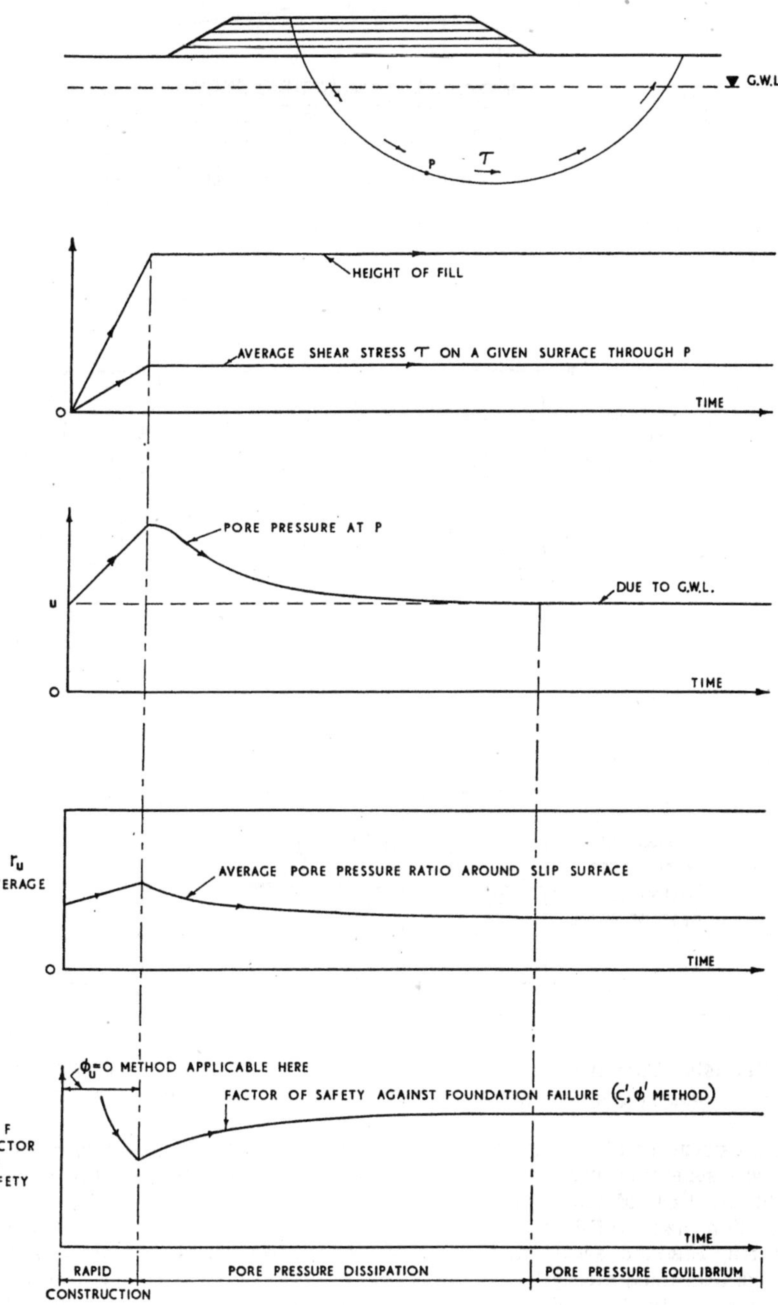

Fig. 14a.—Variation with time of the shear stress, local and average pore pressure, and factor of safety for the saturated clay foundation beneath a fill.

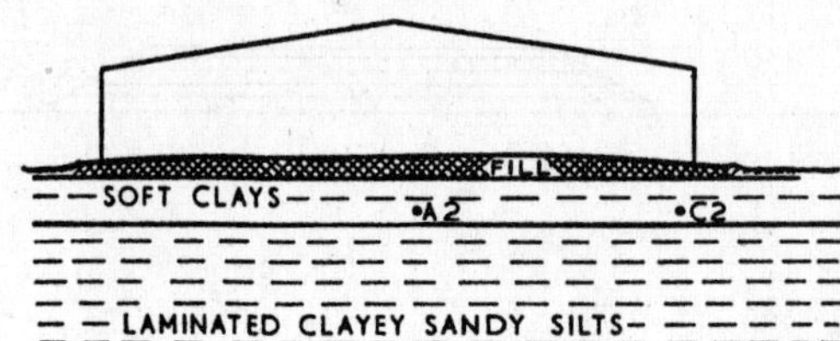

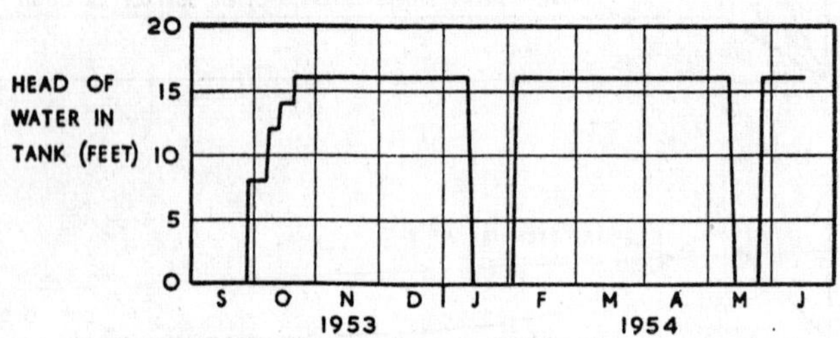

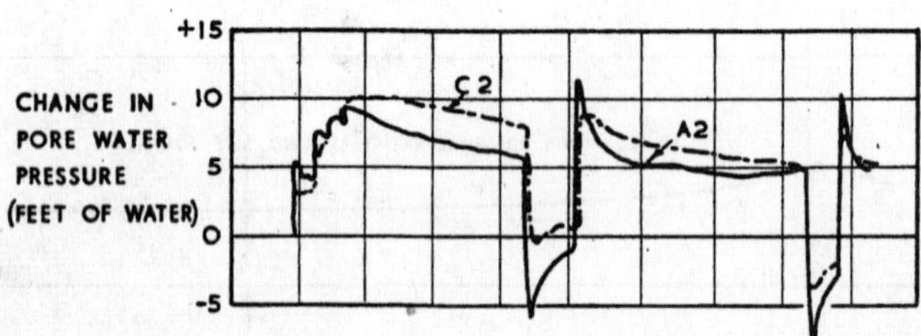

Fig. 14b.—Pore pressure changes in a soft clay foundation on filling and emptying a storage tank (after Gibson and Marsland, 1960).

failure occurred at a height of 25 feet; the factor of safety by the $\Phi_u = 0$ analysis was subsequently found to be 1.08, which can be accepted as agreement to within the limit of experimental accuracy.

In this case the fill was a granular material and its contribution to the shearing resistance was small. In cases where the fill is a cohesive material of high undrained strength, the use of the full value of this strength in the $\Phi_u = 0$ analysis gives misleading results. The explanation appears to be that shear deformations set up in the soft clay foundation under undrained loading

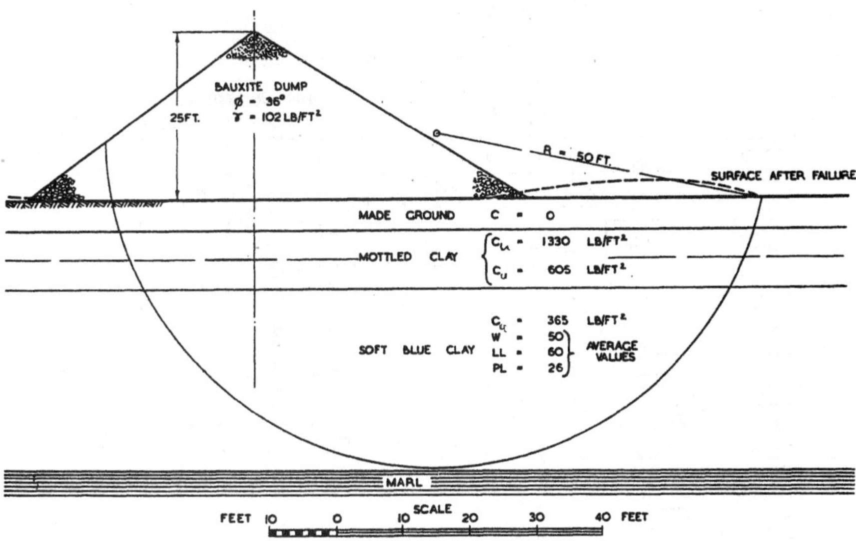

Fig. 15.—Failure of a bauxite dump at Newport (after Skempton and Golder, 1948).

set up tensile stresses in horizontal direction in the more rigid fill above and result in vertical cracks. A description of cracks wider at the bottom than at the top and passing right through the fill is given by Toms (1953a).

As the factor of safety rises with time, no long term failures can be quoted in this category.

The calculation of the ultimate bearing capacity of a structural foundation on a saturated clay is in principle the same as the problem treated above. However, the bearing capacity is not calculated by assuming a circular sliding surface, but is computed from the theory of plasticity for both the total and effective stress analyses, the results being expressed directly as bearing capacity factors.

For large foundations on soft clay the ultimate bearing capacity will increase with time after loading. For small, shallow foundations on stiff clay the ultimate bearing capacity will decrease with time, but in most cases settlement considerations will govern the design.

An example of the $\phi_u = 0$ analysis of an end of construction foundation failure has been given by Skempton (1942). Here a footing 8 feet by 9 feet founded on a soft clay with $c_u = 350$ lb./sq. ft. failed at a nett foundation pressure of 2500 lb./sq. ft. (Fig. 16). Using a bearing capacity factor of 6.7 for this depth to breadth ratio, a factor of safety of 0.95 is obtained.

Two series of loading tests on the stiff fissured London clay may also be mentioned, where the $\phi_u = 0$ analysis has led to factors of safety about 1.02 (Skempton 1959). These tests are particularly interesting as showing that the effect of fissures, which can lead to serious difficulties with end of construction problems in open excavations (see section 6b) does not prevent the successful use of the $\phi_u = 0$ analysis in bearing capacity calculations under end of construction conditions.

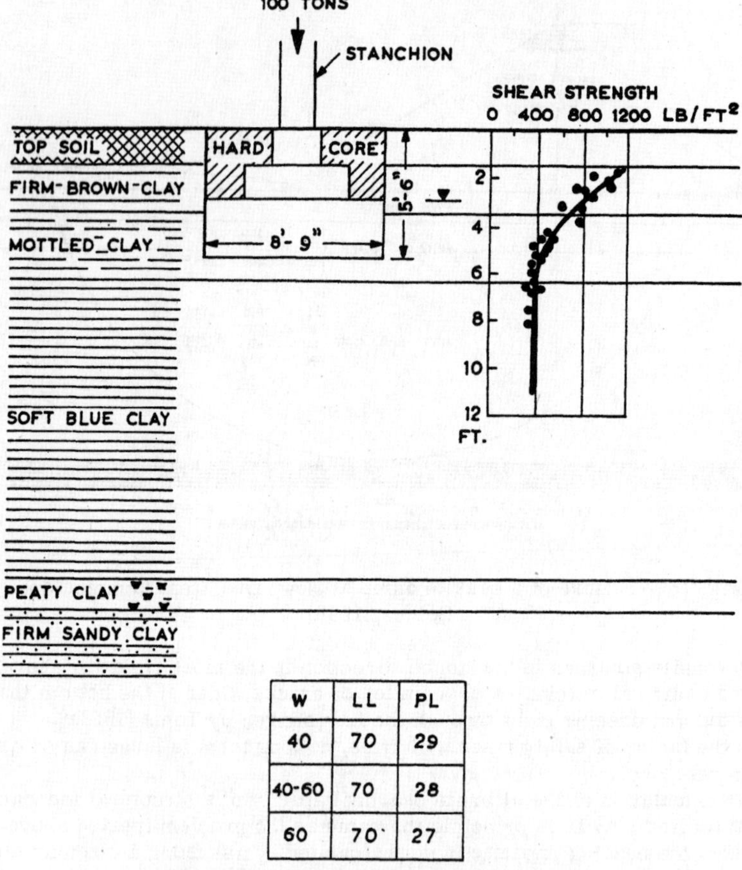

W	LL	PL
40	70	29
40-60	70	28
60	70	27

NETT FOUNDATION PRESSURE AT FAILURE = 2500 LB/FT²
IF q = 6·7 c FACTOR OF SAFETY = 0·95

Fig. 16.—Failure of a foundation on soft clay at Kippen (after Skempton, 1942).

An example of the long term failure of a small heavily loaded foundation of stiff clay is more difficult to find. The long term failure of tunnel arch footings described by Campion (1951) probably falls in this category.

It may however be concluded that with few exceptions the end of construction conditions is the most critical for the stability of foundations and that for saturated clays this may be examined more simply by the $\Phi_u = 0$ analysis. From the field tests and full scale failures tabulated in Table II it is apparent that an accuracy of ± 15% can be expected in the estimate of factor of safety. One of the exceptions is dealt with in section 6(i). Where partial dissipation of pore pressure occurs during construction an analysis in terms of effective stress is used, and examples of the analysis are discussed in section 6(h).

Table II.—End of Construction Failures of Footings and Fills on a Saturated Clay Foundation: $\Phi_u = 0$ Analysis.

1. Footings, loading tests.

Locality	Data of clay					Safety factor $\Phi_u = 0$ analysis	Reference
	W	LL	PL	PI	$\frac{W-PL}{PI}$		
Loading test, Marmorerá	10	35	15	20	-0.25	0.92	Haefeli, Bjerrum
Kensal Green	–	–	–	–	–	1.02	Skempton 1959
Silo, Transcona	50	110	30	80	0.25	1.09	Peck, Bryant 1953
Kippen	50	70	28	42	0.52	0.95	Skempton 1942
Screw pile, Lock Ryan	–	–	–	–	–	1.05	Morgan 1944, Skempton 1950
Screw pile, Newport	–	–	–	–	–	1.07	Wilson 1950
Oil tank, Fredrikstad	45	55	25	30	0.67	1.08	Bjerrum, Øverland 1957
Oil tank A, Shellhaven	70	87	25	62	0.73	1.03	Nixon 1949
Oil tank B, Shellhaven	–	–	–	–	–	1.05	Nixon (Skempton 1951)
Silo, U.S.A.	40	–	–	–	–	0.98	Tschebotarioff 1951
Loading test, Moss	9	–	–	–	–	1.10	NGI
Loading test, Hagalund	68	55	20	35	1.37	0.93	Odenstad 1949
Loading test, Torp	27	24	16	8	1.39	0.96	Bjerrum 1954 c
Loading test, Rygge	45	37	19	18	1.44	0.95	Bjerrum 1954 c

2. Fillings

Locality	W	LL	PL	PI	$\frac{W-PL}{PI}$	Safety factor $\Phi_u = 0$ analysis	Reference
Chingford	90	145	36	109	0.50	1.05	Skempton, Golder 1948
Gosport	56	80	30	50	0.48	0.93	Skempton 1948 d
Panama 2	80	111	45	66	0.53	0.93	Berger 1951
Panama 3	110	125	75	50	0.70	0.98	Berger 1951
Newport	50	60	26	34	0.71	1.08	Skempton, Golder 1948
Bromma II	100	–	–	–	1.00	1.03	Cadling, Odenstad 1950
Bocksjön	100	90	30	60	1.17	1.10	Cadling, Odenstad 1950
Huntington	400	–	–	–	–	0.98	Berger 1951

Table III.—End of Construction Failures in Excavations: $\Phi_u = 0$ Analysis.

Location	Soil type	Data of Clay					Factor of safety: $\Phi_u = 0$ analysis	Reference
		W	LL	PL	PI	$\frac{W-PL}{PI}$		
Huntspill	Intact clay	56	75	28	47	0.6	0.90	Skempton, Golder, 1948
Congress Street	Intact clay	24	33	18	15	0.4	1.10	Ireland, 1954
Skattsmanso I	Intact clay	101	98	39	59	1.05	1.06	Cadling, Odenstad, 1950
Skattmanso II	Intact clay	73	69	24	45	1.09	1.03	Cadling, Odenstad, 1950
Bradwell	Stiff-fissured clay	33	95	32	63	0.02	1.7	Imperial College, 1959 (Skempton, La Rochelle)

(b) The Stability of Cuts and Free-Standing Excavations in Clay.—

The changes in pore pressure and factor of safety during and after the excavation of a cut in clay are illustrated in Fig. 17.

The change in pore pressure can conveniently be expressed by putting B = 1 and re-arranging equ. (4) in the form:

$$\Delta u = [(\Delta\sigma_1 + \Delta\sigma_3)/2] + (A - 1/2)(\Delta\sigma_1 - \Delta\sigma_3) \qquad (23)$$

The reduction in mean principal stress will thus lead to a decrease in pore pressure, and the shear stress term will also lead to a decrease in pore pressure unless A is greater than 1/2, if the unknown effect on pore pressure of changing the directions of the principal stresses is neglected. An estimate of the stress distribution can be made from elastic theory if the initial factor of safety of the slope is high, or from the state of limiting equilibrium round a potential slip surface if the factor of safety is close to 1.

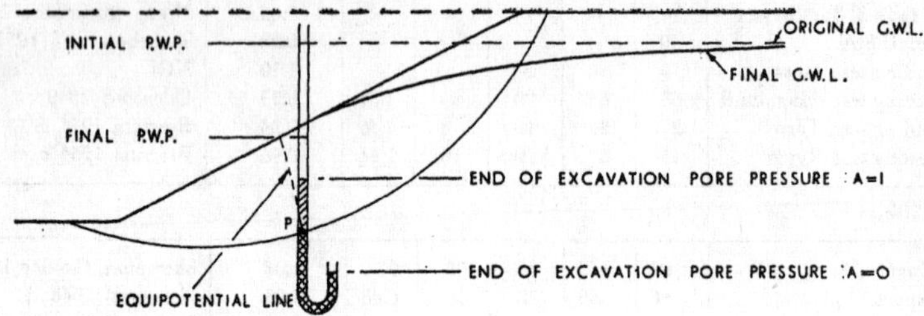

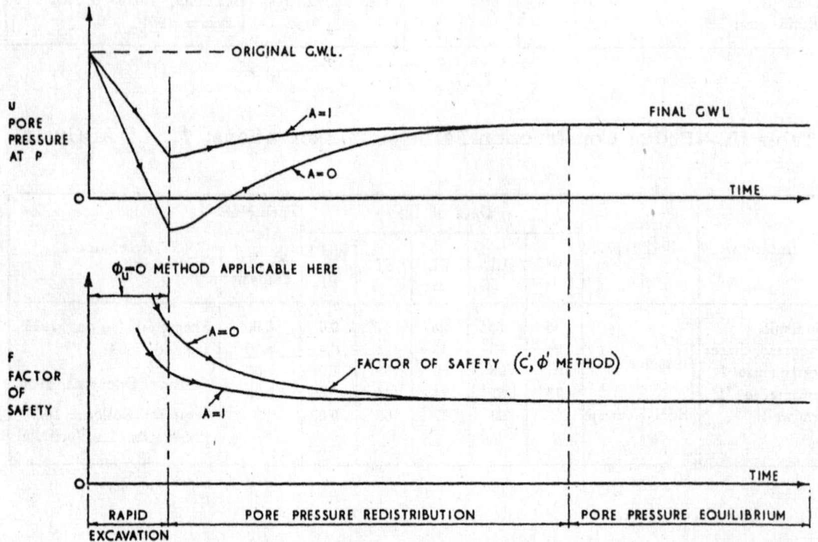

Fig. 17.—The changes in pore pressure and factor of safety during and after the excavation of a cut in clay.

In Fig. 17 the changes in pore pressure at a representative point are shown for the values A = 1 and A = 0. The final equilibrium values of pore pressure are taken from the flow pattern corresponding to steady seepage[s].

Using values of c' and Φ' from drained tests or consolidated-undrained tests expressed in terms of effective stress the factor of safety can be calculated at all stages from equation 14. In the majority of cases, unless special drainage measures are taken to lower the final ground water level, the factor of safety reaches its minimum value under the long term equilibrium pore pressure conditions.

An example of the investigation of a long term failure of a cut in terms of effective stress has been given by Sevaldson (1956). The slide took place in 1954 in a clay slope at Lodalen near Oslo, originally excavated about 30 years earlier (Fig. 18). Since the slide occurred without any apparent change in external loading, it can be considered to be the result of a gradual reduction in the stability of the slope. Extensive field investigations and laboratory studies were carried out to determine the pore pressure in the slope at the time of failure and the shear parameters of the clay.

Triaxial tests gave the values c' = 250 lb./sq. ft. and $\Phi' = 32^o$. An effective stress analysis using equation 14 gave a factor of safety of 1.05, and confirms the validity of the approach to within acceptable limits of accuracy.

Where the final pore pressures are obtained from a flow net not based on field measurements, allowance should be made for the fact that the permeability of a water laid sediment is generally greater in a horizontal direction (Sevaldson 1956). The highest wet season values obviously represent the most critical conditions.

The excavation of cuts in stiff fissured and weathered clays presents some special problems which have been discussed in detail by Terzaghi, 1936 a; Skempton, 1938 c; Henkel and Skempton, 1955; Henkel, 1957; etc. The reduction in stress enables the fissures to open up and they will then represent weak zones which a sliding surface will tend to follow. The fissures will also increase the bulk permeability of the clay (an increase of about 100 times is reported by Skempton and Henkel, 1960) so that the pore pressure rise leading to the long term equilibrium state will occur more rapidly.

The presence of fissures is reflected in the factors of safety obtained using the effective stress analysis with field values of pore pressure and values of c' and Φ' measured in the laboratory on 1-1/2" diameter samples. An analysis of three long-term cutting failures in London clay by Henkel (1957), using the laboratory values of c' = 250 lb./sq. ft. and $\Phi' = 20^o$, gave factors of safety of 1.32, 1.35, and 1.18. Putting c' = 0 gave values of 0.78, 0.81, and 0.82 respectively and obviously underestimated the factor of safety. If the value of c' required to give a factor of safety of 1 is plotted against time after construction (Henkel, 1957; De Lory, 1957), it is found that the value of c' shows a definite correlation with time (Fig. 19) and, as will be seen in section 6(c), appears to approach zero in natural slopes on a geological time scale.

An effective stress analysis based on a value of c' related empirically with time for each clay type will obviously give a close approximation to the correct factor of safety. For large scale work and for remedial measures on active slips where the large strains tend to reduce c', it is prudent to ensure a factor of safety of at least 1 with c' = 0.

s. This method has also been given since 1956 by Skempton in his lectures at Imperial College.

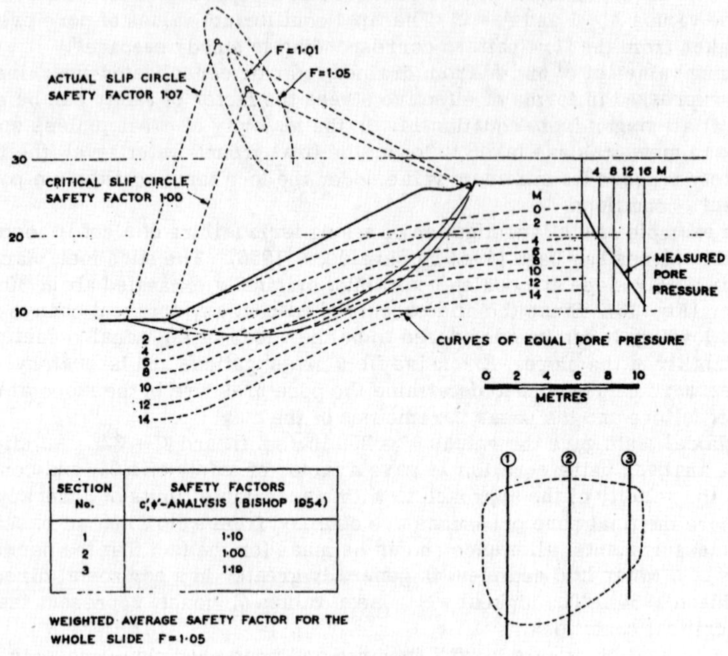

Fig. 18.—Long term failure in a cut at Lodalen (after Sevaldson, 1956).

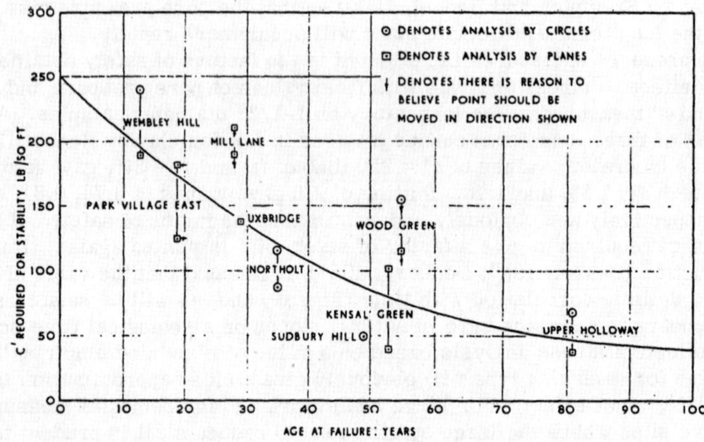

Fig. 19.—Long term failures in stiff-fissured London clay: Correlation of apparent cohesion c' required in effective stress analysis with age of cut at failure (after De Lory, 1957).

The mechanism of the drop in c' before a failure is initiated is not clearly understood, but may be associated with stress concentrations due to the presence of fissures, the progressive spread of an overstressed zone in a soil which tends to dilate and absorb water on shear, and the effect of cyclical fluctuations in effective stress due to seasonal water level changes. On the limited evidence so far available from Lodalen and Selset (see section 6(c)) it does not appear to occur in any marked way in non-fissured clays.

In temporary work, where the end of construction condition is of primary interest, the factor of safety may obviously be calculated by using the $\Phi_u = 0$ analysis and the undrained shear strength. This method may also be used with advantage where it is necessary to check that the initial factor of safety is not lower than the long term value, as it avoids the necessity of explicitly determining the stress distribution and pore pressure values at the end of construction. Four examples of its use are given in Table III.

Also included in Table III is an example of the use of the $\Phi_u = 0$ method for end of construction conditions in a cut in London clay, which led to an over-estimate of the factor of safety by 70%. Whether this is simply a consequence of the opening of fissures due to stress release, or due to changes in pore pressure even in the short period of excavation due to the high bulk permeability, is not yet clear. The reduction in strength at low stresses is apparent in stiff fissured clay even in undrained tests in the triaxial apparatus (Fig. 20), but hardly appears adequate to account for the 70% error. A conservative factor of safety must obviously be used in similar cases, and the rapid adjustment to the equilibrium pore pressure condition must be allowed for in prolonged construction operations.

(c) Natural Slopes.—

Natural slopes represent the ultimate long term equilibrium state of a profile formed by geological processes. The pore pressures are controlled by the prevailing ground water conditions which correspond to steady seepage, subject to minor seasonal variations in ground water level. Natural slopes therefore fall into class (a) in which the pore pressure is an independent variable.

In principle the analysis is the same as that of the long term equilibrium of a cut or excavation. However the pore pressures will have already reached their equilibrium pattern which can be ascertained from piezometer measurements in the field; and the natural processes of softening, leaching etc., will have already reached an advanced stage. An analysis based on laboratory tests of this material would therefore be expected to lead to close agreement with observed slopes in limiting equilibrium.

Relatively few natural slopes in limiting equilibrium have yet been analysed in terms of effective stress, but some representative examples are collected in Table IV. The two cases involving intact clays, Drammen and Selset, are being examined in greater detail, but the preliminary values of factor of safety of 1.15 and 1.03 respectively show that the method can be used with reasonable confidence. Had c' tended to zero the Selset slope would have shown a factor of safety of less than 0.7, for example, which is outside the limit of experimental error.

However, in stiff-fissured clays special account has to be taken of the progressive reduction in the value of c', which appears eventually to approach zero in the failure zone, since the shear strains and water content change associated with failure are very localized and tests on the bulk of the soil do not reveal the decrease in c'.

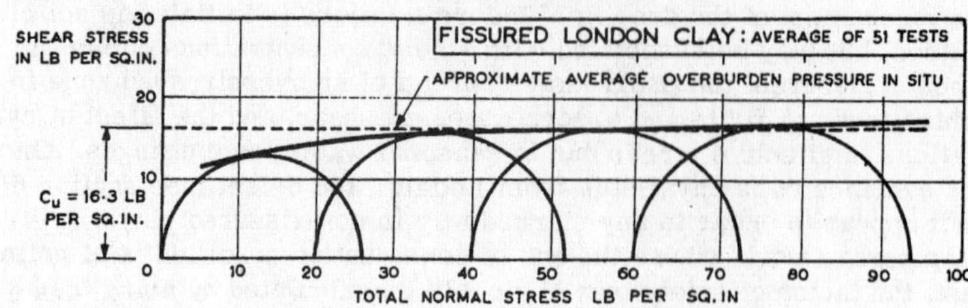

Fig. 20.—The strength of stiff-fissured London clay in undrained tests (after Bishop and Henkel, 1957).

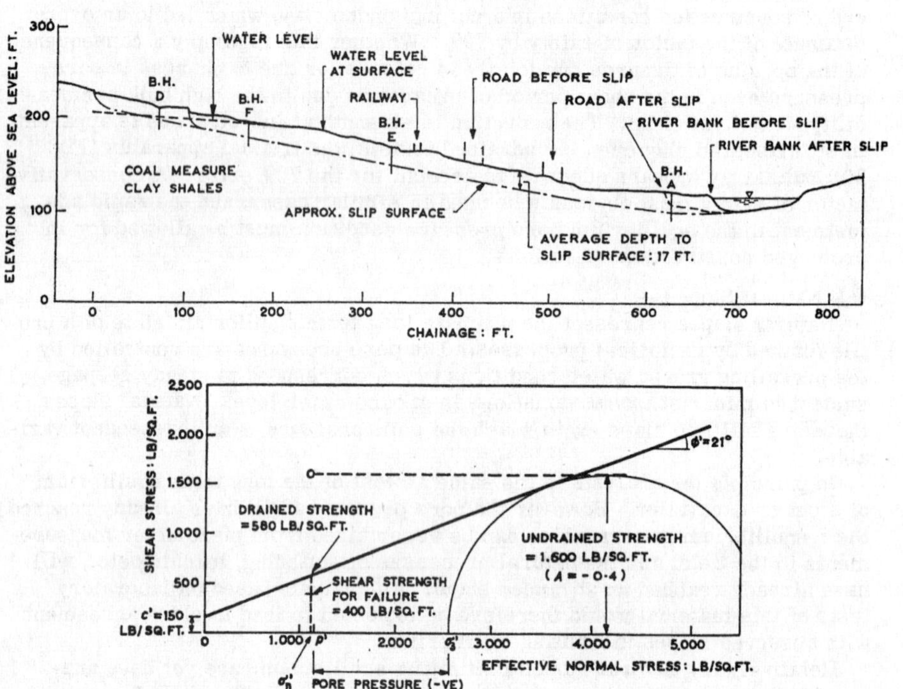

Fig. 21.—Natural slope failure in stiff-fissured clay at Jackfield (after Henkel and Skempton, 1955): Cross-section and shear strength data. Shear strength for limiting equilibrium, 400 lb./sq. ft. (F = 1.0). Undrained shear strength (c' = 150 lb./sq. ft. Φ' = 21° A = -0.4), 1.600 lb./sq. ft. (F = 4.0). Drained shear strength (c' = 150 lb./sq. ft. Φ' = 21°), 580 lb./sq. ft. (F = 1.45). Drained shear strength (c' =0 Φ' = 21°), 430 lb./sq. ft. (F = 1.07).

Table IV.—Failure of Natural Slopes: Effective Stress Analysis.

Location	Soil type	Data of Clay					Factor of safety: c', Φ' analysis	Reference
		W	LL	PL	PI	$\frac{W-PL}{PI}$		
Drammen	Normally Consolidated (Intact)	31	30	19	11	1.09	1.15	Bjerrum and Kjærnsli 1957
Selset	Overconsolidated (Intact)	13	26	13	13	0	1.03	Imperial College
Jackfield	Overconsolidated (stiff-fissured)	20	45	20	25	0	1.45 (1.07 with $c' = 0$)	Henkel and Skempton 1955

Table V.—Long Term Failures in Cuts and Natural Slopes: $\Phi_u = 0$ Analysis (after Bjerrum and Kjaernsli, 1957).

1. Overconsolidated, fissured clays

Locality	Type of slope	Data of clay					Safety factor, $\Phi_u = 0$ analysis	Reference
		W	LL	PL	PI	$\frac{W-PL}{PI}$		
Toddington	Cutting	14	65	27	38	-0.34	20	Cassel, 1948
Hook Norton	Cutting	22	63	33	30	-0.36	8	Cassel, 1948
Folkestone	Nat. slope	20	65	28	37	-0.22	14	Toms, 1953 b
Hullavington	Cutting	19	57	24	33	-0.18	21	Cassel, 1948
Salem, Virginia	Cutting	24	57	27	30	-0.10	3.2	Larew, 1952
Walthamstow	Cutting	–	–	–	–	–	3.8	Skempton, 1942
Sevenoaks	Cutting	–	–	–	–	–	5	Toms, 1948
Jackfield	Nat. slope	20	45	20	25	0.00	4	Henkel/Skempton, 1955
Park Village	Cutting	30	86	30	56	0.00	4	Skempton, 1948 c
Kensal Green	Cutting	28	81	28	53	0.00	3.8	Skempton, 1948 c
Mill Lane	Cutting	–	–	–	–	–	3.1	Skempton, 1948 c
Bearpaw, Canada	Nat. slope	28	110	20	90	0.09	6.3	Peterson, 1952
English Indiana	Cutting	24	50	20	30	0.13	5.0	Larew, 1952
SH 62, Indiana	Cutting	37	91	25	66	0.19	1.9	Larew, 1952

2. Overconsolidated, intact clays

Locality	Type of slope	W	LL	PL	PI	$\frac{W-PL}{PI}$	Safety factor, $\Phi_u = 0$ analysis	Reference
Tynemouth	Nat. slope	–	–	–	–	–	1.6	Imperial College
Frankton, N.Z.	Cutting	43	62	35	27	0.20	1.0	Murphy, 1951
Lodalen	Cutting	31	36	18	18	0.72	1.01	N.G.I.

3. Normally consolidated clays

Locality	Type of slope	W	LL	PL	PI	$\frac{W-PL}{PI}$	Safety factor, $\Phi_u = 0$ analysis	Reference
Munkedal	Nat. slope	55	60	25	35	0.85	0.85	Cadling/Odenstad, 1950
Säve	Nat. slope	–	–	–	–	–	0.80	Cadling/Odenstad, 1950
Eau Brink cut	Cutting	63	55	29	26	1.02	1.02	Skempton, 1945
Drammen	Nat. slope	31	30	19	11	1.09	0.60	N.G.I.

The landslide at Jacksfield provided a good example of the application of the effective stress analysis to a natural slope in stiff-fissured clay (Henkel and Skempton, 1955) and is illustrated in Fig. 21. The slope of the hillside is 10.5°, and when the slip took place in the winter 1951-52, a soil mass 600 feet by 700 feet and 17 feet in thickness moved gradually downward about 100 feet.

The calculated average shear stress in the clay was about 400 lb./sq. ft. Drained tests on undisturbed samples gave $c' = 150$ lb./sq. ft. and $\Phi' = 21°$, which with the observed pore pressures gave a shear strength of 580 lb./sq. ft. and a factor of safety of 1.45. Putting $c' = 0$ gave a shear strength of 430 lb./sq. ft. and a factor of safety of 1.07.

For natural slopes in stiff fissured clays it therefore appears necessary to use $c' = 0$ in the effective stress analysis. This is confirmed by observations made by Skempton and De Lory (1957) on the maximum stable natural slope found in London clay, and by Suklje (1953 a and b), and Nonveiller and Suklje (1955) in other fissured materials. It is interesting to speculate on whether the drop in c' is due to the fissures, or whether both are due to some more fundamental difference in the stress-strain-time relationships between the fissured and intact clays.

A second class of soil which gives rise to special problems includes very sensitive or quick clays. These clays show almost no strength in the remoulded state, and they will therefore tend to flow as a liquid if a slide occurs. A small initial slip in a slope may therefore have catastrophic consequences as the liquified clay will flow away and will not form a support for the exposed clay face, with the result that the whole of an otherwise stable slope may fail in a series of retrogressive slips taking place under undrained conditions.

A factor which affects the quantitative analysis in the case of quick clays is the influence of sample disturbance on the values of c' and Φ' measured in the laboratory. Soft clays of low plasticity are very sensitive to disturbance and reconsolidation in the triaxial test is always accompanied by a reduction in water content. Particularly where the initial water content is above the liquid limit laboratory tests appear to overestimate the value of Φ'. The investigation of a recent slide in quick clay in Norway has given a value of Φ' calculated from the statics of the sliding mass which is less than 50% of the value measured in the triaxial test.

The occurrence of quick clays is limited to certain well defined geological conditions, and where they are encountered special precautions in sampling, testing and analysis are always necessary (Holmsen, 1953; Rosenqvist, 1953; Bjerrum, 1954a and 1955c).

The application of laboratory tests to the stability of natural slopes raises two general matters of principle. The time scales of the load application are so different in the laboratory and in the field that it is perhaps surprising that satisfactory agreement between the results can be obtained at all. Laboratory results quoted by Bishop and Henkel (1957), and Bjerrum, Simons and Torblaa (1958) indicate that under certain conditions Φ' may have a lower value at low rates of loading. This effect may be partly offset by the fact that the worst ground water conditions which touch off the slip are only of seasonal occurrence; and by factors such as the effect of plane strain on the value of Φ' and the omission of 'end effects' in the stability analysis. It is also well known that considerable creep movements occur in slopes still classed at stable. However, with the exceptions noted, the overall correlation between laboratory and field results is quite acceptable from a practical point of view.

Secondly, it may well be asked why the factor of safety cannot be calculated with equal accuracy using the $\Phi_u = 0$ analysis and the undrained strength of samples from the slope where pore pressure and water content equilibrium have been attained. A large number of case records of slides in both natural slopes and cuttings are summarized in Table V and it is evident that as a practical method it is most unreliable, giving values of factor of safety ranging between 0.6 in sensitive clays to 20 in heavily overconsolidated clays.

The fundamental reason for the difference is that in the undrained test the pore pressure is a function of the stress applied during the test, and it is not necessarily equal to the pore pressure in situ. To obtain a factor of safety of 1 for a slope in limiting equilibrium using undrained tests would require that the same pore pressure should be set up in the sample when the in situ normal and shear stresses were replaced. This is in general prevented by the irreversibility of the stress-strain characteristics of the soil and by the changes in the principal stress directions. The latter occur even with in situ tests.

The position is made worse by the fact that the water content changes both in overconsolidated and in sensitive clays are very localized at failure and samples which do not pick up these layers can have little bearing on the stability analysis. A sample from the 2 inch thick slip zone at Jackfield in which large strains had occurred was found to have a water content 10% above the adjacent clay and an undrained strength within 12% of that required for a factor of safety of 1 (Henkel and Skempton, 1955). This layer was difficult to find and sample, and as clay outside the failure zone had an undrained strength nearly four times as great the method has little predictive value.

(d) Base Failure of Strutted Excavations in Clay.—

During excavation in soft clay, base failure sometimes occurs accompanied by settlement of the adjacent ground. Failures of this type[t] have occurred in excavations for basements, in trenches for water and sewage pipes, and in the shafts for deep foundations.

The construction of temporary excavations is generally carried out sufficiently rapidly for pore pressure changes to be ignored. The change in stress thus occurs under undrained conditions and the stability can be calculated using the $\Phi_u = 0$ analysis and undrained tests.

The factor of safety F against base failure can be derived from the familiar bearing capacity theory, considering the excavation as a negative load. This leads to the expression (Bjerrum and Eide, 1956):

$$F = N_c \cdot c_u/(\gamma D + q) \tag{24}$$

where D denotes depth of excavation
- γ » density of the clay
- c_u » the undrained strength of the clay beneath the bottom of the excavation
- q » the surface surcharge (if any)
- N_c » dimensionless bearing capacity factor depending on shape and depth of excavation.

t. Bottom heave failures can also occur in clay if a pervious layer containing water under sufficient head lies close beneath the excavation. For example see Garde-Hansen and Thernöe 1960, and Coates and Slade 1958.

The analysis of the failure of seven excavations is given in Table VI. The results indicate that in practice an accuracy of within ± 20% can be expected.

It should be noted that this type of failure is not caused by inadequate strutting, but the loads and distortion after its occurrence may initiate a more general collapse.

Table VI.—Base Failure of Strutted Excavations in Saturated Clay: $\Phi_u = 0$ Analysis (after Bjerrum and Eide, 1956).

Site	Dimensions $B \times L$: m	Depth D: m	Surcharge p: tons/sq. m	Density γ: tons/cu. m	Shear strength S: tons/sq. m	Sensitivity	B/L	D/B	N_c theoretical	Safety factor F	Average safety factor
1. Pumping station, Fornebu, Oslo	5.0×5.0	3.0	0.0	1.75	0.75	50	1.0	0.60	7.2	1.03	
2. Storehouse, Drammen	4.8× ∞	2.4	1.5	1.90	1.2	5–10	0.0	0.50	5.9	1.16	
3. Pier shaft, Göteborg	ø 0.9	25.0	0.0	1.54	3.5	20–50	1.0	28.0	9.0	0.82	
4. Sewage tank, Drammen	5.5×8.0	3.5	1.0	1.80	1.0	20	0.69	0.64	6.7	0.93	0.96
5. Test shaft (N) Ensjøveien, Oslo	ø 1.5	7.0	0.0	1.85	1.2	140	1.0	4.7	9	0.84	
6. Excavation, Grev Vedels pl., Oslo	5.8×8.1	4.5	1.0	1.80	1.4	5–10	0.72	0.78	7.0	1.08	
7. "Kronibus shaft", Tyholt, Trondheim	2.7×4.4	19.7	0.0	1.80	3.5	40	0.61	7.3	8.5	0.84	

(e) Earth Pressures on Earth Retaining Structures.—

If the displacement of an earth retaining structure is sufficient for the full development of a plastic zone in the soil adjacent to it, the earth pressure will be a function of the shear strength of the soil. This condition is apparently satisfied in much temporary work and in many permanent structures. The distribution of pressure is a function of the deformation of the structure and the soil, and can only be predicted after detailed consideration of the movements involved.

For temporary excavations in intact saturated clay it is generally sufficient to calculate the total earth pressure using the $\Phi_u = 0$ analysis and the undrained shear strength. Justification for this procedure is to be found in the field measurements published by Peck (1942), Skempton and Ward (1952), and Kjaernsli (1958). More recent measurements in soft clay carried out by the Norwegian Geotechnical Institute however indicate that the total earth pressure may exceed the value determined by the $\Phi_u = 0$ analysis and that the ratio of the actual to the calculated load increases with the number of struts used to carry it.

The long term earth pressure is logically computed using the effective stress analysis with values of c' and Φ' taken from drained tests or consolidated-undrained triaxial tests with pore pressure measurements together with the least favourable position of the water table. This will in most cases represent a rise in earth pressure. Few examples are available to confirm this analysis other than of gravity retaining walls which are themselves founded in the same clay stratum. The problem is then in effect one of overall stability since the slip surface passes beneath the wall which is then of little more consequence than one 'slice' in the slices method of analysis (Fig. 22).

A number of failures of this type in stiff fissured London clay have been analysed by Henkel (1957), and here consistent active and passive pressures on the walls have been obtained using the effective stress analysis and observed water levels, together with the reduced values of c' shown in Fig. 19. It should be noted that where the excavation in front of the wall is deep, the presence of the passive pressure is insufficient to prevent the occurrence of progressive softening.

The behaviour of gravity retaining walls can of course throw little light on the end of construction earth pressures in fissured clays. Measurements of strut load have however been made by the Norwegian Geotechnical Institute in a trench in the weathered stiff fissured crust overlying a soft clay stratum (Di Biagio and Bjerrum, 1957; Bjerrum and Kirkedam, 1958). Here the softening appeared to proceed more rapidly than in cuts in London clay, for after only a few months the strut loads corresponded to the value given by the effective stress analysis with c' = 0.

The evidence from this cut and the indirect evidence from the Bradwell slip (section 6(b)) indicates that the $\Phi_u = 0$ analysis does not correctly represent the behaviour of stiff fissured clays under decreasing stresses even shortly after excavation. The rapid dissipation of negative pore pressures due to the presence of open fissures is obviously an important factor in temperate climates. Under long term conditions the $\Phi_u = 0$ method is also inapplicable for the reasons given in sections 6(b) and (c).

(f) The Stability of Earth Dams.—

It is impossible to deal adequately with all the stability problems arising in earth dam construction in one short section. However, the most important

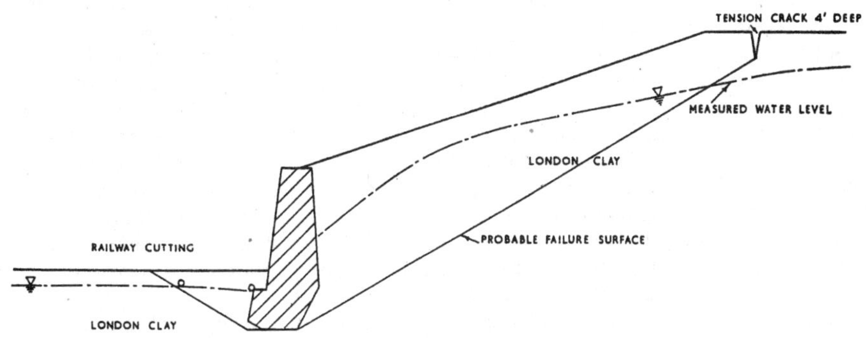

Fig. 22.—Retaining wall failure in stiff-fissured London clay (after Henkel, 1957).

principles may be illustrated by considering the stability of a water retaining dam built mainly of rolled earth fill (Fig. 23).

The stability of the slopes and foundation of an earth dam against shear failure will generally have to be considered under three conditions:

(1) During and shortly after construction,
(2) With the reservoir full (steady seepage), and
(3) On rapid drawdown of the impounded water.

Additional considerations arising from the possibility of failure in a clay foundation stratum are outlined in section 6(h). In this section attention will be limited to the fill.

The stability may be calculated for all three conditions in terms of effective stresses. This involves the measurement of c' and Φ' in the laboratory and an estimate of the pore pressure values at each stage. The use of explicitly determined pore pressures in the analysis enables the field measurements of pore pressure which are made on all important structures to be used as a direct check on stability during and after construction. It also enables the design estimates to be checked against the wealth of pore pressure data now becoming available from representative dams—for example the extensive work of the U.S.B.R. recently summarized by Gould (1959) and special cases such as the Usk dam (Sheppard and Aylen, 1957) and Selset dam (Bishop, Kennard and Penman, 1960).

To work directly with undrained test results expressed in terms of total stress may be unsafe in low dams as it implies dependence on negative pore pressures which will subsequently dissipate; and uneconomical in high dams in wet climates as no account is taken of the dissipation of excess pore pressure during the long construction period.

For the effective stress analysis the values of c' and Φ' are generally obtained from undrained triaxial tests with pore pressure measurement. In earth fill compacted at water contents well above the optimum a series of consolidated-undrained tests with pore pressure measurements may have to be used instead, in order to obtain a sufficient range of effective stresses to define a satisfactory failure envelope.

For the analysis of the condition of long term stability under steady seepage and for the case of rapid drawdown it is necessary to consider the effect of saturation on the values of c' and Φ'. As mentioned in section 4(h), test results show that, in general, the value of Φ' remains almost unchanged. Where c' has an appreciable value in the undrained tests this will decrease. However, tests using the improved techniques described by Bishop (1960) and Bishop, Alpan, Blight, and Donald (1960) have failed to reproduce the high cohesion intercepts previously reported in undrained tests. Provided c' has been accurately measured in either type of test, the differences may only be of significance in important works where the margin of safety is small. Whether c' is likely to become zero in rolled fill on a really long term basis is discussed by Bishop (1958 b) and Terzaghi (1958). Evidence so far does not appear to point to such a reduction.

The principal factors controlling the pore pressure set up during construction are:

(i) The placement moisture content and amount of compaction, and hence the pore pressure parameters;
(ii) The state of stress in the zone of the fill considered, and
(iii) The rate of dissipation of pore pressure during construction.

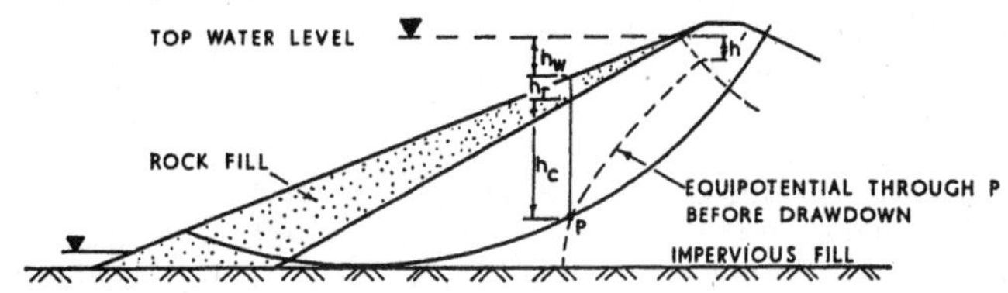

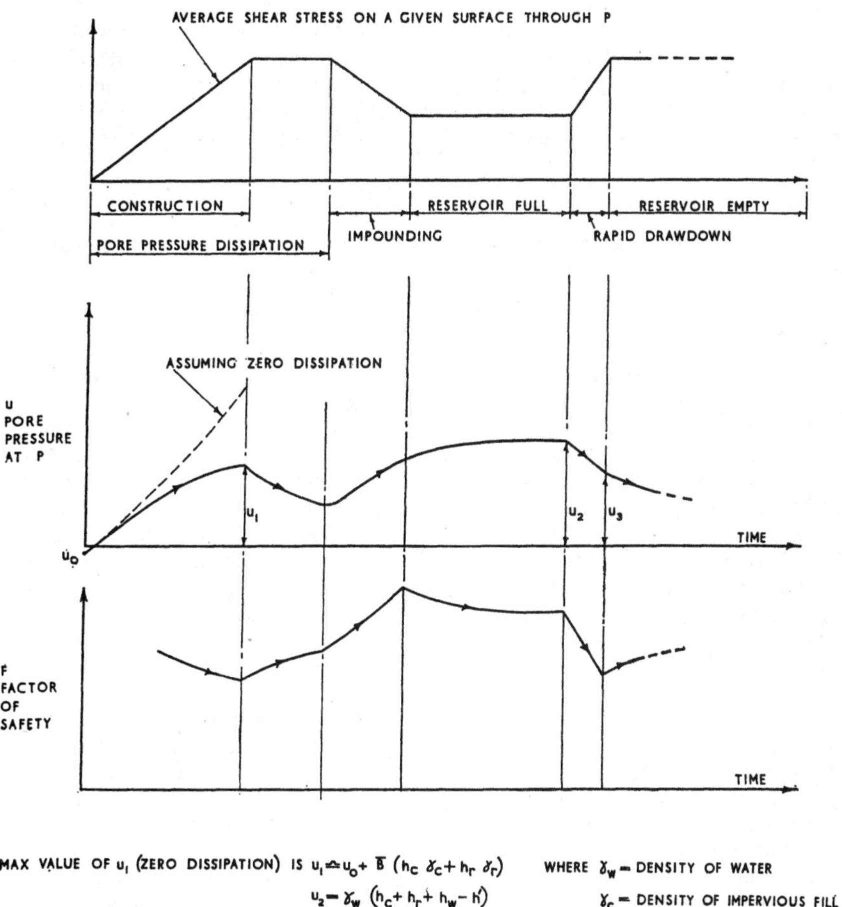

MAX VALUE OF u_1 (ZERO DISSIPATION) IS $u_1 \simeq u_0 + \bar{B}\,(h_c\,\gamma_c + h_r\,\gamma_r)$

$u_2 = \gamma_w\,(h_c + h_r + h_w - h')$

FOR FULL SATURATION ———— $u_3 \simeq \gamma_w\,[h_c + h_r\,(1-n) - h']$

WHERE γ_w — DENSITY OF WATER
γ_c — DENSITY OF IMPERVIOUS FILL
γ_r — DENSITY OF ROCK FILL
n — SPECIFIC POROSITY OF ROCK FILL

Fig. 23.—The changes in shear stress, pore pressure and factor of safety for the upstream slope of an earth dam.

In section 3 it was shown that the pore water pressure set up under undrained conditions can be expressed in the form:

$$u = u_o + \bar{B} . \Delta\sigma_1 \qquad (7)$$

In Fig. 24 the values of u_0 and $\bar{B}$ are plotted against water content for a series of samples prepared with the compactive effort used in the standard compaction test. This clearly shows the sensitivity of the value of the initial pore pressure to the placement water content; the importance of this effect, both in design and construction, cannot be overemphasised.

It is usually assumed that the value of σ_1 is equal to the vertical head of soil γ. h above the point considered, although the direction in which σ_1 acts is not necessarily vertical. This is a reasonably satisfactory assumption when averaged around a complete slip surface, but tends to overestimate the pore pressure in the centre of the dam and underestimate it near the toes (Fig. 25, after Bishop, 1952). It enables the pore pressure ratio required for the stability analysis to be expressed, under undrained conditions as

$$r_u = \bar{B} + u_o/\gamma h \qquad (9)$$

or, at the higher water contents where $\bar{B}$ is large and u_0 small, more simply as

$$r_u = \bar{B} \qquad (10)$$

However, in most earth fills a considerable reduction in the average pore pressure results from dissipation even during the construction period. A numerical method of solving the practical consolidation problem with a moving boundary has been given by Gibson (1958). It should be noted that in many almost saturated soils even a small amount of drainage has a marked effect on the final pore pressure, since it not only reduces the pore pressure already set up, but also reduces the value of $\bar{B}$ under the next increment of load. The theoretical basis of this reduction is discussed by Bishop (1957), and it is confirmed by field results from the Usk dam (Fig. 26a).

As an example of the distribution of pore pressure at the end of construction the contours from the Usk dam are illustrated in Fig. 26b. The effectiveness of the drainage layers placed to reduce the average pore pressure in the fill will be apparent.

The average pore pressure ratio along a potential slip surface at the end of construction may be kept within safe limits either by restricting the size of the impervious zone, by strict placement water content control or by special drainage measures (as at the Usk and Selset dams). Which is the more economical procedure will depend on the climatic conditions and the fill materials available.

For reservoir full conditions the pore pressure distribution may be predicted from the flow net corresponding to steady seepage. Accuracy is difficult to obtain owing to the non-uniformity of the rolled fill and differences in the ratio of horizontal to vertical permeability, so conservative assumptions should be made. However, with properly placed drainage zones the average value of r_u for the downstream slope is generally less than during construction, except in low dams or with rather dry placement.

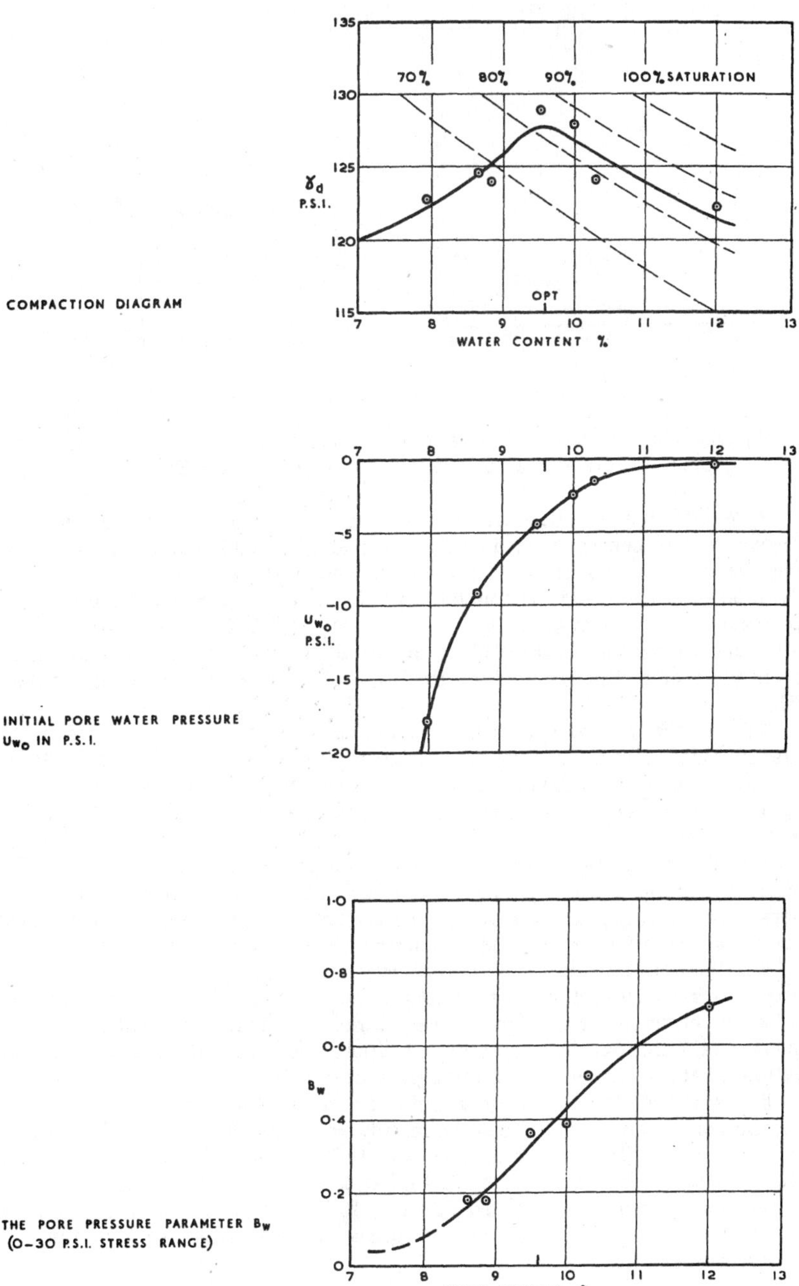

Fig. 24.—Represents Values from Standard Compaction Test under Equal All Round Pressure Increase: Compacted Boulder Clay, Clay Fraction 6%

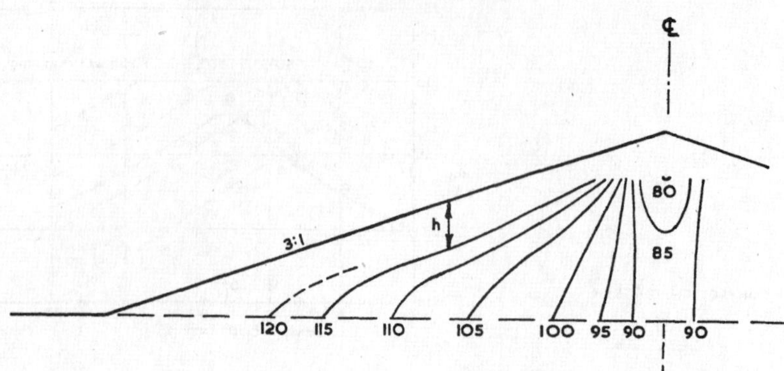

Fig. 25.—Major Principal Stress σ, as a Percentage of γ . h, the Vertical Head of Soil Above the Point Considered (After Bishop, 1952).

A method of predicting the excess pore pressures resulting from rapid drawdown has been proposed by Bishop (1952 and 1954a). In this method the change in pore pressure on drawdown is assumed to take place under undrained conditions and is deduced from the stress change and the pore pressure parameters (see Fig. 23). For saturated fills the value of $\bar{B}$ is taken as 1; the change in the value of σ_1 is due to the removal of the water load from the face of the dam and the drainage of water from the voids of the rockfill.

This method shows reasonable agreement with the results of the field measurements on the Alcova dam (Glover, Gibbs and Daehn, 1948). Two recent cases of very rapid drawdown soon after completion are not in such good agreement, but both involve complicating factors (Bazett, 1958; Paton and Semple, 1960).

Fig. 23 shows diagrammatically the variation in pore pressure and factor of safety for the various phases in the life of the upstream slope of the embankment calculated as described above. The lowest values of factor of safety are usually reached at the end of construction and on rapid drawdown.

For the downstream slope, end of construction and steady seepage are the two critical stages. However, during steady seepage the danger is generally not so much from the pore pressures, which are easily controlled by drainage measures, but from the possibility of piping and internal erosion in the foundation strata, and from crack formation in the fill.

For small earth dams built largely of saturated soft clay, the stability during the construction period can be calculated by the $\Phi_u = 0$ analysis using the undrained strength (Cooling and Golder, 1942). However, where the fill is much stronger than the clay foundation strata satisfactory results are not obtained (for example, Golder and Palmer, 1955) for the reasons given in section 6(a). The long term stability can of course be determined only by the effective stress analysis.

(g) Stability of Slopes in Sand and Gravel on Drawdown.—

In relatively pervious soils of low compressibility the distribution of pore pressure on drawdown is controlled by the rate of drainage of pore water

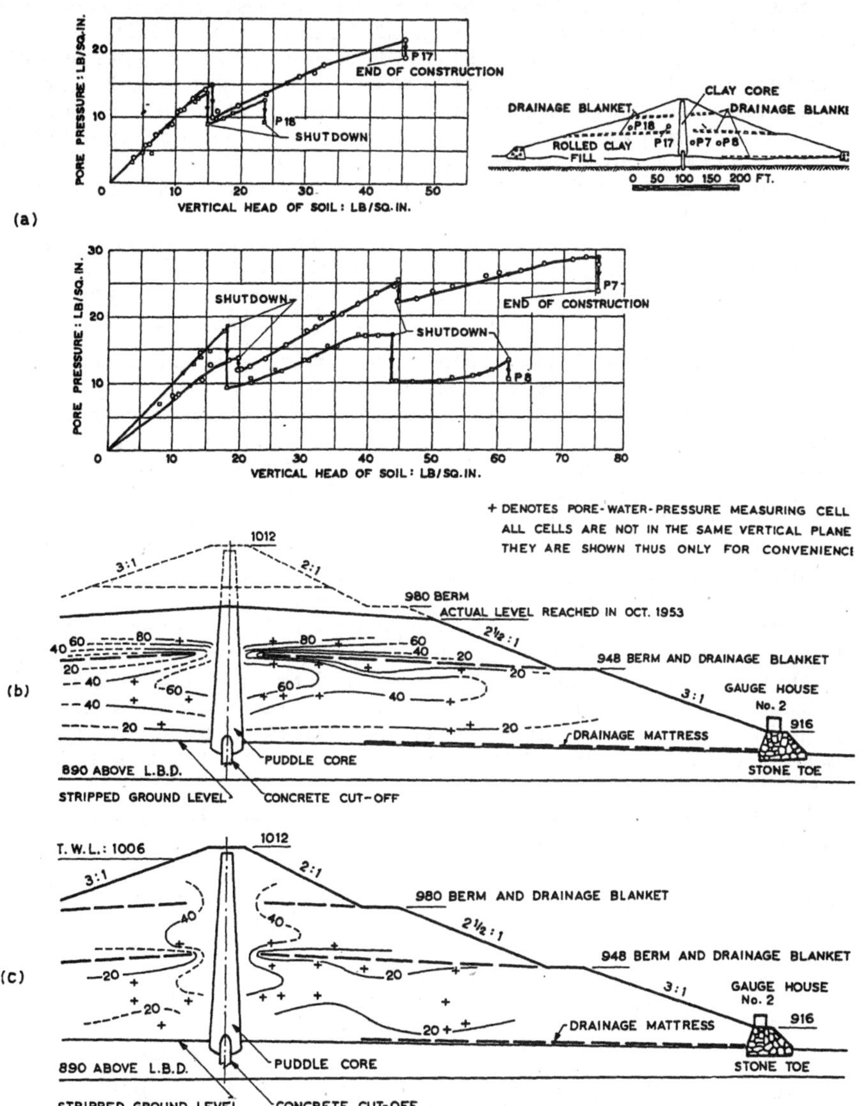

Fig. 26.—Field measurements of pore pressure from the Usk dam: (a) The reduction in pore pressure and in the value of $\bar{B}$ due to pore pressure dissipation (after Bishop, 1957); (b) and (c) contours of pore pressure expressed as a percentage of the vertical head of soil in October, 1953 and October 1954 (after Sheppard and Aylen, 1957).

from the soil. This condition can be represented by a series of flownets with a moving boundary as shown by Terzaghi (1943) and Reinius (1948).

The flow pattern is a function of the ratio of drawdown rate to permeability and the values of the pore pressures to be used in the stability analysis can be taken from the appropriate flow-net. The influence of the greater permeability in the horizontal direction is considerable, but, in one case examined, tended to increase rather than reduce the factor of safety.

The values of c' and Φ' are obtained from drained tests, c' approaching zero for free-draining materials.

An example of a drawdown failure in Thames gravel is shown in Fig. 27. The initial slope of the gravel was 33° and the permeability about 0.05 cm/sec. The value of Φ' in the loose state was 36°. Failure occurred when the pool was lowered at a rate of about 1 foot per day (Bishop, 1952).

(h) The Stability of a Clay Foundation of an Embankment where the Rate of Construction Permits Partial Consolidation.—

It is not uncommon in earth dam construction to encounter geological conditions in which the foundation strata include a soft clay layer at or near the surface, of sufficient extent to be likely to lead to failure in an embankment having conventional side slopes (for example Cooling and Golder, 1942; McLellan, 1945; Bishop, 1948; Skempton and Bishop, 1955; Bishop, Kennard and Penman, 1960). It is then necessary to assess the economics and practicability of a number of alternative solutions. The soft layer may be excavated, if its depth and ground water conditions permit; or an embankment with very flat slopes may be accepted, its factor of safety being calculating using the $\Phi_u = 0$ method which assumes zero drainage. Alternatively, account may be taken of the dissipation of pore pressure which occurs due to natural drainage (for example, Bishop, 1948) or due to special measures, such as vertical sand drains, designed to accelerate consolidation (for example, Skempton and Bishop, 1955; Bishop, Kennard and Penman, 1960). In this case an effective stress analysis is used.

An expression for the initial excess pore pressure in a saturated soft clay layer where B = 1 has been obtained by Bishop (1952):

$$\Delta u = \Delta p + p_o \,.\, [(1 - K)/2] + (2A - 1)\sqrt{p_o^2[(1 - K)/2]^2 + \tau^2} \qquad (25)$$

where Δp denotes change in total vertical stress due to the fill,
τ » shear stress along the layer set up by the fill,
p_o » initial vertical effective stress,
and Kp_o » initial horizontal effective stress in clay layer.

This expression illustrates the dependence of pore pressure on the change in shear stress as well as on the change in vertical stress, though the latter predominates in most practical cases. To avoid the error in the estimate of A which may arise from the change in void ratio on reconsolidation of undisturbed samples, the value of A may be deduced from the relation between the undrained strength and the effective stress envelope, using an assumed value of K. The relationship is given by Bishop (1952).

The estimate of the rate of dissipation of pore pressure is based on the theory of consolidation. It is here that the greatest uncertainity arises, especially in stratified deposits, and field observations of pore pressure are advisable on important works.

Fig. 27.—Drawdown failure in Thames Valley gravel (after Bishop, 1952).

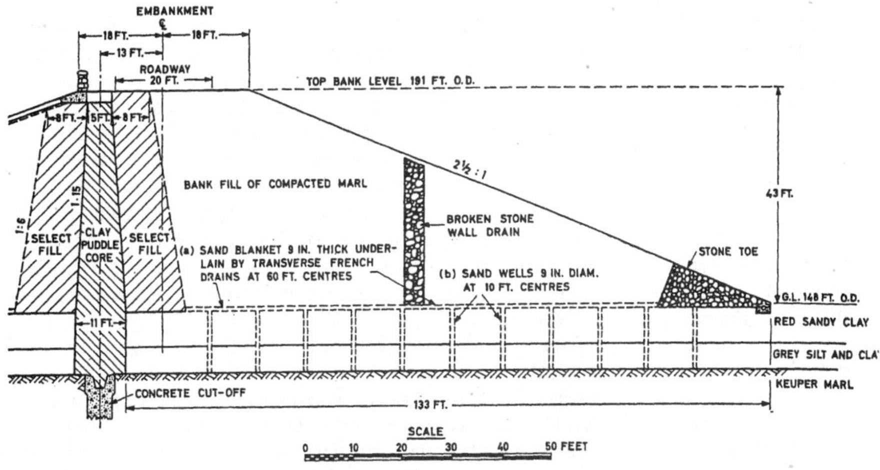

Fig. 28.—Downstream slope of the Chew Stoke dam showing vertical sand drains to accelerate dissipation of pore pressure in soft clay foundation (after Skempton and Bishop, 1955).

The values of c' and Φ' are taken from drained tests or consolidated undrained tests with pore pressure measurement.

The downstream slope of the Chew Stoke dam (Fig. 28) which had a factor of safety against a foundation failure of 0.8 using the $\Phi_u = 0$ analysis was safely constructed using a sand drain spacing designed to give a factor of safety of 1.5 (Skempton and Bishop, 1955). Field observations of pore pressure indicated that the actual factor of safety was rather higher than 1.5 owing to the greater horizontal permeability resulting from stratification of the clay. The Selset dam, founded on a boulder clay with little apparent stratification, showed a smaller difference between predicted and observed pore pressure values (Bishop, Kennard and Penman, 1960).

(i) Some Special Cases.—

In the examples described above the variation in safety factor with time was either a steady increase or a steady decrease during the period from the end of construction until the pore pressures reached an equilibrium condition. The following examples will illustrate that under certain conditions we may temporarily encounter a lower factor of safety at an intermediate stage.

Such cases are obviously very dangerous, as a failure might well occur some weeks or months after the completion of construction, in spite of the fact that it had been ascertained that the factor of safety was adequate in both the initial and final stages. The basic reason in each case is that the redistribution of excess pore pressure which occurs during the consolidation process may lead to a temporary rise in pore pressure outside the zone where the load is applied.

An interesting example is the stability of a river bank in a clay stratum under the action of the excess pore pressure set up by pile driving for a bridge abutment in the vicinity (Bjerrum and Johannessen, 1960). The changes in pore pressure with time are shown diagrammatically in Fig. 29a for two points, one in the centre of the pile group and one outside it, but beneath the slope. The excess pore pressures set up by pile driving will dissipate laterally as well as vertically, particularly in a water-laid sediment where the horizontal permeability tends to be greater than the vertical.

The exact magnitude of the effect is difficult to predict theoretically, and in this case field measurements of pore pressure were used to regulate the progress of pile driving. The most critical distribution of pore pressure occurred shortly after piling was completed (Fig. 29b). The factor of safety of the slope was calculated using the effective stress analysis with c' = 200 lb./sq. ft. and $\Phi' = 27°$, and dropped from an initial value of 1.4 to 1.15 after piling, assuming low water in the river in each case.

A case in which the spread of pore pressure led to an embankment failure some time after construction has been analysed in detail by Ward, Penman and Gibson (1955). The clay foundation on which the embankment was built included two horizontal layers of peat. Due to the relatively higher permeability of the peat, the redistribution of pore pressure after the end of construction resulted in temporary high pore pressures in the peat on both sides of the embankment, where initially the pore pressures were low. Because of the difference between the void ratio pressure curves for consolidation and swelling, the reduction in effective stress due to the presence of an additional volume of pore water is much greater than the increase in effective stress in the zone from which this volume has migrated. An overall decrease in effective stress along a critical slip path can thus occur (Fig. 30) at an

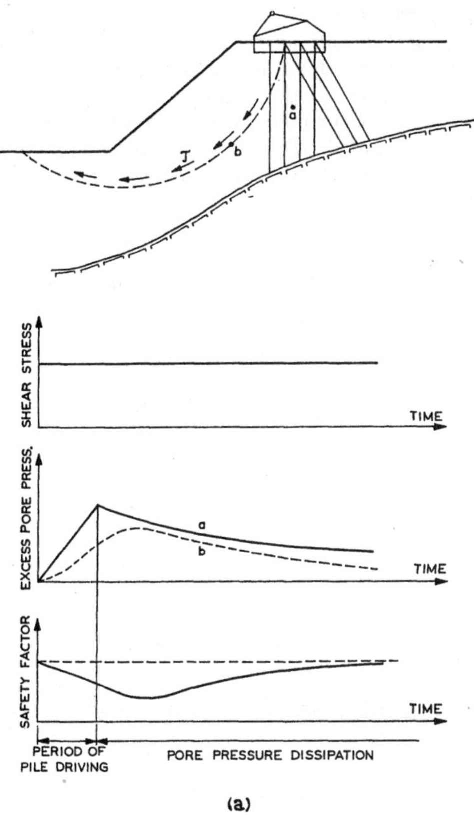

(b)

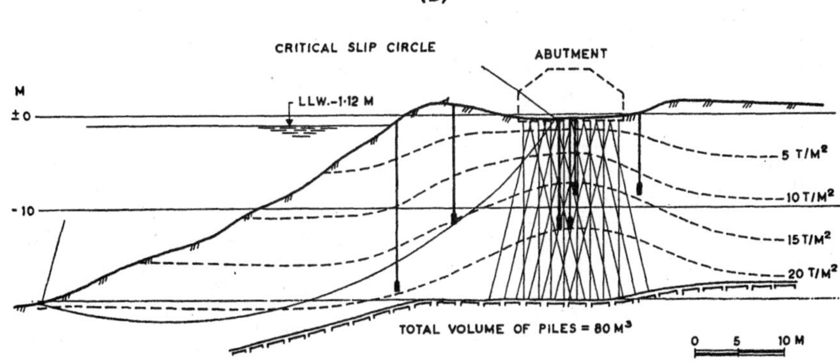

Fig. 29.—The effect of pore pressure set up by pile driving on the stability of an adjacent clay slope: (a) Changes in pore pressure and factor of safety with time: diagrammatic, (b) observed pore pressures just after the completion of pile driving (after Bjerrum and Johannessen, 1960).

intermediate stage, although in the long term equilibrium state the bank foundation would have been stable.[u]

Other engineering operations may result in a similar danger, such as the rapid construction of an embankment or stockpile even some way back from a river bank, cut or quay wall close to limiting equilibrium, and the driving of ordinary piles or screw piles through the clay slopes of rivers or harbours. In such cases an awareness of the danger will either lead to a modification of the operation, or to the use of field measurements of pore pressure as a control while it is carried out.

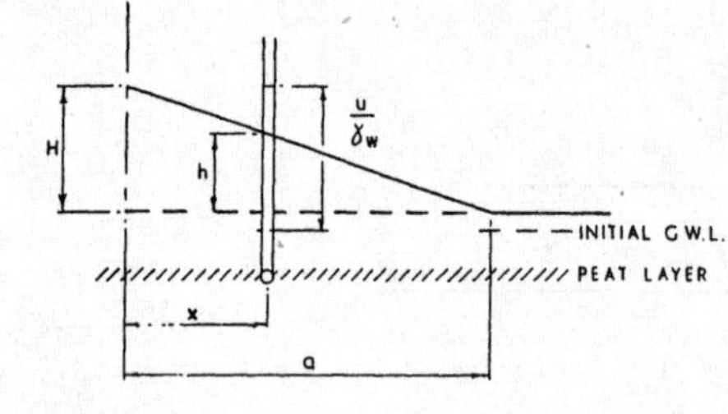

(*a*) Embankment on a clay foundation with a horizontal peat layer–diagrammatic;

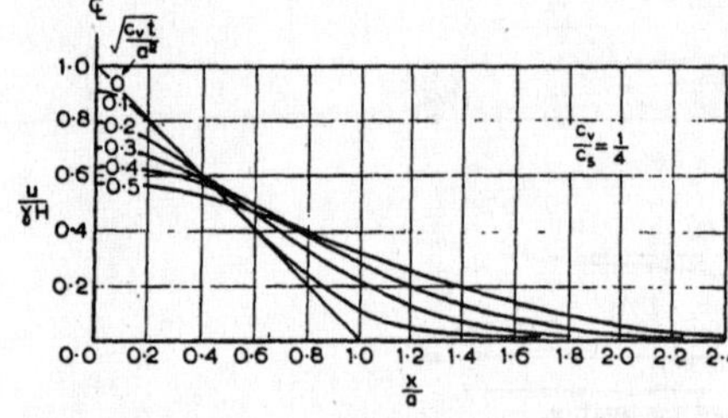

(*b*) changes in pore pressure with time, as a fraction of $\gamma \cdot H$ where γ is density of fill;

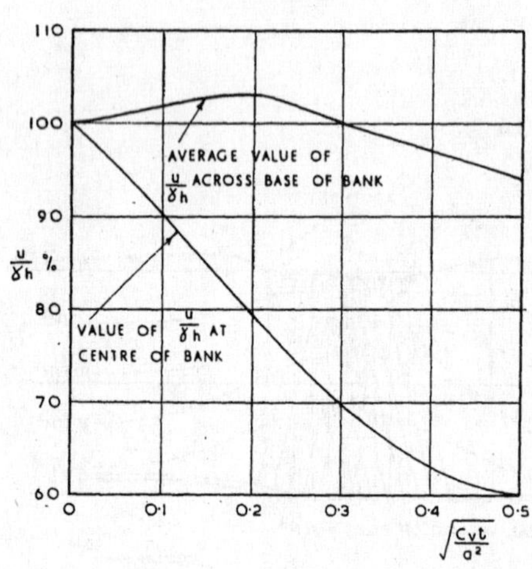

(*c*) effect of redistribution on maximum and average pore pressure for $c_v/c_s = 1/4$ where c_v is coefficient of consolidation and c_s is coefficient of swelling (after Ward, Penman, and Gibson, 1954).

Fig. 30.

u. Other failures of this type are described by Terzaghi and Peck (1948).

It is probable that a number of failures attributed to 'creep' could more correctly be attributed to the redistribution of pore pressure which occurs after construction.

(j) Design in Earthquake Areas.—

The analysis of the stability of a structure or dam in an area subject to earthquakes raises special problems which are outside the scope of this paper.

It is, however, known that a transient load will leave residual excess pore pressures which may be positive or negative depending on the void ratio and stress history of the soil (Bishop and Henkel, 1953). A possible way of evaluating the stability under earthquake conditions may therefore be to use a series of consolidated-undrained tests in which the stress ratio during consolidation is chosen to represent the conditions prevailing in the field before the earthquake. The sample is then subjected to a series of small variations in deviator stress under undrained conditions corresponding to the additional seismic stresses. The magnitude of the residual pore pressure and the additional strain will indicate the likelihood of failure under field conditions.

A discussion of the additional shear stresses likely to be set up in earthquake areas is given by Ambraseys (1959).

7. CONCLUSIONS

The discussion and case records presented in this paper point to four main conclusions:

(I) The effective stress analysis is a generally valid method for analysing any stability problem and is particularly valuable in revealing trends in stability which would not be apparent from total stress methods.

Its application in practice is limited to cases where the pore pressures are known or can be estimated with reasonable accuracy. These include all the class (a) problems, such as long term stability and drawdown in incompressible soils, where the pore pressure is controlled by ground water conditions or by a flow pattern. It is also applicable to both class (a) and class (b) problems where field measurements of pore pressure are available.

Those class (b) problems where the magnitude of the pore pressure has to be estimated from the stress distribution and the measured values of the pore pressure parameters can often be solved more simply by the $\phi_u = 0$ analysis. However, this alternative gives no indication of the long term stability and does not enable account to be taken of dissipation of pore pressure during construction, which may contribute greatly to economy in design.

(II) Where a saturated clay is loaded or unloaded at such a rate that there is no significant dissipation of the excess pore pressures set up, the stability can be determined by the $\phi_u = 0$ analysis, using the undrained strength obtained in the laboratory or from in-situ vane tests.

This method is very simple and reliable if its use is restricted to the conditions specified above. It is essentially an end of construction method, and in the majority of foundation problems, where the factor of safety increases with time, it provides a sufficient check on stability. For cuts, on the other hand, where the factor of safety generally decreases with time, the $\phi_u = 0$ method can be used only for temporary work and the long term stability must be calculated by the effective stress analysis.

(III) The two methods of analysis require the measurement of the shear strength parameters c' and Φ' in terms of effective stress on the one hand and the undrained shear strength c_u under the stress conditions obtaining in the field on the other.

For saturated soils the values of c' and Φ' are obtained from drained tests or consolidated undrained tests with pore pressure measurement, carried out on undisturbed samples. The range of stresses at failure should be chosen to correspond with those in the field. Values measured in the laboratory appear to be in satisfactory agreement with field records with two exceptions. In stiff fissured clays the field value of c' is lower than the value given by standard laboratory tests; in some very sensitive clays the field value of Φ' is lower than the laboratory value.

For partly saturated soils the values of c' and Φ' are obtained from undrained or consolidated-undrained tests with pore pressure measurement, or from drained tests. Provided comparable testing procedures are used the differences between the values of Φ' obtained appear not to be significant from a practical point of view. The values of c' will be slightly influenced by moisture content differences resulting from the different procedures.

The undrained shear strength c_u is obtained from undrained triaxial tests on undisturbed samples (or from unconfined compression tests, except on fissured clays) and from vane tests in situ. It cannot be obtained, without risk of error on the unsafe side, from consolidated-undrained tests where the sample is reconsolidated under the overburden pressure. The error is serious in normally consolidated clays of low plasticity, and though it can be minimised by consolidating under the stress ratio obtaining in the field, the effect of reconsolidation on the void ratio cannot be avoided.

For this same reason it is probably more realistic to calculate the value of the pore pressure parameter A for undisturbed soil from the relationship between the undrained strength of undisturbed samples and the values of c' and Φ', rather than to measure it in a consolidated-undrained test.

(IV) The reliability of any method can ultimately be checked only by making the relevant field measurements when failures occur or when construction operations are likely to bring a soil mass near to limiting equilibrium. The number of published case records in which the data is sufficiently complete for a critical comparison of methods is still regrettably small.

ACKNOWLEDGMENTS

The work of K. Terzaghi, Arthur Casagrande, A. W. Skempton and the late D. W. Taylor has contributed so much to the background of any study of shear strength and stability that specific references in the text are inadequate acknowledgment. The authors would also like to express their gratitude to their colleagues at Imperial College and the Norwegian Geotechnical Institute for valuable comments on the manuscript.

APPENDIX I.—THE USE OF THE PARAMETER Φ_{cu}

In section 4b reference has been made to errors likely to arise in applying in the field the relationship between undrained strength and consolidation pressure obtained in the laboratory from the consolidated-undrained test.

Two inherent errors have been referred to: The effect of reconsolidation after sampling on the void ratio and on the value of the pore pressure parameter A; and the error arising from consolidation under a stress ratio different from that obtaining in the ground. A further error may arise from the way in which the results are introduced into the stability analysis.

This point is illustrated in Fig. 31 (after Bishop and Henkel, 1957). The test is usually performed by consolidating the sample under a cell pressure p, and then causing failure under undrained conditions by increasing the axial stress. The total minor principal stress at failure (σ_3) is thus equal to p; the total major principal stress is $(\sigma_1)_{cu}$. The slope of the envelope to a series of total stress circles obtained in this manner (Fig. 31a) is denoted ϕ_{cu}, the angle of shearing resistance in consolidated undrained tests, and is about one half of the slope of the effective stress envelope (denoted by ϕ') for normally consolidated samples. This relationship between shear strength and total normal stress can only be used in practice if the identity between consolidation pressure and total minor principal stress imposed in the test also

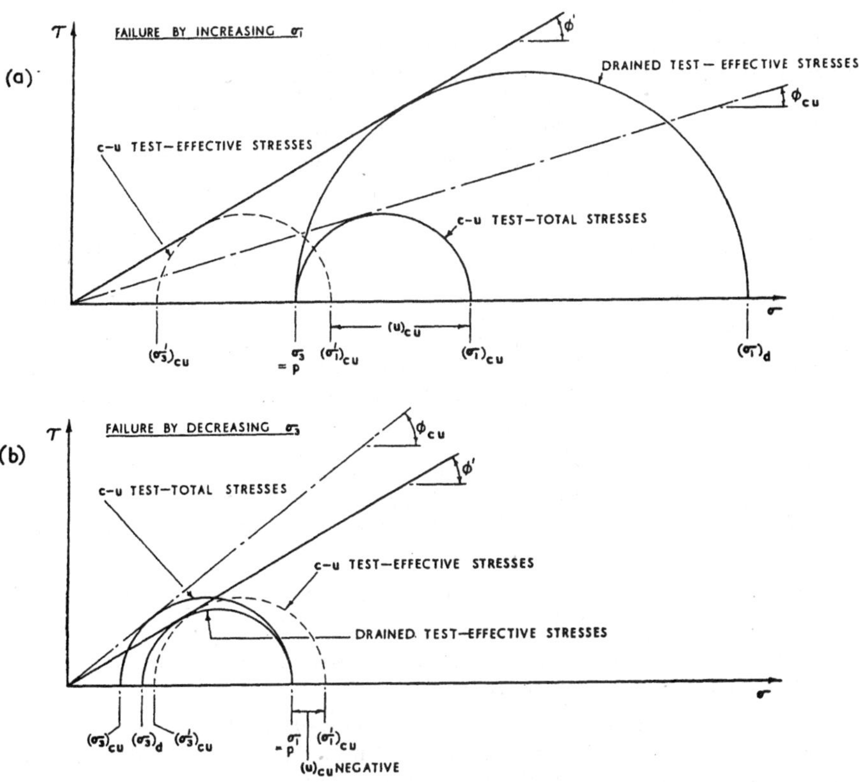

Fig. 31.—The consolidated-undrained test on a saturated cohesive soil in terms of total and effective stresses: (a) Failure by increasing major principal stress σ_1; (b) failure by decreasing minor principal stress σ_3 (after Bishop and Henkel, 1957).

applies around the slip surface considered. Passive earth pressure appears to be the only case in which this is approximately true.

Had the failure been caused by holding σ_1 constant and equal to p and decreasing the total minor principal stress σ_3, the undrained strength would have remained the same, and a radically different value of Φ_{cu} would have been obtained, Fig. 31b. The relationship between shear strength and total normal stress would then approximate to the case of active earth pressure.

The general use of Φ_{cu} (defined in Fig. 31a) as an angle of shearing resistance in conventional stability analyses is therefore likely to lead to very erroneous results, even if the samples are anisotropically consolidated. If σ_3 increases during the undrained loading (as in foundation problems) the factor of safety will be overestimated; if σ_3 decreases, as in the excavation of a cutting, the error may lead to an under-estimate of the factor of safety.

The most logical solution appears to be to plot contours of undrained strength in terms of the consolidation pressure in the ground prior to the undrained loading to be examined, and then to use the $\Phi_u = 0$ analysis. This method is of course limited to the end of construction analysis, in which it is assumed that insufficient time has elapsed for consolidation or swelling to occur.

In rapid drawdown analyses suggested by Terzaghi (1943) and Lowe and Karafiath (1959) the undrained strength is related to the effective normal stress on the potential failure plane before drawdown. However, unless the samples are failed by reducing the stresses, there is a danger of overestimating the undrained strength of compacted samples which are difficult to saturate fully in the laboratory.

APPENDIX II.—BIBLIOGRAPHY ON SHEAR STRENGTH AND STABILITY

1. Ambraseys, N. N. (1959), The seismic stability of earth dams. Thesis. (University of London). London. 2 vol.
2. Bazett, D. J. (1958), Field measurement of pore water pressures. Canadian Soil Mechanics Conference, 12. Saskatoon. Proceedings, p. 2-15.
3. Berger (1951), Unpublished report.
4. Bishop, A. W. (1948), Some factors involved in the design of a large earth dam in the Thames valley. International Conference on Soil Mechanics and Foundation Engineering, 2. Rotterdam. Proceedings, Vol. 2, p. 13-18.
5. Bishop, A. W. (1952), The stability of earth dams. Thesis. (University of London). London. 176 p.
6. Bishop, A. W. (1954 a), The use of pore pressure coefficients in practice. Geotechnique, Vol. 4, No. 4, p. 148-152.
7. Bishop, A. W. (1954 b), The use of the slip circle in the stability analysis of slopes. European Conference on Stability of Earth Slopes, Stockholm. Proceedings, Vol. 1, p. 1-13. Geotechnique, Vol. 5, No. 1, 1955, p. 7-17.
8. Bishop, A. W. (1957), Some factors controlling the pore pressures set up during the construction of earth dams. International Conference on Soil Mechanics and Foundation Engineering, 4. London. Proceedings, Vol. 2, p. 294-300.

9. Bishop, A. W. (1958 a), Test requirements for measuring the coefficient of earth pressure at rest. Brussels Conference on Earth Pressure Problems. Proceedings, Vol. 1, p. 2-14.

10. Bishop, A. W. (1958 b), Discussion on: Terzaghi, K. Design and performance of the Sasumua dam. Institution of Civil Engineers. Proceedings, Vol. 11, November, p. 348-352.

11. Bishop, A. W. (1959), The principle of effective stress. Teknisk ukeblad, Vol. 106, No. 39, p. 859-863. (Norwegian Geotechnical Institute. Publ., 32.)

12. Bishop, A. W. (1960), The measurement of pore pressure in the triaxial test. Pore Pressure and Suction in Soil Conference, London, p. 52-60.

13. Bishop, A. W., Alpan, J., Blight, G. and Donald, V. (1960), Factors controlling the strength of partly saturated soils. Research Conference on Shear Strength of Cohesive Soils. Proceedings.

14. Bishop, A. W. and Eldin, G. (1950), Undrained triaxial tests on saturated sands and their significance in the general theory of shear strength. Geotechnique, Vol. 2, No. 1, p. 13-32.

15. Bishop, A. W. and Eldin, A. K. G. (1953), The effect of stress history on the relation between Φ and porosity in sand. International Conference on Soil Mechanics and Foundation Engineering, 3. Zürich. Proceedings, Vol. 1, p. 100-105.

16. Bishop, A. W. and Henkel, D. J. (1953), Pore pressure changes during shear in two undisturbed clays. International Conference on Soil Mechanics and Foundation Engineering, 3. Zürich. Proceedings, Vol. 1, p. 94-99.

17. Bishop, A. W. and Henkel, D. J. (1957), The measurement of soil properties in the triaxial test. London, Arnold. 190 p.

18. Bishop, A. W., Kennard, M. F. and Penman, A. D. M. (1960), Pore pressure observations at Selset dam. Pore Pressure and Suction in Soil Conference, London, p. 36-47.

19. Bishop, A. W. and Morgenstern, N. (1960), Stability coefficients for earth slopes. In preparation.

20. Bjerrum, L. (1954 a), Geotechnical properties of Norwegian marine clays. Geotechnique, Vol. 4, No. 2, p. 49-69. (Norwegian Geotechnical Institute. Publ., 4).

21. Bjerrum, L. (1954 b), Theoretical and experimental investigations on the shear strength of soils. Thesis. Oslo. 113 p. (Norwegian Geotechnical Institute. Publ., 5).

22. Bjerrum, L. (1954 c), Stability of natural slopes in quick clay. European Conference on Stability of Earth Slopes, Stockholm. Proceedings, Vol. 2, p. 16-40. Geotechnique, Vol. 5, No. 1, 1955, p. 101-119. (Norwegian Geotechnical Institute, Publ., 10).

23. Bjerrum, L. and Eide, O. (1956), Stability of strutted excavations in clay. Geotechnique, Vol. 6, No. 1, p. 32-47. (Norwegian Geotechnical Institute. Publ., 19).

24. Bjerrum, L. and Johannessen, I. (1960), Pore pressures resulting from driving piles in soft clay. Pore Pressure and Suction in Soil Conference, London, p. 14-17.

25. Bjerrum, L. and Kirkedam, R. (1958), Some notes on earth pressure in stiff fissured clay. Brussels Conference on Earth Pressure Problems. Proceedings, Vol. 1, p. 15-27. (Norwegian Geotechnical Institute. Publ., 33.)

26. Bjerrum, L. and Kjaernsli, B. (1957), Analysis of the stability of some Norwegian natural clay slopes. Geotechnique, Vol. 7, No. 1, p. 1-16. (Norwegian Geotechnical Institute. Publ., 24).

27. Bjerrum, L., Simons, N. and Torblaa, I. (1958), The effect of time on the shear strength of a soft marine clay. Brussels Conference on Earth Pressure Problems. Proceedings, Vol. 1, p. 148-158. (Norwegian Geotechnical Institute. Publ., 33.)

28. Bjerrum, L. and Øverland, A. (1957), Foundation failure of an oil tank in Fredrikstad, Norway. International Conference on Soil Mechanics and Foundation Engineering, 4. London. Proceedings, Vol. 1, p. 287-290. (Norwegian Geotechnical Institute. Publ., 26).

29. Bruggeman, J. R., Zangar, C. N. and Brahtz, J. H. A. (1939), Notes on analytic soil mechanics. Denver, Colo. (Department of the Interior, Bureau of Reclamation. Technical memorandum, 592).

30. Cadling, L. and Odenstad, S. (1950), The vane borer. Sthm. 87 p. (Royal Swedish Geotechnical Institute. Proceedings, 2).

31. Campion, F. E. (1951), Part reconstruction of Bo-Peep tunnel at St. Leonards-on-Sea. Institution of Civil Engineers. Journal, Vol. 36, p. 52-75.

32. Casagrande, A. (1934), Discussion of Dr. Jürgenson's papers, entitled "The application of the theory of elasticity and theory of plasticity to foundation problems" and "Research on the shearing resistance of soils." Boston Society of Civil Engineers. Journal, Vol. 21, p. 276-283. Boston Society of Civil Engineers. Contributions to soil mechanics (925-940. Boston 1940, p. 218-225.

33. Casagrande, A. (1949), Soil mechanics in the design and construction of the Logan airport. Boston Society of Civil Engineers. Journal, Vol. 36, p. 192-221. (Harvard University. Graduate School of Engineering. Publ., 467—Soil mechanics series, 33).

34. Casagrande, A. and Albert, S. G. (1930), Research on the shearing resistance of soils. Cambr., Mass. Unpubl. (Massachusetts Institute of Technology. Report).

35. Cassel, F. L. (1948), Slips in fissured clay. International Conference on Soil Mechanics and Foundation Engineering, 2. Rotterdam. Proceedings, Vol. 2, p. 46-50.

36. Coates, R. H. and Slade, L. R. (1958), Construction of circulating-water pump house at Cowes Generating Station, Isle of Wight. Institution of Civil Engineers. Proceedings, Vol. 9, p. 217-232.

37. Cornforth, D. (1960), Thesis. (University of London). In preparation.

38. Cooling, L. F. and Golder, H. Q. (1942), The analysis of the failure of an earth dam during construction. Institution of Civil Engineers. Journal, Vol. 19, p. 38-55.

39. Daehn, W. W. and Hilf, J. W. (1951), Implications of pore pressure in design and construction of rolled earth dams. International Congress on Large Dams, 4. New Delhi. Transactions, Vol. 1, p. 259-270.

40. De Lory, L. A. (1957), Long-term stability of slopes in over-consolidated clays. Thesis. (University of London). London.

41. Di Biagio, E. and Bjerrum, L. (1957), Earth pressure measurements in a trench excavated in stiff marine clay. International Conference on Soil Mechanics and Foundation Engineering, 4. London. Proceedings, Vol. 2, p. 196-202. (Norwegian Geotechnical Institute. Publ., 26).

42. Fraser, A. M. (1957), The influence of stress ratio on compressibility and pore pressure coefficients in compacted soils. Thesis. (University of London). London.

43. Garde-Hansen, P. and Thernöe, S. (1960), Grain silo of 100,000 tons capacity, Mersin, Turkey. CN Post (Cph.), No. 48, p. 14-22.

44. Gibson, R. E. (1958), The progress of consolidation in a clay layer increasing in thickness with time. Geotechnique, Vol. 8, No. 4, p. 171-182.

45. Gibson, R. E. and Marsland, A. (1960), Pore-water observations in a saturated alluvial deposit beneath a loaded oil tank. Pore Pressure and Suction in Soil Conference, London, p. 78-84.

46. Glover, R. E., Gibbs, H. J. and Daehn, W. W. (1948), Deformability of earth materials and its effect on the stability of earth dams following a rapid drawdown. International Conference on Soil Mechanics and Foundation Engineering, 2. Rotterdam. Proceedings, Vol. 5, p. 77-80.

47. Golder, H. Q. and Palmer, D. J. (1955), Investigation of a bank failure at Scrapsgate, Isle of Sheppey, Kent. Geotechnique, Vol. 5, No. 1, p. 55-73.

48. Gould, J. P. (1959), Construction pore pressures observed in rolled earth dams. Denver, Colo. 97 p. (Department of the Interior. Bureau of Reclamation. Technical memorandum, 650).

49. Hamilton, L. W. (1939), The effects of internal hydrostatic pressure on the shearing strength of soils. American Society for Testing Materials. Proceedings, Vol. 39, p. 1100-1121.

50. Hansen, J. B. and Gibson, R. E. (1949), Undrained shear strengths of anisotropically consolidated clays. Geotechnique, Vol. 1, No. 3, p. 189-204.

51. Henkel, D. J. (1957), Investigations of two long-term failures in London clay slopes at Wood Green and Northolt. International Conference on Soil Mechanics and Foundation Engineering, 4. London. Proceedings, Vol. 2, p. 315-320.

52. Henkel, D. J. (1959), The relationships between the strength, pore-water pressure, and volume-change characteristics of saturated clays. Geotechnique, Vol. 9, No. 3, p. 119-135.

53. Henkel, D. J., (1960), The strength of saturated remoulded clay. Research Conference on Shear Strength of Cohesive Soils. Proceedings.

54. Henkel, D. J. and Skempton, A. W. (1955), A landslide at Jackfield, Shropshire, in a heavily over-consolidated clay. Geotechnique, Vol. 5, No. 2, p. 131-137.

55. Hilf, J. W. (1948), Estimating construction pore pressures in rolled earth dams. International Conference on Soil Mechanics and Foundation Engineering, 2. Rotterdam. Proceedings, Vol. 3, p. 234-240.

56. Hilf, J. W. (1956), An investigation of pore-water pressure in compacted cohesive soils. Denver, Colo. 109 p. (Department of the Interior. Bureau of Reclamation. Technical memorandum, 654).

57. Holmsen, P. (1953), Landslips in Norwegian quick-clays. Geotechnique, Vol. 3, No. 5, p. 187-194. (Norwegian Geotechnical Institute. Publ., 2).

58. Hvorslev, M. J. (1937), Über die Festigkeitseigenschaften gestörter bindiger Böden. Kbh., (Gad). 159 p. (Ingeniørvidenskabelige skrifter, A 45).

59. Ireland, H. O. (1954), Stability analysis of the Congress street open cut in Chicago. Geotechnique, Vol. 4, No. 4, p. 163-168.

60. Janbu, N. (1954), Application of composite slip surfaces for stability analysis. European Conference on Stability of Earth Slopes, Stockholm. Proceedings, vol. 3, p. 43-49.

61. Janbu, N. (1957), Earth pressure and bearing capacity by generalized procedure of slices. International Conference on Soil Mechanics and Foundation Engineering, 4. London. Proceedings, Vol. 2, p. 207-212.

62. Kallstenius, J. and Wallgren, A. (1956), Pore water pressure measurement in field investigations. Sthm. 57 p. (Royal Swedish Geotechnical Institute. Proceedings, 13).

63. Kenney, T. C. (1956), An examination of the methods of calculating the stability of slopes. Thesis. (University of London). London.

64. Kjaernsli, B. (1958), Test results, Oslo subway. Brussels Conference on Earth Pressure Problems. Proceedings, Vol. 2, p. 108-117.

65. Larew, H. G. (1952), Analysis of landslides. Wash. D. C. 39 p. (Highway Research Board. Bulletin, 49).

66. Laughton, A. S. (1955), The compaction of ocean sediments. Thesis. (University of Cambridge). Cambr.

67. Little, A. L. and Price, V. E. (1958), The use of an electronic computer for slope stability analysis. Geotechnique, Vol. 8, No. 3, p. 113-120.

68. Lowe, J. and Karafiath, L. (1959), Stability of earth dams upon drawdown. Panamerican Conference on Soil Mechanics and Foundation Engineering, 1. Mexico. Paper 2-A, 15 p.

69. McLellan, A. G. (1945), The Hollowell reservoir scheme for Northampton. Water and water engineering, Vol. 48, p. 7-26.

70. Morgan, H. D. (1944), The design of wharves on soft ground. Institution of Civil Engineers. Journal, Vol. 22, p. 5-25.

71. Murphy, V. A. (1951), A new technique for investigating the stability of slopes and foundations. New Zealand Institution of Engineers. Proceedings, Vol. 37, p. 222-285.

72. Nixon, J. K. (1949), $\phi = 0$ analysis. Geotechnique, Vol. 1, No. 3, 4, p. 208-209, 274-276.

73. Nonveiller, E. and Suklje, L. (1955), Landslide Zalesina. Geotechnique, Vol. 5, No. 2, p. 143-153.

74. Odenstad, S. (1949), Stresses and strains in the undrained compression test. Geotechnique, Vol. 1, No. 4, p. 242-249.

75. Paton, J. and Semple, N. G. (1960), Investigation of the stability of an earth dam subject to rapid drawdown including details of pore pressure recorded during a controlled drawdown test. Pore pressure and Suction in Soil Conference, London, p. 66-71.

76. Peck, R. B. (1942), Earth pressure measurements in open cuts, Chicago (Ill.) subway. American Society of Civil Engineers. Proceedings, Vol. 68, p. 900-928. American Society of Civil Engineers. Transactions, Vol. 108, 1943, p. 1008-1036.

77. Peck, R. B. and Bryant, F. G. (1953), The bearing capacity failure of the Transcona elevator. Geotechnique, Vol. 3, No. 5, p. 201-208.

78. Penman, A. D. M. (1956), A field piezometer apparatus. Geotechnique, Vol. 6, No. 2, p. 57-65.

79. Peterson, R. (1952), Studies—Bearpaw shale at damsite in Saskatchewan. N. Y. 53 p. (American Society of Civil Engineers. Preprint, 52).

80. Reinius, E. (1948), The stability of the upstream slope of earth dams. Sthm. 107 p. (Swedish State Committee for Building Research. Bulletin, 12).

81. Rendulic, L. (1937), Ein Grundgesetz der Tonmechanik und sein experimenteller Beweis. Bauingenieur, Vol. 18, No. 31/32, p. 459-467.

82. Rosenqvist, I. T. (1953), Considerations on the sensitivity of Norwegian quick-clays. Geotechnique, Vol. 3, No. 5, p. 195-200. (Norwegian Geotechnical Institute. Publ., 2).

83. Sevaldson, R. A. (1956), The slide in Lodalen, October 6th, 1954. Geotechnique, Vol. 6, No. 4, p. 1-16. (Norwegian Geotechnical Institute. Publ., 24).

84. Sheppard, G. A. R. and Aylen, L. B. (1957), The Usk scheme for the water supply of Swansea. Institution of Civil Engineers. Proceedings, Vol. 7, paper 6210, p. 246-274.

85. Simons, N. (1958), Discussion on: General theory of earth pressure. Brussels Conference on Earth Pressure Problems. Proceedings, Vol. 3, p. 50-53. (Norwegian Geotechnical Institute. Publ., 33.)

86. Skempton, A. W. (1942), An investigation of the bearing capacity of a soft clay soil. Institution of Civil Engineers. Journal, Vol. 18, p. 307-321.

87. Skempton, A. W. (1945), A slip in the West Bank of the Eau Brink cut. Institution of Civil Engineers. Journal, Vol. 24, p. 267-287.

88. Skempton, A. W. (1948 a), The $\Phi = 0$ analysis of stability and its theoretical basis. International Conference on Soil Mechanics and Foundation Engineering, 2. Rotterdam. Proceedings, Vol. 1, p. 145-150.

89. Skempton, A. W. (1948 b), A study of the immediate triaxial test on cohesive soils. International Conference on Soil Mechanics and Foundation Engineering, 2. Rotterdam. Proceedings, Vol. 1, p. 192-196.

90. Skempton, A. W. (1948 c), The rate of softening in stiff fissured clays, with special reference to London clay. International Conference on Soil Mechanics and Foundation Engineering, 2. Rotterdam. Proceedings, Vol. 2, p. 50-53.

91. Skempton, A. W. (1948 d), The geotechnical properties of a deep stratum of post-glacial clay at Gosport. International Conference on Soil Mechanics and Foundation Engineering, 2. Rotterdam. Proceedings, Vol. 1, p. 145-150.

92. Skempton, A. W. (1950), Discussion on: Wilson, G. The bearing capacity of screw piles and screwcrete cylinders. Institution of Civil Engineers. Journal, Vol. 34, p. 76.

93. Skempton, A. W. (1951), The bearing capacity of clays. Building Research Congress, London. Papers, division 1, part 3, p. 180-189.

94. Skempton, A. W. (1954), The pore pressure coefficients A and B. Geotechnique, Vol. 4, No. 4, p. 143-147.

95. Skempton, A. W. (1959), Cast in-situ bored piles in London clay. Geotechnique, Vol. 9, No. 4, p. 153-173.

96. Skempton, A. W. and Bishop, A. W. (1954), Soils. Building materials, their elasticity and inelasticity. Ed. by M. Reiner with the assistance of A. G. Ward. Amsterdam, North-Holland Publ. Co. Chapter X, p. 417-482.

97. Skempton, A. W. and Bishop, A. W. (1955), The gain in stability due to pore pressure dissipation in a soft clay foundation. International Congress on Large Dams, 5. Paris. Transactions, Vol. 1, p. 613-638.

98. Skempton, A. W. and DeLory, F. A. (1957), Stability of natural slopes in London clay. International Conference on Soil Mechanics and Foundation Engineering, 4. London. Proceedings, Vol. 2, p. 378-381.

99. Skempton, A. W. and Golder, H. Q. (1948), Practical examples of the $\Phi = 0$ analysis of stability of clays. International Conference on Soil Mechanics and Foundation Engineering, 2. Rotterdam. Proceedings, Vol. 2, p. 63-70.

100. Skempton, A. W. and Henkel, D. J. (1960), Field observations on pore pressures in London clay. Pore Pressure and Suction in Soil Conference, London, p. 48-51.

101. Skempton, A. W. and Ward, W. H. (1952), Investigations concerning a deep cofferdam in the Thames estuary clay at Shellhaven. Geotechnique, Vol. 3, No. 3, p. 119-139.

102. Suklje, L. (1953 a), Discussion on: Stability and deformations of slopes and earth dams, research on pore-pressure measurements, groundwater problems. International Conference on Soil Mechanics and Foundation Engineering, 3. Zürich, Proceedings, Vol. 3, p. 211.

103. Suklje, L. (1953 b), Plaz pri Lupoglavu v ecocenskem flisu. (Landslide in the eocene flysch at Lupoglav.) Gradbeni vestnik, Vol. 5, No. 17/18, p. 133-138.

104. Taylor, D. W. (1944), Cylindrical compression research program on stress-deformation and strength characteristics of soils; 10 progress report. Cambr., Mass. 46 p. Publ. by Massachusetts Institute of Technology. Soil Mechanics Laboratory.

105. Taylor, D. W. (1948), Fundamentals of soil mechanics. N. Y., Wiley. 700 p.

106. Terzaghi, K. (1923), Die Berechnung der Durchlässigkeitsziffer des Tones aus dem Verlauf der hydrodynamischen Spannungserscheinungen. Akademie der Wissenschaften in Wien. Mathematisch-naturwissenschaftliche Klasse. Sitzungsberichte. Abteilung II a, Vol. 132, No. 3/4, p. 125-138.

107. Terzaghi, K. (1925), Erdbaumechanik auf bodenphysikalischer Grundlage. Lpz., Deuticke. 399 p.

108. Terzaghi, K. (1932), Tragfähigkeit der Flachgründungen. International Association for Bridge and Structural Engineering. Congress, 1. Paris. Preliminary publ., p. 659-683, final publ., 1933, p. 596-605.

109. Terzaghi, K. (1936 a), Stability of slopes of natural clay. International Conference on Soil Mechanics and Foundation Engineering. 1. Cambr., Mass. Proceedings, Vol. 1, p. 161-165.

110. Terzaghi, K. (1936 b), The shearing resistance of saturated soils and the angle between the planes of shear. International Conference on Soil Mechanics and Foundation Engineering, 1. Cambr., Mass. Proceedings, Vol. 1, p. 54-56.

111. Terzaghi, K. (1943), Theoretical soil mechanics. N. Y., Wiley. 510 p.

112. Terzaghi, K. (1958), Design and performance of the Sasumua dam. Institution of Civil Engineers. Proceedings, Vol. 11, November, p. 360-363.

113. Terzaghi, K. and Peck, R. B. (1948), Soil mechanics in engineering practice. N. Y., Wiley. 566 p.

114. Toms, A. H. (1948), The present scope and possible future development of soil mechanics in British railway civil engineering construction and maintenance. International Conference on Soil Mechanics and Foundation Engineering, 2. Rotterdam. Proceedings, Vol. 4, p. 226-237.

115. Toms, A. H. (1953 a), Discussion on: Conference on the North Sea floods of January 31st-February 1st, 1953. Institution of Civil Engineers, Publ., p. 103-105.

116. Toms, A. H. (1953 b), Recent research into coastal landslides at Folkstone Warren, Kent, England. International Conference on Soil Mechanics and Foundation Engineering, 3. Zürich. Proceedings, Vol. 2, p. 288-293.

117. Tschebotarioff, G. P. (1951), Soil mechanics, foundations, and earth structures; an introduction to the theory and practice of design and construction. N. Y., McGraw-Hill. 655 p.

118. U. S. Department of the Interior. Bureau of Reclamation (1951), Earth manual; a manual on the use of earth materials for foundation and construction purposes. Tentative ed. Denver, Colo. 332 p.

119. Ward, W. H., Penman, A., and Gibson, R. E. (1954), Stability of a bank on a thin peat layer. European Conference on Stability of Earth Slopes, Stockholm. Proceedings, Vol. 1, p. 122-138, Vol. 3, p. 128-129. Geotechnique, Vol. 5, No. 2, 1955, p. 154-163.

120. Waterways Experiment Station, Vicksb., Miss. (1947), Triaxial shear research and pressure distribution studies on soils. Vicksb., Miss. 332 p.

121. Waterways Experiment Station, Vicksb., Miss. (1950), Potamology investigations. Triaxial tests on sands, Reid Bedford Bend, Mississippi river. Vicksb., Miss. 54 p. (Report, 5-3).

122. Wilson, G. (1950), The bearing capacity of screw piles and screwcrete cylinders. Institution of Civil Engineers. Journal, Vol. 34, p. 4-73.

123. Wood, C. C. (1958), Shear strength and volume change characteristics of compacted soils under conditions of plane strain. Thesis. (University of London). London.

Journal of the

SOIL MECHANICS AND FOUNDATIONS DIVISION

Proceedings of the American Society of Civil Engineers

SEEPAGE REQUIREMENTS OF FILTERS AND PERVIOUS BASES

By Harry R. Cedergren,[1] M.ASCE

SYNOPSIS

The water-removing capabilities of two common but distinctly different types of filter designs are analyzed by the flow-net. Typical solutions and numerical examples are presented to emphasize the importance of boundary conditions and permeability upon water-removing capacity.

INTRODUCTION

A rational method is available[2] for the design of base courses that will drain rapidly after becoming flooded by entry of water from the sides, through the pavement, by temporary rises in ground water level, and so forth, and the influence of capillarity can be estimated.[3,4] Up to the present time, however, there has been no rational method for predicting minimum design requirements of filters and pervious bases that will assure the continued removal of infiltering ground water or other steady seepage without excessive build-up of hydrostatic head.

The water-removing capacity of pervious filters and bases varies with boundary conditions, permeability, and the thickness of the water-removing layer. In some types of installation sufficient water-removing capacity is easily attained; however, in others, the discharging seepage is removed with

Note.—Discussion open until March 1, 1961. To extend the closing date one month, a written request must be filed with the Executive Secretary, ASCE. This paper is part of the copyrighted Journal of the Soil Mechanics and Foundations Division, Proceedings of the American Society of Civil Engineers, Vol. 86, No. SM 5, October, 1960.

1 Sr. Materials and Research Engr., Calif. Div. of Highways, Sacramento, Calif.

2 "Base Course Drainage for Airport Pavements," by A. Casagrande and W. L. Shannon, Proceedings, ASCE, Vol. 77, June, 1951, pp. 792-820.

3 "Subsurface Drainage of Highways and Airports," Bulletin 209, Highway Research Board, 1959, pp. 3-6.

4 "Procedures for Testing Soils," American Society for Testing Materials, September, 1944, pp. 56 and 58.

sufficient rapidity only through careful control of the key factors. Flow-net solutions are presented to illustrate a rational approach to filter design and to show the relative importance of some basic factors.

FILTERS MUST SATISFY CONFLICTING REQUIREMENTS

The safe, quick removal of seeping water from engineering structures constructed of earth or founded on earth is often an essential design requirement. Earth dams and levees usually are provided with drains and filters for the removal of seepage caused by the hydraulic conditions that they create (Fig. 1(b)). Airport and highway pavements and roadbeds frequently are protected from the harmful effects of surface infiltration and ground water intrusion by "pervious" bases or filter blankets (Fig. 2). Without adequate drainage many projects would not be economically feasible. In fact, drainage facilities have made many projects possible.

The remarkable benefits possible from filters can be fully assured only if their gradation is carefully adjusted to meet the requirements of the job. Specifications for the grading of granular materials for filters and pervious bases are established by the need for satisfying two conflicting requirements. It must insure permanence of operation. Filters or other porous water-removing layers must be properly graded to prevent the movement of soil particles into or through the pores of the filter. However, it must also permit the rapid removal of water. Filters must be sufficiently more pervious than the protected soil so that the incoming water will be removed without appreciable build-up of hydrostatic head in the filter.

Criteria that have been used since about 1930 for the design of filters for dams and levees, and have been referred at as "Terzaghi Criteria," may be stated as follows:

1. The 15% size of the filter material, D_{15}, must be not more than 4 or 5 times the 85% size, D_{85}, of the protected soil (to prevent piping).
2. The 15% size of the filter material, D_{15}, must be at least 4 or 5 times the 15% size of the protected soil, D_{15}, (to insure adequate permeability).

Experimental work and field observations by a number of investigators[5,6,7,8] has verified the general suitability of the above criteria; however, some modifications have been proposed. Generally the stipulation is made that the grain-size curves of soil and filter shall be somewhat parallel.

If a filter is allowed to be too coarse with respect to the protected soil, and severe discharge conditions develop, serious internal erosion and piping may be expected.

Criterion No. 2 generally has been assumed to assure sufficient permeability to hold seepage forces and gradients in filters to small amounts. It does not

---

5 "An Experimental Investigation of Protective Filters," by G. E. Bertram, Harvard Graduate School of Engineering, Pub. 267, Series 7, 1940.

6 "Design of Drainage Facilities for Airfields," Office of Chief of Engineers, War Department, Engrg. Manual, February, 1943, Ch. XXI.

7 "The Use of Laboratory Tests to Develop Design Criteria for Protective Filters," by K. P. Karpoff, Proceedings, ASTM, Vol. 55, 1955, pp. 1183-1193.

8 "Underdrain Practice of the Connecticut Highway Department," by Philip Keene, Proceedings, Highway Research Board, 1944, pp. 377-389.

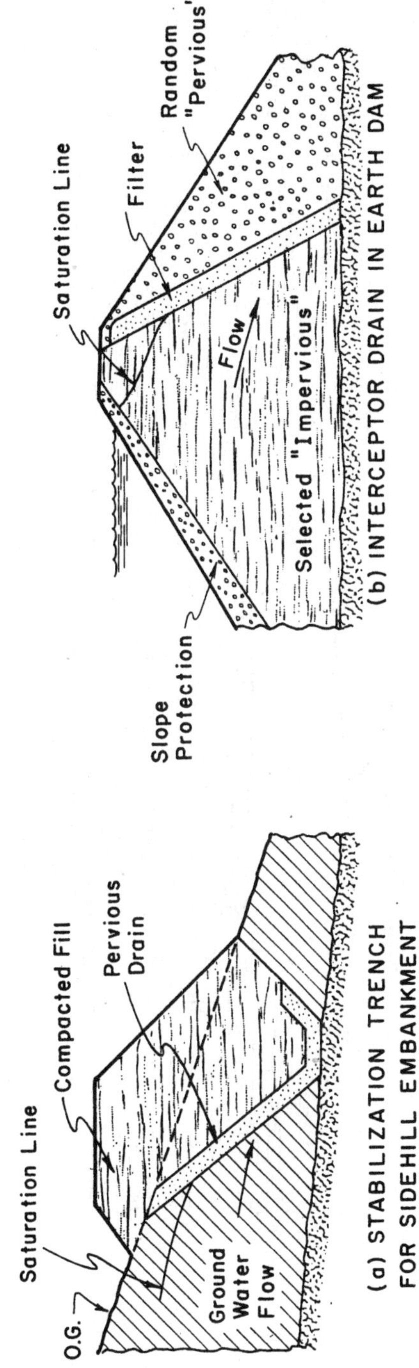

FIG. 1.—EXAMPLES OF ENGINEERING DESIGNS UTILIZING SLOPING PERVIOUS FILTERS OR DRAINS FOR REMOVAL OF STEADY SEEPAGE

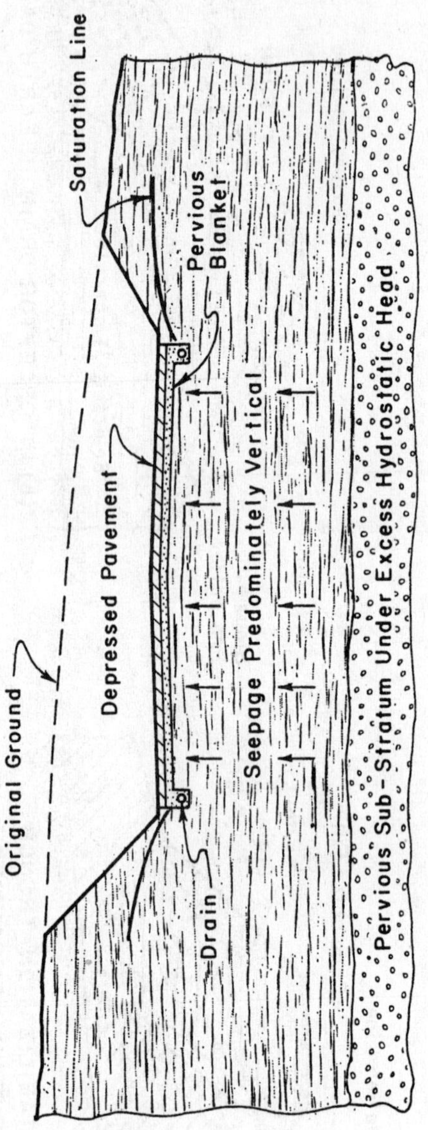

FIG. 2.—EXAMPLE OF ENGINEERING DESIGN UTILIZING RELATIVELY HORIZONTAL DRAINAGE BLANKET FOR REMOVAL OF STEADY SEEPAGE

necessarily prohibit the build-up of head within filters that are required to operate under small gradients.

In general terms, filters often are considered to have adequate permeability if they are "somewhat more pervious" or simply "more pervious" than the protected soil. Such standards (or lack of standards) may not guarantee adequate permeability for removal of seepage. In fact, the required permeability of filters varies substantially with their shape and oreintation. Steeply inclined filters, for example, have relatively high gradients available to hasten the discharge of seepage, whereas nearly flat or horizontal "pervious" bases have relatively minute gradients available.

This paper presents a method for determining the effect of boundary conditions and gradients upon the design requirements of filters subjected to steady seepage. Typical solutions are developed to illustrate a procedure for designing filters and pervious bases that will discharge seepage without excessive build-up of head due to resistance within the water-removing layer.

The condition of "steady seepage" as used herein is not necessarily a year-around condition, but may be of temporary duration. It implies the worst conditions that are likely to prevail for a sufficient length of time to have harmful effects.

PERMEABILITY AND ORIENTATION OF FILTERS CAN BE IMPORTANT

Ordinarily, the permeability of filters and "pervious" bases is not specified, and seldom are hydraulic conditions within such layers analyzed. Under some common usages, the permeability of filters is not particularly critical, and adequate water-removing capacity is easily obtained. Possibly, for this reason, there have been no generally established permeability criteria for filters, pervious bases, and blankets. Casagrande and Shannon[2], however, recognized that the speed with which water can drain from saturated base courses for airfields is directly related to permeability.

The factors influencing the quantity of water flowing through porous media under non-turbulent conditions are expressed by Darcy's law;

$$Q = k\,i\,A \qquad (1)$$

Thus, the seepage quantity, Q, is proportional to the permeability, k, gradient, i, and cross sectional area, A.

If any two of these factors are held constant, the water-removing potential is increased or decreased in direct proportion to changes in the third. Thus, the water-removing capacity of a filter of given cross section and hydraulic gradient, is directly proportional to its permeability. Also, if the permeability and gradient are held constant, the capacity for removing water is directly proportional to the cross sectional area. Recognition of the fact that all three of these factors influence water-removing capacity of filters helps to clarify the basic problem.

Seepage gradients are influenced primarily by hydrostatic head imposed by natural or man-made conditions and boundary conditions established by design. The solutions presented show that the water-removing potential of filters and pervious bases can be increased by providing either greater thickness or greater permeability. The examples are basically "single-layer" filters. If one applies the principle noted above that water-removing capacity is proportional to permeability one might conclude that any quantity of seepage can be removed simply by providing great enough permeability in the filter. Of course, this may lead to situations where the filter would become so coarse as to allow migration of the fines of the protected layer into or through the filter. When sufficient water-removing capacity cannot safely be assured by means of a single-layer filter some other means must be used, such as a multiple-layered system, or graded filter.

ANALYSIS OF HYDRAULIC CONDITIONS WITHIN FILTERS

General Analysis.—The solutions to be presented show the relative influence of seepage gradients as affected by boundary conditions, and the interrelationship between the permeability ratio (ratio of permeability of filter to permeability of soil) and the physical dimensions of water-removing filter layers.

The influence of boundary conditions and gradients will be shown by analysis of two common, but distinctly different kinds of drainage systems: (a) a steeply inclined filter such as might be placed at the boundary between an "impervious" zone of an earth dam or levee and a "pervious" downstream zone, or against the excavated slope of a "stabilization" trench keying a side-hill fill into firm ground and removing harmful ground water (See Fig. 1); and (b) a flat or horizontal filter or pervious base, such as might be placed beneath a highway or other pavement to intercept rising ground water (See Fig. 2).

The two kinds of design were analyzed by means of flow-nets. Fairly broad ranges of conditions were analyzed and summarized in charts.

The shape of the flow pattern characteristic of any given cross section adjusts to the existing physical conditions, as water will always follow the lines of least resistance. When true conditions at a given location can be approximated by simplified cross sections the flow-net offers a means of predicting paths of flow and distribution of hydrostatic head. Commonly, flow-nets are used in analyzing seepage through dams, levees, and foundations; however, the principle had been applied to diverse situations, including problems of steady-state heat flow, electrostatics, current flow through conductors, pressure distribution in spillways, certain cases in the theory of the torsion of elastic rods, and the flow of viscous liquids.[9,10]

In two-dimensional hydraulic flow, the flow-net is a network of two intersecting families of curves. One family is called the stream lines and represents paths of flow, and the other family the equipotential lines or simply the "potential lines." These networks may be obtained from models or by graphical sketching. Those developed in the present study were obtained by the graphical method, which is readily adapted to the problem under discussion.

One must always recognize that flow-net studies are no more accurate than the degree to which the simplified sections represent true conditions, and true conditions may be extremely difficult to evaluate. Once a set of conditions has been set up, however, there is but one solution,[9] regardless of the method used in finding the solution. Some of the most familiar types of flow-nets are for

9 "The Flow of Homogeneous Fluids Through Porous Media," by M. Muskat, McGraw-Hill Book Co., Inc., New York, N. Y., 1937, pp. 139-140.

10 "The Flow-net and the Electric Analogy," by E. W. Lane, F. B. Campbell, and W. H. Price, Civil Engineering, October, 1934, p. 510.

homogeneous cross sections, assumed to be of one permeability throughout;[11,12,13] however, some rather complex steady-state seepage problems have been solved[14,15] and approximate nonsteady conditions[16,17] analyzed by this method. Techniques for constructing flow-nets have been capably outlined in engineering publications.[18]

The case under consideration is that of flow from a soil mass of one permeability (the subsoil, or compacted earth) into a porous zone of higher permeability (the filter or pervious blanket). It may be considered a "two-permeability" case, with an indeterminate saturation line which is located simultaneously with construction of the flow-net.

In preparing to construct a flow-net one must establish the physical conditions defining the cross section to be analyzed, including the shape of the porous masses, their average or effective permeabilities, and hydraulic conditions at the edges of the section, or the "boundary conditions." By graphical means (or models) one obtains for a given section the two sets of intersecting lines that have previously been noted—the "flow-lines" and the "potential lines." For a given cross section the resulting pattern has a very definite shape. By constructing a number of nets for varied conditions the relationship may be established between the distance saturation penetrates a "pervious" filter or blanket and the ratio of the permeability of the soil to the permeability of the pervious layer. This relationship is the basis for the typical design charts presented herein.

For each of the two cases analyzed, flow-nets were constructed for a fairly broad range of conditions, and the results summarized on charts. The detailed conditions assumed are shown on the charts and are described in the text.

Sloping Filters.—The first case to be described is not a steeply inclined filter used with construction of the general types illustrated in Fig. 1. Typical flow-nets for this case are shown in Fig. 3 and the solutions have been plotted in Fig. 4 for slopes from $1\frac{1}{2}$:1 to 6:1. A more detailed chart for a $1\frac{1}{2}$:1 slope is given in Fig. 5.

Key dimensions for the section are the height of the saturated discharge face, H, (see Fig. 4) and the depth of penetration, T, of saturation into the filter. This distance is expressed non-dimensionally as the ratio H/T on Figs. 3, 4, and 5.

The slope of the discharge face is designated as S and is expressed as 1 on S. Thus S = 1.5 means a slope of 1 vertically to 1.5 horizontally. The permeability of the soil is designated as k_s and the permeability of the filter k_f. An

---

11 "Fundamentals of Soil Mecahnics," by D. W. Taylor, John Wiley & Sons, Inc., N. Y., 1948, p. 157.

12 Discussion by A. Casagrande of ASCE Transactions Paper No. 1919, Vol. 100, 1935, p. 1291.

13 "Methods of Analysis of Flow Problems for Highway Subdrainage," by B. McClelland and L. E. Gregg, Proceedings, Highway Research Board, 1944, pp. 364-376.

14 Discussions of ASCE Transactions Paper No. 2270, Vol. 111, 1946, pp. 236 and 241-244.

15 "Use of Flow-net in Earth Dam and Levee Design," by H. R. Cedergren, Proceedings, Second International Conference on Soil Mechanics and Foundation Engineering, June 21-30, 1948, pp. 293-298.

16 Civil Engineering, August, 1941, p. 499.

17 Discussion by H. R. Cedergren of ASCE Transactions Paper No. 2356, Vol. 113, 1948, pp. 1285-1293.

18 "Seepage Through Dams," by A. Casagrande, Journal, New England Water Works Assn., June, 1937.

impervious boundary is assumed at the elevation of the bottom of the filter. Horizontal and vertical permeabilities are assumed to be equal. A gravel pocket surrounds a longitudinal drainage pipe that permits the free discharge of seepage.

Examination of typical flow-nets for this case (Fig. 3) indicates that seepage in the soil is predominately horizontal. At the boundary between the soil

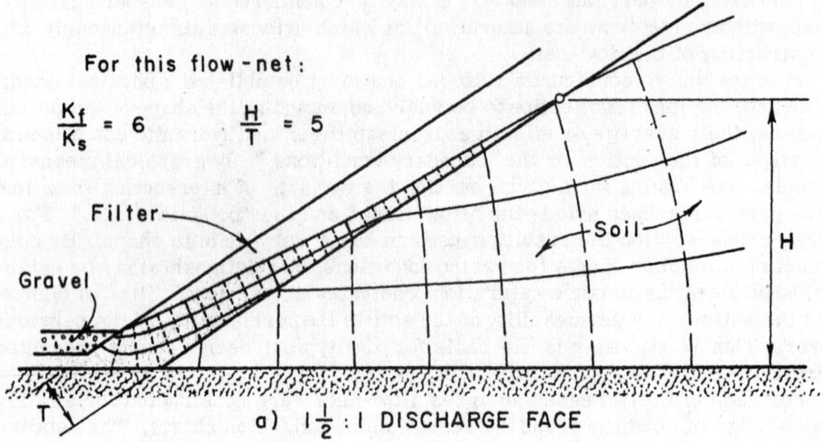

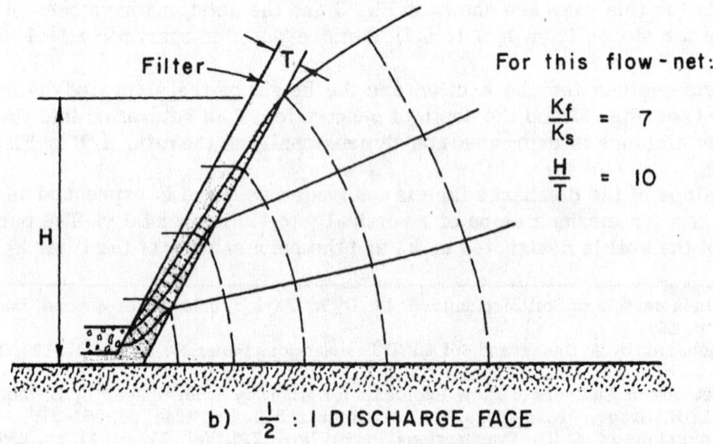

FIG. 3.—TYPICAL FLOW-NETS FOR SEEPAGE INTO SLOPING FILTERS

and the more pervious filter, the flow-lines deflect downward. As seepage accumulates in the filter the thickness of penetration increases to its maximum value (shown as T), which will just contain the zone of saturation. The depth of penetration for a given cross section is dependent upon the ratio of k_f to k_s.

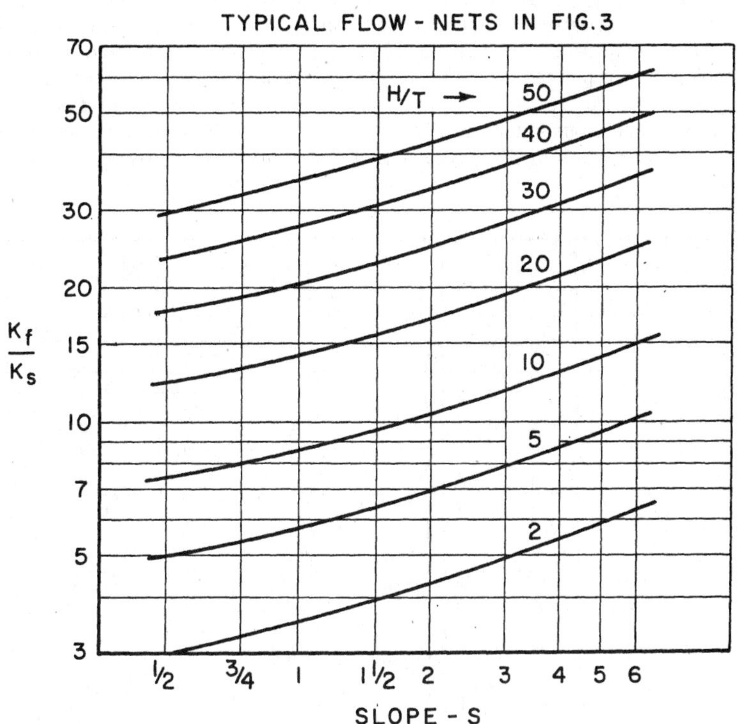

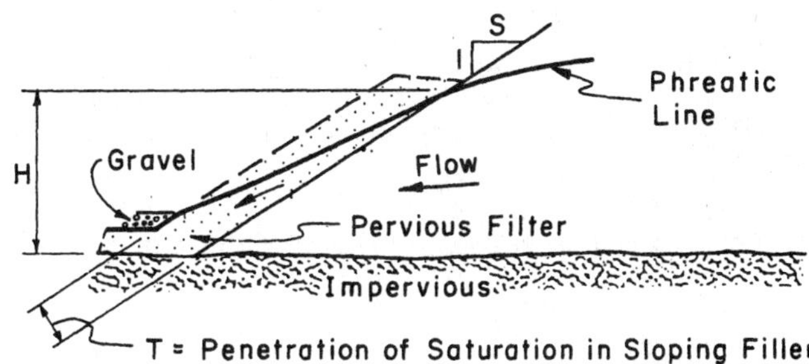

FIG. 4.—FLOW-NET SOLUTION FOR SEEPAGE INTO SLOPING FILTERS ON VARIOUS SLOPES

Graphical checks establish criteria that control the shape of the flow-net and determine the specific ratio k_f/k_s that corresponds to a specific value of T/H. One may select a value of T/H, and construct a balanced flow-net from which the ratio k_f/k_s may be determined or a k_f/k_s ratio may be selected and the depth of penetration determined from the balanced flow-net. In a balanced flow-net the ratio, k_f/k_s is equal to the degree of elongation of the rectangles comprising the net as it passes from the soil into the filter. Numerically, the ratio, k_f/k_s, is equal to the length of a single flow-channel in the filter to its width. Thus, for the conditions shown in Fig. 3(a) a value of H/T of 5 corresponds to a k_f/k_s ratio of 6, and in Fig. 3(b) a H/T ratio of 10 corresponds to a k_f/k_s

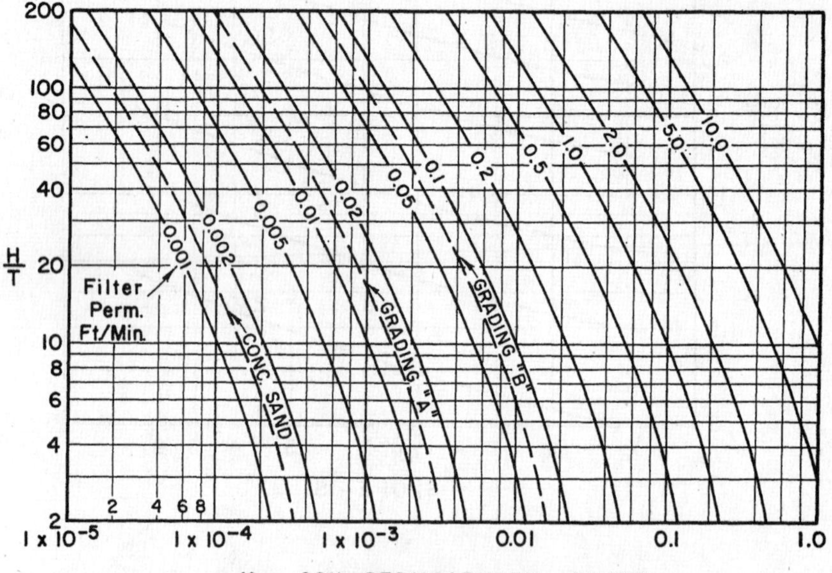

Explanation: This chart was developed from H/T values obtained by flow-net solution of steady seepage for boundary conditions shown in diagram in Fig. 4. Typical Flow-nets in Fig. 3.

FIG. 5.—DESIGN CHART FOR PREVIOUS FILTERS OR BLANKETS ON $1\frac{1}{2}$: 1 SLOPE

ratio of 7. As the ratio k_f/k_s becomes greater (and the permeability of the filter greater with respect to the soil) the depth of penetration of saturation into the filter becomes less. This conclusion is evident from scrutiny of the curves in Figs. 4 and 5.

A number of flow-nets similar to the samples in Fig. 3 were constructed, and the results summarized in Fig. 4. The curves in Fig. 4 give the relationship between slope, S, k_f/k_s ratio, and H/T ratio. From Fig. 4 any one of the three variables may be determined if the other two are known. Thus, if the slope is known, and the permeability of the filter, k_f, and the permeability of the soil, k_s, are known, the ratio H/T may be read from the chart. For the probable height of the saturation face, H, which is estimated from field and soil conditions, the minimum adequate thickness of filter at the bottom of the slope is determined.

To simplify application of the flow-net solutions, detailed charts such as the one reproduced in Fig. 5 may be prepared. Fig. 5 permits direct reading of design ratios (H/T) for given soil permeabilities, k_s, and filter permeabilities, k_f. One chart of the type illustrated in Fig. 5 is required for each slope. Thus, Fig. 5 gives the solution for a $1\frac{1}{2}$:1 slope. It provides the design criteria used in the first of the numerical examples.

Examination of the flow-nets in Fig. 3 for a steeply sloping filter indicates that the total head, H, is, in some degree, available to induce flow within the filter. The significance of this condition on performance will become evident by comparison with the second case.

Horizontal Filters.—The second type of filter installation analyzed by flow-nets is a horizontal "pervious" base for the general type of design illustrated in Fig. 2. Sample flow-nets are presented in Figs. 6 and 7. The pervious blanket is assumed to be horizontal. Water is removed from this layer by a series of shallow, longitudinal drains at a spacing, D. A hydrostatic excess head of 0.2 D is assumed at the top of this aquifer. This head produces an average vertical gradient of 0.4 in the subsoil. The flow-nets give the height of rise of saturation, h, within the filter, for steady seepage excluding capillary rise. The ratio of the distance between drains to the height of rise of saturation due to steady seepage is expressed nondimensionally as D/h. This ratio is related to the permeability ratio (k_f/k_s). Examination of the flow-nets in Figs. 6 and 7 indicates that flow within the subsoil is essentially vertical, while flow in the filter is essentially horizontal. In contrast to the relatively high gradients available in the sloping filter, the horizontal filter has very small gradients available to discharge the incoming seepage.

These small gradients have a retarding effect on seepage discharge that must be compensated by use of filter material of relatively high permeability. This requires the use of relatively coarse and clean material to remove seepage, which in turn means larger voids and greater ease of infiltration. Fortunately, vertical upward flow is the safest from the standpoint of soil migration, since the weight of the soil particles is acting opposite to the direction of flow. An uplift gradient of approximately 1.0 is needed to overcome the downward force exerted by the soil particles, and place the soil in suspension. In the examples, an uplift gradient of 0.4 was assumed.

The relationships derived from these flow-nets are summarized in the chart in Fig. 8. This chart gives the relationship between the permeability ratio (k_f/k_s) and the design ratio, D/h. This chart permits determination of the ratio D/h for a wide range of permeabilities. It permits determination of any one of the variables if the others are known. To facilitate the application of this solution the more detailed plot in Fig. 9 has been prepared. Some computed thicknesses of pervious horizontal bases required for the removal of steady seepage for the assumed boundary conditions are summarized in Table 1.

The practical significance of these solutions will now be illustrated by numerical examples. It might be well to introduce the reminder that the solutions

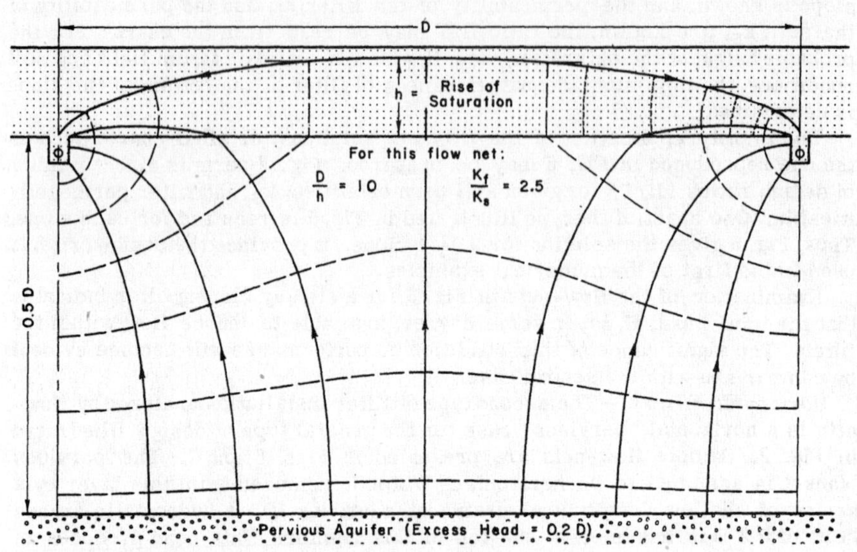

FIG. 6.—TYPICAL FLOW-NET FOR VERTICAL SEEPAGE INTO HORIZONTAL BLANKET

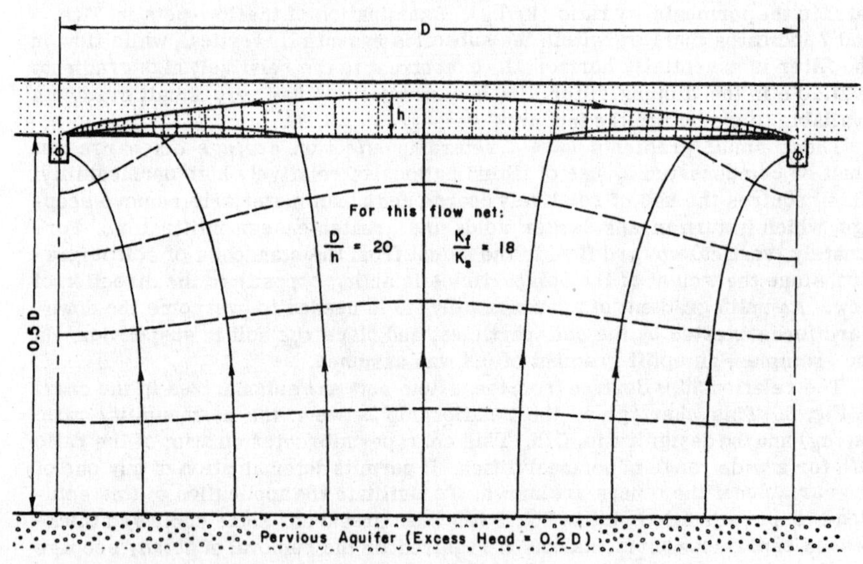

FIG. 7.—TYPICAL FLOW-NET FOR VERTICAL SEEPAGE INTO HORIZONTAL BLANKET

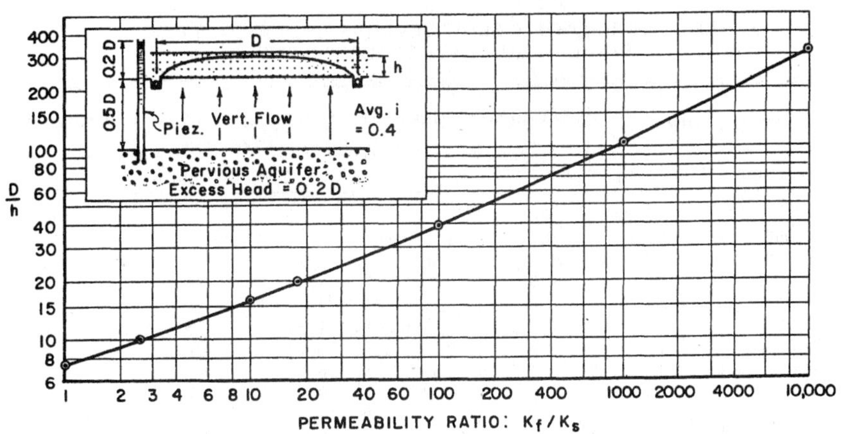

FIG. 8.—FLOW-NET SOLUTION FOR VERTICAL SEEPAGE INTO HORIZONTAL BLANKET

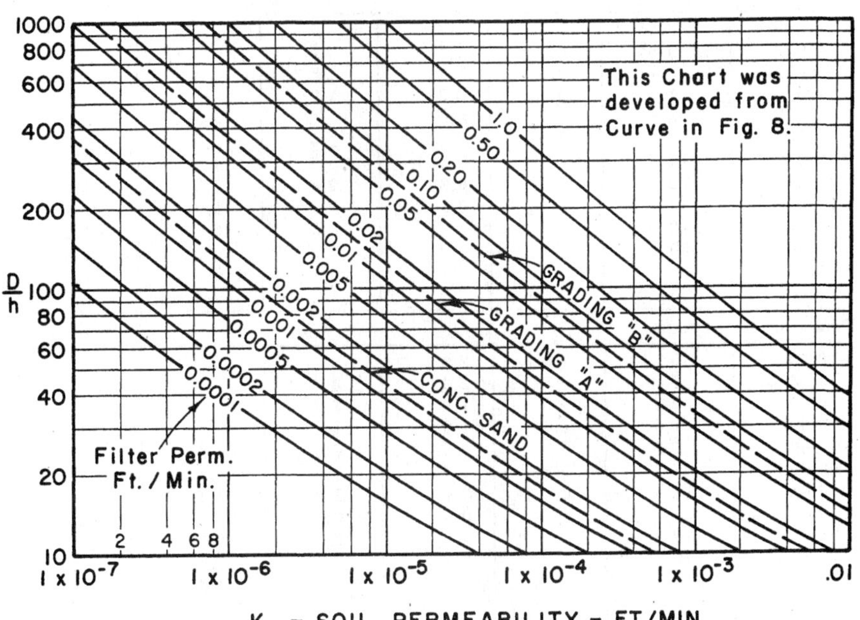

FIG. 9.—DRAIN DESIGN CHART FOR HORIZONTAL BLANKET WITH SHALLOW COLLECTOR DRAINS

TABLE 1.—HEIGHT OF RISE OF SEEPAGE IN PERVIOUS BASE WITH SHALLOW LONGITUDINAL DRAINS[a,b]

Permeability		Height of Rise of Steady Seepage, feet for Effective Soil Perm., ft per min				
ft per min	ft per day	1×10^{-6}	1×10^{-5}	1×10^{-4}	1×10^{-3}	1×10^{-2}
Drains at 20 ft spacing[c]						
0.0001	0.14	0.5	1.3			
0.0002	0.28	0.4	1.0	2.2		
0.0005	0.7	0.3	0.7	1.6		
0.001	1.4	0.2	0.5	1.3		
0.002	2.8	0.1	0.4	1.0	2.2	
0.005	7		0.3	0.7	1.6	
0.01	14		0.2	0.5	1.3	
0.02	28		0.1	0.4	1.0	2.2
0.05	70			0.3	0.7	1.6
0.10	140			0.2	0.5	1.3
0.20	280			0.1	0.4	1.0
0.50	700				0.3	0.7
1.00	1400				0.2	0.5
Drains at 30 ft spacing[c]						
0.0001	0.14	0.8	1.9			
0.0002	0.28	0.6	1.5	3.3		
0.0005	0.7	0.4	1.0	2.4		
0.001	1.4	0.3	0.8	1.9		
0.002	2.8	0.2	0.6	1.5	3.3	
0.005	7	0.1	0.4	1.0	2.4	
0.01	14		0.3	0.8	1.9	
0.02	28		0.2	0.6	1.5	3.3
0.05	70		0.1	0.4	1.0	2.4
0.10	140			0.3	0.8	1.9
0.20	280			0.2	0.6	1.5
0.50	700			0.1	0.4	1.0
1.00	1400				0.3	0.8
Drains at 40 ft spacing[c]						
0.0001	0.14	1.0	2.5			
0.0002	0.28	0.8	2.0	4.4		
0.0005	0.7	0.5	1.4	3.2		
0.001	1.4	0.4	1.0	2.5		
0.002	2.8	0.3	0.8	2.0	4.4	
0.005	7	0.2	0.5	1.4	3.2	
0.01	14	0.1	0.4	1.0	2.5	
0.02	28		0.3	0.8	2.0	4.4
0.05	70		0.2	0.5	1.4	3.2
0.10	140		0.1	0.4	1.0	2.5
0.20	280			0.3	0.8	2.0
0.50	700			0.2	0.5	1.4
1.00	1400			0.1	0.4	1.0

[a] Boundary conditions shown in Fig. 8.
[b] Steady discharge from subgrade under vertical gradient of 0.4.
[c] With pervious base.

TABLE 1.—CONTINUED

Permeability		Height of Rise of Steady Seepage, feet for Effective Soil Perm., ft per min				
ft per min	ft per day	1 x 10^{-6}	1 x 10^{-5}	1 x 10^{-4}	1 x 10^{-3}	1 x 10^{-2}
Drains at 50 ft spacing[c]						
0.0001	0.14	1.3	3.1			
0.0002	0.28	1.0	2.5	5.5		
0.0005	0.7	0.7	1.8	4.0		
0.001	1.4	0.5	1.3	3.1		
0.002	2.8	0.4	1.0	2.5	5.5	
0.005	7	0.3	0.7	1.8	4.0	
0.01	14	0.2	0.5	1.3	3.1	
0.02	28		0.4	1.0	2.5	5.5
0.05	70		0.3	0.7	1.8	4.0
0.10	140		0.2	0.5	1.3	3.1
0.20	280			0.4	1.0	2.5
0.50	700			0.3	0.7	1.8
1.00	1400			0.2	0.5	1.3
Drains at 75 ft spacing[c]						
0.0001	0.14	1.9	4.7			
0.0002	0.28	1.4	3.7	8.3		
0.0005	0.70	1.0	2.5	6.0		
0.001	1.4	0.7	1.9	4.7		
0.002	2.8	0.5	1.4	3.7	8.3	
0.005	7	0.3	1.0	2.5	6.0	
0.01	14	0.2	0.7	1.9	4.7	
0.02	28		0.5	1.4	3.7	8.3
0.05	70		0.3	1.0	2.5	6.0
0.10	140		0.2	0.7	1.9	4.7
0.20	280			0.5	1.4	3.7
0.50	700			0.3	1.0	2.5
1.00	1400			0.2	0.7	1.9
Drains at 100 ft spacing[c]						
0.0001	0.14	2.6	6.3			
0.0002	0.28	1.9	4.9	11.2		
0.0005	0.7	1.3	3.4	8.0		
0.001	1.4	1.0	2.6	6.3		
0.002	2.8	0.7	1.9	4.9	11.2	
0.005	7	0.5	1.3	3.4	8.0	
0.01	14	0.3	1.0	2.6	6.3	
0.02	28		0.7	1.9	4.9	11.2
0.05	70		0.5	1.3	3.4	8.0
0.10	140		0.3	1.0	2.6	6.3
0.20	280			0.7	1.9	4.9
0.50	700			0.5	1.3	3.4
1.00	1400			0.3	1.0	2.6

are illustrative of a rational method for designing filters, and that the charts represent specific solutions.

NUMERICAL EXAMPLES

Description of Filter Materials Used in Examples.—In the examples presented, design requirements will be developed for certain assumed physical conditions. Since the water-removing capacity of pervious filters and bases is related to filter permeability, designs will be developed for filters constructed of materials of several permeabilities. To permit a tie-in with practical experience, several specific filter materials of definite permeability will be considered. The assumed permeabilities and gradients are given in Table 2. The values shown are believed reasonably typical for basically rounded, granular filter materials of durable particles. No implication is made that all filter materials of the indicated gradings will have permeabilities of the values shown,

TABLE 2.—PROPERTIES OF FILTER MATERIALS USED IN EXAMPLES

Material	Percentage Passing Sieve Size or No.						Permeability		
							Approx. Range	Used in Example	
	3/8 in.	4	16	30	50	100	(ft per day)	(ft per day)	(ft per min.)
Concrete Sand	100	95-100	45-80		10-30	2-10	0.01-60	2	0.0014
Grading "A"	100	90-100	30-80	15-50	3-10	0-2[a]	20-400	20	0.014
Grading "B"	100	90-100	25-70	10-40	0-6	0-1[a]	100-1000	100	0.07

[a] To assure cleanness, percent passing No. 100 not to exceed one-fifth of the percent passing No. 50.

because the permeability of soils and filter materials depends not only upon gradation, but also upon density, moisture content during placement, method of placement, and densification, and upon the shape, strength, and surface texture of the soil particles. Wide variations in permeability can be attributed to most of these factors and the only sure way of establishing the permeability of a specific material is by test. The values shown in Table 2 are presented only to permit a demonstration of the influence of permeability upon the design of water-removing, protective filters.

Fine concrete aggregate is considered by many engineers to be an excellent filter material. Concrete sand that complies with AASHO specifications can have a wide range of permeability, depending upon the quantity and properties of the fines, as well as other factors. Relatively clean concrete sand can have permeabilities around 50 ft to 60 ft per day (0.035 ft to 0.042 ft per min), but if it contains the maximum permissible amount of fines (10% of -100 mesh) and is well compacted, its permeability can be in the order of only 0.1 to 0.01 ft per day (0.7×10^{-4} ft to 0.7×10^{-5} ft per min). In the examples, concrete sand is assumed to have a permeability of 2 ft. per day (0.0014 ft per min). Characteristic data for this material, and two progressively cleaner filter materials used in the examples are summarized in Table 2.

Any of the previously described gradings of sound and durable mineral grains, which are placed with reasonable care to avoid segregation, should be satisfactory as a general purpose filter material for protection of many soils under small to moderate seepage gradients without serious danger of infiltration. For high head usages, or installations where infiltration of clogging could have serious consequences, accepted criteria for the prevention of infiltration or clogging should be applied on the basis of the gradation of filter and soil.

Example No. 1 - Sloping Blanket.—A stabilization trench of the type illustrated in Fig. 1(a) is to be constructed in hilly terrain. High ground water imposes a threat to the stability of a highway fill to be constructed at this location. Conditions assumed for the development of the charts in Figs. 4 and 5 are reasonably similar to the conditions at the site. The thickness and permeability of a pervious filter blanket are to be determined from the chart in Fig. 5.

The assumptions are that the height of saturated discharge face, H, is 20 ft, that the slope equals $1\frac{1}{2}$:1 so that S is $1\frac{1}{2}$, and that the soil permeability, k_s, equals 1×10^{-4} ft per min. From Fig. 5 determine required thickness, T, for the given conditions.

(a) Using fine concrete aggregate with a minimum permeability of 0.0014 ft per min (2 ft per day); from Fig. 5, H/T is determined as 16 and required $T = \left(\frac{20}{16}\right) = 1.25$ ft. (b) Using Grading "A", with a minimum k_f of 0.014 ft per min; from Fig. 5, H/T is found to be 140 and required $T = \left(\frac{20}{140}\right) = 0.15$ ft. Under the proposed conditions it is considered that a 2-ft-thick blanket is about as thin as it is practical to construct, hence, fine concrete aggregate, with a minimum permeability of 0.0014 ft per min is satisfactory. A 1-ft blanket of material "A", with a minimum permeability of 0.014 ft per min would have a water-removing capacity substantially greater than that of a 2-ft blanket of fine concrete aggregate of the specified permeability. The thinner blanket would require only half the material needed for the thicker blanket and would provide a much greater discharge capacity. Under conditions where thin blankets are practical, these potentialities are worth considering.

Example No. 2 - Horizontal Blanket.—A freeway through a metropolitan area is to be located in a trench section for several miles. Normal ground water in the area lies several feet above the pavement grade. A pervious sand and gravel substratum gives every indication of sustaining a plentiful supply of ground water. A permanent drainage system for interception of ground water is being considered as a means of keeping excessive water out of the pavement base to prevent uplift, pumping, and general disintegration of the roadway.

Conditions at this location may be reasonably approximated by the solution developed in Figs. 8 and 9.

The assumptions are that the cross section is comparable to Fig. 8; that collector drains are to be at 80-ft intervals, (D = 80 ft); that there will be horizontal top and bottom surfaces of filter (small slope can be neglected); that the blanket thickness is h + 6 in.; and that the soil permeability, k_s, is as in Example 1 equal to 1×10^{-4} ft per min. From Fig. 9 the blanket thickness required for the assumed conditions is determined. (a) Using fine concrete aggregate with minimum k_f equal to 0.0014 ft per min for $k_f = 0.0014$ and $k_s = 1 \times 10^{-4}$ ft per min $D/h = 18$, and $h = D/18 = 80/18 = 4.4$ ft (from Fig. 9). The required blanket thickness = 4.4 + 0.5 = 4.9 ft. The quantity of filter material required per foot of an 80-ft blanket = 4.9 x 80/27 = 14.5 cu yd. (b) Using Grading "A", with minimum k_f = 0.014 ft per min for k_f = 0.014 ft per min, and $k_s = 1 \times 10^{-4}$ ft per min D/h = 44, and h = D/44 = 80/44 = 1.8 ft, (from Fig. 9). The required blanket thickness = 1.8 + 0.5 = 2.3 ft. The quantity of filter material, "A", required per foot of blanket = 2.3 x 80/27 = 6.8 cu yd. (c) Using grading "B", with minimum permeability = 0.7 ft per min for k_f = 0.07 ft per min, and $k_s = 1 \times 10^{-4}$ ft per min D/h = 90, and h = D/90 = 80/90 = 0.9 ft (from Fig. 9). The required blanket thickness = 0.9 + 0.5 = 1.4 ft. The quantity of filter material, "B", required per foot of blanket = 1.4 x 80/27 = 4.2 cu yd.

The practical design of drainage systems should take into consideration the spacing and depth of collector drains as well as the thickness and quality of the drainage layers. Solutions of the kind presented in Fig. 9 can permit an economic evaluation of these various factors.

CONCLUSIONS

The solutions presented illustrate application of the flow-net to the analysis of permeability requirements of filters and pervious bases for the removal of steady seepage. A complete analysis should also take into consideration the possibility of temporary flooding and rise in saturation caused by capillarity.

The numerical examples and flow-net solutions point up the relatively major influence of boundary conditions and seepage gradients upon the minimum permeability that will assure the rapid removal of steady seepage.

Application of these methods depends upon knowledge of the effective or average permeability of the saturated media from which water is being drained. This determination is the most difficult part of the analysis, and should have the benefit of an experienced evaluation of all available information, including permeability data, ground water observations, and records of performance of projects in the area when such records are available. Every effort must be made to detect and evaluate the presence of porous seams and joints that are easily overlooked but that contribute substantially to the flow of water.

Charts that are presented provide flow-net solutions for two common types of filter design. While the purpose of the charts is to demonstrate a design method, rather than to provide general solutions, they should have some practical value for filter design in conditions comparable to those assumed. For conditions substantially different individual solutions would be needed.

The analysis presented leads to the conclusion that relatively high permeability is needed in filter materials when boundary conditions restrict discharge gradients to small values. In such cases, coarser gradings of filters may sometimes be permitted without danger of infiltration and clogging. In cases where a single layer cannot safely remove the incoming water without danger of infiltration or clogging, other designs should be used, such as multiple-layer or graded filters. In such cases, the layer in contact with the soil should be designed to prevent clogging or infiltration, and another layer should be designed to provide the required water-removing capacity.

In the record numerical example, the design thickness of the filter layer was taken as 6 in. greater than the theoretical penetration of saturation caused by steady seepage. This 6 in. thickness is a small allowance for capillarity and "free board." It will be recognized that capillary rise can vary from nearly zero in clean pea gravel to several feet in compacted graded filter material containing appreciable fines. Possible entrapment or "looking" of water beneath pavements by capillarity lends support to the desirability of thinking in terms of materials cleaner than are commonly used in pervious bases.

The solutions point to the desirability of a general appraisal of some commonly accepted specifications for filter materials. Classes of materials should be provided that will not only assure protection against infiltration when this protection is needed, but will have sufficient water-removing capacity to meet the needs of individual installations. Because any analysis of this kind can only approximate true conditions, reasonable margins of safety should be provided. Flow-net solutions can furnish a yardstick that is consistent with available knowledge of seepage gradients and soil conditions.

2626 October, 1960 SM 5

Journal of the
SOIL MECHANICS AND FOUNDATIONS DIVISION
Proceedings of the American Society of Civil Engineers

GENERALIZED SOLUTIONS FOR LATERALLY LOADED PILES

By Hudson Matlock,[1] M. ASCE and Lymon C. Reese,[2] M. ASCE

SYNOPSIS

To reach rational solutions for problems of laterally loaded piles, the nonlinear force-deformation characteristics of the soil must be considered. This may be done by repeated application of elastic theory. Soil modulus constants are adjusted for each successive trial until satisfactory compatibility is obtained in the structure-pile-soil system. The computations are facilitated by non-dimensional solutions.

Basic equations and methods of computation are given for both elastic-pile theory and rigid-pile theory. Several forms of soil modulus variation with depth are considered. Typical solutions are presented and recommendations given for their use in design problems.

INTRODUCTION

The problem of the laterally loaded pile is of particular interest in connection with drilling platforms and other offshore pile-supported industrial and defense installations. Lateral loads from wind and wave are frequently the most critical factor in the design of such structures. Solutions of the general problem also apply to a variety of cases onshore, including power poles, pile-supports for earthquake resistant structures, and pile-supported structures which may be subjected to lateral blast forces.

The problem of laterally loaded piles is closely related to the familiar problem of a beam on an elastic foundation; however, in one respect, it represents a more specialized case. All external forces and moments applied to

Note.—Discussion open until March 1, 1961. To extend the closing date one month, a written request must be filed with the Executive Secretary, ASCE. This paper is part of the copyrighted Journal of the Soil Mechanics and Foundations Division, Proceedings of the American Society of Civil Engineers, Vol. 86, No. SM 5, October, 1960.

[1] Assoc. Prof. of Civ. Engrg., The Univ. of Texas, Austin, Tex.
[2] Assoc. Prof. of Civ. Engrg., The Univ. of Texas, Austin, Tex.

the pile-soil system are introduced through boundary conditions existing at one point, the top of the pile, while loading may be applied at many points along a beam. On the other hand, rational solutions of pile-soil interaction problems require generalization of the beam-on-elastic-foundation theory to account for the non-linear characteristics of real soils.

To account for the non-linearity between pile deflection and soil resistance, the most convenient approach appears to be one in which repeated application of elastic theory is used. Soil resistance moduli are adjusted upon completion of each trial until satisfactory compatibility is obtained between the predicted behavior of the soil and the load-deflection relationships required by an elastic pile.

In the final trial in a series of iterative approximations the final variation in soil modulus may assume any form with respect to distance along the pile. If it is required that the soil modulus values be adjusted independently at each depth considered, then, in practice, a rigorous solution for any particular pile problem requires the use of a digital computer.

Fortunately, the final computed deflections, bending moments, and other quantities are not very sensitive to changes in soil modulus values. Satisfactory results may be obtained for most practical cases with simple forms of soil modulus variation with depth. Adjustments are limited to changes in the values of the coefficients of soil modulus variation.

In most cases, for both clay and sandy soils, the final soil modulus values tend to increase with depth. The principal reasons for this are that (1) soils frequently increase in strength characteristics with depth as the result of overburden pressures and of natural deposition and consolidation processes and (2) pile deflections decrease with depth for any given loading, and the corresponding equivalent elastic moduli of soil reaction tend to increase with decreasing deflection.

Non-dimensional solutions have been presented previously[3] in which the soil modulus, E_s, increases in simple proportion to depth x, or $E_s = k\,x$. While this simple form appears to be applicable to most laterally loaded pile problems, a few cases have been encountered where it would be helpful to use some other soil modulus function. For example, in the interpretation of the results of several series of extensive field tests with an instrumented pile[4,5] other forms have been found desirable. With design problems, however, the uncertainty inherent in estimating soil behavior characteristics from conventional soil tests is usually consistent with the small errors which may be introduced by the use of a simple form of soil modulus-depth function, such as $E_s = k\,x$.

The purpose of the present paper is to consider general solutions for laterally loaded piles which are supported by an elastic medium. Derivations

---

3 "Non-Dimensional Solutions for Laterally Loaded Piles with Soil Modulus Assumed Proportional to Depth," by Lymon C. Reese, Proceedings, Eighth Texas Conf. on Soil Mechanics and Foundation Engrg., Special Publ. No. 29, Bur. of Engrg. Res., The Univ. of Texas, Austin, Tex., September, 1956.

4 "Procedures and Instrumentation for Tests on a Laterally Loaded Pile," by Hudson Matlock and E. A. Ripperger, Proceedings, Eighth Texas Conf. on Soil Mechanics and Foundation Engrg., Special Publ. No. 29, Bur. of Engrg. Res., The Univ. of Texas, Austin, Tex., September, 1956.

5 "Measurement of Soil Pressure on a Laterally Loaded Pile," by Hudson Matlock and E. A. Ripperger, Proceedings, ASTM, Boston, Mass., June, 1958.

and methods are given by which non-dimensional solutions may be computed for any desired form of variation of soil modulus with respect to depth.

Notation.—The letter symbols adopted for use in this paper are defined where they first appear and are arranged alphabetically, for convenience of reference, in Appendix II.

STATEMENT OF PROBLEM

The soil modulus is defined as

$$E_s = \frac{-p}{y} \qquad (1)$$

in which y is the lateral deflection of the pile and p is the soil resistance expressed as force per unit length of pile. The negative sign indicates that the direction of the soil reaction is always opposite to the direction of the pile deflection. The soil reaction need not be a linear function of pile deflection and in general is not. A typical relation between p and y is shown in Fig. 1. The soil modulus, E_s, is the slope of a secant drawn from the origin to any point along the p - y curve. Its units are force × length^{-2}.

Soil reaction-pile deflection relations have been the subject of field and analytical research studies;[6,7,8,9] recommendations have been given for predicting p - y relations from the results of soil investigations,[6,10] and other research is in progress. In time, most of this information probably will be available to the engineering profession.

A typical foundation pile of length, L, and flexural stiffness, E I, is shown in Fig. 2(a). The depth, x, is measured downward from the ground line. In this example, the boundary condition at the top consists of an imposed moment, M_t, and a shear, P_t, and each is shown acting in a positive sense. Other combinations of boundary values would be applicable in variations of this problem.

The variation in soil modulus with depth, corresponding to specific values of the loading P_t and M_t, is indicated in Fig. 2(b). If the loads change in value, a different deflection pattern will be taken by the pile and different values of the soil modulus will result. Considering the non-linearity of p - y relations at various depths, E_s is a function of both x and y. Therefore, the form of the E_s versus depth relationship also will change if the loading is changed. However, it may be assumed temporarily (subject to adjustment of E_s values by trial and error) that the soil modulus is some function of x only, or that

$$E_s = E_s(x) \qquad (2)$$

For solution of the problem the elastic curve y (x) of the pile, shown in Fig. 2(c), must be determined, together with various derivatives which are of interest. These derivatives yield values of slope, moment, shear, and soil re-

[6] "Soil Modulus for Laterally Loaded Piles," by Bramlette McClelland and John A. Focht, Jr., Transactions, ASCE, Vol. 123, 1958, p. 104.

[7] "Static and Cyclic Lateral Loading of and Instrumented Pile," by Hudson Matlock, E. A. Ripperger, and Don P. Fitzgibbon, Report to Shell Oil Co., Austin, Tex., July, 1956.

[8] "Lateral Loading of an Instrumented Pile with Soil Conditions Varied," by Hudson Matlock and E. A. Ripperger, Report to Shell Oil Co., Austin, Tex., June, 1957.

[9] "Theoretical Analysis and Laboratory Studies of Laterally Loaded Model Pile Segments," by Hudson, Matlock and E. A. Ripperger, Report to Shell Oil Co., Austin, Tex., July, 1957.

[10] "Recommendations Pertaining to the Design of Laterally Loaded Piles," by Hudson Matlock, Report to Shell Oil Co., Austin, Tex., November, 1957.

action as functions of depth. The successive curves are shown in Fig. 3 for a typical pile problem. The sign conventions are shown in Fig. 4.

DIMENSIONAL ANALYSIS FOR ELASTIC-PILE THEORY

The principles of dimensional analysis may be used to establish the form of non-dimensional relations for the laterally loaded pile. With the use of model

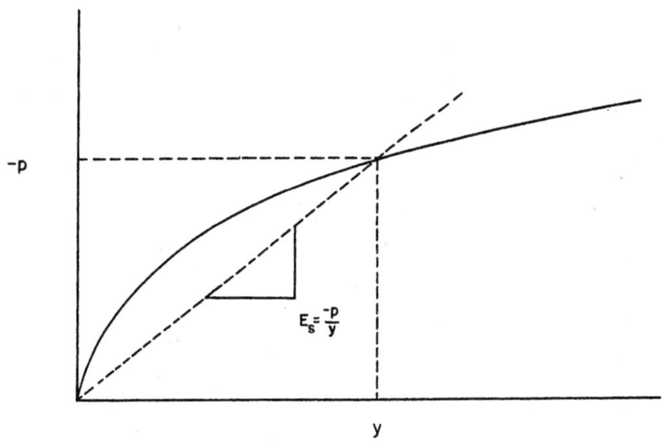

FIG. 1.—TYPICAL RELATION BETWEEN SOIL REACTION p AND PILE DEFLECTION y

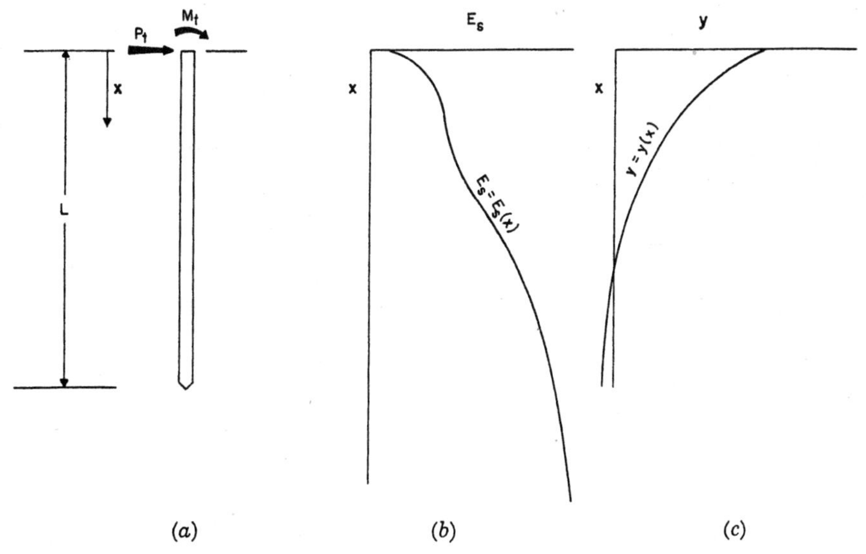

FIG. 2.—A LATERALLY LOADED PILE PROBLEM AND ITS SOLUTION

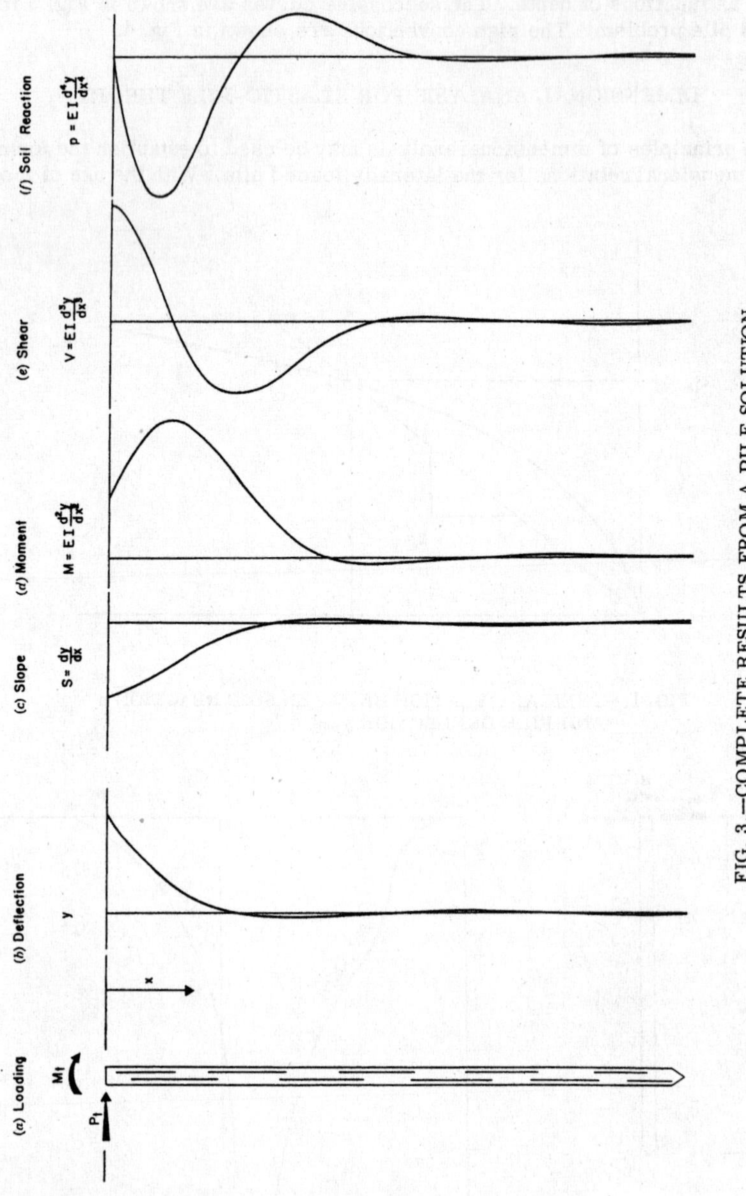

FIG. 3.—COMPLETE RESULTS FROM A PILE SOLUTION

theory the necessary relations will be determined between a "prototype" having any given set of dimensions, and a similar "model" for which a solution may be available. Although the principles of dimensional analysis are usually applied to results obtained from tests on physical models, the method is equally applicable to results obtained by the solution of a mathematical model.

For very long piles, the length, L, loses significance because the deflection may be very nearly zero for much of the length of the pile. It is convenient to introduce some characteristic length dimension as a substitute. A linear dimension, T, is therefore included in the quantities to be considered. T may be defined in any convenient way which will simplify the notation. The specific definition of T will vary with the form of the function for soil modulus versus depth. However, it will be seen later that, for each definition used, T expresses a relation between the stiffness of the soil and the flexural stiffness of the pile. It is therefore given the term "relative stiffness factor."

For the case of an applied shear, P_t, and moment, M_t, the solution for deflections of the elastic curve may include the relative stiffness factor and be expressed as

$$y = y\left(x, T, L, E_s, E I, P_t, M_t\right) \quad \ldots\ldots\ldots\ldots (3)$$

Other boundary values could be substituted for P_t and M_t.

If the assumption of elastic behavior is introduced for the pile, and if deflections remain small, relative to the pile dimensions, the principle of superposition may be applied. Thus the effects of an imposed lateral load, P_t, and imposed moment, M_t, may be considered separately, as shown in Fig. 5. If y_A represents the deflection due to the lateral load, P_t, and if y_B is the deflection caused by the moment, M_t, the total deflection is

$$y = y_A + y_B \quad (4)$$

Furthermore, it is the ratios of y_A to P_t and of y_B to M_t which are sought in reaching generalized elastic-case solutions. Thus, the solutions may be expressed for Case A as

$$\frac{y_A}{P_t} = f_A\left(x, T, L, E_s, E I\right) \quad \ldots\ldots\ldots\ldots (5)$$

and for Case B

$$\frac{y_B}{M_t} = f_B\left(x, T, L, E_s, E I\right) \quad \ldots\ldots\ldots\ldots (6)$$

in which f_A and f_B represent two different functions of the same terms. In each case there are six terms and two dimensions (force and length) involved. There are, therefore, four independent, non-dimensional groups which can be formed. The arrangements chosen are, for Case A,

$$\frac{y_A E I}{P_t T^3}, \frac{x}{T}, \frac{L}{T}, \frac{E_s T^4}{E I} \quad \ldots\ldots\ldots\ldots (7)$$

and for Case B,

$$\frac{y_B E I}{M_t T^2}, \frac{x}{T}, \frac{L}{T}, \frac{E_s T^4}{E I} \quad \ldots\ldots\ldots\ldots (8)$$

To satisfy conditions of similarity, each of these groups must be equal for both model and prototype, or,

$$\frac{x_p}{T_p} = \frac{x_m}{T_m} \quad \ldots\ldots\ldots\ldots\ldots\ldots (9)$$

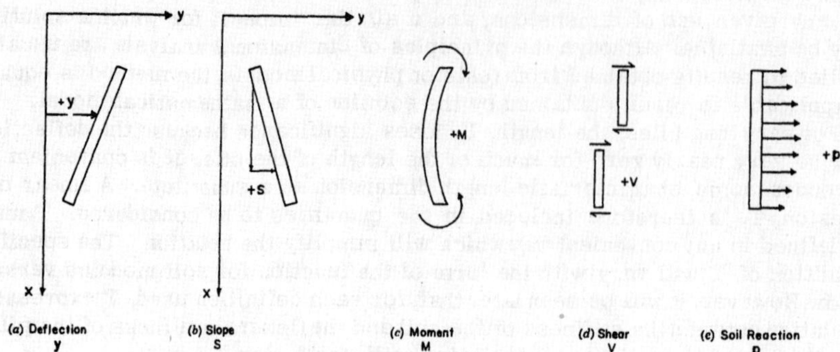

FIG. 4.—SIGN CONVENTIONS

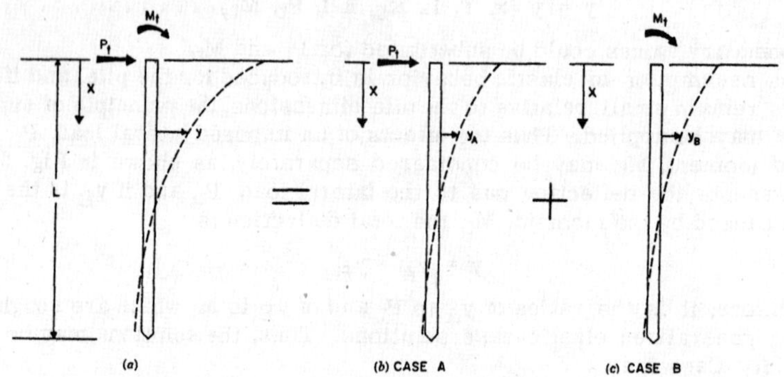

FIG. 5.—APPLICATION OF THE PRINCIPLE OF SUPERPOSITION TO THE LATERALLY LOADED PILE PROBLEM

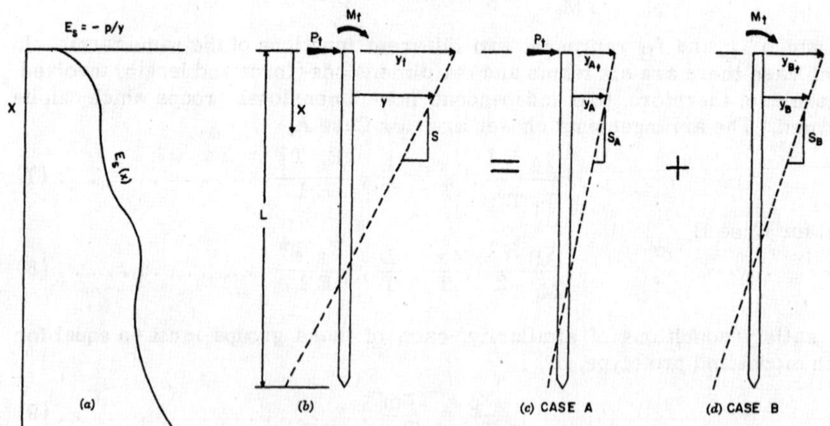

FIG. 6.—THE PROBLEM OF A SHORT, RIGID PILE

$$\frac{L_p}{T_p} = \frac{L_m}{T_m} \qquad (10)$$

$$\frac{E_{s_p}\, T_p{}^4}{E\, I_p} = \frac{E_{s_m}\, T^4}{E\, I_m} \qquad (11)$$

$$\frac{y_{A_p}\, E\, I_p}{P_{t_p}\, T_p{}^2} = \frac{y_{a_m}\, E\, I_m}{P_{t_m}\, T_m{}^3} \qquad (12)$$

and

$$\frac{y_{B_p}\, E\, I_p}{M_{t_p}\, T_p{}^2} = \frac{y_{B_m}\, E\, I_m}{M_{t_m}\, T_m{}^2} \qquad (13)$$

A group of non-dimensional parameters may be defined which will have the same numerical value for any pair of structurally similar cases, or for any model and its prototype. These are

Depth Coefficient,

$$Z = \frac{x}{T} \qquad (14)$$

Maximum Depth Coefficient,

$$Z_{max} = \frac{L}{T} \qquad (15)$$

Soil Modulus Function,

$$\phi\,(Z) = \frac{E_s\, T^4}{E\, I} \qquad (16)$$

Case A Deflection Coefficient,

$$A_y = \frac{y_A\, E\, I}{P_t\, T^3} \qquad (17)$$

Case B Deflection Coefficient,

$$B_y = \frac{y_B\, E\, I}{M_t\, T^2} \qquad (18)$$

Thus, from Eqs. 14 through 18, for (1) similar soil-pile stiffness systems, (2) similar positions along the piles, and (3) similar pile lengths (unless lengths are very great and need not be considered), the solution of the problem can be expressed from Eq. 4, and from Eqs. 17 and 18, as

$$y = \left[\frac{P_t\, T^3}{E\, I}\right] A_y + \left[\frac{M_t\, T^2}{E\, I}\right] B_y \qquad (19)$$

By the same type of reasoning other forms of the solution can be expressed as

Slope,

$$S = S_A + S_B = \left[\frac{P_t\, T^2}{E\, I}\right] A_s + \left[\frac{M_t\, T}{E\, I}\right] \cdot B_s \qquad (20)$$

Moment,

$$M = M_A + M_B = \left[P_t\, T\right] A_m + \left[M_t\right] B_m \qquad (21)$$

Shear,

$$V = V_A + V_B = \left[P_T\right] A_v + \left[\frac{M_t}{T}\right] B_v \dots\dots\dots\dots (22)$$

Soil Reaction,

$$p = p_A + p_B = \left[\frac{P_t}{T}\right] A_p + \left[\frac{M_t}{T^2}\right] B_p \dots\dots\dots\dots (23)$$

It is still necessary to obtain a particular set of A and B coefficients (as functions of the depth parameter Z) by a solution of a particular model. However, Eqs. 20 through 23 are independent of the characteristics of the model except that elastic behavior and small deflections are assumed. Note that T is still an undefined characteristic length dimension and that the variation of E_s with depth, or the corresponding form of ϕ (Z), has not been specified.

The relations previously derived are applicable to step-tapered piles which are conventionally used in offshore construction. However, it is necessary that structural similarity be maintained between the mathematical model and the prototype. This means that the changes in E I must be proportionate and must occur at corresponding values of Z along the lengths of both model and prototype.

THE DIFFERENTIAL EQUATION

From beam theory, the basic equation for an elastic beam is

$$E\,I \frac{d^4y}{dx^4} = p \dots\dots\dots\dots\dots\dots (24)$$

Introducing the definition of $p = -E_s\, y$ of Eq. 1, the basic equation for a beam on an elastic foundation, or for a laterally loaded pile, is

$$\frac{d^4y}{dx^4} + \frac{E_s}{E\,I}\, y = 0 \dots\dots\dots\dots\dots\dots (25)$$

Where an applied lateral load, P_t, and an applied moment, M_t, are considered separately according to the principal of superposition, Eq. 25 becomes, for Case A,

$$\frac{d^4y_A}{dx^4} + \frac{E_s}{E\,I}\, y_A = 0 \dots\dots\dots\dots\dots (26)$$

and for Case B,

$$\frac{d^4y_B}{dx^4} + \frac{E_s}{E\,I}\, y_B = 0 \dots\dots\dots\dots\dots (27)$$

Substituting the definitions of non-dimensional parameters contained in Eqs. 14 through 18, a non-dimensional differential equation can be written for Case A as

$$\frac{d^4A_y}{dZ^4} + \phi\,(Z)\, A_y = 0 \dots\dots\dots\dots\dots (28)$$

and for Case B,

$$\frac{d^4B_y}{dZ^4} + \phi\,(Z)\,B_y = 0 \quad \ldots\ldots\ldots\ldots\ldots \quad (29)$$

SOLUTION OF THE DIFFERENTIAL EQUATIONS

To produce a particular set of non-dimensional A and B coefficients, it is necessary (1) to specify ϕ (Z), including a convenient definition of the relative stiffness factor, T, and (2) to solve the differential equations (Eqs. 28 and 29). The resulting A and B coefficients may then be used, with Eqs. 19 through 23, to compute deflections, slopes, moments, shears, and soil reactions for any pile problem which is similar to the case for which non-dimensional solutions have been obtained.

Based on the boundary conditions, P_t and M_t, and the resulting A and B coefficients, relations may be derived so that problems may be solved for cases in which other boundary conditions are known. These may include a specified deflection, y_t, at the top of the pile or a specified value of slope, S_t. Where structural restraints are involved at the pile-to-structure connection, the boundary conditions may consist of a lateral shear, P_t, and rotational restraint stiffness, M_t/S_t. It is conceivable that still other combinations and ratios may have application in variations of the problem.

Further non-dimensional coefficients could be obtained for these other boundary conditions. However, it appears that any conceivable structure-pile-soil system can be solved by use of the A and B coefficients. Some of these solutions involve a trial and error process in satisfying the boundary conditions.

The only case for which a closed analytical solution is possible is that in which the soil modulus is constant with depth.[11] Series-type solutions have also been developed for the case in which the soil modulus has a linear variation with depth.[12]

Since the available special-case solutions impose limitations on the nature of the soil modulus variation, a more general approach is desirable. The difference-equation method appears to offer the most practical approach to generalized solutions.[13] Procedures have been developed for a once-through computation with successive elimination of unknowns.[14,15] A further extension has been made[3] to enable the introduction of moment and shear as the boundary conditions, and to produce a set of non-dimensional solutions for $E_s = k\,x$.

A computer solution has been developed in which the flexural stiffness E I of the pile may be changed abruptly at points along the length of the pile.[16]

---

11 "Strength of Materials," by S. Timoshenko, Part II, Van Nostrand, New York, 1930.

12 "Beams on Elastic Foundations," by M. Hetenyi, Univ. of Michigan Press, Ann Arbor, Mich., Oxford Univ. Press, London, England, 1946.

13 "Horizontal Pressures on Pile Foundations," by L. A. Palmer and James B. Thompson, Proceedings, Third Internatl. Conf. on Soil Mechanics and Foundation Engrg., Vol. V, Rotterdam, The Netherlands, 1948, pp. 156-161.

14 "Analysis of Laterally Loaded Piles by Difference Equation Solution," by John A. Focht and Bramlette McClelland, The Texas Engineer, Texas Section of ASCE, September, October, November, 1955.

15 "A Numerical Method for Predicting the Behavior of Laterally Loaded Piling," by R. J. Howe, TS Memorandum 9, Shell Oil Co., Houston, Tex., 1953.

16 "Difference Equation Method for Laterally Loaded Piles with Abrupt Changes in Flexural Rigidity," by Lymon C. Reese and A. S. Ginzbarg, EPR Memorandum Report 39, Shell Development Co., Houston, Tex., 1958.

The changes in pile cross section ordinarily encountered in offshore construction produce relatively moderate effects in computed moments. However, the development of a limited number of non-dimensional solutions which include step-changes in E I, may be warranted for a few typical cases.

The difference-equation method is summarized in Appendix I, as it is applied to the case in which the pile stiffness, E I, is constant. This summary includes the equations which are needed to introduce the applied shear, P_t, and moment, M_t, as the boundary conditions.

The summary in Appendix I is given in terms of ordinary parameters having physical dimensions. To produce a set of non-dimensional solutions for any desired soil modulus function, it is possible to use input values from a dimensioned "model" and then to determine the corresponding non-dimensional coefficients by Eqs. 14 through 23. However, values may be computed directly by (1) substituting unit values for P_t and M_t, (2) setting the k coefficients in the soil modulus function $E_s(x)$ to correspond to constants in ϕ (Z) (thus the relative stiffness factor, T, is made equal to unity), and (3) letting the length, L, of the pile be numerically equal to the desired value of Z_{max}. The various results will then be numerically equal to the corresponding non-dimensional coefficients.

LIMITING CASE FOR SHORT RIGID PILES

Piles or posts having relatively shallow imbedment are frequently encountered in practice. Such piles tend to behave as rigid members, and the difference-equation method used in the elastic-theory solutions tends to become inaccurate because of the very small successive differences which are involved. For such cases, a simpler theory is applicable, in which the pile is considered to be a rigid member.

Although computations are much more simple for the rigid-pile case than for the elastic-pile theory, it is still convenient to use generalized solutions and to consider separately the effects of applied lateral load and applied moment. This is particularly true where repeated solutions must be made to obtain compatibility between the behavior of the pile and the predicted characteristics of the soil.

The statement of the rigid-pile problem is illustrated in Fig. 6. For convenience, the notation and format of the development are made to correspond as closely as possible to those of the elastic-pile case.

DIMENSIONAL ANALYSIS FOR RIGID PILES

The pertinent factors in the rigid-pile problem and the principle of superposition are shown in Fig. 6. It is convenient to include an additional term, J, which is later given particular definitions which depend on the form of the soil modulus variation with depth. For the present, J is simply a constant having the same dimensions $(\text{force} \times \text{length}^{-2})$ as the soil modulus, E_s.

For either Case A $(M_t = 0)$ or Case B $(P_t = 0)$ there are a total of six factors to be considered. For Case A,

$$y_A = y_A\left(x, L, E_s, J, P_t\right) \quad \ldots\ldots\ldots\ldots \quad (30)$$

and for Case B,

$$y_B = y_B \left(x, L, E_s, J, M_t\right) \quad \text{(31)}$$

In each trial computation in an actual design problem, the soil is considered to be elastic. Thus, for either Case A or Case B, it is the ratio of deflection to loading which is sought in reaching generalized solutions. This reduces the number of non-dimensional groups to three. For Case A these are

$$\frac{y_A\, J\, L}{P_t}, \quad \frac{x}{L}, \quad \frac{E_s}{J} \quad \text{(32)}$$

and for Case B,

$$\frac{y_B\, J\, L^2}{M_t}, \quad \frac{x}{L}, \quad \frac{E_s}{J} \quad \text{(33)}$$

For similarity between a prototype and a computed model, each non-dimensional group may be defined as a dimensionless parameter. These are as follows:

Depth Coefficient,

$$h = \frac{x}{L} \quad \text{(34)}$$

Soil Modulus Function,

$$\phi\,(h) = \frac{E_s}{J} \quad \text{(35)}$$

Case A Deflection Coefficient,

$$a_y = \frac{y_A\, J\, L}{P_t} \quad \text{(36)}$$

Case B Deflection Coefficient,

$$b_y = \frac{y_B\, J\, L^2}{M_t} \quad \text{(37)}$$

By superposition, the total deflection $y = y_A + y_B$, is

$$y = \left[\frac{P_t}{J\,L}\right] a_y + \left[\frac{M_t}{J\,L^2}\right] b_y \quad \text{(38)}$$

From reasoning similar to the preceding, other forms of the solution can be expressed as:

Slope,

$$S = S_A + S_B = \left[\frac{P_t}{J\,L^2}\right] a_s + \left[\frac{M_t}{J\,L^3}\right] b_s \quad \text{(39)}$$

Moment,

$$M = M_A + M_B = \left[P_t\, L\right] a_m + \left[M_t\right] b_m \quad \text{(40)}$$

Shear,

$$V = V_A + V_B = \left[P_t\right] a_v + \left[\frac{M_t}{L}\right] b_v \quad \text{(41)}$$

Soil reaction,

$$p = p_A + p_B = \left[\frac{P_t}{L}\right] a_p + \left[\frac{M_t}{L^2}\right] b_p \quad \text{(42)}$$

For any given problem the slope $S = dy/dx$ is a constant and all higher derivatives of y are zero. The last three expressions are related to the first two through the relation between soil reaction and pile deflection, $E_s = -p/y$ or, in terms of the non-dimensional coefficients,

$$\phi\,(h) = \frac{-a_p}{a_y} = \frac{-b_p}{b_y} \quad \text{(43)}$$

The preceding dimensional analysis will apply to any form of the soil modulus functions E_s or ϕ (h). The soil modulus constant, J, is to be defined subsequently.

The non-dimensional soil modulus function, ϕ (h), is equivalent to the corresponding function, ϕ (Z), used with the elastic-pile theory except that ϕ (h) is related to the length of the pile rather than to a relative stiffness between the pile and the soil.

For any given ϕ (h), there exists a single set of non-dimensional coefficient curves (for deflection, slope, moment, shear, and soil reaction). Design problems may be solved by essentially the same procedures as for the elastic-pile case. The choice of which theory to use is aided by comparing the results of non-dimensional solutions obtained by the two methods.

EQUATIONS FOR RIGID-PILE SOLUTIONS

The equation for deflection y of a rigid pile is

$$y = y_t + S x \quad \text{. . . . .} \quad (44)$$

in which y_t is the deflection at x = 0 and S is the constant slope of the pile. The soil reaction $p = - E_s\ y$ is

$$p = - E_s\ y_t - E_s\ S\ x \quad \text{. . . . .} \quad (45)$$

A free body of the upper portion of a rigid pile is shown in Fig. 7 with all quantities shown acting in positive directions. By statics, the equation for shear is

$$V = P_t + \int_o^x p\ dx \quad \text{. . . . .} \quad (46)$$

Substituting Eq. 45 into Eq. 46,

$$V = P_t - y_t \int_o^x E_s\ dx - S \int_o^x x\ E_s\ dx \quad \text{. . . . .} \quad (47)$$

The equation for moment is

$$M = M_t + V x - \int_o^x x\ p\ dx \quad \text{. . . . .} \quad (48)$$

or,

$$M = M_t + V x + y_t \int_o^x x\ E_s\ dx + S \int_o^x x^2\ E_s\ dx \quad \text{. . . . .} \quad (49)$$

The shear and moment are zero at the bottom of the pile: at x = L, V = 0, and M = 0. Thus, Eqs. 50 and 51 may be written from Eqs. 47 and 49 so that y_t and S may be evaluated by simultaneous solution.

$$P_t = y_t \int_o^L E_s\ dx + S \int_o^L x\ E_s\ dx \quad \text{. . . . .} \quad (50)$$

and

$$M_t = - y_t \int_o^L x\ E_s\ dx - S \int_o^L x^2\ E_s\ dx \quad \text{. . . . .} \quad (51)$$

The values obtained for y_t and S are then substituted into Eqs. 47 and 49 to complete the solution.

As in the procedure used in the elastic-pile theory, unit values may be introduced into the solution to obtain numerically correct values of the non-dimensional coefficients defined in Eqs. 34 through 37. This amounts to de-

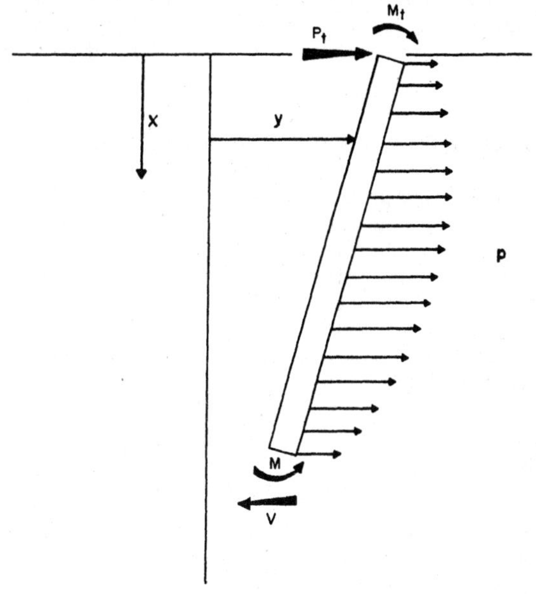

FIG. 7.—FREE BODY DIAGRAM FOR RIGID-PILE THEORY

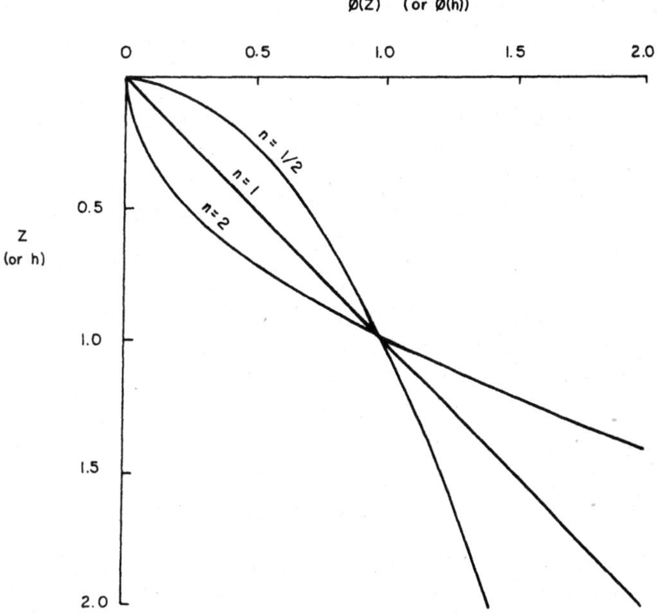

FIG. 8.—TYPICAL POWER FUNCTIONS FOR SOIL MODULUS

termining the non-dimensional coefficients from the results of a numerically convenient model having unit values of L, P_t, and M_t. Coefficients in E_s (x) are chosen to agree with those in the soil modulus function ϕ (h), and J is thus made equal to unity.

FORMS OF SOIL MODULUS VARIATION WITH DEPTH

In solving actual design problems of laterally loaded piles by use of previously computed non-dimensional solutions, the constants in the expressions describing the variation of soil modulus, E_s, with depth, x, are adjusted by trial until reasonable compatibility is obtained. The selected form of the soil modulus relation should be kept as simple as possible so that a minimum number of constants need be adjusted.

Two general forms which are adequate to express any continuous variation with depth are a power form,

$$E_s = k\,x^n \qquad (52)$$

and a polynomial form,

$$E_s = k_o + k_1\,x + k_2\,x^2 \qquad (53)$$

The form $E_s = k\,x$ is seen to be a special case of either of these. A form similar to Eq. 52 has been suggested previously.[17]

The relative stiffness factor, T, of the elastic-pile theory and the soil modulus constant, J, of the rigid-pile theory must be defined for each form of the soil modulus-depth relation. While they may be defined in any way, it is convenient to select definitions that will simplify the corresponding non-dimensional soil modulus functions.

POWER FUNCTIONS; $E_s = k\,x^n$

From the elastic-pile theory, Eq. 16 defining the non-dimensional soil modulus function is $\phi\,(Z) = \dfrac{E_s\,T^4}{E\,I}$.

If the form $E_s = k\,x^n$ is substituted in Eq. 16, the result is

$$\phi\,(Z) = \frac{k}{E\,I}\,x^n\,T^4 \qquad (54)$$

For the elastic-pile case, it is convenient to define the relative stiffness factor, T, by the following expression.

$$T^{n+4} = \frac{E\,I}{k} \qquad (55)$$

Substituting Eq. 55 into Eq. 54 gives

$$\phi\,(Z) = \frac{x^n\,T^4}{T^{n+4}} = \left[\frac{x}{T}\right]^n \qquad (56)$$

---

17 "Piles Subjected to Lateral Thrust Part II - Analysis of Pressure, Deflection, Moment, and Shear by the Method of Difference Equations," by L. A. Palmer and P. P. Brown, Supplement of Symposium on Lateral Load Tests on Piles, ASTM Special Tech. Publication, No. 154-A, 1954, pp. 22-44.

Since $x/T = Z$, the general non-dimensional soil modulus function is

$$\phi\ (Z) = Z^n \qquad (57)$$

Eq. 57 contains only one arbitrary constant, the power n. Therefore, for each value of n which may be selected, one complete set of independent non-dimensional solutions may be obtained from solution of Eqs. 28 and 29. For relatively short elastic piles, separate computations must be made for each Z_{max} considered.

From the rigid-pile theory the soil modulus function has been defined by Eq. 35 as $\phi\ (h) = E_s/J$. If the soil modulus constant, J, is now defined as

$$J = k\ L^n \qquad (58)$$

the corresponding general non-dimensional soil modulus function is

$$\phi\ (h) = \frac{k\ x^n}{k\ L^n} \qquad (59)$$

or, since $h = x/L$,

$$\phi\ (h) = h^n \qquad (60)$$

Only one set of non-dimensional curves will be needed for each selected value of n, regardless of the length L.

In Fig. 8 several typical power functions are shown ($\phi = Z^n$ or $\phi = h^n$). The coordinates are used interchangeably for either the elastic-pile theory or the rigid-pile theory, although values of h greater than unity do not apply in the rigid-pile case.

POLYNOMIAL FUNCTION; $E_s = k_o + k_1\ x + k_2\ x^2$

When a polynomial is used to express the form of the soil modulus variation with depth, the relative stiffness factor, T, or the soil modulus constant, J, may be defined to simplify only one of the terms in the polynomial.

For the elastic-pile case, introducing the polynomial form into Eq. 16 gives

$$\phi\ (Z) = \frac{k_o\ T^4}{E\ I} + \frac{k_1\ T^5}{E\ I}\left[\frac{x}{T}\right] + \frac{k_2\ T^6}{E\ I}\left[\frac{x}{T}\right]^2 \qquad (61)$$

To simplify the second term, as an example, T may be defined by the following expression:

$$T^5 = \frac{E\ I}{k_1} \qquad (62)$$

The resulting soil modulus function is

$$\phi\ (Z) = r_o + Z + r_2\ Z^2 \qquad (63)$$

in which

$$r_o = \frac{k_o}{k_1}\left[\frac{1}{T}\right] \qquad (64)$$

and

$$r_2 = \frac{k_2}{k_1}\left[T\right] \qquad (65)$$

For the rigid pile theory, from Eq. 35,

$$\phi\ (h) = \frac{k_o}{J} + \frac{k_1\ x}{J} + \frac{k_2\ x^2}{J} \qquad (66)$$

To again simplify the second term, J is defined by

$$J = k_1 L \quad \ldots\ldots\ldots\ldots\ldots\ldots\ldots\ldots\ldots \quad (67)$$

and

$$\phi\ (h) = \frac{k_o}{k_1 L} + \frac{k_1 x}{k_1 L} + \frac{k_2 x^2}{k_1 L} \quad \ldots\ldots\ldots \quad (68)$$

or

$$\phi\ (h) = r_o + h + r_2 h^2 \quad \ldots\ldots\ldots\ldots\ldots \quad (69)$$

in which h = x/L and

$$r_o = \frac{k_o}{k_1 L} \quad \ldots\ldots\ldots\ldots\ldots\ldots\ldots \quad (70)$$

and

$$r_2 = \frac{k_2 \cdot L}{k_1} \quad \ldots\ldots\ldots\ldots\ldots\ldots\ldots \quad (71)$$

A separate set of non-dimensional curves would be needed for each desired combination of r-constants. Because of the complexity which otherwise would

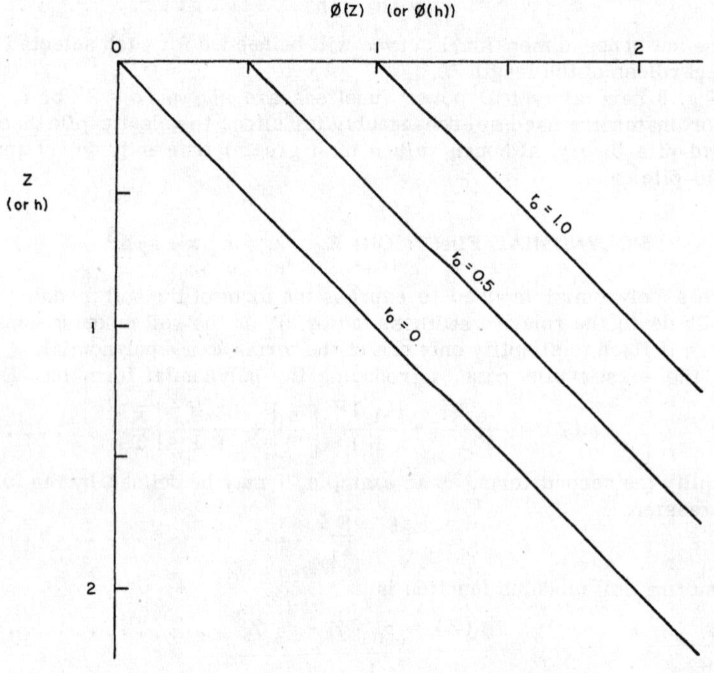

FIG. 9.—EXAMPLES OF SIMPLE POLYNOMIAL SOIL MODULUS FUNCTIONS

result, it does not appear reasonable to vary more than one constant and such forms as those following appear to be about as complicated as should be considered.

$$\phi\ (Z) = r_o + Z \quad \ldots\ldots\ldots\ldots\ldots\ldots \quad (72)$$

and

$$\phi\ (Z) = r_o + Z^2 \quad \ldots\ldots\ldots\ldots\ldots\ldots \quad (73)$$

Curves for the first form (Eq. 72) are shown in Fig. 9 $\left(\phi = r_o + Z \text{ or } \phi = r_o + h\right)$ for three values of r_o. From problems encountered in actual practice, the second form (Eq. 73) appears to offer interesting possibilities, but no solutions are included for this form.

While it would be permissible for some of the r-constants to have negative values, care must be taken that ϕ does not become negative. If this were to occur, some very peculiar results would be possible in the computed solutions.

TYPICAL SOLUTIONS FROM ELASTIC-PILE THEORY

The deflection and moment curves usually are the results of primary interest. These two quantities are used as a basis of comparison of various soil modulus functions in the results presented in the figures which follow. In most of these, results are shown separately for Case A $\left(M_t = 0\right)$ and Case B $\left(P_t = 0\right)$.

Fig. 10 shows the results of elastic-pile theory when various values of the exponent n are used in the form $E_s = k\,x^n$. The three values used correspond to the three curves of ϕ shown in Fig. 8. The solutions in Fig. 10 are applicable to very long piles with Z_{max} equal to approximately 5 or greater.

Although the three curves of Fig. 8 depart widely from each other at values of Z greater than about 1.0, the resulting deflections and moments shown in Fig. 10 are relatively much closer in agreement. This is in accordance with the fact that the behavior of the pile is related, through the relative stiffness factor, T, to the (n + 4) root of the ratio of pile stiffness, E I, to soil modulus constants.

The maximum deflections and moments are seen to increase as the value of n increases. This shows that the soil modulus values in the vicinity of Z less than unity clearly dominate the behavior of the pile.

Similar conclusions can be drawn from solutions shown in Fig. 11. The corresponding soil modulus functions are given in Fig. 9. The function ϕ (Z) = Z is common to both Fig. 10 and Fig. 11 and the corresponding deflection and moment curves may be used as a basis for comparisons between these two figures.

Greater variations are seen among the curves of Fig. 11 than among those in Fig. 10. There is also greater variation in soil modulus values near the top of the pile. This is further confirmation of the importance of soil modulus values at small values of Z.

Additional indication of the importance of the zone near the top is given by the comparison shown in Fig. 12. The constants r_1 in ϕ (Z) = r_1 Z are adjusted to give the same maximum positive moment coefficients as those obtained with the non linear function, ϕ (Z) = Z^n. In this figure, the solid curves have been reproduced from Figs. 8 and 10. They show the forms of soil modulus functions ϕ (Z) = $Z^{1/2}$ and ϕ (Z) = Z^2, and the corresponding moment coefficient curves for a long pile loaded by a lateral force, P_t, but with moment, M_t, equal to zero. The dashed lines in Fig. 12(a) are for soil modulus functions of the form $E_s = k\,x$. The coefficients of Z were determined to give the straight-line variations required to produce the same maximum positive moments as the two non-linear soil modulus functions. The values of Z at the intersections of the curved and straight-line variations are only 0.33 and 0.15 for the two examples considered. The corresponding moment curve comparisons are given in Fig. 12(b).

The comparisons in Fig. 12 suggest that good moment-curve predictions may be made by using $E_s = k\ x$ even though real soil modulus variations

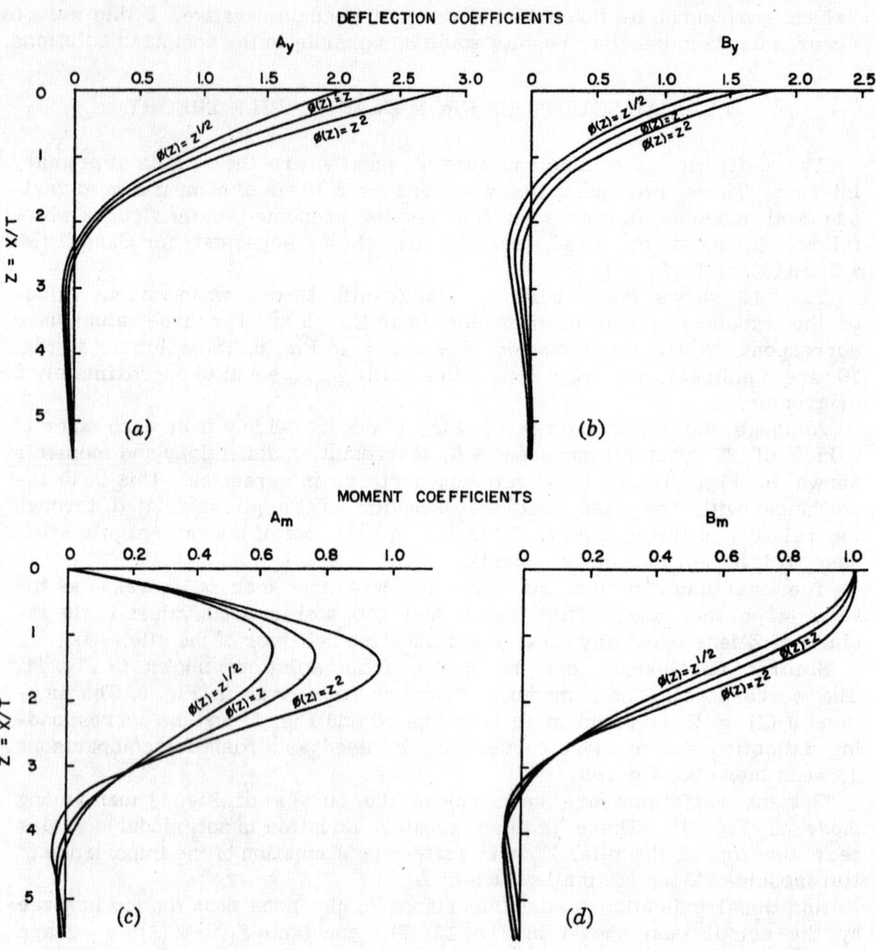

FIG. 10.—TYPICAL RESULTS OF ELASTIC-PILE THEORY WITH $E_s = k\ x^n$, or $\phi(Z) = Z^n$

may be quite non-linear with respect to depth. However, to make such satisfactory approximations, the designer must recognize the relatively great importance of close fitting very near the top of the pile.

TYPICAL SOLUTIONS FROM RIGID-PILE THEORY

The three soil modulus functions shown in Fig. 8 have been used to compute the behavior of a rigid pile also. The resulting deflection and moment curves for both Case A and Case B are shown in Fig. 13.

These results would be applicable to the case of a post or spud pile installed to relatively shallow depths. However, essentially the same trial and error design procedures may be followed as are used for reaching compatibility in the elastic-theory approach. As compared to the theory

DEFLECTION COEFFICIENTS

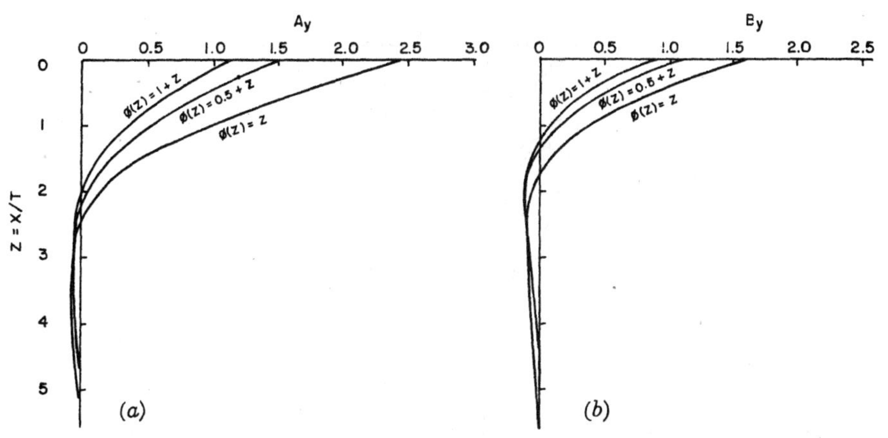

MOMENT COEFFICIENTS

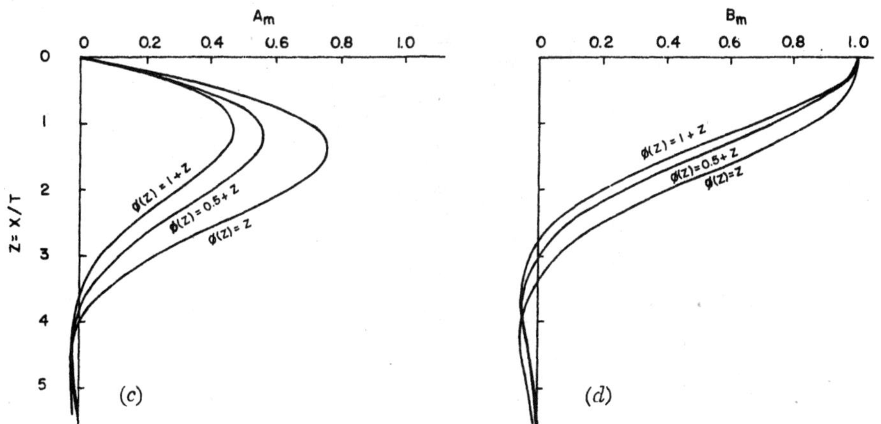

FIG. 11.—TYPICAL RESULTS OF ELASTIC-PILE THEORY WITH $E_s = k_o + k_1 x$, or, $\phi(Z) = r_o + Z$

for very long elastic piles, the major difference is that, for any given soil conditions, the final soil modulus function is related to the length of the pile.

The variations among the moment curves given in Fig. 13 are approximately consistent with those previously shown in Fig. 10 for the elastic-pile theory. However, these elastic-theory results were based on solutions for very long piles and direct comparison between these two figures has limited significance.

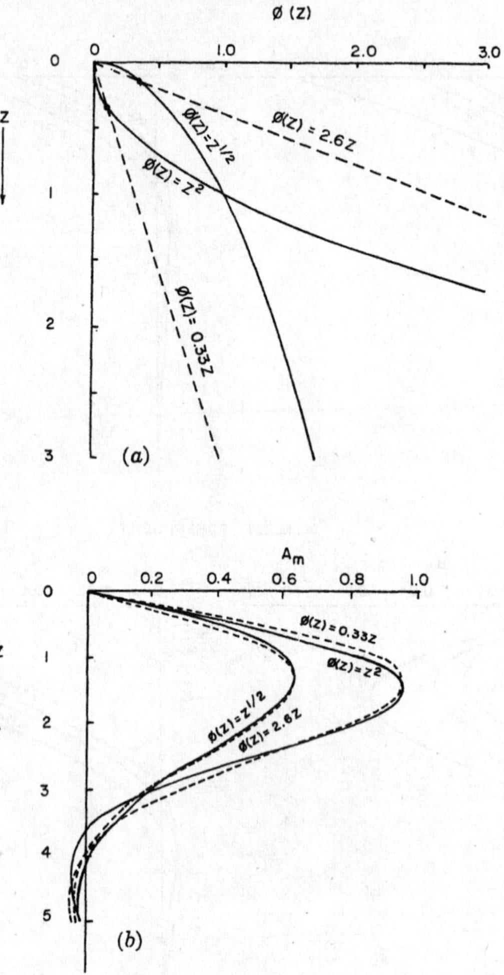

FIG. 12.—COMPARISON OF MOMENT VALUES OF $E_s = k\,x$ and $E_s = k\,x^n$

Considerable variation may be noted in Fig. 13 among the deflection curves for the rigid pile. Examination of the basic equations will show that, for any given length, L, the deflection and slope of the pile will vary according to quantitative variations in Es; but that the moment, shear, and soil reaction depend only on the form of the soil modulus function.

EFFECT OF PILE LENGTH

Results obtained from elastic-pile theory for piles having Z_{max} values of 2, 3, 4, and 10 are shown in Fig. 14 by solid curves. These curves are based

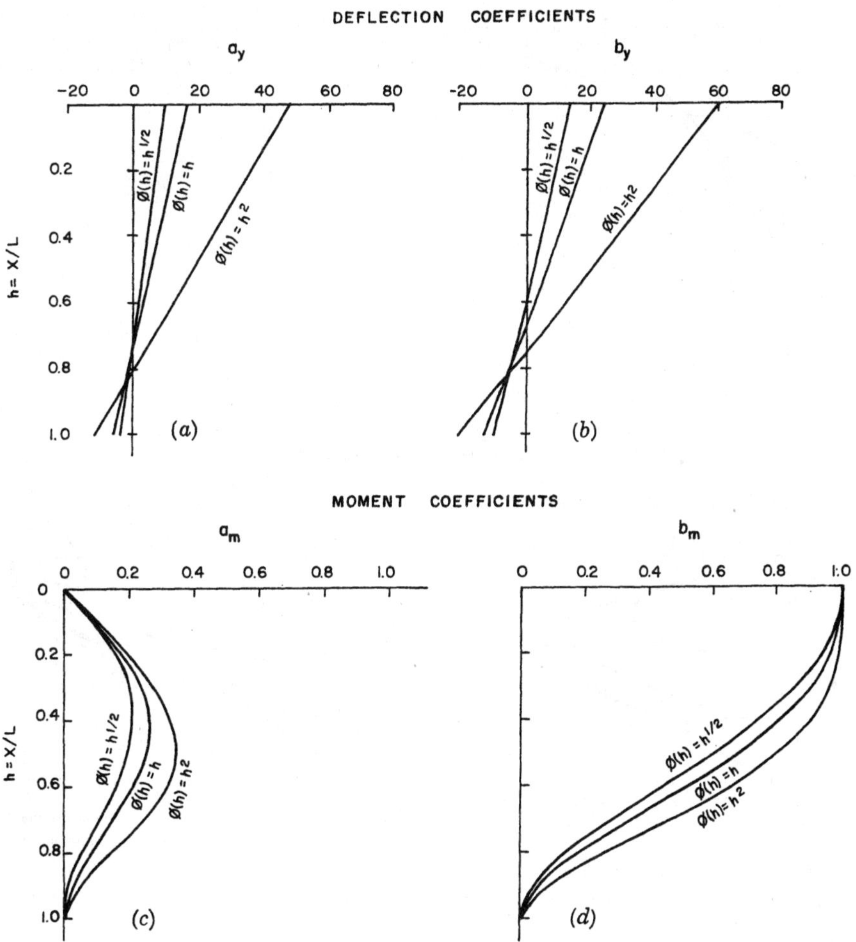

FIG. 13.—TYPICAL SOLUTIONS FROM RIGID-PILE THEORY WITH $E_s = k\,x^n$ $(\phi(Z) = Z^n)$

on a soil modulus which increases in simple proportion to depth, or $E_s = k\,x$ (ϕ = Z or ϕ = h). The corresponding non-dimensional function is ϕ (Z) = Z. Results for the longest pile $(Z_{max} = 10)$ have been shown in previous figures.

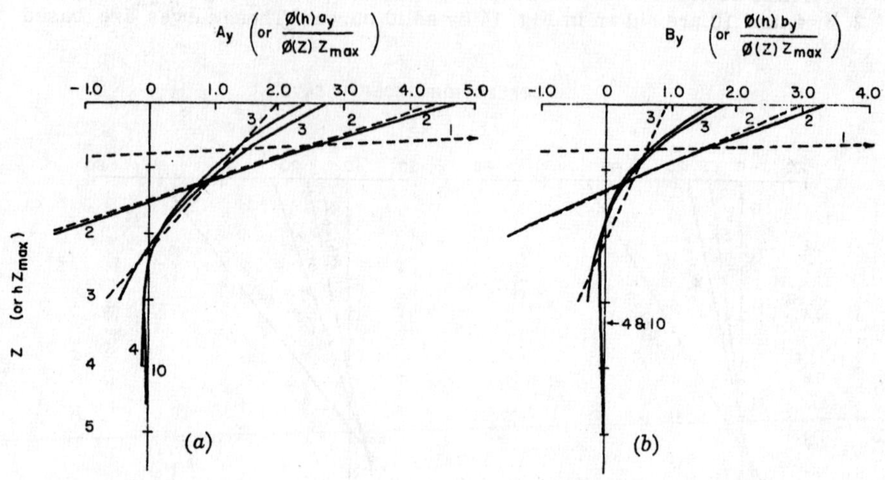

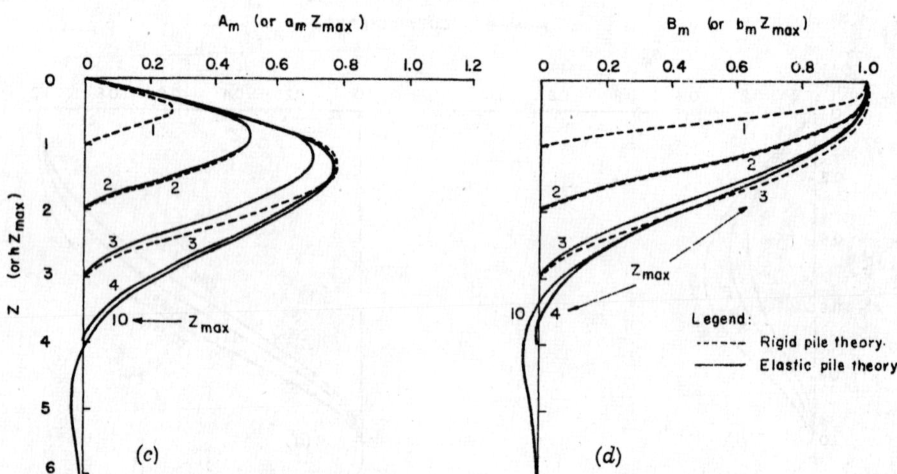

FIG. 14.—EFFECT OF PILE LENGTH AND COMPARISON OF RIGID-PILE THEORY WITH ELASTIC-PILE THEORY

The close agreement between the curves for $Z_{max} = 4$ and those for $Z_{max} = 10$ indicates that for all values of about 5 or greater, an elastic pile will act almost identically to one of infinite length.

The wide differences between the solid curves for $Z_{max} = 2$ and $Z_{max} = 3$ show that more solutions would be needed in this zone for actual design computations. Linear interpolation would be very inaccurate.

Results from the rigid-pile theory, also, are shown in Fig. 14. Dashed curves show rigid-pile deflection and moment coefficient values compared to those from the elastic-pile theory, for Z_{max} values of 2 and 3. Also, results from the rigid-pile theory are shown which correspond to $Z_{max} = 1$.

The method of adapting the rigid-pile results in explained by the following example.

To determine the basis of comparison between the elastic-pile deflection coefficient A_y and the corresponding coefficient a_y of the rigid-pile theory, the expressions for y_A from the two theories may be equated. Thus, from Eqs. 17 and 36,

$$\left[\frac{P_t\ T^3}{E\ I}\right] A_y = \left[\frac{P_t}{J\ L}\right] a_y \qquad (74)$$

Substituting the expression for E I contained in Eq. 16 and the definition of J of Eq. 35,

$$A_y\ P_t\ T^3\ \frac{\phi\ (Z)}{E_s\ T^4} = a_y\ \frac{P_t}{L}\ \frac{\phi\ (h)}{E_s} \qquad (75)$$

Since $Z_{max} = L/T$, this reduces to

$$A_y = \left[\frac{\phi\ (h)}{\phi\ (Z)}\ \frac{1}{Z_{max}}\right] a_y \qquad (76)$$

which is the relation shown in Fig. 14 on the horizontal axis of the graph of A_y versus Z. In a similar manner the relations shown on each of the other graphs have been obtained.

Actually, the rigid-pile is simply a special case of the elastic-pile theory in which the flexural stiffness E I is infinite. By a more general analysis, non-dimensional coefficients could have bee~ defined which would be common to both theories. The arrangement selected is justified on the basis that it is more convenient to use.

By the method of comparison of rigid-pile and elastic-pile results used in Fig. 14, the range of conditions under which each theory is applicable may be determined. The good agreement at $Z_{max} = 2$ indicates that the rigid-pile theory is a satisfactory substitute for the elastic-pile theory for all shorter piles and therefore should be used in this range to avoid the difficulties in difference-equation computations of very short stiff piles. Even at $Z_{max} = 3$, the rigid-pile theory provides a fairly good approximation.

APPLICATION OF GENERALIZED SOLUTIONS TO PILE DESIGN PROBLEMS

Non-dimensional curves for the form of soil modulus variation $E_s = k\ x$ have been available for some time for the elastic pile and have proved to be of value in the design of laterally loaded piles.[3] By use of techniques described in this paper, the reader may recompute these non-dimensional curves or may derive new sets for any desired form of the soil modulus variation. This may be done for both the elastic and the rigid pile.

Solution of actual design problems for laterally loaded piles requires a trial and error adjustment of constants in the soil modulus functions, and perhaps in the form of the function. The steps of this procedure are as follows:

1. A soil modulus variation, usually $E_s = k\,x$, is selected and a trial value of k is assumed.
2. The value of Z_{max} is determined from elastic-pile theory. (If $Z_{max} < 2$, rigid-pile solutions should be used in the following steps).
3. Using non-dimensional coefficients, a trial deflection curve y (x) is computed. For the trials, deflections are needed only in the vicinity of Z less than about 1.0.
4. The computed trial values of deflection y are used as arguments to enter p-y relations which have been previously predicted for the given soil conditions. Revised values of $E_s = -p/y$ are obtained at each depth.
5. A revised value of the constant of soil modulus variation, and possibly a revised form of the soil modulus function itself, are determined by fitting to a plot of the revised E_s values.
6. A new trial is performed. The process is repeated until satisfactory compatibility is reached.

Satisfactory closure is indicated when the soil modulus values resulting from a trial are found to agree closely with the preceding set of values. The necessary degree of precision in soil modulus agreement can be determined by comparing the corresponding changes in values of maximum deflection and bending moment.

It seems reasonable to start the solution of any problem with an assumed soil modulus variation of the form $E_s = k\,x$. A more complex form should be used only after a reasonably good fit has been obtained by the use of $E_s = k\,x$, and then only if much doubt exists as to the correctness of approximating the computed soil modulus variations with the simple straight-line relation. In this connection, the comparisons given herein, among various forms of soil modulus functions, should be helpful as aids to judgment.

APPENDIX I. DIFFERENCE-EQUATION METHOD OF SOLUTION FOR ELASTIC-PILE THEORY

The basic differential equation of the elastic-pile theory, Eq. 25, may be rewritten in difference form as

$$E\,I\left[\frac{y_{i+2} - 4\,y_{i+1} + 6\,y_i - 4\,y_{i-1} + y_{i-2}}{(L/t)^4}\right] = -E_{s_i}\,y_i \quad \ldots\ldots (77)$$

Similarly, expressions may be written as follows for slope, S, moment, M, shear, V, and soil reaction, p.

$$S_i = \frac{1}{2(L/t)}\left(-y_{i+1} + y_{i-1}\right) \quad \ldots\ldots (78)$$

$$M_i = \frac{E\,I}{(L/t)^2}\left(y_{i+1} - 2\,y_i + y_{i-1}\right) \quad \ldots\ldots (79)$$

$$V_i = \frac{E\,I}{2(L/t)^3}\left(-\,y_{i+2} + 2\,y_{i+1} - 2\,y_{i-1} + y_{i-2}\right) \ldots\ldots\ldots \quad (80)$$

and

$$p_i = \frac{E\,I}{(L/t)^4}\left(y_{i+2} - 4\,y_{i+1} + 6\,y_i - 4\,y_{i-1} + y_{i-2}\right) \ldots\ldots \quad (81)$$

The subdivision of the pile into t increments is shown in Fig. 15, together with a summary of equations which are based on Eq. 77 and the Gleser method

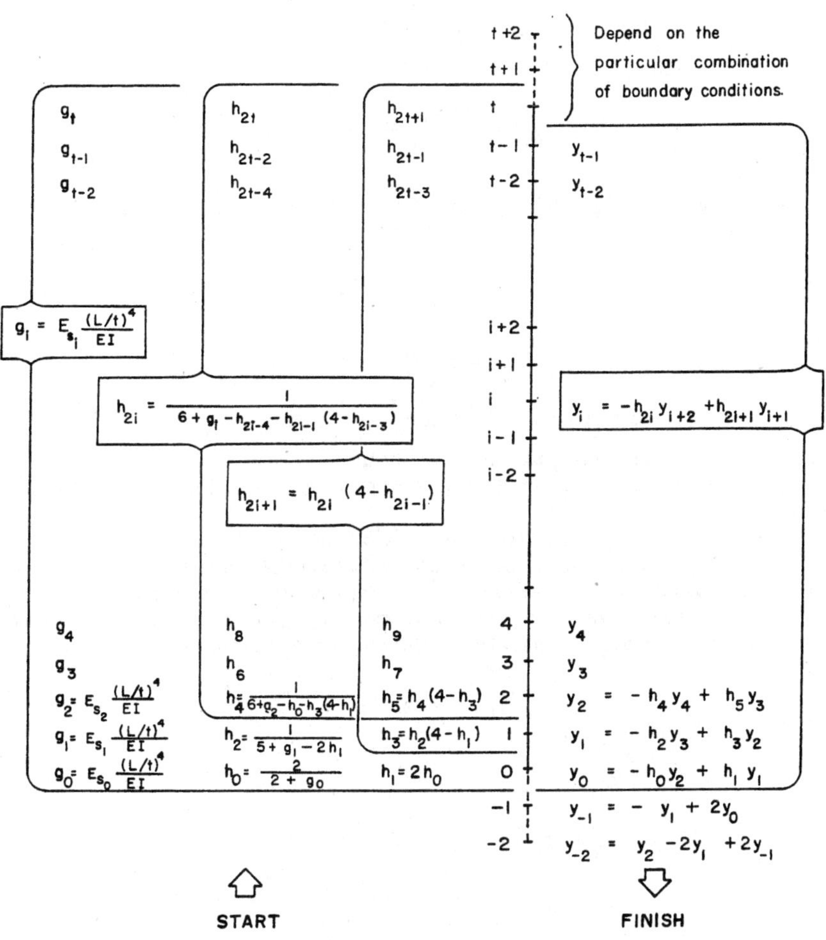

FIG. 15.—SUMMARY OF EQUATIONS AND COMPUTATION FORMAT

of once-through computation.[18] Two imaginary points at the top and two imaginary points at the bottom are added as a device for introducing the

[18] "Lateral Load Tests on Vertical Fixed-Head and Free-Head Piles," by Sol M. Gleser, Symposium on Lateral Load Tests on Piles, ASTM Special Tech. Publ. No. 154, July, 1953, pp. 75-101.

boundary conditions. All necessary equations are given in Fig. 15 except those for y_t, y_{t+1}, and y_{t+2}. The format in Fig. 15 is for a constant E I, see Reese-Ginzbarg for an extension to accommodate step-tapered piles. When the known boundary conditions consist of a lateral load, P_t, and a moment, M_t, these equations are as follows:

$$y_t = \frac{j_4\, h_{2t}(1 - h_{2t-2}) - j_3\left[h_{2t}(2 + 2\, h_{2t-2} - h_{2t-2}\, h_{2t-3}) - h_{2t+1}\right]}{1 - h_{2t-2} - 2\, h_{2t+1} + h_{2t-1}\, h_{2t+1} + h_{2t}(4 - 4\, h_{2t-1} + 4\, h_{2t-2} - h_{2t-4} + h_{2t-1}\, h_{2t-3} - 2\, h_{2t-2}\, h_{2t-3} + h_{2t-2}\, h_{2t-4})} \quad \ldots (82)$$

$$y_{t+1} = \frac{j_3 + y_t\,(2 - h_{2t-1})}{1 - h_{2t-2}} \quad \ldots\ldots\ldots\ldots (83)$$

and

$$y_{t+2} = \frac{h_{2t+1}\, y_{t+1} - y_t}{h_{2t}} \quad \ldots\ldots\ldots\ldots (84)$$

The constants in Eqs. 82, 83, and 84 are as shown in Fig. 15 and by the following two definitions.

$$j_3 = \frac{(L/t)^2}{E\,I}\, M_t \quad \ldots\ldots\ldots\ldots (85)$$

and

$$j_4 = \frac{2(L/t)^3}{E\,I}\, P_t \quad \ldots\ldots\ldots\ldots (86)$$

Following are the steps necessary for computation.

1. Compute a value of g_i at each real point.
2. Proceeding from bottom to top, compute values of the two h-constants at each point.
3. Compute y_t, y_{t+1}, y_{t+2} from Eqs. 82, 83, and 84.
4. Proceeding from top to bottom, compute all other values of deflection y.
5. With y-values known, compute slopes, moments, and shears by Eqs. 78, 79, and 80. Rather than using Eq. 81 to compute soil reactions, it is more convenient and accurate to use the following expression which is based on Eq. 1.

$$p_i = -\,E_{s_i}\, y_i \quad \ldots\ldots\ldots\ldots (87)$$

APPENDIX II. NOTATION

The following symbols are adopted for use in the paper and for the guidance of discussers.

A_y, A_s, A_m, A_v, A_p = non-dimensional coefficients in elastic-pile theory, relating to an applied force P_t, for deflection, slope, moment, shear, and soil reaction respectively;

a_y, a_s, a_m, a_v, a_p = non-dimensional coefficients, same as A-coefficients, except for rigid-pile theory;

B_y, B_s, B_m, B_v, B_p = non-dimensional coefficients in elastic-pile theory, relating to an applied moment M_t, for deflection, slope, moment, shear, and soil reaction respectively;

b_y, b_s, b_m, b_v, b_p = non-dimensional coefficients, same as B-coefficients, except for rigid-pile theory;

E_s = soil modulus (force per unit length of pile and per unit deflection), in pounds per square inches;

E I = flexural stiffness of pile, the product of modulus of elasticity and moment of inertia of pile cross section, in pound inches^2;

g = coefficient in the difference-equation solution;

h = coefficient in the difference-equation solution;

h = depth coefficient in rigid-pile theory, = x/L;

J = soil modulus constant, as defined for each form of E_s (x), in pounds per square inch;

j = coefficient in the difference-equation solution;

k = constant of soil modulus variation in Eq. 52;

k_o, k_1, k_2 = constants of soil modulus variation in Eq. 53;

L = length of pile, in inches;

M = moment, in inch pounds;

M_t = moment at x = 0, in inch pounds;

n = exponent in Eq. 52 or Eq. 57;

P_t = shear at x = 0, in pounds;

p = soil reaction per unit of length of pile, in pounds per inch;

r_o, r_1, r_2 = constants in polynomial soil modulus functions;

S = slope;

T = relative stiffness factor, as defined for each E_s (x), in inches;

t = number of increments in the difference-equation solution;

V = shear, in pounds;

x = depth below groundline, in inches;

y = lateral deflection, in inches;

Z = depth coefficient in elastic-pile theory;

Z_{max} = maximum value of elastic-pile theory depth coefficient L/T;

ϕ (h) = non-dimensional soil modulus function of rigid-pile theory, = E_s/J;

ϕ (Z) = non-dimensional soil modulus function of elastic-pile theory = $E_s\ T^4/(E\ I)$.

Journal of the
SOIL MECHANICS AND FOUNDATIONS DIVISION
Proceedings of the American Society of Civil Engineers

FUNDAMENTAL ASPECTS OF THIXOTROPY IN SOILS

By James K. Mitchell,[1] A. M. ASCE

SYNOPSIS

Thixotropic phenomena are described and previous investigations on the behavior of thixotropic systems are summarized. The complex nature of the phenomena is pointed out, as well as the fact that thixotropy is of quite general occurrence in fine-grained materials

A hypothesis for thixotropic behavior based on initial nonequilibrium of interparticle forces after remolding or compaction, and the effects of this nonequilibrium on subsequent structure changes within the soil, is offered as one possible explanation of the phenomenon. Experimental results are presented which are consistent with the hypothesis.

Some of the practical aspects of thixotropy are pointed out, and it is suggested that thixotropic strength increases warrant consideration in design, provided they are accurately evaluated.

INTRODUCTION

The term "thixotropy" was introduced in 1927 by A. F. Peterfi (1)[2] and used by H. Freundlich (2) to describe the well-known phenomenon of isothermal, reversible gel-sol transformation in colloidal suspensions. J. M. Burgers and G. W. Scott-Blair (3) define thixotropy as a "process of softening caused by remolding, followed by a time-dependent return to the original harder state."

Note.—Discussion open until November 1, 1960. To extend the closing date one month, a written request must be filed with the Executive Secretary, ASCE. This paper is part of the copyrighted Journal of the Soil Mechanics and Foundations Division, Proceedings of the American Society of Civil Engineers, Vol. 86, No. SM 3, June, 1960.

1 Asst. Prof. of Civ. Engrg. and Asst. Research Engr., Inst. of Transp. and Traffic Engrg., Univ. of California, Berkeley, Calif.

2 Numerals in parentheses, thus (1), refer to corresponding items in the Bibliography.

Thixotropic effects in remolded natural clays have been studied by O. Moretto (4) and A. W. Skempton and R. D. Northey (5). These investigations were aimed at determining the extent to which thixotropic hardening could contribute to the sensitivity of clays. H. B. Seed and C. K. Chan (6) have shown that compacted clays may also exhibit appreciable thixotropic strength gain with time. These studies along with a study by P. G. H. Boswell (7) and conclusions reached by H. R. Kruyt (8) suggest that thixotropy may be of general occurrence in the majority of clay-water systems.

Since thixotropic effects can result in strength increases of up to 100% or more after remolding, even in compacted soils, and thus could be of significance from an engineering standpoint, it is useful to examine the phenomenon in some detail. A knowledge of the factors responsible for thixotropic strength increase should aid in further understanding of soil structure and shear strength behavior.

At the present time thixotropy implies slightly different things to different investigators. For example, to the soil engineer, thixotropy signifies strength increase with time after remolding or compaction. He is concerned with water contents generally less than the liquid limit and times measured in weeks, months, or even years. On the other hand to the colloid chemist or rheologist thixotropy immediately brings to mind dilute suspensions, and gel setting times of minutes, hours, or, at most, a few days. In fact it has been suggested that the term thixotropy should not be applied to the phenomena of time-dependent strength gain in soils and a term such as "age hardening" be used. If however, thixotropy is defined as an isothermal, reversible, time-dependent process occurring under conditions of constant composition and volume whereby a material stiffens while at rest and softens or liquifies upon remolding, then the difference between thixotropy in fine-grained soils at water contents less than the liquid limit (concentrated suspensions) and dilute suspensions may be considered one of degree rather than fundamental behavior. The properties of a purely thixotropic material are illustrated in Fig. 1.

In the following two sections, literature that sheds some light on the fundamentals of thixotropy is reviewed. Next, a hypothesis for the cause of thixotropic behavior is presented. Finally, experimental datum is offered which is consistent with the hypothesis.

THIXOTROPY IN DILUTE SUSPENSIONS

Although dilute suspensions bear little resemblance to clay soils in terms of physical properties, it is not unreasonable to suppose that some of the interparticle force mechanisms and principles of aggregation and dispersion are the same as in more concentrated clay-water systems. Freundlich (2) was among the first to present a detailed treatise on the subject of thixotropy. He pointed out that thixotropic behavior undoubtedly depends on the balance of forces acting between particles. The great influence of electrolyte concentration in the system is noted, with an increase in electrolyte content leading to a decrease in setting time for a thixotropic gel.

A tendency for particles of the same substance to adhere upon contact may be important in thixotropy All particles of finely divided matter exhibit an unbalance of forces at their surface. These unsatisfied forces can be satisfied either by contact with particles of the same substance or by adsorption of ions from the adjacent phase. It is possible that particles in a suspension, as

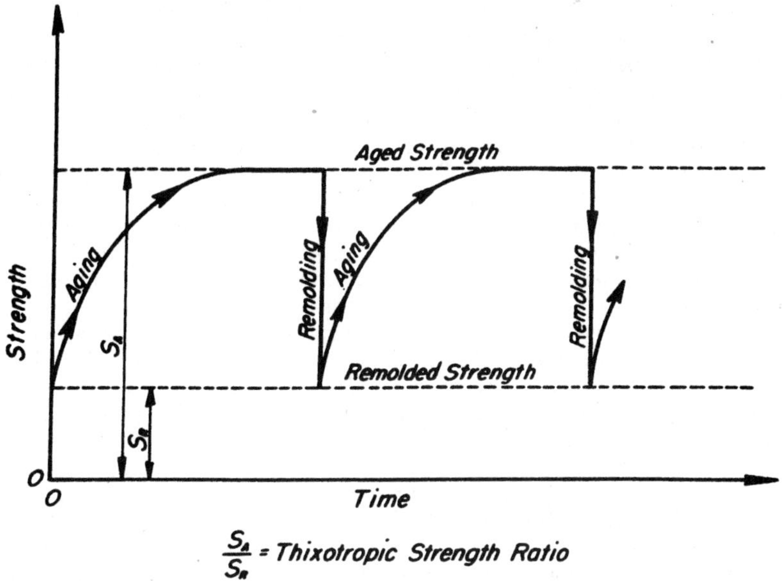

$\frac{S_A}{S_R}$ = Thixotropic Strength Ratio

FIG. 1.—PROPERTIES OF A PURELY THIXOTROPIC MATERIAL.

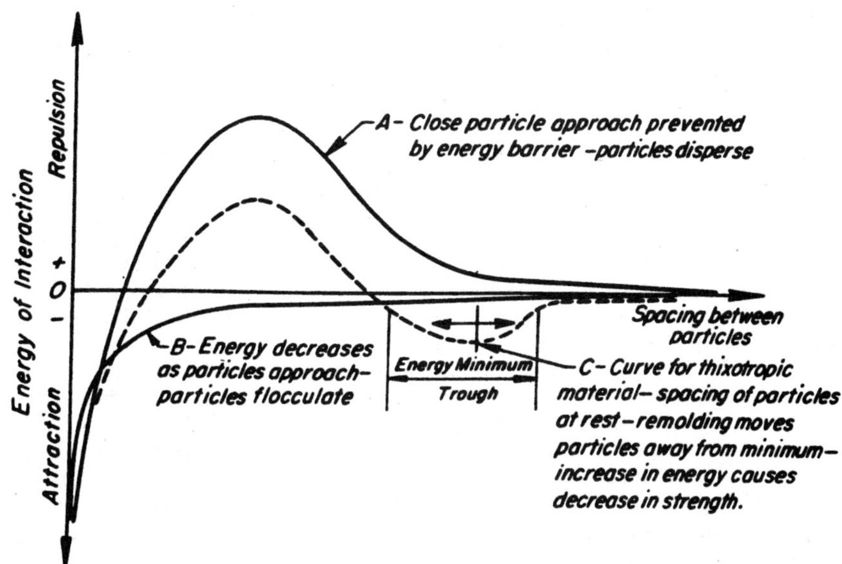

FIG. 2.—ENERGY-DISTANCE CURVES FOR DILUTE SUSPENSIONS OF DISPERSED, FIOCCULATEDAND THIXOTROPIC MATERIALS.

a result of their thermal motion, may be brought into such positions that they adhere, causing solidification of the suspension.

Freundlich also concluded that thixotropy cannot be explained without acknowledging the effects of far-reaching forces of attraction. The explanation of thixotropy on the basis of attraction-repulsion force balance has received wide support by other investigators, such as C. E. Marshall (9), E. J. W. Verwey and J. T. Overbeek (10), and T. W. Lambe (11). This concept is illustrated in Fig. 2 where the ordinates to the curves represent the energy (positive if repulsion and negative if attraction) necessary to bring the particles from an infinite spacing to any given spacing along the abscissa. Curve A represents a stable suspension which exhibits neither flocculation nor thixotropy because of the energy barrier preventing close approach of particles. Curve B represents a condition where particles will spontaneously agglomerate and settle out. The energy minimum indicated on curve C represents the position of particles in a thixotropic gel. Any movement (such as caused by shaking or shearing strains) which tends to change the particle spacing causes an increase in energy of repulsion leading to a more fluid condition.

Curves of the form of those in Fig 2 are usually derived on the assumption of parallel plate-shaped particles charged only on their surface. Lambe(12), J. K. Mitchell (13), H. Van Olphen (14), R. K. Schofield and H. R. Samson (15), and P. A. Theissen (16), among others have shown, however, that clay particles deviate quite appreciably from the assumptions on which the curves are derived. An objection to the situation illustrated in Fig. 2 arises from a consideration of particle orientation in thixotropic systems. According to Van Olphen (14), and R. Fahn, A. Weiss, and H. Hoffman (17), particles in the gelled state are linked in a random array, forming somewhat of a rigid structure. Fig. 2 assumes particles sitting at some distance from each other in a parallel array.

Marshall (9) and Van Olphen (14) present evidence that many times there exist two separate zones of thixotropic behavior for the same material. It was noted from studies of the behavior of suspensions of constant concentration, but increasing electrolyte content, that reversible hardening—liquifaction effects occurred first at the very low salt contents. Slightly higher salt contents led to stable dispersions. Further increase in salt content resulted in the reappearance of thixotropic characteristics. These two regions of observed thixotropic behavior correspond to zones of "non-salt" and "salt" flocculation as described by Lambe (12) and Van Olphen (14).

In non-salt flocculation particles presumably associate in an edge-to-face array, as a result of attractions between negative surfaces and positive edges. Particle associations in salt-flocculated systems are a combination of face-to-face, edge-to-edge and edge-to-face arrangements, that result when the depression of double layers due to increased electrolyte content is sufficiently great that repulsive forces are no longer able to prevent flocculation. Recent light-scattering studies by M. B. Mac Ewen and M. I. Pratt (18) have shown, on the other hand, that for very dilute suspensions of bentonite, the basic structural unit is a linear aggregate one lattice layer thick, in which particles are oriented edge-to-edge in the form of flat ribbons. No evidence of an edge-to-face or face-to-face association was found. Additional studies by Mac Ewen and Mould (19) led to the conclusion that these structures form in zones of potential minima. When the bentonite is dispersed in water the particles are flocculated with edges in close contact; whereas, in the presence of alkali the particles are associated in a minimum such as indicated by the energy trough for curve C in Fig. 2. In this case the edges are separated by a potential barrier preventing close approach, but there is no indication of a parallel face-to-face arrangement. Regardless of the type of particle association there seems ample evidence that thixotropic behavior is associated with a tendency for particles to flocculate, as long as they are free to choose their positions.

H. L. Roeder (20) studied the rheological behavior of thixotropic systems and suggested that thixotropy occurs in systems composed of unstable or partly unstable particles. A system may then be visualized in which the particles have sufficient repulsion forces to resist immediate aggregation but will fall victims to such tendencies in the course of time.

Kruyt (8) points out that thixotropy occurs preferentially, although perhaps not exclusively, in systems with elongated particles, such as clays. The phenomena may be explained by a certain slowness of gel formation, based on either the analogue of slow flocculation or on the rareness of particle encounters due to slow thermal motion of the particles. There may also exit cases where particles are held by long-range forces. Kruyt also noted that thixotropic suspensions of bentonite have been observed wherein the particles come to rest although there is no material contact between them. On the other hand a suspension of graphite in mineral oil is nonconducting when fluid but conducting after gelation, indicating that particles in this system must be in contact and form a continuous network.

THIXOTROPY IN CONCENTRATED SUSPENSIONS

Moretto (4), Skempton and Northey (5), L. Berger and J. Gnaedinger (21), and Seed and Chan (6) all present data which vividly illustrate the increase in strength of clay soils resulting from aging at constant water content. It is apparent from the work of these investigators that this hardening can be of importance in a great variety of soils at water contents of concern to the engineer (plastic limit to liquid limit). Seed and Chan (6) also showed that the hardening effect was completely reversible for a compacted silty clay. Thus the age-hardening phenomena apparently has the characteristics of thixotropy as previously defined. However the mechanism of thixotropic hardening has not been explained. Fig. 3 summarizes the strength gain data as obtained by the above named investigators.

The thixotropic strength ratio, that is, the strength at time t divided by the strength at time equals zero) as suggested by Seed and Chan (6) rather than absolute strength increase is used as a measure of thixotropic effects in order to permit comparisons between soils of different composition, or between samples of the same soil at different water contents. In these cases the absolute strength increase might be misleading. For example a sample with a remolded strength of 1 kg per sq cm might show a strength increase to 1.1 kg per sq cm after a certain time, thus having a thixotropic strength ratio of 1.1. A second sample might have an initial strength of 0.1 kg per sq cm and increase to 0.2 kg per sq cm in the same time. The thixotropic strength ratio for this case would be 2.0. It is felt that for a material to double its strength (even though its initial strength is low) is more significant in terms of thixotropic effects than for a sample to increase its strength by 10%, even though the actual strength increase for the two materials is the same.

Examination of Fig. 3 reveals several points of interest. With the exception of kaolin all the soils showed a significant strength increase with time of aging.

However, tests at the University of California have shown that even kaolinite may be made very thixotropic by the addition of a dispersing agent in order to reduce the degree of flocculation present in the natural material. There appears from Fig. 3 to be no unique relationship between thixotropic strength ratio and time.

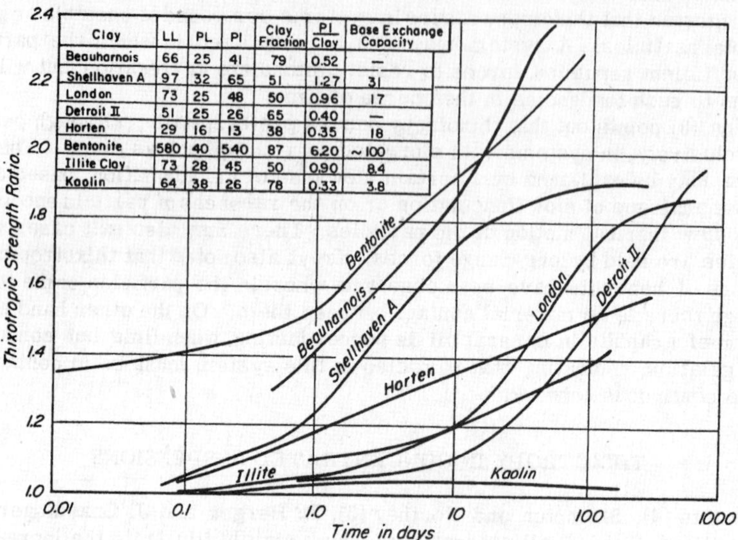

(a) Thixotropic strength gain at water contents equal to the liquid limit.

(Data from A.W. Skempton & R.D. Northey)

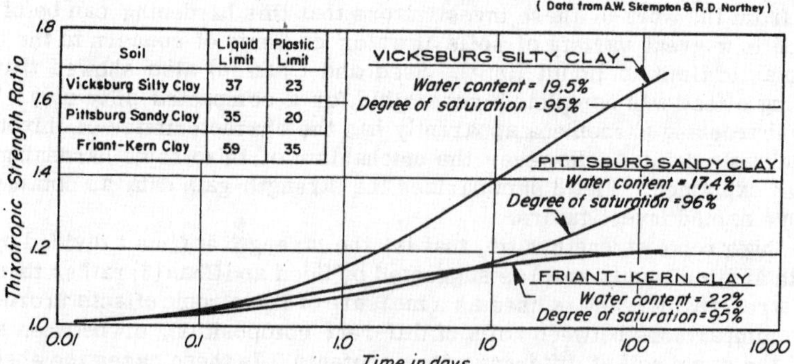

(b) Thixotropic strength gain at water contents approximately equal to the plastic limit.

FIG. 3.—SUMMARY OF THIXOTROPIC STRENGTH GAIN DATA.

The curves in Fig. 3 for the samples tested at water contents near the plastic limit show that thixotropy can be significant at low water contents. In other cases it may be negligible as shown by G. A. Leonards (22), who found no thixotropic tendencies in two compacted soils. The role of water content in determining thixotropic behavior will be considered in more detail subsequently.

Available data all indicate that the lower the strain the more pronounced is the thixotropic effect. This effect is indicated in Fig. 4 from Seed and Chan (6) which shows the thixotropic strength ratio after 1 week as a function of water content for a compacted silty clay. Optimum water content for the compaction procedure used was 17.5%. From Fig. 4 it is seen that the thixotropic effect is small for samples dry of optimum and the effect is the same for all strains. On the other hand, for samples prepared wet of optimum the thixotropy is very pronounced, particularly at low strains. These samples were prepared by kneading compaction. It is believed that a flocculated structure is induced dry of optimum which changes gradually to a dispersed structure wet of optimum by this method of compaction. (See Seed and Chan (23) and Lambe (12) for a complete discussion of the types of structure induced in compacted soils by various methods of compaction and the effect of these structures on properties.

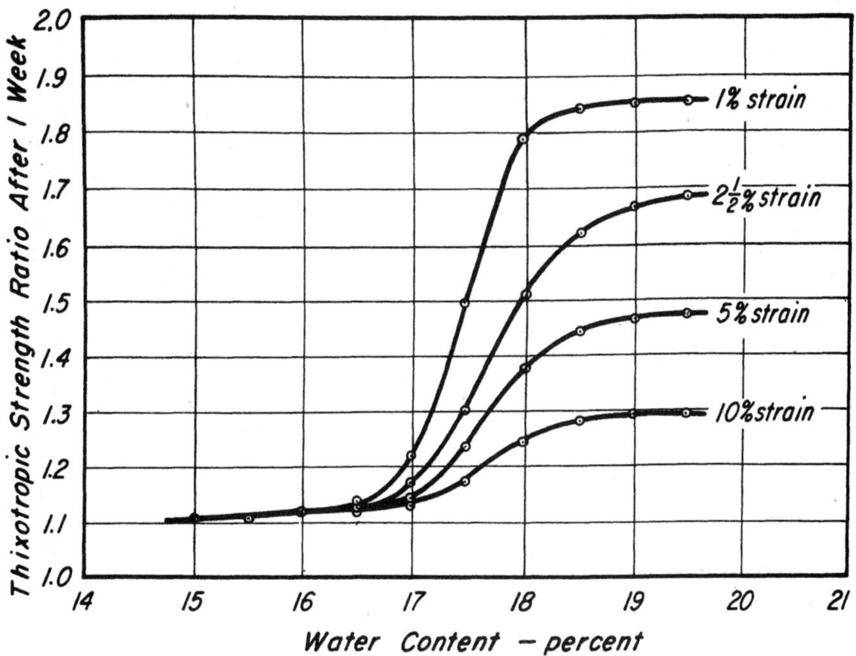

FIG. 4.—THIXOTROPIC STRENGTH RATIOS FOR DIFFERENT VALUES OF AXIAL STRAIN.

The term flocculated structure refers to a structure in which clay particles are arranged in a more or less random array and where interparticle forces were predominantly attractive at the time the soil was compacted. The term dispersed structure refers to a structure in which clay particles are arranged in a more or less parallel array and where interparticle forces were predominantly repulsive at the time the soil was compacted.) The possibility therefore exists that thixotropy depends to some extent on initial structure as well as initial water content. The decrease in thixotropic effect at high strains is to be expencted since shearing tends to remold the soil.

Data obtained by Berger and Gnaedinger (21) indicate that aging may lead to an increase in stiffness without any consequent increase in ultimate strength. It thus becomes apparent that thixotropic effects must be referenced to some criterion of measurement, such as a particular strain, in order to evaluate their true significance.

The influence of thixotropic effects on clay compressibility is not apparent from a study of available literature. It would seem reasonable that if the strength of a soil increases with time after remolding then it might also undergo less volume change, at least under small pressures, during consolidation than if no thixotropic stiffening took place, because strength and resistance to compression both depend in some way on the strength of interparticle bonds.

Berger and Gnaedinger (21) performed consolidation tests on a plastic clay and noted no significant differences in compression curves for samples tested at 0, 6 and 12 months after remolding. On the other hand, P. L. Newland and B. H. Allely (24) found that a sensitivity was developed in a highly plastic clay during consolidation under a given load. After thorough remolding of the consolidated clay and reapplication of the same consolidation pressure, a further decrease in void ratio was observed. It is possible that thixotropy was responsible for this behavior.

Boswell (7) made an examination of the thixotropy of a great number of sedimentary deposits. All the materials examined, with the exception of clean sands, exhibited thixotropic characteristics. Boswell noted further that many of the soils were rheopectic; that is, gentle motion accelerated the rate of thixotropic hardening. In many of the soils at high void ratio syneresis was also noted; the gels begin to split open along irregular cracks with a liberation of water. This phenomenon can be attributed to a contraction of the solid phase structure, due possibly to the formation of more and more interparticle attractive bonds with time.

D. H. Trollope and Chan (25) hypothesize that thixotropic strength gain in compacted soils is due to changes in disposition of the colloidal particles within the soil mass, since according to their hypothesis of soil structure, any variation in the strength of the colloidal matrix in a constant stress environment can only occur if the mean effective distance between particles is changed. These changes, it is suggested, are accomplished through the redistribution of internal stresses. Internal stress redistributions occur because it is unlikely that the most stable colloidal structure will be immediately established everywhere on compaction.

Perhaps the most significant datum available thus far relating to the fundamentals of thixotropy is that reported by P. R. Day (26). The effect of shear on the forces of cohesion between clay particles was studied through measurements of the intensity of the water adsorptive forces. A measure of these forces was obtained by determination of the soil moisture tension (negative pore pressure) using a simple tensiometer. Saturated clay samples at water contents near the liquid limit were allowed to come to equilibrium with the tensiometer; the suction was noted and then the sample was stirred with a motor driven mixer. Changes in tension with time after stirring were noted. Fig. 5, which summarizes some of Day's results, indicates that tension falls to a minimum value after stirring and then gradually returns to nearly the initial value, at which point the process could be repeated. Although no strength values were obtained during these tests, it was noted that the "stirring action produced a decrease in viscosity and that the gel slowly returned to its original viscous state as the tension was restored." Fig. 5 shows that the phenomenon of decreased tension (or increased pore pressure) occurred in a variety of different clay materials, and therefore appears to be of quite general occurrence.

The fact that aging a thixotropic material leads to increased water tension (decreased pore pressure) aids in the fundamental understanding of the phenomena. It is to be expected that an increase in tension, and thus an increase in effective pressure, would lead to an increase in strength. It is the basic cause of this increase in tension that is of interest. According to L. D. Baver (27)

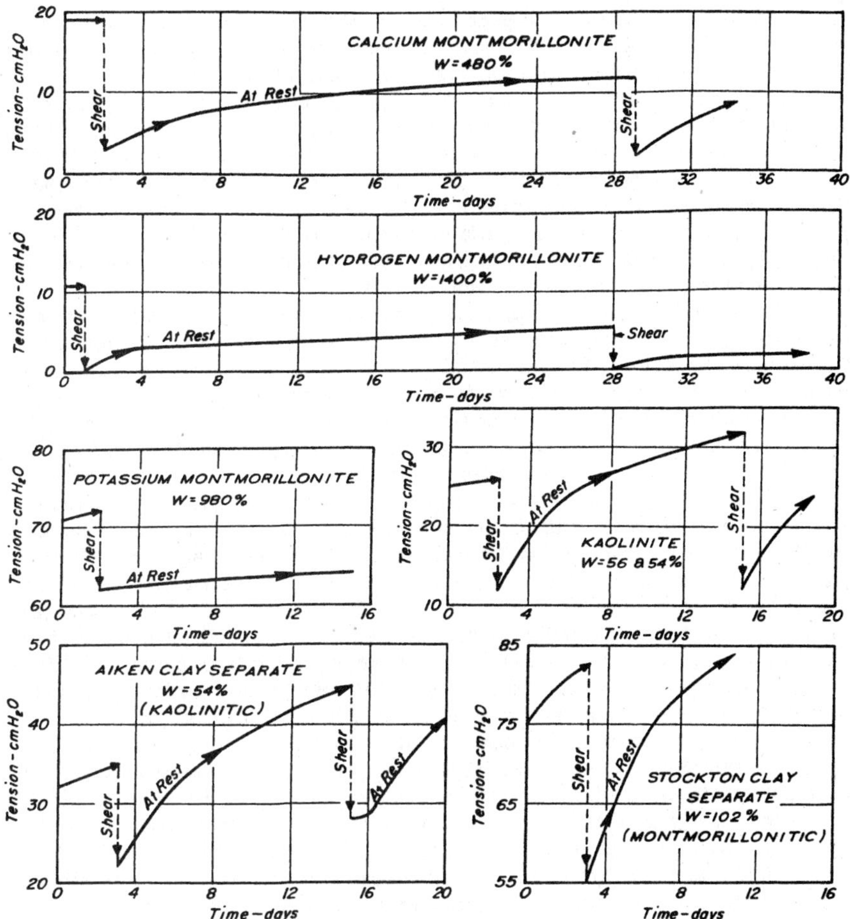

FIG. 5.—THE EFFECT OF SHEAR ON PORE WATER TENSION FOR VARIOUS CLAYS.

the tension in the soil water is a function of four factors (assuming the soil is not subjected to external forces).

1. The total hydrostatic head in the water;
2. The capillary potential due to air-water interfaces;
3. Osmotic pressure differences arising from ionic distributions in the double layers between particles; and
4. Adsorptive forces due to the attraction of water molecules to particle surfaces.

In tests on saturated clays conducted under identical conditions the first two factors are constant, thus time and shear dependent changes in the osmotic or swelling pressure and in water adsorptive forces at particle surfaces become of interest. Although the adsorptive forces at the surfaces of the clay particles may remain constant regardless of whether or not the material is at rest or undergoing stirring, the structural arrangement of the water molecules in response to these forces could conceivably differ. This would be reflected through the tension values measured. Any changes in particle orientation would involve changes in the amounts of double layer interaction between particles which in turn would be reflected through the observed tensions.

That the tension or negative pore pressure increases with time after stirring is of great importance since at constant volume the tension is a direct measure of the free energy of the water. An increase in tension (decrease in pore pressure) is associated with a decrease in free energy and a decrease in tension (increase in pore pressure) is associated with an increase in free energy. Thus, direct evidence is available to indicate that the free energy of the system decreases with time, which means that spontaneous reactions are occurring and the system is seeking equilibrium.

Finally, actual chemical change, for example, time-dependent formation of cementation bonds between particles, must be considered as a possible cause of thixotropic behavior. Iron oxides, carbonates, organic matter, and amorphous silica and alumina are known to exist in varying quantities in most soils. All of these materials could bond particles together. Remolding or compacting a soil could destroy these bonds. Whether or not these bonds would reform with time is not known. The observed reversibility of thixotropic systems argues against the formation of additional cementing materials, particularly in soils which have been in equilibrium with the same water for long periods of time or in dilute suspensions which gel almost instantaneously after shaking. It is not meant to imply however, that chemical changes cannot take place during aging of a moist soil. A. C. Mathers, S. B. Weed, and N. T. Coleman (28) have presented data to show the partial conversion of hydrogen montmorillonoids to the aluminum form in several days when stored both moist and dry at normal temperatures. R. T. Martin (29) found that lithium kaolinite converts to aluminum kaolinite on aging for a period of 100 days in 70 C water.

Thus it may be reasonable to conclude that chemical changes may occur in thixotropic systems and assist strength gain tendencies. There may also be thixotropic systems in which chemical changes are negligible. It should be pointed out, however, that according to the definition of thixotropy presented at the beginning of this paper the process occurs under conditions of constant composition. Therefore any effects due to changes of the type found by Mathers, et al., and Martin would not, strictly speaking, be thixotropic.

A HYPOTHESIS FOR THE CAUSE OF THIXOTROPIC BEHAVIOR

The following hypothesis is offered as one possible explanation of thixotropic effects in soils. The hypothesis is based on the assumption that the internal energy and stress conditions in a thixotropic soil immediately after remolding are not equilibrium conditions. This assumption appears reasonable since, if equilibrium conditions existed, there would be no changes in properties with time. But properties do change, and Day's results (26) show that the material undergoes an energy decrease, which is compatible with spontaneous changes.

This hypothesis is intended to apply to those soils in which thixotropy is primarily a structural effect; that is, property changes with time result from changes in such factors as particle arrangement, adsorbed water structure and distribution of ions in the fluid phase. Any chemical effects must be considered in addition to the mechanism outlined in the following paragraphs.

The structure assumed by a given soil is in direct response to its composition (including pore fluid type and amount) and the environmental conditions at the time of formation, plus changes in structure arising during its history. Such factors as electrolyte content and type, temperature, pH, and subsequent pressures are determinant for sedimented soils. In the case of compacted soils the method of compaction is extremely important. Lambe (12) and Seed and Chan (23) have shown the important role of structure in determining compacted soil properties. Seed and Chan have also convincingly demonstrated the important influence of method of compaction in determining soil structure (23). Furthermore, it has been suggested (12) and effectively shown (23) that the effect of shear strain on a soil is to disperse, that is, to force a more or less parallel arrangement of particles in the shear zone.

Most soils have been found to have a balance of interparticle forces leading, at equilibrium, to a structure somewhere between complete dispersion and complete flocculation. That is, few soils have been encountered that cannot be either further dispersed or further aggregated by appropriate mechanical or chemical treatment. The ease with which such structure change may be accomplished, of course, varies widely for different soils, and at different water contents for the same soil. That a given soil exhibits different thixotropic properties at different water contents has been previously pointed out.

Based on these observations, it may be postulated that a soil will exhibit thixotropic behavior if the following conditions are met:

1. The net interparticle force balance is such that the soil will flocculate if given a change.
2. The flocculation tendency is not strong enough, however, that it cannot be overcome by mechanical action, that is, shearing or stirring the soil.

The mechanism may thus be visualized as follows. When a thixotropic soil is remolded or compacted a part of the externally applied shearing energy is utilized in dispersing the platy clay particles into a uniform, parallel arrangement; that is, the externally applied energy assists the repulsive forces between particles resulting from double layer interaction in producing a dispersed system. The energy of interaction between particles is at a level comcensurate with the externally applied forces, and the adsorbed water layers and ions are distributed in accordance with this high energy level. The net result is a structure similar to that shown schematically in Fig. 6(a). Such a structure is compatible with the conditions during the shearing process.

However, as soon as shearing ceases the externally applied energy (a part of which had been assisting the normal interparticle repulsive forces in creating a disperse structure) drops to zero. Thus the net repulsive force decreases; or, stated another way, the attractive forces now exceed the repulsive forces for the particular arrangement of particles and distribution of water. As a consequence the structure attempts to adjust itself to a new, lower energy condition. The energy dissipation may be accompanied by changes in particle arrangements, adsorbed water structure and distribution of ions. Since structural changes of this type are dependent on actual physical movement of particles, water and ions, they are time-dependent. These internal structural changes

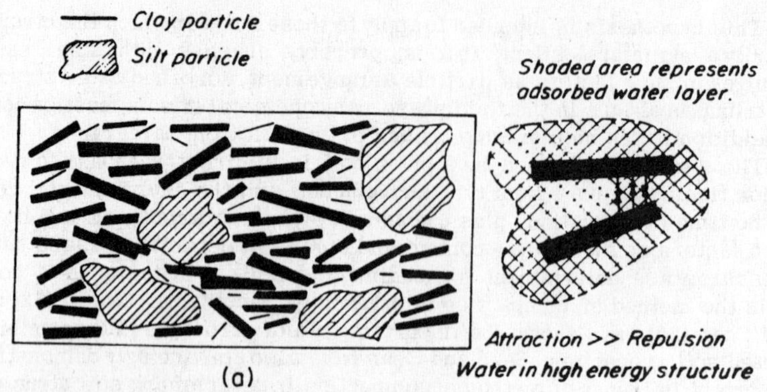

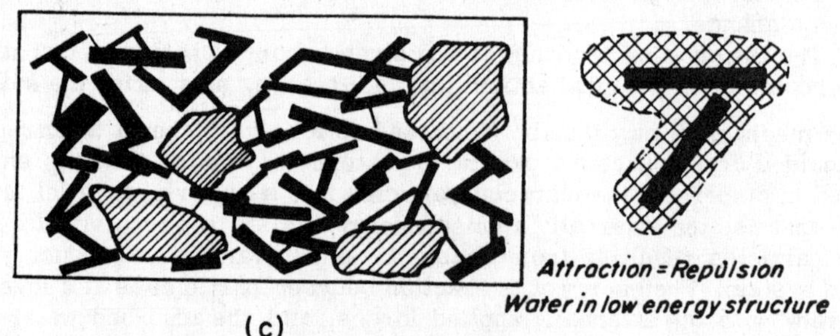

FIG. 6.—SCHEMATIC DIAGRAM OF THIXOTROPIC STRUCTURE CHANGE IN A FINE GRAINED SOIL.

are, in turn, reflected by the physical behavior of the soil (such as increasing strength).

Curves depicting energy of attraction, energy of repulsion, and total energy of interaction between particles for the condition at rest and during remolding are shown in Fig. 7. Although these curves cannot be constructed on a quantitative basis, they are useful qualitatively for the purpose of indicating the energy status within a thixotropic soil. In Fig. 7(a) are shown curves for double layer energy of repulsion and energy of attraction. In addition a line is shown indicating an additional energy of repulsion introduced by the externally applied shearing strains. This energy is shown as increasing with decreasing particle spacing because it is felt that the smaller the particle spacing the greater the percentage of the externally applied energy required to initiate particle movement. At large particle spacings a greater percentage of the applied energy is simply converted to kinetic energy of particle motion. The curve for net energy of interaction is humped, indicating an energy barrier preventing close approach or flocculation of particles. Thus a weak, dispersed structure is formed. When the external strains cease to be applied, however, Fig. 7(b), the energy of repulsion is decreased leading to a total energy of interaction as shown. The energy barrier preventing flocculation is removed.

The system may be visualized as having an excess of internal energy for the new conditions. A spontaneous dissipation of this energy occurs. Since, in the case where attraction exceeds repulsion, minimum energy demands that particles flocculate, the particles will attempt to do so. If this is to happen, however, local redistributions of ions in the double layer and of the double layer water itself are required. Concurrently with these processes a slightly altered adsorbed water structure will form. Since these processes require a dispalcement of particles and a movement of ions and water they are not instantaneous. A viscous flow of water is involved which requires time. Thus the excess energy dissipation is not instantaneous, but occurs over protracted periods. As a result, property changes continue to occur for appreciable periods. A schematic diagram of the structure at some intermediate time after remolding is shown in Fig. 6(b) the final structure is shown in Fig. 6(c). The time required for the achievement of equilibrium will vary with the initial difference between the "as-molded" structure and the equilibrium structure and the mobility of the particles which is related to such factors as water content, particle size distribution, particle shape, the ease of displacement of adsorbed water molecules, and the magnitude of effective stress during aging.

Since the process occurs at constant volume the particle movements during thixotropic hardening are probably small and of a rotational nature. A difference of a few Angstroms spacing between points of closest particle approach could have a pronounced effect on strength, however, since interparticle attractions vary as a negative power of the spacing and only small decreases in spacings at points of closest particle approach could lead to large increases in strength.

Changes in structure of the adsorbed water itself could be a significant contributor to thixotropic effects. That the free energy of the water is diminished with time after remolding is clearly shown by the tension measurements of Day (26) and additional data presented subsequently. Whether this energy decrease is caused by an actual change in the arrangement of water molecules adjacent to the clay surfaces in response to the removal of the remolding energy, or whether it reflects the changes in particle arrangement and spacing is not known. Quite possibly it reflects both factors.

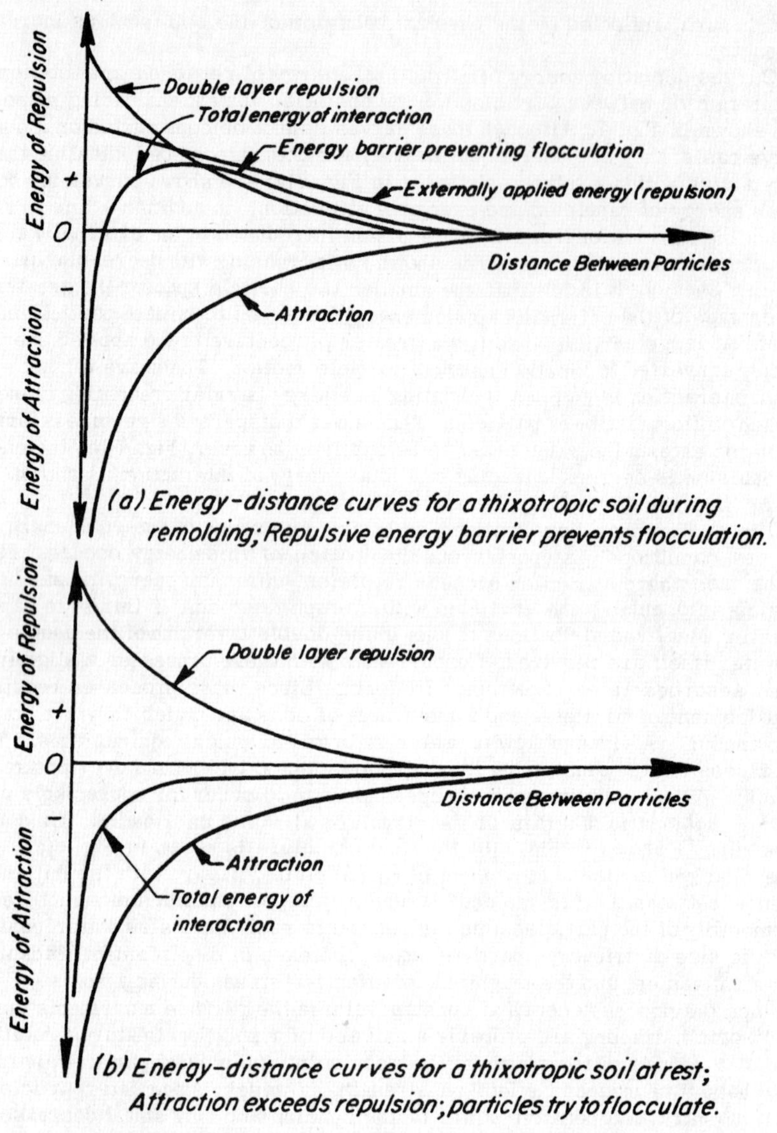

FIG. 7.—ENERGY-DISTANCE CURVES FOR A THIXOTROPIC SOIL.

In summary, thixotropic behavior is the natural response of a soil structure to a change in ambient conditions. When a soil is remolded or compacted a structure is induced which is compatible with the externally applied shearing stresses. When shearing stops the soil is left with an excess of internal energy which is dissipated by means of small particle movements and water redistribution until a structure in equilibrium with the at rest forces is created. Observed behavior of thixotropic soils is considered in the light of this hypothesis in the following sections.

EXPERIMENTAL RESULTS

Reversibility.—It is well known that the sol-gel transformation of dilute thixotropic suspensions is completely reversible. Whenever shaking or stirring of the suspension ceases the material gels; when stirring again commences it liquifies. Since the definition of thixotropy as proposed states that the phenomenon is completely reversible, it is desirable to consider the extent of reversibility at water contents of non practical interest to the engineer. The results of a reversibility study on a silty clay soil from Vicksburg, Miss. (LL = 39%, PL = 25%) are shown in Fig. 8. The aged strength of this compacted silty clay soil after 6 days storage is indicated on the figure and varied from 1.1 to 1.8 times the original as-molded strength, depending on the water content and the strain at which the strength was measured. Values of stress required to cause a given amount of strain as a function of molding water content for samples tested immediately after compaction and for samples aged 6 days, remolded, recompacted, and tested are shown. The data for the as-compacted and the remolded aged samples show no significant variation. It may be concluded, therefore, that the stiffening process was reversible. Additional evidence of complete reversibility is indicated in a later section in connection with the effect of temperature on thixotropy.

The Effect of Initial Structure on the Thixotropic Behavior of Compacted Clays.—If thixotropic effects are caused by dispersion-flocculation phenomena, then it would be expected that the relationship of the structure of a soil immediately after compaction to the structure that would exist if all particles were free to choose their own positions would influence the magnitude of thixotropic property changes. In the case where interparticle forces are such that particles are strongly flocculated and remain flocculated in spite of applied external strains then an initially flocculated structure will result. Negligible thixotropic effects would be expected since the initial structure is one of equilibrium. Similarly if interparticle forces are such that dispersion is the equilibrium condition then applied, external strains will have little additional effect and the soil should show negligible thixotropic behavior. However, if the interparticle forces dictate flocculation, but weak flocculation, and the strains induced during compaction are able to disperse the structure, then, with time after compaction, the structure will seek equilibrium through reflocculation. In this case thixotropic effects should be pronounced.

Vicksburg silty clay which shows rather unique structure changes as a function of water content was investigated. This material when compacted dry of optimum exhibits a flocculated structure, regardless of the method of compaction used, as shown by Seed and Chan (23). At water contents greater than optimum the structure may be dispersed by means of a large shear strain inducing method of compaction (such as kneading) but remains flocculated for nonshear strain inducing methods of compaction (such as static). At considerably

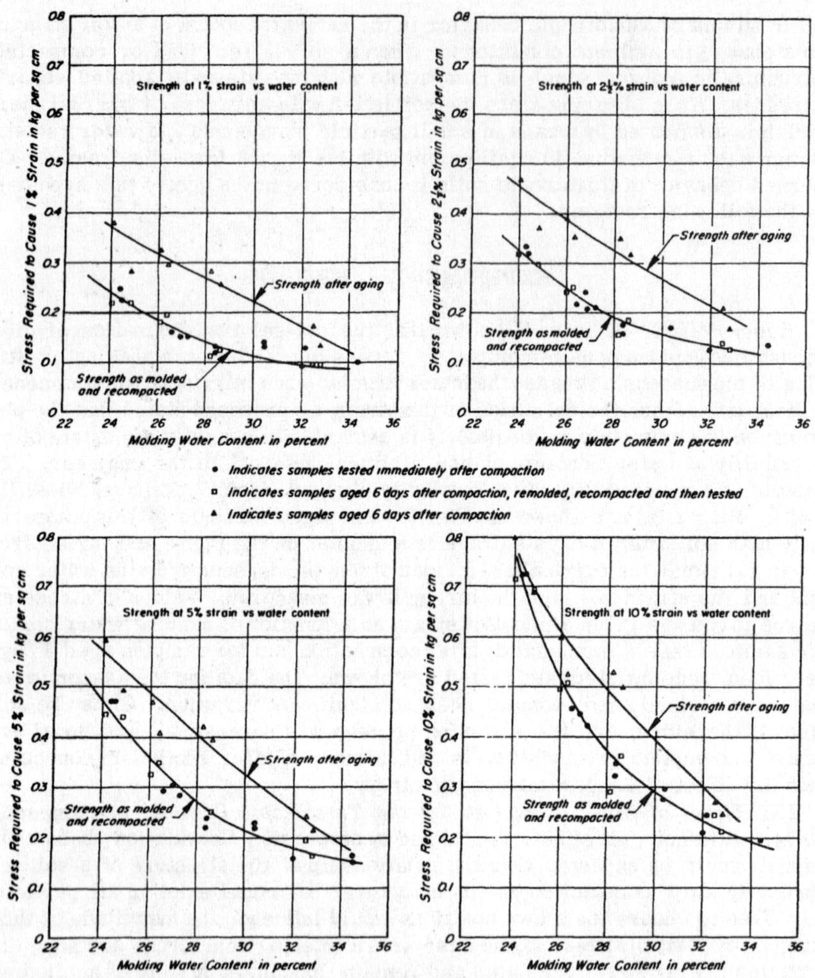

FIG. 8.—COMPARISON OF THE AS-COMPACTED AND THE AGED-REMOLDED-RECOMPACTED STRENGTH OF VICKSBURG SILTY CLAY.

higher water contents this material disperses of its own accord. Wet mechanical analyses of this soil gave identical results with and without a dispersing agent. It thus seems likely that at low water contents the material is strongly flocculated due to double layer water deficiency; a condition described by Lambe (12). At high water contents the material disperses of its own accord from high double layer repulsions. At intermediate water contents the structure may be made to disperse through the application of shear strains. It would therefore be expected that thixotropic effects in this soil would be negligible at low and high water contents and pass through a maximum at some intermediate water content.

Samples were prepared by kneading compaction over the range of water contents from 14% to 35%. Optimum water content for this material is about 17.7% for the compactive effort used. Curves indicating the deviator stress required to cause 1%, 2-1/2%, 5% and 10% strain in triaxial compression tests (quick tests with σ_3 = 1.0 kg per sq cm) as a function of molding water content are shown in Fig. 9, for samples tested immediately after compaction and for samples tested after aging for 6 days. Compaction data for these samples are shown in Fig. 9(e). It may be noted from Fig. 9 that for all strains the curves for both the aged and non-aged samples are essentially parallel up to a water content of about 22%, at which stage they begin to diverge. From 22% to about 29% water content the strength of the aged samples decreases at a slower rate with increasing water content than does the strength of the non-aged samples. At water contents above 29% the strength of the aged samples decreases at a faster rate with increasing water content than does the strength of the non-aged samples. At 34% water, which was the practical upper limit of water content for triaxial testing, the strengths of both types of samples are approximately the same. It is apparent that aging had the greatest effect on strength in the range of water contents of 22% to 29%. It is also evident from Fig. 9 that the aging effect was greatest when measured at low strains.

The thixotropic effect is illustrated more graphically in Fig. 10 which was prepared from the data in Fig. 9. Here, the thixotropic strength ratio (that is, the ratio of the strength of the aged sample to the strength of the non-aged sample for the particular strain under consideration) is plotted as a function of water content. The effect of strain is also readily apparent. These results are consistent with the concepts previously presented: strong flocculation at low water contents, dispersion at high water contents and a weak state of flocculation susceptible to dispersion by shear at intermediate water contents. A decrease in thixotropic effect with increasing strain during shear is to be expected since the strains progressively destroy the thixotropic structure that has formed on aging.

These data also clarify the role of water content in thixotropic behavior. Water content appears to be of primary importance in determining the status of interparticle forces. If a soil can be made to pass from a state of flocculation to dispersion as water content is increased then thixotropic effects will be noted. Thixotropic effects have been found to increase with increasing water content in most past investigations probably because either a sufficiently high water content was not studied to permit partial dispersion under shear strain or the soil remained flocculated regardless of the water content used. However, results reported by Skempton and Northey (5) indicate that for five different clays aged 100 days three show an increase in thixotropy with increasing water content over the range studied; whereas two pass through a maximum at or near the liquid limit and then drop off for further water content increase.

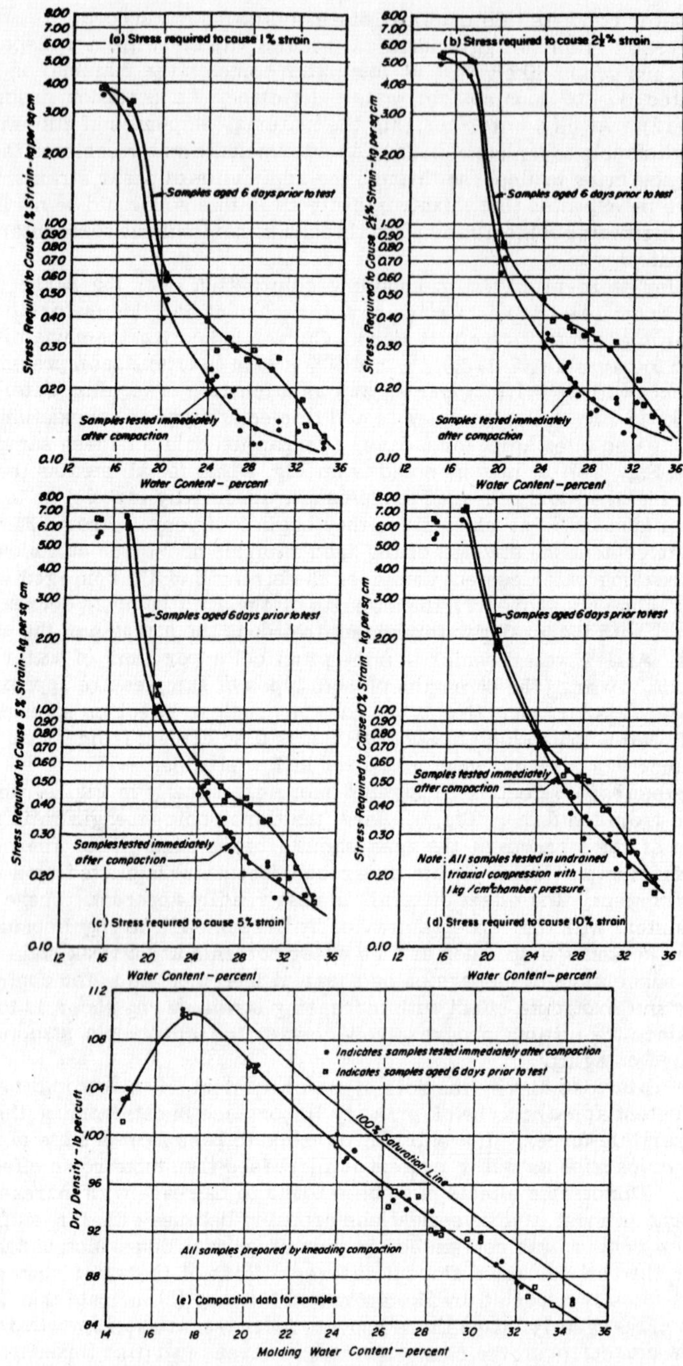

FIG. 9.—STRENGTH OF COMPACTED VICKSBURG SILTY CLAY AS A FUNCTION OF MOLDING WATER CONTENT.

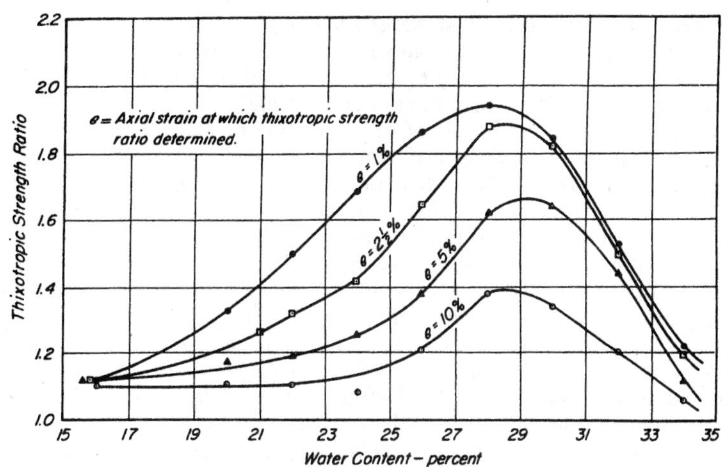

FIG. 10.—THIXOTROPIC STRENGTH RATIO FOR SILTY CLAY AS A FUNCTION OF MOLDING WATER CONTENT.

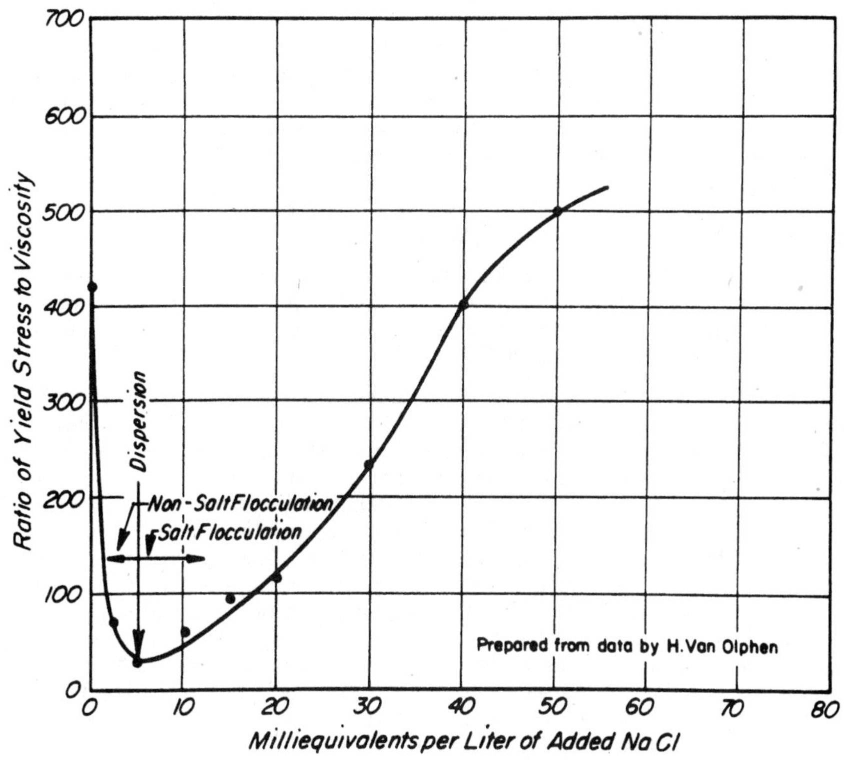

FIG. 11.—EFFECT OF Na Cl CONCENTRATION ON THIXOTROPIC CHARACTERISTICS OF A 3.22% Na-BENTONITE SUSPENSION.

Illustration of the effect of structure on thixotropy in dilute clay suspensions is afforded by data reported by Van Olphen (14). Measurements were made on a 3.22% by weight suspension of pure sodium Wyoming bentonite. Both the yield stress of the gels after a period of rest and the viscosity of the suspensions during shear were measured as a function of added sodium chloride. A measure of the thixotropic effect may be obtained by dividing the gel yield stress by the viscosity. This provides a type of thixotropic strength ratio. Fig. 11 shows the variation in this ratio as a function of added NaCl. It may be seen that with no salt addition the ratio is high, falling to a very low value at 5 m.Eq. NaCl per l and then increasing again rapidly for further addition of salt. Van Olphen was led, by the results of these and other measurements, to the conclusion that at the very low salt concentrations a non-salt flocculation was caused by Coulombic attraction between negative particle surfaces and positive edges. At a salt content of 5m.Eq. NaCl per l the suspension was dispersed and did not gel on standing. With further addition of sodium chloride a salt type flocculation sets in which differs from the non-salt flocculation in that gel times are slower and particles are associated in face-to-face and edge-to-edge associations as well as edge-to-face. (In Van Olphen's view the non-salt flocculated gels are not thixotropic since they set up immediately after stirring. Herein, however, they are considered thixotropic since, according to the definition given in the first part of the paper, no restriction has been put on the time required.)

The importance of this work to the present discussion lies in the fact that only if the suspensions tend to flocculate are they thixotropic. If the suspension remains stably dispersed thixotropic behavior is absent. Thus we have further evidence of thixotropic behavior being caused by a tendency towards flocculation which is over balanced by the dispersing effects of external forces when the material is sheared.

The following quotation from Van Olphen (14) further aids in the understanding of time effects during thixotropic hardening:

> "the thixotropic increase of the modulus of elasticity of a gel with time is a consequence of the increase of the number of links with time after these links have been broken by previous shear. The phenomenon is analogous to slow coagulation."

Effect of Method of Compaction on Thixotropic Behavior.—Since it has been shown by Seed and Chan (23) that method of compaction can have a pronounced influence on the properties of soils compacted wet of optimum, particularly for soils susceptible to a structure change during shear, it would be expected that method of compaction would influence the thixotropic behavior. Samples of Vicksburg silty clay were compacted to the same density at a water content wet of optimum by both static and kneading compaction. Unconsolidated-undrained triaxial tests were performed on these samples at various ages after molding. The stresses required to cause various amounts of strain are shown in Fig. 12 for the two types of compaction. It is immediately evident from this figure that at all ages the static samples are significantly stronger than the kneading samples. This is a logical consequence of the differences in initial structure induced by the two compaction procedures. The differences between the strengths decreases at high value of strain in accordance with the concept of the change of the flocculated structure to a dispersed structure during the test.

The actual strength increase over the 28-day period is slightly greater for the statically compacted samples when measured at low strains and slightly

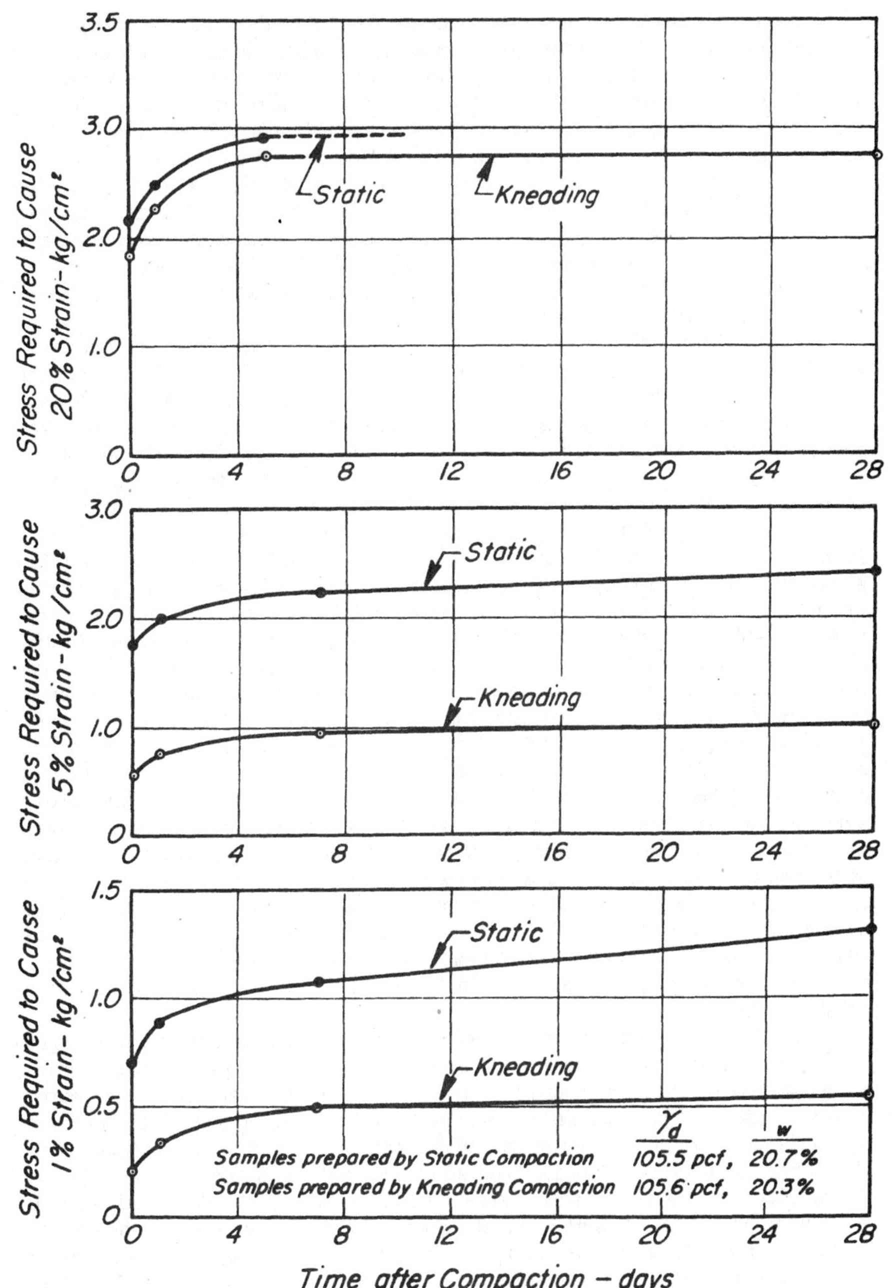

FIG. 12.—EFFECT OF METHOD OF COMPACTION ON THE STRENGTH OF COMPACTED SILTY CLAY.

less when measured at high strains. Quite possibly this greater strength increase in the case of the statically compacted samples is the result of the initial flocculated structure wherein particles are tightly held and very close at so-called "contact points," that is, those points that control interparticle forces. Very slight decreases in spacing in these areas of close approach should result in high strength increases since attraction forces increase rapidly with decreasing spacing at small distances between particles. For the samples compacted by kneading compaction, however, the "effective" initial spacing between particles is somewhat greater due to the more disperse arrangement. At these larger spacings the same decrease in distance would not cause such a great strength increase since the attraction force-distance curve is flatter at larger spacings than for small spacings.

The thixotropic strength ratio is perhaps a more significant indication of thixotropic effects than is the actual strength increase. The ratio reflects the strength change with respect to the initial value and thus offers a more graphic illustration of thixotropic effects for a given initial structure. The thixotropic strength ratio as a function of time is shown in Fig. 13 for the two methods of compaction used. When compared on this basis the samples compacted by kneading are much more thixotropic than those compacted by static compaction. This result might be anticipated because, for the water content used, interparticle forces dictate flocculation in the absence of externally applied shear strains. For samples prepared by static compaction where shear strains are small, a flocculated structure is produced. Since this structure is close to the equilibrium structure for the soil, thixotropic effects are small. On the other hand, the as-compacted structure of samples prepared by kneading compaction is dispersed. As a result the initial structure deviates appreciably from the at-rest equilibrium structure, and thixotropic effects are pronounced.

Effect of Aging on Pore Pressures During Shear.—As noted in a previous section the pressure (or tension) in the pore water of a soil-water system is a measure of the free energy of the water. If free water at atmospheric pressure and 20 C is considered "normal" water then any soil water that registers a pressure less than one atmosphere is at a lower energy level than normal water and any soil water that registers a pressure greater than one atmosphere is at a higher energy level than normal water. Any change in pore water stress during shear or with time reflects a change in the free energy of the system as well as changes in effective stresses. Changes in energy reflect, in turn, changes in particle arrangement, water structure, and electrolyte distribution within the system.

Texts with measured pore pressures were run in order to investigate changes in pore pressure with time and to ascertain whether or not sufficient structural change took place during aging to be reflected by differences in pore water pressures during shear between aged and freshly prepared samples. Seed and Chan (23) have previously shown that dispersed structures develop higher pore pressures during shear than flocculated structures, thus providing a vital and logical link between the effective stress concept of soil strength and the soil structure concept of strength. The structure of the water itself may be an important factor influencing measured pore pressures: a well-ordered water structure leading to lower pressures than a more random arrangement of water molecules.

Typical results of unconsolidated-undrained triaxial compression tests with pore pressure measurements on Vicksburg silty clay are shown in Fig. 14. All points in this figure represent the average of two tests. Four samples

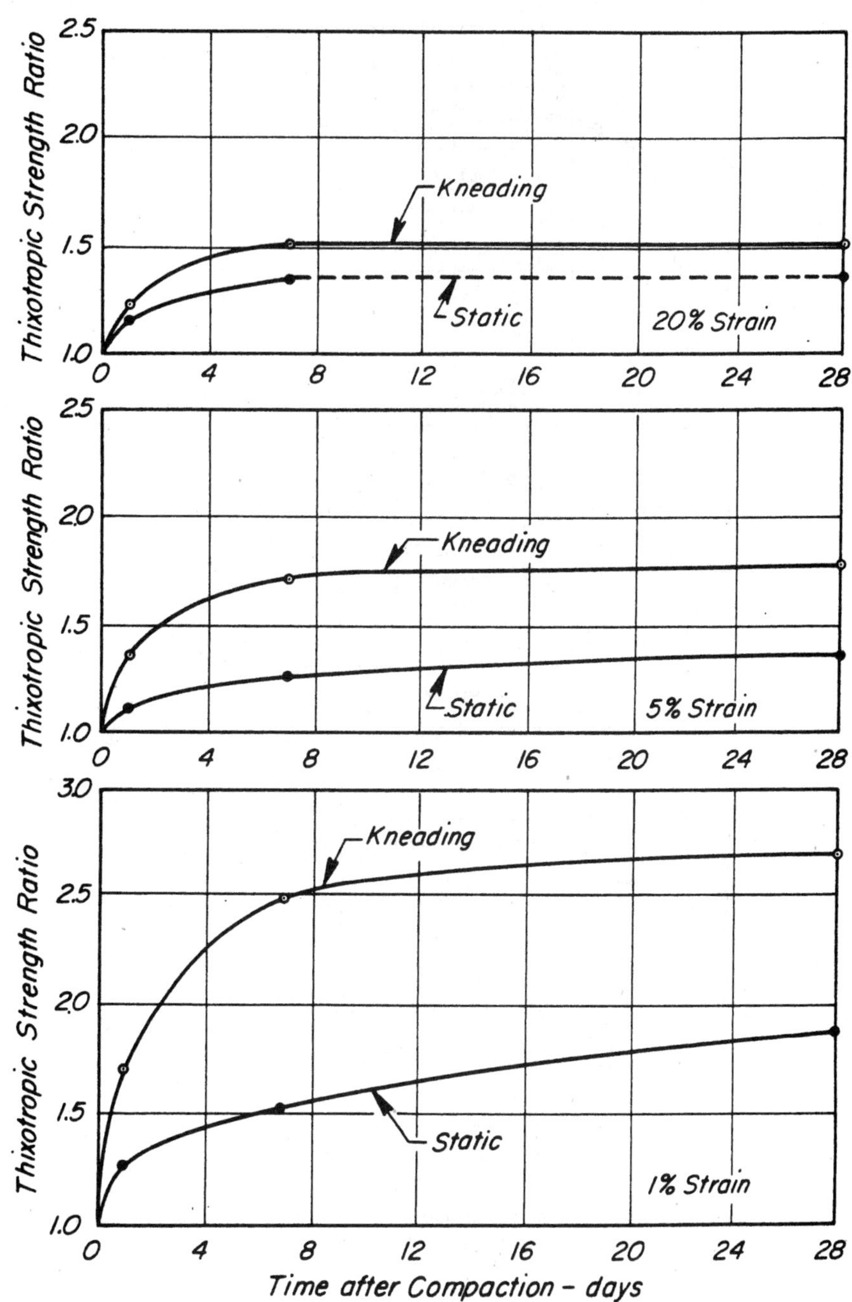

FIG. 13.—EFFECT OF METHOD OF COMPACTION ON THIXOTROPIC CHARACTERISTICS OF COMPACTED SILTY CLAY.

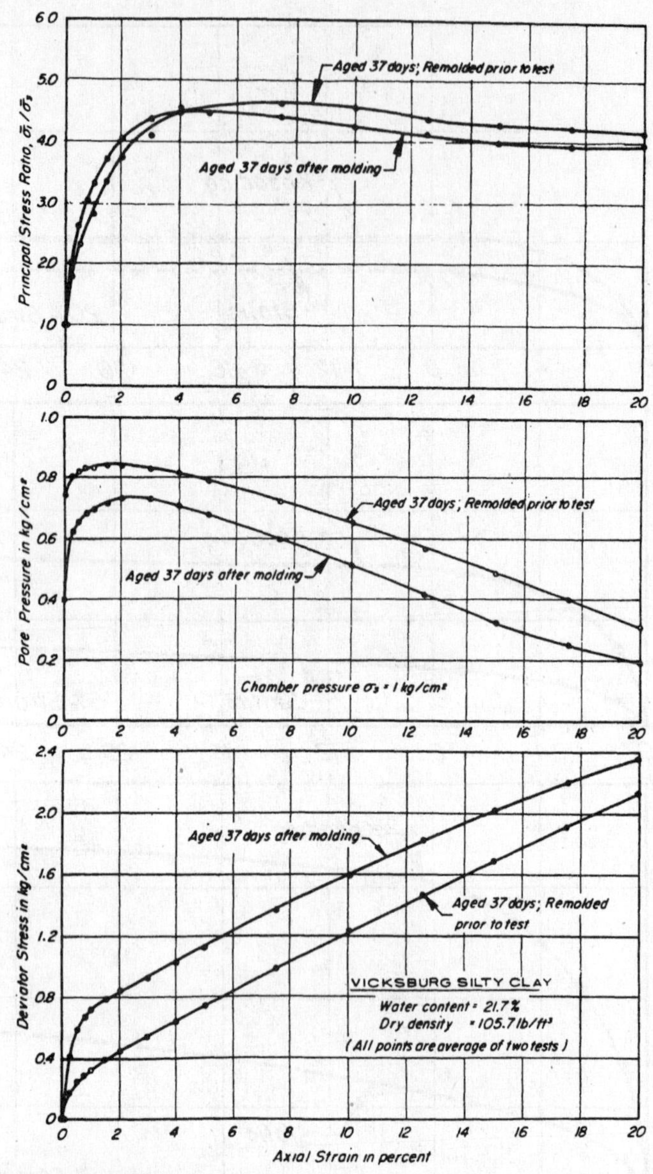

FIG. 14.—THE EFFECT OF AGING ON THE PORE PRESSURES DEVELOPED DURING SHEAR OF COMPACTED VICKSBURG SILTY CLAY.

were compacted by kneading compaction at a water content of 21.7%, wrapped in rubber membranes, and aged for 37 days. Two samples were then tested without disturbance, while the other two were thoroughly remolded, recompacted to the same density and then tested.

The initial pore pressure value shown on the figure (u at 0% strain) is the pressure that had to be applied to the pore water to prevent either water pick-up by the sample or extrusion of water from the sample. Since the initial values of pore pressures are less than 1.0 kg per sq cm—the chamber pressure—it is evident that the initial pore water in the samples before application of the confining pressure was at a lower pressure than normal water, that is, under tension. (That is, were the chamber pressure zero than the measured pore pressure would be negative. Since pressures are more conveniently measured than tensions, the above described method of testing was found most satisfactory.) The difference between the chamber pressure and the pore water pressure at 0% strain is an indication of both the swelling pressure and the intergranular pressure.

Fig. 14 shows that initially the aged sample has a lower pore pressure (0.39 kg per sq cm) than the freshly recompacted sample (0.74 kg per sq cm). This is consistent with the concept of the development of a more flocculated particle orientation and a more ordered water structure with time in a thixotropic soil. During shear the pore pressures developed in the aged sample remain below those developed in the remolded material. The strength, in terms of total stresses (deviator stress), is significantly greater for the aged sample than for the remolded sample, as would be expected. In terms of effective stresses, however, the strengths of the two materials are about the same as evidenced by the plot of effective principal stress ratio versus strain in Fig. 14. These results are in remarkable accord with data previously reported by Seed and Chan (23) concerning the effect of structure on pore pressure and serve to further substantiate the concept that thixotropic effects result primarily from structure changes with time.

Similar results for compacted San Francisco Bay Mud aged 7 days and tested in the same manner as the silty clay are shown in Fig. 15. Here the initial pore pressure difference between aged and remolded soil amounts to 0.25 kg per sq cm. Again it may be seen that the higher pore pressures develop in the remolded material and that the aged samples are stronger in terms of total stresses. The aged specimens of this soil appear only slightly stronger in terms of effective stresses. A final point worthy of note shown by these tests is that, for both soils, with increasing strain the difference between the two pore pressure curves decreases. The curve for the aged specimens shows a more rapid initial build-up of pore pressure with strain than does the curve for the recompacted specimen. This may be attributed to a rapid breakdown of the thixotropic structure during shear, and is an effect similar to decreasing thixotropic strength ratio with increasing strain as shown in previous figures.

Effect of Thixotropy on Compressibility and Permeability.—If the hypothesis for thixotropic behavior is correct then it must necessarily follow that aging a thixotropic soil will have some effect, however small, on the compressibility, because different structures of the same material consolidated from the same initial void ratio exhibit different compression behavior. Flocculent structures offer a greater resistance to compression under low pressures than do disperse structures.

Remolded samples of San Francisco Bay mud were studied at a natural water content of 90.4%. Fig 16(a) shows the thixotropic character of this material

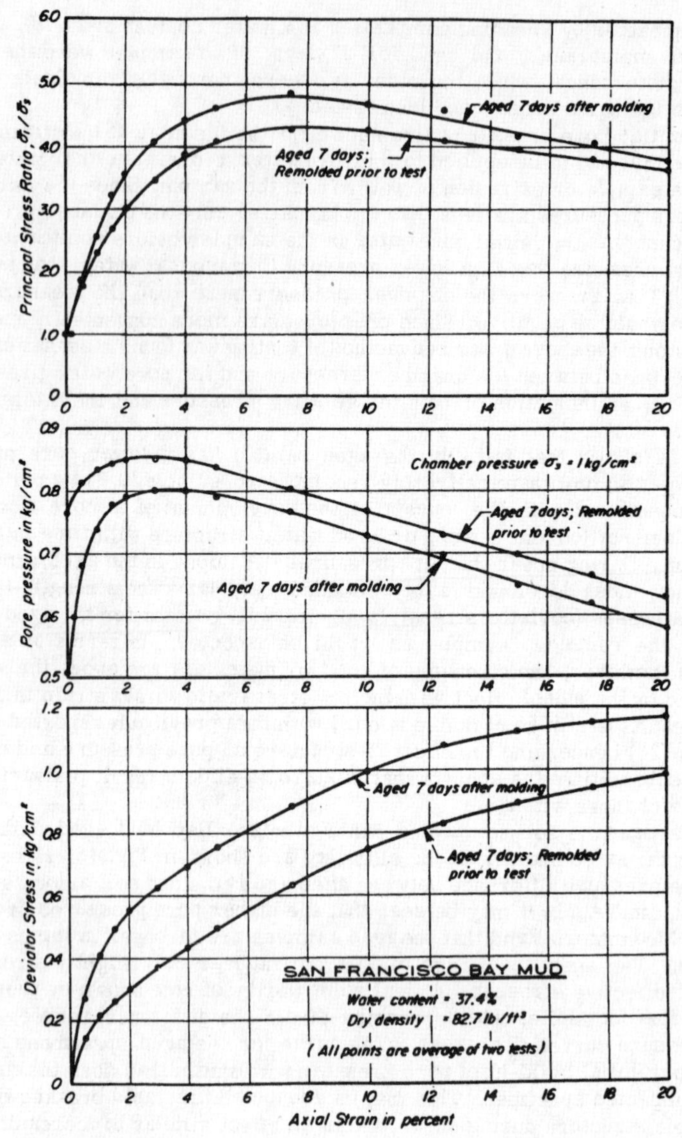

FIG. 15.—THE EFFECT OF AGING ON THE PORE PRESSURES DEVELOPED DURING SHEAR OF COMPACTED SAN FRANCISCO BAY MUD.

as determined by the laboratory vane shear apparatus. Two samples were thoroughly remolded, aged 12 days at constant water content, and tested; two other samples were similarly remolded and aged but again remolded prior to the strength test. Fig. 16(a) indicates significant thixotropic strength gain over a period of 12 days.

Similar samples of this material were tested by means of the standard one-dimensional consolidation test. The void ratio as a function of applied pressure is shown to natural scale in Fig. 16(b) and to log scale in Fig. 16(c). As indicated by Fig. 16 the aged sample has an e-p curve everywhere above that of the remolded sample. That is, aging a thixotropic soil has the effect of in-

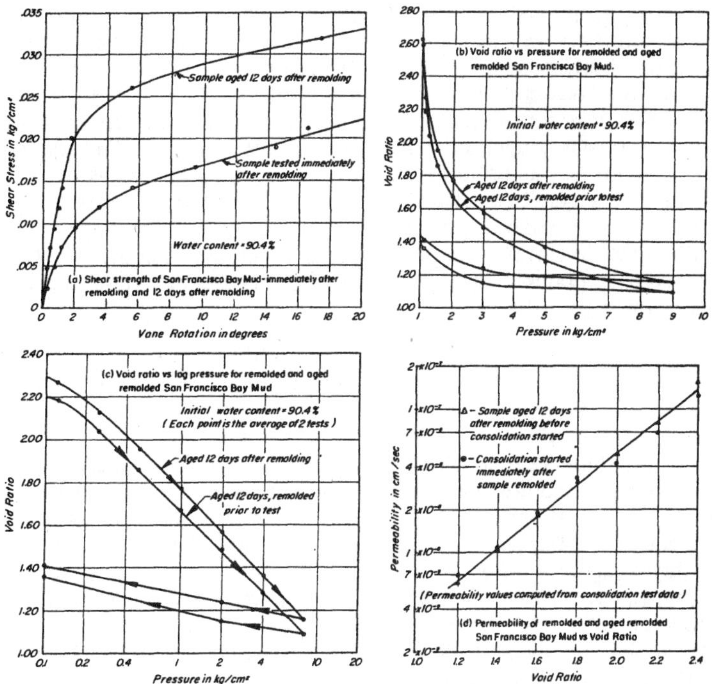

FIG. 16.—EFFECT OF AGING ON THE STRENGTH AND CONSOLIDATION CHARACTERISTIC OF SAN FRANCISCO BAY MUD.

creasing the resistance to compression. Examination of Fig. 16 further indicates that above a pressure of 0.1 kg per sq cm the compression curves are very nearly parallel, although a slight tendency towards convergence at some high pressure may be noted. This means that the effect of aging is exhibited under the initial loading; the increased rigidity of the aged sample being responsible for less compression than in the dispersed remolded structure. But once the applied pressure is great enough to overcome this rigidity, the aged sample behaves similarly to the remolded sample. At the same time, because of the time involved in performing a consolidation test, the remolded sample is gaining thixotropic rigidity, which in turn tends to cause similarity in the curves at the higher pressures.

Permeability values were computed from the coefficient of consolidation values as determined by the logarithm of time fitting method measured during each load increment. These values are plotted in Fig. 16(d). Permeability has been plotted as a function of void ratio rather than pressure as it is felt that void ratio is a more rational basis of comparison. Permeability differences between aged and freshly remolded samples at the same void ratio would be directly attributable to structural differences. No significant differences can be noted between the permeabilities of the aged and remolded samples. There is a suggestion, however, that the permeability of the aged sample is slightly higher at the high void ratio (low pressures) where thixotropy has the greatest effect on structure. This would be expected from considerations of the structures of the two systems; however, the uncertainties involved in the computation of permeability from consolidation data do not warrant a detailed analysis of the results. Similar consolidation and permeability behavior was observed in tests performed on aged and remolded samples of Vicksburg silty clay.

Effect of Temperature on the Rate of Thixotropic Strength Increase.—Altering the temperature in a system containing colloids (such as clay) and water is known to affect the status of interparticle forces, the properties of the water and the rates of any reactions that might occur. Thixotropic behavior appears to be a function of the balance between attractive and repulsive forces and further, in thixotropic systems attraction exceeds repulsion by some small amount. The thickness of the double layer, t which controls interparticle repulsion, is given approximately by the relation

$$t = \sqrt[2]{\frac{E\,k\,T}{8\,n\,e^2\,v^2}} \qquad \text{. . . . . (1)}$$

in which E is the dielectric constant of the pore fluid, k denotes the Boltzmann constant, T is the absolute temperature, n represents the concentration of electrolyte, e is the electronic charge, and v denotes the valence of the exchangeable cation. If temperature alone could be varied, then the thickness of the double layer and therefore the repulsive forces would increase with increasing temperature. However temperature cannot be altered without altering the dielectric constant, E, which decreases with increasing temperature. It may be shown that the overall effect of these changes is a decrease in double layer thickness with increasing temperature. Thus, on this basis, increasing temperature causes decreasing repulsive forces between particles. These principles when applied to clays prepared to identical structures, aged at different temperatures, but tested under identical conditions would lead to the expectation that the higher the aging temperature the higher the rate of strength gain. The reason for this is that increasing the aging temperature would accentuate the excess of attractive over repulsive forces, causing a greater driving force for structure alteration to a more flocculated condition than exists in the as-molded state. In addition higher temperatures would lead to decreased water viscosity and increased thermal activity of dissolved ions and the water molecules themselves, which in turn would allow particle reorientations, water structure changes and ionic redistributions to occur more rapidly.

This expectation was confirmed with the following experiments. A mixture consisting of 25% Wyoming bentonite and 75% sand containing sizes between 100 and 200 mesh by weight was prepared to a water content of 24%. Samples were compacted to a dry density of 100.2 ± 1.0 pcf by kneading compaction. All samples were prepared at a temperature of 70 F, wrapped with two rubber membranes separated by a layer of grease, and stored under water. One-third of the samples were stored at 40 F, a third at 70 F and a third at 140 F. After intervals of storage of 0.29, 1, 3, 7, 14, 24, and 51 days, strength values were measured. All samples were brought to 70 F prior to testing in undrained triaxial compression with a chamber pressure of 1 kg per sq cm. Reliable strength determinations could not be made on the samples aged at 140 F after storage of more than 14 days because the elevated temperature caused deterioration of the membranes followed by adsorption of water by the samples.

Results of these tests are shown in Fig. 17 where the deviator stress required to cause strains of 1%, 2.5%, 5%, and 10% are plotted as a function of time after remolding. The data can be fitted by a straight line on the semi-log plot for times above one day. The rate of strength gain as expressed by the slope of the line for each temperature is as follows:

Temperature in °F	Average rate of strength gain in kg per sq cm per log cycle
40	0.064
70	0.096
140	0.140

It may be seen from the foregoing that the rate of strength gain is more than doubled over the temperature range of 40 F to 140 F.

Strength values were determined on the samples remolded after 51 days storage and are shown in Fig. 17. Due to the water absorption by the samples stored at 140 F meaningful 51-day remolded values could not be determined. However for samples stored at 70 F and 40 F the remolded strength returned almost precisely to the value measured immediately after the initial compaction. This affords additional evidence of the reversible nature of thixotropic effects in soils.

These data are useful in another way. By means of the Arrhenius equation (30) for rate processes the activation energy for the processes involved in the strength changes may be calculated (the activation energy is the energy that any ion, molecule or particle must acquire before it is able to take part in any process, either physical or chemical). The Arrhenius equation is of the form of

$$E_a = \frac{R\, T_1\, T_2}{T_2 - T_1} \ln\left(\frac{k_2}{k_1}\right) \qquad (2)$$

in which E_a is the activation energy in calories per mole, T_1 and T_2 denote the temperatures in °K at which the rates of property changes are measured, R is the gas constant, and k_1 and k_2 are the rate of change of a given property at temperatures T_1 and T_2, respectively.

If Eq. 2 is applied to the rates of strength increase at 40 F and 70 F an activation energy of 3,900 cal per mole is calculated; whereas, if applied to the rates at 70 F and 140 F the activation energy is 1,900 cal per mole. According to S. Glasstone, K. Laidler and H. Eyring (30) activation energies of the order of 4,000 cal per mole at about 70 F and 2,800 cal per mole at 212 F are associated with viscous flow of water. This appears a most reasonable phenomenon to be associated with thixotropic effects since particle movements, water redistribution, etc., must necessarily involve movements in the water phase. On the basis of the data at hand no good explanation for the relatively low value of activation energy at the higher temperature is available. It is known, however, that the effects of clay particle surfaces on the properties of the adjacent water phase may be considerable. The effect of the differences in adsorbed wa-

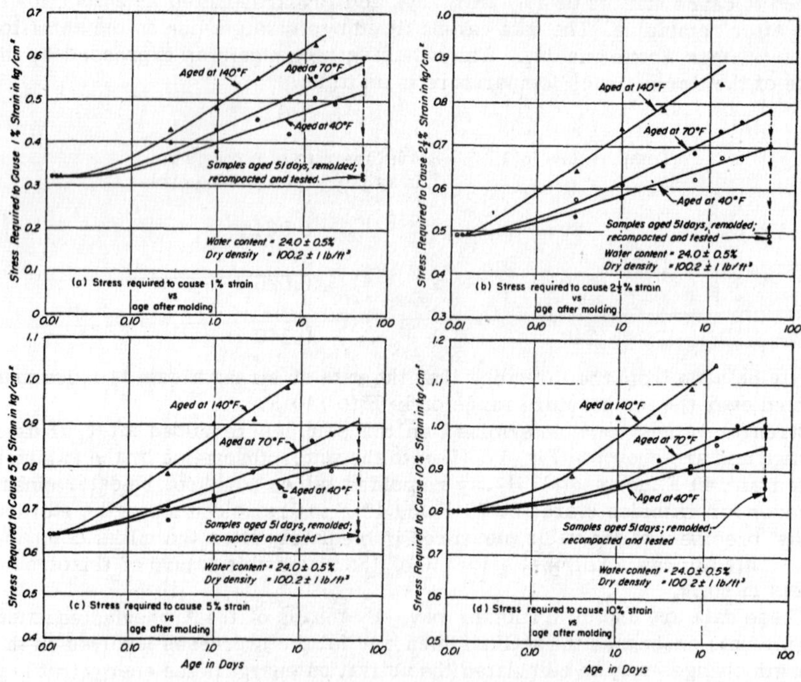

FIG. 17.—EFFECT OF AGING TEMPERATURE ON THE RATE OF THIXOTROPIC STRENGTH INCREASE OF A SAN-BENTONITE MIXTURE.

ter properties from normal water properties on activation energies required for various processes are at this time, unknown.

Although the energy calculations and thermodynamic considerations are somewhat crude when applied to the present data there appear to be great potentialities for the application of such concepts to the study of soil behavior. The feasibility of determining changes in thermodynamic properties associated with such phenomena as creep, secondary compression, and stress-strain behavior has yet to be thoroughly explored; however, should such studies be practical then a much better insight into the fundamentals of soil behavior might result.

It appears that thixotropic effects can cause significant changes in soil properties, at least over some range of water contents that might be encountered in practice. In compacted soils thixotropic effects may be of importance if the material is compacted on the wet side of optimum. In normally consolidated clays thixotropy may be a contributor to sensitivity, and will most certainly be an important factor when dealing with these materials in the remolded state.

Although theoretical thixotropic strength change equations are not available to permit direct accommodation of thixotropic effects into design problems, certain factors appear worthy of note when considering any problem where a thixotropic soil might be encountered. For example, in any laboratory testing program it would be desirable to place strict controls on the time between sample preparation and testing, in order that all tests be equally affected by thixotropy. In the case of compacted soils, comparisons between the as-molded and soaked properties may be influenced by thixotropic effects occurring during the time required for soaking.

The influence of thixotropy on the Mohr failure envelope based on total stresses for a compacted clay was determined by Seed and Chan (6). The effect of aging was to shift the envelope upwards; the strength intercept was increased, whereas the strength angle remained constant. Thus a total stress design based on the immediate strength envelope for a soil of this type will have an added element of safety due to thixotropic effects. It does not necessarily follow, however, that the strength in terms of effective stresses is increased as a result of thixotropic effects. Reference to Figs. 14 and 15 shows that the principal stress ratio curves for aged samples and samples tested immediately after compaction are not significantly different. Thus, it appears possible that the strength in terms of effective stress is unaffected by thixotropic phenomena.

Reference to the moisture tension data obtained by Day (26) and the initial pore pressure values shown in Figs. 14 and 15 show a very significant decrease in pore water pressure with time in a thixotropic soil. Stated another way, thixotropy causes a marked increase in the pore water tension. The effect of this might be to increase the water adsorption potential of the soil, which could be detrimental in practice.

The effect of thixotropy on compressibility has been indicated in Fig. 16. Two parctical implications of these results may be pointed out. An embankment constructed of a thixotropic soil may show less settlement than anticipated as a result of beneficial structure change with time. Thixotropy, in effect, builds precompression into the soil. Secondly, it is doubtful that a "true" consolidation curve can ever be obtained for a thixotropic soil because of the time involved in testing. Thixotropy and secondary compression work at cross purposes. It would seem reasonable that thixotropic effects during consolidation lead to a smaller compression index then would be obtained if there were no thixotropy.

In the filed when dealing with soft clays, thixotropic effects will probably always be beneficial. The effects of disturbance will decrease with time adding some element of safety to any operation wherein remolding cannot be avoided. An example would be the driving of friction piles into soft clay. The build-up of skin friction with time after driving will be assisted by thixotropic hardening as well as the dissipation of excess pore pressure in the distrubed zone.

Finally, there seems to be no fundamental reason why thixotropic effects cannot be included in design whenever it appears that they are of sufficient magnitude to warrant consideration, and carefully controlled tests have been run to assess this magnitude.

SUMMARY AND CONCLUSIONS

Previous investigations of thixotropy in suspensions and in soils have been summarized. These studies have pointed out the variety of factors involved in and the complex nature of the phenomena. The importance of such factors as pore fluid composition, particle size and shape and the balance between interparticle attractions and repulsions have all been recognized. Available evidence indicates that thixotropy is of quite general occurrence in fine-grained materials; however a comprehensive theory for the basic cause of thixotropy has not heretofore been available.

An hypothesis based on initial non-equilibrium of interparticle forces after remolding or compaction, and the effects of this non-equilibrium on subsequent structure changes within the soil has been suggested as one possible explanation of the phenomenon. Data consistent with this hypothesis have been presented. It has been found that thixotropic effects will occur if the initial structure is dispersed artificially to an extent greater than dictated by the interparticle forces. Changes in properties occur in response to the structural changes resulting from a dissipation of excess energy from within the system as the soil adjusts its structure to one compatible with the excess of attractive over repulsive forces that exists after the removal of externally applied loads.

Available data indicate that the process is completely reversible, and for the soils tested chemical effects appear minor. The role of water content in determining thixotropic behavior has been found to be primarily one of altering interparticle forces. If it is of appropriate magnitude so that a dispersed structure can be induced by means of externally applied forces even though the soil particles would tend to flocculate when at rest then the soil will be thixotropic. If the water content and double layer conditions are such that the soil assumes a stable dispersed or flocculated structure independently of applied forces then thixotropic effects are negligible.

Pore pressures have been shown to decrease with the aging of a thixotropic soil. This indicates a change in structure of the system which may be attributed to changes in particle orientation, adsorbed water structure and ionic distributions. The decrease in pore pressure reflects the decrease in free energy of the pore water. The pore pressures during shear are lower for an aged thixotropic soil then for one sheared immediately after molding, which is compatible with previous results for flocculated and dispersed structures. Pore pressure measurements have further shown that while thixotropic effects may cause appreciable increases in strength in terms of total stresses they may cause little or no effect in terms of effective stresses.

The effect of thixotropy on one dimensional compression is to raise the pressure-void ratio curve. The greatest effects are exhibited under low pressures. Thixotropy measurements at different aging temperatures have given results in complete accord with the proposed theory, and have further indicated, according to energy calculations that a viscous flow of water may be involved in thixotropic phenomena. Some of the practical aspects of thixotropy have been pointed out and it is suggested that such effects warrant consideration in design, provided they are accurately evaluated.

ACKNOWLEDGMENTS

The writer is pleased to acknowledge the assistance of those who aided in the conduct of the research and preparation of the paper. H. B. Seed and C. L. Monismith offered helpful suggestions and critically reviewed the manuscript. The writer's colleague, C. K. Chan, was of inestimable assistance in development of experimental procedures. M. Markowicz and D. R. Hooper, Graduate Research Engineers, performed many of the tests. The drawings were prepared by G. Dierking.

APPENDIX I.—BIBLIOGRAPHY

1. "Entwichlungsmech. d. Organism," by Arch. F. Peterfi, Vol. 112, 1927, p. 689.
2. "Thixotropy," by H. Fruendlich, Hermann et Cie, Paris, 1935.
3. "Report on the Principles of Rheological Nomenclature," by J. M. Burgers G. W. Scott Blair, Joint Committee on Rheology of the Internatl. Council of Scientific Unions, Proceedings, Internatl. Rheologic Congress, Amsterdam, 1948.
4. "Effect of Natural Hardening on the Unconfined Compression Strength of Remolded Clays," by O. Moretto, Proceedings, Second Internatl. Conference on Soil Mechanics, Vol. I, 1948.
5. "The Sensitivity of Clays," by A. W. Skempton and R. D. Northey, Geotechnique, Vol. III, No. 1, March, 1952.
6. "Thixotropic Characteristics of Compacted Clays," by H. B. Seed and C. K. Chan, Proceedings, ASCE, Vol. 83, No. SM4, November, 1957.
7. "A Preliminary Examination of the Thixotropy of Some Sedimentary Rocks," by P. G. H. Boswell, Quarterly Journal of Geological Science, Vol. 104, 1949, p. 499.
8. "Colloid Science, I, Irreversible Systems," by H. R. Kruyt, Elsevier Pub. Co., New York, 1952.
9. "The Colloid Chemistry of the Silicate Minerals," by C. E. Marshall, Agronomy, Vol. I, Academic Press, Inc., New York, 1949.
10. "Theory of the Stability of Lyophobic Colloids," by E. J. W. Verwey and J. T. Overbeek, Elsevier Pub. Co., New York, 1948.
11. "The Structure of Inorganic Soil," by T. W. Lambe, ASCE, Speerate 315, October, 1953.
12. "The Structure of Compacted Clay," by T. W. Lambe, Proceedings, ASCE, Vol. 84, No. SM2, May, 1958.
13. "The Importance of Structure to the Engineering Behavior of Clay," by J. K. Mitchell, D. Sc. Dissertation, Massachusetts Inst. of Tech., 1956.

14. "Forces Between Suspended Bentonite Particles," by H. Van Olphen, Clays and Clay Minerals, Proceedings, Fourth Natl. Conference on Clays and Clay Minerals, Publication 456, Natl. Academy of Sciences, Natl. Research Council, 1956.

15. "The Deflocculation of Kaolinite Suspensions and the Accompanying Change-over from Positive to Negative Chloride Adsorption," by R. K. Schofield and H. R. Samson, Clay Minerals Bulletin, Vol. 2, No. 9, British Mineralogical Soc., July, 1953.

16. "Wechselseitige Adsorption von Kolloiden," by P. A. Theissen, Zeitschrift fur Elechtrochemie, Vol. 48, 1942, p. 675.

17. "Thixotropy in Clays," by R. Fahn, A. Weiss and H. Hofmann, Berichte der Deutsche Keramischen Gesellschaft, Vol. 30, 1953, p. 21.

18. "The Gelation of Montmorillonite, Part I—The Formation of a Structural Framework in Sols of Wyoming Bentonite," by M. B. Mac Ewen and M. I. Pratt, Transactions, Faraday Soc., Vol. 53, 1957, p. 535.

19. "The Gelation of Montmorillonite, Part II—The Nature of Interparticles Forces in Sols of Wyoming Bentonite," by M. B. Mac Ewen and D. L. Mould, Transactions, Faraday Soc., Vol. 53, 1957, p. 542.

20. "Rheology of Suspensions: A Study of Dilatency and Thixotropy," by H. L. Roeder, H. J. Paris, Amsterdam, 1939.

21. "Strength Regain of Clays," by L. Berger and J. Gnaedinger, Amer. Soc. of Testing Materials, Bulletin 160, September, 1949, p. 64.

22. "Strength Characteristics of Compacted Clays," by G. A. Leonards, Transactions, ASCE, Vol. 120, 1955.

23. "Structure and Strength Characteristics of Compacted Clays," by H. B. Seed and C. K. Chan, Proceedings, ASCE, Vol. 85, No. SM5, October, 1959.

24. "A Study of the Sensitivity Resulting from Consolidation of a Remolded Clay," by P. L. Newland and B. H. Allely, Proceedings, Fourth Internatl. Conference on Soil Mechanics and Foundation Engrg., Vol. I, 1957, p. 83.

25. "Soil Structure and the Step-Strain Phenomenon," by D. H. Trollope and C. K. Chan, Proceedings, ASCE, Vol. 86, No. SM2, April, 1960.

26. "Effect of Shear on Water Tension in Saturated Clay," by P. R. Day, I and II, Annual Reports, Western Regional Research Proj. W-30, 1954, 1955.

27. "Soil Physics," by L. D. Baver, John Wiley and Sons, New York, 1956.

28. "The Effect of Acid and Heat Treatment on Montmorillonoids," by A. C. Mathers, S. B. Weed and N. T. Coleman, Clays and Clay Minerals, Natl. Academy of Sciences, Natl. Research Council, Publication 395, 1955.

29. "Water Vapor Sorption on Lithium Kaolinite," by R. T. Martin, Clays and Clay Minerals, Natl. Academy of Sciences, Natl. Research Council, Publication 566, 1958, p. 23.

30. "Theory of Rate Processes," by S. Glasstone, K. Laidler and H. Eyring, McGraw-Hill Book Co., New York, 1941.

31. "Physico-Chemical Analysis of the Compressibility of Pure Clays," by G. H. Bolt, Geotechnique, Vol. VI, No. 1, 1956, p. 86.

Journal of the
SOIL MECHANICS AND FOUNDATIONS DIVISION
Proceedings of the American Society of Civil Engineers

PILE-DRIVING ANALYSIS BY THE WAVE EQUATION

By E. A. L. Smith[1]

SYNOPSIS

There are a great many different pile-driving formulas in use, and engineers have never been able to agree as to which one is best. This situation has arisen primarily because, until recently, the mathematics of pile-driving action could not be solved in any practical manner. As a result all pile-driving formulas are partly empirical and, consequently, apply only to certain types or lengths of pile. This paper is presented with the purpose of giving engineers a mathematical method of wider application, depending on the use of electronic computers and numerical integration. The method is also applicable to other impact problems.

INTRODUCTION

Pile-driving formulas are widely used to determine the static bearing capacity of piles. Some of these formulas are also used to determine stresses in the pile during driving.

An astonishing amount of effort and ingenuity has been expended by engineers in the development of pile-driving formulas, with the result that there are a great many different formulas in use, many of which are specified in various building codes throughout the United States and abroad. Robert D. Chellis lists[2] thirty-eight pile driving formulas, and the editors of *Engineering News-Record* have on file four hundred and fifty such formulas.

Note.—Discussion open until January 1, 1961. To extend the closing date one month, a written request must be filed with the Executive Secretary, ASCE. This paper is part of the copyrighted Journal of the Engineering Mechanics Division, Proceedings of the American Society of Civil Engineers, Vol. 86, No. EM 4, August, 1960.

[1] Tucson, Ariz. Formerly Chf. Mech. Engr., Raymond Internatl. Inc., New York, N. Y.

[2] "Pile Foundations," by Robert D. Chellis, McGraw-Hill Book Co., New York, 1951, pp. 28-33, 449-450, 525-538.

The fact that engineers have been unable to agree on any one pile-driving formula is understandable because, until recently, the mathematics of pile-driving action could not be solved in any practical manner. Pile driving is not a simple problem of impact that may be solved directly by Newton's laws.[3] Pile driving is a problem in longitudinal wave transmission that is covered in a general way by the wave equation, which is well known to mathematicians.

Furthermore, pile driving involves many complications such as the use of capblocks, pile caps, cushion blocks, composite piles, and tapered piles, as well as the elastic-plastic action of the ground and other problems in soil mechanics. As a result of these difficulties, all pile-driving formulas are partly empirical and consequently apply only to certain types or lengths of pile. This paper is presented with the purpose of giving engineers a mathematical method of wider application.

So far as the writer is aware, D. V. Isaacs, in 1931, was the first to point out[4] that wave action occurred during the driving of piles. In 1938, E. N. Fox published[5] a solution of the wave equation applied to pile driving, but, as no electronic computers were available at that time, he was forced to use a number of simplifying assumptions that lessened the value of his solution. At the present time (1960), by using the conceptions of the wave equation and resorting to numerical integration and electronic computers, a solution of the pile-driving problem can be obtained that produces mathematical accuracy within about 5%. This degree of accuracy is more than sufficient in view of our present imperfect knowledge of the physical conditions involved.

The mathematical method described herein may, with slight modification, be applied to other impact problems such as the design of a foundation for a forging hammer, or a fendering system for a dock.

OUTLINE OF NUMERICAL METHOD

The hammer, pile, and other parts involved, such as the capblock and pile cap, are represented as a series of weights and springs as shown in Fig. 1, and the time during which the action occurs is divided into small time intervals such as 1/4,000 sec. The action of each weight and each spring is then calculated separately in each and every time interval. In this way a mathematical determination may be made of stresses, and of pile penetration or permanent set per blow, against any amount or kind of ground resistance.

The process may be compared to making drawings for an animated motion picture. The artists who prepare such drawings must take account of the fact that the film will be projected at 24 frames per sec. Each drawing must therefore differ from the preceding drawings by 1/24 sec. In order that the picture may appear realistic when projected, computations must be made to determine how far any moving object will progress in each 1/24 sec. If the motion is uniform the displacements in each succeeding drawing should be uniform; if the motion is uniformly accelerated, as in the case of a falling object, the displacements should differ by increasing amounts readily calculated by using the well-known laws of uniformly accelerated motion.

---

3 "Dynamic Pile Driving Formulas," by A. E. Cummings, Journal of Boston Soc. of Civ. Engrs., January, 1940.

4 "Reinforced Concrete Pile Fourmulae," by D. V. Isaacs, Paper No. 370, Transactions of the Institution of Engrs., Australia, Vol. XII, pp. 312-323, 1931.

5 "An Investigation of the Stresses in Reinforced Concrete Piles during Driving," by W. H. Glanville, G. Grime, E. N. Fox, and W. W. Davies, British Bldg. Research Bd. Tech. Paper No. 20, D.S.I.R., 1938.

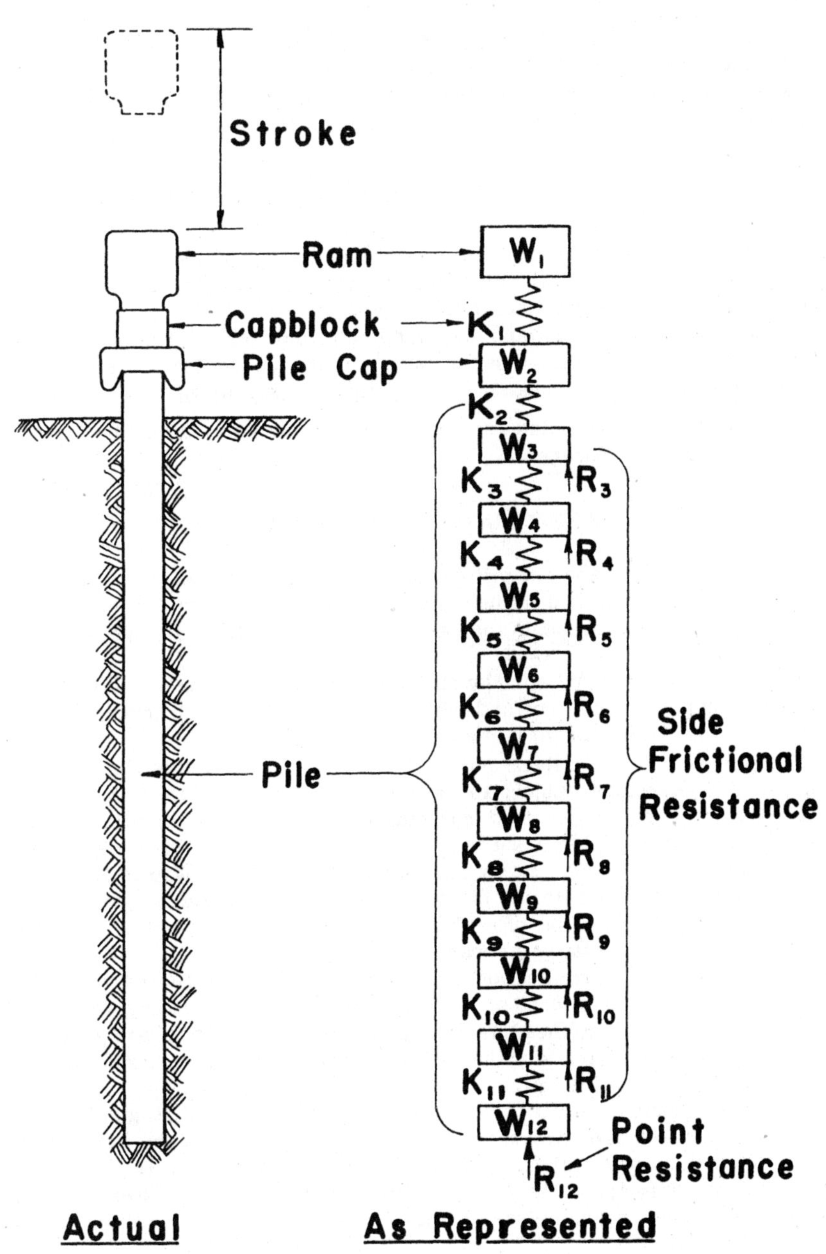

FIG. 1.—METHOD OF REPRESENTING PILE FOR PURPOSE OF CALCULATION

Obviously the rules for making such a set of motion-picture drawings are these:

1. Time is divided into intervals of 1/24 sec each.
2. No motion can be shown in any one drawing.
3. Motion is depicted by making each successive drawing differ from the preceding drawing by just enough to represent the changes occurring during one interval.

The rules for making a numerical pile calculation are almost identical, and may be stated thus:

a. Time is divided into small intervals such as 1/4,000 sec.
b. It is assumed that all velocities, forces, and displacements will have fixed values during any particular interval.
c. The velocities, forces, and displacements for each interval will be computed so as to differ from those existing in the preceding interval by just enough to represent the change occurring during one interval.

If these two sets of rules are compared rule by rule, it will be found that there is no essential difference.

DIAGRAMMATIC REPRESENTATION

The hammer-ram and the pile can are usually short, heavy, rigid objects which may be represented as individual weights without elasticity. In Fig. 1 the first weight W_1 represents the ram, and the second weight W_2 represents the pile cap.

The capblock is a short springy object of wood, plastic, or similar material which is comparatively light and which may therefore be represented by a spring. In Fig. 1 this spring is numbered K_1.

The pile is a heavy object, but is somewhat compressible because of its length. It is therefore subject to wave action under the blow of the hammer. This wave action canbe analyzed mathematically by dividing the pile into short sections or "unit lengths" such as 5 ft or 10 ft in length. The weight of each unit length is represented by an individual weight (W_3 to W_{12} in Fig. 1) and the elasticity of each unit length is represented by an individual spring (K_2 to K_{11} in Fig. 1). The motion of each weight and each spring is then calculated as though each were actually a separate and distinct object.

If the pile is of uniform section, the weights and springs representing it are all alike. If the pile is tapered or is a composite pile, or is in any way irregular in cross section, then these weights and springs differ from one another so as to represent approximately the actual distribution of weight and elasticity throughout the length of the pile.

One may wonder at the fact that comparatively large unit lengths, such as 5 ft or 10 ft, give desired accuracy in computing the action of the impact wave. The following analogy, using water waves, may make the matter clear. Of course impact waves in a pile are longitudinal waves, whereas water waves are transverse waves; nevertheless, water waves can be used to illustrate the principles involved:

Imagine a body of water on which moderately long waves are traveling from right to left. If a long strip of flexible material such as sponge rubber is allowed to float on the surface, it will follow the wave action exactly. If for purposes of mathematical analysis we were to represent this flexible strip by a number of short rigid floats connected together by flexible links, the picture would appear as shown in Fig. 2. This would be an approximation, but the approximation would involve negligible error because the small floats would ride the waves almost exactly like the flexible strip.

But suppose we make the rigid floats comparatively long so that they approach or exceed the length of the water wave, then the picture would look like Fig. 3. These long floats cannot follow the wave form closely. If the floats were made longer still, the resemblance to the true wave form would disappear completely.

From the foregoing, it may be concluded that in dividing a pile into unit lengths for purposes of calculation, the unit lengths chosen must be considerably shorter than the wave length of the stress or impact wave produced by the hammer. Fortunately, a pile-driving impact usually produces a fairly long wave form, therefore division of the pile into lengths of the order of 5 ft to 10 ft will produce acceptable accuracy.

In the special and unusual case where the exact form taken by the impact wave is being investigated, a smaller unit length such as 2 ft or even 1 ft may be advisable. In this case, it may be advisable to divide the ram into unit lengths as well as the pile. In this way a high degree of accuracy may be obtained.

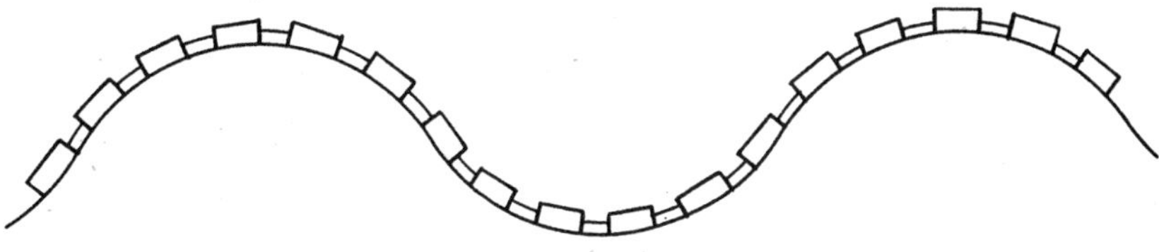

FIG. 2.—SHORT FLOATS

It has been explained previously, that the pile must be divided into unit lengths and that time must be divided into small intervals such as 1/4,000 sec. There is a very important relationship between these two procedures. The smaller the unit lengths chosen, the smaller must be the time interval.

This can be explained in a general way by again considering a motion picture. If the action to be represented is of the normal kind, a 1/24-sec time interval between pictures is suitable. If the action is high speed in character, such as the rotation of a wheel, a much smaller interval is required to show the action accurately. If a 1/24-sec interval is used, the wheel may actually appear to run backward. The same sort of phenomenon occurs in a numerical wave equation calculation because the calculation is a step-by-step process which must keep ahead of the stress wave caused by the hammer blow. The smaller the unit length is made, the smaller must be the time interval. It is also true that especially large or suddenly applied external forces or resistances may require the use of a smaller time interval.

It follows from the foregoing, that whenever a numerical wave equation calculation is made, care must be taken to make certain that the time interval used is not too large. On the other hand, the time interval should not be unnecessarily small, because this would use an unnecessarily large amount of time in computation with little or no increase in accuracy. Consequently, the choice of the correct time interval becomes an important consideration. For pile calculations, a time interval of about 1/4,000 sec will usually be found to be satisfactory. This matter is discussed further in the Appendix.

GENERAL

The foregoing discussion outlines the basic method for obtaining a numerical solution of the pile-driving problem. This method has the outstanding advantage of dividing the problem into a number of simpler problems each of which may be considered more or less independently.

Before going into the mathematics, let us first consider the practical or physical aspects of the problem.

Soil Mechanics.—In order to make a pile calculation, it must be assumed that the soil will act in some particular way. When future investigators develop new facts, the mathematical method explained herein can be modified readily to take account of them, but on the basis of information presently available, the assumptions listed in what follows are recommended.

Resistance At Point Of Pile.—Chellis gives[2] a precedent to follow, which teaches that the ground compresses elastically for a certain distance, (which

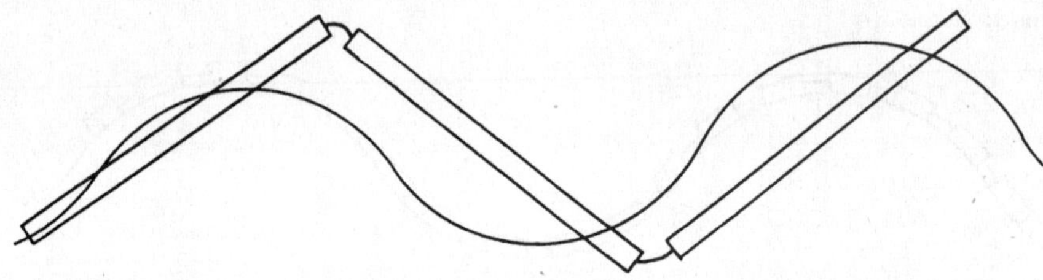

FIG. 3.—LONG FLOATS

Chellis calls C_3, but which herein will be termed Q for "quake") and then fails plastically with constant or "ultimate" resistance R_u. This concept is illustrated in Fig. 4.

Starting at 0, the pile point moves ahead a distance Q (usually assumed to be 0.1 in.) compressing the soil elastically so that at point A the ground resistance has built up to its ultimate value R_u. Plastic failure then occurs and ground resistance remains equal to R_u until the pile point reaches B. Elastic rebound equal to Q then occurs, and motion ceases at point C where all forces are zero. The permanent set of the pile is the distance $s = \overline{OC} = \overline{AB}$.

This conception fails to consider the element of time. Some piles penetrate the ground more rapidly than others. Obviously, the ground will offer more instantaneous resistance to rapid motion than to slow motion. We therefore introduce the additional factor of "viscous damping," which is commonly used in vibration problems.

The numerical wave equation calculation gives the instantaneous velocity of the point of the pile in any time interval. If this instantaneous velocity is called v_p, and if J is a damping constant, then the product J v_p may be used to increase (or decrease) the ground resistance so as to produce damping. Thus at any point x on the line OABC of Fig. 4, the instantaneous damping resistance is J v_p R_x. If J is assumed to have a value of 0.15, and if at point x the calculated value of v_p is 2 fps, the damping resistance is $0.15 \times 2 \times R_x = 0.30\ R_x$, and the total instantaneous resistance is $R_x + 0.30\ R_x = 1.30\ R_x$.

The damping resistance is, of course, temporary or instantaneous, and does not contribute to the bearing capacity of the pile. It should also be noted that the constant J refers only to resistance at the point of the pile, such as R_{12} of Fig. 1.

Thus we see that resistance at the point of the pile is calculated to take account of the following:

1. Elastic ground compression or quake "Q."
2. Ultimate ground resistance "R_u."
3. Viscous damping based on a damping constant "J."

Resistance Alongside The Pile.—Resistance alongside the pile is calculated like resistance at the point of the pile, except that a damping constant J' is used instead of the damping constant J which has already been discussed in connection with resistance at the point. Thus in Fig. 1 J' would apply to re-

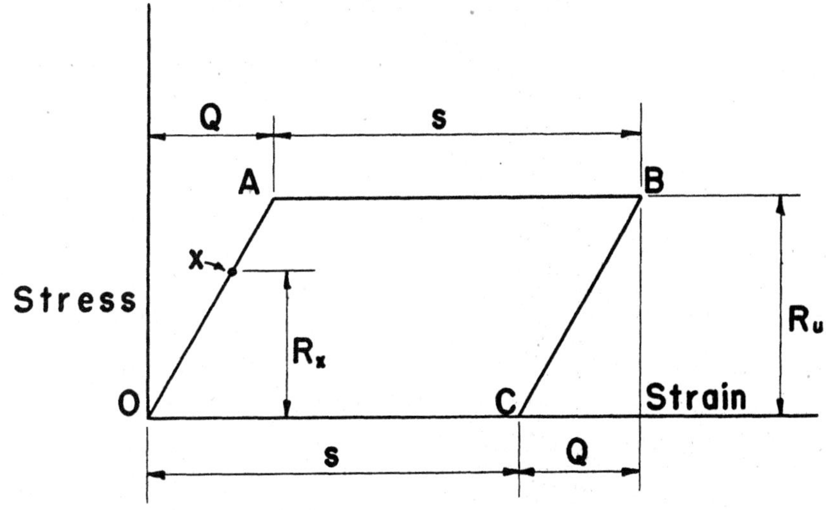

FIG. 4.—STRESS-STRAIN DIAGRAM AT PILE POINT

sistances R_3 to R_{11} inclusive, whereas J would apply only to the point resistance R_{12}.

As the pile is driven downward, the soil under the point of the pile is displaced or caused to flow aside very rapidly. However, the soil alongside the pile is not correspondingly displaced. The value of J' should therefore be smaller than the value of J. A value of J' = 0.05 will be assumed. (This compares with a value of 0.15 assumed for J).

In the numerical wave equation calculation the ground resistance may be distributed over the full length of the pile in any way that is desirable. For simplicity all resistance may be considered to be concentrated at the point of the pile, but the foundation engineer is at liberty to adopt any distribution that in his judgment is best suited to the specific ground conditions as disclosed by borings or other soil investigations.

Physics.—In addition to the action of the soil, we must consider the physical characteristics of the hammer, capblock, pile cap or follower, and pile. Precast concrete piles also require a cushion on the head of the pile. The physical characteristics of this cushion must also be considered. Each of these elements will now be discussed separately.

Hammer.—The hammer ram is ordinarily a short, heavy, rigid object that can be represented by a single weight without elasticity, such as W_1 of Fig. 1. In special cases where the ram is comparatively long and slender, or where an especially accurate stress analysis is required, it may be advisable to divide the ram into a number of weights and springs (as shown in Fig. 10(b) to be presented subsequently).

The velocity of the ram at the instant of impact is needed in order to start the numerical calculation. Ordinarily the rated foot pounds of energy of the hammer is given by the manufacturer. The efficiency is sometimes given and

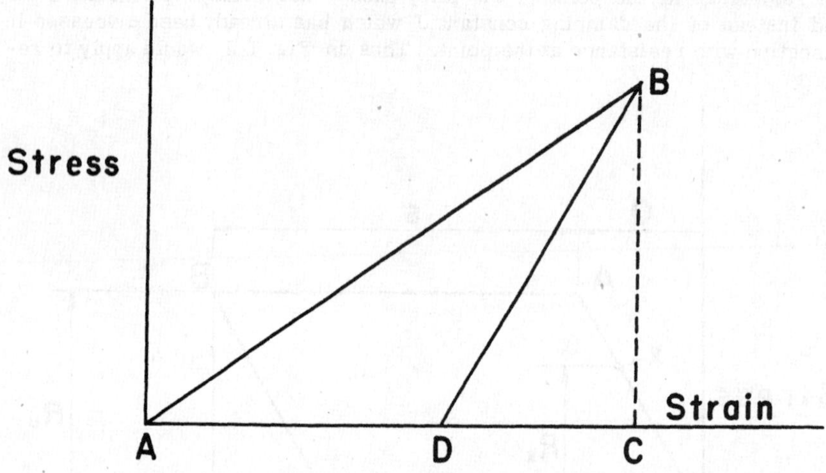

FIG. 5.—STRESS-STRAIN DIAGRAM FOR CAPBLOCK

sometimes must be assumed. From these data the velocity at impact may be calculated by means of the following formula:

$$\text{Velocity At Impact, in fps} = \sqrt{\frac{\text{Rated energy, in ft - lb} \times \text{Efficiency} \times 64.4}{\text{Weight, in lb}}} \quad . . (1)$$

Capblock (Termed "Dolly" in England).—The capblock is represented by spring K_1 in Fig. 1. The form of the stress-strain diagram (or the hysteresis loop) that is produced as the capblock is suddenly compressed and then allowed to reexpand, from information now available may be assumed to be as shown in Fig. 5.

Compression occurs along line AB whose slope is determined by the elastic constant K_1 of the capblock. Restitution occurs first along the line BD and then, because the capblock cannot transmit tension, is completed along line DA, thus forming the hysteresis loop ABDA. The slope of the line BD is automatically determined by the electronic computer so that

$$\frac{\text{Area BCD}}{\text{Area ABC}} = \left(e_1\right)^2 = \frac{\text{Energy output}}{\text{Energy input}} \quad . . . . (2)$$

where e_1 is the coefficient of restitution of the capblock. This is in accord with Newton's laws of impact.

Few tests have been made to determine the elastic characteristics of capblocks under impact conditions. The writer's former employer recently conducted a small number of tests by placing capblocks between two horizontally swinging rams weighing 4,800 lb each. The capblock was struck by one ram with about 15,000 ft-lb of energy, and the subsequent motion of both rams was recorded. A lead pellet was so mounted that it was squashed by the blow and thus measured the maximum compression of the capblock. From these measurements, the spring constant and the coefficient of restitution were determined mathematically.

It was found that the characteristics of a wood capblock vary during driving, but the tests led to the conclusion that in order to be on the conservative side in computing pile penetration per blow, a hardwood capblock with grain vertical, 6 in. in height originally, and with a horizontal area of A square inches, may be assumed to have the following characteristics:

Spring constant, K = 20,000 A lb per in. of compression
Coefficient of restitution, e = 50%

The tests also indicated that a Micarta capblock (Nema Grade "C") 12 in. in height has the following characteristics (which do not vary much during driving):

Spring constant, K = 45,000 A lb per in. of compression
Coefficient of restitution, e = 80%

Pile Cap or Follower or Helmet.—Like the ram of the hammer, the pile cap is ordinarily a short heavy rigid object that can be represented by a single weight without elasticity, such as W_2 of Fig. 1.

If the pile cap is long and slender, as is the case when it is to be used as a follower to drive the piles below ground or below water, then it may have to be represented by a number of weights and springs (as shown in Fig. 10(e), to be presented subsequently). In this case elastic constants must be computed. The elastic constants for any object of uniform cross section are computed by the well-known formula:

$$K = \frac{A\,E}{l} \quad \ldots\ldots\ldots\ldots\ldots\ldots\ldots \quad (3)$$

where A is the sectional area in square inches, E denotes the modulus of elasticity in pounds per square inch, and l is the unit length, in inches, represented by a single spring.

Cushion Blocks (Called "Head Packing" in England).—In Fig. 1, springs K_2 to K11, inclusive, represent the elasticity of the pile itself. However, if a precast concrete pile is being driven, a cushion block must be used under the pile cap W_2 so as to protect the concrete from shattering. In this case, spring K_2 in Fig. 1, would represent the cushion.

Fig. 5 applies to the cushion block as well as to the capblock, and computing methods for the cushion block and capblock are also alike.

Dynamic tests similar to those previously described for capblocks indicate that a wood cushion 4 in. thick composed of pine boards with grain horizontal, as used on top of a precast pile to distribute the blow evenly, may fairly be assumed to have the following characteristics during driving, where A is the area of the head of the pile:

Spring constant, K = 3,480 A lb per in. of compression
Coefficient of restitution, e = 50%

These characteristics vary widely during driving because the pine boards with grain horizontal become very hard and very elastic under continued pounding. The foregoing values are on the conservative side when used for computing pile penetration per blow.

Pile.—In Fig. 1, the ten springs with elastic constants K_2 to K_{11}, inclusive, represent the elasticity of the pile itself.

The experimental data available indicates small hysteresis loss in the pile itself, consequently, the springs representing the pile elasticity, such as K_2 to K_{11} of Fig. 1, may, with negligible error, be considered to be perfectly elastic. (Eq. 24, which is given in the Appendix, makes possible the inclusion of hysteresis loss in the pile itself if in the future this appears to be worthwhile.)

In Fig. 1, springs K_3 to K_{11} can transmit tension; but springs K_1 and K_2 cannot because the ram, the pile cap, and the pile are separate pieces. Therefore the electronic computer is programmed accordingly.

MATHEMATICS

The foregoing material outlines the problem and discusses the physical conditions that must be taken into account. The following gives the mathematics used for the numerical solution of the wave equation, as applied to pile driving.

In order to set up a program for the electronic computer, a number of equations and routines are required, as will be seen.

Notation.—The subscript m denotes the general case; thus W_m denotes any weight as in Fig. 1. For instance W_m might denote W_5, in which case K_m would denote K_5 and R_m would denote R_5, etc. Thus the letter m denotes position in space. The letter n is used to denote position in time.

The capital letters C, D, F, R, V, and Z, refer to instantaneous values for compression, displacement, force, resistance, velocity, and accelerating force, calculated for any time interval n. The corresponding small letters refer to corresponding values in the preceding time interval n-1. The notation d* refers to a displacement value in interval n-2 (two intervals back). Compressions and displacements are totals up to and including the time interval specified:

A = cross sectional area, in square inches;

C = spring compression in time interval n, in inches;

c = spring compression in time interval n-1, in inches;

D = displacement in time interval n, in inches;

d = displacement in time interval n-1, in inches;

d* = displacement in time interval n-2, in inches;

D' = ground plastic displacement in time interval n, in inches;

d' = ground plastic displacement in time interval n-1, in inches;

E = modulus of elasticity, in pounds per square inch;

e = coefficient of restitution;

F = force exerted by spring in time interval n, in pounds;

g = acceleration due to gravity, in feet per second per second;

J = damping constant applicable to resistance at point of pile (R_{12} of Fig. 1);

J' = damping constant applicable to resistance at side of pile (R_3 to R_{11} of Fig. 1);

K = spring constant, in pounds per inch;

K' = spring constant applicable to ground, in pounds per inch;

l = unit length of pile, in inches;

m = subscript denoting the general case;

n = time interval for which calculations are being made;

p = subscript denoting "at point of pile";

Q = quake or maximum elastic ground deformation, in inches;

R = resistance in time interval n, in pounds;

R_u = total ultimate ground resistance to driving, in pounds;

R_{u_m} = portion of R_u applicable to weight W_m;

s = permanent set of pile per blow, in inches;

T_m = critical time interval for spring m, in seconds;

Δt = time interval used for calculation, in seconds;

V = velocity in time interval n, in feet per second;

v = velocity in time interval n-1, in feet per second;

W = weight, in pounds; and

Z = accelerating force in time interval n, in pounds.

Equations.—In a previous paper[6] the writer derived the following five basic equations from the elementary laws of physics. They are based on the assumption that all springs are perfectly elastic, and that the pile is represented typically as shown in Fig. 1:

$$D_m = d_m + v_m (12\ \Delta t) \qquad (4)$$

$$C_m = D_m - D_{m+1} \qquad (5)$$

$$F_m = C_m K_m \qquad (6)$$

$$Z_m = F_{m-1} - F_m - R_m \qquad (7)$$

6 "Impact and Longitudinal Wave Transmission," by E. A. L. Smith, Transactions, ASME, August, 1955.

and

$$V_m = v_m + Z_m \frac{\Delta t\, g}{W_m} \quad \ldots\ldots\ldots\ldots\ldots\ldots (8)$$

Referring to Fig. 1, the velocity of any particular weight in any particular time interval, produces a displacement in the next time interval as given by Eq. 4. The displacements of two adjacent weights produce a compression in the spring between them as given by Eq. 5. This spring compression results in a spring force as given by Eq. 6. The two spring forces and the resistance acting on one particular weight produce a net force as given by Eq. 7. This net force accelerates or decelerates the weight and produces a new velocity as given by Eq. 8. This new velocity, in turn, produces a new displacement in the next succeeding time interval, etc. The process is repeated for each weight and each spring in each time interval, until all downward velocity is lost.

Subsequently these equations will be modified so as to provide for inelastic and plastic characteristics of the capblock, cushion block, and the ground.

If, by methods well known to mathematicians,[7] the wave equation, with resistance included, is converted into a difference equation suitable for numerical computation, the following expression is obtained:

$$D_m = 2\, d_m - d^*_m + \frac{12\, g\, \Delta t^2}{W_m}\left[\left(d_{m-1} - d_m\right)K_{m-1} - \left(d_m - d_{m+1}\right)K_m - R_m\right] \ldots (9)$$

Eq. 9 can also be obtained by combining Eqs. 4 through 8 into a single equation. This shows that these five basic equations are equivalent to the wave equation for purposes of numerical computation.

Eqs. 4 through 8 may be combined in numerous ways. For instance, Z_m can be eliminated by combining Eqs. 7 and 8 thus:

$$V_m = v_m + \left(F_{m-1} - F_m - R_m\right) \frac{\Delta t\, g}{W_m} \quad \ldots\ldots\ldots (10)$$

Similarly, C_m can be eliminated by combining Eqs. 5 and 6 thus:

$$F_m = \left(D_m - D_{m+1}\right) K_m \quad \ldots\ldots\ldots\ldots\ldots (11)$$

The ultimate ground resistance R_u, may be distributed throughout the length of the pile in any way desired by writing

$$R_u = R_{u_3} + R_{u_4} + R_{u_5} + \ldots \quad \ldots\ldots\ldots\ldots\ldots (12)$$

Any of these individual ultimate resistances may be denoted by the general expression R_{u_m} (Fig. 1).

In connection with Fig. 4, it was explained that the ground is assumed to compress elastically a distance Q, after which the ground fails plastically at constant resistance, which for the general case of weight W_m would be called

7 "Methods of Applied Mathematics," by F. B. Hildebrand, Prentice-Hall, Inc., New York, 1952, pp. 322-328.

R_{um}. For computation this may be represented diagrammatically as in Fig. 6.

In Fig. 6, springs K'_{m-1} and K'_m are introduced to represent ground elasticity. Spring K'_m, the general case, must compress a distance Q one, in order to reach the individual ultimate ground resistance R_{um}; therefore, to compute the spring constant for the ground we may write:

$$K'_m = \frac{R_{um}}{Q} \quad \text{..................(13)}$$

In Fig. 6, D_m is the displacement of weight W_m from its original position, as computed by Eq. 4, and D'_m is the measure of the plastic ground failure permitting movement in excess of Q. The amount that ground spring K'_m is compressed may be seen from Fig. 6, to be equal to $D_m - D'_m$. From this we may write an equation for R_m similar to Eq. 11:

$$R_m = \left(D_m - D'_m\right) K'_m \quad \text{..............(14)}$$

In order to include viscous damping as already discussed under the heading "General: Soil Mechanics," Eq. 14 is modified to include a damping constant J or J' multiplied by the instantaneous velocity. For frictional resistance alongside the pile, Eq. 14 thus becomes:

$$R_m = \left(D_m - D'_m\right) K'_m \left(1 + J' \, v_m\right) \text{..........(15)}$$

When applied specifically to resistance at the point of the pile this becomes:

$$R_p = \left(D_p - D'_p\right) K'_p \left(1 + J \, v_p\right) \text{............(16)}$$

Routines.—Evaluating ground displacements D'_m and D'_p as used in Eqs. 15 and 16 involves the use of computer routines rather than equations.

Routine #1.—The displacement D'_m remains constant (starting at zero) unless changed by one of the two following conditions:

1. D'_m cannot be less than $D_m - Q$
2. D'_m cannot be more than $D_m + Q$

The computer makes these comparisons in each time interval and adjusts the value of D'_m accordingly:

Remarks.—Routine #1 applies only to friction alongside the pile, and insures that the compression or extension of spring K'_m cannot exceed Q, the maximum elastic ground deformation. When the upper part of the pile rebounds after the blow, the motion reverses itself; consequently, routine #1 provides for plastic movement D'_m either in the normal downward direction or in an upward direction resulting from pile rebound. This routine is illustrated in Fig. 7.

Routine #2.—The displacement D'_p remains constant (starting at zero) unless changed by the following condition:

D'_p cannot be less than $D_p - Q$

The computer makes this comparison in each time interval and adjusts the value of D'_p accordingly.

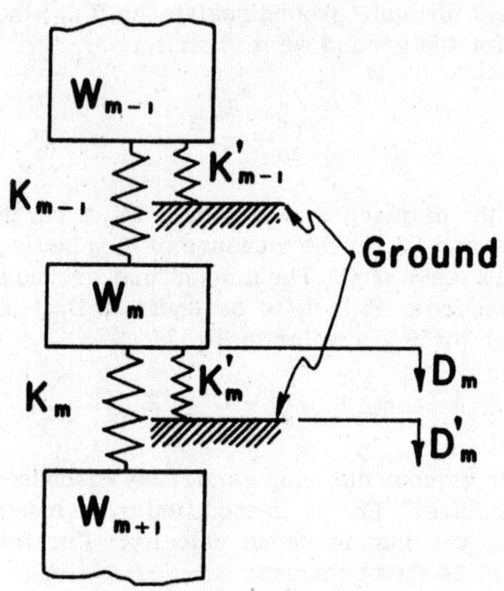

FIG. 6.—GROUND ELASTICITY

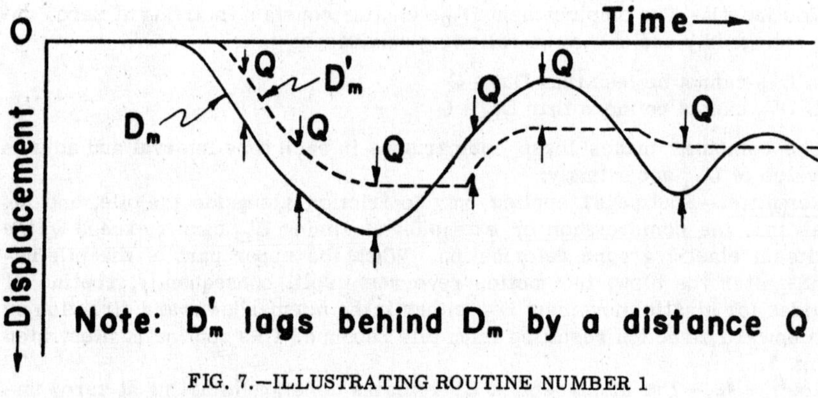

FIG. 7.—ILLUSTRATING ROUTINE NUMBER 1

Remarks.—Routine #2 applies only to the point of the pile. It is similar to routine #1 except that only downward or positive plastic ground movement D'_p is considered. The maximum value of D'_p is the "permanent set" of the pile, which is called s, as shown in Fig. 4. Therefore, it would be illogical to consider reversal of plastic failure of the ground as was done in routine #1. This routine is illustrated in Fig. 8.

Routine #3.—This routine provides for inelasticity in spring K_1 by taking account of its coefficient of restitution e_1, and involves the alternate use of Eqs. 17 and 18 as follows:

$$F_1 = C_1 K_1 \quad \text{.....................(17)}$$

for compression. This equation is identified with the number 17 merely as a matter of convenience. Actually it is nothing but Eq. 6 applied to spring K_1. For restitution

$$F_1 = \left[\frac{K_1}{(e_1)^2}\right] C_1 - \left[\frac{1}{(e_1)^2} - 1\right] K_1 C_{1\,max} \quad \text{........(18)}$$

in which e_1 is the coefficient of restitution for spring K_1 and $C_{1\,max}$ is the maximum value of C_1, used as a constant.

It should be noted that if $e_1 = 1.00$, then Eqs. 17 and 18 become identical and equivalent to Eq. 6.

Eq. 18 is used for restitution. Its derivation is based on the relationships previously explained in connection with Fig. 5, and is not difficult.

Routine #3 is based on a stress-strain (or force-compression) diagram as given in Fig. 9. Letters (a), (b), etc., in Fig. 9, correspond to the correspondingly numbered steps in the routine, as given subsequently.

The actual routine is as follows:

(a) Use Eq. 17 until $C_1 - c_1$ becomes negative. This last value of c_1 is thereafter treated as a constant called C_{1max} for use in Eq. 18.

(b) Use Eq. 18 instead of Eq. 17.

(c) Sometimes recompression occurs; nevertheless, continue using Eq. 18 until $C_1 - C_{1max}$ becomes positive.

(d) Then again use Eq. 17 until $C_1 - c_1$ again becomes negative. This gives a new value for C_{1max} equal to the latest value of c_1.

(e) Then use Eq. 18 with the new value of C_{1max} as a constant. (If additional recompressions occur, steps (c), (d), and (e) are repeated).

(f) F_1 can never be less than zero, that is, can never be negative, because the ram is always a separate piece.

Routine #4.—This routine provides for inelasticity in spring K_2 by taking account of its coefficient of restitution e_2. It uses two equations corresponding to Eqs. 17 and 18, thus:

$$F_2 = C_2 K_2 \quad \text{...................(19)}$$

for compression, and

$$F_2 = \left[\frac{K_2}{(e_2)^2}\right] C_2 - \left[\frac{1}{(e_2)^2} - 1\right] K_2 \, C_{2max} \quad \ldots\ldots\ldots (20)$$

for restitution.

Note that if the coefficient of restitution e_2 = 1.00 Eqs. 19 and 20 become identical and equivalent to Eq. 6.

The routine is the same as routine #3, with subscripts changed from 1 to 2, with the one exception of item (f) of routine 3. For routine #4, item (f) reads as follows:

(f) Alternate No. 1: F_2 cannot be less than zero (that is, cannot be negative).

(f) Alternate No. 2: F_2 can be less than zero (that is, can be negative).

The proper alternate must be specified. Alternate No. 1 applies if K_2 cannot transmit tension. Alternate No. 2 applies if K_2 can transmit tension.

PROGRAM FOR ELECTRONIC DIGITAL COMPUTER

By combining the foregoing formulas and routines a complete program for the computer can be prepared.

The calculation starts with a specified ram velocity V_1 at the "beginning of impact" denoted as time interval 0. All other variables except V_1 start with a value of zero in time interval 0, and remain at zero until modified by one of the following items bearing Roman numerals, which comprise the computer program. These items are listed in the order in which they are calculated by the computer in each and every time interval successively. Time interval #1 is the first time interval for which computations are made by the electronic computer.

Computer Program #1 For Piles With All Resistance At The Point Of The Pile

I Compute D_1 to D_p by means of Eq. 4.

II Compute D'_p by means of routine #2.

III Compute R_p by means of Eq. 16.

IV Compute C_1 to C_{p-1} by means of Eq. 5.

V Compute F_1 by means of routine #3.

VI Compute F_2 by means of routine #4.

VII Compute F_3 to F_{p-1} by means of Eq. 6.

VIII Compute V_1 to V_p by means of Eq. 10.

Computer Program #2 For Piles With Slide Friction As Well As Point Resistance

I to VII inclusive, same as above.

VIII Compute D'_3 to D'_{p-1} by means of routine #1.

IX Compute R_3 to R_{p-1} by means of Eq. 15.

X Compute V_1 to V_p by means of Eq. 10.

In the above programs Eq. 10 has been used instead of Eqs. 7 and 8 because the accelerating force Z is seldom of interest and the use of Eq. 10 saves computer time.

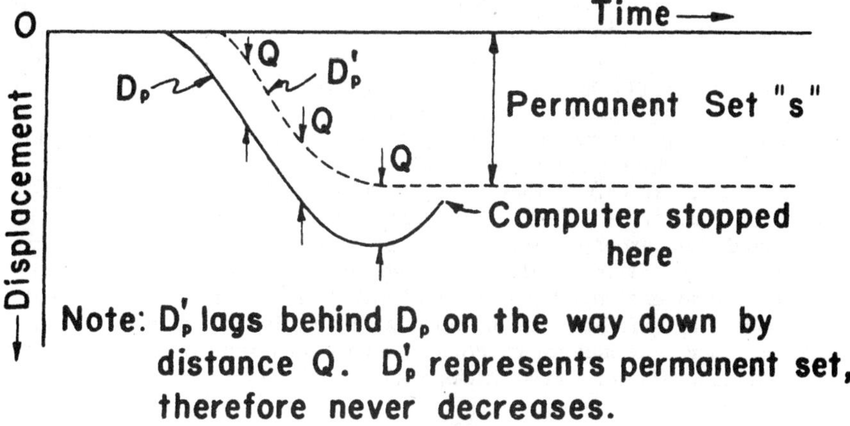

FIG. 8.—ILLUSTRATING ROUTINE NUMBER 2

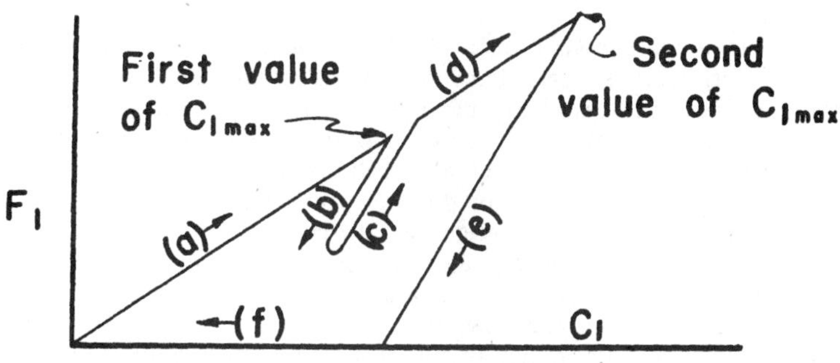

FIG. 9.—STRESS-STRAIN DIAGRAM FOR SPRING K_1

The computer should be programmed to stop automatically when the following two conditions are reached:

(1) $D'_p - d'_p$ equals zero.
(2) V_1 to V_p inclusive, are all simultaneously negative or equal to zero.

This will ordinarily stop the calculation very soon after D'_p reaches a maximum and the point of the pile begins to rebound as indicated in Fig. 8. By this time the major driving force has been expended and only secondary or residual forces are acting. These are of little interest and are difficult to analyze accurately.

RECOMMENDATIONS AND COMMENTS

The mathematical method explained herein was originally developed as a means of computing stresses during driving in capblocks, pile caps, piles, and pile-driving cores or mandrels, in order to insure proper design. Previously all these parts had been designed by trial and error.

During the past 12 yr this mathematical method has been used for over 1,000 calculations, and has been gradually improved and extended to cover pile driving in general. It is now being presented with the hope that it will be useful in determining stresses during driving, and especially stresses in precast concrete piles, so that these piles may be provided with correct reinforcement and correct driving procedures in order to prevent or eliminate cracking or breakage during driving. It is also hoped that in due time the results of many calculations will be correlated with the results of pile-load tests carried to failure, and also with the rapidly increasing knowledge of soil mechanics.

In order to make calculations at all, values must be assigned to certain constants that describe soil action during driving. Up to the present time no instrumented field experiments have been performed to determine these constants accurately. From experience gained through working extensively with this problem, and as a result of making a limited number of comparisons with load tests carried to failure, the writer has personally concluded that certain values are accurate enough for practical use until such time as more accurate values become available. These values are

$$Q = 0.10 \text{ in.}$$
$$J = 0.15$$
$$J' = 0.05$$

The writer is emboldened to make the foregoing recommendations (which may easily be criticized as rash) because he has found from the many calculations already performed that the numerical wave equation solution is not "sensitive," that is, a small change in the value assigned to any constant will produce a smaller change in the calculated results. He therefore believes that the results of calculations are meaningful and worthwhile even though the constants used are not as accurate as might be desired.

To determine a suitable value for R_u to be used in a calculation, the working load to be placed on the pile must be multiplied by a factor such as 2 or 3. However, if soil investigation indicates that the soil is of a type that "sets up" or one that "relaxes" after driving, this factor should be modified accordingly. The numerical wave equation calculation tells what happens during the actual driving of the pile; it cannot predict what will happen a week or a year later. For this information soil mechanics must be consulted.

Modern large electronic computers can complete a calculation such as described in this paper, in a matter of seconds. It is therefore practical to make a large number of calculations and to tabulate (or plot) the results for future reference. Such a tabulation would cover the types of piles and hammers in common use. Special calculations would then need to be made only for unusual types.

The following conclusions come partly from the results of numerical wave equation calculations, but primarily from many years of practical pile-driving experience.

CONCLUSIONS

1. A numerical method suitable for use on modern electronic computers is now available that permits calculation of pile-driving action under any specified set of conditions, and gives permanent set per blow as well as instantaneous stresses, displacements, and velocities.

2. The knowledge of soil mechanics is incomplete, especially the knowledge of soil mechanics under pile-driving action. This offers a fertile field for future investigation, especially now that a mathematical method of calculating driving action is available for checking and analyzing field test results.

3. The knowledge of the physical action of capblocks and cushion blocks under driving conditions could also be improved by dynamic testing. Only a few tests have been made.

4. The numerical wave equation calculation can be used to determine the driving characteristics of various types of piles and hammers. It can also be used to determine the range of application throughout which any particular pile-driving formula may be considered reasonably accurate.

5. It is well known that precast concrete piles are sometimes cracked or broken due to excessive compressive stresses caused by the hammer blow. It is not equally well known that in certain cases excessive tensile stresses may also result from the hammer blow, especially if the pile is long. A wave-equation analysis can be used to determine correct reinforcement and driving procedures to help eliminate these troubles.

6. Many engineers assume that at final penetration, a light pile is easier to drive than a heavy pile of the same overall dimensions. Actually this assumption is correct only for light loading. When loads are heavy and consequently resistance is high, a heavy pile is usually easier to drive than a light pile, because it is stiff and strong and thus is better able to carry large forces down into the ground.

APPENDIX

This appendix is presented primarily for those who actually make calculations, and gives practical details that will be found helpful.

In the writer's paper,[6] a recommendation was made that, in representing the pile, the weight of each unit length should be concentrated at the center. Experience since that paper was written has shown that a preferable procedure is to concentrate the weight of each unit length at the end away from the place where impact occurs, as was done in Fig. 1. This procedure improves accuracy and permits any combination of pile, hammer, etc., to be represented logically as shown in Fig. 10.

The elements shown in Fig. 10 are combined in the following typical ways to form single diagrams similar to Fig. 1:

1. Ram, capblock, pile cap and pile.
2. Ram, capblock and core or mandrel.
3. Ram and pile.
4. Ram, capblock, long follower and pile.
5. Ram, capblock, pile cap, cushion and pile. In this case the spring representing the cushion must be combined with the top spring of the pile by means of Kirchhoff's law

$$\frac{1}{K} = \frac{1}{k_1} + \frac{1}{k_2} \qquad (21)$$

6. Long ram, capblock, pile cap and pile. In this case the spring representing the capblock must be combined with the bottom spring of the ram. (See #5 above).

7. Ram, pile cap or anvil and pile. In this case combine the weights representing the ram and the pile cap, and assign a velocity at impact so that the momentum of the combined weight equals the momentum of the ram at impact.

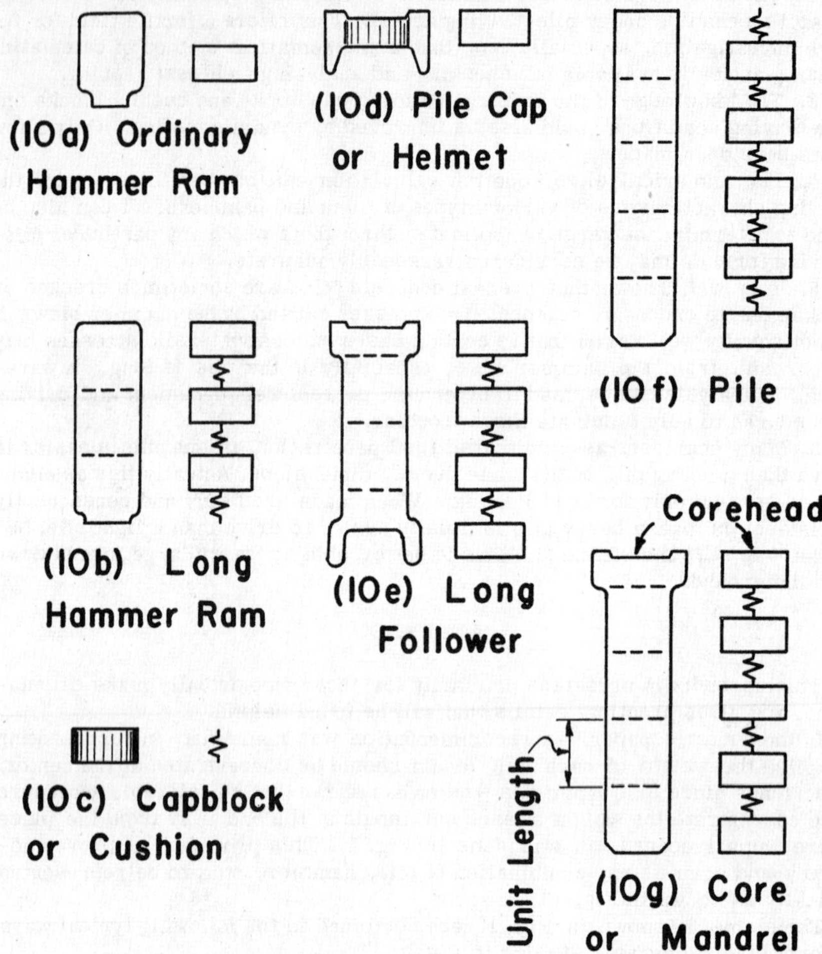

FIG. 10.—REPRESENTATION OF PILE AND DRIVING ELEMENTS

The best time interval Δt, for use in making a pile calculation, may be defined as the largest interval that will produce a completely stable calculation. No simple rule can be given that will cover every possible condition, but the writer has discussed this question and has presented two formulas for the "critical" time interval based on the velocity of travel of the stress-wave as follows: [6]

$$T_m = \frac{1}{1.9648}\sqrt{\frac{W_{m+1}}{K_m}} \quad \text{..............(22)}$$

and

$$T_m = \frac{1}{1.9648}\sqrt{\frac{W_m}{K_m}} \quad \text{...............(23)}$$

The minimum value of T_m that can be obtained by using these formulas is the "critical" one.

The time interval Δt should be about half of this critical value so as to prevent instability from arising due to other factors not included in the above formulas, such as the effects produced by coefficients of restitution e, ground quake Q and damping constants J and J'. As a general rule, the following values for Δt, will be found to be satisfactory for use with piles divided into 8 ft or 10 ft unit lengths:

For steel piles	1/4,000 sec
For wood piles	1/4,000 sec
For concrete piles	1/3,000 sec

If shorter unit lengths are used, the time interval Δt should be reduced proportionately.

An interesting and instructive combination of figures for use with a steel pile of uniform section is to take l = 100 in. (8.33 ft), Δt = 1/4,000 sec, and E = 29,300,000 psi. Sound or stress waves travel at a speed equal to $\sqrt{E/\rho}$ where ρ is the mass per unit volume. Steel weighs 0.283 lb per cu in. and the velocity of sound is 16,660 fps. Thus sound or stress waves will traverse an 8.33 ft unit length in 1/2,000 sec. If 1/4,000 sec is used for Δt, forces will progress from one pile spring to the next (such as from spring K_4 to K_5 of Fig. 1) every second time interval. Similarly velocities will progress from one pile weight to the next (such as W_4 to W_5 of Fig. 1) every second time interval. This may be observed from the numerical wave equation calculation, though the figures will not come out absolutely exact. However, when the calculation has been carried on long enough so that the stress wave is reflected from the point of the pile, the wave action may become harder to observe from the calculated figures, though sometimes the travel of the reflected wave up the pile is easy to follow. If ground friction alongside the pile is included in the calculation, the sound or stress wave will travel at a slower speed.

If the time interval used is too large, instability in the calculation may result. Usually this will force the computer to handle such large numbers that it will give warning. A good plan is to program the computer so that it will give warning if the value of V_2 or V_p, in any time interval, exceeds twice the velocity of the ram at the moment of impact (that is, twice V_1 in time interval 0). This will usually detect instability promptly. It is also a good plan to program the computer to stop the calculation after a maximum of 300 time intervals. This will eliminate any possibility of the computer running indefinitely if the regular program does not stop the calculation at the correct point, as in Fig. 8.

As a check, more than one calculation should be made for any particular pile and hammer combination using varying values of R_u. If four or five such calculations are made, the calculated values of s should plot as a smooth curve. If they do not, it may be concluded that instability was present, or that some error has been made in the calculation. Other methods of checking have been given by the writer.[6]

In rare cases the capblock spring K_1 may be very stiff and the weight W_2 may be very light. This combination tends to produce instability, and if tested according to Eqs. 22 and 23, may call for an excessively small time interval. From a practical standpoint it may be preferable to arbitrarily soften spring K_1 somewhat, or to arbitrarily increase the weight of W_2. Either or both of these changes will permit use of a normal time interval such as 1/4,000 sec, will have only a small effect, and will give results on the safe or conservative side.

It should be noted that if the hammer ram is long and is represented by a number of weights and springs as shown in Fig. 10 (b), then all these weights will have equal initial velocities V_1, V_2, V_3, etc., and the computer must be specially programmed to accept all these velocities as input data.

It should also be noted that the methods described herein may be used to analyze the driving of a pile in which the ram strikes its blow part way down in the pile, or all the way down at the point. Fig. 11 shows, diagrammatically, a pipe pile being driven by a drop hammer operating inside the pile and striking on the point. All equations already given apply to Fig. 11 as well as to Fig. 1 except that the compression of spring K_1 must be computed by the equation

$$C_1 = D_1 - D_{12} \quad \ldots\ldots\ldots\ldots\ldots\ldots (24)$$

the accelerating force for weight W_{12} must be computed by the equation

$$Z_{12} = F_1 + F_{11} - R_{12} \quad \ldots\ldots\ldots\ldots\ldots (25)$$

and the accelerating force for weight W_2 by the equation

$$Z_2 = -F_2 = R_2 \quad \ldots\ldots\ldots\ldots\ldots\ldots (26)$$

These changes are obvious from Fig. 11, and of course require special computer programming.

It is of great interest to note that driving on the point of the pile in this way has been found to be no more effective than driving on the upper end of the pipe. The explanation appears to be that driving on the point of the pile tends to produce high point velocity, which causes high temporary resistance because of soil viscosity. This illustrates the necessity of introducing the damping constant J.

Some types of ground furnish side frictional resistance during driving, but add little or nothing to the permanent bearing capacity of the pile. In this case side resistances may be used in the calculation but ignored in determining the bearing capacity of the pile. Thus in Eq. 12 R_{u_3}, R_{u_4}, etc., would be given values, but only R_{u_p} would be considered to be permanent bearing capacity.

The wrtier has also presented[6] formulas for computing K-values for tapered piles.

It is usual to ignore gravity forces because the pile-bearing capacity wanted is the bearing capacity that the pile has in addition to its own weight. Anyone who wishes to include gravity will find methods of doing so discussed elsewhere by the writer.[6]

Illustrative Problem.—Let Fig. 1 represent a steel pipe or H pile 100 ft long with 15.58 sq in. of area, weighing 53 lb per ft, driven by a #1 Vulcan hammer

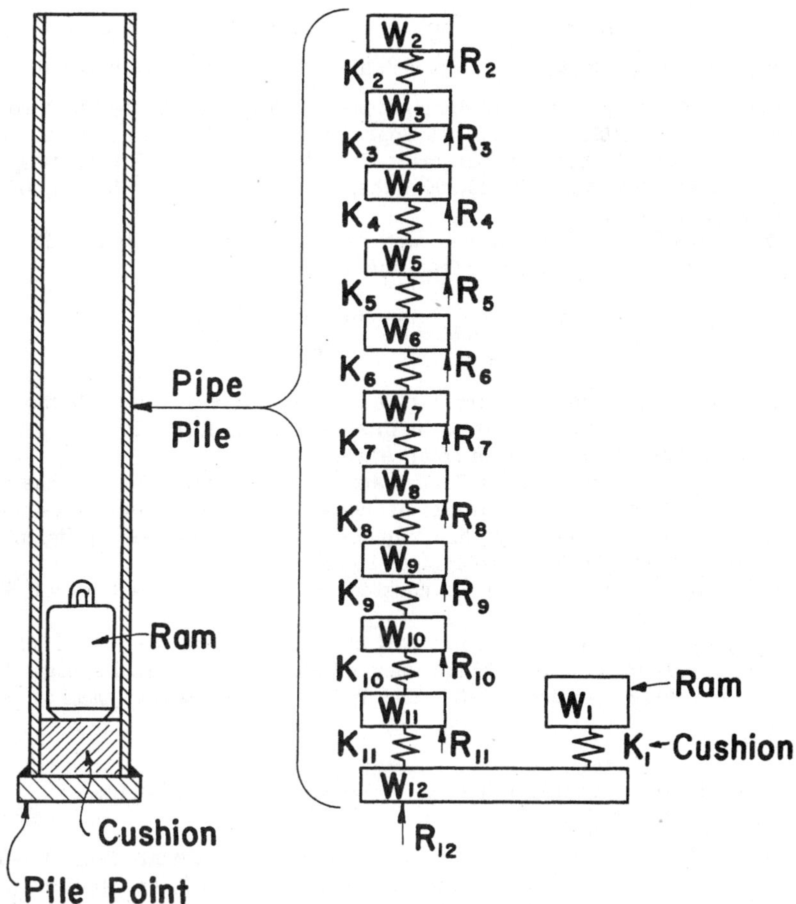

FIG. 11.—PIPE PILE

having a 5,000 lb ram falling 3 ft, using a 700 lb pile cap and a hardwood cap-block 6 in. thick when new and 11-1/2 in. in diameter. A special pile point is specified weighing 100 lb.

Assuming that hammer efficiency is 80%, ground quake Q is 0.1 in., and damping constant J = 0.15, determine the permanent set per blow when driving against an ultimate ground resistance R_u of 200,000 lb concentrated at the point of the pile.

Step No. 1, Choice of Unit Length.—A 10 ft unit length will be adopted. Experience has shown this unit length to be generally satisfactory.

Step No. 2, Choice of Time Interval.—As a general rule, a time interval $\Delta t = 1/4{,}000$ sec will produce a stable calculation using 10 ft unit lengths, and this will be used here.

Step No. 3, Determination of Weights W_m.—

W_1 = the hammer ram = 5,000 lb;

W_2 = the pile cap = 700 lb;

W_3 to W_{11} = 10 ft unit lengths of pile at 53 lb per ft = 530 lb each;

W_{12} includes a special pile point weighing 100 lb, therefore W_{12} = 630 lb.

Step No. 4, Determination of Spring Constants K_m and K'_m.—The term K_1 represents the capblock. This has approximately 100 sq in. of area, and it was stated under the heading "General: Physics: Capblock," that the spring constant for such a capblock = 20,000 A, therefore $K_1 = 20{,}000 \times 100 = 2{,}000{,}000$ lb per in. of compression.

The terms K_2 to K_{11} inclusive, each represent the elasticity of a 10 ft unit length of pile. This is computed from the formula:

$$K = \frac{A\,E}{l} = 15.58 \times \frac{30{,}000{,}000}{10 \times 12} = 3{,}900{,}000$$

Since all resistance is assumed to be at the point of the pile, all values of K'_m except K'_p are equal to zero. In this case K'_p is K'_{12}. From Eq. 13 we get K'_{12} = 2,000,000 lb per in. of compression.

Step No. 5, Determination of Coefficient of Restitution.—The pile has been assumed to be perfectly elastic as discussed under the heading "General: Physics: Pile," consequently, the computer program requires coefficients of restitution only for the first two springs K_1, and K_2.

Spring K_1 represents the wood capblock for which a value of $e_1 = 0.50$ is suitable.

Spring K_2 in this case represents part of the pile, therefore, $e_2 = 1.00$.

Step No. 6, Determination of Ram Impact Velocity.—The rated energy of the hammer is $5{,}000 \times 3 = 15{,}000$ ft-lb. The impact velocity is computed by using Eq. 1:

$$\text{Impact velocity, in fps} = \sqrt{\frac{15{,}000 \times .80 \times 64.4}{5{,}000}} = 12.4 \text{ fps.}$$

Step No. 7, Decide Whether or Not Spring K_2 Can Transmit Tension.—The term W_2 represents the pile cap which merely rests loosely on top of the pile, therefore spring K_2 cannot transmit tension, therefore F_2 can never have a negative value and its minimum value will be 0.

Step No. 8, Summary.—The foregoing values may be summarized for computer input as follows:

$\Delta t = 1/4{,}000$

$W_1 = 5{,}000$, $W_2 = 700$, W_3 to W_{11} = 530 each, $W_{12} = 630$

$K_1 = 2{,}000{,}000$, K_2 to K_{11} = 3,900,000 each, $K'_{12} = 2{,}000{,}000$

$e_1 = 0.50$ $e_2 = 1.00$

Initial value of V_1 = 12.4

Q = 0.1

J = 0.15

F_2 can be equal to or greater than zero, but can never be less than zero.

If computer program #1 is being used the foregoing constitutes the required input data. If computer program #2 is being used, the following must also be specified:

$$K'_1 \text{ to } K'_{11} = 0$$

$$J' = 0$$

Step No. 9, Computation.—The values listed in step 6 are used as computer input for a computer that has been programmed in accordance with Program #1 or #2.

This has already been done for this particular problem. The permanent set s was computed to be 0.20311 in. per blow. (5 blows per in.) The computer was programmed to print maximum compressive and tensile forces in the pile. These were found to be as follows:

		Maximum Forces			
		Compression			Tension
At head of pile	(F_2) =	290,000 lb	=	18,600 psi	0
At middle of pile	(F_6) =	300,000 lb	=	19,250 psi	0
At point of pile	(F_{11}) =	405,000 lb	=	26,000 psi	0

Forces such as these are primarily of interest as a means of determining unit stresses during driving in the case of precast and prestressed concrete piles.

In the above example tension did not occur because R_u = 100 tons which is a moderately high value. Tension in piles occurs primarily during the early stages of driving when R_u is quite small.

The compressive stress at the point of the pile is higher than at the head because the fairly high value of R_u produces a reflected compressive wave which travels back up to the head of the pile. This upward wave overlaps the original downward traveling compressive wave and thus increases the stress at the point of the pile.

The calculated displacements D_1 to D_{12}, which apply to weights W_1 to W_{12}, as well as the plastic ground displacement D'_p are plotted in Fig. 12. Each dot on each curve represents a computed point. Fig. 12 gives a lot of information and is worthy of study.

For The Future.—Future investigators with more complete experimental data available may consider it worthwhile to include hysteresis (internal damping) in the pile as part of the calculation. This can be done by modifying Eq. 6 to read as follows

$$F_m = C_m K_m + B K_m \left(\frac{C_m - c_m}{12 \Delta t}\right) \quad \text{. . . . . (27)}$$

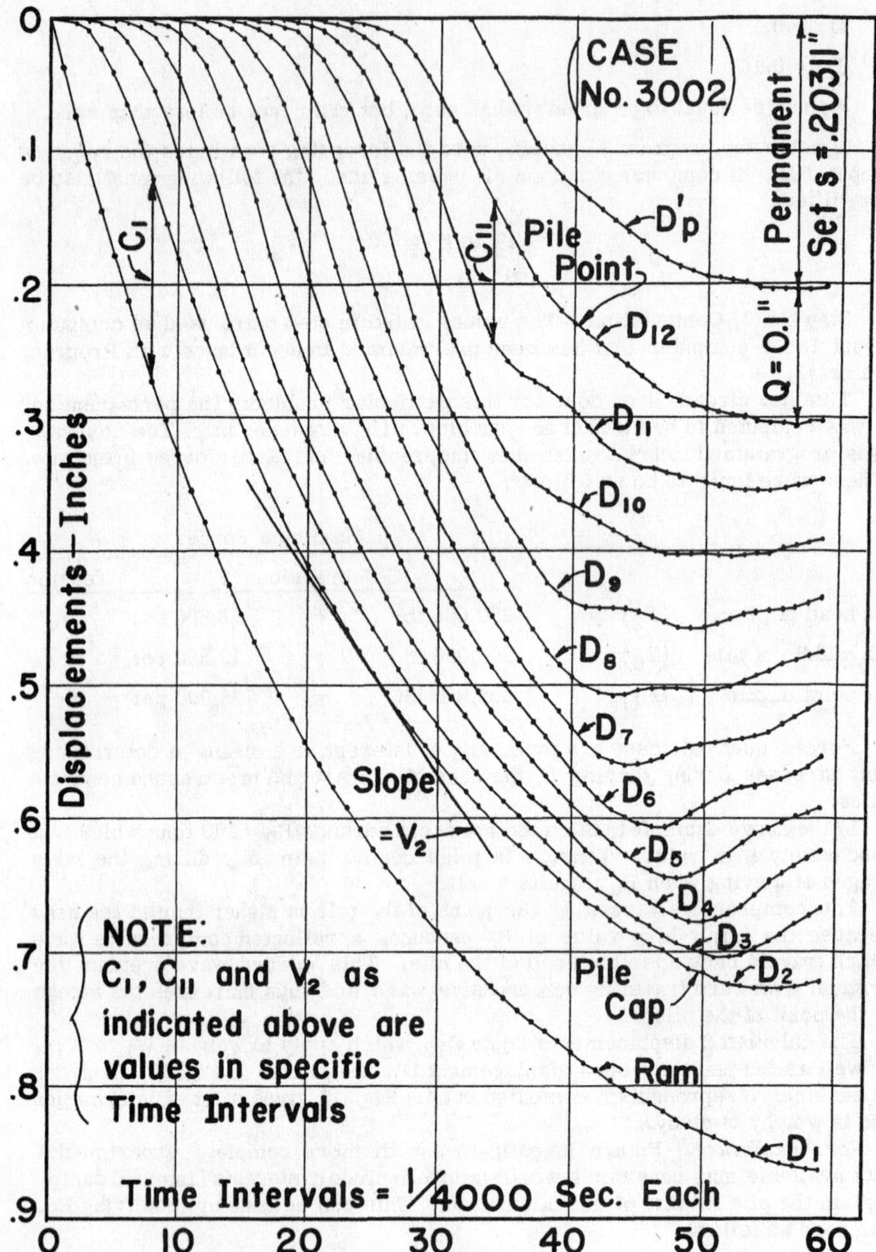

FIG. 12.—ILLUSTRATIVE PROBLEM

where B is a hysteresis or internal damping constant. A small value (such as 0.0002) should be assigned to B so as to produce a narrow hysteresis loop. Actually, the "loop" is a sort of spiral starting at the origin and returning to the origin when both C_m and $C_m - c_m$ become equal to zero.

The expression $\left(\frac{C_m - c_m}{12\,\Delta t}\right)$ in Eq. 24 is the instantaneous rate of spring compression. This corresponds in a general way to v_m and v_p in Eqs. 15, 16, 25, and 26.

A recommendation has been given previously to the effect that the calculation be stopped soon after maximum pile point penetration is reached as indicated in Fig. 8 and actually done in Fig. 12. Future investigators may want to continue beyond this point. If so, they should take note of the fact that Eqs. 15 and 16 give no damping at all when $(D_m - D'_m)$ becomes zero, even though the velocity may be considerable. To overcome this difficulty, Eqs. 15 and 16 can be used until $(D_m - D'_m)$ or $(D_p - D'_p)$ is first equal to Q, which corresponds to point A of Fig. 4. From this point on the following equations may be used instead of Eqs. 15 and 16, respectively:

$$R_m = \left(D_m - D'_m\right) K'_m + J' K'_m Q v_m \quad \ldots\ldots\ldots\ldots (28)$$

and

$$R_p = \left(D_p - D'_p\right) K'_p + J K'_p Q v_p \quad \ldots\ldots\ldots\ldots (29)$$

These two equations produce damping at all times except when v_m (or v_p) is equal to zero. Changing equations in this way is similar to the use of two different equations in routines #3 and #4.

It is also possible that Eq. 24 may in the future be found suitable for use with the capblock and cushion block, and thus replace routines #3 and #4. Before this can be done, suitable values for B in Eq. 24 would have to be determined experimentally.

PAST AND FUTURE OF APPLIED SOIL MECHANICS

By Karl Terzaghi,* *Honorary Member*

(Presented at the annual convention of the ASCE, held in Boston at which Professor Terzaghi was presented the first copy of the book "From Theory to Practice in Soil Mechanics.")

Introduction

I must confess that I have already had a sneak preview of the volume and I was much impressed by the splendid organization of the contents and the esthetic qualities of the book. Hence I wish to express my gratitude to the editors as well as to the publishers, John Wiley and Sons, Inc., for their successful effort to present the essence of the fruits of my labors in a beautiful nutshell.

Once I have published a paper I hardly look at it again, because my thoughts and efforts are already concentrated on other subjects and these are scattered over a large area. Therefore perusing this volume was quite an experience for me. It revealed the history of my gropings through the dark to a clearer understanding of the essential requirements of practicing my chosen profession. Since every one of us has to pass through a painful period of trial and error before he becomes a master in his own house, the story of my own endeavors may be a source of encouragement to those who are still at the foot of the slope.

The Old Code

When I began the ascent, at the beginning of this century, I was equipped with a set of axioms which were then believed, at least by those who taught them to a flock of eager students, to be gospel truth. These were the rules:

(a) At a given load per unit of area the settlement of a spread footing is independent of the area covered by the footing.

(b) The settlement of a pile foundation is equal to the settlement of an individual pile under the same load per pile.

(c) The constants in Coulomb's equation for the shearing resistance of cohesive soils are independent of time.

(d) The earth pressure on lateral supports is independent of the amount of lateral yield of the support.

* Professor of the Practice of Civil Engineering, Emeritus, Harvard University.

(e) The influence of the presence of water on the shearing resistance of soils is caused by the lubricating effect of the water.

Design in earthwork engineering on the basis of these rules required practically no mental effort, but the results were often disappointing. This I found out at an early stage of my professional career, whereupon I started to worry. The following are a few of the incidents which attracted my attention.

In 1907, I designed my first footing foundation, using the rules and procedures which I had been taught; the walls of the structure cracked badly and the case was brought before the courts. Fortunately the misbehavior of the foundation was classified, also in accordance with the prevalent conceptions of those days, as an "Act of God."

In the same year my employer designed and built a low concrete diversion dam with concrete apron in a narrow rock valley in the eastern Alps. The dam was founded on a thick stratum of sand and gravel containing scattered boulders. The textbooks informed their readers that structures of this kind should be protected against piping failure by driving a row of sheetpiles along both the upstream and downstream edge of the base of the structure, but nothing was said about the required depth of penetration. Therefore the piles were simply driven to refusal which was met at some points at a rather shallow depth. When the reservoir was filled for the first time the structure failed by piping and the remnants were buried in sand and gravel. Fortunately this accident too was ascribed by the arbitrator to an "Act of God."

In 1913 the subsoil exploration for the design of the foundations of the new M.I.T. buildings in Cambridge, Mass., was started. According to a very competent geological report prepared by Prof. W. O. Crosby, the bedrock was encountered at a depth of 120 to 135 ft below the ground surface. The materials overlying the bedrock are in succession: boulder clay with a thickness of a few feet to 25 or 30 ft, blue clay with a normal thickness of 80 to 100 ft, glacial sand and gravel from zero to 35 ft, and loose overburden with a thickness of about 20 ft. It was decided to establish the foundations wherever possible on point-bearing piles driven to refusal into the glacial gravel overlying the clay. Where no such refusal could be obtained, the piles were driven through the gravel to a maximum of 50 ft into the clay.

In order to obtain what was expected to be a rational basis for the design of the pile foundation, 80 test piles were driven and on 38 of them complete pile loading tests were performed (Maine and Sawtelle, 1918). On the basis of the test results 13 rather elaborate rules were set up for the purpose of adapting the length of the piles to the variable subsoil conditions and the design load per pile was selected in such a manner that the anticipated settlement would nowhere exceed 1/16 inch.

At the time when these tests were performed, I was associated with the design and construction of pile foundations in Portland, Ore. I already had vague misgivings concerning the soundness of the current procedures regarding the design of such foundations, but I would still have been unable to raise any valid objections to the procedures which were used at the site of M.I.T. In 1925, when I gave the first lectures on soil mechanics on the American continent at M.I.T., the settlement of the library dome of the Institute had passed the 7-inch mark, and was still increasing at a rate of about ½ inch per year, but all I could do at that stage was to offer a rational explanation of what had happened.

Equally unsatisfactory were the conditions at the beginning of the twentieth century in the field of earth-dam design and construction. Earth dams with a height up to 70 ft had already been built in India before the beginning of the Christian Era. During the second half of the 19th century the record height of earth dams gradually increased to 125 ft, but there is hardly one volume of Engineering News or Engineering Record published in those days which does not contain an account of the failure of at least one of these structures. Therefore, as late as 1901 the Board of Consultants of the New York Water Supply arrived at the conclusion that the construction of an earth dam with a height of more than 70 ft cannot be considered advisable. This conclusion was fully justified because at the time when it was made there were not even any generally accepted empirical rules for the design and construction of earth dams. Some engineers believed that the upstream slope of homogeneous earth dams should be flatter than the downstream slope and others felt equally justified in making it steeper. Some engineers insisted on compacting the dam construction materials whereas others insisted that the material should be placed by dumping it into shallow pools to be maintained on the working surface.

The reason for the absence of agreement between engineers concerning the design and construction of earth dams is quite obvious. The relationship between water content, degree of compaction and shearing resistance was still unknown. Hence it was impossible to estimate the factor of safety of the slopes of earth dams with respect to sliding. As a consequence, if a dam failed, it was also impracticable to determine the cause of the failure and the failure was classified as an "Act of God."

One of the most impressive manifestations of the rudimentary state of our knowledge in this field in relatively recent times is a published account of the performance of three homogeneous earth dams in the former Netherlands Indies (Van Es, 1933). Two of the dams failed during construction, when the fill had attained a height of about 33 ft whereas the third one was successfully completed to its final height of 100 ft and has remained stable ever since. The difference between the performance of these dams was ascribed to a difference between the index properties of the construction materials and the placement water content did not receive any attention.

Soil Mechanics Takes Over

As I grew older, the number and size of our projects increased at an accelerated rate whereupon the shortcomings of the traditional procedures in the field of earthwork design became more and more conspicuous. Therefore, in different parts of the world, and almost simultaneously, attempts were made to discover and to eliminate the weak spots in the "Old Code" of the earthwork engineer. In January 1913 the American Society of Civil Engineers appointed a "Special Committee to Codify Present Practice on the Bearing Value of Soils", and in December of the same year the "Geotechnical Commission of the Swedish State Railways" started to develop procedures for determining the factor of safety with respect to sliding of the numerous slopes located along the railroads of southern Sweden. I remained unaware of these developments for a whole decade from 1914 to 1924, on account of the consequences of the first world war. In the meantime, in 1918, I embarked on a quest of my own at American Robert College near Istanbul in Turkey, where I was located at the end of the war. I started at what I considered the root of the evil, the absence of reliable information concerning the physical properties of soils.

I had neither research funds nor a laboratory at my disposal, no private practice, and a very modest salary. Therefore everything had to be improvised at minimum expense. My first earth-pressure apparatus was made out of empty cigar boxes and my loading devices consisted of empty oil cans filled with sand and attached to the ends of wooden beams with homemade knife-edge support (Fig. 1). The

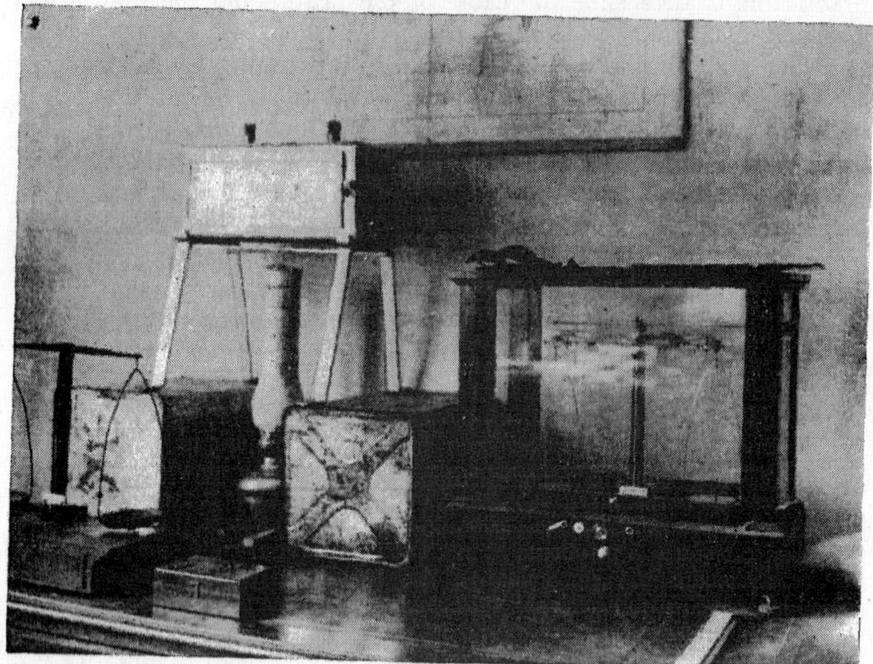

FIGURE 1.—THE AUTHOR'S DRYING OVEN IN HIS LABORATORY AT ROBERT COLLEGE (1918-1925).

drying ovens (Fig. 2) were heated with kerosene lamps and if a lamp started to smoke during the night all the samples were spoiled. Many constituents of my equipment were picked up on the college dump. Hence the prospects for success were not very bright. However in the pioneering stage of any development an individual has very much better prospects for success than a committee because at that stage the development requires years of concentrated and undivided attention. Therefore, it is not surprising that the impetus for a radical revision of the basic principles of earthwork engineering was given by the publication of my single-handed findings in 1925 and not by that of committee reports.

The most important immediate effect of the publication of my test results was the formation of a vacuum created by the demonstration that the old rules for earthwork and foundation design were fallacious. In order to fill the vacancy it became necessary to de-

FIGURE 2.—THE AUTHOR'S CONSOLIDATION APPARATUS AT ROBERT COLLEGE (1920).

velop new techniques for sampling and testing, and for measuring earth and pore-water pressures under field conditions. Simultaneously our knowledge of the physical properties of soils increased rapidly. The shearing resistance of cohesive soils is an impressive example. In 1910 it was believed by everybody, including myself, that the relationship between normal pressure and shearing resistance on plane sections through cohesive soils could be expressed adequately by Coulomb's equation which was derived in the eighteenth century. During recent years the same relationship constituted the sole topic of three important conferences, one of which was held in England and two in the United States. Yet our knowledge of this relationship is still in a state of development. The practical value of the results of some of the investigations concerning shearing resistance is not yet apparent, but once a practical problem requires relevant information it may be too late to get it. Hence, there is no doubt that the fruits of these patient efforts will some day serve a vital purpose.

Satisfactory Performance Forecasts

The practical application of soil mechanics started with attempts to compute the magnitude, distribution, and rate of settlement of structures located above homogeneous clay strata on the basis of the results of soil tests performed in the laboratory, and the results were very encouraging. Equally satisfactory were the forecasts of the performance of homogeneous fills made out of earth prepared and compacted in accordance with rigorous specifications. Earth dams with complete cutoffs, resting on a sand or gravel foundation, belong to this category.

As the years passed, the number of publications describing satisfactory performance forecasts increased steadily and left no more doubt that the fundamental principles of soil mechanics were sound. For this reason extensive subsoil exploration supplemented by soil testing is now compulsory in connection with all earthwork and foundation projects involving important capital investment, and every report dealing with the design of an earth dam is expected to contain a computation of the factor of safety of the slopes with respect to sliding. A departure from this practice is considered negligent. Nevertheless some engineers who have had the benefit of training in soil mechanics and take advantage of modern procedures of sampling and testing occasionally arrive at very erroneous conclusions concerning the future performance of the structures they design. In France, this fact has led to what a French engineer has described as "The Crisis of Confidence in Soil Mechanics (Lossier 1958)."

The examination of case records disclosing flagrant misjudgment in spite of apparently adequate subsoil exploration commonly leads to the following conclusions: Either the designer did not notice significant gaps in the information furnished by the subsurface exploration, or else the sampling and testing operations were carried out by inexperienced personnel. In both instances the structure was designed on the basis of faulty assumptions. Such a procedure can only be classified as a misuse of soil mechanics and leads inevitably to disappointing results. The psychological and technical circumstances responsible for such misuse are examined in the next section.

Misuse of Soil Mechanics

After the beginning of the 19th century, when applied mathematics successfully invaded the field of structural engineering, members of the engineering profession were taught, and even began to believe, that everything can be computed provided our knowledge of the strength of materials is far enough advanced. On the basis of this knowledge they were able to solve most of their problems without any guesswork, while sitting at their desks, and without being compelled to worry about the assumptions. Hence when soil mechanics came into existence it was expected that it would serve the same function in connection with earthwork engineering as applied mechanics does in bridge design.

These expectations were further encouraged by the fact that only successful performance forecasts in the field of earthwork engineering were considered worthy of publication, and in many articles no clear distinction was made between true forecast and forecast by hindsight. Case records of this category diverted attention from the important fact that there are many situations in which the geological conditions preclude the possibility of making an accurate performance forecast. Furthermore, many engineers do not yet realize the importance of the errors which may grow out of inadequate methods of testing. They believe that they have done their duty if they send their samples to the most conveniently located testing laboratory and accept the results at face value. Finally, many engineers do not yet realize that every boring record leaves a wide margin for interpretation unless the geological conditions are exceptionally simple.

Boring records and the report on the results of the soil tests are passed on to the designer, who constructs a physical soil profile on the basis of these data, replaces it by a simplified one, which may involve quite radical modifications, and starts to compute. In many instances the designer has not even seen the site and does not suspect that the constructed soil profile may have little resemblance to the real one. He may also consider it unnecessary to call on the testing laboratory and get first-hand information concerning the equipment which was used in making the soil tests and the qualifications of the technicians who made the tests. Hence it is not surprising that the application of soil mechanics to practical problems occasionally leads to very disappointing results.

When I published the book *Erdbaumechanik* I was not yet aware of the uncertainties associated with the interpretation of boring records and my methods of testing were still very primitive. Therefore I myself passed through a period during which my activities could be described as "misuse of soil mechanics." I am still shocked when I remember the bold conclusions which I drew in those days from the results of tests performed with primitive apparatus on an inadequate number of disturbed samples. However, this period was of short duration. It came to an end soon after 1926 when I settled in the United States. Here I had for the first time an opportunity to examine exposures of sedimentary deposits of very different origin and to secure closely spaced samples from individual drill holes. Through the results of the investigations of my disciples, foremost among them A. Casagrande, I also became acquainted with the importance of the influence of the methods of sampling and testing on the test results. Thus, as time went on and my consulting activities spread over larger and larger areas, I realized that sedimentary deposits which an erratic pattern of stratification are far more common than I had previously suspected and that these patterns preclude accurate performance forecasts.

At the outset these findings gave me a profound shock and I felt temporarily deeply discouraged. However, it did not take long to discover that the capacity for making reasonably accurate forecasts of the performance of structures resting on practically homogeneous strata with large horizontal dimensions is by no means the most important benefit which I derived from the results of my early investigations. Far more important, though much less spectacular are the practical consequences of the newly acquired insight into the mechanics of the processes of settlement and the influence of the porewater pressures on shearing resistance. Equipped with that insight I was now in a position to determine in advance of construction the type and location of potential sources of trouble in the subsoil and to proceed in accordance with the findings.

In order to take full advantage of this possibility, it is sufficient, first, to carry at least one exploratory boring at the site of each new structure to the maximum depth to which the proposed structure may have a significant influence on the stress conditions in the subsoil; and second, to determine the significant properties of the weak strata encountered in the borings by soil tests on representative samples or, in cohesionless strata, by penetration tests in the drill holes. If the subsoil exploration shows that the pattern of stratification of the troublesome strata is erratic, a reasonably accurate performance forecast is impracticable, but for many projects, also superfluous, because the design of a structure cannot be adapted to intricate details of the stratification of the subsoil. What counts under such circumstances are the variations of the average properties of the troublesome strata in horizontal directions. The following case records illustrate the procedure.

Foundation Design

Case 1. A tall and heavy office building was to be built on a thick stratum of sand and gravel, overlain by a 12-foot layer of loose, artificial fill. Along three lines crossing the site the building was located above very thick walls resting on the gravel, which were the remnants of an old fortress. The architects intended to establish the structure directly on the old walls and between them to support it on point-bearing concrete piles driven into the gravel. The settlement of the test piles under the service load assigned to them was 1/16 in. Since the architects had grown up under the influence of the "Old Code," they considered their foundation design satisfactory.

When I was requested to review the project I called attention to the fact that the settlement of the pile-supported portion of the structure would be many times greater than the settlement of the test pile, at equal load per pile, whereas the settlement of the old foundation walls would be negligible. Therefore I requested that the top part of the old walls should be replaced by a compressible cushion and bridged. After the structure was completed it was found that it had settled by amounts up to two inches. Hence, if part of the structure had rested on the old walls it would have been defaced by cracks. A quantitative performance forecast was neither practicable nor essential.

Case 2. On some projects, designers have been misled by the results of exploratory borings which were not deep enough. This possibility is illustrated by the following incident. During the second world war I was requested to review the project for an industrial development on the west coast of North America. The site was located at the shore of a drowned valley, a few feet below high-tide level. It had been explored by borings to a maximum depth of 80 ft. The boring records showed that the subsoil consists of sand, fine silty sand, and clay, resting at a depth of about 70 ft on dense sand and gravel. Therefore, it was intended to cover the site to an elevation of a few feet above high-tide level with a dredged fill and then to establish the factory buildings on point-bearing piles to be driven to refusal into the gravel stratum. (Terzaghi 1953)

Clay strata contained in drowned valley deposits are commonly normally consolidated and some of them are very compressible. On account of the large horizontal dimensions of the area to be covered by the hydraulic fill the weight of the fill alone can produce very important settlement due to the consolidation of clay strata of this kind located at a depth of far more than 80 ft. Therefore I requested that additional holes should be drilled to greater depth. At a depth of about 100 ft these new borings encountered a thick layer of highly compressible clay with a water content close to the liquid limit and an erratic pattern of stratification. A rough estimate based on the results of a few consolidation tests on representative samples showed that the combined weight of the fill and the superimposed loads would have produced unequal settlement of the pile-supported structures by more than one foot and the cost of carrying the foundation to a depth below the base of the newly discovered clay stratum would have been prohibitive. On account of these conditions the site had to be abandoned and the plant was built at another one located close to the outer boundary of the drowned valley deposits, although the new site was much less desirable from the manufacturer's point of view. At a later date an oil tank was erected on point-bearing piles close to the original site and it settled more than one foot.

Before soil mechanics came into existence, a loading test on a pile at the original site would have convinced the designer that the proposed foundation was satisfactory and the subsequent settlement of the structures would have been a surprise, as it was at the site of M.I.T.

Case 3. Another instructive example of the services rendered by soil mechanics without accurate performance forecast is the stabilization of a track with a length of about 1200 ft supporting one end of an ore bridge in a steel plant in eastern Ohio (Terzaghi and Peck, 1957). During the period 1933 to 1952 the cumulative outward movement of the track support was locally 4.5 ft, associated with a heave up to 3 ft (Fig. 3). At times the movement was so large and abrupt as to threaten the continuity of the operation of the plant. Borings showed that the track was underlain by the following strata, enumerated from top to bottom: (1) a 10-ft layer of very stiff clay, (2) a 10-ft layer of saturated, slightly plastic inorganic silt (rock flour), and (3) glacial till. Even between loading seasons the pore water in the silt stratum was locally under a moderate artesian pressure and during the loading seasons, the pore water pressure increased. The displacement of the track support was caused by inadequate shearing resistance of the subsoil along planes located in the silt.

FIGURE 3.—DISPLACEMENT OF ORE BRIDGE TRACK BY HEAVE AND LATERAL MOVEMENT.

It was originally intended to stabilize the track by a combination of vertical and batter piles. A simple computation showed that this procedure was impracticable because locally both sets of piles would have been lifted off their seat. However, the knowledge of the fact disclosed by soil mechanics that the shearing resistance of silt, like that of any other soil, depends on the effective and not on the total stress on the surface of sliding made it possible to stabilize the track without underpinning it. The desired result was accomplished by the installation of a 700-ft row of well points at 5-ft centers driven into the silt stratum. The header was attached to a vacuum pump which operates every year during the entire loading season. The quantity of water pumped out of the saturated silt stratum is not more than about 2 gal./min. Yet the movements have stopped.

Use of Flow Nets

If we face the problem of evaluating the degree of stability of the slopes of an earth dam resting on a permeable deposit, and not provided with a deep cut off, or of a slope on natural ground located downstream from a storage dam, it is necessary to determine the seepage pressures which act on the subsoil in the direction of the flow of the percolating water. This procedure requires the construction of a flow net. In order to construct the flow net the zone of seepage must be divided into several layers each of which is assumed to be perfectly homogeneous and it must be further assumed that the pattern of seepage on all vertical planes in the direction of the flow is identical.

This procedure is universally used, but quite often the constructed flow net has very little resemblance to the real pattern of seepage disclosed by the readings on observation wells after the reservoir is filled. Such discrepancies have even been encountered at the site of dams on strata which are fairly uniformly compressible throughout. Nevertheless I have found that the construction of flow nets can be of inestimable value even at sites where they cannot be expected to furnish any information concerning the real pattern of seepage. This statement is illustrated by Case 4.

Case 4. Fig. 4 is a vertical section across an erratically stratified glacial deposit through which water percolates out of a reservoir over the crest of a buried rock ridge towards a steep slope downstream from the left abutment of a concrete gravity dam. On account of the presence of gradual transitions between some of the strata, of sharp, steep boundaries between others, and of the difficulties involved in distinguishing between till with little cohesion and silty sand and gravel, an accurate interpretation of the results of the boring and sampling operations was impracticable and the constructed soil profile, Fig. 4, cannot be expected to disclose more than the salient features of the stratification. The details are unknown.

Many years before construction was started a large diameter watermain had been installed at mid-height of the slope between the valley floor and the reservoir level and the portion located downstream from the dam formed part of the new water-supply system. Because leakage from the reservoir would compromise the stability of the slope it was decided to intercept the seepage by means of a grout curtain established on or close to the crest of the rock ridge extending from the left abutment of the concrete dam to a distance of about 550 ft. from the abutment. (See Fig. 4). The grout holes

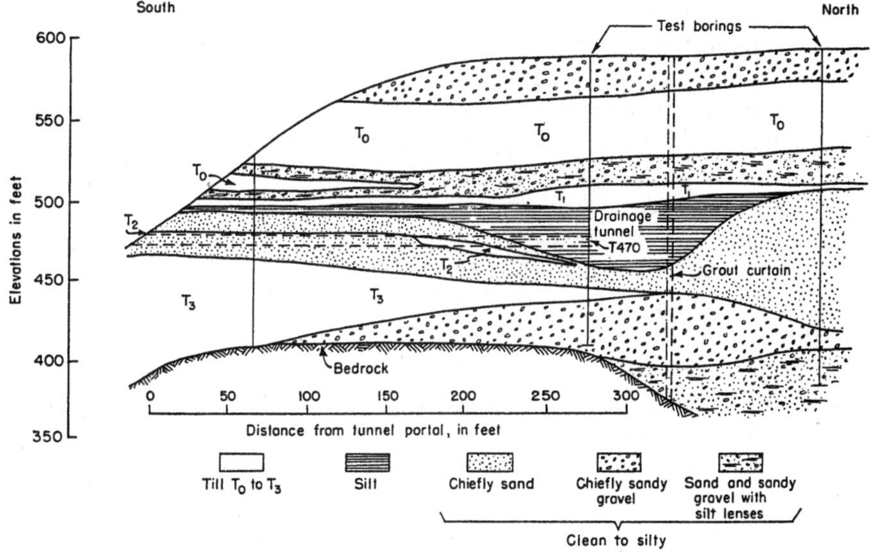

FIGURE 4.—GEOLOGICAL SECTION THROUGH GLACIAL AND FLUVIO-GLACIAL DEPOSITS. The deposits occupy the space between a grout cutoff and a steep slope downstream from left abutment of a concrete gravity dam.

with a depth up to 200 ft. below reservoir level were drilled along a single line and spaced 10 ft. Through the holes a quantity of about 1100 tons of cement was injected into the sediments.

While the reservoir was being filled for the first time, new springs came out of the slope and local subsidence was observed on the road located along the water main. Therefore, it was decided to supplement the grout curtain by drainage provisions in those portions of the slope which gave rise to concern. However, in order to design a drainage system it is necessary to know, at least in a general way, the variations in the permeability of the subsoil between the submerged area and the slope where the water comes out of the ground. At the site under consideration these variations could not even be roughly estimated, and it was not known whether or not the grout curtain served its purpose.

In order to fill the broad gaps in our knowledge 25 observation wells were installed within a distance of 1000 ft. from the left abutment. Some of these are located upstream from the grout curtain and most of them between the grout curtain and the slope where the springs came out. Several of these wells contain multiple piezometers, with perforated sections at different elevations. Observations on these wells showed that the piezometric elevation along vertical lines varies within wide limits. In order to use the observational data as a source of information concerning the variations in the permeability of the sediments acted upon by seepage pressures, the measured piezometric elevations were compared to the values which would have been obtained if the permeability of the subsoil did not change in the direction of the flow. These values can be determined by constructing an ideal flow net on adequate but radically simplifying assumptions.

The assumptions required for constructing the ideal flow net at the site illustrated by Fig. 4 were based on the following facts. The boring records have shown that the permeability of the subsoil below an elevation of about 470 is very low compared to the average permeability of the strata above it. Therefore it was assumed that the base of the ideal substitute for the real pervious layer is located at el. 470. The water in the reservoir does not rise to an elevation of more than about 80 ft. above this base. To a distance of about 700 ft. upstream from the left abutment the submerged outcrops of the pervious deposits are covered with a practically impervious blanket. Therefore the flow lines start upstream from that blanket and the average length of the path of percolation towards the area where the water comes out of the ground is about 1500 ft. Because this distance is very great compared to the vertical distance of 80 ft. between the base of the pervious layer and the water level in the reservoir it was further assumed that all the flow lines are located in horizontal planes. On these assumptions the flow net shown in Fig. 5*a* was obtained. However it was doubtful whether the grout

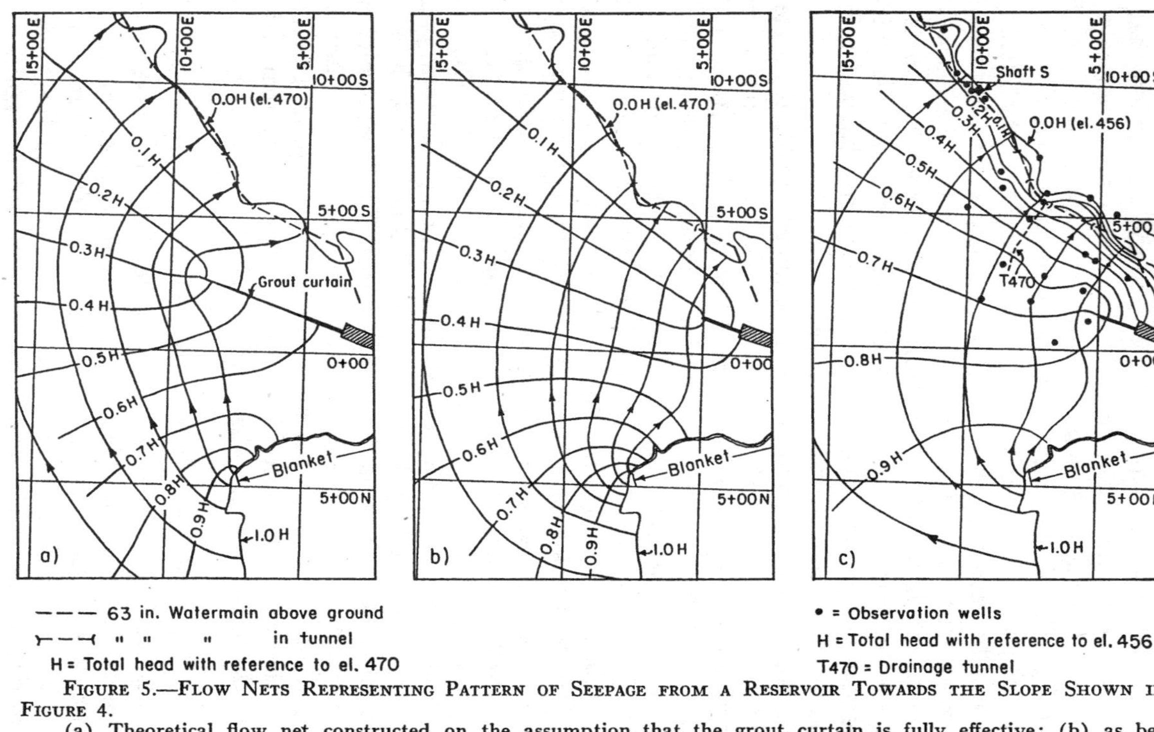

FIGURE 5.—FLOW NETS REPRESENTING PATTERN OF SEEPAGE FROM A RESERVOIR TOWARDS THE SLOPE SHOWN IN FIGURE 4.

(a) Theoretical flow net constructed on the assumption that the grout curtain is fully effective; (b) as before, but grout curtain entirely ineffective; (c) flow net constructed on the basis of the results of readings on observation wells before drains *T*470 and *S* were installed.

curtain shown in the diagram really served its purpose. Therefore a second flow net, Fig. 5*b*, was constructed on the assumption that the grout curtain does not exist. Fig. 5*c* represents approximately the real pattern of seepage. It was obtained by adapting the ideal flow nets to the results of the readings on the observation wells.

By comparing the real pattern of seepage with the ideal patterns it can be seen that the real flow lines cross the site occupied by the grout curtain as if the curtain did not exist. Thus the flow nets lead to the important conclusion that the grout curtain is entirely ineffective. On account of these conditions it became evident that the stability of the slope supporting the vulnerable water main can be assured only by adequate drainage.

The slope was drained by means of a tunnel, *T*470, at an elevation of about 470, at a distance of about 600 ft. downstream from the left abutment and a shaft *S* with collar elevation 470 at a distance of about 1200 ft. downstream from the left abutment, supplemented by horizontal pipe drains. The location of *T*470 and *S* is shown in Figs. 5 and 6 and that of *T*470 also appears in Fig. 4.

On account of the erratic pattern of stratification of the glacial deposit, the installation of the drains called for continuous experimentation in the field and there were various unanticipated incidents which required modification of the original design. Yet the final result was remarkably satisfactory. It is illustrated by Fig. 6. The plain curves are curves of equal lowering of the elevation of the total piezometric head *H* produced by drainage into shaft *S* and the dash curves represent the effect of drainage towards tunnel *T*470 with reference to the elevations which prevailed before the drains were installed. On account of successful drainage combined with local trimming operations the factor of safety of the slope is now higher than it was before the dam was built and the reservoir filled.

Observational Procedures

The case records presented under the heading "Foundation Design" showed that many problems of earthwork engineering can be solved without a detailed and accurate forecast of performance. Satisfactory solutions of such problems can be obtained on the basis of our knowledge of the fundamental principles of soil mechanics supplemented by a moderate amount of boring and testing. However there are others in which the geological conditions preclude the pos-

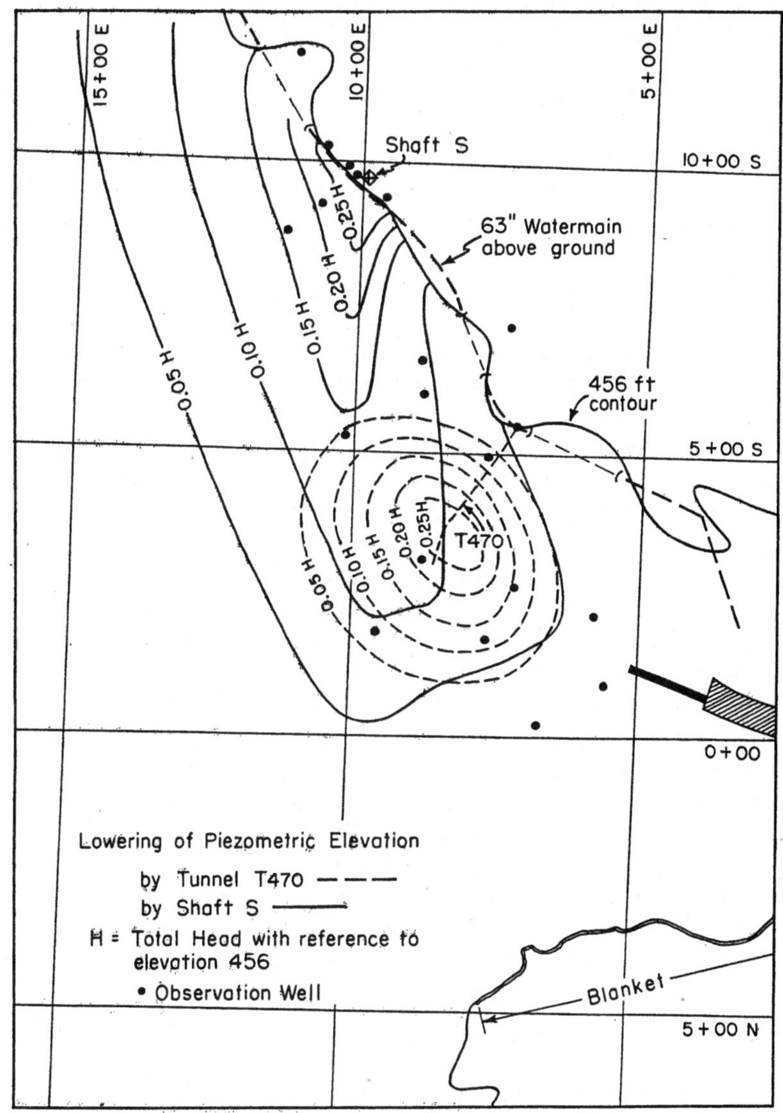

FIGURE 6.—OBSERVED LOWERING OF THE PIEZOMETRIC LEVELS PRODUCED BY DRAINAGE.
The figures, such as 0.2 H, inscribed in the contour lines indicate the decrease of the total head, *H*, produced by the drainage tunnels *T*470 and *S*.

sibility of securing in advance of construction all the essential information required for adequate design. If this condition prevails, sound engineering calls for design on the basis of the most unfavorable assumptions compatible with the results of the subsoil explorations. This rather uneconomical procedure can be avoided only on the condition that the project permits modifications in the design during or after construction in accordance with the results of significant observational data which are secured after construction is started. This can be called an "observational procedure."

Case 5. The observational procedure was used for the first time on a large scale in earthwork engineering between 1912 and 1922 by the administration of the Swedish State Railroads in connection with an investigation of the stability of slopes on glacial deposits, chiefly glacial clays, in railway cuts in southern Sweden, for the purpose of reducing the alarming frequency of catastrophic slides. Statens Järnvägars, 1922).

At an early stage of the investigations Petterson (1916) invented the graphical procedure known as the *Swedish circle method* by means of which the factor of safety of slopes with respect to sliding can be computed if the shearing resistance s of the slope-forming material is known. The Swedish investigators estimated the value s on the basis of the results of field penetration tests in bore holes, combined with cone pentration tests on remolded samples in the laboratory.

Using this procedure and refining the technique for determining the value s, the slopes subject to investigation were classified as safe, possibly dangerous, and obviously dangerous. The dangerous slopes were eliminated by relocation of the line or radical trimming of the slopes. The slopes of dubious stability were rather numerous and a cost estimate showed that the cost of adequately increasing their stability would be prohibitive. Therefore, in these cuts the railroad administration installed automatic devices which stopped oncoming trains at some distance from the ends of the cuts, as soon as the road bed started to move with reference to the firm base of the weak strata.

Case 6. As our knowledge of the significant properties of natural soil deposits increased with intensified subsoil exploration, the field of application of observational procedures expanded. I had for the first time an opportunity to use such a procedure in 1926 in connection with the water-supply reservoir formed by the Granville

Dam near Westfield, Mass. (Terzaghi, 1929). A wedge-shaped portion of the submerged slope adjacent to the right abutment of this homogeneous earth-dam is located on the outcrop of a fluvio-glacial deposit. This deposit occupies the entire space between the reservoir and an adjacent valley at a distance of about 3000 ft. from the reservoir. The surface of the deposit is dotted with kettle holes.

In order to get adequate information concerning the permeability of the strata exposed on the submerged portion of the slope I tested a great number of samples which were recovered from test shafts on profiles I to VIII, Fig. 7. The results of the permeability tests are represented in the figure by the horizontal dimension of the shaded blocks. It can be seen that the coefficient of permeability k of the materials exposed on the slope ranges between a value close to zero and 0.13 cm/sec. The corresponding variations of k in the direction of the flow lines between the submerged slope and their points of exit is unknown.

If it is assumed that k has the same value throughout the length of a flowline as it has at the submerged slope, a seepage computation shows that the loss of water from the reservoir would be excessive. It could be prevented only by covering the outcrops of the pervious strata on the reservoir slope shown in Fig. 7 with an impervious blanket. However, the geological origin of the sediments exposed on the slope suggested that the deposit had a lenticular pattern of stratification. On account of the discontinuity of the most pervious portions the real average permeability of such deposits may be very low. Therefore it was decided to postpone the decision concerning the construction of the blanket until the reservoir was filled for the first time and the loss of water from the reservoir could be measured. Observations after the reservoir had been filled showed clearly that the blanket was unnecessary.

If the estimate of the cost of the blanket had shown that the construction of the blanket would make the project economically unsound, I would have proposed the same procedure, but at the same time I would have notified the owners that the project involved calculated risk, and left it to them to decide whether or not they were willing to take it. Many lawsuits have grown out of the failure of the designers to recognize the existence of such risks or to notify their clients about them.

Case 7. Fig. 8 illustrates the application of observational pro-

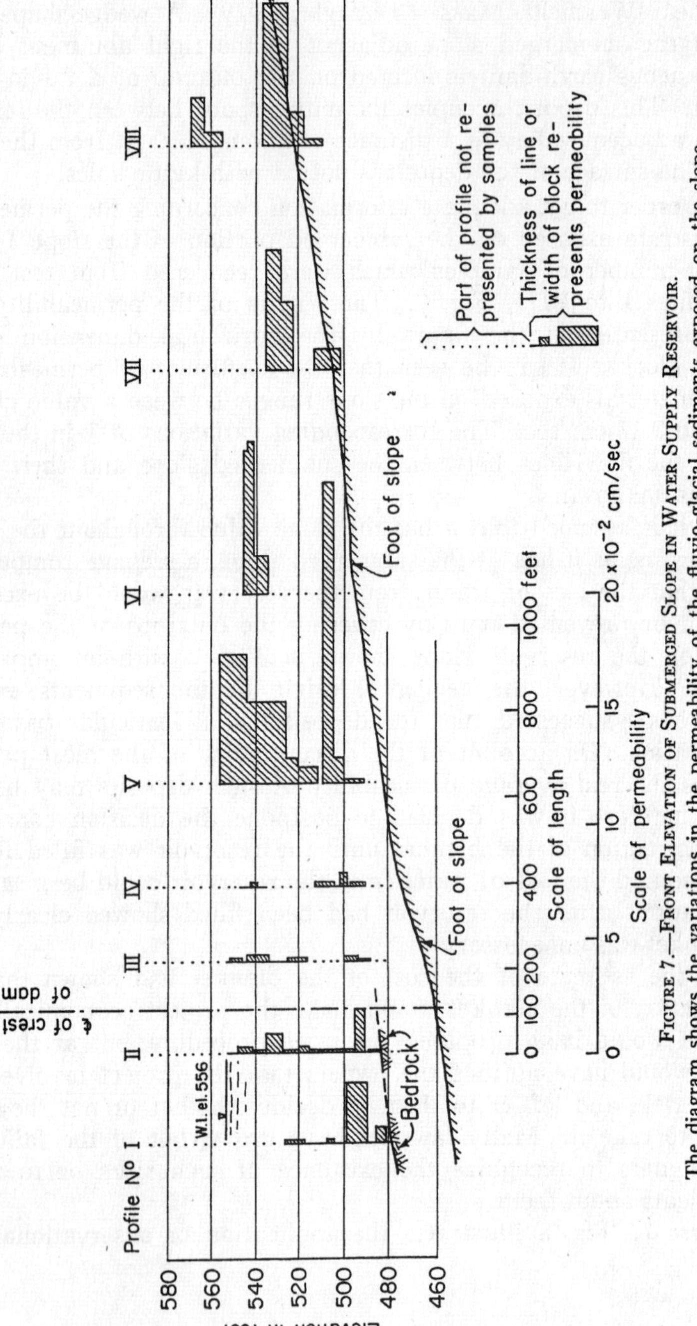

FIGURE 7. FRONT ELEVATION OF SUBMERGED SLOPE IN WATER SUPPLY RESERVOIR.
The diagram shows the variations in the permeability of the fluvio-glacial sediments exposed on the slope.

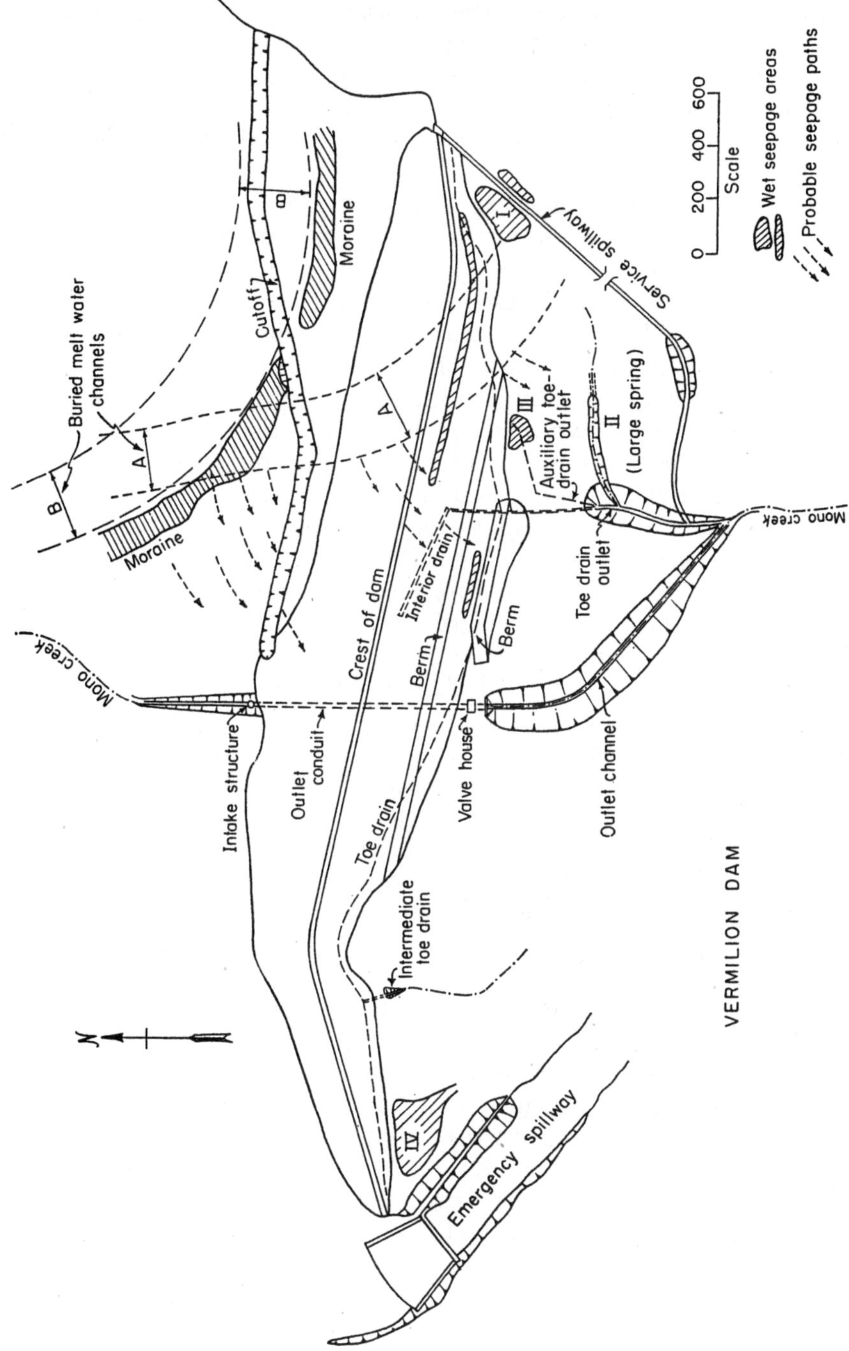

FIGURE 8.—PLAN OF VERMILION DAM.
In areas I to IV, water came out of the ground while the reservoir was being filled for the first time.

cedures to the design of the Vermilion Dam in Southern California with a height of about 150 ft. (Terzaghi and Leps, 1959). The dam shown in the figure rests on thick fluvio-glacial deposits with an erratic pattern of stratification. The deposit was laid down between crescent-shaped terminal moraines and rests on granite at a depth up to 200 ft.

The cost of a cutoff extending down to bedrock would have been prohibitive and a reliable forecast of the pattern of seepage through the fluvio-glacial deposits was impracticable. Yet this pattern has a decisive influence on the factor of safety of the downstream slope with respect to sliding and on the layout of the drainage system required to eliminate the danger of piping. On the other hand, during construction and the first filling of the reservoir, it was possible to detect significant differences between design assumptions and reality, before it was too late to prevent detrimental consequences. Owing to this circumstance, the original design could be based on what appeared to be the most probable, and not the most unfavorable, assumptions compatible with the results of the subsoil exploration. The observational devices required for checking on the original design assumptions were installed as construction proceeded.

The original project included an incomplete cutoff to a maximum depth of 20 ft. on the bottom of the reservoir as shown in Fig. 8, an impervious blanket covering the area between the cutoff and the base of the impervious core of the dam (not shown in the figure), and the installation of a row of bleeder wells (not shown in the figure) with a maximum depth of 100 ft., located along the center-line of the bottom of the toe drain. While the dam was being built all the observation wells were installed which were needed to get a clearer picture of the real pattern of seepage, as in Case 4 illustrated by Figs. 4 to 6. Furthermore filter material was stockpiled on the valley floor downstream from the dam. As the reservoir was being filled it was found that the piezometric elevations were almost everywhere below the elevations which were assumed to prevail when the stability computation for the downstream slope was made. At the one spot where they were slightly higher some additional fill material was added to the slope. It was further found that the bleeder wells located in the proximity of one abutment served no useful purpose. However, water came out of the natural slope in the areas indicated by shading and labeled I to IV in Fig. 8, at distances up to 300 ft.

from the downstream toe of the dam. The danger of piping by subsurface erosion starting from these areas was eliminated at a very moderate expense by covering them with inverted graded filter.

Case 8. An unusual opportunity for the application of the observational procedure was encountered in connection with design and construction of submerged shipways on the Atlantic Coast (Fitz-Hugh, Miller and Terzaghi, 1945). These shipways differ from the conventional type inasmuch as the side walls consist of cellular cofferdams surrounding a thin, but pile-supported concrete floor (Fig. 9). The side walls were acted upon by the lateral pressure exerted by a hydraulic back-fill consisting of marl chunks, and the piles supporting the floor were driven through a layer of marl with a thickness of 20 ft. into an underlying aquifer consisting of silty sand. The side walls had to satisfy the condition that they should remain practically vertical at every stage of operation of the shipways, and the sum of the weight of the shipway floor and of the underlying marl stratum should be at least 1.5 times the hydrostatic uplift on

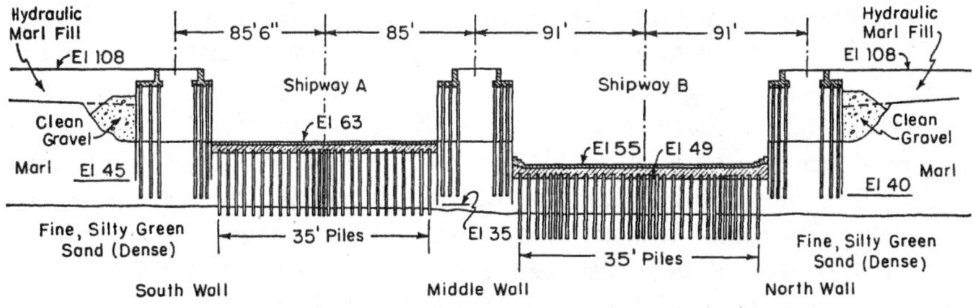

FIGURE 9.—CROSS SECTION THROUGH TWO SUBMERGED SHIPWAYS.

The gravel pockets on both sides of the shipways were placed in dredged cuts after the side walls were built and backfilled, to reduce lateral pressure on the walls.

the base of the marl stratum when the shipways were empty. However, a reliable forecast of the lateral deflection of the side walls was impracticable and the hydrostatic conditions which would prevail in the aquifer during operation of the shipways were unknown. Therefore, the following procedure was adopted.

During the back-filling operations the increase of the lateral deflection of the side walls was measured. When it was found, by extrapolation from the observational data, that the lateral deflection of the side walls would be excessive, a wedge-shaped portion of the hydraulically placed marl fill was removed by dredging in the area adjacent to the outer face of the cellular sidewalls of the shipways. This material was replaced by gravel communicating with the shipways through pipes across the side walls. Thus the lateral pressure on the side walls was reduced to a nominal value.

In order to get information concerning the hydrostatic pressure conditions in the aquifer underlying the shipway floors, piezometric tubes were installed with their open ends located in the aquifer. While the shipways were being pumped out for the first time, the piezometric elevations were measured. When it was found that the factor of safety of the shipway floors was locally smaller than 1.5 the uplift pressure was reduced within these areas by the installation of relief wells. The cost of the modifications of the design during construction, amounting to about $250,000, were very small compared to the savings realized by the radical departure from the conventional design of submerged shipways.

The preceding case records are representative of the manifold practical applications of observational procedures in earthwork engineering. In every case the procedure was used to compensate for the inevitable uncertainties involved in the interpretation of the results of the subsoil exploration.

There are no case records which demonstrate more impressively the existence of these uncertainties, and the means for coping with them, than those describing the successful use of observational procedures. Yet there are still many engineers engaged in earthwork design who are not yet fully aware of the existence and the importance of these uncertainties. Therefore the publication of such case records serves a vital educational purpose, by counterbalancing the psychological effect of the records of reasonably accurate performance forecasts which can be expected only under exceptionally favorable conditions.

Role of Geological Considerations in Earthwork Engineering

The nature and importance of the uncertainties involved in the interpretation of the results of subsoil exploration are essentially determined by the geological history of the site. Therefore I have acquired the habit of starting the investigation of a new site by collecting information regarding the geological characteristics of the site from resident geologists and from published sources, supplemented by a painstaking examination of the site. The interpretation of the findings is based on my previously acquired empirical knowledge of the engineering properties of the products of the different geological processes such as glaciation or stream action. The knowledge was obtained with the assistance of soil mechanics by correlating the processes with the information obtained by experimental determination of the engineering properties of their products and with the pattern of stratification of the resulting deposits.

The results of the geological inquiry combined with the general layout for the project determine the maximum depth at which the seat of potential trouble may be located. Depending on the nature of the project the seat may consist of one or more strata with high compressibility, of an aquifer containing water under high artesian pressure, avenues for the escape of water from a reservoir and the like. This procedure for estimating the maximum depth for the first exploratory drillholes was illustrated by case 2 under the heading "Foundation Design."

The next step is to determine the horizontal and vertical boundaries of the seat of potential trouble by means of exploratory borings and the pattern of stratification of the deposits involved. In some instances this pattern can be predicted on the basis of the results of the geological reconnaissance. Varved clay deposits in an undisturbed state are always practically homogeneous in horizontal directions because they were deposited in still water. On the other hand the index properties of clay located at the base of drowned-valley deposits are likely to vary over short distances in both vertical and horizontal directions, because the currents responsible for their transport prior to sedimentation shifted throughout the year. In doubtful cases information concerning the degree of homogeneity of a seat of potential trouble can be obtained by a moderate amount of testing of samples recovered from the exploratory borings. The degree of homogeneity determines subsequent procedures.

If the seat of potential difficulties is relatively homogeneous, an elaborate subsoil exploration involving the testing of numerous undisturbed samples may be justified. On the basis of the results of the investigations, a settlement or stability computation can be made.[1] The prerequisites for the success of such operations are satisfied

[1] Quite recently I encountered a deposit of glacial lake clay which proved to be far more compressible than the consolidation tests on undisturbed samples indicated, but this deposit is unique in my experience.

throughout large portions of the coastal plains, or on lake deposits such as those underlying some of the plains northwest of the Great Lakes Region or the valley of Mexico. On the other hand, if the much more common condition of nonhomogeneity is encountered, very little can be gained by accumulating test data in addition to those which disclosed the absence of homogeneity. In this case the third step is an evaluation of the inevitable uncertainties involved in the interpretation of the results of the subsoil exploration. Design has then to be based on the most unfavorable possibilities compatible with the known features of the subsoil conditions. However, if the project permits modification of the design during construction, this uneconomical method can be avoided by using the observational procedure described in the preceding section of this paper (Case 5 to 8).

If this general procedure is followed, developments conspicuously at variance with what is anticipated can be due only to significant geological details located between drillholes or between the points where samples were recovered. The possibility of such developments came to my attention shortly after I published the book *Erdbaumechanik* (1925) when I was asked to review the design for the substructure of a steam power station. The site for the station covered an area of about 200 by 100 ft. and it was explored by borings spaced 50 ft. both ways, to a depth of 60 ft. below the ground surface. The borings showed that the site was located on a stratum of waterbearing sand and gravel with a thickness of 15 ft. resting on a remarkably homogeneous deposit of stiff clay. The station was to be provided with a sub basement on a heavily reinforced concrete mat, resting at a depth of about 20 ft. on the stiff clay.

Prior to starting the excavation the site for the subbasement was surrounded by a row of steel sheet piles driven to a depth of about two feet into the stiff clay. After the excavation was completed no water entered the pit except for the water which leaked out of the sand through the interlocks of the sheetpiles.

When I visited the site on the evening of the day when the excavation was completed and the laborers had left, the night watchman informed us that a "blow" had occurred in the pit, a few minutes earlier. When we arrived at the pit we saw a jet of water shooting out of the clay close to the center of the pit. The point of emergence of the jet was buried beneath a cone-shaped and rapidly growing accumulation of clean sand.

To stop the flow of sand we replaced the cone-shaped sand deposit by an empty cement barrel and filled the barrel with concrete aggregate to act as a filter. As soon as this operation was completed a second blow occurred at some distance from the first one, which was then treated in the same manner. During the next few hours a whole string of barrels was installed above springs on a meandering line which did not approach any of the points where test-borings had been made. Subsequent investigations showed that the springs came out of a narrow belt of clean, waterbearing sand located at a depth of a few feet below the bottom of the excavation. Before the reinforcement for the concrete floor of the subbasement was placed, the cavities formed by the discharge of sand from the belt were plugged by grouting.

Buildings on spread footings may be—and have been—severely damaged on account of a few footings resting on exceptionally compressible material contained in pockets located entirely between the points where test borings had been made. Slopes of earth dikes or dams on stiff clay have failed on account of the clay containing a few thin layers of exceptionally weak material such as bentonite located between the elevations at which samples were recovered from the bore holes. Fortunately such incidents are rare, but inevitable.

Under normal conditions the general procedure described in this paper protects the engineer in charge of an earthwork project against the serious danger of underestimating the uncertainties involved in the results of his subsoil exploration. However no hard-and-fast rules can be established concerning the details of practicing the procedure, because at every new site the designer is likely to encounter some conditions which are without any precedent in his own experience or in published case records. This fact points to an important difference between structural engineering and applied soil mechanics.

Many problems of structural engineering can be solved solely on the basis of information contained in textbooks, and the designer can start using this information as soon as he has formulated his problem. By contrast, in applied soil mechanics a large amount of original brain work has to be performed before the procedures described in the textbooks can safely be used. If the engineer in charge of earthwork design does not have the required geological training, imagination and common sense, his knowledge of soil mechanics may do more harm than good. Instead of using soil mechanics he will abuse it.

Conclusions and Outlook

In 1936, when the First International Conference on Soil Mechanics and Foundation Engineering convened at Harvard, I no longer had any illusions concerning the limits nature has set to the degree of accuracy which can be achieved in the prediction of performance in the realm of earthwork engineering. Hence at that time I started to promote comparison between forecasts and observed performance and correlation between case records and the geological characteristics of the site. I expected that once practicing engineers had learned to evaluate the inevitable uncertainties associated with performance forecasts at given sites, the misuse of soil mechanics would automatically stop. This, however, turned out to be a wish-dream.

We are now living in the year 1960. The generation of engineers which still believes that the settlement of a pile-supported structure can be predicted from the results of a loading test on a single pile is gradually dying out. However, the misuse of soil mechanics is still practiced all over the globe, because the members of our profession are spoiled almost beyond recovery by the success of applied mathematics in other fields of civil engineering. Once an engineer has left his alma mater his mind is likely to become dogmatic. Therefore he must be warned against the misuse of soil mechanics at an early date.

Most of the obstacles to a rigorous treatment of the problems of earthwork engineering grow out of the geological history of the soil deposits underlying our sites. Therefore, I used to call the course in engineering geology I gave at Harvard an antidote to "Theoretical Soil Mechanics" and I taught it in this spirit. However, so far very few courses in engineering geology accomplish their vital mission because in most of them the subject is still presented as an accumulation of facts leaving it to the student to discover that these facts are essential ingredients in engineering reasoning and the basis for observational procedures. That discovery rarely ensues. Therefore the misuse of soil mechanics will continue unless the center of gravity of the courses in engineering geology is shifted from the geological facts to their engineering consequences. Among the most important of these are the limits which nature has set to the degree of reliability of the information which can be obtained by presently available means for subsurface exploration. (Terzaghi, 1961).

References

Bjerrum, L., Casagrande, A., Peck, R. B., and Skempton, A. W. (eds.). 1960. From Theory to Practice in Soil Mechanics. Selections from the writings of Karl Terzaghi with bibliography and contributions on his life and achievements. John Wiley and Sons, Inc., New York. Pp. 425.

FitzHugh, M. M., Miller, J. S., and Terzaghi, K. 1945. "Shipways with Cellular Walls on a Marl Foundation." Trans. A.S.C.E. Vol. 112 (1947), pp. 298-324.

Lossier, H. 1958. La Crise de Confiance de la Mécanique des sols. Le Génie Civil. July 1958. Paris.

Main, Charles T., and Sawtell, H. E. 1918. "Foundations of the New Buildings of the Massachusetts Institute of Technology, Cambridge, Mass." Journal of the Boston Society of Civil Engineers, Vol. V, Jan. 1918, pp. 1-34. Discussion by L. M. Hastings, pp. 35-38.

Petterson, K. E. 1916. "Kajraseti Göteborg den 5te Mars 1916." Teknisk Tiøsskrift 1916, V. U. Vol. 30, pp. 281-287, and Vol. 31, pp. 289-291.

Statens Järnvägars Geotekniska Kommission. 1922. "Slutbetänkande avgiet den Maj 31, 1922." Stockholm.

Terzaghi, K. 1925. "Erdbaumechanik." Wien.

Terzaghi, K. 1929. "Soil Studies for the Granville Dam at Westfield, Mass." Journal of the New England Water Works Association, Vol. 43, pp. 191-213.

Terzaghi, K. 1936. "Relation Between Soil Mechanics and Foundation Engineering." Presidential Address, Proc. First Intern. Conf. on Soil Mech. and Found. Eng., Cambridge, Mass., Vol. 1, pp. 54-56.

Terzaghi, K. 1953. "Fifty Years of Subsoil Exploration." Proc. Third Intern. Conf. on Soil Mech. and Found. Eng., Zürich, Vol. 3, pp. 227-237.

Terzaghi, K., and Peck, R. B. 1957. Stabilization of an Ore Pile by Drainage. Proc. A.S.C.E., Vol. 83, No. SM 1, Paper 1144.

Terzaghi, K., and Leps, T. M. 1958. Design and Performance of Vermilion Dam, California. Proc. A.S.C.E., Paper 1728, SM, pp. 1-30.

Terzaghi, K. 1961. Engineering Geology on the Job and in the Classroom. Journal of the Boston Society of Civil Engineers, Vol. 48, pp. 97-109.

Van Es, L. J. C. 1933. Das Untersuchungsverfahren über die Eignung von Bodenarten für den Bau von Staudämmen mit Hilfe der Konsistenzwerte von Atterberg. 1. Erster Intern. Talsperren-Kongress, Stockholm, 1933. Question 2 and Rep. No. 21.

PAST AND FUTURE OF APPLIED SOIL MECHANICS*

DISCUSSION

By Ralph B. Peck†

One paragraph in Dr. Terzaghi's stimulating paper deserves to be reproduced and to hang in the office of every subsurface engineer:

"Many problems of structural engineering can be solved solely on the basis of information contained in textbooks, and the designer can start using this information as soon as he has formulated his problem. By contrast, in applied soil mechanics, a large amount of original brain work has to be performed before the procedures described in the textbooks can safely be used. If the engineer in charge of earthwork design does not have the required geological training, imagination and common sense, his knowledge of soil mechanics may do more harm than good. Instead of using soil mechanics he will abuse it."

The abuse of soil mechanics has received attention in several publications by Dr. Terzaghi and others. It is a matter of serious concern to all those engaged in the practice of subsurface engineering. Abuse may occur to different degrees and on different levels. This discussion is concerned with the possible abuse of soil mechanics in the observational procedure.

The observational procedure so well described in Dr. Terzaghi's paper is one of the most notable and useful contributions of applied soil mechanics. It often permits a very economical solution of a difficult problem at an adequate and known factor of safety. As the method is popularly understood by subsurface engineers, it consists of basing a design or construction procedure on average conditions, or on conditions considered to be most probable, rather than on the most unfavorable conditions that are compatible with the conditions at the site. Field observations are then devised to give adequate warning regarding undesirable deviations from the assumed conditions. If necessary, modifications or corrective measures are introduced as indicated by the observations.

---

* Paper printed in April, 1961 JOURNAL, by Dr. Karl Terzaghi.
† Professor of Foundation Engineering, University of Illinois, Urbana.

One essential feature of the observational method is omitted in this conception. The engineer must not only assess the average or most probable conditions, but he must also carefully assess the most unfavorable conceivable conditions. Moreover, he must have a sound plan for coping with the unfavorable conditions if they should actually develop. This plan may consist, as Dr. Terzaghi suggested in one of his examples, of making known to the owner the consequences of exceptionally unfavorable conditions, because the owner may be willing to take the risk. Under many circumstances, however, the unfavorable conditions, if they should prevail, would require radical changes in design or in construction procedure. If such changes are required, but if the engineer has not planned in advance what they should be, the consequences may be serious indeed. The writer suspects that some designs and construction procedures that have been advocated in the name of the observational procedure have been proposed without giving proper attention to the steps that would be taken if unfavorable conditions developed. The laws of probability indicate that such designs or procedures would be adequate in most instances, whereupon the designer gets the credit for proceeding in an economical fashion. Such a modus operandi is an abuse of soil mechanics and should not be tolerated.

A second requirement in planning an observational approach to a design is that field measurements can be planned and executed that will disclose significant unfavorable developments. Considerable ingenuity may often be required in determining what measurements should be made and how to make them. The nature of some problems, however, may preclude measurements that will give the necessary information in time to permit appropriate action. Under these circumstances the observational procedure should not be applied.

As an example, the foundations for two grain elevators, A and B, may be considered. The soil conditions at both sites were virtually identical; indeed the two sites were in close proximity to each other. Both structures carried approximately the same loads. Both were supported on vertical piles driven through loose fill and relatively soft cohesive deposits to a firm base. Along one side of each structure was a sheet pile bulkhead; against the opposite side was a fill supporting railroad tracks. Lateral movements of structure A became perceptible after about 20 years of service and were measured from time to time. The amount of movement was found to be doubling at about five year intervals, and reached a total of more than one foot. Because of the acceleration, the interested parties became alarmed and provided additional support for the structure before any serious consequences developed.

On the other hand, structure B gave no visible evidence of appreciable lateral movement. Therefore, it was considered to be more stable than structure A and no periodic observations or remedial measures were considered necessary. Nevertheless, after comparatively minor incidents that may have slightly altered the loads on the structure or its foundations, the structure suddenly underwent very large lateral deformations and within a period of one or two days suffered a complete collapse and foundation failure.

The principal difference between the two structures was the presence of a relatively small number to timber batter piles beneath structure B. These piles had evidently supplied sufficient resistance to prevent the development of the large lateral movements observed at structure A. In resisting the movements, however, they were very highly stressed. When some relatively minor incident caused one or a few of these piles to fail, the subsoil could not quickly accommodate itself to the changed conditions and an almost immediate collapse occurred.

Had the writer been requested, prior to the collapse, to investigate the stability of structure B, he would quite possibly have suggested refined observations, such as careful measurements of lateral movements and the installation of slope indicators, to detect the beginnings of movement. He would probably have recommended that observations be made at frequent intervals and that no remedial measures should be taken unless the structure began to accelerate. Had this prescription been advocated and followed, it is quite certain that the collapse would have occurred precisely as it did.

The defect in the use of the observational procedure, as suggested at site B, lies in the presence of what might be termed brittle and rigid elements in the foundation. These elements had sufficient rigidity to prevent a movement in the early stages but not sufficient strength to withstand the forces that developed as time went by. It is doubtful that suitable deformation measurements could have been devised to give adequate warning of the failure that occurred, because the brittle load-carrying component could not have deformed perceptibly before failure. On the other hand, a structural analysis of the behavior of the foundation, even from a qualitative point of view, would have led to the conclusion that the batter piles must have been highly loaded and that they constituted danger spots. A proper analysis and assessment of the conditions would, in other words, have led the engineer to the conclusion that the usual observational procedures might be misleading and unconservative in this instance.

This illustration points out that the subsurface engineer dealing with the observational method needs an additional ingredient besides a knowledge of the geology of the site and a knowledge of the soil conditions. He also needs a good conception of structural behavior and of the interaction between soil and structure. He must be especially wary of elements that may carry dangerously high loads without being able to deflect appreciably in comparison with the deflection required to mobilize the strength of the other elements.

A second type of problem in which current observational procedures may not lead to a sound conclusion involves the stability of slopes. During the construction of earth dams, or during the placement of fills, it has become rather customary to install pore pressure devices in the fill and in the foundation materials. One purpose of the pore pressure devices is to determine if, or when, the rate of filling must be decreased in order to eliminate the possibility of a slope or base failure. Under some simple circumstances, for example, if the piezometer is located beyond the toe of the embankment and registers a pressure approaching the weight of the overburden, a fairly clear cut statement may be made regarding the factor of safety. In most instances, however, the relationships between the measured pore pressures and the factor of safety against failure are by no means simple. The pore pressure observations unquestionably lead to a better understanding of the subsurface conditions and of the stability of the slope. They may not, however, lead directly to an indication of the factor of safety. If they are made, but if inadequate thought has been given to the way in which they are to be interpreted, reliance on them as part of an observational procedure may also be considered an abuse of soil mechanics.

In short, the observational procedure must be planned with great care. The engineer must be certain that he can cope with the consequences of learning that the real conditions are extremely unfavorable. He must also assure himself that the observations he intends to make will actually disclose the state of affairs in sufficent time and with sufficient clarity to indicate the degree of safety. If the engineer overlooks these matters, the observational procedure may lead to a catastrophe made all the less defensible because the techniques at first glance might appear to represent the latest developments in the application of soil mechanics.

AMERICAN SOCIETY OF CIVIL ENGINEERS

Founded November 5, 1852

TRANSACTIONS

Paper No. 3351
(Vol. 127, 1962, Part I)

FOUNDATION VIBRATIONS

By F. E. Richart, Jr.,[1] F. ASCE

With discussion by Messrs. D. F. Coates; I. Alpan; Joseph G. Perri; A. A. Eremin; Bobby Ott Hardin; V. A. Smoots, J. F. Stickel and J. A. Fischer; S. D. Wilson and R. P. Miller; and F. E. Richart, Jr.

SYNOPSIS

Theoretical solutions for the vibrations of foundations resting on soil have been studied and evaluated. For vertical vibrations of a machine foundation, it is usually necessary to use the theory for an oscillator resting on an elastic semi-infinite body because of the important effects introduced by oscillations of the soil. This effect becomes of less importance if the force of the oscillator varies with the frequency of rotation. The use of Sung's theory for vertical foundation vibrations is outlined and curves are presented which may be used for analysis or design.

By comparison of the ordinary theory of vibration, which considers the spring to be weightless, to the theory for an oscillator on a semi-infinite elastic body, it is shown that the ordinary theory is usually satisfactory for determining the natural frequency for the rocking and sliding modes of vibration of the foundation.

In order to evaluate the dynamic characteristics of a foundation, it was found necessary to determine the shear modulus and Poisson's ratio of the underlying soil. Typical values of these soil constants are given and methods for establishing them at a particular site are described. The shear modulus of a granular soil should increase approximately with the 1/3 power of the confining pressure.

Two examples are included to show the details of application of the theoretical methods. A diagram is also included which illustrates the possibility of

Note.—Published essentially as printed here, in August, 1960, in the Journal of the Soil Mechanics and Foundations Division, as Proceedings Paper 2564. Positions and titles given are those in effect when the paper or discussion was approved for publication in Transactions.

[1] Prof. of Civ. Engrg., Univ. of Florida, Gainesville, Fla.

stiffening the foundation, by the use of point-bearing piles to rock, in order to increase the natural frequency of vertical vibration.

INTRODUCTION

Oscillating or rotating machine components are a source of periodic impulses. Because large machines are usually supported directly on the soil in a manner that permits a direct transmission of these periodic impulses into the soil, the design of the machine foundation involves the problem of soil dynamics. When the natural frequency of a soil-foundation system happens to coincide with the frequency of the exciting forces generated by the machine, excessive vibration amplitudes may occur which can lead to structural damage or operational failure of the machine. Periodic or intermittent wind forces on tall structures may also set up vibrations of the foundations for a sufficient time to cause damage.

The type of foundation considered in this study consists of a massive block resting directly on the soil. The block is freely supported and has six degrees of freedom in vibration: It may vibrate in vertical, longitudinal, or lateral translation, or it may oscillate in the rotational modes of rocking, pitching, or yawing. These motions are illustrated in Fig. 1. In general, the vertical translation and the rocking types of vibration occur most often in machine foundations.

The first approximation to the study of foundation vibrations considered the vibrating system to behave as a single mass supported by a weightless spring and subjected to a viscous damping. As a result of an extensive series of tests made by DEGEBO (Deutsche Ferschungagesellschaft für Bodenmechanik) it was found that the simple damped mass-spring theory was not adequate to explain the test results obtained from vertical motions of an oscillator resting on soil. A. Gertwig, G. Früh, H. Lorenz, and co-workers found (1)[2] that it was necessary to include consideration of an "in-phase" mass of soil which oscillated with the foundation to made the simple theory agree with test results. Unfortunately, test data prove that the computed in-phase soil mass is not a constant quantity, even for a particular foundation, because it varies with both the static and dynamic loading. In the intervening years since these test results became available, repeated attempts have been made to provide rules for determining the weight of this in-phase mass, with negligible success.

In 1936, E. Reissner presented (2) an analytical solution to the problem of vertical motion of an oscillator resting on an elastic semi-infinite body. His solution included the dynamic behavior of the elastic body, and represented the oscillator by a pulsating pressure uniformly distributed over a circular area of contact. More recently, both T. Y. Sung (3) and P. M. Quinlan (4) have extended Reissner's treatment to cover different contact pressure distributions between the oscillator and elastic body, and R. N. Arnold, G. N. Bycroft, and G. B. Warburton (5), and I. Toriumi (6) have considered different modes of vibration. The theories based on the treatment of an elastic semi-infinite body have shown relatively good agreement with test results.

It is the purpose of this paper to examine briefly some of the theoretical procedures which are available for analysis and design of foundations subjected to vibratory loads. Sand compaction by vibrations and the resulting settlements of foundations is a separate problem which will not be considered here.

---

2 Numerals in parentheses, thus (1), refer to corresponding items in the Bibliography.

Notation.—The letters and symbols used in the text are defined as they appear and are assembled in Appendix I.

LIMITATIONS OF VIBRATION AMPLITUDE

It is obviously impossible to eliminate the oscillating motion completely from a foundation which is subjected to significant periodic impulses. The designer can only attempt to reduce the foundation vibration to a magnitude which is tolerable at the operating frequency for the design conditions. In general, the permissible amplitude of vibration is decreased as the frequency is increased. Thus, no value of allowable amplitude should be considered as a design criterion unless the frequency is also specified. In cases where the design specifications do not include permissible amplitudes of vibration, the

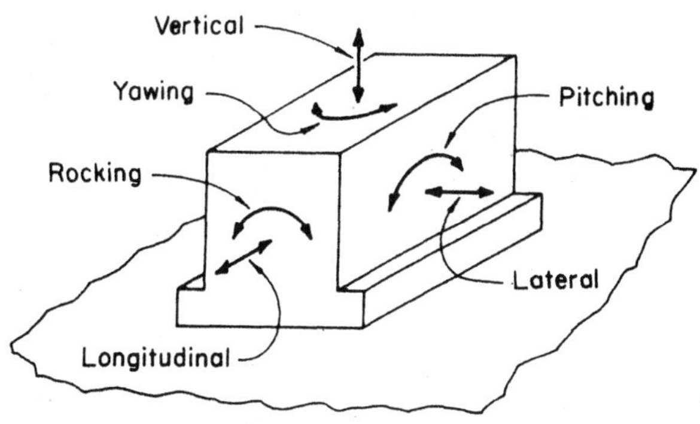

Translational Modes	Rotational Modes
Vertical	Rocking
Longitudinal	Pitching
Lateral	Yawing

FIG. 1.—SIX MODES OF VIBRATION FOR A FOUNDATION

information given in Fig. 2 may be used as a guide. It is important to note from Fig. 2 that the vibration amplitude which is designated as the limit for machines and machine foundations is approximately one hundred times that which is barely noticeable to persons.

DAMPED FORCED VIBRATIONS FOR ONE DEGREE OF FREEDOM

It is useful to review briefly the simple case of damped forced vibrations of a single mass supported by a weightless spring. As shown in Fig. 3 the mass has a weight, W_o, which is set into oscillation by a periodic force Q. The motion is restrained by the inertia of the mass, by the reaction of the spring, and by the viscous damping force. The periodic exciting force is defined by

$$Q = Q_1 \sin\left(2 \pi f_1 t\right) \quad (1)$$

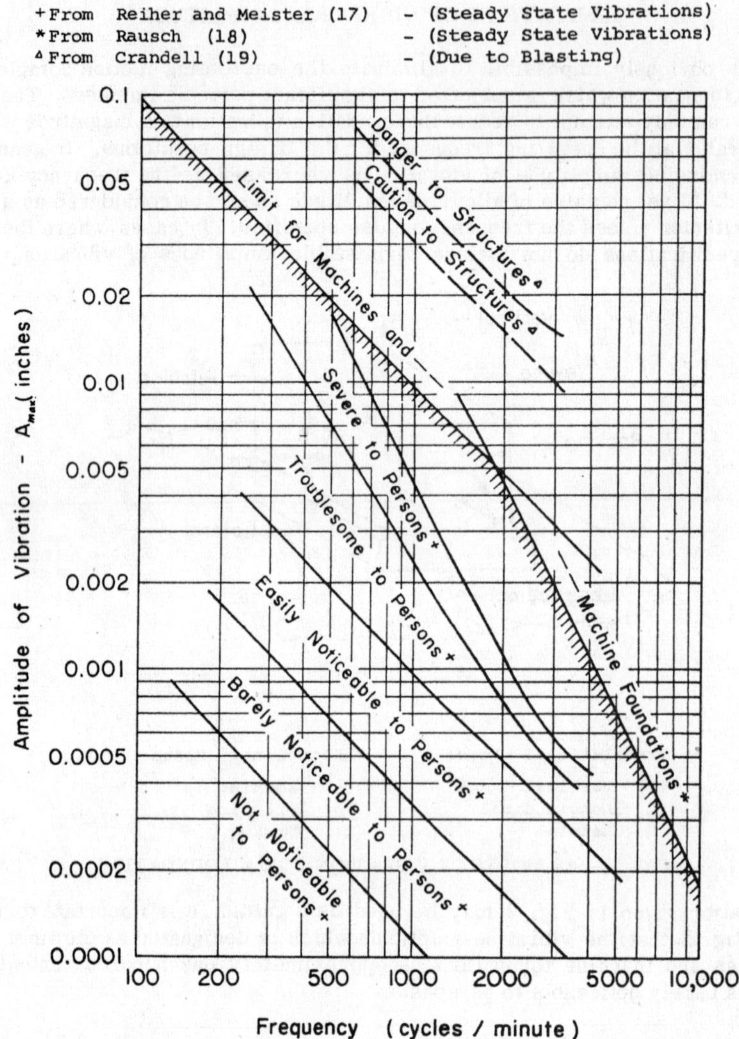

FIG. 2.—ALLOWABLE VERTICAL VIBRATION AMPLITUDE FOR A PARTICULAR FREQUENCY OF VIBRATION

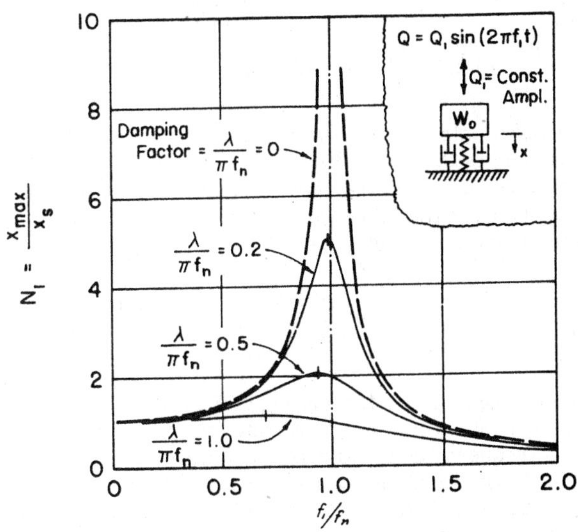

(a) For Constant Amplitude Exciting Force

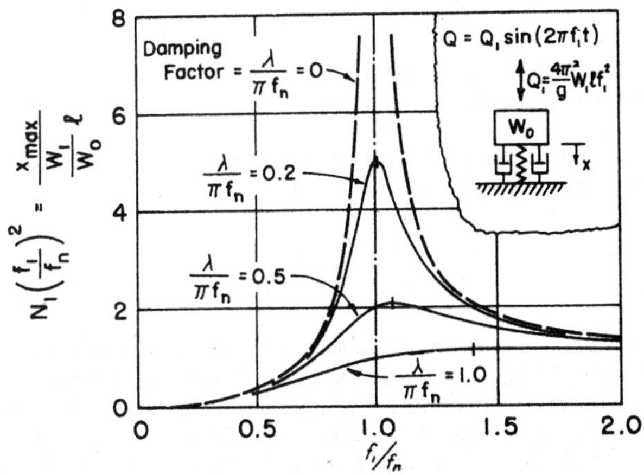

(b) For Exciting Force Dependent Upon Exciting Frequency

FIG. 3.—AMPLITUDE—FREQUENCY RELATIONS FOR DAMPED FORCED VIBRATION OF A MASS—SPRING SYSTEM

in which Q_1 may be either a constant or a function of the exciting frequency, f_1, and t denotes time. For the case where Q_1 is a constant and thus independent of the exciting frequency, the resulting displacement of the mass is given by

$$x = x_s N_1 \sin\left[(2\pi f_1 t) - \alpha\right] \quad (2)$$

in which α is the phase angle between the exciting force and the resulting motion of the mass, and x_s is the static displacement which would be produced by a force of magnitude Q_1 acting on the spring. The force required to compress the spring a unit distance defines the spring constant, c_s, which is then used to evaluate the static displacement by

$$x_s = \frac{Q_1}{c_s} \quad (3)$$

The displacement magnification factor, N_1, in Eq. 2 is shown as the ordinate in Fig. 3(a) and the curves on this diagram show the manner in which N_1 varies as a function of the frequency ratio and of the damping factor, $\lambda/(\pi f_n)$. The frequency ratio is the ratio of the exciting frequency, f_1, to the natural frequency, f_n, of free vibration of the mass-spring system. The natural frequency of free vibration without damping of the mass-spring system is given by

$$f_n = \frac{1}{2\pi}\sqrt{\frac{c_s g}{W_o}} \quad (4)$$

in which g is the acceleration of gravity. The derivation of Eq. 2 and a discussion of the preparation of Fig. 3(a) is readily available in any book on elementary vibrations.

For rotating machinery with unbalanced weights, the exciting force is a function of the exciting frequency, f_1, or

$$Q_1 = \frac{W_1}{g}\, l\, 4\pi^2 f_1^2 \quad (5)$$

in which W_1 is the weight of the unbalanced rotating part and l is the eccentric radius from the center of gravity of the part to the center of rotation. By introducing the value of Q_1 as defined by Eq. 5, the solution for the displacement, x, can again be determined. The results are shown in Fig. 3(b), in which the ordinate is equal to the ordinate of Fig. 3(a) multiplied by $(f_1/f_n)^2$. The value of maximum vibration amplitude is

$$x_{max} = N_1 \left(\frac{f_1}{f_n}\right)^2 \frac{W_1}{W_o}\, l \quad (6)$$

in which W_o represents the static weight of the mass. It should be noted that on Fig. 3(b) the curves all approach an ordinate value of 1.0 as the frequency ratio becomes large. Thus, for large values of the frequency ratio, the force transmitted through the spring to the rigid base is

$$(F)_{f_1/f_n \to \infty} = c_s\, x_{max} = c_s \frac{W_1}{W_0} l \ldots\ldots\ldots\ldots (7)$$

Another important quantity which can be defined by consideration of Figs. 3(a) and 3(b) is the term "resonance," used to denote the condition of vibration corresponding to a large increase in amplitude. For undamped forced vibrations this condition occurs at $f_1/f_n = 1.0$. When the exciting force has a constant amplitude regardless of frequency, the curves representing damped forced vibrations in Fig. 3(a) show the maximum amplitude magnification at values of f_1/f_n less than 1.0. For small values of damping the amplitude peaks occur at frequency ratios so close fo $f_1/f_n = 1.0$ that the difference is usually ignored. However, when the damping factor, $\lambda/(\pi\, f_n)$, equals 1.0, the peak is at $f_1/f_n = 0.7$. The other condition, for which the exciting force is a function of the exciting frequency, establishes peak amplitudes at a value of f_1/f_n greater than 1.0, which for the damping factor of 1.0 results in a peak at $f_1/f_n = 1.4$. This shows the effect of damping in shifting the frequency for maximum amplitude of vibration away from the "natural" frequency of the foundation. It also demonstrates that if Fig. 3(a) is used to determine the frequency for peak amplitude when the exciting force is actually frequency dependent, the selected frequency may be as low as one half the correct value.

OSCILLATORS ACTING ON THE SURFACE OF A SEMI-INFINITE ELASTIC BODY

Elastic Waves.—The vibrations examined previously have considered a closed system in which the periodic motion was confined to the mass and spring, with the spring supported by an immovable base. An actual foundation which has been set into vibratory motion by a periodic force becomes the source of periodic impulses which proceed into the subgrade in radial directions, similar to a sound wave. In the course of transmitting waves the particles of the subgrade also undergo periodic motion, but only at particular locations do these motions correspond to the motion of the foundation. Directly below the foundation base the subgrade material moves with the foundation and is "in phase" with the foundation motion. At a greater distance a zone of the subgrade moves opposite to the foundation motion and is designated as "180° out of phase." Fig. 4 illustrates the concept of phase relations of zones of the subgrade with the shaded areas representing the "in phase" zones. The spacing between the centers of the shaded zones is determined by the wave length, which, in turn, is established by the velocity of propagation of the elastic wave in the subgrade and the frequency of the load applications.

In an infinite, elastic, isotropic, homogeneous body disturbances may be propagated by waves of volume change, designated as the compressive wave, push wave, or P-wave, and by waves of distortion of constant volume designated as the shear wave, transverse wave, or S-wave. The compression and shear waves also transmit disturbances throughout the interior of a semi-infinite elastic, isotropic, homogeneous body, but, because of the free surface, a third type of wave appears. This surface wave has been designated as the

Rayleigh wave or R-wave. Additional waves appear in non-isotropic or layered materials and are described, for example, by L. D. Leet (10) and W. M. Ewing, W. S. Jardetzky, and F. Press (11). The velocity of propagation of the shear wave is given by

$$v_s = \sqrt{\frac{G\,g}{\gamma}} = \sqrt{\frac{G}{\rho}} \qquad (8)$$

in which G is the shear modulus of elasticity, γ is the unit weight, and ρ is the mass density of the subgrade material. The velocities of the compression and Rayleigh waves are dependent on Poisson's ratio and are related to the shear wave velocities as shown in Fig. 5.

For actual materials, the different wave velocities may be determined by seismic methods in the field or in the laboratory. The significant variables which affect these velocities are the modulus of elasticity, the unit weight, and Poisson's ratio, as shown by Eq. 8 and Fig. 5. For soils, the water content and confining pressure contribute to changes in these variables and, as a consequence, seasonal variations in water content, for example, will cause seasonal changes in the wave velocities at a particular location. The compressional wave velocity for granular materials varies approximately with the 1/6 power of the confining pressure as shown by the test data assembled in Fig. 6.

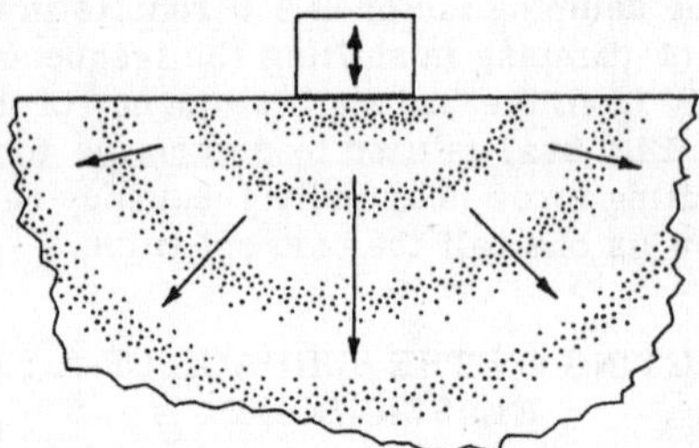

FIG. 4.—PROPAGATION OF ELASTIC WAVES INTO THE SOIL BENEATH AN OSCILLATOR

Vertical Impulses Acting at the Surface.—H. Lamb (17) studied the effects produced by a single impulse which acted at the surface of a semi-infinite, isotropic, homogeneous, elastic solid. He considered primarily the effects produced at the surface and found that the disturbance produced by the impulse spreads over the surface in the form of a symmetrical annular wave system. His theory showed nothing corresponding to the long succession of to and fro vibrations which are characteristic of the seismograph records.

Lamb also studied the effects produced by periodic vertical and horizontal forces applied at a point or distributed along a line on the surface of the semi-infinite solid. By integration of the effects of the periodic vertical point load over a circular area, Reissner obtained (2) a solution which could be used to approximate the behavior of a vibrating mass resting on the surface of a semi-infinite solid. Reissner's solution considered the pressure to be uniformly distributed over the circular base area. Later, Sung (3) and Quinlan (4) studied the effects resulting from uniform, parabolic, and rigid base types of pressure distributions over the circular loading area.

The vertical periodic force which acts on the surface of the semi-infinite solid can be generated by a vibrator having a force output independent of frequency, or by a mechanical vibrator consisting of rotating eccentric weights.

The oscillator shown in Fig. 7(a) develops the same force amplitude Q_1 at all frequencies. The oscillator shown in Fig. 7(b) produces a force depending on the angular velocity of two unbalanced masses m_1. The radius of rotation of the center of gravity of each mass is designated by l and the masses rotate at an angular velocity, ω, in opposing directions to balance the horizontal components of the rotating forces. The unbalanced vertical force has an amplitude, Q_1, determined by

$$Q_1 = 2\, m_1\, l \omega^2 \quad \text{.................. (9)}$$

In their theoretical studies of the problem, Reissner, Sung, and Quinlan considered the contact area between the oscillator and semi-infinite body to be

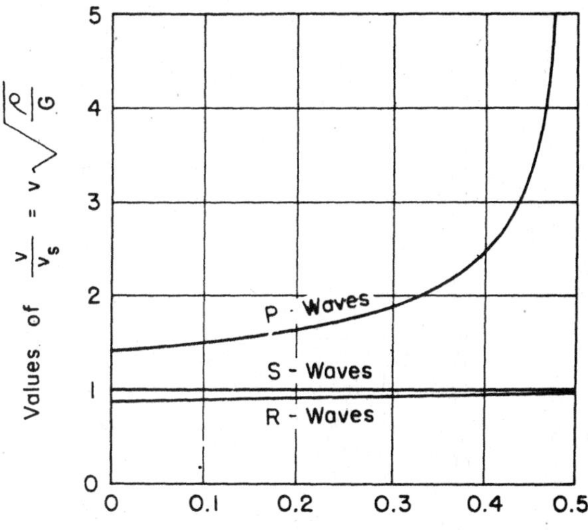

FIG. 5.—RELATION BETWEEN POISSON'S RATIO, μ, AND VELOCITIES OF PROPAGATION OF COMPRESSION (P), SHEAR (S), AND RAYLEIGH (R) WAVES IN A SEMI-INFINITE ELASTIC MEDIUM

circular and of radius r_o. The dynamic characteristics of the vibrator-soil system were found to depend on the parameters: (a) the radius r_o of the loading area, (b) the total mass m_o (= W_o/g) of the oscillator, (c) the amplitude of the dynamic force applied (Q_1), (d) the distribution of the contact pressure on the circular oscillator base, and (e) the Poisson's ratio, μ, mass density, ρ, and shear modulus, G, of the foundation material. A significant term used in the analysis is the dimensionless quantity

$$b = \frac{m_o}{\rho\, r_o^3} = \frac{W_o}{\gamma\, r_o^3} = \frac{\pi\, p_o}{\gamma\, r_o} \quad \text{.............(10)}$$

which denotes the mass ratio of the system. The quantity b can be interpreted as the ratio between the static mass of the oscillator and the mass of a cylinder of the elastic half-space having a radius r_o and a height r_o/π. Note tnat

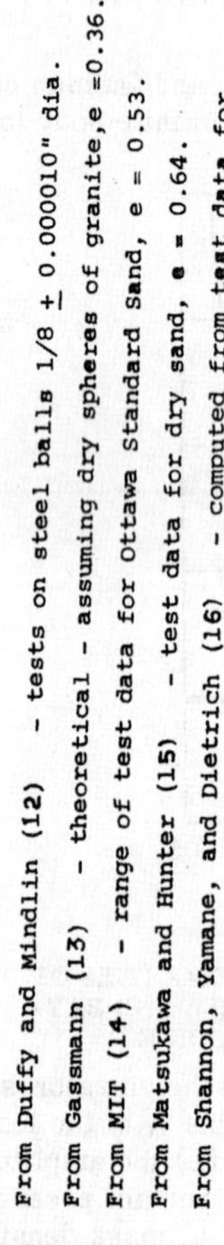

From Duffy and Mindlin (12) - tests on steel balls 1/8 ± 0.000010" dia.
From Gassmann (13) - theoretical - assuming dry spheres of granite, e = 0.36.
From MIT (14) - range of test data for Ottawa Standard Sand, e = 0.53.
From Matsukawa and Hunter (15) - test data for dry sand, e = 0.64.
From Shannon, Yamane, and Dietrich (16) - computed from test data for Ottawa Standard Sand, e = 0.61.

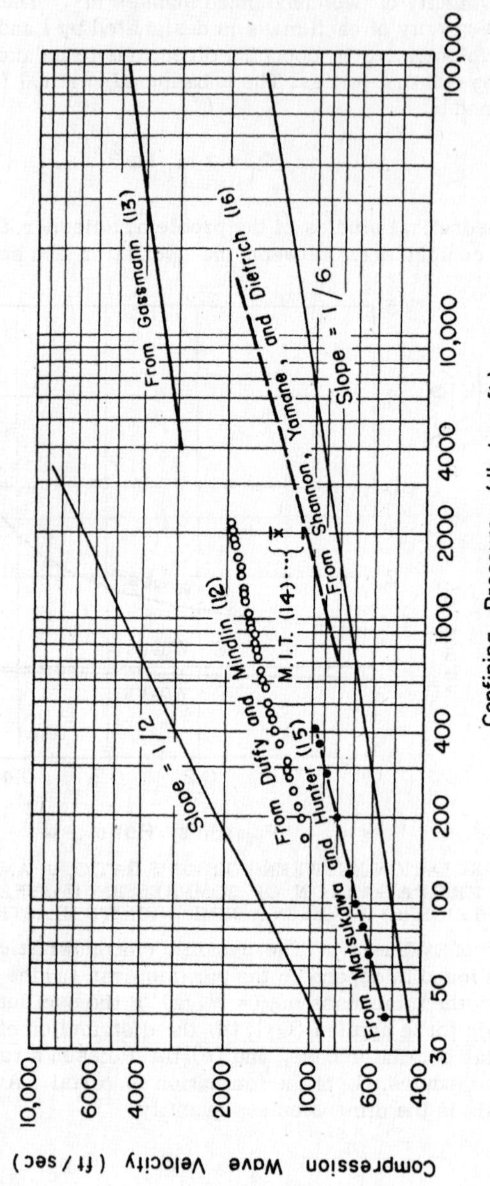

FIG. 6.—EFFECT OF CONFINING PRESSURE ON THE COMPRESSION WAVE VELOCITY IN GRANULAR MATERIALS

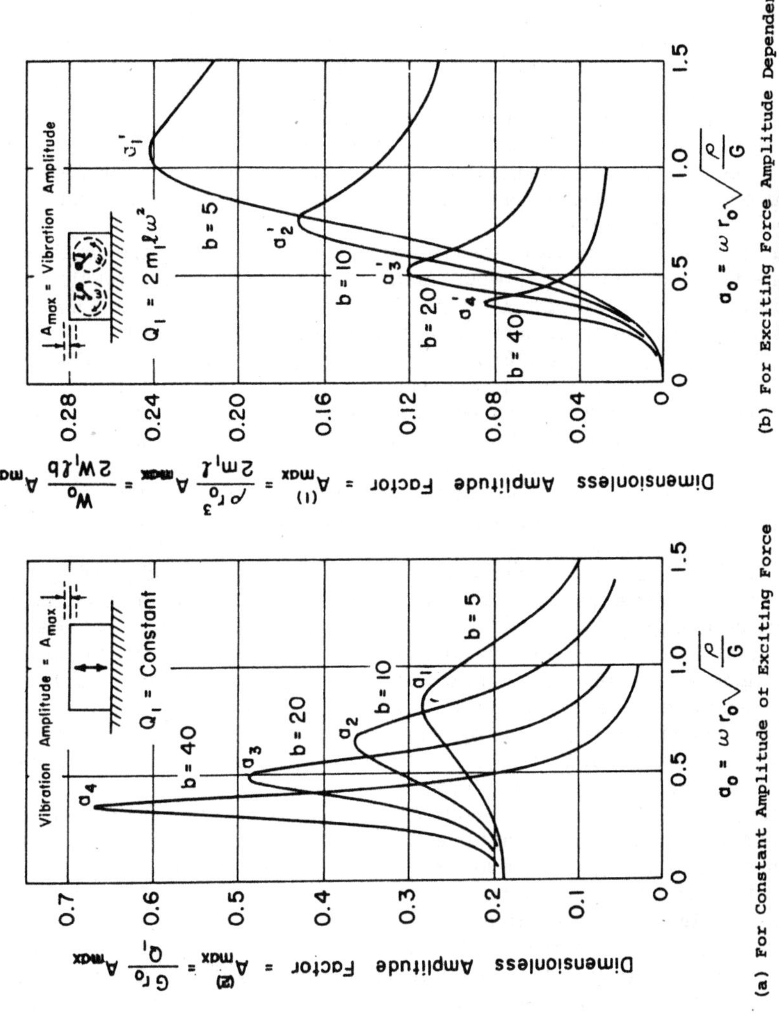

FIG. 7.—AMPLITUDE VS. FREQUENCY RELATIONS FOR RIGID BASE OSCILLATOR RESTING ON ELASTIC SEMI-INFINITE BODY ($\mu = 0.25$)

for submerged soils the total saturated density is used in evaluating ρ. A dimensionless frequency term is also utilized which has the form

$$a_o = \omega r_o \sqrt{\frac{\rho}{G}} = 2 \pi f_o r_o \sqrt{\frac{\rho}{G}} \qquad (11)$$

in which f_o is the resonant frequency for the forced vibration.

By using these dimensionless mass ratio and frequency factors along with the elastic characteristics of the subgrade, a theoretical solution for the amplitude of oscillation was obtained. For convenience in further computations, the actual amplitude was converted into a dimensionless amplitude factor. In Fig. 7 the variation of this amplitude factor, with changes in the frequency factor, are shown for several constant values of the mass ratio, b. The diagrams of Fig. 7 result from the assumptions of Poisson's ratio of 0.25 for the subgrade material and a distribution of contact pressure corresponding to that produced by a rigid circular base.

Fig. 7 (a) shows the amplitude-frequency factor curves which are developed by a constant force oscillator, whereas the curves on Fig. 7(b) result from consideration of the rotating mass type of oscillator. The curves in Fig. 7(a) and 7(b) are similar in appearance to those obtained for damped forced vibrations as shown in Figs. 3(a) and 3(b) for the corresponding types of oscillators. The value of a_o corresponding to the peak amplitude for each curve in Fig. 7 may be used to determine the resonant frequency for a particular oscillator-soil system. Thus, by assembling the values of a_o corresponding to the peak amplitude for a particular value of b, the relations between a_o and b may be established as shown in Fig. 8(a). Also, by taking the value of the peak amplitude factor and plotting it against the corresponding value of b, the curves of Fig. 8(b) were obtained. Thus, the points designated a_1, a_2, a_3, ... a'_1, a'_2, a'_3, ... in Fig. 7 determine the points in Fig. 8 indicated by the same symbols. In this manner, Sung has condensed the results of voluminous computations into graphical form, which may be used readily for analysis or design. His diagrams show the relations between b, Poisson's ratio, the pressure distribution, and the maximum power input required, and their influence on the dimensionless frequency factor, a_o, and the maximum amplitude of oscillation. The computations made by Sung were used by the writer to prepare Fig. 8, which is applicable for the condition of a rigid base type pressure distribution on the circular oscillator base.

The curves in Fig. 8 may be used to determine the resonant frequency and maximum amplitude of oscillation for an idealized equivalent of the actual machine foundation. It is necessary that the designer obtain data including (a) the unbalanced forces and operating frequencies of the machines, and (b) at least approximate values for the shear modulus, G, the unit weight, γ, and effective Poisson's ratio of the supporting soil. Then the designer has a choice of the footing-soil contact area (which establishes the radius, r_o, of an equivalent circular area of contact) and the static weight, W_o, of the machine and foundation. These quantities may be varied to provide acceptable dynamic characteristics for the installation. A detailed illustration of the use of the curves in Fig. 8 for the study of a machine foundation is presented subsequently.

Rocking Oscillation.—The rocking type of oscillation of a mass resting on the surface of an elastic semi-infinite body consists of a periodic rotation about a horizontal line passing through the base of the mass (Fig. 9). Arnold, Bycroft, and Warburton (5) have given a solution to this problem for the conditions of a cylindrical mass resting on an elastic semi-infinite body which has

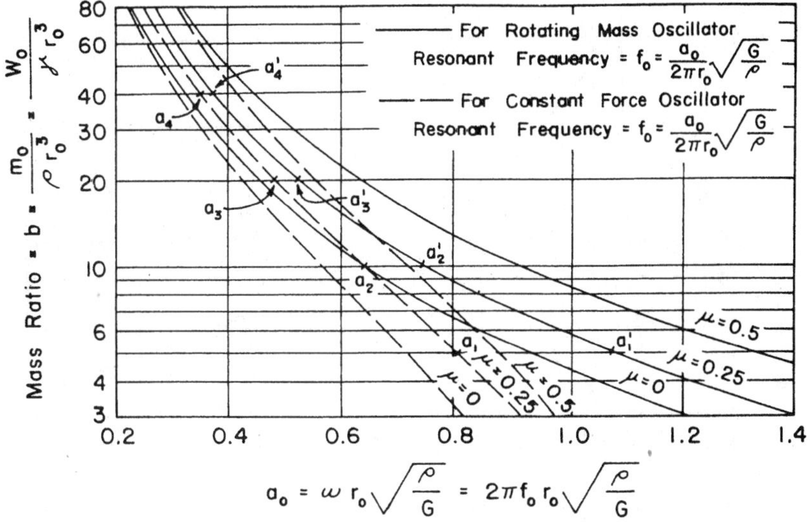

(a) Mass Ratio vs. Frequency Factor Relations

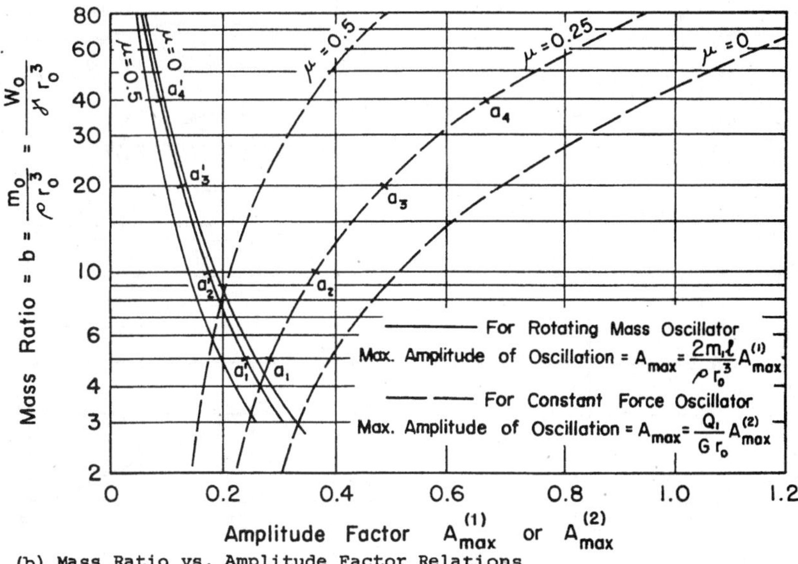

(b) Mass Ratio vs. Amplitude Factor Relations

FIG. 8.—CHARACTERISTICS OF VERTICAL OSCILLATION FOR AN OSCILLATOR WITH A RIGID CIRCULAR BASE RESTING ON A SEMI-INFINITE ELASTIC MEDIUM

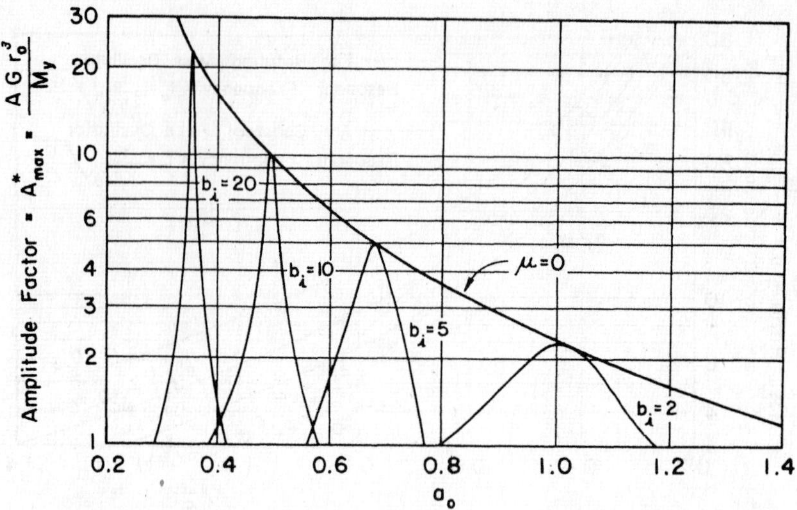

(a) Amplitude Factor vs. Frequency Factor.

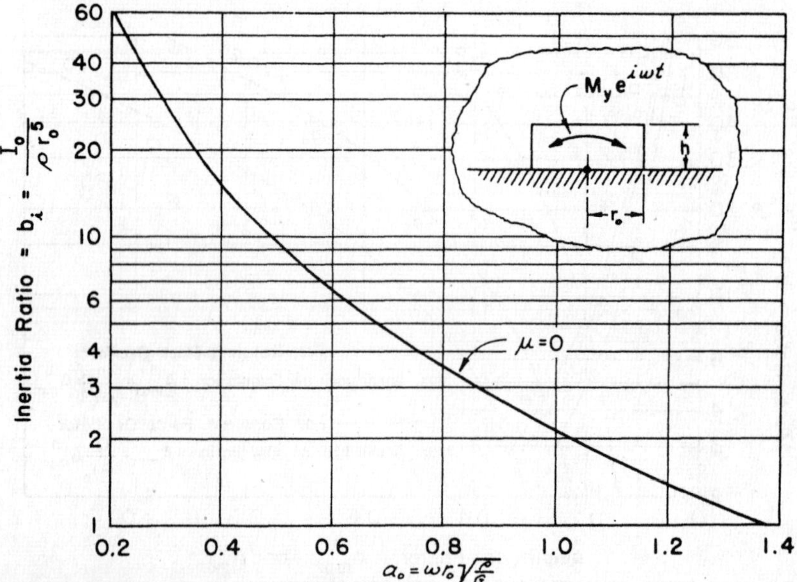

(b) Inertia Ratio vs. Frequency Factor.

FIG. 9.—CHARACTERISTICS OF ROCKING OSCILLATION FOR AN OSCILLATOR WITH A RIGID CIRCULAR BASE RESTING ON A SEMI-INFINITE ELASTIC MEDIUM

Poisson's ratio equal to zero. The amplitude of the periodic moment, M_y, was considered to be independent of frequency and the distribution of pressure at the circular base of the oscillator was assumed as the rigid base distribution.

For the rocking condition, the significant parameter is the "inertia ratio," b_i, which is defined by

$$b_i = \frac{I_o}{\rho\, r_o^5} \qquad (12)$$

in which I_o is the mass moment of inertia of the oscillator. For a cylindrical oscillator as shown in Fig. 9, which has a uniform distribution of its mass, the expression for I_o is

$$I_o = \frac{W_o}{g}\left(\frac{r_o^2}{4} + \frac{h^2}{3}\right) \qquad (13)$$

In Fig. 9(a) the amplitude versus frequency curves are shown for b_i = 2, 5, 10, and 20. Also shown is the envelope curve which is tangent to each curve of amplitude versus frequency for the different values of b_i. The point of tangency is close to the point of maximum amplitude, particularly for larger values of b_i, and this tangent curve is used to define the relation between frequency at maximum amplitude and the value of the inertia ratio, b_i, which is shown in Fig. 9(b).

In analyzing a given oscillator-soil combination, the value of b_i may be computed from the known physical constants of the system. Then by using Fig. 9(b), the value of a_o corresponding to this value of b_i is found directly. The resonant frequency can be computed directly from the value of a_o determined previously and the value of amplitude of oscillation at resonance may be determined from Fig. 9(a). It should be noted that the amplitude versus frequency curves for a given value of b_i are much narrower for the rocking case (Fig. 9) than for the case of vertical oscillation (Fig. 7). This means that the range of frequency which will excite resonant or near resonant amplitudes of oscillation is much smaller for the rocking case, and that a relatively small change of frequency will reduce the vibration amplitude to small values.

Sliding Type Oscillation.—Arnold, et al (5) also treated the problem of horizontal oscillation of a cylindrical mass which rests on the semi-infinite body. They evaluated the results only for the case of Poisson's ratio equal to zero. In Fig. 10(a) the amplitude versus frequency relations are shown, and from this information the curves for the mass ratio versus frequency were established as shown in Fig. 10(b).

A foundation which tends to oscillate horizontally will undoubtedly have both the sliding and rocking types of oscillation. The possibilities of resonance can be estimated for each condition independently, and then the resulting motion determined by superposition. Superposition of these motions will usually establish a center of rotation of the foundation at some distance into the subgrade.

Interference Effects.—The theoretical solutions consider that only one oscillator is acting on the surface of the semi-infinite body. In many practical cases, several machines are placed within the same building on adjacent foundations. The question becomes: "What is the effect of one machine installation

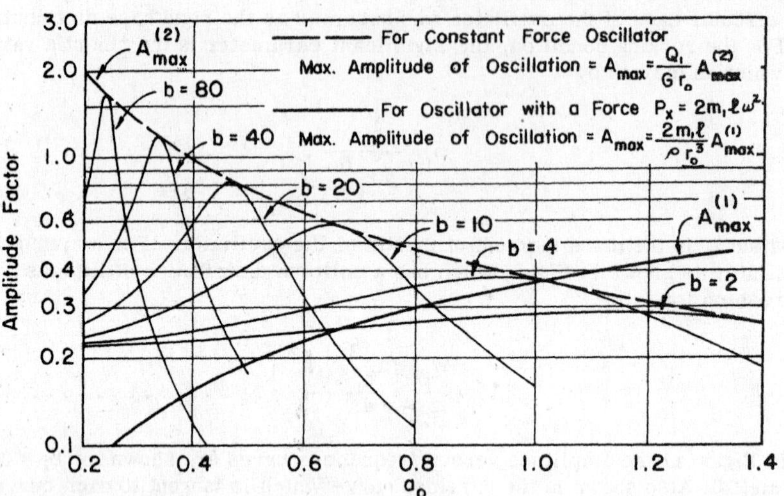

(a) Amplitude Factor vs. Frequency Factor.

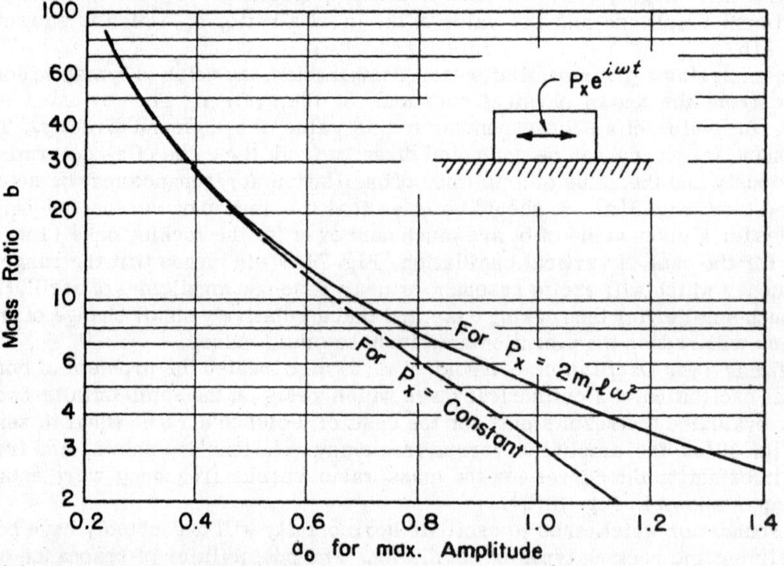

(b) Mass Ratio vs. Frequency Factor.

FIG. 10.—CHARACTERISTICS OF SLIDING OSCILLATION FOR AN OSCILLATOR WITH A RIGID CIRCULAR BASE RESTING ON A SEMI-INFINITE ELASTIC MEDIUM

upon the dynamic behavior of a neighboring installation?" The variables involved in this problem include the weight of the foundation blocks and machinery, the shape and elevation of the block bases, and the dynamic characteristics of the subgrade. In order to study the interference between two rectangular footings, J. W. Clawson (18) mounted two rectangular "feet," each having plan dimensions of 3.5 in. by 11.25 in., beneath a single rotating mass oscillator. The center line spacings of these rectangular feet were varied from 4 in. to 20 in., or from about 1 to 6 footing widths. Preliminary tests, using only dense sand as the subgrade material, indicated that for spacings from 1 to approximately 4 footing widths, the footings behaved as a single unit and their behavior could be treated by use of Fig. 8. For spacings greater than approximately 4 footing widths they acted as independent footings. Further tests using different ratios of length to width for rectangular footings, and different subgrades must be carried out before definite relations can be established for these interference effects.

Effect of a Layered Subgrade.—Vertical oscillations of a mass at the surface of a single elastic layer resting on a rigid medium, were treated by G. B. Warburton (19) (20). He considered the rigid base type of pressure distribution to exist at the circular oscillator base. The additional parameter which becomes significant in the study of oscillators on a layered subgrade is the ratio, R, of the thickness of the layer to the radius, r_o, of the oscillator base.

For different values of b (defined by Eq. 10), it was found that the amplitude of oscillation was many times larger than that occurring for the same value of b and a semi-infinite medium. This is caused by the reflection of the elastic waves from the rigid surface underlying the elastic layer. However, for values of R of approximately 6 or greater (that is, for a layer more than 3 oscillator base diameters in thickness) it was indicated that the dynamic behavior would correspond to that for a semi-infinite medium.

The amplitude-magnifying effect of a rigid surface underlying the elastic layer should not constitute a practical difficulty. If a rigid layer (rock) is within 3 foundation diameters of the design base elevation, it would be logical to support the block directly on piles to rock.

In the intermediate case the subgrade modulus increases with depth below the ground surface. For a subgrade consisting primarily of clean sand, the increase in stiffness with depth will approximate a continuous function. An average value of the elastic soil constants must be estimated for the zone of the subgrade which is effective in supporting the foundation block. Then these values can be used with the theory of homogeneous subgrades (Fig. 8) to evaluate the dynamic characteristics of the foundation. If several distinct layers of marked difference in stiffness exist in the subgrade, the use of the homogeneous subgrade theory could give only a very crude approximation to the correct values.

Oscillator on a Weightless Subgrade.—The similarity between the curves of Figs. 3 and 7 has led to attempts to establish a correlation between the results from the theory of damped forced vibrations and the theory considering the elastic semi-infinite body. Reissner (2) found that a correlation was possible for any specific case, but that a general relation could not be established which would permit extrapolation from one b-value to another. The extrapolation from one value of b corresponding to a model to another value of b corresponding to the prototype is necessary if this kind of procedure is to be of any use in design.

Another approach is to eliminate any attempts to determine a damping constant and to compute the natural frequency of free vibration of a weight supported by a weightless, undamped spring. This procedure will give no information about the amplitude of resonant vibration but will give a value of frequency. It is of interest to evaluate the approximation involved.

For the case of a weight having a rigid circular base resting on the surface of a weightless elastic medium, the spring constant for vertical motion is

$$c_s = \frac{P_z}{z} = \frac{4\,G\,r_o}{(1-\mu)} \quad \text{.................(14)}$$

By substituting this value of c_s into Eq. 4, the natural frequency of vertical oscillation is determined as

$$f_n = \frac{1}{\pi}\sqrt{\frac{G\,r_o\,g}{W_o\,(1-\mu)}} \quad \text{.................(15)}$$

or

$$f_n^2 = \frac{G\,r_o\,g}{\pi^2\,W_o\,(1-\mu)} \quad \text{.................(16)}$$

In order to evaluate the difference between the natural frequency as given by Eq. 15 and the resonant frequency given by the theory which includes the subgrade weight (as in Fig. 7), R. Jones (21) noted that the parameters b and a_o (Eqs. 10 and 11) contain many of the terms included in Eq. 16. That is,

$$b\,a_o^2 = \frac{W_o}{\gamma\,r_o^3}\;\frac{4\,\pi^2\,f_o^2\,r_o^2\,\gamma}{g\,G} = \frac{4\,\pi^2\,f_o^2\,W_o}{G\,r_o\,g} \quad \text{.......(17)}$$

or

$$f_o^2 = \frac{G\,r_o\,g}{\pi^2\,W_o}\;\frac{b\,a_o^2}{4} \quad \text{...............(18)}$$

Eqs. 16 and 18 will be equal when

$$\frac{b\,a_o^2}{4} = \frac{1}{1-\mu} \quad \text{...................(19)}$$

The relation given in Eq. 19 was used to prepare the curves shown in Fig. 11, which show the approximations involved in the use of the weightless subgrade procedure.

The solid curves in Fig. 11 result from excitation provided by a rotating mass oscillator and the dashed curves result from excitation by a constant force oscillator. Previous comparisons made by Reissner (2), Arnold et al (5), and Jones (21), using diagrams similar to Fig. 11 have considered only the constant-force oscillator and curves corresponding to the dashed curves in Fig. 11 have been obtained. Consequently, it is a little surprising to find the close agreement between the weightless-subgrade results and those obtained from consideration of the rotating mass oscillator. Differences of less than 10% are unimportant when compared to the probable errors in estimating the elastic constants for the subgrade.

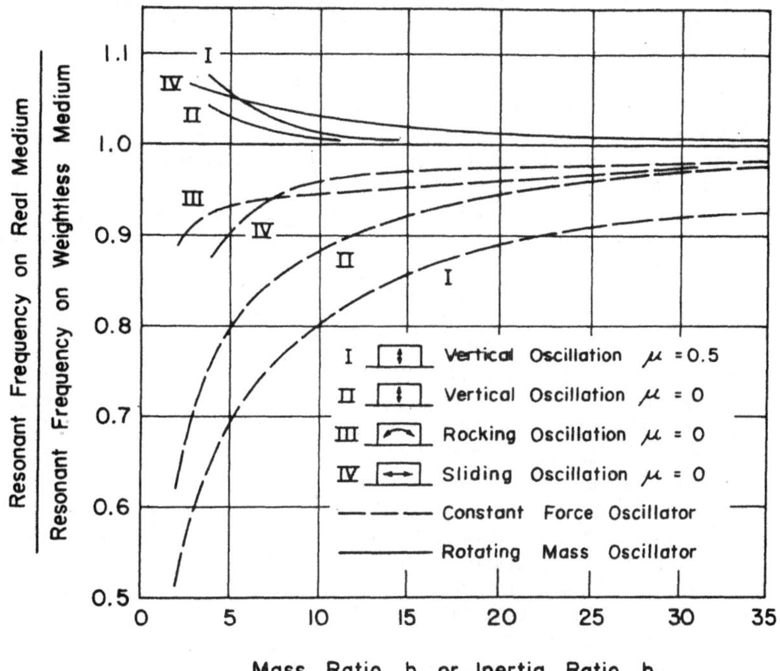

FIG. 11.—COMPARISON OF THE RESONANT FREQUENCY OBTAINED BY CONSIDERING THE SOIL AS AN ELASTIC SEMI-INFINITE BODY HAVING WEIGHT TO THAT OBTAINED BY CONSIDERING THE MEDIUM TO BE WEIGHTLESS (FOR A RIGID BASE TYPE OF PRESSURE DISTRIBUTION ON THE CIRCULAR OSCILLATOR BASE)

When the foundation has a base shape which can be approximated by a circle, the dynamic characteristics of the installation can be determined directly by use of Fig. 8. If the foundation plan deviates markedly from the circular shape, or if the foundation is embedded to an appreciable depth, the subgrade can be treated as a weightless spring for computation of the natural frequency. Then, from the computed value of the mass ratio, b, the probable deviations from the correct resonant frequency can be estimated with the aid of Fig. 11.

VALUES FOR THE SHEAR MODULUS OF SOILS

For actual soils the shear modulus depends on the magnitude of the static and dynamic loads as well as on the water content, which may have a seasonal variation near the surface. Consequently, it is often difficult to determine a G-value for a particular soil which is actually a constant. However, a range of G-values can usually be established which permits a reasonable estimate of the foundation frequency and amplitude of vibration.

The preferable methods of estimating the soil constants involve the use of data obtained at the actual site. These data may be obtained by evaluating the dynamic behavior of a small oscillator or they may be obtained by seismic methods. If borings show the subgrade to be reasonably uniform, then the use of an oscillator operating at frequencies and amplitudes comparable to those anticipated for the prototype foundation will give useful design information. If the oscillator test data include the resonant frequency plus either the amplitude of vibration or the power input to the drive motor, the method of determining both G and π described by Sung may be used. The values of G and μ determined from the model test may then be used with the prototype foundation data to determine the resonant frequency and expected amplitudes of vibration.

Use of Oscillator Test Data.—Table 1 has been prepared to show typical values of the shear modulus and amplitudes of vibration obtained from test data of oscillators acting on soils. Except for the two cases from Toriumi (6) the data refer to vertical oscillations. It should be noted that most of the values listed in Cols. 6, 7, and 9 of Table 1 were computed from test data including oscillator weight, base plate area, resonant frequency, and eccentric weight. It was arbitrarily assumed that Poisson's ratio was 0.25 (but 0 for horizontal oscillation) and that a rigid base type pressure distribution existed.

Col. 9 was included in order to estimate the probability of the oscillator "jumping" free of the ground when the oscillator force acts upward. When the quantity $A\omega^2$ equals 1.0 g, no contact pressure exists, and when this quantity becomes greater than 1.0 g, a jump occurs. It would be anticipated that the pressure distribution at the oscillator base would approach a parabolic distribution as the ratio of maximum to minimum average base contact pressure increases. Under these conditions, failure of the soil would undoubtedly occur near the edges of the base plate where the confining pressure was low. This condition is indicated by test No. 206 from the DEGEBO tests listed in Table 1, for which the computed amplitude of oscillation was 0.0064 in. (rigid-base assumption, with $\mu = 0.25$) as contrasted to a measured amplitude of 0.0208 in. A computed amplitude of 0.015 in. is obtained by using the assumptions of $\mu = 0.25$ and a parabolic base pressure distribution in conjunction with the appropriate curves given by Sung (3). For the higher vibration amplitudes, the oscillator acts as a compactor and develops a harder "lump" of sand directly below the center of the oscillator (22), which tends to change the pressure distribution at the oscillator base. Test data given by A. Pauw (22) and the writer (24) have pointed out that it is very likely that the base pressure distribution changes as the vibration amplitude becomes more violent.

From a careful consideration of the data given in Col. 9 of Table 1 and the brief examination in the preceding paragraph, it should be evident that oscillator test data can be quite misleading for design purposes if the test conditions are not comparable to the design conditions. The right hand boundary of the limiting curve from E. Rausch (8) which is shown in Fig. 2 corresponds to values of $A\omega^2$ equal to 0.50 g. Thus at the maximum upward position of the foundation during oscillation, the soil pressure is reduced to one-half the static value, and at the maximum downward position the soil pressure is 1.5 times the static value. K. Terzaghi has pointed out (25) the variations in the "hysteresis" modulus of soils which would be anticipated for differences in the range of the

TABLE 1.—OSCILLATOR TEST DATA

Material (1)	Test No. (2)	Oscillator Base Area, in sq ft (3)	Static Contact Pressure in psf (4)	Resonant Frequency, in cps (5)	Vibration Amplitude, in in. (6)	Computed Value of G in psi (7)	Assumed Poisson's Ratio (8)	A ω^2 (9)	Type of Oscillator (10)[a]	Motion (11)[b]	Reference in Bibliography (12)
Berlin	52	2.69	1,110	21.9	0.041[c]	2,360	0.25	2.01 g			
sand,	54	2.69	1,110	22.8	0.020[c]	2,560	0.25	1.09 g			
e = 0.53	56	2.69	1,110	27.5	0.007[c]	3,720	0.25	0.77 g			
Micaceous fine sand	209	10.76	315	29.5	0.008[c]	2,330	0.25	0.67 g	R	V	(1)
Loamy sand	206	10.76	315	23.2	0.0064[c] 0.0208[d]	1,430	0.25	1.15 g			
Dry silty clay	A	0.175	430	108		4,520	0.50				
		0.175	920	77		4,660	0.50				
		0.175	1,520	59		4,550	0.50				
Wet, soft silty clay	B	0.175	430	58		1,300	0.50				
		0.175	920	44.5		1,560	0.50				
		0.175	1,520	38		1,890	0.50				
Wet, soft silty clay	C	0.785	73	110		2,270	0.50		E	V	(21)
		0.785	430	50		2,000	0.50				
	C	0.136	410	77.5		1,800	0.50				
		0.136	2,470	41		3,040	0.50				
Dry, soft silty clay	D	0.349	216	96		2,570	0.50				
		0.349	760	52		2,400	0.50				
	D	0.087	825	71.5		2,410	0.50				
		0.087	3,014	43		3,010	0.50				
Clean dense quartz sand		0.136	420	46.8	0.0022[c]	1,830	0.25	0.49 g			
		0.136	660	39	0.0018[c]	1,990	0.25	0.28 g	R	V	(18)
		0.136	960	33	0.0013[c]	2,060	0.25	0.14 g			
	1	686	1,230	3.75	0.0077[c] 0.0059[d]	6,700	0		R	H	(6)
	3	576	1,080	4.17	0.0091[c] 0.0075[d]	3,400	0				

[a] R denotes rotating mass, E denotes electromagnetic. [b] V denotes vertical motion, H is horizontal. [c] Computed, using Fig. 8 or Fig. 11. [d] Measured.

maximum to minimum stresses. This factor should be considered in planning the oscillator tests.

G. Values from Elastic Wave Velocities.—Evaluation of the soil parameters can also be estimated from tests to determine the velocities of the elastic waves through the soil. Generally, this is carried out at or near the ground surface and it is necessary to evaluate the wave velocities for the zone of soil which will be effective in supporting the foundation. The theories, procedures, and equipment involved in the determinations of elastic wave velocities in soils have been considered by R. K. Bernhard and J. Finelli (26) and Bernhard (27). Test data obtained by Jones (21) at sites A and B permitted evaluation of the shear modulus at the two sites as 5,750 psi and 1,870 psi, respectively. These values compare favorably with the values listed in Table 1, which were obtained by oscillator tests at the same sites. Toriumi (6) also used the elastic-wave data to evaluate the shear modulus at the sites indicated as tests 1 and 3 in Table 1. By this method he obtained G = 6,060 psi for test 1 and G = 3,400 psi for test 3, values which compare well with 6,700 psi and 3,400 psi as shown in Table 1.

The velocity of propagation of elastic waves varies with the type of soil and its water content. For granular soils, the velocity of propagation of the longitudinal wave also varies approximately to proportion to the 1/6 power of the confining pressure. The curves shown in Fig. 6 indicate the relations between the compression wave velocity and confining pressure for several series of tests on granular materials. It would be anticipated that the shear wave velocity would vary in a corresponding manner. Test data obtained by K. Iida (28) (29) have also shown a marked change in both the compression and shear wave velocities in granular materials with an increase in water content. However, these tests were run on small vertical cylinders confined only by a thin cellophane shell, with the result that the confining pressure at any point in the cylinder was caused only by the weight of the overlying material. Probably the effect of water content in drastically reducing the shear modulus would become of less importance as the confining pressure is increased.

The use of elastic wave velocities for evaluation of soil properties at specific locations is becoming a more common practise in recent years. In interpretation and use of the data it is necessary to recognize, as is also the case for concrete, that the elastic modulus determined from the wave velocities is likely to be higher than that which will occur under the operating conditions. This increase exists because only very small amplitudes of motion occur during the wave propagation whereas motions of a higher order of magnitude are induced by a vibrating foundation. The use of both a small oscillator and the elastic wave procedure should permit determination of soil constants which would "bracket" the effective values.

USE OF PILES

For high-speed machinery having operating speeds greater than 1,000 rpm, it is usually satisfactory to provide a foundation which has a natural frequency of one-half to one-third of the operating frequency. During starting and stopping, the machine will operate briefly at the resonant frequency of the foundation. Thus, the designer must estimate the probable amplitudes of vibration both at resonance and at the operating frequency, which may be done with the aid of Figs. 3 and 8.

At the other extreme, heavy, low-speed machinery requires a foundation which has a natural frequency of at least twice the operating speed of the machine. When the foundation rests directly on the soil the natural frequency of the unit may be increased by reducing the total static weight of the assembly, by increasing the base area of the foundation, or by increasing the G-value of the soil by compaction or by chemical injection. In certain installations, adjustment of these variables within the allowable limits will not increase the frequency to a satisfactory value. Often the use of piles will provide the required stiffness.

When point-bearing piles to rock are installed, an estimate of the possible increase in stiffness of the foundation due to the pile support may be obtained through the theory of longitudinal elastic wave propagation through bars. Even when the pile is unloaded a longitudinal exciting force will cause the pile to vibrate longitudinally at its natural frequency. This phenomenon has been well understood by engineers concerned with pile driving, and studies of this problem have been presented by A. E. Cummings (30), and E. A. Smith (31), among others. A weight added to the upper end of the pile lowers the natural frequency of longitudinal oscillation. From the theory given by S. P. Timoshenko (32), Fig. 12 was prepared to show the influence of pile length and working stress on its natural frequency under the conditions of absolute fixity at the rock surface. Note that the theoretical value is the maximum frequency that could be expected from the use of a point bearing pile. The actual frequency would be somewhat lower because the point would cause an elastic compression of the rock at the point of contact. Fig. 12 shows quickly whether an increase in resonant frequency of the foundation can be provided economically by the use of point bearing piles which carry a negligible load by skinfriction.

USE OF THE THEORETICAL PROCEDURES

To demonstrate the use of some of the theoretical methods described previously, two typical block type foundations for machines are analyzed. Before the dynamic behavior of the foundation can be treated, it is necessary to evaluate the unbalanced forces which exist in the particular type of machine. The manufacturer's data may state the unbalanced forces explicitly, or it may be necessary to compute these forces from the weights and dimensions of the moving parts. W. K. Newcomb (33) has presented a description of the unbalanced forces in multicylinder engines which may be used, or it may be preferable to refer to the basic principles of forces developed in crank-type engines as given, for instance, by J. P. Den Hartog (34).

In addition, an evaluation of reasonable values for the shear modulus, G, and Poisson's ratio, μ, of the soil must be made by any one of the procedures indicated previously.

Example A - Foundation for a Single-Cylinder Engine.—Fig. 13 shows a hollow concrete foundation block which was designed to rest on the subgrade just below the basement level and to support the engine at the first-floor level. The subgrade is a sandy clay for which the soil constants were estimated to be G = 2,500 psi, μ = 0.50, and γ = 95 lb per cu ft.

The single-cylinder engine operates at 1,800 rpm, at which speed the vertical primary unbalanced force is 3,450 lb and the secondary unbalanced force

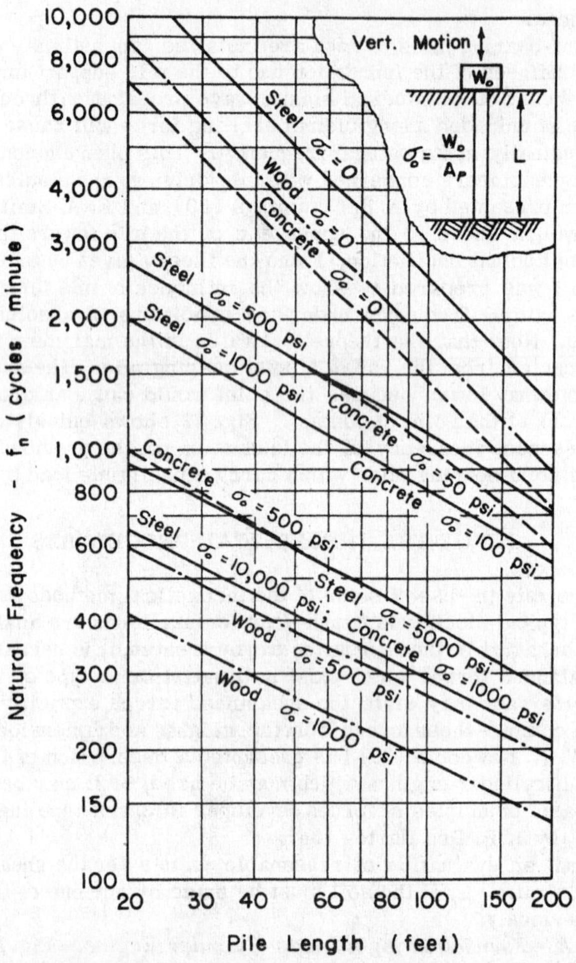

FIG. 12.—RESONANT FREQUENCY OF VERTICAL OSCILLATION FOR A POINT-BEARING PILE CARRYING A STATIC LOAD W_0. (LOADED STRATUM IS RIGID)

is 1,075 lb. The expression for the acceleration of the piston

$$\frac{d^2 x_p}{dt^2} = r_1 \omega^2 \left(\cos \omega t + \frac{r_1}{r_2} \cos 2\omega t \right) \quad \ldots\ldots\ldots\ldots (20)$$

establishes the fact that the primary unbalanced force pulsates at the same frequency as the crankshaft rotation, whereas the secondary unbalanced force oscillates at twice the crankshaft frequency. The terms r_1 and r_2 in Eq. 20 refer to the crank throw and connecting rod length, respectively, as shown in Fig. 13.

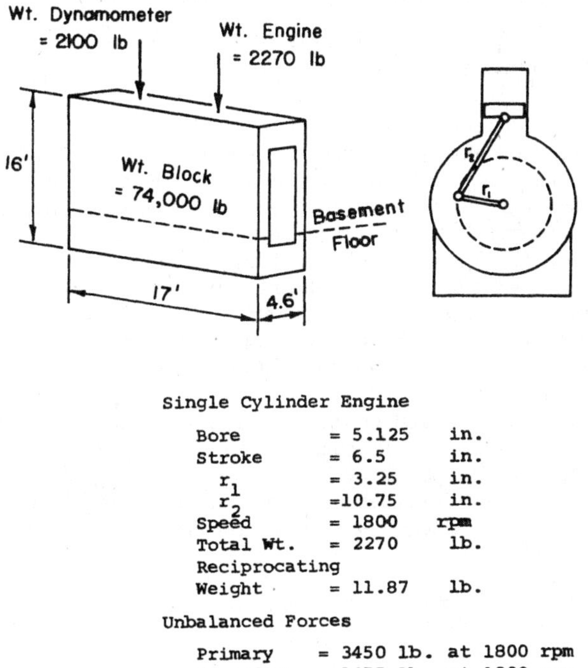

FIG. 13.—MACHINE AND FOUNDATION DATA FOR EXAMPLE A

The design conditions for this installation require that the resonant frequency of vertical oscillation be less than 900 rpm (half the operating frequency) and that the amplitude of vibration be less than ± 0.0005 in.

For the block without any footing extension, the static soil pressure is 0.5 tons per sq ft, which is less than one quarter of the estimated allowable bearing pressure for the sandy clay. The pulsating vertical motion caused by the unbalanced engine forces may be considered in two parts, one a vertical load applied directly above the centroid of the foundation base, and a moment which tends to cause a pitching motion of the block. For a first approximation to the vertical vibration characteristics of the foundation block, consider a circular footing having a radius r_o determined by

$$r_o = \sqrt{\frac{\text{base area}}{\pi}} = \sqrt{\frac{78.2}{\pi}} = 5.0 \text{ ft}$$

Then the mass ratio b is

$$b = \frac{W_o}{\gamma r_o^3} = \frac{78,370}{95 \times 125} = 6.6$$

From Fig. 8(a), $a_o = 1.14$, and the resonant frequency is

$$f_o = \frac{a_o}{2 \pi r_o} \sqrt{\frac{G g}{\gamma}} = \frac{1.14}{10 \pi} \sqrt{\frac{2500 \times 144 \times 32.17}{95}} = 12.7 \text{ cps or } 760 \text{ rpm}$$

Note that for this example G and μ were given as 2,500 psi and 0.50, respectively. In case the test data are not sufficient to establish a value of μ, but require that μ be assumed before G can be evaluated, the frequency and amplitude corresponding to the range between $\mu = 0.25$ and $\mu = 0.50$ must be evaluated. From Fig. 8 it is seen that the highest frequency, for a given value of b, occurs for $\mu = 0.50$, and the greatest amplitude of oscillation occurs for $\mu = 0$. For materials other than hard rock, μ is generally greater than 0.20, with the result that soils which would normally support block-type foundations exhibit effective Poisson's ratios between 0.25 and 0.50, for conditions involving small strains.

Because the rotating speed of the machine will pass through the resonant frequency during starting and stopping, it is necessary to evaluate the amplitude of vibration at 760 rpm as well as for the operating speed of 1,800 rpm. The greatest vertical force occurs in each cycle when the primary and secondary maximum forces act together, which for this example produces a total force at 760 rpm of

$$F_{760} = 807 \text{ lb} = 2 m_1 l \omega^2$$

and

$$2 m_1 l = 0.127 \text{ lb sec}^2$$

From Fig. 8(b), $A^{(1)}_{max} = 0.170$ for b = 6.6, or

$$A_{max,760} = \frac{2 m_1 l}{\rho r_o^3} A^{(1)}_{max.} = \frac{0.127 \times 0.170 \times 32.17}{95 \times 125} = 0.000058 \text{ ft or } 0.00070 \text{ in.}$$

The operating speed of 1,800 rpm is 2.37 times the resonant frequency of 760 rpm. Consequently, the use of Figs. 3 and 11 will permit an estimate of the amplitude of vibration expected at 1,800 rpm. The actual resonant frequency will be approximately 1.04 times that predicted by the weightless subgrade theory, which gives f_1/f_n = 2.37 x 1.04 = 2.45. For this frequency ratio the quantity $N_1\ (f_1/f_n)^2$ is 1.20 and with the foundation and machine data of W_o = 78,370 lb, W = 11.87 lb, and 2 m_1 l = 0.127 lb sec^2, the computed amplitude of vibration at 1,800 rpm is

$$x_{max} = A_{1800} = 1.20\ \frac{0.127 \times 386}{78{,}370} = 0.00074 \text{ in.}$$

The value of A_{1800} should be less than $A_{max,760}$ but the computed value is slightly higher in this case because of the approximate method used.

Because both values of amplitude are higher than the design limit, but the frequency is satisfactory, it is necessary to increase the effective circular radius, r_o, without decreasing the value of the parameter b appreciably. This means that sufficient weight must be added to maintain the ratio b approximately constant as r_o is increased. If the foundation base width is increased to 6.0 ft, without changing the length, the new radius of the equivalent circular base area is r_{o_2} = 5.7 ft. The ratio of new to old radius, cubed, is

$$\left(\frac{r_{o_2}}{r_o}\right)^3 = \left(\frac{5.7}{5.0}\right)^3 = 1.48$$

Consequently, to maintain the value of b constant, the total static weight of the foundation block must be increased by the factor 1.48; from 78,370 lb to 116.000 lb. The values of amplitude now become, $A_{max,760}$ = 0.00047, and x_{max} = A_{1800} = 0.00050 in., which correspond to the design requirements. If a factor of safety is required, the base area and weight could be increased still more; the size of the factor of safety would depend on the accuracy of determination of the soil and machine constants and the consequences resulting from larger vibration amplitudes.

The unbalanced periodic moment tends to cause pitching of the block in the direction of the long axis. Because of the great stability of the foundation in that direction, this motion should be insignificant. Both the pitching and rocking oscillations of the foundation may be studied by the procedure given in Example B.

Example B - Foundation for a Horizontal Piston-Type Compressor.—To study the dynamic behavior of a foundation for a low-speed machine, consider the installation of a motor-compressor unit on soil having the same characteristics as that for Example A. The air compressors consist of horizontally opposed cylinders; large unbalanced horizontal forces are produced but the vertical unbalanced forces are negligible. The important characteristics of the machine and foundation unit are shown in Fig. 14.

The mass moment of inertia of the foundation block and machinery, about a line through the center of the base, is

$$I_o = \frac{W_o}{3\,g}\left(h^2 + a^2\right) + \frac{W_{mach}}{g}\,(14)^2 = 3.60 \times 10^6 \text{ ft-lb sec}^2$$

A uniform circular cylinder, 10 ft high, which would have the same value of I_o, would have a radius of r_o = 13.3 ft. For this equivalent body

$$b_i = \frac{I_o}{\rho \, r_o^5} = \frac{2.92 \times 10^6 \times 32.17}{95 \times (13.3)^5} = 2.9$$

From Fig. 9, a_o = 0.87, A^*_{max} = 3.0, and the first estimate of frequency is

$$f_o = \frac{a_o}{2 \pi r_o} \sqrt{\frac{G}{\rho}} = 3.63 \text{ cps or } 218 \text{ rpm}$$

The amplitude of rotational oscillation at this frequency is

$$A_{240} = \frac{A^*_{max} \, M_y}{G \, r_o^3} = \frac{3.0 \times 14 \times 5730 \left(\frac{218}{277}\right)^2}{2500 \times 144 \times (13.3)^3} = 0.00017 \text{ rad.}$$

or at a distance of 14 ft above the assumed point of rotation, the horizontal amplitude of vibration is 0.030 in.

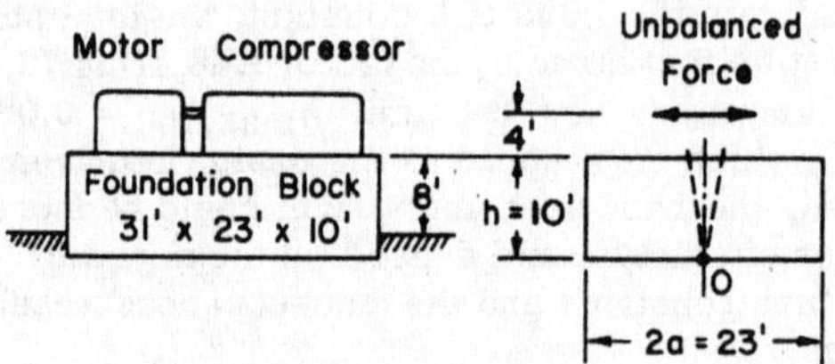

WEIGHTS

Motor	=	40,000 lb.
Compressor	=	156,000 lb.
Foundation Block	=	1,000,000 lb.

MACHINE CHARACTERISTICS

4000 H.P. at 277 rpm

Unbalanced Forces:

Primary	=	4680 lb. at 277 rpm
Secondary	=	1050 lb. at 277 rpm

FIG. 14.—MACHINE AND FOUNDATION DATA FOR EXAMPLE B

These values of resonant frequency and amplitude of oscillation were obtained using the curves from Fig. 9, which were obtained on the basis of assuming Poisson's ratio equal to zero. From a consideration of Fig. 8, it would be expected that for a value of μ = 0.5 that the resonant frequency would be higher and the amplitude of vibration would be lower than the value computed on the basis of $\mu = 0$.

As a check on the resonant frequency value of 240 rpm the constant for the soil, considered as a weightless spring, may be determined and the natural frequency computed from Eq. 4. From these computations it was found that $f_n = 290$ cpm when $\mu = 0$ and $f_n = 410$ cpm when $\mu = 0.5$. Thus the natural frequency was increased by a factor of 1.4 by changing μ from 0 to 0.5.

Because the operating frequency for the machine is 277 rpm, the resonant frequency of the foundation should be at least two times the frequency of the secondary force (554 rpm). Consequently, the block as shown in Fig. 14 does not represent a foundation with satisfactory dynamic characteristics. The best way to improve the dynamic behavior of this foundation is to reduce the height of the machinery above the base. If this cannot be done, a reduction in the weight of the block and spreading of the footing would help some, but it would be more advantageous to stiffen the support. If rock is within a reasonable distance below the base elevation, point-bearing piles may be used. From Fig. 12 the materials, lengths, and working stresses can be selected which will provide a natural frequency of 1,100 cpm or greater. If rock is beyond the reach of bearing piles, friction piles may be used to transfer the foundation load to the more rigid underlying soil. In this case care must be taken during construction to insure that additional settlements of the friction piles do not occur when the vibratory operating loads are applied. C. F. Dodge and W. F. Swiger (35) have shown the importance of these additional settlements of friction piles caused by vibratory loads.

CONCLUSIONS

From a consideration of the usual damped-forced vibration theory, and the theory for an oscillator resting on the surface of an elastic subgrade having weight, it was demonstrated that the value of resonant frequency was markedly higher when an oscillator had a force amplitude output increasing with rotational frequency than when the oscillator force amplitude was constant. The difference in frequency can be quite significant for design purposes. It was also shown that the natural frequency computed for an oscillator resting on a weightless subgrade was significantly higher than the frequency determined for a subgrade, having weight, which was excited by a constant-force oscillator. However, the resonant frequency for a rotating mass oscillator acting on the actual subgrade was very close to the corresponding value of natural frequency computed for a weightless subgrade. The weightless-subgrade procedure gives no value for the amplitude of vibration at the natural frequency. The theory for the elastic medium having weight must be used for amplitude determinations. These conclusions regarding resonant frequencies and amplitude of oscillation apply to vertical, sliding, and rocking modes of oscillation of the foundation.

Some of the test results previously obtained from oscillators resting directly on soil have indicated a non-linear behavior. This has caused concern about the usefulness of elastic theories for predicting the dynamic behavior of foundations on soil. However, from computed and measured amplitudes of motion of test oscillators, it was found that the dynamic soil pressures were a much higher proportion of the static soil pressure than could be tolerated in a machine foundation. Some of the test oscillators must have even jumped free of the ground each cycle, thereby acting as a compactor. A change in pressure distribution because of the increase in amplitude of vibration could account for much of the non-linear behavior found in some tests. This emphasizes that oscillator test data may be misleading if the vibration parameters are quite different from the prototype design values.

Oscillator test data as well as seismic methods have been used successfully to establish soil constants at particular locations. The use of these experimentally determined soil constants with the theory has given reasonably good predictions for the dynamic behavior of the prototype foundation. Data were also given to show typical values of the shear modulus for both clays and granular soils. Test data for velocities of the compression wave in granular soils show that the compression modulus of elasticity varies approximately with the 1/3 power of the confining pressure. The shear modulus should vary in a similar manner.

An analysis of two example machine foundations has shown that the theoretical methods gave reasonable results for the resonant frequencies and amplitudes of vibration of these block type foundations. The theoretical procedures demonstrate the importance of varying each of the parameters which affect the dynamic characteristics of a particular foundation. A diagram was also included to show the effectiveness of point bearing piles to rock in stiffening a foundation and thereby increasing its natural frequency.

ACKNOWLEDGMENTS

This study was completed during the writer's tenure as a Science Faculty Fellow, sponsored by the National Science Foundation. The support provided by this fellowship is gratefully acknowledged. The writer wishes to express his appreciation to Harvard University, and particularly to A. Casagrande, Professor of Soil Mechanics, Harvard University, for providing office and library facilities during this tenure. Finally, the writer wishes to acknowledge the helpful comments and suggestions for improving the text which were contributed by R. C. Hirschfeld, Instructor in Soil Mechanics, at Harvard University.

APPENDIX I. —NOTATION

The following symbols have been adopted for use in the paper and are presented herewith for the guidance of discussers:

A^*_{max}, A_{max} = amplitude of oscillation (maximum);

$A^{(1)}_{max}$ = dimensionless amplitude factor corresponding to amplitude produced by a rotating mass oscillator;

$A^{(2)}_{max}$ = dimensionless amplitude factor corresponding to amplitude produced by a constant force oscillator;

$\bar{A}$ = area of foundation base;

A_p = area of pile cross section;

2 a	= width of a strip footing or width of a rectangular base;
a_o	= dimensionless frequency factor, or velocity ratio, defined by Eq. 11;
b	= mass ratio for vertical or sliding oscillation, defined by Eq. 10;
b'	= mass ratio for vertical oscillation of an infinite strip footing;
b_i	= inertia ratio for rocking oscillation, defined by Eq. 12;
c_s	= spring constant of subgrade, or coefficient of vertical subgrade reaction;
f_o	= resonant frequency for forced vibration;
f_1	= frequency of the exciting force;
f_n	= natural frequency of free vibration;
G	= shear modulus of elasticity;
g	= acceleration of gravity;
h	= height of the foundation block;
I_o	= rotational mass moment of inertia;
l	= radius from center of gravity of rotating part to center of rotation;
M_y	= external moment causing rocking of the foundation;
m_o	= total static mass of oscillator = W_o/g;
m'_o	= static mass per unit of length of infinite strip oscillator;
m_1	= unbalanced rotating mass = W_1/g;
N_1	= displacement magnification factor;
P_x	= horizontal force;
P_z	= vertical force;
P_o	= average soil pressure at the oscillator base;
Q	= periodic force;
Q_1	= amplitude of periodic force;
R	= ratio of thickness of elastic layer to radius of oscillator base;
r_o	= radius of circular base of oscillator;
r_1	= crank throw;
r_2	= connecting rod length;
t	= time;
v	= velocity of elastic wave;
v_s	= velocity of the shear wave;
W_o	= total static weight of the oscillator;
W_1	= weight of unbalanced rotating or reciprocating machine part;

x = displacement;

x_s = static displacement, defined by Eq. 3;

x_p = displacement of piston;

z = vertical displacement;

α = phase angle;

γ = unit weight of elastic body or soil;

μ = Poisson's ratio of elastic body or soil;

ρ = mass density of elastic body or soil = γ/g;

σ_o = nominal compression stress in pile caused by static load;

ω = angular velocity; and

λ = damping coefficient.

APPENDIX II. - BIBLIOGRAPHY

1. "Die Ermittlung der für das Bauwesen wichtigsten Eigenschaften des Bodens durch erzwungene Schwingungen," by A. Hertwig, G. Früh, and H. Lorenz, DEGEBO, No. 1, J. Springer, Berlin, 1933.

2. "Stationäre, axialsymmetrische durch eine schüttelnde Masse erregte Schwingungen eines homogenen elastischen Halbraumes," by E. Reissner, Ingenieur-Archiv, Vol. 7, Pt. 6, December, 1936, pp. 381-396.

3. "Vibrations in Semi-Infinite Solids due to Periodic Surface Loading," by T. Y. Sung, ASTM Special Technical Publication No. 156, "Symposium on Dynamic Testing of Soils, 1953, pp. 35-64.

4. "The Elastic Theory of Soil Dynamics," by P. M. Quinlan, ASTM Special Technical Publication No. 156, "Symposium on Soil Dynamics," 1953, pp. 3-34.

5. "Forced Vibrations of a Body on an Infinite Elastic Solid," by R. N. Arnold, G. N. Bycroft, and G. B. Warburton, Journal of Applied Mechanics, Transactions, ASME, Vol. 77, 1955, pp. 391-401.

6. "Vibrations in Foundations of Machines," by I. Toriumi, Technical Report of Osaka University, Vol. 5, No. 146, Osaka, Japan, March, 1955, pp. 103-126.

7. "Die Empfindlichkeit des Menschen gegen Erschutterungen," by H. Reiher and F. J. Meister, Forsch, Gebiete Ingenieurwesen, Vol. 2, No. 11. 1931, pp. 381-386.

8. "Maschinenfundamente und andere dynamische Bauaufgaben," by E. Rausch, Vertrieb VDI, Verlag G.M.B.H., Berlin, N.W.7, 1943, 3 parts.

9. "Ground Vibration due to Blasting and its Effect on Structures," by F. J. Crandell, Journal, Boston Society of Civil Engineers, April, 1949. Also, "Contributions to Soil Mechanics 1940-1953," B.S.C.F.

10. "Earth Waves," by L. D. Leet, Harvard Univ. Press and J. Wiley & Sons, 1950.

11. "Elastic Waves in Layered Media," by W. M. Ewing, W. S. Jardetzky, and F. Press, McGraw-Hill Book Co., 1957.

12. "Stress-Strain Relations of a Granular Medium," by J. Duffy and R. D. Mindlin, Journal of Applied Mechanics, ASME, December, 1957, p. 585.

13. "Elastic Waves Through a Packing of Spheres," by F. Gassman, Geophysics, Vol. 16, No. 4, October, 1951, pp. 673-685.

14. "The Behavior of Soils Under Dynamic Loadings," 3, MIT, Final Report on Laboratory Studies, Dept. of Civ. and Sanitary Engrg., Soil Mech. Lab., AFSWP-118, Report to Office of the Chf. of Engrs., 1954.

15. "The Variation of Sound Velocity with Stress in Sand," by E. Matsukawa and A. N. Hunter, Proceedings of The Physical Society, Sect. B, Vol. 69, Part 8, No. 440.B, August, 1956, pp. 847-848.

16. "Dynamic Triaxial Tests on Sand," by W. L. Shannon, G. Yamane, and R. J. Dietrich, Proceedings, 1st Panamerican Conf. on Soil Mech. and Foundation Engrg., Mexico City, September, 1959.

17. "On the Propagation of Tremors Over the Surface of an Elastic Solid," by H. Lamb, Philosophical Transactions, Royal Society, A. Vol. 203, London, 1904, pp. 1-42.

18. "The Effect of Footing Shape on Foundation Vibrations," by J. W. Clawson, thesis presented to the Univ. of Florida, at Gainesville, in 1959, in partial fulfilment of the requirements for the degree of Master of Science.

19. "Forced Vibration of a Body on a Stratum of Soil," by G. B. Warburton, 9th Internatl. Cong. of Appl. Mech, Vol. 7, Brussels, 1956, pp. 137-142.

20. "Forced Vibration of a Body on an Elastic Stratum," by G. M. Warburton, Journal of Applied Mechanics, Vol. 24, 1957, pp. 55-58.

21. "In-Situ Measurement of the Dynamic Properties of Soil by Vibration Methods," by R. Jones, Geotechnique, Vol. VIII, No. 1, March, 1958, pp. 1-21.

22. "Compaction of Sands at Resonant Frequency," by F. J. Converse, ASTM Special Technical Publication No. 156, "Symposium on Dynamic Testing of Soils," 1953, pp. 124-137.

23. Discussion of Sung, by A. Pauw, 1953, pp. 67-68.

24. Discussion of Sung, by F. E. Richart, Jr., 1953, pp. 64-68.

25. "Principles of Soil Mechanics," by K. Terzaghi, A Summary of Experimental Studies of Clay and Sand, Reprinted from Engineering News-Record, 1925.

26. "Pilot Studies on Soil Dynamics," by R. K. Bernhard and J. Finelli, ASTM Special Technical Publication No. 156, "Symposium on Dynamic Testing of Soils," 1953, pp. 211-253.

27. "Microseisms," by R. K. Bernhard, ASTM Special Technical Publication No. 206, "Papers on Soils," 1957, pp. 83-102.

28. "The Velocity of Elastic Waves in Sand," by K. Iida, Tokyo Imperial Univ., Bulletin of the Earthquake Research Institute, Vol. 16, 1938, pp. 131-144.

29. "On the Elastic Properties of Soil, Particularly in Relation to its Water Content," by K. Iida, Tokyo Imperial Univ., Bulletin of the Earthquake Institute, Vol. 18, 1940, pp. 675-690.

30. "Dynamic Pile Driving Formulas," by A. E. Cummings, Journal, Boston Society of Civil Engineers, January, 1940. Also, "Contributions to Soil Mechanics 1925-1930," B.S.C.E.

31. "Pile Calculation by the Wave Equation," by E. A. Smith, Concrete and Constructional Engineering, Vol. 53, No. 6, London, June, 1958.

32. "Vibration Problems in Engineering," by S. P. Timoshenko, 3d Ed., D. Van Nostrand Co., Inc., New York, 1955.

33. "Principles of Foundation Design for Engines and Compressors," by W. K. Newcomb, Transactions, ASME, Vol. 73, 1951, pp. 307-312. Discussion 313-318.

34. "Mechanical Vibrations," by J. P. Den Hartog, 4th Ed., McGraw-Hill Book Co., Inc., New York, 1956.

35. "Vibration Testing of Friction Piles," by C. F. Dodge and W. F. Swiger, Engineering News-Record, May 13, 1948, pp. 84-88.

36. "Performance Records of Engine Foundations," by G. P. Tschebotarioff, ASTM Special Technical Publication No. 156, "Symposium on Dynamic Testing of Soils," 1953, pp. 163-168.

37. "Effect of Consolidation Pressure on Elastic and Strength Properties of Clay," by S. Wilson and R. Dietrich, ASCE Research Conference on Shear Strength of Cohesive Soils, 1960.

38. "Soil Mechanics, Foundations and Earth Structures," by G. Tschebotarioff, McGraw-Hill Book Co., Inc., New York, N.Y., 1951, p. 585.

39. "Der mitschwingende Baugrund bei dynamisch belasteten Systemen," by K. Polz, Die Bautechnik, 33 Jgg, Heft 6, 1956.

40. "The Dynamic Principles of Machine Foundations and Ground," by J. H. A. Crockett and R. E. R. Hammond, Procedings, Inst., Mech. Engrg., Vol. 160, No. 4, London, 1949.

41. "Geophysical Investigations concerning the Seismic Resistance of Earth Dams," by C. A. Heiland, A. I. M. M. E., T. P., No. 1054, New York, 1939.

42. "Vibrations in Foundations," by W. Eastwood, J. Inst. Struct., Engrg., Vol. 31, No. 3, London, 1953.

43. "Der Baugrund als Federung in schwingenden Systemen," by G. Ehlers, Beton u. Eisen, Bd. 41 pp. 197-203, 1942.

44. "Der Baugrund als Federung und Dämpfung schwingender Körper," by H. Lorenz, Der Bauingenieur, 25. Jgg., Heft 10, 1950.

45. "Elasticity and Damping Effects of Oscillating Bodies on Soil," by H. Lorenz, A. S. T. M, Sp. Tech. Bulletin, No. 156, 1953.

46. "Grundbau-Dynamik," by H. Lorenz, Springer Verlag, Berlin, 1960.

47. "Ueber die Nichtlinearität der Vertikalschwingungen von starren Körpern auf dem Baugrunde," by M. Novak, Acta Technica, Vol. 2, Praha, 1957.

48. "Principles of Foundation Design for Engines and Compressors," by W. K. Newcomb, Transactions, A. S. M. E., Vol. 73, No. 3, 1951.

49. "Grouting to Prevent Vibration of Machinery Foundations," by J. P. Gnaedinger, Presented at the June, 1960 Meeting of A. S. C. E., Reno, Nev.

50. "Field Investigations of the Theory of Vibrations of Massive Foundation Under Machines," by D. D. Barkam, Proceedings, Internatl. Conference of Soil Mechanics and Foundations Engrg., Vol. 11, 1936, p. 285.

51. "Determination of Elastic Constants of Soil by Means of Vibration Methods," by A. Ishinoto, and K. Iida, Bulletin of the Earthquake Research Inst., Vol., XIV, 1936, part 1; also Vol XV, 1937, part 11.

52. "Analytical and Experimental Methods in Engineering Seismology," by M. Biot, Proceedings, ASCE, January, 1942.

53. "Mathematical Methods in Engineering," by Th. Von Kármán, and M. A. Biot, McGraw Hill Book Co., Inc., New York, 1940, p. 404.

54. "Elasticity and Internal Friction in a Long Column of Granite," by F. Birch and D. Bandcroft, Bulletin, Seis, Soc. of America, Vol. 28, October, 1938.

55. "Geophysical Exploration," by C. A. Heiland, 1946, Prentice-Hall, Inc., New York.

56. "Performance Records of Engine Foundations," by Gregory P. Tschebotarioff, ASTM Special Technical Publication No. 156, "Symposium on Dynamic Testing of Soils," July 2, 1953, pp. 163-168.

57. "Principles of Foundation Design for Engines and Compressors," by W. K. Newcomb, Transactions, ASME, April, 1951.

58. "Vibrations in Foundations of Machines," by Isao Toriumi, Tech. Reports of the Osaka Univ., Vol. 5, No. 146.

59. "Stationäre, axialsymmetrische, durch eine schüttelnde Masse erregte Schwingunger eines homogenen elastischen Halbraumes," by E. Reissner, Ingenieur Archiv, Vol. 7, 1936.

60. "Vibrations in Semi-Infinite Solids Due to Periodic Surface Loading," by T. Y. Sung, ASTM Special Technical Publication No. 156, "Symposium on Dynamic Testing of Soils," 1953, pp. 35-64.

61. "A Dynamic Analogy for Foundation-Soil Systems," by Adrian Pauw, ASTM Special Technical Publication No. 156, "Symposium on Dynamic Testing of Soils," July 2, 1953, pp. 90-112.

62. "Effect of Consolidation Pressure on Elastic and Strength Properties of Clay," by S. D. Wilson and R. J. Dietrich, preprint from the June, 1960 ASCE Research Conference on Shear Strength of Cohesive Soils, at Boulder, Colo.

63. "Engineering Vibrations," by L. S. Jacobsen and R. S. Ayre, McGraw-Hill Book Co., Inc., New York, 1958.

64. "Dimensional Analysis in Soil Mechanics," by H. Lundgren, Acta Polytechnica, Civil Engineering and Building Construction Series, Vol. 4, No. 10, 1957.

65. "Principles of Soil Mechanics - A Summary of Experimental Studies of Sand and Clay," by K. Terzaghi. Engineering News Record, 1925.

66. "Sound Absorbing Material," by C. Zwikker and C. W. Kosten, Elsevier Press, New York.

67. "Theory of Propagation of Elastic Waves in a Fluid-Saturated Porous Solid. I. Low Frequency Range," by M. A. Biot, Journal of the Acoustical Society of America, Vol. 28, 1956, p. 168.

68. "Study of Elastic Wave Propagation and Damping in Granular Materials," by Bobby Ott Hardin, thesis presented to the University of Florida in Gainesville, Fla., in August, 1961, in partial fulfillment of the requirements for the degree of Doctor of Philosophy.

69. "Freie und erzwungene Torsionsschwingungen des elastischen Halbraumes," by E. Reissner; Ingenieur Archiv, Vol. VIII, No. 4, p. 229.

70. "Forced Torsional Oscillations of an Elastic Half-Space," Part I by E. Reissner and H. F. Sagoci; Part II by H. F. Sagoci Journal of Applied Physics, Vol. 15, No. 9, p. 652.

3507 May, 1963 SM 3

Journal of the
SOIL MECHANICS AND FOUNDATIONS DIVISION
Proceedings of the American Society of Civil Engineers

STRESS-DILATANCY, EARTH PRESSURES, AND SLOPES

By P. W. Rowe[1]

SYNOPSIS

Soils consist of individual particles in contact and the physical laws that govern inter-particle sliding are stated. A few examples serve to illustrate how these laws fit, and yet qualify, current thought on deformation and failure, energy corrections, and the concept of a yield surface.

The general equation for the stability of earth masses is developed and applied to the cases of active and passive pressure, slope stability, and bearing capacity. The equations, which include terms for work done both internally and externally during volume change, yield solutions that are identical to those based on the Coulomb equation in the special case of no volume-change rate during failure.

Much more attention is required to the particulate nature of soils if fundamental progress in soil mechanics is to be achieved. This paper shows that such treatment can still lead to simple practical applications along lines similar to those in present use.

STRESS-DILATANCY RELATION

Notation.—The letter symbols adopted for use in this paper are defined where they first appear and are arranged alphabetically in the Appendix.

Note.—Discussion open until October 1, 1963. To extend the closing date one month, a written request must be filed with the Executive Secretary, ASCE. This paper is part of the copyrighted Journal of the Soil Mechanics and Foundations Division, Proceedings of the American Society of Civil Engineers, Vol. 89, No. SM3, May, 1963.

1 Prof. of Soil Mechanics, Manchester Univ., England.

General.—During the course of a series of experiments on sheet-pile walls[2,3] (between 1947 and 1957), the importance of soil deformations prior to failure became increasingly evident. Visual observations on sands showed that each increment of effective stress caused small inter-particle slides that were progressively arrested by additional particle contacts. This process was described as an increase in interlocking with increase in mobilized Coulomb ϕ and an hypothesis was suggested that the basic laws which governed the failure condition also applied throughout deformation.[4]

In order to understand the detailed mechanism of the deformation and failure of simple arrangements of particles in contact, a series of studies were made of the stress-strain-volume change relations for ideal assemblies of rods and spheres. The work was extended, by means of the principle of minimum energy, to include the case of a random mass of irregular particles.[5] (Although the following material summarizes the essential findings of the work, for continuity it will be helpful if the reader is conversant with this previous publication.) It was established that, at any stage of deformation of any ideal packing, the effective-stress ration in a triaxial test is given by

$$\frac{\sigma'_1}{\sigma'_3} = \tan \alpha \tan \left(\phi_\mu + \beta\right) \quad \ldots\ldots\ldots\ldots (1)$$

in which α is the average inclination to the major plane at which the particles are interlocked and β is the inclination to the minor principal plane at which the particles tend instantaneously to slide (Fig. 1). The angle ϕ_μ is the physical angle of friction between the particle surfaces for the material of which the assembly is composed.

During a small increase in the effective-stress ratio, there is a small change in the major and minor principal strain rates denoted by $\dot{\epsilon}_1$ and $\dot{\epsilon}_3$, respectively. If the corresponding volume change is denoted by $\dot{dv}$, it can be shown by geometry that

$$\frac{1}{2}\left[1 + \frac{\dot{dv}}{v\,\dot{\epsilon}_1}\right] = \frac{\dot{\epsilon}_3}{\dot{\epsilon}_1} = \frac{1}{2} \tan \alpha \tan \beta \quad \ldots\ldots (2)$$

in which $\dot{dv}$ and $\dot{\epsilon}_3$ are positive on expansion and $\dot{\epsilon}_1$ is positive on compression. Eqs. 1 and 2 combine to give the ratio

$$\dot{E} = \frac{\sigma'_1 \dot{\epsilon}_1}{2\sigma'_3 \dot{\epsilon}_3} = \frac{\tan\left(\phi_\mu + \beta\right)}{\tan \beta} \quad \ldots\ldots\ldots\ldots (3)$$

The term $\dot{E}$ states the ratio of the instantaneous rate of work done on the sample by σ'_1 to that done by the sample against σ'_3. This latter term properly

2 "A Theoretical and Experimental Analysis of Sheet-Pile Walls," by P. W. Rowe, Proceedings, Inst. of Civ. Engrs., London, Vol. 4, No. 1, 1955, p. 32.

3 "Sheet-Pile Walls in Clay," by P. W. Rowe, Proceedings, Inst. of Civ. Engrs., Vol. 7, July, 1957, p. 629.

4 "A Stress-Strain Theory for Cohesionless Soil with Application to Earth Pressures at Rest and Moving Walls," by P. W. Rowe, Géotechnique, Vol. IV, No. 2, 1954, p. 70.

5 "The Stress-Dilatancy Relation for Static Equilibrium of an Assembly of Particles in Contact," by P. W. Rowe, Proceedings, Royal Soc., London, Series A, Vol. 269, p. 500.

includes any additional internal work that arises directly as a result of the rate of volume change against the cell pressure.

The equations also apply to any two irregular particles in contact, and because the angle of interlocking of the particles, α, is missing from Eq. 3, the possibility arose that throughout the mass of particles there was a preferential inter-particle sliding direction β such that the rate of internal work absorbed in frictional heat loss was a minimum.

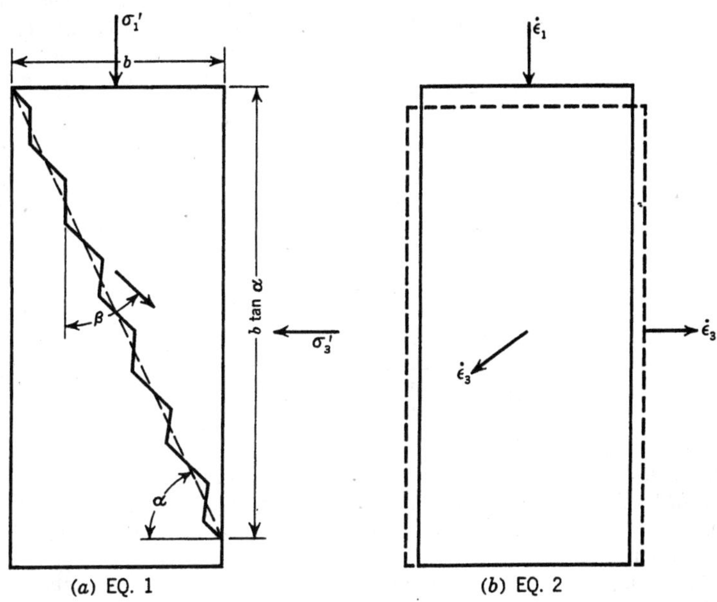

FIG. 1.—STRESS AND STRAIN RATIO RELATIONSHIPS

The increment of internal work absorbed is

$$dW = \sigma_1' \dot{\epsilon}_1 - 2\sigma_3' \dot{\epsilon}_3 = \sigma'_1 \dot{\epsilon}_1 \left[1 - \frac{1}{\dot{E}}\right] \quad \ldots\ldots\ldots (4)$$

Denoting the instantaneous slope of the effective stress ratio-strain graph as λ,

$$dW = \sigma_1' \frac{\Delta\left(\frac{\sigma_1'}{\sigma_3'}\right)}{\lambda} \left[1 - \frac{1}{\dot{E}}\right] \quad \ldots\ldots\ldots\ldots (5)$$

At a given stage of test, σ_1' and λ have fixed values during a small increment and for a chosen increment of effective-stress ratio, dW is a minimum when $\dot{E}$ is a minimum.

The ratio E is a minimum when in Eq. 3

$$\frac{d\dot{E}}{d\beta} = 0 \qquad (6)$$

This gives

$$\beta = \left(45° - \frac{\phi_\mu}{2}\right) \qquad (7)$$

as the preferential direction of sliding. Substituting this preferential value into Eqs. 1, 2, and 3, the minimum energy criterion is satisfied when

$$\frac{\sigma'_1}{\sigma'_3} = \tan\alpha \tan\left(45° + \frac{\phi_\mu}{2}\right) \qquad (8)$$

in which

$$\tan\alpha = \sqrt{\frac{\sigma'_1}{\sigma'_3}\left(1 + \frac{d\dot{v}}{v\,\dot{\epsilon}_1}\right)} \qquad (9)$$

and

$$\frac{\sigma'_1}{\sigma'_3\left[1 + \frac{d\dot{v}}{v\,\dot{\epsilon}_1}\right]} = \tan^2\left(45° + \frac{\phi_\mu}{2}\right) \qquad (10)$$

Eqs. 8, 9, and 10 are the basic stress-dilatancy relations. The ratio tan α in a mass of irregular particles equals the ratio of the number of particle contacts per unit area on the major principal plane to that number on the minor principal plane.

A series of triaxial compression tests and direct measurements of ϕ_μ, on quartz, glass, and steel, in which all the terms of Eq. 10 were measured independently, established that Eq. 10 was valid for dense, overconsolidated or reloaded soils throughout deformation, where the strains to the peak were of the order of a few percentages.

On the other hand, loose, normally consolidated soils on first loading suffered additional loss of energy in friction due to deviations of β from the preferred direction, associated with a process of rearragning the particle assembly during deformation. By the time failure was reached, the loose assemblies were completely remolded to the critical voids ratio condition and this requires the maximum additional internal work done in friction, associated with maximum work done in remolding.

Accepting the form of Eq. 10 and substituting the observed values of σ'_1/σ'_3 and $1 + (d\dot{v}/v\,\dot{\epsilon}_1)$, a value of ϕ_f must be substituted in place of ϕ_μ in order to satisfy the equation for any packing other than very dense. In Fig. 2, the familiar relations between the effective-stress ratio and strain have been replotted as the mobilized Coulomb ϕ_m versus the major principal strain using

$$\frac{\sigma'_1}{\sigma'_3} = \tan^2\left(45° + \frac{\phi_m}{2}\right) \qquad (11)$$

The shaded area in Fig. 2 shows the variation of ϕ_f with strain, writing

$$\frac{\sigma_1'}{\sigma_3'\left[1+\frac{d\dot{v}}{v\,\dot{\epsilon}_1}\right]} = \tan^2\left(45^\circ + \frac{\phi_f}{2}\right) \quad \ldots\ldots\ldots (12)$$

Special care in the measurement of strain and volume change is necessary, together with the use of silicone grease and rubber-membrane type frictionless end plattens to ensure absence of end restraint; otherwise, the values of ϕ_f appear to be several degrees higher. If, for example, a slip line forms,

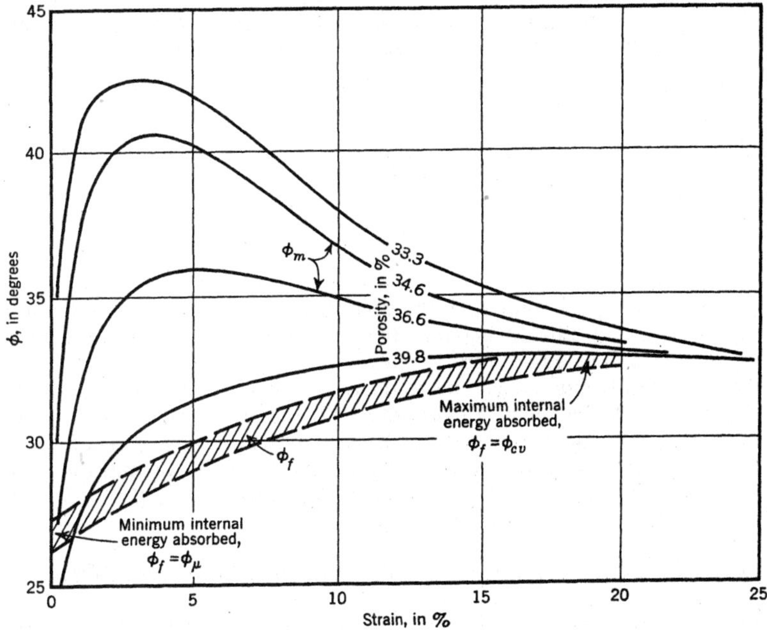

FIG. 2.—UNIQUE RELATIONSHIP BETWEEN ϕ AND STRAIN

measured volume changes $d\dot{v}$ are associated with values of v much smaller than the sample volume so that $(d\dot{v}/v)$-values for the entire sample are too small and ϕ_f is too high. It has been shown[5] experimentally, and can be proved theoretically, that slip lines are the result of failure rather than associated with the cause of failure and do not feature in the stress-dilatancy treatment of soils.

It is seen in Fig. 2 that, under minimum energy conditions, at small strains, $\phi_f = \phi_\mu$, whereas under maximum energy conditions at large strains, when the soil is completely remolded, ϕ_f may be 6° higher than ϕ_μ.

Eq. 10 states that, in the absence of remolding and with complete correction for volume change, the stress ratio throughout deformation and failure is identical to that of an ideal frictional material which obeys the Mohr-Coulomb criterion at constant volume with $\phi = \phi_\mu$, the physical angle of friction of the material.

In Eq. 12, the work done in remolding is expressed by elevating ϕ_μ to ϕ_f as in the past the work done in both remolding and volume change has been expressed, in effect, by elevating ϕ_μ to ϕ_{max} and writing $\dot{v} = 0$. However, with increase in strains to failure, from dense to loose packings, ϕ_{max} decreases, whereas ϕ_f increases. The lowest value of ϕ_f $(= \phi_\mu)$ strictly applies only during negligible strains, as when a soil is being reloaded, but $\phi_f \simeq \phi_\mu$ for dense or overconsolidated packings during initial low strains.

Postulating the existence of a cohesion c_μ between particles in terms of the average over the sample area, and modifying it to c_f to allow for remolding, it will be shown[6] that Eq. 12 becomes extended to the form

$$\frac{\sigma'_1}{\sigma'_3\left[1 + \frac{d\dot{v}}{v\,\dot{\epsilon}_1}\right]} = \tan^2\left(45° + \frac{\phi_f}{2}\right) + \frac{2\,c_f}{\sigma'_3}\tan\left(45° + \frac{\phi_f}{2}\right) \quad \ldots\ldots (13)$$

Initial experiments[6] indicate that the strength of clays can be studied to advantage by means of Eq. 13 because the components of packing and particle surface properties are separated.

Volume changes due to change in mean principal stress are additive to those in Eq. 13. At the peak, the effective stresses remain constant and the volume-change rate is entirely of a dilatant nature. Having measured σ'_1/σ'_3 and $d\dot{v}/v\;\dot{\epsilon}_1$ at the peak for a series of values of σ'_3, the parameters ϕ_f and c_f are determined from Eq. 13 and α is determined from Eq. 9.

STRESS-DILATANCY RELATION AND CURRENT CONCEPTS

At this point, it may be useful to give a few examples of how the treatment of soil as particulate matter fits into currnt thought.

For the special case of $d\dot{v}/v = 0$, the form of the solution is identical to that based on Mohr-Coulomb. The α-line takes the place of the "slip line," but, otherwise, forces must be integrated over this line in the normal manner. If $d\dot{v}/v = 0$, then sliding also occurs along the α-line, as previously assumed, but if $d\dot{v}/v \neq 0$, particle sliding occurs on β-planes, which diverge from the direction of the α-line.

K. H. Roscoe, A. N. Schofield, and C. P. Wroth, in dealing with granular materials, applied[7] the Taylor-Bishop energy correction to the observed Coulomb ϕ_{max} to obtain a "corrected" ϕ_r (expressed in terms of a reduced

6 "The Stress-Dilatancy Performance of Two Clays," by P. W. Rowe, D. B. Oates, and N. A. Skermer, Symposium on Laboratory Shear Testing of Soils, Ottawa, Canada, September, 1963.

7 "On the Yielding of Soils," by K. H. Roscoe, A. N. Schofield, and C. P. Wroth, Géotechnique, Vol. 8, No. 1, 1958, p. 22.

deviator stress q_i). In the case of their tests on steel balls, the value of ϕ_r happened to be constant and equal to the ultimate residual Coulomb ϕ_{cv} at the critical voids ratio. In their (q_i, p, n)-plot, a unique yield or state boundary surface was proved to exist.

Their energy correction did not account for additional internal work done due to dilatancy,[8,9] and making the proper allowance, the yield surface may be transposed to a ϕ_f-strain plot when it becomes identical to the shaded area in Fig. 2. Here it is seen that ϕ_f increases a few degrees progressively with large strains and it may well be that there is a fundamental relationship between the amount of strain, the associated remolding, and the value of ϕ_f. In this respect and with the foregoing qualifications, there is agreement between two lines of thought which germinated in the realms of very small and very large movements.

The use of stress-dilatancy requires the separation of effective stresses and pore-water pressure in analysis and this procedure is widely accepted. However, in the special case of studies of the undrained strength of saturated, normally consolidated, clays, Roscoe, et al, state[10] that the maximum deviator stress is the significant failure point, whereas the peak effective-stress ratio is irrelevant. As implied by Roscoe, this may be true in the laboratory, where water flow is artifically prohibited and where pore-water pressure changes are determined entirely by the soil porosity, consolidation pressure, and stress path. Under field conditions, it is readily accepted that the presence of silt layers in varved clays or anisotropic permeability in the case of laminated clays can lead to a local spread of pore-water pressure and a breakdown of strict undrained conditions. However, even with an intact homogeneous clay, water migrations can occur between zones subject to different pore pressures. In Fig. 3(a), line AB is the locus of elements subject to peak effective-stress ratio. With further increase in compression stress, the pore pressures must decrease on this line from value 1 to value 2, Fig. 3(b). However, the pore pressures in adjacent, less highly stressed zones are higher so that water can migrate toward the failure zone. As a result, the effective-stress ratio at failure differs from that in the undrained test. Clearly, zone ruptures are less critical than line ruptures in this respect and, also, foundations subject to "immediate" loading will satisfy undrained conditions of intact clays. A useful consideration of total and effective strength criteria has been given by R. V. Whitman.[11]

L. Bjerrum, et al, have shown[12] that, in the undrained case, when $du/d\epsilon \neq 0$, failure does not coincide with the peak effective-stress ratio. Their detailed results for sands have been reanalyzed and found to lie on a stress-dilatancy line. This finding may be transposed to the statement that failure can occur at α-values less than the drained maximum. The rate $du/d\epsilon$ is a

8 "Volume Changes in Drained Triaxial Tests on Granular Materials," by P. L. Newland and B. H. Alleley, Géotechnique, Vol. 9, No. 1, 1957, p. 17.

9 Discussion of Session I by P. W. Rowe, Proceedings, 5th Internatl. Conf. on Soil Mechanics, Paris, 1961, Vol. III, p. 137.

10 Discussion of "The Shear Strength Properties of Calcium Illite," by K.H. Roscoe, A. N. Schofield, C. P. Wroth, and A. Thurairajah, Géotechnique, Vol. XIII, No. 3, 1962, p. 246.

11 Discussion of Session 4 by R. V. Whitman, ASCE Research Conf. on Shear Strength of Cohesive Soils, Colorado, July, 1960, p. 1069.

12 "The Shear Strength of a Fine Sand," by L. Bjerrum, S. Kringstad, and O. Kummeneje, Proceedings, 5th Internatl. Conf. on Soil Mechanics, Paris, 1961, Vol. I, p. 29.

function of $\dot{dv}/(v\,\dot{\epsilon}_1)$ which, in turn, is related to the mobilized ratio σ_1'/σ_3'. Consequently, stress-dilatancy may eventually enchance research into sensitive clays.

In the past, the consideration of an energy "correction" has not led to any application to stability problems. The present objective is, therefore, to take a first look at a possible treatment of retaining walls and slopes, using the stress-dilatancy equations, which account for energy spent in volume change.

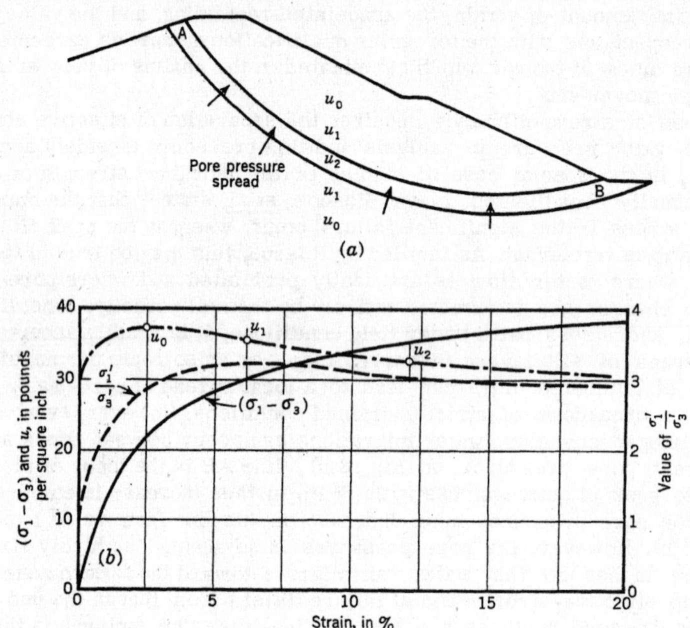

FIG. 3.—USE OF THE PEAK EFFECTIVE STRESS RATIO

The special cases are taken in which the boundary stresses are defined without reference to boundary strain.

APPLICATION TO THE STABILITY OF EARTH MASSES

There are two main approaches to earth-pressure analysis. The first is to consider the differential equilibrium equations of an element, followed by integration throughout the soil mass such that the boundary equilibria are satisfied. This has provided "rigorous" total-stress solutions for ideal uniform soils. The second is to divide the mass within a chosen boundary into a series of vertical slices, to consider the equilibrium of each slice, and to integrate over the boundary, as described by A. W. Bishop[13] and N. Janbu.[14] This flexible analysis, which is readily applicable to the practical situation of variable strata and water pressures, is adopted herein with the following differences:

1. The boundary line is an α-line rather than a slip line.
2. In certain cases, only one α-line need be considered as the minimum condition is already written into the equations.
3. Boundary directions of principal stress must be considered as is usual in the case of the "rigorous" analytical methods.

The general equation for the stability of an elemental slice may be obtained by considering first the equilibrium of two blocks in contact (Fig. 4)

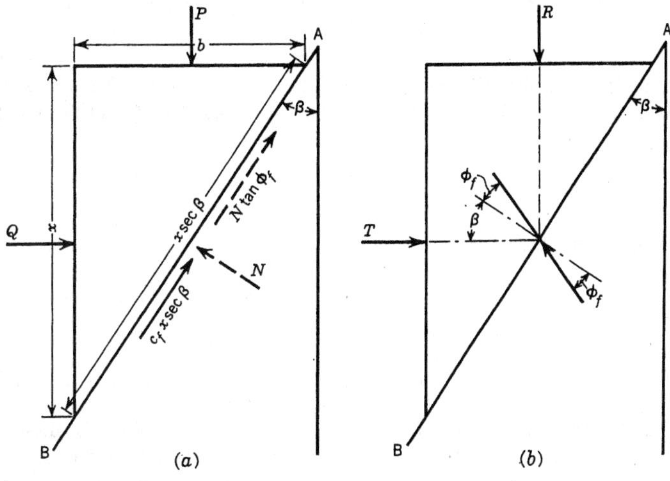

FIG. 4.—BASIC EQUILIBRIUM DIAGRAM

where the plane of sliding, AB, differs from the direction of either P or Q; c_f and ϕ_f are the shearing parameters on the interface.

Resolving the cohesion forces in the P and Q directions, Fig. 4(a), the resultant forces to be balanced by friction are

$$P - c_f x \sec \beta \cos \beta = R \qquad (14)$$

and

$$Q + c_f x \sec \beta \sin \beta = T \qquad (15)$$

13 "The Use of the Slip Circle in the Stability Analysis of Slopes," by A. W. Bishop, Géotechnique, Vol. V, No. 1, 1955, p. 7.

14 "Earth Pressures and Bearing Capacity Calculations by Generalized Procedure of Slices," by N. Janbu, Proceedings, 4th Internatl. Conf. on Soil Mechanics, London, 1957, Vol. II, p. 207.

Considering friction, Fig. 4(b),

$$\frac{R}{T} = \tan \phi = \tan \left(\phi_f + \beta\right) \quad \ldots\ldots\ldots\ldots (16)$$

Substituting for R and T,

$$\frac{P - c_f x}{Q + c_f x \tan \beta} = \tan \left(\phi_f + \beta\right) \quad \ldots\ldots\ldots (17)$$

from which

$$P = Q \tan \left(\phi_f + \beta\right) + \frac{c_f x \sec^2 \beta}{1 - \tan \phi_f \tan \beta} \quad \ldots\ldots\ldots (18)$$

In this equation, sliding occurs in the β-direction along AB but P and Q have not necessarily been obtained by integrating over AB. Considering the case of $c_f = 0$,

$$P = Q \tan \left(\phi_f + \beta\right) \quad \ldots\ldots\ldots\ldots\ldots\ldots (19)$$

With reference to Fig. 1,

$$P = \sigma'_1 \; b \quad \ldots\ldots\ldots\ldots\ldots\ldots\ldots (20)$$

and

$$Q = \sigma'_3 \; b \tan \alpha \quad \ldots\ldots\ldots\ldots\ldots\ldots (21)$$

Substituting these values into Eq. 19, the form of Eq. 1 is obtained. Therefore, Eq. 18 becomes of general application provided P and Q are obtained by integrating pressures over the α-plane. In the special case in which

$$\alpha = 90° - \beta \quad \ldots\ldots\ldots\ldots\ldots\ldots (22)$$

sliding occurs along the α-plane, $d\dot{v} = 0$, and the plane is identical to the Mohr-Coulomb "slip plane." The same is true when $c_f \neq 0$ and writing

$$x = b \tan \alpha \quad \ldots\ldots\ldots\ldots\ldots\ldots (23)$$

in Eq. 18, then

$$P = Q \tan \left(\phi_f + \beta\right) + \frac{c_f \, b \tan \alpha \sec^2 \beta}{\left(1 - \tan \beta \tan \phi_f\right)} \quad \ldots\ldots (24)$$

for which

$$\sigma'_1 = \sigma'_3 \tan \alpha \tan \left(\phi_f + \beta\right) + \frac{c_f \tan \alpha \sec^2 \beta}{\left(1 - \tan \beta \tan \phi_f\right)} \quad \ldots (25)$$

Dividing Eq. 25 by Eq. 2, differentiating with respect to β and substituting the preferential β-value gives Eq. 13. Eq. 18 is the general expression, whereas Eqs. 24 and 25 apply only where P and Q are principal stress directions.

In general, the principal stresses change direction throughout the earth mass, whereas the known gravitational force P is vertical and the required restraining force Q is horizontal. Figs. 5 illustrate the rotation of the direc-

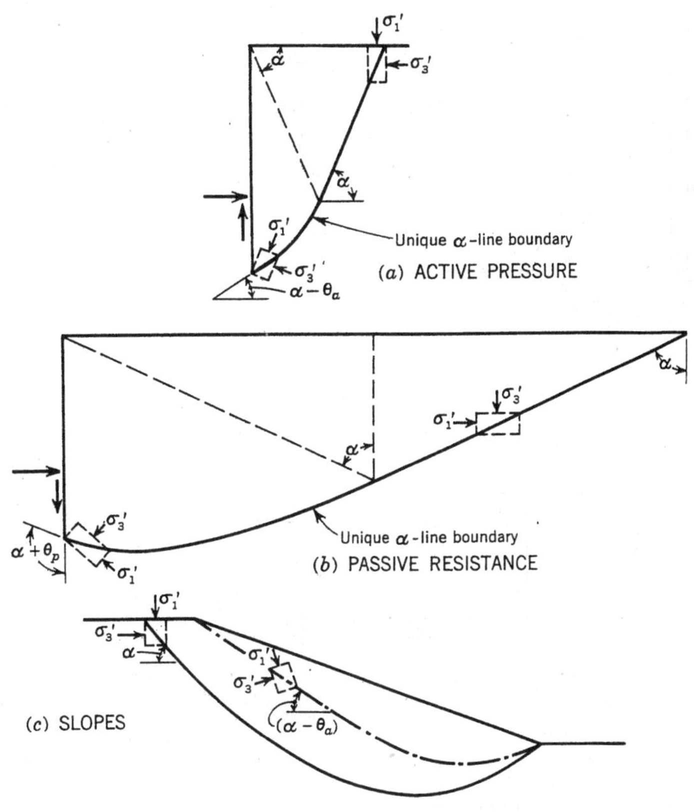

FIG. 5.—TYPICAL α-LINE ROTATIONS

tion of the principal stresses for three common engineering cases. It is, therefore, necessary to consider the case of an element rotated through θ_a° relative to the vertical direction of P, as in Fig. 6(b), or through θ_p° relative to the horizontal direction of Q at the passive end of an α-line, as in Fig. 6(a).

The direction of particle movement, previously at $(45^\circ - \phi_f/2)$ to the major principal stress, must lie at $(45^\circ - \phi_f/2 + \theta_a)$ to the vertical. To distinguish between the special case of the vertical element and the general case, the

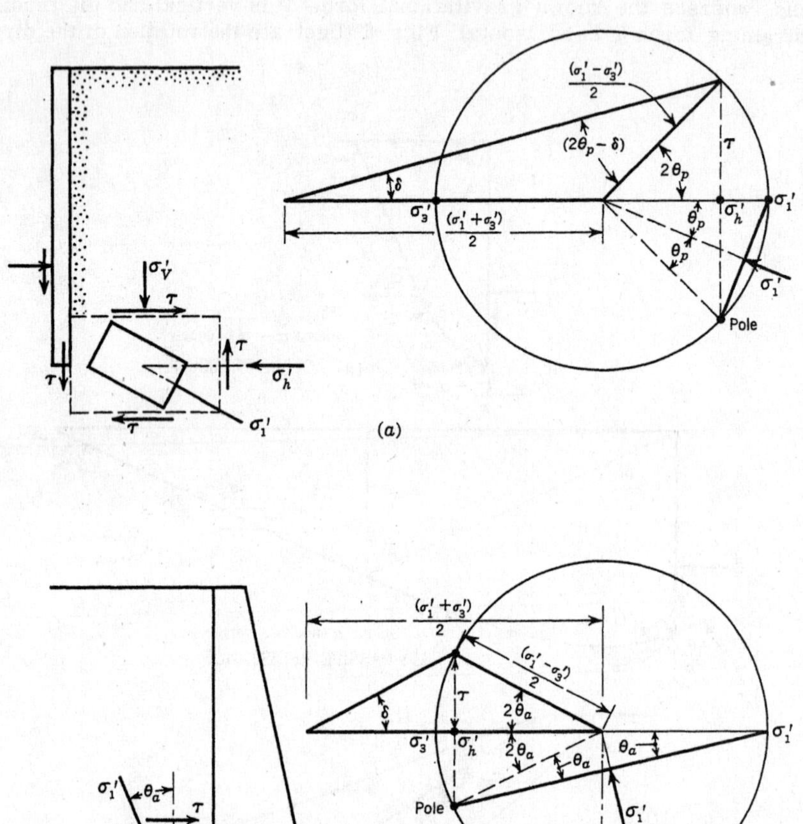

FIG. 6.—ROTATION OF PRINCIPAL STRESSES

particle movement is taken at ω to the horizontal and the α-line at η to the horizontal, from which

$$\omega = \left(45° + \frac{\phi_f}{2} - \theta_a\right) \quad \text{.............(26a)}$$

and

$$\eta = \left(\alpha - \theta_a\right) \quad \text{...............(26b)}$$

Substituting $(90° - \omega)$ for β and $x = b \tan \eta$ in Eq. 18, with rearrangement, one obtains

$$Q = P \tan\left(\omega - \phi_f\right) - c_f\, b \frac{\tan \eta}{\tan \omega} \frac{\sec^2 \omega}{\left(1 + \tan \omega \tan \phi_f\right)} \quad \text{.... (27)}$$

Considering a vertical slice of soil (width b) with total internal boundary forces as shown in Fig. 7(b) and with a water pressure u at the base of the

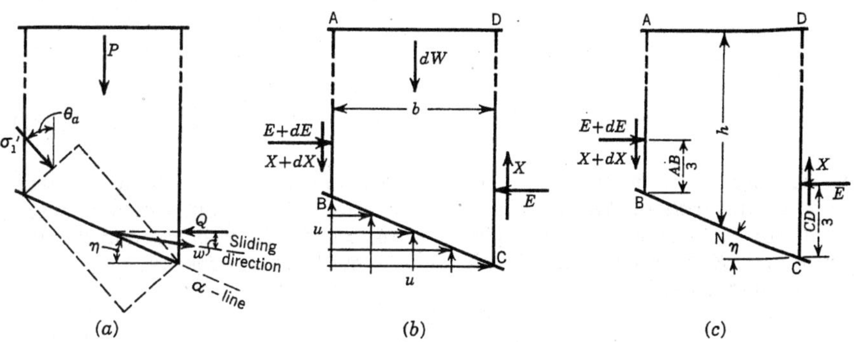

FIG. 7.—FORCES ON ELEMENTAL SLICE

slice, the effective forces acting on the α-line boundary are found from Figs. 7(a) and 7(b):

$$P = dW + dX - u\,b \quad \text{.............. (28a)}$$

$$Q = -dE - u\,b \tan \eta \quad \text{.............. (28b)}$$

Substituting these values into Eq. 27, rearranging and summating elements,

$$\sum_{\theta=0}^{\theta=\theta_{max}} (-dE - u\,b \tan \eta) = \sum (dW + dX - u\,b) \tan \omega$$

$$- \sum \left[dW + dX - u\,b \tan \phi_f + \phi_f + c_f b \frac{\tan \eta}{\tan \omega} \right]$$

$$\frac{\sec^2 \omega}{(1 + \tan \omega \tan \phi_f)} \quad \ldots\ldots\ldots\ldots\ldots\ldots \quad (29)$$

The difficulty associated with the general application of Eq. 29 is that the distribution of internal shear over the various slices cannot be directly determined. However, there is a direct relationship between the principal-stress-direction change θ and the boundary forces on a vertical element. Using Mohr's circle of stress for the boundary passive case, Fig. 6(a), it follows that

$$\frac{\sigma_1' + \sigma_3'}{\sigma_1' - \sigma_3'} = J = \sin \frac{(2\,\theta_p - \delta)}{\sin \delta} \quad \ldots\ldots \quad (30)$$

in which

$$\tan \delta = \frac{\tau}{\sigma_h'} \quad \ldots\ldots\ldots\ldots\ldots \quad (31)$$

at the base of the slice and θ_p is the rotation of the major principal stress from the horizontal. Assuming that each vertical slice is subject to gravitational forces increasing linearly with depth, tan δ may be equated to the average ratio X/E, where X and E are the internal boundary forces.

Eq. 30 then becomes,

$$E = X \left(J + \cos 2\,\theta_p \right) \operatorname{cosec} 2\,\theta_p \quad \ldots\ldots\ldots \quad (32)$$

It is also necessary to satisfy moment equilibrium for each slice. Assuming that the forces normal to each slice act at the one-third points, Fig. 7(c), and taking moments about N, the mid-point of the base of each slice, it is found that

$$X + \frac{dX}{2} = \frac{2}{3} E \tan \eta + \frac{dE}{3} \left(\frac{h}{b} + \tan \eta \right) \quad \ldots\ldots \quad (33)$$

Differentiating Eq. 32,

$$dE = \left(J + \cos 2\,\theta_p \right) \operatorname{cosec} 2\,\theta_p \, dX - 2\,X \left(J \cos 2\,\theta_p + 1 \right)$$

$$\operatorname{cosec}^2 2\,\theta_p \, d\theta \quad \ldots\ldots\ldots\ldots \quad (34)$$

Substituting for dE (Eq. 34) and E (Eq. 32) into Eq. 33 and rearranging, the following relation is obtained between the shape of the α-line and the distribution of the X-forces:

$$\frac{dX}{X} = \frac{2\left[1 - \frac{2}{3}\tan\eta\left(J + \cos 2\theta_p\right)\operatorname{cosec} 2\theta_p + \frac{2\,d\theta}{3}\left(\frac{h}{b} + \tan\eta\right)\frac{\left(J\cos 2\theta_p + 1\right)}{\sin^2 2\theta_p}\right]}{\left[\frac{2}{3}\left(\frac{h}{b} + \tan\eta\right)\left(J + \cos 2\theta_p\right)\operatorname{cosec} 2\theta_p - 1\right]} \quad \text{. . (35)}$$

The importance of the distribution of X is most for passive-pressure calculations and least for active pressure and, for this reason, Eq. 35 has been developed in terms of θ_p rather than θ_a. The direction of the α-line is determined exactly at the wall where the value of δ is known and at a point some distance from the wall where the shear on a vertical slice is zero. Where model observations reveal a slip line after failure with the shape of a logarithmic spiral, this implies that the direction of principal stresses, and therefore the α-line rotates between limits over a spiral path. Accordingly, Eqs. 29 and 35 have been solved numerically using log-spiral α-lines for uniform granular materials to give the earth pressure coefficients shown in Fig. 8 for the case of zero pore pressure.

In order to be able to deal rapidly with more complex site conditions, fairly simple direct solutions can be obtained in the following way for the cases of active pressure and slope stability using slight approximations of no practical significance. The minimum-energy criterion eliminates, or greatly reduces, the present need to investigate a large number of possible slip surfaces to find the minimum stability condition.

ACTIVE PRESSURE COEFFICIENTS

Frictionless Wall.—In Fig. 9(a) there is shown a frictionless vertical wall, behind which there is a level fill of dry or submerged cohesionless soil. Referring to Eq. 29, u = 0, $c_f = 0$, X at wall = 0, and $\theta_a = 0$. Therefore,

$$-dE = +E = dW\tan\omega - dW\tan\phi_f\,\frac{\sec^2\omega}{1 + \tan\omega\tan\phi_f} \quad \text{. . (36)}$$

From which the active force,

$$+E = dW\tan\left(\omega - \phi_f\right) \quad \text{. . . . . (37)}$$

writing

$$\omega = 45° + \frac{\phi_f}{2} \quad \text{. . . . . (38)}$$

and

$$dW = \frac{1}{2}\gamma H^2\cot\alpha \quad \text{. . . . . (39)}$$

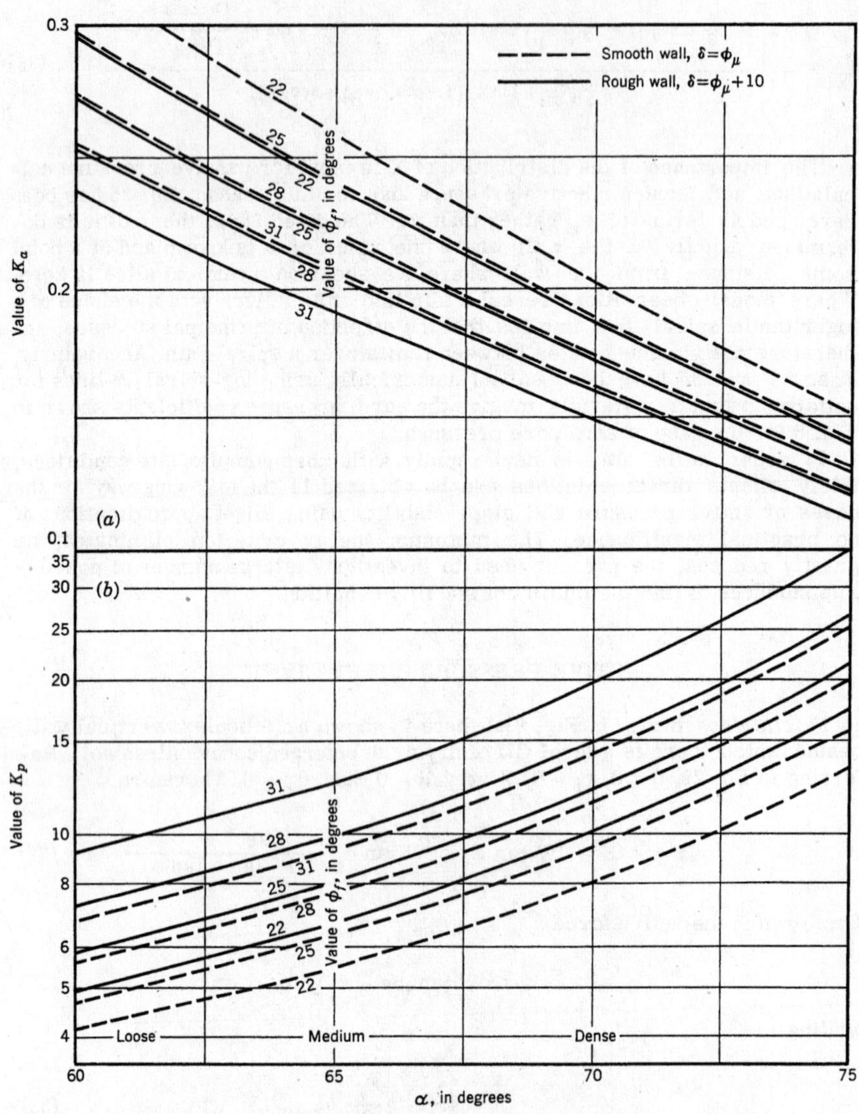

FIG. 8.—EARTH PRESSURE COEFFICIENTS

and substituting Eq. 39 into Eq. 37 leads to

$$E = \frac{1}{2} \gamma H^2 K_a \quad \ldots\ldots\ldots\ldots\ldots\ldots \quad (40)$$

in which

$$K_a = \cot \alpha \tan \left(45^\circ - \frac{\phi_f}{2}\right) \quad \ldots\ldots\ldots\ldots \quad (41)$$

This solution demonstrates that the force on a retaining wall supporting dense sand is less than that supporting loose sand, not because the angle of friction is less for loose sand (for rather ϕ_f is larger) but because the weight of the wedge bearing directly on the wall is less for dense sand. This is entirely due to the greater degree of interlocking of dense sand. The solu-

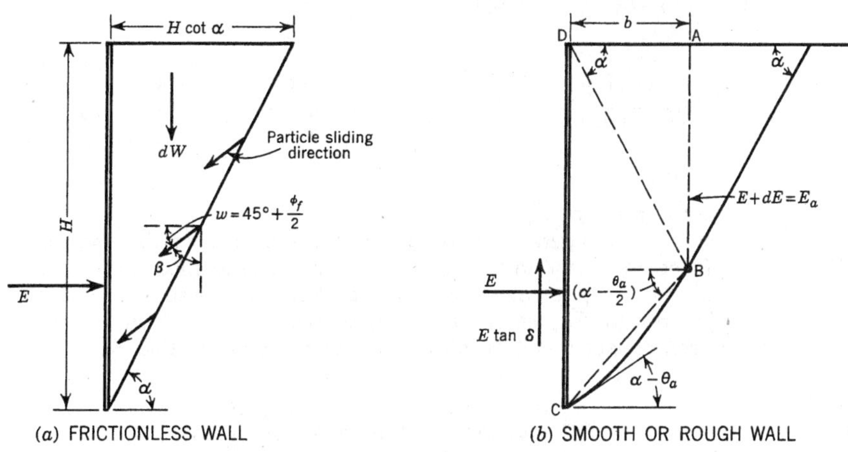

FIG. 9.—ACTIVE PRESSURE CALCULATION

tion is numerically identical to that using Coulomb's equation because, whatever the theory, the ratio of the lateral to vertical force is directly related to the ratio of the principal stresses measured in the laboratory for this simple case with X equal to zero.

It is noted that this particular solution may be obtained from first principles by drawing a line at slope α behind the wall and considering the equilibrium of the weight of the mass bounded by this line and the wall, with respect to sliding in the preferential β-direction. Thus, in Fig. 9(a),

$$\frac{dW}{E} = \tan\left(\phi_f + \beta\right) = \tan\left(45^\circ + \frac{\phi_f}{2}\right) \quad \ldots\ldots\ldots \quad (42)$$

leads to

$$E = \frac{1}{2}\gamma H^2 \cot\alpha \tan\left(45^\circ - \frac{\phi_f}{2}\right) \quad \ldots\ldots\ldots (43)$$

Rough Wall.—The conditions in Fig. 9(b) are the same as those for Fig. 9(a) except that the wall is rough with friction angle $\delta = \phi_w + r$. The angle ϕ_w is the friction angle between the soil grains and the wall material, and is generally close to ϕ_μ. For a smooth wall, $r = 0$ and $\delta = \phi_w$. The angle δ equals zero only for a shearless wall. The present tendency to express δ as a ratio of the Coulomb ϕ_{max} is unsound because ϕ_{max} empirically includes strength due to α, which does not appear to affect δ.

At the wall, the major principal stress and the α-line turn through θ_a from the vertical, Fig. 9(b). Inserting $\theta_p = 90^\circ - \theta_a$ into Eq. 30 or using Mohr's circle of stress, Fig. 6(b), together with the use of the form of Eq. 4, and writing ϕ_f in place of ϕ_μ leads to

$$\frac{\sin\left(2\theta_a + \delta\right)}{\sin\delta} = J = \frac{\tan\alpha \tan\left(45^\circ + \frac{\phi_f}{2}\right) + 1}{\tan\alpha \tan\left(45^\circ + \frac{\phi_f}{2}\right) - 1} \quad \ldots\ldots (44)$$

which provides the value of θ_a.

In Fig. 9(b), the conjugate α-line DB provides a boundary beyond which all vertical planes above and to the right of DB are free of shear stress. The active force on AB, E_a, is therefore given by Eq. 41. Between B and C, the α-line rotates through θ_a, which is usually on the order of 10° in the active case. It is, therefore, sufficient to integrate over the straight line BC:

$$DC = b\left[\tan\alpha + \tan\left(\alpha - \frac{\theta_a}{2}\right)\right] \quad \ldots\ldots\ldots\ldots (45)$$

and

$$E = E_a + \sum - dE = E_a + \sum (dW + dX)\tan\left(\omega - \phi_f\right) = \frac{1}{2}\gamma H^2 K_a \quad \ldots (46)$$

in which

$$dW = \frac{1}{2}\gamma\, b\,(H + b\tan\alpha) \quad \ldots\ldots\ldots\ldots (47a)$$

$$\omega = \left(45^\circ + \frac{\phi_f}{2} - \frac{\theta_a}{2}\right) \quad \ldots\ldots\ldots\ldots (47b)$$

$$dX = X = -E\tan\delta \quad \ldots\ldots\ldots\ldots (47c)$$

and

$$E_a = \frac{1}{2} \gamma b^2 \tan^2 \alpha \cot \alpha \tan\left(45° - \frac{\phi_f}{2}\right) \quad \ldots\ldots (47d)$$

These equations lead to the following expression for the coefficient K_a:

$$K_a = \frac{\tan \alpha \tan\left(45° - \frac{\phi_f}{2}\right) + \left[2 \tan \alpha + \tan\left(\alpha - \frac{\theta_a}{2}\right)\right] \tan\left(45° - \frac{\phi_f}{2} - \frac{\theta_a}{2}\right)}{\left[\tan \alpha + \tan\left(\alpha - \frac{\theta_a}{2}\right)\right]^2 \left[1 + \tan \delta \tan\left(45° - \frac{\phi_f}{2} - \frac{\theta_a}{2}\right)\right]} \quad \ldots (48)$$

For chosen values of α, ϕ_f, and δ, the values of θ_a may be obtained from Eq. 44 and inserted into Eq. 48.

The values of the active-pressure coefficient obtained (in this way) from Eq. 48 with assumed straight α-lines agree to within 2% with the values shown in Figs. 8 using Eqs. 29 and 35 with curved α-lines.

SLOPE-STABILITY EQUATION

Multiplying both sides of Eq. 29 by $\cos \eta$ and adding $-dX \sin \eta$ to both sides,

$$\sum(-dE \cos\eta - dX \sin \eta) = \sum dW \sin \eta \frac{\tan \omega}{\tan \eta} + \sum(dX - u\,b) \sin \eta \left(\frac{\tan \omega}{\tan \eta} - 1\right)$$

$$- \sum \left[(dW + dX - u\,b) \tan \phi_f + c_f\, b \frac{\tan \eta}{\tan \omega}\right] \frac{\sec^2 \omega \cos \eta}{1 + \tan \omega \tan \phi_f} \quad \ldots\ldots\ldots\ldots (49)$$

The left side of Eq. 49 represents the summation of the internal forces resolved parallel to the α-line. If this line has a circular path, and no external forces act at the boundaries, each term has an approximately constant radius moment arm about the center of the circle and, therefore, the summation is proportional to the moment of the internal forces about an external point, which is zero. For noncircular lines, where a circle forms a close approximation, the summation may also be taken as zero for practical purposes. In addition, for the Coulomb case of no dilatancy, $\omega = \eta$ and the following equilibrium condition is then found:

$$\sum dW \sin \eta = \sum\left[(dW + dX - u\,b) \tan(\phi_f) + c_f\, b\right] \frac{\sec \eta}{1 + \tan \eta \tan \phi_f} \quad \ldots\ldots (50)$$

This is the form of Bishop's stability equation for slopes using the Coulomb type of parameters. In Eq. 50 it has also been found that, because $\Sigma dX = 0$, the summation $\Sigma \frac{dX \tan \phi_f \sec \eta}{1 + \tan \eta \tan \phi_f}$ is near zero and very small compared with the remaining terms in Eq. 50.

The pore pressure may be expressed by the coefficient

$$r_u = \frac{u\,b}{dW} \quad \ldots\ldots\ldots\ldots\ldots\ldots\ldots \quad (51)$$

A special case of interest is that in which this ratio is constant. It has also been shown[6] that c_f may be approximately equal to zero for certain

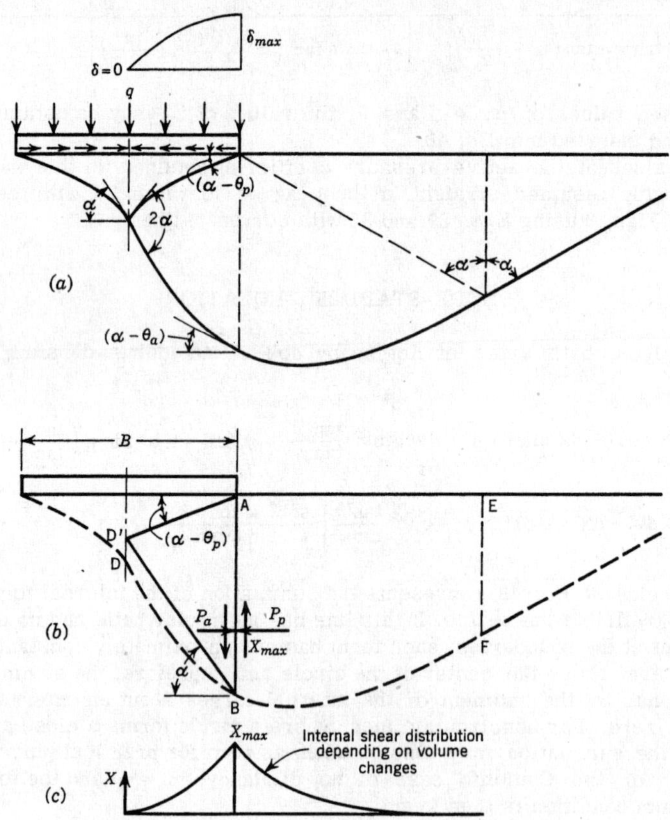

FIG. 10.—BEARING CAPACITY CALCULATION

types of clay. For these conditions, and neglecting the terms dX, Eq. 49, may be written in the form

$$r_u = \frac{\sum dW \cos\eta \tan(\omega - \phi_f)}{\sum dW \cos\eta \left[\tan(\omega - \phi_f) - \tan\eta\right]} \quad \ldots\ldots\ldots \quad (52)$$

which states the average pore-pressure coefficient at failure for a given α-line boundary and values of ϕ_f.

BEARING CAPACITY OF SHALLOW FOUNDATIONS

The general shape of the α-line boundary for a surface infinite strip foundation is shown in Fig. 10(a) and the maximum bearing capacity q may be found by the method of slices in which the foundation load is added to the weight of the slice. However, both the size and the distribution of the internal dX-forces are important and cannot be determined precisely without considering boundary and internal deformations. The following simplified solution appears to involve no error of practical importance.

Consider a vertical plane AB at the foundation edge [Fig. 10(b)]. This may be treated as a rough wall subject to an active force P_a and a passive resistance P_p. The internal vertical shear is zero on CD and EF and reaches a maximum on AB. The distribution between CD and AB for the active case and AB and EF for the passive case has already been considered in determining the values of K_a and K_p. Then,

$$P_a = \frac{1}{2} \gamma H^2 K_a + q_s H K_a - c_f H K_{ac} \quad \text{........ (53)}$$

$$P_p = \frac{1}{2} \gamma H^2 K_p + q_s H K_p + c_f H K_{pc} \quad \text{........ (54)}$$

and

$$P_a = P_p \quad \text{.................... (55)}$$

from which

$$q = \frac{1}{2} \gamma B N_\gamma + q_s N_q + c_f N_c \quad \text{........... (56)}$$

in which

$$N_\gamma = \frac{H}{B} \left[\frac{K_p}{K_a} - 1 \right] \quad \text{............(57a)}$$

$$N_q = \frac{K_p}{K_a} \quad \text{................(57b)}$$

and

$$N_c = \frac{K_{pc} + K_{ac}}{K_a} \quad \text{..............(57c)}$$

The present purpose is to calculate the value of N_γ. This depends on the ratio H/B, which is derived from the geometry of the α-lines, and on the ratio X/P_p or tan δ on AB, which controls the values of K_p and K_a.

In Fig. 10(b), the actual (dashed) α-lines AD and DB are curved, but the depth H is closely found by substituting the straight lines AD' and D'B at angles $(\alpha - \theta_p)$ and α to the horizontal, respectively. These are the directions of the α-lines at A and at D. With this approximation,

$$\frac{H}{B} = \frac{1}{2}\left[\tan\left(\alpha - \theta_p\right) + \tan\alpha\right] \quad \ldots\ldots\ldots (58)$$

The equivalent value of δ on AB is not necessarily the maximum value of δ, for AB is an imaginary line through the soil mass that is not necessarily everywhere at failure. According to Fig. 1, particles slide at angle $(\alpha + \beta - 90°)$ from the α-line, which becomes $(\alpha - \phi_f/2 - 45°)$. If AB were a conjugate α-line, the maximum wall roughness angle would state the deviation of particles from AB and

$$r_{max} = \left(\alpha - \frac{\phi_f}{2} - 45°\right) \quad \ldots\ldots\ldots\ldots (59)$$

therefore,

$$\delta_{max} = \phi_f + r = \left(\alpha + \frac{\phi_f}{2} - 45°\right) \quad \ldots\ldots\ldots (60)$$

Inserting this value into Eq. 30 and writing

$$\frac{\sigma_1'}{\sigma_3'} = \tan\alpha\,\tan\left(45° + \frac{\phi_f}{2}\right) \quad \ldots\ldots\ldots (61)$$

it is found that angle ABF = 2α and, therefore, AB is indeed a conjugate α-line.

Although Eq. 60 gives the maximum value of δ, it does not follow that this is reached. With a dense sand, external expansion may be expected to mobilize high δ-values with small movements, whereas much larger movements may be necessary with loose sand. There is no reason to assume that failure is reached on AB, and solutions have been made for values of δ such that

$$\delta = \delta_{max} - \psi \quad \ldots\ldots\ldots\ldots\ldots\ldots (62)$$

For chosen values of α, ϕ_f, and ψ, using Eqs. 62, 60, 30, 58, and 57a and Fig. 8 for K_p and K_a, values of N_γ are plotted in Fig. 11. Experimental results[15]

15 "Etude experimentale de la Capacite portante du Sable sous des Fondations directes etablies en Surface," by E. E. De Beer and A. B. Vesic, Annales Des Travaux Publics de Belgique, No. 3, 1958.

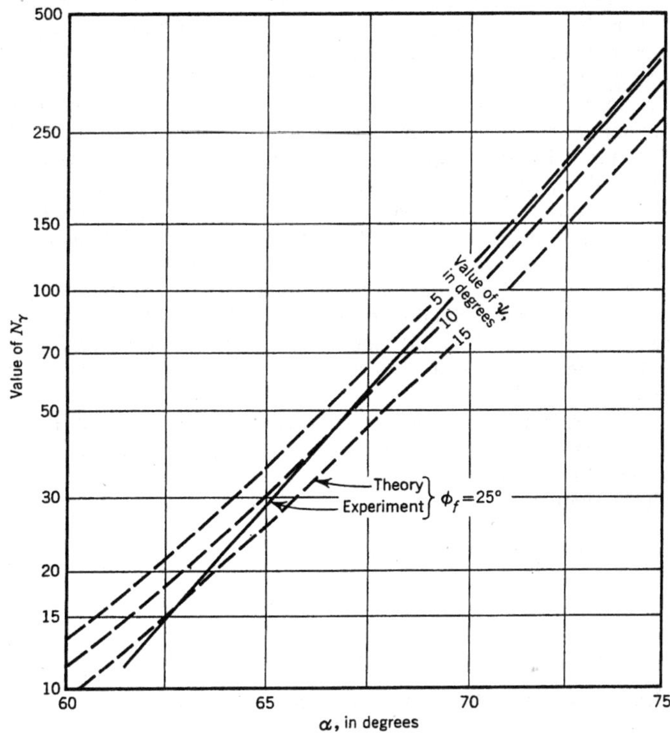

FIG. 11.—BEARING CAPACITY COEFFICIENT N_γ

have been added in which the chosen α and ϕ_f values fit their published values of ϕ_{max}. It is seen that a fit between theory and experiments requires an increasing proportional mobilization of internal shear with increase in relative density, as would be expected.

CONCLUSIONS

These examples serve to illustrate the way in which the familiar types of solution may be obtained by beginning with stress-dilatancy instead of the Mohr-Coulomb criterion. For the engineering purpose of computing the stability of slopes and gravity retaining walls, the new calculations yield solutions of no practical difference from those based on the Mohr-Coulomb criterion. For these cases, especially for those involving clays, the no-dilatancy assumption can be inserted into the equations to provide the familiar solutions in terms of Coulomb parameters.

However, the use of Coulomb's equation as the entire basis for teaching and research imposes a severe restriction on the development of soil mechanics. Clearly, it is inconsistent, particularly in teaching, to begin with the Mohr-Coulomb criterion, which has nothing to say concerning volume change, when it is necessary to emphasize that volume change in shear is one of the most important properties indigenous to soil materials. The solution of deformation problems such as flexible walls and foundation settlement requires consideration of particle movement before failure and compatibility with the structure at the boundary. In problems of this type, large differences appear to exist between observations and calculations that ignore dilatancy. In the classical earth-pressure problem, the angle of wall friction plays an important part and its value requires study in terms of the roughness angle of various types of wall surface. The increasing studies of the chemistry of clay mineral surfaces and the structure of clay matrix require the facility of properly dissociating the inter-particle shear strength denoted by c_f and ϕ_f from that due to packing denoted by α. In the last resort, a correlation between observations made using apparatus such as the triaxial cell, plane strain, and direct shear apparatus can only be achieved by considering particle movement.

Much more attention to the particulate nature of soils is essential if fundamental progress in soil mechanics is to be achieved. The paper shows that a particulate treatment of soils can already be applied to the more common practical problems.

APPENDIX.—NOTATION

The following symbols have been adopted for use in this paper:

b = width of basic wedge;

c_f = cohesion intercept in stress-dilatancy plot;

E = internal earth force;

$\dot{E}$ = ratio of instantaneous work done on sample to that done by sample;

h = height of elemental slice;

K_a, K_{ac} = active pressure coefficients;

K_p, K_{pc} = passive pressure coefficients;

N_γ, N_q, N_c = bearing capacity coefficients;

P = vertical force on basic wedge;

P_a = active earth force;

P_p = passive earth force;

Q = horizontal force on basic wedge;

R, T = resultant forces on basic wedge in equilibrium with friction;

r = wall roughness angle;

u = pore pressure;

$\frac{d\dot{v}}{v}$ = increment of unit volume change;

W = weight of soil;

X = internal shear force on a vertical slice;

x = height of basic wedge;

α = angle of interlocking with reference to major principal plane;

β = direction of particle movement with reference to minor principal plane;

γ = density;

δ = angle of wall friction;

ϵ_1, ϵ_3 = principal strains;

$\dot{\epsilon}_1, \dot{\epsilon}_3$ = increments of principal strain;

η = deviation of the α-line from the horizontal;

θ = angle of rotation of principal stresses;

θ_a = value of θ measured from the vertical;

θ_p = value of θ measured from the horizontal;

λ = instantaneous slope of the effective stress ratio-strain graph;

σ_1', σ_3' = principal effective stresses;

σ_h', σ_v' = horizontal and vertical effective stresses, respectively;

τ = shear stress;

ϕ_{cv} = ultimate Coulomb ϕ at critical voids ratio;

ϕ_f = angle of inter-particle friction modified for remolding;

ϕ_m = mobilized Coulomb ϕ;

ϕ_{max} = maximum Coulomb ϕ;

ϕ_r = Taylor-Bishop modified ϕ;

ϕ_w = true friction angle between grains and material of wall;

ϕ_μ = true friction angle between soil grains; and

ω = deviation of the direction of particle movement from the horizontal.

Bearing Capacity of Deep Foundations in Sand

ALEKSANDAR B. VESIĆ, Associate Professor of Civil Engineering, Georgia Institute of Technology Soil Mechanics Laboratory

Large-scale model experiments have been made to provide information on factors which influence bearing capacity of deep foundations in sand. Cylindrical and prismatical foundations of various sizes resting at different depths in homogeneous sand masses of different relative densities were loaded statically to failure. Special loading cells permitted separate registration of point and skin loads throughout the tests. Additional tests with models of sand colored in layers were made to study the mechanism of shear failure in the soil mass. The model experiments were accompanied by standard laboratory tests for determination of physical characteristics of the soils used.

An analysis of shear patterns observed indicates that, depending on relative density of sand, all three types of failure previously described in the literature may occur at shallow depths: general shear failure, local shear failure and punching shear failure. However, at greater depths only punching shear failure occurs, irrespective of the relative density of sand.

The unit point and skin resistances of the foundation increase linearly with depth only at shallow depths. At greater depths, both resistances show a hyperbolic increase and reach asymptotically constant final values. These final values are independent of overburden pressure and appear to be functions of relative density of sand only. This is explained by the "arching" of sand above the foundation base.

Analyses of observed ultimate loads indicate that a fair estimate of bearing capacity can be made by assuming failure surfaces in accordance with observed shear patterns.

•FROM the point of view of soil mechanics, there are two general types of deep foundations. The first type may be represented by a foundation installed by some process of excavation or drilling which does not induce significant changes in density or structure of the bearing soil. Practically all piers and caissons and some piles belong to this type. The other type may be represented by a deep foundation forced into the ground by driving or a similar operation, that induces significant changes in adjacent soil. Most piles belong to this second type.

SCOPE

The present paper is an investigation of ultimate bearing capacity of deep foundations in homogeneous masses of sand. This problem is considered to be fundamental. Only after a thorough understanding of principles of behavior of deep foundations in homogeneous masses of soil will it be possible properly to see and treat the problems of different types of deep foundations encountered in engineering practice.

Paper sponsored by Committee on Stress Distribution in Earth Masses.

Theoretical Considerations

The basic problem of bearing capacity of deep foundations in sand can be formulated as follows: A rigid foundation of known shape and dimensions is placed at a depth D in a homogeneous mass of sand of defined physical properties (Fig. 1). A static, vertical, central load is applied on the top. What is the ultimate load Q that this foundation can support?

The load is generally transmitted partially along the foundation shaft or skin, partially at the foundation base or point. The two bearing components of the load, the skin load Q_s and the base or point load Q_p are usually considered separately. The total ultimate load is then expressed as the sum of these two components:

$$Q = Q_p + Q_s = p_o A_p + s_o A_s \tag{1}$$

Here p_o represents the unit base resistance and s_o unit skin resistance of the foundation (psi or kg/cm²). A_p and A_s are, respectively, bearing areas of the base and the skin.

The solution of the problem of bearing capacity of the base has been sought in the past primarily by an approach base on the classical work by Prandtl (1, 2), and Reissner (3). They presented a solution of the problem of penetration of a rigid stamp into an incompressible (rigid-plastic) solid (Fig. 2). That solution, first applied to the problem of bearing capacity of soils by Caquot (4) and Buisman (5) is usually written in the following general form (6):

$$p_o = c\, N_c\, \zeta_c + q\, N_q\, \zeta_q + \tfrac{1}{2}\, \gamma\, B\, N_\gamma\, \zeta_\gamma \tag{2}$$

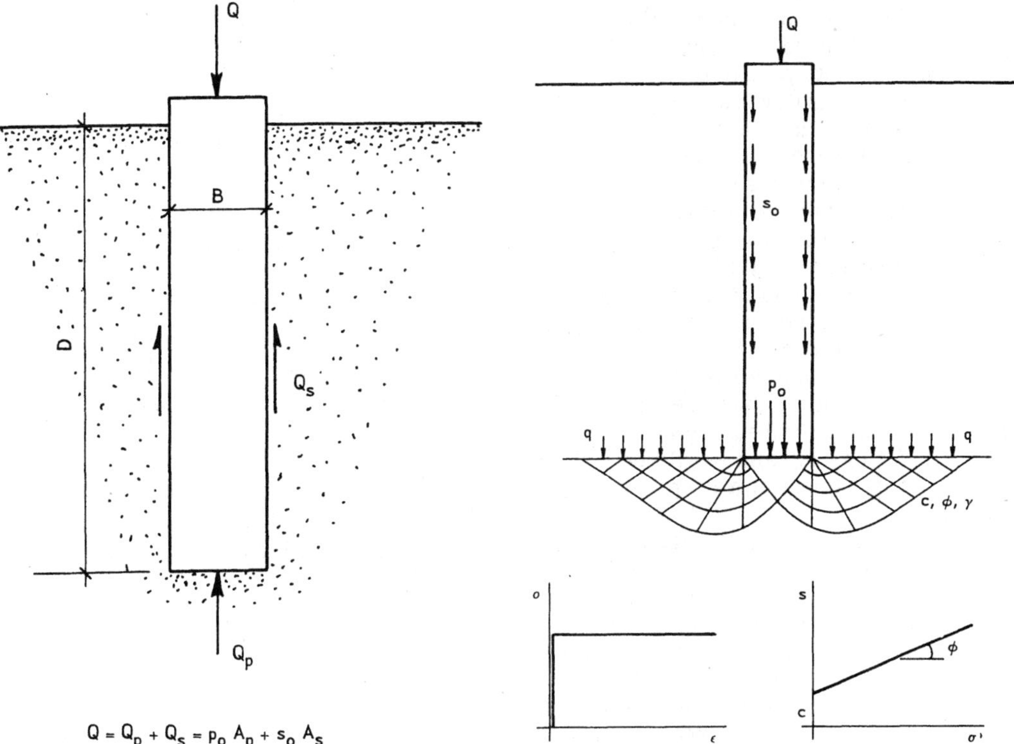

Figure 1. Basic problem of bearing capacity of a deep foundation.

Figure 2. Prandtl-Reissner solution as applied by Caquot and Buisman.

In Eq. 2, c represents the shear strength intercept (cohesion) of the soil, q the overburden pressure, γ the unit weight of the soil involved in shear and B the foundation width. N_c, N_q, N_γ are bearing capacity factors for a strip foundation and ζ_c, ζ_q, ζ_γ are shape factors. Both N and ζ factors are, generally, dimensionless functions of the angle of shearing resistance φ.

In the case of foundations in sand c = 0 and

$$p_o = q\, N_q\, \zeta_q + \tfrac{1}{2}\, \gamma\, B\, N_\gamma\, \zeta_\gamma \qquad (3)$$

Using the same general approach but different shear patterns with rupture lines reverting to the shaft (Fig. 3) De Beer (7, 8) and Jaky (9) have obtained another solution of the same problem with considerably higher bearing capacity factors. Their work has been further extended by Meyerhof (10, 11), and others.

All of the previously mentioned solutions were obtained by considering a plane problem (long rectangular foundation). Some work in developing solutions for an axially symmetrical problem (circular foundation) has also been done, primarily in the USSR (12). However, there is still a general tendency to determine the shape factors empirically.

Another possible approach to the problem of base bearing capacity originated in the work by Bishop, Hill and Mott (13), and Skempton, Yassin and Gibson (14), who have considered the problem of expansion of a spherical or cylindrical cavity inside an infinite mass of an ideal solid. In such a case there exists around the cavity a highly stressed zone where the material, by assumption, behaves as a rigid-plastic solid. Outside that zone it behaves as an ideal elastic (or linearly deformable) solid.

The solutions of this kind have been applied to the problem of bearing capacity of deep foundations by Gibson (15), Skempton, Yassin and Gibson (14) and Ladanyi (16). According to these solutions the bearing capacity p_o can be computed by an expression analogous to Eq. 2 without the third term, however. In conditions of infinite mass or of very great depth this third term becomes negligible compared with other two terms. Consequently, both theoretical approaches indicate that at greater depths the bearing capacity of the base should be practically independent of its size and proportional to the overburden pressure q. Based on this conclusion and some limited experimental evidence, it has been generally admitted that the point bearing capacity of pile foundations in sand should be equal to the point resistance of a deep cone penetrometer.

Values of bearing capacity factors N_q found by different theoretical solutions mentioned are shown in Figure 4 left. The diagram on the right shows the corresponding factors $\zeta_q N_q = N_q'$ for circular foundations, as proposed by different investigators. This diagram contains also the empirical curves for N_q recommended by Brinch Hansen (17, 19) and Caquot and Kerisel (18). There is an appreciable difference in proposed numerical values.

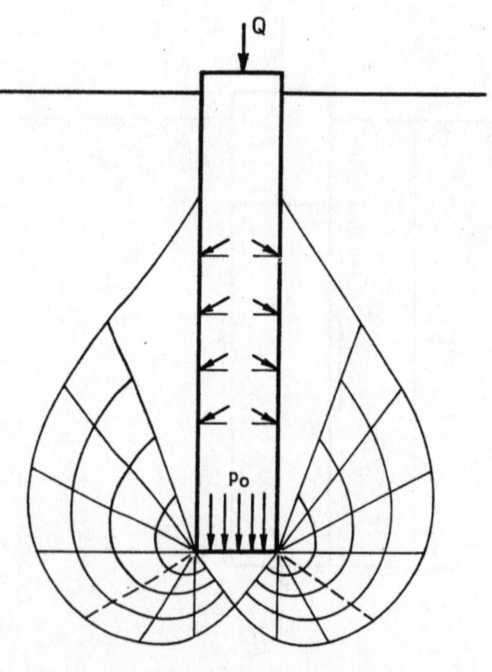

Figure 3. Shear pattern with rupture lines reverting to the shaft.

The unit skin resistance s_o consists of two terms: one representing adhesion c_a; the other, friction along the skin:

$$s_o = c_a + p_s \tan \delta \qquad (4)$$

The term δ denotes the angle of skin friction and p_s average normal pressure on the skin. This pressure is generally assumed to be proportional to the corresponding average overburden pressure along the skin q_s. In sand $c_a = 0$ and

$$p_s = K_s\, q_s \qquad (5)$$

K_s is a dimensionless number, which can be called coefficient of skin pressure. With this, the following expression for skin resistance in sand is obtained:

$$s_o = K_s \tan \delta\, q_s \qquad (6)$$

Eq. 6 suggests that, in homogeneous soil conditions, unit skin resistance s_o should be proportional to the average overburden pressure q_s.

Previous Experiments

Among numerous experimental studies related to the problem, comparatively few have been of sufficiently general nature to permit drawing definite conclusions concerning the influence of the different parameters involved.

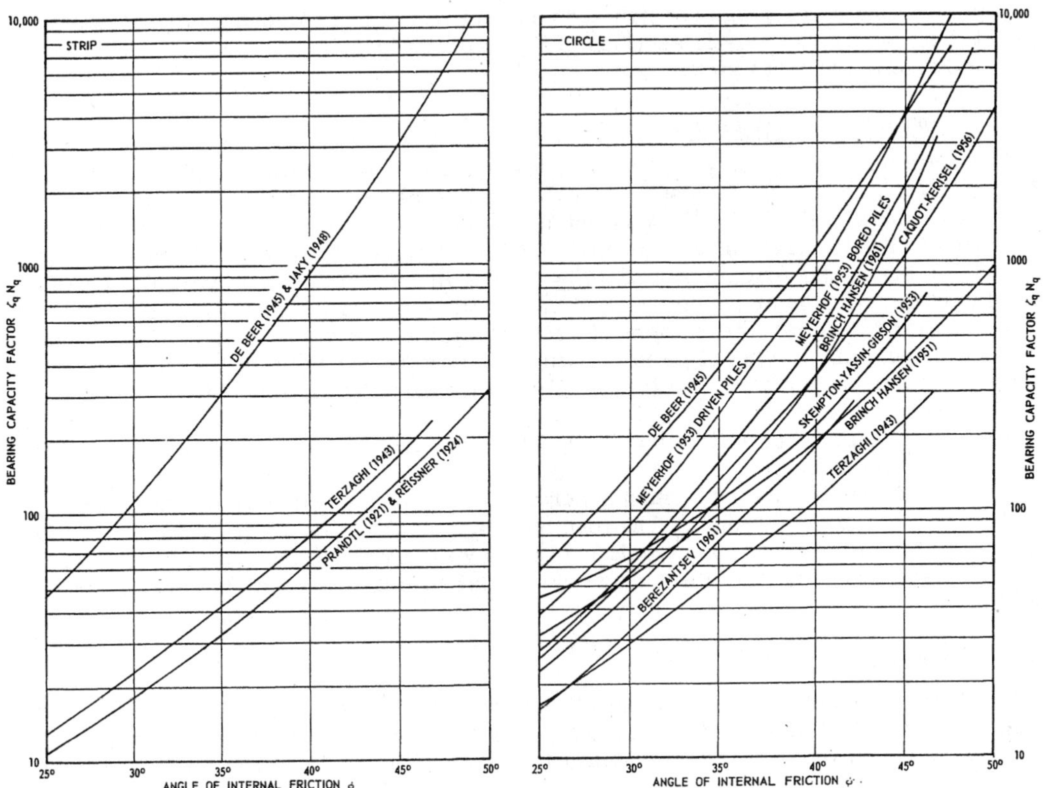

Figure 4. Theoretical bearing capacity factors.

Successful field experiences with predetermination of ultimate bearing capacity of piles by means of deep cone penetration tests (5, 7, 20, 21, 22, 23, 24, 25) have built a certain confidence in the general validity of the theoretical approaches described. It has been found in very many instances that the point bearing capacity of driven piles was indeed comparable to that of a deep cone penetrometer. One of the solutions proposed was being used with apparent success for evaluation of shear strength of sand in situ (8). Also, small-scale model tests (10, 15) as well as several well-documented full-scale tests on piles and piers (32) indicated N_q values in the wide general range predicted by the theories.

Some experiences, however, were not that encouraging. For instance, scale effects of a nature opposite to those predicted by the theories have been reported (26, 27). Observations of shear patterns in sand around deep foundations (28, 29, 30, 31) showed failure surfaces to be localized to the immediate vicinity of the foundation base. To the author's knowledge, not a single test ever indicated failure surfaces reverting to the shaft. The latest large-scale experiments, undertaken by the Institut de Recherches Appliquees du Beton Arme (IRABA) near Paris, led Kerisel (32) to conclude, without a rational explanation, that both foundation depth and size significantly influence the bearing capacity factor N_q. He suggested that N_q was not a unique function of φ but a complex function of φ, D/B and B. This conclusion will be discussed subsequently.

Investigations of the problem were undertaken several years ago with the general aim to contribute, if possible, to the understanding of phenomena occurring beneath and around deep foundations in sand. Both laboratory and field studies were envisaged in different phases of work. As a pilot study, it was proposed to perform large-scale model tests with foundations buried and driven to various depths in uniform masses of sand of different densities. This paper reports on the first phase of the work: tests with buried foundations.

TEST APPARATUS AND MATERIALS

A pile testing facility was constructed adjacent to the Soil Mechanics Laboratory. Figure 5 shows characteristic sections of this facility in different phases of operation. Figure 6 shows a general view.

Test Pit and Loading Equipment

The main feature is a large cylindrical test pit 8 ft 4 in. in diameter and 22 ft deep (Fig. 5), in which models of deep foundations can be placed in any kind of soil under controlled conditions. The pit is connected to a 12-in. sump that allows regulation and control of water level in the models. A 200-ton capacity reaction frame permits vertical or horizontal loading of models by means of corresponding hydraulic jacks. An adjustable A-frame at the upper level serves as support for pile-driving equipment as well as for miscellaneous equipment used in placing and excavating sand for models. This frame permits driving of piles vertically or at any batter up to 3:1 by means of a drop-hammer sliding along the leads. The entire facility is served by a 1.5-ton service crane.

For small-scale tests a steel box 50 by 50 in. square 70 in. deep with a 5-ton loading frame was constructed.

Loading of models was performed generally by means of hydraulic jacks of appropriate capacity (up to 200 tons). The load measurements were made by a corresponding set of proving rings and electronic load cells with a precision of less than 1 percent. Displacement measurements were made by ordinary micrometer dial gages (0.0001-in. precision).

Model Foundations

Two types of foundations were generally built: cylindrical, with circular bases 2.13-, 4- and 6.75-in. diameter, and prismatical, with 2.44- by 12.44-in. rectangular bases. The lengths varied according to the foundation depths (Table 1) between 10 and 113 in.

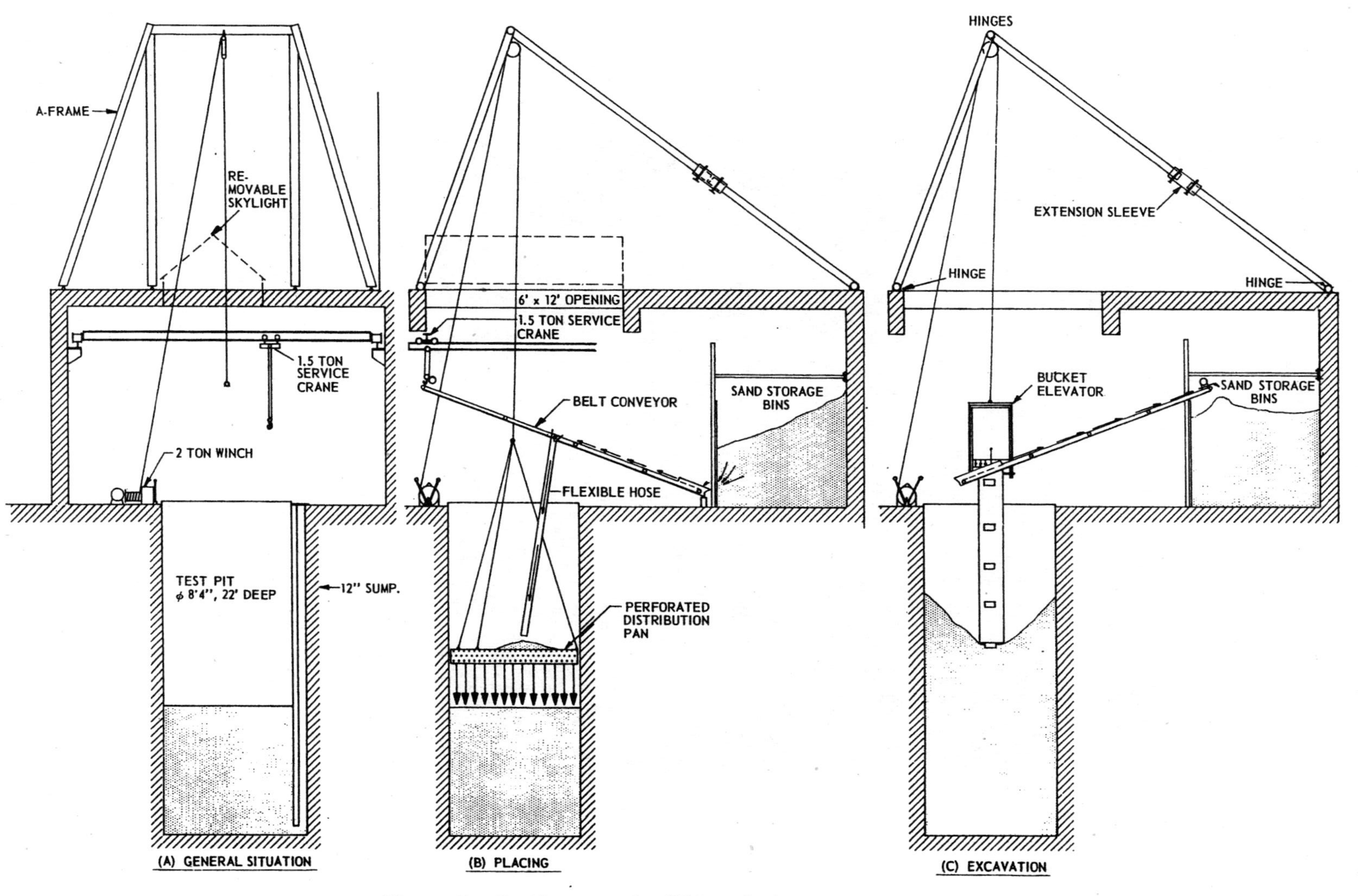

Figure 5. Testing area in different phases of operation.

Figure 6. Testing area.

TABLE 1
SUMMARY OF LOADING TESTS

Series	Test Numbers	Foundation			Different Sand Densities	Container Size (in.)
		Shape	Size (in.)	Depth (in.)		
A[a]	21 - 40	Circular	2.13	0, 10, 20 30, 40	4	50 × 50 × 70
B[b]	1 - 20	Rectangular	2 × 12 2.44 × 12.44	0 10, 20 30, 40	4	50 × 50 × 70
C	41 - 52	Circular	3.94 4	0 40	4	50 × 50 × 70
				80	3	100ϕ, 264
D	61 - 70	Circular	6 6.75	0 60, 100	4 3	100ϕ, 264
E	81 - 84	Circular	8	0	4	100ϕ, 264

[a]Skin diameter, 2 in.
[b]Skin dimensions, 2.25 × 12.25 in.

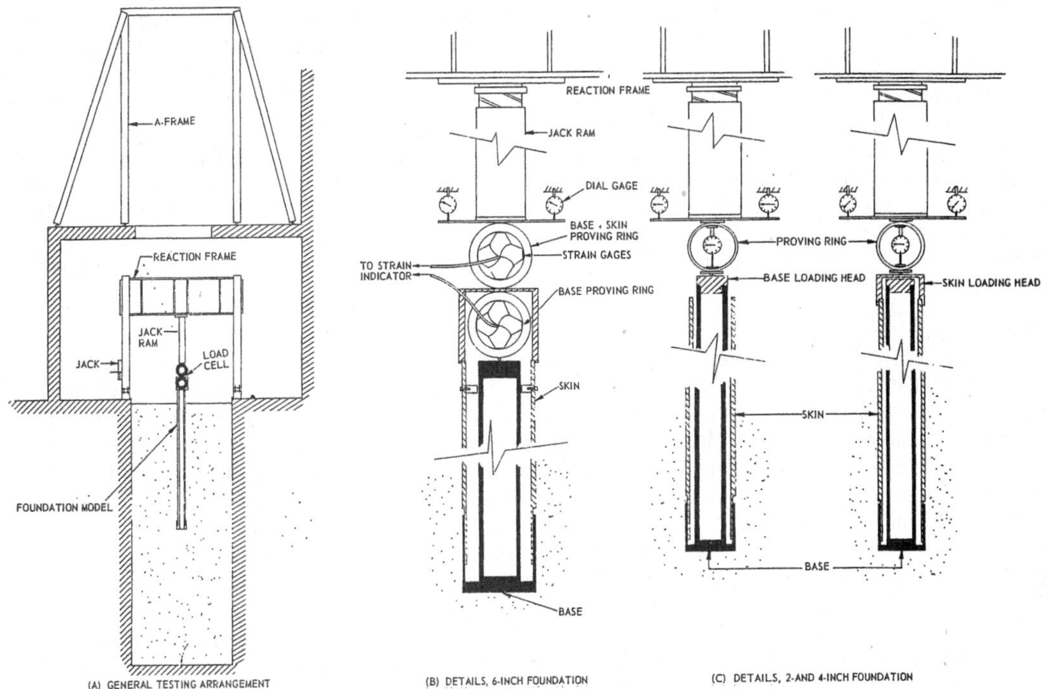

Figure 7. Model foundations.

Figure 8. Test in progress.

The foundations were constructed on a principle similar to that of a deep cone penetrometer. They consist of a steel casing inside which a steel shaft independently connects the loading head with the base. In the case of 2- and 4-in. diameter foundations (Fig. 7) separate loading of base and skin of the foundation is possible through exchange of the loading head. In the case of 6-in. diameter foundations (Fig. 7b) the loading head is constructed so as to allow separate registration on the strain-indicator of base and total (base + skin) load. Figure 8 shows an outside view of this special head.

The flat bearing surfaces of the bases are covered by sandpaper to assure perfect roughness. To prevent caving-in of sand while the base only is pushed, the latter is protected by a cap tightly fitted with the bottom of the casing. To minimize friction between the cap and casing, the contact surface is kept clean and perfectly lubricated.

Properties of Sand

All the tests in this investigation were performed with a medium sand originating from the Chattahoochee River, near Atlanta. The sand was sieved through a window screen (equivalent to 1.44-mm sieve opening). The grain size distribution curve (Fig. 9) and microscopic examinations indicate a medium, uniform sand composed mostly of subangular quartz particles, but rich in mica. The material has been air-dried prior to use in tests. The water content, controlled throughout the investigation, varied between 0.2 and 0.3 percent.

Maximum and minimum densities of this sand, as determined by standard procedures are given in Table 2, which also contains corresponding minimum and maximum porosities n and void ratios e.

Shear strength characteristics of the sand were determined by standard triaxial tests (constant cell pressure and positive deviator stress). A total of 54 air-dry samples 2.8 in. in diameter and approximately 6 in. high were prepared at four different densities, and tested using cell pressures varying from 5 to 80 psi. A strain-controlled loading machine was used for all the tests with the axial strain rate 0.02 in. per min (Table 3).

Assuming the Coulomb-Mohr criterion of failure to be valid, an ordinary plot of these results in τ vs σ or $\sigma_1 - \sigma_3$ vs $\sigma_1 + \sigma_3$ presentation can be made. Such a plot indicated that the strength envelopes of the sand in question are slightly curved. For better insight into the nature of this curvature the results are also plotted as $\sigma_1 - \sigma_3$ vs $1/\sigma_3$, a plot proposed by Hansen and Odgaard (33). In such a presentation a straight-line strength envelope appears as a straight line having the equation

$$\frac{\sigma_1 - \sigma_3}{\sigma_3} = \frac{2 \sin \varphi}{1 - \sin \varphi} + \frac{2 \cos \varphi}{1 - \sin \varphi} \frac{c}{\sigma_3} \qquad (7)$$

Figure 10 shows that a reasonably good straight-line approximation of the actually curved envelopes can be obtained by separately considering tow ranges of confining pressures σ_3, namely, $\sigma_3 < 10$ psi and $\sigma_3 > 10$ psi.

For $\sigma_3 < 10$ psi ($1/\sigma_3 > 0.10$ psi) the $\sigma_1 - \sigma_3/\sigma_3$ values are practically independent of $1/\sigma_3$, which means that the shear strength intercept c_o is zero. The angles of internal friction φ_o corresponding to observed shear strengths are given in Table 2 (Col. 9) and plotted in Figure 11, which shows that φ_o can be expressed as a function of e approximately by

$$\tan \varphi_o = \frac{0.68}{e} \qquad (8)$$

For 80 psi $> \sigma_3 >$ 10 psi the $\sigma_1 - \sigma_3/\sigma_3$ values can be approximated by linear functions of $1/\sigma_3$; which indicates that the shear strength intercept c_1 is different from zero. Table 2 (Cols. 10 and 11) gives values in-

TABLE 2

MINIMUM AND MAXIMUM DENSITIES OF CHATTAHOOCHEE RIVER SAND

Density	Dry Unit Weight (pcf)	Void Ratio, e	Porosity n (%)
Minimum	79.0	1.10	52.4
Maximum	102.5	0.615	38.1

TABLE 3

TRIAXIAL TEST RESULTS

Test No.	γ_d (pcf)	Void Ratio, e	Mean Void Ratio	Cell Pressure, σ_3 (psi)	Stress Difference at Failure, $\sigma_1 - \sigma_3$ (psi)	Axial Strain at Failure (%)	$\frac{\sigma_1 - \sigma_3}{\sigma_3}$	Angle of Internal Friction, φ_0	Strength Intercept, c_0 (psi)	Angle of Internal Friction, φ_0	Strength Intercept, c_1 (psi)
(1)	(2)	(3)	(4)	(5)	(6)	(7)	(8)	(9)	(10)	(11)	(12)
1	84.6	0.957		5	13.3	2.7	2.65				
2	84.6	0.952		10	27.4	7.6	2.74				
3	84.6	0.957		10	31.0	4.6	4.10				
4	84.4	0.961		20	49.1	6.6	2.46				
6	84.0	0.970	0.957	40	90.4	11.9	2.26	35° 20'	0	32°	1.08
8	84.6	0.957		80	182.0	8.1	2.28				
34	85.1	0.945		10	24.5	13.2	2.45				
40	84.6	0.957		40	96.0	7.4	2.40				
46	84.4	0.96		7	18.3	3.8	2.62				
47	84.6	0.956		7	20.6	2.8	2.94				
48	84.7	0.955		5	14.5	2.4	2.90				
5	90.5	0.830		20	62.2	4.6	3.11				
7	90.1	0.838		40	117.3	9.3	2.93				
11	89.6	0.847		5	16.9	2.9	3.39				
12	90.1	0.838		10	32.6	4.5	3.26				
13	90.1	0.838	0.836	20	62.4	8.2	3.12	38° 40'	0	35° 10'	1.55
14	90.1	0.838		40	108.6	10.7	2.72				
15	90.5	0.830		10	40.5	4.4	4.05				
31	91.3	0.813		20	60.1	6.4	3.00				
32	90.3	0.833		40	118.3	9.6	2.96				
33	90.5	0.830		55	156.2	9.4	2.84				
35	90.7	0.827		10	34.0	7.0	3.40				
38	89.0	0.859		40	115.7	6.4	2.89				
49	90.1	0.838		7	22.8	4.0	3.26				
50	90.2	0.836		7	25.0	3.4	3.57				
51	90.1	0.838		5	15.9	3.0	3.18				
16	94.8	0.747		5	22.5	3.0	4.51				
17	94.5	0.751		10	43.1	4.5	4.31				
18	95.0	0.742		20	78.3	6.7	3.92				
19	95.4	0.737		40	142.0	7.1	3.55				
20	95.4	0.737	0.745	75	247.9	9.2	3.30	42° 50'	0	38° 20'	2.42
39	94.3	0.755		35	125.7	4.6	3.60				
52	94.8	0.747		7	28.3	3.8	4.04				
53	94.9	0.745		7	31.0	3.0	4.44				
54	95.0	0.742		5	20.5	3.1	4.10				
21	97.9	0.691		5	24.3	5.0	4.87				
22	98.8	0.676		10	48.2	4.3	4.82				
23	98.7	0.678		20	79.4	5.4	3.97				
24	98.5	0.681		35	140.9	7.9	4.02				
25	98.7	0.678	0.681	70	250.7	9.8	3.58	45°	0	40°	2.85
26	98.6	0.680		40	146.1	8.9	3.65				
36	98.2	0.685		10	48.7	4.8	4.87				
41	98.5	0.681		5	23.5	3.0	4.87				
42	98.7	0.678		10	47.0	3.2	4.70				
43	98.5	0.681		20	83.2	6.7	4.16				
44	98.6	0.680		7	32.6	3.6	4.66				
45	98.4	0.682		7	34.1	3.9	4.87				

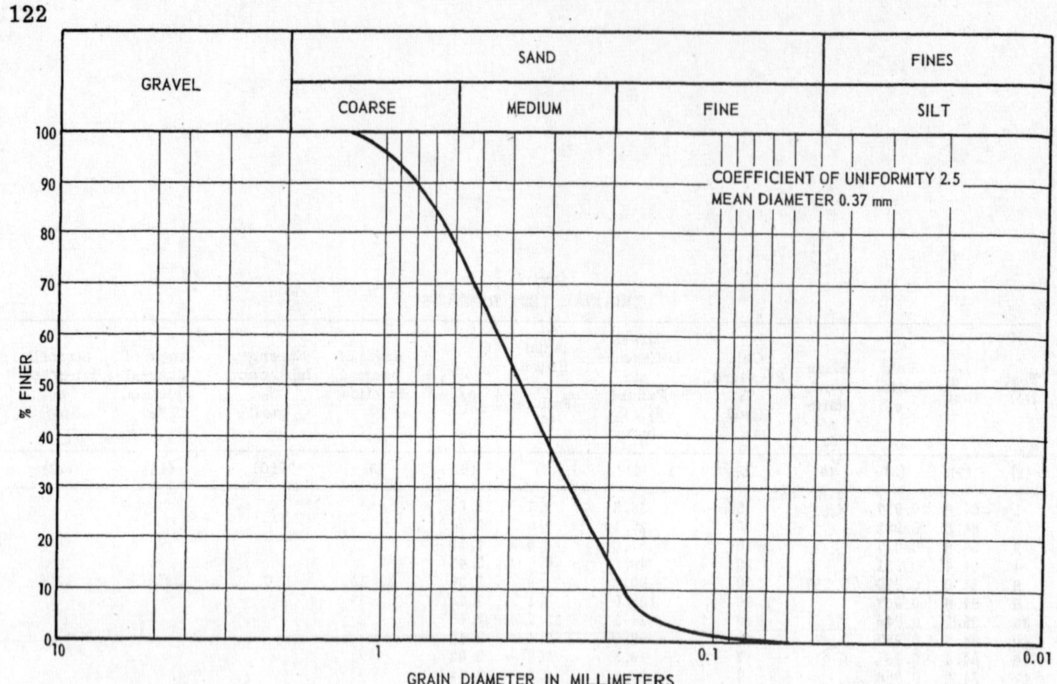

Figure 9. Grain-size distribution curve.

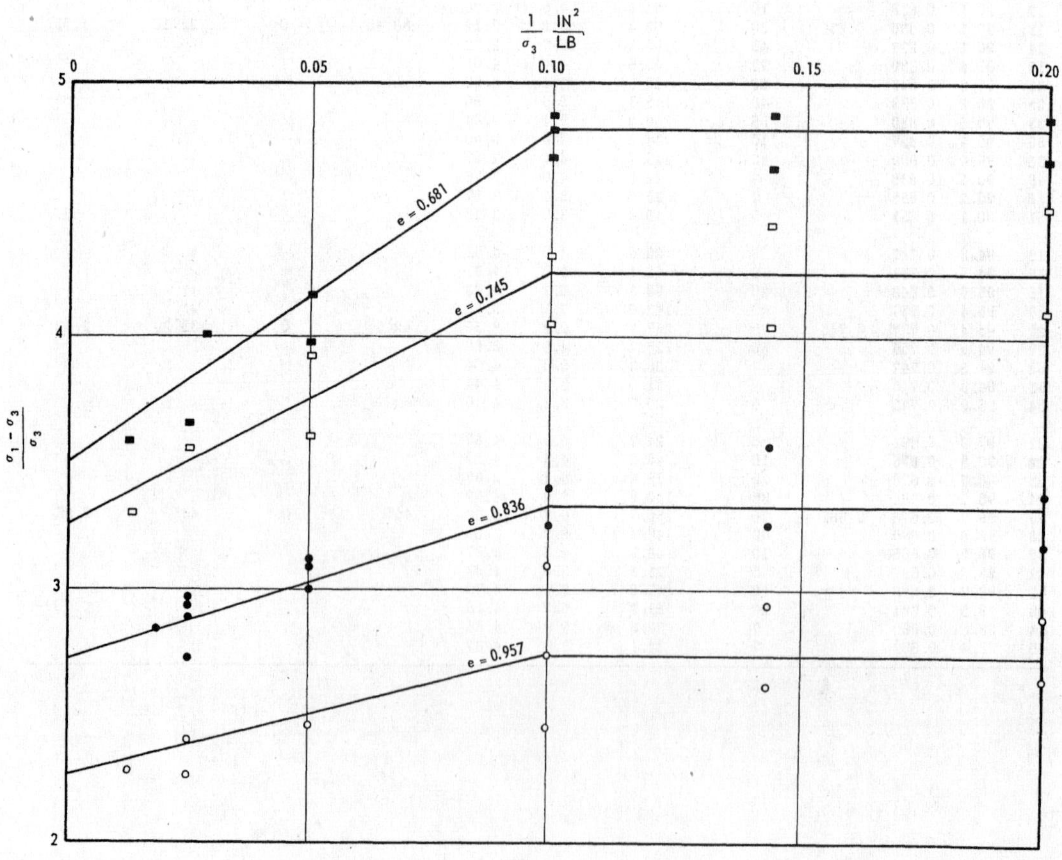

Figure 10. Triaxial test results.

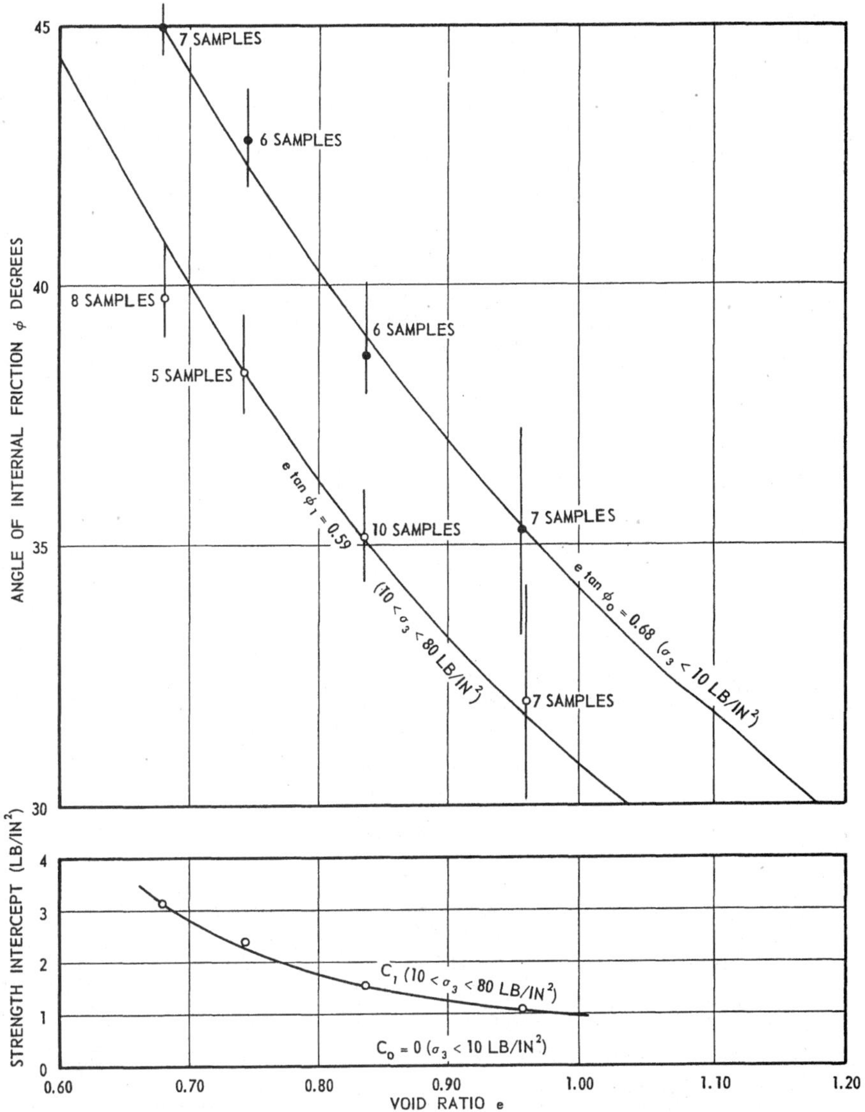

Figure 11. Shear strength parameters as a function of initial void ratio.

dicated by Figure 10 (left). The same c_1 and φ_1 values as functions of initial void ratio e are shown in Figure 11. The following analytical expression gives a good approximation of φ_1 as a function of e:

$$\tan \varphi_1 = \frac{0.59}{e} \tag{9}$$

DESCRIPTION OF TESTS

Placing of Sand and Control of Density

All the models for this investigation have been constructed in the following way. First, sand of desired uniform density was placed up to the planned elevation of the foundation base. The model foundation was then brought to full contact with the care-

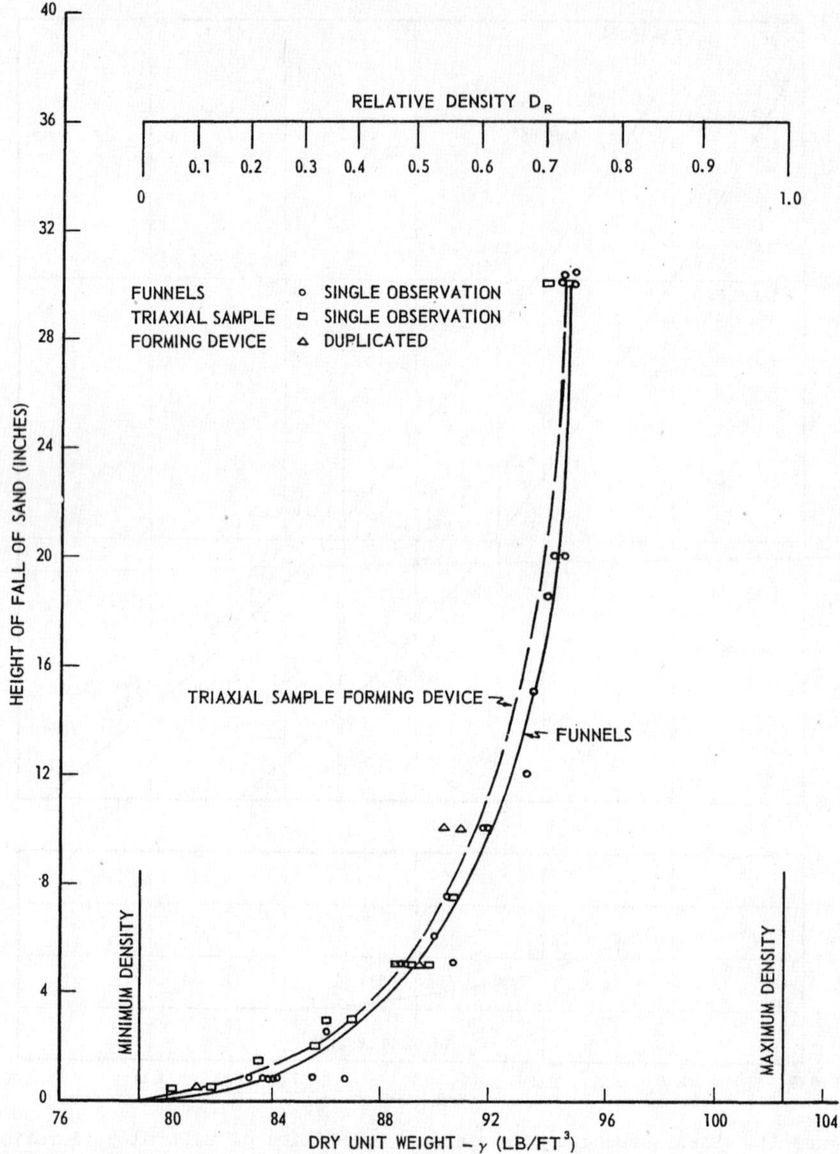

Figure 12. Relative density of sand as function of height of fall.

fully prepared horizontal sand surface and fixed in place so that it could not move during the subsequent operation of filling the sand above the base of the footing. Following former experience (34), exceptionally uniform loose- and medium-dense sand models (D_R < 0.70) were built by pouring sand from containers with perforated bottoms (Fig. 5b). It was confirmed again that the density of sand models so built is a unique function of the height of free fall of sand as long as other variables (rate of flow) remain the same (Fig. 12).

Denser sand models (D_R > 0.70) were built by surface vibration of 4-in. thick sand layers obtained by pouring sand 30 in. from the perforated container. Electric vibrators with a frequency of 3,600 cpm, attached to steel plates of appropriate shape, were used for surface vibration. Lead surcharge was added as necessary to achieve maximum compaction. All possible care was exercised to obtain uniform density throughout a model.

The homogeneity of sand in models was checked by penetrometer soundings. A simple static-cone micropenetrometer was constructed. This device has a ½-in. point diameter and a ⅜-in. shaft diameter of the casing. The assembly can be pushed into the sand by means of a screw jack at about 4 in. per min. Total resistance (point + skin) was recorded in several positions across the model and plotted against depth for each test. To convert the measurements of this kind into density, an empirical relationship was established between total resistance reduced to unit area of the point end dry unit weight of the material (Fig. 13). This was achieved by sounding sand models in a 24- by 16- by 60-in. box placed on a scale and filled by the same methods as used for building larger models.

Review of Tests Performed

Following the previously outlined program, six series of tests were performed. The main characteristics of five series of regular loading tests are given in Table 1.

The sixth series of tests, Series M, numbered 101 to 105, was devoted to the study of failure phenomena under foundations. Models of soil were built of distinct layers of sand to which cements of two different colors were added (10% by weight). After performing the loading test in the usual way, water was added to the models to cause setting of the layered mass. A few days later, the hardened block was cut through characteristic sections where shear patterns at failure were visible for observation and analysis.

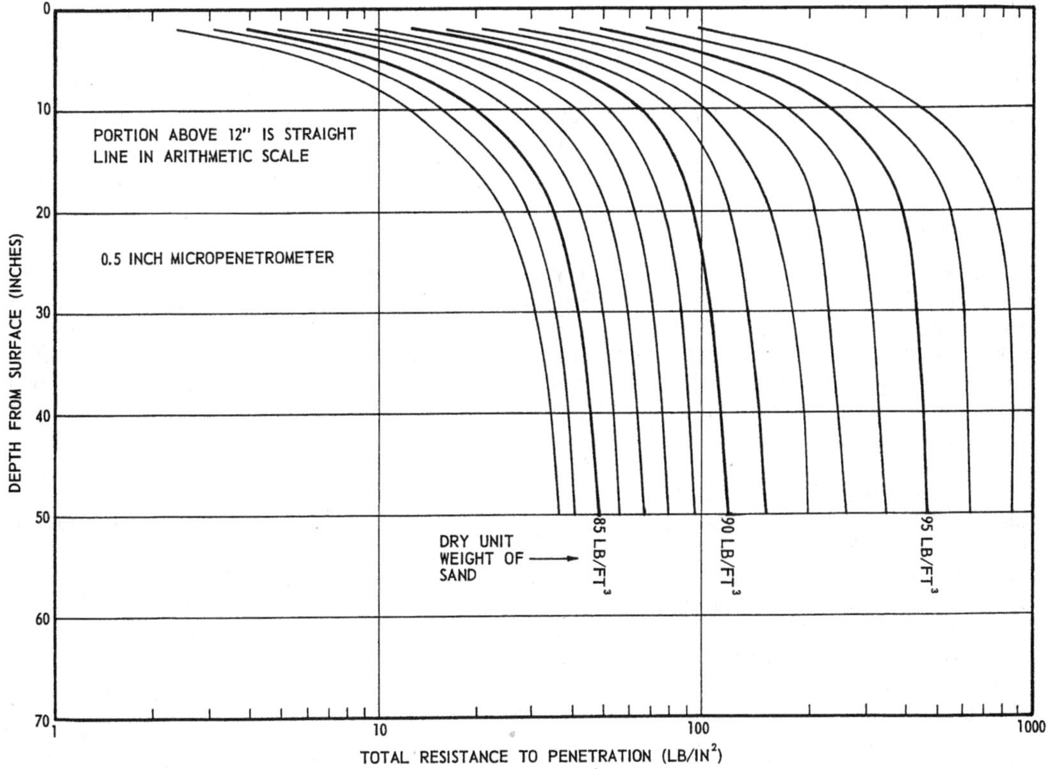

Figure 13. Relationship between depth and total penetration resistance for different sand densities.

Loading Procedure

The loading procedure for tests at the surface was similar to that followed in ordinary plate load tests. The load was applied in increments of about $\frac{1}{20}$ of the estimated failure load at 1-min intervals.

The loading procedure for tests beneath the surface was, in principle, the same. However, three separate loading stages existed in each test of series A, B and C. First, the foundation base was pushed until failure was reached; second, the same procedure was repeated with the foundation shaft; and third, after the shaft reached the base, both were pushed together and the total resistance was recorded.

Inasmuch as the loading head of the 6.75-in. foundations was of different construction (Fig. 7), the loading stages in D-tests differed somewhat from those just described. By pushing the foundation base to failure, the loading head was brought to contact with the skin and the entire foundation was forced into the soil. Special proving rings (Fig. 7) registered base and skin loads separately during this second stage.

Displacements of foundation base and skin were recorded at 1-min intervals by two micrometer dial gages placed near the loading head (Fig. 7). Also, in tests of series D, displacements of the sand surface were measured at different locations around the foundation.

TEST RESULTS

Significant results of the loading tests performed are given in Tables 4 through 9. Characteristic load-settlement diagrams of surface tests are shown in Figures 14 through 16. The black points indicate ultimate and first failure loads. The criterion by which these loads were established will be discussed subsequently.

Characteristic load-settlement diagrams of base and skin loading tests at greater depth are shown in Figures 17 through 19 and 20 through 22. Black points indicate ultimate loads.

Characteristic failure patterns at greater depth obtained in tests with colored sand are shown in Figures 23 through 25. Figure 23 shows what happens when a circular shaft penetrates through dense sand ($D_R \sim 0.9$) from a relative depth of D/B = 10 to a relative depth of D/B = 19. Figure 24 shows the analogous phenomenon for a rectangular foundation penetrating from D/B = 10 to D/B = 11, and Figure 25 for a rectangular foundation penetrating from D/B = 5 to D/B = 6.

Types of Failure

Three characteristic types of failure were observed in surface tests. Foundations on relatively dense sand ($D_R > 0.70$) fail suddenly with very pronounced peaks of base resistance (Fig. 26a) when the settlement reaches about 7 percent of the foundation width. The failure is accompanied by the appearance of failure surfaces at the sand surface and by considerable bulging of sheared mass of sand. The phenomenon corresponds exactly to that described earlier by Terzaghi (6) as "general shear failure."

Foundations on sand of medium density ($0.35 < D_R < 0.70$) do not show a sudden failure. As the settlements exceed about 8 percent of the foundation width small sudden shears within the sand mass are apparent from observations of load and settlement gages. Simultaneously, bulging of the sand surface starts. At settlements of about 15 percent of foundation width, a visible boundary of sheared zone at the sand surface appears. However, the peak of base resistance may never be reached.

The phenomenon is of the same nature as that described by Terzaghi (6) and by De Beer and Vesić (34) as "local shear failure" (rupture par refoulement incomplet). In the latter investigation, however, the tests were stress-controlled, so that the beginning of large shears in the soil mass was much more pronounced and was always recorded as the first failure of the foundation.

Finally, foundations on relatively loose sand ($D_R < 0.35$) penetrate into the soil without any bulging of the sand surface (Fig. 26c). The base resistance steadily increases as the settlement progresses. The rate of settlement, however, increases and reaches a maximum at a settlement of about 15 to 20 percent of foundation width.

TABLE 4

SIGNIFICANT RESULTS OF LOADING TESTS WITH CIRCULAR PLATES AT THE SURFACE

Test No.	Plate Diameter, B (in.)	Dry Unit Weight of Sand, γ_d (pcf)	Ultimate Pressure, p_o (psi)	$\frac{p_o}{\frac{1}{2}\gamma B}$	Ultimate Settlement, w (psi)	$\frac{w}{B}$ (%)	Type of Failure[a]
(1)	(2)	(3)	(4)	(5)	(6)	(7)	(8)
34	2.13	96.0	33.8	572	0.126	5.9	G
21	2.13	93.0	10.2 (7.6)[b]	173 (129)	0.272	12.8	L
22	2.13	89.8	7.4 (3.9)	134 (70.6)	0.414	19.5	L
23	2.13	82.6	2.4 (0.8)	48.2 (15.0)	0.463	21.8	P
44	3.94	96.7	53.8	432	0.227	5.8	G
41	3.94	93.8	19.8 (13.4)	186 (126)	0.446	11.3	L
42	3.94	91.0	13.0 (6.7)	126 (64.7)	0.621	19.2	L
43	3.94	83.3	3.1 (1.6)	32.5 (16.3)	0.582	14.8	P
61	6.00	96.2	73.4	432	0.425	7.1	G
62	6.00	93.0	31.0 (23.0)	193 (143)	0.869	14.5	L
63	6.00	91.7	19.3 (13.6)	121 (85.5)	0.875	14.6	L
64	6.00	95.0	5.3 (3.3)	35.9 (22.1)	0.852	14.2	P
84	8.00	96.2	79.2	432	0.667	8.3	G
81	8.00	95.2	55.9	262	0.737	9.2	G
82	8.00	95.2	42.1	193	0.695	8.7	G
83	8.00	88.0	9.0 (7.0)	44.2 (34.4)	1.08	13.5	L

TABLE 5

SIGNIFICANT RESULTS OF LOADING TESTS WITH RECTANGULAR PLATES AT THE SURFACE

Test No.	Plate Size (in.)	Dry Unit Weight of Sand, γ_d (pcf)	Ultimate Pressure, p_o (psi)	$\frac{p_o}{\frac{1}{2}\gamma B}$	Ultimate Settlement, w (in.)	$\frac{w}{B}$	Type of Failure[a]
(1)	(2)	(3)	(4)	(5)	(6)	(7)	(8)
16	2 × 12	96.4	47.0	842	0.212	10.6	G
1	2 × 12	93.6	22.1 (14.8)[b]	268	0.397	19.9	L
2	2 × 12	91.9	11.6 (8.3)	147	0.403	20.1	L
3	2 × 12	84.0	2.1 (1.4)	28.8	0.429	21.4	P

[a]G = general shear; L = local shear; P = punching shear.
[b]Numbers in parentheses refer to first failure.

TABLE 6

SIGNIFICANT TEST RESULTS – CIRCULAR DEEP FOUNDATIONS[a]

Test No.	Depth, D (in.)	Dry Unit Weight of Sand, γ_d (pcf)	Ultimate Base Resistance, p_o (psi)	Ultimate Settlement, w (in.)	$\frac{w}{B}$ (%)	Ultimate Skin Resistance, s_o (psi)	Ultimate Skin Displacement (in.)	Total Ult. Load (lb)	Ult. Displacement for Total Load (in.)
(1)	(2)	(3)	(4)	(5)	(6)	(7)	(8)	(9)	(10)
37		96.5	238.0	0.315	14.8	0.582	0.352	850	0.22
24	10	92.0	62.5	0.560	26.3	–	–	–	–
25		90.0	41.1	0.574	27.0	0.332	0.115	173	0.13
26		83.5	14.1	0.434	20.4	0.267	0.249	57	0.06
38		96.4	298.0	0.447	21.0	1.020	0.420	1,340	0.39
27	20	93.8	99.2	0.633	29.8	0.342	0.352	477	0.18
28		91.5	54.1	0.625	29.4	0.341	0.385	290	0.28
29		82.8	14.3	0.425	20.0	0.314	0.344	93	0.10
39		96.2	329.0	0.429	20.2	1.618	0.704	1,730	0.28
30	30	94.0	112.5	0.703	33.1	0.412	0.330	540	0.20
31		91.3	61.6	0.616	29.0	0.350	0.294	287	0.17
32		84.2	20.2	0.483	22.7	0.264	0.270	126	0.07
40		95.9	302.0	0.459	21.6	1.653	0.757	1,670	0.18
33	40	93.3	113.8	0.623	29.3	0.366	0.362	572	0.22
34		90.9	73.1	0.561	26.4	0.331	0.335	323	0.21
35		82.4	17.4	0.474	22.3	0.238	0.315	144	0.10

[a]Base diameter, 2.13 in.; skin diameter, 2.00 in.

TABLE 7

SIGNIFICANT TEST RESULTS – CIRCULAR DEEP FOUNDATIONS[a]

Test No.	D (in.)	γ_d (pcf)	p_o (psi)	w (in.)	$\frac{w}{B}$ (%)	Ult. Skin Resistance, s_o (psi)	Ult. Skin Displacement (in.)	Total Ult. Load (lb)	Ult. Displacement for Total Load (in.)
(1)	(2)	(3)	(4)	(5)	(6)	(7)	(8)	(9)	(10)
48		95.8	202.0	0.933	23.3	1.002	0.251	3,370	0.30
45	40	94.6	114.9	0.971	24.3	0.484	0.348	1,990	0.30
46		91.1	62.0	1.126	28.2	0.465	0.362	1,010	0.23
47		83.8	26.3	1.174	29.4	0.400	0.462	515	0.22
51	80	94.8	184.0	0.941	23.5	2.260	0.267	4,260	0.30
49		93.8	130.8	1.417	35.4	0.793	0.300	2,340	0.19
50		82.3	27.8	1.073	26.8	0.454	0.376	800	0.39

[a]Base and skin diameter, 4.00 in.

TABLE 8

SIGNIFICANT TEST RESULTS – CIRCULAR DEEP FOUNDATION[a]

Test No.	D (in.)	γ_d (pcf)	p_o (psi)	w (in.)	w/B (%)	Ult. Skin Resistance, s_o (psi)	Ult. Skin Displacement (in.)	Total Ult. Load (lb)
(1)	(2)	(3)	(4)	(5)	(6)	(7)	(8)	(9)
68		96.5	341.6	2.17	32.2	1.700	0.350	14,000
69	60	94.0	98.8	1.88	27.8	0.491	0.330	3,950
70		84.1	30.6	1.26	18.7	0.430	0.380	1,570
67	110	95.6	271.0	2.20	32.6	1.750	0.421	12,800
65	113	91.8	96.4	2.90	43.0	0.683	0.350	4,700
66	110	85.8	55.8	1.81	26.8	–	–	2,830

[a]Base and skin diameter, 6.75 in.

TABLE 9

SIGNIFICANT TEST RESULTS — RECTANGULAR DEEP FOUNDATIONS[a]

Test No.	Depth, D (in.)	Dry Unit Weight of Sand, γ_d (pcf)	Ult. Pressure, p_o (psi)	Ult. Displacement, w (in.)	$\frac{w}{B}$ (%)	Ult. Skin Resistance, s_o (psi)	Ult. Skin Displacement (in.)	Total Ult. Load (lb)	Ult. Displacement for Total Load (in.)
(1)	(2)	(3)	(4)	(5)	(6)	(7)	(8)	(9)	(10)
17		95.0	126.0	0.605	24.4			3,140	0.23
4	10	94.0	52.8	0.600	24.6			1,550	0.18
5		91.1	32.4	0.712	29.2			1,050	0.10
6		83.8	8.8	0.594	24.3			420	0.10
18		95.1	159.8	0.626	25.7	0.299	0.222	4,810	0.41
7	20	93.8	79.5	0.766	31.4	0.168	0.268	2,610	0.25
8		90.9	39.4	0.689	28.2	0.174	0.165	1,570	0.18
9		82.0	11.1	0.657	26.9	0.124	0.124	560	0.18
19		96.4	181.5	0.715	29.3	0.484	0.235	6,190	0.40
10	30	94.2	84.2	0.740	30.3	0.282	0.197	3,000	0.27
11		91.8	48.4	0.759	31.1	0.208	0.185	2,000	0.29
12		82.0	11.9	0.566	22.8	0.153	0.170	705	0.22
20		96.5	188.6	0.600	24.6	0.673	0.152	6,500	0.40
13	40	94.1	85.2	0.700	28.7	0.270	0.118	3,320	0.40
14		90.7	45.0	0.700	28.7	0.210	0.125	1,980	0.22
15		82.0	11.5	0.373	15.3	0.188	0.162	735	0.08

[a]Base width, 2.44 in.; base length, 12.44 in.; skin width, 2.25 in.; skin length, 12.25 in.

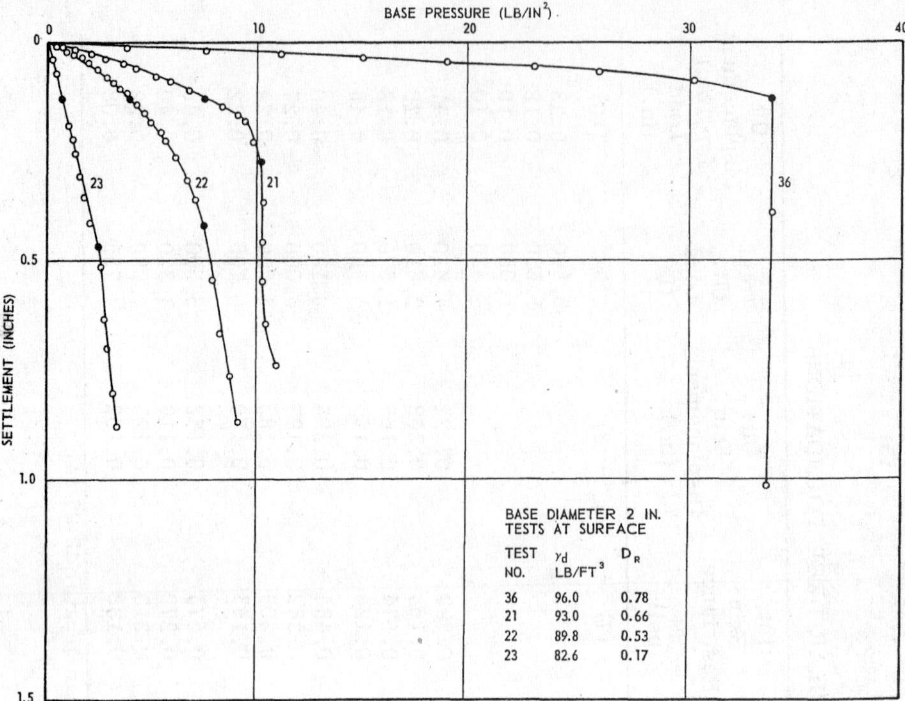

Figure 14. Typical results of surface tests.

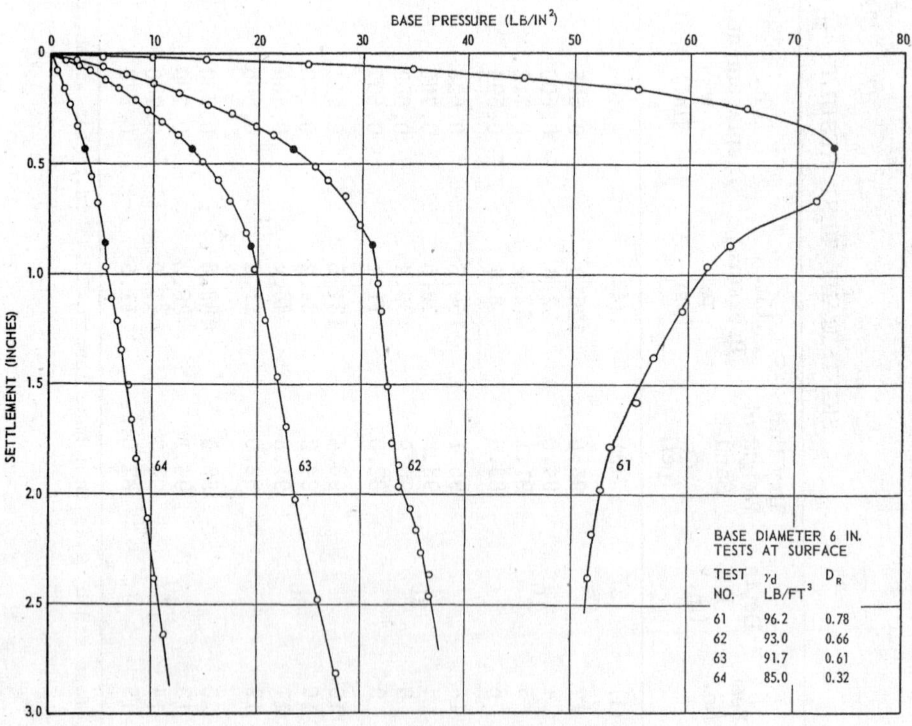

Figure 15. Typical results of surface tests.

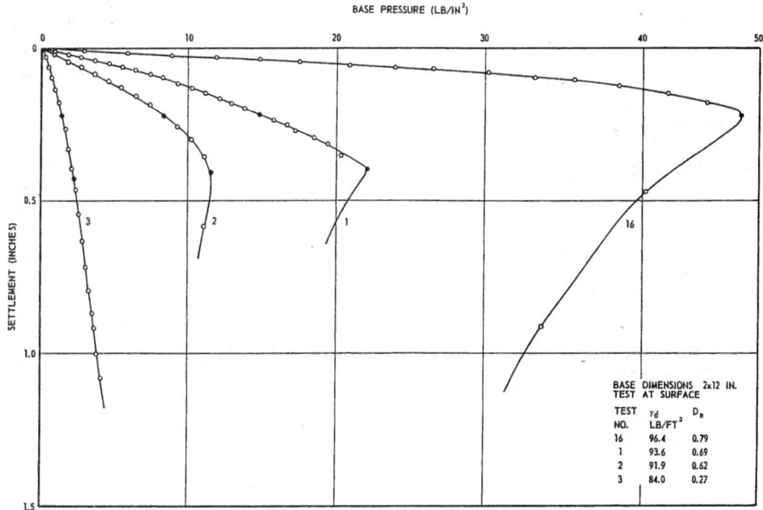

Figure 16. Typical results of surface tests.

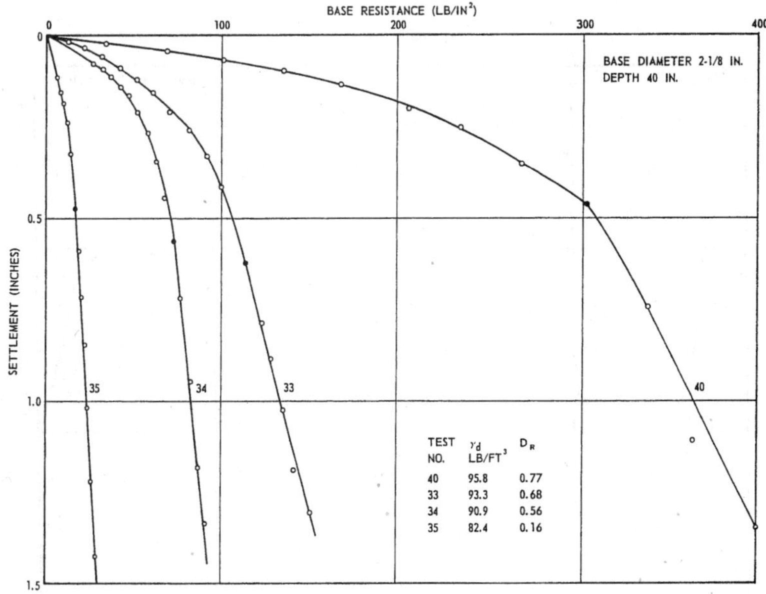

Figure 17. Typical results of base loading tests at greater depth.

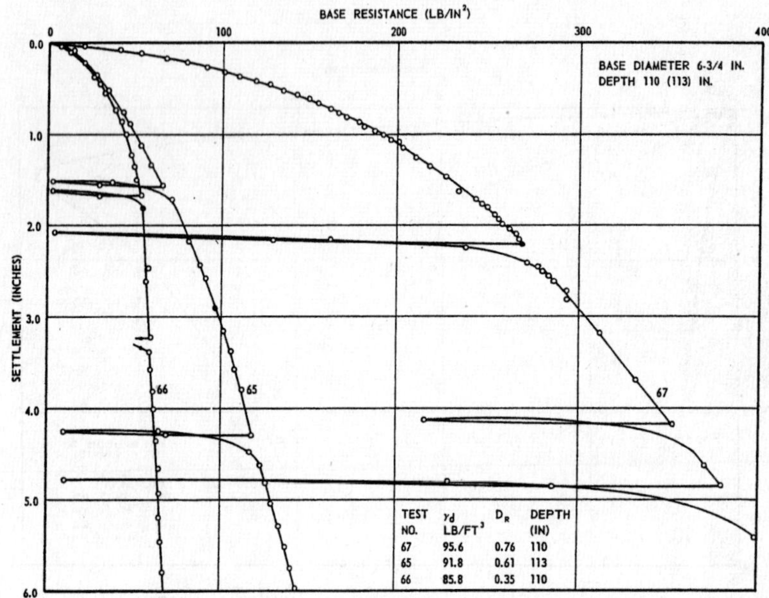

Figure 18. Typical results of base loading tests at greater depth.

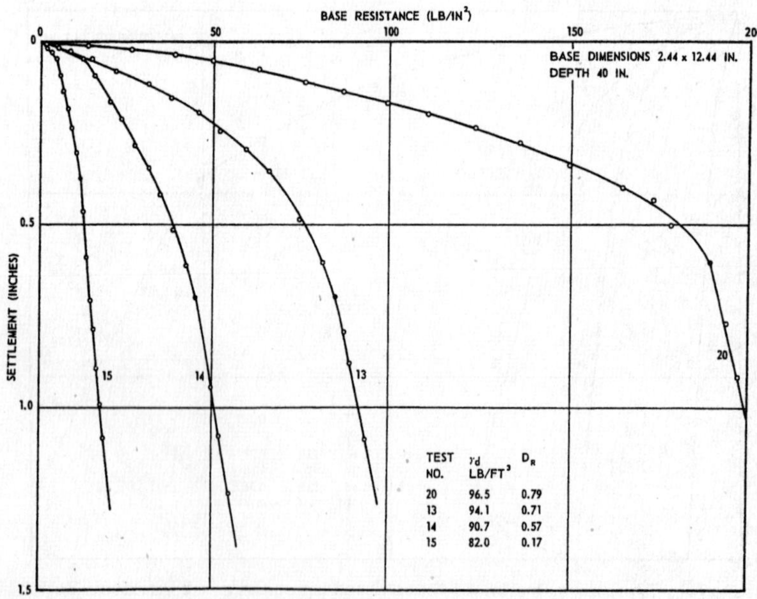

Figure 19. Typical results of base loading tests at greater depth.

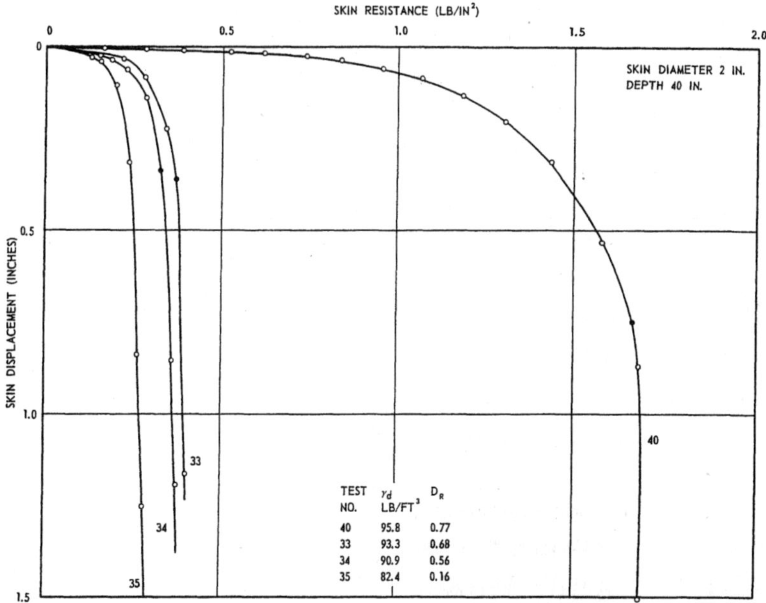

Figure 20. Typical results of skin loading tests.

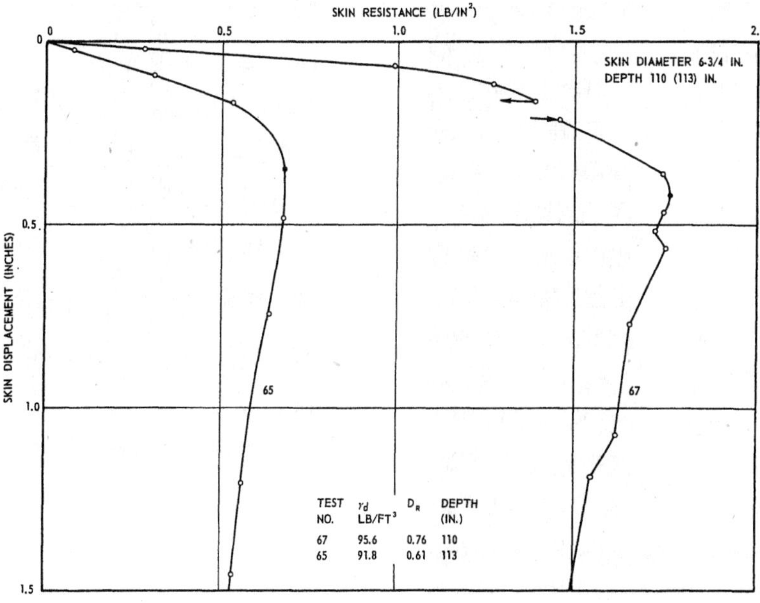

Figure 21. Typical results of skin loading tests.

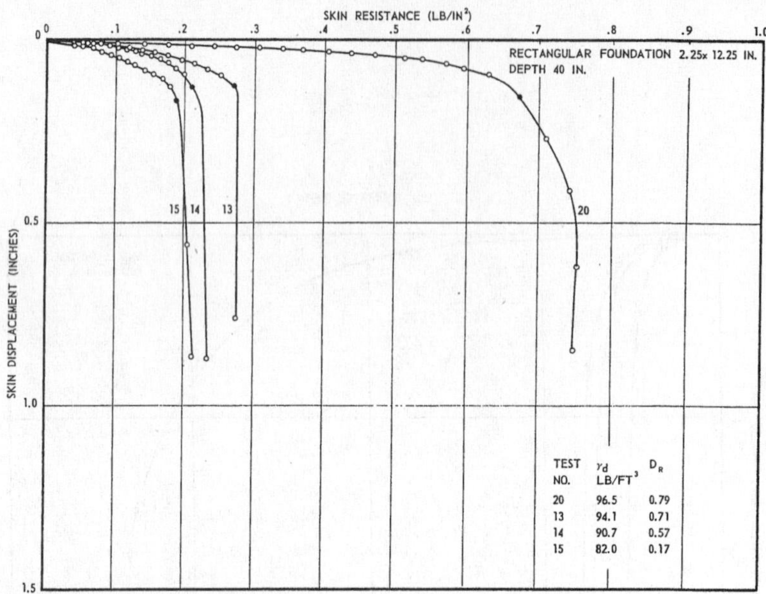

Figure 22. Typical results of skin loading tests.

TEST NO. 104 D/B=10 CIRCULAR FOUNDATION B=1 IN. D_R~0.9

Figure 23. Shear pattern under a circular foundation placed at greater depth in very dense sand.

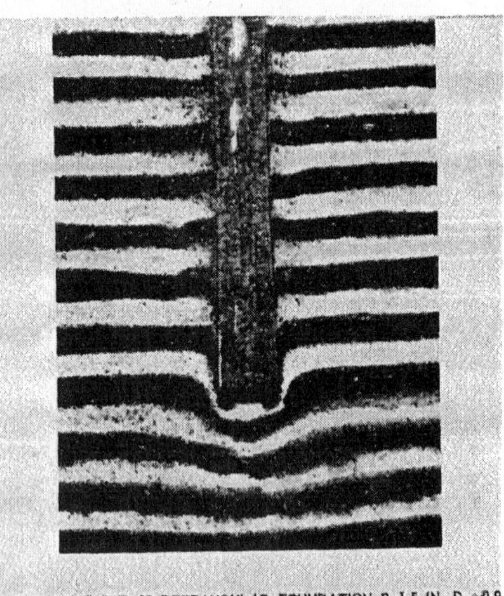

TEST NO. 101 D/B=10 RECTANGULAR FOUNDATION B=1.5 IN. D_R~0.9

Figure 24. Shear pattern under a rectangular foundation placed at greater depth in very dense sand.

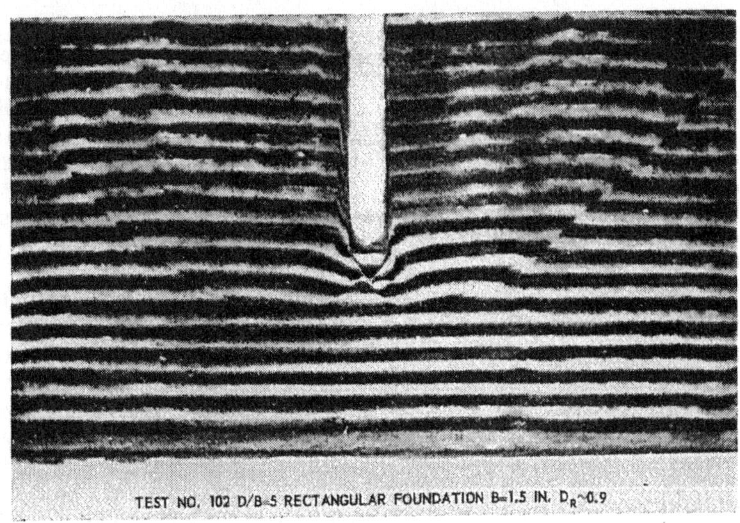

Figure 25. Shear pattern under a rectangular foundation placed at shallow depth in very dense sand.

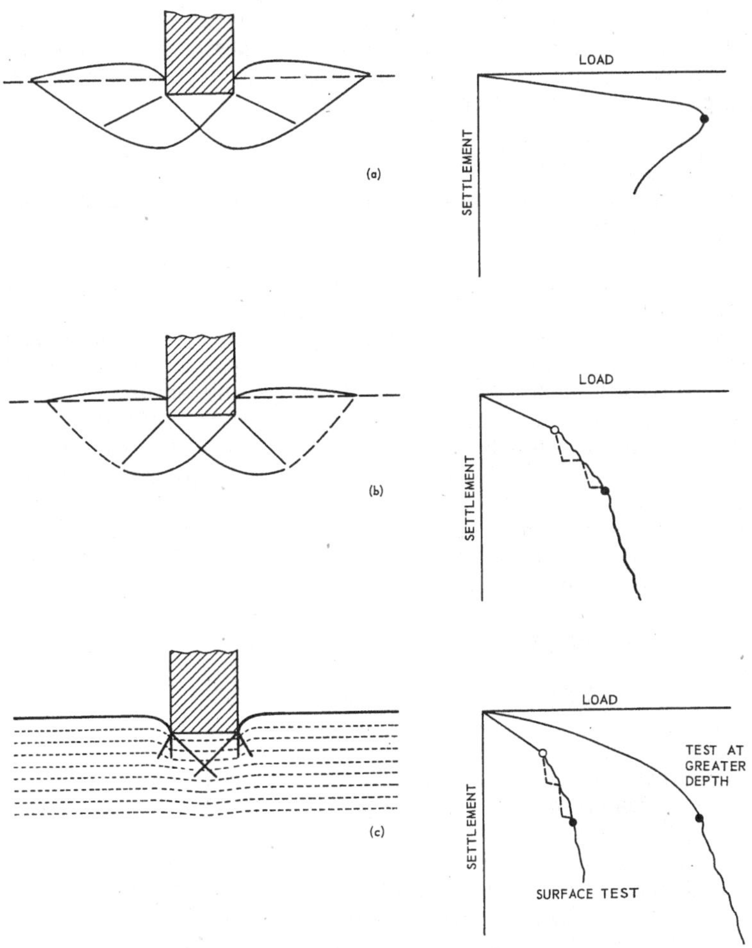

Figure 26. Types of failure.

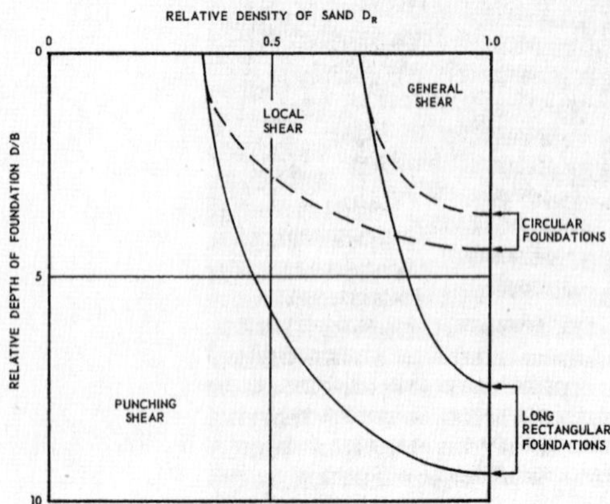

Figure 27. Types of failure at different relative depth D/B of foundations in sand.

Sudden shears can be observed in sequence as soon as the settlement reaches about 6 to 8 percent of foundation width. The failure surface, which is vertical or slightly inclined and follows the perimeter of the base, never reaches the sand surface. The phenomenon is essentially "punching shear failure," as described by De Beer and Vesić (34).

The same three characteristic types of failure are observed at shallow depths. However, as the relative depth D/B increases, the limiting relative densities at which failure types change increase. The approximate limits of types of failure to be expected as relative depth D/B and relative density of sand D_R vary are shown in Figure 27. There is a critical relative depth below which only punching shear failure occurs. For circular foundations this critical relative depth seems to be around D/B = 4, and for long rectangular foundations around D/B = 8.

It is important to note that the limits of types of failure depend on the compressibility of the material. More compressible materials will generally have lower critical relative depths. Following this trend, it is not difficult to explain why some materials may exhibit punching shear failure only.

Criterion of Failure or Ultimate Load

In accordance with observations just described the following criteria of failure or ultimate load were established:

1. In the case of general shear failure, the criterion is very clear: a peak of base resistance is always reached, corresponding to the appearance of failure surfaces at the sand surface, and to an abrupt change of rate of settlement from positive to negative.
2. In the case of local shear failure, there is not always a peak of base resistance, however, the rate of settlement reaches a maximum at the same load at which failure becomes visible at the surface. This load is considered as ultimate. In addition, first failure, clearly distinguishable only in stress-controlled tests, can be noted when settlements reach magnitudes at which the general shear failure occurs in dense sand (34).
3. In the case of punching shear failure, there is no peak of base resistance nor any appearance of failure surfaces. However, a peak of settlement rate can be noted. The corresponding load is considered as ultimate load.

Analogous criteria are adopted for skin loading tests.

DISCUSSION OF TEST RESULTS

Foundations at Surface

Figure 28 compares observed bearing capacities in surface tests with corresponding theoretical values. Measured values of $p/\frac{1}{2}\gamma B$ (col. 5, Tables 4 and 5) are shown as a function of dry unit weight of sand γ_d or relative density D_R. To have a better basis for comparison, the ultimate pressures of rectangular foundation have been multiplied by a shape factor 0.60 (a value recently confirmed by very extensive experiments, 35). Both first and ultimate failure pressures are shown for medium and loose sands. Figure 28 also shows theoretical bearing capacity factor N_γ after Caquot and Kerisel (36)

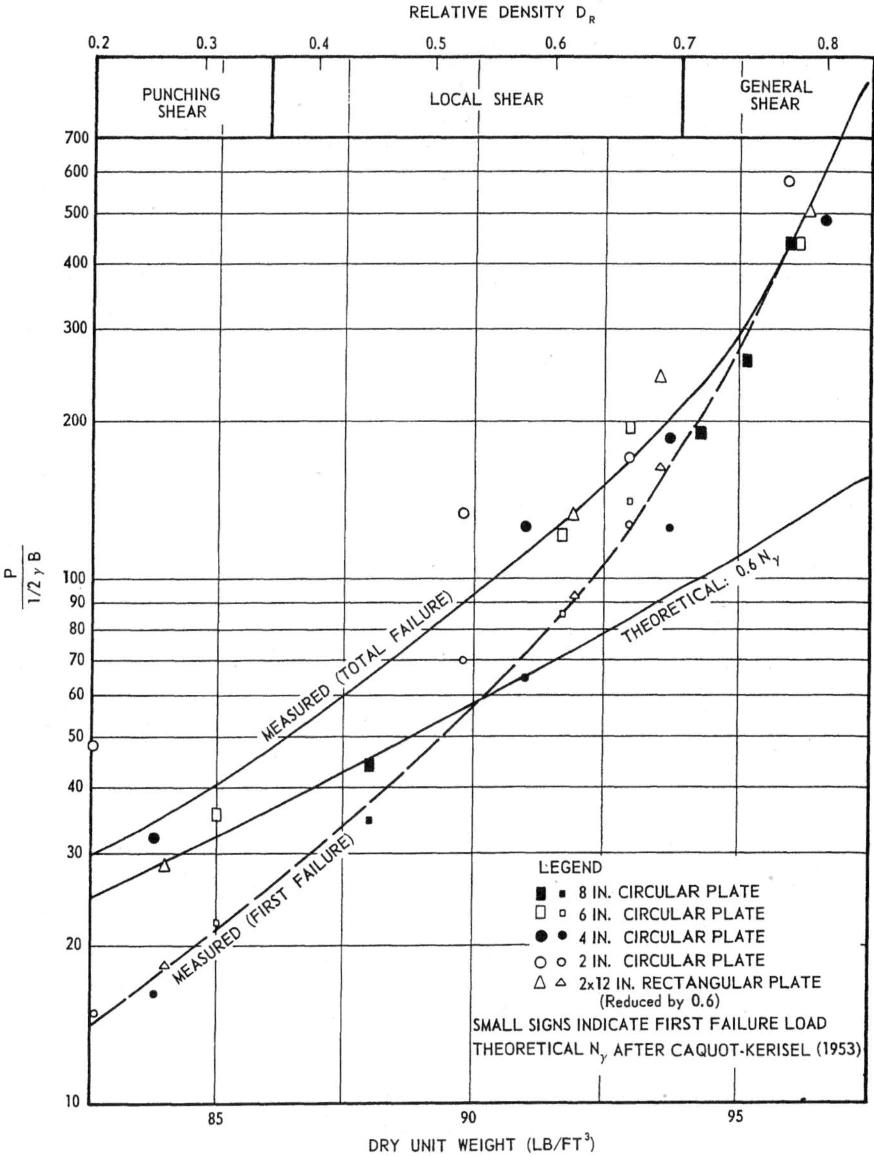

Figure 28. Observed bearing capacities of foundations at surface.

multiplied by shape factor 0.60. To present the factor N_γ as a function of dry unit weight or void ratio, the experimentally established relationship (Eq. 8) between the angle of internal friction φ and void ratio e is used.

Figure 28 shows that the observed ultimate bearing capacities are generally 1.2 to 4 times higher than corresponding theoretical values. This is in general agreement with findings of earlier experiments of similar nature (34, 33). A fully satisfactory explanation of this phenomenon has not yet been found.

The ranges of relative densities in which different types of failure occur (Fig. 28, top) also agree well with those found in an earlier investigation (34). It seems that the conventional classification of sands by relative density into loose ($D_R < 0.33$), medium ($0.33 < D_R < 0.67$) and dense ($D_R > 0.67$) has a certain meaning concerning the type of failure of shallow foundations on such materials.

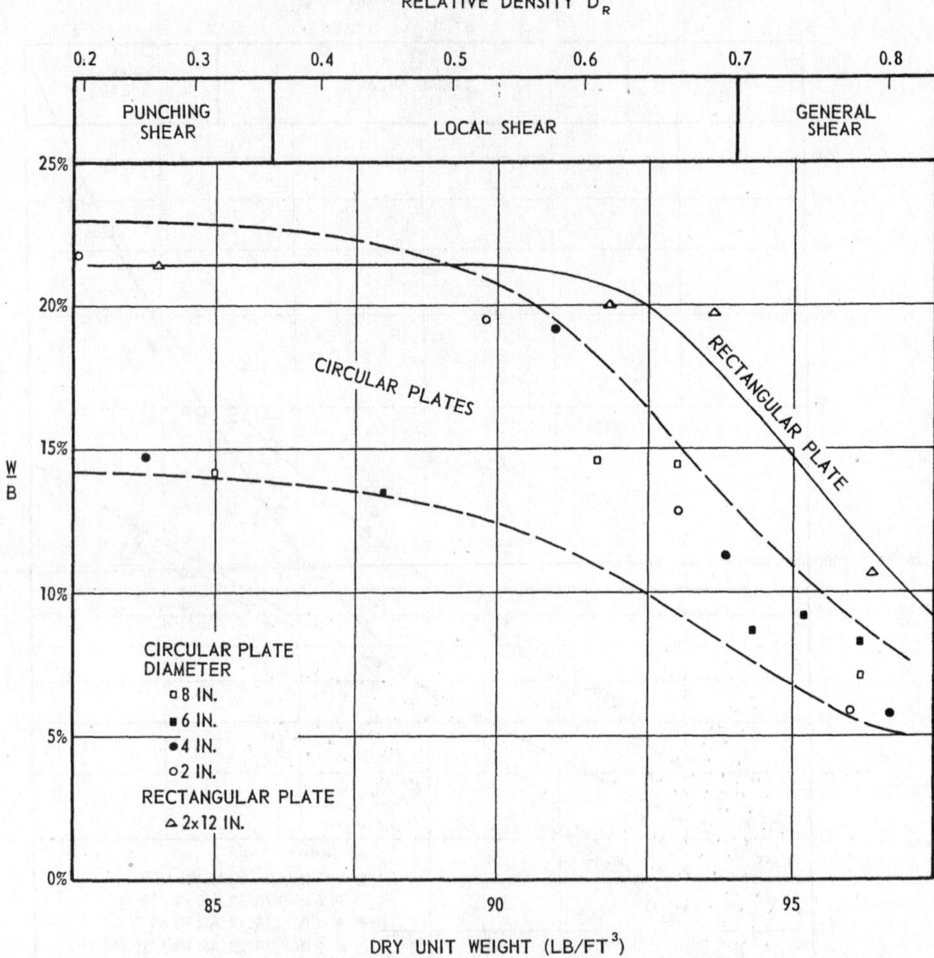

Figure 29. Settlement at failure for surface foundations.

Figure 29 shows the settlements at which ultimate loads were recorded, expressed as percentage of the foundation width. General shear failure usually occurs at settlements not exceeding 10 percent of foundation width; the other failure types take place at settlements of about 15 to 20 percent of the foundation width. This is in general agreement with former observations. However, slightly higher relative settlements at failure of rectangular foundations do not conform with some earlier findings (35). A similar trend was observed in tests with deep foundations (Fig. 32).

Base Resistance of Deep Foundations

Figures 30 and 31 show the general trend of increase in bearing capacity of the base with increase of foundation depth. Figure 30 shows the ultimate base resistance of 2.13-in. circular foundations as a function of foundation depth D. Figure 31 is an analogous plot for 2.44- by 12.44-in. rectangular foundations. A practically linear increase of bearing capacity with depth can be observed only at shallow depths, not exceeding approximately D/B = 4 for circular and D/B = 6 for rectangular foundations. As the foundation depth increases further, the rate of increase of bearing capacity with depth decreases. At a relative depth of approximately D/B = 15 the bearing capacity reaches asymptotically final values which appear to be functions of sand density only.

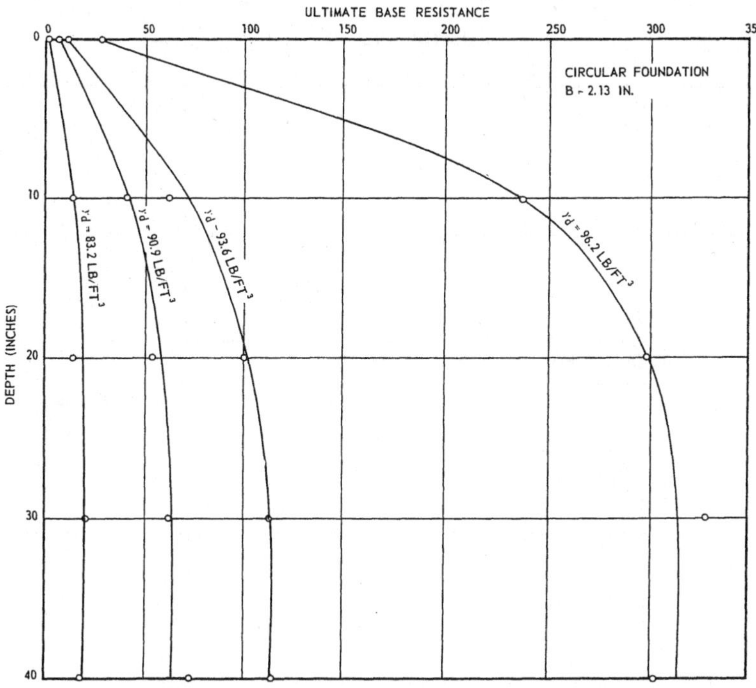

Figure 30. Measured bearing capacities of base—circular foundation, B = 2.13 in.

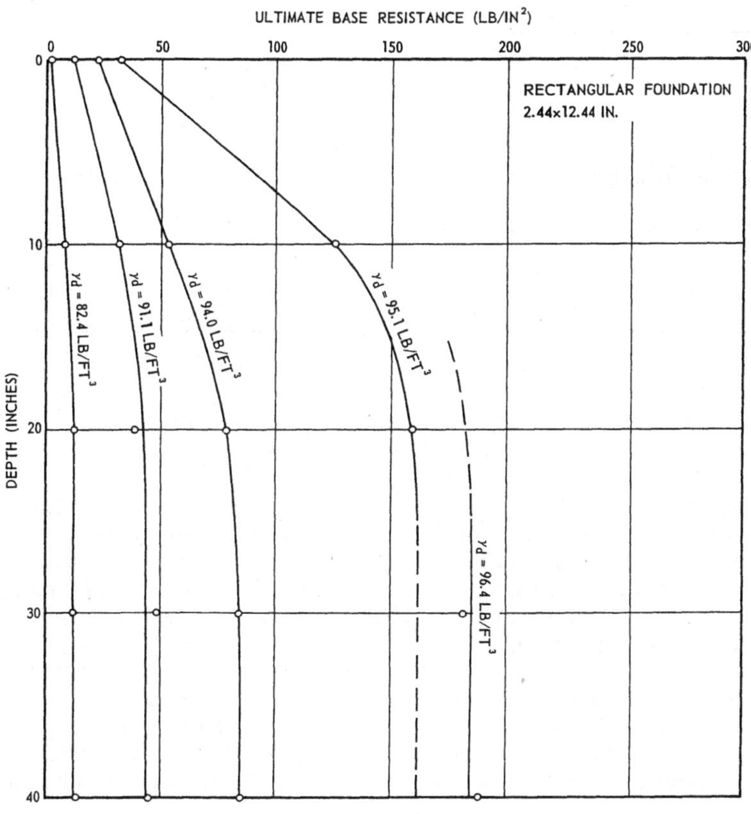

Figure 31. Measured bearing capacities of base—rectangular foundation, 2.44 x 12.44 in.

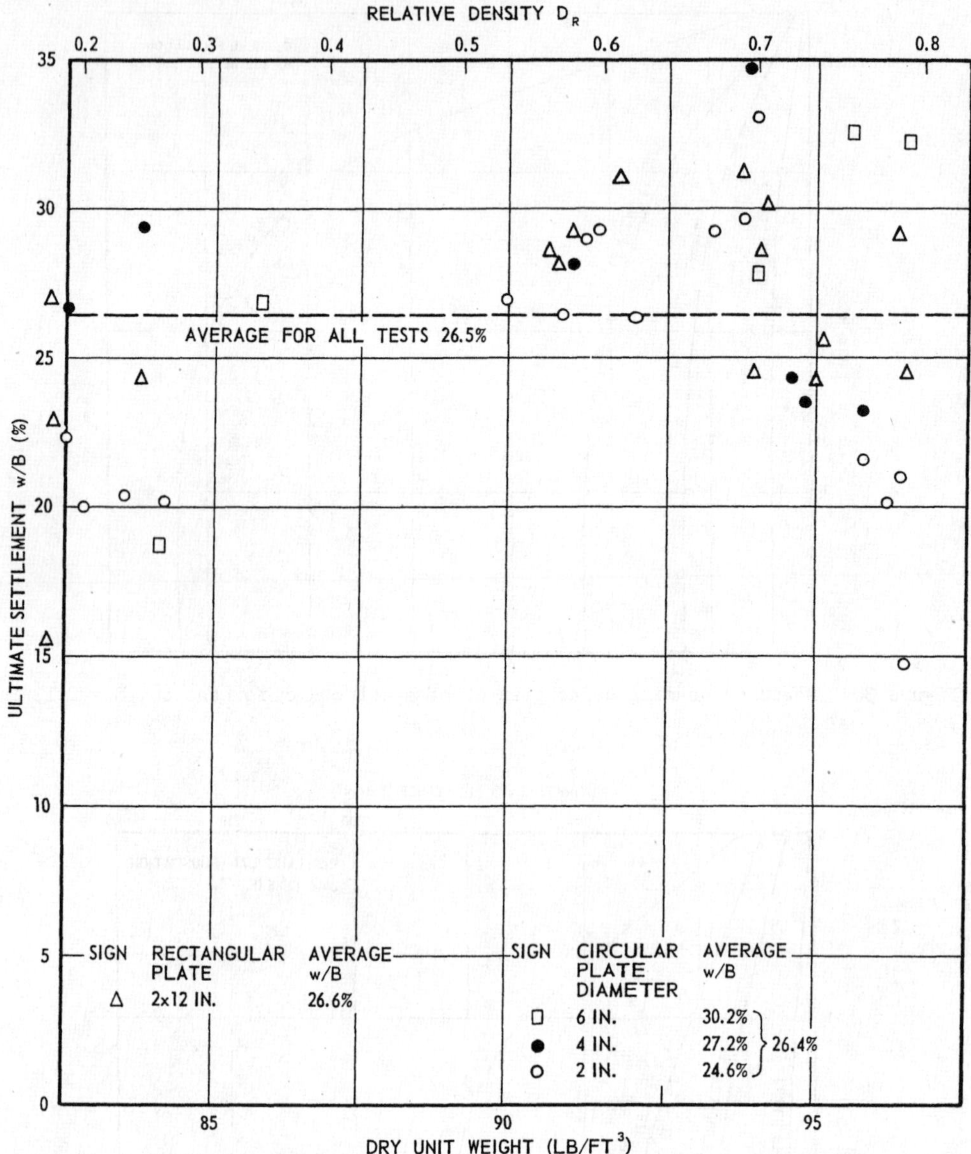

Figure 32. Ultimate settlement of deep foundations.

Base displacements or settlements needed to reach the ultimate loads are shown in Figure 32. Apparently, there is a tendency of ultimate settlements to increase with both foundation size and depth; however, this tendency is not pronounced. It may be stated that, in the range of foundations sizes and depth used in this investigation, ultimate loads are reached at settlements of about 20 to 30 percent of foundation depth. Figure 32 is in general agreement with isolated former observations.

A comparison of final bearing capacities of circular and long rectangular foundations indicates that the former are approximately 1.50 times higher. Figure 33 shows the average final bearing capacity of the base as a function of dry unit weight of sand, with bearing capacities of rectangular foundations multiplied by a shape factor of 1.50. Final bearing capacities observed in tests with 4- and 6.75-in. circular foundations are

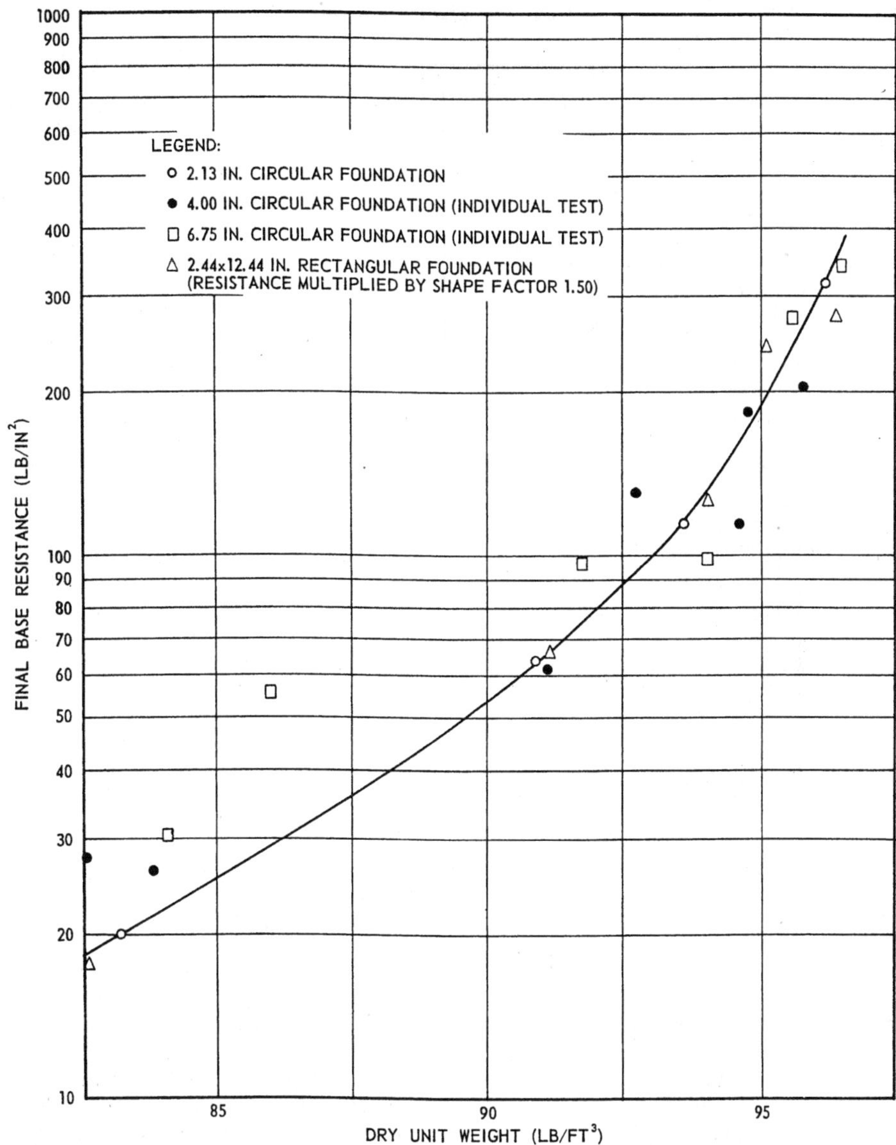

Figure 33. Ultimate base resistance at greater depth.

also plotted. The final bearing capacities are apparently independent of foundation size, at least for dense and medium dense sands.

A similar conclusion can be reached by studying Figure 6 of Kerisel's paper (32), although the numerical values obtained by the two investigations are not directly comparable due to differences in experimental approach and sand properties.

Skin Resistance of Deep Foundations

The variation of ultimate skin resistance s_o with depth for 2-in. circular foundations is shown in Figure 34. For models in dense, vibrated sand, a long initial linear increase of s_o with depth (up to D/B = 15) is followed by a sharp turn into a final skin resistance which remains constant as the depth increases further. For models in loose and medium-dense sand the shape of the initial part of the s_o curve is not quite clear.

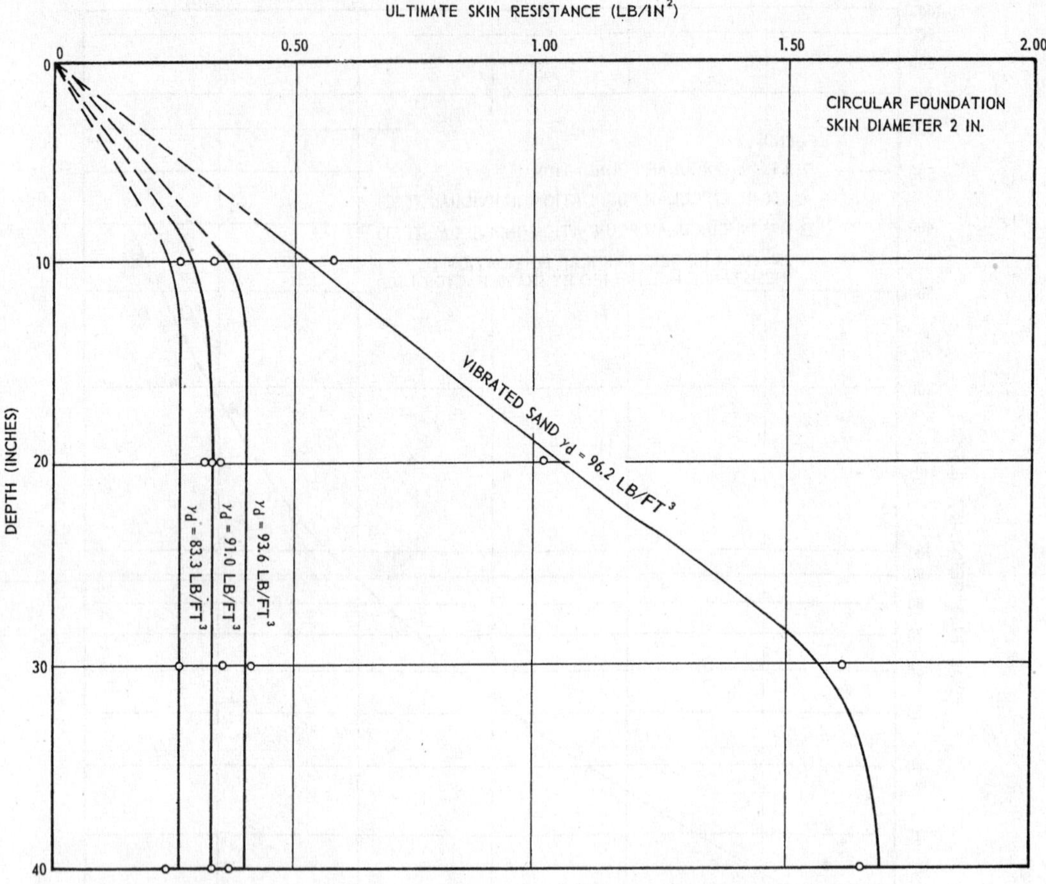

Figure 34. Ultimate skin resistance—circular foundation, 2 in.

It appears that there is also an initial linear increase limited to a depth of about four diameters. Beyond this depth the skin resistance turns sharply into a practically constant final value, varying with sand density only.

Figure 35 shows analogous diagrams for 2.25- by 12.25-in. rectangular foundations. The trend is similar, however, the initial part along which s_0 increases linearly with depth seems to be longer. The slope of the initial linear part is approximately three to four times less than the corresponding slope in the case of circular foundations. The final skin resistance, however, appears to be approximately 1.5 times lower for the rectangular shape.

Figure 36 shows the final average skin resistance as a function of dry unit weight of sand. Skin resistances of rectangular foundations multiplied by a shape factor of 1.5 are also plotted. The curve takes into account the fact that, due to method of placing of medium-dense sand, the density in the immediate vicinity of the skin was lower than the average density of the entire model, particularly in the case of 2-in. foundations.

The general shape of the s_0 curves found in the present investigation differs from that found in IRABA tests (32). However, it agrees well with numerous former observations on full-scale piers and caissons, which usually show a linear increase of s_0 at shallow depths, but a practically constant s_0 at greater depths.

Figure 37 shows skin displacements needed to reach ultimate skin resistance. It appears that these displacements are not dependent on foundation width and depth nor on sand density. For circular foundations they vary in the range of 0.30 to 0.40 in.; for rectangular foundations they are about one-half that magnitude. This finding con-

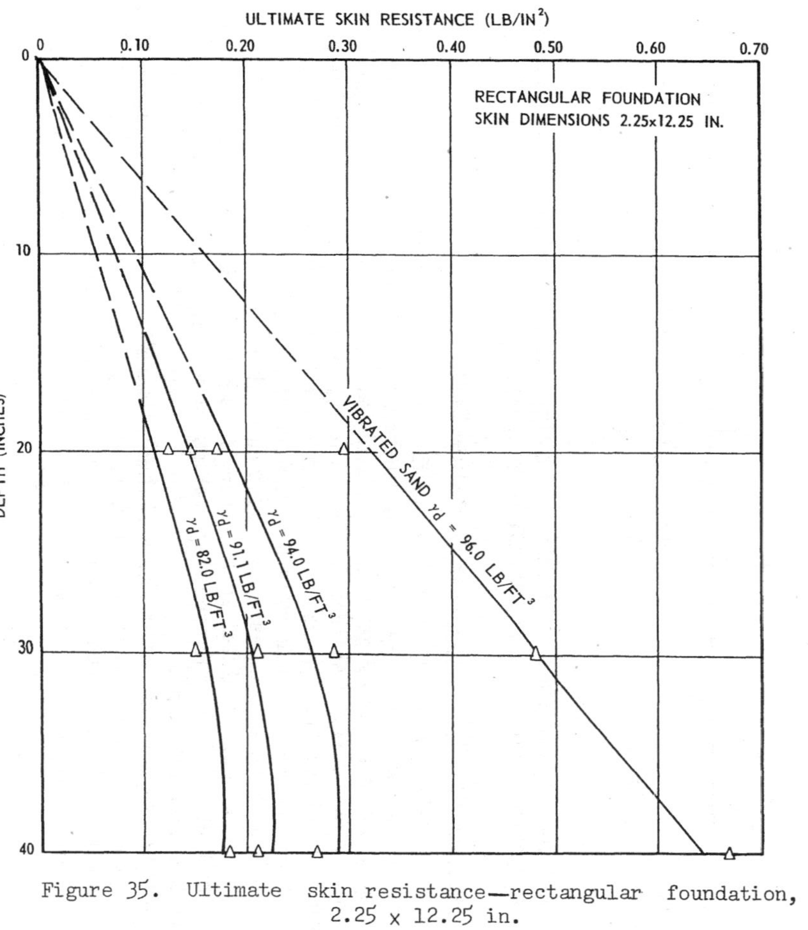

Figure 35. Ultimate skin resistance—rectangular foundation, 2.25 x 12.25 in.

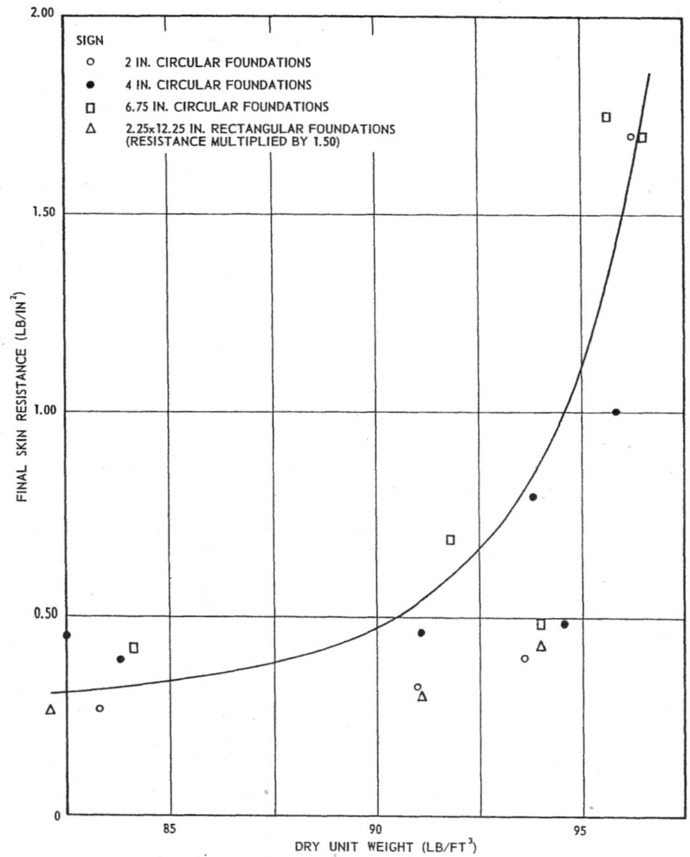

Figure 36. Final skin resistance at greater depth.

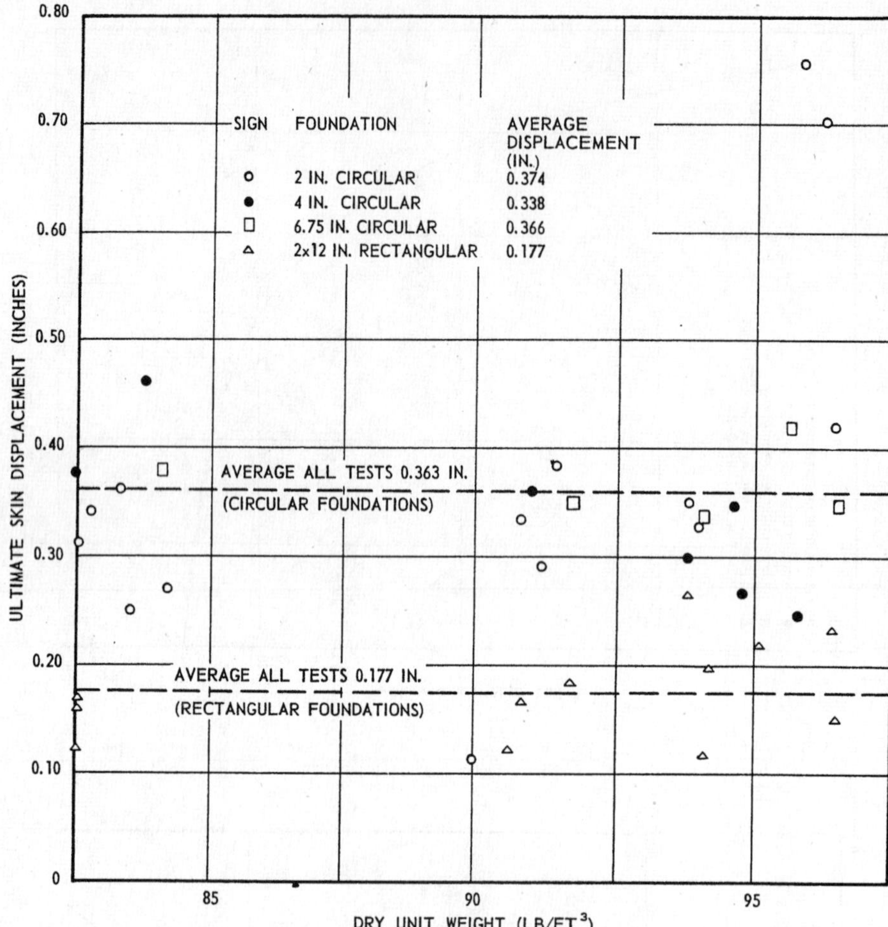

Figure 37. Ultimate skin displacement of deep foundations.

firms previously expressed beliefs that the mobilization of shear strength along a fixed rupture surface is governed by absolute displacement along that surface.

Bearing Capacity and Shape Factors at Shallow Depth

As previously mentioned, at shallow depths not exceeding $D/B = 4$ the increase of bearing capacity with depth appears to be linear as proposed by Eq. 3. Therefore the initial slopes of curves in Figures 30 and 31 indicate the experimental values of bearing capacity factor N_q at shallow depths. The N_q factors evaluated from these slopes are shown in Figure 38. To take into account the effect of shape, the depth term of circular foundations was reduced by an assumed shape factor of $\xi_q = 2.00$. Good agreement resulting from such an assumption indicates that shape factor ξ_q for a circular foundation in sand cannot differ greatly from 2. Terzaghi (6) proposed for that factor a value of 1.30 and Brinch Hansen (19), values increasing with φ from about 1.30 for $\varphi = 35°$ to 2.20 for $\varphi = 45°$.

As a basis for comparison, a curve of theoretical N_q values after Prandtl-Reissner is also shown (Fig. 38). To trace this curve the relationship (Eq. 8) between φ and e was assumed valid. This is justified by the probability that the average normal stress along a rupture surface under foundations does not exceed 10 percent of the foundation pressure.

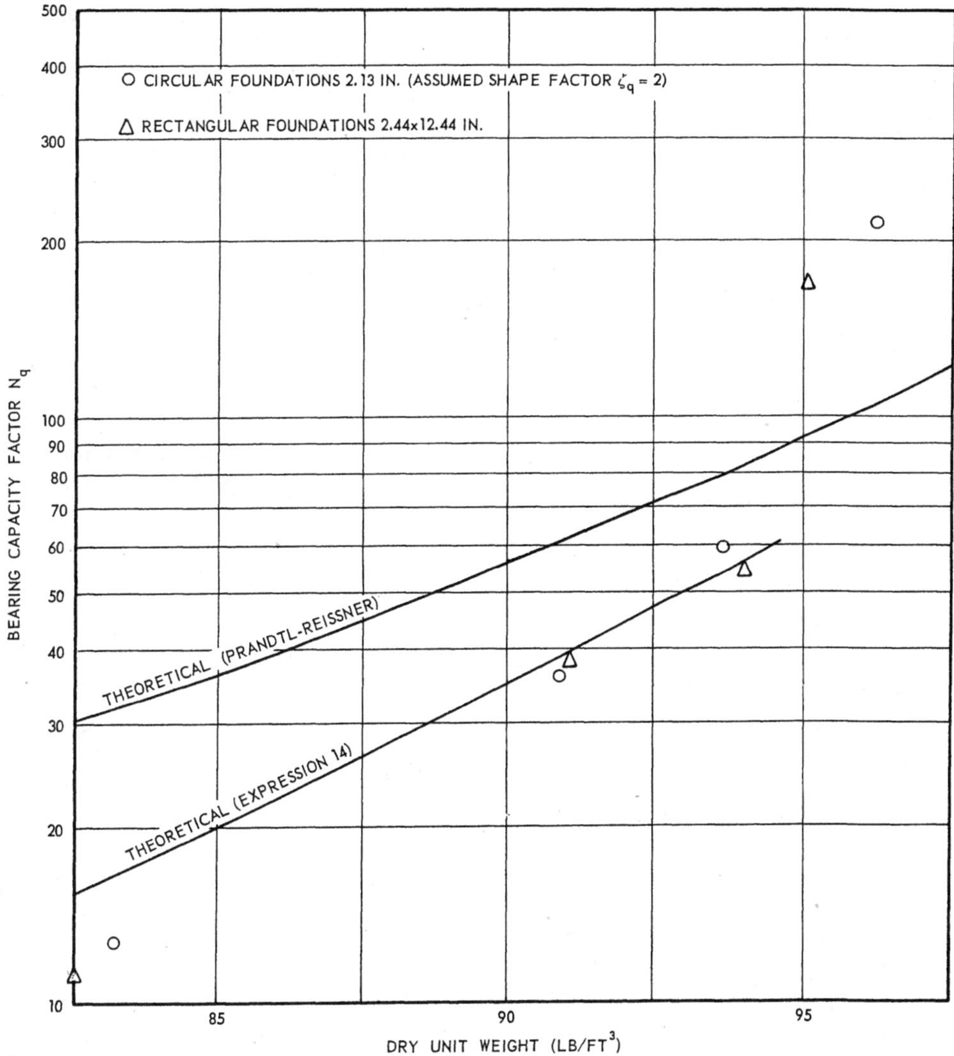

Figure 38. Measured bearing capacity factors N_q at shallow depth.

Higher N_q values for two very dense models can easily be explained. At high relative densities general shear failure still occurs at shallow depths (Figs. 25 and 27). As failure surfaces extend above foundation level, bearing capacity must be higher than indicated by Prandtl-Reissner theory which neglects shear resistance of the overburden (Fig. 2). If this resistance is taken into account a depth factor of approximately 2 should be introduced for $D/B = 4$ and $\varphi = 42°$ (10, 19). Therefore, excellent agreement of existing theory and experiments can be stated if the sand is dense. However, lower N_q values observed for medium and loose sand models cannot be explained by the existing theories, which consider general shear failure only.

Bearing Capacity in Local Shear Failure

To evaluate the bearing capacity factor N_q in the case of local or punching shear failure of a long rectangular foundation, a shear pattern based on observations on colored sand models will be considered (Fig. 39). It consists of an elastic zone ABC with two adjoining plastic zones BCD. The extent of development of these zones is determined by the angle θ at the apex.

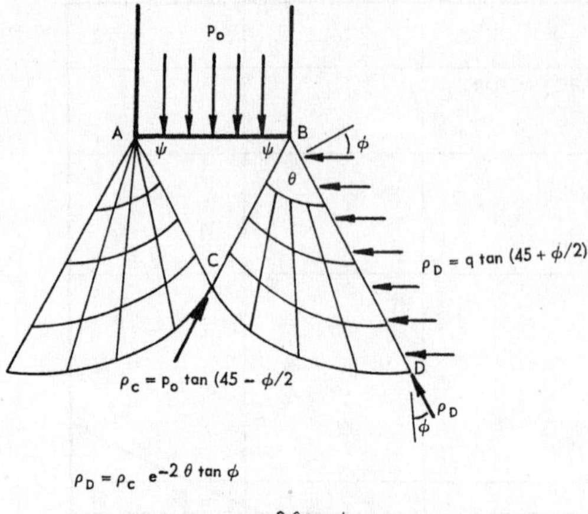

Figure 39. Analysis of punching or local shear failure.

It will be assumed that the overburden pressure q is great enough to allow neglecting the soil's own weight γ. Under such circumstances, solutions for weightless soil (3) can be applied to analyze stress conditions along CD. It is easy to show that the stress ρ_D acting on rupture line at D and the analogous stress ρ_C at C are connected by

$$\rho_D = \rho_C\ e^{-2\theta \tan \varphi} \tag{10}$$

However,

$$\rho_C = \rho_A = p_o \tan (45 - \varphi/2) \tag{11}$$

Also, assuming that the minor principal stress along BD is equal to overburden pressure q,

$$\rho_D = q \tan (45 + \varphi/2) \tag{12}$$

Eliminating ρ_C and ρ_D from Eqs. 10, 11, and 12,

$$p_o = q \tan^2 (45 + \varphi/2)\ e^{2\theta \tan \varphi} \tag{13}$$

By introducing $\theta = 1.9\ \varphi$, on the basis of observations, the following expression for N_q is obtained:

$$N_q = e^{3.8\ \varphi \tan \varphi} \tan^2 (45 + \varphi/2) \tag{14}$$

Numerical values for different angles φ are given in Table 10. They are lower than classical Prandtl-Reissner values. Reasonable agreement between N_q values computed by Eq. 14 and observed experimentally are shown in Figure 38.

TABLE 10

BEARING CAPACITY FACTOR N_q IN THE CASE OF LOCAL OR PUNCHING SHEAR FAILURE

Angle of Internal Friction, φ (deg)	Bearing Capacity Factor, N_q
0	1.0
5	1.2
10	1.6
15	2.2
20	3.3
25	5.3
30	9.5
35	18.7
40	42.5
45	115
50	422

Bearing Capacity at Greater Depth

Earlier discussion of base and skin resistances p_o and s_o has shown (Figs. 30 through 36) that, beyond some limiting relative depth D/B, the increase of p_o and s_o with depth is not linear. As D/B increases over 15, p_o and s_o do not increase any more. Final values of p_o and s_o appear to be functions of density of sand only (Figs. 33 and 36).

These observations seem to contradict the fundamental structure of bearing capacity Eq. 3 and 6 derived by using theories of plastic or elastic-plastic equilibrium. As previously mentioned, similar observations made recently by Kerisel (32) have led him to conclude that the bearing capacity factor N_q is a complex function of ω D/B and B. As long as no explanation of these findings is offered this appears to be the only possible conclusion.

However, attempts to explain the obtained results by an appropriate rational analysis leave serious doubts as to the correctness of the conclusion. No matter how limited an extent of plastic zone adjacent to the foundation base is assumed, there still must be a certain increase of p_o as overburden pressure increases. Therefore, N_q cannot be zero for any increment of loading as long as the same material is dealt with. When loosening of sand structure or significant crushing of sand grains occurs, there is still a lower limit of angle of internal friction of the newly formed material. Consequently, sooner or later, there must be an increase of p_o if overburden pressure continues to increase.

On the basis of these and other considerations, the conclusion was reached that constant values of p_o at greater depth do not result from decrease in N_q alone as suggested.

The explanation of the phenomena observed must, therefore, be sought through the assumption that q is not proportional to initial overburden pressure, as conventionally assumed. In connection with this, the true meaning of q in different theories should be remembered. In Prandtl-Reissner theory q is defined as normal stress at failure on horizontal plane of the foundation base (Fig. 2). In De Beer-Jaky-Meyerhof theories it is defined as normal stress at failure on the lower portion of the foundation shaft (Fig. 3). There is no good reason to take these stresses a priori equal or proportional to the initial overburden pressure, if foundation is deeply embedded in sand.

To demonstrate the meaning of the results, it is assumed that both p_o and s_o increase linearly with q as indicated by Eq. 3 and 6, but that q is strictly q_f or effective normal stress at failure acting on an elemental horizontal plane next to the foundation (Fig. 40). In a plane problem, or a rectangular foundation placed at greater depth, Eq. 3 can be rewritten in the following form:

$$p_o = q_f N_q \tag{15}$$

$$s_o = q_f K_s \tan \delta \tag{16}$$

Eliminating q_f

$$N_q = \frac{p_o}{s_o} K_s \tan \delta \tag{17}$$

In analogous way the following expressions can be written for a circular foundation:

$$p_o = q_f N_q \zeta_q \tag{18}$$

$$s_o = q_f K_s \tan \delta \, \zeta_s \tag{19}$$

$$N_q = \frac{p_o}{s_o} K_s \tan \delta \, \frac{\zeta_s}{\zeta_q} \tag{20}$$

Thus, it is possible to evaluate N_q from results of tests at greater depth under mentioned assumptions without really knowing q_f. $K_s \tan \delta$ or $K_s \tan \delta \, \zeta_s$ can be evaluated from the initial straight-line part of the s_o line.

The results of such an evaluation are shown in Figure 41, where individual results from tests at greater depth with 2- 4- and 6.75-in. circular foundations as well as with 2- by 12-in. rectangular foundations are plotted. To plot comparable magnitudes, both shape factors ζ_s and ζ_q for circular foundations were taken equal to 3 although ζ_s appeared to slightly higher. Figure 41 also shows, for comparison, bearing capacity factor N_q after Prandtl-Reissner and after Eq. 14. It is evident that experimental values obtained by using Eqs. 17 and 20 are primarily functions of sand density and that they are independent of absolute magnitude of q_f. Their numerical values in dense sand are reasonably well estimated by using the Prandtl-Reissner classical expression for N_q with a shape factor of approximately 3 for circular foundations. In medium and loose sands experimental values are lower, and comparable to those estimated by Eq. 14.

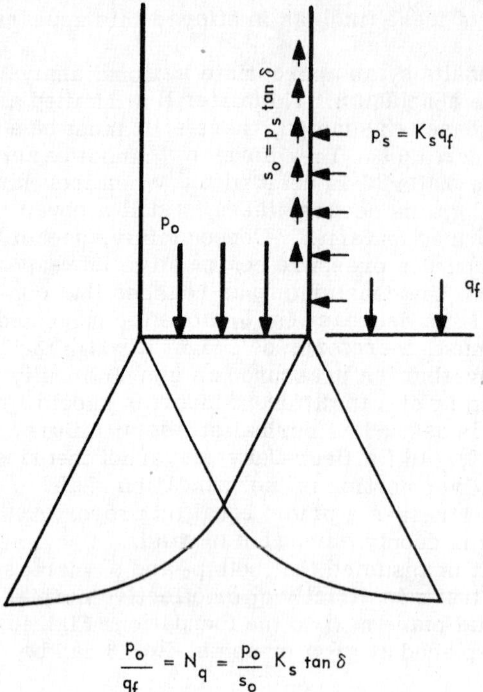

Figure 40. Stress conditions in vicinity of base of deep foundation.

Some details of the analysis presented undoubtedly need further clarification, particularly the choice of shape factors which, due to limited number of tests performed, could not have been determined very accurately. It appears certain, however, that both base resistance p_o and skin resistance s_o are linear functions of vertical stress at failure, q_f. This stress is not necessarily equal nor proportional to the overburden pressure q. The nonlinear increase of base or skin resistance with depth can be explained by a similar increase of q_f with depth. If the base and skin resistance reach constant values at greater depth, it is because q_f also becomes constant at greater depth.

Analysis of Vertical Stress Around the Foundation

According to the preceding discussion, curves in Figures 30 and 31 indicate the nature of variation with depth of vertical stress at the base level, q_f. At shallow depths ($D/B < 4$), q_f is equal to the overburden stress q; at greater depths ($D/B > 15$), q_f reaches a constant value independent of overburden stress.

In a similar way, curves in Figures 34 and 35 indicate the variation of average vertical stress along foundation shaft q_s with foundation depth. From the shape of these curves it may be concluded that the distribution of vertical stresses q_z at any point z along the shaft must follow a curve similar to that in Figure 42c; namely, there should be a linear increase of q_z along a certain depth z_o, followed by a peak and gradual decrease to the final magnitude q_f. It is to be understood, however, that the foundation depth, sand density and some other factors may have influence on the shape of curves in question. Therefore, the peak mentioned may be more or less pronounced, or even nonexisting, leading to q_z curves of shapes between those in Figures 42c and 42b.

Looking for an explanation of this general trend of variation of q_z with depth, it was concluded that the nonlinear increase of bearing capacity with depth could be attributed to "arching" in sand above the foundation base. There exists, indeed, a striking similarity between curves in Figure 42 and curves of vertical pressure in a mass of sand above a yielding horizontal support (6, Fig. 18d).

On the basis of all the observations made, the following explanation of stress conditions around a deep foundation is suggested: When the foundation is loaded (Fig. 42a) the mass of sand beneath is compressed downward. At the same time sand around the foundation tends to follow the general downward movement of the mass. As a consequence of this, the originally horizontal stresses on a vertical plane n-n at a certain distance from the foundation become inclined. The inclination of these stresses is a function of the amount of displacement w of the foundation and of the distance z′ from the base level. If the foundation depth D is great enough, and if the base displacement w remains limited, there may be a distance z_o' beyond which the effect of downward movement is not felt any more. Above that distance stresses on vertical planes may remain horizontal, and the vertical stress q_z may be equal to overburden stress q.

The following arguments can be added in support of the explanation:

1. Measurements of displacements of sand surface during loading tests on foundations placed at greater depth ($D/B > 8$) showed downward movement of soil adjacent to the foundation.

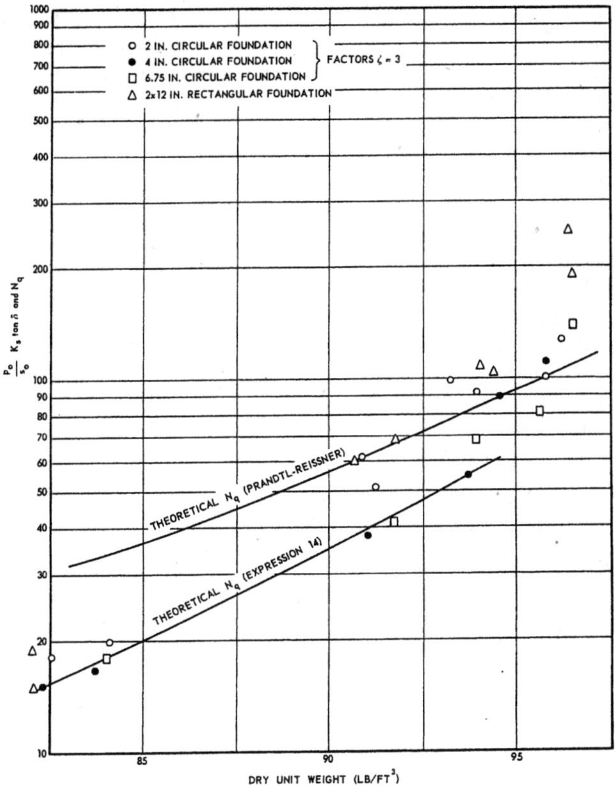

Figure 41. Bearing capacity factors at greater depth.

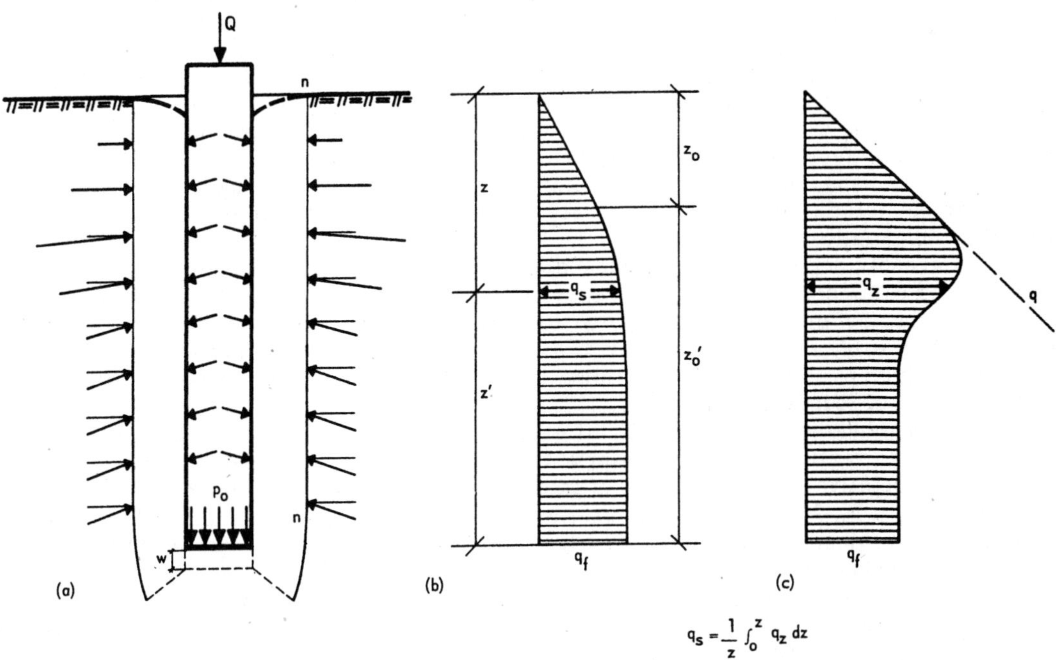

$$q_s = \frac{1}{z}\int_0^z q_z\,dz$$

Figure 42. Stress distribution around a deep foundation in sand.

2. Measurements of sand density around the foundation after load testing to failure indicated, in the case of models made of dense sand, considerable loosening in a zone immediately above base level, but a slight densification below that level.

3. Final resistances of the base and skin of rectangular foundations are found to be 1.50 times lower than corresponding resistances of circular foundations of the same diameter. However, measurements of real shape factor for base and skin resistances at greater depth indicate values of approximately 3. This leads to the conclusion that the final vertical stresses q_f around rectangular foundations are two times higher than corresponding stresses around circular foundations. The same ratio of final vertical stresses is found in the case of rectangular versus circular bins of the same diameter, where a similar phenomenon of arching occurs.

4. Under similar conditions there is less arching under smaller foundations, resulting in higher bearing capacity for the same relative depth D/B, because base displacements at failure increase proportionally to the foundation width. Displacement of a larger foundation will mobilize friction along relatively longer distance z_o'.

5. Measurements of skin resistance along model foundations in sand have indicated distributions similar to curves in Figure 42 (38, 39).

Numerous other observations on actual deep foundations as well as on models can be cited in favor of the explanation offered. However, it should be realized that the problem in question is very complex, and that variables such as roughness of foundation skin or method of construction may also be of significant influence.

A special study of this problem, including the development of a method for rational analysis of stresses around deep foundations in a homogeneous sand stratum is in progress.

CONCLUSIONS

1. Shear patterns observed underneath buried model foundations in sand indicate that, depending on relative density of sand, all three types of failure previously described in the literature may occur at shallow depth: general shear failure, local shear failure and punching shear failure. However, at greater depths only punching shear failure occurs, irrespective of relative density of sand. Limits of types of failure to be expected vary with relative density or compressibility as well as with relative depth D/B of the foundation (Fig. 27).

2. At shallow depths, not exceeding four foundation widths ($D/B < 4$) the increase of point bearing capacity with depth is linear. In dense sand ($D_R > 0.70$) the bearing capacity factor N_q can be estimated with sufficient accuracy using an analysis based on conventional theory of general shear failure in a rigid-plastic solid. In loose or medium-dense sand, failure surfaces being localized, better agreement with test results may be stated if an expression for N_q derived under the assumption of local or punching shear failure is used.

3. At greater depths, generally exceeding 15 foundation widths, both base resistance p_o and skin resistance s_o reach constant final values. These values are independent of overburden pressure q and appear to be functions of relative density of sand only. This is explained by the arching of sand above the foundation base. It is demonstrated that both p_o and s_o are proportional to the effective vertical stress at failure, q_f at the level of foundation base.

4. The bearing capacity factory N_q at greater depth, defined as the ratio of base resistance p_o to vertical stress q_f, is practically independent of foundation size and is a function of relative density or angle of internal friction of sand. Observed bearing capacity factors N_q for long rectangular foundations at greater depth do not differ from those at shallow depth. However, the shape factor for circular foundations appears to be somewhat higher at greater depths.

5. Skin resistance along the foundation shaft is not necessarily increasing linearly with depth. Instead, it is proportional to the vertical stress q_z at the corresponding elevation. Vertical stress increases linearly only at shallow depths. If the foundation is deeply embedded in sand, the distribution of vertical stress, as well as of skin resistance, is likely to be similar to that in Figure 42.

6. The fundamental fallacy of conventional analyses of bearing capacity of deep foundations in sand consists in the assumption that q is always equal to the initial overburden stress at the level of foundation base. This may be correct if a deep foundation penetrates only slightly into a sand stratum overlain by compressible soil. However, it may be entirely wrong if a deep foundation is completely embedded in sand.

ACKNOWLEDGMENTS

The tests described in this paper have been performed in the Soil Mechanics Laboratory of Georgia Institute of Technology, Atlanta, Georgia. The investigation is being sponsored by the State Highway Department of Georgia and the U. S. Bureau of Public Roads. Thanks are due to officials of these institutions for their interest and support. The pile testing area has been constructed and equipped using funds allocated by the Engineering Experiment Station, Georgia Institute of Technology. The contributions of McKinney Drilling Company and Armco Drainage & Metal Products, Inc., are gratefully acknowledged.

The author is indebted to James M. Duncan, Don C. Banks, and Ray L. Holmes, Graduate Research Assistants, who assisted in construction of equipment and performed the loading tests. Special credit is due to Guillermo Restrepo, Graduate Student, for preparation and testing of two models with colored sand. Graduate Research Assistants, O. S. Lord, R. C. Williams, and W. L. Boyd, as well as Student Assistants, T. W. Russell, D. E. Hodges, and Houshang Rahimzadeh, also gave much valuable help in different phases of the project.

Finally, the author wishes to express his thanks to Robert E. Stiemke, Director of the Engineering Experiment Station, for his help and encouragement in the early stages of the project. He is also grateful to Mrs. Milena A. Sedmak Vesić for her interest and assistance in preparation of diagrams.

REFERENCES

1. Prandtl, L., "Uber die Harte plastisher Korper." Nachrichten Kon. Gesell. der Wissenschaffen, Math. Phys. Klasse, 74-85, Gottingen (1920).
2. Prandtl, L., "Uber die Eindringungsfestigkeit plastisher Baustoffe und die Festigkeit von Schneiden." Zeitschrift fur Angewandte Mathematik und Mechanik, 1: 1, 15-20 (1921).
3. Reissner, H., "Zum Erddruckproblem." Proc. First Intern. Conf. Applied Mech., 295-311, Delft (1924).
4. Caquot, A., "Equilibre des massifs a frottement interne." Gauthier-Villars, Paris (1934).
5. Buisman, A. S. K., "De weerstand van paalpunten in zand." De Ingenieur, 50: 25-28, 31-35 (1935).
6. Terzaghi, K., "Theoretical Soil Mechanics." Wiley (1943).
7. De Beer, E. E., "Donnees concernant la resistance au cisaillement deduites des essais de penetration en profondeur." Geotechnique, 1: 22-40 (1948).
8. De Beer, E. E., " Etude des fondations sur pilotis et des fondations directes." Annales des Travaux Publics de Belgique, 46: 1-78 (1945).
9. Jaky, J., "On the Bearing Capacity of Piles." Proc., Sec. Intern. Conf. Soil Mech. and Found. Engg., 1: 100-103, Rotterdam (1948).
10. Meyerhof, G. G., "The Ultimate Bearing Capacity of Foundations." Geotechnique, 2: 301-332 (1951).
11. Meyerhof, G. G., "Recherches sur la force portante des pieux." Annales de l' Institut Technique du Batiment et des Travaux Publics, 6: 63-64, 371-374 (1953).
12. Berezantsev, V. G., "Osesimetrichnaia zadacha teorii predel 'nogo ravnovesia sypuchei sredy." Gostechizdat, Moscow (1952).
13. Bishop, R. F., Hill, R., and Mott, N. F., "The Theory of Indentation and Hardness Test." Proc. of the Physical Soc., 57: 147-159 (1945).
14. Gibson, R. E., Discussion. Jour. Instit. of Civil Engineers, 34: 382 (1950).

15. Skempton, A. W., Yassin, A. A., and Gibson, R. E., "Theorie de la force portante des pieux." Annales de l'Institut Rechnique du Batiment et des Travaux Publics, 6: 63-64, 285-290 (1953).
16. Ladanyi, B., "Etude theorique et experimentale de l'expansion dans un sol pulverulent d'une cavite presentant une symetrie spherique ou cylindrique." Annales des Travaux Publics de Belgique, 62: 105-148, 365-406 (1961).
17. Brinch Hansen, J., "Simple Statical Computation of Permissible Pile Loads." Christiani and Nielsen Post, 14-17 (1951).
18. Caquot, A., and Kerisel, J., "Traite de Mecanique des Sols." 3rd edit. Gauthier-Villars, Paris (1956).
19. Brinch Hansen, J., "A General Formula for Bearing Capacity." Danish Technical Institute, Bull. 11, Copenhagen (1961).
20. Delft Soil Mechanics Laboratory, "The Predetermination of the Required Length and the Prediction of the Toe Resistance of Piles." Proc. Intern. Conf. on Soil Mech. and Found. Engg., I: 181-184, Cambridge, Mass. (1936).
21. Plantema, G., "Results of a Special Loading Test on a Reinforced Concrete Pile." Proc. Sec. Intern. Conf. Soil Mech. Found. Engg., IV: 112-118, Rotterdam (1948).
22. Huizinga, T. K., "Application of Results of Deep Penetration Tests to Foundation Piles." Building Research Congress, 173-179, London (1951).
23. Van der Veen, C., "The Bearing Capacity of a Pile." Proc. Third Int. Conf. Soil Mech. Found. Engg., II: 84-90, Zurich (1953).
24. Van der Veen, C., and Boersma, L., "The Bearing Capacity of a Pile Predetermined by a Cone Penetration Test." Proc. Fourth Intern. Confer. on Soil Mech. and Found. Engg., II: 76-78, London (1957).
25. Bogdanovic, Lj., "The Use of Penetration Tests for Determining the Bearing Capacity of Piles." Proc. Fifth Intern. Confer. Soil Mech. Found. Engg., II: 17-22, Paris (1961).
26. Geuze, E. C. W. A., "Resultats d'essais de penetration en profondeur et de mise en charge de pieux." Annales de l'Institut Technique du Batiment et des Travaux Publics 6: 63-64, 313-319 (1953).
27. De Beer, E. E., "Les essais de penetration en profondeur." Unpublished lecture, Yugoslavia (1955).
28. Kahl, H. and Muhs, H., "Uber die Untersuchung des Baugrundes mit einer Spitzendrucksonde." Die Bautechnik, 29: 81-88 (1952).
29. Kerisel, J., "Deformations et contraintes au voisinage des pieux." Bull. 10, A. B. E. M., Brussels (1953).
30. L'Herminier, R., "Remarques sur le poinconnement continu des sables et graviers." Annales de l'Institut Technique du Batiment et des Travaux Publics, 6: 63-64, 371-374 (1953).
31. Berezantsev, V. G., "Raschet prochnosti osnovania sooruzhenii." Gosstroiizdat, Ch. III, Moscow (1960).
32. Kerisel, J., "Foundations profondes en milieu sableux." Proc., Fifth Intern. Confer. Soil Mech. Found. Engg., II: 73-83, Paris (1961).
33. Hansen, Bent, and Odgaard, D., "Bearing Capacity Tests on Circular Plates on Sand." Danish Technical Institute, Bull. 8, 19 pp., Copenhagen (1960).
34. De Beer, E. E., and Vesić, A. B., "Etude experimentale de la capacite portante du sable sous des foundations directes etablies en surface." Annales des Travaux Publics de Belgique, 59: 3, 5-58 (1958).
35. De Beer, E. E., and Ladanyi, B., "Etude Experimentale de la capacite portante du sable sous des fondations circulaires etablies en surface." Proc. Fifth Intern. Confer. Soil Mech. Found. Engg., I: 577-585, Paris (1961).
36. Caquot, A., and Kerisel, J., "Sur le terme de surface dans le calcul des fondations en milieu pulverulent." Proc. Third Intern. Confer. Soil Mech. Found. Engg., I: 336-337, Zurich (1953).
37. L'Herminier, R., Habib, P., Tcheng, Y., and Bernede, J., "Fondations superficielles." Proc. Fifth Intern. Confer. Soil Mech. Found. Engg., I: 713-717, Paris (1961).

38. Zweck, H., "Mesures sur modeles reduits du frottement lateral et de la resistance de pointe des pieux." Annales de l'Institut Technique du Batiment et des Travaux Publics, 6: 63-64, 367-370 (1953).

39. Habib, P., "Essais de charge portante de pieux en modele reduit." Annales de l'Institut Technique du Batiment et des Travaux Publics, 6: 63-64, 361-366 (1953).

Journal of the
SOIL MECHANICS AND FOUNDATIONS DIVISION
Proceedings of the American Society of Civil Engineers

FOUNDATION BEHAVIOR OF IRON ORE STORAGE YARDS[a]

By Ralph B. Peck,[1] F. ASCE, and Tonis Raamot,[2] M. ASCE

SYNOPSIS

Iron ore storage yards constitute some of the largest, most heavily loaded soil masses to be found; moreover, the load usually varies widely over an annual cycle. The behavior of twelve such yards above plastic clay deposits is examined.

The relation between lateral deformation and loading is studied to ascertain the extent to which the clay behaves elastically, the possible existence of a threshold stress at which progressive nonrecoverable movements are initiated, and the influence of the cyclic character of the loading. The conditions are also investigated under which the clays gain strength or fail to do so as a consequence of the ore loading.

The influence is evaluated of such structural features as tierods, piles, sheetpile cells, and other means used to increase the capacity of the yards.

The relative stability of the various yards is investigated by an adaptation of the theory of the squeeze test on a purely cohesive material. Some insight is provided into the most suitable distribution of the ore load for stability.

Note.—Discussion open until October 1, 1964. To extend the closing date one month, a written request must be filed with the Executive Secretary, ASCE. This paper is part of the copyrighted Journal of the Soil Mechanics & Foundations Division, Proceedings of the American Society of Civil Engineers, Vol. 90, No. SM3, May, 1964.

[a] Presented at the 1963 Terzaghi lecture on October 8, 1963 during the annual meeting of the American Society of Civil Engineers, San Francisco, Calif.

[1] Prof. of Foundation Engrg., Univ. of Illinois, Urbana, Ill.

[2] Asst. Prof. in Civ. Engrg., Newark Coll. of Engrg., Newark, N. J.

DESCRIPTION AND SIGNIFICANCE OF ORE YARDS

Introduction.—Some of the largest, most heavily loaded masses of soil to be found in the world are undoubtedly those beneath the storage areas for iron ore adjacent to blast furnaces. A modern blast furnace consumes some 1500 tons of iron ore each day. To assure no interruption in its operation, an ample reserve of ore is needed close at hand. Moreover, along the Great Lakes, which are closed to traffic by ice during almost half the year, each furnace must be provided with enough ore during the shipping season to last until shipping can be resumed the next year. To satisfy this requirement, storage areas several hundred feet wide and often on the order of 1000 ft long are customarily filled in the fall to heights of roughly 50 ft with ore having a unit weight of about 160 lb per cu ft. If the subsoil contains strata of clay, the corresponding unit load of about 8000 psf is likely to produce substantial displacements.

Plastic clays constitute the subsoil for many of the ore yards in the Great Lakes region and for several on the Atlantic seaboard. Moreover, since the ore is usually delivered by ship, the loaded area is bordered by a channel that increases the tendency for displacements. A typical arrangement of the facilities is shown in Fig. 1. The ore is unloaded onto the forward part of the dock, usually in a trough, by rapid unloaders. It is subsequently placed in a suitable portion of the main storage area by a clamshell bucket operated from a gantry crane, or ore bridge, usually supported by rails on retaining walls on both sides of the storage area. In several instances, these retaining walls are connected at their bases by steel tierods extending across the ore yard.

Most of the ore yards in the United States were established in the early 1900's. At the time of the two world wars, increased demands for steel led to substantial increases in loading and usually to significant movements of dock and retaining structures; in a few instances outright failure occurred. Numerous additions have also been made from time to time; almost every ore yard has been enlarged on several occasions.

The necessity for adjustments to keep the unloader and ore bridge tracks reasonably straight, level, and parallel gave rise to numerous level and alignment surveys. Each steel company has extensive records of such surveys and ajdustments, as well as much supplementary data concerning behavior and soil conditions. In many instances, the yards have been reinforced by tierods, vertical and batter pile foundations, or other structural systems. The records of behavior constitute a considerable body of information about the reaction of the subsoil to heavy, variable, and often cyclic loads, and about the interaction between soil and structure.

This paper is based on the behavior of twelve ore yards with clay subsoils. Six of these (Yards A to F, Fig. 2), all located in Cleveland, Ohio transfer the ore load directly to the soil. Three other yards (G, H, and I, Fig. 3) are located in other localities bordering the lakes and similarly transfer the load directly to the soil. The remaining three (J, K, and L, Fig. 4) include features providing substantial structural support for the ore. Two of these are in the Great Lakes area and one is on the Atlantic seaboard. In addition to a description of the behavior of the clays and of the structures, the paper contains an attempt to correlate the behavior and the properties of the subsoil on an analytical basis.

History of Yard A.—Each of the ore yards differs in many respects from the others; consequently, no one of them can be considered typical. As a sample, however, Yard A is possibly the most illuminating. It is structurally one of the simplest, as the ore rests directly on the ground and there is no structural floor, front retaining wall, or tierod system across the yard as a whole. It has experienced movements larger than any of the others except for those that have failed outright; therefore, its behavior can be described quantitatively despite the low accuracy of some of the observations. The history and construction details of this yard are outlined in the following paragraphs to demonstrate the complexity of the information that must be assembled and considered before the behavior of any ore yard can be characterized.

The first blast furnaces at the site were built in 1902 and 1909. Originally the ore was brought in by rail, but in 1907 an ore bridge was constructed. The bridge tracks were laid directly on the ground. According to the original plan the ore was to be piled to a height of 46.5 ft, but the actual height in the earliest years is not known. For many years the normal maximum height above yard level has been about 40 ft to 45 ft.

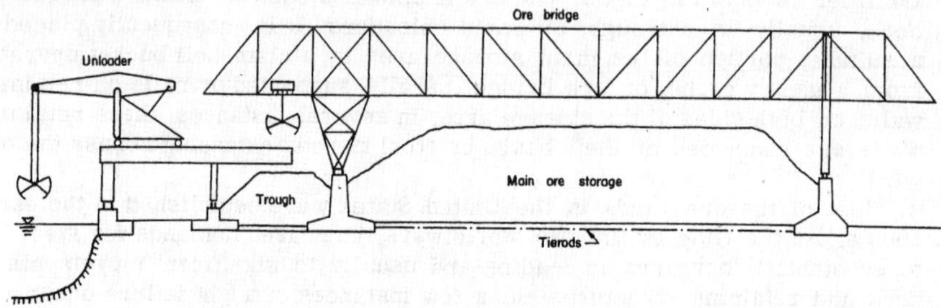

FIG. 1.—TYPICAL ARRANGEMENT OF ORE STORAGE YARD

The details of the earliest riverfront construction are unknown. In 1918-19, a reinforced concrete dock wall was constructed between Stas. 0+00 and 10+04.6 on 45-ft timber piles. The location is shown on the plan, Fig. 5a, and in section in Fig. 5b. Immediately thereafter, a concrete trough 500 ft long was built opposite the north half of the new dock wall. The trough consisted of two L-shaped cantilever retaining walls connected by a concrete floor. The landward retaining wall was founded on 45-ft piles, some vertical and some on batter. Steel tierods connected this wall to the dock. The riverward wall of the trough was not supported on piles.

In 1943 the trough was extended from Sta. 5+00 to Sta. 7+21.6. The construction was similar to that of the earlier trough except that one row of batter piles and one row of vertical piles were driven beneath the riverward retaining wall. Tierods were established between the new trough and the dock wall. The trough and tieback systems were extended in 1943 and 1944 from Sta. 7+65 to Sta. 10+69, leaving a gap to avoid the intake tunnel and screen house. At the same time the dock was extended from Sta. 10+04.6 to Sta. 10+69.

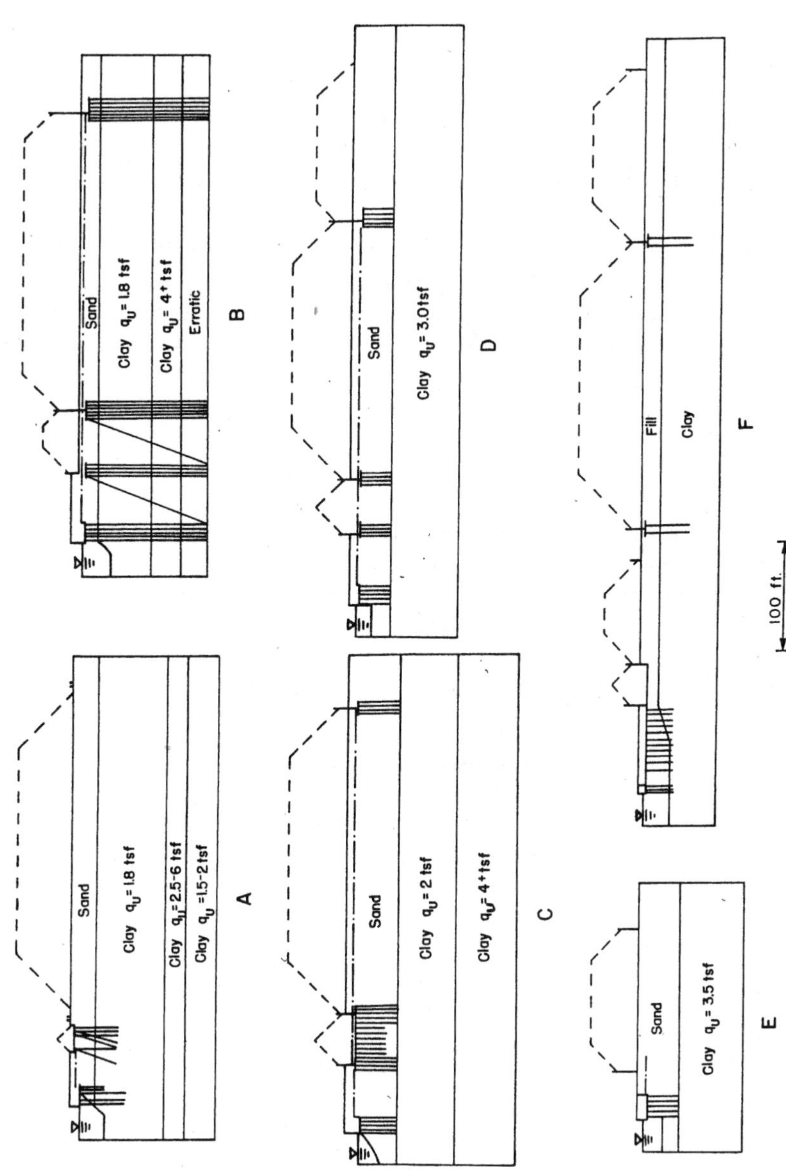

FIG. 2.—SIMPLIFIED CROSS SECTIONS, YARDS A-F

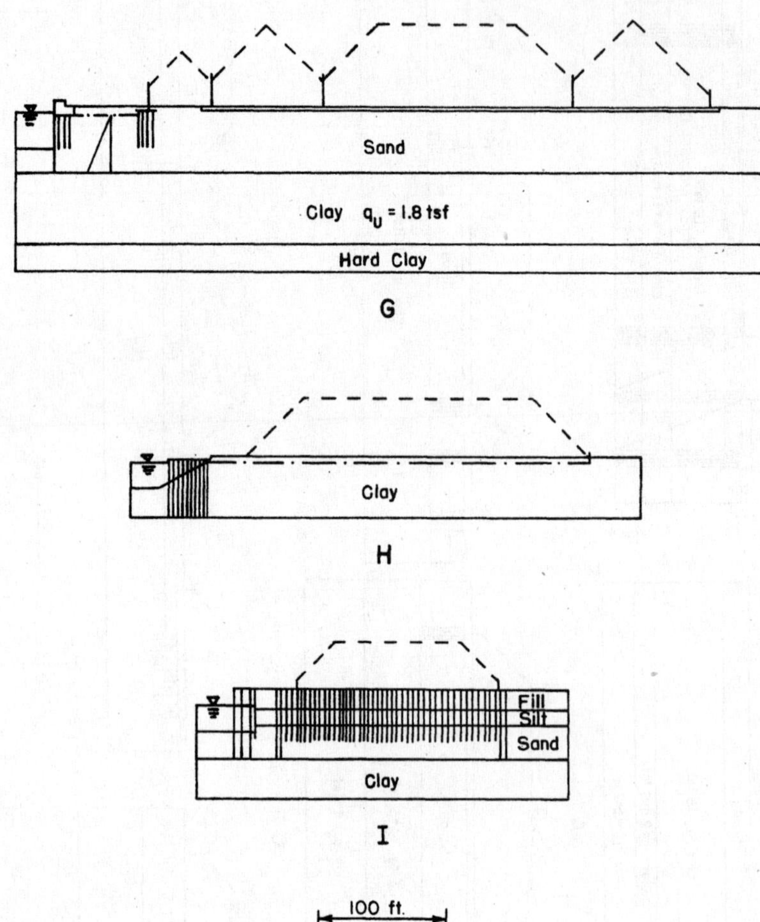

FIG. 3.—SIMPLIFIED CROSS SECTIONS, YARDS G-I

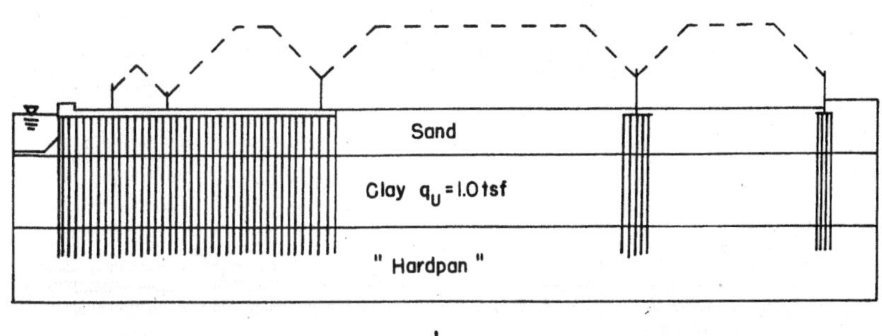

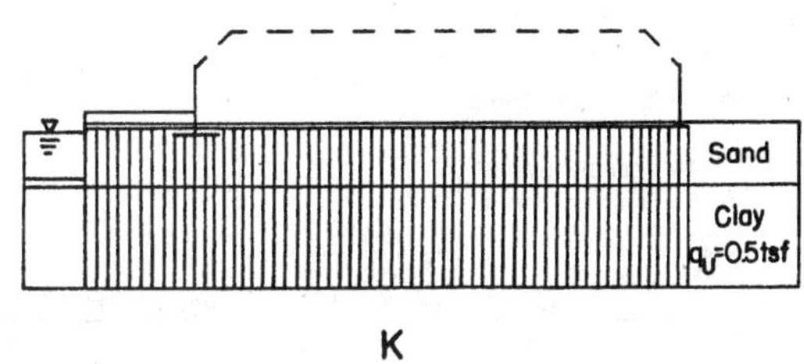

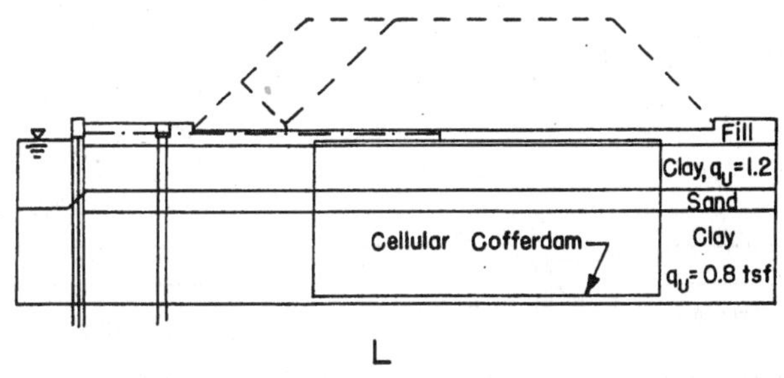

100 ft.

FIG. 4.—SIMPLIFIED CROSS SECTIONS, YARDS J-L

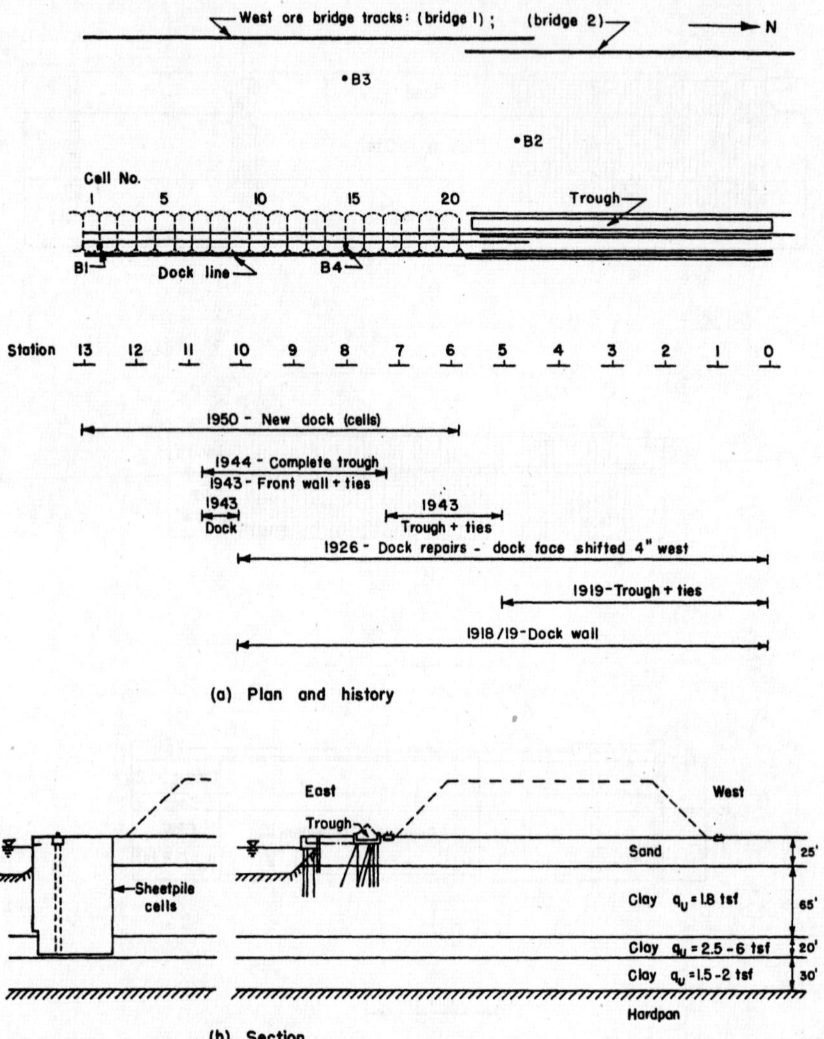

FIG. 5.—HISTORY AND CONSTRUCTION DETAILS, YARD A

In 1950 the old dock and trough were removed south of Sta. 5+00 and replaced by new construction consisting of 20 sheetpile cells (Fig. 5b). The sheets have lengths up to 110 ft. The new dock was extended south to Sta. 13+00.

The subsurface conditions beneath Yard A are indicated in condensed and simplified fashion in Fig. 5b. The most significant feature is the deposit of plastic clay, about 65 ft thick, with an unconfined compressive strength of about 1.8 tons per sq ft. The deposit was preconsolidated during the glacial epoch by an overburden of deltaic materials, since removed by erosion. The preconsolidation load is estimated to be about 5 tons per sq ft (equivalent to about 60 ft of ore) above the present overburden pressures. The clay is overlain by about 25 ft of sand upon which the ore is piled. It is apparent that the 45-ft timber piles beneath the 1919 and 1943 construction penetrated only a short distance into the clay beneath the sand.

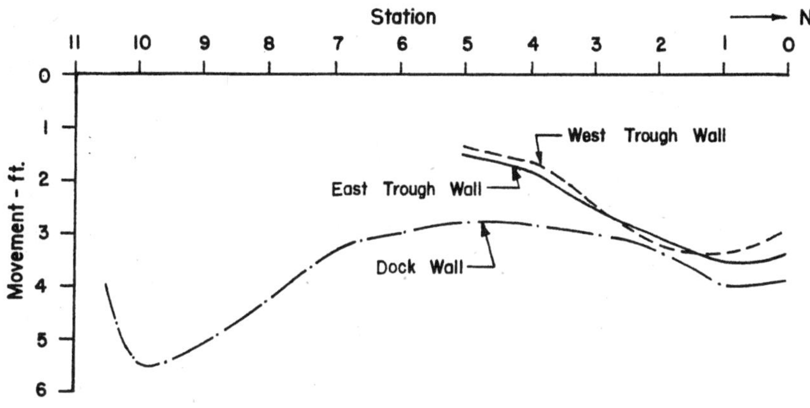

FIG. 6.—CUMULATIVE MOVEMENTS, YARD A, 1919-1948

The thick clay deposit is underlain by a variable but much harder clay till, underlain in turn by somewhat softer clays. It will be shown later that these lower deposits do not enter significantly into the movements of the yard. The succession of clays rests upon a very hard clay, locally designated hardpan, which is underlain by shale bedrock.

During the years 1919-1948 the dock wall and ore trough progressively moved toward the river. The accumulated movements, shown in Fig. 6, amounted to as much as 5 ft. The annual increase, which appears to have been roughly constant, was about 2 in. Between about Stas. 1 and 3 the dock wall and trough moved through approximately the same distances; hence, the tierods probably remained intact. South of about Sta. 3, however, the dock wall moved much farther than the trough and the tierods unquestionably failed, probably because of local overloading. Settlements of the trough and of the dock wall (as indicated by the riverward rail of the unloader) increased during the same period to amounts ranging from 1.5 ft to 3 ft.

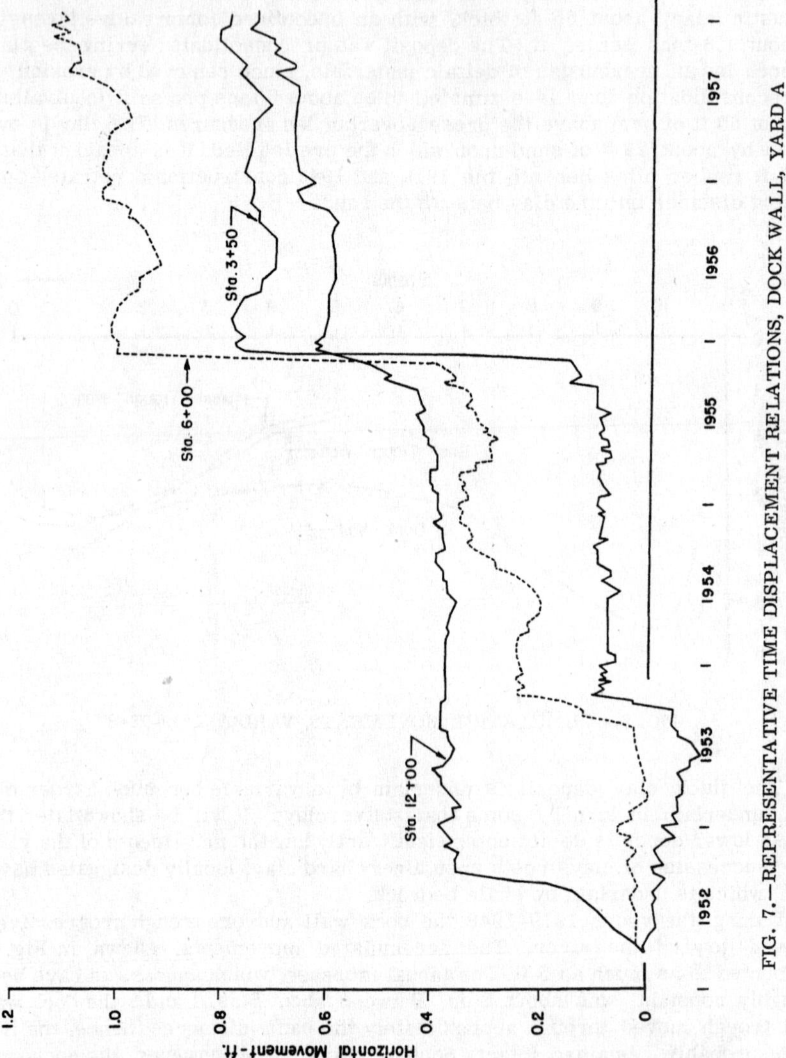

FIG. 7.—REPRESENTATIVE TIME DISPLACEMENT RELATIONS, DOCK WALL, YARD A

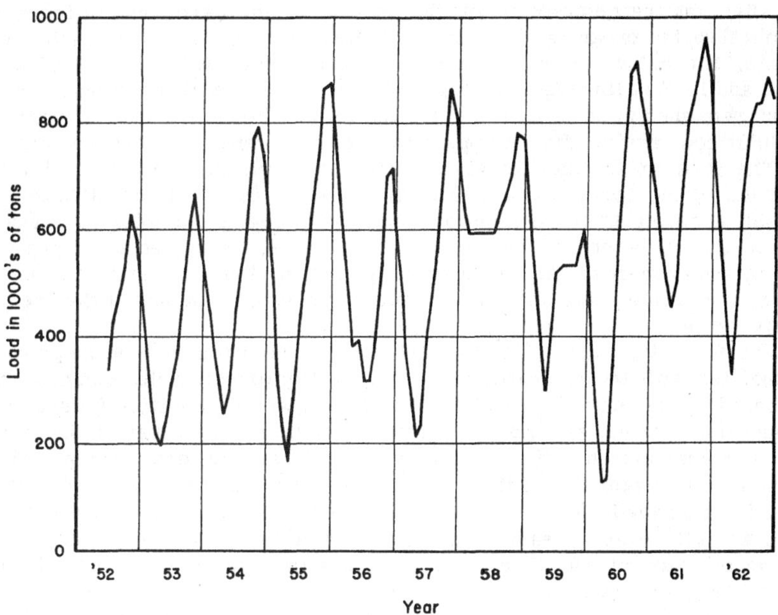

FIG. 8.—TOTAL LOAD ON YARD A BY MONTHS

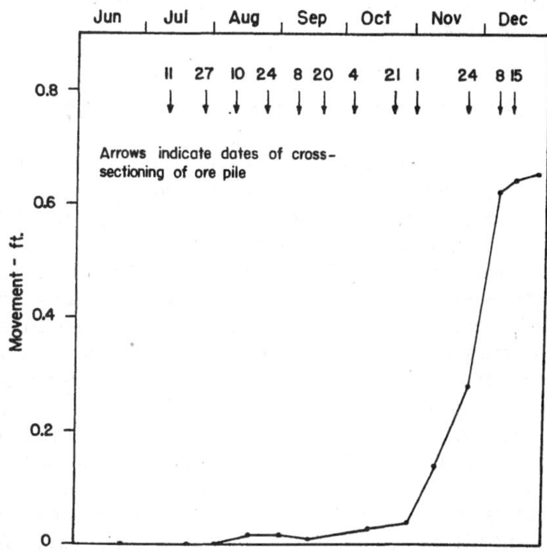

FIG. 9.—1955 MOVEMENT AT STATION 4, YARD A

After the reconstruction of the portion of the yard south of Sta. 5+00 a comprehensive program of observations was initiated to determine the movements. The nature of the horizontal displacements at the dock line between 1952 and 1957 is illustrated in Fig. 7. The average annual increment of movement remained at about 2 in. per yr and, as will be considered presently, the cellular construction did not appear to alter the trend of the movements.

The load on the storage area fluctuates seasonally. The record of total load during the decade 1952-1962 is shown in Fig. 8. The information is based on shipping weights of incoming ore boats and charging loads for the furnaces; the accuracy is not of high order. The ore is segregated according to its origin and properties. As a consequence, its distribution over the area of the yard is extremely variable. From time to time the contours of the ore piles are surveyed.

The foregoing study of Yard A shows the general nature of the problems associated with the ore yards and the type of information that can usually be gathered from the records of the steel companies. To avoid failure and to reduce movements, various ways have been tried to strengthen the storage areas. A comparison of the information about the several ore yards permits at least a crude evaluation of the relative effectiveness of the different systems for adding strength. In connection with a few of the yards, much more precise data are available, at least for certain intervals of time. As various aspects of the behavior of subsoil and structure are introduced, pertinent data from the different ore yards will be presented.

The data will be examined first to gain insight into the behavior of large clay masses under load. The effectiveness of strengthening measures and questions of soil-structure interaction will then be considered. Finally, an attempt will be made to compare the behavior of the ore yards by means of an analytical procedure based on modifications of existing solutions of problems in the theory of plasticity.

BEHAVIOR OF SOIL MASS UNDER LOAD

Pertinent Fields of Investigation.—In the normal operation of the ore yards along the Great Lakes it has been customary to attempt to store as much ore as possible at the end of each shipping season. The limiting load has usually been determined over the years by adding ore until the movements of the retaining walls or dock structures were considered excessive. At some of the yards, the capacity established in this manner a number of years ago has not been exceeded and movements at present are inconsequential. At a few, the experimental procedure is still going on and it is possible from the observational data to learn about the nature of the load-deformation characteristics of the clay masses and about the increase in their load-carrying capacity as the years go by. As field evidence concerning such phenomena as the existence of a threshold stress and the recovery of deformation upon unloading is rather uncommon, these matters will be examined in some detail.

Elastic and Inelastic Distortions.—The relations between ore load and horizontal displacements have been studied in detail for Yards A and B. The behavior of Yard A is the more instructive because there are no batter piles through the clay and no tierods or other structural features to introduce complications.

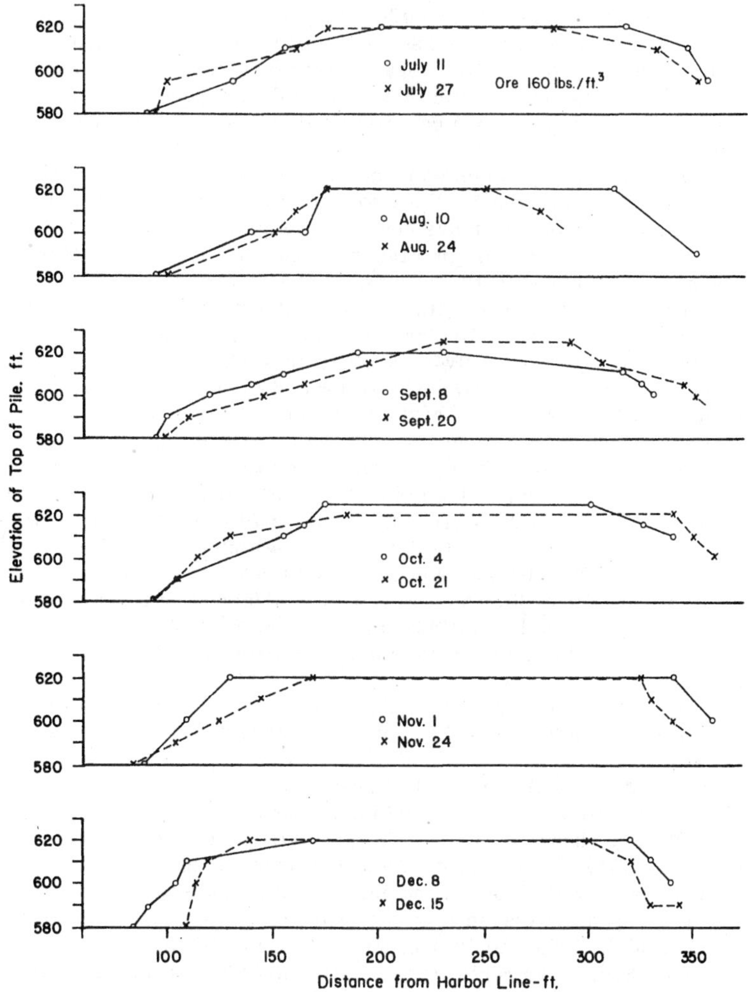

FIG. 10.—LOADING AT STATION 4, 1955 SEASON, YARD A

Each loading season the dock wall advances toward the river. Surveys have shown that the horizontal movement is essentially the same at a given station whether measured at the dock line or at any distance up to at least 80 ft from the dock; hence, it appears that the block of soil including the dock structure moves or distorts as a unit under the influence of the ore. The general pattern of the movements is represented by Fig. 7 which covers a 6-yr period from 1952 to 1957. It may be observed that major movements, if any, occur suddenly near the end of a loading season. A small percentage of such movements, up to an inch, may be recoverable, but recovery does not necessarily occur upon each unloading. The elastic or recoverable component of the deformation is therefore very small when compared with the total magnitude of ore yard movement.

During 1955, when the displacements of the ore yard were a maximum and when the total load exceeded all previous maxima, relatively frequent observations were made of both movement and the contours of the ore pile. These data are particularly useful in reviewing the behavior of the storage area.

The maximum movements occurred in the neighborhood of Sta. 4+00. The development of movements with time is shown in Fig. 9. Near the end of November the increase in displacement was very abrupt. The increment, on the order of 0.4 ft, obviously took place in such a short time that the total load on the storage area could not have changed by an appreciable percentage. Nevertheless, the distribution of the load in the critical area did change significantly. This fact is shown in Fig. 10 in which are plotted the cross sections of the ore near Sta. 4+00 at successive dates from July 11 to December 15. Although there was a large quantity of ore in storage, to heights exceeding 40 ft, at various times between July 11 and October 21, the riverward face of the ore pile had a relatively flat slope. Beginning in November, however, the front face of the pile was steepened and the contours of the storage area were filled out. The contours on November 1, December 8, and December 15 all indicate only small additions in the total quantity of ore in storage at Sta. 4+00, but substantial increases in the amount stored on the river side of the ore pile.

It is quite apparent that extraordinarily large displacements are initiated by comparatively minor additions of load near the riverward face of the ore pile.

From similar detailed records, the increase in deflection for each year from 1952 to 1962 has been plotted for points along the entire storage area. The results are shown in Fig. 11. The greatest general movements occurred in 1955 and 1962; in contrast, almost no movement took place in 1956 or 1959. The total weight of stored material for the same years is shown in Fig. 8. It is apparent that in 1955 the load exceeded previous maxima. However, the load in 1960 and 1961 exceeded that in 1962 by a small margin but the corresponding movements were greatest in 1962. Hence, there is no precise correlation between average load on the entire yard and average deflection, although there may be a general relation.

Fig. 12 is a plot of a deflection index versus maximum total load for each year between 1952 and 1962. The deflection index is a measure of the total movement that occurred during any one year and is proportional to the area under the deflection curves shown in Fig. 11. The total load corresponding to a deflection index is the peak load for the particular year taken from Fig. 8. While the scatter of points in Fig. 12 is considerable, the results indicate

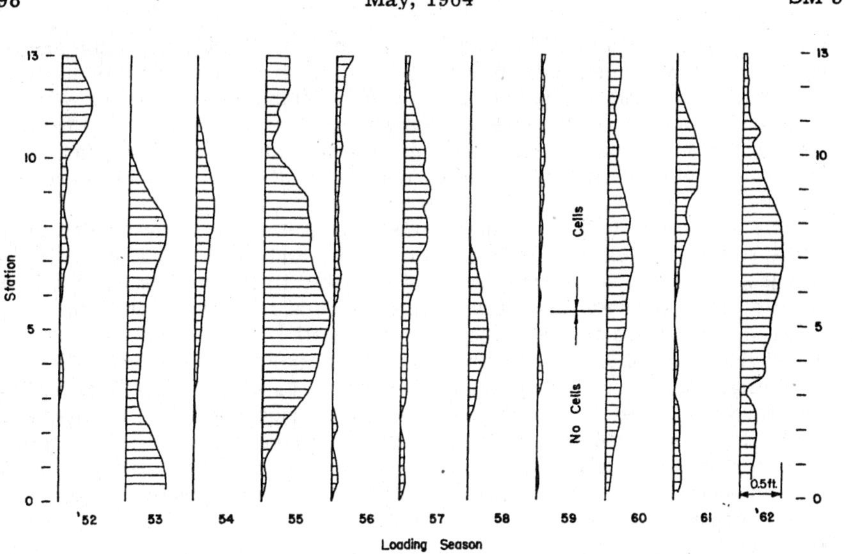

FIG. 11.—ANNUAL MOVEMENTS OF YARD A

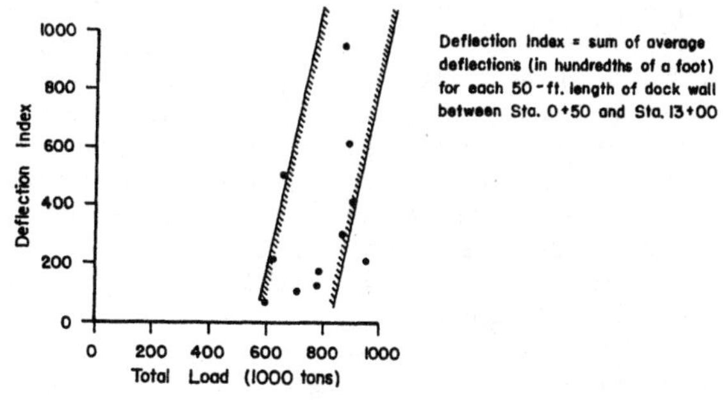

FIG. 12.—DEFLECTION INDEX AS TOTAL LOAD, YARD A

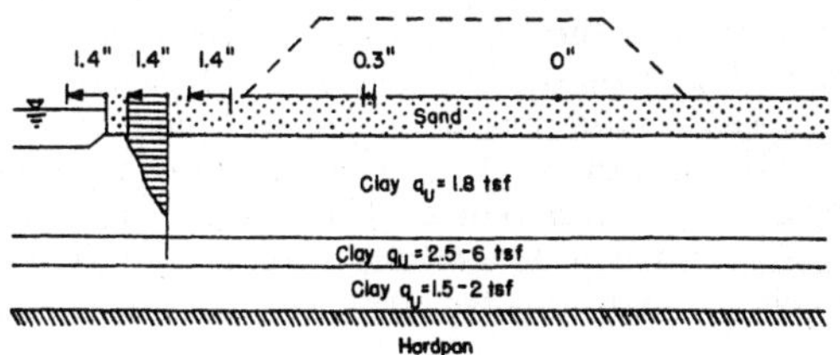

FIG. 13.—APPROXIMATE HORIZONTAL MOVEMENTS 1956-58, STATION 3, YARD A

that a threshold load in the neighborhood of 600,000 tons may exist and that considerable movement may be expected once this threshold load is exceeded. The total magnitude of the permanent deformations, once initiated, may be a function of the period of time the yard is subjected to the critical load, although there are no data to substantiate this statement. It is apparent that extremely detailed observations are needed for a complete understanding of the movements.

The foregoing study has demonstrated that the recoverable displacements of the soil are small, usually less than an inch, and that at a rather delicately balanced arrangement and magnitude of load the displacements increase very greatly. This behavior indicates a threshold value of stress initiating large progressive displacements. It is also indicated that, at Yard A at least, the threshold stress did not increase perceptibly during the period 1955-1962.

Similar increases in displacement under apparently negligible changes in loading were noted by K. Terzaghi in connection with Yard B. They were attributed in his 1948 report[3] to breakage of tierods connecting the retaining walls enclosing the storage area. Although there is no doubt that tierod breakage did occur and may have been reflected in such movements, it also seems possible that the cause of the movements may have been similar to those at Yard A where no tierods existed to complicate the situation. Both yards are located in the same geological setting above preconsolidated clays.

No comparable data are available for the other ore yards, especially with respect to the distribution of the loads. Therefore, it is not known whether the yards underlain by overconsolidated clays behave differently than the others with respect to a threshold stress.

The preceding analysis refers only to the movement of the surface of the soil on the river side of the storage pile. Fig. 13 gives a general conception of the distortions at Yard A as gained from surface observations in other parts of the yard and from slope indicator measurements. Although the data are not complete, they may serve as a basis of comparison between Yard A, where no soil-structure interaction is involved, and other yards provided with piles, tierods, or other strengthening devices. It may also be noticed that the lower stratum of plastic clay, beneath the intermediate hard layer, has not participated in the distortions.

Increase of Capacity under Ore Load.—It is striking that ore is piled to about the same height in almost all the soil-supported yards in spite of the fact that some of them are located in regions of very weak clays. Although a few yards have failed, such as Yards H and I, others have survived and today support loads as great as those on initially much stronger clays. Because the geometry of most of the yards is rather similar, it seems evident that the underlying clays must either have possessed initially or have developed a certain strength not too different for the different yards.

A question of considerable practical importance is whether a natural deposit of preconsolidated clay, when reloaded to an intensity equal to or slightly less than the previous maximum load, experiences a significant gain in shearing strength. Evidence concerning this point, as well as on the behavior of more nearly normally loaded clays, will be examined in this section.

---

[3] Terzaghi, K., "Final Report on the Performance of the Ore Yard of the RFC Plancor 257 during the Service Period 1943 to 1948," From Theory to Practice in Soil Mechanics, John Wiley and Sons, Inc., New York, N. Y., 1960.

Direct information concerning the present strength of the clays beneath storage piles is fragmentary. The principal data pertain to Yards G, A and C. Yard G is located in an area not previously consolidated under a pressure in excess of the present ore loads, whereas at Yards A and C the preconsolidation pressure exceeded the present ore loads.

At Yard G, three 2-in. thin-walled tube-sample borings were made at the locations shown in Fig. 14 after the ore yard had been in service some 50 yr. Boring 1 was located about 220 ft outside the limits of the ore yard beneath an area that had been filled with sand to a level about 15 ft above the base level of the ore pile. Boring 3, located between two of the three main ore piles, penetrated clay that had long been subjected to the influence of the ore, although not to the maximum intensity of the ore load. The clay in Boring 5, at the dock line about 50 ft from the ore trough, should have been only slightly influenced by the ore load. The values of unconfined compressive strength and natural water content from the three borings within the stratum of plastic clay are compared in Fig. 14. The average unconfined compressive strengths for Borings 1, 3, and 5 were respectively about 1.0, 1.7, and 1.3 tons per sq ft. Moreover, numerous other borings have been made in the vicinity. They indicate that the unconfined compressive strength, in areas not affected by loading, ranges from about 0.8 to 1.2 tons per sq ft.

At Yard A, four borings were made in 1949 and 1950 that furnished 3-in. undisturbed samples for unconfined compression tests. Boring 1, Fig. 5a, was made near the dock at a point about 200 ft south of the area that up to that time had been subjected to ore load; Borings 2 to 4 were made beneath or near the sides of the ore pile. Because of the scattering of the results, a statistical study was made to determine whether the variations in the average values between depths of about 25 ft and 80 ft below lake level were significant. The results indicated that the average strengths (from 1.9 to 2.0 tons per sq ft) of the samples taken from Borings 2 to 4 did not differ significantly among themselves, but that they were significantly greater than the strength (1.3 tons per sq ft) of the samples taken from Boring 1. The strength of the samples from Boring 1 seems to be anomalously low even in consideration of the distant location of the boring from a previously loaded area, inasmuch as other borings in the locality generally exhibit strengths ranging from 1.5 to 1.8 tons per sq ft.

At Yard C two borings were made in 1943 for the purpose of comparing the strengths beneath loaded and unloaded areas; one was located in the central part of the main storage area, and the other 200 ft from the end of the yard. The complexity of the glacial stratigraphy in the Cleveland area, especially in the upper clays,[4] was not appreciated at the time. Only one stratum, between 59 ft and 65 ft below lake level, could be said with assurance to be identified at comparable depths in both borings. The material was a lean blue clay with silt partings at 1/4-in. intervals. The water contents from the two borings were approximately the same, but the unconfined compressive strengths of 2-in. tube samples were 2.0 tons per sq ft and 1.5 tons per sq ft beneath and outside the loaded areas, respectively.

Hence, the fragmentary data from the strength tests are rather inconclusive. At Yards A and C, where the clays were originally preconsolidated,

---

[4] Peck, Ralph B., "Foundation Conditions in the Cuyahoga River Valley," Proc. Sep. No. 513, ASCE, Vol. 80, October, 1954.

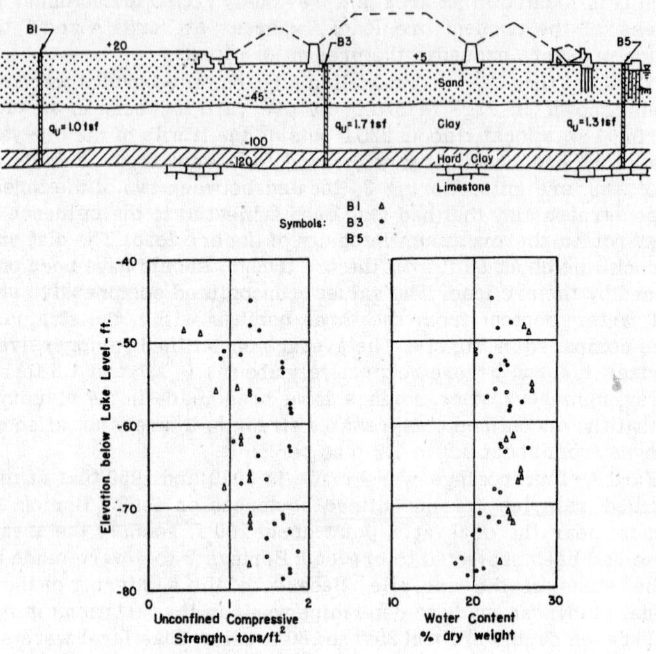

FIG. 14.—SOIL PROPERTIES BENEATH YARD G, 1956

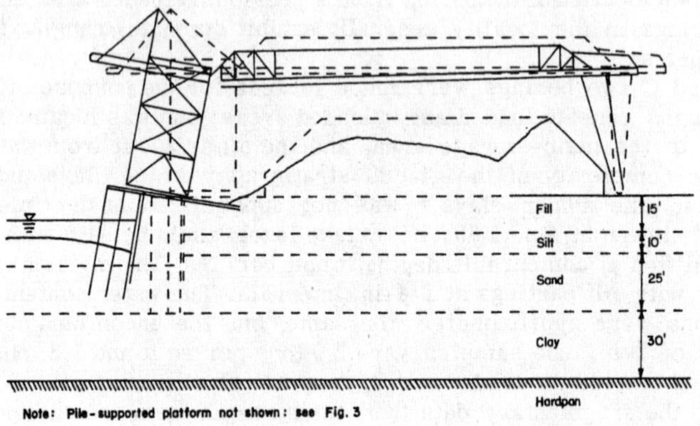

FIG. 15.—APPROXIMATE CONDITIONS OF YARD I AFTER 1912 FAILURE

there is at least an indication that the present strengths beneath the ore yards somewhat exceed those of the clays near the periphery of the yards. However, the variability of the natural deposits, especially near Yard C, renders this indication rather uncertain. If the indication is reliable, it can logically support either of two hypotheses: (1) The strength of the peripheral soil corresponds to the undisturbed condition and the ore load has increased by consolidation the strength of the underlying soil, or (2) the underlying soil has not been strengthened by consolidation, and shearing distortions with associated porepressure changes have slightly weakened the peripheral soils. Of course, any combination of these two mechanisms is possible. Conversely, at Yard G, there is no doubt that the subsoil has been strengthened by consolidation under the influence of the ore load. Nevertheless, even this conclusion cannot be directly supported by observations of settlement or of decrease in water content because of limitations imposed by the quality of the data or by the variability of the subsoil.

Almost all the settlement records, for example, refer to the retaining walls and dock structures rather than to the loaded areas proper. Undoubtedly, only part of the settlements of the dock structures is the result of consolidation; part is a consequence of shear distortions. All the loaded areas are underlain by sands that have probably contributed substantially to the total settlement. For these reasons no valid conclusions regarding changes in soil strength can be drawn from the settlement data. Furthermore, the natural variations in water content of the clay deposits have been found to be so great as to preclude definite conclusions concerning a systematic change as a consequence of long-term loading.

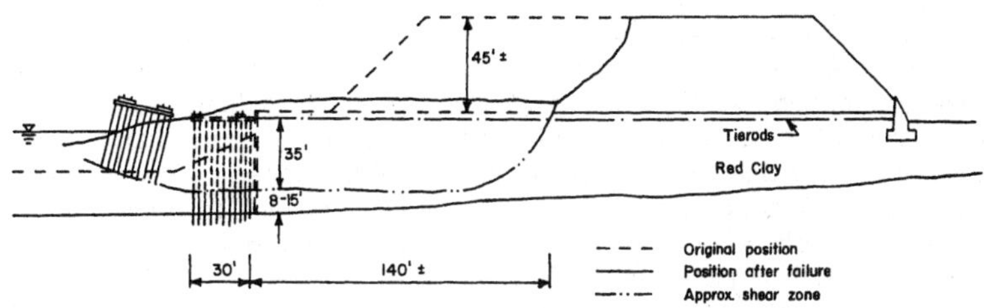

FIG. 16.—APPROXIMATE CONDITIONS OF YARD H AFTER 1913 FAILURE

That the weight of the ore was able to increase the strength of the clays in areas of low preconsolidation pressure is demonstrated indirectly by the survival of some of these yards under today's high ore loads. There is no question that the present ore loads could not have been placed on these yards early in their history. This is attested by failures such as that experienced by Yard I, Fig. 3. This yard was constructed in 1909. The original portion of the yard was covered by a hardwood floor underlain by about 2 ft of soil beneath which were driven 40-ft timber piles on about 4-ft centers. Inasmuch as the piles were not connected to the floor and did not penetrate into the clay deposit, the load may be regarded as being carried directly by the clay. Failure occurred suddenly near the close of the 1912 shipping season, when the ore load was at a record intensity. The configuration after failure, determined from photographs, is shown in Fig. 15. The photographs also indicate the height of the pile to have been about 35 ft to 40 ft.

From numerous borings made near the old ore yard the soil profile is fairly well known. No tests have been made to determine the strength of the clay. However, the standard penetration resistance is about 5 blows per foot. Correlations indicate that the unconfined compressive strength of unloaded deposits of the clay in the locality can hardly exceed 0.8 ton per sq ft and is more probably about 0.6 ton per sq ft.

The failure demonstrates that consolidation was not adequate to increase the strength from its initially low value to an amount capable of withstanding a large increase in load to 35 or 40 ft in a single loading season following a history of moderate loading.

The same conclusion can possibly be drawn from the failure of Yard H. This yard, Fig. 3, was constructed in 1901. It was subjected to moderate loadings of unknown magnitude until 1913. At the end of the loading season in 1913, as the last ore boat was being unloaded, an extensive section of the dock failed. According to contemporary accounts, the failure took place very suddenly and along a surface of sliding that must have included a long horizontal section (Fig. 16). The clay is known to be varved. According to N. D. Lea,[5] the unconfined compressive strength from a site immediately adjacent to the reconstructed section was found to be about 0.7 ton per sq ft; the corresponding strength based on the results of vane tests was about 1.0 ton per sq ft. Because the 1913 loading was substantially greater than that of preceding years, it is likely that the clay would not have developed by consolidation the requisite unconfined compressive strength to prevent failure. The failure may have taken place by spreading, as a consequence of high pore pressures in some of the silt or sand layers in the varves.

In contrast to Yards H and I, other yards such as G did not fail, although the initial strength of the clay was quite low and the load is now very heavy. Further evidence is provided by the part of Yard J (Fig. 4) not supported by piles: This yard also carries a high load but the strength of the virgin clay in the neighborhood is only about 0.8 ton per sq ft.

It seems clear, in view of the foregoing evidence, that ore loads greater than the preconsolidation pressure either produced failure if applied too rapidly or produced consolidation and strengthening if increased slowly enough. The behavior of the preconsolidated clays in the Cleveland area, however, is not so clear. Additional insight can be gained by a study of Yard B, the only yard on which observations have been made systematically from the first loading season.

Yard B was completed and first loaded in 1942. In his report of 1948, Terzaghi[3] investigated the relation between unit load on the ore yard and lateral movement of the riverward group of walls as a possible means to ascertain whether the ultimate capacity increased under continued or increasing loading. M. I. Esrig[6] recalculated the data on a slightly modified basis and extended it through the first sixteen loading seasons. According

5 Lea, N. D., "Performance of a Steel Sheet Piling Bulkhead," Proceedings, 3rd Interntl. Conf. on Soil Mechanics and Foundations, Vol. 2, Zurich, 1953, p. 180.

6 Esrig, M. I., "The Load-Deflection Behavior of a Heavily Loaded Area in Cleveland, Ohio," thesis presented to the University of Illinois, at Urbana, Ill., in 1961, in partial fulfilment of the requirements for the degree of Doctor of Philosophy.

to Esrig's study, the lateral movements correlated best with the average unit load on a strip bordering the retaining wall and having half the width of the ore yard. Fig. 17 shows the load-deflection diagram for the riverward retaining wall. The envelope to the diagram has the typical shape of a load-deformation curve for a purely cohesive soil, with an ultimate capacity of about 3 tons per sq ft.

The envelope to Fig. 17 does not in itself indicate an increase in capacity with time; it might be interpreted to suggest that in its earliest years the yard was not loaded to capacity. Nevertheless, the movements during the first few years under low loadings caused concern that failure might be imminent. It is conceivable that the stiffness, as measured by the slope of the stress-deformation curve, rather than the strength of the clay may have increased with repetition of loading. Esrig has investigated this possibility by a detailed study of the load-deflection curves of which Fig. 17 is representative. The stiffness has been defined arbitrarily as the increase in deflection, during a given loading season, corresponding to the increment of load from 1.0 ton per sq ft to 1.5 tons per sq ft. The slopes are plotted in Fig. 18 as a function of the cycle of loading. The low accuracy of the observations leads to a large scatter as can be seen from Fig. 17, but there appears to be a trend toward an increase in stiffness during the first few cycles, a behavior that might be expected from soil under repeated loading.

A set of piezometers was installed at Yard B prior to the first loading season. The installation and the results of the observations have been described in detail by Terzaghi.[3] Some of the results are represented, after slight refinements in the loading used by Terzaghi, in Figs. 19a and b. It is evident that the pore pressures within the clay have faithfully reflected changes in the ore load in the vicinity of the piezometers, and that there was a general decrease of pore pressure with respect to the applied pressure by the end of the second loading season.

A more detailed study of piezometer reaction is shown in Fig. 19c, in which the difference between applied stress and pore pressure is plotted for all four piezometers. The plot discloses a considerable irregularity in the reaction of the individual piezometers. According to Fig. 19c, virtually no permanent decrease in pore pressure took place during the first loading season. After the second season, however, all piezometers registered a considerable pressure decrease amounting to 20 ft of water or more. Piezometers Ib and IIa reacted in a similar fashion, whereas IIb and III exhibited pronounced relative pressure decreases during the early part of both loading seasons. Furthermore, the response of Ib and IIa seems to differ during the second season.

These inconsistent reactions can be explained in part by the presence or absence of seams and pockets of coarse material that were encountered during boring operations and that contained water under an initial artesian pressure. Inaccuracies in determining the pressure exerted by the ore at any time may also account for some of the apparent discrepancies in the reactions of the piezometers. Shear distortion may also have had an influence. In overconsolidated clays such as those underlying Yard B, a decrease in pore pressures as a result of shear deformations can be anticipated. As shown in the small scale diagram accompanying Fig. 19, Gages IIb and III are at locations through which potential surfaces of failure may be drawn. The soil mass in the vicinity of gages IIa and Ib would have less opportunity

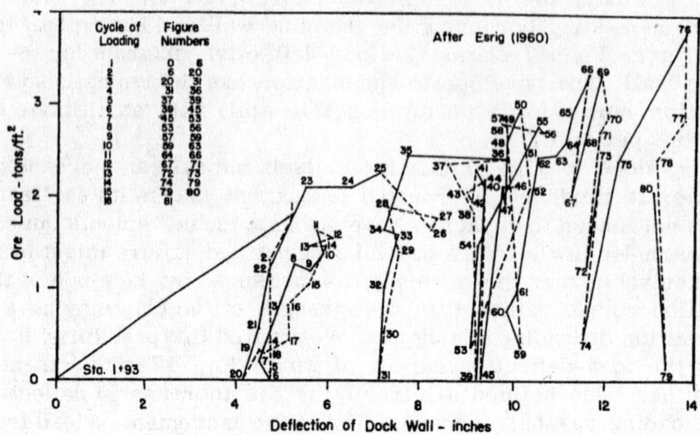

FIG. 17.—LOAD-DEFLECTION RELATION FOR DOCK WALL AT STATION 1+93, YARD B

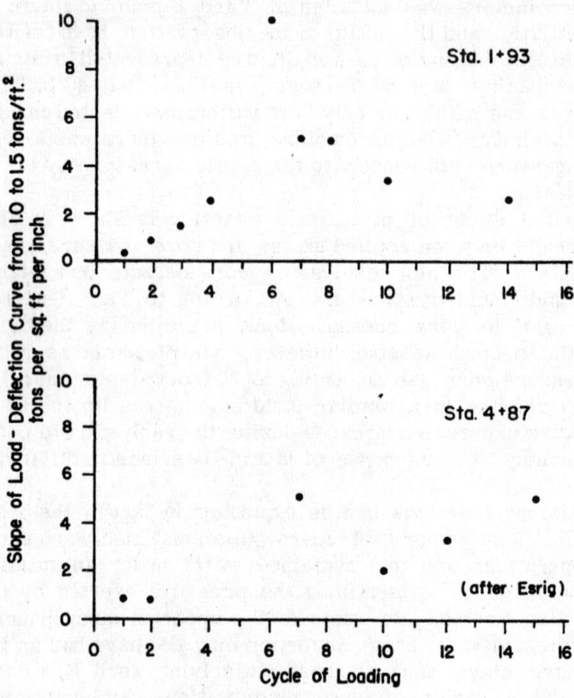

FIG. 18.—RELATION BETWEEN SLOPE OF LOAD DEFLECTION CURVE AND NUMBER OF LOADING CYCLES, YARD B

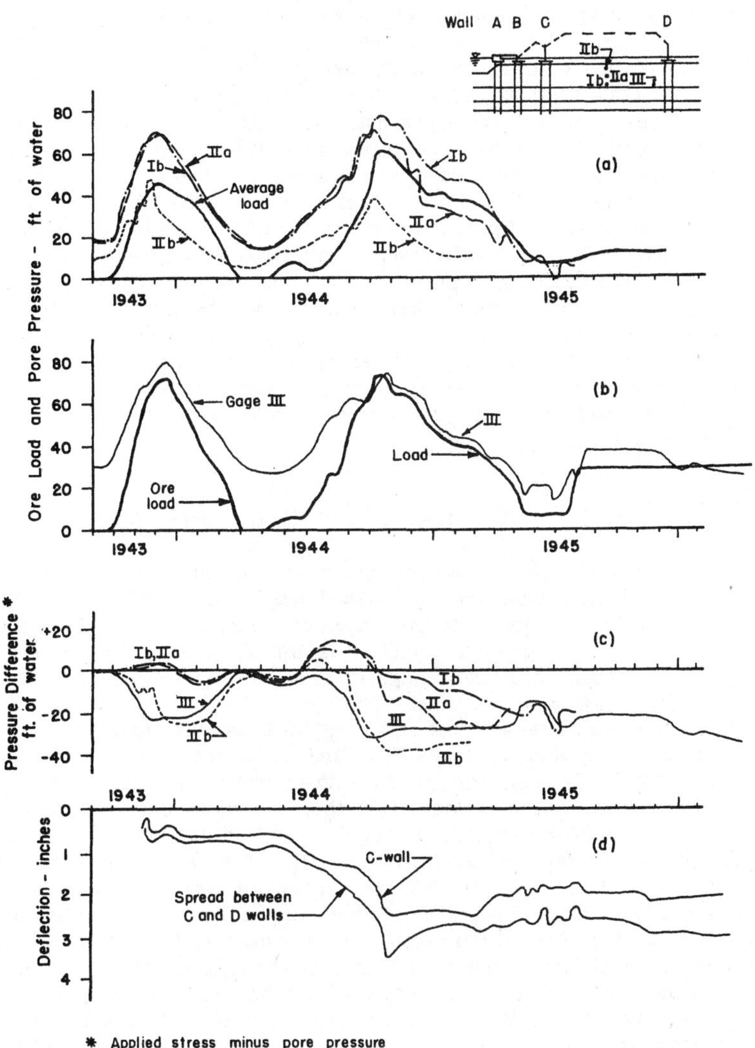

FIG. 19.—RESULTS OF PORE PRESSURE OBSERVATIONS, YARD B

to undergo large shear distortions. In Fig. 19d the movement of walls C and D in the vicinity of the gages is shown. It is evident that the rapid decrease of pore pressure at locations of potential failure surfaces coincides with permanent outward movement of the retaining walls. The evidence is not conclusive, but it may be argued that observed decreases in pore pressure are not necessarily caused by consolidation.

The information in this section suggests that ore yards located above deposits of clay not previously loaded to pressures in excess of the present ore loads have gained considerable strength, probably by consolidation, until an equilibrium condition has been achieved under the prevailing loads. If, however, the annual increment of load was at some time excessively large compared to previous maxima or was applied too rapidly, the yard failed. On the other hand, where the preconsolidation load exceeded the intensity of the prevailing ore loads the initial strength was already large enough to prevent failure under the customary intensities of ore loading. The observed pore pressure behavior can be interpreted in different ways and does not shed much light on the behavior of the clay mass. Whether the ore load significantly altered the strength of the clay, either upward or downward, is debatable. The evidence suggests, however, that the first few cycles of loading involved larger deformations within a given range of stress than developed later in the life of the ore yard.

INSIGHT INTO SOIL-STRUCTURE INTERACTION

Methods for Improving Behavior under Load.—In most ore yards the tracks for the ore bridge are established on retaining walls; the walls, in turn, are usually provided with pile support. In many of the yards, for instance C, D, F, G, the piles can have little influence on behavior except for improving local support of the bridge tracks. No further consideration will be given to these features.

In several yards, however, the retaining walls are tied together by means of steel rods. These include Yards B, C, and D. Yard C provides evidence of the behavior of the tierods. The three retaining walls were built in 1908. The main retaining walls were connected at their bases by pairs of rods, 2-7/8 in. in diameter, on 24-ft centers. During the period 1908-1927 the load on the ore yard increased, especially in 1916 and 1918. The peak tonnage during the period was stored in 1918. By 1927 the riverward retaining wall had moved toward the river about 14-1/2 in., whereas the landward wall had moved about 1-3/4 in. in the opposite direction. In the winter of 1927-28 a second set of tierods was installed in the south half of the ore yard. The new rods were pairs of billets, 3 by 3 in., placed at 24-ft centers to stagger the old pairs. The old rods were found to be intact but with a fairly sharp downward bend at the edge of the wall footings and a sag in the center of the yard of 2 ft to 3 ft. Most of the curvature was in the 50 ft near each wall.

During the following loading season the walls in the northern half of the yard spread about 1 in.; the following winter new tierods were also placed in the north half. Measurements through 1933 indicated no further progressive movement of the walls. The measurements were discontinued in 1933 and the reference points lost. In 1960 the reference points were reestablished. Although new base lines were used, the surveys indicate that very little movement had taken place between 1933 and 1960, despite the fact that an all-time maximum load was placed on the yard in 1937 and, since 1936, annual peak loads have been generally slightly more than the maximum before 1927.

There can be little doubt that the riverward wall was moving toward the river about 1 in. or 2 in. per loading season during the period 1918-1927, under a maximum average unit load on the yard of about 2.4 tons per sq ft, and that the supplementary tierods were adequate to stop this movement. Whether the new rods were also responsible for the ability of the yard to carry up to 3.2 tons per sq ft with negligible movement in subsequent years, or whether strengthening of the soil must be given at least part of the credit, cannot be determined directly. The yield capacity of the original tierods was about 10 tons per ft, and of the combined sets about 23 tons per ft of length of the ore yard. It seems likely that the rods could have provided only a small but nevertheless important margin.

In Yard B the behavior of the tierods was overshadowed, especially as deflections increased, by the influence of the piles supporting the various retaining walls. Except for a line of 16-in. pipe piles at 3.5-ft centers along the river side of the dock wall, all the piles were concrete-filled steel pipes 10-3/4 in. in diameter with lengths of about 120 ft. They were driven to refusal on the shale bedrock. Wall A was supported by a total of 0.7 10-in. piles per ft arranged in 3 rows; wall B by 1.1 piles per ft in 5 rows; wall C by 1.4 piles per ft in 8 rows; and wall D by 1.75 piles per ft in 7 rows. One of the rows in walls B and C was battered at 1:4 toward the channel. The tops of the piles were embedded in the heavily reinforced walls.

On the basis of extensive observations, Terzaghi constructed Fig. 20 representing the maximum displacements of the walls up to 1960. Because of the restraint of the tierods, the main retaining walls C and D tilt toward the ore pile when the yard is full. When the yard is emptied each year the walls assume a more vertical position and the horizontal displacements reduce generally on the order of 0.5 in. The additional permanent horizontal displacement is on the order of 0.3 in. per yr, only about 1/5 of that of Yard A. At Yard B the maximum average intensity of ore loading has been restricted to about 3.0 tons per sq ft; that at Yard A has been about 3.2 tons per sq ft. Thus the smaller annual permanent deformations at Yard B are probably the result of smaller loading, not encroaching on the threshold stress. It is probable that only a part of the elastic recovery is the result of the strain energy in the tierods and piles because the recovery is at least as great at Yard A which does not possess these features.

Although the piles and tierods at Yard B must have added at least slightly to the elasticity and capacity of the system, they have been subjected to severe stresses. The bending stresses in the piles at their points of embedment in the walls are especially large and may eventually limit the loading on the yard by requiring a limitation on horizontal deflection. The row of 16-in. piles near the front of the dock wall was stiff enough to cause the concrete to fail and to release the heads of the piles.

The outright failure of several of the early ore yards was always accompanied by displacements toward the channel. This characteristic evidently led to the conception of providing pile support for a platform to carry the

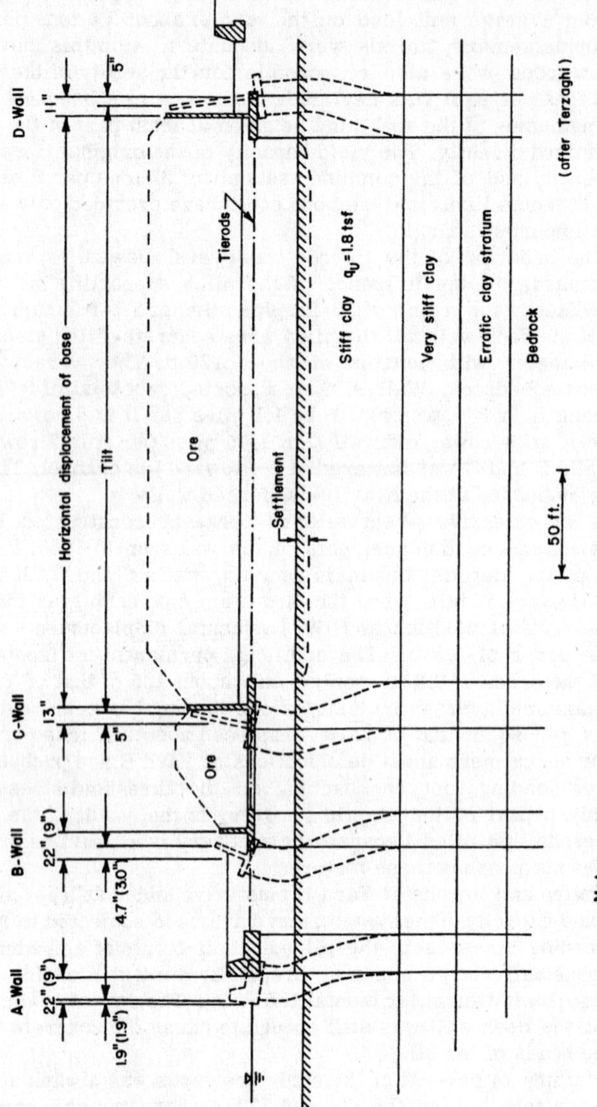

FIG. 20.—APPROXIMATE DISPLACEMENTS OF RETAINING WALLS, 1942–1960, YARD B

ore in at least the riverward portions of the yards. According to E. J. Fucik,[7] the first construction of this sort appears to have been done for the Inland Steel Company at Indiana Harbor in 1914, after failure of 500 ft of the yard the preceding winter. The idea has been extended in some instances to the provision of pile support beneath the entire storage area. Yard J is very similar to that of the Inland Steel Company, whereas Yard K carries the conception to a completely pile-supported platform.

Yard J was constructed in five stages, as indicated in Fig. 21. In the 1917 section the ore is stored directly on the subsoil. In all the additions a concrete floor has been provided between the dock wall and wall No. 4. The floor is supported in the 1924 extension by 25-ft piles, in the 1929 extension by 70 ft

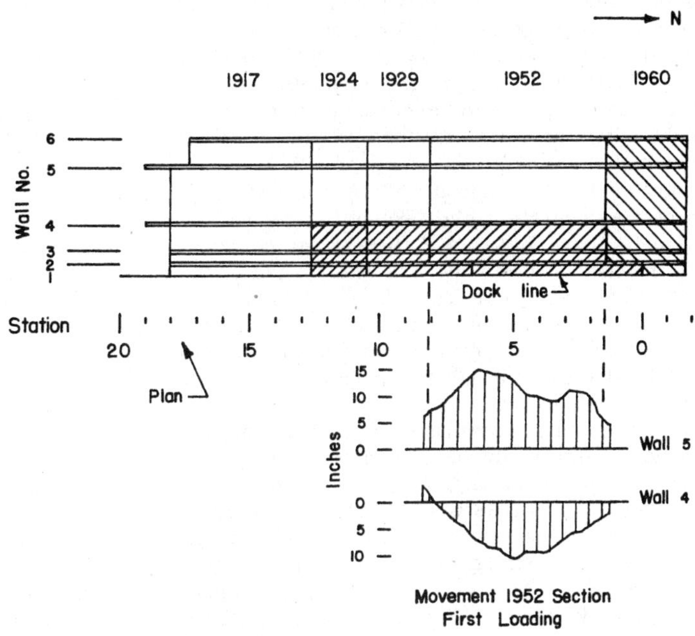

FIG. 21.—CONSTRUCTION HISTORY AND 1952 MOVEMENTS, YARD J

piles, and in the 1952 addition by piles about 100 ft long to hardpan. In the 1960 addition the floor extends across the entire yard and is supported by cylindrical 30-in. piers to hardpan. All the walls in each addition are supported by piles or piers similar to those beneath the floor.

The behavior of the 1917, 1924, and 1929 sections is not known in detail. There is no doubt that movements occurred, but the magnitudes were not considered troublesome. Nevertheless, it was evidently believed desirable to

[7] Fucik, E. J., "The Main Ore Unloading Dock Failures and their Correction 1909-1925, Great Lakes Region," Proceedings, 4th Conf. on Coastal Engrg., Chicago, Ill., 1954, p. 316.

provide positive support for the slab in 1952 by driving 10-3/4 in. steel piles, with upper shell sections, to high resistance in the hardpan. The piles were spaced at 4 ft by 5.5 ft centers and were assigned a load of 50 tons each.

During the first application of the ore load to the 1952 extension, walls 4 and 5, even though located on the land side where no weakening influence of the channel could be felt, was quite substantial. Obviously, the pile support beneath the walls and platforms did not eliminate the movement of walls 1-4 toward the channel. Most striking, however, was the presence of transverse cracks in the floors and walls. The cracks indicate that the 1952 addition increased in length by about 1 ft. The transverse reinforcement, calculated to withstand the lateral pressure of the ore, proved adequate, but the longitudinal steel was nominal and failed.

In the 1960 addition a pier-supported concrete platform, heavily reinforced for the ore pressure in both the longitudinal and transverse directions, was built over the entire area to be subjected to ore load. Such construction, of course, eliminated the strength and compressibility of the subsoil from consideration. However, it was necessary to restrict the load at the junction between the 1952 and 1960 additions to prevent pushing the entire platform longitudinally.

In contrast to pile-supported platforms, which place no reliance on the strength of the underlying clay except to prevent local buckling of the piles, sheetpile cells have been used to restrict horizontal displacements of the foundation materials under the ore load. Yard L includes this feature. Unfortunately, the history of Yard L is complex, the behavior has also been complex, and the history of loading is not accurately known. Nevertheless, considerable insight into the behavior of the system can be obtained by a review of the details although the benefits of the sheetpile cells cannot be assessed with confidence.

The yard was built in several increments. The first 600 ft, constructed in 1909, consisted essentially of a concrete sheetpile bulkhead and tracks for the unloaders and ore bridge. A second, similar section was constructed in 1916 (see Fig. 22). In 1938 the second 600 ft, built in 1916, were reinforced by a steel sheetpile cell system, and in 1948 the first 600 ft were similarly altered. In 1952 an extension 1000 ft long was constructed with cells; ore had been piled in this area as far as Sta. 15+00 since 1937, and to Sta. 25+00 since 1941, as indicated in Fig. 22. The 1952 addition included a dock wall of steel instead of concrete sheetpiles. In the older sections dredging at various times had required increasing the depth and strength of the dock wall.

Little information exists regarding the behavior of the older portions of the yard before 1938. As a consequence, no direct comparison can be made of performance before and after installation of the cells. Outward movements of the dock wall were prevalent. These were principally a consequence of various inadequacies in the tiebacks and anchorages. More accurate data are available concerning the 1952 addition, from which the movements of the cells can be judged. The pertinent information is contained in Fig. 23.

This figure shows that the dock wall is anchored by long tierods to a reinforced concrete anchor wall that rests on the clay within the cells but is restrained from slipping with respect to the crosswalls of the cells by the crenulations of the sheetpiles. Between January, 1953 and December, 1957 the dock wall moved out by amounts varying from 3 in. to 14 in., as shown in Fig. 23. When the sheetpile cells were driven in 1952, the locations of the

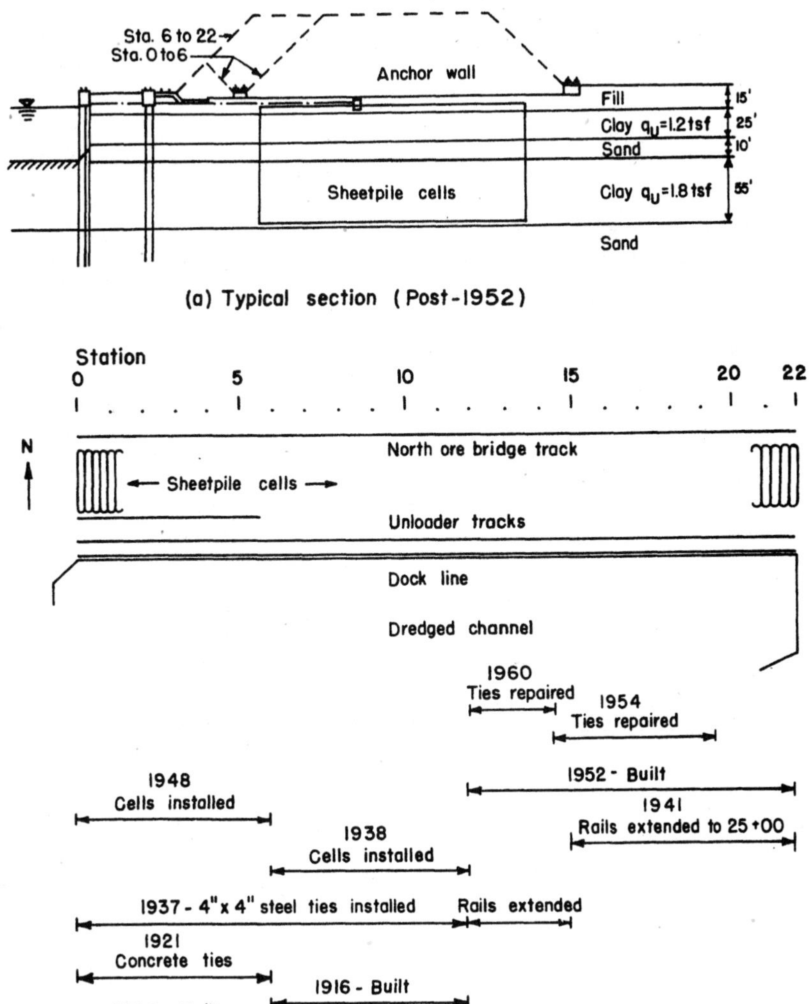

FIG. 22.—HISTORY OF CONSTRUCTION, YARD L

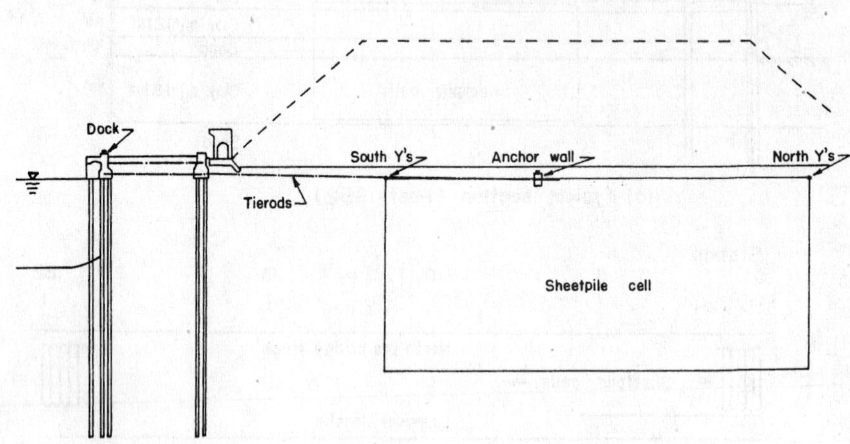

Location of observation points

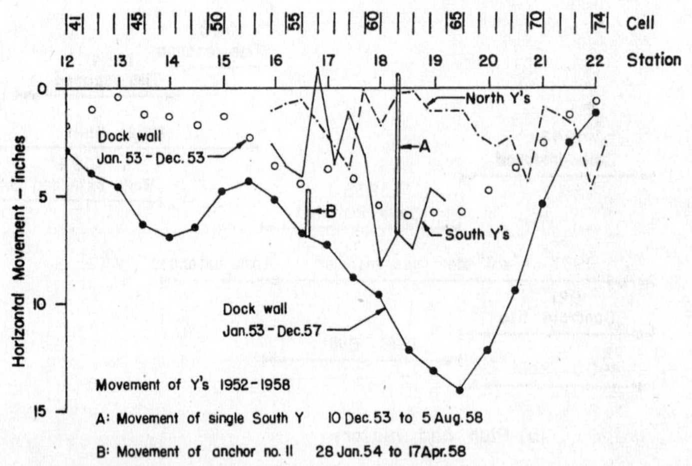

FIG. 23.—STUDY OF MOVEMENTS 1952-1958, YARD L

Y-piles were accurately measured. In 1958 the cells were uncovered and the Y-piles again surveyed. Both the riverward and landward Y-piles had moved toward the river but the movement of the riverward line was the greater by about 0 in. to 7 in. Hence, the cells elongated at their tops by about this amount. Inspection of the tops of the cells showed no signs of rupture at the interlocks. In addition, at one station the movement of a specific Y-pile on the riverward side was measured between December 10, 1953 and August 5, 1958. This measurement (A in Fig. 23) indicated 7 in., in agreement with the observations mentioned previously. Finally, at another station the absolute movement of the anchor wall was measured between January 28, 1954 and April 17, 1958 (B in Fig. 23). This movement was found to be 2 in. At this location the total movement of the bulkhead was appreciably smaller than at the location where the movement of the Y-pile was measured and an anchorage movement of 2 in. is quite compatible with the observed movement of 7 in. of the Y-pile.

These results indicate that the ore yard has experienced a general movement toward the channel. The movement at the anchorage may have ranged from a very small value to as much as 4 in. or 5 in. in a 5-yr period. The maximum rate of movement of the anchorage is, therefore, on the order of 1 in. per year. Such a movement is compatible with a general stretching of the sheetpile cells of up to 7 in. and a corresponding distortion of the clay within the cells.

Much larger movements of the dock walls have been observed from time to time. These have been found to be a consequence of tierod breakage.

The intensity of loading on Yard L has been unusually great, reaching an average of 4.2 tons per sq ft over large areas in some loading seasons. Unconfined compression tests on the underlying clays indicate a strength of about 1.2 tons per sq ft in the upper part of the deposit and about 1.8 tons per sq ft in the lower part. In both parts there is considerable scattering of the results; most of the data pertain to areas not influenced or only slightly influenced by the ore loads. Considering the high intensities of loading and comparatively low initial strengths, the present annual movement of about 1 in. per yr at the anchorage (slightly greater at the dock line) appears rather small and suggests that the cells have had a beneficial influence.

In contrast, cells driven as retaining walls near the front of an ore pile, as those in Yard A, Fig. 5, have unquestionably proven ineffective. The data concerning horizontal movement of the dock wall at Yard A, Fig. 11, show not the slightest difference in performance on either side of the junction between the cellular section and the older unreconstructed portion.

RELATIVE STABILITY OF ORE YARDS

Purpose of Analysis.—The case histories have disclosed that the most significant manifestations of unfavorable ore-yard behavior are large permanent displacements and occasional failures by sliding or rupture. It is apparent that the factor of safety of all the yards is quite low in comparison to the values adopted in the design of ordinary structures. Calculations of the factor of safety based on customary methods of slope stability analysis are of limited interest because they do not readily disclose the relative importance of the significant variables. A means for comparing theoretically the relative stability of many similar structures on the basis of the principal variables would be useful, especially since several examples of rather unsatisfactory behavior have been recorded in the case histories.

The relative stability of the yards is investigated on the assumption that the clay may be regarded as a perfectly plastic medium with a strength corresponding to the ultimate undrained shear strength of the clay.

Development of Analysis.—Solutions for the bearing capacity of a material with a yield value or cohesion c have been obtained by L. Prandtl and by R. Hill,[8] who both arrived at the same ultimate pressure of 5.14 times the yield value for a load applied at the surface of the material. This value corresponds to the bearing capacity of a strip footing resting directly on a saturated cohesive soil under short term loading (the $\phi = 0$ case). Laboratory and field observations, although limited in number, have generally confirmed the predictions based on this theory.

The applicability of the ordinary bearing capacity theory to ore yards is somewhat questionable. Even though it is possible to construct a Hill-type slip field, Fig. 24a, underneath the edge of an ore yard, the sloping face and the horizontal force within the ore pile remain unconsidered. A sloping face tends to decrease the shearing stress in the subsoil and hence it increases the value of the ultimate vertical stress that can be applied at the surface. On the other hand, according to V. V. Sokolovski,[9] the ultimate vertical stress decreases with increasing horizontal component of an applied inclined stress. Thus the existence of a horizontal force within an ore pile and within any sand strata overlying the clay has an unfavorable influence on the ultimate bearing capacity of the ore yard.

Another theoretical solution of interest, first investigated by Prandtl,[8] is the squeezing of a perfectly plastic material between two rough plates. The practical application of this solution to problems in soil mechanics was recognized by L. Jurgenson,[10,11] who recommended the theory for the design of embankments resting on relatively thin clay strata. Jurgenson also showed the applicability of the general principles of the squeezing theory when a horizontal force exists within the embankment, but he did not consider end conditions in detail. The characteristic feature of the squeezing theory is a linear increase of the allowable vertical stress with increasing horizontal distance as shown in Fig. 24b. The squeezing theory has several advantages as compared to the bearing capacity theory: There is an infinite number of potential failure surfaces that can be related to actual sliding surfaces in the field; it is possible to calculate readily the magnitude of the horizontal stress at any point within the clay stratum; and furthermore, consideration can be given to the influence of structural features on the magnitude of the horizontal stress within the clay and on the ultimate load-carrying capacity of the clay. The magnitude of the horizontal stress plays an important role in the comparison of the relative stability of the ore yards.

---

[8] Hill, R., "The Mathematical Theory of Plasticity," Oxford, London, Eng., 1950.
[9] Sokolovski, V. V., "Statics of Soil Media," London, 1960.
[10] Jurgenson, L., "The Application of Theories of Elasticity and Plasticity to Foundation Problems," Journal, Boston Soc. of Civ. Engrs., Vol. 21, 1934, p. 206.
[11] Jurgenson, L., "Stability of Earth and Foundation Works and of Natural Slopes," Proceedings, Internatl. Conf. on Soil Mechanics and Foundations, Vol. 2, Cambridge, Mass., 1936, p. 194.

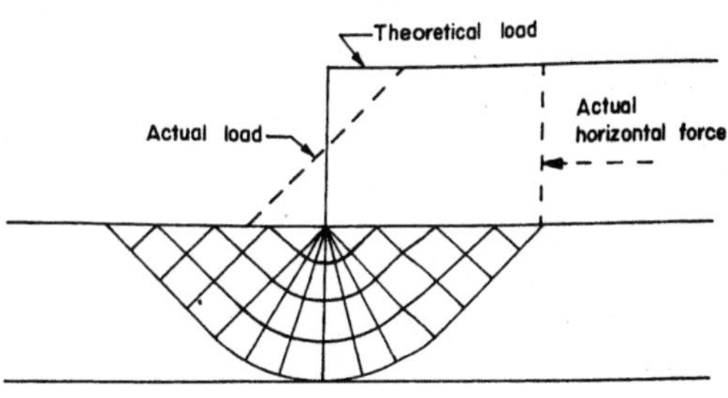

(a) Bearing Capacity Problem

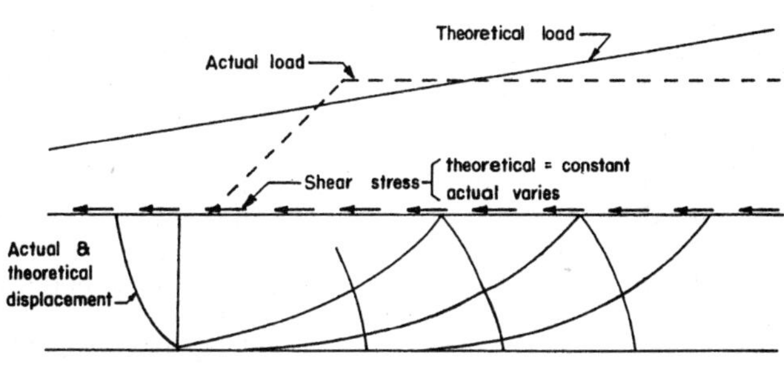

(b) Squeezing Problem

FIG. 24.—FAILURE PATTERNS IN COHESIVE MATERIAL

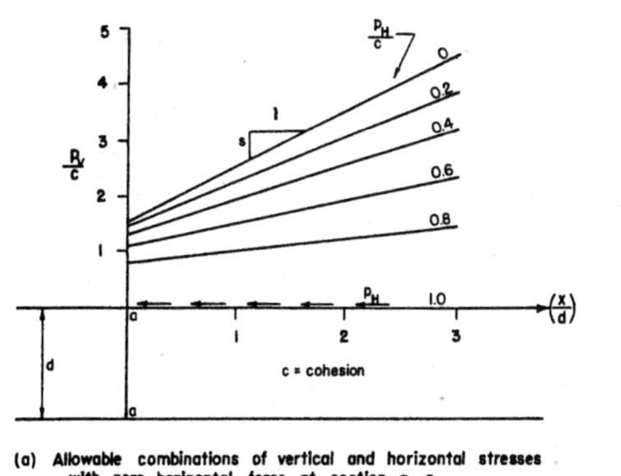

(a) Allowable combinations of vertical and horizontal stresses with zero horizontal force at section a - a.

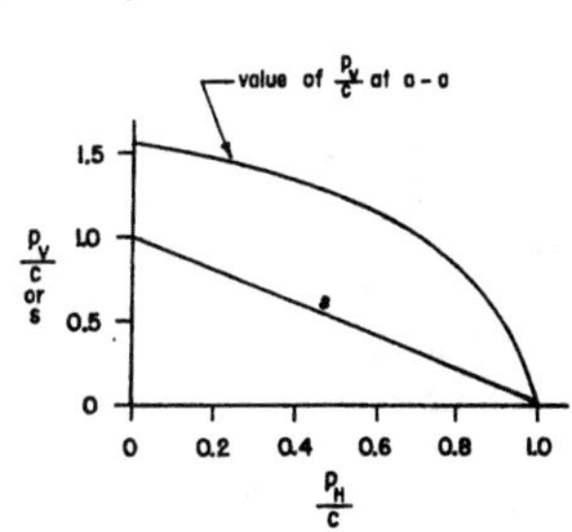

(b) Values of squeezing problem parameters

FIG. 25.—LOAD LINES FOR APPLICATION OF SQUEEZING PROBLEM

An extension of the results of the squeezing problem is shown in Fig. 25a.[12] This graph demonstrates the effect of several variables on the load-carrying capacity of a plastic ($\phi = 0°$) medium of thickness d subjected at its upper surface to a uniformly distributed horizontal stress as might be caused by the horizontal force within the ore pile. The ordinates are expressed in terms of the dimensionless ratio of the vertical applied stress to the cohesion. Similarly, the horizontal applied stress is expressed as the dimensionless ratio p_H/c. The abscissas represent the dimensionless distance x/d. Any given p_H/c line represents the magnitude of the dimensionless vertical pressure that can be applied at the surface of the material so that the average horizontal pressure on a-a is zero. Such a line of constant p_H/c is termed a load line. All the load lines in Fig. 25a may be considered to represent the same factor of safety, which is considerably above unity because the strength of the material to the left of section a-a has not been taken into account. It is readily noted that the magnitude of the applied horizontal stress p_H significantly affects the ultimate vertical stress. An increase in the ordinates of any load line by a constant vertical pressure p_v/c will produce an equal average horizontal pressure p_H/c at section a-a. Specific values of the initial magnitude of the dimensionless vertical stress at section a-a and its allowable rate of increase s per unit thickness of the clay stratum are shown in Fig. 25b as a function of the ratio p_H/c.

The load lines shown in Fig. 25a do not correspond to ultimate failure unless the material to the left of section a-a has zero resistance. They merely correspond to a plastic zone underneath the loaded area. In order to develop a rupture line to the left of section a-a, some horizontal stress must be applied at section a-a. No rigorous theoretical solution exists regarding the magnitude of the horizontal stress required; the value should range between 2 times and 3 times the magnitude of the cohesion.

Because of the considerable width of most ore yards as compared to the thickness of the underlying clay stratum, the squeezing problem is convenient as a basis for comparing the stability of ore yards and for reviewing the probable effect of the variables on behavior. The results are not accurate in the absolute sense but are nevertheless useful for comparisons of the ore yards and evaluations of design features. In this respect they serve the same function as properly used conventional stability analyses.

In the application of the analysis to an ore yard, the horizontal force at an appropriate vertical section within the ore pile and underlying sand stratum is first estimated. The horizontal pressure has been taken for all the yards as equal to 0.4 times the vertical pressure at any point; however, it can be shown that other reasonable ratios have little effect on comparisons of the relative stability of the yards, provided the applied horizontal stress does not approach the shear strength of the clay. The horizontal force determined in this fashion is then assumed to be uniformly distributed over the horizontal distance between the chosen vertical section and the bulkhead a-a. The slope of the load line and the initial value of vertical loading corresponding to zero horizontal force at the bulkhead can be determined from Fig. 25.

The appropriate vertical section for the calculation is chosen by substituting for the actual cross section of the ore yard an equivalent line parallel to the

[12] Raamot, T., "The Foundation Behavior of Ore Yards," thesis presented to the University of Illinois, at Urbana, Ill., in 1962, in partial fulfilment of the requirements for the degree of Doctor of Philosophy.

load line. Because the crest of most ore piles is level and because the load line invariably has a slope, it is necessary to choose a horizontal distance over which the effects of the real and idealized cross sections are the same. It can be shown that the horizontal length of each cycloidal shear line between its beginning and its point of tangency at the bottom of the stratum must be less than 2.57 times the thickness of the stratum; the exact value depends on the ratio p_H/c. In order to develop at least one such shear line, the horizontal length of loading has, therefore, been taken as twice the thickness of the clay stratum. Point A, Fig. 26, located a distance d equal to the thickness of the clay deposit from the break of the slope of the ore pile, is therefore a reasonable point of intersection of the actual crest line and its theoretical approximation. It determines the vertical section at which the horizontal force within the ore pile and the sand stratum is calculated. Although the results of the calculations are fairly sensitive to the choice of the position of point A, the relative stability of the various ore yards is not significantly affected if a consistent procedure is used for all the yards. The ore yards are compared on the basis of the average theoretical horizontal pressure developed at the bulkhead. To permit a direct comparison of numerical values, this pressure is expressed as the nondimensional ratio of the average horizontal stress to the cohesion of the soil.

The analysis cannot be applied to all the ore yards. Yard E is not wide enough. Yard K possesses a pile-supported relieving platform that extends the whole width of the yard. At Yard F the consistency of the clay is not known. As indicated previously, Yard L is too complex for a rational analysis. There remain 8 ore yards to be compared, two of which have experienced large-scale failures and one of which is exhibiting substantial progressive movements. Because the typical loading conditions are not known accurately in most cases, the most appropriate method of rating the relative stability of the ore yards is to assume the unit weight of the ore to be 160 lb per cu ft and to consider the height of the ore pile as a variable.

The ore yards are compared in Figs. 27 and 28. The abscissas represent the height of the ore pile at the critical section. The ordinates are the ratio of theoretical horizontal pressure at the bulkhead to shear strength of the clay. Increasing values of this ratio correspond to decreasing stability. Figure 27 is a comparison of the ore yards based on the strength of the clay alone; the effect of tierods, piles, or passive pressure in sand layers has been ignored. The general trend of relative safety reflects the shear strength of the clay strata. The safest is ore yard D, which is underlain by clay with an unconfined compressive strength of 3.0 tons per sq ft. Five ore yards above clays with unconfined compressive strengths of 1.8 tons per sq ft to 2.0 tons per sq ft rate quite closely together, whereas two on softer clays could be expected to carry considerably lower loads. As a matter of fact, these two have failed.

The analytical procedure permits at least a crude evaluation of the theoretical effect of the various structural features incorporated in some of the yards, as well as of the influence of the passive pressure that might develop in sand deposits on the channel side of the bulkheads. For this evaluation, the tierods are assumed to be stressed to their yield point at 35,000 psi and the total force carried by the tierods is subtracted from the horizontal pressure within the ore piles. The passive resistance of sand at the bottom of the channel is evaluated on the basis of a passive pressure coefficient of 3.0,

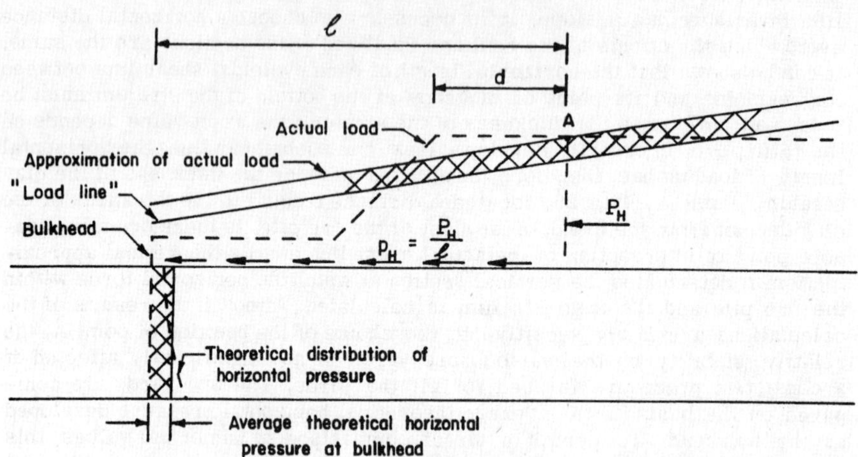

FIG. 26.—APPLICATION OF SQUEEZING PROBLEM TO EVALUATION OF ACTUAL ORE YARD

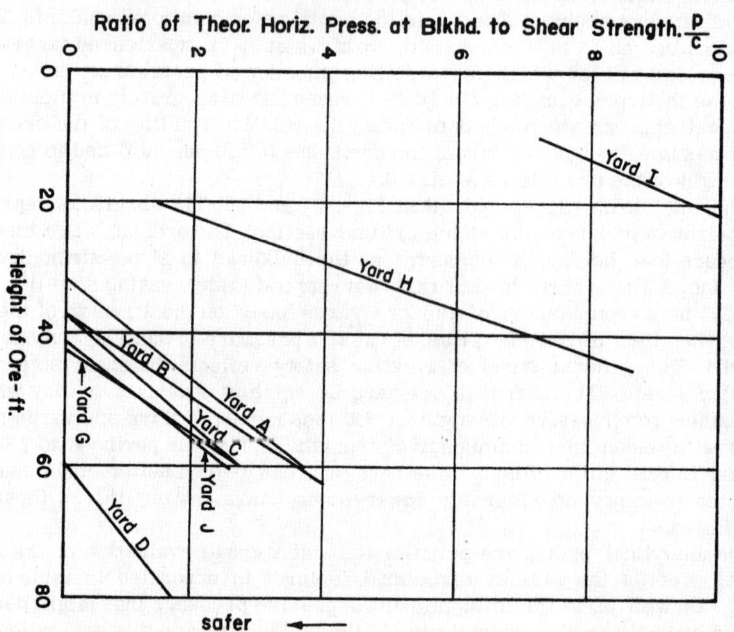

FIG. 27.—RELATIVE STABILITY COMPARISON OF ORE YARDS BASED ON $c = q_u/2$

and an equivalent force is subtracted from the average horizontal pressure within the clay, where appropriate. A reasonable value of the pressure resisted by piles is also subtracted.

Five ore yards represented in Fig. 27 have very similar safety ratios. This group is of special interest because it includes Yard A which still experiences appreciable progressive movements. The behavior of the latter yard has been examined previously in some detail; it was noted that significant movements start when the height of ore is about 40 ft. Yard A is also unique because for much of its length it depends on soil strength alone. Fig. 28a is an enlargement of the group from Fig. 27. Only the strength of the soil is considered. On this basis, an ore height of 40 ft at Yard A is comparable to 43 ft at Yard B, 45 ft at Yard C, and 49 ft at Yard G and the soil-supported portion of Yard J.

In Fig. 28b the tierods are assumed to be stressed to their yield point. In addition, in Fig. 28c passive pressure in the sand on the channel side of the bulkhead is included as well as a resistance of 10 kips per pile in the case

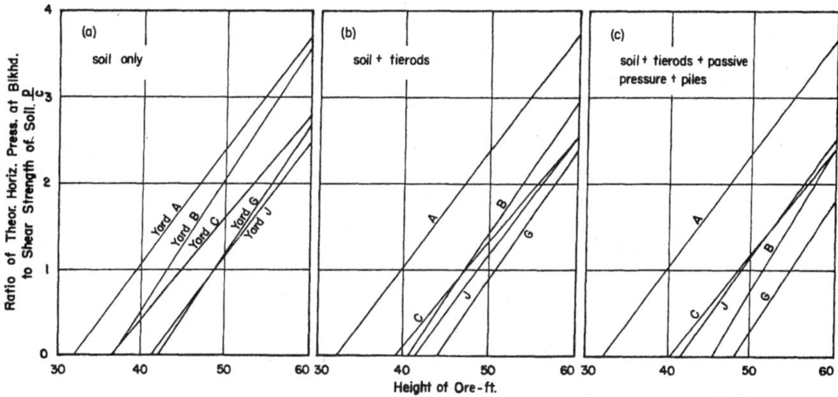

FIG. 28.—STABILITY STUDY OF FIVE ORE YARDS

of Yard B. Since Yard A includes no structural features, the gap between this and the other ore yards widens with each assumption. The last graph indicates that the other ore yards could be loaded to heights of about 50 ft before their factors of safety would become comparable to that of Yard A loaded to a height of 40 ft. This prediction has not been verified because the actual ore storage height at Yards B and C has never been much in excess of 40 ft. It should be noted that actual ore heights of 40 ft at Yards B and C result in no positive theoretical horizontal pressure at the bulkhead.

An interesting conclusion to be drawn from the theoretical studies is that the load could be distributed on most ore storage areas in a manner much more favorable to stability than that beneath the usual ore pile having a trapezoidal cross section with a level crest. It has been shown that steepening the front slope of the pile or building its face out toward the river, was responsible for movements even when the total load on the yard was not unusually great. The analysis, in agreement with these observations, indicates that the loading at these locations determines the magnitude of the theoretical horizontal pressure at the bulkhead and thus the safety of the ore yard. It appears, therefore, that unsatisfactory behavior of the existing ore yards is initiated by loads on a relatively small portion of the total width of the yard. On the other hand, the theoretical load lines indicate that some of the potential capacity of the ore yards is not used. This additional capacity exists in the form of potential storage at low heights between the main ore pile and the bulkhead and at greater heights near the center of the main storage area.

The geometric arrangement of some ore yards appears to be inherently superior to that on others. Figure 28a indicates that Yard G could carry an ore pile 9 ft higher than Yard A. In the comparison, the contribution of structural elements has not been considered, and identical strengths have been assigned to the clay underlying both yards. A minor part of the difference results from the slightly thinner clay stratum at Yard G. The major difference, however, can be attributed to the favorable load distribution on Yard G. This yard has a trough at a reasonable distance from the channel, followed by a narrower storage area that carries ore at a smaller height than the main storage area. Behind the main pile is again a smaller ore pile. This type of loading is more in accordance with the theoretical load lines than the loading at Yard A, which rises abruptly and causes an overload along the front edge of the ore pile. Because the total width of Yard G exceeds that of Yard A by about 110 ft, the maximum height of ore in Yard A can never equal the height of ore in Yard G.

The possibility of failure on the rear side of an ore yard should not be overlooked. Although no failures at that side have occurred, several ore yards, for example Yards B and J, have experienced landward movement of the rear elements. An extension of the analytical approach to the stability of the rear of the ore yards has not been made because of the complications introduced by the rather thick sand stratum overlying the clay in nearly all cases, and because the boat channel along the front of the pile constitutes a distinct weakness in the system, both in terms of increased shear stresses and decreased resistance.

CONCLUSIONS

The consequences of the high intensity and cyclic character of the loading on ore storage areas above plastic clay soils constitute valuable data concerning the behavior of the masses of clay.

With increasing ore load, small elastic lateral movements may develop. At a critical stage of loading, large progressive nonrecoverable movements seem to be initiated by very small increases in the load or changes in its distribution. This conclusion is based on observations on clays preconsolidated to pressures above the maximum ore pressure and may not be valid with respect to deposits where the intensity of ore loading exceeds the preconsolidation stress.

Where the ore loading has exceeded the preconsolidation stress the clay has consolidated and gained strength. In a few instances the ore load was increased rapidly and stability failures ensued; otherwise the strength increased until the yard was capable of supporting the customary load of 40 ft to 50 ft of ore (3 tons to 4 tons per sq ft). Conversely, the strength of the preconsolidated deposits appears not to have increased as a consequence of the ore loading. There is some evidence that with successive applications of load at low stress levels the deformation at a given level of stress becomes smaller.

Tierods between the main retaining walls are capable of increasing the capacity of a yard, but by a small amount. Structural platforms supported by piles or piers have been successful but only when full consideration has been given to the shearing stresses exerted on the slab by the ore pile, and to the lateral forces that may be exerted on the platform by soil-supported ore piles in adjacent areas. Sheet pile cells beneath the central part of the ore pile seem to have been beneficial but the data are insufficient to permit a reliable assessment of their value.

The theory of a perfectly cohesive material squeezed between two rough rigid plates, when adapted to the ore storage problem, provides a means for evaluating the relative stability of the yards. The results are in reasonable agreement with experience. The influence of tie rods and other structural features can be taken into account. It is found that for maximum over-all stability the surface of the ore pile should slope upward from an initial height near the dock at a rate that depends on the thickness and strength of the clay. Some existing ore yards are shown to satisfy these conditions more closely than others, and to have experienced more satisfactory behavior.

Journal of the
SOIL MECHANICS AND FOUNDATIONS DIVISION
Proceedings of the American Society of Civil Engineers

THE TERZAGHI LECTURE[a]

ROLE OF THE "CALCULATED RISK" IN EARTHWORK AND FOUNDATION ENGINEERING

By Arthur Casagrande,[1] Hon. M. ASCE

FOREWORD

I am deeply grateful to my colleagues who elected me to present the 1964 Terzaghi Lecture. There is, indeed, no better way for me to repay all that I owe to Karl Terzaghi than by lecturing in his honor, by keeping alive in my lectures his memory, and by letting my younger colleagues and my students partake in what he has taught me.

On October 21, 1963 I spoke with Karl Terzaghi for the last time. I called on him after returning from several weeks of travel which included attendance at the ASCE convention in San Francisco. I found him still at work, but very weak and suffering severe physical pains. He was anxious to hear about the San Francisco meeting and the first Terzaghi lecture delivered by Ralph Peck, which was so well received by the large audience. With sincere interest he read every line on the plaque of my Terzaghi Award, which I had brought along for him to see. Then he asked me to report about the construction progress of the South Saskatchewan River dam on which he had been serving as consultant. Finally he gave me, for my review, another part of the last paper on which he was still working. Three days later I returned

Note.—Discussion open until December 1, 1965. To extend the closing date one month, a written request must be filed with the Executive Secretary, ASCE. This paper is part of the copyrighted Journal of the Soil Mechanics and Foundations Division, Proceedings of the American Society of Civil Engineers, Vol. 91, No. SM4, July, 1965.

[a] The Karl Terzaghi Award and Lecture were established by the Soil Mechanics and Foundations Division of the Society by gifts from the many friends and admirers of Karl Terzaghi, Hon. M. ASCE. A distinguished engineer is invited periodically to deliver a "Terzaghi Lecture" at an appropriate meeting of the Society. This, the Second Terzaghi Lecture, was presented during the annual Meeting of ASCE on October 21, 1964.

[1] Prof., Soil Mechanics and Foundation Engrg., Harvard Univ., Cambridge, Mass.

with my completed review. Usually I saw him at work silhouetted against the lake behind his home as soon as I looked through the glass panels of the front door; but this time his chair was empty. I gave the manuscript to Mrs. Terzaghi. The next morning at eight, October 25, 1963, she telephoned and said that Karl had passed away an hour earlier.

ON THE MEANING OF "CALCULATED RISK" AND OTHER TYPES OF RISK

Terzaghi's great accomplishment was to replace in earthwork and foundation engineering the large conglomeration of great "unknown risks" of the past in part by rational analyses which are based on the principles of soil mechanics that he developed, and in part by "calculated risks" that we can estimate with the help of soil mechanics, and judgment.

The purpose of this presentation is to discuss all types of risks which are inherent in earthwork and foundation engineering, the extent to which we are justified or forced to assume such risks, and how to cope with them. The title of my paper does not fully cover my subject because I intend to speak not only about calculated risks, but also about risks due to unknown factors, which I shall term "unknown risks," and about uncertainties introduced by human failings which may be termed "human risks."

For an orderly discussion I should start with a definition of the expression "calculated risk." You have all heard it many times; and probably most of you have used it. Its meaning has puzzled me for many years because I observed that it was used particularly when it was not possible to calculate anything, when instead it was necessary to rely on experience and judgment. Eventually, I found in a dictionary that as an adjective, "calculated" means "estimated" rather than "computed"; and that a calculated risk is an estimated risk. Therefore, the often-heard joking remark that a calculated risk is the type of risk that nobody knows how to calculate, is really a play of words on the ambiguity of the adjective "calculated." (I mention in passing that it is difficult to translate correctly "calculated risk" into other languages. I have seen it translated in professional literature into just the opposite of its real meaning.)

Webster's 1961 International Dictionary contains the following two definitions for calculated risk:

1. "A hazard or chance of failure whose degree of probability has been reckoned or estimated before some undertaking is entered upon.
2. "An undertaking or the actual or possible product of an undertaking whose chance of failure has been previously estimated."

Use of this term in other activities is illustrated by the following examples: The May 7, 1963, issue of Engineering News-Record had an editorial advocating that helicopters be permitted to land on the rooftop of the Pan Am Building in New York. It started with this paragraph: "There's A CALCULATED RISK in crossing a street in the busy centers of our cities. There's a calculated risk in living in a flight path of our major airports. There's a calculated risk in flying helicopters to rooftops in cities."

In a 1963 issue of TIME magazine, in the section on "WORLD BUSINESS," there was an article entitled "Calculated Risks" in which the eagerness with which west-European businessmen are extending long-term credits to the Soviet Union was discussed; how they are taking calculated risks in order to gain a toehold in the potentially enormous market in Russia and its European satellites.

Although the term calculated risk is used extensively by engineers, I have not found a published definition specifically pertaining to its use in engineering. Therefore, I have asked a number of colleagues whether or not they have used it and with what meaning. A minority of the replies stated that it is a misleading term, that it can be easily misunderstood and should be avoided. The majority replied that they have been using it, and their definitions agreed in substance and may be summarized as follows: The taking of a carefully considered risk which is based largely on an analysis of factors that require experience and judgment for their evaluation.

This survey and my own observations show that the term "calculated risk" is indeed widely used in engineering, if not somewhat loosely; and that usage and most suggested definitions have in common a meaning that includes the following two distinct steps:

(a) The use of imperfect knowledge, guided by judgment and experience, to estimate the probable ranges for all pertinent quantities that enter into the solution of a problem.

(b) The decision on an appropriate margin of safety, or degree of risk, taking into consideration economic factors and the magnitude of losses that would result from failure.

I intend to use the term "calculated risk" in this paper with this combined meaning. Both steps are inherent in a calculated risk, as illustrated by the following fictitious example.

An embankment is to be built on a clay foundation. From his investigations the designer concludes that the in situ shear strength of this clay may range between 1.0 and 2.0 tons per sq ft. The upper limit is derived from conventional laboratory strength tests on undisturbed samples and on in situ vane tests. The lower limit is based on the designer's experience and judgement of the possible combined effects of (1) lateral transmission of pore pressures due to the stratified character of the clay stratum, which would reduce the average shearing resistance along a potential sliding surface; and (2) the reduction in long-term strength when this clay is subjected to shear deformation at constant water content.

After establishing the controlling range for shearing resistance, the designer selects an allowable (design) value to be used in his stability analyses. If this project is an important dam whose failure would cause catastrophic losses, he might decide to use the very conservative design value of 0.6 tons per sq ft. Thus, he would protect himself against his wide range of uncertainty by an ample margin of safety, i.e. with a factor of safety ranging between the limits of about 1.6 and 3.3. To effect greater economy without compromising safety, he might elect to install numerous piezometers in the clay stratum and to adopt an initial design with a much smaller safety margin. Thus he would utilize the project as a full-scale test, and on the basis of the piezometer observations he would modify the design during construction if required.

If the project were a long highway embankment for which a partial failure would cause only a modest economic loss, the designer would try to achieve greater economy by allowing a greater risk of failure. Therefore, he might use a design value of 1.2 tons per sq ft. Using piezometer observations as a guide, he would add stabilizing berms only if needed. Thus, his initial design would allow a certain probability of failure which the designer hopes to control within tolerable limits with the help of piezometer installations. He may choose to go one step further and deliberately produce failures by constructing full-scale test sections. Thereby he would succeed in reducing to a narrow range the uncertainty concerning the controlling shearing resistance of the clay stratum.

The alternatives in the preceding example not only illustrate the two steps that enter in the evaluation of a calculated risk, but they also demonstrate that the meaning of a statement such as "the designer had to cope with a large calculated risk" is not clear. Does this statement imply (a) a wide range of uncertainty about the strength, or (b) a great risk of failure? With no other information I would assume that it implies a combination of both.

EXAMPLES

By means of several examples I shall illustrate the nature and importance of risks in earthwork and foundation engineering. So many interesting case records in my files clamored for attention that it was difficult for me to settle on only seven examples. I have selected the majority from my own practice, in part because I know them so well, and in part because it is easier for me to be frank about risks for which I have had to accept a major responsibility than for me to discuss risks taken by others.

Unfortunately it is common practice in publications describing projects in applied soil mechanics to present a rationalized picture that has little resemblance to the truth about the actual approach followed in arriving at major design decisions. If judgment based on empirical knowledge played an important role in such decisions, why not admit it? If unknown risks or calculated risks were involved because of the large gaps in our understanding of the mechanics of soils, why not admit it? Authors who, a posteriori, try to rationalize their decisions or make theories fit the facts, merely reveal their limited grasp of the realities of applied soil mechanics. With distorted presentation of important case records they hinder rather than promote progress; and they mislead their younger and less experienced colleagues.

(1) *Piers for Oakland Bay Bridge.*—For my first example, I quote from memory what the late Daniel E. Moran related in a lecture to my students in the spring of 1935:

> "In the course of early investigations for the San Francisco-Oakland Bay bridge, my partner Carlton Proctor was in San Francisco and sent me a cable to Bermuda where I was vacationing. The cable read: 'Can one sink open caissons to a depth of 350 ft?' Open caissons had never before been sunk to such a great depth, and no one knew whether it could be done. There was only one answer I could give. I cabled back one

word—YES. I still don't know whether it can be done, because it developed that we could stop the caissons at considerably shallower depths."

His one-word reply was not a wild guess, but the reply of the man who knew most about that subject and who was endowed with brilliant intuition. In his judgment it was feasible, and he had confidence in his own ability to conquer any difficulties. I consider this an example of calculated risk par excellence.

(2) *Panama Canal Slopes in Cucaracha Shale*.—My second example is the great unknown risk which the builders of the Panama Canal faced where the Canal cuts through the bentonitic Cucaracha shale formation. By continual removal of material from these cut slopes over a period of several decades, in an effort to stay ahead of major slides, the slopes have become extremely flat. Stability analyses which were made in connection with the 1946-47 Sea-level Canal Studies[2] showed that the effective shearing resistance of the shale had decreased from an average value of approximately 2 to 3 tons per sq ft in 1912 to as little as 0.4 to 0.5 tons per sq ft in 1947. This final value is approximately equivalent to a friction angle of 10° with zero cohesion, which was also obtained from laboratory direct shear tests on artificial slickensides made by polishing blocks of the shale.[3]

At the time of these investigations the conventional belief was that the shale would have retained a greater long-term strength if the slopes had been cut at the start with an appropriate factor of safety. However, I was inclined to the opinion[4] that this procedure would not have prevented a gradual reduction of the strength to the same very small residual value that we observe today; and that even if the designers had known that eventually the present extremely flat slopes would develop, the empirical procedure of flattening the slopes in stages was basically a satisfactory and economical approach.

(3) *Fort Peck Dam*.—Unknown risks also caused the failure of a portion of the Fort Peck dam in 1938. Using the hydraulic-fill method, the dam was constructed of river sands and finer-grained alluvial soils on a foundation of alluvial sands, gravels and clays with a total thickness up to 130 ft. Beneath this river alluvium is the Bearpaw clay-shale which contains layers of bentonite.

The failure occurred on September 22, 1938, when the dam was almost completed and the reservoir was partially filled. It affected a 1700 ft long section of the upstream shell next to the right (east) abutment where the underlying alluvium consists chiefly of sands. Comparison of the aerial views in Figs. 1 and 2, taken shortly after the slide, with the plan before and after the slide in Fig. 3 and the typical cross-section in Fig. 4, shows the significant and unusual features of the topography of the slide mass. The movement began by a bulging out of the western portion of the affected upstream slope with simultaneous subsidence of the core pool. Then a transverse crack developed at the western end which widened rapidly into a deep gap while the

---

[2] Binger, W. V., and Thompson, T. F., "Excavation Slopes," Panama Canal Symposium, Proceedings, ASCE, Vol. 74, No. 4, April, 1948, pp. 570-590.

[3] Binger, W. V., "Analytical Studies of Panama Canal Slides," Proceedings, 2nd Internatl. Conf. on Soil Mechanics and Foundation Engrg., Rotterdam, June, 1948, Vol. II, pp. 54-60.

[4] Casagrande, A., discussion of "Excavation Slopes," Panama Canal Symposium, by W. V. Binger and T. F. Thompson, Proceedings, ASCE, Vol. 74, No. 4, April, 1948, pp. 870-874.

FIG. 1.—OBLIQUE AERIAL VIEW OF SLIDE IN FORT PECK DAM

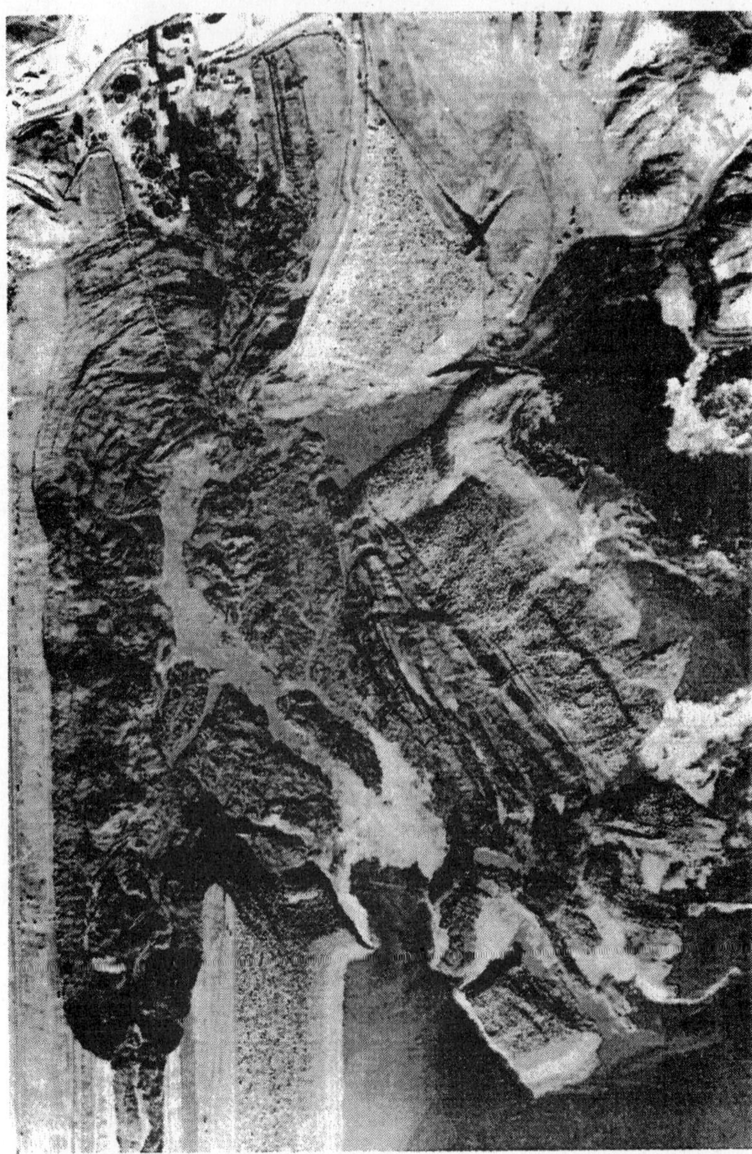

FIG. 2.—AERIAL VIEW OF SLIDE IN FORT PECK DAM

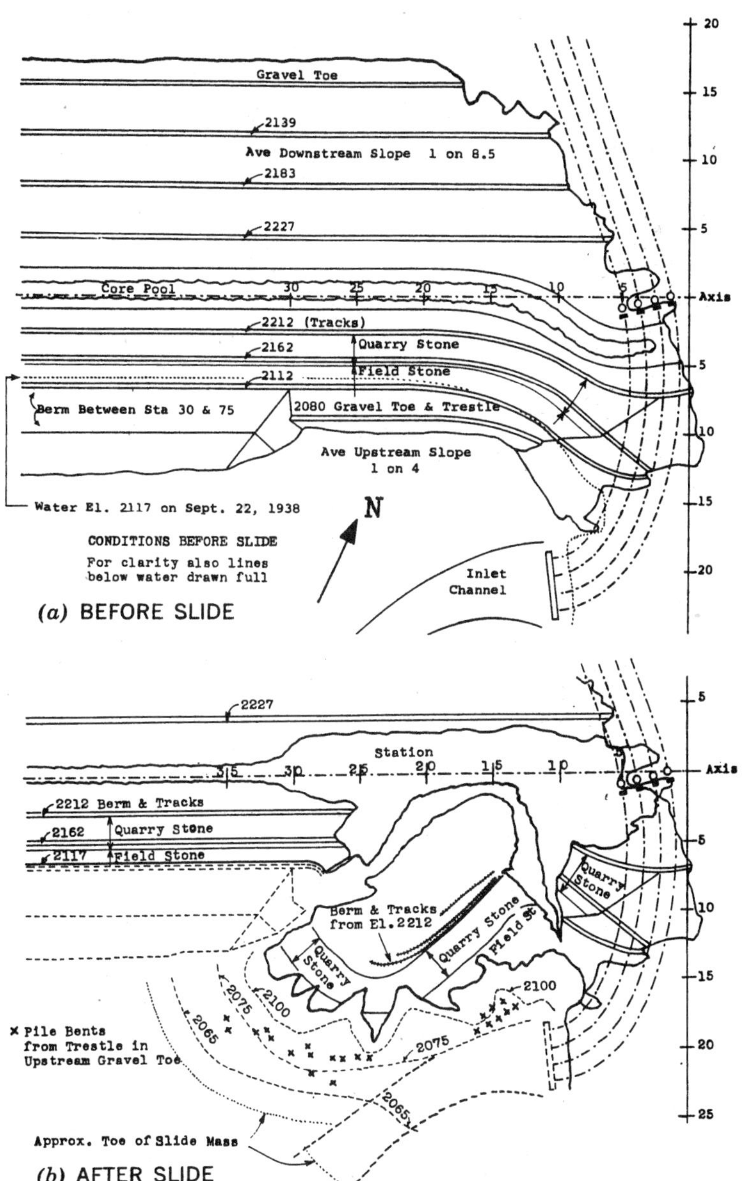

FIG. 3.—PLAN OF FORT PECK DAM SLIDE AREA

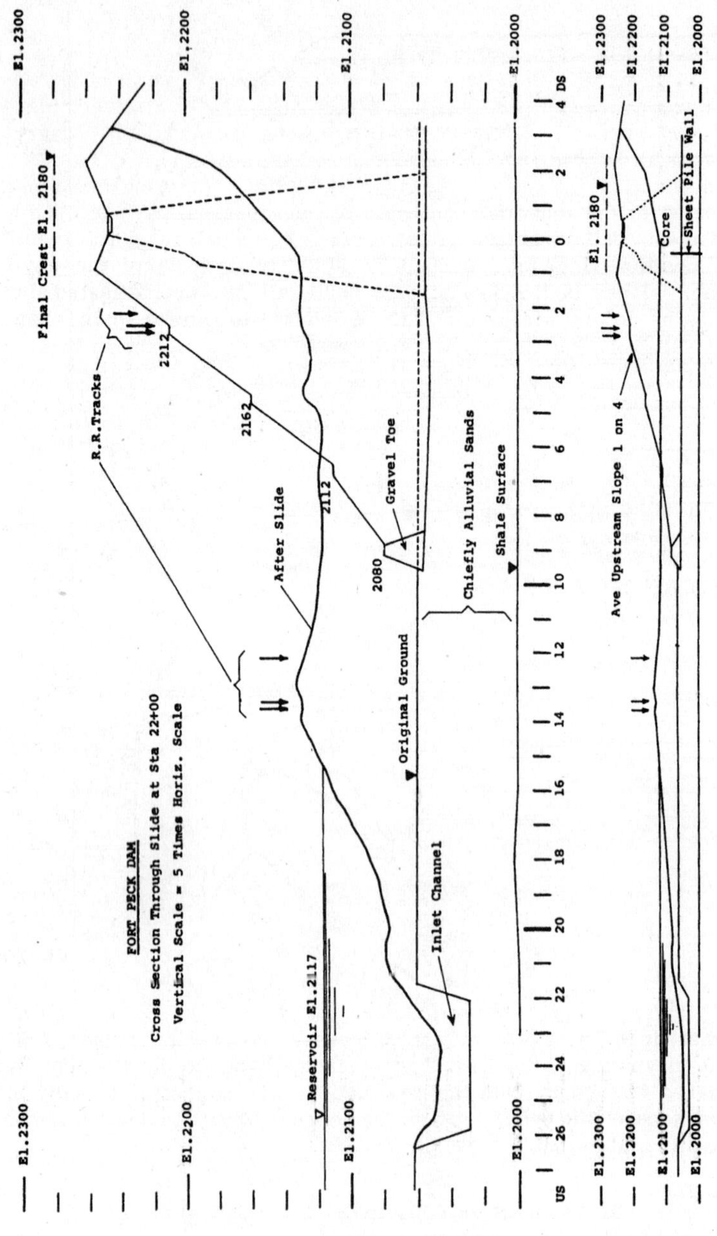

FIG. 4.—CROSS-SECTION OF SLIDE AT STATION 22+00

moving portion of the slope started to swing in a rotational movement as if hinged at the abutment. Through this gap the core pool drained with enormous speed. The western portion which was moving out faster and further, broke into several large blocks and came to rest in the fan-shaped pattern seen in the aerial photographs, Figs. 1 and 2. Next to the abutment, at the fulcrum of the rotational movement, a major wedge-shaped break developed.

By reference to the stations and ranges in Fig. 3 one can see how far some parts moved; e.g. the railroad tracks from the berm at El. 2212; the upper and lower boundaries of the quarry stone riprap; the trestle pile bents which were embedded in the gravel toe 900 ft upstream of the axis, many of which were moved to the locations indicated by the crosses. In very few minutes some of these well identified features moved as much as 1500 ft from their original position. Including secondary sliding, particularly the caving of steep back slopes in the downstream shell, all movements lasted about 10 minutes. The lower section in Fig. 4, plotted to natural scale, shows the almost level surface of the slide mass when it came to rest. The underwater slopes averaged 1 on 20, and the material which flowed into the inlet channel came to rest with an average slope of about 1 on 30.

The slide removed from the original section a volume of 5.2 million cu yd, and this agreed, within the accuracy of the surveys, with the volume of slide material deposited outside the original section. A rough estimate of the total volume of material that participated in the slide is 10 million cu yd.

As a member of the nine-man board appointed to investigate the slide, I inspected the site conditions about ten days after the event. It appeared to me that practically intact blocks of the upstream slope were floating like islands in a mass of thoroughly disturbed material. The level areas between and inside these islands were in a dangerously quick condition and dotted with sand volcanos of different sizes, some still discharging water and sand. Several other members of the board including Glennon Gilboy shared my first impression that only a liquefaction of major proportions could have produced such end results.

On the basis of the subsequent comprehensive investigations, which are summarized in the Corps of Engineers' report[5] on the slide, a majority of the Board concluded that the slide was due to shear failure of the shale foundation, and that "the extent to which the slide progressed upstream may have been due, in some degree, to a partial liquefaction of the material in the slide." This was a compromise wording to bridge the wide gap in the views of the consultants who signed the report. The views of those board members who did not believe that liquefaction was a major factor are discussed in detail in the paper by T. A. Middlebrooks[6] and in several discussions of his paper. In his discussion, G. Gilboy[7] emphasized the views of the minority who concluded that liquefaction was triggered by shear failure in the shale, and that the great magnitude of the failure was principally due to liquefaction. Gilboy and I shared the opinion that the liquefaction was centered principally in the fine sand zone of the shell next to the core, and that liquefaction may have

5 "Report on the Slide of a Portion of the Upstream Face of the Fort Peck Dam," Corps of Engrs., U. S. Dept. of the Army, U. S. Govt. Printing Office, Washington, D. C., July, 1939.

6 Middlebrooks, T. A., "Fort Peck Slide," Transactions, Vol. 107, 1942, pp. 723-764.

7 Ibid., pp. 725-755.

spread into the underlying heavily loaded foundation sands. Except for my verbal statements before the board meetings in 1938/39, and the written comments in Gilboy's discussion, the arguments in support of the liquefaction hypothesis have not been presented to the profession. As I see them, they may be summarized as follows:

(1) From the topography of the slide mass after the failure one can conclude that the shearing resistance in the foundation must have dropped to a very small value (a friction angle of very few degrees and zero cohesion), i.e., to a small fraction of the resistance that existed at the start of the slide. In contrast, experience with slides in clay-shales show that their resistance drops relatively little during the sliding movement, and that the movement occurs at a relatively slow rate. To obtain a quantitative estimate of the friction angle during the movement, Harold M. Westergaard was engaged as special consultant for an investigation of the dynamics of the slide. In his report to the Fort Peck District Engineer, dated January 28, 1939, he presented theoretical analyses and arrived at the following conclusions:

> "All of the computations that have been made point to the same conclusion applying to the partial failure of the Fort Peck Dam: The average coefficient of overall friction was approximately equal to the slope of the line connecting the centers of gravity before and after the motion."

The coordinates of these centers of gravity were as follows:

Center of Gravity of Volume Missing from Original Section		Center of Gravity of Volume Deposited Outside Original Section	
Sta.	17+34.4	Sta.	22+27.1
Range	1+11.5 U	Range	14+34.5 U
El.	2194.9	El.	2087.7

The slope of the line connecting these centers of gravity is 0.076, i.e., equivalent to an average friction angle of 4°20'. Because much larger friction angles must have been effective during the initial stages of the movement, and because there is no doubt that over some area near the abutment, where the displacements were relatively small, shear failure developed through the foundation materials without liquefaction, it is probable that within the liquefied mass the effective friction angle was even less than four degrees during the major portion of the movement.

(2) The rate and distance of movement were greatest at the west end if the slide area, Sta. 20 to Sta. 25, where the thickness of the foundation sand was the greatest. At the eastern end where there was little sand, the surface of the shale was higher and the factor of safety against sliding in the shale was much smaller; yet there the failure was induced as a secondary effect.

(3) If the strains produced by a local shear failure in the shale had not triggered any liquefaction, speed and distance of movement would have been much more limited and the slide mass would have remained much more coherent. Furthermore, arching within the overlying sands might have relieved the stresses in the weakened zone and arrested the movement.

(4) To start liquefaction of alluvial sand, the mass must be subjected to considerable strains. For a number of hours prior to the slide, some movements in the surface of the fill were detected, indicating that shear failure in the underlying shale, and therefore strains in the sand mass were already in progress. Then almost suddenly the rapid movements started.

(5) The slide mass in the outer reaches was stretched horizontally in the direction of movement as well as laterally. This is evidence that the movement was due to an underlying liquefied layer which was getting thinner as the movement progressed.

(6) The fact that many of the undisturbed samples obtained after the slide showed shear planes was interpreted as the strongest evidence against the hypothesis that liquefaction had occurred. However, it is likely that the liquefied layer had flowed out almost completely from under the large blocks of the upstream shell that had been moved bodily on that layer; probably most of the liquified material ended up in the submerged outer zones of the slide mass. With only a few undisturbed sample borings, it would have been very difficult to find any thin remnants of the liquefied layer. Furthermore, the existence of shear planes in a sample does not necessarily prove that this material had not liquified. As the liquefied layer decreased in thickness during the movement, drainage must have caused a decrease in pore pressures. During the last moment of movement the increasing effective stresses could have caused the material to "freeze in its tracks," so to speak, while it still had sufficient momentum to produce shear planes with small displacements.

(7) It was concluded from triaxial compression tests performed after the slide that the shell material could not have liquefied. However, based on present-day knowledge I consider it impossible to determine the susceptibility to liquefaction from the type of triaxial tests which were made at that time. Unfortunately even today we have no laboratory tests that can measure reliably the susceptibility of a sand to liquefaction.

Although the controversy between the board members concerning the mechanism of the failure was never resolved, the majority of the members were able to agree on the redesign of the dam which included a large upstream berm along the entire length of the dam. In addition, all material used for the repair of the slide area and in the new berm, including the hydraulically placed sand, was compacted.

The experience at Fort Peck has influenced the design of other large dams on the Missouri River which are also underlain by alluvial sands and clay-shales. For these later projects the unknown risks of the Fort Peck dam had become calculated risks of which we were well aware, but for which we were not yet able to develop quantitative design procedures. We resorted to two main types of defense against these risks: (1) the use of very flat average slopes, to keep the induced shear stresses in the foundation sands and in the clay-shale very small; and (2) the use of such materials and methods of compaction so that all possibility of liquefaction within the dam itself was eliminated.

In retrospect, we may now ask the question, what information is available today to the designer who has to cope with risks involving the strength of clay-shales and the stability of saturated loose sand deposits?

Concerning the strength of clay-shales, extensive empirical knowledge has accumulated during construction of the Oahe Dam in South Dakota, of the

South Saskatchewan River Dam in Canada, and of Waco Dam in Texas. In addition, the designer who today faces stability problems in highly overconsolidated clays and clay-shales should refer to the recent proposals by Borowicka[8] and Skempton[9] for determination of a lower limit for the shear strength by repetitive direct shear tests.

Concerning the phenomenon of liquefaction of loose, saturated sand deposits, it would be a rewarding project to reanalyze the great amount of data on the Fort Peck slide in the light of other empirical knowledge which has become available since that event. For example, extensive investigations on the liquefaction of sand along the Mississippi have been carried out and reported on by the Waterways Experiment Station. In such an effort experience gained from older hydraulic fill dams should not be overlooked. In particular I call attention to the remarkable paper by Allen Hazen on "Hydraulic-Fill Dams,"[10] in which he described the failure of a portion of the upstream shell of the Calaveras Dam in California, a failure which strikingly resembles the Fort Peck slide, even to the extent that witnesses used almost identical words to describe the initial swinging movement of the affected portion of the upstream shell. In both cases the reservoir was partially full so that the lower zone of the upstream shell was fully saturated. Allen Hazen also concluded that only in a liquefied condition could the shell material have come to rest with almost a level surface. He proposed an explanation for the mechanism of liquefaction which even in the light of present-day knowledge is remarkably clear and accurate. (To persuade the reader to acquaint himself with Hazen's explanations, which were well ahead of his time, they are appended to this paper.) He was also the first one to recommend compaction of the materials in the shells of hydraulic fill dams as a defense against the possibility of liquefaction.

(4) *Logan Airport.*—My next example is the calculated risk in the design and construction of Logan Airport in Boston. In 1943 Governor Saltonstall of Massachusetts was faced with an important decision—whether to enlarge the existing very small airport in the Boston Harbor, which was close to the center of the city, or to establish the airport about 15 miles inland. He strongly favored the harbor location. Because of my familiarity with the properties of the soft clay underlying the greater Boston area, and because I was then particularly active in military airfield design problems, Governor Saltonstall asked me to review the feasibility of building the airport in the Boston harbor on a hydraulic fill dredged from the same clay. He had been advised that this would not be possible because such a hydraulic clay fill would not be strong enough. He wanted my answer in a hurry because New York had already started construction of Idlewild, and he was concerned that Boston would lose its place in the competition for the future overseas air traffic.

From personal observation of hydraulic dredging of clay for construction of levees in the Atchafalaya Basin, I knew that clay discharges from the pipe in the form of well-rounded clay pebbles in a matrix of slippery mud. How

---

[8] Borowicka, H., "Vienna Method of Shear Testing, Laboratory Shear Testing of Soils," Technical Publication No. 361, ASTM, Philadelphia, Pa., 1963, pp. 306-314.

[9] Skempton, A. W., "Long-Term Stability of Clay Slopes," Géotechnique, Institution of Civ. Engrs., London, England, June, 1964, pp. 77-101.

[10] Hazen, Allen, "Hydraulic Fill Dams," Transactions, ASCE, Vol. 83, 1920, pp. 1713-1745.

quickly would such a mass consolidate to achieve the required strength? There was no time for large-scale tests, which would probably have been the only reliable method for answering this important question. I estimated that the voids between the clay pebbles would render the entire mass sufficiently pervious so that it would consolidate within a reasonable time to the strength of the clay pebbles themselves, i.e. the strength of a medium-soft clay. On the basis of this judgment the decision was made to go ahead with the Boston harbor site.

The consensus of several experienced dredging contractors, who were consulted during the design stage, was that the hydraulic clay fill would develop side slopes of about 1 on 20. Also, they were of the opinion that it would not

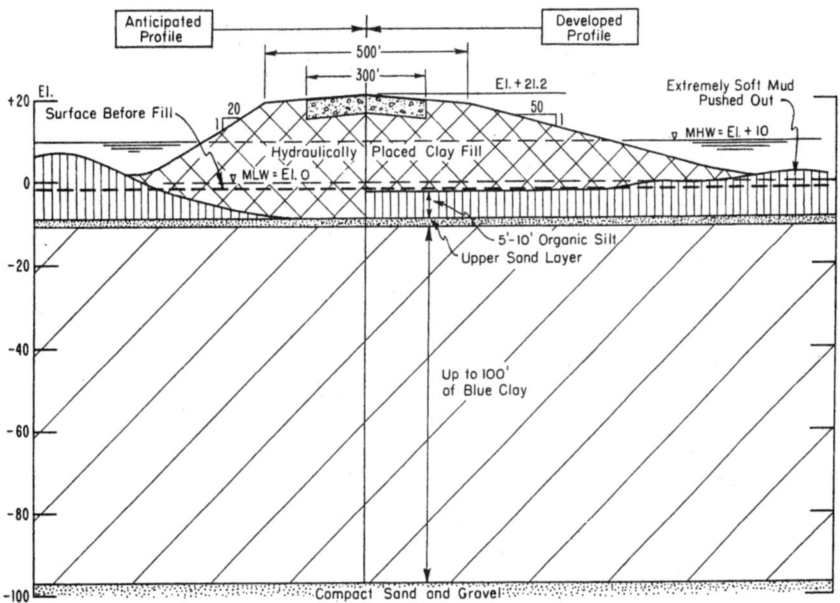

FIG. 5.—COMPARISON OF PROPOSED AND DEVELOPED PROFILE THROUGH HYDRAULIC FILL FOR RUNWAYS AT LOGAN AIRPORT

be necessary to remove the 5 to 10 ft thick layer of very soft organic silt overlying the clay, because it would be squeezed out by the weight of the clay fill. Therefore, the engineers decided not to remove the organic silt. However, the hydraulic clay fill actually assumed average side slopes of 1 on 50, and the organic silt was not displaced by the clay fill. The mixture of clay pebbles and mud had such a small shearing resistance that it simply spread out over the silt almost like a liquid and caused hardly any shear failures within the silt layer. On the left side in Fig. 5 are shown our good intentions and on the right side the actual results of this filling operation.

Much more serious were the implications of the results of the first load tests which we performed as soon as a sufficient fill area became available.

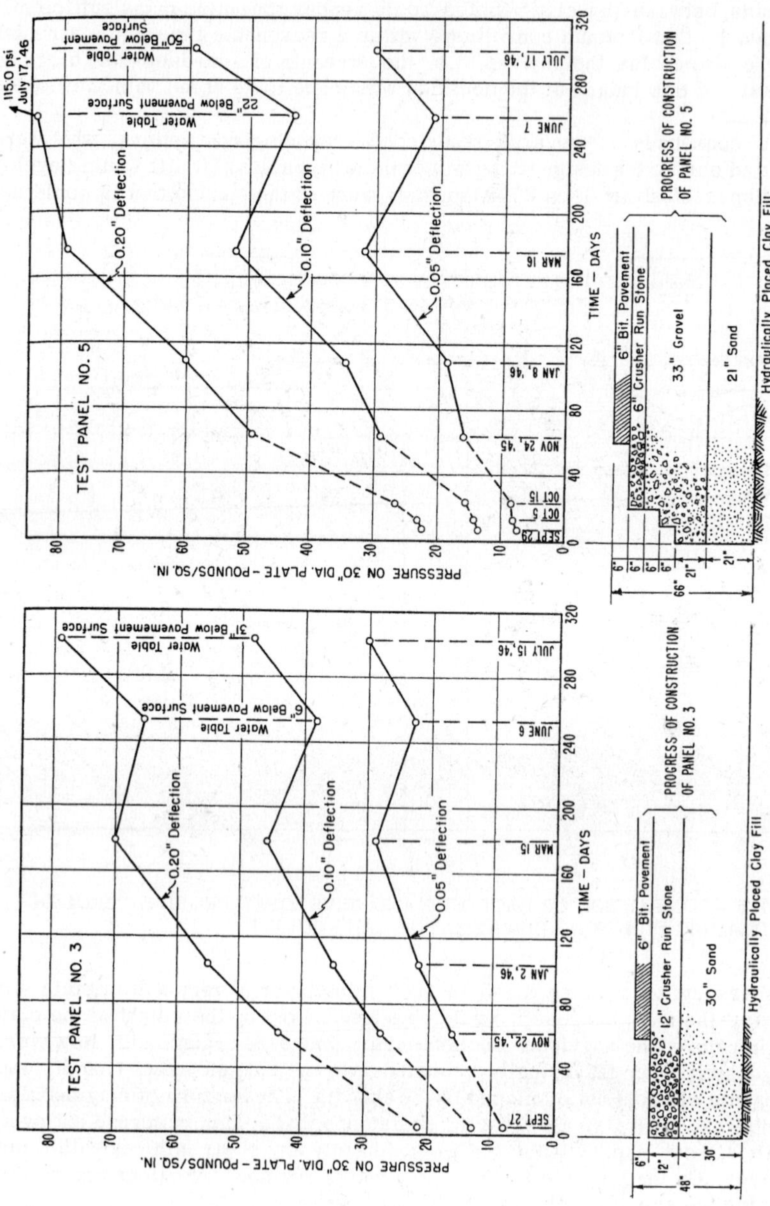

FIG. 6.—INCREASE OF BEARING CAPACITY WITH TIME

A number of test sections were constructed with different combinations of base courses and pavement, and they were subjected to 30 in. diameter plate load tests after several elapsed times. A deflection of 0.2 in. was considered the maximum permissible value, in accordance with the results of extensive investigations by the Corps of Engineers on the behavior of clay subgrades beneath airfield pavements. The initial allowable bearing capacities obtained from these load tests were discouragingly small, as shown by the first load test results in Fig. 6. Fortunately, the capacity kept increasing over a much longer period than I had anticipated. (Note: The reduction at an elapsed time between 180 and 250 days was due to saturation of the base course from rains. As soon as proper drainage was provided the strength immediately started to rise again.) By extrapolation we concluded that for a 5 ft combined thickness of base course and asphalt pavement, the pavement could carry a 65,000 lb single wheel load within one year after completion of the pavement, and a 125,000 lb single wheel load within five years after completion.

Today no one denies that construction of a hydraulic clay fill was the right solution for this project. But this decision had to be based on a major calculated risk concerning the rate of strength increase in the hydraulic clay fill after completion of the pavements. If I had to do this job again, I would insist on only one change—removal of the organic silt by dredging before placing the hydraulic clay fill. This would have substantially reduced the total and differential settlements after completion.[11,12]

(5) *Great Salt Lake Railroad Fill.*—An outstanding example of a calculated risk is the design and construction of Southern Pacific's railroad embankment across the Great Salt Lake to replace the 12 mile long timber trestle which was built at the beginning of this century, Fig. 7. The trestle was constructed because the original efforts to build an embankment across the lake had failed. Construction had proceeded only a few miles from each end when the fill simply kept disappearing in the soft clay. This undertaking was an enormous unknown risk that defeated the railway engineers. Undaunted, they started driving piles for a timber trestle which they completed in about one year—in itself a remarkable achievement.

The new fill, which was built in the years 1956 to 1959, starts from the existing old fill sections in the form of flat S curves and then runs parallel to the trestle 1500 ft to the north, to ensure that construction of the fill would not endanger the trestle. (This margin of safety proved to be ample for all but one human risk. One dark night a tug captain fell asleep, and his tug pushed a loaded 2000 cu yd barge straight through the trestle, slicing off the husky piles as if they were matchsticks. When last seen, the captain was heading across country without having bothered to pick up his paycheck.)

The International Engineering Company carried out the design of the embankment and Morrison-Knudsen was the contractor. They were assisted by a consulting board consisting of F. H. Kellogg, R. R. Philippe, S. D. Wilson and myself. For all practical purposes E. E. Mayo and W. M. Jaekle of the Southern Pacific Company, and L. D. Wilbur of the International Engineering

---

11 Casagrande, A., "Soil Mechanics in the Design and Construction of the Logan Airport," Journal, Boston Soc. of Civ. Engrs., Boston, Mass., April, 1949, pp. 192-221.

12 Gould, J. P., "Analysis of Pore Pressure and Settlement Observations at Logan International Airport," Harvard Soil Mechanics Series No. 34, Harvard Univ., Cambridge, Mass., December, 1949.

Company, were acting as members of the board by virtue of their close and very effective cooperation with the board.

The board started its investigation in 1955 and quickly realized that a design based on laboratory strength tests on undisturbed samples of the soft and sensitive clay, which underlies the lake to a great depth, would involve

FIG. 7.—AERIAL VIEW OF GREAT SALT LAKE RAILROAD FILL, LOOKING WEST

great uncertainties. Therefore, the board first recommended construction of full-scale test sections. However, we soon learned that they could be built only by mobilizing most of the very expensive equipment needed for construction of the entire embankment. Therefore, for the initial design we had to

FIG. 8.—SWEDISH FOIL SAMPLER OPERATING IN GREAT SALT LAKE FROM A PORTABLE TRIPOD

FIG. 9.—UNDISTURBED SPECIMENS OF SOFT GREAT SALT LAKE CLAY

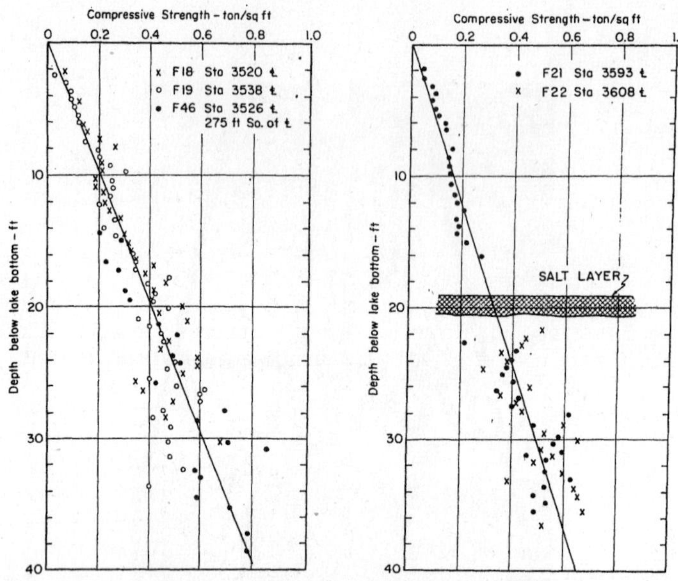

FIG. 10.—TYPICAL STRENGTH VERSUS DEPTH PROFILES FOR NORMALLY CONSOLIDATED GREAT SALT LAKE CLAY

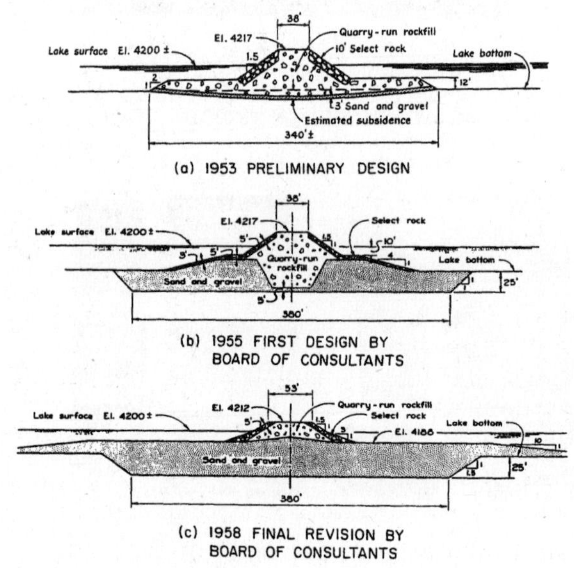

FIG. 11.—EVOLUTION OF GREAT SALT LAKE EMBANKMENT SECTION ON SOFT CLAY

rely chiefly on conventional stability analyses based on the strength of undisturbed specimens of the clay. We were encouraged by the excellent quality of the undisturbed samples obtained from numerous borings by means of a Swedish foil sampler, which is seen in Fig. 8 operating from a tripod that could easily be moved about.

Fig. 9 shows several undisturbed clay specimens. On the right is shown a section through an unconfined compression test specimen after failure, with the shear plane faintly visible. The failure strains were generally very small, as low as 1% or 2%. The liquid limit ranged from 50 to 150, with the points on the plasticity chart lined up along the A-line. Oven-drying caused the plasticity index to drop by about one-half. The natural water content ranged close to the liquid limit. The appearance varied from fairly homogeneous to intensely color-stratified. However, large differences in color of adjoining layers usually were not reflected by any significant variations in water content or limits.

Typical results of unconfined compression tests as a function of depth, for the deeper area of the lake, are plotted in Fig. 10 for two groups of borings about 1.5 miles apart. While the results are consistent within each group, the results plotted on the right, for the group of borings closer to the middle of the lake, show a smaller rate of strength increase with depth. These plots show that the clay is normally consolidated because the strength is practically zero at lake bottom and increases linearly with depth. In contrast, the clay in the shallower areas of the lake and near the shore was found to be overconsolidated as a result of drying at a time when the lake levels were lower.

The salt layer which underlies the fill for many miles, with its upper surface at a depth of 20 to 30 ft below lake bottom, was found to vary greatly in thickness and it seriously complicated the design and construction of the fill.

Several design stages for the cross-section on soft clay (no salt layer) are shown in Fig. 11. Although the 1955 design by the Board had a reasonable factor of safety based on the laboratory strength of the undisturbed samples and conventional stability analyses, the full-scale test sections showed it to be far too weak. We had anticipated the possibility that stabilizing berms might be required, but none so massive as those shown in Fig. 11(c) were envisioned. Even our most pessimistic assumptions did not prepare us for the very low in situ strength of the soft clay which we derived from analysis of the failures of test sections and of other sections of the fill. It also developed that the rock core which was incorporated in the 1955 design, Fig. 11(b), for the purpose of reducing the lateral earth pressure within the fill, did not function as intended. Therefore, the large volume of high-cost rockfill was soon replaced, as shown in Fig. 11(c), by low-cost sand and gravel fill. The evolution of this section and the construction procedure have been explained elsewhere.[13] During construction many variations from the sections shown in Fig. 11(b) and (c) were developed, depending on the progress of dredging and underwater filling that was already accomplished by the time the final design decisions were made. A variety of other sections were designed for areas underlain by stiffer clays or by various thicknesses of the salt layer.

To achieve an economical design it was necessary to build full-scale test fills and to produce failures. Particularly impressive were the failures of

13 Casagrande, A., "An Unsolved Problem of Embankment Stability on Soft Ground," Proceedings, 1st Panamerican Conf. on Soil Mechanics and Foundation Engrg., Mexico City, Mexico, September, 1959, Vol. II, pp. 721-746.

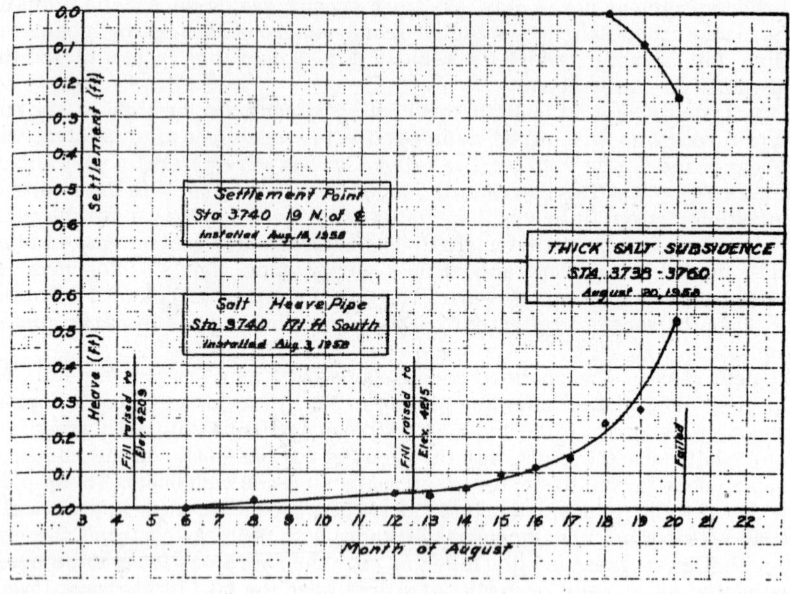

FIG. 12.—HEAVE PIPE OBSERVATIONS PRECEDING THE "MIDLAKE FAILURE"

FIG. 13.—AERIAL VIEW OF "MIDLAKE FAILURE"

FIG. 14.—AERIAL VIEW OF RECONSTRUCTION OF "MIDLAKE FAILURE"

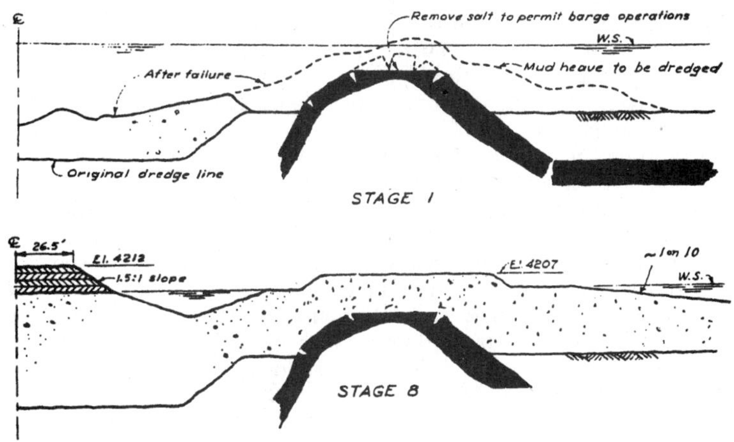

FIG. 15.—CROSS-SECTION THROUGH "MIDLAKE FAILURE" ILLUSTRATING METHOD OF RECONSTRUCTION

fill founded on the salt layer. For practical purposes the salt had to carry the entire lateral thrust of the fill; and when the salt buckled, the fill would sink into the soft clay with extraordinary speed. Here I cannot resist relating an anecdote as it was told to me. Upon completion of one of the 500 ft long test fills on salt, owners' and contractors' engineers standing on it were so impressed by its solid feel that they concluded: This proves that Casagrande is too conservative. The next day this entire test fill disappeared so fast that a tug captain swore he could look down into the hole because the lake could not flow together fast enough. We soon learned that buckling failure of the salt layer was preceded by a slow heaving adjacent to the fill which could be observed easily by means of "heave observation pipes." It became standard procedure to install such pipes where indicated, and to observe them daily. Usually the addition of a small berm would be sufficient to stop any tendency to heave.

There was one section of completed fill in the middle of the lake underlain by a 10 ft thick layer of salt which everyone considered reasonably safe, but which nevertheless was instrumented with heave pipes. Unfortunately, for a short period the results of the heave observations along that section remained in the notebook of an overworked surveyor, instead of being plotted and examined promptly by others. On August 20, 1958, while the surveyor was just completing another heave measurement, he suddenly observed the fill subsiding. A 2000 ft section disappeared in a few minutes in an almost symmetrical failure, referred to as "the midlake failure." The measurements plotted in Fig. 12 speak for themselves. The salt had buckled on both sides at a distance of 180 to 200 ft from the center line, as seen in the aerial view in Fig. 13. Reconstruction followed a procedure developed in connection with the repair of several much smaller subsidences. For example, in the foreground of Fig. 7 one can see on the right of the fill a balancing berm that was built for the purpose of stopping a subsidence in its early stages. Reconstruction of the midlake failure is shown in progress in Fig. 14, and the procedure is explained in Fig. 15. After removal of the salt and clay heaves by dredging to permit barge operations, the two underwater berms were constructed with extremely flat outside slopes. Then truck dumping was used to raise the two berms and the central fill above water, at first maintaining almost exact balance, and later very slowly increasing the differential weight between the central and the berm fills.

Principally on an empirical basis which was developed by extensive full-scale testing and observations, and necessitating many important design changes during construction, the project was completed successfully one year ahead of schedule, and in my judgment with almost optimum economy. Paradoxically, such economy would not have been achieved had we known at the start the strength of the clay that controlled the stability of the embankment during construction. It is even probable that this project would not have been authorized, for such knowledge, combined with the application of a conventional factor of safety, would have forced us to design a fill for which the cost estimate would have been far in excess of our 1955 estimate shown in Table 1. Thus the $50 million limit which Southern Pacific's Board of Directors had established as the maximum expenditure that would be economically justified for this project would have been greatly exceeded; and instead of building the embankment crossing, the deck of the timber trestle would probably have been renewed in accordance with an alternate plan.

I do not wish to leave the impression that the calculated risk in the design of this project also involved risks concerning the long-term safety of this railroad crossing. A fill built on normally consolidated clay has its lowest factor of safety against foundation failure during construction or immediately after its completion. Subsequently, progress of consolidation steadily increases the safety against failure. Therefore, a low initial factor of safety is justified for such a project. (An engineer who had extensive experience in building long highway embankments on soft clay, once remarked that if no failure occurs it proves to him that the embankment was overdesigned.)

Since the summer of 1959, when this new railroad crossing was put in operation, we relied chiefly on settlement observations as a check on the performance of the foundations. The rate of settlements is gradually decreasing in a consistent pattern, reflecting a steadily increasing strength of the clay.

This project is a good example of what Terzaghi liked to call the "observational approach," i.e. the continuous evaluation of observations and new information for redesigning as needed while construction is in progress. It also

TABLE 1.—COST ESTIMATES FOR GREAT SALT LAKE RAILROAD FILL

Date	Job Status	Volume of Fill, in millions of cubic yards	Cost, in millions of dollars
1953	I.E.Co's Preliminary Design	21	$30
1955	Board of Consultants' First Design	32	$49
1956	Start of Construction		
1959	Completion (one year ahead of schedule)	44	$50.5

illustrates another statement that Terzaghi made many times: "In applied soil mechanics a design is not completed until the construction is successfully completed."

In common with the Fort Peck Dam, our experience with the Great Salt Lake Fill has greatly increased our ability to cope with certain problems in applied soil mechanics on an empirical basis. However, in spite of the great effort that went into soil testing, field observations and stability analyses before and throughout the construction of the Great Salt Lake Fill, we have not succeeded in making reliable determinations of the following important quantities that govern the stability of this embankment:

(1) the long-term in situ strength of the clay at unchanged water content;

(2) the pore pressures induced by the fill load in the clay outside a zone where the fill load produces significant normal stresses. (Note: Although piezometers commonly provide reliable measurements of pore pressures, most of the piezometers installed on this project became inoperative soon after installation because of salt deposition. Some of those that worked temporarily

showed that remarkably high pore pressures were transmitted through the salt layer for many hundreds of feet from the fill.); and

(3) the magnitude of the earth pressure within the fill.

This account of the calculated risks involved in this project would not be complete without mentioning two more calculated risks. The "sand and gravel" used for the main body of the underwater fill was largely a silty sand. The question of its stability under dynamic stresses was of serious concern to me.

FIG. 16.—AERIAL VIEW OF BALDWIN HILLS RESERVOIR ON THE DAY OF FAILURE (LOOKING NORTH)

Blasting tests in steep underwater slopes of this material proved that the good gradation and angularity of the grains (chiefly quartz) rendered the mass remarkably stable to shock. This conclusion was substantiated later when on August 30, 1962, the fill was exposed to a severe earthquake which caused only slight local movements. The other risk concerns the relationship between crest elevation and lake level. The lake level fluctuated between a maximum of about El. 4216 almost 100 years ago and a recent minimum of about El. 4196. The design was based on an assumed lake El. 4200 which is approximately the average for the preceding 30 years. There is a long-term trend toward lower lake levels. For this reason, and also to keep the stresses in the clay as small as possible, the relatively low crest El. 4212 was elected for the final design.

Should you some day travel on or fly over this Great Salt Lake Fill, remember that what you see of the entire long railroad embankment is only a very small portion of the total volume of this fill, even a smaller ratio than for the proverbial iceberg; and that this ratio is a monument to a calculated risk which almost defeated those responsible for the design. And when you observe in the middle of the lake three parallel embankments almost a half-mile long, remember that this section of the embankment stands as a monument not only to a calculated risk but also to a human risk.

No other project has caused me so many sleepless nights as this one.

(6) Baldwin Hills Reservoir.—My next example is the Baldwin Hills Reservoir in Los Angeles which failed in December 1963. Essential data, including Figs. 16, 17 and 21, were obtained from the published report prepared by the

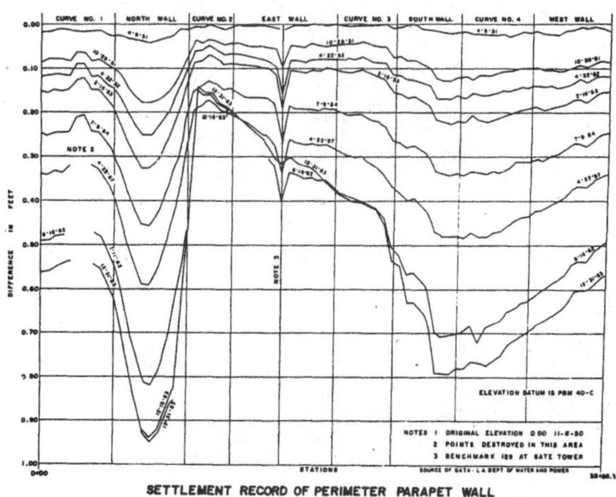

FIG. 17.—SETTLEMENT RECORD OF PERIMETER PARAPET WALL, BALDWIN HILLS RESERVOIR

State of California Engineering Board of Inquiry.[14] Fig. 16 is an aerial view while the water was still pouring through the crevasse that had developed in the natural ground forming the right abutment of the earth dam. The reservoir had been carved from the sides of a valley and closed off by an earth dam with a maximum height of about 130 ft. An open and active geologic fault crosses the reservoir in north-south direction in a direct line with the crevasse.

In Fig. 17 are plotted the settlements along the entire crest of the reservoir. The progress of the settlements with time was quite regular. During the 12.5 year life of the reservoir, the maximum settlement of almost one foot developed under the highest portion of the dam, which in this plot is designated "North Wall." The crevasse developed in the right abutment portion, designated

14 "Investigation of Failure of Baldwin Hills Reservoir," California Dept. of Water Resources, Sacramento, Calif., April, 1964.

FIG. 18.—RIGHT (EAST) SIDE OF CREVASSE, BALDWIN HILLS RESERVOIR

FIG. 19.—LEFT (WEST) SIDE OF CREVASSE, BALDWIN HILLS RESERVOIR

FIG. 20.—LARGEST SINK HOLE ALONG FAULT MERGED WITH CREVASSE

FIG. 21.—ILLUSTRATION OF ERODIBILITY OF INGLEWOOD FORMATION IN WHICH THE BALDWIN RESERVOIR WAS CONSTRUCTED (GULLY IS ABOUT 40 FT DEEP)

"Curve No. 2" in Fig. 17, which had settled a total of about 0.2 ft. The sharp increase in settlement between this abutment and the dam was chiefly due to the steep original ground slope. Thus, the stage was set for tension forces between the abutment and the dam, a tendency that was aggravated by the fact that this junction of dam and abutment was located in a curve which arched outward, such that the water load also tended to produce tension forces.

The entire reservoir is underlain by very weakly cemented silts, sands and gravels. The abutment in which the crevasse developed, Figs. 18 and 19, consisted chiefly of loose, fine sands and silts that are very easily eroded. Some of the fine sand layers had so little cohesion that one could dislodge grains by blowing at them. To prevent seepage from getting into these dangerous foundation and abutment soils, the reservoir was lined with a 10 ft thick clay layer which tapered to 5 ft at the top of the slope. Beneath the clay was a 4 in. drainage layer consisting of cemented pea gravel, i.e. a porous concrete to drain the seepage water. Just below the pea gravel a system of 4 in. tile drains was installed to collect any seepage passing through the clay lining. Beneath the pea gravel was a 1/4 in. thick asphaltic layer intended to prevent seepage from reaching the foundation.

After the failure a number of sink holes were found in the reseroir bottom along the fault line. The largest one, Fig. 20, merged with the crevasse. Along the fault the offset on top of the clay lining averaged 2 to 3 in. In the course of detailed exploration beneath the sink holes, several large cavities were found in the natural foundation materials. Obviously underground erosion had been in progress for a long time.

From the State Report[14] I reproduce Fig. 21, which illustrates how easily the natural materials that underlie the reservoir can be eroded. From the State Report[14] I quote also the last paragraph of the conclusions which summarizes very well the salient aspects:

> "Sitting on the flank of the sensitive Newport-Inglewood fault system with its associated tectonic restlessness, at the rim of a rapidly depressing subsidence basin, on a foundation adversely influenced by water, this reservoir was called upon to do more than it was able to do."

In addition to the detailed investigations by the State Board, an investigation was carried out by the Board of Inquiry appointed by the Mayor of Los Angeles. The report of this board contained the following statements:

> "The Baldwin Hills Reservoir was built across a series of minor geological faults . . . accurately located during construction of the reservoir. These minor faults were reported as active by the Department of Water & Power geologist who prepared the final geological report on the reservoir site.
>
> "A Board of Consultants consisting of a geologist and two engineers which was appointed by the Department of Water & Power in connection with the design and construction of the reservoir was aware of this system of minor faults in the reservoir area, and reported in 1948 that it was 'very unlikely that any appreciable movement will again occur along these auxiliary faults.' That Board of Consultants also in 1948 approved the 'feasibility of the proposed Baldwin Hills Project, the suitability of the site, and the plans of design.'

"Movements took place along the faults below the reservoir, probably beginning shortly after completion. These movements increased at an accelerated rate after December of 1957.

"Movements on the faults fractured the drainage system underneath the reservoir and ruptured the asphalt membrane which was intended to prevent leakage of reservoir water into the underlying soil and rock. These fractures probably began in 1951 at several places along the faults.

"The breaks in the asphalt membrane permitted drainage water from the reservoir in the pea gravel to flow into the underlying natural soil or rock along the local faults with the result that piping (channeling) took place in the soil or rock at various locations."

The designers of this project and their consultants were well aware that the City of Los Angeles must have within its vast limits strategically located reservoirs to ensure a dependable water supply and protection in case of large fires. The Baldwin Hills are the only elevated area in southwestern Los Angeles which appeared suitable for a distribution reservoir. Were these persuasive arguments responsible for taking a grave risk, or did the designers underestimate the risk?

(7) Foundations for Brookhaven Synchrotron.—My last example is the foundation design for the 33 billion electron volt synchrotron at Brookhaven National Laboratory on Long Island, New York. The principal element is an 850 ft diameter concrete conduit, called the tunnel, which is covered with a substantial mound of earth. In this tunnel there is a stainless steel tube in which protons are accelerated by means of hundreds of magnets weighing about 20 tons each. Stone & Webster, who were engaged for the design, and I as consultant, were presented by the physicists with the requirement that relative to each other these magnets must stay aligned with an accuracy of plus or minus 0.005 in.—an unheard of accuracy for foundations.

Fortunately we had good foundation soils, the Long Island sand which extends to great depth. However, the original topography of the area was far from level; there were groundwater fluctuations to consider; there was the weight of the mound of earth over the tunnel; there was the target building with very heavy but movable shielding loads; etc. It seemed almost an impossible task. Bill Swiger of Stone & Webster and I soon realized that even if the magnets are supported on piles, secondary compression in the sand may cause excessive differential movements. Large-area load tests were made to obtain as much information on long-term time-effects in sand as was possible in the available time. With strict temperature control in the tunnel and other refinements, we finally arrived at what we believed was a satisfactory solution. Then came the day when we were invited to answer questions. The auditorium at Brookhaven was filled with physicists, and there I was trying to explain how we foundation engineers extraploate from crude tests and using simple computations the probable movements of their vital magnets. I could see in the eyes of Dr. G.K. Green, Brookhaven's Associate Director who led this inquisition, and also from the expressions on the faces of many in the audience, what they were thinking. I vividly remember the scene. After I had finished my explanations, Dr. Green said: "Tell me frankly, you haven't just pulled your figures out of the air?" He accompanied his question with a sweeping motion of his arm, demonstrating how I might have caught my numbers skillfully in the air as they were flying by. I replied: "No, Dr. Green, we foundation engineers do not pull figures out of the air; we pull them out of the ground." And stooping down I demonstrated with a similar sweeping motion how we do it. After the laughter had died down, I had the nerve to ask him whether he could explain just how he arrived at the five mils limitation. He hesitated a bit and said: "Well, my colleagues and I made our computations, and we arrived at the conclusion that we could stand 20 mils. But we did not trust our computations, so we cut it down to five mils to be on the safe side." I thanked Dr. Green and his colleagues for showing so much confidence in the ability of foundation engineers to control settlements to such accuracy, when they had so little confidence in their own computations.

In several years of operation, except for temporary new construction in the area and for changes in the arrangement of shielding loads in the target area, the relative movements between magnets have not exceeded 10 mils. Furthermore, once when some construction was in progress near the tunnel, a settlement of about 100 mils occurred locally; but the machine continued to operate for one week without causing the operators to notice or object.

Of course we could not be certain about our prediction. But we knew that at worst the magnets would have had to be realigned more frequently than the owners desired; (and the shaky confidence of Brookhaven's physicists in soil mechanics would have suffered a mortal blow).

DISCUSSION

Throughout the history of engineering, progress required bold advances beyond the limits of knowledge, often at great risks. What should the engineer's attitude be toward such risks? In part, I should like to answer this question in Terzaghi's words, by quoting from a letter he wrote to André Coyne immediately after the failure of the Malpasset Dam:

"When I read in the papers about the failure of the Malpasset Dam, my thoughts turned immediately to you and to the terrible shock you must have experienced when the sad news reached you. In situations of this kind it is at the outset impossible to divorce the technical aspects of the event from the human tragedies involved. Yet every fair-minded engineer will remember that failures of this kind are, unfortunately, essential and inevitable links in the chain of progress in the realm of engineering, because there are no other means for detecting the limits to the validity of our concepts and procedures. I have witnessed the shocking manifestations of this painful process during the First World War in the field of aviation, when we tried to proceed in a few years from the primitive types of airplanes to larger and more elaborate ones, and in the field of dam construction the price of our lessons is equally high.

"Having known you well for many years, I feel confident that the failure was not a consequence of an error in your design. Therefore, it will serve the vital purpose of disclosing a factor which in the past has not received the attention which it requires. The fact that its implications became manifest on one of your jobs is not your fault, because the occurrence of failures at the borderline of our knowledge is governed by the laws of statistics, and these laws hit at random. None of us is immune. You as an individual, and the equally innocent victims of the failure have paid one of the many fees which nature has stipulated for the advancement in the realm of dam construction. Therefore, the torments which you experienced should at least be tempered by the knowledge that the sympathies of your colleagues in the engineering profession will be coupled with their gratitude for the benefits which they have derived from your bold poineering."

When referring to the pioneer days of military aviation (he was in charge of an aviation research establishment during the first World War), Terzaghi might well have added that the days of risk in designing airplanes have by no means passed. Even today, after a new design is developed and the airplane is carefully constructed of materials manufactured to precise specifications, it must be subjected to many months of exhaustive testing before it is finally certified for service. And even then one can not be certain that all weaknesses have been discovered.

By comparison, in foundation and earthwork engineering we often deal with very erratic, natural materials whose properties are in part far from clearly understood. Then we build upon such materials and with such materials, huge dams which hold back more potential destruction than man could create by any other peacetime activity. Therefore it seems hardly necessary to stress the great importance of including in the education of those who specialize in soil mechanics, a realistic approach to the risks involved in earthwork and foundation engineering.

Classification of Risks and Potential Losses.—To facilitate discussion, I will use the following rough classification of risks and potential losses in earthwork and foundation engineering:

Classification of Risks

A. Engineering Risks
   1. Unknown risks.
   2. Calculated risks.

B. Human Risks

   Most human risks, both unknown and calculated, fall into the following:

   1. Unsatisfactory organization, including division of responsibility between design and supervision of construction.
   2. Unsatisfactory use of available knowledge and judgment.
   3. Corruption.

Often there is no sharp line of demarkation between these three groups of human risks. In particular, division of responsibility is frequently the cause of insufficient use of available knowledge and judgment, and it can also facilitate corrupt practices.

Classification of Potential Losses

I. Catastrophic loss of lives and property.
II. Heavy loss of lives and property.
III. Serious financial loss; probably no loss of lives.
IV. Tolerable financial loss; no loss of lives.

Engineering Risks.

Unknown Engineering Risks.—By definition such risks cannot be identified until they reveal themselves by a failure or other event that can be observed and investigated. I hope that I am not too optimistic when I state my belief that soil mechanics has advanced far enough to permit at least a qualitative estimate of the response of all soils and rocks on our globe when subjected to our conventional engineering activities; in other words, that we are not likely to encounter any major unknown engineering risks. However, there may well be some unknown risks in store for us in connection with the application of nuclear energy to large-scale excavations, e.g. in materials such as the Cucaracha formation in the Panama Canal Zone.

Calculated Engineering Risks.—I list below some of the more important calculated risks in applied soil mechanics for which we still depend largely on crude empirical knowledge and judgment because quantitative analyses are either non-existent or of very doubtful validity:

1. Liquefaction slides in granular soils.
2. Liquefaction slides in extremely sensitive clays.
3. Stress-deformation and strength characteristics of coarse-granular materials, including rockfills, under very high confining pressures.
4. Long-term stress-deformation and strength properties of clays at constant water content.
5. Stability characteristics of highly plastic, stiff clays and clay-shales.
6. Control of transverse and longitudinal cracks in the core of high rockfill dams.
7. Effects of earthquakes on high earth and rockfill dams.

In passing, I mention that significant progress in our ability to cope with these problems will depend largely on expensive investigations carried out in connection with major projects.

The margin of safety that we incorporate into our structures should bear a direct relationship to the magnitude of potential losses, and it must also take into account the range of uncertainty involved. If this range is very small, then we are approaching problems in structural engineering which are resolved by the use of a conventional factor of safety. When the range is large, we cannot usually express the results in terms of a numerical factor of safety; we then speak of a margin of safety which is based largely on experience and judgment.

Projects in Category I, i.e. those involving potentially catastrophic losses, are almost always undertaken by owners with ample financial resources who are fully aware of their great responsibility. Therefore, the best knowledge and judgment are mobilized to ensure the best possible design and construction. On such projects the designer will resort to a liberal margin of safety and/or to independent lines of defense whenever he deals with aspects that he cannot evaluate with any degree of certainty. For example, in the case of very high dams the designer will resort to particularly conservative foundation and abutment treatments; he may resort to additional freeboard in order to allow for possible slumping during earthquakes; he will pay special attention to the transition zones between the core and rock shells to ensure that any transverse cracks in the core will be self-healing; he may resort to arching of a high dam between steep abutment slopes; he will require the best possible compaction of the shell zones to reduce the magnitude of horizontal and vertical deflections during the application of the water load; and finally, he will use very conservative slopes. Earth and rockfill dams that have been designed and built in accordance with such practices are considered safe beyond human doubt. However, there is a great need for a thorough investigation of dams which were designed and built at a time when earth dam engineering still involved many unknown risks, and when greater chances were taken. This need is now generally recognized and several federal and state agencies are in the process of implementing such efforts.

As we progress through the categories of potential loss to Category IV, the probability of failure due to calculated risks becomes significant. One is, of course, justified in taking greater chances when only financial losses are involved.

Human Risks.

Unsatisfactory Organization.—Division of responsibility between design and supervision of construction is one of the most frequent causes of trouble in earthwork and foundation engineering. If the designer has no control over the execution, and particularly if he lacks confidence in those who will supervise and execute the construction, should he introduce an extra margin of safety in his design, or select an entirely different and costlier design which would be less liable to become unsafe through poor execution? How can the designer protect himself if he not only has no control over construction, but if he is not even informed when those in charge of construction introduce changes which they consider still within the designers intention, but which may actually have dangerous consequences? There is no satisfactory solution to such problems except elimination of the basic cause.

Division of responsibility can also adversely affect the design itself. This happens particularly in the case of important dam projects when the owners wish to reduce their risks by having an independent review made, or when a governmental agency may be obliged by law to obtain an independent review of the design. Such reviews are sometimes carried out after the work of the designers is already well advanced, or when construction has actually started. Almost invariably such independent reviews lead to very awkward situations. It should not come as a surprise when many consulting engineers refuse to accept such assignments. In response high-level administrators have replied: "Surely the owner is entitled to an independent opinion"; or "When the safety of a major dam is questioned, the engineering profession owes it to the public to provide the best possible advice"; or "How else could a Government exercise its power to police design and construction of dams that potentially are an enormous hazard to life and property?"

If experience had proven that an independent review of a design contributes to a better and safer project, consultants would have to accept the unpleasant aspects of such assignments. However, on the basis of my long experience as an individual consultant and as a member of consulting boards for many earth and rockfill dams, I conclude that an independent review of the design invariably leads to a division of responsibility and to serious controversies between opposing experts, and which in the end may be harmful to the design and construction of the project. In part the reason is a human weakness that hardly needs to be emphasized. Even a brilliant man can be very sensitive when his carefully prepared design, on which he may have worked for years, is attacked by someone who on the basis of a brief review believes that he has good reasons to criticize the design. Even if the criticism were justified, it would be presented under such unfavorable conditions that effective cooperation between the designer and the reviewer would be very difficult.

In my experience the only approach for avoiding such situations is to have only one board of consultants which is appointed jointly by all parties concerned with the design, i.e., usually by (1) the owners, (2) the engineers (when the designers are an independent engineering organization), and (3) the state or governmental supervisory agency if approval of the final design has to be obtained from such authority. Furthermore, it should be understood that each of these organizations, as well as the board of consultants, will have the right to suggest the addition of other experts to that board, if the need for additional specialized advice should develop. However, when a new consultant is added to a board at an advanced stage of design or during construction, it is essential that he be given sufficient time to study, thoroughly, all available information before he is asked to express his views.

When advocating a single consulting board, I have been asked: Why should membership on the same board automatically assure elimination of serious controversies? Certainly it is not automatic. But again and again I have been pleasantly surprised to observe how experts who disagree, when working together in a cooperative effort, by exchange of their knowledge and by mutual stimulation, will succeed in developing a solution to a vexing problem which is significantly better than any of the solutions proposed individually; and that a consulting board can agree on a solution even when important differences in viewpoints between members could not be reconciled.

To show that a single board can arrive at satisfactory solutions even when its internal stresses and strains are great, I present two cases which have been most difficult because of the wide divergence of opinions between members of the same board:

Case (1).—The Fort Peck Dam failure was probably the first case when a major failure was investigated not by a consulting board or panel appointed entirely new for this purpose, but rather by simply adding several new members to the original board that was responsible for the design. In some quarters this approach was criticized because the original members were considered to be biased. However, I believe that creation of a single board was the only satisfactory approach for investigating the failure and for preparing a revised design. By working together as one board and by learning from each other, a substantial majority, 7 out of 9, were able to agree on the redesign of the dam. Two of the original board members wrote minority reports. One recommended that the project be abandoned; and the other proposed a radically different method for reconstruction of the dam.[5]

Case (2).—Three eminent engineers were appointed as the consulting board for a high rockfill dam. They were so far apart in their views that a compromise appeared impossible. The principal engineer of the design organization became impatient and locked the three board members into a room with the assurance that he would not let them out until they agreed. They finally reached a solution which in my judgment was better than any one of the original proposals.

In this last example, assume that one of the three experts were to serve as principal consultant to the designers, the second one were to be appointed later by the owners for a review of the design, and still later the third one were to be appointed by the government's supervisory agency to carry out an independent review of the design. It is almost certain that the design would be changed twice, and that the views of the consultant to the government agency would prevail, but not necessarily because his proposals had greater merit. In any case, no matter what solution is finally reached, it is likely that it would be inferior to the solution that would have been adopted if these three men had worked together as members of the same board.

In the very rare case when a conflict on a vital issue cannot be resolved within a board of consultants, or when the organization charged with the design (either an independent engineering firm or the owners' design department) disagrees with the recommendations of the consultants, the final decision must rest with the owner or their engineers. In such cases, or when an independent review by a governmental authority enforces a change with which the board of consultants cannot agree, the board would have no choice but to resign.

For the current high standard of engineering practice in the United States, the probability that any division of responsibility would contribute to failure of a project in Category I (potentially catastrophic losses) is extremely small. However, as we move down the scale of potential losses, this probability increases rapidly.

Unsatisfactory Use of Available Knowledge and Judgment.—Under this heading I include all cases for which insufficient engineering knowledge and judgment are used in design and construction. They range from honest mistakes or lack of knowledge to the other extreme when a consultant is merely used as window dressing. In the latter case he may be made the scapegoat for anything that goes wrong on a project even though his advice was entirely satisfactory.[15]

The engineer who is ultimately responsible for design and/or construction, is dependent on many subordinates whose work he cannot personally check. Even with the best system of controls and checks, errors in judgment and evaluation can creep into some part of design or construction.

I include under this heading the error in judgment by owners and engineers when they force the contractor to take risks that they may not dare to assume themselves. Contractors are used to inequities in contracts. But when inequities become iniquities then a responsible contractor has no choice but to refuse to bid. Shortly before this paper went to print, my attention was called to a paper by C. P. Dunn.[16] It contains a competent and timely discussion of the reasons for the excessively heavy burden of risks which some owners and engineers force upon contractors, and of the resulting conflicts which cause " . . . the lines of communication between contractors and engineers to break down to the point where a smooth running job is a rare event, and the usual situation is mutual distrust and misunderstanding." Sometimes an engineer learns only by bitter experience that he cannot absolve himself from responsibility, by a simple statement in the contract, for the insufficient or unreliable information he furnishes to the bidders; and that in the end the engineer will achieve for the owner a better project at a more equitable cost if he furnishes

15 Terzaghi, Karl, "Consultants, Clients and Contractors," Journal, Boston Soc. of Civ. Engrs., Boston, Mass., January, 1958, pp. 1-41; closure, July, 1958, pp. 255-277.

16 Dunn, C. P., "Contractors versus Engineers-The Worsening Conflict," Western Construction, King Publications, San Francisco, Calif., April, 1965.

the bidders with reliable information on the subsoil conditions rather than letting the contractors guess and gamble. Gradually it is becoming accepted that the engineer should also reduce the contractor's risks by undertaking the design of such vital construction details as sheeting and bracing systems and cofferdams. This procedure protects the owner against a contractor who is willing to gamble the most, who gets the job, but who in the end merely gets himself and the owner into trouble by causing delays and additional costs which render the project more expensive than if it had been given to one of the higher but more responsible bidders.

The possibility of catastrophic failures resulting from unsatisfactory use of available knowledge and judgment is probably remote. However, for projects for which the potential losses are of much smaller magnitude, and particularly for projects of Category IV, failures due to errors caused by such human risks are relatively frequent.

Corruption.—The importance of a realistic attitude toward all types of human risks in earthwork and foundation engineering, including the possibility of corruption, cannot be overemphasized. Although found most frequently on the level of inspection, often initiated by dishonest contractors, occasionally corruption will reach into much higher echelons. Even the most experienced designer who can cope well with engineering risks may see his career ruined by human risks, particularly by corruption. What is the engineer to do when he knows or suspects that corruption may endanger the safety of a project? On a broader level, it is an age-old problem. About 2500 years ago two Chinese philosophers held opposite views. Lao-Tze recommended that during periods of corrupt government the good men should stay away from public service, while Confucius urged that they should devote their efforts to reforming the government.

CLOSING REMARKS

While reviewing a final draft for publication, General James H. Stratton suggested that a better title for this paper would have been: "Value Judgments Versus Calculated Risks in Civil Engineering." Implied in this suggestion is a separation of the two aspects that are included in the term calculated risk as described in the introduction and as used in this paper. If the term value judgment were used to describe the evaluation of engineering problems which at present defy quantitative analysis, it might be better to use a term other than calculated risk to designate the process of choosing an appropriate margin of safety.

In the course of my preparation for this Terzaghi lecture I became convinced that the task I intended to perform is too large and too difficult for one man to accomplish. The best I can hope for is that this presentation will serve as a catalyst to promote a more comprehensive study by a group of engineers and contractors who have extensive experience in dealing with risks in earthwork and foundation engineering. In addition, owners of large dams and governmental authorities that have police power over such dams should be represented by knowledgeable administrators who have accumulated the wisdom that is so necessary for guiding difficult human affairs. I envision that such a group would start its efforts by an exchange of letters which would define the main questions and present factual examples. This would be followed by group conferences. I hope that eventually such efforts will crystallize into a set of quite informal but very useful guidelines for our profession for dealing with risks in civil engineering.

APPENDIX.—HAZEN ON HYDRAULIC FILL DAMS

During meetings of the consulting board for the Fort Peck Dam failure, 1938-39, I presented quotations from Allen Hazen's outstanding paper on "HYDRAULIC FILL DAMS,"[10] and I referred to him as one of the early pioneers of modern soil mechanics. This statement so aroused the ire of a senior member of the board (who was already well known for his low regard for the contributions by soil mechanics to earthwork and foundation engineering) that he considered it his duty to defend the memory of Allen Hazen against my "insulting suggestion."

The following chapter from Hazen's paper (pp. 1739-1745) is reprinted here not only because of its bearing on a topic in my paper, but also because I believe that every student of soil mechanics would profit by reading it. Hazen's clear description of the physical causes of liquefaction of sand, i.e. of the manner in which the relative amount of stress carried by the grain skeleton and by the pore water influences the strength of soils, may well be unsurpassed; and it has lost none of its value as a result of the more scientific explanation of this phenomenon that was developed after his death.

"QUICKSAND CONDITIONS IN DAMS"

"When a granular material has its pores completely filled with water and is under pressure, two conditions may be recognized. In the first or normal case, the whole of the pressure is communicated through the material from particle to particle by the bearings of the edges and points of the particles on each other. The water in the pores is under no pressure that interferes with this bearing. Under such condition the frictional resistance of the material against sliding on itself may be assumed to be the same, or nearly the same, as it would be if the pores were not filled with water. In the second case, the water in the pores of the material is under pressure. The pressure of the water on the particles tends to hold them apart; and part of the pressure is transmitted through the water. To whatever extent this happens, the pressure transmitted by the edges and points of the particles is reduced. As water pressure is increased, the pressure on the edges is reduced and the friction resistance of the material becomes less. If the pressure of the water in the pores is great enough to carry all the load, it will have the effect of holding the particles apart and of producing a condition that is practically equivalent to that of quicksand.

"An extra pressure in the water in the pores of such a material may be produced by a sudden blow or shock which tends to compress the solid material by crushing the edges and points where they bear, or by causing a re-arrangement of particles with smaller voids. An illustration of this can be seen in the sand on the seashore. Such sand, comparable to dune sand in size, is usually found to be saturated with water for a certain distance above the water level. This condition is maintained by capillarity. If a weight is slowly placed on this saturated sand, there is a slight settlement, the grains of sand coming to firmer bearings, and the weight is carried. A sharp blow, as with the foot, however, liquefies a certain volume and makes quicksand. The condition of quicksand lasts for only a few seconds until the surplus water can find its way out. When this happens the grains again come to solid bearings and stability is restored. During a few seconds after the sand is struck, however, it is almost liquid, and is capable of moving or flowing or of transmitting pressure in the same measure as a liquid.

"Fine-grained sand in which this condition exists is called quicksand. The properties of quicksand are well known. Fine-grained sand saturated with water and then mixed with an additional 5% of water acts practically as a liquid. It will flow through small orifices or in a pipe at a 5% slope. The sand in a mechanical filter in process of washing is a typical case. When the sand is drained, in the filter, it forms a bed as hard as any sand bed. When subjected to a reverse current of water strong enough to lift it slightly, however, the volume is increased by perhaps 5%, and it becomes liquid. An object can then be pushed through it with scarcely more resistance than would be offered by a liquid of high specific gravity.

"The conditions that control the stability or lack of stability of quicksand may also control the stability or lack of stability of materials in dams.

"The puddle clay core of an hydraulic-fill dam is physically like quicksand, but with particles one hundred times smaller in diameter and a million times smaller in weight. It has the instability of quicksand in full measure and it retains it for a long time, or perhaps indefinitely.

"With the coarse-grained part of an hydraulic-fill dam, that is to say, with rock toes,[a] there may be also, at times, a similar condition. Generally speaking, it would be expected that such coarse-grained materials would have sufficient drainage to let out surplus water and prevent the possibility of an excess sufficient to destroy its stability. With hard-grained materials from glacial drift of New England and the Northern States, it is hard to conceive of a lack of drainage in gravel that would permit the accumulation of an excess of water. With the softer materials of the Pacific Coast, however, the conditions may be different. In the first place, these soft rocks by partial crushing under pressure produce fine material which tends to fill the remaining spaces and to reduce the drainage capacity.

"On the other hand, each increment of loading applied to soft-grained material produces a certain compression and settlement; and with it a reduction in voids. This may happen to a toe of soft rock on the upstream side of a dam against which water is being stored during construction. There is first an open condition with ample voids and ample drainage. As the dam is built higher, pressure increases; there is compression and reduction in porosity. Each additional increment of loading and compression means that a certain quantity of water representing the difference between the old voids and the new voids must be expelled. As long as the material remains sufficiently pervious to carry off this excluded water, the process of compression is harmless. The sur-

---

[a] In Fig. 3 of Hazen's paper he uses the term "toe" for the zones of a dam outside the core, which today are usually referred to as "shells."

plus water is pushed back into the reservoir and stability is retained. There may come a time, however, when the compression goes forward so rapidly that the surplus water cannot be carried off fast enough. When that point is reached (if it is reached), there will tend to be an excess of water in the interstices and that excess will transmit some of the pressure that was before carried only by the bearings of solid particles, and the frictional resistance of the material will be less, and perhaps much less, than it was before.

"The thought has occurred to the writer, in looking at the material that slid in the Calaveras Dam, that something of this kind may have happened on a large scale—800,000 cu yd of fill flowed for a brief space, and then became solid. It was, in fact, so solid that in examining it afterward, by samples and by borings, it was difficult to see how the material could have flowed—as it certainly did flow.

"It may be that after the first movement there was some readjustment of the material in the toe which resulted in producing temporarily this condition of quicksand, and which destroyed for a moment the stability of the material and facilitated the movement that took place.

"This will not account for the initial movement; but the initial movement of some part of the material might result in accumulating pressure, first on one point, and then on another, successively, as the early points of concentration were liquefied and in that way a condition comparable to quicksand in a large mass of material may have been produced."

From the "Summary" of his paper, the following paragraph is particularly pertinent:

"Stability is increased by compactness. It is worth while to watch voids closely, and to make every effort to hold them at a minimum. The extra weight is advantageous, but security against compression and re-arrangement with resulting temporary quicksand conditions, can be best reached in this way."

4271 March, 1965 SM 2

Journal of the
SOIL MECHANICS AND FOUNDATIONS DIVISION
Proceedings of the American Society of Civil Engineers

SHALLOW FOUNDATIONS[a]

By George G. Meyerhof,[1] F. ASCE

SYNOPSIS

A review of the design and performance of spread footings and rafts in relation to the prediction and control of settlement reveals that allowable bearing pressures for a given settlement of shallow foundations on sand and gravel, when estimated from standard or static cone penetration tests, are rather conservative. Based on the present analysis, the bearing pressures could safely be increased by 50%, when the corresponding predicted settlements would vary from about 0.8 to 2 times the observed values. The allowable bearing pressures for shallow foundations on clay, when based on the sum of immediate and final consolidation settlements with an allowance for any lateral consolidation and the effects of shearing stresses, usually give results of sufficient accuracy. More field observations are needed on the contact pressures and stresses in raft foundations to develop a rational method of their design.

INTRODUCTION

In the past, the design of shallow foundations was mainly empirical and it was based largely on allowable bearing pressures for various soil types. The study of foundation behavior in the field, amplified by laboratory research has led to a more rational approach. It is now recognized that the results of a

Note.—Discussion open until August 1, 1965. To extend the closing date one month, a written request must be filed with the Executive Secretary, ASCE. This paper is part of the copyrighted Journal of the Soil Mechanics and Foundations Division, Proceedings of the American Society of Civil Engineers, Vol. 91, No. SM2, March, 1965.

[a] This is one of the "state-of-the-art" papers presented at the ASCE Soil Mechanics and Foundations Division Conf. on "Design of Foundations for Control of Settlements" held at Northwestern Univ., Evanston, Ill., June, 1964; the compiled papers were presented in the September, 1964, Division Journal.

[1] Dean, Nova Scotia Tech. Coll., Halifax, N. S., Canada.

site exploration and soil tests form the basis of an analysis of the stability and probable movements of the proposed foundation and superstructure.

If the stability of the foundation is ensured with an adequate margin of safety, the design of foundations is governed largely by the requirement that the foundation movements must be within limits that can be tolerated by the superstructure. Moreover, where the foundation is located below the depth at which the soil is subjected to climatic influences, erosion, or other extraneous disturbances, the movements caused by structural loads form the main criterion for the allowable bearing pressures of shallow foundations. The magnitude of the probable movements and even more their time rate and their distribution over the foundation area depend on so many factors that they can only be estimated within fairly wide limits.

Some of these problems are briefly summarized herein in relation to the design and performance of shallow foundations, namely, spread footings and rafts, on sands and clays. Considerations of foundation movements resulting from swelling clays, collapsible soils, or vibrations are outside the scope of the material covered herein.

FOUNDATIONS ON SAND

Spread footings of reinforced concrete or steel grillage are the simplest types and, if practicable, they are usually also the most economical. Because settlements of footings having a constant bearing pressure on sand increase roughly with the square root of the base width, the footings should be kept as small as the allowable bearing pressure will permit. Rafts of reinforced concrete spread the load over the maximum area of ground and thereby decrease the bearing pressure to a minimum. Rafts also reduce the differential settlements, especially if they are made fairly rigid or if the depth of any basements is roughly varied in proportion to the applied loads.

Settlements of foundations on sands and gravels, not underlain by softer strata, are nearly complete at the end of the construction period. Allowable bearing pressures for a given settlement on sand are commonly estimated from dynamic or static penetration tests and sometimes also from plate loading tests. These tests are conducted within the significant depth or seat of settlement below the level of the foundation. For sands and gravels this depth varies roughly from one to two times the foundation width, the lower limit applying to a square foundation and the upper limit to a strip foundation.

Standard penetration tests are widely used to estimate the relative density of sand and they have been correlated with the results of plate loading tests and settlement observations on structures.[2] On this basis the allowable bearing pressure, in tons per square foot, of spread footings on sand can be written, approximately,[3,4]

---

[2] Terzaghi, Karl, and Peck, R. B., "Soil Mechanics in Engineering Practice," John Wiley & Sons, Inc., New York, N. Y., 1948.

[3] Meyerhof, George G., "Penetration Tests and Bearing Capacity of Cohesionless Soils," Journal of the Soil Mechanics and Foundations Division, ASCE, Vol. 82, No. SM1, Proc. Paper 866, 1956.

[4] Meyerhof, George G., closure to "Penetration Tests and Bearing Capacity of Cohesionless Soils," Journal of the Soil Mechanics and Foundations Division, ASCE, Vol. 83, No. SM1, Proc. Paper 1155, 1957.

$$p_a = \frac{N\ S_a}{8} \text{ for } B \leq 4 \text{ ft} \quad \ldots\ldots\ldots\ldots\ldots \quad (1)$$

or $$p_a = \frac{N\ S_a}{12}\left(\frac{B+1}{B}\right)^2 \text{ for } B > 4 \text{ ft} \quad \ldots\ldots\ldots\ldots \quad (2)$$

and for rafts on sand

$$p_a = \frac{N\ S_a}{12} \quad \ldots\ldots\ldots\ldots\ldots\ldots \quad (3)$$

in which B is the foundation width, in feet, N is the standard penetration resistance, in blows per foot (corrected for compact and dense, submerged silty sand) and S_a is the allowable settlement, in inches. This relationship is sensibly independent of the shape of the foundation, and for a foundation depth approaching the width B, an increase of the allowable pressure by approximately one-third is sometimes made.

The standard penetration resistance depends, however, not only on the relative density of the soil but also on the effective overburden pressure, ground-water conditions, and other factors.[5] Even without making any allowance for the ground-water conditions, which are already reflected in the measured values of N, as suggested previously,[3,4] the method still furnishes conservative bearing pressures. This is shown by the analysis of settlement observations on buildings on sand[6,7,8] summarized in Table 1. A comparison between these observed and estimated settlements using Eqs. 1 to 3 is given in Fig. 1, which shows that the estimated movements vary from approximately 1.5 to 3 times the observed values. In most cases the foundations were fully submerged and if an allowance for submergence had been made in the present estimates, overconservative allowable bearing pressures would have been obtained. This is confirmed by an analysis of plate loading tests above and below the ground-water table.[9]

On the other hand, comparative plate loading tests on several sites[10,11] indicate that the standard penetration resistance does not fully account for the

---

5 Gibbs, H. J., and Holtz, W. G., "Research on Determining the Density of Sands by Spoon Penetration Testing," Proceedings, 4th Internatl. Conf. on Soil Mechanics and Foundation Engrg., London, 1957, Vol. 1, p. 35.

6 Rios, L., and Silva, F. P., "Foundations in Downtown Sao Paulo (Brazil)," Proceedings, 2nd Internatl. Conf. on Soil Mechanics and Foundation Engrg., Rotterdam, 1948, Vol. 4, p. 69.

7 Vargas, M., "Building Settlement Observations in Sao Paulo," Proceedings, 2nd Internatl. Conf. on Soil Mechanics and Foundation Engrg., Rotterdam, 1948, Vol. 4, p. 13.

8 Vargas, M., "Foundations of Tall Buildings on Sand in Sao Paulo (Brazil)," Proceedings, 5th Internatl. Conf. on Soil Mechanics and Foundation Engrg., Paris, 1961, Vol. 1, p. 841.

9 Sutherland, H. B., "The Use of In-Situ Tests to Estimate the Allowable Bearing Pressure of Cohesionless Soils," Structural Engineer, London, 1963, Vol. 41, p. 85.

10 Meigh, A. C., and Nixon, I. K., "Comparison of In-Situ Tests for Granular Soils," Proceedings, 5th Internatl. Conf. on Soil Mechanics and Foundation Engrg., Paris, 1961, Vol. 1, p. 499.

11 Rodin, S., "Experiments with Penetrometers with Particular Reference to the Standard Penetration Test," Proceedings, 5th Internatl. Conf. on Soil Mechanics and Foundation Engrg., Paris, 1961, Vol. 1, p. 517.

effect of the soil grain properties on the compressibility characteristics o granular soils. Thus, from the penetration resistance reasonably safe allow· able bearing pressure were obtained for fine and medium sands, whereas fo: sandy gravels the estimated values were generally less than one-half of thos‹ indicated by plate loading tests. However, the analysis of recent settlemen observations of structures on rafts on sandy gravel[12] given in Table 1 an‹ plotted in Fig. 1 shows that estimates of allowable bearing pressures on grave using the values of N and Eq. 3 are no more conservative than those for sand

TABLE 1.—COMPARISON OF OBSERVED AND ESTIMATED SETTLEMENTS OF SHALLOW FOUNDATIONS ON SAND AND GRAVEL

Structure	Reference	Foundation width, in feet	Soil type	Average standard penetration resistance in number of blows per foot	Bearing pressure in tons per square foot	Maximum Settlement		Ratio estimated t‹ observe settlement
						Observed, in inches	Estimated, in inches	
T. Edison, Sao Paulo	6	60	Fine clayey sand	15	2.4	0.6	1.9	3.2
Banco do Brasil, Sao Paulo	6,8	75	Fine clayey sand	18	2.5	1.1	1.7	1.5
Iparanga, Sao Paulo	7	30	Fine-medium clayey sand	9	2.3	1.4	3.1	2.2
C.B.I. Esplanada, Sao Paulo	8	48	Fine-medium clayey sand	22	4.0	1.1	2.2	2.0
Riscala, Sao Paulo	7	13	Fine-medium clayey sand	c.20	2.4	0.5	c.1.2	c.2.4
Thyssen, Dusseldorf	12	74	Sandy gravel	25	2.5	0.95	1.2	1.3
Ministry, Dusseldorf	12	52	Coarse gravel	20	2.3	0.85	1.4	1.6
Chimney, Cologne	12	67	Sandy gravel	10	1.8	0.4	2.1	5.3

Static cone penetration (deep-sounding) tests have been used to estimat‹ a compressibility index of sand[13]

12 Schultze, E., "Probleme bei der Auswertung von Setzungsmessungen," Proceedings, Baugrundtagung, Essen, Germany, 1962, p. 343.

13 DeBeer, E., "Settlement Records on Bridges Founded on Sand," Proceedings 2nd Internatl. Conf. on Soil Mechanics and Foundation Engrg., Rotterdam, 1948, Vol. 2, p. 111.

$$C = \frac{1.5\ q_c}{p_o} \qquad (4)$$

in which q_c is the cone resistance at a depth z with an effective overburden pressure p_o. According to this method the in-situ coefficient of volume change is

$$m_v = \frac{2}{3\ q_c} \qquad (5)$$

and the settlement can be estimated from the expression

$$S = \int_o^{\infty} \frac{1}{C} \log_e \frac{p_o + \Delta_p}{p_o} \qquad (6)$$

or
$$S = \int_o^{\infty} m_v\ \Delta_p\ dz \qquad (7)$$

in which Δ_p is the increase in vertical pressure at depth z resulting from the foundation load using the elastic theory.

A comparison of this method with observations on bridges on wide shallow strip footings on generally compact fine-medium sand[13,14,15] is shown in Fig. 1. It is found that the calculated settlements using Eqs. 4 and 6 are within a range of approximately 1.2 to 2.5 times the observed maximum settlements with an average ratio of calculated to observed settlements of approximately 1.9. It may, therefore, be concluded that both of the preceding methods of estimating settlements and the corresponding allowable bearing pressures for spread foundations on sand are rather conservative. On the basis of the present analysis to 50% greater bearing pressures could safely be used in conjunction with both methods, which would give sufficiently accurate results for practical purposes and lead to a range of predicted settlements between approximately 0.8 to 2 times the observed values with an average of approximately 1.3.

Attempts have also been made to estimate the total settlement of shallow foundations on granular soils either from recompression curves in consolidation tests or from drained triaxial compression tests with constant principal stress ratio.[12] However, the different stress conditions under laboratory and field conditions make this procedure rather uncertain, unless the compression characteristics of the soil are established from an analysis of settlement observations of structures in a given region of similar soil conditions.[8,16]

---

14 Marivoet, L., "Settlement Observations on Bridges Built on Shallow Foundations," Proceedings, 3rd Internatl. Conf. on Soil Mechanics and Foundation Engrg., Zurich, 1953, Vol. 1, p. 418.

15 DeBeer, E., and Martens, A., "Method of Computation of an Upper Limit for the Influence of Heterogeneity of Sand Layers on the Settlement of Bridges," Proceedings, 4th Internatl. Conf. on Soil Mechanics and Foundation Engrg., London, 1957, Vol. 1, p. 275.

16 Muhs, H., "Ergebnisse der Setzungsmessungen an den Hochhäusern im Hansaviertel in Berlin," Mitteilungen Degebo, Berlin, Germany, 1961, No. 15, p. 1.

The movements which can be tolerated by a structure should preferably be decided in each particular case on the basis of a structural analysis. For this purpose a method has been developed[17,18] in which the foundation and settlement characteristics of the soil at the site are interrelated with the loading, layout and rigidity of the superstructure. However, the composite action between the framework of buildings and their cladding as well as the creep of structural materials under slow rates of strain make it difficult at present to predict allowable foundation movements from stress calculations for structures. Consequently a statistical approach based on the performance of structures under settlement becomes necessary.

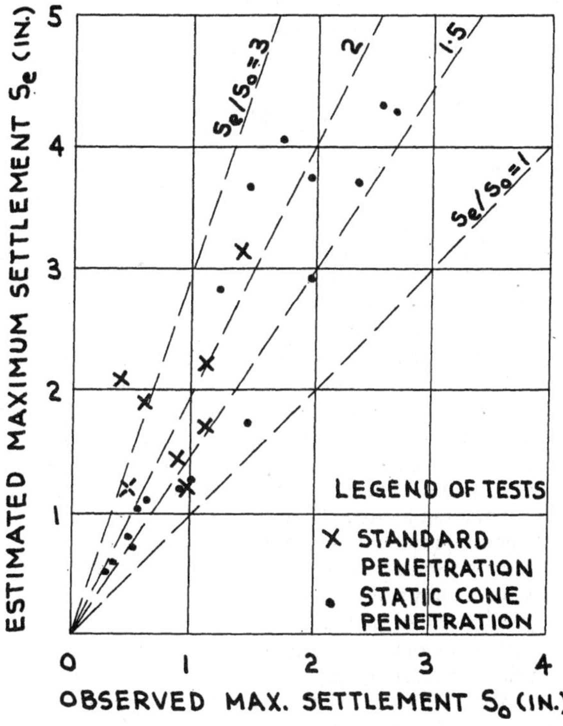

FIG. 1.—COMPARISON OF OBSERVED AND ESTIMATED SETTLEMENTS OF SHALLOW FOUNDATIONS ON SAND AND GRAVEL

Analysis of field observations indicates[19] that for traditional encased steel and reinforced concrete buildings an angular distortion exceeding approxi-

[17] Meyerhof, George G., "The Settlement Analysis of Building Frames," Structural Engineer, London, 1947, Vol. 25, p. 369.

[18] Meyerhof, George G., "Some Recent Foundation Research and its Application to Design," Structural Engineer, London, 1953, Vol. 31, p. 151.

[19] Skempton, A. W., "The Allowable Settlement of Buildings," Proceedings, Institution of Civ. Engrs., London, 1956, Part 3, Vol. 5, p. 727.

mately 1/150 will usually lead to structural damage; an angular distortion exceeding approximately 1/300 will generally lead to undesirable cracking of walls although such architectural damage is rather a subjective criterion. The angular distortion of buildings can be statistically related to the maximum differential and total settlements. For traditional forms of construction on sand an allowable differential settlement of 3/4 in. and an allowable total settlement of 1 in. to 2 in. has been suggested for shallow foundations, the lower limit applying to footings and the upper limit to rafts.[2] For a rigid structure the differential settlements are governed by the allowable tilting of the order of 1/200. In the case of non-traditional buildings, bridges and industrial structures the allowable differential and total settlements have to be assessed in each given case on the basis of an analysis and past experience.

FOUNDATIONS ON CLAY

The allowable bearing pressure of foundations on clay is generally controlled by the settlement of the structure, which depends to a considerable extent on the net pressure under the foundation. This pressure can be reduced by excavation and the construction of basements, each of which reduces the net pressure on the soil by approximately 1/2 ton per square feet and is equivalent to the weight of about 5 storys of office or apartment buildings.

The settlement of spread footings resting on clay increases roughly in direct proportion to the base width, and because of overlapping of the soil stresses from adjacent footings uniformly loaded buildings tend to settle more at the center than at the edges. The resulting differential settlements can conveniently be reduced by using a greater bearing pressure for the edge footings than for the center footings of the structure, provided the safe bearing capacity of the soil is not exceeded. The differential movements can be further reduced by providing a raft, especially if the substructure is made rigid. The resulting bending moments in a rigid substructure can be decreased by using a greater net pressure at the edges of the building than at the center. This can readily be achieved by providing basements with a smaller excavation around the perimeter of the building than near the middle.

However, on sites underlain by clay the removal of soil produces a heave of the bottom of the excavation, and the resultant settlement of the structure will be roughly equal to the sum of the heave and the settlement under the net foundation pressure. The advantage of compensated foundations may also partly be offset by the cost of deep excavations. The magnitude and distribution of the heave can be estimated as for settlement to be indicated subsequently, but using rebound triaxial compression tests for the elastic component and expansion consolidometer tests for the swelling component of clays. Field observations have shown that the amount of heave is usually overestimated unless excellent undisturbed samples are used in carefully conducted tests.

The settlement of the foundation resulting from replacement of the overburden load is similarly estimated, but using recompression tests. The subsequent settlement under the net foundations loads can be divided into (1) immediate settlement caused by deformation of the soil without volume change, and (2) consolidation settlement caused by closer packing of the grains. The immediate settlement occurs mainly on application of the load and consists of an elastic (recoverable) part, which can be estimated from elastic theory and a plastic (irrecoverable) part, which can be estimated from plastic theory.[20] In practice it is, however, frequently sufficiently accurate to predict the whole immediate settlement from elastic theory allowing for the shape, depth, and rigidity of the foundation and the thickness of a compressible layer with a mean secant modulus of deformation without the working range of stresses and the seat of settlement.[21]

Comparison between calculated and observed net immediate settlements of structures on spread foundations on clay[22] shows that the ratio of calculated to observed settlements varied between approximately 0.8 and 1.4 with an average of approximately 1.2. The values for normally consolidated clays were somewhat greater than for highly over-consolidated clays because of the difficulty of obtaining truly undisturbed samples for determination of the modulus of deformation of sensitive clays.

The magnitude and time-rate of consolidation settlements can be estimated from consolidation theory for the primary compression part of a thin bed of clay between two sand strata or a large loaded area in relation to the thickness of the compressible stratum, although the drainage boundaries are sometimes difficult to assess. For a thick clay stratum or a foundation resting directly on clay an allowance should be made for lateral drainage and the pore pressures resulting from shearing stresses from the foundation loads.[23] Otherwise the magnitude of the estimated settlements will usually be too great and the time-rate will be too small. This has been confirmed by a comparison between calculated and observed net consolidation settlements deduced from the difference between total and immediate settlements.[22,24] It was found that the present methods give calculated net settlements of approximately 0.7 to 1.5 times the observed final settlements with an average ratio close to unity. The agreement appears to be somewhat better for normally consolidated and lightly over-consolidated clays than for highly consolidated clays.

The rate of settlement is, however, usually underestimated even where some allowance is made for lateral consolidation. The maximum time to be considered in this connection depends on the economic life of the structure and for practical purposes the final movements are mainly required. The secondary compression of clays can only be assessed roughly for the condition of a laterally confined clay layer using the logarithmic time relationship obtained from consolidation tests. For thick clay layers the time relationship from drained triaxial compression tests might be preferable for this purpose.

---

20 Meyerhof, George G., "The Tilting of a Large Tank on Soft Clay," Proceedings, South Wales Institution of Engrs., 1951, Vol. 67, p. 53.

21 Skempton, A. W., "The Bearing Capacity of Clays," Proceedings, Bldg. Research Congress, London, 1951, Vol. 1, p. 180.

22 Macdonald, D. H., and Skempton, A. W., "A Survey of Comparisons Between Calculated and Observed Settlements of Structures on Clay," Proceedings, Conf. on Correlation Between Calculated and Observed Stresses and Displacements in Structures, London, 1955, Vol. 1, p. 318.

23 Skempton, A. W., and Bjerrum, L., "A Contribution to the Settlement Analysis of Foundations on Clay," Geotechnique, Institution of Civ. Engrs., London, 1957, Vol. 7, p. 168.

24 Skempton, A. W., Peck, R. B., and Macdonald, D. H., "Settlement Analyses of Six Structures in Chicago and London," Proceedings, Institution of Civ. Engrs., London, 1955, Part 1, Vol. 4, p. 525.

Fortunately, secondary consolidation settlements are usually of little structural significance.

For shallow foundations on clay the pattern of stratification and the variation of the applied loads and of the compressibility of the soil in plan make it difficult to give general values of the allowable differential and total settlements. However, a range of approximately 1 to 2 times the corresponding values suggested for sand may be used for preliminary estimates of traditional buildings on clay,[19] the lower limit applying to foundations on a shallow and relatively thin stratum and the upper limit applying to a deep-seated or thick bed. In the case of a deep-seated compressible stratum the footings of a structure may also be subjected to horizontal movements associated with the settlement due to the bending of the natural raft formed by the overlying firm stratum[25] and these movements should be considered in the design.

Horizontal movements and tilting of individual footings must also be expected where they are subjected to a horizontal thrust and bending moment in addition to a vertical load, as in the case of arched and framed structures and retaining walls. Approximate expressions have been developed to estimate the lateral and rotational movements of footings under eccentric and inclined loads and to predict their effect on simple superstructures.[18] Some support for this analysis has been obtained from the results of loading tests on model portal frames and full-sized experiments, which have shown that customary spread footings should generally be considered as hinged bases, and for fixity conditions either wide or deep foundations are required. However, no field observations appear to be available (as of 1965) to access the allowable horizontal movements and tilting of footings for different types of structures.

STRESSES IN FOUNDATIONS

In practice, spread footings can be considered as perfectly rigid and the contact pressure between the base and soil under working loads is, therefore, nonuniform. While the distribution of the contact pressure has no appreciable effect on the total settlement of footings, the corresponding bending stresses in the footings differ from those calculated from the usual assumption of a uniform pressure distribution. Measurements of contact pressures under rigid spread foundations[26] reported in 1961 have confirmed earlier model experiments that except near the surface of cohesionless soils the contact pressure under a rigid base at working load is generally a maximum near the edges and a minimum near the middle of the foundation, similar to the theoretical distribution given by Boussinesq. In cohesive soils consolidation and plastic flow gradually change the contact pressure to a more uniform distribution. On the basis of the most unfavourable observed pressure distributions in which the maximum contact pressure was approximately twice the average pressure under the base and the minimum pressure was approximately one-half of the average, the maximum bending stresses in a centrally loaded foot-

---

25 Hardy, R. M., Ripley, C. F., and Lee, K. L., "Horizontal Movements Associated with Vertical Settlements," Proceedings, 5th Internatl. Conf. on Soil Mechanics and Foundation Engrg., Paris, 1961, Vol. I, p. 665.

26 Schultze, E., "Distribution of Stress Beneath a Rigid Foundation," Proceedings, 5th Internatl. Conf. on Soil Mechanics and Foundation Engrg., Paris, 1961, Vol. 1, p. 807.

ing are approximately one-third greater than given by the customary assumption of a uniform contact pressure. A linear contact pressure distribution is, therefore, sufficiently accurate for the structural design of spread footings.

The contact pressure distribution and stresses in rafts can, however, vary within wide limits and depend mainly on the magnitude and distribution of the external loading, the thickness and the relative stiffness of the raft and superstructure to that of the soil, and the size of the raft in relation to the thickness of the compressible stratum on which it rests. Thus, for a uniformly loaded raft on a comparatively thin stratum the contact pressure is fairly uniform and the bending stresses are small irrespective of the stiffness of the raft. For a stratum thickness exceeding approximately the width of the raft the contact pressure is less uniform and the corresponding bending stresses increase rapidly with the relative stiffness of the raft.[18]

Although the assumption of a uniform contact pressure distribution under rafts is frequently safe for structural design purposes, it is found that the bending stresses are sensitive to changes in the contact pressure distribution. For fairly flexible rafts carrying concentrated loads it is believed that reasonable estimates of the contact pressure distribution and the maximum bending stresses can be obtained on the basis of the coefficient of subgrade reaction. On the other hand, for relatively rigid rafts the distribution of the contact pressure and the bending stresses depend to a considerable extent on the variation of the compressibility of the soil in plan. This is shown by observations on a large raft foundation on clay, which also indicated a change of the contact pressure with time.[27] Because these variations are difficult to predict, several limiting pressure distributions may have to be considered to determine the worst conditions for the design of different portions of the raft. However, until field observations on the contact pressures and stresses in rafts of various stiffnesses and on different soils are available, the structural design of rafts remains one of the least satisfactory aspects in the design of shallow foundations.

CONCLUSIONS

Allowable bearing pressures for a given settlement of shallow foundations on sand and gravel, when estimated from standard or static cone penetration tests, are rather conservative, even without making an allowance for groundwater conditions in the standard penetration resistance. On the basis of the present analysis the bearing pressures estimated by both methods could safely be increased by 50%, when the corresponding predicted settlements would vary from approximately 0.8 to 2 times the observed values.

The allowable bearing pressures for shallow foundations on clay, when based on the sum of immediate and final consolidation settlements with an allowance for any lateral consolidation and the effects of shearing stresses usually give results of sufficient accuracy for practical purposes. The allowable lateral and rotational movements of spread footings cannot be assessed at present.

---

[27] Teng, W. C., "Determination of the Contact Pressure Against a Large Raft Foundation," Geotechnique, Institution of Civ. Engrs., London, 1949, Vol. 1, p. 222.

Whereas the bending stresses in spread footings can be determined from the customary assumption of a linear contact pressure distribution with sufficient accuracy, the stresses in rafts are sensitive to changes in the contact pressure distribution and the variation of the compressibility of the soil in plan. It is hoped that more field observations will be made on the contact pressures and stresses in raft foundations so that a rational method of their design can be developed.

Geotechnical Special Publications

1 *Terzaghi Lectures*
2 *Geotechnical Aspects of Stiff and Hard Clays*
3 *Landslide Dams: Processes, Risk, and Mitigation*
4 *Tiebacks for Bulkheads*
5 *Settlement of Shallow Foundation on Cohesionless Soils: Design and Performance*
6 *Use of In Situ Tests in Geotechnical Engineering*
7 *Timber Bulkheads*
8 *Foundations for Transmission Line Towers*
9 *Foundations & Excavations in Decomposed Rock of the Piedmont Province*
10 *Engineering Aspects of Soil Erosion, Dispersive Clays and Loess*
11 *Dynamic Response of Pile Foundations–Experiment, Analysis and Observation*
12 *Soil Improvement: A Ten Year Update*
13 *Geotechnical Practice for Solid Waste Disposal '87*
14 *Geotechnical Aspects of Karst Terrains*
15 *Measured Performance Shallow Foundations*
16 *Special Topics in Foundations*
17 *Soil Properties Evaluation from Centrifugal Models*
18 *Geosynthetics for Soil Improvement*
19 *Mine Induced Subsidence: Effects on Engineered Structures*
20 *Earthquake Engineering & Soil Dynamics II*
21 *Hydraulic Fill Structures*
22 *Foundation Engineering*
23 *Predicted and Observed Axial Behavior of Piles*
24 *Resilient Moduli of Soils: Laboratory Conditions*
25 *Design and Performance of Earth Retaining Structures*
26 *Waste Containment Systems: Construction, Regulation, and Performance*
27 *Geotechnical Engineering Congress*
28 *Detection of and Construction at the Soil/Rock Interface*
29 *Recent Advances in Instrumentation, Data Acquisition and Testing in Soil Dynamics*
30 *Grouting, Soil Improvement and Geosynthetics*
31 *Stability and Performance of Slopes and Embankments II*
32 *Embankment of Dams–James L. Sherard Contributions*
33 *Excavation and Support for the Urban Infrastructure*
34 *Piles Under Dynamic Loads*
35 *Geotechnical Practice in Dam Rehabilitation*
36 *Fly Ash for Soil Improvement*
37 *Advances in Site Characterization: Data Acquisition, Data Management and Data Interpretation*
38 *Design and Performance of Deep Foundations: Piles and Piers in Soil and Soft Rock*
39 *Unsaturated Soils*
40 *Vertical and Horizontal Deformations of Foundations and Embankments*
41 *Predicted and Measured Behavior of Five Spread Footings on Sand*
42 *Serviceability of Earth Retaining Structures*
43 *Fracture Mechanics Applied to Geotechnical Engineering*
44 *Ground Failures Under Seismic Conditions*
45 *In Situ Deep Soil Improvement*
46 *Geoenvironment 2000*
47 *Geo-Environmental Issues Facing the Americas*
48 *Soil Suction Applications in Geotechnical Engineering*
49 *Soil Improvement for Earthquake Hazard Mitigation*
50 *Foundation Upgrading and Repair for Infrastructure Improvement*
51 *Performance of Deep Foundations Under Seismic Loading*
52 *Landslides Under Static and Dynamic Conditions–Analysis, Monitoring, and Mitigation*
53 *Landfill Closures–Environmental Protection and Land Recovery*
54 *Earthquake Design and Performance of Solid Waste Landfills*
55 *Earthquake-Induced Movements and Seismic Remediation of Existing Foundations and Abutments*
56 *Static and Dynamic Properties of Gravelly Soils*
57 *Verification of Geotechnical Grouting*
58 *Uncertainty in the Geologic Environment*
59 *Engineered Contaminated Soils and Interaction of Soil Geomembranes*
60 *Analysis and Design of Retaining Structures Against Earthquakes*

- 61 *Measuring and Modeling Time Dependent Soil Behavior*
- 62 *Case Histories of Geophysics Applied to Civil Engineering and Public Policy*
- 63 *Design with Residual Materials: Geotechnical and Construction Considerations*
- 64 *Observation and Modeling in Numerical Analysis and Model Tests in Dynamic Soil-Structure Interaction Problems*
- 65 *Dredging and Management of Dredged Material*
- 66 *Grouting: Compaction, Remediation and Testing*
- 67 *Spatial Analysis in Soil Dynamics and Earthquake Engineering*
- 68 *Unsaturated Soil Engineering Practice*
- 69 *Ground Improvement, Ground Reinforcement, Ground Treatment: Developments 1987-1997*
- 70 *Seismic Analysis and Design for Soil-Pile-Structure Interactions*
- 71 *In Situ Remediation of the Geoenvironment*
- 72 *Degradation of Natural Building Stone*
- 73 *Innovative Design and Construction for Foundations and Substructures Subject to Freezing and Frost*
- 74 *Guidelines of Engineering Practice for Braced and Tied-Back Excavations*
- 75 *Geotechnical Earthquake Engineering and Soil Dynamics III*
- 76 *Geosynthetics in Foundation Reinforcement and Erosion Control Systems*
- 77 *Stability of Natural Slopes in the Coastal Plain*
- 78 *Filtration and Drainage in Geotechnical/Geoenvironmental Engineering*
- 79 *Recycled Materials in Geotechnical Applications*
- 80 *Grouts and Grouting: A Potpourri of Projects*
- 81 *Soil Improvement for Big Digs*
- 82 *Risk-Based Corrective Action and Brownfields Restorations*
- 83 *Design and Construction of Earth Retaining Systems*
- 84 *Effects of Construction on Structures*
- 85 *Application of Geotechnical Principles in Pavement Engineering*
- 86 *Big Digs Around the World*
- 87 *Jacked Tunnel Design and Construction*
- 88 *Analysis, Design, Construction, and Testing of Deep Foundations*
- 89 *Recent Advances in the Characterization of Transportation Geo-Materials*
- 90 *Geo-Engineering for Underground Facilities*
- 91 *Special Geotechnical Testing: Central Artery/Tunnel Project in Boston, Massachusetts*
- 92 *Behavioral Characteristics of Residual Soils*
- 93 *National Geotechnical Experimentation Sites*
- 94 *Performance Confirmation of Constructed Geotechnical Facilities*
- 95 *Soil-Cement and Other Construction Practices in Geotechnical Engineering*
- 96 *Numerical Methods in Geotechnical Engineering: Recent Developments*
- 97 *Innovations and Applications in Geotechnical Site Characterization*
- 98 *Pavement Subgrade, Unbound Materials, and Nondestructive Testing*
- 99 *Advances in Unsaturated Geotechnics*
- 100 *New Technological and Design Developments in Deep Foundations*
- 101 *Slope Stability 2000*
- 102 *Trends in Rock Mechanics*
- 103 *Advances in Transportation and Geoenvironmental Systems Using Geosynthetics*
- 104 *Advances in Grouting and Ground Modification*
- 105 *Environmental Geotechnics*
- 106 *Geotechnical Measurements: Lab & Field*
- 107 *Soil Dynamics and Liquefaction 2000*
- 108 *Use of Geophysical Methods in Construction*
- 109 *Educational Issues in Geotechnical Engineering*
- 110 *Computer Simulation of Earthquake Effects*
- 111 *Judgment and Innovation: The Heritage and Future of the Geotechnical Engineering Profession*
- 112 *Soft Ground Technology*
- 113 *Foundations and Ground Improvement*
- 114 *Soils Magic*
- 115 *Expansive Clay Soils and Vegetative Influence on Shallow Foundations*
- 116 *Deep Foundations 2002: An International Perspective on Theory, Design, Construction, and Performance*
- 117 *Discrete Element Methods: Numerical Modeling of Discontinua*
- 118 *A History of Progress: Selected U.S. Papers in Geotechnical Engineering*